Second Edition

HANDBOOK OF Physical-Chemical Properties and Environmental Fate for Organic Chemicals

Volume I
Introduction and Hydrocarbons

Second Edition

Handbook of Physical-Chemical Properties and Environmental Fate for Organic Chemicals

Volume I
Introduction and Hydrocarbons

Donald Mackay
Wan Ying Shiu
Kuo-Ching Ma
Sum Chi Lee

Taylor & Francis
Taylor & Francis Group

Boca Raton   London   New York

A CRC title, part of the Taylor & Francis imprint, a member of the
Taylor & Francis Group, the academic division of T&F Informa plc.

Published in 2006 by
CRC Press
Taylor & Francis Group
6000 Broken Sound Parkway NW, Suite 300
Boca Raton, FL 33487-2742

© 2006 by Taylor & Francis Group, LLC
CRC Press is an imprint of Taylor & Francis Group

No claim to original U.S. Government works
Printed in the United States of America on acid-free paper
10 9 8 7 6 5 4 3 2 1

International Standard Book Number-10: 1-56670-687-4 (Hardcover)
International Standard Book Number-13: 978-1-56670-687-2 (Hardcover)
Library of Congress Card Number 2005051402

This book contains information obtained from authentic and highly regarded sources. Reprinted material is quoted with permission, and sources are indicated. A wide variety of references are listed. Reasonable efforts have been made to publish reliable data and information, but the author and the publisher cannot assume responsibility for the validity of all materials or for the consequences of their use.

No part of this book may be reprinted, reproduced, transmitted, or utilized in any form by any electronic, mechanical, or other means, now known or hereafter invented, including photocopying, microfilming, and recording, or in any information storage or retrieval system, without written permission from the publishers.

For permission to photocopy or use material electronically from this work, please access www.copyright.com (http://www.copyright.com/) or contact the Copyright Clearance Center, Inc. (CCC) 222 Rosewood Drive, Danvers, MA 01923, 978-750-8400. CCC is a not-for-profit organization that provides licenses and registration for a variety of users. For organizations that have been granted a photocopy license by the CCC, a separate system of payment has been arranged.

Trademark Notice: Product or corporate names may be trademarks or registered trademarks, and are used only for identification and explanation without intent to infringe.

Library of Congress Cataloging-in-Publication Data

Handbook of physical-chemical properties and environmental fate for organic chemicals.--2nd ed. / by Donald Mackay ... [et al.].
    p. cm.
      Rev. ed. of: Illustrated handbook of physical-chemical properties and environmental fate for organic chemicals / Donald Mackay, Wan Ying Shiu, and Kuo Ching Ma. c1992-c1997.
    Includes bibliographical references and index.
    ISBN 1-56670-687-4 (set : acid-free paper)
    1. Organic compounds--Environmental aspects--Handbooks, manuals, etc. 2. Environmental chemistry--Handbooks, manuals, etc.
I. Mackay, Donald, 1936- II. Mackay, Donald, 1936- Illustrated handbook of physical-chemical properties and environmental fate for organic chemicals.

TD196.O73M32 2005
628.5'2--dc22
                                                2005051402

Taylor & Francis Group is the Academic Division of T&F Informa plc.

Visit the Taylor & Francis Web site at
http://www.taylorandfrancis.com

and the CRC Press Web site at
http://www.crcpress.com

Preface

This handbook is a compilation of environmentally relevant physical-chemical data for similarly structured groups of chemical substances. These data control the fate of chemicals as they are transported and transformed in the multimedia environment of air, water, soils, sediments, and their resident biota. These fate processes determine the exposure experienced by humans and other organisms and ultimately the risk of adverse effects. The task of assessing chemical fate locally, regionally, and globally is complicated by the large (and increasing) number of chemicals of potential concern; by uncertainties in their physical-chemical properties; and by lack of knowledge of prevailing environmental conditions such as temperature, pH, and deposition rates of solid matter from the atmosphere to water, or from water to bottom sediments. Further, reported values of properties such as solubility are often in conflict. Some are measured accurately, some approximately, and some are estimated by various correlation schemes from molecular structures. In some cases, units or chemical identity are wrongly reported. The user of such data thus has the difficult task of selecting the "best" or "right" values. There is justifiable concern that the resulting deductions of environmental fate may be in substantial error. For example, the potential for evaporation may be greatly underestimated if an erroneously low vapor pressure is selected.

To assist the environmental scientist and engineer in such assessments, this handbook contains compilations of physical-chemical property data for over 1000 chemicals. It has long been recognized that within homologous series, properties vary systematically with molecular size, thus providing guidance about the properties of one substance from those of its homologs. Where practical, plots of these systematic property variations can be used to check the reported data and provide an opportunity for interpolation and even modest extrapolation to estimate unmeasured properties of other substances. Most handbooks treat chemicals only on an individual basis and do not contain this feature of chemical-to-chemical comparison, which can be valuable for identifying errors and estimating properties. This most recent edition includes about 1250 compounds and contains about 30 percent additional physical-chemical property data. There is a more complete coverage of PCBs, PCDDs, PCDFs, and other halogenated hydrocarbons, especially brominated and fluorinated substances that are of more recent environmental concern. Values of the physical-chemical properties are generally reported in the literature at a standard temperature of 20 or 25°C. However, environmental temperatures vary considerably, and thus reliable data are required on the temperature dependence of these properties for fate calculations. A valuable enhancement to this edition is the inclusion of extensive measured temperature-dependent data for the first time. The data focus on water solubility, vapor pressure, and Henry's law constant but include octanol/water and octanol/air partition coefficients where available. They are provided in the form of data tables and correlation equations as well as graphs.

We also demonstrate in Chapter 1 how the data may be taken a stage further and used to estimate likely environmental partitioning tendencies, i.e., how the chemical is likely to become distributed between the various media that comprise our biosphere. The results are presented numerically and pictorially to provide a visual impression of likely environmental behavior. This will be of interest to those assessing environmental fate by confirming the general fate characteristics or behavior profile. It is, of course, only possible here to assess fate in a "typical" or "generic" or "evaluative" environment. No claim is made that a chemical will behave in this manner in all situations, but this assessment should reveal the broad characteristics of behavior. These evaluative fate assessments are generated using simple fugacity models that flow naturally from the compilations of data on physical-chemical properties of relevant chemicals. Illustrations of estimated environmental fate are given in Chapter 1 using Levels I, II, and III mass balance models. These and other models are available for downloading gratis from the website of the Canadian Environmental Modelling Centre at Trent University (www.trent.ca/cemc).

It is hoped that this new edition of the handbook will be of value to environmental scientists and engineers and to students and teachers of environmental science. Its aim is to contribute to better assessments of chemical fate in our multimedia environment by serving as a reference source for environmentally relevant physical-chemical property data of classes of chemicals and by illustrating the likely behavior of these chemicals as they migrate throughout our biosphere.

Acknowledgments

We would never have completed the volumes for the first and second editions of the handbook and the CD-ROMs without the enormous amount of help and support that we received from our colleagues, publishers, editors, friends, and family. We are long overdue in expressing our appreciation.

We would like first to extend deepest thanks to these individuals: Dr. Warren Stiver, Rebecca Lun, Deborah Tam, Dr. Alice Bobra, Dr. Frank Wania, Ying D. Lei, Dr. Hayley Hung, Dr. Antonio Di Guardo, Qiang Kang, Kitty Ma, Edmund Wong, Jenny Ma, and Dr. Tom Harner. During their past and present affiliations with the Department of Chemical Engineering and Applied Chemistry and/or the Institute of Environment Studies at the University of Toronto, they have provided us with many insightful ideas, constructive reviews, relevant property data, computer know-how, and encouragement, which have resulted in substantial improvements to each consecutive volume and edition through the last fifteen years.

Much credit goes to the team of professionals at CRC Press/Taylor & Francis Group who worked on this second edition. Especially important were Dr. Fiona Macdonald, Publisher, Chemistry; Dr. Janice Shackleton, Input Supervisor; Patrica Roberson, Project Coordinator; Elise Oranges and Jay Margolis, Project Editors; and Marcela Peres, Production Assistant.

We are indebted to Brian Lewis, Vivian Collier, Kathy Feinstein, Dr. David Packer, and Randi Cohen for their interest and help in taking our idea of the handbook to fruition.

We also would like to thank Professor Doug Reeve, Chair of the Department of Chemical Engineering and Applied Chemistry at the University of Toronto, as well as the administrative staff for providing the resources and assistance for our efforts.

We are grateful to the University of Toronto and Trent University for providing facilities, to the Natural Sciences and Engineering Research Council of Canada and the consortium of chemical companies that support the Canadian Environmental Modelling Centre for funding of the second edition. It is a pleasure to acknowledge the invaluable contributions of Eva Webster and Ness Mackay.

Biographies

Donald Mackay, born and educated in Scotland, received his degrees in Chemical Engineering from the University of Glasgow. After working in the petrochemical industry he joined the University of Toronto, where he taught for 28 years in the Department of Chemical Engineering and Applied Chemistry and in the Institute for Environmental Studies. In 1995 he moved to Trent University to found the Canadian Environmental Modelling Centre. Professor Mackay's primary research is the study of organic environmental contaminants, their properties, sources, fates, effects, and control, and particularly understanding and modeling their behavior with the aid of the fugacity concept. His work has focused especially on the Great Lakes Basin; on cold northern climates; and on modeling bioaccumulation and chemical fate at local, regional, continental and global scales.

His awards include the SETAC Founders Award, the Honda Prize for Eco-Technology, the Order of Ontario, and the Order of Canada. He has served on the editorial boards of several journals and is a member of SETAC, the American Chemical Society, and the International Association of Great Lakes Research.

Wan-Ying Shiu is a Senior Research Associate in the Department of Chemical Engineering and Applied Chemistry, and the Institute for Environmental Studies, University of Toronto. She received her Ph.D. in Physical Chemistry from the Department of Chemistry, University of Toronto, M.Sc. in Physical Chemistry from St. Francis Xavier University, and B.Sc. in Chemistry from Hong Kong Baptist College. Her research interest is in the area of physical-chemical properties and thermodynamics for organic chemicals of environmental concern.

Kuo-Ching Ma obtained his Ph.D. from Florida State University, M.Sc. from The University of Saskatchewan, and B.Sc. from The National Taiwan University, all in Physical Chemistry. After working many years in the aerospace, battery research, fine chemicals, and metal finishing industries in Canada as a Research Scientist, Technical Supervisor/Director, he is now dedicating his time and interests to environmental research.

Sum Chi Lee received her B.A.Sc. and M.A.Sc. in Chemical Engineering from the University of Toronto. She has conducted environmental research at various government organizations and the University of Toronto. Her research activities have included establishing the physical-chemical properties of organochlorines and understanding the sources, trends, and behavior of persistent organic pollutants in the atmosphere of the Canadian Arctic.

Ms. Lee also possesses experience in technology commercialization. She was involved in the successful commercialization of a proprietary technology that transformed recycled material into environmentally sound products for the building material industry. She went on to pursue her MBA degree, which she earned from York University's Schulich School of Business. She continues her career, combining her engineering and business experiences with her interest in the environmental field.

Contents

Volume I
Chapter 1 Introduction 1
Chapter 2 Aliphatic and Cyclic Hydrocarbons 61
Chapter 3 Mononuclear Aromatic Hydrocarbons 405
Chapter 4 Polynuclear Aromatic Hydrocarbons (PAHs) and Related Aromatic Hydrocarbons 617

Volume II
Chapter 5 Halogenated Aliphatic Hydrocarbons 921
Chapter 6 Chlorobenzenes and Other Halogenated Mononuclear Aromatics 1257
Chapter 7 Polychlorinated Biphenyls (PCBs) 1479
Chapter 8 Chlorinated Dibenzo-p-dioxins 2063
Chapter 9 Chlorinated Dibenzofurans 2167

Volume III
Chapter 10 Ethers 2259
Chapter 11 Alcohols 2473
Chapter 12 Aldehydes and Ketones 2583
Chapter 13 Carboxylic Acids 2687
Chapter 14 Phenolic Compounds 2779
Chapter 15 Esters 3023

Volume IV
Chapter 16 Nitrogen and Sulfur Compounds 3195
Chapter 17 Herbicides 3457
Chapter 18 Insecticides 3711
Chapter 19 Fungicides 4023

Appendix 1 4133
Appendix 2 4137
Appendix 3 4161

1 Introduction

CONTENTS

1.1	The Incentive	2
1.2	Physical-Chemical Properties	3
	1.2.1 The Key Physical-Chemical Properties	3
	1.2.2 Partitioning Properties	3
	1.2.3 Temperature Dependence	5
	1.2.4 Treatment of Dissociating Compounds	7
	1.2.5 Treatment of Water-Miscible Compounds	8
	1.2.6 Treatment of Partially Miscible Substances	8
	1.2.7 Treatment of Gases and Vapors	8
	1.2.8 Solids, Liquids and the Fugacity Ratio	9
	1.2.9 Chemical Reactivity and Half-Lives	10
1.3	Experimental Methods	11
	1.3.1 Solubility in Water and pK_a	11
	1.3.2 Vapor Pressure	12
	1.3.3 Octanol-Water Partition Coefficient K_{OW}	13
	1.3.4 Henry's Law Constant	13
	1.3.5 Octanol-Air Partition Coefficient K_{OA}	14
1.4	Quantitative Structure-Property Relationships (QSPRs)	14
	1.4.1 Objectives of QSPRs	14
	1.4.2 Examples of QSARs and QSPRs	15
1.5	Mass Balance Models of Chemical Fate	18
	1.5.1 Evaluative Environmental Calculations	18
	1.5.2 Level I Fugacity Calculations	19
	1.5.3 Level II Fugacity Calculations	22
	1.5.4 Level III Fugacity Calculations	23
1.6	Data Sources and Presentation	28
	1.6.1 Data Sources	28
	1.6.2 Data Presentation	29
1.7	Illustrative QSPR Plots and Fate Calculations	29
	1.7.1 QSPR Plots for Mononuclear Aromatic Hydrocarbons	29
	1.7.2 Evaluative Calculations for Benzene	32
	1.7.3 QSPR Plots for Chlorophenols and Alkylphenols	36
	1.7.4 Evaluative Calculations for Pentachlorophenol	39
1.8	References	49

1.1 THE INCENTIVE

It is believed that there are some 50,000 to 100,000 chemicals currently being produced commercially in a range of quantities with approximately 1000 being added each year. Most are organic chemicals, and many are pesticides and biocides designed to modify the biotic environment. Of these, perhaps 1000 substances are of significant environmental concern because of their presence in detectable quantities in various components of the environment, their toxicity, their tendency to bioaccumulate, their persistence and their potential to be transported long distances. Some of these chemicals, including pesticides, are of such extreme environmental concern that international actions have been taken to ensure that all production and use should cease, i.e., as a global society we should elect not to synthesize or use these chemicals. They should be "sunsetted." PCBs, "dioxins" and DDT are examples. A second group consists of less toxic and persistent chemicals which are of concern because they are used or discharged in large quantities. They are, however, of sufficient value to society that their continued use is justified, but only under conditions in which we fully understand and control their sources, fate and the associated risk of adverse effects. This understanding is essential if society is to be assured that there is negligible risk of adverse ecological or human health effects. Other groups of more benign chemicals can presumably be treated with less rigor.

A key feature of this "cradle-to-grave" approach to chemical management is that society must improve its skills in assessing chemical fate in the environment. We must better understand where chemicals originate, how they migrate in, and between, the various media of air, water, soils, sediments and their biota which comprise our biosphere. We must understand how these chemicals are transformed by chemical and biochemical processes and, thus, how long they will persist in the environment. We must seek a fuller understanding of the effects that they will have on the multitude of interacting organisms that occupy these media, including ourselves.

It is now clear that the fate of chemicals in the environment is controlled by a combination of three groups of factors. First are the prevailing environmental conditions such as temperatures, flows and accumulations of air, water and solid matter and the composition of these media. Second are the properties of the chemicals which influence partitioning and reaction tendencies, i.e., the extent to which the chemical evaporates or associates with sediments, and how fast the chemical is eventually destroyed by conversion to other chemical species. Third are the patterns of use, into which compartments the substance is introduced, whether introduction is episodic or continuous and in the case of pesticides how and with which additives the active ingredient is applied.

In recent decades there has emerged a discipline within environmental science concerned with increasing our understanding of how chemicals behave in our multimedia environment. It has been termed environmental chemistry or "chemodynamics." Practitioners of this discipline include scientists and engineers, students and teachers who attempt to measure, assess and predict how this large number of chemicals will behave in laboratory, local, regional and global environments. These individuals need data on physical-chemical and reactivity properties, as well as information on how these properties translate into environmental fate. This handbook provides a compilation of such data and outlines how to use them to estimate the broad features of environmental fate. It does so for classes or groups of chemicals, instead of the usual approach of treating chemicals on an individual basis. This has the advantage that systematic variations in properties with molecular structure can be revealed and exploited to check reported values, interpolate and even extrapolate to other chemicals of similar structure.

With the advent of inexpensive and rapid computation there has been a remarkable growth of interest in this general area of quantitative structure-property relationships (QSPRs). The ultimate goal is to use information about chemical structure to deduce physical-chemical properties, environmental partitioning and reaction tendencies, and even uptake and effects on biota. The goal is far from being fully realized, but considerable progress has been made. In this series of handbooks we have adopted a simple and well-tried approach of using molecular structure to deduce a molar volume, which in turn is related to physical-chemical properties. In the case of pesticides, the application of QSPR approaches is complicated by the large number of chemical classes, the frequent complexity of molecules and the lack of experimental data. Where there is a sufficient number of substances in each class or homologous series QSPRs are presented, but in some cases there is a lack of data to justify them. QSPRs based on other more complex molecular descriptors are, of course, widely available, especially in the proceedings of the biennial QSAR conferences.

Regrettably, the scientific literature contains a great deal of conflicting data, with reported values often varying over several orders of magnitude. There are some good, but more not-so-good reasons for this lack of accuracy. Many of these properties are difficult to measure because they involve analyzing very low concentrations of 1 part in 10^9 or 10^{12}. For many purposes an approximate value is adequate. There may be a mistaken impression that if a vapor pressure is low, as is the case with DDT, it is not important. DDT evaporates appreciably from solution in water, despite its low vapor pressure, because of its low solubility in water. In some cases the units are reported incorrectly. There may be uncertainties about temperature or pH. In other cases the chemical is wrongly identified. Errors tend to be perpetuated

Introduction

by repeated citation. The aim of this handbook is to assist the user to identify such problems, provide guidance when selecting appropriate values and where possible determine their temperature dependence.

The final aspect of chemical fate treated in this handbook is the depiction or illustration of likely chemical fate. This is done using multimedia "fugacity" models as described later in this chapter. The aim is to convey an impression of likely environmental partitioning and transformation characteristics, i.e., a "behavior profile." A fascinating feature of chemodynamics is that chemicals differ so greatly in their behavior. Some, such as chloroform, evaporate rapidly and are dissipated in the atmosphere. Others, such as DDT, partition into the organic matter of soils and sediments and the lipids of fish, birds and mammals. Phenols and carboxylic acids tend to remain in water where they may be subject to fairly rapid transformation processes such as hydrolysis, biodegradation and photolysis. By entering the physical-chemical data into a model of chemical fate in a generic or evaluative environment, it is possible to estimate the likely general features of the chemical's behavior and fate. The output of these calculations can be presented numerically and pictorially.

In summary, the aim of this series of handbooks is to provide a useful reference work for those concerned with the assessment of the fate of existing and new chemicals in the environment.

1.2 PHYSICAL-CHEMICAL PROPERTIES

1.2.1 THE KEY PHYSICAL-CHEMICAL PROPERTIES

In this section we describe the key physical-chemical properties and discuss how they may be used to calculate partition coefficients for inclusion in mass balance models. Situations in which data require careful evaluation and use are discussed.

The major differences between behavior profiles of organic chemicals in the environment are attributable to their physical-chemical properties. The key properties are recognized as solubility in water, vapor pressure, the three partition coefficients between air, water and octanol, dissociation constant in water (when relevant) and susceptibility to degradation or transformation reactions. Other essential molecular descriptors are molar mass and molar volume, with properties such as critical temperature and pressure and molecular area being occasionally useful for specific purposes. A useful source of information and estimation methods on these properties is the handbook by Boethling and Mackay (2000).

Chemical identity may appear to present a trivial problem, but most chemicals have several names, and subtle differences between isomers (e.g., cis and trans) may be ignored. The most commonly accepted identifiers are the IUPAC name and the Chemical Abstracts System (CAS) number. More recently, methods have been sought of expressing the structure in line notation form so that computer entry of a series of symbols can be used to define a three-dimensional structure. For environmental purposes the SMILES (Simplified Molecular Identification and Line Entry System, Anderson et al. 1987) is favored, but the Wismesser Line Notation is also quite widely used.

Molar mass or molecular weight is readily obtained from structure. Also of interest for certain purposes are molecular volume and area, which may be estimated by a variety of methods.

When selecting physical-chemical properties or reactivity classes the authors have been guided by:

1. The acknowledgment of previous supporting or conflicting values,
2. The method of determination,
3. The perception of the objectives of the authors, not necessarily as an indication of competence, but often as an indication of the need of the authors to obtain accurate values, and
4. The reported values for structurally similar, or homologous compounds.

The literature contains a considerable volume of "calculated" data as distinct from experimental data. We have generally not included such data because they may be of questionable reliability. In some cases an exception has been made when no experimental data exist and the calculation is believed to provide a useful and reliable estimate.

1.2.2 PARTITIONING PROPERTIES

Solubility in water and vapor pressure are both "saturation" properties, i.e., they are measurements of the maximum capacity that a solvent phase has for dissolved chemical. Vapor pressure P (Pa) can be viewed as a "solubility in air," the corresponding concentration C (mol/m^3) being P/RT where R is the ideal gas constant (8.314 J/mol.K) and T is absolute temperature (K). Although most chemicals are present in the environment at concentrations well below saturation, these concentrations are useful for estimating air-water partition coefficients as ratios of saturation values. It is usually assumed that the same partition coefficient applies at lower sub-saturation concentrations. Vapor pressure and solubility thus provide estimates of the air-water partition coefficient K_{AW}, the dimensionless ratio of concentration in air (mass/volume) to that in water. The related Henry's law constant H (Pa.m³/mol) is the ratio of partial pressure in air (Pa) to the concentration in water (mol/m³). Both express the relative air-water partitioning tendency.

When solubility and vapor pressure are both low in magnitude and thus difficult to measure, it is preferable to measure the air-water partition coefficient or Henry's law constant directly. It is noteworthy that atmospheric chemists frequently use K_{WA}, the ratio of water-to-air concentrations. This may also be referred to as the Henry's law constant.

The octanol-water partition coefficient K_{OW} provides a direct estimate of hydrophobicity or of partitioning tendency from water to organic media such as lipids, waxes and natural organic matter such as humin or humic acid. It is invaluable as a method of estimating K_{OC}, the organic carbon-water partition coefficient, the usual correlation invoked being that of Karickhoff (1981)

$$K_{OC} = 0.41\ K_{OW}$$

Seth et al. (1999) have suggested that a better correlation is

$$K_{OC} = 0.35\ K_{OW}$$

and that the error limits on K_{OC} resulting from differences in the nature of organic matter are a factor of 2.5 in both directions, i.e. the coefficient 0.35 may vary from 0.14 to 0.88.

K_{OC} is an important parameter which describes the potential for movement or mobility of pesticides in soil, sediment and groundwater. Because of the structural complexity of these agrochemical molecules, the above simple relationship which considers only the chemical's hydrophobicity may fail for polar and ionic compounds. The effects of pH, soil properties, mineral surfaces and other factors influencing sorption become important. Other quantities, K_D (sorption partition coefficient to the whole soil on a dry weight basis) and K_{OM} (organic matter-water partition coefficient) are also commonly used to describe the extent of sorption. K_{OM} is often estimated as 0.56 K_{OC}, implying that organic matter is 56% carbon.

K_{OW} is also used to estimate equilibrium fish-water bioconcentration factors K_B, or BCF using a correlation similar to that of Mackay (1982)

$$K_B = 0.05\ K_{OW}$$

where the term 0.05 corresponds to a lipid content of the fish of 5%. The basis for this correlation is that lipids and octanol display very similar solvent properties, i.e., K_{LW} (lipid-water) and K_{OW} are equal. If the rate of metabolism is appreciable, equilibrium will not apply and the effective K_B will be lower to an extent dictated by the relative rates of uptake and loss by metabolism and other clearance processes. If uptake is primarily from food, the corresponding bioaccumulation factor also depends on the concentration of the chemical in the food.

For dissociating chemicals it is essential to quantify the extent of dissociation as a function of pH using the dissociation constant pK_a. The parent and ionic forms behave and partition quite differently; thus pH and the presence of other ions may profoundly affect chemical fate. This is discussed later in more detail in Section 1.2.4.

The octanol-air partition coefficient K_{OA} was originally introduced by Paterson et al. (1991) for describing the partitioning of chemicals from the atmosphere to foliage. It has proved invaluable for this purpose and for describing partitioning to aerosol particles and to soils. It can be determined experimentally using the technique devised by Harner and Mackay (1995). Although there are fewer data for K_{OA} than for K_{OW}, its use is increasing and when available, data are included in this handbook. K_{OA} has been applied to several situations involving partitioning of organic substances from the atmosphere to solid or liquid phases. Finizio et al. (1997) have shown that K_{OA} is an excellent descriptor of partitioning to aerosol particles, while McLachlan et al. (1995) and Tolls and McLachlan (1994) have used it to describe partitioning to foliage, especially grasses. Hippelein and McLachlan (1998) have used K_{OA} to describe partitioning between air and soil.

An attractive feature of K_{OA} is that it can replace the liquid or supercooled liquid vapor pressure in a correlation. K_{OA} is an experimentally measurable or accessible quantity, whereas the supercooled liquid vapor pressure must be estimated from the solid vapor pressure, the melting point and the entropy of fusion. The use of K_{OA} thus avoids the potentially erroneous estimation of the fugacity ratio, i.e., the ratio of solid and liquid vapor pressures. This is especially important for solutes with high melting points and, thus, low fugacity ratios.

Introduction

The availability of data on K_{AW}, K_{OW} and K_{OA} raises the possibility of a consistency test. At first sight it appears that K_{OA} should equal K_{OW}/K_{AW}, and indeed this is often approximately correct. The difficulty is that in the case of K_{AW}, the water phase is pure water, and for K_{OA} the octanol phase is pure "dry" octanol. For K_{OW}, the water phase inevitably contains dissolved octanol, and the octanol phase contains dissolved water and is thus not "dry." Beyer et al. (2002) and Cole and Mackay (2000) have discussed this issue.

If the partition coefficients are regarded as ratios of solubilities S (mol/m³)

$$K_{AW} = S_A/S_W \text{ or } \log K_{AW} = \log S_A - \log S_W$$

$$K_{OA} = S_O/S_A \text{ or } \log K_{OA} = \log S_O - \log S_A$$

$$K_{OW} = S_{OW}/S_{WO} \text{ or } \log K_{OW} = \log S_{OW} - \log S_{WO}$$

where subscript A applies to the gas phase or air, W to pure water, O to dry octanol, OW to "wet" octanol and WO to water saturated with octanol. It follows that the assumption that K_{OA} is K_{OW}/K_{AW} is essentially that

$$(\log S_{OW} - \log S_O) - (\log S_{WO} - \log S_W) = 0$$

$$\text{or } S_{OW} S_W/(S_O \cdot S_{WO}) \text{ is } 1.0$$

This is obviously satisfied when S_{OW} equals S_O and S_{WO} equals S_W, but this is not necessarily valid, especially when K_{OW} is large.

There are apparently two sources of this effect. The molar volume of water changes relatively little as a result of the presence of a small quantity of dissolved octanol, however the quantity of dissolved water in the octanol is considerable, causing a reduction in molar volume of the octanol phase. The result is that even if activity coefficients are unaffected, $\log S_O/S_W$ will be about 0.1 units less than that of $\log K_{OW}$. Effectively, the octanol phase "swells" as a result of the presence of water, and the concentration is reduced. In addition, when $\log K_{OW}$ exceeds 4.0 there is an apparent effect on the activity coefficients which causes $\log (S_O/S_W)$ to increase. This increase can amount to about one log unit when $\log K_{OW}$ is about 8. A relatively simple correlation based on the analysis by Beyer et al. (2002) (but differing from their correlation) is that

$$\log K_{OA} = \log (K_{OW}/K_{AW}) - 0.10 + [0.30 \log K_{OW} - 1.20]$$

when $\log K_{OW}$ is 4 or less the term in square brackets is ignored
when $\log K_{OW}$ is 4 or greater that term is included

1.2.3 Temperature Dependence

All partitioning properties change with temperature. The partition coefficients, vapor pressure, K_{AW} and K_{OA}, are more sensitive to temperature variation because of the large enthalpy change associated with transfer to the vapor phase. The simplest general expression theoretically based temperature dependence correlation is derived from the integrated Clausius-Clapeyron equation, or van't Hoff form expressing the effect of temperature on an equilibrium constant K_p,

$$R \cdot \ln K_p = A_o - B/T$$

which can be rewritten as

$$\ln (\text{Property}) = A - \Delta H/RT$$

where A_o, B and A are constants, ΔH is the enthalpy of the phase change, i.e., evaporation from pure state for vapor pressure, dissolution from pure state into water for solubility, and for air-water transition in the case of Henry's law constant.

The fit is improved by adding further coefficients in additional terms. The variation of these equilibrium constants with temperature can be expressed by (Clarke and Glew 1966),

$$R \cdot \ln K_p(T) = A + B/T + C \cdot \ln T + DT + ET^2 + FT^3 + \ldots\ldots$$

where A, B, C, D, E, F are constants.

There have been numerous approaches to describing the temperature dependence of the properties. For aqueous solubility, the most common expression is the van't Hoff equation of the form (Hildebrand et al. 1970):

$$d(\ln x)/d(1/T) = -\Delta_{sol}H/R$$

where x is the mole fraction solubility, T is the temperature in K, R is the ideal gas constant, and $\Delta_{sol}H$ is the enthalpy of solution of the solute. The enthalpy of solution can be considered as the sum of various contributions such as cavity formation and interactions between solute-solute or solute-solvent as discussed by Bohon and Claussen (1951), Arnold et al. (1958), Owen et al. (1986) and many others. Assuming the enthalpy of solution is constant over a narrow temperature range, integrating gives,

$$\ln x = -\Delta_{sol}H/RT + C$$

where C is a constant.

The relation between aqueous solubility and temperature is complicated because of the nature of the interactions between the solute and water structure. The enthalpy of solution can vary greatly with temperature, e.g., some liquid aromatic hydrocarbons display a minimum solubility corresponding to zero enthalpy of solution between 285 and 320 K. For instance, benzene has a minimum solubility at 291 K (Bohon and Claussen 1951, Arnold et al. 1958, Shaw 1989a) and alkylbenzenes display similar behavior (Shaw 1989a,b, Owens 1986). As is illustrated later in chapter 3, solid aromatic hydrocarbons show a slight curvature in plots of logarithm of mole fraction solubility versus reciprocal absolute temperature. For narrow ranges in environmental temperatures, the enthalpy of solution may be assumed to be constant, and the linear van't Hoff plot of $\ln x$ versus $1/T$ is often used (Dickhut et al. 1986). Other relationships such as quadratic or cubic equations have been reported (May et al. 1978), and polynomial series (Clarke and Glew 1966, May et al. 1983, Owens et al. 1986) have been used when the data justify such treatment.

Equations relating vapor pressure to temperature are usually based on the two-parameter Clausius-Clapeyron equation,

$$d(\ln P^S)/dT = \Delta_{vap}H/RT^2$$

where P^S is vapor pressure, $\Delta_{vap}H$ is the enthalpy of vaporization. Again assuming $\Delta_{vap}H$ is constant over a narrow range of temperature, this gives,

$$\ln P^S = -\Delta_{vap}H/RT + C$$

which can be rewritten as the Clapeyron equation

$$\log P^S = A - B/T$$

This can be empirically modified by introducing additional parameters to give the three-parameter Antoine equation by replacing T with $(T + C)$, where C is a constant, which is the most common vapor pressure correlation used to represent experimental data (Zwolinski and Wilhoit 1971, Boublik et al. 1984, Stephenson and Malanowski 1987, and other handbooks).

$$\log P^S = A - B/(t + C)$$

where A, B and C are constants and t often has units of °C.

Other forms of vapor pressure equations, such as Cox equation (Osborn and Douslin 1974, Chao et al. 1983), Chebyshev polynomial (Ambrose 1981), Wagner's equation (Ambrose 1986), have also been widely used. Although

Introduction

the enthalpy of vaporization varies with temperature, for the narrow environmental temperature range considered in environmental conditions, it is often assumed to be constant, for example, for the more volatile monoaromatic hydrocarbons and the less volatile polynuclear aromatic hydrocarbons.

The van't Hoff equation also has been used to describe the temperature effect on Henry's law constant over a narrow range for volatile chlorinated organic chemicals (Ashworth et al. 1988) and chlorobenzenes, polychlorinated biphenyls, and polynuclear aromatic hydrocarbons (ten Hulscher et al. 1992, Alaee et al. 1996). Henry's law constant can be expressed as the ratio of vapor pressure to solubility, i.e., p/c or p/x for dilute solutions. Note that since H is expressed using a volumetric concentration, it is also affected by the effect of temperature on liquid density whereas k_H using mole fraction is unaffected by liquid density (Tucker and Christian 1979), thus

$$ln\ (k_H/Pa) = ln\ [(P^S/Pa)/x];$$

or,

$$ln\ (H/Pa \cdot m^3 \cdot mol^{-1}) = ln\ [(P^S/Pa)/(C^S_W/mol \cdot m^{-3})];$$

where C^S_W is the aqueous solubility.

By substituting equations for vapor pressure and solubility, the temperature dependence equation for Henry's law constant can be obtained, as demonstrated by Glew and Robertson (1956), Tsonopoulos and Wilson (1983), Heiman et al. (1985), and ten Hulscher et al. (1991).

Care must be taken to ensure that the correlation equations are applied correctly, especially since the units of the property, the units of temperature and whether the logarithm is base e or base 10. The equations should not be used to extrapolate beyond the stated temperature range.

1.2.4 TREATMENT OF DISSOCIATING COMPOUNDS

In the case of dissociating or ionizing organic chemicals such as organic acids and bases, e.g., phenols, carboxylic acids and amines, it is desirable to calculate the concentrations of ionic and non-ionic species, and correct for this effect. A number of authors have discussed and reviewed the effect of pH and ionic strength on the distribution of these chemicals in the environment, including Westall et al. (1985), Schwarzenbach et al. (1988), Jafvert et al. (1990), Johnson and Westall (1990) and the text by Schwarzenbach, Gschwend and Imboden (1993).

A simple approach is suggested here for estimating the effect of pH on properties and environmental fate using the phenols as an example. A similar approach can be used for bases. The extent of dissociation is characterized by the acid dissociation constant, K_a, expressed as its negative logarithm, pK_a, which for most chloro-phenolic compounds range between 4.75 for pentachlorophenol and 10.2 to phenol, and between 10.0 and 10.6 for the alkylphenols. The dissolved concentration in water is thus the sum of the undissociated, parent or protonated compound and the dissociated phenolate ionic form. When the pK_a exceeds pH by 2 or more units, dissociation is 1% or less and for most purposes is negligible. The ratio of ionic to non-ionic or dissociated to undissociated species concentrations is given by,

$$ionic/non\text{-}ionic = 10^{(pH-pKa)} = I$$

The fraction ionic x_I is $I/(1 + I)$. The fraction non-ionic x_N is $1/(1 + I)$. For compounds such as pentachlorophenol in which pH generally exceeds pK_a, I and x_I can be appreciable, and there is an apparently enhanced solubility (Horvath and Getzen 1985, NRCC 1982, Yoshida et al. 1987, Arcand et al. 1995, Huang et al. 2000). There are other reports of pH effects on octanol-water partition coefficient (Kaiser and Valdmanis 1982, Westall et al. 1985, Lee et al. 1990, Smejtek and Wang 1993), soil sorption behavior (Choi and Amoine 1974, Lee et al. 1990, Schellenberg et al. 1984, Yoshida et al. 1987, Lee et al. 1990), bioconcentration and uptake kinetics to goldfish (Stehly and Hayton 1990) and toxicity to algae (Smith et al. 1987, Shigeoka et al. 1988).

The following treatment has been suggested by Shiu et al. (1994) and is reproduced briefly below. The simplest, "first-order" approach is to take into account the effect of dissociation by deducing the ratio of ionic to non-ionic species I, the fraction ionic x_I and the fraction non-ionic x_N for the chemical at both the pH and temperature of experimental data determination (I_D, x_{ID}, x_{ND}) and at the pH and temperature of the desired environmental simulation (I_E, x_{IE}, x_{NE}). It is assumed that dissociation takes place only in aqueous solution, not in air, organic carbon, octanol or lipid phases. Some ions and ion pairs are known to exist in the latter two phases, but there are insufficient data to justify a general procedure for estimating the quantities. No correction is made for the effect of cations other than H^+. This approach must be regarded as merely a first correction for the dissociation effect. An accurate evaluation should preferably be based on experimental determinations. The reported solubility C mol/m³ and K_{OW} presumably refer to the total of ionic and non-ionic forms, i.e., C_T and $K_{OW,T}$, at the pH of experimental determination, i.e.,

$$C_T = C_N + C_I$$

The solubility and K_{OW} of the non-ionic forms can be estimated as

$$C_N = C_T \cdot x_{ND}; \qquad K_{OW,N} = K_{OW,T}/x_{ND}$$

Vapor pressure P^S is not affected, but the apparent Henry's law constant H_T, must also be adjusted to H_T/x_N, being P^S/C_N or $P^S/(C_T \cdot x_N)$.

C_N and $K_{OW,N}$ can be applied to environmental conditions with a temperature adjustment if necessary. Values of I_E, x_{Ix} and x_{NE} can be deduced from the environmental pH and the solubility and K_{OW} of the total ionic and non-ionic forms calculated.

In the tabulated data presented in this handbook the aqueous solubilities selected are generally those estimated to be of the non-ionic form unless otherwise stated.

1.2.5 TREATMENT OF WATER-MISCIBLE COMPOUNDS

In the multimedia models used in this series of volumes, an air-water partition coefficient K_{AW} or Henry's law constant (H) is required and is calculated from the ratio of the pure substance vapor pressure and aqueous solubility. This method is widely used for hydrophobic chemicals but is inappropriate for water-miscible chemicals for which no solubility can be measured. Examples are the lower alcohols, acids, amines and ketones. There are reported "calculated" or "pseudo-solubilities" that have been derived from QSPR correlations with molecular descriptors for alcohols, aldehydes and amines (by Leahy 1986; Kamlet et al. 1987, 1988 and Nirmalakhandan and Speece 1988a,b). The obvious option is to input the H or K_{AW} directly. If the chemical's activity coefficient γ in water is known, then H can be estimated as $v_W \gamma P_L^S$, where v_W is the molar volume of water and P_L^S is the liquid vapor pressure. Since H can be regarded as P_L^S/C_L^S, where C_L^S is the solubility, it is apparent that $(1/v_W\gamma)$ is a "pseudo-solubility." Correlations and measurements of γ are available in the physical-chemical literature. For example, if γ is 5.0, the pseudo-solubility is 11100 mol/m³ since the molar volume of water v_W is 18×10^{-6} m³/mol or 18 cm³/mol. Chemicals with γ less than about 20 are usually miscible in water. If the liquid vapor pressure in this case is 1000 Pa, H will be 1000/11100 or 0.090 Pa·m³/mol and K_{AW} will be H/RT or 3.6×10^{-5} at 25°C. Alternatively, if H or K_{AW} is known, C_L^S can be calculated. It is possible to apply existing models to hydrophilic chemicals if this pseudo-solubility is calculated from the activity coefficient or from a known H (i.e., C_L^S, P_L^S/H or P_L^S or $K_{AW} \cdot RT$). This approach is used here. In the fugacity model illustrations all pseudo-solubilities are so designated and should not be regarded as real, experimentally accessible quantities.

1.2.6 TREATMENT OF PARTIALLY MISCIBLE SUBSTANCES

Most hydrophobic substances have low solubilities in water, and in the case of liquids, water is also sparingly soluble in the pure substance. Some substances such as butanols and chlorophenols display relatively high mutual solubilities. As temperature increases, these mutual solubilities increase until a point of total miscibility is reached at a critical solution temperature. Above this temperature, no mutual solubilities exist. A simple plot of solubility versus temperature thus ends at this critical point. At low temperatures near freezing, the phase diagram also become complex. Example of such systems have been reported for *sec*-butyl alcohol (2-butanol) by Ochi et al. (1996) and for chlorophenols by Jaoui et al. (1999).

1.2.7 TREATMENT OF GASES AND VAPORS

A volatile substance may exist in one of three broad classes that can be loosely termed gases, vapors and liquids.

A *gaseous* substance such as oxygen at normal environmental conditions exists at a temperature exceeding its critical temperature of 155 K. No vapor pressure can be defined or measured under this super-critical condition, thus no Henry's law constant can be calculated. Empirical data are required.

A substance such as propane with a critical temperature of 370 K has a measurable vapor pressure of 998000 Pa, or approximately 10 atm at 27°C, which exceeds atmospheric pressure of 101325 Pa, the boiling point being –42°C or 231 K. It is thus a *vapor* at normal temperatures and pressures. A Henry's law constant can be calculated from this vapor pressure and a solubility as described earlier.

Introduction

Most substances treated in this handbook are liquids or solids at environmental conditions; thus their boiling points exceed 25°C. Benzene, for example, has a critical temperature of 562 K, a boiling point of 80°C and a vapor pressure of 12700 Pa at 25°C.

When a solubility in water is measured and reported for gases and vapors an ambiguity is possible. For gases the solubility and the corresponding partial or total pressure in the gas phase must be reported since the solubility is dependent on this pressure as dictated by Henry's Law. For liquids and solids the solubility is presumably measured under conditions when the partial pressure equals the vapor pressure. For vapors such as propane the solubility can be measured either at a specified pressure (usually 1 atmosphere) or under high-pressure conditions (e.g., 10 atm) when the substance is a liquid. When calculating H or K_{AW} it is essential to use the correct pressure corresponding to the solubility measurement. Care must be exercised when treating substances with boiling points at or below environmental temperatures to ensure that the solubility is interpreted and used correctly.

1.2.8 Solids, Liquids and the Fugacity Ratio

Saturation properties such as solubility in water and vapor pressure can be measured directly for solids and liquids. For certain purposes it is useful to estimate the solubility that a solid substance would have if it were liquid at a temperature below the melting point. For example, naphthalene melts at 80°C and at 25°C the solid has a solubility in water of 33 g/m^3 and a vapor pressure of 10.9 Pa. If naphthalene was a liquid at 25°C it is estimated that its solubility would be 115 g/m^3 and its vapor pressure 38.1 Pa, both a factor of 3.5 greater. This ratio of solid to liquid solubilities or vapor pressures is referred to as the fugacity ratio. It is 1.0 at the melting point and falls, in this case at lower temperatures to 0.286 at 25°C.

Solubilities and vapor pressures of a solid substance in the liquid state are often reported for the following four reasons.

Measurements of gas chromatographic retention time are often used as a fast and easy method of estimating vapor pressure. These estimated pressures are related to the gas/substrate partition coefficient, which can be regarded as a ratio of solubility of the substance in the gas to that in the substrate, both solubilities being of the substance in the liquid state. As a result the estimated vapor pressures are of the liquid state. To obtain the solid vapor pressure requires multiplication by the fugacity ratio. It is important to establish if the estimated and reported property is of the vapor or liquid.

QSPRs in which solubilities and vapor pressures are correlated against molecular structure are done exclusively using the liquid state property. This avoids the complication introduced by the effect of fugacity ratio or melting point on the solid state property.

When a solid is in liquid solution it behaves according to its liquid state properties because it is in a liquid mixture. When applying Raoult's Law or similar expressions, the pure substance property is that of the liquid. Liquids such as crude oils and PCB mixtures consist largely of solid substances, but they are in the liquid state and generally unable to precipitate as solid crystals because of their low individual concentrations.

When estimating air-aerosol partitioning of gas phase substances such as PAHs, most of which are solids, it is usual to use the liquid state vapor pressure as the correlating parameter. This is because the PAH is effectively in a liquid-like state on or in the aerosol particle. It does not exist in crystalline form.

When calculating partition coefficients such as K_{AW}, K_{OW} or K_{OA} from solubilities it is immaterial if the values used are of solids or liquids, but it is erroneous to mix the two states, e.g., a solid solubility and a liquid vapor pressure.

The fugacity ratio F can be estimated at temperature T (K) from the expression

$$\ln F = -\Delta S\,(T_M - T)/RT$$

where ΔS is the entropy of fusion, T_M is the melting point, and R is the gas constant. ΔS is related to the measurable enthalpy of fusion ΔH at the melting point as $\Delta H/T_M$. The reader should use experimental data for ΔH, ΔS and melting point whenever possible. The most reliable method is to measure ΔH calorimetrically, calculate ΔS and use this value to estimate F. Only in the absence of ΔH data should a QSPR be used or Walden's Rule applied that ΔS is approximately 56.5 J/mol K. This assumption leads to the equations

$$F = \exp(-6.79(T_M/T - 1))$$

$$\log F = -0.01(T_M - 298)$$

F is thus 1.0 at the melting point, with lower values at lower temperatures. It is not applied at temperatures exceeding T_M. This issue is discussed by Mackay (2001), Tesconi and Yalkowsky (2000), Yalkowsky and Banerjee (1992) and Chickos et al. (1999).

1.2.9 Chemical Reactivity and Half-Lives

Characterization of chemical reactivity presents a challenging problem in environmental science in general and especially in handbooks. Whereas radioisotopes have fixed half-lives, the half-life of a chemical in the environment depends not only on the intrinsic properties of the chemical, but also on the nature of the environmental compartments. Factors such as sunlight intensity, hydroxyl radical concentration and the nature of the microbial community, as well as temperature, affect the chemical's half-life so it is impossible (and misleading) to document a single reliable half-life. We suggest that the best approach is to suggest a semi-quantitative classification of half-lives into groups or ranges, assuming average environmental conditions to apply. Obviously, a different class will generally apply between compartments such as in air and bottom sediment. In this compilation we use the following class ranges for chemical reactivity in a single medium such as water.

These times are divided logarithmically with a factor of approximately 3 between adjacent classes. With the present state of knowledge it is probably misleading to divide the classes into finer groupings; indeed, a single chemical is likely to experience half-lives ranging over three classes, depending on season. These half-lives apply to the reaction of the parent substance. Often a degradation product or metabolite is formed that is of environmental concern. Since it has different properties it requires separate assessment. The ultimate degradation to inorganic species may require a much longer time than is indicated by the initial half-life.

class	mean half-life (hours)	range (hours)
1	5 < 10	
2	17 (~ 1 day)	10–30
3	55 (~ 2 days)	30–100
4	170 (~ 1 week)	100–300
5	550 (~ 3 weeks)	300–1,000
6	1700 (~ 2 months)	1,000–3,000
7	5500 (~ 8 months)	3,000–10,000
8	17000 (~ 2 years)	10,000–30,000
9	55000 (~ 6 years)	30,000–100,000
10	> 11 years	> 100,000

When compiling the suggested reactivity classes, the authors have examined the available information on reaction rates of the chemical in each medium by all relevant processes. These were expressed as an overall half-life for transformation. The product of the half-life and the corresponding rate constant is ln2 or 0.693. For example, a chemical may be subject to biodegradation with a half-life of 20 days or 480 hours (rate constant 0.0014 h^{-1}) and simultaneous photolysis with a rate constant of 0.0011 h^{-1} (half-life 630 hours). The overall rate constant is thus 0.0025 h^{-1} and the half-life is 277 hours or 12 days. Data for homologous chemicals have also been compiled, and insights into the reactivity of various functional groups considered. In most cases a single reaction class is assigned to the series; in the above case, class 4 with a mean half-life of 170 hours would be chosen. These half-lives must be used with caution, and it is wise to test the implications of selecting longer and shorter half-lives.

The most reliable kinetic data are for atmospheric oxidation by hydroxyl radicals. These data are usually reported as second-order rate constants applied to the concentration of the chemical and the concentration of hydroxyl radicals (usually of the order of 10^6 radicals per cm^3). The product of the assumed hydroxyl radical concentration and the second-order rate constant is a first-order rate constant from which a half-life can be deduced.

Extensive research has been conducted into the atmospheric chemistry of organic chemicals because of air quality concerns. Recently, Atkinson and coworkers (1984, 1985, 1987, 1988, 1989, 1990, 1991), Altshuller (1980, 1991) and Sabljic and Güsten (1990) have reviewed the photochemistry of many organic chemicals of environmental interest for their gas phase reactions with hydroxyl radicals (OH), ozone (O_3) and nitrate radicals (NO_3) and have provided detailed information on reaction rate constants and experimental conditions, which allowed the estimation of atmospheric lifetimes. Klöpffer (1991) has estimated the atmospheric lifetimes for the reaction with OH radicals to range from 1 hour to 130 years, based on these reaction rate constants and an assumed constant concentration of OH

Introduction

radicals in air. As Atkinson (1985) has pointed out, the gas phase reactions with OH radicals are the major tropospheric loss process for the alkanes, haloalkanes, the lower alkenes, the aromatic hydrocarbons, and a majority of the oxygen-containing organics. In addition, photooxidation reactions with O_3 and NO_3 radicals can result in transformation of these compounds. The night-time reaction with NO_3 radicals may also be important (Atkinson and Carter 1984, Sabljic and Güsten 1990).

There are fewer studies on direct or indirect photochemical degradation in the water phase; however, Klöpffer (1991) had pointed out that the rate constant or lifetimes derived from these studies "is valid only for the top layer or surface waters." Mill (1982, 1989, 1993) and Mill and Mabey (1985) have estimated half-lives of various chemicals in aqueous solutions from their reaction rate constants with singlet oxygen, as well as photooxidation with hydroxyl and peroxy radicals. Buxton et al. (1988) gave a critical review of rate constants for reactions with hydrated electrons, hydrogen atoms and hydroxyl radicals in aqueous solutions. Mabey and Mill (1978) also reviewed the hydrolysis of organic chemicals in water under environmental conditions. Recently, Ellington and coworkers (1987a,b, 1988, 1989) also reported the hydrolysis rate constants in aqueous solutions for a variety of organic chemicals.

In most cases, a review of the literature suggested that reaction rates in water by chemical processes are 1 to 2 orders of magnitude slower than in air, but with biodegradation often being significant, especially for hydrocarbons and oxygen-containing chemicals. Generally, the water half-life class is three more than that in air, i.e., a factor of about 30 slower. Chemicals in soils tend to be shielded from photolytic processes, and they are less bioavailable, thus the authors have frequently assigned a reactivity class to soil of one more than that for water. Bottom sediments are assigned an additional class to that of soils largely on the basis that there is little or no photolysis, there may be lack of oxygen, and the intimate sorption to sediments renders the chemicals less bioavailable.

Because of the requirements of regulations for certain chemicals such as pesticides, extensive data usually exist on partitioning properties and reactivity or half-lives of active ingredients. In some cases these data have been peer-reviewed and published in the scientific literature, but often they are not generally available. A reader with interest in a specific pesticide can often obtain additional data from manufacturers or from registration literature, including accounts of chemical fate under field application conditions. Frequently these data are used as input to pesticide fate models, and the results of these modeling exercises may be available or published in the scientific literature.

The chemical reactivity of these substances is a topic which continues to be the subject of extensive research; thus there is often detailed, more recent information about the fate of chemical species which are of particular relevance to air or water quality. The reader is thus urged to consult the original and recent references because when considering the entire multimedia picture, it is impossible in a volume such as this to treat this subject in the detail it deserves.

1.3 EXPERIMENTAL METHODS

1.3.1 SOLUBILITY IN WATER AND pK_a

Most conventional organic contaminants are fairly hydrophobic and thus exhibit a low but measurable solubility in water. Solubility is often used to estimate the air-water partition coefficient or Henry's law constant, but this is not possible for miscible chemicals; indeed the method is suspect for chemicals of appreciable solubility in water, i.e., exceeding 1 g/100 g. Direct measurement of the Henry's law constant is thus required.

The conventional method of preparing saturated solutions for the determination of solubility is batch equilibration. An excess amount of solute chemical is added to water and equilibrium is achieved by shaking gently (generally referred as the "shake flask method") or slow stirring with a magnetic stirrer. The aim is to prevent formation of emulsions or suspensions and thus avoid extra experimental procedures such as filtration or centrifuging which may be required to ensure that a true solution is obtained. Experimental difficulties can still occur with sparingly soluble chemicals such as longer chain alkanes and polycyclic aromatic hydrocarbons (PAHs) because of the formation of emulsion or microcrystal suspensions. An alternative approach is to coat a thin layer of the chemical on the surface of the equilibration flask before water is added. An accurate "generator column" method is also used (Weil et al. 1974, May et al. 1978a,b) in which a column is packed with an inert solid support, such as glass beads and then coated with the solute chemical. Water is pumped through the column at a controlled, known flow rate to achieve saturation.

The method of concentration measurement of the saturated solution depends on the solute solubility and its chemical properties. Some common methods used for solubility measurement are listed below.

1. Gravimetric or volumetric methods (Booth and Everson 1948)
    An excess amount of solid compound is added to a flask containing water to achieve saturation solution by shaking, stirring, centrifuging until the water is saturated with solute and undissolved solid or liquid

residue appears, often as a cloudy phase. For liquids, successive known amounts of solute may be added to water and allowed to reach equilibrium, and the volume of excess undissolved solute is measured.
2. Instrumental methods
   a. UV spectrometry (Andrews and Keefer 1950, Bohon and Claussen 1951, Yalkowsky and Valvani 1976);
   b. Gas chromatographic analysis with FID, ECD or other detectors (McAuliffe 1966, Mackay et al. 1975, Chiou et al. 1982, Bowman and Sans 1983);
   c. Fluorescence spectrophotometry (Mackay and Shiu 1977);
   d. Interferometry (Gross and Saylor 1931);
   e. High-pressure liquid chromatography (HPLC) with I.R., UV or fluorescence detection (May et al. 1978a,b, Wasik et al. 1983, Shiu et al. 1988, Doucette and Andren 1988a);
   f. Liquid phase elution chromatography (Schwarz 1980, Schwarz and Miller 1980);
   g. Nephelometric methods (Davis and Parke 1942, Davis et al. 1942, Hollifield 1979);
   h. Radiotracer or liquid scintillation counting (LSC) method (Banerjee et al. 1980, Lo et al. 1986).

For most organic chemicals the solubility is reported at a defined temperature in distilled water. For substances which dissociate (e.g., phenols, carboxylic acids and amines) it is essential to report the pH of the determination because the extent of dissociation affects the solubility. It is common to maintain the desired pH by buffering with an appropriate electrolyte mixture. This raises the complication that the presence of electrolytes modifies the water structure and changes the solubility. The effect is usually "salting-out." For example, many hydrocarbons have solubilities in seawater about 75% of their solubilities in distilled water. Care must thus be taken to interpret and use reported data properly when electrolytes are present.

The dissociation constant K_a or its commonly reported negative logarithmic form pK_a is determined in principle by simultaneous measurement or deduction of the ionic and non-ionic concentrations and the pH of the solution.

The most common problem encountered with reported data is inaccuracy associated with very low solubilities, i.e., those less than 1.0 mg/L. Such solutions are difficult to prepare, handle and analyze, and reported data often contain appreciable errors.

As was discussed earlier, care must be taken when interpreting solubility data for gases, i.e., substances for which the temperature exceeds the boiling point. Solubility then depends on the pressure which may be atmospheric or the higher vapor pressure.

1.3.2 Vapor Pressure

In principle, the determination of vapor pressure involves the measurement of the saturation concentration or pressure of the solute in a gas phase. The most reliable methods involve direct determination of these concentrations, but convenient indirect methods are also available based on evaporation rate measurements or chromatographic retention times. Some methods and approaches are listed below.

   a. Static method, the equilibrium pressure in a thermostatic vessel is directly measured by use of pressure gauges: diaphragm gauge (Ambrose et al. 1975), Rodebush gauge (Sears and Hopke 1947), inclined-piston gauge (Osborn and Douslin 1975);
   b. Dynamic method (or boiling point) for measuring relatively high vapor pressure, eg., comparative ebulliometry (Ambrose 1981);
   c. Effusion methods, torsion and weight-loss (Balson 1947, Bradley and Cleasby 1953, Hamaker and Kerlinger 1969, De Kruif 1980);
   d. Gas saturation or transpiration methods (Spencer and Cliath 1970, 1972, Sinke 1974, Macknick and Prausnitz 1979, Westcott et al. 1981, Rordorf 1985a,b, 1986);
   e. Dynamic coupled-column liquid chromatographic method- a gas saturation method (Sonnefeld et al. 1983);
   f. Calculation from evaporation rates and vapor pressures of a reference compound (Gückel et al. 1974, 1982, Dobbs and Grant 1980, Dobbs and Cull 1982);
   g. Calculation from GC retention time data (Hamilton 1980, Westcott and Bidleman 1982, Bidleman 1984, Kim et al. 1984, Foreman and Bidleman 1985, Burkhard et al. 1985a, Hinckley et al. 1990).

The greatest difficulty and uncertainty arises when determining the vapor pressure of chemicals of low volatility, i.e., those with vapor pressures below 1.0 Pa. Vapor pressures are strongly dependent on temperature, thus accurate temperature control is essential. Data are often regressed against temperature and reported as Antoine or Clapeyron constants. Care

Introduction

must be taken if the Antoine or other equations are used to extrapolate data beyond the temperature range specified. It must be clear if the data apply to the solid or liquid phase of the chemical.

1.3.3 Octanol-Water Partition Coefficient K_{OW}

The experimental approaches are similar to those for solubility, i.e., employing shake flask or generator-column techniques. Concentrations in both the water and octanol phases may be determined after equilibration. Both phases can then be analyzed by the instrumental methods discussed above and the partition coefficient is calculated from the concentration ratio C_O/C_W. This is actually the ratio of solute concentration in octanol saturated with water to that in water saturated with octanol.

As with solubility, K_{OW} is a function of the presence of electrolytes and for dissociating chemicals it is a function of pH. Accurate values can generally be measured up to about 10^7, but accurate measurement beyond this requires meticulous technique. A common problem is the presence of small quantities of emulsified octanol in the water phase. The high concentration of chemical in that emulsion causes an erroneously high apparent water phase concentration.

Considerable success has been achieved by calculating K_{OW} from molecular structure; thus, there has been a tendency to calculate K_{OW} rather than measure it, especially for "difficult" hydrophobic chemicals. These calculations are, in some cases, extrapolations and can be in serious error. Any calculated log K_{OW} value above 7 should be regarded as suspect, and any experimental or calculated value above 8 should be treated with extreme caution.

For many hydrophilic compounds such as the alcohols, K_{OW} is low and can be less than 1.0, resulting in negative values of log K_{OW}. In such cases, care should be taken when using correlations developed for more hydrophobic chemicals since partitioning into biota or organic carbon phases may be primarily into aqueous rather than organic media.

Details of experimental methods are described by Fujita et al. (1964), Leo et al. (1971), Hansch and Leo (1979), Rekker (1977), Chiou et al. (1977), Miller et al. (1984, 1985), Bowman and Sans (1983), Woodburn et al. (1984), Doucette and Andren (1987), and De Bruijn et al. (1989).

1.3.4 Henry's Law Constant

The Henry's law constant is essentially an air-water partition coefficient which can be determined by measurement of solute concentrations in both phases. This raises the difficulty of accurate analytical determination in two very different media which usually requires different techniques. Accordingly, effort has been devoted to devising techniques in which concentrations are measured in only one phase and the other concentration is deduced from a mass balance. These methods are generally more accurate. The principal difficulty arises with hydrophobic, low-volatility chemicals which can establish only very small concentrations in both phases.

Henry's law constant can be regarded as a ratio of vapor pressure to solubility, thus it is subject to the same effects that electrolytes have on solubility. Temperature affects both properties. Some methods are as follows:

   a. Volatility measurement of dilute aqueous solutions (Butler et al. 1935, Burnett 1963, Buttery et al. 1969);
   b. Multiple equilibration method (McAuliffe 1971, Munz and Roberts 1987);
   c. Equilibrium batch stripping (Mackay et al. 1979, Dunnivant et al. 1988, Betterton and Hoffmann 1988, Zhou and Mopper 1990);
   d. GC-determined distribution coefficients (Leighton and Calo 1981);
   e. GC analysis of both air/water phases (Vejrosta et al. 1982, Jönsson et al. 1982);
   f. EPICS (Equilibrium Partitioning In Closed Systems) method (Lincoff and Gossett 1984, Gossett 1987, Ashworth et al. 1988);
   g. Wetted-wall column (Fendinger and Glotfelty 1988, 1989, 1990);
   h. Headspace analyses (Hussam and Carr 1985);
   i. Calculation from vapor pressure and solubility (Mackay and Shiu 1981);
   j. GC retention volume/time determined activity coefficient at infinite dilution γ^∞ (Karger et al. 1971a,b, Sugiyama et al. 1975, Tse et al. 1992).

When using vapor pressure and solubility data, it is essential to ensure that both properties apply to the same chemical phase, i.e., both are of the liquid, or of the solid. Occasionally, a solubility is of a solid while a vapor pressure is extrapolated from higher temperature liquid phase data.

As was discussed earlier under solubility, for miscible chemicals it is necessary to determine the Henry's law constant directly, since solubilities are not measurable.

1.3.5 Octanol-Air Partition Coefficient K_{OA}

As was discussed earlier the octanol-air partition coefficient is increasingly used as a descriptor of partitioning between the atmosphere and organic phases in soils and vegetation. A generator column technique is generally used in which an inert gas is flowed through a column containing a substance dissolved in octanol. The concentration in the equilibrated gas leaving the column is then measured (Harner and Mackay 1995). More recent methods have been described by Harner and Bidleman (1996) and Shoeib and Harner (2002). Su et al (2002) have described a GC retention time method.

1.4 QUANTITATIVE STRUCTURE-PROPERTY RELATIONSHIPS (QSPRs)

1.4.1 Objectives of QSPRs

Because of the large number of chemicals of actual and potential concern, the difficulties and cost of experimental determinations, and scientific interest in elucidating the fundamental molecular determinants of physical-chemical properties, considerable effort has been devoted to generating quantitative structure-property relationships (QSPRs). This concept of structure-property relationships or structure-activity relationships (QSARs) is based on observations of linear free-energy relationships, and usually takes the form of a plot or regression of the property of interest as a function of an appropriate molecular descriptor which can be calculated using only a knowledge of molecular structure or a readily accessible molecular property.

Such relationships have been applied to solubility, vapor pressure, K_{OW}, K_{AW}, K_{OA}, Henry's law constant, reactivities, bioconcentration data and several other environmentally relevant partition coefficients. Of particular value are relationships involving various manifestations of toxicity, but these are beyond the scope of this handbook. These relationships are valuable because they permit values to be checked for "reasonableness" and (with some caution) interpolation is possible to estimate undetermined values. They may be used (with extreme caution!) for extrapolation.

A large number of descriptors have been, and are being, proposed and tested. Dearden (1990) and the compilations by Karcher and Devillers (1990) and Hermens and Opperhuizen (1991) give comprehensive accounts of descriptors and their applications.

A valuable source of up-to-date information is the proceedings of the biennial QSAR conferences. The QSAR 2002 conference proceedings have been edited by Breton et al. (2003). A set of critical reviews has been edited by Walker (2003). Of particular note is the collection of estimation methods developed by the Syracuse Research Corporation with US EPA support and available on the internet at www.syrres.com under "estimation methods."

Among the most commonly used molecular descriptors are molecular weight and volume, the number of specific atoms (e.g., carbon or chlorine), surface areas (which may be defined in various ways), refractivity, parachor, steric parameters, connectivities and various topological parameters. Several quantum chemical parameters can be calculated from molecular orbital calculations including charge, electron density and superdelocalizability. It is likely that existing and new descriptors will continue to be tested, and that eventually a generally preferred set of readily accessible parameters will be adopted for routine use for correlating purposes.

From the viewpoint of developing quantitative correlations it is desirable to seek a linear relationship between descriptor and property, but a nonlinear or curvilinear relationship is adequate for illustrating relationships and interpolating purposes. In this handbook we have elected to use the simple descriptor of molar volume at the normal boiling point as estimated by the Le Bas method (Reid et al. 1987). This parameter is very easily calculated and proves to be adequate for the present purposes of plotting property versus relationship without seeking linearity.

The Le Bas method is based on a summation of atomic volumes with adjustment for the volume decrease arising from ring formation. The full method is described by Reid et al. (1987), but for the purposes of this compilation, the volumes and rules as listed in Table 1.3.1 are used.

Example: The experimental molar volume of chlorobenzene 115 cm³/mol (Reid et al. 1987). From the above rules, the Le Bas molar volume for chlorobenzene (C_6H_5Cl) is:

$$V = 6 \times 14.8 + 5 \times 3.7 + 24.6 - 15 = 117 \text{ cm}^3/\text{mol}$$

Accordingly, plots are presented at the end of each chapter for solubility, vapor pressure, K_{OW}, and Henry's law constant versus Le Bas molar volume.

Introduction

TABLE 1.3.1
Le Bas molar volume

	increment, cm³/mol
Carbon	14.8
Hydrogen	3.7
Oxygen	7.4
In methyl esters and ethers	9.1
In ethyl esters and ethers	9.9
Join to S, P, or N	8.3
Nitrogen	
Doubly bonded	15.6
In primary amines	10.5
In secondary amines	12.0
Bromine	27.0
Chlorine	24.6
Fluorine	8.7
Iodine	37.0
Sulfur	25.6
Rings	
Three-membered	–6.0
Four-membered	–8.5
Five-membered	–11.5
Six-membered	–15.0
Naphthalene	–30.0
Anthracene	–47.5

As was discussed earlier in Section 1.2.8 a complication arises in that two of these properties (solubility and vapor pressure) are dependent on whether the solute is in the liquid or solid state. Solid solutes have lower solubilities and vapor pressures than they would have if they had been liquids. The ratio of the (actual) solid to the (hypothetical supercooled) liquid solubility or vapor pressure is termed the fugacity ratio F and can be estimated from the melting point and the entropy of fusion. This "correction" eliminates the effect of melting point, which depends on the stability of the solid crystalline phase, which in turn is a function of molecular symmetry and other factors. For solid solutes, the correct property to plot is the calculated or extrapolated supercooled liquid solubility. This is calculated in this handbook using where possible a measured entropy of fusion, or in the absence of such data the Walden's Rule relationship suggested by Yalkowsky (1979) which implies an entropy of fusion of 56 J/mol·K or 13.5 cal/mol·K (e.u.)

$$F = C_S^S/C_L^S = P_S^S/P_L^S = \exp\{6.79(1 - T_M/T)\}$$

where C^S is solubility, P^S is vapor pressure, subscripts S and L refer to solid and liquid phases, T_M is melting point and T is the system temperature, both in absolute (K) units. The fugacity ratio is given in the data tables at 25°C, the usual temperature at which physical-chemical property data are reported. For liquids, the fugacity ratio is 1.0.

The usual approach is to compile data for the property in question for a series of structurally similar molecules and plot the logarithm of this property versus molecular descriptors, on a trial-and-error basis seeking the descriptor which best characterizes the variation in the property. It may be appropriate to use a training set to obtain a relationship and test this relationship on another set. Generally a set of at least ten data points is necessary before a reliable QSPR can be developed.

1.4.2 Examples of QSARs and QSPRs

There is a continuing effort to extend the long-established concept of quantitative-structure-activity-relationships (QSARs) to quantitative-structure-property relationships (QSPRs) to compute all relevant environmental physical-chemical properties (such as aqueous solubility, vapor pressure, octanol-water partition coefficient, Henry's law constant, bioconcentration factor (BCF), sorption coefficient and environmental reaction rate constants from molecular structure).

Examples are Burkhard (1984) and Burkhard et al. (1985a), who calculated solubility, vapor pressure, Henry's law constant, K_{OW} and K_{OC} for all PCB congeners. Hawker and Connell (1988) also calculated log K_{OW}; Abramowitz and Yalkowsky (1990) calculated melting point and solubility for all PCB congeners based on the correlation with total surface area (planar TSAs). Doucette and Andren (1988b) used six molecular descriptors to compute the K_{OW} of some chlorobenzenes, PCBs and PCDDs. Mailhot and Peters (1988) employed seven molecular descriptors to compute physical-chemical properties of some 300 compounds. Isnard and Lambert (1988, 1989) correlated solubility, K_{OW} and BCF for a large number of organic chemicals. Nirmalakhandan and Speece (1988a,b, 1989) used molecular connectivity indices to predict aqueous solubility and Henry's law constants for 300 compounds over 12 logarithmic units in solubility. Kamlet and co-workers (1986, 1987, 1988) have developed the "solvatochromic" parameters with the intrinsic molar volume to predict solubility, log K_{OW} and toxicity of organic chemicals. Warne et al. (1990) correlated solubility and K_{OW} for lipophilic organic compounds with 39 molecular descriptors and physical-chemical properties. Atkinson (1987, 1988) has used the structure-activity relationship (SAR) to estimate gas-phase reaction rate constants of hydroxyl radicals for organic chemicals. Mabey et al. (1984) have reviewed the estimation methods from SAR correlation for reaction rate constants and physical-chemical properties in environmental fate assessment. Other correlations are reviewed by Lyman et al. (1982) and Yalkowsky and Banerjee (1992). As Dearden (1990) has pointed out, "new parameters are continually being devised and tested, although the necessity of that may be questioned, given the vast number already available." It must be emphasized, however, that regardless of how accurate these predicted or estimated properties are claimed to be, ultimately they have to be confirmed or verified by experimental measurement.

A fundamental problem encountered in these correlations is the mismatch between the accuracy of experimental data and the molecular descriptors which can be calculated with relatively high precision, usually within a few percent. The accuracy may not always be high, but for correlation purposes precision is more important than accuracy. The precision and accuracy of the experimental data are often poor, frequently ranging over a factor of two or more. Certain isomers may yield identical descriptors, but have different properties. There is thus an inherent limit to the applicability of QSPRs imposed by the quality of the experimental data, and further efforts to improve descriptors, while interesting and potentially useful, may be unlikely to yield demonstrably improved QSPRs.

One of the most useful and accessible set of QSARs is that developed primarily by Howard and Meylan at the Syracuse Research Corporation, NY. These estimation methods are available as the EPISuite set from their website at www.syrres.com.

For correlation of solubility, the correct thermodynamic quantities for correlation are the activity coefficient γ, or the excess Gibbs free energy ΔG, as discussed by Pierotti et al. (1959) and Tsonopoulos and Prausnitz (1971). Examples of such correlations are given below.

1. Carbon number or carbon plus chlorine number (Tsonopoulos and Prausnitz 1971, Mackay and Shiu 1977);
2. Molar volume cm³/mol
   a. Liquid molar volume - from density (McAuliffe 1966, Lande and Banerjee 1981, Chiou et al. 1982, Abernethy et al. 1988, Wang et al. 1992);
   b. Molar volume by additive group contribution method, e.g., Le Bas method, Schroeder method (Reid et al. 1987, Miller et al. 1985);
   c. Intrinsic molar volume, V_I, cm³/mol - from van der Waals radius with solvatochromic parameters α and β (Leahy 1986, Kamlet et al. 1987, 1988);
   d. Characteristic molecular volume, m³/mol (McGowan and Mellors 1986);
3. Group contribution method (Irmann 1965, Korenman et al. 1971, Polak and Lu 1973, Klopman et al. 1992);
4. Molecular volume - Å³/molecule (cubic Angstrom per molecule)
   a. van der Waals volume (Bondi 1964);
   b. Total molecular volume (TMV) (Pearlman et al. 1984, Pearlman 1986);
5. Total surface area (TSA) - Å²/molecule (Hermann 1971, Amidon et al. 1975, Yalkowsky and Valvani 1976, Yalkowsky et al. 1979, Iwase et al. 1985, Pearlman 1986, Andren et al. 1987, Hawker and Connell 1988, Dunnivant et al. 1992);
6. Molecular connectivity indices (MCI) or χ (Kier and Hall 1976, Andren et al. 1987, Nirmalakhandan and Speece 1988b, 1989);
7. Boiling point (Almgren et al. 1979);
8. Melting point (Amidon and Williams 1982);
9. Melting point and TSA (Abramowitz and Yalkowsky 1990);
10. High-pressure liquid chromatography (HPLC) - retention data (Locke 1974, Whitehouse and Cooke 1982, Brodsky and Ballschmiter 1988);

Introduction

11. Adsorbability index (AI) (Okouchi et al. 1992);
12. Fragment solubility constants (Wakita et al. 1986).

Several workers have explored the linear relationship between octanol-water partition coefficient and solubility as a means of estimating solubility.

Hansch et al. (1968) established the linear free-energy relationship between aqueous and octanol-water partition of organic liquid. Others, such as Tulp and Hutzinger (1978), Yalkowsky et al. (1979), Mackay et al. (1980), Banerjee et al. (1980), Chiou et al. (1982), Bowman and Sans (1983), Miller et al. (1985), Andren et al. (1987) and Doucette and Andren (1988b) have all presented similar but modified relationships.

The UNIFAC (UNIQUAC Functional Group Activity Coefficient) group contribution (Fredenslund et al. 1975, Kikic et al. 1980, Magnussen et al. 1981, Gmehling et al. 1982 and Hansen et al. 1991) is widely used for predicting the activity coefficient in nonelectrolyte liquid mixtures by using group-interaction parameters. This method has been used by Kabadi and Danner (1979), Banerjee (1985), Arbuckle (1983, 1986), Banerjee and Howard (1988) and Al-Sahhaf (1989) for predicting solubility (as a function of the infinite dilution activity coefficient, γ^∞) in aqueous systems. Its performance is reviewed by Yalkowsky and Banerjee (1992).

HPLC retention time data have been used as a pseudo-molecular descriptor by Whitehouse and Cooke (1982), Hafkenscheid and Tomlinson (1981), Tomlinson and Hafkenscheid (1986) and Swann et al. (1983).

The octanol-water partition coefficient K_{OW} is widely used as a descriptor of hydrophobicity. Variation in K_{OW} is primarily attributable to variation in activity coefficient in the aqueous phase (Miller et al. 1985); thus, the same correlations used for solubility in water are applicable to K_{OW}. Most widely used is the Hansch-Leo compilation of data (Leo et al. 1971, Hansch and Leo 1979) and related predictive methods. Examples of K_{OW} correlations are:

1. Molecular descriptors
   a. Molar volumes: Le Bas method; from density; intrinsic molar volume; characteristic molecular volume (Abernethy et al. 1988, Chiou 1985, Kamlet et al. 1988, McGowan and Mellors 1986);
   b. TMV (De Bruijn and Hermens 1990);
   c. TSA (Yalkowsky et al. 1979, 1983, Pearlman 1980, 1986, Pearlman et al. 1984, Hawker and Connell 1988);
   d. Molecular connectivity indices (Doucette and Andren 1988b);
   e. Molecular weight (Doucette and Andren 1988b).
2. Group contribution methods
   a. π-constant or hydrophobic substituent method (Hansch et al. 1968, Hansch and Leo 1979, Doucette and Andren 1988b);
   b. Fragment constants or f-constant (Rekker 1977, Yalkowsky et al. 1983);
   c. Hansch and Leo's f-constant (Hansch and Leo 1979; Doucette and Andren 1988b).
3. From solubility - K_{OW} relationship
4. HPLC retention data
   a. HPLC-k' capacity factor (Könemann et al. 1979, McDuffie 1981);
   b. HPLC-RT retention time (Veith et al. 1979, Rapaport and Eisenreich 1984, Doucette and Andren 1988b);
   c. HPLC-RV retention volume (Garst 1984);
   d. HPLC-RT/MS HPLC retention time with mass spectrometry (Burkhard et al. 1985c).
5. Reversed-phase thin-layer chromatography (TLC) (Ellgehausen et al. 1981, Bruggeman et al. 1982).
6. Molar refractivity (Yoshida et al. 1983).
7. Combination of HPLC retention data and molecular connectivity indices (Finizio et al. 1994).
8. Molecular orbital methods (Reddy and Locke 1994).

As with solubility and octanol-water partition coefficient, vapor pressure can be estimated with a variety of correlations as discussed in detail by Burkhard et al. (1985a) and summarized as follows:

1. Interpolation or extrapolation from equation for correlating temperature relationships, e.g., the Clausius-Clapeyron, Antoine equations (Burkhard et al. 1985a);
2. Carbon or chlorine numbers (Mackay et al. 1980, Shiu and Mackay 1986);
3. Le Bas molar volume (Shiu et al. 1987, 1988);
4. Boiling point T_B and heat of vaporization ΔH_v (Mackay et al. 1982);
5. Group contribution method (Macknick and Prausnitz 1979);

6. UNIFAC group contribution method (Jensen et al. 1981, Yair and Fredenslund 1983, Burkhard et al. 1985a, Banerjee et al.1990);
7. Molecular weight and Gibbs' free energy of vaporization ΔG_v (Burkhard et al. 1985a);
8. TSA and ΔG_v (Amidon and Anik 1981, Burkhard et al. 1985a, Hawker 1989);
9. Molecular connectivity indices (Kier and Hall 1976, 1986, Burkhard et al. 1985a);
10. Melting point T_M and GC retention index (Bidleman 1984, Burkhard et al. 1985a);
11. Solvatochromic parameters and intrinsic molar volume (Banerjee et al. 1990).

As described earlier, Henry's law constants can be calculated from the ratio of vapor pressure and aqueous solubility. Henry's law constants do not show a simple linear pattern as solubility, K_{OW} or vapor pressure when plotted against simple molecular descriptors, such as numbers of chlorine or Le Bas molar volume, e.g., PCBs (Burkhard et al. 1985b), pesticides (Suntio et al. 1988), and chlorinated dioxins (Shiu et al. 1988). Henry's law constants can be estimated from:

1. UNIFAC-derived infinite dilution activity coefficients (Arbuckle 1983);
2. Group contribution and bond contribution methods (Hine and Mookerjee 1975, Meylan and Howard 1991);
3. Molecular connectivity indices (Nirmalakhandan and Speece 1988b, Sabljic and Güsten 1989, Dunnivant et al. 1992);
4. Total surface area - planar TSA (Hawker 1989);
5. Critical reviews by Mackay and Shiu 1981, Shiu and Mackay 1986 and Suntio et al. 1988.

For water-miscible compounds the use of aqueous solubility data is obviously impossible.

Bioconcentration Factors:

1. Correlation with K_{OW} (Neely et al. 1974, Könemann and van Leeuwen 1980, Veith et al. 1980, Chiou et al. 1977, Mackay 1982, Briggs 1981, Garten and Trabalka 1983, Davies and Dobbs 1984, Zaroogian et al. 1985, Oliver and Niimi 1988, Isnard and Lambert 1988);
2. Correlation with solubility (Kenaga 1980, Kenaga and Goring 1980, Briggs 1981, Garten and Trabalka 1983, Davies and Dobbs 1984, Isnard and Lambert 1988);
3. Correlation with K_{OC} (Kenaga 1980, Kenaga and Goring 1980, Briggs 1981);
4. Calculation with HPLC retention data (Swann et al. 1983);
5. Calculation with solvatochromic parameters (Hawker 1989, 1990b).

Sorption Coefficients:

1. Correlation with K_{OW} (Karickhoff et al. 1979, Schwarzenbach and Westall 1981, Mackay 1982, Oliver 1984);
2. Correlation with solubility (Karickhoff et al. 1979);
3. Molecular connectivity indices (Gerstl and Helling 1984; Sabljic 1984, 1987, Bahnick and Doucette 1988, Sabljic et al. 1989, Meylan et al. 1992);
4. Estimation from molecular connectivity index/fragment contribution method (Meylan et al. 1992, Lohninger 1994);
5. From HPLC retention data (Swann et al. 1983, Szabo et al. 1990).
6. Molecular orbital method (Reddy and Locke 1994).

Octanol-Air Partition coefficient.
The molecular descriptors used for K_{OW}, solubility in water and vapor pressure can potentially be applied to K_{OA}.

1.5 MASS BALANCE MODELS OF CHEMICAL FATE

1.5.1 Evaluative Environmental Calculations

When conducting assessments of how a chemical is likely to behave in the environment and especially how different chemicals behave in the same environment, there is incentive to standardize the evaluations using "evaluative" environmental models. The nature of these calculations has been described in a series of papers, notably Mackay (1979),

Introduction

Paterson and Mackay (1985), Mackay and Paterson (1990, 1991), and a recent text (Mackay 2001). Only the salient features are presented here. Three evaluations are completed for each chemical, namely the Level I, II and III fugacity calculations. These calculations can also be done in concentration format instead of fugacity, but for this type of evaluation the fugacity approach is simpler and more instructive. The mass balance models of the types described below can be downloaded for the web site www.trentu.ca/cemc

1.5.2 Level I Fugacity Calculations

The Level I calculation describes how a given amount of chemical partitions at equilibrium between six media: air, water, soil, bottom sediment, suspended sediment and fish. No account is taken of reactivity. Whereas most early evaluative environments have treated a one square kilometre region with about 70% water surface (simulating the global proportion of ocean surface), it has become apparent that a more useful approach is to treat a larger, principally terrestrial area similar to a jurisdictional region such as a US state. The area selected is 100,000 km^2 or 10^{11} m^2, which is about the area of Ohio, Greece or England. This environment was used in previous editions of this Handbook and is identical to the EQC or Equilibrium Criterion model described by Mackay et al. (1996).

The atmospheric height is selected as an arbitrary 1000 m reflecting that region of the troposphere which is most affected by local air emissions. A water surface area of 10% or 10,000 km^2 is used, with a water depth of 20 m. The water volume is thus 2×10^{11} m^3. The soil is viewed as being well mixed to a depth of 10 cm and is considered to be 2% organic carbon. It has a volume of 9×10^9 m^3. The bottom sediment has the same area as the water, a depth of 1 cm and an organic carbon content of 4%. It thus has a volume of 10^8 m^3.

For the Level I calculation both the soil and sediment are treated as simple solid phases with the above volumes, i.e., the presence of air or water in the pores of these phases is ignored.

Two other phases are included for interest. Suspended matter in water is often an important medium when compared in sorbing capacity to that of water. It is treated as having 20% organic carbon and being present at a volume fraction in the water of 5×10^{-6}, i.e., it is about 5 to 10 mg/L. The volume is thus 10^6 m^3. Fish is also included at an entirely arbitrary volume fraction of 10^{-6} and are assumed to contain 5% lipid, equivalent in sorbing capacity to octanol. The volume is thus 2×10^5 m^3. These two phases are small in volume and rarely contain an appreciable fraction of the chemical present, but it is in these phases that the highest concentration of chemical often exists.

Another phase which is introduced later in the Level III model is aerosol particles with a volume fraction in air of 2×10^{-11}, i.e., approximately 30 μg/m^3. Although negligible in volume, an appreciable fraction of the chemical present in the air phase may be associated with aerosols. Aerosols are not treated in Level I or II calculations because their capacity for the chemical at equilibrium is usually negligible when compared with soil.

These dimensions and properties are summarized in Tables 1.5.1 and 1.5.2. The user is encouraged to modify these dimensions to reflect conditions in a specific area of interest.

The amount of chemical introduced in the Level I calculation is an arbitrary 100,000 kg or 100 tonnes. If dispersed entirely in the air, this amount yields a concentration of 1 μg/m^3 which is not unusual for ubiquitous contaminants such as hydrocarbons. If dispersed entirely in the water, the concentration is a higher 500 μg/m^3 or 500 ng/L, which again is reasonable for a well-used chemical of commerce. The corresponding value in soil is about 0.0046 μg/g. Clearly for restricted chemicals such as PCBs, this amount is too large, but it is preferable to adopt a common evaluative amount

TABLE 1.5.1
Compartment dimensions and properties for Levels I and II calculations

Compartment	Air	Water	Soil	Sediment	Suspended sediment	Fish
Volume, V (m^3)	10^{14}	2×10^{11}	9×10^9	10^8	10^6	2×10^5
Depth, h (m)	1000	20	0.1	0.01	—	—
Area, A (m^2)	100×10^9	10×10^9	90×10^9	10×10^9	—	—
Fraction OC	—	—	0.02	0.04	0.2	—
Density, ρ (kg/m^3)	1.2	1000	2400	2400	1500	1000
Adv. Residence Time, t (hours)	100	1000	—	50,000	—	—
Adv. flow, G (m^3/h)	10^{12}	2×10^8	—	2000	—	—

TABLE 1.5.2
Bulk compartment dimensions and volume fractions (v) for Level III calculations

Compartment		Volume
Air	Total volume	10^{14} m^3 (as above)
	Air phase	10^{14} m^3
	Aerosol phase	2000 m^3 (v = 2 × 10^{-11})
Water	Total volume	2×10^{11} m^3
	Water phase	2×10^{11} m^3 (as above)
	Suspended sediment phase	10^6 m^3 (v = 5 × 10^{-6})
	Fish phase	2×10^5 m^3 (v = 1 × 10^{-6})
Soil	Total volume	18×10^9 m^3
	Air phase	3.6×10^9 m^3 (v = 0.2)
	Water phase	5.4×10^9 m^3 (v = 0.3)
	Solid phase	9.0×10^9 m^3 (v = 0.5) (as above)
Sediment	Total volume	500×10^6 m^3
	Water phase	400×10^6 m^3 (v = 0.8)
	Solid phase	100×10^6 m^3 (v = 0.2) (as above)

TABLE 1.5.3
Equations for phase Z values used in Levels I, II and bulk phase values used in Level III

Compartment	Z values
Air	$Z_1 = 1/RT$
Water	$Z_2 = 1/H = C^S/P^S$
Soil	$Z_3 = Z_2 \cdot \rho_3 \cdot \phi_3 \cdot K_{OC}/1000$
Sediment	$Z_4 = Z_2 \cdot \rho_4 \cdot \phi_4 \cdot K_{OC}/1000$
Suspended Sediment	$Z_5 = Z_2 \cdot \rho_5 \cdot \phi_5 \cdot K_{OC}/1000$
Fish	$Z_6 = Z_2 \cdot \rho_6 \cdot L \cdot K_{OW}/1000$
Aerosol	$Z_7 = Z_1 \cdot 6 \times 10^6/P_L^S$ or $0.1 \, Z_1 \, K_{OA}$

where
R = gas constant (8.314 J/mol·K)
T = absolute temperature (K)
C^S = solubility in water (mol/m^3)
P^S = vapor pressure (Pa)
H = Henry's law constant (Pa·m^3/mol)
P_L^S = liquid vapor pressure (Pa)
K_{OA} = octanol-air partition coefficient
K_{OW} = octanol-water partition coefficient
ρ_i = density of phase i (kg/m^3)
ϕ_i = mass fraction organic-carbon in phase i (g/g)
L = lipid content of fish

Note for solids $P_L^S = P_S^S/\exp\{6.79(1 - T_M/T)\}$, where T_M is melting point (K) of the solute and T is 298 K. An experimental entropy of fusion should be used if available.

for all substances. No significance should, of course, be attached to the absolute values of the concentrations which are deduced from this arbitrary amount. Only the relative values have significance.

The Level I calculation proceeds by deducing the fugacity capacities or Z values for each medium (see Table 1.5.3), following the procedures described by Mackay (2001). These working equations show the necessity of having data on molecular mass, water solubility, vapor pressure, and octanol-water partition coefficient. The fugacity f (Pa) common to all media is deduced as

$$f = M/\Sigma V_i Z_i$$

Introduction

where M is the total amount of chemical (mol), V_i is the medium volume (m^3) and Z_i is the corresponding fugacity capacity for the chemical in each medium. It is noteworthy that Z values contain all the necessary partition information. The partition coefficient K_{12} is simply the ratio of Z values, i.e., Z_1/Z_2. Definition of the Z values starts in the air compartment then proceeds to other compartments using the appropriate partition coefficients.

The molar concentration C (mol/m^3) can then be deduced as Zf mol/m^3 or as WZf g/m^3 or 1000 WZf/ρ μg/g, where ρ is the phase density (kg/m^3) and W is the molecular mass (g/mol). The amount m_i in each medium is $C_i V_i$ mol, and the total in all media is M mol. The information obtained from this calculation includes the concentrations, amounts and distribution.

Note that this simple treatment assumes that the soil and sediment phases are entirely solid, i.e., there are no air or water phases present to "dilute" the solids. Later in the Level III calculation these phases and aerosols are included (see Table 1.5.4).

Correction for Dissociation

As discussed earlier in Section 1.2.4, for dissociating or ionizing organic chemicals in aqueous solution, it is necessary to consider the effect of pH and thus the degree of dissociation, and to calculate the concentrations of both ionic and non-ionic species. The EQC model does not address dissociation.

The Z values are calculated using the conventional equations at the pH of the experimental data (i.e., the system pH). The total Z value in water is then separated into its ionic and non-ionic contributions, i.e., fractions of I/(I + 1) and 1/(I + 1). The Z value for the non-ionic form in water is assumed to apply at all pHs i.e., including the environmental pH, but an additional and possibly different ionic Z value in water is deduced at the environmental pH using I calculated at that pH. The total Z values in water are then calculated. Z values in other media are unaffected.

The calculation is illustrated in Table 1.5.5 for pentachlorophenol. The experimental aqueous solubility is 14.0 g/m^3 at a pH of 5.1. The environmental pH is 7. Higher environmental pH increases the extent of dissociation, thus increasing the Z value in water, increasing the apparent solubility, decreasing the apparent K_{OW} and Henry's law constant and the air-water partition coefficient, and decreasing the soil-water partition coefficient.

Note: At pH of 5.1, K_{OW} is 112200 and is the ratio of concentration in octanol to total concentration in water comprising fractions 1/(1 + I) or 1/(1 + 2.29) or 0.304 of neutral and 0.696 of ionic species. K_{OW} is thus 112200/0.304 or 369000 for the neutral species and zero for the ionic species. For the neutral species K_{OC} is assumed to be 0.41·K_{OW} or 151300, thus K_P is 151300 × 0.02 L/kg, i.e., 3027 for a soil of 2% organic carbon. K_{SW} is thus 3027 × 2.4 where 2.4 is the solid density (kg/L) or 7265. Z_S for the neutral species is thus 7265 × Z_W or 27970. At pH of 7, the neutral species Z values are unaffected, but the Z value for water increases to 704 because of the greater extent of dissociation. K_{SW} thus decreases to 27970/704 or 39.72.

TABLE 1.5.4
Bulk phase Z values, Z_{Bi} deduced as $\Sigma v_i Z_i$, in which the coefficients, e.g., 2×10^{-11}, are the volume fractions v_i of each pure phase as specified in Table 1.5.2

Compartment	Bulk Z values	
Air	$Z_{B1} = Z_1 + 2 \times 10^{-11} Z_7$	(approximately 30 μg/m^3 aerosols)
Water	$Z_{B2} = Z_2 + 5 \times 10^{-6} Z_5 + 1 \times 10^{-6} Z_6$	(5 ppm solids, 1 ppm fish by volume)
Soil	$Z_{B3} = 0.2 Z_1 + 0.3 Z_2 + 0.5 Z_3$	(20% air, 30% water, 50% solids)
Sediment	$Z_{B4} = 0.8 Z_2 + 0.2 Z_4$	(80% water, 20% solids)

TABLE 1.5.5
Calculated Z values at different experimental and environmental pHs of pentachlorophenol. Z values at 25°C, log K_{OW} is 5.05, pK_a 4.74, at data pH of 5.1 and environmental pH of 7.0 for air, water and soil of fraction organic carbon 0.02 and density of soil 2.4 kg/L

	At data pH of 5.1 (I = 2.29)			At environ. pH of 7 (I = 182)		
	Neutral	Ionic	Total	Neutral	Ionic	Total
Air	4.03 × 10^{-4}	0	4.03 × 10^{-4}	4.03 × 10^{-4}	0	4.03 × 10^{-4}
Water	3.85	8.82	12.67	3.85	700.4	704.2
Soil solids	27970	0	27970	27970	0	27970

TABLE 1.5.6
Calculated Z_W values and some partition coefficients at different environmental pHs for pentachlorophenol (PCP), 2,4-dichlorophenol (2,4-DCP), 2,4,6-trichlorophenol (2,4,6-TCP) and p-cresol at 25°C. K_{AW} is the air-water partition coefficient and K_{SW} is the soil-water partition coefficient

	Z values in water				Partitioning properties			
At pH	Neutral	Ionic	Total Z_W	Fraction x_N	S_T g/m³	H_T Pa·m³/mol	K_{AW}	K_{SW}
PCP								
4	3.849	0.7004	4.549	0.846	16.55	0.224	8.9×10^{-5}	6147
6	3.849	70.04	73.89	0.052	268.8	0.0135	5.46×10^{-6}	378.5
7	3.849	700.4	704.2	0.0055	2562	0.00142	5.73×10^{-7}	39.7
2,4,6-TCP								
4	1.7677	0.0140	1.7817	0.992	434	0.5612	2.26×10^{-4}	105.2
6	1.7677	1.4041	3.1718	0.557	772	0.315	1.272×10^{-4}	59.09
7	1.7677	14.041	15.8088	0.118	3644	0.172	6.945×10^{-5}	11.86
2,4-DCP								
4	3.063	0.000386	3.063	1.0	6000	0.326	0.000132	31.24
6	3.063	0.0386	3.101	0.988	6073	0.322	0.000130	30.85
7	3.063	0.386	3.448	0.888	6760	0.290	0.000117	27.75
p-Cresol								
4	11.97	0	11.948	1.0	22000	0.0836	3.37×10^{-5}	1.968
7	11.97	0.0066	11.975	1.0	22000	0.0836	3.35×10^{-5}	1.968

This is further demonstrated in Table 1.5.6 which shows the effects of environmental pH on the partitioning behavior of 2,4-dichlorophenol (pK_a = 7.90, solubility of 6000 g/m³ at pH of 5.1 and log K_{OW} = 3.20), 2,4,6-trichlorophenol (pK_a = 6.10, solubility of 430 g/m³ at pH of 5.1 and log K_{OW} = 3.69), pentachlorophenol (pK_a = 4.74, solubility of 14.0 g/m³ at pH of 5.1 and log K_{OW} = 5.05) and p-cresol (pK_a = 10.26, a solubility of 22000 g/m³ and log K_{OW} = 2.0) in the multimedia environment at 25°C. For environmental pH from 4 to 7, there is no significant effect for p-cresol (or for chemicals for which $pK_a \gg pH$), very little effect for 2,4-dichlorophenol (and chemicals with pK_a ranging between 7–10). There is some effect on 2,4,6-trichlorophenol (and chemicals with pK_a of 6–7) and a large effect for pentachlorophenol.

A similar treatment can be applied to other dissociating compounds such as the carboxylic acids, nitrophenols. For bases such as amines the pK_a is defined as (14 - pK_b), and the extent of dissociation is estimated as above.

1.5.3 Level II Fugacity Calculations

The Level II calculation simulates a situation in which a chemical is continuously discharged into the multimedia environment and achieves a steady-state and equilibrium condition, at which input and output rates are equal. The task is to deduce the rates of loss by reaction and advection and the prevailing concentrations and masses.

The reaction rate data developed for each chemical in the tables are used to select a reactivity class as described earlier, and hence a first-order rate constant for each medium. Often these rates are in considerable doubt; thus the quantities selected should be used with extreme caution because they may not be widely applicable. The rate constants k_i h⁻¹ are used to calculate reaction D values for each medium D_{Ri} as $V_i Z_i k_i$. The rate of reactive loss is then $D_{Ri} f$ mol/h.

For advection, it is necessary to select flow rates. This is conveniently done in the form of advective residence times, t in hour (h); thus the advection rate G_i is V_i/t m³/h for each medium. For air, a residence time of 100 hours is used (approximately 4 days), which is probably too long for the geographic area considered, but shorter residence times tend to cause air advective loss to be a dominant mechanism. For water, a figure of 1000 hours (42 days) is used, reflecting a mixture of rivers and lakes. For sediment burial (which is treated as an advective loss), a time of 50,000 hours or 5.7 years is used. Only for very persistent, hydrophobic chemicals is this process important. No advective loss from soil is included. The D value for loss by advection D_{Ai} is $G_i Z_i$, and the rates are $D_{Ai} f$ mol/h.

Introduction

There may thus be losses caused by both reaction and advection D values for the four primary media. These loss processes are not included for fish or suspended matter. At steady-state and equilibrium conditions, the input rate E mol/h can be equated to the sum of the output rates, from which the common fugacity can be calculated as follows

$$E = f \cdot \Sigma D_{Ai} + f \cdot \Sigma D_{Ri}$$

thus,

$$f = E/(\Sigma D_{Ai} + \Sigma D_{Ri})$$

The common assumed emission rate is 1000 kg/h or 1 tonne/h. To achieve an amount equivalent to the 100 tonnes in the Level I calculation requires an overall residence time of 100 hours. Again, the concentrations and amounts m_i and Σm_i or M can be deduced, as well as the reaction and advection rates. These rates obviously total to give the input rate E. Of particular interest are the relative rates of these loss processes, and the overall persistence or residence time, which is calculated as

$$t_O = M/E$$

where M is the total amount present. It is also useful to calculate a reaction and an advection persistence t_R and t_A as

$$t_R = M/\Sigma D_{Ri}f \qquad t_A = M/\Sigma D_{Ai}f$$

Obviously,

$$1/t_O = 1/t_R + 1/t_A$$

These persistences indicate the likelihood of the chemical being lost by reaction as distinct from advection. The percentage distribution of chemical between phases is identical to that in Level I. A pie chart depicting the distribution of losses can be drawn.

1.5.4 Level III Fugacity Calculations

Whereas the Levels I and II calculations assume equilibrium to prevail between all media, this is recognized as being excessively simplistic and even misleading. In the interests of algebraic simplicity, only the four primary media are treated for this level. The task is to develop expressions for intermedia transport rates by the various diffusive and non-diffusive processes as described by Mackay (2001). This is done by selecting values for 12 intermedia transport velocity parameters which have dimensions of velocity (m/h or m/year), are designated as U_i m/h and are applied to all chemicals. These parameters are used to calculate seven intermedia transport D values.

It is desirable to calculate new "bulk phase" Z values for the four primary media which include the contribution of dispersed phases within each medium as described by Mackay and Paterson (1991) and as listed earlier. The air is now treated as an air-aerosol mixture, water as water plus suspended particles and fish, soil as solids, air and water, and sediment as solids and porewater. The Z values thus differ from the Level I and Level II "pure phase" values. The necessity of introducing this complication arises from the fact that much of the intermedia transport of the chemicals occurs in association with the movement of chemical in these dispersed phases. To accommodate this change the same volumes of the soil solids and sediment solids are retained, but the total phase volumes are increased. These Level III volumes are also given in Table 1.5.2. The reaction and advection D values employ the generally smaller bulk phase Z values but the same residence times; thus the G values are increased and the D values are generally larger.

Intermedia D Values

The justification for each intermedia D value follows. It is noteworthy that, for example, air-to-water and water-to-air values differ because of the presence of one-way non-diffusive processes. A fuller description of the background to these calculations is given by Mackay (2001).

1. Air to Water (D_{12})

Four processes are considered: diffusion (absorption), dissolution in rain of gaseous chemical, and wet and dry deposition of particle-associated chemical.

For diffusion, the conventional two-film approach is taken with water-side (k_W) and air-side (k_A) mass transfer coefficients (m/h) being defined. Values of 0.05 m/h for k_W and 5 m/h for k_A are used. The absorption D value is then

$$D_{VW} = 1/[1/(k_A A_W Z_1) + 1/(k_W A_W Z_2)]$$

where A_W is the air-water area (m²) and Z_1 and Z_2 are the pure air and water Z values. The velocities k_A and k_W are designated as U_1 and U_2.

For rain dissolution, a rainfall rate of 0.876 m/year is used, i.e., U_R or U_3 is 10^{-4} m/h. The D value for rain dissolution D_{RW} is then

$$D_{RW} = U_R A_W Z_2 = U_3 A_W Z_2$$

For wet deposition, it is assumed that the rain scavenges Q (the scavenging ratio) or about 200,000 times its volume of air. Using a particle concentration (volume fraction) v_Q of 2×10^{-11}, this corresponds to the removal of Qv_Q or 4×10^{-6} volumes of aerosol per volume of rain. The total rate of particle removal by wet deposition is then $Qv_Q U_R A_W$ m³/h, thus the wet "transport velocity" $Qv_Q U_R$ is 4×10^{-10} m/h.

For dry deposition, a typical deposition velocity U_Q of 10 m/h is selected yielding a rate of particle removal of $U_Q v_Q A_W$ or $2 \times 10^{-10} A_W$ m³/h corresponding to a transport velocity of 2×10^{-10} m/h. Thus,

$$U_4 = Qv_Q U_R + U_Q v_Q = v_Q(QU_R + U_Q)$$

The total particle transport velocity U_4 for wet and dry deposition is thus 6×10^{-10} m/h (67% wet and 33% dry) and the total D value D_{QW} is

$$D_{QW} = U_4 A_W Z_7$$

where Z_7 is the aerosol Z value.

The overall D value is given by

$$D_{12} = D_{VW} + D_{RW} + D_{QW}$$

2. Water to Air (D_{21})

Evaporation is treated as the reverse of absorption; thus D_{21} is simply D_{VW} as before.

3. Air to Soil (D_{13})

A similar approach is adopted as for air-to-water transfer. Four processes are considered with rain dissolution (D_{RS}) and wet and dry deposition (D_{QS}) being treated identically except that the area term is now the air-soil area A_S.

For diffusion, the approach of Jury et al. (1983, 1984a,b,c) is used as described by Mackay and Stiver (1991) and Mackay (1991) in which three diffusive processes are treated. The air boundary layer is characterized by a mass transfer coefficient k_S or U_7 of 5 m/h, equal to that of the air-water mass transfer coefficient k_A used in D_{12}.

For diffusion in the soil air-pores, a molecular diffusivity of 0.02 m²/h is reduced to an effective diffusivity using a Millington-Quirk type of relationship by a factor of about 20 to 10^{-3} m²/h. Combining this with a path length of 0.05 m gives an effective air-to-soil mass transfer coefficient k_{SA} of 0.02 m/h, which is designated as U_5.

Similarly, for diffusion in water a molecular diffusivity of 2×10^{-6} m²/h is reduced by a factor of 20 to an effective diffusivity of 10^{-7} m²/h, which is combined with a path length of 0.05 m to give an effective soil-to-water mass transfer coefficient of k_{SW} 2×10^{-6} m/h.

It is probable that capillary flow of water contributes to transport in the soil. For example, a rate of 7 cm/year would yield an equivalent water velocity of 8×10^{-6} m/h, which exceeds the water diffusion rate by a factor of four. For illustrative purposes we thus select a water transport velocity or coefficient U_6 in the soil of 10×10^{-6} m/h, recognizing that this will vary with rainfall characteristics and soil type. These soil processes are in parallel with boundary layer diffusion in series, so the final equations are

$$D_{VS} = 1/[1/D_S + 1/(D_{SW} + D_{SA})]$$

Introduction

where

$$D_S = U_7 A_S Z_1 \quad (U_7 = 5 \text{ m/h})$$

$$D_{SW} = U_6 A_S Z_2 \quad (U_6 = 10 \times 10^{-6} \text{ m/h})$$

$$D_{SA} = U_5 A_S Z_1 \quad (U_5 = 0.02 \text{ m/h})$$

where A_S is the soil horizontal area.

Air-soil diffusion thus appears to be much slower than air-water diffusion because of the slow migration in the soil matrix. In practice, the result will be a nonuniform composition in the soil with the surface soil (which is much more accessible to the air than the deeper soil) being closer in fugacity to the atmosphere.

The overall D value is given as

$$D_{13} = D_{VS} + D_{QS} + D_{RS}$$

4. Soil to Air (D_{31})

Evaporation is treated as the reverse of absorption, thus the D value is simply D_{VS}.

5. Water to Sediment (D_{24})

Two processes are treated, diffusion and deposition.

Diffusion is characterized by a mass transfer coefficient U_8 of 10^{-4} m/h, which can be regarded as a molecular diffusivity of 2×10^{-6} m²/h divided by a path length of 0.02 m. In practice, bioturbation may contribute substantially to this exchange process, and in shallow water current-induced turbulence may also increase the rate of transport. Diffusion in association with organic colloids is not included. The D value is thus given as $U_8 A_W Z_2$.

Deposition is assumed to occur at a rate of 5000 m³/h, which corresponds to the addition of a depth of solids of 0.438 cm/year; thus 43.8% of the solids resident in the accessible bottom sediment is added each year. This rate is about 12 cm³/m²·day, which is high compared to values observed in large lakes. The velocity U_9, corresponding to the addition of 5000 m³/h over the area of 10^{10} m², is thus 5×10^{-7} m/h.

It is assumed that of this 5000 m³/h deposited, 2000 m³/h or 40% is buried (yielding the advective flow rate in Table 1.5.1), 2000 m³/h or 40% is resuspended (as discussed later) and the remaining 20% is mineralized organic matter. The organic carbon balance is thus only approximate.

The transport velocities are thus:

deposition U_9 5.0×10^{-7} m/h or 0.438 cm/y

resuspension U_{10} 2.0×10^{-7} m/h or 0.175 cm/y

burial U_B 2.0×10^{-7} m/h or 0.175 cm/y

(included as an advective residence time of 50,000 h)

The water-to-sediment D value is thus

$$D_{24} = U_8 A_W Z_2 + U_9 A_W Z_5$$

where Z_5 is the Z value of the particles in the water column.

6. Sediment to Water (D_{42})

This is treated similarly to D_{24} giving:

$$D_{42} = U_8 A_W Z_2 + U_{10} A_W Z_4$$

where U_{10} is the sediment resuspension velocity of 2.0×10^{-7} m/h and Z_4 is the Z value of the sediment solids.

7. Sediment Advection or Burial (D_{A4})

This D value is $U_B A_W Z_4$, where U_B, the sediment burial rate, is 2.0×10^{-7} m/h. It can be viewed as $G_B Z_{B4}$, where G_B is the total burial rate specified as V_S/t_B where t_B (residence time) is 50,000 h, and V_S (the sediment volume) is the product of sediment depth (0.01 cm) and area A_W. Z_4, Z_{B4} are the Z values of the sediment solids and of the bulk sediment, respectively. Since there are 20% solids, Z_{B4} is about $0.2 Z_4$. There is a slight difference between these approaches because in the advection approach (which is used here) there is burial of water as well as solids.

8. Soil to Water Run-Off (D_{32})

It is assumed that there is run-off of water at a rate of 50% of the rain rate, i.e., the D value is

$$D = 0.5\, U_3 A_S Z_2 = U_{11} A_S Z_2$$

thus the transport velocity term U_{11} is $0.5 U_3$ or 5×10^{-5} m/h.

For solids run-off it is assumed that this run-off water contains 200 parts per million by volume of solids; thus the corresponding velocity term U_{12} is $200 \times 10^{-6} U_{11}$, i.e., 10^{-8} m/h. This corresponds to the loss of soil at a rate of about 0.1 mm per year. If these solids were completely deposited in the aquatic environment (which is about 1/10th the soil area), they would accumulate at about 0.1 cm per year, which is about a factor of four less than the deposition rate to sediments. The implication is that most of this deposition is of naturally generated organic carbon and from sources such as bank erosion.

Summary

The twelve intermedia transport parameters are listed in Table 1.5.7 and the equations are summarized in Table 1.5.8.

Algebraic Solution

Four mass balance equations can be written, one for each medium, resulting in a total of four unknown fugacities, enabling simple algebraic solution as shown in Table 1.5.9. From the four fugacities, the concentration, amounts and rates of all transport and transformation processes can be deduced, yielding a complete mass balance.

The new information from the Level III calculations are the intermedia transport data, i.e., the extent to which chemical discharged into one medium tends to migrate into another. This migration pattern depends strongly on the proportions of the chemical discharged into each medium; indeed, the relative amounts in each medium are largely a reflection of the locations of discharge. It is difficult to interpret these mass balance diagrams because, for example, chemical depositing from air to water may have been discharged to air, or to soil from which it evaporated, or even to water from which it is cycling to and from air.

To simplify this interpretation, it is best to conduct three separate Level III calculations in which unit amounts (1000 kg/h) are introduced individually into air, soil and water. Direct discharges to sediment are unlikely and are not

TABLE 1.5.7
Intermedia transport parameters

U		m/h	m/year
1	Air side, air-water MTC*, k_A	5	43,800
2	Water side, air-water MTC, k_W	0.05	438
3	Rain rate, U_R	10^{-4}	0.876
4	Aerosol deposition	6×10^{-10}	5.256×10^{-6}
5	Soil-air phase diffusion MTC, k_{SA}	0.02	175.2
6	Soil-water phase diffusion MTC, k_{SW}	10×10^{-6}	0.0876
7	Soil-air boundary layer MTC, k_S	5	43,800
8	Sediment-water MTC	10^{-4}	0.876
9	Sediment deposition	5.0×10^{-7}	0.00438
10	Sediment resuspension	2.0×10^{-7}	0.00175
11	Soil-water run-off	5.0×10^{-5}	0.438
12	Soil-solids run-off	10^{-8}	8.76×10^{-5}

*MTC is mass transfer coefficient. Scavenging ratio Q is 2×10^5, dry deposition velocity U_Q is 10 m/h and sediment burial rate U_B is 2.0×10^{-7} m/h

Introduction

TABLE 1.5.8
Intermedia transport D value equations

Air-Water	$D_{12} = D_{VW} + D_{RW} + D_{QW}$
	$D_{VW} = A_W/(1/U_1Z_1 + 1/U_2Z_2)$
	$D_{RW} = U_3A_WZ_2$
	$D_{QW} = U_4A_WZ_7$
Water-Air	$D_{21} = D_{VW}$
Air-Soil	$D_{13} = D_{VS} + D_{RS} + D_{QS}$
	$D_{VS} = 1/(1/D_S + 1/(D_{SW} + D_{SA}))$
	$D_S = U_7A_SZ_1$
	$D_{SA} = U_5A_SZ_1$
	$D_{SW} = U_6A_SZ_2$
	$D_{RS} = U_3A_SZ_2$
	$D_{QS} = U_4A_SZ_7$
Soil-Air	$D_{31} = D_{VS}$
Water-Sediment	$D_{24} = U_8A_WZ_2 + U_9A_WZ_5$
Sediment-Water	$D_{42} = U_8A_WZ_2 + U_{10}A_WZ_4$
Soil-Water	$D_{32} = U_{11}A_SZ_2 + U_{12}A_SZ_3$

TABLE 1.5.9
Level III solutions to mass balance equations

Compartment	Mass balance equations
Air	$E_1 + f_2D_{21} + f_3D_{31} = f_1D_{T1}$
Water	$E_2 + f_1D_{12} + f_3D_{32} + f_4D_{42} = f_2 D_{T2}$
Soil	$E_3 + f_1D_{13} = f_3D_{T3}$
Sediment	$E_4 + f_2D_{24} = f_4D_{T4}$
where	E_i is discharge rate, E_4 usually being zero.
	$D_{T1} = D_{R1} + D_{A1} + D_{12} + D_{13}$
	$D_{T2} = D_{R2} + D_{A2} + D_{21} + D_{23} + D_{24}$, ($D_{23} = 0$)
	$D_{T3} = D_{R3} + D_{A3} + D_{31} + D_{32}$, ($D_{A3} = 0$)
	$D_{T4} = D_{R4} + D_{A4} + D_{42}$
Solutions:	
	$f_2 = [E_2 + J_1J_4/J_3 + E_3D_{32}/D_{T3} + E_4D_{42}/D_{T4}]/(D_{T2} - J_2J_4/J_3 - D_{24} \cdot D_{42}/D_{T4})$
	$f_1 = (J_1 + f_2J_2)/J_3$
	$f_3 = (E_3 + f_1D_{13})/D_{T3}$
	$f_4 = (E_4 + f_2D_{24})/D_{T4}$
where	
	$J_1 = E_1/D_{T1} + E_3D_{31}/(D_{T3} \cdot D_{T1})$
	$J_2 = D_{21}/D_{T1}$
	$J_3 = 1 - D_{31} \cdot D_{13}/(D_{T1} \cdot D_{T3})$
	$J_4 = D_{12} + D_{32} \cdot D_{13}/D_{T3}$

considered here. These calculations show clearly the extent to which intermedia transport occurs. If, for example, the intermedia D values are small compared to the reaction and advection values, the discharged chemical will tend to remain in the discharge or "source" medium with only a small proportion migrating to other media. Conversely, if the intermedia D values are relatively large, the chemical becomes very susceptible to intermedia transport. This behavior is observed for persistent substances such as PCBs, which have very low rates of reaction.

A direct assessment of multimedia behavior is thus possible by examining the proportions of chemical found at steady state in the "source" medium and in other media. For example, when discharged to water, an appreciable fraction of the benzene is found in air, whereas for atrazine, only a negligible fraction of atrazine reaches air.

Linear Additivity or Superposition of Results

Because these equations are entirely linear, the solutions can be scaled linearly. The concentrations resulting from a discharge of 2000 kg/h are simply twice those of 1000 kg/h. Further, if discharge of 1000 kg/h to air causes 500 kg in water and discharge of 1000 kg/h to soil causes 100 kg in water, then if both discharges occur simultaneously, there will be 600 kg in water. If the discharge to soil is increased to 3000 kg/h, the total amount in the water will rise to (500 + 300) or 800 kg. It is thus possible to deduce the amount in any medium arising from any combination of discharge rates by scaling and adding the responses from the unit inputs. This "linear additivity principle" is more fully discussed by Stiver and Mackay (1989).

The persistence or residence time of the chemical is independent of the emission rate, but it does depend on the "mode of entry, i.e., into which compartment the chemical is emitted."

In the diagrams presented later, these three-unit (1000 kg/h) responses are given. Also, an illustrative "three discharge" mass balance is given in which a total of 1000 kg/h is discharged, but in proportions judged to be typical of chemical use and discharge to the environment. For example, benzene is believed to be mostly discharged to air with minor amounts to soil and water.

Also given in the tables are the rates of reaction, advection and intermedia transport for each case.

The reader can deduce the fate of any desired discharge pattern by appropriate scaling and addition. It is important to emphasize that because the values of transport velocity parameters are only illustrative, actual environmental conditions may be quite different; thus, simulation of conditions in a specific region requires determination of appropriate parameter values as well as the site-specific dimensions, reaction rate constants and the physical-chemical properties which prevail at the desired temperature.

In total, the aim is to convey an impression of the likely environmental behavior of the chemical in a readily assimilable form.

1.6 DATA SOURCES AND PRESENTATION

1.6.1 DATA SOURCES

Most physical properties such as molecular weight (MW, g/mol), melting point (m.p., °C), boiling point (b.p., °C), and density have been obtained from commonly used handbooks such as the *CRC Handbook of Chemistry and Physics* (Weast 1972, 1982; Lide 2003), Lange's *Handbook of Chemistry* (Dean 1979, 1985, 1992), Dreisbach's *Physical Properties of Chemical Compounds*, Vol. I, II and III (1955, 1959, 1961), Organic Solvents, Physical Properties and Methods of Purification (Riddick et al. 1986), *The Merck Index* (Windholz 1983, Budavari 1989) and several handbooks and compilations of chemical property data for pesticides. Notable are the text by Hartley and Graham-Bryce (1980), the *Agrochemicals Handbook* (Hartley and Kidd 1987), the *Pesticide Manual* (Worthing and co-workers 1983, 1987, 1991, Tomlin 1994), the *CRC Handbook of Pesticides* (Milne 1995), the *Agrochemicals Desk Reference* (Montgomery 1993) and the SCS/ARS/CES Pesticide Properties Database by Wauchope and co-workers (Wauchope et al. 1992, Augustijn-Beckers et al. 1994, Hornsby et al. 1996). Other physical-chemical properties such as aqueous solubility, vapor pressure, octanol-water partition coefficient, Henry's law constant, bioconcentration factor and sorption coefficient have been obtained from scientific journals or other environmental handbooks, notably Verschueren's *Handbook of Environmental Data on Organic Chemicals* (1977, 1983) and Howard and co-workers' *Handbook of Environmental Fate and Exposure Data*, Vol. I, II, III and IV (1989, 1990, 1991 and 1993). Other important sources of vapor pressure are the *CRC Handbook of Chemistry and Physics* (Weast 1972, 1982), Lange's *Handbook of Chemistry* (Dean 1992), the *Handbook of Vapor Pressures and Heats of Vaporization of Hydrocarbons and Related Compounds* (Zwolinski and Wilhoit 1971), the *Vapor Pressure of Pure Substances* (Boublik et al. 1973, 1984), the *Handbook of the Thermodynamics of Organic Compounds* (Stephenson and Malanowski 1987). For aqueous solubilities, valuable sources include the *IUPAC Solubility Data Series* (Barton 1984, Horvath and Getzen 1985, Shaw 1989a,b) and Horvath's *Halogenated Hydrocarbons, Solubility-Miscibility with Water* (Horvath 1982). Octanol-water partition coefficients are conveniently obtained from the compilation by Leo et al. (1971), Hansch and Leo (1979), Hansch et al. (1995), and Sangster (1989, 1993), or can be calculated from molecular structure by the methods of Hansch and Leo (1979) or Rekker (1977). Lyman et al. (1982) and Boethling and Mackay (2000) also outline methods of estimating solubility, K_{OW}, vapor pressure, and the bioconcentration factor for organic chemicals. The recent *Handbook of Environmental Degradation Rates* by Howard et al. (1991) is a valuable source of rate constants and half-lives.

The most reliable sources of data are the original citations of valuable experimental data in the reviewed scientific literature. Particularly reliable are those papers which contain a critical review of data from a number of sources as well as independent experimental determinations. Calculated or correlated values are viewed as being less reliable. The aim

Introduction

in this work has been to gather sufficient experimental data with a list of citations to interpret them and select a "best" or "most likely" value.

1.6.2 Data Presentation

Chemical Properties.

The emphasis in this handbook is on experimentally determined values rather than estimated values. The latter are included when there is a lack of experimental data. Included in the experimental data are indirect measurements using GC or HPLC retention times.

The names, formula, melting and boiling point and density data are self-explanatory.

The molar volumes are in some cases at the stated temperature and in other cases at the normal boiling point. Certain calculated molecular volumes are also used; thus the reader is cautioned to ensure that when using a molar volume in any correlation, it is correctly selected. In the case of polynuclear aromatic hydrocarbons, the Le Bas molar volume is regarded as suspect because of the compact nature of the multi-ring compounds. It should thus be regarded as merely an indication of relative volume, not an absolute volume.

Heats of fusion, ΔH_{fus}, are generally expressed in kcal/mol or kJ/mol and entropies of fusion, ΔS_{fus} in cal/mol·K (e.u. or entropy unit) or J/mol·K. The fugacity ratio F, as discussed in Section 1.2.8, is used to calculate the supercooled liquid vapor pressure or solubility for correlation purposes. In the case of liquids such as benzene, it is 1.0. For solids it is a fraction representing the ratio of solid-to-liquid solubility or vapor pressure.

A wide variety of solubilities (in units of g/m^3 or the equivalent mg/L) have been reported. Experimental data have the method of determination indicated. In other compilations of data the reported value has merely been quoted from another secondary source. In some cases the value has been calculated. The abbreviations are generally self-explanatory and usually include two entries, the method of equilibration followed by the method of determination. From these values a single value is selected for inclusion in the summary data table. Vapor pressures and octanol-water partition coefficients are selected similarly.

The reader is advised to consult the original reference when using these values of bioconcentration factors (BCF), bioaccumulation factors (BAF), K_{OC} and K_{OM}, to ensure that conditions are as close as possible to those of specific interest.

The "Environmental Fate Rate Constants" refer to specific degradation processes rather than media. As far as possible the original numerical quantities are given and thus there is a variety of time units with some expressions being rate constants and others half-lives. The conversion is that the rate constant k is $0.693/t_{1/2}$ where $t_{1/2}$ is the half-life.

From these data a set of medium-specific degradation reaction half-lives is selected for use in Levels II and III calculations. Emphasis is placed on the fastest and the most plausible degradation process for each of the environmental compartments considered. Instead of assuming an equal half-life for both the water and soil compartment as suggested by Howard et al. (1991), a slower active class (in the reactivity table described earlier) was assigned for soil and sediment compared to that of the water compartment. This is in part because the major degradation processes are often photolysis (or photooxidation) and biodegradation. There is an element of judgment in this selection, and it is desirable to explore the implications of selecting other values.

The "Half-life in the Environment" data reflect observations of the rate of disappearance of the chemical from a medium, without necessarily identifying the cause of mechanism of loss. For example, loss from water may be a combination of evaporation, biodegradation and photolysis. Clearly these times are highly variable and depend on factors such as temperature, meteorology and the nature of the media. Again, the reader is urged to consult the original references.

1.7 ILLUSTRATIVE QSPR PLOTS AND FATE CALCULATIONS

Illustrative QSPR plots and their interpretation are given in this section, followed by examples of Levels I, II and III fate calculations. A relatively simple evaluation of benzene is given first followed by the more complex evaluation of pentachlorophenol.

1.7.1 QSPR Plots for Mononuclear Aromatic Hydrocarbons

The physical-chemical data for mononuclear aromatics are plotted in the appropriate QSPR plots on Figures 1.7.1 to 1.7.5 (which are also Figures 3.2.1 to 3.2.5 for the mononuclear aromatic hydrocarbons in Chapter 3). These plots show that the data are relatively "well-behaved," there being consistency among the reported values for this homologous series. In the case of benzene this QSPR plot is of little value because this is a well-studied chemical, but for other less-studied chemicals the plots are invaluable as a means of checking the reasonableness of data. The plots can also be used, with appropriate caution, to estimate data for untested chemicals. We do not develop linear regressions of these data since we suggest that the plots be used directly for data estimation purposes. This enables the user to assess into account the values of similarly structured compounds and it gives a direct impression of likely error. We discuss, below, the general nature of the relationships and in particular the slopes of the QSPR plots.

Figures 1.7.1 to 1.7.4 show the dependence of the physical-chemical properties on Le Bas molar volume. Figure 1.7.1 shows that the solubilities of the monoaromatics decrease steadily with increasing molar volume. The vapor pressure data in Figure 1.7.2 are similar, but log K_{OW} in Figure 1.7.3 increases with increasing molar volume also in a linear fashion.

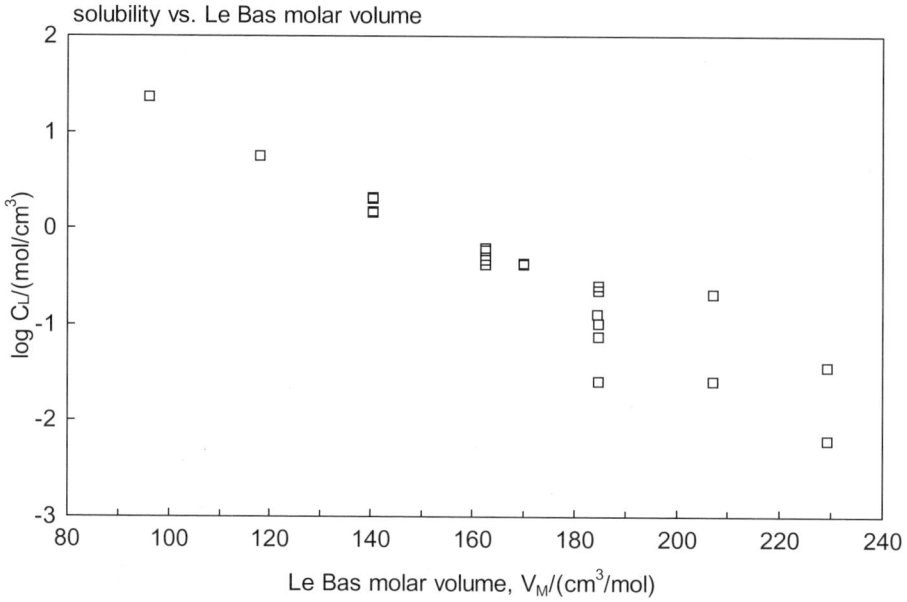

FIGURE 1.7.1 Molar solubility (liquid or supercooled liquid) versus Le Bas molar volume for mononuclear aromatic hydrocarbons.

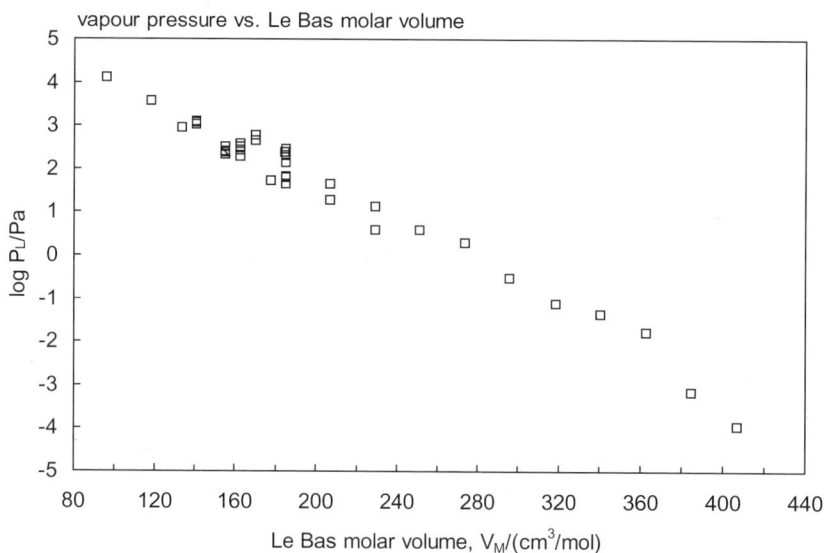

FIGURE 1.7.2 Vapor pressure (liquid or supercooled liquid) versus Le Bas molar volume for mononuclear aromatic hydrocarbons.

Introduction

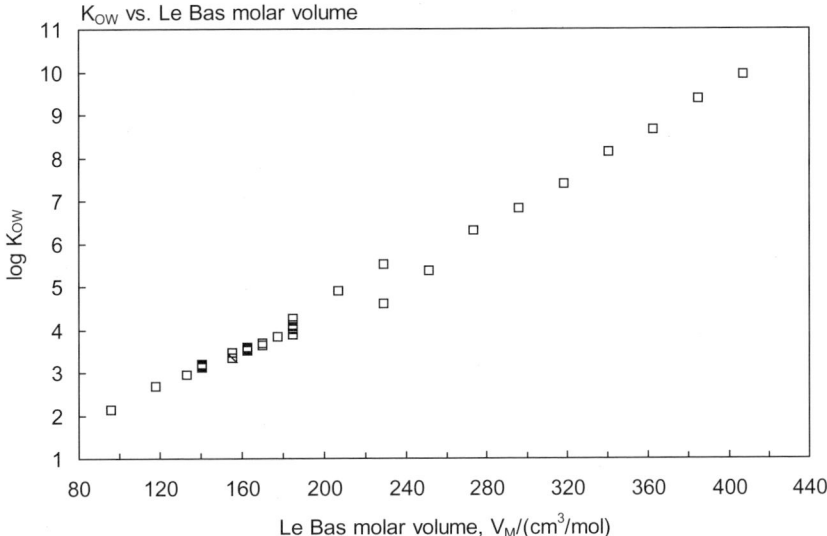

FIGURE 1.7.3 Octanol-water partition coefficient versus Le Bas molar volume for mononuclear aromatic hydrocarbons.

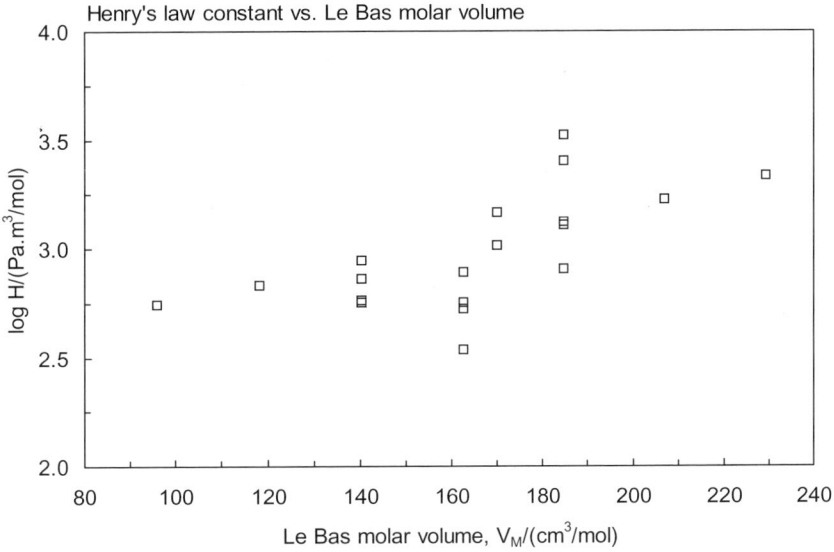

FIGURE 1.7.4 Henry's law constant versus Le Bas molar volume for mononuclear aromatic hydrocarbons.

The plot between Henry's law constant and molar volume (Figure 1.7.4) is more scattered. Figure 1.7.5 shows the often-reported inverse relationship between octanol-water partition coefficient and the supercooled liquid solubility.

The QSPR plots show that an increase in molar volume by 100 cm³/mol generally causes:

(i) A decrease in log solubility by 2.5 units, i.e., a factor of $10^{2.5}$ or 316;
(ii) A decrease in log vapor pressure by 2.2 units, i.e., a factor of $10^{2.2}$ of 159;
(iii) An increase in log Henry's law constant of 0.3 (i.e., 2.5 − 2.2) or a factor or $10^{0.3}$ or 2.0;
(iv) An increase in log K_{OW} by 2.0 units, i.e., a factor of 100.

The plot of log K_{OW} versus log solubility thus has a slope of approximately 2.0/2.5 or 0.8. This slope of less than 1.0 has been verified experimentally by Chiou et al. (1982) and Bowman and Sans (1983). Its theoretical basis has been discussed in detail by Miller et al. (1985).

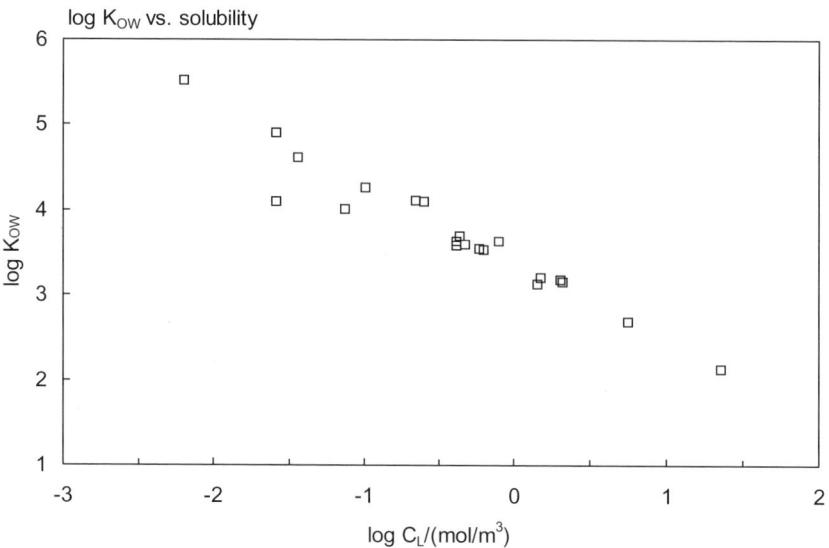

FIGURE 1.7.5 Octanol-water partition coefficient versus molar solubility (liquid or supercooled liquid) for mononuclear aromatic hydrocarbons.

Similar inferences can be made for other homologous series such as the chlorobenzenes and PCBs. In such cases the property change caused by substitution of one chlorine can be deduced as is illustrated later for chlorophenols.

The "Half-life in the Environment" and "Environmental Fate Rate Constants" are medium-specific degradation reaction half-lives selected for use in Level II and Level III calculations. As discussed earlier, emphasis was based on the fastest and the most plausible degradation process for each of the environmental compartments considered.

In summary, the physical-chemical and environmental fate data listed result in the tabulated selected values of solubility, vapor pressure, K_{OW}, dissociation constant where appropriate and reaction half-lives at the end of each chapter. These values are used in the evaluative environmental calculations.

1.7.2 Evaluative Calculations for Benzene

The illustrative evaluative environmental calculations described here are presented in the following format. Levels I, II and III diagrams are assigned to separate pages, and the physical-chemical properties are included in the Level I diagram. Two types of Level III diagrams are given; one depicts the transport processes and the other the distribution among compartments.

Level I

The Level I calculation suggests that if 100,000 kg (100 tonnes) of benzene are introduced into the 100,000 km² environment, 99% will partition into air at a concentration of 9.9×10^{-7} g/m³ or about 1 µg/m³. The water will contain nearly 1% at a low concentration of 4 µg/m³ or equivalently 4 ng/L. Soils would contain 5×10^{-6} µg/g and sediments about 9.7×10^{-6} µg/g. These values would normally be undetectable as a result of the very low tendency of benzene to sorb to organic matter in these media. The fugacity is calculated to be 3.14×10^{-5} Pa. The dimensionless soil-water and sediment-water partition coefficients or ratios of Z values are 2.6 and 5.3 as a result of a K_{OC} of about 55 and a few percent organic carbon in these media. There is little evidence of bioconcentration with a very low fish concentration of 3.0×10^{-5} µg/g. The pie chart in Figure 1.7.6 clearly shows that air is the primary medium of accumulation.

Level II

The Level II calculation includes the half-lives of 17 h in air, 170 h in water, 550 h in soil and 1700 h in sediment. No reaction is included for suspended sediment or fish. The input of 1000 kg/h results in an overall fugacity of 6×10^{-6} Pa, which is about 20% of the Level I value. The concentrations and amounts in each medium are thus about 20% of the Level I values. The relative mass distribution is identical to Level I. The primary loss mechanism is reaction in air, which accounts for 802 kg/h or 80.2% of the input. Most of the remainder is lost by advective outflow. The water, soil and sediment loss processes are unimportant largely because so little of the benzene is present in these media, but also

Introduction

Chemical name: Benzene
Fugacity Level I calculations: (six-compartment model)

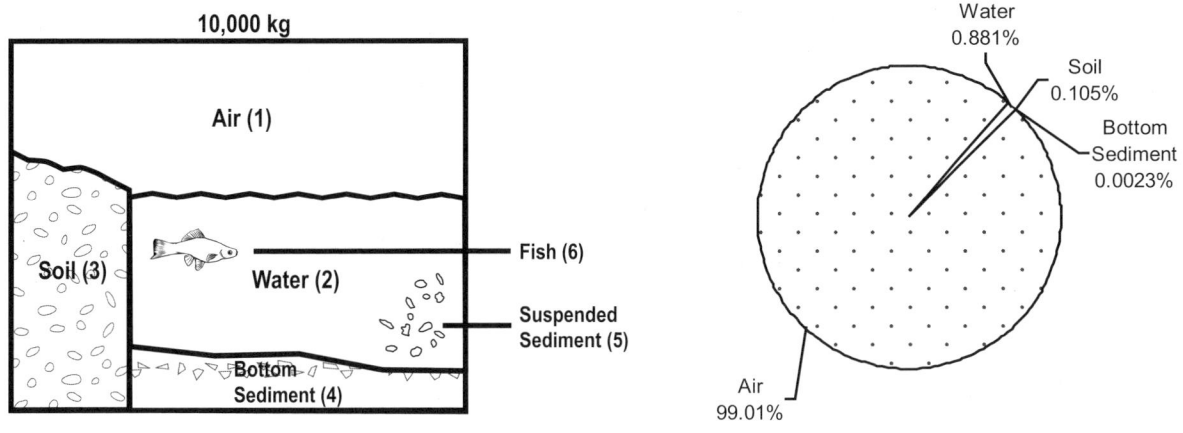

Compartment	Z	Concentration			Amount	Amount
	mol/m^3·Pa	mol/m^3	g/m^3	µg/g	kg	%
Air (1)	4.034E-04	1.268E-08	9.901E-07	8.251E-04	9.901E+04	9.901E+01
Water (2)	1.794E-03	5.638E-08	4.404E-06	4.404E-06	8.808E+02	8.808E-01
Soil (3)	4.764E-03	1.497E-07	1.169E-05	4.871E-06	1.052E+02	1.052E-01
Bottom sediment (4)	9.527E-03	2.994E-07	2.338E-05	9.743E-06	2.338E+00	2.338E-03
Suspended sediment (5)	2.977E-02	9.355E-07	7.307E-05	4.871E-05	7.307E-02	7.307E-05
Biota (6)	1.210E-02	3.803E-07	2.970E-05	2.970E-05	5.941E-03	5.941E-06
Total					1.000E+05	1.000E+02

Fugacity, f = 3.142E-05

FIGURE 1.7.6 Level I fugacity calculations for benzene in a generic environment.

because of the slower reaction and advection rates. The overall residence time is 19.9 h; thus, there is an inventory of benzene in the system of 19.9 × 1000 or 19900 kg. The pie chart in Figure 1.7.7 illustrates the dominance of air reaction and advection.

If the primary loss mechanism of atmospheric reaction is accepted as having a 17h half-life, the D value is 1.6 × 10^9 mol/Pa·h. For any other process to compete with this would require a value of at least 10^8 mol/Pa·h. This is achieved by advection (4 × 10^8), but the other processes range in D value from 19 (advection in bottom sediment) to 1.5 × 10^6 (reaction in water) and are thus a factor of over 100 or less. The implication is that the water reaction rate constant would have to be increased 100-fold to become significant. The soil rate constant would require an increase by 10^4 and the sediment by 10^6. These are inconceivably large numbers corresponding to very short half-lives, thus the actual values of the rate constants in these media are relatively unimportant in this context. They need not be known accurately. The most sensitive quantity is clearly the atmospheric reaction rate.

The amounts in the compartments can be calculated easily from the total amount and the percentages of mass distribution in Level I. For example, the amount in water is 0.881% of 19877 kg or 175 kg.

Level III

The Level III calculation includes an estimation of intermedia transport. Examination of the magnitude of the intermedia D values given in the fate diagram (Figure 1.7.8) suggests that air-water and air-soil transport are most important with water-sediment and soil-water transport being negligible in potential transfer rate. The magnitude of these larger intermedia

Chemical name: Benzene
Fugacity Level II calculations: (six-compartment model)

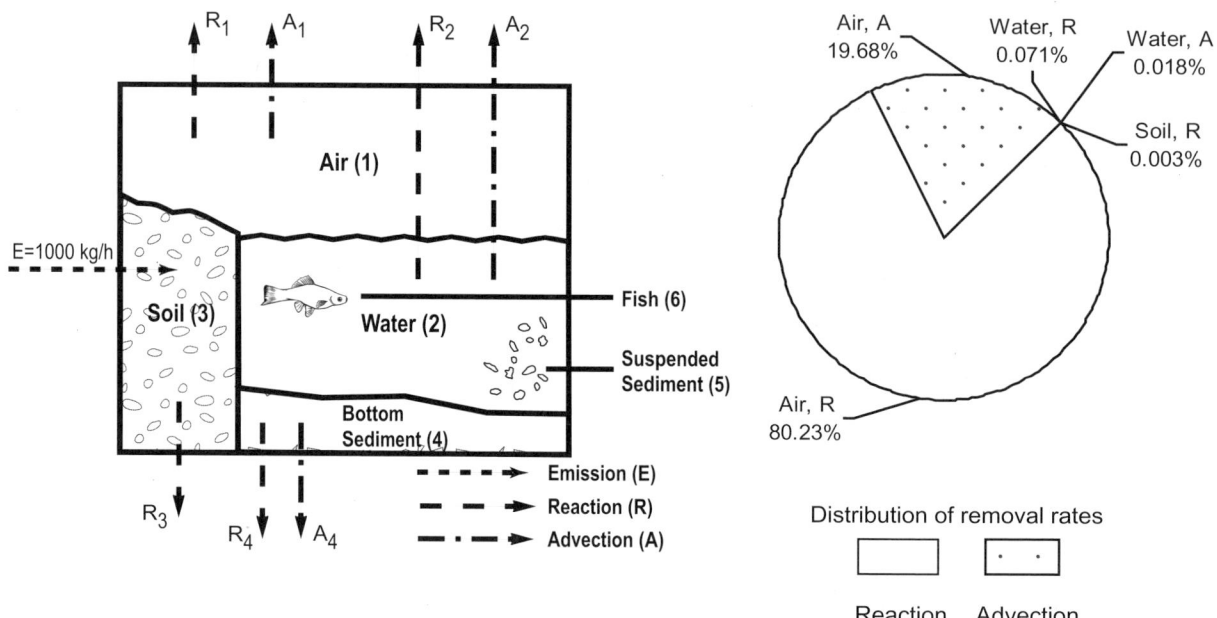

Compartment	Half-life h	D Value Reaction mol/Pa.h	D Value Advection mol/Pa.h	Concentration mol/m³	Loss Reaction kg/h	Loss Advection kg/h	Total Removal %
Air (1)	17	1.645E+09	4.034E+08	2.520E-09	8.023E+02	1.968E+02	9.991E+01
Water (2)	170	1.463E+06	3.589E+05	1.121E-08	7.137E-01	1.751E-01	8.888E-02
Soil (3)	550	5.402E+04	-	2.975E-08	2.635E-02	-	2.635E-03
Bottom sediment (4)	1700	3.884E+02	1.905E+01	5.950E-08	1.895E-04	9.296E-06	1.988E-05
Suspended sediment (5)	170	1.214E+02	2.977E+01	1.859E-07	5.921E-05	1.452E-05	7.373E-06
Biota (6)	170	9.867E+00	2.421E+00	7.559E-08	4.814E-06	1.181E-06	5.995E-07

Fugacity, f 6.246E-06 Pa
Total amount, M 2.545E+05 mol
Total amount 1.988E+04 kg
Total reaction D value, D_R 1.646E+09 mol/Pa.h
Total advection D value, D_A 4.038E+08 mol/Pa.h
Total D value, D_T 2.050E+09 mol/Pa.h
Total loss by reaction 8.030E+02 kg/h
Total loss by advection 1.970E+02 kg/h
Total loss 1.000E+03 kg/h
Reaction residence time, t_R 2.475E+01 h
Advection residence time, t_A 1.009E+02 h
Overall residence time, t_O 1.988E+01 h

FIGURE 1.7.7 Level II fugacity calculations for benzene in a generic environment.

transport D values (approximately 10^6 mol/Pa·h) compared to the atmospheric reaction and advection values of 10^8 to 10^9 suggests that reaction and advection will be very fast relative to transport.

The bulk Z values are similar for air and water to the values for the "pure" phases in Levels I and II, but they are lower for soil and sediment because of the "dilution" of the solid phase with air or water.

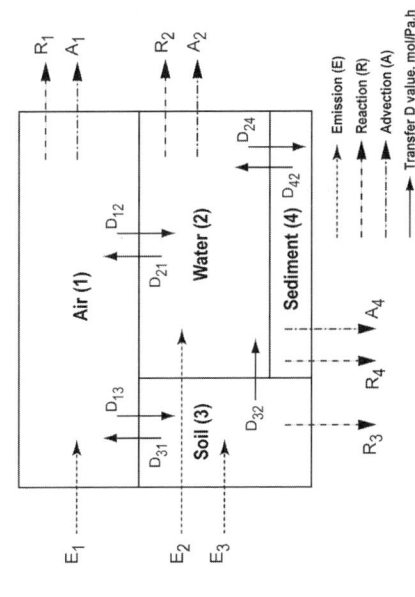

FIGURE 1.7.8 Level III fugacity calculations for benzene in a generic environment.

The first row describes the condition if 1000 kg/h is emitted into the air. The result is similar to the Level II calculation with 19700 kg in air, 57 kg in water, 24 kg in soil and only 0.2 kg in sediment. It can be concluded that benzene discharged to the atmosphere has very little potential to enter other media. The rates of transfer from air to water and air to soil are both only about 0.4 kg/h. Even if the transfer coefficients were increased by a factor of 10, the rates would remain negligible. The reason for this is the value of the mass transfer coefficients which control this transport process. The overall residence time is 19.8 hours, similar to Level II.

If 1000 kg/h of benzene is discharged to water, as in the second row, there is predictably a much higher concentration in water (by a factor of over 2000). There is reaction of 546 kg/h in water, advective outflow of 134 kg/h and transfer to air of 320 kg/h with negligible loss to sediment. The amount in the water is 134000 kg; thus the residence time in the water is 134 h and the overall environmental residence time is a longer 140 hours. The key processes are thus reaction in water (half-life 170 h), evaporation (half-life 290 h) and advective outflow (residence time 1000 h). The evaporation half-life can be calculated as (0.693 × mass in water)/rate of transfer, i.e., (0.693 × 133863)/320 = 290 h. Clearly, competition between reaction and evaporation in the water determines the overall fate. Ninety-five percent of the benzene discharged is now found in the water, and the concentration is a fairly high 6.7×10^{-4} g/m^3, or 670 ng/L.

The third row shows the fate if discharge is to soil. The amount in soil is 67460 kg, reflecting an overall 87 h residence time. The rate of reaction in soil is only 85 kg/h and there is no advection; thus, the primary loss mechanism is transfer to air (T_{31}) at a rate of 905 kg/h, with a relatively minor 10 kg/h to water by run-off. The net result is that the air concentrations are similar to those for air discharge and the soil acts only as a reservoir. The soil concentration of 3.75×10^{-3} g/m^3 or 2.5×10^{-3} µg/g or 2.5 ng/g is controlled almost entirely by the rate at which the benzene can evaporate.

The net result is that benzene behaves entirely differently when discharged to the three media. If discharged to air it reacts rapidly and advects with a residence time of 20 h with little transport to soil or water. If discharged to water it reacts and evaporates to air with a residence time of 140 h. If discharged to soil it mostly evaporates to air with a residence time in soil of 53 h.

The final scenario is a combination of discharges, 600 kg/h to air, 300 kg/h to water, and 100 kg/h to soil. The concentrations, amounts and transport and transformation rates are merely linearly combined versions of the three initial scenarios. For example, the rate of reaction in air is now 632 kg/h. This is 0.6 of the first (air emission) rate of 803 kg/h, i.e., 482 kg/h, plus 0.3 of the second (water emission) rate of 257 kg/h, i.e., 77 kg/h and 0.1 of the third (soil emission) rate of 729 kg/h, i.e., 73 kg yielding a total of (482 + 77 + 73) or 632 kg/h. It is also apparent that the amount in the air of 15500 kg causing a concentration of 0.155 µg/m^3 is attributable to emissions to air (0.6 × 0.197 or 0.118 µg/m^3), emissions to water (0.3 × 0.063 or 0.019 µg/m^3) and emissions to soil (0.1 × 0.179 or 0.018 µg/m^3). The concentration in water of 2.0×10^{-4} g/m^3 or 202 µg/m^3 or ng/L is largely attributable to the discharges to water, which alone cause 0.3 × 669 or 200 µg/m^3. Although more is emitted to air, it contributes less than 1 µg/m^3 to the water with soil emissions accounting for about 1 µg/m^3. Similarly, the prevailing soil concentration is controlled by the rate of discharge to the soil.

In this multimedia discharge scenario the overall residence time is 59 hours, which can be viewed as 60% of the air residence time of 19.7 h, 30% of the water residence time of 140 h and 10% of the overall soil residence time of 53 h. The overall amount in the environment of 59,000 kg is thus largely controlled by the discharges to water, which account for (0.3 × 133863) or 40,000 kg.

Figure 1.7.9 shows the distributions of mass and removal process rates for these four scenarios. Clearly, when benzene is discharged into a specific medium, most of the chemical is found in that medium. Only in the case of discharges to soil is an appreciable fraction found in another compartment, namely air. This is because benzene evaporates fairly rapidly from soil without being susceptible to reaction or advection.

Finally, it is interesting to note that the fugacity in this final case (in units of mPa) are for the four media 5.0×10^{-3}, 1.4, 1.6 and 1.1. The soil, sediment and water are fairly close to equilibrium, with the air notably "under-saturated" by a factor of about 200. This is the result of the rapid loss processes from air.

1.7.3 QSPR Plots for Chlorophenols and Alkylphenols

These QSPR (quantitative structure-property relationship) plots display the usual approximately linear relationships similar to those of the alkyl and chlorinated aromatic hydrocarbons.

Most acid dissociation constants pK_a exceed environmental pH values, the exceptions being the highly chlorinated phenols. As a result, these substances tend to have higher apparent solubilities in water because of dissociation. The structure-property relationships apply to the un-ionized or protonated species; thus, experimental data should preferably be "corrected" to eliminate the effect of ionization, thus eliminating pH effects.

Introduction

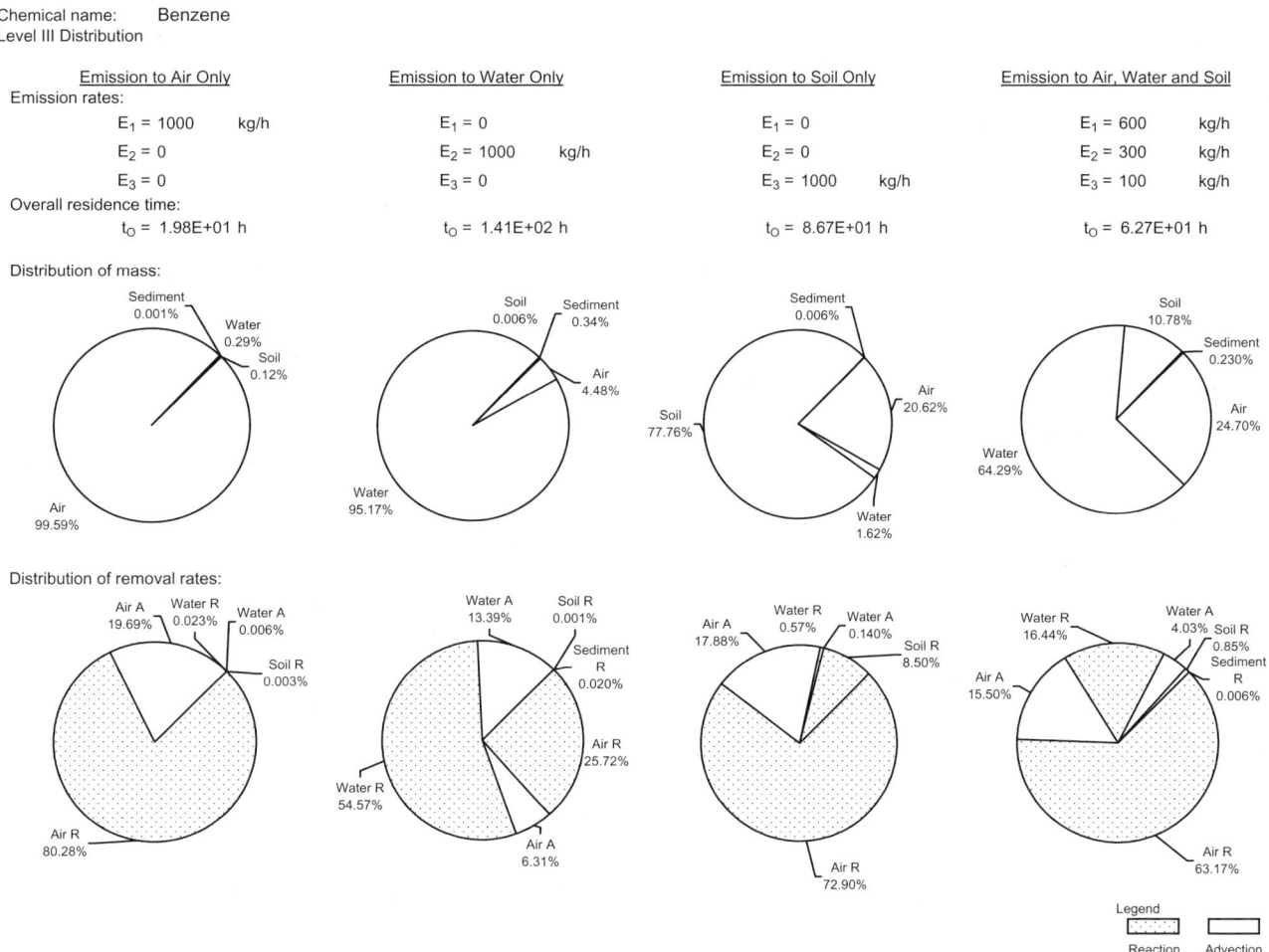

FIGURE 1.7.9 Level III fugacity distributions of benzene for four emission scenarios.

Figure 1.7.10 shows that the chlorophenol solubilities behave similarly to other chemical series with slopes of about 0.62 log units per 20.9 cm^3/mol, which is the volume difference resulting from substitution by one chlorine. The result is a factor of $10^{0.62}$ or 4.2 drop in solubility per chlorine. The alkylphenols have a lower slope of about 0.5 per CH$_2$ and usually have higher solubilities at the same molar volume. The two sets of data are, however, generally similar.

The vapor pressure data in Figure 1.7.11 show a slope of about 0.60 log units per 20.9 cm^3/mol (i.e., a factor or 4.0) per chlorine. There is a lower slope for the alkylphenols, and they usually have higher vapor pressures, especially for the larger molecules.

The K_{OW} data in Figure 1.7.12 show that the chlorophenols and alkylphenols differ in properties, there being more uncertainty about the K_{OW} of the longer-chain phenols. The chlorophenols tend to partition more into octanol at the same molar volume and are thus expected to be more bioaccumulative. The slope of the chlorophenol line is about 0.78 log units per chlorine or a factor of 6.0. The alkylphenol slope is lower and about 0.36 log units per CH$_2$, i.e., a factor of 2.3.

The Henry's law constant data calculated as the ratio of vapor pressure to solubility in Figure 1.7.13 are quite scattered. There is little systematic variation with molar volume. Most values of log H lie between –0.1 to –0, i.e., H lies between 0.8 and 0.08, and the resulting air-water partition coefficient K_{AW} or H/RT thus lies between 3×10^{-4} and 3×10^{-5}.

Figure 1.7.14, the plot of log K_{OW} versus log solubility, shows a relatively high slope of 1.25 for the chlorophenols and a lower slope of 0.70 for the alkylphenols.

Addition of a chlorine causes a drop in chlorophenol solubility in water by about 0.62 log units, and K_{OW} increases by about 0.78 log units. For the alkylphenols, addition of a methylene causes about a 0.50 log unit drop in solubility in water, and K_{OW} increases by only about 0.36 log units. The slope of the log K_{OW} versus solubility lines are thus about 0.78/0.62 or 1.25 for the chlorophenols and 0.36/0.5 or 0.72 for the alkylphenols. An implication is that since K_{OW} can

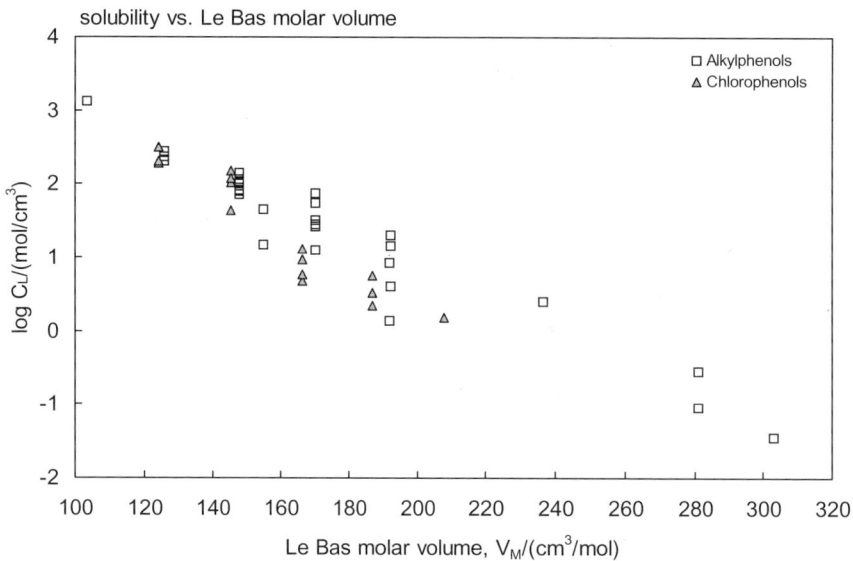

FIGURE 1.7.10 Molar solubility (liquid or supercooled liquid) versus Le Bas molar volume for alkylphenols and chlorophenols.

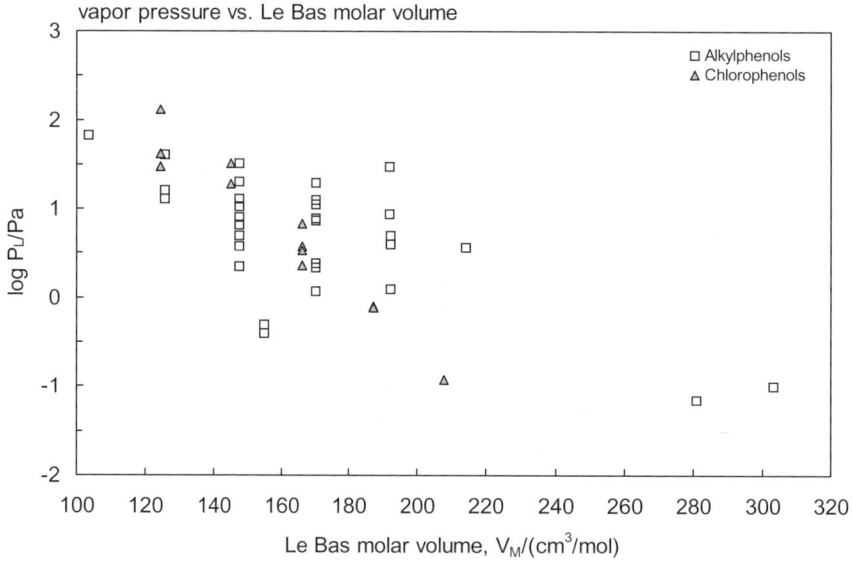

FIGURE 1.7.11 Vapor pressure (liquid or supercooled liquid) versus Le Bas molar volume for alkylphenols and chlorophenols.

be viewed as a ratio of "solubility" in octanol and solubility in water, the solubility of the chlorophenols in octanol increases by (0.78 – 0.62) or 0.16 log unit per chlorine, while for the alkylphenols the corresponding change is (0.36 – 0.50) or –0.14 log unit, or a decrease of a factor of 1.4. The reasons for this difference are not known. The chlorophenols thus appear to have an unusually strong tendency to partition into octanol. Whether or not this tendency applies to lipid phases in biota or to organic carbon is not certain, but such a tendency is obviously of considerable interest when interpreting the toxicity and fate of these chemicals.

These data show clearly that the structure-property relationships which apply to hydrophobic organic chemicals such as the chloro- and alkyl-aromatics also apply to the phenols, but the relationships are more scattered and less well defined. The absolute values of properties differ greatly. This scatter is probably attributable, in part, to insufficient experimental data or errors in experimental measurements, to dissociation and to the greater polar character of these chemicals. It is not recommended that correlations developed for non-polar organic chemicals be applied to the phenols. Separate treatment of each homologous series is required.

Introduction

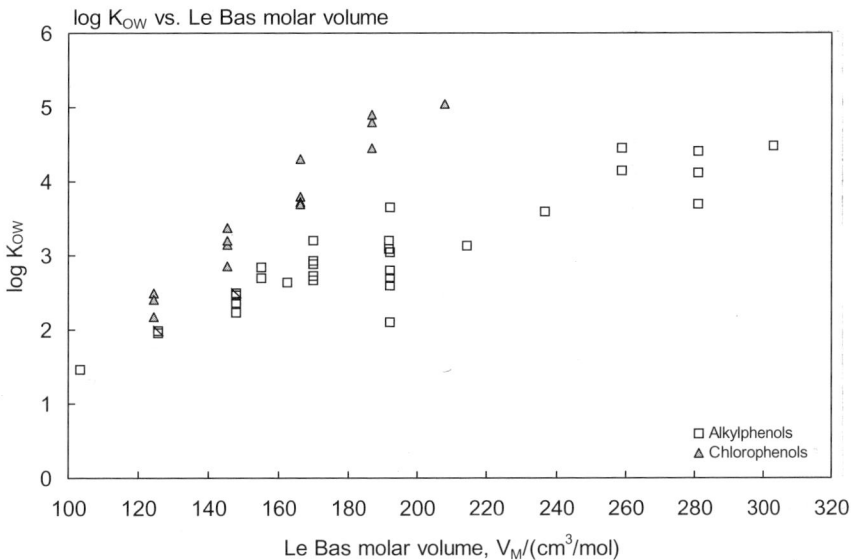

FIGURE 1.7.12 Octanol-water partition coefficient versus Le Bas molar volume for alkylphenols and chlorophenols.

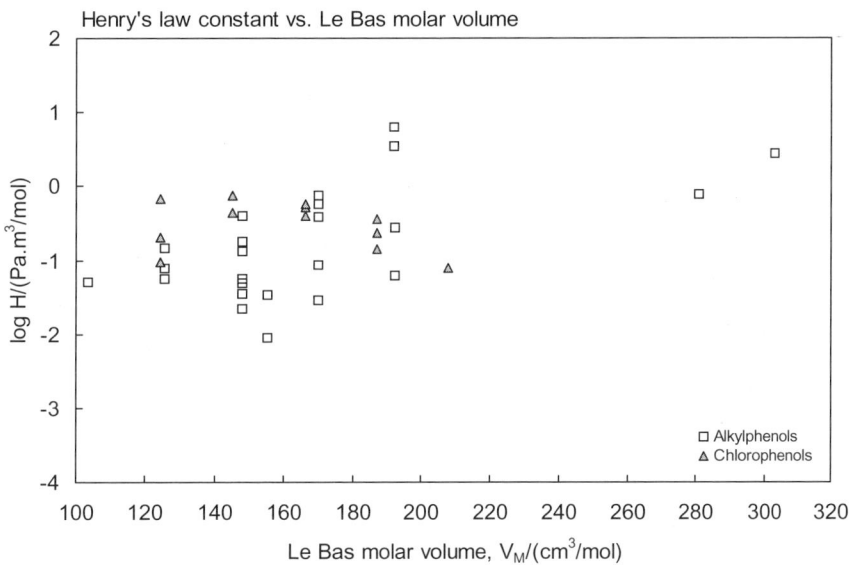

FIGURE 1.7.13 Henry's law constant versus Le Bas molar volume for alkylphenols and chlorophenols.

1.7.4 Evaluative Calculations for Pentachlorophenol

For dissociating compounds the environmental pH is specified and the calculation of Z values has been modified to include ionic species as discussed in Section 1.2.4. Generally, if discharge is to a compartment such as water, most chemical will be found in that compartment, and will react there, but a quantity does migrate to other compartments and is lost from these media. Three pie charts corresponding to discharges of 1000 kg/h to air, water and soil are included. The percentage emission in each medium in this case has been selected to be 5, 25 and 70% discharged to air, water and soil, respectively. A fourth pie chart with discharges to all three compartments is also given. This latter chart is in principle the linear sum of the first three, but since the overall residence times differ, the diagram with the longer residence time, and greater resident mass, tends to dominate.

Figures 1.7.15 to 1.7.18 show the mass distributions obtained in Level I calculations and the removal distribution from Level II fugacity calculation of pentachlorophenol (PCP) at two different environmental pHs for the generic

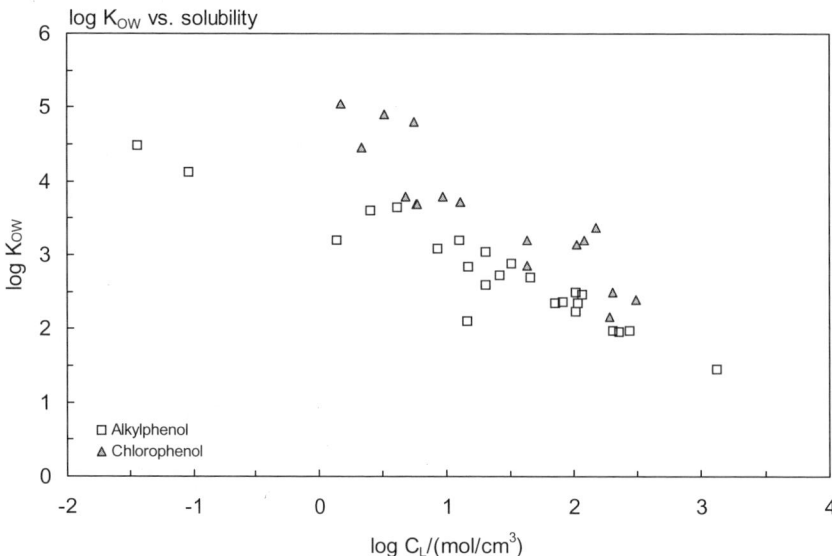

FIGURE 1.7.14 Octanol-water partition coefficient versus molar solubility (liquid or supercooled liquid) for alkylphenols and chlorophenols.

environment. Figures 1.7.19 to 1.7.22 show the corresponding Level III fugacity calculations. Both mass and removal distributions are shown in these figures for the four scenarios of discharges to air, water, soil, and mixed compartments.

Level I

The Level I calculations for environmental pHs of 5.1 and 7 suggest that if 100,000 kg (100 tonnes) of pentachlorophenol (PCP) are introduced into the 100,000 km^2 environment, most PCP will tend to be associated with soil. This is especially the case at low pH when the protonated form dominates. Very little partitions into air and only about 1% partitions into water. Soil contains most of the PCP. Sediments contain about 2%. There is evidence of bioconcentration with a rather high fish concentration. Note that only four media (air, water, soil and bottom sediment) are depicted in the pie chart; therefore, the sum of the percent distribution figures is slightly less than 100%. The air-water partition coefficient is very low. As pH increases, dissociation increases and there is a tendency for partitioning to water to become more important. Essentially, the capacity of water for the chemical increases. Partitioning to air is always negligible.

Level II

The Level II calculations at pH 5.1 include the reaction half-lives of 550 h in air, 550 h in water, 1700 h in soil and 5500 h in sediment. No reaction is included for suspended sediment or fish. The steady-state input of 1000 kg/h results in an overall fugacity of 3.43×10^{-8} Pa, which is about 24 times the Level I value. The concentrations and amounts in each medium are thus about 24 times the Level I values. The relative mass distribution is identical to Level I. The primary loss mechanism is reaction in soil, which accounts for 936 kg/h, or 94% of the input. Most of the remainder is lost by reaction and advection in water. The air and sediment loss processes are unimportant largely because so little of the PCP is present in these media. The overall residence time is 2373 h; thus, there is an inventory of PCP in the system of 2373 × 1000 or 2,373,000 kg.

The primary loss mechanism of soil reaction has a D value of 1.03×10^{11}; thus, for any other process to compete with this would require a D value of at least 10^{10} mol/Pa·h. The next largest D values are 3.19×10^9 and 2.53×10^9 for reaction and advection in water, which are about a factor of 30 smaller. Only if the water advection or reaction rates are increased by about this factor will these processes become significant. As pH increases, reaction in, and advection from, water increase in importance.

Level III

The Level III diagrams (Figures 1.7.19 to 1.7.22 for the two pHs) are regarded as the most realistic depictions of chemical fate.

This calculation includes an estimation of intermedia transport. Examination of the magnitude of the intermedia D values given in the fate diagrams suggest that water-sediment and air-soil transport are most important, with soil-water, and air-water exchange being slower. This chemical tends to be fairly immobile in terms of intermedia transport.

Introduction

Chemical name: Pentachlorophenol
Fugacity Level I calculations: (six-compartment model) at data pH of 5.1

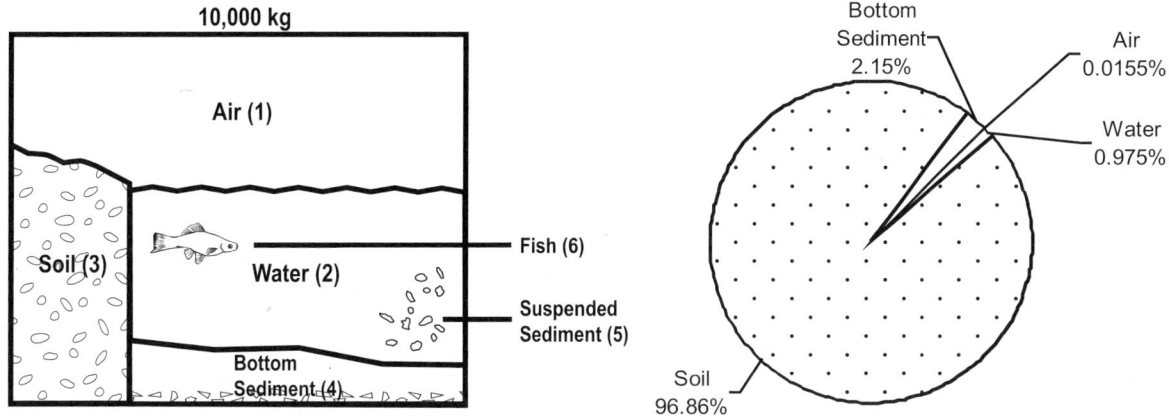

Physical-chemical properties:		Partition coefficients:		Z values in water:		
Molecular weight (g/mol)	266.34	Henry's law constant	7.90E-02	at data pH		5.1
Melting point (°C)	174	Air/water	3.19E-05	neutral		3.849
Solubility (g/m³)	14	Organic carbon, K_{OC}	4.60E+04	ionic		8.817
Vapor pressure (Pa)	4.15E-03	Bioconcentration factor, BCF	5.61E+03	total		12.666
log K_{OW}	5.05	Soil/water	2.21E+03	at environ. pH		5.1
Fugacity ratio, F	3.36E-02	Sediment/water	4.42E+03	neutral		3.849
Dissociation const, pKa	4.74	Suspended sediment/water	1.38E+04	ionic		8.817
		Aerosol/water	4.86E+07	total		12.666

	Fugacity, f =	1.44E-09				
Compartment	Z	Concentration			Amount	Amount
	mol/m³·Pa	mol/m³	g/m³	µg/g	kg	%
Air (1)	4.03E-04	5.82E-13	1.55E-10	1.31E-07	1.55E+01	1.55E-02
Water (2)	1.27E+01	1.83E-08	4.87E-06	4.87E-06	9.74E+02	9.74E-01
Soil (3)	2.80E+04	4.04E-05	1.08E-02	4.48E-03	9.68E+04	9.68E+01
Bottom sediment (4)	5.59E+04	8.80E-05	2.15E-02	8.96E-03	2.15E+03	2.15E+00
Suspended sediment (5)	1.75E+05	2.52E-04	6.72E-02	4.48E-02	6.72E+01	6.72E-02
Biota (6)	7.11E+04	1.03E-04	2.73E-02	2.73E-02	5.46E+00	5.46E-03
Total					1.00E+05	1.00E+02

FIGURE 1.7.15 Level I fugacity calculations for PCP at data determination pH of 5.1.

The bulk Z values are similar for air and water to the values for the "pure" phases in Level I and II, but they are lower for soil and sediment because of the "dilution" of the solid soil and sediment phases with air or water.

The complete discussion of PCP fate as deduced in these calculations is beyond our scope, but to assist the reader we describe the behavior at a pH of 5.1 in some detail below.

These tabulated data are given in numerical and pictorial form in Figures 1.7.19 to 1.7.22. The first row of figures at the foot of Figure 1.7.19 describes the condition if 1000 kg/h is emitted to the air. The result is similar to the Level II calculation with 65780 kg in air, 21070 kg in water, 504700 kg in soil and only 40800 kg in sediment. It can be concluded that PCP discharged to the atmosphere has fairly high potential to enter other media. The rate of transfer from air to

Chemical name: Pentachlorophenol
Fugacity Level II calculations: (six-compartment model) at data pH of 5.1

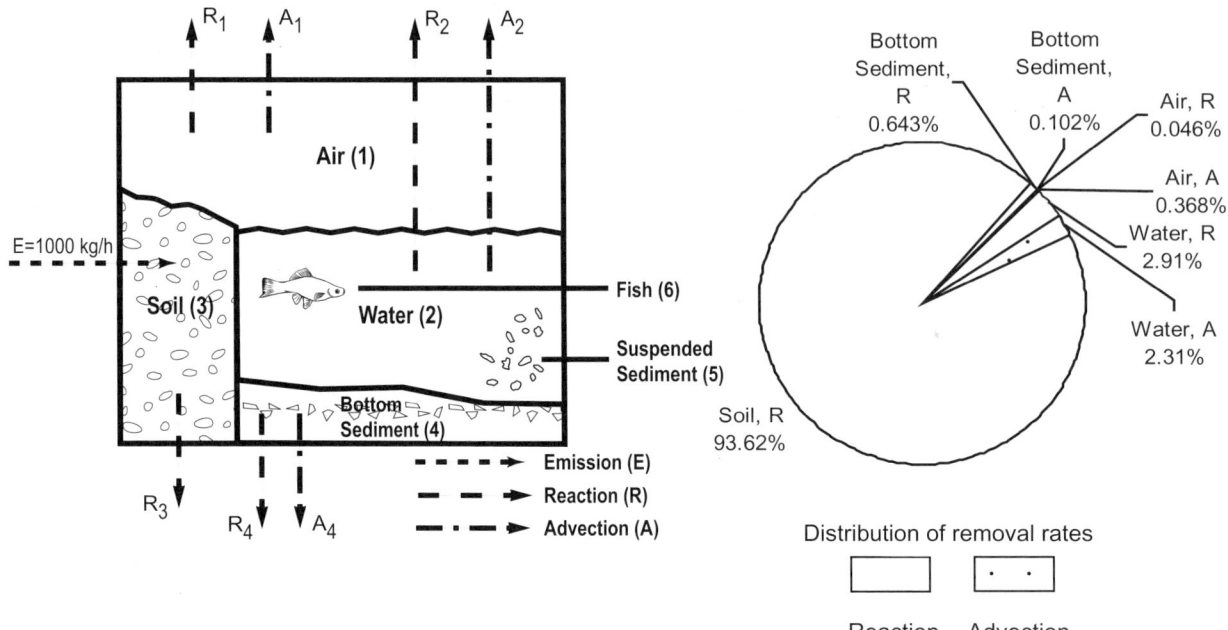

Compartment	Half-life h	D Value Reaction mol/Pa.h	D Value Advection mol/Pa.h	Concentration mol/m^3	Loss Reaction kg/h	Loss Advection kg/h	Total Removal %
Air (1)	550	5.08E+07	4.03E+08	1.38E-11	4.64E-01	3.68E+00	4.14E-01
Water (2)	550	3.19E+09	2.53E+09	4.34E-07	2.91E+01	2.31E+01	5.22E+00
Soil (3)	1700	1.03E+11	-	9.58E-04	9.36E+02	-	9.36E+01
Bottom sediment (4)	5500	7.05E+08	1.12E+08	1.92E-03	6.43E+00	1.02E+00	7.45E-01
Suspended sediment (5)	-	-	-	5.99E-03	-	-	-
Biota (6)	-	-	-	2.43E-03	-	-	-

Fugacity, f	3.43E-08 Pa
Total amount, M	8.91E+06 mol
Total amount	2.37E+06 kg
Total reaction D value, D_R	1.06E+11 mol/Pa.h
Total advection D value, D_A	2.94E+09 mol/Pa.h
Total D value, D_T	1.09E+11 mol/Pa.h
Total loss by reaction	9.72E+02 kg/h
Total loss by advection	2.78E+01 kg/h
Total loss	1.00E+03 kg/h
Reaction residence time, t_R	2.44E+03 h
Advection residence time, t_A	8.53E+04 h
Overall residence time, t_O	2.37E+03 h

FIGURE 1.7.16 Level II fugacity calculations for PCP at data determination pH of 5.1.

water (T_{12}) is about 54 kg/h and that from air to soil (T_{13}) 206 kg/h. The reason for this is the value of the mass transfer coefficients which control this transport process. The overall residence time is 632 h.

If 1000 kg/h of PCP is discharged to water, as in the second row, there is, as expected, a much higher concentration in water. There is reaction of 494 kg/h in water, advective outflow of 392 kg/h and transfer to air (T_{21}) of 2.90 kg/h with

Introduction

Chemical name: Pentachlorophenol
Fugacity Level I calculations: (six-compartment model) at environmental pH of 7

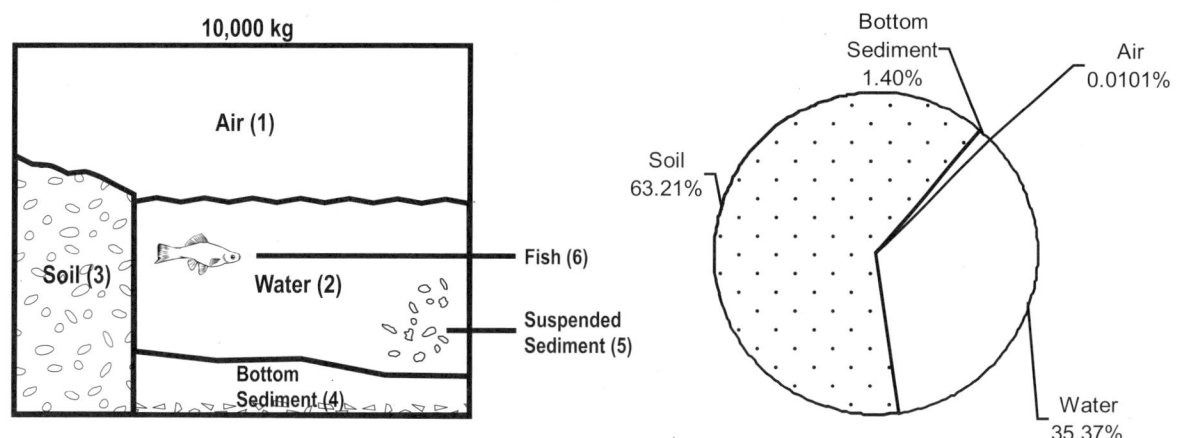

Physical-chemical properties:		Partition coefficients:		Z values in water:	
Molecular weight (g/mol)	266.34	Henry's law constant	1.42E-03	at data pH	5.1
Melting point (°C)	174	Air/water	5.73E-07	neutral	3.849
Solubility (g/m³)	14	Organic carbon, K_{OC}	4.60E+04	ionic	8.817
Vapor pressure (Pa)	4.15E-03	Bioconcentration factor, BCF	5.61E+03	total	12.666
log K_{OW}	5.05	Soil/water	3.97E+01	at environ. pH	7
Fugacity ratio, F	3.36E-02	Sediment/water	7.94E+01	neutral	3.849
Dissociation const, pKa	4.74	Suspended sediment/water	2.48E+02	ionic	700.379
		Aerosol/water	4.86E+07	total	704.228

	Fugacity, f =	9.43E-10				
Compartment	Z	Concentration			Amount	Amount
	mol/m³.Pa	mol/m³	g/m³	µg/g	kg	%
Air (1)	4.03E-04	3.80E-13	1.01E-10	8.54E-08	1.01E+01	1.01E-02
Water (2)	7.04E+02	6.64E-07	1.77E-04	1.77E-04	3.54E+04	3.54E+01
Soil (3)	2.80E+04	2.64E-05	7.02E-03	2.93E-03	6.32E+04	6.32E+01
Bottom sediment (4)	5.59E+04	5.27E-05	1.40E-02	5.85E-03	1.40E+03	1.40E+00
Suspended sediment (5)	1.75E+05	1.65E-04	4.39E-02	2.93E-02	4.39E+01	4.39E-02
Biota (6)	7.11E+04	6.70E-05	1.78E-02	1.78E-02	3.57E+00	3.57E-03
Total					1.00E+05	1.00E+02

FIGURE 1.7.17 Level I fugacity calculations for PCP at environmental pH of 7.

substantial loss of 128 kg/h to sediment. The amount in the water is 392,200 kg; thus, the residence time in the water is 392 h, and the overall environmental residence time is a longer 1153 h. The key processes are thus reaction in water (half-life 550 h) and advective outflow (residence time 1000 h). The evaporation half-life can be calculated as (0.693 × mass in water)/rate of transfer, i.e., (0.693 × 392,200)/2.90 = 93700 h. Clearly competition between advection and reaction in the water determines the overall fate. Thirty-four percent of the PCP discharged is now found in the water and the concentration is fairly high, namely 1.96×10^{-3} g/m³ or 1.96 µg/L.

The third row shows the fate if PCP is discharged to soil. The amount in soil is 245100 kg, with only 7.43 kg in air. The overall residence time is 2452 hours, which is largely controlled by the reaction rate in soil. The rate of reaction in soil is 999 kg/h and there is no advection; thus, the other loss mechanism is transfer to air (T_{31}) at a rate of 0.11 kg/h, with a relatively minor 0.8 kg/h to water by run-off. The soil concentration of 0.136 g/m³ is controlled almost entirely by the rate at which the PCP reacts.

Chemical name: Pentachlorophenol
Fugacity Level II calculations: (six-compartment model) at environmental pH of 7

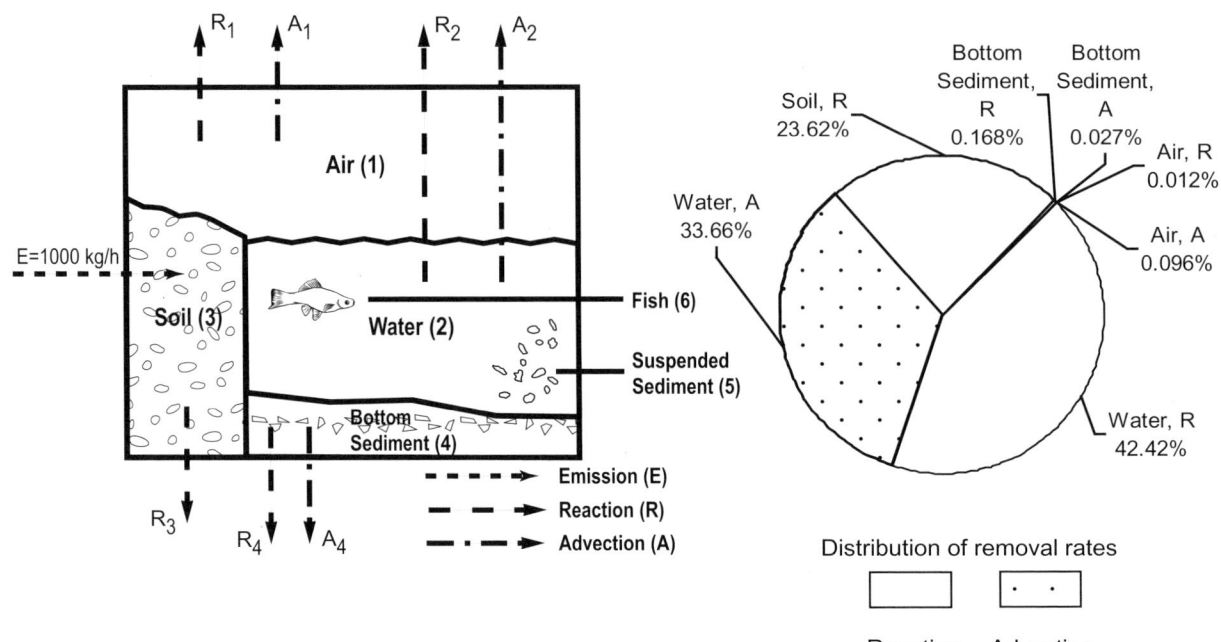

Compartment	Half-life h	D Value Reaction mol/Pa.h	D Value Advection mol/Pa.h	Concentration mol/m^3	Loss Reaction kg/h	Loss Advection kg/h	Total Removal %
Air (1)	550	5.08E+07	4.03E+08	3.59E-12	1.20E-01	9.56E-01	1.08E-01
Water (2)	550	1.77E+11	1.41E+11	6.26E-06	4.20E+02	3.34E+02	7.54E+01
Soil (3)	1700	1.03E+11	-	2.49E-04	2.34E+02	-	2.43E+01
Bottom sediment (4)	5500	7.05E+08	1.12E+08	4.97E-04	1.67E+00	2.65E-01	1.93E-01
Suspended sediment (5)	-	-	-	1.55E-03	-	-	-
Biota (6)	-	-	-	6.32E-04	-	-	-

Fugacity, f 8.89E-09 Pa
Total amount, M 3.54E+06 mol
Total amount 9.44E+05 kg
Total reaction D value, D_R 2.80E+11 mol/Pa.h
Total advection D value, D_A 1.41E+11 mol/Pa.h
Total D value, D_T 4.21E+11 mol/Pa.h
Total loss by reaction 6.65E+02 kg/h
Total loss by advection 3.35E+02 kg/h
Total loss 1.00E+03 kg/h
Reaction residence time, t_R 1.42E+03 h
Advection residence time, t_A 2.82E+03 h
Overall residence time, t_O 9.44E+02 h

FIGURE 1.7.18 Level II fugacity calculations for PCP at environmental pH of 7.

The net result is that PCP behaves entirely differently when discharged to the three media. If discharged to air, it advects rapidly and reacts with a residence time of 632 h or about 26.3 days, with substantial transport to soil or water. If discharged to water, it reacts and evaporates to air with a residence time of 1153 h or 48 days. If discharged to soil, it mostly reacts with an overall residence time of about 2452 h or 102 days.

Chemical name: Pentachlorophenol
Fugacity Level III calculations: (four-compartment model) at data pH of 5.1

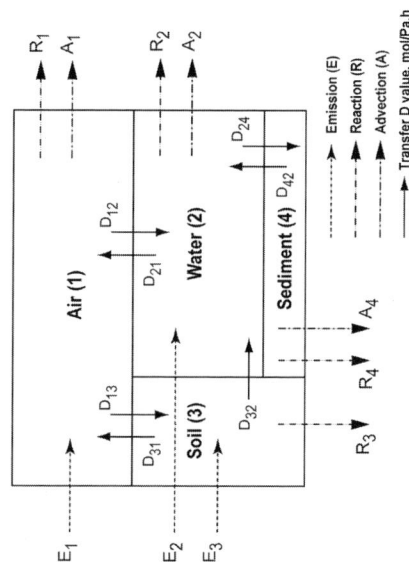

Phase Properties, Compositions, and Transport and Transformation Rates:

Emission Scenario	Emission (kg/h)			Fugacity (Pa)				Concentration (g/m³)			
	E_1	E_2	E_3	f_1	f_2	f_3	f_4	C_1	C_2	C_3	C_4
(A)ir Only	1000	0	0	6.116E-06	2.907E-08	7.526E-09	2.736E-08	6.578E-07	1.054E-04	2.804E-02	8.160E-02
(W)ater Only	0	1000	0	1.772E-08	5.410E-07	2.180E-11	5.092E-07	1.905E-09	1.961E-03	8.122E-05	1.519E+00
(S)oil Only	0	0	1000	6.909E-10	4.359E-10	3.655E-08	4.103E-10	7.430E-11	1.580E-06	1.362E-01	1.224E-03
A+W+S	50	250	700	3.107E-07	1.370E-07	2.597E-08	1.290E-07	3.342E-08	4.966E-04	9.674E-02	3.846E-01

Emission Scenario	Emission (kg/h)			Loss by Reaction (kg/h)				Loss by Advection (kg/h)				Amount (kg)				Total Amount (kg)
	E_1	E_2	E_3	R_1	R_2	R_3	R_4	A_1	A_2	A_3	A_4	w_1	w_2	w_3	w_4	
(A)ir Only	1000	0	0	8.288E+01	2.655E+01	2.06E+02	5.141E+00	6.578E+02	2.107E+01		8.160E-01	6.578E+04	2.107E+04	5.047E+05	4.080E+04	6.324E+05
(W)ater Only	0	1000	0	2.401E-01	4.942E+02	5.96E-01	9.567E+01	1.905E+00	3.922E+02		1.519E+01	1.905E+02	3.922E+05	1.462E+03	7.593E+05	1.153E+06
(S)oil Only	0	0	1000	9.362E-03	3.982E-01	9.99E+02	7.709E-02	7.430E-02	3.160E-01		1.224E-02	7.430E+00	3.160E+02	2.451E+06	6.118E+02	2.452E+06
A+W+S	50	250	700	4.211E+00	1.252E+02	7.10E+02	2.423E+01	3.342E+01	9.933E+01		3.846E+00	3.342E+03	9.933E+04	1.741E+06	1.923E+05	2.036E+06

Phase Properties and Rates:

Compartment	Bulk Z mol/m³·Pa	Half-life h	D Value Reaction mol/Pa·h	D Value Advection mol/Pa·h
Air (1)	4.038E-04	5.500E+02	5.09E+07	4.04E+08
Water (2)	1.361E+01	5.500E+02	3.43E+09	2.72E+09
Soil (3)	1.399E+04	1.700E+03	1.03E+11	-
Sediment (4)	1.120E+04	5.500E+02	7.05E+08	1.12E+08

Residence time (h)	Emission to			
	(A)ir Only	(W)ater Only	(S)oil Only	A+W+S
Overall residence time	6.324E+02	1.153E+03	2.452E+03	2.036E+03
Reaction residence time	1.974E+03	1.952E+03	2.453E+03	2.358E+03
Advection residence time	9.304E+02	2.817E+03	6.090E+06	1.491E+04

Emission Scenario	Emission (kg/h)			Intermedia Rate of Transport (kg/h)							
	E_1	E_2	E_3	T_{12}	T_{21}	T_{13}	T_{31}	T_{32}	T_{24}	T_{42}	
(A)ir Only	1000	0	0	5.358E+01	1.557E-01	2.059E+02	2.278E-02	1.647E-01	6.864E+00	9.076E-01	
(W)ater Only	0	1000	0	1.552E-01	2.897E+00	5.965E-01	6.599E-05	4.770E-04	1.278E+02	1.689E+01	
(S)oil Only	0	0	1000	6.052E-03	2.334E-03	2.326E-02	1.106E-01	7.999E-01	1.029E-01	1.361E-02	
A+W+S	50	250	700	2.722E+00	7.337E-01	1.046E+01	7.861E-02	5.863E-01	3.235E+01	4.277E+00	

FIGURE 1.7.19 Level III fugacity calculations for PCP at pH of 5.1.

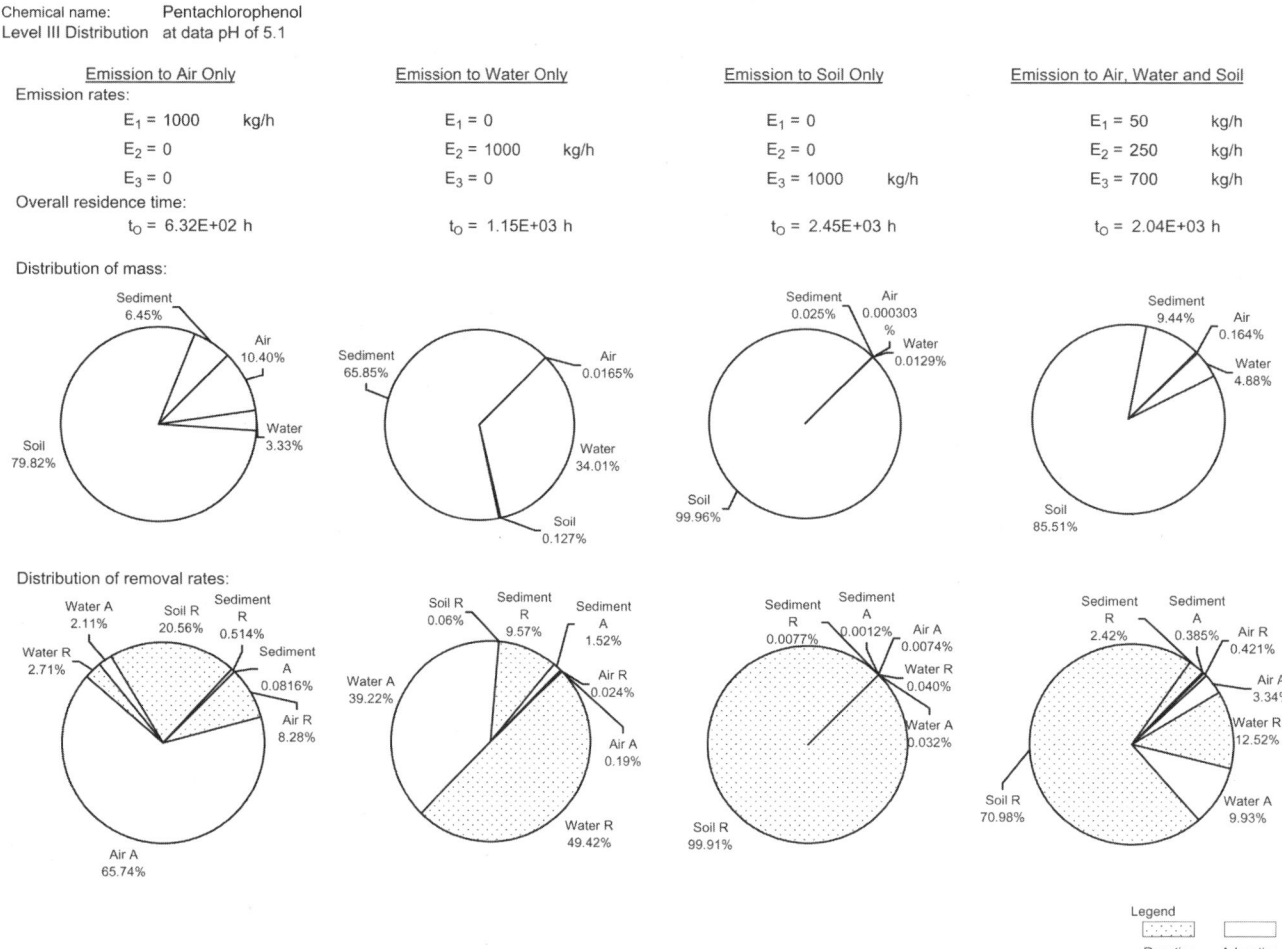

FIGURE 1.7.20 Level III fugacity distributions of PCP for four emission scenarios at pH of 5.1.

The final scenario is a combination of discharges, 50 kg/h to air, 250 kg/h to water, and 700 kg/h to soil (which are different from the often assumed equal emissions). The concentrations, amounts and transport and transformation rates are merely linearly combined versions of the three initial scenarios. For example, the rate of reaction in air is now 4.21 kg/h. This is 0.05 of the first (air emission) rate of 82.9 kg/h, i.e., 4.14 kg/h, plus 0.25 of the second (water emission) rate of 0.24 kg/h, i.e., 0.06 kg/h and 0.7 of the third (soil emission) rate of 0.0094 kg/h, i.e., 0.0066 kg/h yielding a total of (4.14 + 0.06 + 0.0066) or 4.21 kg/h. It is also apparent that the amount in the air of 3342 kg causing a concentration of 3.342×10^{-8} g/m^3 or 33 ng/m^3 is attributable to emissions to air (0.05×658 or 33 ng/m^3), emissions to water (0.25×1.9 or 0.5 ng/m^3) and emissions to soil (0.7×0.0743 or 0.052 µg/m^3). The concentration in water of 4.97×10^{-4} g/m^3, or 497 ng/L, is largely attributable to the discharges to water, which alone cause $0.25 \times 1.96 \times 10^{-3}$ g/m^3 or 4.9×10^{-4} g/m^3 or 490 µg/m^3, or 490 ng/L. Although more is emitted to soil, it contributes only about 1.1 µg/m^3 to the water with air emissions accounting for about 5.27 µg/m^3. Similarly, the prevailing soil concentration is controlled by the rate of discharge to the soil.

In this multimedia discharge scenario the overall residence time is 2036 h, which can be viewed as the sum of 5% of the air emission residence time of 632 h, 25% of the water emission residence time of 1153 h and 70% of the soil emission residence time of 2452 h. The overall amount in the environment of 2.04×10^6 kg is thus largely controlled by the discharges to soil and water.

Finally, it is interesting to note that the fugacities in this final case (in units of µPa) are for the four media: 0.31 (air), 0.137 (water), 0.026 (soil) and 0.129 (sediment). The media are fairly close to equilibrium, i.e., within a factor of about 5 of the average value.

At pH 7, Figure 1.7.21, the capacity of water for PCP increases; thus, the water compartment becomes more important as do intermedia transport processes involving water such as wet deposition in dissolved form and run-off

Introduction

Chemical name: Pentachlorophenol
Fugacity Level III calculations: (four-compartment model) at environmental pH of 7

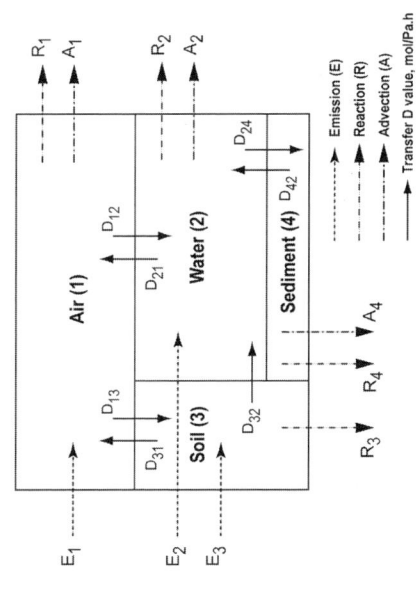

Phase Properties and Rates:

Compartment	Bulk Z mol/m³·Pa	Half-life h	D Value Reaction mol/Pa.h	Advection mol/Pa.h
Air (1)	4.038E-04	5.500E+02	5.09E+07	4.04E+08
Water (2)	7.052E+02	5.500E+02	1.78E+11	1.41E+11
Soil (3)	1.420E+04	1.700E+03	1.04E+11	-
Sediment (4)	1.175E+04	5.500E+03	7.40E+08	1.18E+08

	Emission to			
Residence time (h)	(A)ir Only	(W)ater Only	(S)oil Only	A+W+S
Overall residence time	2.074E+03	4.588E+02	2.393E+03	1.894E+03
Reaction residence time	2.319E+03	8.218E+02	2.426E+03	2.164E+03
Advection residence time	1.961E+04	1.039E+03	1.805E+05	1.515E+04

Phase Properties, Compositions, and Transport and Transformation Rates:

Emission Scenario	Emission (kg/h)			Fugacity (Pa)				Concentration (g/m³)				Amount (kg)				Total Amount (kg)
	E_1	E_2	E_3	f_1	f_2	f_3	f_4	C_1	C_2	C_3	C_4	w_1	w_2	w_3	w_4	
(A)ir Only	1000	0	0	4.907E-07	1.408E-09	2.958E-08	1.328E-09	5.278E-08	2.645E-04	1.118E-01	4.158E-03	5.278E+03	5.290E+04	2.013E+06	2.078E+03	2.074E+06
(W)ater Only	0	1000	0	3.097E-11	1.175E-08	1.867E-12	1.108E-08	3.331E-12	2.207E-03	7.060E-06	3.467E-02	3.331E-01	4.413E+05	1.271E+02	1.733E+04	4.588E+05
(S)oil Only	0	0	1000	6.453E-10	3.510E-10	3.497E-08	3.309E-10	6.940E-11	6.592E-05	1.322E-01	1.036E-03	6.940E+00	1.318E+04	2.380E+06	5.178E+02	2.393E+06
A+W+S	50	250	700	2.500E-08	3.253E-09	2.596E-08	3.067E-09	2.688E-09	6.110E-04	9.814E-02	9.600E-03	2.688E+02	1.222E+05	1.766E+06	4.800E+03	1.894E+06

Emission Scenario	Emission (kg/h)			Loss by Reaction (kg/h)				Loss by Advection (kg/h)				Intermedia Rate of Transport (kg/h)							
	E_1	E_2	E_3	R_1	R_2	R_3	R_4	A_1	A_2	A_3	A_4	T_{12}	T_{21}	T_{13}	T_{31}	T_{24}	T_{32}	T_{42}
(A)ir Only	1000	0	0	6.650E+00	6.665E-02	8.21E+02	2.618E-01	5.278E-01	5.290E+01	4.156E-02	4.156E-02	9.470E-01	7.565E-03	8.470E-02	1.112E+00	5.920E-01	2.517E+01	2.886E-01
(W)ater Only	0	1000	0	4.197E-04	5.561E+02	5.18E-02	2.184E+00	3.331E-03	4.413E+02	3.467E-02	3.467E-02	5.977E-02	6.312E-02	5.346E-02	7.020E-05	4.939E+00	1.589E-03	2.408E+00
(S)oil Only	0	0	1000	8.745E-03	1.661E+01	9.70E+02	6.525E-02	6.940E-02	1.318E+01	1.036E-02	1.036E-02	1.245E-01	1.885E-01	1.114E+00	1.315E+00	1.475E-01	2.975E+01	7.193E-02
A+W+S	50	250	700	3.387E-01	1.540E+02	7.20E+02	6.048E-01	2.688E+00	1.222E+02	9.600E-02	9.600E-02	4.823E+00	1.748E+00	4.314E+01	9.758E-01	1.368E+00	2.208E+01	6.667E-01

FIGURE 1.7.21 Level III fugacity calculations for PCP at pH of 7.

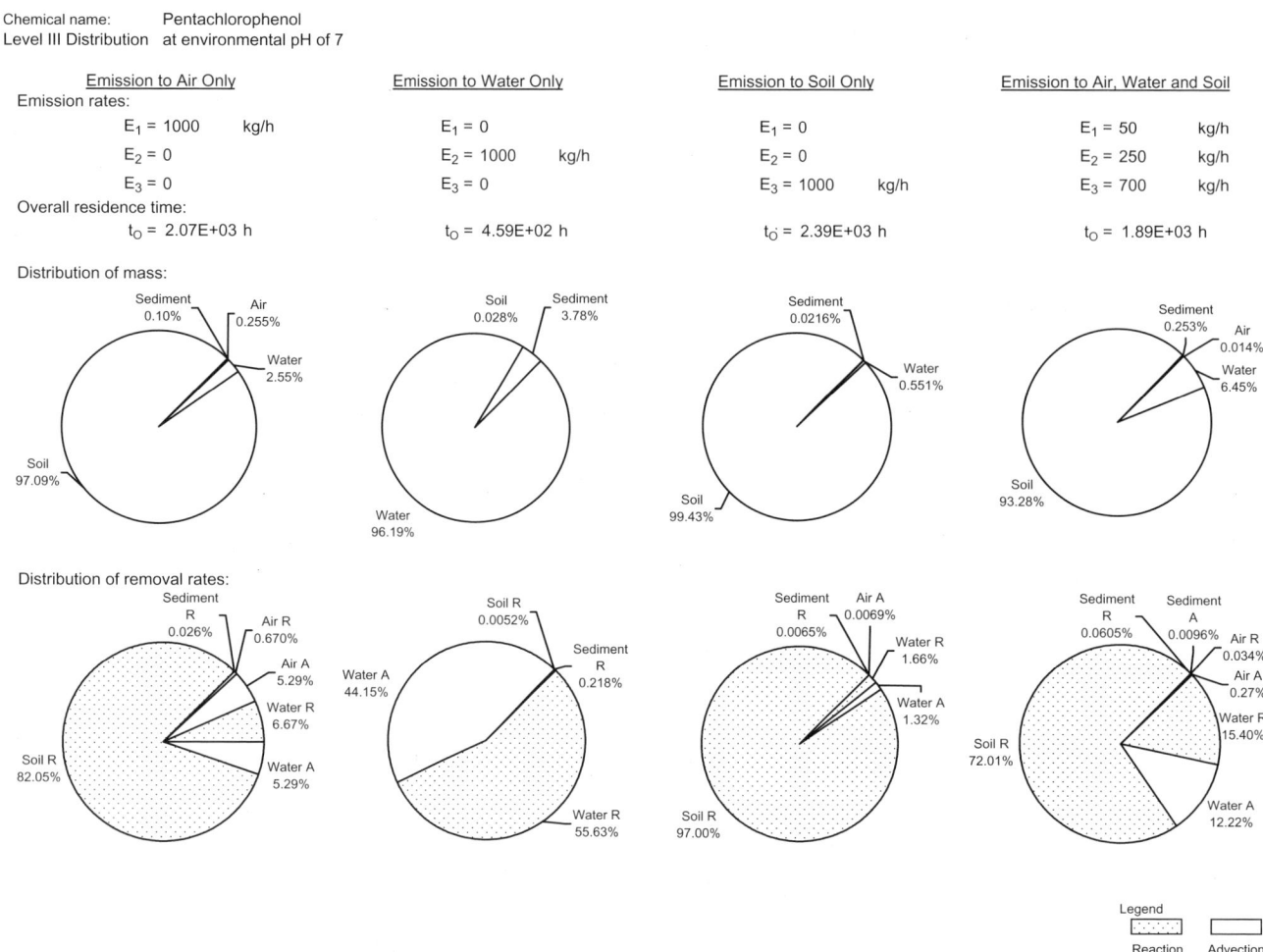

FIGURE 1.7.22 Level III fugacity distributions of PCP for four emission scenarios at pH of 7.

from soil to water. The net effect is that if discharged to air, the amounts transferred to soil and water increase as does the overall residence time. If discharged to water, there is less water to sediment transfer because of the reduced apparent hydrophobicity, and the residence time decreases. If discharged to soil, there is little effect of the pH increase because the PCP tends to remain there.

Similar diagrams could be prepared for other phenolic compounds at a range of pH values. The results suggest that the same broad patterns of behavior apply as for PCP but the residence times are generally shorter because of reduced hydrophobicity and more rapid reactions. The lower chlorinated phenols are relatively short-lived and are not subject to appreciable intermedia transport, i.e., when discharged to a medium they tend to remain there until degraded or advected. The longest persistence occurs when the chemical is present in soils.

Such simulations suggest that because of their relatively high water solubility which in combination with low vapor pressure causes low air-water partition coefficients, the phenols tend to remain in water or in soil and show little tendency to evaporate. Their environmental fate tends to be dominated by reaction in soil and water, and for the more sorptive species, in sediments. Their half-lives are relatively short, because of their susceptibility to degradation.

It is believed that examining these three behavior profiles, and their combination in the fourth, illustrate and explain the environmental fate characteristics of this and other chemicals. Important intermedia transport processes and levels in various media that arise from discharges into other media become clear. It is believed that the broad characteristics of environmental fate as described in the generic environment are generally applicable to other environments, albeit with differences attributable to changes in volumes, temperature, flow rates and compartment compositions.

1.8 REFERENCES

Abernethy, S., Mackay, D., McCarty, L. S. (1988) Volume fraction correlation for narcosis in aquatic organisms: The key role of partitioning. *Environ. Toxicol. Chem.* 7, 469–481.

Abramowitz, R., Yalkowsky, S. H. (1990) Estimation of aqueous solubility and melting point of PCB congeners. *Chemosphere* 21, 1221–1229.

Alaee, M., Whittal, R.M., Strachan, W.M.J. (1996) The effect of water temperature and composition on Henry's law constant for various PAH's. *Chemosphere* 32, 1153–1164.

Almgren, M., Grieser, F., Powell, J. R., Thomas, J. K. (1979) A correlation between the solubility of aromatic hydrocarbons in water and micellar solutions, with their normal boiling points. *J. Chem. Eng. Data* 24, 285–287.

Al-Sahhaf, T. A. (1989) Prediction of the solubility of hydrocarbons in water using UNIFAC. *J. Environ. Sci. Health* A24, 49–56.

Altshuller, A. P. (1980) Lifetimes of organic molecules in the troposphere and lower stratosphere. *Adv. Environ. Sci. Technol.* 10, 181–219.

Altshuller, A. P. (1991) Chemical reactions and transport of alkanes and their products in the troposphere. *J. Atmos. Chem.* 12, 19–61.

Ambrose, D. (1981) Reference value of vapor pressure. The vapor pressures of benzene and hexafluorobenzene. *J. Chem. Thermodyn.* 13, 1161–1167.

Ambrose, D., Lawrenson, L. J., Sprake, C. H. S. (1975) The vapour pressure of naphthalene. *J. Chem. Thermodyn.* 7, 1173–1176.

Amidon, G. L., Yalkowsky, S. H., Anik, S. T., Leung, S. (1975) Solubility of nonelectrolytes in polar solvents. V. Estimation of the solubility of aliphatic monofunctional compounds in water using a molecular surface area approach. *J. Phys. Chem.* 9, 2239–2245.

Amidon, G. L., Anik, S. T. (1981) Application of the surface area approach to the correlation and estimation of aqueous solubility and vapor pressure. Alkyl aromatic hydrocarbons. *J. Chem. Eng. Data* 26, 28–33.

Amidon, G. L., Williams, N. A. (1982) A solubility equation for non-electrolytes in water. *Intl. J. Pharm.* 11, 156–249.

Anderson, E., Veith, G. D., Weininger, D. (1987) *SMILES: A Line Notation and Computerized Interpreter for Chemical Structures.* EPA Environmental Research Brief, U.S. EPA, EPA/600/M-87/021.

Andren, A. W., Doucette, W. J., Dickhut, R. M. (1987) Methods for estimating solubilities of hydrophobic organic compounds: Environmental modeling efforts. In: *Sources and Fates of Aquatic Pollutants.* Hites, R. A., Eisenreich, S. J., Eds., pp. 3–26, Advances in Chemistry Series 216, American Chemical Society, Washington, D.C.

Arbuckle, W. B. (1983) Estimating activity coefficients for use in calculating environmental parameters. *Environ. Sci. Technol.* 17, 537–542.

Arbuckle, W. B. (1986) Using UNIFAC to calculate aqueous solubilities. *Environ. Sci. Technol.* 20, 1060–1064.

Arcand, Y., Hawari, J., Guiot, S. R. (1995) Solubility of pentachlorophenol in aqueous solutions: the pH effect. *Water Res.* 29, 131–136.

Arnold, D.S., Plank, C.A., Erickson, E.E., Pike, F.P. (1958) Solubility of benzene in water. *Chem. Eng. Data Ser.* 3, 253–256.

Ashworth, R. A., Howe, G. B., Mullins, M. E., Roger, T. N. (1988) Air-water partitioning coefficients of organics in dilute aqueous solutions. *J. Hazard. Materials* 18, 25–36.

Atkinson, R. (1985) Kinetics and mechanisms of the gas phase reaction of hydroxyl radicals with organic compounds under atmospheric conditions. *Chem. Rev.* 85, 69–201.

Atkinson, R. (1987) A structure-activity relationship for the estimation of the rate constants for the gas phase reactions of OH radicals with organic compounds. *Int. J. Chem. Kinetics* 19, 790–828.

Atkinson, R. (1989) Kinetics and mechanisms of the gas-phase reactions of the hydroxyl radical with organic compounds. *J. Phys. Chem. Ref. Data Monograph No. 1.* 1–246.

Atkinson, R. (1990) Gas-phase tropospheric chemistry of organic compounds, a review. *Atmos. Environ.* 24A, 1–41.

Atkinson, R. (1991) Kinetics and mechanisms of the gas-phase reactions of the NO_3 radicals with organic compounds. *J. Phys. Chem. Data* 20, 450–507.

Atkinson, R., Carter, W. L. (1984) Kinetics and mechanisms of the gas-phase reactions of ozone with organic compounds under atmospheric conditions. *Chem. Rev.* 84, 437–470.

Augustijn-Beckers, P. W. M., Hornsby, A. G., Wauchope, R. D. (1994) The SCS/ARS/CES pesticide properties database for environmental decision making. II. Additional compounds. *Rev. Environ. Contam. Toxicol.* 137, 1–82.

Bahnick, D. A., Doucette, W. J. (1988) Use of molecular connectivity indices to estimate soil sorption coefficients for organic chemicals. *Chemosphere* 17, 1703–1715.

Balson, E. W. (1947) Studies in vapour pressure measurement. Part III. An effusion manometer sensitive to 5×10^{-6} millimetres of mercury: vapour pressure of D.D.T. and other slightly volatile substances. *Trans. Farad. Soc.* 43, 54–60.

Banerjee, S. (1985) Calculation of water solubility of organic compounds with UNIFAC-derived parameters. *Environ. Sci. Technol.* 19, 369–370.

Banerjee, S., Howard, P. H. (1988) Improved estimation of solubility and partitioning through correction of UNIFAC-derived activity coefficients. *Environ. Sci. Technol.* 22, 839–841.

Banerjee, S., Howard, P. H., Lande, S. S. (1990) General structure-vapor pressure relationships for organics. *Chemosphere* 21, 1173–1180.

Banerjee, S., Yalkowsky, S. H., Valvani, S. C. (1980) Water solubility and octanol/water partition coefficients of organics. Limitations of the solubility-partition coefficient correlation. *Environ. Sci. Technol.* 14, 1227–1229.

Barton, A.F.M. (1984) IUPAC Solubility Data Series Vol. 15. *Alcohols With Water.* Pergamon Press Inc., Oxford, England, U.K.

Betterton, E. A., Hoffmann, M. R. (1988) Henry's law constants of some environmentally important aldehydes. *Environ. Sci. Technol.* 22, 1415–1418.

Beyer, A., Mackay, D., Matthies, M., Wania, F., Webster, E. (2000) Assessing long-range transport potential of persistent organic pollutants. *Environ. Sci. Technol.* 34,699–703.

Bidleman, T. F. (1984) Estimation of vapor pressures for nonpolar organic compounds by capillary gas chromatography. *Anal. Chem.* 56, 2490–2496.

Boethling, R.S., Mackay, D. Eds. (2000) Handbook of Property Estimation Methods: Environmental and Health Sciences. CRC Press, Boca Raton, FL.

Bohon, R. L., Claussen, W. F. (1951) The solubility of aromatic hydrocarbons in water. *J. Am. Chem. Soc.* 73, 1571–1576.

Bondi, A. (1964) van der Waals volumes and radii. *J. Phys. Chem.* 68, 441–451.

Booth, H. S., Everson, H. E. (1948) Hydrotropic solubilities: solubilities in 40 percent sodium xylenesulfonate. *Ind. Eng. Chem.* 40, 1491–1493.

Boublik, T., Fried, V., Hala, E. (1973) *The Vapor Pressure of Pure Substances*, Elsevier, Amsterdam, The Netherlands.

Boublik, T., Fried, V., Hala, E. (1984) *The Vapor Pressure of Pure Substances*, 2nd revised Edition, Elsevier, Amsterdam, The Netherlands.

Bowman, B. T., Sans, W. W. (1983) Determination of octanol-water partitioning coefficient (K_{OW}) of 61 organophosphorus and carbamate insecticides and their relationship to respective water solubility (S) values. *J. Environ. Sci. Health* B18, 667–683.

Bradley, R. S., Cleasby, T. G. (1953) The vapour pressure and lattice energy of some aromatic ring compounds. *J. Chem. Soc.* 1953, 1690–1692.

Breton, R., Schuurmann, G., Purdy, R. (2003) Proceedings 10th International Quantitative Structure Activity Relationships workshop. *QSAR and Combinatorial Science* 22, 1–409.

Briggs, G. G. (1981) Theoretical and experimental relationships between soil adsorption, octanol-water partition coefficients, water solubilities, bioconcentration factors, and the parachor. *J. Agric. Food Chem.* 29, 1050–1059.

Brodsky, J., Ballschmiter, K. (1988) Reversed phase liquid chromatography of PCBs as a basis for the calculation of water solubility and log K_{OW} for polychlorobiphenyls. *Fresenius Z. Anal. Chem.* 331, 295–301.

Bruggeman, W. A., van der Steen, J., Hutzinger, O. (1982) Reversed-phase thin-layer chromatography of polynuclear aromatic hydrocarbons and chlorinated biphenyls. Relationship with hydrophobicity as measured by aqueous solubility and octanol-water partition coefficient. *J. Chromatogr.* 238, 335–346.

Budavari, S., Editor (1989) *The Merck Index. An Encyclopedia of Chemicals, Drugs and Biologicals*. 11th Edition, Merck & Co., Rahway, New Jersey.

Burkhard, L.P. (1984) Physical-Chemical Properties of the Polychlorinated Biphenyls: Measurement, Estimation, and Application to Environmental Systems. Ph.D. Thesis, University of Wisconsin-Madison, Wisconsin.

Burkhard, L. P., Andren, A. W., Armstrong, D. E. (1985a) Estimation of vapor pressures for polychlorinated biphenyls: A comparison of eleven predictive methods. *Environ. Sci. Technol.* 19, 500–507.

Burkhard, L. P., Armstrong, D. E., Andren, A. W. (1985b) Henry's law constants for polychlorinated biphenyls. *Environ. Sci. Technol.* 590–595.

Burkhard, L. P., Kuehl, D. W., Veith G. D. (1985c) Evaluation of reversed phase liquid chromatograph/mass spectrometry for estimation of n-octanol/water partition coefficients of organic chemicals. *Chemosphere* 14, 1551–1560.

Burnett, M. G. (1963) Determination of partition coefficients in infinite dilution by the gas chromatographic analysis of the vapor above dilute solutions. *Anal. Chem.* 35, 1567–1570.

Butler, J. A. V., Ramchandani, C. N., Thomson, D. W. (1935) The solubility of non-electrolytes. Part I. The free energy of hydration of some aliphatic alcohols. *J. Chem. Soc.* 280–285.

Buttery, R. B., Ling, L. C., Guadagni, D. G. (1969) Volatilities of aldehydes, ketones, and esters in dilute water solution. *J. Agric. Food Chem.* 17, 385–389.

Buxton, G. V., Greenstock, G. L., Helman, W. P., Ross, A. B. (1988) Critical review of rate constants for reactions of hydrated electrons, hydrogen atoms and hydroxyl radicals in aqueous solutions. *J. Phys. Chem. Ref. Data* 17, 513–886.

Chao, J., Lin, C.T., Chung, T.H. (1983) Vapor pressure of coal chemicals. *J. Phys. Chem. Ref. Data* 12, 1033–1063.

Chickos, J.S., Acree, W.E., Jr., Liebman, J.F. (1999) Estimating solid-liquid phase change enthalpies and entropies. *J. Phys. Chem. Ref. Data* 28, 1535–1673.

Chiou, C. T. (1985) Partition coefficients of organic compounds in lipid-water systems and correlations with fish bioconcentration factors. *Environ. Sci. Technol.* 19, 57–62.

Chiou, C. T., Freed, V. H., Schmedding, D. W. (1977) Partition coefficient and bioaccumulation of selected organic chemicals. *Environ. Sci. Technol.* 11, 475–478.

Chiou, C. T., Schmedding, D. W., Manes, M. (1982) Partitioning of organic compounds in octanol-water system. *Environ. Sci. Technol.* 16, 4–10.

Choi, J., Amoine, S. (1974) Adsorption of pentachlorophenol by soils. *Soil Sci. Plant Nutr.* 20, 135–144.

Clark, E.C.W., Glew, D.N. (1966) Evaluation of thermodynamic functions from equilibrium constants. *Trans. Farad. Soc.* 62, 539–547.

Cole, J.G., Mackay, D. (2000) Correlating environmental partitioning properties of organic compounds: The "three solubility approach". *Environ. Toxicol. Chem.* 19, 265–270.

Davies, R. P., Dobbs, A. J. (1984) The prediction of bioconcentration in fish. *Water Res.* 18, 1253–1262.

Introduction

Davis, W. W., Krahl, M. E., Clowes, G. H. (1942) Solubility of carcinogenic and related hydrocarbons in water *J. Am. Chem. Soc.* 64, 108–110.

Davis, W. W., Parke, Jr., T. V. (1942) A nephelometric method for determination of solubilities of extremely low order. *J. Am. Chem. Soc.* 64, 101–107.

Dean, J. D., Ed. (1979) *Lange's Handbook of Chemistry.* 12th Edition, McGraw-Hill, New York.

Dean, J. D., Ed. (1985) *Lange's Handbook of Chemistry.* 13th Edition, McGraw-Hill, New York.

Dean, J. D., Ed. (1992) *Lange's Handbook of Chemistry.* 14th Edition, McGraw-Hill, New York.

Dearden, J. C. (1990) Physico-chemical descriptors. In: *Practical Applications of Quantitative Structure-Activity Relationships (QSAR) in Environmental Chemistry and Toxicology.* Karcher, W. and Devillers, J., Eds., pp. 25–60. Kluwer Academic Publisher, Dordrecht, The Netherlands.

De Bruijn, J., Busser, G., Seinen, W., Hermens, J. (1989) Determination of octanol/water partition coefficient for hydrophobic organic chemicals with the "slow-stirring" method. *Environ. Toxicol. Chem.* 8, 499–512.

De Bruijn, J., Hermens, J. (1990) Relationships between octanol/water partition coefficients and total molecular surface area and total molecular volume of hydrophobic organic chemicals. *Quant. Struct.-Act. Relat.* 9, 11–21.

De Kruif, C. G. (1980) Enthalpies of sublimation and vapour pressures of 11 polycyclic hydrocarbons. *J. Chem. Thermodyn.* 12, 243–248.

Dobbs, A. J., Grant, C. (1980) Pesticide volatilization rate: a new measurement of the vapor pressure of pentachlorophenol at room temperature. *Pestic. Sci.* 11, 29–32.

Dobbs, A. J., Cull, M. R. (1982) Volatilization of chemical relative loss rates and the estimation of vapor pressures. *Environ. Pollut. Ser.B.*, 3, 289–298.

Doucette, W. J., Andren, A. W. (1987) Correlation of octanol/water partition coefficients and total molecular surface area for highly hydrophobic aromatic compounds. *Environ. Sci. Technol.* 21, 521–524.

Doucette, W. J., Andren, A. W. (1988a) Aqueous solubility of selected biphenyl, furan, and dioxin congeners. *Chemosphere* 17, 243–252.

Doucette, W. J., Andren, A. W. (1988b) Estimation of octanol/water partition coefficients: Evaluation of six methods for highly hydrophobic aromatic hydrocarbons. *Chemosphere* 17, 345–359.

Dreisbach, R. R. (1955) *Physical Properties of Chemical Compounds.* No. 15 of the Adv. in Chemistry Series, American Chemical Society, Washington, D.C.

Dreisbach, R. R. (1959) *Physical Properties of Chemical Compounds-II.* No. 22, Adv. in Chemistry Series, American Chemical Society, Washington, D.C.

Dreisbach, R. R. (1961) *Physical Properties of Chemical Compounds-III.* No. 29, Adv. in Chemistry Series, American Chemical Society, Washington, D.C.

Dunnivant, F. M., Coate, J. T., Elzerman, A. W. (1988) Experimentally determined Henry's law constants for 17 polychlorobiphenyl congeners. *Environ. Sci. Technol.* 22, 448–453.

Dunnivant, F. M., Elzerman, A. W., Jurs, P. C., Hansen, M. N. (1992) Quantitative structure-property relationships for aqueous solubilities and Henry's law constants of polychlorinated biphenyls. *Environ. Sci. Technol.* 26, 1567–1573.

Ellgehausen, H., D'Hondt, C., Fuerer, R. (1981) Reversed-phase chromatography as a general method for determining octan-1-ol/water partition coefficients. *Pestic. Sci.* 12, 219–227.

Ellington, J. J. (1989) *Hydrolysis Rate Constants for Enhancing Property-Reactivity Relationships.* U.S. EPA, EPA/600/3-89/063, Athens, GA.

Ellington, J. J., Stancil, Jr., F. E., Payne, W. D. (1987a) *Measurements of Hydrolysis Rate Constant for Evaluation of Hazardous Land Disposal: Volume I. Data on 32 Chemicals* U.S. EPA, EPA/600/3-86/043, Athens, GA.

Ellington, J. J., Stancil, Jr., F. E., Payne, W. D., Trusty, C. D. (1987b) *Measurements of Hydrolysis Rate Constant for Evaluation of Hazardous Land Disposal: Volume II. Data on 54 Chemicals.* U.S. EPA, EPA/600/3-88/028, Athens, GA.

Ellington, J. J., Stancil, Jr., F. E., Payne, W. D., Trusty, C. D. (1988) *Interim Protocol for Measurement Hydrolysis Rate Constants in Aqueous Solutions.* USEPA, EPA/600/3-88/014, Athens, GA.

Fendinger, N. J., Glotfelty, D. E. (1988) A laboratory method for the experimental determination of air-water Henry's law constants for several pesticides. *Environ. Sci. Technol.* 22, 1289–1293.

Fendinger, N. J., Glotfelty, D. E. (1989) A comparison of two experimental techniques for determining air-water Henry's laws constants. *Environ. Sci. Technol.* 23, 1528–1531.

Fendinger, N. J., Glotfelty, D. E. (1990) Henry's law constants for selected pesticides, PAHs and PCBs. *Environ. Toxicol. Chem.* 9, 731–735.

Finizio, A., Di Guardo, A., Vighi, M. (1994) Improved RP-HPLC determination of K_{OW} for some chloroaromatic chemicals using molecular connectivity indices. *SAR & QSAR in Environ. Res.* 2, 249–260.

Finizio, A., Mackay, D., Bidleman, T.F., Harner, T. (1997) Octanol-air partition coefficient as a predictor of partitioning of semi-volatile organic chemicals to aerosols. *Atmos. Environ.* 31, 2289–2296.

Foreman, W. T., Bidleman, T. F. (1985) Vapor pressure estimates of individual polychlorinated biphenyls and commercial fluids using gas chromatographic retention data. *J. Chromatogr.* 330, 203–216.

Fredenslund, A., Jones, R. L., Prausnitz, J. M. (1975) Group-contribution estimation of activity coefficients in nonideal liquid mixtures. *AIChE J.* 21, 1086–1099.

Fujita, T., Iwasa, J., Hansch, C. (1964) A new substituent constant, "pi" derived from partition coefficients. *J. Am. Chem. Soc.* 86, 5175–5180.

Garst, J.E. (1984) Accurate, wide-range, automated, high-performance chromatographic method for the estimation of octanol/water partition coefficients. II: Equilibrium in partition coefficient measurements, additivity of substituent constants, and correlation of biological data. *J. Pharm. Sci.* 73, 1623–1629.

Garten, C. T., Trabalka, J. R. (1983) Evaluation of models for predicting terrestrial food chain behavior of xenobiotics. *Environ. Sci. Technol.* 17, 590–595.

Gerstl, Z., Helling, C. S. (1984) Evaluation of molecular connectivity as a predictive method for the adsorption of pesticides by soils. *J. Environ. Sci. Health* B22, 55–69.

Glew, D.N., Roberson, R.E. (1956) The spectrophotometric determination of the solubility of cumene in water by a kinetic method. *J. Phys. Chem.* 60, 332–337.

Gmehling, J., Rasmussen, P., Fredenslund, A. (1982) Vapor-liquid equilibria by UNIFAC group contribution. Revision and extension. 2 *Ind. Eng. Chem. Process Des. Dev.* 21, 118–127.

Gossett, R. (1987) Measurement of Henry's law constants for C_1 and C_2 chlorinated hydrocarbons. *Environ. Sci. Technol.* 21, 202–208.

Gross, P. M., Saylor, J. H. (1931) The solubilities of certain slightly soluble organic compounds in water. *J. Am. Chem. Soc.* 1931, 1744–1751.

Gückel, W., Rittig, R., Synnatschke, G. (1974) A method for determining the volatility of active ingredients used in plant protection. II. Application to formulated products. *Pestic. Sci.* 5, 393–400.

Gückel, W., Kästel, R., Lawerenz, J., Synnatschke, G. (1982) A method for determining the volatility of active ingredients used in plant protection. Part III: The temperature relationship between vapour pressure and evaporation rate. *Pestic. Sci.* 13, 161–168.

Hafkenscheid, T. L., Tomlinson, E. (1981) Estimation of aqueous solubilities of organic non-electrolytes using liquid chromatographic retention data. *J. Chromatogr.* 218, 409–425.

Hamaker, J. W., Kerlinger, H. O. (1969) Vapor pressures of pesticides. *Adv. Chem. Ser.* 86, 39–54.

Hamilton, D. J. (1980) Gas chromatographic measurement of volatility of herbicide esters. *J. Chromatogr.* 195, 75–83.

Hansch, C., Leo, A. (1979) *Substituent Constants for Correlation Analysis in Chemistry and Biology*. Wiley-Interscience, New York.

Hansch. C., Leo, A. J., Hoekman, D. (1995) *Exploring QSAR, Hydrophobic, Electronic, and Steric Constants*. ACS Professional Reference Book, American Chemical Society, Washington, D.C.

Hansch, C., Quinlan, J. E., Lawrence, G. L. (1968) The linear-free energy relationship between partition coefficient and aqueous solubility of organic liquids. *J. Org. Chem.* 33, 347–350.

Hansen, H. K., Schiller, M., Gmehling, J. (1991) Vapor-liquid equilibria by UNIFAC group contribution. 5. Revision and extension. *Ind. Eng. Chem. Res.* 30, 2362–2356.

Harner, T. and Bidleman, T.F. (1996) Measurements of octanol-air partition coefficients for polychlorinated biphenyls. *J. Chem. Eng. Data* 41, 895–899.

Harner T. and Mackay, D. (1995) Measurement of octanol-air partition coefficients for chlorobenzenes, PCBs, and DDT. *Environ. Sci. Technol.* 29, 1599–1606.

Hartley, G. S., Gram-Bryce, I. J. (1980) *Physical Principles of Pesticide Behavior, the Dynamics of Applied Pesticides in the Local Environments in Relation to Biological Response*. Vol. 2, Academic Press, London.

Hartley, D., Kidd, H., Eds., (1987) *The Agrochemical Handbook*, 2nd ed., The Royal Society of Chemistry, The University of Nottingham, England.

Hawker, D. W. (1989) The relationship between octan-1-ol/water partition coefficient and aqueous solubility in terms of solvatochromic parameters. *Chemosphere* 19, 1586–1593.

Hawker, D. W. (1990a) Vapor pressures and Henry's law constants of polychlorinated biphenyls. *Environ. Sci. Technol.* 23, 1250–1253.

Hawker, D. W. (1990b) Description of fish bioconcentration factors in terms of solvatochromic parameters. *Chemosphere* 20, 267–477.

Hawker, D. W., Connell, D. W. (1988) Octanol-water partition coefficients of polychlorinated biphenyl congeners. *Environ. Sci. Technol.* 22, 382–387.

Heidman, J. L., Tsonopoulos, C., Brady, C. J., Wilson, G. M. (1985) High-temperature mutual solubilities of hydrocarbons and water. *AIChE J.* 31, 376–384.

Hermann, R. B. (1971) Theory of hydrophobic bonding. II. The correlation of hydrocarbon solubility in water with solvent cavity surface area. *J. Phys. Chem.* 76, 2754–2758.

Hermens, J. L. M., Opperhuizen, A., Eds. (1991) *QSAR in Environmental Toxicology IV*. Elsevier, Amsterdam, The Netherlands. Also published in *Sci. Total Environ.* vol. 109/110.

Hildebrand,, J. H., Prausnitz, J. M., Scott, R. L. (1970) *Regular and Related Solutions. The Solubility of Gases, Liquids, and Solids*. Van Nostrand Reinhold. Co., New York.

Hinckley, D. A., Bidleman, T.F., Foreman, W.T. (1990) Determination of vapor pressures for nonpolar and semipolar organic compounds from gas chromatographic retention data. *J. Chem. Eng. Data* 35, 232–237.

Hine, J., Mookerjee, P. K. (1975) The intrinsic hydrophilic character of organic compounds. Correlations in terms of structural contributions. *J. Org. Chem.* 40, 292–298.

Hippelein, M. and McLachlan, M.S. (1998) Soil air partitioning of semivolatile organic chemicals. 1. Method development and influence of physical chemical properties. *Environ. Sci. Technol.* 32, 310–316.

Hollifield, H. C. (1979) Rapid nephelometric estimate of water solubility of highly insoluble organic chemicals of environmental interest. *Bull. Environ. Contam. Toxicol.* 23, 579–586.

Hornsby, A. G., Wauchope, R. D., Herner, A. E. (1996) *Pesticide Properties in the Environment.* Springer-Verlag, New York.

Horvath, A. L. (1982) *Halogenated Hydrocarbons, Solubility - Miscibility with Water.* Marcel Dekker, New York.

Horvath, A.L., Getzen, F.W. Eds. (1985) IUPAC Solubility Data Series: Vol. 20. *Halogenated Benzenes, Toluenes and Phenols with Water.* Pergamon Press, Oxford.

Howard, P. H., Ed. (1989) *Handbook of Fate and Exposure Data for Organic Chemicals. Vol. I. Large Production and Priority Pollutants.* Lewis Publishers, Chelsea, MI.

Howard, P. H., Ed. (1990) *Handbook of Fate and Exposure Data for Organic Chemicals. Vol. II. Solvents.* Lewis Publishers, Chelsea, MI

Howard, P. H., Ed. (1991) *Handbook of Fate and Exposure Data for Organic Chemicals. Vol. III. Pesticides.* Lewis Publishers, Chelsea, MI.

Howard, P. H., Ed. (1993) *Handbook of Fate and Exposure Data for Organic Chemicals. Vol. IV. Solvents 2.* Lewis Publishers, Chelsea, MI.

Howard, P. H., Boethling, R. S., Jarvis, W. F., Meylan, W. M., Michalenko, E. M. (1991) *Handbook of Environmental Degradation Rates.* Lewis Publishers, Chelsea, MI.

Huang, G. -L., Xiao, H., Chi, J., Shiu, W. -Y., Mackay, D. (2000) Effects of pH on the aqueous solubility of selected chlorinated phenols. *J. Chem. Eng. Data* 45, 411–414.

Hussam, A., Carr, P. W. (1985) A study of a rapid and precise methodology for the measurement of vapor-liquid equilibria by headspace gas chromatography. *Anal. Chem.* 57, 793–801.

Irmann, F. (1965) Eine einfache korrelation zwishen wasserlöslichkeit und strucktur von kohlenwasserstoffen und halogenkohlenwasserstoffen. *Chem.-Ing.-Techn.* 37, 789–798.

Isnard, P., Lambert, S. (1988) Estimating bioconcentration factors for octanol-water partition coefficient and aqueous solubility. *Chemosphere* 17, 21–34.

Isnard, P., Lambert, S. (1989) Aqueous solubility/n-octanol-water partition coefficient correlations. *Chemosphere* 18, 1837–1853.

IUPAC Solubility Data Series (1984) Vol. 15: *Alcohols with Water.* Barton, A. F. M., Ed., Pergamon Press, Oxford, England.

IUPAC Solubility Data Series (1985) Vol. 20: *Halogenated Benzenes, Toluenes and Phenols with Water.* Horvath, A. L., Getzen, F. W., Eds., Pergamon Press, Oxford, England.

IUPAC Solubility Data Series (1989a) Vol. 37: *Hydrocarbons (C_5 - C_7) with Water and Seawater.* Shaw, D. G., Ed., Pergamon Press, Oxford, England.

IUPAC Solubility Data Series (1989b) Vol. 38: *Hydrocarbons (C_8 –C_{36}) with Water and Seawater.* Shaw, D. G., Ed., Pergamon Press, Oxford, England.

Iwase, K., Komatsu, K., Hirono, S., Nakagawa, S., Moriguchi, I. (1985) Estimation of hydro-phobicity based on the solvent-accessible surface area of molecules. *Chem. Pharm. Bull.* 33, 2114–2121.

Jafvert, C. T., Westall, J. C., Grieder, E., Schwarzenbach, P. (1990) Distribution of hydrophobic ionogenic organic compounds between octanol and water: organic acids. *Environ. Sci. Technol.* 24, 1795–1803.

Jaoui, M., Luszczyk, M., Rogalski, M. (1999) Liquid-liquid and liquid-solid equilibria of systems containing water and selected chlorophenols. *J. Chem. Eng. Data* 44, 1269–1272.

Jensen, T., Fredenslund, A., Rasmussen, P. (1981) Pure-compound vapor pressures using UNIFAC group contribution. *Ind. Eng. Chem. Fundam.* 20, 239–246.

Johnson, C. A., Westall, J. C. (1990) Effect of pH and KCl concentration on the octanol-water distribution of methylanilines. *Environ. Sci. Technol.* 24, 1869–1875.

Jönsson, J. A., Vejrosta, J., Novak, J. (1982) Air/water partition coefficients for normal alkanes (n-pentane to n-nonane). *Fluid Phase Equil.* 9, 279–286.

Jury, W. A., Spencer, W. F., Farmer, W. J. (1983) Behavior assessment model for trace organics in soil: I. Model description. *J. Environ. Qual.* 12, 558–566.

Jury, W. A., Farmer, W. J., Spencer, W. F. (1984a) Behavior assessment model for trace organics in soil: II. Chemical classification and parameter sensitivity. *J. Environ. Qual.* 13, 567–572.

Jury, W. A., Farmer, W. J., Spencer, W. F. (1984b) Behavior assessment model for trace organics in soil: III. Application of screening model. *J. Environ. Qual.* 13, 573–579.

Jury, W. A., Spencer, W. F., Farmer, W. J. (1984c) Behavior assessment model for trace organics in soil: IV. Review of experimental evidence. *J. Environ. Qual.* 13, 580–587.

Kabadi, V. N., Danner, R. P. (1979) Nomograph solves for solubilities of hydrocarbons in water. *Hydrocarbon Processing*, 68, 245–246.

Kaiser, K. L. E., Valdmanis, I. (1982) Apparent octanol/water partition coefficients of pentachlorophenol as a function of pH. *Can. J. Chem.* 61, 2104–2106.

Kamlet, M. J., Doherty, R. M., Veith, G. D., Taft, R. W., Abraham, M. H. (1986) Solubility properties in polymers and biological media. 7. An analysis toxicant properties that influence inhibition of bioluminescence in *Photobacterium phosphoreum* (the Microtox test). *Environ. Sci. Technol.* 20, 690–695.

Kamlet, M. J., Doherty, R. M., Abraham, M. H., Carr, P. W., Doherty, R. F., Raft, R. W. (1987) Linear solvation energy relationships. Important differences between aqueous solubility relationships for aliphatic and aromatic solutes. *J. Phys. Chem.* 91, 1996–2004.

Kamlet, M. J., Doherty, R. M., Carr, P. W., Mackay, D., Abraham, M. H., Taft, R. W. (1988) Linear solvation energy relationships. 44. Parameter estimation rules that allow accurate prediction of octanol/water partition coefficients and other solubility and toxicity properties of polychlorinated biphenyls and polycyclic aromatic hydrocarbons. *Environ. Sci. Technol.* 22, 503–509.

Karcher, W., Devillers, J., Eds., (1990) *Practical Applications of Quantitative-Structure-Activity Relationships (QSAR) in Environmental Chemistry and Toxicology.* Kluwer Academic Publisher, Dordrecht, The Netherlands.

Karger, B. L., Castells, R. C., Sewell, P. A., Hartkopf, A. (1971a) Study of the adsorption of insoluble and sparingly soluble vapors at the gas-liquid interface of water by gas chromatography. *J. Phys. Chem.* 75, 3870–3879.

Karger, B. L., Sewell, P. A., Castells, R. C., Hartkopf, A. (1971b) Gas chromatographic study of the adsorption of insoluble vapors on water. *J. Colloid Interface Sci.* 35(2), 328–339.

Karickhoff, S. W. (1981) Semiempirical estimation of sorption of hydrophobic pollutants on natural sediments and soil. *Chemosphere* 10, 833–846.

Karickhoff, S. W., Brown, D. S., Scott, T. A. (1979) Sorption of hydrophobic pollutants on natural water sediments. *Water Res.* 13, 241–248.

Kenaga, E. E. (1980) Predicted bioconcentration factors and soil sorption coefficients of pesticides and other chemicals. *Ecotox. Environ. Saf.* 4, 26–38.

Kenaga, E. E., Goring, C. A. I. (1980) Relationship between water solubility, soil sorption, octanol-water partitioning, and concentration of chemicals in biota. In: *Aquatic Toxicology.* ASTM STP 707, Eaton, J. G., Parrish, P. R., Hendrick, A. C., Eds., pp. 78–115, Am. Soc. for Testing and Materials, Philadelphia, PA.

Kier, L. B., Hall, L. H. (1976) Molar properties and molecular connectivity. In: *Molecular Connectivity in Chemistry and Drug Design.* Medicinal Chem. Vol. 14, pp. 123–167, Academic Press, New York.

Kier, L. B., Hall, L. H. (1986) *Molecular Connectivity in Structure-Activity Analysis.* Wiley, New York.

Kikic, I., Alesse, P., Rasmussen, P., Fredenslund, A. (1980) On the combinatorial part of the UNIFAC and UNIQUAC models. *Can. J. Chem. Eng.* 58, 253–258.

Kim, Y.-H., Woodrow, J. E., Seiber, J. N. (1984) Evaluation of a gas chromatographic method for calculating vapor pressures with organophosphorus pesticides. *J. Chromatogr.* 314, 37–53.

Klöpffer, W. (1991) Photochemistry in environmental research: Its role in abiotic degradation and exposure analysis. *EPA Newsletter* 41, 24–39.

Klopman, G., Wang, S., Balthasar, D. M. (1992) Estimation of aqueous solubility of organic molecules by the group contribution approach. Application to the study of biodegradation. *J. Chem. Inf. Comput. Sci.* 32, 474–482.

Könemann, H., van Leeuewen, K. (1980) Toxicokinetics in fish: accumulation of six chloro-benzenes by guppies. *Chemosphere* 9, 3–19.

Könemann, H., Zelle, R., Busser, F. (1979) Determination of log P_{oct} values of chloro-substituted benzenes, toluenes and anilines by high-performance liquid chromatography on ODS-silica. *J. Chromatogr.* 178, 559–565.

Korenman, I.M., Gur'ev, I.A., Gur'eva, Z.M. (1971) Solubility of liquid aliphatic compounds in water. *Russ. J. Phys. Chem.* 45, 1065–1066.

Lande, S. S., Banerjee, S. (1981) Predicting aqueous solubility of organic nonelectrolytes from molar volume. *Chemosphere* 10, 751–759.

Leahy, D. E. (1986) Intrinsic molecular volume as a measure of the cavity term in linear solvation energy relationships: octanol-water partition coefficients and aqueous solubilities. *J. Pharm. Sci.* 75, 629–636.

Lee, L. S., Rao, P. S. C., Nkedi-Kizza, P., Delfino, J. (1990) Influence of solvent and sorbent characteristics on distribution of pentachlorophenol in octanol-water and soil-water systems. *Environ. Sci. Technol.* 24, 654–661.

Leighton, Jr., D. T., Calo, J. M. (1981) Distribution coefficients of chlorinated hydrocarbons in dilute air-water systems for groundwater contamination applications. *J. Chem. Eng. Data* 26, 382–385.

Leo, A., Hansch, C., Elkins, D. (1971) Partition coefficients and their uses. *Chem. Rev.* 71, 525–616.

Lide, D.R., Ed. (2003) *CRC Handbook of Chemistry and Physics*, 84th edition. CRC Press, Boca Raton, FL.

Lincoff, A. H., Gossett, J. M. (1984) The determination of Henry's law constants for volatile organics by equilibrium partitioning in closed systems. In: *Gas Transfer at Water Surfaces.* Brutsaert, W., Jirka, G. H., Eds., pp. 17–26, D. Reidel Publishing Co., Dordrecht, The Netherlands.

Lo, J. M., Tseng, C. L., Yang, J. Y. (1986) Radiometric method for determining solubility of organic solvents in water. *Anal. Chem.* 58, 1596–1597.

Locke, D. (1974) Selectivity in reversed-phase liquid chromatography using chemically bonded stationary phases. *J. Chromatogr. Sci.* 12, 433–437.

Lohninger, H. (1994) Estimation of soil partition coefficients of pesticides from their chemical structure. *Chemosphere* 29, 1611–1626.

Lyman, W. J., Reehl, W. F., Rosenblatt, D. H. (1982) *Handbook of Chemical Property Estimation Methods.* McGraw-Hill, New York.

Mabey, W. J., Mill, T., Podoll, R. T. (1984) *Estimation Methods for Process Constants and Properties used in Fate Assessment.* USEPA, EPA-600/3-84-035, Athens, GA.

Mackay, D. (1979) Finding fugacity feasible. *Environ. Sci. Technol.* 13, 1218–1223.

Mackay, D. (1982) Correlation of bioconcentration factors. *Environ. Sci. Technol.* 16, 274–278.

Mackay, D. (1991) *Multimedia Environmental Models. The Fugacity Approach.* Lewis Publishers, Chelsea, MI.

Mackay, D. (2001) *Multimedia Environmental Models: The Fugacity Approach.* 2nd edition, Lewis Publishers, CRC Press, Boca Raton, FL.

Mackay, D., Bobra, A. M., Shiu, W.-Y., Yalkowsky, S. H. (1980) Relationships between aqueous solubility and octanol-water partition coefficient. *Chemosphere* 9, 701–711.

Mackay, D., Bobra, A. M., Chan, D. W., Shiu, W.-Y. (1982) Vapor pressure correlation for low-volatility environmental chemicals. *Environ. Sci. Technol.* 16, 645–649.

Mackay, D., Di Guardo, A., Paterson, S., Cowan, C.E. (1996) Evaluating the environmental fate of a variety of types of chemicals using the EQC model. *Environ. Toxicol. Chem.* 15, 1627–1637.

Mackay, D., Paterson, S. (1990) Fugacity models. In: *Practical Applications of Quantitative Structure-Activity Relationships (QSAR) in Environmental Chemistry and Toxicology.* Karcher, W., Devillers, J., Eds., pp. 433–460, Kluwer Academic Publishers, Dordrecht, The Netherlands.

Mackay, D., Paterson, S. (1991) Evaluating the multimedia fate of organic chemicals: A Level III fugacity model. *Environ. Sci. Technol.* 25, 427–436.

Mackay, D., Shiu, W. Y. (1977) Aqueous solubility of polynuclear aromatic hydrocarbons. *J. Chem. Eng. Data* 22, 339–402.

Mackay, D., Shiu, W. Y. (1981) A critical review of Henry's law constants for chemicals of environmental interest. *J. Phys. Chem. Ref. Data* 11, 1175–1199.

Mackay, D., Shiu, W. Y., Sutherland, R.P. (1979) Determination of air-water Henry's law constants for hydrophobic pollutants. *Environ. Sci. Technol.* 13, 333–337.

Mackay, D., Shiu, W. Y., Wolkoff, A. W. (1975) Gas chromatographic determination of low concentration of hydrocarbons in water by vapor phase extraction. In: *Water Quality Parameters. ASTM STP 573*, pp. 251–258, American Society for Testing and Materials, Philadelphia, PA.

Mackay, D., Stiver, W. H. (1991) Predictability and environmental chemistry. In: *Environmental Chemistry of Herbicides.* Vol. II, Grover, R., Cessna, A. J., Eds., pp. 281–297, CRC Press, Boca Raton, FL.

Macknick, A. B., Prausnitz, J. M. (1979) Vapor pressure of high-molecular weight hydrocarbons. *J. Chem. Eng. Data* 24, 175–178.

Magnussen, T., Rasmussen, P., Fredenslund, A. (1981) UNIFAC parameter table for prediction of liquid-liquid equilibria. *Ind. Eng. Chem. Process Des. Dev.* 20, 331–339.

Mailhot, H., Peters, R. H. (1988) Empirical relationships between the 1-octanol/water partition coefficient and nine physicochemical properties. *Environ. Sci. Technol.* 22, 1479–1488.

May, W. E., Wasik, S. P., Freeman, D. H. (1978a) Determination of the aqueous solubility of polynuclear aromatic hydrocarbons by a coupled-column liquid chromatographic technique. *Anal. Chem.* 50, 175–179.

May, W. E., Wasik, S. P., Freeman, D. H. (1978b) Determination of the solubility behavior of some polycyclic aromatic hydrocarbons in water. *Anal. Chem.* 50, 997–1000.

May, W. E., Wasik, S. P., Miller, M. M., Tewari Y. B., Brown-Thomas, J. M., Goldberg, R. N. (1983) Solution thermodynamics of some slightly soluble hydrocarbons in water. *J. Chem. Eng. Data* 28, 197–200.

McAuliffe, C. (1966) Solubility in water of paraffin, cycloparaffin, olefin, acetylene, cycloolefin and aromatic hydrocarbons. *J. Phys. Chem.* 76, 1267–1275.

McAuliffe, C. (1971) GC determination of solutes by multiple phase equilibration. *Chem. Tech.* 1, 46–51.

McDuffie, B. (1981) Estimation of octanol/water partition coefficient for organic pollutants using reversed phase HPLC. *Chemosphere* 10, 73–83.

McLachlan, M.S., Welsch-Pausch, K. and Tolls, J. (1995) Field validation of a model of the uptake of gaseous SOC in *Lolium multiflorum* (Rye Grass). *Environ. Sci. Technol.* 29, 1998–2004.

McGowan, J.C., Mellors, A. (1986) *Molecular Volumes in Chemistry and Biology-Applications including Partitioning and Toxicity.* Ellis Horwood Limited, Chichester, England.

Meylan, W. M., Howard, P. H. (1991) Bond contribution method for estimating Henry's law constants. *Environ. Toxicol. Chem.* 10, 1283–1293.

Meylan, W. M., Howard, P. H., Boethling, R. S. (1992) Molecular topology/fragment contribution for predicting soil sorption coefficient. *Environ. Sci. Technol.* 26, 1560–1567.

Mill, T. (1982) Hydrolysis and oxidation processes in the environment. *Environ. Toxicol. Chem.* 1, 135–141.

Mill, T. (1989) Structure-activity relationships for photooxidation processes in the environment. *Environ. Toxicol. Chem.* 8, 31–45.

Mill, T. (1993) Environmental chemistry. In: *Ecological Risk Assessment.* Suter, II, G.W., Ed., pp. 91–127, Lewis Publishers, Chelsea, MI.

Mill, T., Mabey, W. (1985) Photodegradation in water. In: *Environmental Exposure from Chemicals.* Vol. 1. Neely, W. B., Blau, G. E., Eds., pp. 175–216, CRC Press, Boca Raton, FL.

Miller, M. M., Ghodbane, S., Wasik, S. P., Tewari, Y. B., Martire, D. E. (1984) Aqueous solubilities, octanol/water partition coefficients and entropies of melting of chlorinated benzenes and biphenyls. *J. Chem. Eng. Data* 29, 184–190.

Miller, M. M., Wasik, S. P., Huang, G.-L., Shiu, W.-Y., Mackay, D. (1985) Relationships between octanol-water partition coefficient and aqueous solubility. *Environ. Sci. Technol.* 19, 522–529.

Milne, G. W. A., Editor (1995) *CRC Handbook of Pesticides.* CRC Press, Boca Raton, FL.

Montgomery, J. H. (1993) *Agrochemicals Desk Reference. Environmental Data.* Lewis Publishers, Chelsea, MI.

Munz, C., Roberts, P. V. (1987) Air-water phase equilibria of volatile organic solutes. *J. Am. Water Works Assoc.* 79, 62–69.

Neely, W. B., Branson, D. R., Blau, G. E. (1974) Partition coefficient to measure bioconcentration potential of organic chemicals in fish. *Environ. Sci. Technol.* 8, 1113–1115.

Nirmalakhandan, N. N., Speece, R. E. (1988a) Prediction of aqueous solubility of organic chemicals based on molecular structure. *Environ. Sci. Technol.* 22, 328–338.

Nirmalakhandan, N. N., Speece, R. E. (1988b) QSAR model for predicting Henry's law constant. *Environ. Sci. Technol.* 22, 1349–1357.

Nirmalakhandan, N. N., Speece, R. E. (1989) Prediction of aqueous solubility of organic chemicals based on molecular structure. 2. Application to PNAs, PCBs, PCDDs, etc. *Environ. Sci. Technol.* 23, 708–713.

NRCC (1982) *Chlorinated Phenols: Criteria for Environmental Quality.* National Research Council Canada, Publication No. 18578, Ottawa, Canada.

Ochi, K., Saito, T., Kojima, K. (1996) Measurement and correlation of mutual solubilities of 2-butanol + water. *J. Chem. Eng. Data* 41, 361–364.

Okouchi, H., Saegusa, H., Nojima, O. (1992) Prediction of environmental parameters by adsorbability index: water solubilities of hydrophobic organic pollutants. *Environ. Intl.* 18, 249–261.

Oliver, B. G. (1984) The relationship between bioconcentration factor in rainbow trout and physical-chemical properties for some halogenated compounds. In: *QSAR in Environmental Toxicology.* Kaiser, K. L. E., Ed., pp. 300–317, D. Reidel Publishing, Dordrecht, The Netherlands.

Oliver, B. G., Niimi, A. J. (1988) Trophodynamic analysis of polychlorinated biphenyl congeners and other chlorinated hydrocarbons in the Lake Ontario ecosystem. *Environ. Sci. Technol.* 22, 388–397.

Osborn, A.G., Douslin, D.R. (1974) Vapor-pressure relations of 15 hydrocarbons. *J. Chem. Eng. Data* 19, 114–117.

Osborn, A. G., Douslin, D. R. (1975) Vapor pressures and derived enthalpies of vaporization of some condensed-ring hydrocarbons. *J. Chem. Eng. Data* 20, 229–231.

Owens, J.W., Wasik, S.P., DeVoe, H. (1986) Aqueous solubilities and enthalpies of solution of n-alkylbenzenes. *J. Chem. Eng. Data* 31, 47–51.

Paterson, S., Mackay, D. (1985) The fugacity concept in environmental modelling. In: *The Handbook of Environmental Chemistry.* Vol. 2/Part C, Hutzinger, O., Ed., pp. 121–140, Springer-Verlag, Heidelberg, Germany.

Paterson, S., Mackay, D., Bacci, E. and Calamari, D. (1991) Correlation of the equilibrium and kinetics of leaf-air exchange of hydrophobic organic chemicals. *Environ. Sci. Technol.* 25, 866–871.

Pearlman, R. S. (1980) Molecular surface areas and volumes and their use in structure/activity relationships. In: *Physical Chemical Properties of Drugs.* Yalkowsky, S.H., Sinkula, A.A., Valvani, S.C., Eds., *Medicinal Research Series,* Vol. 10, pp. 321–317, Marcel Dekker, New York.

Pearlman, R. S. (1986) Molecular surface area and volume: Their calculation and use in predicting solubilities and free energies of desolvation. In: *Partition coefficient, Determination and Estimation.* Dunn, III, W. J., Block, J. H., Pearlman R. S., Eds., pp. 3–20, Pergamon Press, New York.

Pearlman, R. S., Yalkowsky, S. H., Banerjee, S. (1984) Water solubilities of polynuclear aromatic and heteroaromatic compounds. *J. Phys. Chem. Ref. Data* 13, 555–562.

Pierotti, C., Deal, C., Derr, E. (1959) Activity coefficient and molecular structure. *Ind. Eng. Chem. Fundam.* 51, 95–101.

Polak, J., Lu, B. C. Y. (1973) Mutual solubilities of hydrocarbons and water at 0° and 25°C. *Can. J. Chem.* 51, 4018–4023.

Rapaport, R. A., Eisenreich, S. J. (1984) Chromatographic determination of octanol-water partition coefficients (K_{OW}'s) for 58 polychlorinated biphenyl congeners. *Environ. Sci. Technol.* 18, 163–170.

Reddy, K. N., Locke, M. A. (1994) Relationships between molecular properties and log p and soil sorption (K_{OC}) of substituted phenylureas: QSAR models. *Chemosphere* 28, 1929–1941.

Reid, R. C., Prausnitz, J. M., Polling, B. E. (1987) *The Properties of Gases and Liquids.* 4th Edition, McGraw-Hill, New York.

Rekker, R. F. (1977) *The Hydrophobic Fragmental Constant.* Elsevier, Amsterdam/New York.

Riddick, J. A., Bunger, W. B., Sakano, T. K. (1986) *Organic Solvents, Physical Properties and Methods of Purification.* 4th Edition, Wiley-Science Publication, John Wiley & Sons, New York.

Rordorf, B. F. (1985a) Thermodynamic and thermal properties of polychlorinated compounds: the vapor pressures and flow tube kinetic of ten dibenzo-*p*-dioxins. *Chemosphere* 14, 885–892.

Rordorf, B. F. (1985b) Thermodynamic properties of polychlorinated compounds: the vapor pressures and enthalpies of sublimation of ten dibenzo-*p*-dioxins. *Thermochimica Acta* 85, 435–438.

Rordorf, B. F. (1986) Thermal properties of dioxins, furans and related compounds. *Chemosphere* 15, 1325–1332.

Sabljic, A. (1984) Predictions of the nature and strength of soil sorption of organic pollutants by molecular topology. *J. Agric. Food Chem.* 32, 243–246.

Sabljic, A. (1987) On the prediction of soil sorption coefficients of organic pollutants from molecular structure: Application of molecular topology model. *Environ. Sci. Technol.* 21, 358–366.

Sabljic, A., Lara, R., Ernst, W. (1989) Modelling association of highly chlorinated biphenyls with marine humic substances. *Chemosphere* 19, 1665–1676.

Sabljic, A., Güsten, H. (1989) Predicting Henry's law constants for polychlorinated biphenyls. *Chemosphere* 19, 1503–1511.

Sabljic, A., Güsten, H. (1990) Predicting the night-time NO_3 radical reactivity in the troposphere. *Atmos. Environ.* 24A, 73–78.

Sangster, J. (1989) Octanol-water partition coefficients of simple organic compounds. *J. Phys. Chem. Ref. Data* 18, 1111–1230.

Sangster, J. (1993) LOGKOW databank. Sangster Research Laboratory, Montreal, Quebec, Canada.

Schellenberg, K., Leuenberger, C., Schwarzenbach, R. P. (1984) Sorption of chlorinated phenols by natural sediments and aquifer materials. *Environ. Sci. Technol.* 18, 652–657.

Introduction

Schwarz, F. P. (1980) Measurement of the solubilities of slightly soluble organic liquids in water by elution chromatography. *Anal. Chem.* 52, 10–15.

Schwarz, F. P., Miller, J. (1980) Determination of the aqueous solubilities of organic liquids at 10.0, 20.0, 30.0 °C by elution chromatography. *Anal. Chem.* 52, 2162–2164.

Schwarzenbach, R. P., Gschwend, P. M., Imboden, D. M. (1993) *Environmental Organic Chemistry.* John Wiley & Sons, New York.

Schwarzenbach, R. P., Stierli, R., Folsom, B. R., Zeyer, J. (1988) Compound properties relevant for assessing the environmental partitioning of nitrophenols. *Environ. Sci. Technol.* 22, 83–92.

Schwarzenbach, R. P., Westall, J. (1981) Transport of nonpolar compounds from surface water to groundwater. Laboratory sorption studies. *Environ. Sci. Technol.* 11, 1360–1367.

Sears, G. W., Hopke, E. R. (1947) Vapor pressures of naphthalene, anthracene and hexachlorobenzene in a low pressure region. *J. Am. Chem. Soc.* 71, 1632–1634.

Seth, R., Mackay, D., Munthe, J. (1999) Estimation of organic carbon partition coefficient and its variability for hydrophobic chemicals. *Environ, Sci. Technol.* 33, 2390–2394.

Shaw, D.G., Ed. (1989a) IUPAC Solubility Data Series: Vol. 37. *Hydrocarbons (C_5-C_7) with Water and Seawater.*, Pergamon Press, Oxford, England.

Shaw, D.G., Ed. (1989b) IUPAC Solubility Data Series: Vol. 38. *Hydrocarbons (C_8-C_{36}) with Water and Seawater.* Pergamon Press, Oxford, England.

Shigeoka, T., Sato, Y., Takeda, Y. (1988) Acute toxicity of chlorophenols to green algae, *Selenastrum capricornutum* and *Chlorella vulgaris*, and quantitative structure-activity relationships. *Environ. Toxicol. Chem.* 7, 847–854.

Shiu, W.-Y., Ma, K.-C., Varhanickova, D., Mackay, D. (1994) Chlorophenols and alkylphenols: A review and correlation of environmentally relevant properties and fate in an evaluative environment. *Chemosphere* 29(6), 1155–1224.

Shiu, W.-Y., Mackay, D. (1986) A critical review of aqueous solubilities, vapor pressures, Henry's law constants, and octanol-water partition coefficients of the polychlorinated biphenyls. *J. Phys. Chem. Ref. Data* 15, 911–929.

Shiu, W.-Y., Gobas, F. A. P. C., Mackay, D. (1987) Physical-chemical properties of three congeneric series of chlorinated aromatic hydrocarbons. In: *QSAR in Environmental Toxicology II.* Kaiser, K. L. E., Ed., pp. 347–362, D. Reidel Publishing, Dordrecht, The Netherlands.

Shiu, W.-Y., Doucette, W., Gobas, F. A. P. C., Mackay, D., Andren, A. W. (1988) Physical-chemical properties of chlorinated dibenzo-p-dioxins. *Environ. Sci. Technol.* 22, 651–658.

Shoeib, M., Harner,T. (2002) Using measured octanol-air partition coefficients to explain environmental partitioning of organochlorine pesticides. *Environ. Toxicol. Chem.* 21,984–990.

Sinke, G. C. (1974) A method for measurement of vapor pressures of organic compounds below 0.1 torr. Naphthalene as reference substance. *J. Chem. Thermodyn.* 6, 311–316.

Smejtek, P., Wang, S. (1993) Distribution of hydrophobic ionizable xenobiotics between water and lipid membranes: pentachlorophenol and pentachlorophenate. A comparison with octanol-water partition. *Arch. Environ. Contam. Toxicol.* 25, 394–404.

Smith, P.D., Brockway, D.L., Stancil, Jr., F.E. (1987) Effect of hardness, alkalinity and pH on toxicity of pentachlorophenol to *senastrum capricornutum* (printz). *Environ. Toxicol. Chem.* 6, 891–990.

Sonnefeld, W. J., Zoller, W. H., May, W. E. (1983) Dynamic coupled-column liquid chromatographic determination of ambient temperature vapor pressures of polynuclear aromatic hydrocarbons. *Anal. Chem.* 55, 275–280.

Spencer, W. F., Cliath, M. M. (1970) Vapor density and apparent vapor pressure of lindane (γ-BHC). *J. Agric. Food Chem.* 18, 529–530.

Spencer, W. F., Cliath, M. M. (1972) Volatility of DDT and related compounds. *J. Agric. Food Chem.* 20, 645–649.

Stehly, G. R., Hayton, W. L. (1990) Effect of pH on the accumulation kinetics of pentachlorophenol in goldfish. *Arch. Environ. Contam. Toxicol.* 19, 464–470.

Stephenson, R. M., Malanowski, A. (1987) *Handbook of the Thermodynamics of Organic Compounds.* Elsevier, New York.

Stiver, W., Mackay, D. (1989) The linear additivity principle in environmental modelling: Application to chemical behaviour in soil. *Chemosphere* 19, 1187–1198.

Su,Y., Lei, D.L. Daly, G.,Wania, F. (2002) Determination of octanol-air partition coefficient (K_{OA}) values for chlorobenzenes and polychlorinated naphthalenes from gas chromatographic retention times. *J. Chem. Eng. Data,* 47, 449–455.

Sugiyama, T., Takeuchi, T., Suzuki, Y. (1975) Thermodynamic properties of solute molecules at infinite dilution determined by gas-liquid chromatography. I. Intermolecular energies of *n*-alkane solutes in C_{28} - C_{36} *n*-alkane solvents. *J. Chromatogr.* 105, 265–272.

Suntio, L. R., Shiu, W.-Y., Mackay, D. (1988) Critical review of Henry's law constants for pesticides. *Rev. Environ. Contam. Toxicol.* 103, 1–59.

Swann, R. L., Laskowski, D. A., McCall, P. J., Vander Kuy, K., Dishburger, H. J. (1983) A rapid method for the estimation of the environmental parameters octanol/water partition coefficient, soil sorption constant, water to air ratio, and water solubility. *Res. Rev.* 85, 17–28.

Szabo, G., Prosser, S., Bulman, R. A. (1990) Determination of the adsorption coefficient (K_{OC}) of some aromatics for soil by RP-HPLC on two immobilized humic acid phases. *Chemosphere* 21, 777–788.

ten Hulscher, Th.E.M., van der Velde, Bruggeman, W. A. (1992) Temperature dependence of Henry's law constants for selected chlorobenzenes, polychlorinated biphenyls and polycyclic aromatic hydrocarbons. *Environ. Toxicol. Chem.* 11, 1595–1603.

Tesconi, M., Yalkowsky, S.H. (2000) Melting Point, Chapter 1 in Boethling, R.S. and Mackay, D. (Eds). *Handbook of Property Estimation Methods: Environmental and Health Sciences*, CRC Press, Boca Raton, FL.

Tolls, J. and McLachlan, M.S. (1994) Partitioning of semivolatile organic compounds between air and *Lolium multiflorum* (Welsh Rye Grass). *Environ. Sci. Technol.* 28, 159–166.

Tomlin, C., Ed. (1994) *The Pesticide Manual (A World Compendium)*, 10th Ed., Incorporating the Agrochemicals Handbook, The British Crop Protection Council and The Royal Society of Chemistry, England.

Tomlinson, E., Hafkenscheid, T. L. (1986) Aqueous solution and partition coefficient estimation from HPLC data. In: *Partition Coefficient, Determination and Estimation.* Dunn, III, W. J., Block, J. H., Pearlman, R. S., Eds., pp. 101–141, Pergamon Press, New York.

Tse, G., Orbey, H., Sandler, S. I. (1992) Infinite dilution activity coefficients and Henry's law coefficients for some priority water pollutants determined by a relative gas chromatographic method. *Environ. Sci. Technol.* 26, 2017–2022.

Tsonopoulos, C., Prausnitz, J. M. (1971) Activity coefficients of aromatic solutes in dilute aqueous solutions. *Ind. Eng. Chem. Fundam.* 10, 593–600.

Tsonopoulos, C., Wilson, G.M.W. (1983) High-temperature mutual solubilities of hydrocarbons and water. *AIChE. J.* 29, 990–999.

Tucker, E.E., Christian, S.D. (1979) A prototype hydrophobic interaction. The dimerization of benzene in water. *J. Phys. Chem.* 83, 426–427.

Tulp, M. T. M., Hutzinger, O. (1978) Some thoughts on the aqueous solubilities and partition coefficients of PCB, and the mathematical correlation between bioaccumulation and physico-chemical properties. *Chemosphere* 7, 849–860.

Veith, G. D., Austin, N. M., Morris, R. T. (1979) A rapid method for estimating log P for organic chemicals. *Water Res.* 13, 43–47.

Veith, G. D., Macek, K. J., Petrocelli, S. R., Caroll, J. (1980) An evaluation of using partition coefficients and water solubilities to estimate bioconcentration factors for organic chemicals in fish. In: *Aquatic Toxicology.* ASTM ATP 707, Eaton, J. G., Parrish, P. R., Hendrick, A.C., Eds, pp. 116–129, Am. Soc. for Testing and Materials, Philadelphia, PA.

Vejrosta, J., Novak, J., Jönsson, J. (1982) A method for measuring infinite-dilution partition coefficients of volatile compounds between the gas and liquid phases of aqueous systems. *Fluid Phase Equil.* 8, 25–35.

Verschueren, K. (1977) *Handbook of Environmental Data on Organic Chemicals.* Van Nostrand Reinhold, New York.

Verschueren, K. (1983) *Handbook of Environmental Data on Organic Chemicals.* 2nd Edition, Van Nostrand Reinhold, New York.

Wakita, K., Yoshimoto, M., Miyamoto, S., Watanabe, H. (1986) A method for calculations of the aqueous solubility of organic compounds by using new fragment solubility constants. *Chem. Pharm. Bull.* 34, 4663–4681.

Walker, J.D. (Ed.) 2003 Annual review quantitative structure-activity relationships. *Environ. Toxicol. Chem.* 22, 1651–1935.

Wang, L., Zhao, Y., Hong, G. (1992) Predicting aqueous solubility and octanol/water partition coefficients of organic chemicals from molar volume. *Environ. Chem.* 11, 55–70.

Warne, M., St. J., Connell, D. W., Hawker, D. W. (1990) Prediction of aqueous solubility and the octanol-water partition coefficient for lipophilic organic compounds using molecular descriptors and physicochemical properties. *Chemosphere* 16, 109–116.

Wasik, S. P., Miller, M. M., Tewari, Y. B., May, W. E., Sonnefeld, W. J., DeVoe, H., Zoller, W. H. (1983) Determination of the vapor pressure, aqueous solubility, and octanol/water partition coefficient of hydrophobic substances by coupled generator column/liquid chromatographic methods. *Res. Rev.* 85, 29–42.

Wauchope, R. D., Buttler, T. M., Hornsby, A. G., Augustijn-Beckers, P. W. M., Burt, J. P. (1992) The SCS/ARS/CES pesticide properties database for environmental decision making. *Rev. Environ. Contam. Toxicol.* 123, 1–156.

Weast, R., Ed. (1972–73) *Handbook of Chemistry and Physics.* 53th Edition, CRC Press, Cleveland, OH.

Weast, R., Ed. (1982–83) *Handbook of Chemistry and Physics.* 64th Edition, CRC Press, Boca Raton, FL.

Weil, L., Dure, G., Quentin, K. L. (1974) Solubility in water of insecticide, chlorinated hydrocarbons and polychlorinated biphenyls in view of water pollution. *Z. Wasser Abwasser Forsch.* 7, 169–175.

Westall, J. C., Leuenberger, C., Schwarzenbach, R. P. (1985) Influence of pH and ionic strength on the aqueous-nonaqueous distribution of chlorinated phenols. *Environ. Sci. Technol.* 19, 193–198.

Westcott, J. W., Bidleman, T. F. (1982) Determination of polychlorinated biphenyl vapor pressures by capillary gas chromatography. *J. Chromatogr.* 210, 331–336.

Westcott, J. W., Simon, J. J., Bidleman, T. F. (1981) Determination of polychlorinated biphenyl vapor pressures by a semimicro gas saturation method. *Environ. Sci. Technol.* 15, 1375–1378.

Whitehouse, B. G., Cooke, R. C. (1982) Estimating the aqueous solubility of aromatic hydrocarbons by high performance liquid chromatography. *Chemosphere* 11, 689–699.

Windholz, M., Ed. (1983) *The Merck Index, An Encyclopedia of Chemicals, Drugs and Biologicals.* 10th Edition, Merck & Co. Rahway, NJ.

Woodburn, K. B., Doucette, W. J., Andren, A. W. (1984) Generator column determination of octanol/water partition coefficients for selected polychlorinated biphenyl congeners. *Environ. Sci. Technol.* 18, 457–459.

Worthing, C. R., Walker, S. B., Eds. (1983) *The Pesticide Manual (A World Compendium)*, 7th Edition, The British Crop Protection Council, Croydon, England.

Worthing, C. R., Walker, S. B., Eds. (1987) *The Pesticide Manual (A World Compendium)*, 8th Edition, The British Crop Protection Council, Croydon, England.

Worthing, C. R., Hance, R. J., Eds. (1991) *The Pesticide Manual (A World Compendium)*, 9th Edition, The British Crop Protection Council, Croydon, England.

Yair, O. B., Fredenslund, A. (1983) Extension of the UNIFAC group-contribution method for the prediction of pure-component vapor pressure. *Ind. Eng. Chem. Fundam. Des. Dev.* 22, 433–436.

Yalkowsky, S. H. (1979) Estimation of entropies of fusion of organic compounds. *Ind. Eng. Chem. Fundam.* 18, 108–111.

Yalkowsky, S. H., Banerjee, S. (1992) *Aqueous Solubility, Methods of Estimation for Organic Compounds*. Marcel Dekker, New York.

Yalkowsky, S. H., Valvani, S. C. (1976) Partition coefficients and surface areas of some alkylbenzenes. *J. Med. Chem.* 19, 727–728.

Yalkowsky, S. H., Valvani, S. C. (1979) Solubility and partitioning. I: Solubility of nonelectrolytes in water. *J. Pharm. Sci.* 69, 912–922.

Yalkowsky, S. H., Orr, R. J., Valvani, S. C. (1979) Solubility and partitioning. 3. The solubility of halobenzenes in water. *Ind. Eng. Chem. Fundam.* 18, 351–353.

Yalkowsky, S. H., Valvani, S. S., Mackay, D. (1983) Estimation of the aqueous solubility of some aromatic compounds. *Res. Rev.* 85, 43–55.

Yoshida, K., Shigeoka, T., Yamauchi, F. (1983) Relationship between molar refraction and n-octanol/water partition coefficient. *Ecotox. Environ. Saf.* 7, 558–565.

Yoshida, K., Shigeoka, T., Yamauchi, F. (1987) Evaluation of aquatic environmental fate of 2,4,6-trichlorophenol with a mathematical model. *Chemosphere* 16, 2531–2544.

Zhou, X., Mopper, K. (1990) Apparent partition coefficients of 15 carbonyl compounds between air and seawater and between air and freshwater: Implications for air-sea exchange. *Environ. Sci. Technol.* 24, 1864–1869.

Zaroogian, G.E., Heltshe, J. F., Johnson, M. (1985) Estimation bioconcentration in marine species using structure-activity models. *Environ. Toxicol. Chem.* 4, 3–12.

Zwolinski, B. J., Wilhoit, R. C. (1971) *Handbook of Vapor Pressures and Heats of Vaporization of Hydrocarbons and Related Compounds*. API-44, TRC Publication No. 101, Texas A&M University, College Station, TX.

2 Aliphatic and Cyclic Hydrocarbons

CONTENTS

2.1 List of Chemicals and Data Compilations 64
    2.1.1 Saturated Hydrocarbons 64
        2.1.1.1 Alkanes 64
            2.1.1.1.1 Isobutane (2-Methylpropane) 64
            2.1.1.1.2 2,2-Dimethylpropane (Neopentane) 67
            2.1.1.1.3 n-Butane 70
            2.1.1.1.4 2-Methylbutane (Isopentane) 73
            2.1.1.1.5 2,2-Dimethylbutane 77
            2.1.1.1.6 2,3-Dimethylbutane 79
            2.1.1.1.7 2,2,3-Trimethylbutane 83
            2.1.1.1.8 n-Pentane 85
            2.1.1.1.9 2-Methylpentane (Isohexane) 93
            2.1.1.1.10 3-Methylpentane 98
            2.1.1.1.11 2,2-Dimethylpentane 101
            2.1.1.1.12 2,4-Dimethylpentane 103
            2.1.1.1.13 3,3-Dimethylpentane 105
            2.1.1.1.14 2,2,4-Trimethylpentane (Isooctane) 109
            2.1.1.1.15 2,3,4-Trimethylpentane 112
            2.1.1.1.16 n-Hexane 114
            2.1.1.1.17 2-Methylhexane (Isoheptane) 123
            2.1.1.1.18 3-Methylhexane 125
            2.1.1.1.19 2,2,5-Trimethylhexane 127
            2.1.1.1.20 n-Heptane 129
            2.1.1.1.21 2-Methylheptane 137
            2.1.1.1.22 3-Methylheptane 139
            2.1.1.1.23 n-Octane 141
            2.1.1.1.24 4-Methyloctane 150
            2.1.1.1.25 n-Nonane 152
            2.1.1.1.26 n-Decane 159
            2.1.1.1.27 n-Undecane 164
            2.1.1.1.28 n-Dodecane 167
            2.1.1.1.29 n-Tridecane 172
            2.1.1.1.30 n-Tetradecane 175
            2.1.1.1.31 n-Pentadecane 179
            2.1.1.1.32 n-Hexadecane 183
            2.1.1.1.33 n-Heptadecane 187
            2.1.1.1.34 n-Octadecane 190
            2.1.1.1.35 n-Eicosane 194
            2.1.1.1.36 n-Tetracosane 201
            2.1.1.1.37 n-Hexacosane 206

	2.1.1.2	Cycloalkanes	211
		2.1.1.2.1 Cyclopentane	211
		2.1.1.2.2 Methylcyclopentane	217
		2.1.1.2.3 1,1,3-Trimethylcyclopentane	219
		2.1.1.2.4 *n*-Propylcyclopentane	221
		2.1.1.2.5 Pentylcyclopentane	223
		2.1.1.2.6 Cyclohexane	224
		2.1.1.2.7 Methylcyclohexane	233
		2.1.1.2.8 1,2,-*cis*-Dimethylcyclohexane	240
		2.1.1.2.9 1,4-*trans*-Dimethylcyclohexane	245
		2.1.1.2.10 1,1,3-Trimethylcyclohexane	247
		2.1.1.2.11 Ethylcyclohexane	249
		2.1.1.2.12 Cycloheptane	254
		2.1.1.2.13 Cyclooctane	258
		2.1.1.2.14 Decalin	263
2.1.2	Unsaturated Hydrocarbons		270
	2.1.2.1	Alkenes	270
		2.1.2.1.1 2-Methylpropene	270
		2.1.2.1.2 1-Butene	273
		2.1.2.1.3 2-Methyl-1-butene	276
		2.1.2.1.4 3-Methyl-1-butene	280
		2.1.2.1.5 2-Methyl-2-butene	283
		2.1.2.1.6 1-Pentene	288
		2.1.2.1.7 *cis*-2-Pentene	292
		2.1.2.1.8 2-Methyl-1-pentene	295
		2.1.2.1.9 4-Methyl-1-pentene	297
		2.1.2.1.10 1-Hexene	299
		2.1.2.1.11 1-Heptene	304
		2.1.2.1.12 1-Octene	308
		2.1.2.1.13 1-Nonene	311
		2.1.2.1.14 1-Decene	314
	2.1.2.2	Dienes	317
		2.1.2.2.1 1,3-Butadiene	317
		2.1.2.2.2 2-Methyl-1,3-butadiene (Isoprene)	322
		2.1.2.2.3 2,3-Dimethyl-1,3-butadiene	328
		2.1.2.2.4 1,4-Pentadiene	330
		2.1.2.2.5 1,5-Hexadiene	334
		2.1.2.2.6 1,6-Heptadiene	337
	2.1.2.3	Alkynes	338
		2.1.2.3.1 1-Butyne	338
		2.1.2.3.2 1-Pentyne	340
		2.1.2.3.3 1-Hexyne	342
		2.1.2.3.4 1-Heptyne	344
		2.1.2.3.5 1-Octyne	346
		2.1.2.3.6 1-Nonyne	348
	2.1.2.4	Cycloalkenes	349
		2.1.2.4.1 Cyclopentene	349
		2.1.2.4.2 Cyclohexene	352
		2.1.2.4.3 1-Methylcyclohexene	357
		2.1.2.4.4 Cycloheptene	359
		2.1.2.4.5 Cyclooctene	361
		2.1.2.4.6 1,4-Cyclohexadiene	364
		2.1.2.4.7 Cycloheptatriene	367

Aliphatic and Cyclic Hydrocarbons

		2.1.2.4.8	*dextro*-Limonene [(*R*)-(+)-Limonene]	371
		2.1.2.4.9	α-Pinene	373
		2.1.2.4.10	β-Pinene	379
2.2	Summary Tables and QSPR Plots			383
2.3	References			395

2.1 LIST OF CHEMICALS AND DATA COMPILATIONS

2.1.1 Saturated Hydrocarbons

2.1.1.1 Alkanes

2.1.1.1.1   Isobutane (2-Methylpropane)

Common Name: Isobutane
Synonym: 2-methylpropane
Chemical Name: 2-methylpropane
CAS Registry No: 75-28-5
Molecular Formula: C_4H_{10}
Molecular Weight: 58.122
Melting Point (°C):
    –159.4 (Weast 1984; Lide 2003)
Boiling Point (°C):
    –11.73 (Lide 2003)
Density (g/cm³ at 20°C):
    0.5490 (Weast 1984)
    0.5571, 0.5509 (20°C, 25°C, Riddick et al. 1986)
Molar Volume (cm³/mol):
    105.9 (calculated-density, Stephenson & Malanowski 1987; Ruelle & Kesselring 1997)
    96.2 (calculated-Le Bas method at normal boiling point)
Enthalpy of Vaporization, ΔH_V (kJ/mol):
    19.121, 21.297 (25°C, bp, Riddick et al. 1986)
Enthalpy of Fusion, ΔH_{fus} (kJ/mol):
    4.540 (Dreisbach 1959; Riddick et al. 1986)
Entropy of Fusion, ΔS_{fus} (J/mol K):
Fugacity Ratio at 25°C, F: 1.0

Water Solubility (g/m³ or mg/L at 25°C):
    48.9 (shake flask-GC at atmospheric pressure, McAuliffe 1963, 1966)

Vapor Pressure (Pa at 25°C or as indicated and reported temperature dependence equations. Additional data at other temperatures designated * are compiled at the end of this section.):
    101783* (–11.609°C, static method-manometer, measured range –85.5 to –11.609°C, Aston et al. 1940)
    101325* (–11.7°C, summary of literature data, temp range –109.2 to –11.7°C, Stull 1947)
    348100 (calculated from determined exptl. data, Dreisbach 1959)
    $\log(P/\text{mmHg}) = 6.74808 - 882.80/(240.0 + t/°C)$; temp range –75 to 30°C (Antoine eq. for liquid state, Dreisbach 1959)
    357000 (interpolated-Antoine eq., temp range –86.57 to 18.88°C, Zwolinski & Wilhoit 1971)
    $\log(P/\text{mmHg}) = 6.91048 - 946.35/(246.68 + t/°C)$; temp range –86.57 to 18.88°C (Antoine eq., Zwolinski & Wilhoit 1971)
    $\log(P/\text{mmHg}) = [-0.2185 \times 5084.4/(T/K)] + 7.250$; temp range –115 to –34°C (Antoine eq., Weast 1972–73)
    $\log(P/\text{mmHg}) = [-0.2185 \times 5416.2/(T/K)] + 7.349085$; temp range –109.5 to 137.5°C (Antoine eq., Weast 1972–73)
    312486, 313702* (21.07, 21.22°C, vapor-liquid equilibrium, measured range 4.580–71.17°C, Steele et al. 1976)
    356600 (extrapolated-Antoine eq., temp range –87 to 7°C, Dean 1985, 1992)
    $\log(P/\text{mmHg}) = 6.90148 - 946.35/(246.68 + t/°C)$; temp range –87 to 7°C (Antoine eq., Dean 1985, 1992)
    $\log(P/\text{kPa}) = 6.00272 - 947.54/(248.87 + t/°C)$; temp range not specified (Antoine eq., Riddick et al. 1986)
    351130 (interpolated-Antoine eq.-III, Stephenson & Malanowski 1987)

Aliphatic and Cyclic Hydrocarbons

log (P_L/kPa) = 6.03538 − 946.35/(−26.47 + T/K); temp range 186–280 K (Antoine eq.-I, Stephenson & Malanowski 1987)

log (P_L/kPa) = 7.83572 − 1470.08/(3.99 + T/K); temp range 121–187 K (Antoine eq.-II, Stephenson & Malanowski 1987)

log (P_L/kPa) = 5.93028 − 907.164/(−30.14 + T/K); temp range 263–306 K (Antoine eq.-III, Stephenson & Malanowski 1987)

log (P_L/kPa) = 6.26924 − 1102.296/(−2.12 + T/K); temp range 301–366 K (Antoine eq.-IV, Stephenson & Malanowski 1987)

log (P_L/kPa) = 6.95371 − 1648.648/(77.939 + T/K); temp range 361–408 K (Antoine eq.-V, Stephenson & Malanowski 1987)

log (P/mmHg) = 31.2541 − 1.9532 × 10³/(T/K) − 8.806·log (T/K) + 8.9246 × 10⁻¹¹·(T/K) + 5.7501 × 10⁻⁶·(T/K)²; temp range 114–408 K (vapor pressure eq., Yaws 1994)

Henry's Law Constant (Pa m³/mol at 25°C):

118640	(converted from $1/K_{AW} = C_W/C_A$ reported as exptl., Hine & Mookerjee 1975)
100980, 22090	(calculated-group contribution, bond contribution, Hine & Mookerjee 1975)
120000	(calculated-P/C, Mackay & Shiu 1981)
116700	(calculated-vapor-liquid equilibrium (VLE) data, Yaws et al. 1991)

Octanol/Water Partition Coefficient, log K_{OW}:

2.76	(shake flask-GC, Leo et al. 1975)
2.76	(recommended, Sangster 1993)
2.76	(recommended, Hansch et al. 1995)

Octanol/Air Partition Coefficient, log K_{OA}:

Bioconcentration Factor, log BCF:

Sorption Partition Coefficient, log K_{OC}:

Environmental Fate Rate Constants, k and Half-Lives, $t_{1/2}$:

Volatilization:

Photolysis:

Oxidation: rate constant k, for gas-phase second order rate constants, k_{OH} for reaction with OH radical, k_{NO3} with NO_3 radical and k_{O3} with O_3 or as indicated, *data at other temperatures and/or the Arrhenius expression see reference:

k_{OH}*(exptl) = 1.42 × 10¹² cm³ mol⁻¹ s⁻¹, k_{OH}(calc) = 1.31 × 10¹² cm³ mol⁻¹ s⁻¹ at 297 K, measured range 297–499 K (flash photolysis-kinetic spectroscopy, Greiner 1970)

k_{OH} = 1.28 × 10⁹ L mol⁻¹ s⁻¹ at 300 K (Greiner 1967; quoted, Altshuller & Bufalini 1971)

k_{OH} = (2.52 ± 0.05) × 10⁻¹² cm³ molecule⁻¹ s⁻¹ at 300 K (relative rate method, Darnall et al. 1978)

k_{O3} = 2.0 × 10⁻²³ cm³ molecule⁻¹ s⁻¹ at 298 K, temp range 298–323 K (Atkinson & Carter 1984)

k_{OH} = (2.29 ± 0.06) × 10⁻¹² cm³ molecule⁻¹ s⁻¹ at room temp. (relative rate, Atkinson et al. 1984c)

k_{OH} = (2.34 ± 0.33) × 10⁻¹² cm³ molecule⁻¹ s⁻¹ at (24.6 ± 0.4)°C (Edney et al. 1986)

k_{OH}* = 2.34 × 10⁻¹² cm³ molecule⁻¹ s⁻¹ at 298 K (recommended, Atkinson 1989)

k_{OH} = 2.34 × 10⁻¹² cm³ molecule⁻¹ s⁻¹ at 298 K, k_{NO3} = 9.7 × 10⁻¹⁷ cm³ molecule⁻¹ s⁻¹ at 296 K (recommended, Atkinson 1990)

k_{NO3}* = (1.10 ± 0.2) × 10⁻¹⁶ cm³ molecule⁻¹ s⁻¹ at 298 K, measured range 298–523 K, atmospheric $t_{1/2}$ = 1750 h during the night at room temp. (discharge flow system, Bagley et al. 1990)

k_{OH} = 7.38 × 10⁻¹³ cm³ molecule⁻¹ s⁻¹, k_{NO3} = 6.50 × 10⁻¹⁷ cm³ molecule⁻¹ s⁻¹ (Sabljic & Güsten 1990)

k_{NO3} = 9.8 × 10⁻¹⁷ cm³ molecule⁻¹ s⁻¹ at 296 K (Atkinson 1991)

k_{OH} = 2.34 × 10⁻¹² cm³ molecule⁻¹ s⁻¹ and an estimated lifetime was 59 h (Altshuller 1991)

k_{NO3}(exptl) = 9.8 × 10⁻¹⁷ cm³ molecule⁻¹ s⁻¹, k_{NO3}(recommended) = 9.9 × 10⁻¹⁷ cm³ molecule⁻¹ s⁻¹, k_{NO3}(calc) = 7.90 × 10⁻¹⁷ cm³ molecule⁻¹ s⁻¹, at 296 ± 2 K (relative rate method, Aschmann & Atkinson 1995)

k_{OH}* = 2.19 × 10⁻¹² cm³ molecule⁻¹ s⁻¹, k_{NO3}* = 10.6 × 10⁻¹⁷ cm³ molecule⁻¹ s⁻¹ at 298 K (recommended, Atkinson 1997)

Hydrolysis:

Biodegradation:
Biotransformation:
Bioconcentration, Uptake (k_1) and Elimination (k_2) Rate Constants or Half-Lives:

Half-Lives in the Environment:

Air: atmospheric $t_{1/2}$ = 1750 h due to reaction with NO_3 radical during the night at room temp., and $t_{1/2}$ = 82 h for reaction with OH radical (Bagley et al. 1990);
atmospheric lifetime was estimated to be 59 h, based on a photooxidation rate constant k = 2.34 × 10^{-12} cm^3 $molecule^{-1}$ s^{-1} in summer daylight with OH radical (Altshuller 1991).

TABLE 2.1.1.1.1.1
Reported vapor pressures of isobutane (2-methylpropane) at various temperatures and the coefficients for the vapor pressure equations

$\log P = A - B/(T/K)$ (1) $\ln P = A - B/(T/K)$ (1a)
$\log P = A - B/(C + t/°C)$ (2) $\ln P = A - B/(C + t/°C)$ (2a)
$\log P = A - B/(C + T/K)$ (3)
$\log P = A - B/(T/K) - C \cdot \log (T/K)$ (4)

Aston et al. 1940		Stull 1947		Steele et al. 1976	
static method-manometer		summary of literature data		vapor-liquid equilibrium	
t/°C	P/Pa	t/°C	P/Pa	t/°C	P/Pa
−85.5	1516	−109.2	133.3	4.580	183196
−71.704	4261	−91.1	666.6	4.620	185019
−56.431	11579	−86.4	1333	21.07	312486
−44.107	23233	−77.9	2666	21.22	313702
−27.576	52132	−68.4	5333	37.55	497811
−22.071	66405	−62.4	7999	37.62	498721
−18.761	76349	−51.4	13332	54.43	762470
−13.328	95476	−41.5	26664	54.50	763484
−11.609	101783	−27.1	53329	71.06	1112244
		−11.7	101325	71.17	1114778
mp/°C	−145	mp/°C	−145		

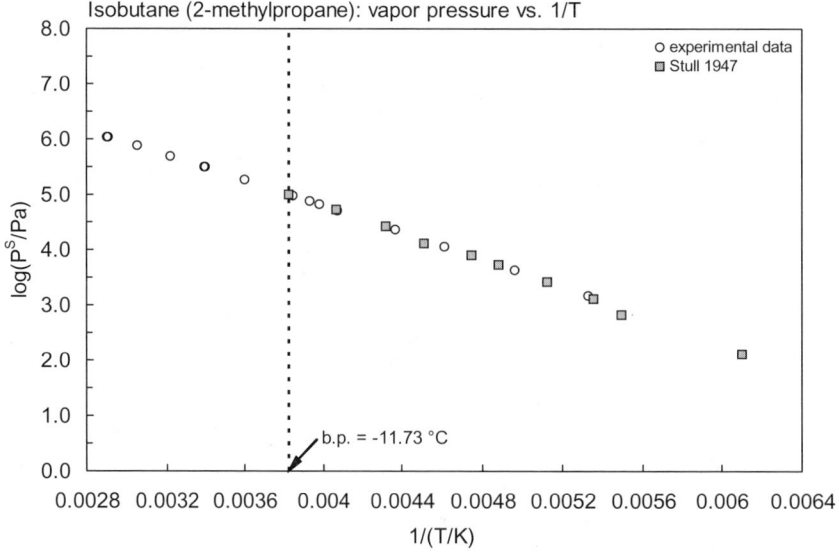

FIGURE 2.1.1.1.1.1 Logarithm of vapor pressure versus reciprocal temperature for isobutane.

2.1.1.1.2 *2,2-Dimethylpropane (Neopentane)*

Common Name: 2,2-Dimethylpropane
Synonym: neopentane, tetramethylmethane
Chemical Name: 2,2-dimethylpropane
CAS Registry No: 463-82-1
Molecular Formula: C_5H_{12}
Molecular Weight: 72.149
Melting Point (°C):
    –16.4   (Lide 2003)
Boiling Point (°C):
    9.503   (Dreisbach 1959; Stephenson & Malanowski 1987)
    9.48   (Lide 2003)
Density (g/cm³ at 20°C):
    0.5910, 0.5852   (20°C, 25°C, Dreisbach 1959; Riddick et al. 1986)
Molar Volume (cm³/mol):
    122.10   (20°C, calculated-density, McAuliffe 1966; Wang et al. 1992)
    117.6   (20°C, Stephenson & Malanowski 1987)
    118.4   (calculated-Le Bas method at normal boiling point)
Enthalpy of Vaporization, ΔH_V (kJ/mol):
    21.778, 22.753   (25°C, bp, Riddick et al. 1986)
Enthalpy of Fusion, ΔH_{fus} (kJ/mol):
    3.255   (Dreisbach 1959; Chickos et al. 1999)
    3.146   (Riddick et al. 1986)
Entropy of Fusion, ΔS_{fus} (J/mol K):
    12.69   (exptl., Chickos et al. 1999)
Fugacity Ratio at 25°C, F: 1.0

Water Solubility (g/m³ or mg/L at 25°C):
    33.2   (shake flask-GC at 1 atmospheric pressure, McAuliffe 1966)

Vapor Pressure (Pa at 25°C or as indicated and reported temperature dependence equations. Additional data at other temperatures designated * are compiled at the end of this section.):
    177930*   (Antoine eq. regression, temp range –112 to 9.8°C, Stull 1947)
    171350   (calculated from determined data, Dreisbach 1959)
    $\log (P/\text{mmHg}) = 6.73812 - 950.84/(237.0 + t/°C)$; temp range –60 to 55°C (Antoine eq. for liquid state, Dreisbach 1959)
    171586*   (derived from compiled data, temp range –13.729 to 29.914°C, Zwolinski & Wilhoit 1971)
    $\log (P_S/\text{mmHg}) = 7.2034 - 1020.7/(230.0 + t/°C)$; temp range –52 to 17.6°C (Antoine eq., solid, Zwolinski & Wilhoit 1971)
    $\log (P_L/\text{mmHg}) = 6.60427 - 883.42/(227.782 + t/°C)$; temp range –13.729 to 29.914°C (Antoine eq., liquid, Zwolinski & Wilhoit 1971)
    $\log (P/\text{mmHg}) = [-0.2185 \times 5648.6/(T/K)] + 7.263947$; temp range –102 to 152.5°C (Antoine eq., Weast 1972–73)
    169019*   (24.56°C, ebulliometry, measured range –5.128 to 40°C, Osborn & Douslin 1974)
    $\log (P/\text{mmHg}) = 6.60427 - 883.42/(227.78 + t/°C)$; temp range –14 to 29°C (Antoine eq., Dean 1985, 1992)
    171300   (selected, Riddick et al. 1986)
    $\log (P/\text{kPa}) = 6.89316 - 938.234/(235.249 + t/°C)$; temp range not specified (Antoine eq., Riddick et al. 1986)
    171520, 171450   (interpolated-Antoine eq.-III and IV, Stephenson & Malanowski 1987)
    $\log (P_S/\text{kPa}) = 6.3283 - 1020.7/(-43.15 + T/K)$; temp range 223–256 K (Antoine eq-I., Stephenson & Malanowski 1987)

log $(P_S/kPa) = 7.07825 - 1372.459/(-8.39 + T/K)$; temp range 223–256 K (Antoine eq.-II, Stephenson & Malanowski 1987)

log $(P_L/kPa) = 5.76532 - 900.545/(-43.111 + T/K)$; temp range 268–313 K (Antoine eq.-III, Stephenson & Malanowski 1987)

log $(P_L/kPa) = 5.83935 - 937.641/(-38.071 + T/K)$; temp range 257–315 K (Antoine eq.-IV, Stephenson & Malanowski 1987)

log $(P_L/kPa) = 6.08953 - 1080.237/(-17.896 + T/K)$; temp range 312–385 K (Antoine eq.-V, Stephenson & Malanowski 1987)

log $(P_L/kPa) = 7.26795 - 2114.713/(128.175 + T/K)$; temp range 382–433 K (Antoine eq-VI, Stephenson & Malanowski 1987)

log $(P/mmHg) = 26.6662 - 1.9307 \times 10^3/(T/K) - 7.0448 \cdot \log(T/K) + 7.4104 \times 10^{-9} \cdot (T/K) + 3.9463 \times 10^{-6} \cdot (T/K)^2$; temp range 257–434 K (vapor pressure eq., Yaws 1994)

Henry's Law Constant (Pa m^3/mol at 25°C):

221000	(calculated as $1/K_{AW}$, C_W/C_A, reported as exptl., Hine & Mookerjee 1975)
125640, 334380	(calculated-group contribution, calculated-bond contribution, Hine & Mookerjee 1975)
373000	(calculated-P/C, Mackay & Shiu 1981)
213250	(calculated-vapor-liquid equilibrium (VLE) data, Yaws et al. 1991)

Octanol/Water Partition Coefficient, log K_{OW}:

3.11	(shake flask-GC, Leo et al. 1975; Leo et al. 1971; Hansch & Leo 1979)
2.95, 3.41, 3.22	(calculated-fragment const., Rekker 1977)
3.30, 3.08	(calculated-MO, calculated-π const., Bodor et al. 1989)
3.11	(recommended, Sangster 1989, 1993)
2.98	(calculated-V_M, Wang et al. 1992)
3.11	(recommended, Hansch et al. 1995)

Octanol/Air Partition Coefficient, log K_{OA}:

Bioconcentration Factor, log BCF:

Sorption Partition Coefficient, log K_{OC}:

Environmental Fate Rate Constant and Half-Lives:
  Volatilization:
  Photolysis:
  Oxidation: rate constant k, for gas-phase second order rate constants, k_{OH} for reaction with OH radical, k_{NO3} with NO_3 radical and k_{O3} with O_3 or as indicated, *data at other temperatures and/or the Arrhenius expression see reference:

   k_{OH}*(exptl) = 6.50×10^{12} cm^3 mol^{-1} s^{-1}, k_{OH}(calc) = 5.27×10^{12} cm^3 mol^{-1} s^{-1} at 298 K, measured range 298–493 K (flash photolysis-kinetic spectroscopy, Greiner 1970)
   k_{OH}(exptl) = 6.50×10^{11} cm^3 mol^{-1} s^{-1}, k_{OH}(calc) = 5.37×10^{11} cm^3 mol^{-1} s^{-1} at 298 K (Greiner 1970)
   $k_{O(3P)}$ = 5.50×10^{-15} cm^3 molecule^{-1} s^{-1} for the reaction with O(^{3}P) atom at room temp. (Herron & Huie 1973)
   k_{OH} = $(1.04 \pm 0.17) \times 10^{-12}$ cm^3 molecule^{-1} s^{-1} at 300 K (relative rate method, Darnall et al. 1978)
   k_{OH} = 9.30×10^{-13} cm^3 molecule^{-1} s^{-1} at room temp. (Atkinson et al. 1979)
   k_{OH} = 9.0×10^{-13} cm^3 molecule^{-1} s^{-1} (Winer et al. 1979)
   k_{OH}* = 8.49×10^{-13} cm^3 molecule^{-1} s^{-1} at 298 K (recommended, Atkinson 1989, 1990, 1991)
   k_{OH}* = 8.48×10^{-13} cm^3 molecule^{-1} s^{-1} at 298 K (recommended, Atkinson 1997)
  Hydrolysis:
  Biodegradation:
  Biotransformation:
  Bioconcentration, Uptake (k_1) and Elimination (k_2) Rate Constants or Half-Lives:

Half-Lives in the Environment:

Aliphatic and Cyclic Hydrocarbons

TABLE 2.1.1.1.2.1
Reported vapor pressures of 2,2-dimethylpropane (neopentane) at various temperatures and the coefficients for the vapor pressure equations

$$\log P = A - B/(T/K) \quad (1) \qquad \ln P = A - B/(T/K) \quad (1a)$$
$$\log P = A - B/(C + t/°C) \quad (2) \qquad \ln P = A - B/(C + t/°C) \quad (2a)$$
$$\log P = A - B/(C + T/K) \quad (3)$$
$$\log P = A - B/(T/K) - C \cdot \log(T/K) \quad (4)$$
$$\log P = A[1 - \Phi/(T/K)] \quad (5) \qquad \text{where } \log A = a + b(T/K) + c(T/K)^2$$

Stull 1947		Zwolinski & Wilhoit 1971				Osborn & Douslin 1974	
summary of literature data		selected values				ebulliometry	
t/°C	P/Pa	t/°C	P/Pa	t/°C	P/Pa	t/°C	P/Pa
			solid		liquid		
−102.0	133.3	−52.0	4000	−13.729	39997	−5.128	57818
−85.4	666.6	−48.0	5333	−7.047	53329	−0.301	70121
−76.7	1333	−45.0	6666	−1.570	66661	4.577	84666
−67.2	2666	−41.9	7999	3.112	79993	9.500	101325
−56.1	5333	−37.4	10666	7.224	93326	14.472	120793
−49.0	7999	−33.8	13332	7.991	95992	19.492	143246
−39.1	13332	−27.0	19998	8.742	98659	24.560	169019
−23.7	26667	−21.8	26664	9.112	101325	29.675	198488
−7.10	53329	−17.6	33331	10.199	103991	34.838	232017
9.50	101325			10.906	106658	40.048	270022
		bp/°C	9.478	14.251	119990		
mp/°C	−16.6	Antoine eq.		17.324	133322	Cox eq.	
		eq. 2	P/mmHg	22.829	159987	eq. 5	P/mmHg
		A	7.2034	29.914	199984	Φ	282.650
		B	1020.7	25.0	171586	a	0.802264
		C	230.0			b	−6.70026 × 10⁻⁴
		temp range −52 to −17.6°C		Antoine eq.		c	11.22918 × 10⁻⁷
		pres. range 30–250 mmHg		eq. 2	P/mmHg	for temp range 268–314 K	
				A	6.60427		
		ΔH$_V$/(kJ mol⁻¹) =		B	883.42		
		at 25°C	21.85	C	227.782		
		at bp	22.75	temp range −14 to 30°C			
				pressure 300–1800 mmHg			

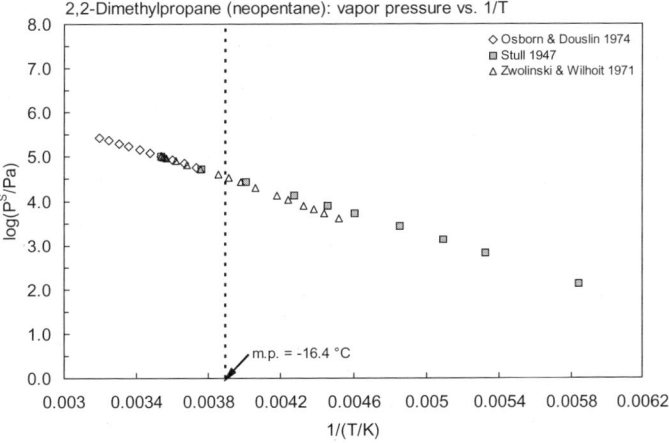

FIGURE 2.1.1.1.2.1 Logarithm of vapor pressure versus reciprocal temperature for 2,2-dimethylpropane.

2.1.1.1.3 n-Butane

Common Name: *n*-Butane
Synonym: 1-butane
Chemical Name: *n*-butane
CAS Registry No: 106-97-8
Molecular Formula: C_4H_{10}
Molecular Weight: 58.122
Melting Point (°C):
    –138.3   (Lide 2003)
Boiling Point (°C):
    –0.50   (Dreisbach 1959; Stephenson & Malanowski 1987; Lide 2003)
Density (g/cm^3 at 20°C):
    0.5788, 0.5730   (20°C, 25°C, Dreisbach 1959)
    0.5786, 0.5729   (20°C, 25°C, Riddick et al. 1986)
Molar Volume (cm^3/mol):
    100.45, 101.45   (20°C, 25°C, calculated-density)
    96.2   (calculated-Le Bas method at normal boiling point)
Enthalpy of Vaporization, ΔH_V (kJ/mol):
    21.066, 22.393   (25°C, bp, Riddick et al. 1986)
Enthalpy of Fusion, ΔH_{fus} (kJ/mol):
    4.393   Parks & Huffman 1931
    4.661   (Dreisbach 1959; Riddick et al. 1986; Chickos et al. 1999)
Entropy of Fusion, ΔS_{fus} (J/mol K):
    34.56   (exptl., Chickos et al. 1999)
Fugacity Ratio at 25°C, F: 1.0

Water Solubility (g/m^3 or mg/L at 25°C):
    65.6   (shake flask-UV, Morrison & Billett 1952)
    67.0   (shake flask-UV, Claussen & Polglase 1952)
    72.7   (shake flask-GC, Franks et al. 1966)
    61.4   (shake flask-GC, McAuliffe 1963, 1966)
    61.66   (shake flask-GC, Coates et al. 1985)

Vapor Pressure (Pa at 25°C or as indicated and reported temperature dependence equations):
    288200   (extrapolated-Antoine eq. regression, temp range –101.5 to –0.5°C, Stull 1947)
    243050   (calculated from determined data, Dreisbach 1959)
    $\log (P/\text{mmHg}) = 6.83029 - 945.9/(240.0 + t/°C)$; temp range –60 to 30°C (Antoine eq. for liquid state, Dreisbach 1959)
    242647   (derived from compiled data, temp range –77.62 to 18.88°C, Zwolinski & Wilhoit 1971)
    $\log (P/\text{mmHg}) = 6.80896 - 935.56/(238.73 + t/°C)$; temp range –77.62 to 18.88°C (Antoine eq., Zwolinski & Wilhoit 1971)
    242840   (extrapolated-Antoine eq., temp range –77 to 19°C, Dean 1985, 1992)
    $\log (P/\text{mmHg}) = 6.80896 - 935.86/(238.73 + t/°C)$; temp range –77 to 19°C (Antoine eq., Dean 1985, 1992)
    243000   (lit. average, Riddick et al. 1986)
    $\log (P/\text{kPa}) = 5.93266 - 935.773/(238.789 + t/°C)$; temp range not specified (Antoine eq., Riddick et al. 1986)
    242810   (interpolated-Antoine eq.-III, Stephenson & Malanowski 1987)
    $\log (P_L/\text{kPa}) = 5.93386 - 935.86/(-34.52 + T/K)$; temp range 195–292 K (Antoine eq.-I, Stephenson & Malanowski 1987)
    $\log (P_L/\text{kPa}) = 7.3327 - 1409.73/(T/K)$; temp range 135–213 K (Antoine eq.-II, Stephenson & Malanowski 1987)
    $\log (P_L/\text{kPa}) = 6.07512 - 1007.247/(-25.272 + T/K)$; temp range 273–321 K (Antoine eq.-III, Stephenson & Malanowski 1987)

Aliphatic and Cyclic Hydrocarbons

log (P_L/kPa) = 6.32267 − 1161.1/(−3.107 + T/K); temp range 316–383 K (Antoine eq.-IV, Stephenson & Malanowski 1987)

log (P_L/kPa) = 7.04942 − 1770.348/(84.979 + T/K); temp range 375–425 K (Antoine eq.-V, Stephenson & Malanowski 1987)

log (P/mmHg) = 27.0441 − 1.9049 × 10^3/(T/K) − 7.1805·log (T/K) − 6.6845 × 10^{-11}·(T/K) + 4.219 × 10^{-6}·(T/K)^2; temp range 135–425 K (vapor pressure eq., Yaws 1994)

592440   (57.01°C, vapor-liquid equilibrium VLE data, Pasanen et al. 2004)

Henry's Law Constant (Pa m^3/mol at 25°C):

94240   (calculated-1/K_{AW}, C_W/C_A, reported as exptl., Hine & Mookerjee 1975)
82080, 22100   (calculated-group contribution, calculated-bond contribution, Hine & Mookerjee 1975)
95900   (calculated-P/C, Mackay & Shiu 1975; Mackay 1981; Mackay & Shiu 1981)
80210   (calculated-MCI χ, Nirmalakhandan & Speece 1988)
92910   (calculated-vapor-liquid equilibrium (VLE) data, Yaws et al. 1991)

Octanol/Water Partition Coefficient, log K_{OW}:

2.89   (shake flask-GC, Leo et al. 1975)
2.89   (concn. ratio, Cramer 1977)
2.46, 2.84, 2.96   (calculated-f const., Rekker 1977)
2.89, 2.76   (Hansch & Leo 1979)
2.79   (calculated-hydrophobicity const., Iwase et al. 1985)

Octanol/Air Partition Coefficient, log K_{OA}:

1.53   (calculated-measured γ° in pure octanol and vapor pressure P, Abraham et al. 2001)

Bioconcentration Factor, log BCF:

Sorption Partition Coefficient, log K_{OC}:

Environmental Fate Rate Constants, k, and Half-Lives, $t_{1/2}$:

Volatilization:
Photolysis:
Oxidation: rate constant k, for gas-phase second order rate constants, k_{OH} for reaction with OH radical, k_{NO3} with NO_3 radical and k_{O3} with O_3 or as indicated, *data at other temperatures and/or the Arrhenius expression see reference:

k_{O3} = 9.8 × 10^{-24} cm^3 molecule^{-1} s^{-1} at 298 K, measured range 298–323 K (Schubert & Pease 1956)

$k_{O(3P)}$ = 3.1 × 10^{-14} cm^3 molecule^{-1} s^{-1} for the reaction with O(^{3}P) atom (Herron & Huie 1973)

k_{OH} = 3.0 × 10^{-12} cm^3 molecule^{-1} s^{-1} (Atkinson et al. 1979)

k_{OH}*(exptl) = 1.66 × 10^{12} cm^3 mol^{-1} s^{-1}, k_{OH}(calc) = 1.54 × 10^{12} cm^3 mol^{-1} s^{-1} at 298 K, measured range 298–495 K (flash photolysis-kinetic spectroscopy, Greiner 1970)

k_{OH} = 1.8 × 10^9 L mol^{-1} s^{-1} with atmospheric $t_{1/2}$ = 2.4 to 24 h at 300 K (Darnall et al. 1976)

k_{OH} = 1.8 × 10^9 M^{-1} s^{-1} in polluted atmosphere at 305 ± 2 K (relative rate method, Lloyd et al. 1976)

k_{OH} = 3.0 × 10^{-12} cm^3 molecule^{-1} s^{-1} with a loss rate at 0.11 d^{-1} for the reaction with OH radical and an average OH concn of 1.2 × 10^6 molecules/cm^3 (Zafonte & Bonamassa 1977)

k_{OH}(exptl) = (2.57 − 4.22) × 10^{-12} cm^3 molecule^{-1} s^{-1}, k_{OH}(calc) = 2.71 × 10^{-12} cm^3 molecule^{-1} s^{-1} at at atmospheric pressure and 300 K (Darnall et al. 1978)

k_{OH} = 2.58 × 10^{-12} cm^3 molecule^{-1} s^{-1} at 299 ± 2 K (relative rate method, Atkinson et al. 1982a, 1984c)

k_{O3} < 10^{-23} cm^3 molecule^{-1} s^{-1} with a loss rate of < 6 × 10^{-7} d^{-1}, k_{OH} = 2.6 × 10^{-12} cm^3 molecule^{-1} s^{-1} with a loss rate of 0.2 d^{-1} and k_{NO3} = 3.6 × 10^{-17} cm^3 molecule^{-1} s^{-1} with a loss rate of 0.0007 d^{-1} (Atkinson & Carter 1984; Atkinson 1985)

k_{OH}* = 2.54 × 10^{-12} cm^3 molecule^{-1} s^{-1} at 298 K (recommended, Atkinson 1989)

k_{OH}* = 2.54 × 10^{-12} cm^3 molecule^{-1} s^{-1} at 298 K, k_{NO3} = 6.5 × 10^{-17} cm^3 molecule^{-1} s^{-1} at 296 ± 2 K (Atkinson 1990; Altshuller 1991).

k_{NO_3}* = (0.45 ± 0.06) × 10^{-16} cm³ molecule⁻¹ s⁻¹ at 298 K, measured range 298–523 K, atmospheric $t_{½}$ = 4300 h during the night at room temp. (discharge flow system, Bagley et al. 1990)

atmospheric lifetime was estimated to be 54 h, based on the photooxidation reaction rate constant with OH radical during summer daylight hours (Altshuller 1991)

k_{NO_3} = (≤ 2.0 – 6.6) × 10^{-17} cm³ molecule⁻¹ s⁻¹ at 296–298 K (Atkinson 1991)

k_{NO_3}(exptl) = 6.70 × 10^{-17} cm³ molecule⁻¹ s⁻¹, k_{NO_3}(recommended) = 4.3 × 10^{-17} cm³ molecule⁻¹ s⁻¹ at 296 ± 2 K (relative rate method, Aschmann & Atkinson 1995)

k_{OH}* = 2.44 × 10^{-12} cm³ molecule⁻¹ s⁻¹, k_{NO_3}* = 4.59 × 10^{-17} cm³ molecule⁻¹ s⁻¹ at 298 K (recommended, Atkinson 1997)

k_{OH}* = 2.37 × 10^{-12} cm³ molecule⁻¹ s⁻¹ at 298 K, measured range 230–400 K (relative rate method, DeMore & Bayes 1999)

Hydrolysis:

Biodegradation:

Biotransformation:

Bioconcentration, Uptake (k_1) and Elimination (k_2) Rate Constants or Half-Lives:

Half-Lives in the Environment:

Air: $t_{½}$ = 6.5 h in ambient air based on reaction with OH radicals at 300 K (Doyle et al. 1975);

photolysis $t_{½}$ = 2.4 to 24 h (Darnall et al. 1976);

atmospheric lifetimes τ(calc) = 4 × 10^7 h for reaction with O_3, τ = 107 h with OH radical and τ = 32150 h with NO_3 radical based on reaction rate constants and environmental concentrations of OH, NO_3 radicals and O_3 in the gas phase (Atkinson & Carter 1984);

atmospheric lifetimes τ(calc) = 222 h for the reaction with OH radical, τ = 4 × 10^7 h with O_3 and τ = 32150 h with NO_3 radical based on the rate constants and environmental concentrations of OH, NO_3 radicals and O_3 in the gas phase (Atkinson 1985);

atmospheric $t_{½}$ = 4300 h due to reaction with NO_3 during the night at room temp., and $t_{½}$ = 77 h for reaction with OH radical (Bagley et al. 1990);

atmospheric lifetime τ ~ 54 h based on a photooxidation reaction rate constant of 2.54 × 10^{-12} cm³ molecule⁻¹ s⁻¹ with OH radicals during summer daylight hours (Altshuller 1991).

Aliphatic and Cyclic Hydrocarbons

2.1.1.1.4   2-Methylbutane (Isopentane)

Common Name: 2-Methylbutane
Synonym: Isopentane
Chemical Name: 2-methylbutane
CAS Registry No: 78-78-4
Molecular Formula: C_5H_{12}; $CH_3CH(CH_3)CH_2CH_3$
Molecular Weight: 72.149
Melting Point (°C):
    –159.77   Lide 2003)
Boiling Point (°C): 27.875
    27.88   (Lide 2003)
Density (g/cm³ at 20°C):
    0.6197, 0.6146   (20°C, 25°C, Dreisbach 1959)
    0.6193, 0.6142   (20°C, 25°C, Riddick et al. 1986)
Molar Volume (cm³/mol):
    116.5, 117.47   (20°C, 25°C, calculated-density)
    118.4   (calculated-Le Bas method at normal boiling point)
Enthalpy of Fusion, ΔH_{fus} (kJ/mol):
    5.1505   (Dreisbach 1959, Riddick et al. 1986)
    5.13   (Chickos et al. 1999)
Entropy of Fusion, ΔS_{fus} (J/mol K):
    45.23, 43.35 (exptl., calculated-group additivity method, Chickos et al. 1999)
Fugacity Ratio at 25°C, F: 1.0

Water Solubility (g/m³ or mg/L at 25°C or as indicated. Additional data at other temperatures designated * are compiled at the end of this section.):
    47.8   (shake flask-GC, McAuliffe 1963,1966)
    46.9*   (20°C, shake flask-GC, measured range 20–60°C, Pavlova et al. 1966)
    72.4; 49.6, 55.2   (0, 25°C, shake flask-GC, calculated-group contribution, Polak & Lu 1973)
    48.0   (shake flask-GC, Price 1976)
    48.0   (selected, Riddick et al. 1986)
    48.5*   (IUPAC recommended best value, temp range 0–60°C, Shaw 1989)
    52.11*   (calculated-liquid-liquid equilibrium LLE data, temp range 273.2–323.2 K, Mączyński et al. 2004)

Vapor Pressure (Pa at 25°C or as indicated and reported temperature dependence equations. Additional data at other temperatures designated * are compiled at the end of this section.):
    82790   (22.04°C, Schumann et al. 1942)
    83720*   (22.44°C, manometer, temp range 16.291–28.587°C, Willingham et al. 1945)
    $\log (P/\text{mmHg}) = 6.87372 - 1075.816/(233.259 + t/°C)$; temp range 16.291–28.587°C (Antoine eq. from exptl. data, manometer, Willingham et al. 1945)
    99550*   (Antoine eq. regression, temp range –82.9 to 27.8°C, Stull 1947)
    91740   (calculated from determined data, Dreisbach 1959)
    $\log (P/\text{mmHg}) = 6.78967 - 1020.012/(223.097 + t/°C)$; temp range –45 to 75°C (Antoine eq. for liquid state, Dreisbach 1959)
    91646*   (extrapolated-Antoine eq, temp range –67.03 to 49.14°C, Zwolinski & Wilhoit 1971)
    $\log (P/\text{mmHg}) = 6.83315 - 1040.73/(235.455C + t/°C)$; temp range –67.03 to 49.14°C (Antoine eq., Zwolinski & Wilhoit 1971)
    $\log (P/\text{mmHg}) = [-0.2185 \times 6470.8/(T/K)] + 7.544680$; temp range: –82.9 to 180.3°C, (Antoine eq., Weast 1972–73)
    91730, 92100 (interpolated, Antoine equations, Boublik et al. 1984)

log (P/kPa) = 6.04913 – 1081.748/(239.817 + t/°C); temp range –56 to 22.4°C (Antoine eq. from reported exptl. data of Schumann et al. 1942, Boublik et al. 1984)

log (P/kPa) = 5.9333 – 1029.602/(234.294 + t/°C); temp range 16.29–28.59°C (Antoine eq. from reported exptl. data of Willingham et al. 1945, Boublik et al. 1984)

91660 (interpolated-Antoine eq., temp range –87 to 7°C, Dean 1985)

log (P/mmHg) = 6.91048 – 946.35/(246.68 + t/°C); temp range –87 to 7°C (Antoine eq., Dean 1985, 1992)

91700 (quoted, Riddick et al. 1986)

log (P/kPa) = 5.92023 – 1022.88/(233.460 + t/°C), temp range not specified (Antoine eq., Riddick et al. 1986)

91640 (interpolated-Antoine eq., Stephenson & Malanowski 1987)

log (P_L/kPa) = 5.95805 – 1040.73/(–37.705 + T/K); temp range 216–323 K (Antoine eq-I., Stephenson & Malanowski 1987)

log (P_L/kPa) = 6.32287 – 1279.08/(–4.481 + T/K); temp range 300–460 K (Antoine eq-II., Stephenson & Malanowski 1987)

log (P_L/kPa) = 6.39629 – 1325.048/(1.244 + T/K); temp range 320–391 K (Antoine eq-III., Stephenson & Malanowski 1987)

log (P_L/kPa) = 6.22589 – 1212.803/(–12.958 + T/K); temp range 385–416 K (Antoine eq-IV., Stephenson & Malanowski 1987)

log (P_L/kPa) = 8.09160 – 3167.07/(233.708 + T/K); temp range 412–460 K (Antoine eq-V., Stephenson & Malanowski 1987)

log (P/mmHg) = 29.2963 – 2.1762 × 10^3/(T/K) – 7.883·log (T/K) – 4.6512 × 10^{-11}·(T/K) + 3.8997 × 10^{-6}·(T/K)^2; temp range 113–460 K (vapor pressure eq., Yaws 1994)

Henry's Law Constant (Pa m^3/mol at 25°C):
140000 (calculated-P/C, Mackay et al. 1979; Mackay 1981)
138000; 140000, 139000, 134700 (recommended; calculated-P/C, Mackay & Shiu 1981))
138210 (selected, Mills et al. 1982)
138290 (calculated-vapor-liquid equilibrium (VLE) data, Yaws et al. 1991)

Octanol/Water Partition Coefficient, log K_{OW}:
2.30 (calculated-π constant, Hansch et al. 1968)
2.41 (calculated-MCI χ, Murray et al. 1975)
2.83 (calculated-molar volume V_M, Wang et al. 1992)
2.4698 (calculated-UNIFAC group contribution, Chen et al. 1993)

Octanol/Air Partition Coefficient, log K_{OA}:

Bioconcentration Factor, log BCF:

Sorption Partition Coefficient, log K_{OC}:

Environmental Fate Rate Constants, k, and Half-Lives, $t_{1/2}$:
Volatilization:
Photolysis:
Oxidation: rate constant k, for gas-phase second order rate constants, k_{OH} for reaction with OH radical, k_{NO3} with NO$_3$ radical and k_{O3} with O$_3$ or as indicated, *data at other temperatures and/or the Arrhenius expression see reference:

k_{OH} = 2.0 × 10^{12} cm^3 mol^{-1} s^{-1} with atmospheric $t_{1/2}$ = 2.4–24 h (Lloyd 1976, Darnall et al. 1976)

k_{OH} = (3.78 ± 0.07) × 10^{-12} cm^3 molecule^{-1} s^{-1} at 300 K (relative rate method, Darnall et al. 1978)

k_{OH} = (3.97 ± 0.11) × 10^{-12} cm^3 molecule^{-1} s^{-1} at room temp. (relative rate, Atkinson et al. 1984c)

k_{OH} = 3.9 × 10^{-12} cm^3 molecule^{-1} s^{-1} at 298 K (recommended, Atkinson 1989)

k_{OH} = 3.9 × 10^{-12} cm^3 molecule^{-1} s^{-1} at 298 K (Atkinson 1990, 1991; Altshuller 1991)

k_{NO3}* = (1.60 ± 0.2) × 10^{-16} cm^3 molecule^{-1} s^{-1} at 298 K, measured range 298–523 K, atmospheric $t_{1/2}$ = 1200 h during the night at room temp. (discharge flow system, Bagley et al. 1990)

k_{OH} = 3.9 × 10^{-12} cm^3 molecule^{-1} s^{-1} at 298 K, estimated atmospheric lifetime of 36 h (Altshuller 1991)

k_{NO_3}(exptl) = 1.56 × 10⁻¹⁶ cm³ molecule⁻¹ s⁻¹, k_{NO_3}(calc) = 2.49 × 10⁻¹⁶ cm³ molecule⁻¹ s⁻¹ at 296 ± 2 K (relative rate method, Aschmann & Atkinson 1995)

k_{OH} = 3.7 × 10⁻¹² cm³ molecule⁻¹ s⁻¹, k_{NO_3}* = 1.62 × 10⁻¹⁶ cm³ molecule⁻¹ s⁻¹ at 298 K (recommended, Atkinson 1997)

Hydrolysis:

Biodegradation:

Biotransformation:

Bioconcentration, Uptake (k_1) and Elimination (k_2) Rate Constants or Half-Lives:

Half-Lives in the Environment:

Air: photooxidation reaction rate constant of 2.0 × 10¹² cm³ mol⁻¹ s⁻¹ for the reaction with hydroxyl radical in air (Darnall et al. 1976; Lloyd et al. 1976) with atmospheric $t_{1/2}$ = 2.4–24 h (Darnall et al. 1976); atmospheric $t_{1/2}$ = 1200 h due to reaction with NO_3 radical during the night at room temp., and $t_{1/2}$ = 50 for reaction with OH radical (Bagley et al. 1990); atmospheric lifetime of 36 h, based on rate constant of 3.90 × 10⁻¹² cm³ molecule⁻¹ s⁻¹ for the reaction with OH radicals during summer daylight (Altshuller 1991).

TABLE 2.1.1.1.4.1
Reported aqueous solubilities of 2-methylbutane (isopentane) at various temperatures

Pavlova et al. 1966		Polak & Lu 1973		Shaw 1989a		Mączyński et al. 2004	
in IUPAC 1989		shake flask-GC/FID		IUPAC recommended		calc-recommended LLE data	
t/°C	S/g·m⁻³	t/°C	S/g·m⁻³	t/°C	S/g·m⁻³	t/°C	S/g·m⁻³
20	46.9	0	72.4	0	72	0	68.14
40	57.7	25	49.6	20	47	20	56.12
50	70.1			25	48.5	25	52.11
60	79.3			30	51	40	52.11
				40	58	50	56.12
				50	70		
				60	79		

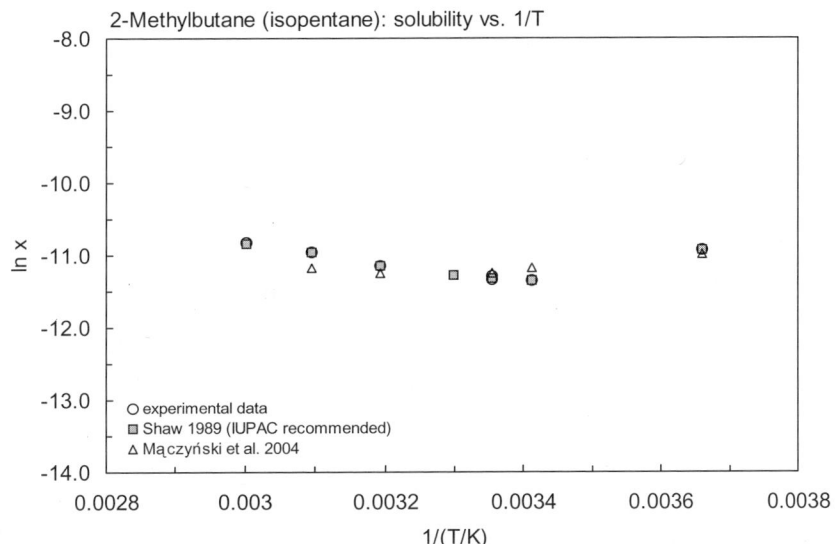

FIGURE 2.1.1.1.4.1 Logarithm of mole fraction solubility (ln x) versus reciprocal temperature for 2-methylbutane.

TABLE 2.1.1.1.4.2
Reported vapor pressures of 2-methylbutane (isopentane) at various temperatures and the coefficients for the vapor pressure equations

$\log P = A - B/(T/K)$ (1) $\qquad \ln P = A - B/(T/K)$ (1a)

$\log P = A - B/(C + t/°C)$ (2) $\qquad \ln P = A - B/(C + t/°C)$ (2a)

$\log P = A - B/(C + T/K)$ (3) $\qquad \ln P = A - B/(C + T/K)$ (3a)

$\log P = A - B/(T/K) - C \cdot \log (T/K)$ (4)

$\ln (P/P_{ref}) = [1 - (T_{ref}/T)] \cdot \exp(a + bT + cT^2)$ (5)

Willingham et al. 1945		Stull 1947		Zwolinski & Wilhoit 1971			
ebulliometry		summary of literature data		selected values			
t/°C	P/Pa	t/°C	P/Pa	t/°C	P/Pa	t/°C	P/Pa
16.291	66760	−82.9	133.3	−57.03	1333	25.517	93326
22.435	83722	−65.8	666.6	−47.32	2666	26.320	95992
26.773	97608	−57.0	1333	−41.14	4000	27.156	98659
27.24	99207	−47.3.	2666	−36.49	5333	27.875	101325
27.673	100700	−36.5	5333	−32.74	6666	25.0	91646
28.16	102402	−29.6	7999	−29.56	7999		
28.587	103922	−20.2	13332	−24.35	10666	eq. 2	P/mmHg
		−5.90	26664	−20.11	13332	A	6.83315
bp/°C	27.852	10.5	53329	−11.97	19998	B	1040.73
		27.8	101325	−5.81	26664	C	235.445
eq. 2	P/mmHg			−0.79	33331	bp/°C	28.875
A	6.78967	mp/°C	−159.7	3.47	39997	$\Delta H_V/(kJ\ mol^{-1}) =$	
B	1020.012			10.53	53329	at 25°C	24.84
C	233.097			16.293	66661	at bp	24.69
				21.208	79993		

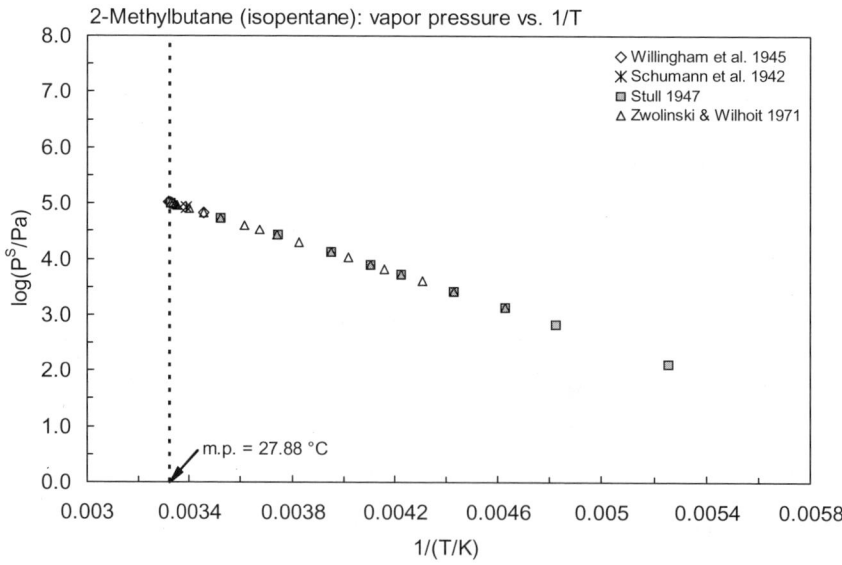

FIGURE 2.1.1.1.4.2 Logarithm of vapor pressure versus reciprocal temperature for 2-methylbutane.

Aliphatic and Cyclic Hydrocarbons

2.1.1.1.5 2,2-Dimethylbutane

Common Name: 2,2-Dimethylbutane
Synonym: neohexane, dimethylpropylmethane
Chemical Name: 2,2-dimethylbutane
CAS Registry No: 75-83-2
Molecular Formula: C_6H_{14}; $CH_3CH(CH_3)_2CH_2CH_3$
Molecular Weight: 86.175
Melting Point (°C):
    –99.8   (Lide 2003)
Boiling Point (°C):
    49.73   (Lide 2003)
Density (g/cm³ at 20°C):
    0.6492, 0.6445 (20°C, 25°C, Dreisbach 1959; Riddick et al. 1986)
Molar Volume (cm³/mol):
    132.74, 133.71   (20°C, 25°C, calculated-density)
    140.6   (calculated-Le Bas method at normal boiling point)
Enthalpy of Fusion, ΔH_{fus} (kJ/mol):
    5.791   (Dreisbach 1959; Riddick et al. 1986)
    5.4, 0.28, 0.58   (–146.35, –132.35, –98.95°C, Chickos et al. 1999)
Entropy of Fusion, ΔS_{fus} (J/mol K):
    45.88, 42.6 (exptl., calculated-group additivity method, total phase change entropy, Chickos et al. 1999)
Fugacity Ratio at 25°C, F: 1.0

Water Solubility (g/m³ or mg/L at 25°C or as indicated):
    18.4   (shake flask-GC, McAuliffe 1963,1966)
    39.4, 23.8 (0, 25°C, shake flask-GC, Polak & Lu 1973)
    21.2   (shake flask-GC, Price 1976)
    21.2   (shake flask-GC, Krzyzanowska, Szeliga 1978)
    18.0   (selected Riddick et al. 1986)
    21.0   (IUPAC recommended best value, Shaw 1989)
    19.63   (calculated-recommended liquid-liquid equilibrium LLE data, Mączyński et al. 2004)

Vapor Pressure (Pa at 25°C or as indicated and reported temperature dependence equations):
    43320   (24.47°C, manometer, temp range 15.376–50.529°C, Willingham et al. 1945)
    log (P/mmHg) = 6.75483 – 1081.176/(229.343 + t/°C); temp range 15.376–50.529°C (Antoine eq. from exptl. data, manometer, Willingham et al. 1945)
    43480   (Antoine eq. regression, temp range –69.3 to 49.7°C, Stull 1947)
    42570   (25°C, Nicolini & Laffitte 1949)
    42540   (calculated from determined data, Dreisbach 1959; quoted, Hine & Mookerjee 1975)
    log (P/mmHg) = 6.75483 – 1081.176/(229.343 + t/°C); temp range –25 to 95°C (Antoine eq. for liquid state, Dreisbach 1959)
    42543   (interpolated-Antoine eq., temp range –41.5 to 72.8°C, Zwolinski & Wilhoit 1971)
    log (P/mmHg) = 6.75483 – 1081.176/(229.343 + t/°C); temp range –41.5 to 72.8°C (Antoine eq., Zwolinski & Wilhoit 1971)
    log (P/mmHg) = [–0.2185 × 7271.0/(T/K)] + 7.84130; temp range –69.3 to 49.7°C (Antoine eq., Weast 1972–73)
    42585, 42550   (interpolated-Antoine equations, Boublik et al. 1984)
    log (P/kPa) = 5.88698 – 1085.038/(229.817 + t/°C); temp range 15.376–50.53°C (Antoine eq. from reported exptl. data of Willingham et al. 1945, Boublik et al. 1984)
    log (P/kPa) = 5.87001 – 1080.723/(229.842 + t/°C); temp range 0–45°C (Antoine eq. from reported exptl. data of Nicolini & Laffitte 1949, Boublik et al. 1984)

42540 (interpolated-Antoine eq., temp range −42 to 73°C, Dean 1985, 1992)
log (P/mmHg) = 6.75483 − 1081.176/(229.34 + t/°C); temp range −42 to 73°C (Antoine eq., Dean 1985, 1992)
42700 (lit. average, Riddick et al. 1986)
log (P/kPa) = 5.87963 − 1081.14/(229.349 + t/°C); temp range not specified (Antoine eq., Riddick et al. 1986)
42560 (interpolated-Antoine eq., Stephenson & Malanowski 1987)
log (P_L/kPa) = 5.87731 − 1079.789/(−43.978 + T/K); temp range 293–324 K (Antoine eq., Stephenson & Malanowski 1987)
log (P/mmHg) = 33.1285 − 2.4527 × 10^3/(T/K) − 9.2016·log (T/K) − 4.7077 × 10^{-10}·(T/K) + 4.1755 × 10^{-6}·(T/K)^2; temp range 174–489 K (vapor pressure eq., Yaws 1994)

Henry's Law Constant (Pa m^3/mol at 25°C):
173000 (calculated-P/C, Mackay & Shiu 1981)
196800 (calculated as 1/K_{AW}, C_W/C_A, reported as exptl., Hine & Mookerjee 1975)
196800, 49430 (calculated-group contribution, calculated-bond contribution, Hine & Mookerjee 1975)
173180 (calculated-P/C, Eastcott et al. 1988)
188040 (calculated-MCI χ, Nirmalakhandan & Speece 1988)
153890 (calculated-vapor-liquid equilibrium (VLE) data, Yaws et al. 1991)

Octanol/Water Partition Coefficient, log K_{OW}:
3.82 (calculated-fragment const., Valvani et al. 1981)
3.25 (calculated-V_M, Wang et al. 1992)
3.82 (recommended, Sangster 1993)
3.82 (Hansch et al. 1995)

Octanol/Air Partition Coefficient, log K_{OA}:

Bioconcentration Factor, log BCF:

Sorption Partition Coefficient, log K_{OC}:

Environmental Fate Rate Constants, k, and Half-Lives, $t_{1/2}$:
  Volatilization:
  Photolysis:
  Oxidation: rate constant k, for gas-phase second order rate constants, k_{OH} for reaction with OH radical, k_{NO3} with NO_3 radical and k_{O3} with O_3 or as indicated, *data at other temperatures and/or the Arrhenius expression see reference:
    k_{OH} = (2.66 ± 0.08) × 10^{-12} cm^3 molecule^{-1} s^{-1} at room temp. (relative rate, Atkinson et al. 1984c)
    k_{OH} = (2.59 - 6.16) × 10^{-12} cm^3 molecule^{-1} s^{-1} at 297–299 K (Atkinson 1985)
    k_{OH}* = (2.22 - 0.36) × 10^{-12} cm^3 molecule^{-1} s^{-1} at 299 K, measured range 245–328 K (relative rate method, Harris & Kerr 1988; Atkinson 1989)
    k_{OH}* = 2.32 × 10^{-12} cm^3 molecule^{-1} s^{-1} at 298 K (recommended, Atkinson 1989, 1990)
    k_{OH}* = 2.34 × 10^{-12} cm^3 molecule^{-1} s^{-1} at 298 K (recommended, Atkinson 1997)
  Hydrolysis:
  Biodegradation:
  Biotransformation:
  Bioconcentration, Uptake (k_1) and Elimination (k_2) Rate Constants or Half-Lives:

Half-Lives in the Environment:

Aliphatic and Cyclic Hydrocarbons

2.1.1.1.6   2,3-Dimethylbutane

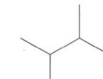

Common Name: 2,3-Dimethylbutane
Synonym: diisopropyl
Chemical Name: 2,3-dimethylbutane
CAS Registry No: 79-29-8
Molecular Formula: C_6H_{14}; $CH_3CH(CH_3)CH(CH_3)CH_3$
Molecular Weight: 86.175
Melting Point (°C):
    –128.10   (Lide 2003)
Boiling Point (°C):
    57.93   (Lide 2003)
Density (g/cm³ at 20°C):
    0.6616, 0.6570   (20°C, 25°C, Dreisbach 1959; Riddick et al. 1986)
    0.6616   (Weast 1984)
Molar Volume (cm³/mol):
    130.25, 131.16   (20°C, 25°C, calculated-density)
    140.6   (calculated-Le Bas method at normal boiling point)
Enthalpy of Vaporization, ΔH_V (kJ/mol):
    29.12, 27.275   (25°C, bp, Dreisbach 1959)
    29.125, 27.276   (25°C, bp, Riddick et al. 1986)
Enthalpy of Fusion, ΔH_{fus} (kJ/mol):
    0.812   (Dreisbach 1959)
    0.7991   (Riddick et al. 1986)
    6.43, 2.37, 0.79 (–137.05, –166.15, –127.95°C, Chickos et al. 1999)
Entropy of Fusion, ΔS_{fus} (J/mol K):
    52.96, 37.6 (exptl., calculated-group additivity method, total phase change entropy, Chickos et al. 1999)
Fugacity Ratio at 25°C, F: 1.0

Water Solubility (g/m³ or mg/L at 25°C or as indicated. Additional data at other temperatures designated * are compiled at the end of this section.):
    32.9, 22.5   (0, 25°C, shake flask-GC, calculated-group contribution, Polak & Lu 1973)
    19.1*   (shake flask-GC, measured range 25–149.5°C, Price 1976)
    11.0   (selected, Riddick et al. 1986)
    21.0*   (IUPAC tentative value, temp range 0 – 150°C, Shaw 1989a)
    18.67*   (calculated-liquid-liquid equilibrium LLE data, temp range 273.2–422.7 K, Mączyński et al. 2004)

Vapor Pressure (Pa at 25°C or as indicated and reported temperature dependence equations. Additional data at other temperatures designated * are compiled at the end of this section.):
    28955*   (23.10°C, ebulliometry, measured range: 14.256–58.789°C, Willingham et al. 1945)
    log (P/mmHg) = 6.80983 – 1127.187/(228.900 + t/°C); temp range 14.256–58.789°C (Antoine eq. from exptl. data, ebulliometry, Willingham et al. 1945)
    31204*   (calculated-Antoine eq. regression, temp range –63.6 to 58°C, Stull 1947)
    31280   (calculated from determined data, Dreisbach 1959)
    log (P/mmHg) = 6.90983 – 1127.187/(228.9 + t/°C); temp range –20 to 100°C (Antoine eq. for liquid state, Dreisbach 1959)
    31277*   (Antoine eq., temp range –34.9 to 81.3°C, Zwolinski & Wilhoit 1971)
    log (P/mmHg) = 6.80983 – 1127.187/(228.900 + t/°C); temp range –34.9 to 81.3°C (Antoine eq., Zwolinski & Wilhoit 1971)
    log (P/mmHg) = [–0.2185 × 7120.0/(T/K)] + 7.536008; temp range –63.6 to 225.5°C (Antoine eq., Weast 1972–73)

31280 (interpolated-Antoine equations, Boublik et al. 1984)

log (P/kPa) = 5.594371 – 1132.099/(229.494 + t/°C); temp range 14.256–58.8°C (Antoine eq. from reported exptl. data of Willingham et al. 1945, Boublik et al. 1984)

32010 (interpolated-Antoine eq., temp range –35 to 81°C, Dean 1985, 1992)

log (P/mmHg) = 6.80983 – 1127.83/(228.90 + t/°C); temp range –35 to 81°C (Antoine eq., Dean 1985, 1992)

31300 (lit. average, Riddick et al. 1986)

log (P/kPa) = 5.93941 – 1129.73/(229.215 + t/°C); temp range not specified (Antoine eq., Riddick et al. 1986)

31290 (interpolated-Antoine eq., Stephenson & Malanowski 1987)

log (P_L/kPa) = 5.95181 – 1136.355/(–43.159 + T/K); temp range 278–322 K (Antoine eq., Stephenson & Malanowski 1987)

log (P/mmHg) = 33.6319 – 2.5524 × 10^3/(T/K) – 9.3142·log (T/K) + 1.4759 × 10^{-10}·(T/K) + 3.914 × 10^{-6}·(T/K)^2; temp range 145–500 K (vapor pressure eq., Yaws 1994)

Henry's Law Constant (Pa m^3/mol at 25°C):

130000 (recommended, Mackay & Shiu 1981)
141000 (calculated-P/C, Mackay & Shiu 1981)
131190 (calculated-vapor-liquid equilibrium (VLE) data, Yaws et al. 1991)

Octanol/Water Partition Coefficient, log K_{OW}:

3.85 (shake flask, Hansch & Leo 1979)
3.85 (calculated-fragment const., Valvani et al. 1981)
3.85 (recommended, Sangster 1989, 1993)
2.42; 2.63 (calculated-S, calculated-molar volume, Wang et al. 1992)
3.42 (recommended, Hansch et al. 1995)

Octanol/Air Partition Coefficient, log K_{OA}:

Bioconcentration Factor, log BCF:

Sorption Partition Coefficient, log K_{OC}:

Environmental Fate Rate Constants, k, and Half-Lives, $t_{1/2}$:

Volatilization:

Photolysis:

Photooxidation: rate constant k, for gas-phase second order rate constants, k_{OH} for reaction with OH radical, k_{NO3} with NO_3 radical and k_{O3} with O_3 or as indicated, *data at other temperatures and/or the Arrhenius expression see reference:

k_{OH}*(exptl) = 5.16 × 10^{12} cm^3 mol^{-1} s^{-1}, k_{OH}(correlated) = 4.49 × 10^{12} cm^3 mol^{-1} s^{-1} at 300 K, measured range 300–498 K (flash photolysis-kinetic spectroscopy, Greiner 1970)

$k_{O(3P)}$ = 2.0 × 10^{-13} cm^3 molecule^{-1} s^{-1} for the reaction with O(^{3}P) atom at room temp. (Herron & Huie 1973)

k_{OH} = 5.50 × 10^{-12} cm^3 molecule^{-1} s^{-1} at room temp. (Atkinson et al. 1979)

k_{OH} = 5.50 × 10^{-12} cm^3 molecule^{-1} s^{-1}; $k_{O(3P)}$ = 2.0 × 10^{-12} cm^3 molecule^{-1} s^{-1} for reaction with O(^{3}P) atom at room temp. (abstraction mechanism, Gaffney & Levine 1979)

k_{OH} = (5.67 ± 0.29) × 10^{-12} cm^3 molecule^{-1} s^{-1} at 300 K (relative rate method, Darnall et al. 1978)

k_{OH} = (6.26 ± 0.06) × 10^{-12} cm^3 molecule^{-1} s^{-1} at room temp. (relative rate, Atkinson et al. 1984c)

k_{NO3} = 4.06 × 10^{-16} cm^3 molecule^{-1} s^{-1} at 296 K (relative rate method, Atkinson et al. 1988)

k_{OH}* = (5.90 - 0.23) × 10^{-12} cm^3 molecule^{-1} s^{-1} at 295 K, measured range 247–327 K (relative rate method, Harris & Kerr 1988)

k_{OH} = 6.2 × 10^{-12} cm^3 molecule^{-1} s^{-1} at 298 K (recommended, Atkinson 1989)

k_{OH} = 6.3 × 10^{-12} cm^3 molecule^{-1} s^{-1} with an estimated atmospheric lifetime of 22 h in air during summer daylight (Altshuller 1991)

k_{OH} = 6.30 × 10^{-12} cm^3 molecule^{-1} s^{-1}, k_{NO3} = 4.06 × 10^{-16} cm^3 molecule^{-1} s^{-1} at 298 K (Atkinson 1990)

k_{OH} = 19.0 × 10^{-12} cm^3 molecule^{-1} s^{-1}, k_{NO3} = 4.06 × 10^{-16} cm^3 molecule^{-1} s^{-1} at 298 K (Sabljic & Güsten 1990)

k_{NO3} = (4.04 - 5.34) × 10^{-16} cm^3 molecule^{-1} s^{-1} at 296 K (review, Atkinson 1991)

k_{OH} = 6.3 × 10^{-12} cm^3 molecule^{-1} s^{-1} at 298 K, k_{NO3} = 40.6 × 10^{-17} cm^3 molecule^{-1} s^{-1} at 296 K (Atkinson 1990)

k_{NO3}(exptl) = 4.08 × 10^{-16} cm^3 molecule^{-1} s^{-1}, k_{NO3}(recommended) = 2.55 × 10^{-16} cm^3 molecule^{-1} s^{-1}, k_{NO3}(calc) = 2.55 × 10^{-16} cm^3 molecule^{-1} s^{-1} at 296 ± 2 K (relative rate method, Aschmann & Atkinson 1995)

k_{OH}* = 5.78 × 10^{-12} cm^3 molecule^{-1} s^{-1}, k_{NO3} = 4.40 × 10^{-16} cm^3 molecule^{-1} s^{-1} at 298 K (recommended, Atkinson 1997)

Hydrolysis:

Biodegradation:

Biotransformation:

Bioconcentration, Uptake (k_1) and Elimination (k_2) Rate Constants or Half-Lives:

Half-Lives in the Environment:

Air: photooxidation reaction rate constant of 6.30 × 10^{-12} cm^3 molecule^{-1} s^{-1} with hydroxyl radicals and an estimated atmospheric lifetime of 22 h during summer daylight (Altshuller 1991).

TABLE 2.1.1.1.6.1
Reported aqueous solubilities of 2,3-dimethylbutane at various temperatures

Polak & Lu 1973		Price 1976		Shaw 1989a		Mączyński et al. 2004	
shake flask-GC		shake flask-GC/FID		IUPAC tentative values		calc-recommended LLE data	
t/°C	S/g·m^{-3}	t/°C	S/g·m^{-3}	t/°C	S/g·m^{-3}	t/°C	S/g·m^{-3}
0	32.9	25	19.1	0	33	25	18.67
25	22.5	40.1	19.2	25	21	40.1	18.19
		55.1	23.7	30	19	55.1	20.11
		99.1	40.1	41	19	99.1	39.74
		1213	56.8	50	21	121.3	67.03
		137.3	97.9	70	28	137.3	100.5
		149.5	171	90	35	149.5	138.8
				110	46		
				130	75		
				150	180		

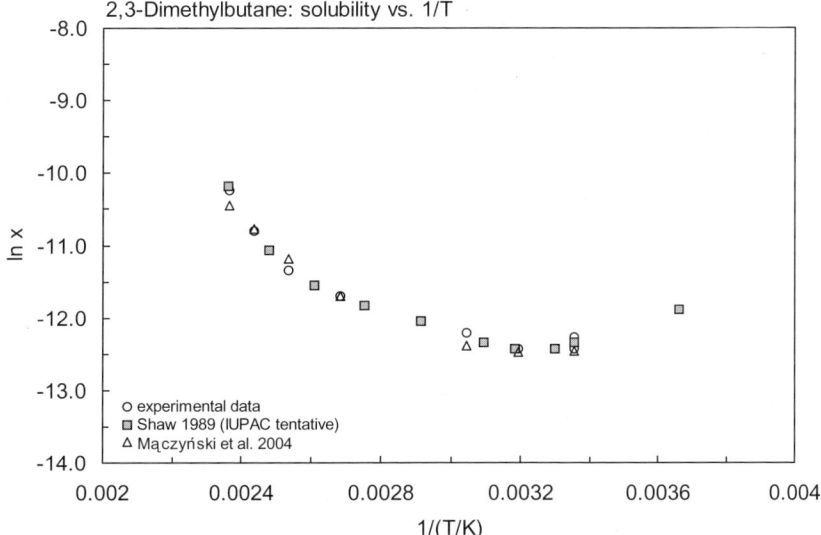

FIGURE 2.1.1.1.6.1 Logarithm of mole fraction solubility (ln x) versus reciprocal temperature for 2,3-dimethylbutane.

TABLE 2.1.1.1.6.2
Reported vapor pressures of 2,3-dimethylbutane at various temperatures and the coefficients for the vapor pressure equations

$$\log P = A - B/(T/K) \quad (1)$$
$$\log P = A - B/(C + t/°C) \quad (2)$$
$$\log P = A - B/(C + T/K) \quad (3)$$
$$\log P = A - B/(T/K) - C \cdot \log (T/K) \quad (4)$$
$$\ln P = A - B/(T/K) \quad (1a)$$
$$\ln P = A - B/(C + t/°C) \quad (2a)$$

Willingham et al. 1945		Stull 1947		Zwolinski & Wilhoit 1971			
ebulliometry		summary of literature data		selected values			
t/°C	P/Pa	t/°C	P/Pa	t/°C	P/Pa	t/°C	P/Pa
14.256	19921	−63.6	133.3	−34.9	1333	57.988	101325
18.044	23451	−44.5	666.6	−24.3	2666	25.0	234.6
23.099	28955	−34.9	1333	−17.5	4000		
27.746	34897	−24.1	2666	−12.5	5333	eq. 2	P/mmHg
33.357	43320	−12.4	5333	−8.4	6666	A	6.80983
39.15	53654	−4.9	7999	−4.9	7999	B	1127.187
45.339	66756	5.40	13332	0.82	10666	C	228.900
52.06	83718	21.1	26664	5.34	13332		
56.806	97604	39.0	53329	14.36	19998	bp/°C	57.998
57.317	99201	58.0	101325	21.097	26664	ΔH_V/(kJ mol^{-1}) =	
57.79	100693			26.588	33331	at 25°C	29.12
58.32	102390	mp/°C	−128.2	31.257	39997	at bp	27.28
58.789	103907			38.982	53329		
				45.927	66661		
bp/°C	57.988			50.682	79993		
eq. 2	P/mmHg			55.404	93326		
A	6.80983			56.283	95992		
B	1127.187			57.145	98659		
C	228.900			57.568	99992		

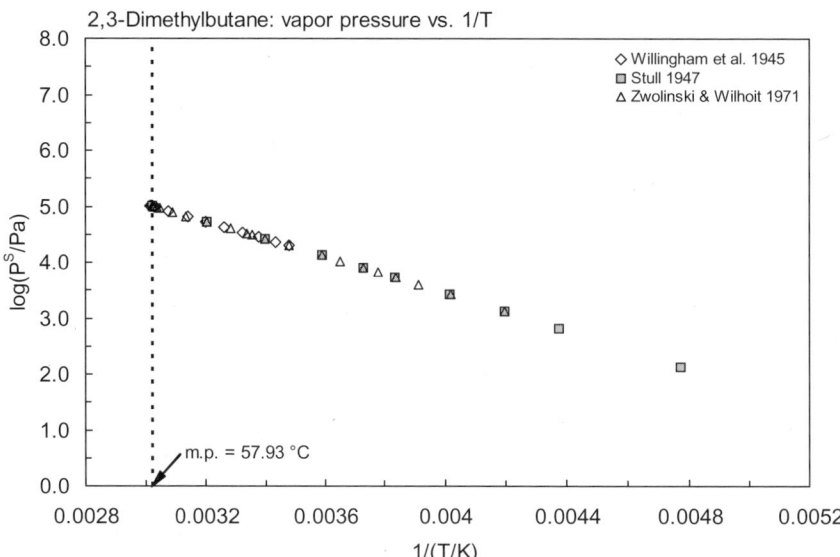

FIGURE 2.1.1.1.6.2 Logarithm of vapor pressure versus reciprocal temperature for 2,3-dimethylbutane.

Aliphatic and Cyclic Hydrocarbons

2.1.1.1.7   2,2,3-Trimethylbutane

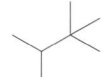

Common Name: 2,2,3-Trimethylbutane
Synonym: triptene
Chemical Name: 2,2,3-trimethylbutane
CAS Registry No: 464-06-2
Molecular Formula: C_7H_{16}; $CH_3CH(CH_3)_2CH(CH_3)_2$
Molecular Weight: 100.202
Melting Point (°C):
    –24.6   (Lide 2003)
Boiling Point (°C):
    80.86   (Lide 2003)
Density (g/cm³ at 20°C):
    0.6901, 0.6859   (20°C, 25°C, Dreisbach 1959)
    0.6901   (Weast 1984)
Molar Volume (cm³/mol):
    145.2, 146.1 (20°C, 25°C, calculated-density)
    162.8   (calculated-Le Bas method at normal boiling point)
Enthalpy of Vaporization, ΔH_V (kJ/mol):
    32.037, 28.94   (25°C, at normal bp, Dreisbach 1959)
Enthalpy of Fusion, ΔH_{fus} (kJ/mol):
    2.259   (at mp, Dreisbach 1959; Chickos et al. 1999)
    2.36, 2.2   (–152.15, –25.45°C, Chickos et al. 1999)
Entropy of Fusion, ΔS_{fus} (J/mol K):
    28.53, 36.7 (exptl., calculated-group additivity method, total phase change entropy, Chickos et al. 1999)
Fugacity Ratio at 25°C, F: 1.0

Water Solubility (g/m³ or mg/L at 25°C):
    4.38   (estimated-nomograph of Kabadi & Danner 1979; Brookman et al. 1985)
    7.09   (calculated-molar volume, mp and mobile order thermodynamics, Ruelle & Kesselring 1997)

Vapor Pressure (Pa at 25°C or as indicated and reported temperature dependence equations):
    13840   (25.3°C, ebulliometry, measured range 12.6-81.8°C, Forziati et al. 1949)
    log (P/mmHg) = 6.79230 – 1200.563/(226.650 + t/°C); temp range 12.6–81.8°C (Antoine eq., ebulliometry-manometer measurements, Forziati et al. 1949)
    13650   (calculated from determined data, Dreisbach 1959)
    log (P/mmHg) = 6.79230 – 1200.563/(226.05 + t/°C); temp range 0 – 125°C (Antoine eq. for liquid state, Dreisbach 1959)
    13652   (derived from compiled data, temp range –18.8–205.94°C, Zwolinski & Wilhoit 1971)
    log (P/mmHg) = 6.79230 – 1200.563/(226.050 + t/°C); temp range –18.8–205.94°C (Antoine eq., Zwolinski & Wilhoit 1971)
    13650, 13660   (interpolated-Antoine equations, Boublik et al. 1984)
    log (P/kPa) = 5.92699 – 11206.087/(226.731 + t/°C); temp range 22.7–105.6°C (Antoine eq. from reported exptl. data, Boublik et al. 1984)
    log (P/kPa) = 5.92037 – 1202.337/(226.256 + t/°C); temp range 12.6–81.77°C (Antoine eq. from reported exptl. data of Forziati et al. 1949, Boublik et al. 1984)
    13650   (interpolated-Antoine eq., temp range –19 to 106°C, Dean 1985, 1992)
    log (P/mmHg) = 6.79230 – 1200.563/(226.05 + t/°C); temp range –19 to 106°C (Antoine eq., Dean 1985, 1992)
    13650   (interpolated-Antoine eq., Stephenson & Malanowski 1987)

log $(P_L/kPa) = 5.9181 - 1201.098/(-47.026 + T/K)$; temp range 284–355 K (Antoine eq.-I, Stephenson & Malanowski 1987)

log $(P_L/kPa) = 6.18145 - 1390.726/(-20.97 + T/K)$; temp range 353–483 K (Antoine eq.-II, Stephenson & Malanowski 1987)

log $(P/mmHg) = 32.3633 - 2.6614 \times 10^3/(T/K) - 8.7743 \cdot \log(T/K) - 7.687 \times 10^{-10} \cdot (T/K) + 3.2006 \times 10^{-6} \cdot (T/K)^2$; temp range 249–531 K (vapor pressure eq., Yaws 1994)

Henry's Law Constant (Pa m^3/mol at 25°C):
    241010   (calculated-vapor-liquid equilibrium (VLE) data, Yaws et al. 1991)

Octanol/Water Partition Coefficient, log K_{OW}:

Octanol/Air Partition Coefficient, log K_{OA}:

Bioconcentration Factor, log BCF:

Sorption Partition Coefficient, log K_{OC}:

Environmental Fate Rate Constants, k, and Half-Lives, $t_{½}$:
    Volatilization:
    Photolysis:
    Oxidation: rate constant k, for gas-phase second order rate constants, k_{OH} for reaction with OH radical, k_{NO3} with NO_3 radical and k_{O3} with O_3 or as indicated, *data at other temperatures and/or the Arrhenius expression see reference:
        k_{OH}*(exptl) = 3.84×10^{12} cm^3 mol^{-1} s^{-1}, k_{OH}(correlated) = 3.15×10^{12} cm^3 mol^{-1} s^{-1} at 296 K, measured range 296–498 K (flash photolysis-kinetic spectroscopy, Greiner 1970)
        k_{OH} = (3.6 - 5.05) $\times 10^{-12}$ cm^3 molecule^{-1} s^{-1} at 296–305 K (Darnall et al. 1978)
        k_{OH} = (4.21 ± 0.08) $\times 10^{-12}$ cm^3 molecule^{-1} s^{-1} at room temp. (relative rate, Atkinson et al. 1984c)
        k_{OH} = 5.23×10^{-12} cm^3 molecule^{-1} s^{-1} at 296 K, k_{OH} = 4.09×10^{-12} cm^3 molecule^{-1} s^{-1} at 297 K (Atkinson 1985)
        k_{OH}* = 4.23×10^{-12} cm^3 molecule^{-1} s^{-1} at 298 K (recommended, Atkinson 1989, 1990)
        k_{NO3}(exptl) = 2.23×10^{-16} cm^3 molecule^{-1} s^{-1}, k_{NO3}(calc) = 1.31×10^{-16} cm^3 molecule^{-1} s^{-1} at 296 ± 2 K (relative rate method, Aschmann & Atkinson 1995)
        k_{OH}* = 4.24×10^{-12} cm^3 molecule^{-1} s^{-1}, k_{NO3}* = 2.4×10^{-16} cm^3 molecule^{-1} s^{-1} at 298 K (recommended, Atkinson 1997)
    Hydrolysis:
    Biodegradation:
    Biotransformation:
    Bioconcentration, Uptake (k_1) and Elimination (k_2) Rate Constants or Half-Lives:

Aliphatic and Cyclic Hydrocarbons

2.1.1.1.8   n-Pentane

Common Name: *n*-Pentane
Synonym: pentane
Chemical Name: *n*-pentane
CAS Registry No: 109-66-0
Molecular Formula: C_5H_{12}; $CH_3(CH_2)_3CH_3$
Molecular Weight: 72.149
Melting Point (°C):
    –129.67   (Lide 2003)
Boiling Point (°C):
    36.06   (Lide 2003)
Density (g/cm³ at 20°C):
    0.6262, 0.6214   (20°C, 25°C, Dreisbach 1959; Riddick et al. 1986)
Molar Volume (cm³/mol):
    115.22, 116.1   (20° C, 25°C, calculated-density)
    118.4   (calculated-Le Bas method at normal boiling point)
Enthalpy of Vaporization, ΔH_V (kJ/mol):
    26.42, 25.77   (25°C bp, Dreisbach 1959)
    26.427, 25.786   (25°C, bp, Riddick et al. 1986)
Enthalpy of Fusion, ΔH_{fus} (kJ/mol):
    8.393   (Dreisbach 1959; Riddick et al. 1986)
    8.4   (Chickos et al. 1999)
Entropy of Fusion, ΔS_{fus} (J/mol K):
    58.59, 63.2   (exptl., calculated-group additivity method, Chickos et al. 1999)
Fugacity Ratio at 25°C, F: 1.0

Water Solubility (g/m³ or mg/L at 25°C or as indicated and reported temperature dependence equations. Additional data at other temperatures designated * are compiled at the end of this section.):

360	(16°C, cloud point, Fühner 1924)
120	(radiotracer method, Black et al. 1948)
38.5	(shake flask-GC, McAuliffe 1963, 1966)
49.7	(vapor saturation-GC, Barone et al. 1966)
40.0	(Baker 1967)
40.3*	(shake flask-GC, measured range 4 – 30°C, Nelson & De Ligny 1968)
11.8*	(shake flask-GC, measured range 5 – 35°C, Pierotti & Liabastre 1972)
65.7; 47.6, 44.6	(0, 25°C, shake flask-GC, calculated-group contribution, Polak & Lu 1973)
39.5*	(shake flask-GC, measured range 25 – 149.5°C, Price 1976)
39.0	(shake flask-GC, Kryzanowska & Szeliga 1978)
40.0	(partition coefficient, Rudakov & Lutsyk 1979)
40.75	(generator column-GC, Tewari et al. 1982a)
36.9	(calculated-activity coeff. γ and K_{OW}, Tewari 1982b)
40.6*	(vapor saturation-GC, measured range 15 – 40°C, Jönsson et al. 1982)
38.9	(shake flask-GC, Coates et al. 1985)
38.0	(selected, Riddick et al. 1986)
42.0*	(IUPAC recommended best value, temp range 0 – 90°C, Shaw 1989)

$\ln x = -333.59719 + 14358.472/(T/K) + 47.97436 \cdot \ln(T/K)$; temp range 290–400 K (eq. derived from literature calorimetric and solubility data, Tsonopoulos 1999)
    44.09*   (calculated-liquid-liquid equilibrium LLE data, temp range 273.2 – 422.7 K, Mączyński et al. 2004)

Vapor Pressure (Pa at 25°C or as indicated and reported temperature dependence equations. Additional data at other temperatures designated * are compiled at the end of this section.):

- 68213* (24.828°C, static method, measured range –65 to 25°C, Messerly & Kennedy 1940)
- 66760* (24.37°C, ebulliometry, measured range 13.282–36.818°C, Willingham et al. 1945)
- log (P/mmHg) = 6.87372 – 1075.816/(233.369 + t/°C); temp range 13.282–36.818°C (Antoine eq. from exptl. data, ebulliometry-manometer, Willingham et al. 1945)
- 71050* (interpolated-Antoine eq. regression, temp range –76.6 to 36.1°C, Stull 1947)
- 68330 (calculated from determined data, Dreisbach 1959)
- log (P/mmHg) = 6.85221 – 1064.63/(232.0 + t/°C); temp range –35 to 80°C (Antoine eq. for liquid state, Dreisbach 1959)
- 68368* (interpolated- Antoine eq., temp range –50.14 to 57.53°C, Zwolinski & Wilhoit 1971)
- log (P/mmHg) = 6.87632 – 1075.78/(233.205 + t/°C); temp range –50.14 to 57.53°C (Antoine eq., Zwolinski & Wilhoit 1971)
- log (P/mmHg) = [–0.2185 × 6595.1/(T/K)] + 7.489673; temp range –76.6 to 191°C (Antoine eq., Weast 1972–73)
- 3400* (–30.86°C, gas saturation, measured range –129.54 to –30.86°C, Carruth & Kobayashi 1973)
- 57820* (20.57°C, ebulliometric method, measured range –4.4 to 68.218°C, Osborn & Douslin 1974)
- 68330 (interpolated-Antoine eq, Boublik et al. 1984)
- log (P/kPa) = 6.12545 – 1132.518/(239.074 + t/°C); temp range –65.2 to 24.83°C (Antoine eq. from reported exptl. data, Boublik et al. 1984)
- log (P/kPa) = 5.96982 – 1060.916/(231.577 + t/°C); temp range 13.28–36.82°C (Antoine eq. from reported exptl. data of Willingham et al. 1945, Boublik et al. 1984)
- log (P/kPa) = 5.99028 – 1071.187/(232.766 + t/°C); temp range –4.4 to 68.21°C (Antoine eq. from reported exptl. data of Osborn & Douslin 1974, Boublik et al. 1984)
- 70915 (interpolated-Antoine eq., temp range –50 to 58°C, Dean 1985. 1992)
- log (P/mmHg) = 6.85296 – 1064.84/(233.01 + t/°C); temp range –50 to 58°C (Antoine eq., Dean 1985, 1992)
- 69810, 68880, 68330 (headspace-GC, correlated, Antoine eq., Hussam & Carr 1985)
- 68330 (lit. average, Riddick et al. 1986)
- log (P/kPa) = 5.97786 – 1064.84/(232.012 + t/°C); temp range not specified (Antoine eq., Riddick et al. 1986)
- 68355 (interpolated-Antoine eq., Stephenson & Malanowski 1987)
- log (P$_L$/kPa) = 7.6922 – 1686.65/(T/K); temp range 143–233 K (Antoine eq.-I, Stephenson & Malanowski 1987)
- log (P$_L$/kPa) = 5.99466 – 1073.139/(–40.188 + T/K); temp range 223–352 K (Antoine eq.-II, Stephenson & Malanowski 1987)
- log (P$_L$/kPa) = 5.98799 – 1070.14/(–40.485 + T/K); temp range 269–335 K (Antoine eq.-III, Stephenson & Malanowski 1987)
- log (P$_L$/kPa) = 6.28417 – 1260.973/(–14.031 + T/K); temp range 350–422 K (Antoine eq.-IV, Stephenson & Malanowski 1987)
- log (P$_L$/kPa) = 7.47436 – 2414.137/(141.919 + T/K); temp range 418–470 K (Antoine eq., Stephenson & Malanowski 1987)
- 68340* (recommended, Ruzicka & Majer 1994)
- ln [(P/kPa)/(P$_o$/kPa)] = [1 – (T$_o$/K)/(T/K)]·exp{2.73425 – 1.988544 × 10^{-3}·(T/K) + 2.408406 × 10^{-6}·(T/K)^2}; reference state at P$_o$ = 101.325 kPa, T$_o$ = 309.209 K (Cox equation, Ruzicka & Majer 1994)
- log (P/mmHg) = 33.3239 – 2.4227 × 10^3/(T/K) – 9.2354·log (T/K) + 9.0199 × 10^{-11}·(T/K) + 4.105 × 10^{-6}·(T/K)^2; temp range 143–470 K (vapor pressure eq., Yaws 1994)

Henry's Law Constant (Pa m^3/mol at 25°C or as indicated. Additional data at other temperatures designated * are compiled at the end of this section.):

- 127050 (calculated as 1/K$_{AW}$, C$_W$/C$_A$, reported as exptl., Hine & Mookerjee 1975)
- 115900, 33430 (calculated-group contribution, calculated-bond contribution, Hine & Mookerjee 1975)
- 128000 (calculated-P/C, Mackay & Shiu 1975, 1990; Bobra et al. 1979; Mackay et al. 1979; Mackay 1981)
- 125000 (recommended, Mackay & Shiu 1981)
- 128000, 125000, 123000, 122200, 10370 (calculated-P/C, Mackay & Shiu 1981)
- 78050, 99075, 126800, 144020, 174480 (14.8, 20.05, 25.1, 30.1, 34.92°C, equilibrium cell-concentration ratio-GC, Jönsson et al. 1982)
- 121410* (calculated-temp dependence eq. derived from exptl data, measured range 15–35°C. Jönsson et al. 1982)

ln $(1/K_{AW})$ = 19237.8/(T/K) + 53.671·ln (T/K) – 372.214; temp range: 15 – 35°C (least-square regression of equilibrium cell-concn ratio-GC measurements, Jönsson et al. 1982)

127670	(selected, Mills et al. 1982)
120970	(calculated-P/C, Eastcott et al. 1988)
100980	(calculated-MCI χ, Nirmalakhandan & Speece 1988)
128050	(calculated-vapor-liquid equilibrium (VLE) data, Yaws et al. 1991)

Octanol/Water Partition Coefficient, log K_{OW}:

2.50	(shake flask-GC, Hansch et al. 1968)
3.39	(shake flask-GC, Leo et al. 1975)
2.99, 3.42, 3.48	(calculated-fragment const., Rekker 1977)
3.23	(Hansch & Leo 1979)
3.64	(calculated-activity coeff. γ, Wasik et al. 1981,1982)
3.62	(generator column-GC, Tewari et al. 1982a,b)
2.37	(HPLC-k$'$ correlation, Coates et al. 1985)
3.62, 3.60	(generator column-GC, calculated-activity coeff. γ, Schantz & Martire 1987)
3.45	(recommended, Sangster 1989, 1993)
3.39	(recommended, Hansch et al. 1995)

Octanol/Air Partition Coefficient, log K_{OA} at 25°C or as indicated:

2.05*	(20.29°C, from GC-determined γ^∞ in octanol, measured range 20.29 – 50.28°C, Gruber et al. 1997)
1.95	(calculated-measured γ^∞ in pure octanol and vapor pressure P, Abraham et al. 2001)

Bioconcentration Factor, log BCF:

Sorption Partition Coefficient, log K_{OC}:

Environmental Fate Rate Constants, k, and Half-Lives, $t_{1/2}$:

Volatilization:

Photolysis:

Oxidation: rate constant k, for gas-phase second order rate constants, k_{OH} for reaction with OH radical, k_{NO3} with NO_3 radical and k_{O3} with O_3 or as indicated *data at other temperatures see reference:

$k_{O(3P)} = 5.8 \times 10^{-14}$ cm^3 molecule^{-1} s^{-1} for the reaction with O(^{3}P) (Herron & Huie 1973)

$k_{OH} = (3.74 \pm 0.13) \times 10^{-12}$ cm^3 molecule^{-1} s^{-1} at 300 K (relative rate method, Darnall et al. 1978)

$k_{OH} = 5.0 \times 10^{-12}$ cm^3 molecule^{-1} s^{-1} (Atkinson et al. 1979)

$k_{OH} = 5.0 \times 10^{-12}$ cm^3 molecule^{-1} s^{-1}; $k_{O(3P)} = 5.8 \times 10^{-14}$ cm^3 molecule^{-1} s^{-1} room temp. (abstraction mechanism, Gaffney & Levine 1979)

$k_{OH} = 4.13 \times 10^{-12}$ cm^3 molecule^{-1} s^{-1} at 299 ± 2 K (relative rate method, Atkinson et al. 1982a)

$k_{NO3} = (8.1 \pm 1.7) \times 10^{-17}$ cm^3 molecule^{-1} s^{-1} at 296 K (relative rate method, Atkinson et al. 1984a; Atkinson 1991)

$k_{OH} = (4.13 \pm 0.08) \times 10^{-12}$ cm^3 molecule^{-1} s^{-1} at room temp. (relative rate, Atkinson et al. 1984c)

$k_{OH} = 4.06 \times 10^{-12}$ cm^3 molecule^{-1} s^{-1} at 297 K (Atkinson 1986; quoted, Edney et al. 1986)

$k_{OH} = 4.29 \times 10^{-12}$ cm^3 molecule^{-1} s^{-1} at 312 K in Smog chamber (Nolting et al. 1988)

$k_{OH}* = 3.94 \times 10^{-12}$ cm^3 molecule^{-1} s^{-1} at 298 K (recommended, Atkinson 1989, 1991)

$k_{OH} = 4.06 \times 10^{-12}$ cm^3 molecule^{-1} s^{-1}, $k_{NO3} = 8.0 \times 10^{-17}$ cm^3 molecule^{-1} s^{-1} (Sabljic & Güsten 1990)

$k_{NO3} = 8.1 \times 10^{-17}$ cm^3 molecule^{-1} s^{-1}, $k_{OH} = 3.94 \times 10^{-12}$ cm^3 molecule^{-1} s^{-1} at 298 K with summer daylight atmospheric lifetime τ = 35 h (Altshuller 1991)

$k_{OH}* = 3.94 \times 10^{-12}$ cm^3 molecule^{-1} s^{-1} at 298 K, $k_{NO3} = 9.0 \times 10^{-17}$ cm^3 molecule^{-1} s^{-1} at 296 K (Atkinson 1990)

k_{NO3}(exptl) = 8.2 × 10^{-17} cm^3 molecule^{-1} s^{-1}, k_{NO3}(calc) = 7.7 × 10^{-17} cm^3 molecule^{-1} s^{-1} at 296 ± 2 K (relative rate method, Aschmann & Atkinson 1995)

$k_{OH}* = 4.00 \times 10^{-12}$ cm^3 molecule^{-1} s^{-1}, $k_{NO3} = 8.7 \times 10^{-17}$ cm^3 molecule^{-1} s^{-1} at 298 K (recommended, Atkinson 1997)

$k_{OH}* = 3.70 \times 10^{-12}$ cm^3 molecule^{-1} s^{-1} at 298 K, measured range 230–400 K (relative rate method, DeMore & Bayes 1999)

Hydrolysis:
Biodegradation:
Biotransformation:
Bioconcentration, Uptake (k_1) and Elimination (k_2) Rate Constants or Half-Lives:

Half-Lives in the Environment:

Air: photooxidation reaction rate constant of 3.94×10^{-12} cm^3 molecule^{-1} s^{-1} with hydroxyl radical and an estimated atmospheric lifetime $\tau = 35$ h (Altshuller 1990).

TABLE 2.1.1.1.8.1
Reported aqueous solubilities of *n*-pentane at various temperatures

1.

Nelson & De Ligny 1968		Pierotti & Liabastre 1972		Polak & Lu 1973		Price 1976	
shake flask-GC		shake flask-GC		shake flask-GC		shake flask-GC	
t/°C	S/g·m^{-3}	t/°C	S/g·m^{-3}	t/°C	S/g·m^{-3}	t/°C	S/g·m^{-3}
4.0	40.9	5.11	10.94	0	65.7	25	39.6
10	42.9	15.21	11.80	25	47.6	40.1	39.8
20	39.3	25.11	11.20			55.7	41.8
25	40.5	35.21	10.89			99.1	69.4
30	40.5					121.3	110
						137.3	210
						149.5	298

2.

Jonsson et al. 1982		Shaw 1989a		Mączyński et al. 2004	
vapor saturation-GC		IUPAC recommended		calc-recommended LLE data	
t/°C	S/g·m^{-3}	t/°C	S/g·m^{-3}	t/°C	S/g·m^{-3}
15	42.9	0	66	0	56.12
20	41.4	10	43	4	52.11
25	40.6	20	40	10	48.1
30	40.3	25	42	15	48.1
40	40.6	30	41	20	44.09
		40	40	25	44.09
		50	41	30	44.09
		60	43	35.1	44.09
		70	46	40.1	44.09
		80	50	55.7	48.1
		90	58	99.1	88.18
		110	86	121.3	136.3
		130	150	137.3	192.4
		150	300	149.5	260.5

Aliphatic and Cyclic Hydrocarbons

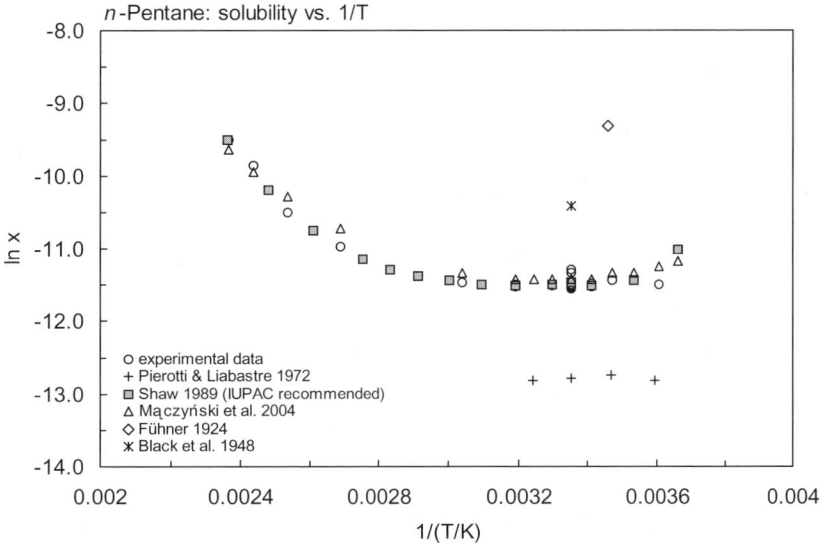

FIGURE 2.1.1.1.8.1 Logarithm of mole fraction solubility (ln x) versus reciprocal temperature for n-pentane.

TABLE 2.1.1.1.8.2
Reported vapor pressures of *n*-pentane at various temperatures and the coefficients for the vapor pressure equations

$\log P = A - B/(T/K)$ (1) $\quad\quad \ln P = A - B/(T/K)$ (1a)

$\log P = A - B/(C + t/°C)$ (2) $\quad\quad \ln P = A - B/(C + t/°C)$ (2a)

$\log P = A - B/(C + T/K)$ (3)

$\log P = A - B/(T/K) - C \cdot \log (T/K)$ (4)

$\log (P/P_\Phi) = A[1 - \Phi/(T/K)]$ (5) $\quad\quad$ where $\log A = a + b(T/K) + c(T/K)^2$ - Cox eq. I

$\ln [(P/kPa)/(P_o/kPa)] = [1 - (T_o/K)/(T/K)] \cdot \exp\{A_0 - A_1 \cdot (T/K) + A_2 \cdot (T/K)^2\}$ (6) $\quad\quad$ - Cox eq. II

1.

Messerly & Kennedy 1940		Willingham et al. 1945		Stull 1947		Zwolinski & Wilhoit 1971	
static method-manometer		ebulliometry		summary of literature data		selected values	
t/°C	P/Pa	t/°C	P/Pa	t/°C	P/Pa	t/°C	P/Pa
–65.178	411	13.282	43322	–76.6	133.3	–50.14	1333
–48.811	1604	18.647	53657	–62.5	666.6	–40.25	2666
–39.537	2818	24.371	66756	–50.1	1333	–33.86	4000
–27.420	5951	30.592	83719	–40.1	2666	–29.24	5333
–17.476	10340	34.981	97604	–29.2	5333	–25.42	6666
–9.793	15341	35.453	99201	–22.2	7999	–22.19	7999
–2.359	21933	35.89	100694	–12.6	13332	–16.98	10666
3.618	28767	36.379	102393	1.9	26664	–12.59	13332
9.621	37303	36.818	103911	18.5	53329	–4.33	19998
14.653	45947			36.1	101325	1.92	26664
18.613	53773	bp/°C	36.073			7.01	33331
21.679	60539			mp/°C	–129.7	11.34	39997

(*Continued*)

TABLE 2.1.1.1.8.2 (*Continued*)

Messerly & Kennedy 1940		Willingham et al. 1945		Stull 1947		Zwolinski & Wilhoit 1971	
static method-manometer		ebulliometry		summary of literature data		selected values	
t/°C	P/Pa	t/°C	P/Pa	t/°C	P/Pa	t/°C	P/Pa
24.828	68213	eq. 2	P/mmHg			18.48	53329
		A	6.87372			24.322	66661
		B	1076.816			29.297	79993
		C	233.359			33.657	93326
						34.469	95992
						35.264	98659
						36.042	101325
						25.0	68368
						bp/°C	36.042
						eq. 2	P/mmHg
						A	6.87632
						B	1075.78
						C	233.205
						$\Delta H_V/(kJ\ mol^{-1})=$	
						at 25°C	26.43
						at bp	25.77

2.

Carruth & Kobayashi 1973		Osborn & Douslin 1974		Ruzicka & Majer 1994	
gas saturation		ebulliometry		recommended	
t/°C	P/Pa	t/°C	P/Pa	T/K	P/Pa
–129.54	0.0809	–4.40	19933	144.82	0.1
–122.64	0.3893	0.512	25023	157.65	1
–115.87	0.7666	5.464	31177	173.73	10
–108.74	2.6	10.458	38656	193.17	100
–93.55	19.2	15.495	47375	219.13	1000
–83.75	56.93	20.572	57818	255.05	10000
–67.19	285.3	25.698	70121	309.21	101325
–54.18	926.6	30.86	84533	298.15	68350
–45.94	1733	36.068	101325		
–30.86	3400	41.32	120792	data calc from Cox eq.	
		46.613	143245	eq. 6	P/kPa
mp/°C	–129.7	51.951	169019	A_0	2.73425
		57.331	198487	A_1	1.966544×10^{-3}
eq. 1a	P/mmHg	62.755	232016	A_2	2.408406×10^{-6}
A	19.7269	68.218	270022	with reference state at	
B	3899.67			P_0/kPa	101.325
		Cox eq.		T_0/K	309.21
		eq. 5	P/mmHg		
		Φ	309.218		
		a	0.81357		
		b	-7.73685×10^{-4}		
		c	9.06731×10^{-7}		
		P_Φ	760		
		temp range: 268–342 K			

Aliphatic and Cyclic Hydrocarbons

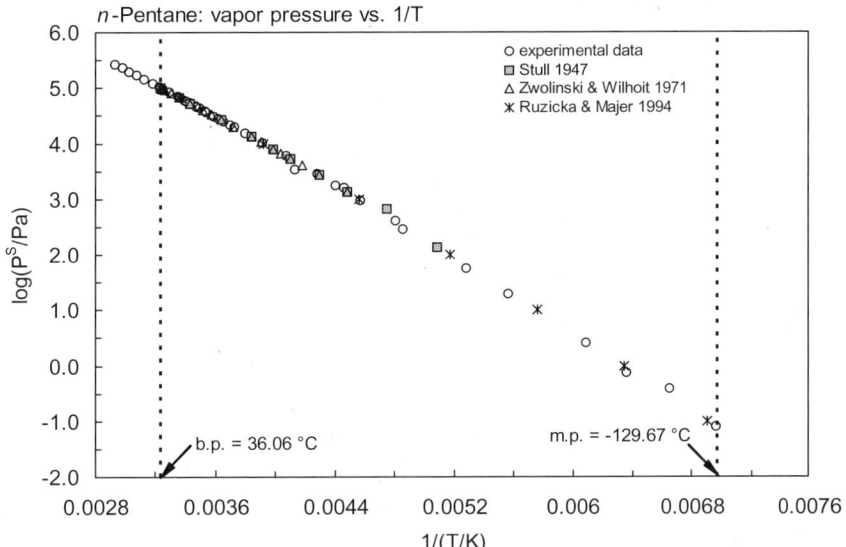

FIGURE 2.1.1.1.8.2 Logarithm of vapor pressure versus reciprocal temperature for *n*-pentane.

TABLE 2.1.1.1.8.3
Reported Henry's law constants and octanol-air partition coefficients of *n*-pentane at various temperatures and temperature dependence equations

$\ln K_{AW} = A - B/(T/K)$	(1)	$\log K_{AW} = A - B/(T/K)$	(1a)
$\ln (1/K_{AW}) = A - B/(T/K)$	(2)	$\log (1/K_{AW}) = A - B/(T/K)$	(2a)
$\ln (k_H/atm) = A - B/(T/K)$	(3)		
$\ln [H/(Pa\,m^3/mol)] = A - B/(T/K)$	(4)	$\ln [H/(atm\cdot m^3/mol)] = A - B/(T/K)$	(4a)
$K_{AW} = A - B\cdot(T/K) + C\cdot(T/K)^2$	(5)		

Henry's law constant		log K_{OA}	
Jönsson et al. 1982		Gruber et al. 1997	
equilibrium cell-GC		GC det'd activity coefficient	
t/°C	H/(Pa m³/mol)	t/°C	log K_{OA}
15.0	78035#	20.29	2.053
15.1	77905	30.3	1.911
15.1	77905	40.4	1.781
20.0	98674	50.28	1.670
20	100713#		
20	97490		
25.0	121511		
25	125828		
25	127774		
30.0	146533#		
30	144022		
35.0	173105#		
35	175477		

interpolated from exptl data

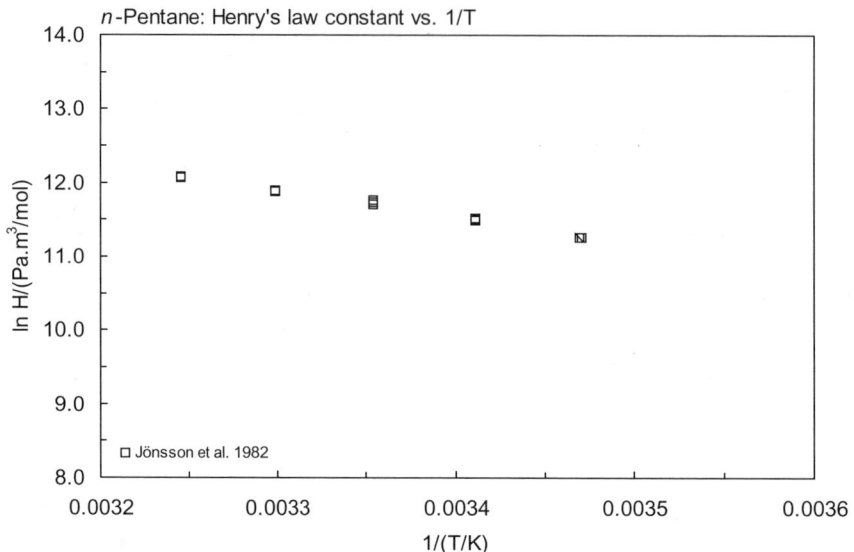

FIGURE 2.1.1.1.8.3 Logarithm of Henry's law constant versus reciprocal temperature for *n*-pentane.

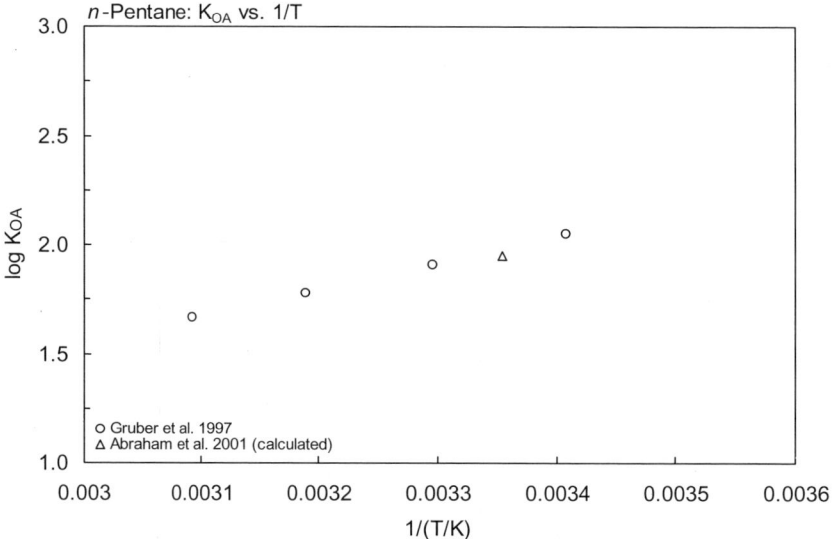

FIGURE 2.1.1.1.8.4 Logarithm of K_{OA} versus reciprocal temperature for *n*-pentane.

Aliphatic and Cyclic Hydrocarbons

2.1.1.1.9   2-Methylpentane (Isohexane)

Common Name: 2-Methylpentane
Synonym: isohexane
Chemical Name: 2-methylpentane
CAS Registry No: 107-83-5
Molecular Formula: C_6H_{14}; $CH_3CH(CH_3)CH_2CH_2CH_3$
Molecular Weight: 86.175
Melting Point (°C):
    –153.6   (Lide 2003)
Boiling Point (°C):
    60.26   (Lide 2003)
Density (g/cm³ at 20°C):
    0.6322, 0.6485 (20°C, 25°C, Dreisbach 1959; Riddick et al. 1986)
Molar Volume (cm³/mol):
    136.31, 132.88 (20°C, 25°C, calculated-density)
    140.6   (calculated-Le Bas method at normal boiling point)
Enthalpy of Vaporization, ΔH_V (kJ/mol):
    29.87, 27.79 (25°C, bp, Riddick et al. 1986)
Enthalpy of Fusion, ΔH_{fus} (kJ/mol):
    6.203   (Dreisbach 1959)
    6.268   (Riddick et al. 1986)
    6.27   (Chickos et al. 1999)
Entropy of Fusion, ΔS_{fus} (J/mol K):
    52.43, 50.6 (exptl., calculated-group additivity method, Chickos et al. 1999)
Fugacity Ratio at 25°C, F: 1.0

Water Solubility (g/m³ or mg/L at 25°C or as indicated. Additional data at other temperatures designated * are compiled at the end of this section.):
    13.8   (shake flask-GC, McAuliffe 1963, 1966)
    16.21   (vapor saturation-GC, Barone et al. 1966)
    19.45; 15.7, 15.6   (0, 25°C, shake flask-GC, calculated-group contribution, Polak & Lu 1973)
    14.2   (shake flask-GC, Leinonen & Mackay 1973)
    13.0*   (shake flask-GC, measured range 25 – 149.5°C, Price 1976)
    13.7*   (recommended best value, temp range 25 – 150°C, IUPAC Solubility Data Series, Shaw 1989a)
    13.88*   (calculated-liquid-liquid equilibrium LLE data, temp range 273.2–422.7 K, Mączyński et al. 2004)

Vapor Pressure (Pa at 25°C or as indicated and reported temperature dependence equations. Additional data at other temperatures designated * are compiled at the end of this section.):
    29040*   (25.64°C, ebulliometry, measured temp range 12.758–61.066°C, Willingham et al. 1945)
    log (P/mmHg) = 6.83910 – 1135.410/(226.572 + t/°C); temp range 12.758–61.066°C (Antoine eq. from exptl. data, ebulliometry-manometer, Willingham et al. 1945)
    27820*   (calculated-Antoine eq. regression, temp range –60 to 60.3°C, Stull 1947)
    28240   (calculated from determined data, Dreisbach 1959)
    log (P/mmHg) = 6.83910 – 1135.410/(226.572 + t/°C); temp range –15 to 100°C (Antoine eq. for liquid state, Dreisbach 1959)
    28238; 28200* (derived from compiled, interpolated-Antoine eq., Zwolinski & Wilhoit 1971)
    log (P/mmHg) = 6.83910 – 1135.410/(226.572 + t/°C); temp range –32.1 to 83.4°C (Antoine eq., Zwolinski & Wilhoit 1971)
    27780   (interpolated-Antoine eq., Weast 1972-73)

log (P/mmHg) = [–0.2185 × 7676.6/(T/K)] + 7.944630; temp range: –60.9 to 60.3°C (Antoine eq., Weast 1972–73)
28240 (interpolated-Antoine eq., Boublik et al. 1984)
log (P/kPa) = 6.86839 – 1151.401/(228.477 + t/°C); temp range 12.78 – 61°C (Antoine eq. from reported exptl. data of Willingham et al. 1945, Boublik et al. 1984)
28230 (interpolated-Antoine eq., temp range –32 to 83°C, Dean 1985, 1992)
log (P/mmHg) = 6.83910 – 1135.41/(226.57 + t/°C); temp range –32 to 83°C (Antoine eq., Dean 1985, 1992)
28300 (selected lit., Riddick et al. 1986)
log (P/kPa) = 5.98850 – 1148.74/(228.166 + t/°C); temp range not specified (Antoine eq., Riddick et al. 1986)
28250 (calculated-Antoine eq., Stephenson & Malanowski 1987)
log (P_L/kPa) = 5.97783 – 1142.922/(–45.657 + T/K); temp range 293–335 K (Antoine eq., Stephenson & Malanowski 1987)
log (P/mmHg) = 30.7477 – 2.4888 × 10^3/(T/K) –8.2295·log (T/K) – 2.3723 × 10^{-11}·(T/K) + 3.2402 × 10^{-6} · $(T/K)^2$; temp range 120–498 K (vapor pressure eq., Yaws 1994)
72190 (50°C, vapor-liquid equilibria VLE data, Horstmann et al. 2004)
log (P/kPa) = 5.99313 – 1151.40/(T/K – 44.673); temp range not specified (Antoine eq., Horstmann et al. 2004)

Henry's Law Constant (Pa m^3/mol at 25°C or as indicated and reported temperature dependence equations. Additional data at other temperatures designated * are compiled at the end of this section.):
170000 (recommended; Mackay & Shiu 1981)
175490 (calculated as 1/K_{AW}, C_W/C_A, reported as exptl., Hine & Mookerjee 1975)
196800, 49430 (calculated-group contribution, calculated-bond contribution, Hine & Mookerjee 1975)
176160 (calculated-P/C, Eastcott et al. 1988)
83590* (EPICS-GC, measured range 10 – 30°C, Ashworth et al. 1988)
ln [H/(atm·m^3/mol)] = 2.959 – 957.2/(T/K); temp range 10 – 30°C (EPICS measurements, Ashworth et al. 1988)
176280 (calculated-vapor-liquid equilibrium (VLE) data, Yaws et al. 1991)
746 (20°C, selected from reported experimental determined values, Staudinger & Roberts 1996, 2001)
log K_{AW} = 2.470 – 288/(T/K) (summary of literature data, Staudinger & Roberts 2001)

Octanol/Water Partition Coefficient, log K_{OW}:
2.80 (calculated-π constant, Hansch et al. 1968)
2.85 (calculated-MCI χ, Murray et al. 1975)
3.23 (calculated-molar volume V_M, Wang et al. 1992)
3.74 (calculated-fragment constant, Müller & Klein 1992)

Octanol/Air Partition Coefficient, log K_{OA}:

Bioconcentration Factor, log BCF:

Sorption Partition Coefficient, log K_{OC}:

Environmental Fate Rate Constants, k, and Half-Lives, $t_{1/2}$:
Volatilization:
Photolysis:
Oxidation: rate constant k, for gas-phase second order rate constants, k_{OH} for reaction with OH radical, k_{NO3} with NO$_3$ radical and k_{O3} with O$_3$ or as indicated, *data at other temperatures see reference:
k_{OH}(exptl) = (3.2 ± 0.6) × 10^9 M^{-1} s^{-1} at 305 ± 2 K (relative rate method, Lloyd et al. 1976, Darnall et al. 1976)
k_{OH} = (5.68 ± 0.24) × 10^{-12} cm^3 molecule^{-1} s^{-1} at room temp. (relative rate, Atkinson et al. 1984c)
k_{OH} = 5.6 × 10^{-12} cm^3 molecule^{-1} s^{-1} at 298 K (recommended, Atkinson 1989, 1990)
k_{OH} = 2.34 × 10^{-12} cm^3 molecule^{-1} s^{-1} at 298 K, k_{NO3} = 9.7 × 10^{-17} cm^3 molecule^{-1} s^{-1} at 296 K (Atkinson 1990)
k_{OH} = 5.6 × 10^{-12} cm^3 molecule^{-1} s^{-1}, estimated atmospheric lifetime was 25 h during summer daylight hours (Altshuller 1991)
k_{NO3}(exptl) = 1.71 × 10^{-16} cm^3 molecule^{-1} s^{-1}, k_{NO3}(calc) = 1.83 × 10^{-16} cm^3 molecule^{-1} s^{-1} at 296 ± 2 K (relative rate method, Aschmann & Atkinson 1995)
k_{OH} = 5.3 × 10^{-12} cm^3 molecule^{-1} s^{-1}, k_{NO3} = 1.8 × 10^{-16} cm^3 molecule^{-1} s^{-1} at 298 K (recommended, Atkinson 1997)

Aliphatic and Cyclic Hydrocarbons

Hydrolysis:
Biodegradation:
Biotransformation:
Bioconcentration, Uptake (k_1) and Elimination (k_2) Rate Constants or Half-Lives:

Half-Lives in the Environment:
  Air: half-life of 2.4–24 h based on photooxidation rate constant of 3.2×10^{12} cm^3 mol^{-1} s^{-1} for the gas-phase reaction with hydroxyl radical (Darnall et al. 1976);
    atmospheric lifetime was estimated to be 25 h during summer daylight, based on photooxidation rate constant of 5.6×10^{-12} cm^3 molecule^{-1} s^{-1} for the reaction with hydroxyl radical in air (Altshuller 1991).

TABLE 2.1.1.1.9.1
Reported aqueous solubilities of 2-methylpentane at various temperatures

Polak & Lu 1973		Price 1976		Shaw 1989a		Mączyński et al. 2004	
shake flask-GC		shake flask-GC		IUPAC recommended		calc-recommended LLE data	
t/°C	S/g·m^{-3}	t/°C	S/g·m^{-3}	t/°C	S/g·m^{-3}	t/°C	S/g·m^{-3}
0	19.45	25	13	25	13.7	0	18.19
25	15.7	40.1	13.8	30	13	25	13.88
		55.7	15.7	40	14	40	13.88
		99.1	27.1	50	15	55.7	15.32
		118	44.9	70	17	99.1	30.64
		137.3	86.8	90	23	118	46.92
		149.5	113	110	36	137.3	76.6
				130	68	145.9	110.12
				150	120		

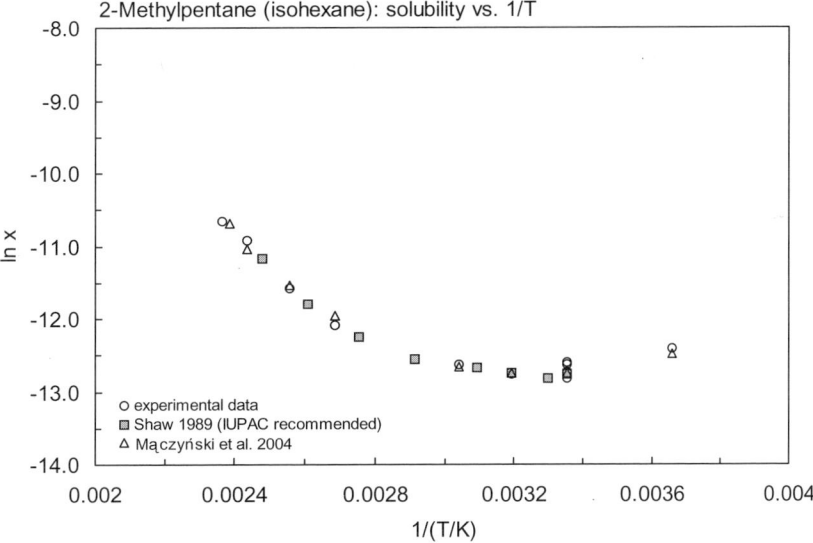

FIGURE 2.1.1.1.9.1 Logarithm of mole fraction solubility (ln x) versus reciprocal temperature for 2-methylpentane.

TABLE 2.1.1.1.9.2
Reported vapor pressures and Henry's law constants of 2-methylpentane at various temperatures and the coefficients for the vapor pressure equations

$\log P = A - B/(T/K)$ (1) $\qquad \ln P = A - B/(T/K)$ (1a)

$\log P = A - B/(C + t/°C)$ (2) $\qquad \ln P = A - B/(C + t/°C)$ (2a)

$\log P = A - B/(C + T/K)$ (3) $\qquad \ln P = A - B/(C + T/K)$ (3a)

$\log P = A - B/(T/K) - C \cdot \log (T/K)$ (4)

$\ln (P/P_{ref}) = [1 - (T_{ref}/T)] \cdot \exp(a + bT + cT^2)$ (5)

Vapor pressure						Henry's law constant	
Willingham et al. 1945		Stull 1947		Zwolinski & Wilhoit 1971		Ashworth et al. 1988	
ebulliometry		summary of literature data		selected values		EPICS-GC	
t/°C	P/Pa	t/°C	P/Pa	t/°C	P/Pa	t/°C	H/(Pa m³/mol)
12.758	16620	−60.9	133.3	−32.1	1333	10	70624
16.82	19921	−41.7	666.6	−21.6	2666	15	70320
20.584	23451	−32.1	1333	−14.8	4000	20	89470
25.617	29038	−214	2666	−9.8	5333	25	83593
30.237	34896	−9.7	5333	−5.7	6666	30	85924
35.810	43322	−1.9	7999	−2.2	7999		
41.507	53656	8.1	13332	3.45	10666	$\ln H = A - B/(T/K)$	
47.714	66757	24.1	26664	8.06	13332	H/(atm m³/mol)	
54.388	83718	41.5	53329	16.92	19998	A	2.959
59.099	97605	60.3	101325	23.624	26664	B	957.2
59.607	99204			29.084	33331		
60.074	100694	mp/°C	−154	33.724	39997		
60.602	102394			41.40	53329		
61.066	103913			47.672	66661		
				53.020	79993		
bp/°C	60.271			57.706	93326		
				58.579	95992		
eq. 2	P/mmHg			59.434	98659		
A	6.93910			60.271	101325		
B	1135.410			25.0	28238		
C	226.572						
				eq. 2	P/mmHg		
				A	6.83910		
				B	1135.410		
				C	226.572		
				bp/°C	60.271		
				$\Delta H_V/(kJ\,mol^{-1})$ =			
				at 25°C	29.76		
				at bp	27.79		

Aliphatic and Cyclic Hydrocarbons

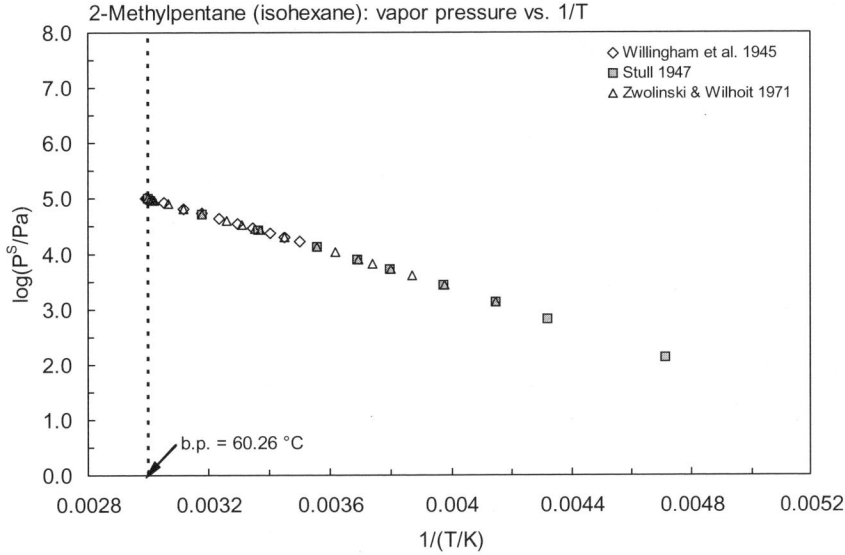

FIGURE 2.1.1.1.9.2 Logarithm of vapor pressure versus reciprocal temperature for 2-methylpentane.

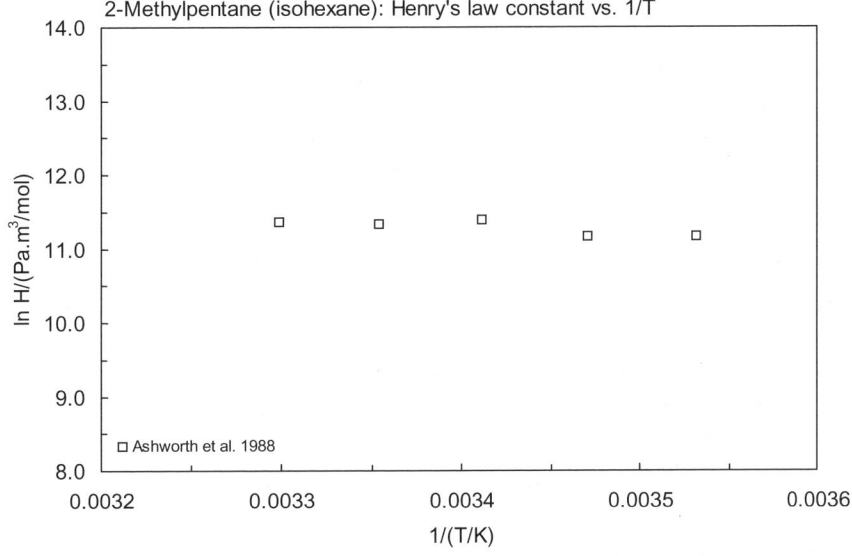

FIGURE 2.1.1.1.9.3 Logarithm of Henry's law constant versus reciprocal temperature for 2-methylpentane.

2.1.1.1.10   3-Methylpentane

Common Name: 3-Methylpentane
Synonym: diethylmethylmethane
Chemical Name: 3-methylpentane
CAS Registry No: 96-14-0
Molecular Formula: C_6H_{14}; $CH_3CH_2(CH_3)CHCH_2CH_3$
Molecular Weight: 86.175
Melting Point (°C):
    –162.9   (Lide 2003)
Boiling Point (°C):
    63.27   (Lide 2003)
Density (g/cm³ at 25°C):
    0.66431, 0.65976 (20°C, 25°C, Dreisbach 1959; Riddick et al. 1986)
Molar Volume (cm³/mol):
    129.7, 130.6 (20°C, 25°C, calculated-density)
    129.7   (20°C, calculated-density, McAuliffe 1966; Stephenson & Malanowski 1987)
    140.6   (calculated-Le Bas method at normal boiling point)
Enthalpy of Vaporization, ΔH_V (kJ/mol):
    30.28, 28.08 (25°C, bp, Riddick et al. 1986)
Enthalpy of Fusion, ΔH_{fus} (kJ/mol):
    5.035   (Riddick et al. 1986)
    5.31   (Chickos et al. 1999)
Entropy of Fusion, ΔS_{fus} (J/mol K):
    48.17, 50.6 (exptl., calculated-group additivity method, Chickos et al. 1999)
Fugacity Ratio at 25°C, F: 1.0

Water Solubility (g/m³ or mg/L at 25°C or as indicated):
    12.8   (shake flask-GC, McAuliffe 1966)
    21.5; 17.9, 17.2 (0; 25°C, shake flask-GC, calculated-group contribution, Polak & Lu 1973)
    13.1   (shake flask-GC, Price 1976)
    12.9   (partition coefficient-GC, Rudakov & Lutsyk 1979)
    13.0   (selected, Riddick et al. 1986)
    12.9   (recommended best value, IUPAC Solubility Data Series, Shaw 1989a)
    22.02, 16.76 (0, 25°C, calculated-recommended liquid-liquid equilibrium LLE data, Mączyński et al. 2004)

Vapor Pressure (Pa at 25°C or as indicated and reported temperature dependence equations. Additional data at other temperatures designated * are compiled at the end of this section.):
    29040   (23.2°C, ebulliometry, measured range 15.29–64.083°C, Willingham et al. 1945)
    $\log (P/\text{mmHg}) = 6.84887 - 1152.368/(227.129 + t/°C)$; temp range 15.29–64.083°C (Antoine eq. from exptl. data, ebulliometry-manometer, Willingham et al. 1945)
    24970   (Antoine eq. regression, temp range –59 to 63.3°C, Stull 1947)
    25300   (calculated from determined data, Dreisbach 1959)
    $\log (P/\text{mmHg}) = 6.84887 - 1152.368/(227.129 + t/°C)$; temp range –15 to 105°C (Antoine eq. for liquid state, Dreisbach 1959)
    25305, 25300 (derived from compiled data, interpolated-Antoine eq., Zwolinski & Wilhoit 1971)
    $\log (P/\text{mmHg}) = 6.84887 - 1152.368/(227.129 + t/°C)$; temp range –30.1 to 86.6°C (Antoine eq., Zwolinski & Wilhoit 1971)
    24940   (interpolated-Antoine eq., Weast 1972–73)
    $\log (P/\text{mmHg}) = [-0.2185 \times 7743.9/(T/K)] + 7.947042$; temp range –59 to 63.3°C (Antoine eq., Weast 1972–73)

Aliphatic and Cyclic Hydrocarbons

25310 (interpolated-Antoine eq, Boublik et al. 1984)

log (P/kPa) = 6.36895 – 1997.558/(202.608 + t/°C); temp range 69.2–271.1°C (Antoine eq. from reported exptl. data, Boublik et al. 1984)

25300 (interpolated-Antoine eq., temp range –30 to 87°C, Dean 1985, 1992)

log (P/mmHg) = 6.84887 – 1152.368/(227.13 + t/°C); temp range –30 to 87°C (Antoine eq., Dean 1985, 1992)

25300 (lit. average, Riddick et al. 1986)

log (P/kPa) = 5.98356 – 11133.52/(224.944 + t/°C); temp range not specified (Antoine eq., Riddick et al. 1986)

25320 (interpolated-Antoine eq., Stephenson & Malanowski 1987)

log (P$_L$/kPa) = 5.97897 – 1155.28/(–45.659 + T/K); temp range 293–338 K (Antoine eq., Stephenson & Malanowski 1987)

log (P/mmHg) = 35.2848 – 2.6773 × 10^3/(T/K) – 9.8546·log (T/K) + 2.2352 × 10^{-11}·(T/K) + 4.0277 × 10^{-6}·(T/K)^2; temp range 110–504 K (vapor pressure eq., Yaws 1994)

Henry's Law Constant (Pa m^3/mol at 25°C):

172000 (recommended; Mackay & Shiu 1981)
171490 (calculated as 1/K$_{AW}$, C$_W$/C$_A$, reported as exptl., Hine & Mookerjee 1975)
196800, 49430 (calculated-contribution, calculated-bond contribution, Hine & Mookerjee 1975)
170210 (calculated-P/C, Eastcott et al. 1988)
139390 (calculated-MCI χ, Nirmalakhandan & Speece 1988)
113670 (calculated-vapor-liquid equilibrium (VLE) data, Yaws et al. 1991)

Octanol/Water Partition Coefficient, log K$_{OW}$:

2.80 (calculated-π const., Hansch et al. 1968; Hansch & Leo 1979)
2.88 (calculated-MCI χ, Murray et al. 1975)
3.81 (calculated-intrinsic molar volume V$_I$ and solvatochromic parameters, Leahy 1986)
3.60 ± 0.20 (recommended, Sangster 1989)
3.18 (calculated-molar volume V$_M$, Wang et al. 1992)
3.74 (calculated-fragment constant, Müller & Klein 1992)
2.9168 (calculated-UNIFAC group contribution, Chen et al. 1993)
3.60 (recommended, Sangster 1993)

Octanol/Air Partition Coefficient, log K$_{OA}$:

Bioconcentration Factor, log BCF:

Sorption Partition Coefficient, log K$_{OC}$:

Environmental Fate Rate Constants, k, and Half-Lives, t$_{½}$:

Volatilization:

Photolysis:

Oxidation: rate constant k, for gas-phase second order rate constants, k$_{OH}$ for reaction with OH radical, k$_{NO3}$ with NO$_3$ radical and k$_{O3}$ with O$_3$ or as indicated, *data at other temperatures see reference:

k$_{OH}$ = 1.28 × 10^9 L mol^{-1} s^{-1} at 300 K (Greiner 1967; quoted, Altshuller & Bufalini 1971)

k$_{OH}$ = (4.3 ± 0.9) × 10^9 M^{-1} s^{-1} at 305 ± 2 K (relative rate method, Lloyd et al. 1976)

k$_{OH}$ = 6.82 × 10^{-12} cm^3 molecule^{-1} s^{-1} at atmospheric pressure and 305 K (Darnall et al. 1978)

k$_{OH}$ = (5.78 ± 0.11) × 10^{-12} cm^3 molecule^{-1} s^{-1} at room temp. (relative rate, Atkinson et al. 1984c)

k$_{OH}$ = 5.7 × 10^{-12} cm^3 molecule^{-1} s^{-1} at 298 K (recommended, Atkinson 1989, 1990, 1991)

k$_{OH}$ = 5.7 × 10^{-12} cm^3 molecule^{-1} s^{-1} at 298 K, estimated atmospheric lifetime 25 h, during summer daylight hours (Altshuller 1991)

k$_{NO3}$(exptl) = 2.04 × 10^{-16} cm^3 molecule^{-1} s^{-1}, k$_{NO3}$(calc) = 2.53 × 10^{-16} cm^3 molecule^{-1} s^{-1} at 296 ± 2 K (relative rate method, Aschmann & Atkinson 1995)

k$_{OH}$ = 5.4 × 10^{-12} cm^3 molecule^{-1} s^{-1}, k$_{NO3}$* = 2.2 × 10^{-16} cm^3 molecule^{-1} s^{-1} at 298 K (recommended, Atkinson 1997)

Hydrolysis:
Biodegradation:
Biotransformation:
Bioconcentration, Uptake (k_1) and Elimination (k_2) Rate Constants or Half-Lives:

Half-Lives in the Environment:

Air: photooxidation reaction rate constant of 4.30×10^{12} cm^3 mol^{-1} s^{-1} with hydroxyl radical with half-life of 2.4–24 h (Darnall et al. 1976; Lloyd et al. 1976);
rate constant of 5.7×10^{-12} cm^3 molecule$^{-1}\cdot$s^{-1} for the reaction with OH radical with an estimated atmospheric lifetime of 25 h during summer daylight (Altshuller 1991).

Aliphatic and Cyclic Hydrocarbons

2.1.1.1.11   2,2-Dimethylpentane

Common Name: 2,2-Dimethylpentane
Synonym:
Chemical Name: 2,2-dimethylpentane
CAS Registry No: 590-35-2
Molecular Formula: C_7H_{16}; $CH_3C(CH_3)_2CH_2CH_2CH_3$
Molecular Weight: 100.202
Melting Point (°C):
    −123.7   (Lide 2003)
Boiling Point (°C):
    79.2   (Lide 2003)
Density (g/cm³ at 20°C):
    0.6739   (Weast 1984)
    0.6739, 0.6695   (20°C, 25°C, Dreisbach 1959)
Molar Volume (cm³/mol):
    148.69   (20°C, calculated from density)
    162.8   (calculated-Le Bas method at normal boiling point)
Enthalpy of Fusion, ΔH_{fus} (kJ/mol):
    5.812   (Dreisbach 1959)
    5.86   (Chickos et al. 1999)
Entropy of Fusion, ΔS_{fus} (J/mol K):
    39.55, 49.74 (exptl., calculated-group additivity method, Chickos et al. 1999)
Fugacity Ratio at 25°C, F: 1.0

Water Solubility (g/m³ or mg/L at 25°C):
    4.40   (shake flask-GC, Price 1976)
    4.90   (calculated-recommended liquid-liquid equilibrium LLE data, Mączyński et al. 2004)

Vapor Pressure (Pa at 25°C or as indicated and reported temperature dependence equations):
    13824   (24.67°C, manometer, measured range 15.325–80.05°C, Willingham et al. 1945)
    log (P/mmHg) = 6.81509 − 1190.298/(223.343 + t/°C); temp range 15.325–80.05°C (Antoine eq. from exptl. data, ebulliometry-manometer measurements, Willingham et al. 1945)
    13850   (24.708°C, ebulliometry, Forziati et al. 1949)
    log (P/mmHg) = 6.81479 − 1190.033/(223.303 + t/°C); temp range 12.188–80.074°C (Antoine eq., ebulliometry-manometer measurements, Forziati et al. 1949)
    14030   (calculated from determined data, Dreisbach 1959)
    log (P/mmHg) = 6.81480 − 1190.033/(223.303 + t/°C); temp range 0–115°C (Antoine eq. for liquid state, Dreisbach 1959)
    14026   (interpolated-Antoine eq., temp range −18.6 to 103.75°C, Zwolinski & Wilhoit 1971)
    log (P/mmHg) = 6.81480 − 1190.033/(223.303 + t/°C); temp range −18.6 to 103.75°C (Antoine eq., Zwolinski & Wilhoit 1971)
    13500   (interpolated-Antoine eq., temp range −69.3 to 49.7°C, Weast 1972–72)
    log (P/mmHg) = [−0.2185 × 7271.0/(T/K)] + 7.841340; temp range −69.3 to 49.7°C (Antoine eq., Weast 1972–73)
    14030   (interpolated-Antoine eq., Boublik et al. 1984)
    log (P/kPa) = 5.93788 − 1189.09/(223.198 + t/°C); temp range 15.32-80.05°C (Antoine eq. from reported exptl. data of Willingham et al. 1945, Boublik et al. 1984)
    14030   (interpolated-Antoine eq., temp range −18 to 103°C, Dean 1985, 1992)
    log (P/mmHg) = 6.84180 − 1190.033/(223.3 + t/°C); temp range −18 to 103°C (Antoine eq., Dean 1985, 1992)
    14010   (interpolated-Antoine eq.-I, Stephenson & Malanowski 1987)

log (P_L/kPa) = 5.93117 − 1185.576/(−50.37 + T/K); temp range 277-354 K (Antoine eq.-I, Stephenson & Malanowski 1987)

log (P_L/kPa) = 6.2280 − 1399.333/(−20.934 + T/K); temp range 353–483 K (Antoine eq.-II, Stephenson & Malanowski 1987)

log $(P/mmHg)$ = 6.2875 − 2.1682 × 10^3/(T/K) + 2.6936·log (T/K) − 1.5525 × 10^{-2}·(T/K) + 1.0917 × 10^{-5}· (T/K)^2; temp range 149–527 K (vapor pressure eq., Yaws 1994)

Henry's Law Constant (Pa m^3/mol at 25°C):

    318000   (calculated-P/C, Mackay & Shiu 1981)
    319200   (selected, Mills et al. 1982)
    319420   (calculated-vapor-liquid equilibrium (VLE) data, Yaws et al. 1991)

Octanol/Water Partition Coefficient, log K_{OW}:

    3.10   (calculated-π substituent constant, Hansch et al. 1968)
    3.62   (calculated-V_M, Wang et al. 1992)
    4.14   (calculated-fragment constant, Müller & Klein 1992)

Octanol/Air Partition Coefficient, log K_{OA}:

Bioconcentration Factor, log BCF:

Sorption Partition Coefficient, log K_{OC}:

Environmental Fate Rate Constants, k, and Half-Lives, $t_{1/2}$:

    Volatilization:

    Photolysis:

    Oxidation: rate constant k, for gas-phase second order rate constants, k_{OH} for reaction with OH radical, k_{NO3} with NO$_3$ radical and k_{O3} with O$_3$ or as indicated, *data at other temperatures see reference:

        k_{OH} = (2.66 ± 0.08) × 10^{-12} cm^3 molecule^{-1} s^{-1} at room temp. (relative rate, Atkinson et al. 1984c)
        k_{OH} = 3.37 × 10^{-12} cm^3 molecule^{-1} s^{-1} at 300 K (Atkinson 1989)
        k_{OH} = 3.4 × 10^{-12} cm^3 molecule^{-1} s^{-1} at 298 K (Atkinson 1990)
        k_{OH} = 2.34 × 10^{-12} cm^3 molecule^{-1} s^{-1} at 298 K, k_{NO3} = 9.7 × 10^{-17} cm^3 molecule^{-1} s^{-1} at 296 K (Atkinson 1990)
        k_{OH} = 3.4 × 10^{-12} cm^3 molecule^{-1} s^{-1} at 298 K (recommended, Atkinson 1997)

    Hydrolysis:

    Biodegradation:

    Biotransformation:

    Bioconcentration, Uptake (k_1) and Elimination (k_2) Rate Constants or Half-Lives:

Half-Lives in the Environment:

Aliphatic and Cyclic Hydrocarbons

2.1.1.1.12   2,4-Dimethylpentane

Common Name: 2,4-Dimethylpentane
Synonym: diisopropylmethane
Chemical Name: 2,4-dimethylpentane
CAS Registry No: 108-08-7
Molecular Formula: C_7H_{16}; $CH_3CH(CH_3)CH_2CH(CH_3)CH_3$
Molecular Weight: 100.202
Melting Point (°C):
    –119.5   (Lide 2003)
Boiling Point (°C):
    80.49   (Lide 2003)
Density (g/cm³ at 20°C):
    0.6727, 0.6683   (20°C, 25°C, Dreisbach 1959; Riddick et al. 1986)
Molar Volume (cm³/mol):
    148.95, 149.9   (20°C, 25°C, calculated-density)
    162.8   (calculated-Le Bas method at normal boiling point)
Enthalpy of Vaporization, ΔH_V (kJ/mol):
    32.89, 29.501   (25°C, bp, Riddick et al. 1986)
Enthalpy of Fusion, ΔH_{fus} (kJ/mol):
    6.840   (Dreisbach 1959)
    6.845   (Riddick et al. 1986)
    6.85   (Chickos et al. 1999)
Entropy of Fusion, ΔS_{fus} (J/mol K):
    44.46, 44.7 (exptl., calculated-group additivity method, Chickos et al. 1999)
Fugacity Ratio at 25°C, F: 1.0

Water Solubility (g/m³ or mg/L at 25°C or as indicated):
    3.62   (shake flask-GC, McAuliffe 1963)
    4.06   (shake flask-GC, McAuliffe 1966; quoted, Hermann 1972; Price 1976)
    6.50; 5.50   (0; 25°C, shake flask-GC, Polak & Lu 1973)
    4.41   (shake flask-GC, Price 1976)
    4.20   (recommended best value, IUPAC Solubility Data Series, Shaw 1989a)
    6.12, 4.45   (0, 25°C, calculated-recommended liquid-liquid equilibrium LLE data, Mączyński et al. 2004)

Vapor Pressure (Pa at 25°C or as indicated and reported temperature dependence equations):
    12635   (Antoine eq. regression, temp range –48 to 80.5°C, Stull 1947)
    11730   (22.54°C, ebulliometry, measured range 13.714–81.374°C, Forziati et al. 1949)
    $\log (P/\text{mmHg}) = 6.82621 - 1192.041/(221.634 + t/°C)$; temp range 13.714–81.374°C (Antoine eq. from exptl. data, ebulliometry-manometer, Forziati et al. 1949)
    13120   (calculated from determined data, Dreisbach 1959)
    $\log (P/\text{mmHg}) = 6.82621 - 1192.041/(221.634 + t/°C)$; temp range 0 – 115°C (Antoine eq. for liquid state, Dreisbach 1959)
    13119   (interpolated-Antoine eq, temp range –17.0 to 104.94°C, Zwolinski & Wilhoit 1971)
    $\log (P/\text{mmHg}) = 6.82621 - 1192.041/(221.634 + t/°C)$; temp range –17.0 to 104.94°C (Antoine eq., Zwolinski & Wilhoit 1971)
    12620   (interpolated-Antoine eq., temp range –48.0 to 80.5°C, Weast 1972–73)
    $\log (P/\text{mmHg}) = [-0.2185 \times 8167.4/(T/K)] + 7.961374$; temp range –48.0 to 80.5°C (Antoine eq., Weast 1972–73)
    13120   (interpolated-Antoine eq., temp range 13.7–81.37°C, Boublik et al. 1984)
    $\log (P/\text{kPa}) = 5.95675 - 1195.154/(221.992 + t/°C)$; temp range 13.7–81.37°C (Antoine eq. from reported exptl. data, Boublik et al. 1984)

15450 (interpolated-Antoine eq., temp range –17 to 105°C, Dean 1985, 1992)

log (P/mmHg) = 6.82621 – 1192.04/(225.32 + t/°C); temp range –17 to 105°C (Antoine eq., Dean 1985, 1992)

13000 (selected, Riddick et al. 1986)

log (P/kPa) = 5.94917 – 1191.06/(221.540 + t/°C); temp range not specified (Antoine eq., Riddick et al. 1986)

13125 (interpolated-Antoine eq., Stephenson & Malanowski 1987)

log (P_L/kPa) = 5.95921 – 1196.516/(–50.993 + T/K); temp range 284–355 K (Antoine eq., Stephenson & Malanowski 1987)

log (P/mmHg) = 35.9436 – 2.846 × 10^3/(T/K) – 9.9938·log (T/K) + 8.0613 × 10^{-11}·(T/K) + 3.6419 × 10^{-6}·(T/K)^2; temp range 154–520 K (vapor pressure eq., Yaws 1994)

Henry's Law Constant (Pa m^3/mol at 25°C):

319300 (calculated–1/K_{AW}, C_W/C_A, reported as exptl., Hine & Mookerjee 1975)
326600, 73120 (calculated-group contribution, calculated-bond contribution, Hine & Mookerjee 1975)
300000 (recommended; Mackay & Shiu 1981)
297300 (calculated-P/C, Eastcott et al. 1988)
160050 (calculated-molecular connectivity index MCI χ, Nirmalakhandan & Speece 1988)
298050 (calculated-vapor-liquid equilibrium (VLE) data, Yaws et al. 1991)

Octanol/Water Partition Coefficient, log K_{OW}:

3.10 (calculated-π constant, Hansch et al. 1968)
3.17 (calculated-MCI χ, Murray et al. 1975)
3.66 (calculated-molar volume V_M, Wang et al. 1992)
4.14 (calculated-f constant, Müller & Klein 1992)

Octanol/Air Partition Coefficient, log K_{OA}:

Bioconcentration Factor, log BCF:

Sorption Partition Coefficient, log K_{OC}:

Environmental Fate Rate Constants, k, and Half-Lives, $t_{1/2}$:

Volatilization:
Photolysis:
Oxidation: rate constant k, for gas-phase second order rate constants, k_{OH} for reaction with OH radical, k_{NO3} with NO_3 radical and k_{O3} with O_3 or as indicated, *data at other temperatures see reference:

k_{OH} = (5.26 ± 0.11) × 10^{-12} cm^3 molecule^{-1} s^{-1} at room temp. (relative rate, Atkinson et al. 1984c)
k_{OH} = 5.10 × 10^{-12} cm^3 · molecule^{-1} s^{-1} at 298 K (Atkinson 1990)
k_{NO3}(exptl) = 1.44 × 10^{-16} cm^3 molecule^{-1} s^{-1}, k_{NO3}(calc) = 2.89 × 10^{-16} cm^3 molecule^{-1} s^{-1} at 296 ± 2 K (relative rate method, Aschmann & Atkinson 1995)
k_{OH} = 5.0 × 10^{-12} cm^3 molecule^{-1} s^{-1}, k_{NO3}* = 1.5 × 10^{-16} cm^3 molecule^{-1} s^{-1} at 298 K (recommended, Atkinson 1997)

Hydrolysis:
Biodegradation:
Biotransformation:
Bioconcentration, Uptake (k_1) and Elimination (k_2) Rate Constants or Half-Lives:

Half-Lives in the Environment:

Air: photooxidation reaction rate constant of 5.10 × 10^{-12} cm^3 molecule^{-1} s^{-1} with hydroxyl radicals and an estimated lifetime of 27 h during summer daylight (Altshuller 1991).

Aliphatic and Cyclic Hydrocarbons

2.1.1.1.13   3,3-Dimethylpentane

Common Name: 3,3-Dimethylpentane
Synonym:
Chemical Name: 3,3-dimethylpentane
CAS Registry No: 562-49-2
Molecular Formula: C_7H_{16}; $CH_3CH_2(CH_3)_2CH_2CH_3$
Molecular Weight: 100.202
Melting Point (°C):
    –134.4   (Lide 2003)
Boiling Point (°C):
    86.06   (Lide 2003)
Density (g/cm^3 at 20°C):
    0.6933, 0.6891   (20°C, 25°C, Dreisbach 1959)
Molar Volume (cm^3/mol):
    144.5   (20°C, calculated-density, Stephenson & Malanowski 1987)
    162.8   (calculated-Le Bas method at normal boiling point)
Enthalpy of Vaporization, ΔH_V (kJ/mol):
    33.02, 29.74   (25°C, normal bp, Dreisbach 1959)
Enthalpy of Fusion, ΔH_{fus} (kJ/mol):
    7.067   (Dreisbach 1959)
    7.07   (Chickos et al. 1999)
Entropy of Fusion, ΔS_{fus} (J/mol K):
    51.16, 49.7 (exptl., calculated-group additivity method, Chickos et al. 1999)
Fugacity Ratio at 25°C, F: 1.0

Water Solubility (g/m^3 or mg/L at 25°C. Additional data at other temperatures designated * are compiled at the end of this section.):
    5.94*   (shake flask-GC, measured range 25–150.4°C, Price 1976)
    6.68*   (calculated-liquid-liquid equilibrium LLE data, temp range 298.2–423.2 K, Mączyński et al. 2004)

Vapor Pressure (Pa at 25°C or as indicated and reported temperature dependence equations. Additional data at other temperatures designated * are compiled at the end of this section.):
    10600*   (Antoine eq. regression, temp range –45.9 to 86.1°C Stull 1947)
    10316*   (23.521°C, ebulliometry, measured range –13.5 to 82°C Forziati et al. 1949)
    $\log (P/mmHg) = 6.82668 - 1228.063/(225.316 + t/°C)$; temp range 13.484–80.962°C (Antoine eq., ebulliometry-manometer measurements, Forziati et al. 1949)
    11044   (calculated from determined data, Dreisbach 1959)
    $\log (P/mmHg) = 6.82667 - 1228.663/(225.316 + t/°C)$; temp range 5–130°C (Antoine eq. for liquid state, Dreisbach 1959)
    11039*   (interpolated-Antoine eq., temp range –14.4 to 111.25°C, Zwolinski & Wilhoit 1971)
    $\log (P/mmHg) = 6.82667 - 1228.663/(225.316 + t/°C)$; temp range –14.4 to 111.25°C (Antoine eq., Zwolinski & Wilhoit 1971)
    10600   (interpolated-Antoine eq., temp range –45.9 to 86.1°C, Weast 1972–73)
    $\log (P/mmHg) = [-0.2185 \times 8145.4/(T/K)] + 7.869254$; temp range –45.9 to 86.1°C (Antoine eq., Weast 1972–73)
    11045   (interpolated-Antoine eq., temp range 13.484–86.96°C, Boublik et al. 1984)
    $\log (P/kPa) = 5.95327 - 1229.625/(225.427 + t/°C)$; temp range 13.484–86.96°C (Antoine eq. from reported exptl. data, Boublik et al. 1984)
    11045   (interpolated-Antoine eq., temp range –14 to 112°C, Dean 1985, 1992)

log (P/mmHg) = 6.82667 – 1228.663/(225.32 + t/°C); temp range –14 to 112°C (Antoine eq., Dean 1985, 1992)
11150, 11040 (interpolated-Antoine eq.-I, III, Stephenson & Malanowski 1987)
log (P_L/kPa) = 5.95139 – 1228.138/(–47.819 + T/K); temp range 285–360 K (Antoine eq.-I, Stephenson & Malanowski 1987)
log (P_L/kPa) = 6.35011 – 1415.316/(–31.302 + T/K); temp range 213–281 K (Antoine eq.-II, Stephenson & Malanowski 1987)
log (P_L/kPa) = 5.94685 – 1225.973/(–48.144 + T/K); temp range 280–360 K (Antoine eq.-III, Stephenson & Malanowski 1987)
log (P/mmHg) = 30.2570 – 2.6313 × 10^3/(T/K) – 7.9839·log (T/K) + 4.6848 × 10^{-13}·(T/K) + 2.717 × 10^{-6}·(T/K)^2; temp range 139–536 K (vapor pressure eq., Yaws 1994)

Henry's Law Constant (Pa m^3/mol at 25°C):
    186000   (calculated-P/C, Mackay & Shiu 1981)
    186305   (calculated-vapor-liquid equilibrium (VLE) data, Yaws et al. 1991)

Octanol/Air Partition Coefficient, log K_{OA}:

Octanol/Water Partition Coefficient, log K_{OW}:

Bioconcentration Factor, log BCF:

Sorption Partition Coefficient, log K_{OC}:

Environmental Fate Rate Constants, k and Half-Lives, $t_{1/2}$:

Half-Lives in the Environment:

TABLE 2.1.1.1.13.1
Reported aqueous solubilities of 3,3-dimethylpentane at various temperatures

Price 1976		Mączyński et al. 2004	
shake flask-GC		calc-recommended LLE data	
t/°C	S/g·m^{-3}	t/°C	S/g·m^{-3}
25	5.92	25	6.68
40.1	6.78	40	6.68
55.7	8.17	56	7.79
69.7	10.3	70	8.91
99.1	15.8	99	16.14
118	27.3	118	26.17
140.4	67.3	150	66.81
150.4	86.1		

Aliphatic and Cyclic Hydrocarbons

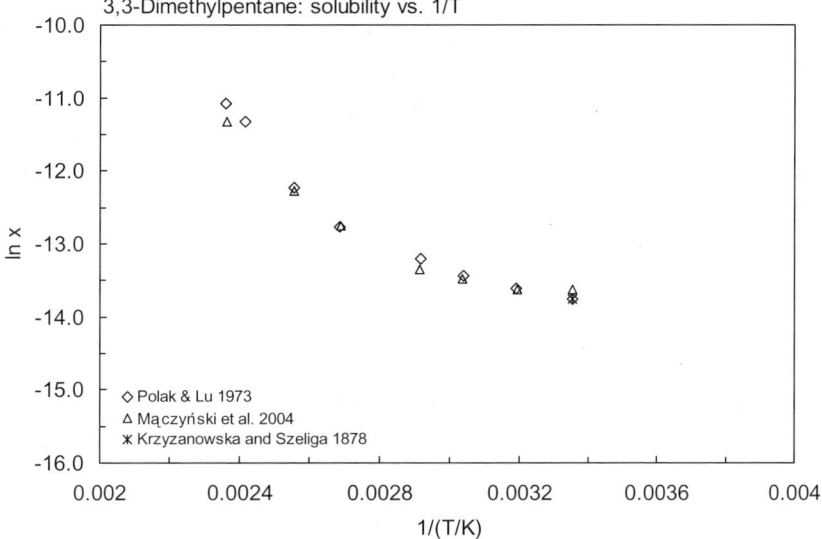

FIGURE 2.1.1.1.13.1 Logarithm of mole fraction solubility (ln x) versus reciprocal temperature for 3,3-dimethylpentane.

TABLE 2.1.1.1.13.2
Reported vapor pressures of 3,3-dimethylpentane at various temperatures and the coefficients for the vapor pressure equations

$\log P = A - B/(T/K)$ (1)　　　　$\ln P = A - B/(T/K)$ (1a)

$\log P = A - B/(C + t/°C)$ (2)　　$\ln P = A - B/(C + t/°C)$ (2a)

$\log P = A - B/(C + T/K)$ (3)　　$\ln P = A - B/(C + T/K)$ (3a)

$\log P = A - B/(T/K) - C \cdot \log (T/K)$ (4)

$\ln (P/P_{ref}) = [1 - (T_{ref}/T)] \cdot \exp(a + bT + cT^2)$ (5)

Stull 1947		Forziati et al. 1949				Zwolinski & Wilhoit 1971	
summary of literature data		ebulliometry				selected values	
t/°C	P/Pa	t/°C	P/Pa	t/°C	P/Pa	t/°C	P/Pa
		set 1		set 2			
–45.9	133.3	13.443	6383	17.163	7658	–14.41	1333
–25.0	666.6	17.231	7677	20.484	8969	–2.96	2666
–14.4	1333	30.533	8985	34.355	16623	4.36	4000
–2.90	2666	23.521	10316	38.804	19926	9.852	5333
9.90	5333	26.318	11704	53.385	34902	14.297	6666
18.1	7999	30.086	13826	59.444	43327	18.055	7999
29.3	13332	34.406	16613	84.792	976–7	24.231	10666
46.2	26664	38.790	19917	85.344	99201	29.241	13332
65.5	53329	48.72	23438	85.854	100696	38.880	19998
86.1	101325	48.34	28954	86.429	102396	46.173	26664
		53.37	34893			52.114	33331
mp/°C	–135	59.44	43322			57.164	39997
		65.697	53645			65.519	53329
		72.384	66733			72.347	66661
		79.651	83682			78.168	79993
		84.78	97575	bp/°C	86.069	83.272	93326
		85.335	99206			84.222	95992

(Continued)

TABLE 2.1.1.1.13.2 (Continued)

Stull 1947 summary of literature data		Forziati et al. 1949 ebulliometry				Zwolinski & Wilhoit 1971 selected values	
t/°C	P/Pa	t/°C	P/Pa	t/°C	P/Pa	t/°C	P/Pa
		86.421	102377	eq. 2	P/mmHg	85.153	98659
		86.928	103895	A	6.81813	86.064	101325
				B	1223.543	25.0	11039
				C	224.687		
						eq. 2	P/mmHg
						A	6.82667
						B	1228.663
						C	225.316
						bp/°C	86.064
						$\Delta H_V/(kJ\ mol^{-1})$ =	
						at 25°C	33.02
						at bp	29.65

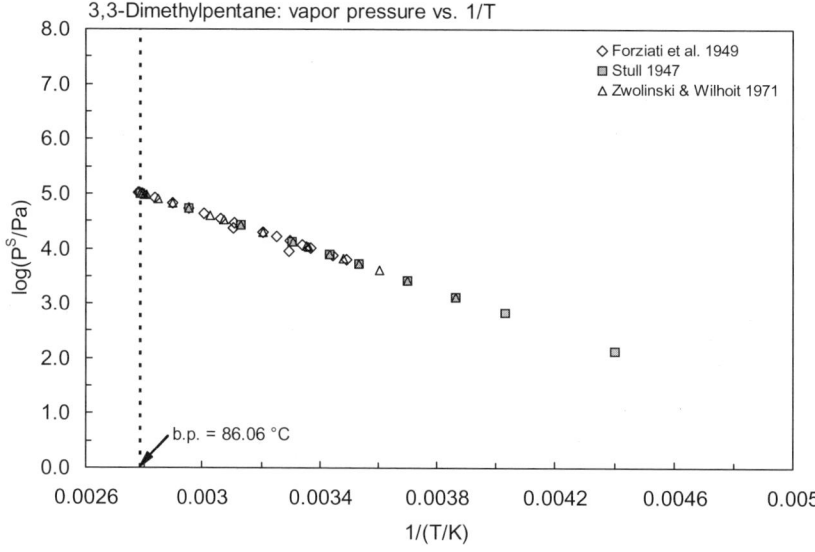

FIGURE 2.1.1.1.13.2 Logarithm of vapor pressure versus reciprocal temperature for 3,3-dimethylpentane.

2.1.1.1.14   2,2,4-Trimethylpentane (Isooctane)

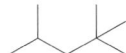

Common Name: 2,2,4-Trimethylpentane
Synonym: isooctane, isobutyltrimethylmethane
Chemical Name: 2,2,4-trimethylpentane
CAS Registry No: 504-84-1
Molecular Formula: C_8H_{18}; $CH_3C(CH_3)_2CH_2CH(CH_3)CH_3$
Molecular Weight: 114.229
Melting Point (°C):
    –107.30   (Stull 1947; Lide 2003)
Boiling Point (°C):
    99.22   (Lide 2003)
Density (g/cm³ at 20°C):
    0.6919, 0.6878   (20°C, 25°C, Dreisbach 1959; Riddick et al. 1986)
Molar Volume (cm³/mol):
    165.1   (20°C, calculated-density, McAuliffe 1966; Stephenson & Malanowski 1987)
    185.0   (calculated-Le Bas method at normal boiling point)
Enthalpy of Vaporization, ΔH_V (kJ/mol):
    36.92, 32.01   (25°C, bp, Riddick et al. 1986)
Enthalpy of Fusion, ΔH_{fus} (kJ/mol):
    9.196   (Riddick et al. 1986)
    9.27   (Chickos et al. 1999)
Entropy of Fusion, ΔS_{fus} (J/mol K):
    55.52, 43.8   (exptl., calculated-group additivity method, Chickos et al. 1999)
Fugacity Ratio at 25°C, F: 1.0

Water Solubility (g/m³ or mg/L at 25°C or as indicated):
    2.44   (shake flask-GC, McAuliffe 1963,1966)
    2.46; 2.05   (0; 25°C, shake flask-GC, Polak & Lu 1973)
    1.14   (shake flask-GC, Price 1976)
    2.50, 2.0, 2.20   (0, 20, 25°C, IUPAC recommended values, Shaw 1989b)
    2.30, 1.65, 1.65   (0. 20, 25°C, calculated-liquid-liquid equilibrium LLE data, Mączyński et al. 2004)

Vapor Pressure (Pa at 25°C or as indicated and reported temperature dependence equations):
    6371   (24.4°C, ebulliometry, measured temp range 24.4–100.13°C, Willingham et al. 1945)
    log (P/mmHg) = 6.81189 – 1257.840/(220.735 + t/°C); temp range 24.4–100.13°C (Antoine eq., ebulliometry-manometer measurements, Willingham et al. 1945)
    6250   (calculated-Antoine eq. regression, temp range –36.4 to 99°C, Stull 1947)
    1739, 6573, 19528   (0, 25, 50°C, static method, vapor-liquid equilibrium VLE data, Kretschmer et al. 1948)
    6580   (calculated from determined data, Dreisbach 1959; quoted, Hine & Mookerjee 1975)
    log (P/mmHg) = 6.81189 – 1257.840/(220.735 + t/°C); temp range 15 – 120°C (Antoine eq. for liquid state, Dreisbach 1959)
    6573   (interpolated-Antoine eq., temp range –4.3 to 125.22°C, Zwolinski & Wilhoit 1971)
    log (P/mmHg) = 6.81189 – 1257.840/(220.735 + t/°C); temp range –4.3 to 125.22°C (Antoine eq., Zwolinski & Wilhoit 1971)
    6570   (interpolated-Antoine eq., temp range 24.6–100.13°C, Boublik et al. 1973, 1984)
    log (P/mmHg) = 6.80304 – 1252.59/(220.119 + t/°C), temp range 24.6–100.13°C (Antoine eq. from reported exptl. data of Willingham et al. 1945, Boublik et al. 1984)
    6240   (interpolated-Antoine eq., temp range –36.5 to 99.2°C, Weast 1972–73)
    log (P/mmHg) = [–0.2185 × 8548.0/(T/K)] + 7.934852; temp range –36.5 to 99.2°C (Antoine eq., Weast 1972–73)

110 Handbook of Physical-Chemical Properties and Environmental Fate for Organic Chemicals

log (P/kPa) = 5.92751 – 1252.348/(220.09 + t/°C); temp range 69.2 – 271.1°C (Antoine eq. from reported exptl. data of Willingham et al. 1945, Boublik et al. 1984)

6580 (interpolated-Antoine eq., temp range: 24 – 100°C, Dean 1985, 1992)

log (P/mmHg) = 6.81189 – 1257.84/(220.74 + t/°C), temp range: 24 – 100°C (Antoine eq., Dean 1985, 1992)

6500 (quoted lit., Riddick et al. 1986)

log (P/kPa) = 5.92885 – 1153.36/(220.241 + t/°C), temp range not specified (Antoine eq., Riddick et al. 1986)

6580 (interpolated, Antoine eq., Stephenson & Malanowski 1987)

log (P_L/kPa) = 5.93934 – 1254.146/(–52.831 + T/K), temp range: 297–314 K, (Antoine eq.-I, Stephenson & Malanowski 1987)

log (P_L/kPa) = 6.44016 – 1650.17/(T/K); temp range 423–523 K (Antoine eq.-II, Stephenson & Malanowski 1987)

log (P_L/kPa) = 6.33252 – 1441.485/(–36.695 + T/K); temp range 194–299 K (Antoine eq.-III, Stephenson & Malanowski 1987)

log (P_L/kPa) = 6.97534 – 1283/067/(–40.166 + T/K); temp range 372–416 K (Antoine eq.-IV, Stephenson & Malanowski 1987)

log (P_L/kPa) = 6.26002 – 1501.036/(–19.15 + T/K); temp range 413–494 K (Antoine eq.-V, Stephenson & Malanowski 1987)

log (P_L/kPa) = 7.76427 – 3268.783/(206.659 + T/K); temp range 490–544 K (Antoine eq.-VI, Stephenson & Malanowski 1987)

log (P/mmHg) = 35.954 – 3.0569 × 10^3/(T/K) – 9.8896·log (T/K) – 7.2916 × 10^{-11}·(T/K) + 3.1060 × 10^{-6}·(T/K)^2; temp range 161–564 K (vapor pressure eq., Yaws 1994)

Henry's Law Constant (Pa m^3/mol at 25°C):

308030 (calculated-P/C, Mackay & Leinonen 1975)
330000; 308000, 365000 (recommended; calculated-P/C, Mackay & Shiu 1981)
304960 (calculated as 1/K_{AW}, C_W/C_A, reported as exptl., Hine & Mookerjee 1975)
472090, 110700 (calculated-group contribution, calculated-bond contribution, Hine & Mookerjee 1975)
314110 (calculated-P/C, Lyman et al. 1982)
350140 (calculated-MCI χ, Nirmalakhandan & Speece 1988)
327200 (calculated-P/C, Eastcott et al. 1988)
338270 (calculated-vapor-liquid equilibrium (VLE) data, Yaws et al. 1991)

Octanol/Water Partition Coefficient, log K_{OW}:

5.83 (estimated-HPLC/MS, Burkhard et al. 1985)
4.54 (calculated-fragment const., Burkhard et al. 1985)
5.02 (calculated-regression eq. from Lyman et al. 1982, Wang et al. 1992)
4.06 (calculated-molar volume V_M, Wang et al. 1992)

Octanol/Air Partition Coefficient, log K_{OA}:

Bioconcentration Factor, log BCF:

Sorption Partition Coefficient, log K_{OC}:

Environmental Fate Rate Constants, k, and Half-Lives, $t_{1/2}$:

Volatilization: volatilization $t_{1/2}$ = 5.5 h from a water column 1 m^2 in cross section of depth 1 m (Mackay & Leinonen 1975);

$t_{1/2}$ ~ 3.1 h at 20°C in a river 1 m deep flowing at 1 m/s with a wind velocity of 3 m/s (estimated, Lyman et al. 1982).

Photolysis:

Oxidation: rate constant k, for gas-phase second order rate constants, k_{OH} for reaction with OH radical, k_{NO3} with NO$_3$ radical and k_{O3} with O$_3$ or as indicated, *data at other temperatures and/or the Arrhenius expression see reference:

k_{OH}*(exptl) = 2.83 × 10^{12} cm^3 mol^{-1} s^{-1}, k_{OH}(calc) = 2.35 × 10^{12} cm^3 mole^{-1} s^{-1} at 298 K, measured range 298–493 K (flash photolysis-kinetic spectroscopy, Greiner 1970)

k_{OH} = 3.73 × 10^{-12} cm^3 molecule^{-1} s^{-1} at room temp. (Darnall et al. 1978)

$k_{O(3P)} = 9.10 \times 10^{-14}$ cm^3 molecule^{-1} s^{-1} for the reaction with O(^{3}P) atom at room temp. (Herron & Huie 1973)

$k_{OH} = 3.7 \times 10^{-12}$ cm^3·molecule^{-1} s^{-1} at room temp. (Atkinson et al. 1979)

$k_{O3}^* = 2.0 \times 10^{-23}$ cm^3 molecule^{-1} s^{-1} at 298 K, measured range 298–323 K (Atkinson & Carter 1984)

$k_{OH} = (3.66 \pm 0.16) \times 10^{-12}$ cm^3 molecule^{-1} s^{-1} at room temp. (relative rate, Atkinson et al. 1984c)

$k_{OH} = 3.90 \times 10^{-12}$ cm^3 molecule^{-1} s^{-1} at 298 K and 3.56×10^{-12} cm^3 molecule^{-1} s^{-1} at 297 K (Atkinson 1985)

$k_{OH} = 3.68 \times 10^{-12}$ cm^3 molecule^{-1} s^{-1} at 298 K (recommended, Atkinson 1989, 1990)

k_{NO3}(exptl) $= 7.5 \times 10^{-17}$ cm^3 molecule^{-1} s^{-1}, k_{NO3}(calc) $= 1.65 \times 10^{-16}$ cm^3 molecule^{-1} s^{-1} at 296 ± 2 K (relative rate method, Aschmann & Atkinson 1995)

$k_{OH}^* = 3.57 \times 10^{-12}$ cm^3 molecule^{-1} s^{-1}, $k_{NO3} = 9.0 \times 10^{-17}$ cm^3 molecule^{-1} s^{-1} at 298 K (recommended, Atkinson 1997)

Hydrolysis:

Biodegradation:

Biotransformation:

Bioconcentration, Uptake (k_1) and Elimination (k_2) Rate Constants or Half-Lives:

Half-Lives in the Environment:

Air: atmospheric lifetime was estimated to be 16 h, based on the photooxidation reaction rate constant of 3.68×10^{-12} cm^3 molecule^{-1} s^{-1} with OH radical in air during summer daylight (Altshuller 1991).

Surface water: volatilization $t_{1/2} = 5.5$ h from a water column 1 m^2 in cross section of depth 1-m (Mackay & Leinonen 1975);

estimated $t_{1/2} = 3.1$ h at 20°C in a river 1 m deep flowing at 1 m/s with a wind velocity of 3 m s^{-1} (Lyman et al. 1982).

2.1.1.1.15 2,3,4-Trimethylpentane

Common Name: 2,3,4-Trimethylpentane
Synonym:
Chemical Name: 2,3,4-trimethylpentane
CAS Registry No: 565-75-3
Molecular Formula: C_8H_{18}; $CH_3CH(CH_3)CH(CH_3)CH(CH_3)CH_3$
Molecular Weight: 114.229
Melting Point (°C):
    –109.2   (Lide 2003)
Boiling Point (°C):
    113.5   (Lide 2003)
Density (g/cm³ at 20°C):
    0.7191   (Weast 1984)
    0.7191, 0.7191   (20°C, 25°C, Dreisbach 1959)
Molar Volume (cm³/mol):
    158.9   (calculated-density, Stephenson & Malanowski 1987)
    185.0   (calculated-Le Bas method at normal boiling point)
Enthalpy of Vaporization, ΔH_V (kJ/mol):
    38.0, 32.67 (25°C, at normal bp, Dreisbach 1959)
Enthalpy of Fusion, ΔH_{fus} (kJ/mol):
    9.27   (Chickos et al. 1999)
Entropy of Fusion, ΔS_{fus} (J/mol K):
    56.65, 38.8 (exptl., calculated-group additivity method, Chickos et al. 1999)
Fugacity Ratio at 25°C, F: 1.0

Water Solubility (g/m³ or mg/L at 25°C or as indicated):
    2.34, 2.30   (0; 25°C, shake flask-GC, Polak & Lu 1973)
    1.36   (shake flask-GC, Price 1976)
    2.30, 1.36   (quoted lit., IUPAC Solubility Data Series, Shaw 1989)
    2.86, 2.03   (0, 25°C, calculated-recommended liquid-liquid equilibrium LLE data, Mączyński et al. 2004)

Vapor Pressure (Pa at 25°C or as indicated and reported temperature dependence equations):
    $\log (P/mmHg) = 6.85396 - 1315.034/(217.526 + t/°C)$; temp range 36.568–114.381°C (Antoine eq. from exptl. data, ebulliometry-manometer, Willingham et al. 1945)
    3431   (Antoine eq. regression, temp range –26 to 113.5°C Stull 1947)
    3600   (calculated from determined data, Dreisbach 1959)
    $\log (P/mmHg) = 6.85396 - 1315.084/(217.526 + t/°C)$; temp range 25 – 150°C (Antoine eq. for liquid state, Dreisbach 1959)
    3600   (interpolated-Antoine eq., temp range 7.1–140.0°C, Zwolinski & Wilhoit 1971)
    $\log (P/mmHg) = 6.85396 - 1315.084/(217.526 + t/°C)$; temp range 7.1–140.0°C (Antoine eq., Zwolinski & Wilhoit 1971)
    3430   (interpolated-Antoine eq., temp range –26.3 to 113.5°C, Weast 1972–73)
    $\log (P/mmHg) = [-0.2185 \times 8988.2/(T/K)] + 7.997094$; temp range –26.3 to 113.5°C (Antoine eq., Weast 1972–73)
    1179   (4.912°C, static method-inclined piston manometer, measured range –50.325 to 4.912°C, Osborn & Douslin 1974)
    3600   (extrapolated-Antoine eq., Boublik et al. 1984)

log (P/kPa) = 5.98137 – 1316.608/(217.70 + t/°C); temp range 69.2–271.1°C (Antoine eq. from reported exptl. data of Willingham et al. 1945, Boublik et al. 1984)

3600 (extrapolated-Antoine eq., temp range 36–114°C, Dean 1985, 1992)

log (P/mmHg) = 6.85396 – 1315.08/(217.53 + t/°C); temp range 36–114°C (Antoine eq., Dean 1985, 1992)

3610 (interpolated-Antoine eq.-I, Stephenson & Malanowski 1987)

log (P_L/kPa) = 6.00347 – 1330.047/(–53.921 + T/K); temp range 288–400 K (Antoine eq.-I, Stephenson & Malanowski 1987)

log (P_L/kPa) = 6.37038 – 1511.86/(–38.054 + T/K); temp range 223–289 K (Antoine eq.-II, Stephenson & Malanowski 1987)

log (P/mmHg) = 35.1565 – 3.0232×10^3/(T/K) – 9.2267·log (T/K) + 2.7691×10^{-11}·(T/K) + 2.7828×10^{-6}·(T/K)^2; temp range 164–566 K (vapor pressure eq., Yaws 1994)

Henry's Law Constant (Pa m^3/mol at 25°C):

190000; 302000, 179000 (recommended; calculated-P/C, Mackay & Shiu 1981)

178700 (calculated-vapor-liquid equilibrium (VLE) data, Yaws et al. 1991)

Octanol/Water Partition Coefficient, log K_{OW}:

Octanol/Air Partition Coefficient, log K_{OA}:

Bioconcentration Factor, log BCF:

Sorption Partition Coefficient, log K_{OC}:

Environmental Fate Rate Constants, k, and Half-Lives, $t_{1/2}$:

Volatilization:

Photolysis:

Oxidation: rate constant k, for gas-phase second order rate constants, k_{OH} for reaction with OH radical, k_{NO3} with NO$_3$ radical and k_{O3} with O$_3$ or as indicated, *data at other temperatures and/or the Arrhenius expression see reference:

k_{OH}* = 6.99×10^{-12} cm^3 molecule^{-1} s^{-1} at 295 K, measured range 243–313 K (relative rate method with reference to *n*-hexane, Harris & Kerr 1988; Atkinson 1989)

k_{OH} = 7.0×10^{-12} cm^3 molecule^{-1} s^{-1} at 298 K (recommended, Atkinson 1990)

k_{OH} = 7.0×10^{-12} cm^3 molecule^{-1} s^{-1} at 298 K and the atmospheric lifetime was estimated to be 20 h during summer daylight hours (Altshuller 1991)

k_{OH} = 7.1×10^{-12} cm^3 molecule^{-1} s^{-1} at 298 K (recommended, Atkinson 1997)

Hydrolysis:

Biodegradation:

Biotransformation:

Bioconcentration, Uptake (k_1) and Elimination (k_2) Rate Constants or Half-Lives:

Half-Lives in the Environment:

Air: atmospheric lifetime was estimated to be 20 h, based on the photooxidation reaction rate constant of 7.0×10^{-12} cm^3 molecule^{-1} s^{-1} with OH radicals in air during summer daylight (Altshuller 1991).

2.1.1.1.16   n-Hexane

Common Name: *n*-Hexane
Synonym: hexane
Chemical Name: *n*-hexane
CAS Registry No: 110-54-3
Molecular Formula: C_6H_{14}; $CH_3(CH_2)_4CH_3$
Molecular Weight: 86.175
Melting Point (°C):
    –95.35   (Lide 2003)
Boiling Point (°C):
    68.73   (Lide 2003)
Density (g/cm³ at 20°C):
    0.6593, 0.6548   (20°C, 25°C, Riddick et al. 1986)
Molar Volume (cm³/mol):
    130.7; 131.6 (20°C, 25°C, calculated-density)
    140.6   (calculated-Le Bas method at normal boiling point)
Enthalpy of Vaporization, ΔH_V (kJ/mol):
    31.552, 28.853   (25°C, bp, Riddick et al. 1986)
    31.52   (298.15 K, recommended, Ruzicka & Majer 1994)
Enthalpy of Fusion, ΔH_{fus} (kJ/mol):
    13.028   (Dreisbach 1959)
    13.079   (Riddick et al. 1986)
    13.08   (Chickos et al. 1999)
Entropy of Fusion, ΔS_{fus} (J/mol K):
    73.22, 72.5   (exptl., calculated-group additivity method, Chickos et al. 1999)
Fugacity Ratio at 25°C, F: 1.0

Water Solubility (g/m³ or mg/L at 25°C or as indicated and reported temperature dependence equations. Additional data at other temperatures designated * are compiled at the end of this section.):

140	(15.5°C, shake flask-cloud point, Fühner 1924)
< 262	(shake flask-residue volume, Booth & Everson 1948)
36.0	(shake flask-cloud point, Durand 1948)
120	(shake flask-cloud point, McBain & Lissant 1951)
9.50	(shake flask-GC, McAuliffe 1963, 1966;)
16.2	(vapor saturation-GC, Barone et al. 1966)
18.3*	(shake flask-GC, measured range 4–55°C, Nelson & De Ligny 1968)
12.3	(shake flask-GC, Leinonen & Mackay 1973)
12.4*	(shake flask-GC, Polak & Lu 1973)
13.0	(shake flask-GC, Krasnoshchekova & Gubertritis 1973)
16.2	(shake flask-GC, Mackay et al. 1975)
9.47*	(shake flask-GC, measured range 25–151.8°C, Price 1976)
12.3	(shake flask-GC, Aquan-Yuen et al. 1979)
10.09*	(vapor saturation-GC, measured range 15.5–40°C, Jönsson et al. 1982)
12.24	(generator column-GC, Tewari et al. 1982a; Wasik et al. 1982; Miller et al. 1985)
14.1	(calculated-activity coeff. γ and K_{OW}, Tewari et al. 1982b)
11.4*	(37.78°C, shake flask-GC, measured 37.78–200°C, Tsonopoulos & Wilson 1983)
9.55	(shake flask-GC, Coates et al. 1985)
14.0	(shake flask-purge and trap-GC, Coutant & Keigley 1988)
9.8*	(recommended best value, IUPAC Solubility Data Series, temp range 0–140°C, Shaw 1989)

ln x = –374.90804 + 16327.128/(T/K) + 53.89582·ln (T/K); temp range 290–400 K (eq. derived from literature calorimetric and solubility data, Tsonopoulos 1999)
    11.5*   (calculated-liquid-liquid equilibrium LLE data, temp range 273.2–425 K, Mączyński et al. 2004)

Aliphatic and Cyclic Hydrocarbons

Vapor Pressure (Pa at 25°C or as indicated and reported temperature dependence equations. Additional data at other temperatures designated * are compiled at the end of this section.):

24811*   (30°C, static-manometer, measured range 30–60°C, Smyth & Engel 1929)
19920*   (24.7°C, ebulliometry, measured range 13.033–69.541°C Willingham et al. 1945)
log (P/mmHg) = 6.87776 − 1171.530/(224.366 + t/°C); temp range 13.033–69.541°C (Antoine eq. from exptl. data, ebulliometry, Willingham et al. 1945)
19700*   (calculated-Antoine eq. regression, temp range −53.9 to 68.7°C, Stull 1947)
20170   (calculated from determined data, Dreisbach 1959)
log (P/mmHg) = 6.87776 − 1171.53/(224.366 + t/°C); temp range −10 to 110°C (Antoine eq. for liquid state, Dreisbach 1959)
20198*   (interpolated-Antoine eq., temp range −25.1 to 92.1°C, Zwolinski & Wilhoit 1971)
log (P/mmHg) = 6.87776 − 1171.530/(224.366 + t/°C); temp range −25.1 to 92.1°C (Antoine eq., Zwolinski & Wilhoit 1971)
3120*   (−0.51°C, gas saturation, measured range −95.44 to −0.51°C, Carruth & Kobayashi 1973)
20130   (Campbell et al. 1968)
20124, 20141 (static method-differential pressure gauge, Bissell & Williamson 1975)
22090*   (27.1°C, Letcher & Marsicano 1974)
21809, 57929, 130189 (26.85, 51.85, 76.85°C, vapor-liquid equilibrium VLE data, Gutsche & Knapp 1982)
20160   (interpolated-Antoine eq., Boublik et al. 1984)
log (P/kPa) = 6.01098 − 1176.102/(224.899 + t/°C); temp range 13.033–69.54°C (Antoine eq. from reported exptl. data of Willingham et al. 1945, Boublik et al. 1984)
log (P/kPa) = 5.72763 − 1031.938/(208.304 + t/°C); temp range 27.11–45.11°C (Antoine eq. from reported exptl. data of Letcher & Marsicano 1974, Boublik et al. 1984)
20190   (interpolated-Antoine eq., temp range −25 to 92°C, Dean 1985, 1992)
log (P/mmHg) = 6.87601 − 1171.17/(224.41 + t/°C); temp range −25 to 92°C (Antoine eq., Dean 1985, 1992)
20700, 20180, 20160 (headspace-GC, correlated, Antoine eq., Hussam & Carr 1985)
20170   (lit. average, Riddick et al. 1986)
log (P/kPa) = 6.00091 − 1171.91/(224.408 + t/°C), temp range not specified (Antoine eq., Riddick et al. 1986)
20180, 20300, 20165 (interpolated-Antoine equations I, II and IV, Stephenson & Malanowski 1987)
log (P_L/kPa) = 6.00431 − 1172.04/(−48.747 + T/K); temp range 293–343 K (Antoine eq.-I, Stephenson & Malanowski 1987)
log (P_L/kPa) = 6.15142 − 1224.492/(−45.358 + T/K); temp range 238–298 K (Antoine eq.-II, Stephenson & Malanowski 1987)
log (P_L/kPa) = 8.47892 − 1800.89/(−4.115 + T/K); temp range 189–259 K (Antoine eq.-III, Stephenson & Malanowski 1987)
log (P_L/kPa) = 5.99521 − 1167.388/(−49.272 + T/K); temp range 298–341 K (Antoine eq.-IV, Stephenson & Malanowski 1987)
log (P_L/kPa) = 5.91942 − 1123.687/(−54.776 + T/K); temp range 341–377 K (Antoine eq.-V, Stephenson & Malanowski 1987)
log (P_L/kPa) = 6.4106 − 1469.286/(−7.702 + T/K); temp range 374–451 K (Antoine eq.-VI, Stephenson & Malanowski 1987)
log (P_L/kPa) = 7.30814 − 2367.155/(111.016 + T/K); temp range 445–508 K (Antoine eq.-VII, Stephenson & Malanowski 1987)
10854   (calculated-UNIFAC activity coeff., Banerjee et al. 1990)
20136*   (24.96°C, Hg manometer, measured range 9.95–49.97°C, Gracia et al. 1992)
20180*   (recommended, Ruzicka & Majer 1994)
ln [(P/kPa)/(P_o/kPa)] = [1 − (T_o/K)/(T/K)]·exp{2.79797 − 2.022083 × 10^{-3}·(T/K) + 2.287564 × 10^{-6}·(T/K)^2}; reference state at P_o = 101.325 kPa, T_o = 341.863 K (Cox equation, Ruzicka & Majer 1994)
log (P/mmHg) = 69.7378 − 3.6278 × 10^3/(T/K) − 23.927·log (T/K) + 1.281 × 10^{-2}·(T/K) − 1.6844 × 10^{-13}·(T/K)^2; temp range 178–507 K (vapor pressure eq., Yaws 1994)
24938, 53982 (30, 50°C, VLE equilibrium data, Carmona et al. 2000)

Henry's Law Constant (Pa m^3/mol at 25°C or as indicated and reported temperature dependence equations. Additional data at other temperatures designated * are compiled at the end of this section.):

183690   (calculated-1/K_{AW}, C_W/C_A, reported as exptl., Hine & Mookerjee 1975)

160000, 50590   (calculated-group contribution, calculated-bond contribution, Hine & Mookerjee 1975)
190000   (calculated-P/C, Mackay & Shiu 1975,1981; Bobra et al. 1979)
177060   (equilibrium cell-concn ratio-GC, Vejrosta et al. 1982)
173340*   (equilibrium cell-concentration ratio-GC, measured range 14.52–34.9°C Jönsson et al. 1982)
103262, 135400, 172140, 215420, 261420   (15, 20, 25, 30, 35°C, calculated-temp dependence eq. derived from exptl data, Jönsson et al. 1982)
$\ln (1/K_{AW}) = 21493.1/(T/K) + 59.299 \cdot \ln (T/K) - 414.193$; temp range 15–35°C (least-square regression of equilibrium cell-concn ratio-GC measurements, Jönsson et al. 1982)
77820*   (EPICS-GC/FID, measured range 10–30°C, Ashworth et al. 1988)
$\ln [H/(atm\_m^3/mol)] = 25.25 - 7530/(T/K)$; temp range 10–30°C (EPICS measurements, Ashworth et al. 1988)
130790   (calculated-vapor-liquid equilibrium (VLE) data, Yaws et al. 1991)
18600*   (40°C, equilibrium headspace-GC, measured range 40–70°C, Kolb et al. 1992)
$\ln (1/K_{AW}) = -28.60 + 8375/(T/K)$; temp range 40–70°C (equilibrium headspace-GC measurements, Kolb et al. 1992)
163812   (EPICS-GC, Ryu & Park 1999)
65318   (20°C, selected from reported experimental values, Staudinger & Roberts 1996, 2001)
$\log K_{AW} = 12.150 - 3143/(T/K)$ (summary of literature data, Staudinger & Roberts 2001)

Octanol/Water Partition Coefficient, log K_{OW}:
3.00   (calculated-π constant, Hansch et al. 1968; Hansch & Leo 1979)
3.90   (shake flask-concn. ratio-GC, Platford 1979; Platford 1983)
4.20   (calculated-activity coeff. γ, Wasik et al. 1981, 1982)
4.11   (generator column-GC, Tewari et al. 1982a,b)
2.90   (HPLC-k′ correlation, Coates et al. 1985)
4 25   (calculated-activity coeff., Berti et al. 1986)
4.16   (generator column-GC, Schantz & Martire 1987)
4.00   (recommended, Sangster 1989, 1993)
4.29   (calculated-activity coeff., Tse & Sandler 1994)
3.90   (recommended, Hansch et al. 1995)

Octanol/Air Partition Coefficient, log K_{OA} at 25°C or as indicated. Additional data at other temperatures designated * are compiled at the end of this section:
2.55*   (20.29°C, from GC-determined γ° in octanol, measured range 20.29–50.28°C Gruber et al. 1997)
2.44   (calculated-measured γ° in pure octanol and vapor pressure P, Abraham et al. 2001)

Bioconcentration Factor, log BCF:

Sorption Partition Coefficient, log K_{OC}:

Environmental Fate Rate Constants, k, and Half-Lives, $t_{1/2}$:
  Volatilization:
  Photolysis:
  Oxidation: rate constant k, for gas-phase second order rate constants, k_{OH} for reaction with OH radical, k_{NO3} with NO_3 radical and k_{O3} with O_3 or as indicated *data at other temperatures and/or the Arrhenius expression see reference:
    $k_{O(3P)} = 9.30 \times 10^{-14}$ cm^3 molecule^{-1} s^{-1} for reaction with O(^{3}P) atoms (Herron & Huie 1973)
    $k_{OH} = (3.8 \pm 0.8) \times 10^9$ L mol^{-1} s^{-1} at 305 K (relative rate method, Llyod et al. 1976)
    $k_{OH} = 5.90 \times 10^{-12}$ cm^3 molecule^{-1} s^{-1} at room temp. (Atkinson et al. 1979)
    $k_{OH} = (6.1 - 6.8) \times 10^{-12}$ cm^3 molecule^{-1} s^{-1} 292–303 K, k_{OH} (calc) = 6.96×10^{-12} cm^3 molecule^{-1} s^{-1} at 300 K (Darnall et al. 1978)
    $k_{OH} = 5.90 \times 10^{-12}$ cm^3 molecule^{-1} s^{-1}; $k_{O(3P)} = 9.3 \times 10^{-14}$ cm^3 molecule^{-1} s^{-1} for reaction with O(^{3}P) atoms, at room temp. (abstraction mechanism, Gaffney & Levine 1979)
    $k_{OH} = (5.71 \pm 0.09) \times 10^{-12}$ cm^3 molecule^{-1} s^{-1} at 299 ± 2 K (relative rate method, Atkinson et al. 1982a)
    $k_{NO3} = (1.05 \pm 0.2) \times 10^{-16}$ cm^3 molecule^{-1} s^{-1} at 296 K (Atkinson et al. 1984a, Atkinson 1991)
    $k_{OH} = (5.70 \pm 0.09) \times 10^{-12}$ cm^3 molecule^{-1} s^{-1} at room temp. (relative rate, Atkinson et al. 1984c)

k_{OH} = 5.21 × 10^{-12} cm^3 molecule^{-1} s^{-1} at room temp. (relative rate, Atkinson & Aschmann 1984)

k_{OH} = 6.20 × 10^{-12} cm^3 molecule^{-1} s^{-1} at 312 K in smog chamber (Nolting et al. 1988)

k_{OH}* = 5.61 × 10^{-12} cm^3 molecule^{-1} s^{-1} at 298 K (recommended, Atkinson 1989, 1991, Altshuller 1991)

k_{OH} = 5.58 × 10^{-12} cm^3 molecule^{-1} s^{-1}, k_{NO3} = 1.05 × 10^{-16} cm^3 molecule^{-1} s^{-1} (Sabljic & Güsten 1990)

k_{OH} = 5.61 × 10^{-12} cm^3 molecule^{-1} s^{-1} at 298 K, k_{NO3} = 10.5 × 10^{-17} cm^3 molecule^{-1} s^{-1} at 296 K (Atkinson 1990)

k_{NO3} = 1.05 × 10^{-16} cm^3 molecule^{-1} s^{-1} at 296 ± 2 K (Atkinson 1991)

k_{NO3}(exptl) = 1.06 × 10^{-16} cm^3 molecule^{-1} s^{-1}, k_{NO3}(calc) = 1.11 × 10^{-16} cm^3 molecule^{-1} s^{-1} at 296 ± 2 K (relative rate method, Aschmann & Atkinson 1995)

k_{OH}* = 5.45 × 10^{-12} cm^3 molecule^{-1} s^{-1}, k_{NO3} = 11 × 10^{-17} cm^3 molecule^{-1} s^{-1} at 298 K (recommended, Atkinson 1997)

k_{OH}* = 5.19 × 10^{-12} cm^3 molecule^{-1} s^{-1} at 298 K, measured range 230–400 K (relative rate method, DeMore & Bayes 1999)

Hydrolysis:

Biodegradation:

Biotransformation:

Bioconcentration, Uptake (k_1) and Elimination (k_2) Rate Constants or Half-Lives:

Half-Lives in the Environment:

Air: photooxidation reaction $t_{1/2}$ = 2.4–24 h in air, based on reaction rate constant of 3.8 × 10^9 L mol^{-1} s^{-1} for the reaction with hydroxyl radical (Darnall et al. 1976);

atmospheric lifetime ~ 25 h, based on a rate constant of 5.61 × 10^{-12} cm^3 molecule^{-1} s^{-1} for the reaction with OH radicals in summer daylight (Altshuller 1991).

TABLE 2.1.1.1.16.1
Reported aqueous solubilities of *n*-hexane at various temperatures

1.

Nelson & De Ligny 1968		Polak & Lu 1973		Price 1976		Jönsson et al. 1982	
shake flask-GC		shake flask-GC		shake flask-GC		vapor saturation-GC	
t/°C	S/g·m^{-3}	t/°C	S/g·m^{-3}	t/°C	S/g·m^{-3}	t/°C	S/g·m^{-3}
4.0	16.4	0	16.5	25.0	9.47	15	10.72
14.0	15.2	25	12.4	40.1	10.1	20	10.32
25.0	18.3			55.7	13.2	25	10.09
35.0	12.9			69.7	15.4	30	10.02
45.0	22.2			69.7	15.2	35	10.10
55.0	21.2			99.1	22.4		
				114.4	29.2		
				121.3	37.6		
				137.3	56.9		
				151.8	106.0		

(Continued)

TABLE 2.1.1.1.16.1 (*Continued*)

2.

Tsonopoulos & Wilson 1983		Shaw 1989a		Mączyński et al. 2004	
shake flask-GC		IUPAC recommended		calc-recommended LLE	
t/°C	S/g·m^{-3}	t/°C	S/g·m^{-3}	t/°C	S/g·m^{-3}
37.78	11.4	0	17	0	15.8
93.33	27.4	20	12	4	14.36
94.4	25.6	24	11	13	12.45
100	29.7	30	9.9	15	12.45
148.09	130	40	11	20	12.0
150	162	50	12	25	11.5
200	885	60	13.6	30	11.5
		70	15.7	35	11.5
		80	18.5	40.1	11.5
		100	27	55.7	12.93
		120	45	69.7	15.32
		140	80	99.1	26.33
				113.8	37.34
				121.3	44.52
				137.2	67.02
				151.8	105.3

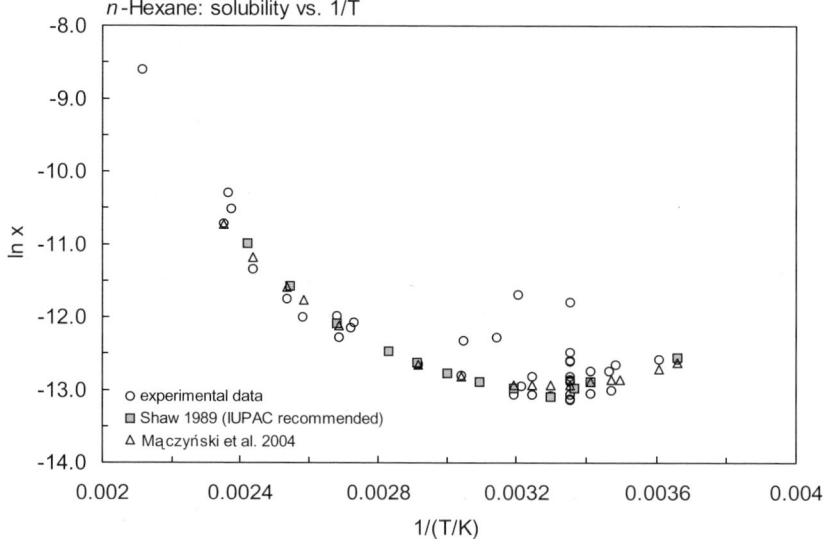

FIGURE 2.1.1.1.16.1 Logarithm of mole fraction solubility (ln x) versus reciprocal temperature for *n*-hexane.

TABLE 2.1.1.1.16.2
Reported vapor pressures of *n*-hexane at various temperatures and the coefficients for the vapor pressure equations

$$\log P = A - B/(T/K) \quad (1) \qquad \ln P = A - B/(T/K) \quad (1a)$$
$$\log P = A - B/(C + t/°C) \quad (2) \qquad \ln P = A - B/(C + t/°C) \quad (2a)$$
$$\log P = A - B/(C + T/K) \quad (3) \qquad \ln P = A - B/\{(T/K) - C\} \quad (3a)$$
$$\log P = A - B/(T/K) - C \cdot \log (T/K) \quad (4)$$
$$\ln [(P/kPa)/(P_0/kPa)] = [1 - (T_0/K)/(T/K)] \cdot \exp\{A_0 - A_1 \cdot (T/K) + A_2 \cdot (T/K)^2\} \quad (5)\text{ - Cox eq.}$$

1.

Smyth & Engel 1929		Willingham et al. 1945		Stull 1947		Zwolinski & Wilhoit 1971	
static-manometer		ebulliometry		summary of literature data		selected values	
t/°C	P/Pa	t/°C	P/Pa	t/°C	P/Pa	t/°C	P/Pa
30	24811	13.033	11698	−53.9	133.3	−25.19	1333
40	36757	16.576	13819	−34.5	666.6	−14.36	2666
50	53409	20.618	16620	−25.0	1333	−7.50	4000
60	75727	24.717	19920	−14.1	2666	−2.36	5333
		28.528	23451	−2.30	5333	1.79	6666
bp/°C	68.8	33.631	28956	5.40	7999	5.31	7999
		38.311	34896	15.8	13332	11.08	10666
		43.967	43322	31.6	26664	15.762	13332
		49.803	53655	49.8	53329	24.764	19998
		56.030	66757	68.7	101325	31.572	26664
		62.785	83719			37.114	33331
		67.554	97605	mp/°C	−95.3	41.824	39997
		68.067	99201			49.612	53329
		68.540	100694			55.973	66661
		69.081	102394			61.394	79993
		69.541	103913			66.145	93326
						67.030	95992
		bp/°C	68.740			67.896	98659
						68.744	101325
		eq. 2	P/mmHg				
		A	6.87776			eq. 2	P/mmHg
		B	1171.530			A	6.87024
		C	224.366			B	118.72
		temp range: 13–69.5°C				C	224.210
						bp/°C	68.744
						$\Delta H_V/(kJ\ mol^{-1}) =$	
						at 25°C	31.55
						at bp	28.85

(*Continued*)

TABLE 2.1.1.1.16.2 (Continued)

2.

Carruth & Kobayashi 1973		Letcher & Marsicano 1974		Gracia et al. 1992		Ruzicka & Majer 1994	
gas saturation		static method-manometer		Hg manometer		recommended	
t/°C	P/Pa	t/°C	P/Pa	t/°C	P/Pa	T/K	P/Pa
–95.44	1.360	27.11	22091	9.95	10081	162.54	0.1
–95.13	1.373	29.31	24151	14.98	12818	176.70	1.0
–91.70	2.266	33.81	29197	19.97	16136	193.88	10
–88.93	3.213	36.44	32437	24.96	20136	215.51	100
–83.99	5.973	39.88	37117	29.96	24949	243.81	1000
–75.94	13.60	42.45	40996	34.95	30577	282.95	10000
–61.25	64.13	45.15	45289	39.93	37185	341.863	101325
–53.29	128	48.11	50476	44.96	45000	298.15	20180
–34.84	656			49.97	54028		
–24.91	1264	Antoine eq.				data calc. from Cox eq.	
–14.39	2560	eq. 3	P/mmHg			eq. 5	P/kPa
–0.51	3120	A	6.6298	Antoine eq.		A_0	2.73425
		B	1050.38	eq. 3a	P/kPa	A_1	2.02283×10^{-3}
mp/°C	–95.36	C	210.477	A	13.74029	A_2	2.287564×10^{-6}
				B	2654.670	with reference state at	
eq. 1a	P/mmHg	$\Delta H_V/(kJ\ mol^{-1}) = 30.84$		C	50.869	P_o/kPa	101.325
A	19.5553					T_o/K	341.863
B	4292.8						

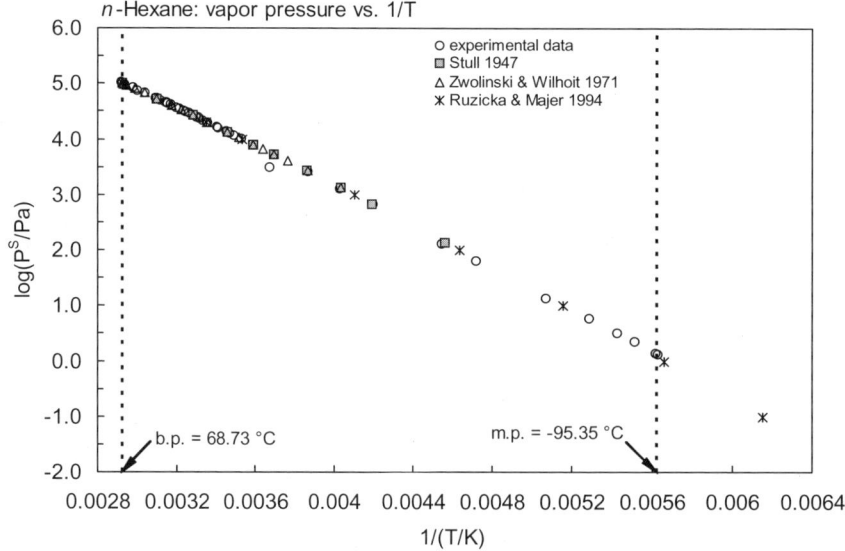

FIGURE 2.1.1.1.16.2 Logarithm of vapor pressure versus reciprocal temperature for *n*-hexane.

Aliphatic and Cyclic Hydrocarbons

TABLE 2.1.1.1.16.3
Reported Henry's law constants and octanol-air partition coefficients of *n*-hexane at various temperatures and temperature dependence equations

$$\ln K_{AW} = A - B/(T/K) \quad (1)$$
$$\ln (1/K_{AW}) = A - B/(T/K) \quad (2)$$
$$\ln (k_H/\text{atm}) = A - B/(T/K) \quad (3)$$
$$\ln [H/(\text{Pa m}^3/\text{mol})] = A - B/(T/K) \quad (4)$$
$$K_{AW} = A - B \cdot (T/K) + C \cdot (T/K)^2 \quad (5)$$

$$\log K_{AW} = A - B/(T/K) \quad (1a)$$
$$\log (1/K_{AW}) = A - B/(T/K) \quad (2a)$$
$$\log [H/(\text{atm} \cdot \text{m}^3/\text{mol})] = A - B/(T/K) \quad (4a)$$

Henry's law constant						log K_{OA}	
Jönsson et al. 1982		Ashworth et al. 1988		Kolb et al. 1992		Gruber et al. 1997	
equilibrium cell-GC		EPICS-GC		equilibrium headspace-GC		GC det'd activity coefficient	
t/°C	H/(Pa m³/mol)	t/°C	H/(Pa m³/mol)	t/°C	H/(Pa m³/mol)	t/°C	log K_{OA}
14.52	103536	10	24115	40	18600	20.29	2.548
14.52	103090	15	41847	60	64410	30.3	2.386
15.0	103262	20	89470	70	237750	40.4	2.216
20.05	135426	25	77818			50.28	2.090
20.05	136182	30	158067	eq. 2	1/K_{AW}		
20.0	135403			A	−28.60		
25	174565	eq. 4	H/(atm m³/mol)	B	−8375		
25	172140	A	25.25				
25	172140	B	7530				
30.2	215560						
30.2	211937						
30.0	215418						
34.9	264034						
34.9	256113						
35.0	261424						

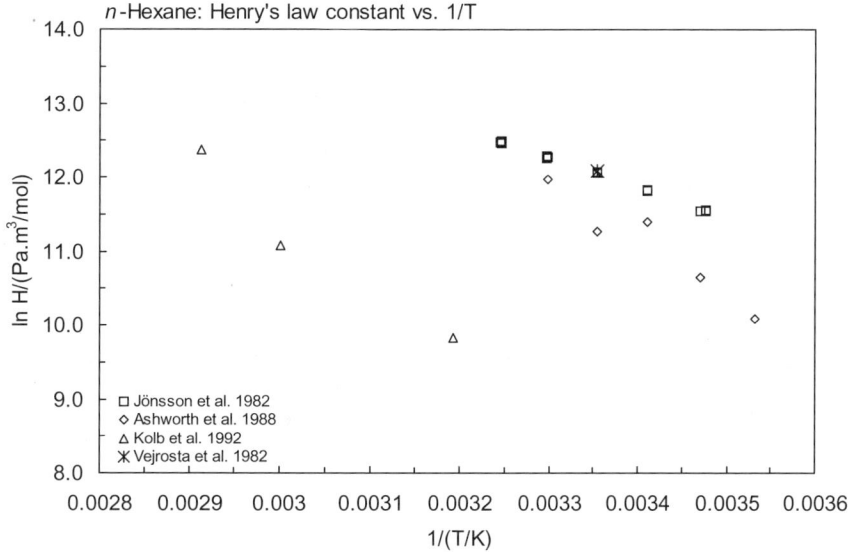

FIGURE 2.1.1.1.16.3 Logarithm of Henry's law constant versus reciprocal temperature for *n*-hexane.

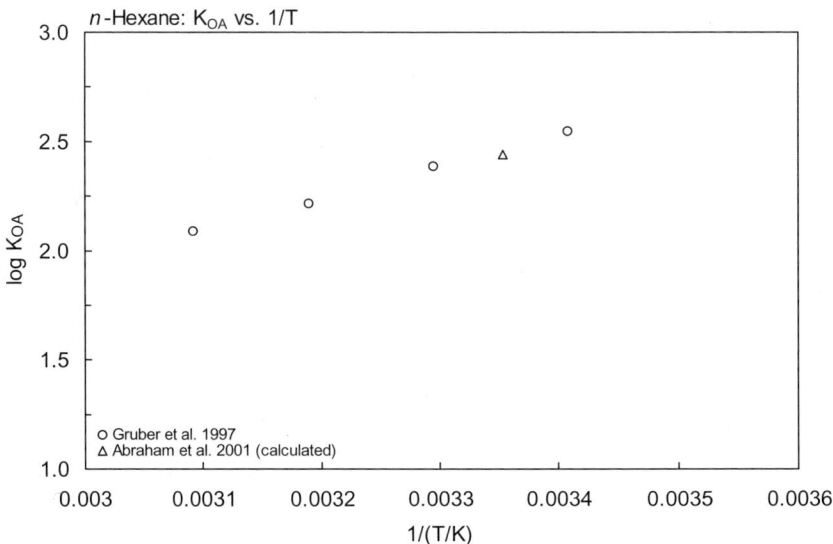

FIGURE 2.1.1.1.16.4 Logarithm of K_{OA} versus reciprocal temperature for *n*-hexane.

Aliphatic and Cyclic Hydrocarbons

2.1.1.1.17   2-Methylhexane (Isoheptane)

Common Name: 2-Methylhexane
Synonym: isoheptane, ethylisobutylmethane
Chemical Name: 2-methylhexane
CAS Registry No: 591-76-4
Molecular Formula: C_7H_{16}; $CH_3CH(CH_3)CH_2(CH_2)_2CH_3$
Molecular Weight: 100.202
Melting Point (°C):
    –118.2   (Lide 2003)
Boiling Point (°C):
    90.04   (Lide 2003)
Density (g/cm³ at 20°C):
    0.6786, 0.6744   (20°C, 25°C, Dreisbach 1959)
    0.6786, 0.6548   (20°C, 25°C, Riddick et al. 1986)
Molar Volume (cm³/mol):
    147.66, 148.58   (20°C, 25°C, calculated-density)
    162.8   (calculated-Le Bas method at normal boiling point)
Enthalpy of Vaporization, ΔH_V (kJ/mol):
    34.807, 30.669   (25°C, bp, Riddick et al. 1986)
Enthalpy of Fusion, ΔH_{fus} (kJ/mol):
    9.184   (Dreisbach 1959; Riddick et al. 1986)
    9.18   (Chickos et al. 1999)
Entropy of Fusion, ΔS_{fus} (J/mol K):
    59.29, 57.8   (exptl., calculated-group additivity method, Chickos et al. 1999)
Fugacity Ratio at 25°C, F: 1.0

Water Solubility (g/m³ or mg/L at 25°C):
    2.54   (shake flask-GC, Price 1976)
    3.51   (calculated-recommended liquid-liquid equilibrium LLE data, Mączyński et al. 2004)

Vapor Pressure (Pa at 25°C or as indicated and reported temperature dependence equations):
    8380   (calculated-Antoine eq. regression, temp range –40.4 to 90°C, Stull 1947)
    9000   (25.518°C, ebulliometry, measured range 18.5–90.9°C, Forziati et al. 1949)
    log (P/mmHg) = 6.87319 – 1236.026/(219.545 + t/°C); temp range (Antoine eq., ebulliometry-manometer measurements, Forziati et al. 1949)
    8780   (calculated from determined data, Dreisbach 1959)
    log (P/mmHg) = 6.87318 – 1236.026/(219.545 + t/°C); temp range 10–110°C (Antoine eq. for liquid state, Dreisbach 1959)
    8780   (interpolated-Antoine eq., Zwolinski & Wilhoit 1971)
    log (P/mmHg) = 5.87318 – 1236.026/(219.545 + t/°C); temp range –9.10 to 114.78°C (Antoine eq., Zwolinski & Wilhoit 1971)
    8370   (interpolated-Antoine eq., Weast 1972–73)
    log (P/mmHg) = [–0.2185 × 8538.7/(T/K)] + 8.055523; temp range –40.4 to 90°C (Antoine eq., Weast 1972–73)
    87778, 8785 (interpolated-Antoine eq., Boublik et al. 1984)
    log (P/kPa) = 6.0031 – 1238.614/(219.867 + t/°C); temp range 18.5–90.9°C (Antoine eq. from reported exptl. data of Forziati et al. 1949, Boublik et al. 1984)
    log (P/kPa) = 6.06712 – 1270.535/(222.971 + t/°C); temp range 0–45°C (Antoine eq. from reported exptl. data, Boublik et al. 1984)
    8790   (interpolated-Antoine eq., temp range –9 to 115°C, Dean 1985, 1992)
    log (P/mmHg) = 6.87318 – 1236.026/(219.55 + t/°C); temp range –9 to 115°C (Antoine eq., Dean 1985, 1992)

8800 (lit. average, Riddick et al. 1986)
log (P/kPa) = 5.99612 − 1235.10/(219.469 + t/°C); temp range not specified (Antoine eq., Riddick et al. 1986)
8790 (interpolated-Antoine eq., Stephenson & Malanowski 1987)
log (P_L/kPa) = 6.00513 − 1240.11/(−53.123 + T/K); temp range 296–365 K (Antoine eq., Stephenson & Malanowski 1987)
log (P/mmHg) = 54.1075 − 3.785 × 10^3/(T/K) − 17.547·log (T/K) + 8.2594 × 10^{-3}·(T/K) − 3.4967 × 10^{-14}·(T/K)^2; temp range 155–530 K (vapor pressure eq., Yaws 1994)

Henry's Law Constant (Pa m^3/mol at 25°C or as indicated and reported temperature dependence equations. Additional data at other temperatures designated * are compiled at the end of this section.):
346000 (calculated-P/C, Mackay & Shiu 1981, Eastcott et al. 1988)
346500 (calculated-vapor-liquid equilibrium (VLE) data, Yaws et al. 1991)
51878* (26.9°C, EPICS-GC, measured range 26.9–45°C, Hansen et al. 1993)
ln [H/(kPa·m^3/mol)] = 3608/(T/K) − 8.0; temp range 26.9–45°C (EPICS-GC, Hansen et al. 1993)
63856 (20°C, selected from reported experimental determined values, Staudinger & Roberts 2001)
log K_{AW} = −4.274 + 1669/(T/K) (summary of literature data, Staudinger & Roberts 2001)

Octanol/Water Partition Coefficient, log K_{OW}:

Bioconcentration Factor, log BCF:

Sorption Partition Coefficient, log K_{OC}:

Environmental Fate Rate Constants, k, and Half-Lives, $t_{1/2}$:
Volatilization:
Photolysis:
Oxidation: rate constant k, for gas-phase second order rate constants, k_{OH} for reaction with OH radical, k_{NO3} with NO_3 radical and k_{O3} with O_3 or as indicated, *data at other temperatures see reference:
k_{OH} = 6.80 × 10^{-12} cm^3 molecule^{-1} s^{-1} (Atkinson 1990, 1991)
k_{OH} = 6.80 × 10^{-12} cm^3 molecule^{-1} s^{-1} with atmospheric lifetime of 25 h in summer daylight (Altshuller 1990)
Hydrolysis:
Biodegradation:
Biotransformation:
Bioconcentration, Uptake (k_1) and Elimination (k_2) Rate Constants or Half-Lives:

Half-Lives in the Environment:
Air: photooxidation reaction rate constant of 6.80 × 10^{-12} cm^3 molecule^{-1} s^{-1} with hydroxyl radicals and an estimated lifetime of 25 h in summer daylight (Altshuller 1990).

TABLE 2.1.1.1.17.1
Reported Henry's law constants of 2-methylhexane at various temperatures

Hansen et al. 1993	
EPICS-GC	
t/°C	H/(kPa m^3/mol)
26.9	51.878
35.0	31.512
45.0	25.939
ln [H/(Pa m3/mol)] = A − B/(T/K)	
eq. 4	H/(kPa m^3/mol)
A	−8 ± 3.53
B	−3608 ± 1088

Aliphatic and Cyclic Hydrocarbons

2.1.1.1.18   3-Methylhexane

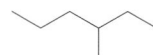

Common Name: 3-Methylhexane
Synonym: ethylmethylpropylmethane
Chemical Name: 3-methylhexane
CAS Registry No: 589-34-4
Molecular Formula: C_7H_{16}; $CH_3CH_2CH(CH_3)(CH_2)_2CH_3$
Molecular Weight: 100.202
Melting Point (°C):
   –119.4   (Riddick et al. 1986; Lide 2003)
Boiling Point (°C):
   92   (Lide 2003)
Density (g/cm³ at 20°C):
   0.6860   (Weast 1984)
   0.6871, 0.6830   (20°C, 25°C, Dreisbach 1959; Riddick et al. 1986)
Molar Volume (cm³/mol):
   145.83, 146.71   (calculated-density)
   162.8   (calculated-Le Bas method at normal boiling point)
Enthalpy of Vaporization, ΔH_V (kJ/mol):
   35.087, 30.79 (25°C, bp, Riddick et al. 1986)
Enthalpy of Fusion, ΔH_{fus} (kJ/mol):
Entropy of Fusion, ΔS_{fus} (J/mol K):
Fugacity Ratio at 25°C, F: 1.0

Water Solubility (g/m³ or mg/L at 25°C or as indicated):
   4.95   (shake flask-GC, Polak & Lu 1973)
   2.64   (shake flask-GC, Price 1976)
   3.80   (suggested IUPAC tentative value, Shaw 1989)
   6.12, 4.29   (0. 25°C, calculated-recommended liquid-liquid equilibrium LLE data, Mączyński et al. 2004)

Vapor Pressure (Pa at 25°C or as indicated and reported temperature dependence equations):
   7782   (Antoine eq. regression, temp range –39 to 91.9°C Stull 1947)
   7702   (23.662°C, ebulliometry, measured range 19.915–92.737°C, Forziati et al. 1949)
   log (P/mmHg) = 6.86754 – 1240.196/(219.223 + t/°C); temp range 19.9–92.7°C (Antoine eq., ebulliometry-manometer measurements, Forziati et al. 1949)
   8210   (calculated from determined data, Dreisbach 1959)
   log (P/mmHg) = 6.86764 – 1240.196/(219.223 + t/°C); temp range 10–130°C (Antoine eq. for liquid state, Dreisbach 1959)
   8213   (interpolated-Antoine eq., temp range –7.90 to 116.73°C, Zwolinski & Wilhoit 1971)
   log (P/mmHg) = 6.86764 – 1240.196/(219.223 + t/°C); temp range –7.90 to 116.73°C (Antoine eq., Zwolinski & Wilhoit 1971)
   7772   (interpolated-Antoine eq., temp range –39.0 to 91.9°C, Weast 1972–73)
   log (P/mmHg) = [–0.2185 × 8596.3/(T/K)] + 8.065472; temp range –39.0 to 91.9°C (Antoine eq., Weast 1972–73)
   8212   (interpolated-Antoine eq., Boublik et al. 1984)
   log (P/kPa) = 5.99489 – 1241.528/(219.375 + t/°C); temp range 20–92.74°C (Antoine eq. from reported exptl. data of Forziati et al. 1949, Boublik et al. 1984)
   8210   (interpolated-Antoine eq., temp range –8 to 117°C Dean 1985, 1992)
   log (P/mmHg) = 6.86764 – 1240.196/(219.22 + t/°C); temp range –8 to 117°C (Antoine eq., Dean 1985, 1992)
   8300   (lit. average, Riddick et al. 1986)
   log (P/kPa) = 5.98993 – 1238.88/(219.10 + t/°C); temp range not specified (Antoine eq., Riddick et al. 1986)
   8215   (interpolated-Antoine eq., Stephenson & Malanowski 1987)

log (P_L/kPa) = 5.9926 – 1239.57/(–53.979 + T/K); temp range 289–366 K (Antoine eq., Stephenson & Malanowski 1987)

log (P/mmHg) = 35.2535 – 2.931 × 10^3/(T/K) – 9.667·log (T/K) – 5.2026 × 10^{-11}·(T/K) + 3.2107 × 10^{-6}·(T/K)^2; temp range 154–535 K (vapor pressure eq., Yaws 1994)

Henry's Law Constant (Pa m^3/mol at 25°C):

172000	(recommended, Mackay & Shiu 1981)
312170	(calculated-P/C, Eastcott et al. 1988)
311620	(calculated-vapor-liquid equilibrium (VLE) data, Yaws et al. 1991)

Octanol/Water Partition Coefficient, log K_{OW}:

Octanol/Air Partition Coefficient, log K_{OA}:

Bioconcentration Factor, log BCF:

Sorption Partition Coefficient, log K_{OC}:

Environmental Fate Rate Constants, k, and Half-Lives, $t_{1/2}$:
    Volatilization:
    Photolysis:
    Oxidation: rate constant k, for gas-phase second order rate constants, k_{OH} for reaction with OH radical, k_{NO3} with NO$_3$ radical and k_{O3} with O$_3$ or as indicated, *data at other temperatures see reference:
        k_{OH} = 7.20 × 10^{-12} cm^3 molecule^{-1} s^{-1} at 298 K (Atkinson 1990,1991)
        k_{OH} = 7.20 × 10^{-12} cm^3 molecule^{-1} s^{-1} with a calculated atmospheric lifetime of 20 h during summer daylight (Altshuller 1991)
    Hydrolysis:
    Biodegradation:
    Biotransformation:
    Bioconcentration, Uptake (k_1) and Elimination (k_2) Rate Constants or Half-Lives:

Half-Lives in the Environment:
    Air: atmospheric $t_{1/2}$ ~ 2.4–24 h for the reaction with hydroxyl radical, based on the EPA Reactivity Classification of Organics (Darnall et al. 1976);
        photooxidation reaction rate constant of 7.20 × 10^{-12} cm^3 molecule^{-1} s^{-1} with hydroxyl radical and an estimated lifetime of 20 h during summer daylight (Altshuller 1991).

2.1.1.1.19   2,2,5-Trimethylhexane

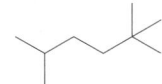

Common Name: 2,2,5-Trimethylhexane
Synonym:
Chemical Name: 2,2,5-trimethylhexane
CAS Registry No: 3522-94-9
Molecular Formula: C_9H_{20}, $(CH_3)_3CCH_2CH_2CH(CH_3)_2$
Molecular Weight: 128.255
Melting Point (°C):
    –105.7   (Lide 2003)
Boiling Point (°C):
    124.09   (Lide 2003)
Density (g/cm³ at 25°C):
    0.7072   (Weast 1984)
    0.7072, 0.7032   (20°C, 25°C, Dreisbach 1959; Riddick et al. 1986)
Molar Volume (cm³/mol):
    181.3, 182.4   (20°C, 25°C, calculated-density)
    207.2   (calculated-Le Bas method at normal boiling point)
Enthalpy of Vaporization, ΔH_V (kJ/mol):
    40.175, 33.76 (25°C, bp, Riddick et al. 1986)
Enthalpy of Fusion, ΔH_{fus} (kJ/mol):
    6.192   (Riddick et al. 1986)
Entropy of Fusion, ΔS_{fus} (J/mol K):
Fugacity Ratio at 25°C, F: 1.0

Water Solubility (g/m³ or mg/L at 25°C or as indicated):
    1.15   (shake flask-GC, McAuliffe 1966)
    0.79; 0.54, 0.54   (0; 25°C, shake flask-GC, calculated-group contribution, Polak & Lu 1973)
    0.80, 0.80   (0, 25°C, IUPAC recommended best value, Shaw1989b)
    0.613   (calculated-recommended liquid-liquid equilibrium LLE data, Mączyński et al. 2004)

Vapor Pressure (Pa at 25°C or as indicated and reported temperature dependence equations):
    log (P/mmHg) = 6.83532 – 1324.049/(210.737 + t/°C); temp range 46.1–125.0°C (Antoine eq., ebulliometry-manometer measurements, Forziati et al. 1949)
    2212   (extrapolated-Antoine eq., Dreisbach 1959)
    log (P/mmHg) = 6.83531 – 1324.059/(210.737 + t/°C); temp range 35–145°C (Antoine eq. for liquid state, Dreisbach 1959)
    2210   (interpolated-Antoine eq., temp range 16.17–151.1°C, Zwolinski & Wilhoit 1971)
    log (P/mmHg) = 6.83531 – 1324.049/(210.737C + t/°C); temp range 16.17–151.1°C (Antoine eq., Zwolinski & Wilhoit 1971)
    2216   (static method-inclined piston manometer, measured range –35 to 30°C, Osborn & Douslin 1964)
    2207   (interpolated-Antoine eq., Boublik et al. 1984)
    log (P/kPa) = 5.96385 – 1326.27/(212.991 + t/°C); temp range 46.14–126.05°C (Antoine eq. from reported exptl. data of Forziati et al. 1949, Boublik et al. 1984)
    2207   (extrapolated-Antoine eq., temp range 46 to 125°C Dean 1985, 1992)
    log (P/mmHg) = 6.83775 – 1325.54/(210.91 + t/°C); temp range 46–125°C (Antoine eq., Dean 1985, 1992)
    2216   (lit. average, Riddick et al. 1986)
    log (P/kPa) = 5.96021 – 1324.049/(210.737 + t/°C); temp range not specified (Antoine eq., Riddick et al. 1986)
    2218   (interpolated-Antoine eq.-I, Stephenson & Malanowski 1987)
    log (P_L/kPa) = 6.99253 – 1243.85/(–60.158 + T/K); temp range 288–399 K (Antoine eq.-I, Stephenson & Malanowski 1987)

log (P_L/kPa) = 6.25179 − 1471.621/(−48.901 + T/K); temp range 238–293 K (Antoine eq.-II, Stephenson & Malanowski 1987)

log $(P/mmHg)$ = 7.8816 − 2.6422 × 10^3/(T/K) + 23.902·log (T/K) − 1.5376 × 10^{-2}·(T/K) + 7.7931 × 10^{-6}·(T/K)^2; temp range 167–568 K (vapor pressure eq., Yaws 1994)

Henry's Law Constant (Pa m^3/mol at 25°C):
- 350000 (recommended, Mackay & Shiu 1981)
- 523760 (calculated-vapor-liquid equilibrium (VLE) data, Yaws et al. 1991)

Octanol/Water Partition Coefficient, log K_{OW}:
- 4.63 (calculated-regression eq. from Lyman et al. 1982, Wang et al. 1992)
- 4.46 (calculated-molar volume V_M, Wang et al. 1992)

Octanol/Air Partition Coefficient, log K_{OA}:

Bioconcentration Factor, log BCF:

Sorption Partition Coefficient, log K_{OC}:

Environmental Fate Rate Constants and Half-Lives:

Half-Lives in the Environment:

Aliphatic and Cyclic Hydrocarbons

2.1.1.1.20 n-*Heptane*

Common Name: *n*-Heptane
Synonym: heptane
Chemical Name: *n*-heptane
CAS Registry No: 142-82-5
Molecular Formula: C_7H_{16}, $CH_3(CH_2)_5CH_3$
Molecular Weight:100.202
Melting Point (°C):
    –90.55   (Lide 2003)
Boiling Point (°C):
    98.4   (Dreisbach 1959; Weast 1972–73; Lide 2003)
Density (g/cm³ at 20°C):
    0.6837   (Weast 1972–73)
    0.6837, 0.6795   (20°C, 25°C, Dreisbach 1959; Riddick et al. 1986)
Molar Volume (cm³/mol):
    146.6, 147.5 (20°C, 25°C, calculated-density)
    162.8   (calculated-Le Bas method at normal boiling point)
Enthalpy of Vaporization, ΔH_V (kJ/mol):
    36.55, 31.7   (25°C, bp, Riddick et al. 1986)
    36.57   (298.15 K, recommended, Ruzicka & Majer 1994)
Enthalpy of Fusion, ΔH_{fus} (kJ/mol):
    14.037   (Riddick et al. 1986)
    14.04   (Chickos et al. 1999)
Entropy of Fusion, ΔS_{fus} (J/mol K):
    76.9, 81.8   (exptl., calculated-group additivity method, Chickos et al. 1999)
Fugacity Ratio at 25°C, F: 1.0

Water Solubility (g/m³ or mg/L at 25°C or as indicated and reported temperature dependence equations. Additional data at other temperatures designated * are compiled at the end of this section.):
    50.0   (15.5°C, shake flask-cloud point, Fühner 1924)
    150   (radiotracer method, Black et al. 1948)
    15.0   (16°C, shake flask-cloud point, Durand 1948)
    2.93   (shake flask-GC, McAuliffe 1963,1966)
    11.0*   (22°C, cloud point, measured range 295–355 K at 17–55 MPa, Connolly 1966)
    2.66*   (shake flask-GC, measured range 4.3–45°C, Nelson & De Ligny 1968)
    4.39; 3.37*   (0, 25°C, shake flask-GC, Polak & Lu 1973)
    2.57   (shake flask-GC, Krasnoshchekova & Gubertritis 1975)
    2.24*   (shake flask-GC, measured range 25–150.4°C, Price 1976)
    3.70   (shake flask-GC, Bittrich et al. 1979)
    2.90   (partition coefficient-GC, Rudakov & Lutsyk 1979)
    3.58   (generator column-GC, Tewari et al. 1982a; Wasik et al. 1982)
    4.62   (calculated-activity coeff. γ and K_{OW}, Tewari et al. 1982b)
    2.51*   (vapor saturation-GC, measured range 15–40°C, Jönsson et al. 1982)
    2.95   (shake flask-GC, Coates et al. 1985)
    3.57   (lit. average, Riddick et al. 1986)
    2.90   (shake flask-purge and trap-GC, Coutant & Keigley 1988)
    2.40*   (recommended best value, IUPAC Solubility Data Series, temp range 0–140°C, Shaw 1989)
    $\ln x = -396.93979 + 17232.298/(T/K) + 56.95927 \cdot \ln (T/K)$, temp range 290–400 K (eq. derived from literature calorimetric and solubility data, Tsonopoulos 1999)
    2.95*   (calculated-liquid-liquid equilibrium LLE data, temp range 273.2–423.6 K, Mączyński et al. 2004)

Vapor Pressure (Pa at 25°C or as indicated and reported temperature dependence equations. Additional data at other temperatures designated * are compiled at the end of this section.):

5520* (22.7°C, static-manometer, measured range 22.7–98.4°C, Smyth & Engel 1929)
6370* (25.9°C, ebulliometry, measured range 25.9–99.3°C, Willingham et al. 1945)
log (P/mmHg) = 6.90342 – 1268.636/(216.951 + t/°C); temp range 25.9–99.3°C (Antoine eq. from exptl. data, ebulliometry-manometer, Willingham et al. 1945)
5795* (calculated-Antoine eq. regression, temp range –34 to 98.4°C Stull 1947)
6425* (26.039°C, ebulliometry, measured range 26–99.3°C Forziati et al. 1949)
log (P/mmHg) = 6.90027 – 1296.871/(216.757 + t/°C); temp range 26.0–99.3°C (Antoine eq., ebulliometry-manometer measurements, Forziati et al. 1949)
6110 (calculated from determined data, Dreisbach 1959)
log (P/mmHg) = 6.90240 – 1268.115/(216.90 + t/°C); temp range 15–155°C (Antoine eq. for liquid state, Dreisbach 1959)
6105 (Harris & Dunlop 1970)
6113* (interpolated-Antoine eq., temp range –2.10 to 123.41°C, Zwolinski & Wilhoit 1971)
log (P/mmHg) = 6.90240 – 1268.115/(216.900 + t/°C); temp range –2.10 to 123.41°C (Antoine eq., Zwolinski & Wilhoit 1971)
log (P/mmHg) = [–0.2185 × 8409.6/(T/K)] + 7.786586; temp range –34.0 to 247.5°C (Antoine eq., Weast 1972–73)
5080* (22.46°C, gas saturation, measured range –87.85 to 22.46°C, Carruth & Kobayashi 1973)
6037, 6057 (static method-differential pressure gauge, Bissell & Williamson 1975)
7826, 18929, 40469, 57059(30, 50, 70, 80°C, vapor-liquid equilibrium VLE data, Gutsche & Knapp 1982)
6090 (interpolated-Antoine eq., Boublik et al. 1984)
log (P/kPa) = 6.02701 – 1167.592/(216.796 + t/°C); temp range 25.92–99.3°C (Antoine eq. from reported exptl. data of Willingham et al. 1945, Boublik et al. 1984)
log (P/kPa) = 4.38001 – 668.768/(159.522 + t/°C); temp range –87.85 to 22.4°C (Antoine eq. from reported exptl. data of Carruth & Kobayashi 1973, Boublik et al. 1984)
6110, 5958, 6090 (headspace-GC, correlated, Antoine eq., Hussam & Carr 1985)
6090 (interpolated-Antoine eq., temp range –2 to 124°C Dean 1985. 1992)
log (P/mmHg) = 6.98677 – 1264.90/(216.54 + t/°C); temp range –2 to124°C (Antoine eq., Dean 1985, 1992)
6090 (literature average, Riddick et al. 1986)
log (P/kPa) = 6.02167 – 1264.90/(216.544 + t/°C); temp range not specified (Antoine eq., Riddick et al. 1986)
6110 (interpolated-Antoine eq., Stephenson & Malanowski 1987)
log (P_L/kPa) = 6.02633 – 1268.583/(–56.054 + T/K); temp range 297–375 K (Antoine eq., Stephenson & Malanowski 1987)
6102* (recommended, Ruzicka & Majer 1994)
ln [(P/kPa)/(P_o/kPa)] = [1 – (T_o/K)/(T/K)]·exp{2.86470 – 2.113204 × 10^{-3}·(T/K) + 2.250991 × 10^{-6}·(T/K)^2}; reference state at P_o = 101.325 kPa, T_o = 371.552 K (Cox equation, Ruzicka & Majer 1994)
log (P/mmHg) = 65.0257 – 3.8188 × 10^3/(T/K) –21.684·log (T/K) + 1.0387 × 10^{-2}·(T/K) + 1.0206 × 10^{-14}·(T/K)^2; temp range 183–540 K (vapor pressure eq., Yaws 1994)
12309 (40°C, average value, vapor-liquid equilibrium VLE data, Rhodes et al. 1997)
7811, 18869 (30, 50°C, VLE equilibrium data, Carmona et al. 2000)
62503* (355.899 K, ebulliometry, measured range 355.899–503.406 K, Weber 2000)

Henry's Law Constant (Pa m^3/mol at 25°C or as indicated and reported temperature dependence equations. Additional data at other temperatures designated * are compiled at the end of this section.):

206200 (calculated-1/K_{AW}, C_W/C_A, reported as exptl., Hine & Mookerjee 1975)
226000, 73120 (calculated-group contribution, calculated-bond contribution, Hine & Mookerjee 1975)
230000 (recommended, Mackay & Shiu 1981)
250420* (25.04°C, equilibrium cell-concentration ratio-GC, measured range 15.3–35.05°C Jönsson et al. 1982)
136120, 184640, 243020, 315050, 394150(15, 20, 25, 30, 35°C, calculated-temp dependence eq. derived from exptl data, Jönsson et al. 1982)

Aliphatic and Cyclic Hydrocarbons

ln $1/K_{AW}$ = 23748.4/(T/K) + 64.927·ln (T/K) – 454.172; temp range 15–35°C (least-square regression of equilibrium cell-concn ratio-GC measurements, Jönsson et al. 1982)

273400 (calculated-vapor-liquid equilibrium (VLE) data, Yaws et al. 1991)

91294* (26°C, EPICS-GC, measured range 26–45°C, Hansen et al. 1993)

ln [H/(kPa·m^3/mol)] = –3730/(T/K) + 17.0; temp range 26–45°C (EPICS-GC, Hansen et al. 1993)

220825 (EPICS-GC, Ryu & Park 1999)

68000 (20°C, selected from reported experimental determined values, Staudinger & Roberts 1996, 2001)

log K_{AW} = 6.532 – 1491/(T/K) (summary of literature data, Staudinger & Roberts 2001)

Octanol/Water Partition Coefficient, log K_{OW}:

3.50 (calculated-π constant, Hansch et al. 1968, Hansch & Leo 1979)
4.76 (calculated-activity coeff. γ, Wasik et al. 1981,1982)
4.66 (generator column-GC, Tewari et al. 1982a,b)
3.44 (HPLC-k′ correlation, Coates et al. 1985)
4.48 (Berti et al. 1986)
4.66, 4.72 (generator column-GC, calculated-activity coeff. γ, Schantz & Martire 1987)
4.50 ± 0.25 (recommended, Sangster 1989)
4.66 (recommended, Sangster 1993)
4.99 (calculated-activity coefficients, Tse & Sandler 1994)
4.66 (recommended, Hansch et al. 1995)

Octanol/Air Partition Coefficient, log K_{OA} at 25°C or as indicated. Additional data at other temperatures designated * are compiled at the end of this section.):

3.05* (20.29°C, from GC-determined γ° in octanol, measured range 20.29–50.28°C Gruber et al. 1997)
2.95 (calculated-measured γ° in pure octanol and vapor pressure P, Abraham et al. 2001)

Bioconcentration Factor, log BCF:

Sorption Partition Coefficient, log K_{OC}:

Environmental Fate Rate Constants, k, and Half-Lives, $t_{1/2}$:

Volatilization:
Photolysis:
Oxidation: rate constant k, for gas-phase second order rate constants, k_{OH} for reaction with OH radical, k_{NO3} with NO$_3$ radical and k_{O3} with O$_3$ or as indicated, *data at other temperatures and/or the Arrhenius expression see reference:

k_{OH} = (7.30 ± 0.17) × 10^{-12} cm^3 molecule^{-1} s^{-1} at 299 ± 2 K (relative rate method, Atkinson et al. 1982a, 1984c)

k_{NO3} = 1.36 × 10^{-16} cm^3 molecule^{-1} s^{-1} at 296 K (Atkinson 1988,1990)

k_{OH} = 7.52 × 10^{-12} cm^3 molecule^{-1} s^{-1} at 312 K in smog chamber (Nolting et al. 1988)

k_{OH}* = 7.15 × 10^{-12} cm^3 molecule^{-1} s^{-1} at 298 K (recommended, Atkinson 1989, 1991)

k_{OH} = 7.15 × 10^{-12} cm^3 molecule^{-1} s^{-1} at 298 K, k_{NO3} = 1.36 × 10^{-16} cm^3 molecule^{-1} s^{-1} at 296 K (Atkinson 1990)

k_{OH} = 7.19 × 10^{-12} cm^3 molecule^{-1} s^{-1}, k_{NO3} = 1.36 × 10^{-16} cm^3 molecule^{-1} s^{-1} (Sabljic & Güsten 1990)

k_{OH} = 7.15 × 10^{-12} cm^3 molecule^{-1} s^{-1} at 296 K, with a calculated atmospheric lifetime of 19 h during summer daylight hours (Altshuller 1991)

k_{NO3} = (1.34 – 1.37) × 10^{-16} cm^3 molecule^{-1} s^{-1} at 296 K (Atkinson 1991)

k_{NO3}(exptl) = (1.36, 1.38) × 10^{-16} cm^3 molecule^{-1} s^{-1}, k_{NO3}(recommended) = 1.37 × 10^{-16} cm^3 molecule^{-1} s^{-1}, k_{NO3}(calc) = 1.45 × 10^{-16} cm^3 molecule^{-1} s^{-1} at 296 ± 2 K (relative rate method, Aschmann & Atkinson 1995)

k_{OH}* = 7.02 × 10^{-12} cm^3 molecule^{-1} s^{-1}, k_{NO3} = 15 × 10^{-17} cm^3 molecule^{-1} s^{-1} at 298 K (recommended, Atkinson 1997)

Hydrolysis:
Biodegradation:

Biotransformation:

Bioconcentration, Uptake (k_1) and Elimination (k_2) Rate Constants or Half-Lives:

Half-Lives in the Environment:

  Air: photooxidation reaction rate constant of 7.15×10^{-12} cm^3 molecule^{-1} s^{-1} with hydroxyl radicals with an estimated lifetime of 19 h in summer daylight (Altshuller 1991).

TABLE 2.1.1.1.20.1
Reported aqueous solubilities of *n*-heptane at various temperatures

1.

Connolly 1966		Nelson & De Ligny 1968		Polak & Lu 1973	
shake flask-cloud point		shake flask-GC		shake flask-GC	
t/°C	S/g·m^{-3}	t/°C	S/g·m^{-3}	t/°C	S/g·m^{-3}
22	1.1	4.3	1.95	0	4.39
57	3.3	13.5	2.02	25	3.37
77	3.7	25.0	2.66		
82	10.3	35.0	2.27		

2.

Price 1976		Jonsson et al. 1982		Shaw 1989a		Mączyński et al. 2004	
shake flask-GC		vapor saturation-GC		IUPAC "tentative" best values		calc-recommended LLE data	
t/°C	S/g·m^{-3}	t/°C	S/g·m^{-3}	t/°C	S/g·m^{-3}	t/°C	S/g·m^{-3}
25.0	2.24	15	2.67	0	3.0	0	4.18
40.1	2.63	20	2.57	10	2.0	15	3.23
55.7	3.11	25	2.51	20	2.4	20	3.06
99.1	5.60	30	2.49	25	2.4	25	2.95
118.0	11.4	35	2.52	30	2.40	30	2.90
136.6	27.3			40	2.5	35	2.90
150.4	43.7			50	2.9	40.1	2.95
				60	3.3	45	3.01
				90	3.9	55.7	3.34
				100	5.8	99.1	7.79
				120	13	118	12.8
				140	31	136.6	22.8
						150.4	36.7
				ΔH_{sol}/(kJ mol^{-1}) = –2.43			

Aliphatic and Cyclic Hydrocarbons

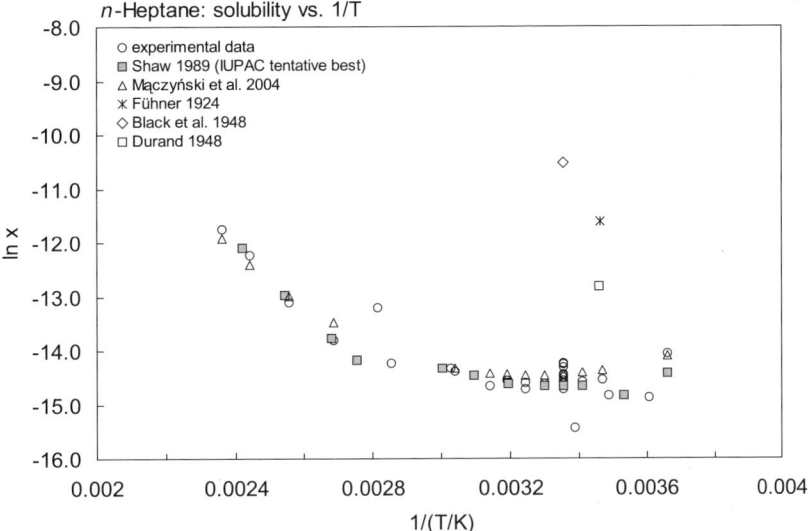

FIGURE 2.1.1.1.20.1 Logarithm of mole fraction solubility (ln x) versus reciprocal temperature for *n*-heptane.

TABLE 2.1.1.1.20.2
Reported vapor pressures of *n*-heptane at various temperatures and the coefficients for the vapor pressure equations

$\log P = A - B/(T/K)$ (1) $\ln P = A - B/(T/K)$ (1a)
$\log P = A - B/(C + t/°C)$ (2) $\ln P = A - B/(C + t/°C)$ (2a)
$\log P = A - B/(C + T/K)$ (3)
$\log P = A - B/(T/K) - C \cdot \log (T/K)$ (4)
$\ln [(P/kPa)/(P_0/kPa)] = [1 - (T_0/K)/(T/K)] \cdot \exp\{A_0 - A_1 \cdot (T/K) + A_2 \cdot (T/K)^2\}$ (5) - Cox eq.

1.

Smyth & Engel 1929		Willingham et al. 1945		Stull 1947		Forziati et al. 1949	
static method-manometer		ebulliometry		summary of literature data		ebulliometry	
t/°C	P/Pa	t/°C	P/Pa	t/°C	P/Pa	t/°C	P/Pa
22.7	5520	25.925	6370	−34.0	133.3	26.030	6425
30.0	7759	29.699	7665	−12.7	666.6	29.813	7717
30.3	7839	33.024	8977	−2.1	1333	33.108	9013
38.4	11466	36.017	10311	9.5	2666	36.105	10352
50.0	18812	38.822	11700	22.3	5333	38.901	11740
51.2	19998	42.599	13821	30.6	7999	42.680	13866
61.2	29331	46.929	16604	41.8	13332	46.987	16659
69.5	39703	51.320	19920	58.7	26664	51.373	19966
70.0	40183	55.394	23443	78.0	53329	55.442	23499
79.5	55955	60.862	28950	98.4	101325	60.902	29003
98.4	101325	65.882	34897			65.916	34948
		71.930	43326	mp/°C	−90.6	71.966	43386
mp/°C	−90.5	78.160	53647			78.202	53270
bp/°C	98.4	84.823	66749			84.856	66815
		92.053	83706			92.678	83769
		97.154	97590			97.180	97663
		97.702	99186			97.728	99257
		98.207	100689			98.237	100758

(Continued)

TABLE 2.1.1.1.20.2 *(Continued)*

Smyth & Engel 1929		Willingham et al. 1945		Stull 1947		Forziati et al. 1949	
static method-manometer		ebulliometry		summary of literature data		ebulliometry	
t/°C	P/Pa	t/°C	P/Pa	t/°C	P/Pa	t/°C	P/Pa
		98.773	102393			98.813	102476
		99.285	103907			99.322	104018
		bp/°C	98.426			bp/°C	98.427
		eq. 2	P/mmHg				
		A	6.90342			eq. 2	P/mmHg
		B	1268.636			A	6.90027
		C	216.951			B	1266.871
		temp range: 25.9–99.3°C				C	216.757
						temp range: 26.0–99.3°C	

2.

Zwolinski & Wilhoit 1971		Carruth & Kobayashi 1973		Ruzicka & Majer 1994		Weber 2000	
selected values		gas saturation		recommended		ebulliometry	
t/°C	P/Pa	t/°C	P/Pa	t/°C	P/Pa	T/K	P/Pa
–2.22	1333	–87.85	0.288	–40	76.3	355.899	62503
9.43	2666	–80.41	0.867	–20	393	359.210	69482
16.78	4000	–68.21	4.240	0	1520	363.492	79475
22.29	5333	–55.70	16.40	20	4730	367.925	89500
26.75	6666	–42.11	57.46	25	6102	370.925	99446
30.519	7999	–27.95	196	40	12300	374.219	109546
36.706	10666	–14.0	595	60	28100	377.268	119434
41.723	13332	1.01	1667	80	57100	380.130	129452
51.366	19998	11.92	2986	100	106000	382.836	139321
58.656	26664	22.46	5080			386.617	154371
64.591	33331			OR		390.085	169321
69.632	39997	mp/°C	–90.60	T/K	P/Pa	394.397	189252
77.966	53329			179.40	0.1	335.188	30303
84.772	66661	eq. 1a	P/mmHg	194.52	1.0	339.317	35308
90.571	79993	A	20.1590	212.98	10	342.996	40310
95.650	93326	B	4852.65	236.18	100	349.358	50269
96.597	95992			266.53	1000	352.183	55286
97.523	98659			308.48	10000	354.804	60285
98.429	101325			371.55	101325	357.252	65269
				298.15	6102	359.559	70277
eq. 2	P/mmHg					363.894	60452
A	6.89385			data calc. from Cox eq.		367.733	90423
B	1264.37			eq. 5	P/kPa	371.255	100423
C	216.636			A_0	2.86470	374.510	110415
bp/°C	98.429			A_1	2.113204×10^{-3}	377.544	120394
ΔH_V/(kJ mol^{-1}) =				A_2	2.250991×10^{-6}	380.390	130382
at 25°C	36.55			with reference state at		383.069	140381
at bp	31.70			P_o/kPa	101.325	386.826	155364
				T_o/K	371.552	390.313	170377
						394.609	190385
						to	
						503.406	1597746
						see ref. for complete set of data	

Aliphatic and Cyclic Hydrocarbons

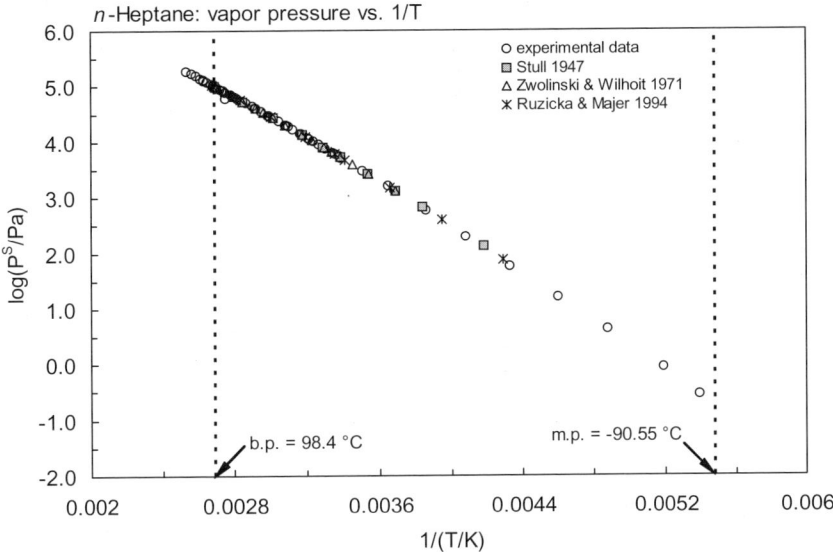

FIGURE 2.1.1.1.20.2 Logarithm of vapor pressure versus reciprocal temperature for *n*-heptane.

TABLE 2.1.1.1.20.3
Reported Henry's law constants and octanol-air water partition coefficients of *n*-heptane at various temperatures and temperature dependence equations

$$\ln K_{AW} = A - B/(T/K) \quad (1)$$
$$\ln (1/K_{AW}) = A - B/(T/K) \quad (2)$$
$$\ln (k_H/atm) = A - B/(T/K) \quad (3)$$
$$\ln [H/(Pa\ m^3/mol)] = A - B/(T/K) \quad (4)$$
$$K_{AW} = A - B \cdot (T/K) + C \cdot (T/K)^2 \quad (5)$$

$$\log K_{AW} = A - B/(T/K) \quad (1a)$$
$$\log (1/K_{AW}) = A - B/(T/K) \quad (2a)$$

$$\ln [H/(atm \cdot m^3/mol)] = A - B/(T/K) \quad (4a)$$

Henry's law constant				log K_{OA}	
Jönsson et al. 1982		Hansen et al. 1993		Gruber et al. 1997	
equilibrium cell-GC		EPICS-GC		GC det'd activity coefficient	
t/°C	H/(Pa m³/mol)	t/°C	H/(Pa m³/mol)	t/°C	log K_{OA}
15.0	136118*	26.0	91294	20.29	3.047
15.3	140244	35.8	121083	30.3	2.839
15.3	136260	45.0	193024	40.4	2.654
20.0	184640*			50.28	2.510
20.05	190443	eq. 4	H/(kPa m³/mol)		
20.05	187513	A	17 ± 2.22		
25.0	243021*	B	3730 ± 686		
25.05	193690				
25.05	190710				
25.04	250419				
29.8	302857				
29.8	316821				
30.0	315049*				
35.05	389419				
34.83	405792				
34.83	379903				
35.0	394148*				

* interpolated data

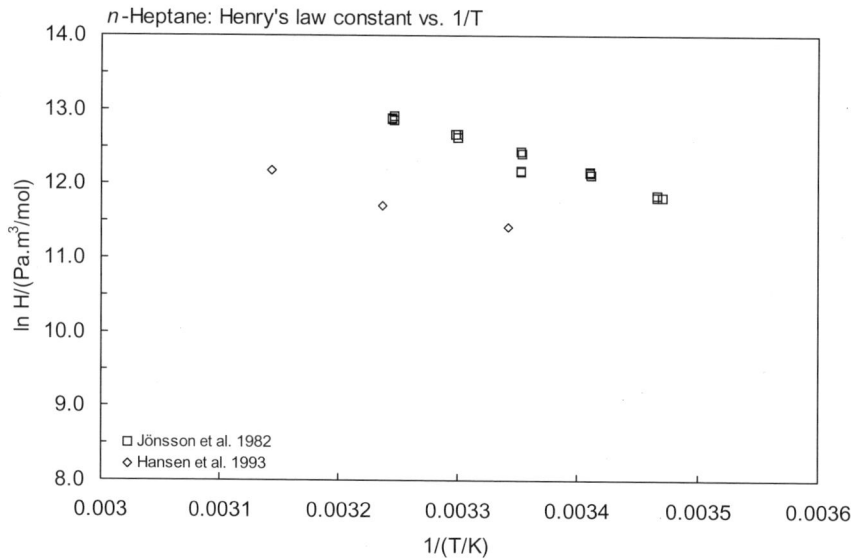

FIGURE 2.1.1.1.20.3 Logarithm of Henry's law constant versus reciprocal temperature for *n*-heptane.

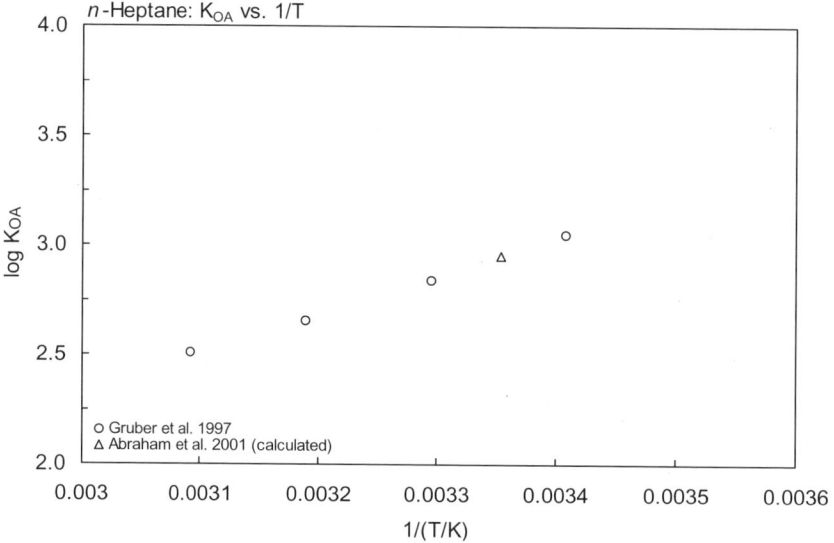

FIGURE 2.1.1.1.20.4 Logarithm of K_{OA} versus reciprocal temperature for *n*-heptane.

Aliphatic and Cyclic Hydrocarbons

2.1.1.1.21   2-Methylheptane

Common Name: 2-Methylheptane
Synonym:
Chemical Name: 2-methylheptane
CAS Registry No: 592-27-8
Molecular Formula: C_8H_{18}; $CH_3CH(CH_3)(CH_2)_4CH_3$
Molecular Weight: 114.229
Melting Point (°C):
  −109.02   (Lide 2003)
Boiling Point (°C):
  117.66   (Lide 2003)
Density (g/cm³ at 20°C):
  0.698   (Lide 2003)
Molar Volume (cm³/mol):
  163.7   (20°C, calculated-density, Stephenson & Malanowski 1987; Ruelle & Kesselring 1997)
  185.0   (calculated-Le Bas method at normal boiling point)
Enthalpy of Vaporization, ΔH_V (kJ/mol):
  39.68, 33.60 (25°C, normal bp, Dreisbach 1959)
Enthalpy of Fusion, ΔH_{fus} (kJ/mol):
  11.92   (Chickos et al. 1999)
Entropy of Fusion, ΔS_{fus} (J/mol K):
  72.62, 64.9   (exptl., calculated-group additivity method, Chickos et al. 1999)
Fugacity Ratio at 25°C, F: 1.0

Water Solubility (g/m³ or mg/L at 25°C):
  0.95, 4.55   (quoted, calculated-molar volume V_M, Wang et al. 1992)
  2.84; 5.94   (quoted exptl.; calculated-group contribution method, Kühne et al. 1995)
  0.95; 1.61, 1.61   (quoted exptl.; calculated-molar volume, mp and mobile order thermodynamics, Ruelle & Kesselring 1997)

Vapor Pressure (Pa at 25°C and reported temperature dependence equations):
  log (P/mmHg) = 6.91737 − 1337.468/(213.693 + t/°C); temp range 41.7–118.5°C (Antoine eq. from exptl. data, ebulliometry-manometer, Willingham et al. 1945)
  2620   (calculated-Antoine eq. regression, temp range −21 to 117.6°C Stull 1947)
  6386   (23.4°C, Nicolini & Laffitte 1949)
  2748   (extrapolated-Antoine eq., Dreisbach 1959)
  log (P/mmHg) = 6.91735 − 1337.468/(213.693 + t/°C); temp range 35–150°C (Antoine eq. for liquid state, Dreisbach 1959)
  2748   (interpolated-Antoine eq., temp range 12.3–143.8°C, Zwolinski & Wilhoit 1971)
  log (P/mmHg) = 6.91735 − 1337.468/(213.693 + t/°C); temp range 12.3–143.8°C (Antoine eq., Zwolinski & Wilhoit 1971)
  2620   (interpolated-Antoine eq., temp range −21–117.6°C, Weast 1972–73)
  log (P/mmHg) = [−0.2185 × 9362.0/(T/K)] + 8.154424; temp range −21–117.6°C (Antoine eq., Weast 1972–73)
  2732, 6850 (calculated-Antoine eq., Boublik et al. 1973, 1984)
  log (P/mmHg) = 6.88814 − 1319.539/(211.625 + t/°C); temp range 41–118.5°C (Antoine eq. from reported exptl. data of Willingham et al. 1945, Boublik et al. 1973)
  log (P/mmHg) = 6.85999 − 1313.125/(230.02 + t/°C); temp range 23.4–75°C (Antoine eq. from reported exptl. data of Nicolini & Laffitte 1949, Boublik et al. 1973)
  1161   (10°C, static method-inclined piston manometer, measured range −40 to 10°C, Osborn & Douslin 1974)

2750   (extrapolated-Antoine eq., temp range 42–119°C, Dean 1985, 1992)

log (P/mmHg) = 6.91735 – 1337.47/(213.69 + t/°C); temp range 42–119°C (Antoine eq., Dean 1985, 1992)

2750   (interpolated-Antoine eq., Stephenson & Malanowski 1987)

log (P_L/kPa) = 6.05858 – 1346.996/(–30.648 + T/K); temp range 285–392 K (Antoine eq.-I, Stephenson & Malanowski 1987)

log (P_L/kPa) = 6.81199 – 1703.6/(–30.648 + T/K); temp range 233–286 K (Antoine eq.-II, Stephenson & Malanowski 1987)

log (P/mmHg) = 37.693 – 3.2611 × 10^3/(T/K) – 10.391·log (T/K) – 1.0524 × 10^{-12}·(T/K) + 3.056 × 10^{-6}·(T/K)^2; temp range 164–560 K (vapor pressure eq., Yaws 1994)

Henry's Law Constant (Pa m^3/mol at 25°C):

369880   (calculated-vapor-liquid equilibrium (VLE) data, Yaws et al. 1991)

Octanol/Water Partition Coefficient, log K_{OW}:

3.91   (calculated-regression eq. from Lyman et al. 1982, Wang et al. 1992)
4.04   (calculated-V_M, Wang et al. 1992)

Octanol/Air Partition Coefficient, log K_{OA}:

Bioconcentration Factor, log BCF:

Sorption Partition Coefficient, log K_{OC}:

Environmental Fate Rate Constants, k, and Half-Lives, $t_{½}$:

Volatilization:
Photolysis:
Oxidation: rate constant k, for gas-phase second order rate constants, k_{OH} for reaction with OH radical, k_{NO3} with NO_3 radical and k_{O3} with O_3 or as indicated, *data at other temperatures see reference:
k_{OH}(calc) = 8.20 × 10^{-12} cm^3 molecule^{-1} s^{-1} (SAR structure reactivity relationship, Atkinson 1987)
k_{OH} = 8.20 × 10^{-12} cm^3 molecule^{-1} s^{-1} with a calculated atmospheric lifetime τ = 17 h during summer daylight hours (Altshuller 1991).
Hydrolysis:
Biodegradation:
Biotransformation:
Bioconcentration, Uptake (k_1) and Elimination (k_2) Rate Constants or Half-Lives:

Half-Lives in the Environment:

Air: photooxidation reaction rate constant of 8.20 × 10^{-12} cm^3 molecule^{-1} s^{-1} with hydroxyl radical and an estimated lifetime of 17 h during summer daylight (Altshuller 1991).

Aliphatic and Cyclic Hydrocarbons

2.1.1.1.22   3-Methylheptane

Common Name: 3-Methylheptane
Synonym:
Chemical Name: 3-methylheptane
CAS Registry No: 589-81-1
Molecular Formula: C_8H_{18}; $CH_3CH_2CH(CH_3)(CH_2)_3CH_3$
Molecular Weight: 114.229
Melting Point (°C):
    –120.48   (Lide 2003)
Boiling Point (°C):
    118.9   (Lide 2003)
Density (g/cm³ at 20°C):
    0.7075   (Weast 1984)
    0.7058, 0.7018   (20°C, 25°C, Dreisbach 1959)
Molar Volume (cm³/mol):
    161.4   (20°C, calculated-density, Stephenson & Malanowski 1987)
    185.0   (calculated-Le Bas method at normal boiling point)
Enthalpy of Vaporization, ΔH_V (kJ/mol):
    39.83, 34.08   (25°C, normal bp, Dreisbach 1959)
Enthalpy of Fusion, ΔH_{fus} (kJ/mol):
    11.38   (Dreisbach 1959)
    11.7   (Chickos et al. 1999)
Entropy of Fusion, ΔS_{fus} (J/mol K):
    76.6, 64.9 (exptl., calculated-group additivity method, Chickos et al. 1999)
Fugacity Ratio at 25°C, F: 1.0

Water Solubility (g/m³ or mg/L at 25°C):
    0.792   (shake flask-GC, Price 1976; quoted, Mackay & Shiu 1981; Shaw 1989; Myrdal et al. 1992)
    0.850   (estimated-nomograph, Brookman et al. 1985)
    1.015   (calculated-recommended liquid-liquid equilibrium LLE data, Mączyński et al. 2004)

Vapor Pressure (Pa at 25°C or as indicated and reported temperature dependence equations):
    log (P/mmHg) = 6.89945 – 1331.530/(212.414 + t/°C); temp range 42.7–119.8°C (Antoine eq. from exptl. data, ebulliometry, manometer, Willingham et al. 1945)
    2610   (calculated from determined data, Dreisbach 1959)
    log (P/mmHg) = 6.89944 – 1331.530/(212.414 + t/°C); temp range 30–150°C (Antoine eq. for liquid state, Dreisbach 1959)
    2600   (interpolated-Antoine eq., temp range 13.3–145.2°C, Zwolinski & Wilhoit 1971)
    log (P/mmHg) = 6.89944 – 1331.530/(212.414 + t/°C); temp range 13.3–145.2°C (Antoine eq., Zwolinski & Wilhoit 1971)
    2466   (interpolated-Antoine eq., temp range –19.8 to 118.9°C, Weast 1972–73)
    log (P/mmHg) = [–0.2185 × 9432.0/(T/K)] + 8.179407; temp range –19.8 to 118.9°C (Antoine eq., Weast 1972–73)
    1486   (15°C, static method-inclined piston manometer, measured range –35 to 15°C, Osborn & Douslin 1974)
    2600, 3232 (quoted, calculated-bp, Mackay et al. 1982)
    2600   (extrapolated-Antoine eq., Boublik et al. 1984)
    log (P/kPa) = 6.01647 – 1326.329/(211.776 + t/°C); temp range 42.67–119.8°C (Antoine eq. from reported exptl. data of Willingham et al. 1945, Boublik et al. 1984)
    2605   (extrapolated, Antoine eq., temp range 43–120°C Dean 1985, 1992)
    log (P/mmHg) = 6.89944 – 1331.53/(212.41 + t/°C); temp range 43–120°C (Antoine eq., Dean 1985, 1992)

2630 (interpolated-Antoine eq.-I, Stephenson & Malanowski 1987)

log $(P_L/kPa) = 6.02047 - 1329.42/(-60.945 + T/K)$; temp range 286–393 K (Antoine eq.-I, Stephenson & Malanowski 1987)

log $(P_L/kPa) = 6.50909 - 1567.45/(-40.786 + T/K)$; temp range 238–286 K (Antoine eq.-II, Stephenson & Malanowski 1987)

log $(P/mmHg) = 52.8828 - 3.6231 \times 10^3/(T/K) - 16.804 \cdot \log(T/K) + 7.1828 \times 10^{-3} \cdot (T/K) + 7.4077 \times 10^{-14} \cdot (T/K)^2$; temp range 153–564 K (vapor pressure eq., Yaws 1994)

Henry's Law Constant (Pa m^3/mol):
- 376000 (calculated-P/C, Mackay & Shiu 1981)
- 375900 (selected, Mills et al. 1982)
- 375800 (calculated-vapor-liquid equilibrium (VLE) data, Yaws et al. 1991)

Octanol/Water Partition Coefficient, log K_{OW}:

Octanol/Air Partition Coefficient, log K_{OA}:

Bioconcentration Factor, log BCF:

Sorption Partition Coefficient, log K_{OC}:

Environmental Fate Rate Constants, k, and Half-Lives, $t_{½}$:

Volatilization:

Photolysis:

Oxidation: rate constant k, for gas-phase second order rate constants, k_{OH} for reaction with OH radical, k_{NO3} with NO$_3$ radical and k_{O3} with O$_3$ or as indicated, *data at other temperatures see reference:

k_{OH}(calc) = 8.90×10^{-12} cm^3 molecule^{-1} s^{-1} at room temp. (SAR, Atkinson 1987)

k_{OH} = 8.90×10^{-12} cm^3 molecule^{-1} s^{-1} with a calculated atmospheric lifetime τ = 16 h during summer daylight hours (Altshuller 1991)

Hydrolysis:

Biodegradation:

Biotransformation:

Bioconcentration, Uptake (k_1) and Elimination (k_2) Rate Constants or Half-Lives:

Half-Lives in the Environment:

Air: photooxidation reaction rate constant of 8.90×10^{-12} cm^3 molecule^{-1} s^{-1} with hydroxyl radical with an estimated lifetime of 16 h during summer daylight (Altshuller 1991).

Aliphatic and Cyclic Hydrocarbons

2.1.1.1.23 n-Octane

Common Name: *n*-Octane
Synonym: octane
Chemical Name: *n*-octane
CAS Registry No: 111-65-9
Molecular Formula: C_8H_{18}; $CH_3(CH_2)_6CH_3$
Molecular Weight: 114.229
Melting Point (°C):
    –56.82 (Lide 2003)
Boiling Point (°C):
    125.67 (Lide 2003)
Density (g/cm³ at 20°C):
    0.7027, 0.6886 (20°C, 25°C, Riddick et al. 1986)
    0.70256 (20°C, digital precision densimeter, Dejoz et al. 1996)
Molar Volume (cm³/mol):
    162.6, 165.8 (20°C, 25°C, calculated-density)
    185.0 (calculated-Le Bas method at normal boiling point)
Enthalpy of Vaporization, ΔH_V (kJ/mol):
    41.49, 34.431 (25°C, bp, Riddick et al. 1986)
    41.56 (298.15 K, recommended, Ruzicka & Majer 1994)
Enthalpy of Fusion, ΔH_{fus} (kJ/mol):
    20.74 (Dreisbach 1959; Riddick et al. 1986; Chickos et al. 1999)
Entropy of Fusion, ΔS_{fus} (J/mol K):
    95.86, 91.1 (exptl., calculated-group additivity method, Chickos et al. 1999)
Fugacity Ratio at 25°C, F: 1.0

Water Solubility (g/m³ or mg/L at 25°C or as indicated and reported temperature dependence equations. Additional data at other temperatures designated * are compiled at the end of this section.):
    14.0 (cloud point, Fühner 1924, quoted, Deno & Berkheimer 1960)
    0.66 (shake flask-GC, McAuliffe 1963, 1966)
    0.493 (radiotracer method, Baker 1967)
    0.880* (shake flask-GC, measured range 5–25°C, Nelson & De Ligny 1968)
    0.700 (shake flask-GC, Krzsnoshchekova & Gubergrits 1973)
    1.35; 0.85 (0, 25°C, shake flask-GC, Polak & Lu 1973)
    0.431* (shake flask-GC, measured range 25149.5°C, Price 1976)
    0.615* (vapor saturation-GC, measured range 15–35°C, Jönsson et al. 1982)
    1.103 (generator column-GC, Tewari et al. 1982a; Wasik et al. 1982)
    1.56 (calculated-activity coeff. and K_{OW}, Tewari et al. 1982b)
    0.615 (vapor saturation-partition coefficient-GC, Jönsson et al. 1982)
    0.660 (shake flask-GC, Coates et al. 1985)
    0.762* (37.75°C, shake flask-GC, measured range 37.75–280°C, pressure range 0.0103–8.86 MPa, Heidmen et al. 1985)
    $\ln x = -343.1497 + 13862.49/(T/K) + 49.24600 \cdot \ln (T/K)$; temp range 37.75–280°C (shake flask-GC, Heidman et al. 1985)
    0.884, 0.949 (20°C, shake flask-GC, Burris & MacIntyre 1986)
    1.250 (shake flask-purge and trap-GC, Coutant & Keigley 1988)
    0.71* (recommended, temp range 0–100°C, IUPAC Solubility Data Series, Shaw 1989)
    $\ln x = -415.7563 + 17975.386/(T/K) + 59.55451 \cdot \ln (T/K)$; temp range 290–400 K (eq. derived from literature calorimetric and solubility data, Tsonopoulos 1999)
    0.774* (29.9°C, shake flask-solid extraction-GC/FID, measured range 29.9–183°C, Marche et al. 2003)
    $\ln x = -362.618 + 14904.474/(T/K) + 52.067 \cdot \ln (T/K)$: temp range 29.9–183°C (shake flask-solid extraction-GC/FID measurements, Marche et al. 2003)

0.635*   (calculated-liquid-liquid equilibrium LLE data, temp range 288.2–36.1 K, Mączyński et al. 2004)
0.807   (24.9°C, generator column-GC/FID, measured range 10–45°C, Sarraute et al. 2004)

Vapor Pressure (Pa at 25°C or as indicated and reported temperature dependence equations. Additional data at other temperatures designated * are compiled at the end of this section.):

486.6*   (3.7°C, static-McLeod gauge, measured range –9.31 to 3.7°C, Linder 1931)
7670*   (52.972°C, ebulliometry, measured range 52.972–126.570°C, Willingham et al. 1945)
log (P/mmHg) = 6.92377 – 1355.126/(209.517 + t/°C); temp range 52.9–126.6°C (Antoine eq. from exptl. data, ebulliometry-manometer, Willingham et al. 1945)
1777*   (calculated-Antoine eq. regression, temp range –14 to 125.6°C, Stull 1947)
1870   (calculated from determined data, Dreisbach 1959)
log (P/mmHg) = 6.92377 – 1355.126/(209.517 + t/°C); temp range 40–155°C (Antoine eq. for liquid state, Dreisbach 1959)
1885*   (interpolated-Antoine eq., temp range 19.2–152.1°C, Zwolinski & Wilhoit 1971
log (P/mmHg) = 6.92377 – 11355.126/(209.517 + t/°C); temp range 19.2–152.1°C (Antoine eq., Zwolinski & Wilhoit 1971)
1825   (interpolated-Antoine eq., temp range –14 to 281.4°C, Weast 1972–73)
log (P/mmHg) = [–0.2185 × 9221.0/(T/K)] + 7.894018; temp range –14 to 281.4°C (Antoine eq., Weast 1972–73)
1573*   (23.96°C, gas saturation, measured range –56.35 to 23.96°C, Carruth & Kobayashi 1973)
1885, 2060 (quoted, calculated-bp, Mackay et al. 1982)
1860   (extrapolated-Antoine eq., Boublik et al. 1984)
log (P/kPa) = 6.04394 – 1351.938/(209.12 + t/°C); temp range 52.93–126.57°C (Antoine eq. from reported exptl. data of Willingham et al. 1945, Boublik et al. 1984)
1860   (interpolated-Antoine eq., temp range 19–152°C, Dean 1985, 1992)
log (P/mmHg) = 6.91868 – 1351.99/(209.15 + t/°C); temp range 19–152°C (Antoine eq., Dean 1985, 1992)
1870   (lit. average, Riddick et al. 1986)
log (P/kPa) = 6.04358 – 1351.99/(209.155 + t/°C); temp range not specified (Antoine eq., Riddick et al. 1986)
1854, 1814, 1854 (headspace-GC, correlated, Antoine eq., Hussam & Carr 1985)
1862   (interpolated-Antoine eq., Stephenson & Malanowski 1987)
log (P_L/kPa) = 6.04231 – 1351.491/(–64.014 + T/K); temp range 297–400 K (Antoine eq.-I, Stephenson & Malanowski 1987)
log (P_L/kPa) = 7.90115 – 2238.9/(–4.53 + T/K); temp range 216–278 K (Antoine eq.-II, Stephenson & Malanowski 1987)
log (P_L/kPa) = 6.16936 – 1440.32/(–52.894 + T/K); temp range 396–432 K (Antoine eq.-III, Stephenson & Malanowski 1987)
log (P_L/kPa) = 6.23406 – 1492.068/(–45.851 + T/K); temp range 428–510 K (Antoine eq.-IV, Stephenson & Malanowski 1987)
log (P_L/kPa) = 7.66614 – 7.66614/(159.091 + T/K); temp range 506–569 K (Antoine eq.-V, Stephenson & Malanowski 1987)
1872*   (recommended, Ruzicka & Majer 1994)
ln [(P/kPa)/(P_o/kPa)] = [1 – (T_o/K)/(T/K)]·exp{2.90150 – 2.046204 × 10^{-3}·(T/K) + 2.010759 × 10^{-6}·(T/K)^2}; reference state at P_o = 101.325 kPa, T_o = 398.793 K (Cox equation, Ruzicka & Majer 1994)
log (P/mmHg) = 29.0948 – 3.0114 × 10^3/(T/K) –7.2653·log (T/K) – 2.2696 × 10^{-11}·(T/K) + 1.468 × 10^{-6}·(T/K)^2; temp range 216–569 K (vapor pressure eq., Yaws 1994)
1820*   (24.6°C, ebulliometry, measured range 291.25–409.95 K, Dejoz et al. 1996)
ln (P/Pa) = 13.9183 – 3114.43/[(T/K) – 63.9225]; temp range 291–409 K (ebulliometry, Dejoz et al. 1996)
6727*   (50.43°C, comparative ebulliometry, measured range 323–563 K, data fitted to Wagner eq., Ewing & Ochoa 2003)

Henry's Law Constant (Pa m^3/mol at 25°C or as indicated and reported temperature dependence equations. Additional data at other temperatures designated * are compiled at the end of this section.):

326800   (calculated-1/K_{AW}, C_W/C_A, reported as exptl., Hine & Mookerjee)
311900, 110670   (calculated-group contribution, calculated-bond contribution, Hine & Mookerjee 1975)
325300   (calculated-P/C, Mackay & Leinonen 1975; Mackay & Shiu 1990)
323200   (calculated-P/C, Bobra et al. 1979; Mackay et al. 1979; Mackay 1981)

Aliphatic and Cyclic Hydrocarbons

325000 (calculated-P/C, Mackay & Shiu 1981)

355500* (25.1°C, equilibrium cell-concentration ratio-GC, measured range 14.8–34.92°C, Jönsson et al. 1982)

180130, 253880, 344280, 458250, 595800 (15, 20, 25, 30, 35°C, calculated-temp dependence eq. derived from exptl. data, Jönsson et al. 1982)

$\ln(1/K_{AW}) = 26003.7/(T/K) + 70.571 \cdot \ln(T/K) - 494.151$; temp range: 15–35°C (least-square regression of equilibrium cell-concn ratio-GC measurements, Jönsson et al. 1982)

324200 (calculated-P/C, Lyman et al. 1982)

$\ln(k_H/MPa) = 357.733 - 19363.1/(T/K) - 9.04865 \cdot (T/K)^2 - 49.5296 \cdot \ln(T/K)$; maximum $k_H = 7.836 \times 10^4$ MPa at 372.1 K (Heidman et al. 1985)

314700 (calculated-P/C, Eastcott et al. 1988)

326200 (Valsaraj 1988)

201500 (calculated-MCI χ, Nirmalakhandan & Speece 1988)

499500 (calculated-vapor-liquid equilibrium (VLE) data, Yaws et al. 1991)

39213* (27.9°C, EPICS-GC, measured range 27.9–45°C, Hansen et al. 1993)

$\ln[H/(kPa \cdot m^3/mol)] = -8014/(T/K) + 30.0$; temp range: 27.9–45°C (EPICS-GC, Hansen et al. 1993)

297951 (EPICS-GC, Ryu & Park 1999)

21838 (20°C, selected from reported experimental determined values, Staudinger & Roberts 1996, 2001)

$\log K_{AW} = 12.08 - 3263/(T/K)$ (summary of literature data, Staudinger & Roberts 2001)

257797 (24.9°C, calculated-P/C, Sarraute et al. 2004)

Octanol/Water Partition Coefficient, log K_{OW}:

4.0 (calculated-π substituent const., Hansch et al. 1968; Hansch & Leo 1979)

5.29 (calculated-activity coefficient γ, Wasik et al. 1981, 1982)

5.18 (generator column-GC, Tewari et al. 1982a,b)

4.0 (HPLC-k' correlation, Coates et al. 1985)

5.24 (generator column-GC, calculated-activity coefficient γ, Schantz & Martire 1987)

5.15 ± 0.45 (recommended, Sangster 1989; 1993)

5.61 (calculated-activity coefficients, Tse & Sandler 1994)

5.18 (recommended, Hansch et al. 1995)

Octanol/Air Partition Coefficient, log K_{OA} at 25°C or as indicated. Additional data at other temperatures designated * are compiled at the end of this section:

3.55* (20.29°C, from GC-determined γ° in octanol, measured range 20.29–50.28°C, Gruber et al. 1997)

3.30 (calculated-measured γ° in pure octanol and vapor pressure P, Abraham et al. 2001)

Bioconcentration Factor, log BCF:

Sorption Partition Coefficient, log K_{OC}:

Environmental Fate Rate Constants, k, and Half-Lives, $t_{1/2}$:

Volatilization: $t_{1/2}$ = 5.55 h for a water column of 1 m² minimum cross section of depth 1 m (Mackay & Leinonen 1975)

estimated $t_{1/2}$ ~ 3.1 h at 20°C in a river 1 m deep flowing at 1 m s⁻¹ and with a wind velocity of 3 m s⁻¹ (Lyman et al. 1982).

Photolysis:

Oxidation: rate constant k, for gas-phase second order rate constants, k_{OH} for reaction with OH radical, k_{NO3} with NO_3 radical and k_{O3} with O_3 or as indicated, *data at other temperatures and/or the Arrhenius expression see reference:

k_{OH}*(exptl) = 6.03 × 10¹² cm³ mol⁻¹ s⁻¹, k_{OH}(calc) = 2.35 × 10¹² cm³ mol⁻¹ s⁻¹ at 296 K, measured range 296–497 K (flash photolysis-kinetic spectroscopy, Greiner 1970)

$k_{O(3P)}$ = 1.70 × 10⁻¹³ cm³ molecule⁻¹ s⁻¹ for the reaction with O(³P) atom at room temp. (Herron & Huie 1973)

k_{OH}(exptl) = 8.42 × 10⁻¹² cm³ molecule⁻¹ s⁻¹ at 295 K, k_{OH}(calc) = 7.35 × 10⁻¹² cm³ molecule⁻¹ s⁻¹ at 300 K (Darnall et al. 1978)

k_{OH} = (9.01 ± 0.19) × 10^{-12} cm^3 $molecule^{-1}$ s^{-1} at 299 ± 2 K (relative rate method, Atkinson et al. 1982a, 1984c)

k_{OH} = 8.80 × 10^{-12} cm^3 $molecule^{-1}$ s^{-1} at 312 K in smog chamber (Nolting et al. 1988)

k_{OH}* = 8.68 × 10^{-12} cm^3 $molecule^{-1}$ s^{-1} at 298 K (recommended, Atkinson 1989)

k_{OH} = 8.68 × 10^{-12} cm^3 $molecule^{-1}$ s^{-1}, k_{NO3} = 1.81 × 10^{-16} cm^3 $molecule^{-1}$ s^{-1} at 296 K (Atkinson 1990)

k_{OH} = 8.71 × 10^{-12} cm^3 $molecule^{-1}$ s^{-1}, k_{NO3} = 1.81 × 10^{-16} cm^3 $molecule^{-1}$ s^{-1} (Sabljic & Güsten 1990)

k_{OH} = 8.68 × 10^{-12} cm^3 $molecule^{-1}$ s^{-1}, estimated atmospheric lifetime of 16 h (Altshuller 1991)

k_{NO3} = 1.82 × 10^{-16} cm^3 $molecule^{-1}$ s^{-1} at 296 ± 2 K (Atkinson 1991)

k_{OH} = 8.68 × 10^{-12} cm^3 $molecule^{-1}$ s^{-1} (Paulson & Seinfeld 1992)

k_{NO3}(exptl) = 1.84 × 10^{-16} cm^3 $molecule^{-1}$ s^{-1}, k_{NO3}(calc) = 1.79 × 10^{-16} cm^3 $molecule^{-1}$ s^{-1} at 296 ± 2 K (relative rate method, Aschmann & Atkinson 1995)

k_{OH}* = 8.71 × 10^{-12} cm^3 $molecule^{-1}$ s^{-1}, k_{NO3}* = 1.9 × 10^{-16} cm^3 $molecule^{-1}$ s^{-1} at 298 K (recommended, Atkinson 1997)

Hydrolysis:

Biodegradation:

Biotransformation:

Bioconcentration, Uptake (k_1) and Elimination (k_2) Rate Constants or Half-Lives:

Half-Lives in the Environment:

Air: atmospheric $t_{1/2}$ ~ 2.4–24 h for C_4H_{10} and higher paraffins, based on the EPA Reactivity Classification of Organics (Darnall et al. 1976);
photooxidation reaction k = 8.68 × 10^{-12} cm^3 $molecule^{-1}$ s^{-1} with OH radicals with an estimated lifetime of 16 h in air during summer daylight (Altshuller 1991).

Surface water: volatilization $t_{1/2}$ = 5.55 h for a water column of 1 m^2 minimum cross section of depth 1 m (Mackay & Leinonen 1975); estimated volatilization $t_{1/2}$ = 3.1 h at 20°C in a river 1 m deep flowing at 1 m/s and with a wind velocity of 3 m/s (Lyman et al. 1982).

Ground water:

Sediment:

Soil:

Biota:

TABLE 2.1.1.1.23.1
Reported aqueous solubilities of *n*-octane at various temperatures

1.

Nelson & De Ligny 1968		Polak & Lu 1973		Price 1976		Jonsson et al. 1982	
shake flask-GC		shake flask-GC/FID		shake flask-GC		vapor saturation-GC	
t/°C	S/g·m^{-3}	t/°C	S/g·m^{-3}	t/°C	S/g·m^{-3}	t/°C	S/g·m^{-3}
5	1.65	0	1.35	25.0	0.431	15	0.653
15	0.89	25	0.85	40.1	0.524	20	0.628
25	1.84			69.7	0.907	25	0.615
				99.1	1.12	30	0.612
				121.3	4.62	35	0.620
				136.6	8.52		
				149.5	11.80		

Aliphatic and Cyclic Hydrocarbons

TABLE 2.1.1.1.23.1 (*Continued*)

2.

Heidman et al. 1985		Shaw 1989a		Marche et al. 2003		Mączyński et al. 2004	
shake flask-GC		IUPAC "tentative" best values		shake flask-GC		calc-recommended LLE data	
t/°C	S/g·m⁻³	t/°C	S/g·m⁻³	t/°C	S/g·m⁻³	t/°C	S/g·m⁻³
37.75	0.76	0	0.14	29.9	0.774	15	0.698
93.35	3	20	0.63	30.3	0.762	20	0.698
148.85	24	25	0.71	69.2	1.517	25	0.635
206.35	25	30	0.61	70	2.234	30	0.635
262.35	220	40	0.8	99.9	3.046	35	0.635
279.65	380	50	1	101.1	3.414	37.7	0.635
		60	1.2	124	7.806	40.1	0.952
pressures range from 0.0103 to 8.86 MPa		70	1.5	124	9.011	69.7	4.00
		80	2	131	11.23	121.3	6.98
		90	2.7	131	12.25	136.6	12.06
$\Delta H_{sol}/(kJ\ mol^{-1}) = 13.3$ 25°C		100	3.7	151.2	24.88	149.5	22.21
		120	7.2	165.1	43.79	262.9	50.77
		140	15	165.4	43.28		
		160	22	183	84.85		

$$\ln x = A + B/T + C \cdot \ln T$$
T in K
A   −362.618
B   14904.474
C   52.067

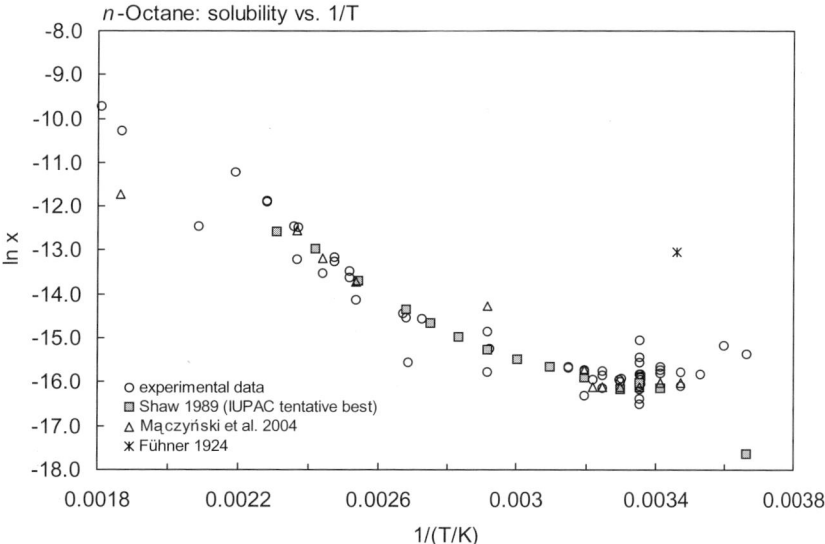

FIGURE 2.1.1.1.23.1 Logarithm of mole fraction solubility (ln *x*) versus reciprocal temperature for *n*-octane.

TABLE 2.1.1.1.23.2
Reported vapor pressures of *n*-octane at various temperatures and the coefficients for the vapor pressure equations

$\log P = A - B/(C + T/K)$ (1) $\ln P = A - B/(T/K)$ (1a)

$\log P = A - B/(C + t/°C)$ (2) $\ln P = A - B/(C + t/°C)$ (2a)

$\log P = A - B/(C + T/K)$ (3) $\ln P = A - B/(C + T/K)$ (3a)

$\log P = A - B/(T/K) - C \cdot \log (T/K)$ (4)   (4a)

$\ln [(P/kPa)/(P_o/kPa)] = [1 - (T_o/K)/(T/K)] \cdot \exp\{A_0 - A_1 \cdot (T/K) + A_2 \cdot (T/K)^2\}$ (5) - Cox eq.

1.

Linder 1931		Willingham et al. 1945		Stull 1947		Zwolinski & Wilhoit 1971	
McLeod gauge		ebulliometry		summary of literature data		selected values	
t/°C	P/Pa	t/°C	P/Pa	t/°C	P/Pa	t/°C	P/Pa
–9.31	196.0	52.972	7670	–14.0	133.3	19.03	1333
–3.0	309.3	56.456	8979	8.30	666.6	39.06	2666
3.7	486.6	59.615	10314	10.2	1333	39.10	4000
		62.592	11702	31.5	2666	44.95	5333
		66.587	13823	45.1	5333	49.68	6666
		71.163	16600	53.8	7999	53.67	7999
		75.820	19918	65.7	13332	60.24	10666
		80.134	23441	83.6	26664	65.56	13332
		85.916	28952	104.0	53329	75.79	19998
		91.230	34894	125.6	101325	83.52	26664
		97.635	43326			89.813	33331
		104.233	53646	mp/°C	–56.8	95.158	39997
		111.277	66742			103.991	53329
		118.924	83695			111.204	66661
		124.319	97584			117.349	79993
		124.809	99161			122.731	93326
		125.433	100662			123.734	95992
		126.035	102386			124.715	98659
		126.570	103900			125.675	101325
		bp/°C	123.665				
		eq. 2	P/mmHg			eq. 2	P/mmHg
		A	6.92377			A	6.90940
		B	1355.126			B	1349.82
		C	209.517			C	209.385
		temp range: 52.9–126.6°C				bp/°C	125.675
						$\Delta H_V/(kJ\,mol^{-1})$	
						at 25°C	41.49
						at bp	34.41

Aliphatic and Cyclic Hydrocarbons

TABLE 2.1.1.1.23.2 *(Continued)*

2.

Carruth & Kobayashi 1973		Ruzicka & Majer 1994		Dejoz et al. 1996		Ewing & Ochoa 2003	
gas saturation		recommended		vapor-liquid equilibrium		comparative ebulliometry	
t/°C	P/Pa	T/K	P/Pa	t/°C	P/Pa	t/°C	P/Pa
−56.35	2.40	194.67	0.1	18.1	1250	50.03	6727
−54.64	2.746	210.84	1.0	20.7	1450	51.811	7640
−49.92	4.40	230.53	10	24.69	1820	60.642	10766
−42.73	9.706	255.25	100	27.4	2130	67.754	14507
−35.08	20.13	287.52	1000	29.6	2410	73.195	18016
−23.06	59.33	332.03	10000	31.2	2630	79.794	23174
−13.75	127.7	398.79	101325	33.8	3020	88.950	32265
5.26	532.0	298.15	1872	36.8	3520	93.009	37123
16.41	1047			42.4	4670	99.293	45789
23.96	1573	data calc. from Cox eq.		48.6	6290	104.234	53686
		eq. 5	P/kPa	51.4	7150	110.649	65523
mp/°C	−56.81	A_0	2.90150	55.7	8690	115.266	75253
		A_1	2.046204×10^{-3}	58.9	10000	118.867	83613
eq. 1a	P/mmHg	A_2	2.010759×10^{-6}	61.4	11130	121.958	91358
A	20.3621	with reference state at		64.0	12410	125.705	101486
B	5294.36	P_o/kPa	101.325	66.7	13890	128.458	109480
		T_o/K	398.793	68.3	14820	131.853	130296
				71.8	17050	134.957	138.490
				74.9	19240		
				77.6	21330	Antoine eq. for P < 145 kPa	
				80.7	23900	in the following form	
				85.3	28330	ln P/kPa = (A−B)/(T− C)	
				90.05	34030	with	
				96.1	41180	A	13.91204
				100.5	47680	B	3110.704
				102.2	50300	C	64.086
				eq. 3(a)	kPa	for temp 323–563 K	
				A	13.9183	or pressure 6.7–2301 kPa	
				B	3114.43	data fitted to Wagner eq.	
				C	−63.925		
				for temp 291–409 K			

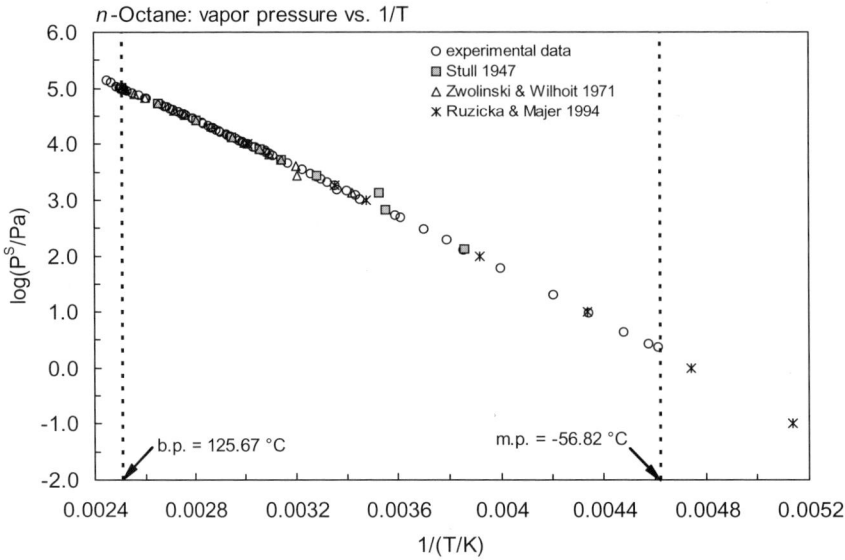

FIGURE 2.1.1.1.23.2 Logarithm of vapor pressure versus reciprocal temperature for *n*-octane.

TABLE 2.1.1.1.23.3
Reported Henry's law constants and octanol-air partition coefficients of *n*-octane at various temperatures and temperature dependence equations

$$\ln K_{AW} = A - B/(T/K) \quad (1)$$
$$\ln (1/K_{AW}) = A - B/(T/K) \quad (2)$$
$$\ln (k_H/atm) = A - B/(T/K) \quad (3)$$
$$\ln [H/(Pa\ m^3/mol)] = A - B/(T/K) \quad (4)$$
$$K_{AW} = A - B \cdot (T/K) + C \cdot (T/K)^2 \quad (5)$$

$$\log K_{AW} = A - B/(T/K) \quad (1a)$$
$$\log (1/K_{AW}) = A - B/(T/K) \quad (2a)$$

$$\ln [H/(atm \cdot m^3/mol)] = A - B/(T/K) \quad (4a)$$

Henry's law constant				log K_{OA}	
Jönsson et al. 1982		Hansen et al. 1993		Gruber et al. 1997	
equilibrium cell-GC		EPICS-GC		GC det'd activity coefficient	
t/°C	H/(Pa m³/mol)	t/°C	H/(Pa m³/mol)	t/°C	log K_{OA}
14.8	171001	27.9	39213	20.29	3.554
14.8	176031	35.0	93827	30.3	3.302
15.0	180126*	45.0	167693	40.4	3.089
20.0	253880*			50.28	2.927
20.05	259326	eq. 4	H/(kPa m³/mol)		
20.05	256596	A	30 ± 5.25		
25.0	344280*	B	8014 ± 1617		
25.1	357298				
25.1	353730				
30.0	458253*				
30.1	452643				
34.92	587452				
35.0	595804*				

* interpolated data

Aliphatic and Cyclic Hydrocarbons

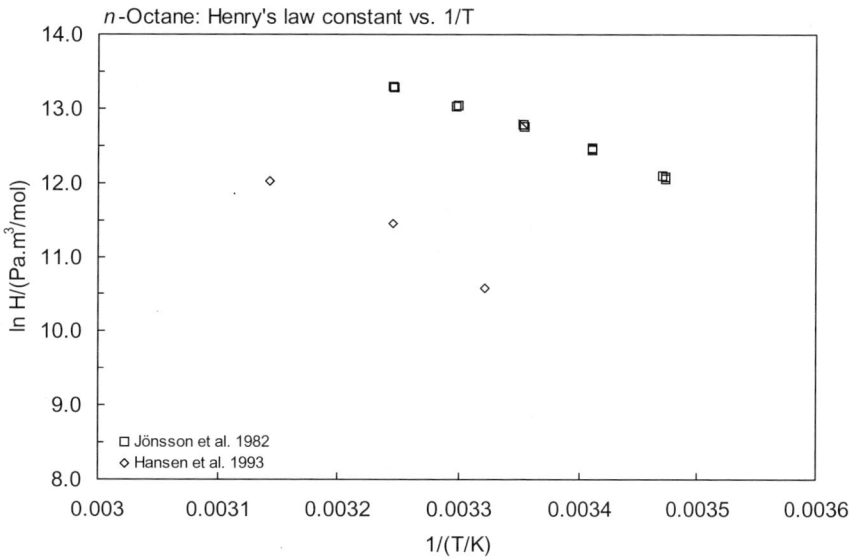

FIGURE 2.1.1.1.23.3 Logarithm of Henry's law constant versus reciprocal temperature for *n*-octane.

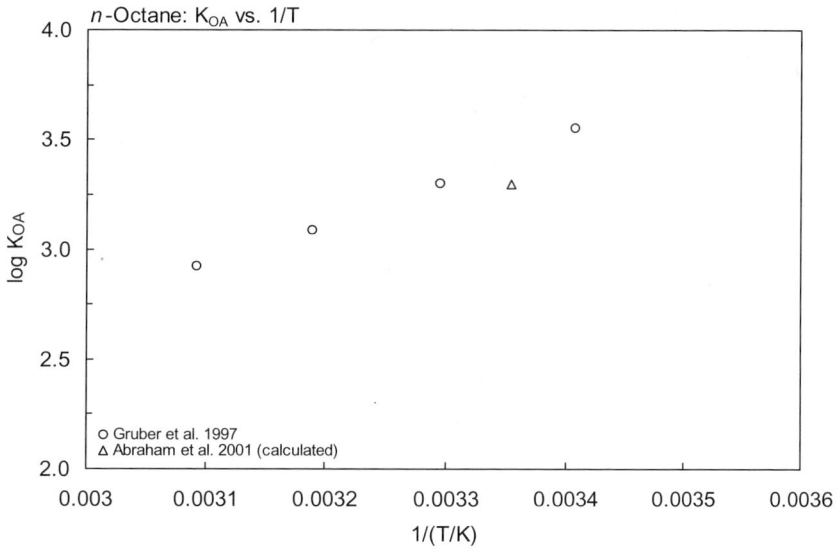

FIGURE 2.1.1.1.23.4 Logarithm of K_{OA} versus reciprocal temperature for *n*-octane.

2.1.1.1.24   4-Methyloctane

Common Name: 4-Methyloctane
Synonym:
Chemical Name: 4-methyloctane
CAS Registry No: 2216-34-4
Molecular Formula: C_9H_{20}; $CH_3(CH_2)_2CH(CH_3)(CH_2)_3CH_3$
Molecular Weight: 128.255
Melting Point (°C):
    −113.3   (Lide 2003)
Boiling Point (°C):
    142.4   (Lide 2003)
Density (g/cm³ at 20°C):
    0.7199, 0.7169   (20°C, 25°C, Dreisbach 1959)
Molar Volume (cm³/mol):
    178.2   (20°C, calculated-density)
    207.2   (calculated-Le Bas method at normal boiling point)
Enthalpy of Vaporization, ΔH_V (kJ/mol):
    44.75, 36.60 (25°C, normal bp, Dreisbach 1961)
Enthalpy of Fusion, ΔH_{fus} (kJ/mol):
Entropy of Fusion, ΔS_{fus} (J/mol K):
Fugacity Ratio at 25°C, F: 1.0

Water Solubility (g/m³ or mg/L at 25°C):
    0.115   (shake flask-GC, Price 1976; quoted, Shaw 1989)

Vapor Pressure (Pa at 25°C and reported temperature dependence equations):
    901   (extrapolated-Antoine eq., Dreisbach 1959)
    $\log (P/\text{mmHg}) = 6.9155 - 1406.0/(206.0 + t/°C)$; temp range 50–165°C (Antoine eq. for liquid state, Dreisbach 1959)
    933   (extrapolated-Antoine eq., temp range 32–170°C, Zwolinski & Wilhoit 1971)
    $\log (P/\text{mmHg}) = 6.90318 - 1399.12/(205.41 + t/°C)$; temp range 32–170°C (Antoine eq., Zwolinski & Wilhoit 1971)
    $\log (P/\text{mmHg}) = 11.2012 - 2.9467 \times 10^3/(T/K) + 1.2133 \cdot \log (T/K) - 1.4423 \times 10^{-2} \cdot (T/K) + 9.177 \times 10^{-6} \cdot (T/K)^2$; temp range 160–588 K (vapor pressure eq., Yaws 1994)

Henry's Law Constant (Pa m³/mol at 25°C):
    1010000; 1000000   (calculated-P/C, recommended, Mackay & Shiu 1981)
    1007000   (selected, Mills et al. 1982)
    1007000   (calculated-vapor-liquid equilibrium (VLE) data, Yaws et al. 1991)

Octanol/Water Partition Coefficient, log K_{OW}:

Octanol/Air Partition Coefficient, log K_{OA}:

Bioconcentration Factor, log BCF:

Sorption Partition Coefficient, log K_{OC}:

Environmental Fate Rate Constants, k, and Half-Lives, $t_{½}$:
    Volatilization:
    Photolysis:

Aliphatic and Cyclic Hydrocarbons

Oxidation: rate constant k, for gas-phase second order rate constants, k_{OH} for reaction with OH radical, k_{NO3} with NO_3 radical and k_{O3} with O_3 or as indicated, *data at other temperatures see reference:

$k_{OH} = 9.72 \times 10^{-12}$ cm^3 molecule^{-1} s^{-1} at 300 K (Atkinson 1989)

$k_{OH} = 9.7 \times 10^{-12}$ cm^3 molecule^{-1} s^{-1} at 298 K (recommended, Atkinson 1997)

Hydrolysis:

Biodegradation:

Biotransformation:

Bioconcentration, Uptake (k_1) and Elimination (k_2) Rate Constants or Half-Lives:

Half-Lives in the Environment:

2.1.1.1.25   n-Nonane

Common Name: *n*-Nonane
Synonym: nonane
Chemical Name: *n*-nonane
CAS Registry No: 111-84-2
Molecular Formula: C_9H_{20}; $CH_3(CH_2)_7CH_3$
Molecular Weight: 128.255
Melting Point (°C):
    –53.46   (Lide 2003)
Boiling Point (°C):
    150.82   (Lide 2003)
Density (g/cm³ at 20°C):
    0.7176, 0.7138   (20°C, 25°C, Dreisbach 1959)
    0.7177, 0.7138   (20°C, 25°C, Riddick et al. 1986)
Molar Volume (cm³/mol):
    178.7, 179.7   (20°C, 25°C, calculated-density)
    207.2   (calculated-Le Bas method at normal boiling point)
Enthalpy of Vaporization, ΔH_V (kJ/mol):
    44.442, 36.915   (25°C, bp, Riddick et al. 1986)
    46.55   (298.15 K, recommended, Ruzicka & Majer 1994)
Enthalpy of Fusion, ΔH_{fus} (kJ/mol):
    15.468   (Dreisbach 1959; Riddick et al. 1986)
    6.28, 15.38   (–55.95, –53.45°C, Chickos et al. 1999)
Entropy of Fusion, ΔS_{fus} (J/mol K):
    99.2, 100.5   (exptl., calculated-group additivity method, total phase change entropy, Chickos et al. 1999)
Fugacity Ratio at 25°C, F: 1.0

Water Solubility (g/m³ or mg/L at 25°C or as indicated and reported temperature dependence equations. Additional data at other temperatures designated * are compiled at the end of this section.):
    0.220   (shake flask-GC, McAuliffe 1969;)
    0.098   (Baker 1967)
    0.071   (shake flask-GC, Krasnoshchekova & Gubertritis 1973)
    0.122*   (shake flask-GC, measured range 25–136.6°C, Price 1976)
    0.289, 0.272 (15, 20°C, vapor saturation-GC, Jönsson et al. 1982)
    0.219   (shake flask-GC, Coates et al. 1985)
    1.70*   (tentative best value, temp range 20–130°C, IUPAC Solubility Data Series, Shaw 1989)
    $\ln x = -433.434 + 18767.82/(T/K) + 61.940 \cdot \ln (T/K)$; temp range 290–400 K (eq. derived from lit. calorimetric and solubility data, Tsonopoulos 1999)
    0.135, 0.477   (25, 100.1°C, calculated-liquid-liquid equilibrium LLE data, Mączyński et al. 2004)

Vapor Pressure (Pa at 25°C or as indicated and reported temperature dependence equations. Additional data at other temperatures designated * are compiled at the end of this section.):
    6349*   (70.127°C, ebulliometry, measured range 70.127–151.764°C, Willingham et al. 1945)
    $\log (P/\text{mmHg}) = 6.94495 - 1435.158/(202.331 + t/°C)$; temp range 70.1–151.8°C (Antoine eq. from exptl. data, ebulliometry-manometer, Willingham et al. 1945)
    623*   (calculated-Antoine eq. regression, temp range 1.4–149.5°C, Stull 1947)
    6405*   (70.343°C, ebulliometry-manometer, measured range 70.343–151.786°C, Forziati et al. 1949)
    $\log (P/\text{mmHg}) = 6.93513 - 1428.811/(201.619 + t/°C)$; temp range 70.343–151.786°C (Antoine eq., ebulliometry-manometer measurements, Forziati et al. 1949)
    580   (extrapolated-Antoine eq., Dreisbach 1959)
    $\log (P/\text{mmHg}) = 6.93513 - 1428.811/(201.619 + t/°C)$; temp range 60–185°C (Antoine eq. for liquid state, Dreisbach 1959)

Aliphatic and Cyclic Hydrocarbons

571* (extrapolated-Antoine eq., temp range 39.32–178.48°C, Zwolinski & Wilhoit 1971)

log (P/mmHg) = 6.93513 – 1428.811/(201.619 + t/°C); temp range 39.32–178.48°C (Antoine eq., Zwolinski & Wilhoit 1971)

log (P/mmHg) = [–0.2185 × 10456.9/(T/K)] + 8.332532; temp range 2.4–149.5°C (Antoine eq., Weast 1972–73)

404.0* (20.99°C, gas saturation, measured range –53.49 to 34.59°C, Carruth & Kobayashi 1973)

570, 713 (extrapolated-Antoine eq., Boublik et al. 1984)

log (P/kPa) = 6.0628 – 1430.638/(201.827 + t/°C), temp range: 70.434–151.8°C (Antoine eq. from reported exptl. data of Willingham et al. 1945, Boublik et al. 1984)

log (P/kPa) = 6.0847 – 1439.2/(205.962 + t/°C), temp range: 66.61–147.86°C (Antoine eq. from reported exptl. data Forziati et al. 1949, Boublik et al. 1984)

571 (extrapolated-Antoine eq., temp range 39–179°C, Dean 1985, 1992)

log (P/mmHg) = 6.93893 – 1431.82/(202.01 + t/°C); temp range 39–179°C (Antoine eq., Dean 1985, 1992)

570 (lit. average, Riddick et al. 1986)

log (P/kPa) = 6.06383 – 1431.82/(202.011 + t/°C); temp range not specified (Antoine eq., Riddick et al. 1986)

517; 571 (extrapolated-Antoine eq.-I; interpolated-Antoine eq.-II, Stephenson & Malanowski 1987)

log (P_L/kPa) = 6.0593 – 1429.46/(–71.33 + T/K); temp range 344–426 K (Antoine eq.-I, Stephenson & Malanowski 1987)

log (P_L/kPa) = 8.17855 – 2523.8/(T/K); temp range 219–308 K (Antoine eq.-II, Stephenson & Malanowski 1987)

580.7* (recommended, Ruzicka & Majer 1994)

ln [(P/kPa)/(P_o/kPa)] = [1 – (T_o/K)/(T/K)]·exp{2.94690 – 2.061933 × 10^{-3}·(T/K) + 1.903683 × 10^{-6}·(T/K)^2}; reference state at P_o = 101.325 kPa, T_o = 423.932 K (Cox equation, Ruzicka & Majer 1994)

log (P/mmHg) = 8.8817 – 2.8042 × 10^3/(T/K) + 1.5262·log (T/K) – 1.0464 × 10^{-2}·(T/K) + 5.7972 × 10^{-6}·(T/K)^2; temp range 220–596 K (vapor pressure eq., Yaws 1994)

Henry's Law Constant (Pa m^3/mol at 25°C or as indicated and reported temperature dependence equations. Additional data at other temperatures designated * are compiled at the end of this section.):

601000, 748000, 333000; 500000 (calculated-P/C values, recommended, Mackay & Shiu 1981)

147400*, 372820 (14.8, 20.05°C, equilibrium cell-concentration ratio-GC, Jönsson et al. 1982)

173520, 348200 (15, 20°C, calculated-temp dependence eq. derived from exptl data, Jönsson et al. 1982)

ln (1/K_{AW}) = 28259/(T/K) + 76.183·ln (T/K) – 524.13; temp range 15–35°C (least-square regression of equilibrium cell-concn ratio-GC measurements, Jönsson et al. 1982)

41950* (EPICS-GC/FID, measured range 10–30°C, Ashworth et al. 1988)

ln [H/(atm m^3/mol)] = –0.1847 + 202.1/(T/K); temp range 10–30°C (EPICS measurements, Ashworth et al. 1988)

599600 (calculated-P/C, Eastcott et al. 1988)

601000 (calculated-vapor-liquid equilibrium (VLE) data, Yaws et al. 1991)

459668 (EPICS-GC, Ryu & Park 1999)

42164 (20°C, selected from reported experimental determined values, Staudinger & Roberts 1996, 2001)

log K_{AW} = 1.104 + 39/(T/K) (summary of literature data, Staudinger & Roberts 2001)

Octanol/Water Partition Coefficient, log K_{OW}:

4.51 (estimated-HPLC-k' correlation, Coates et al. 1985)

5.65 ± 0.60 (recommended, Sangster 1989)

5.42 (recommended, Sangster 1993)

Octanol/Air Partition Coefficient, log K_{OA}:

Bioconcentration Factor,

Sorption Partition Coefficient, log K_{OC}:

Environmental Fate Rate Constants, k, and Half-Lives, $t_{1/2}$:
Volatilization:
Photolysis:
Oxidation: rate constant k, for gas-phase second order rate constants, k_{OH} for reaction with OH radical, k_{NO3} with NO$_3$ radical and k_{O3} with O$_3$ or as indicated, *data at other temperatures see reference:

$k_{OH} = (10.7 \pm 0.4) \times 10^{-12}$ cm^3 molecule^{-1} s^{-1} at 299 ± 2 K (relative rate method, Atkinson et al. 1982a, 1984c)
$k_{OH} = 10.2 \times 10^{-12}$ cm^3 molecule^{-1} s^{-1} at 312 K in smog chamber (Nolting et al. 1988)
$k_{OH} = 1.02 \times 10^{-11}$ cm^3 molecule^{-1} s^{-1} at 298 K (recommended, Atkinson 1989)
$k_{OH} = 1.02 \times 10^{-11}$ cm^3 molecule^{-1} s^{-1} at 298 K, $k_{NO3} = 2.39 \times 10^{-16}$ cm^3 molecule^{-1} s^{-1} at 296 K (Atkinson 1990)
$k_{OH} = 1.02 \times 10^{-11}$ cm^3 molecule^{-1} s^{-1} estimated atmospheric lifetime of 14 h (Altshuller 1991)
$k_{OH} = 1.0 \times 10^{-11}$ cm^3 molecule^{-1} s^{-1} at 298 K, $k_{NO3} = 2.30 \times 10^{-16}$ cm^3 molecule^{-1} s^{-1} (Sabljic & Güsten 1990)
$k_{NO3} = 2.41 \times 10^{-16}$ cm^3 molecule^{-1} s^{-1} at 296 ± 2 K (Atkinson 1991)
k_{NO3}(exptl) = (1.92, 2.59) × 10^{-16} cm^3 molecule^{-1} s^{-1}, k_{NO3}(calc) = 2.47 × 10^{-16} cm^3 molecule^{-1} s^{-1} at 296 ± 2 K (relative rate method, Aschmann & Atkinson 1995)
k_{OH}* = 10.0 × 10^{-12} cm^3 molecule^{-1} s^{-1}, k_{NO3} = 2.3 × 10^{-16} cm^3 molecule^{-1} s^{-1} at 298 K (recommended, Atkinson 1997)

Hydrolysis:
Biodegradation:
Biotransformation:
Bioconcentration, Uptake (k_1) and Elimination (k_2) Rate Constants or Half-Lives:

Half-Lives in the Environment:

Air: atmospheric $t_{1/2}$ ~ 2.4–24 h for C$_4$H$_{10}$ and higher paraffins for the reaction with hydroxyl radical, based on the EPA Reactivity Classification of Organics (Darnall et al. 1976); photooxidation reaction rate constant of 1.02 × 10^{-11} cm^3 molecule^{-1} s^{-1} with OH radical with an estimated lifetime τ = 14 h during summer daylight (Altshuller 1991).

TABLE 2.1.1.1.25.1
Reported aqueous solubilities of *n*-nonane at various temperatures

Price 1976		Jonsson et al. 1982		Shaw 1989a	
shake flask-GC		vapor saturation-GC		IUPAC "tentative" best	
t/°C	S/g·m^{-3}	t/°C	S/g·m^{-3}	t/°C	S/g·m^{-3}
25.0	0.122	15	0.289	20	0.29
69.7	0.309	20	0.272	25	0.17
99.1	0.420			30	0.14
121.3	1.70			40	0.17
136.6	5.07			50	0.22
				60	0.26
				70	0.31
				80	0.34
				90	0.37
				100	0.42
				110	0.80
				120	1.60
				130	3.20

Aliphatic and Cyclic Hydrocarbons

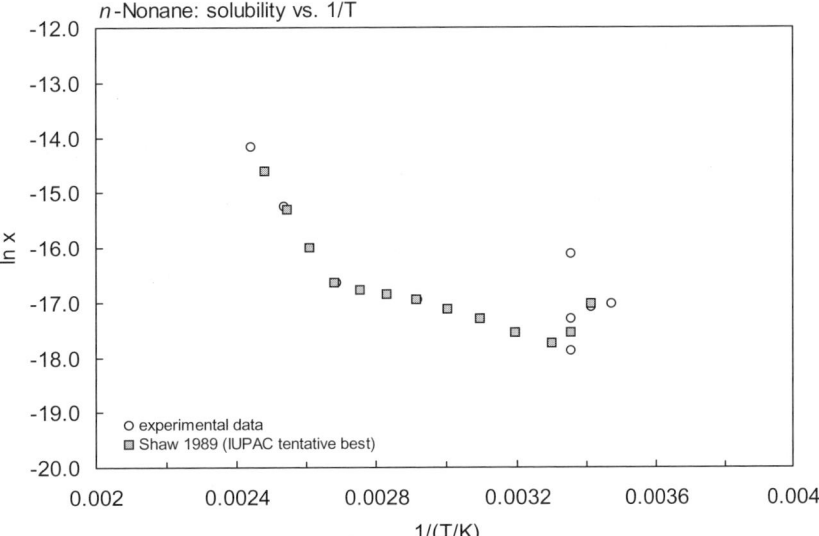

FIGURE 2.1.1.1.25.1 Logarithm of mole fraction solubility (ln x) versus reciprocal temperature for *n*-nonane.

TABLE 2.1.1.1.25.2
Reported vapor pressures of *n*-nonane at various temperatures and the coefficients for the vapor pressure equations

$\log P = A - B/(T/K)$ (1) $\quad\quad \ln P = A - B/(T/K)$ (1a)
$\log P = A - B/(C + t/°C)$ (2) $\quad\quad \ln P = A - B/(C + t/°C)$ (2a)
$\log P = A - B/(C + T/K)$ (3)
$\log P = A - B/(T/K) - C \cdot \log (T/K)$ (4)
$\ln [(P/kPa)/(P_o/kPa)] = [1 - (T_o/K)/(T/K)] \cdot \exp\{A_0 - A_1 \cdot (T/K) + A_2 \cdot (T/K)^2\}$ (5) - Cox eq.

1.

Willingham et al. 1945		Stull 1947		Forziati et al. 1949		Zwolinski & Wilhoit 1971	
ebulliometry-manometer		summary of literature data		ebulliometry-manometer		selected values	
t/°C	P/Pa	t/°C	P/Pa	t/°C	P/Pa	t/°C	P/Pa
70.127	6349	2.4	133.3	70.343	6405	39.06	1333
74.388	7650	26.3	666.6	74.546	7697	51.93	2666
78.097	8955	38.0	1333	78.219	8995	60.12	4000
81.458	10294	51.0	2666	81.548	10335	66.25	5333
84.582	11687	65.6	5333	84.658	11724	71.216	6666
88.801	13822	74.1	7999	88.864	13847	75.409	7999
93.610	16611	86.0	13332	93.601	16639	82.293	10666
98.491	19913	104.7	26664	98.545	19949	87.873	13332
103.047	23451	126.8	53329	103.072	23479	98.593	19998
109.115	28952	149.5	101325	109.136	28984	106.694	26664
114.684	34892			114.712	34930	113.284	33331
121.399	43319	mp/°C	−53.7	121.433	43366	118.882	39997
128.329	53653			128.357	53703	128.131	53329
136.721	66757			135.741	66801	135.680	66661
143.738	83721			143.751	83755	142.110	79993
149.394	97609			149.409	97652	147.741	93326

(Continued)

TABLE 2.1.1.1.25.2 *(Continued)*

Willingham et al. 1945		Stull 1947		Forziati et al. 1949		Zwolinski & Wilhoit 1971	
ebulliometry-manometer		summary of literature data		ebulliometry-manometer		selected values	
t/°C	P/Pa	t/°C	P/Pa	t/°C	P/Pa	t/°C	P/Pa
150.002	99207			150.017	99241	148.790	95992
150.565	100833			150.579	100734	149.816	98659
151.195	102401			151.222	102467	150.321	101325
151.764	103921			151.786	104006		
						eq. 2	P/mmHg
bp/°C	150.796			bp/°C	150.798	A	6.93440
						B	1508.75
eq. 2	P/mmHg			eq. 2	P/mmHg	C	195.374
A	6.94445			A	6.93513	bp/°C	150.821
B	1435.158			B	1428.811	ΔH_v/(kJ mol^{-1})	
C	202.331			C	201.619	at 25°C	36.92
temp range: 70.1–151.8°C				temp range: 70.3–151.8°C		at bp	46.44

2.

Carruth & Kobayashi 1973		Ruzicka & Majer 1994	
gas saturation		recommended	
t/°C	P/Pa	T/K	P/Pa
–53.49	0.7586	209.23	0.1
–49.4	0.825	226.30	1.0
–48.26	0.925	247.08	10
–36.32	3.240	273.13	100
–27.45	7.866	307.09	1000
–9.97	37.33	353.86	10000
2.60	100.4	423.392	101325
20.99	404.0	298.15	580.7
26.9	602.6		
34.59	722.6	data calc. from Cox eq.	
		eq. 5	P/kPa
mp/°C	–53.7	A_0	2.94690
		A_1	2.051933 × 10^{-3}
eq. 1a	P/mmHg	A_2	1.903683 × 10^{-6}
A	20.8468	with reference state at	
B	–5811.26	P_o/kPa	101.325
		T_o/K	423.392

Aliphatic and Cyclic Hydrocarbons

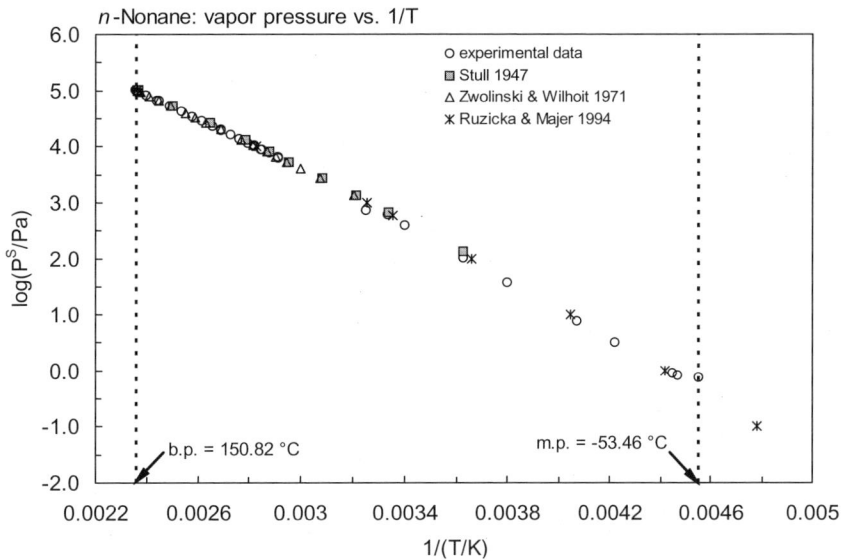

FIGURE 2.1.1.1.25.2 Logarithm of vapor pressure versus reciprocal temperature for *n*-nonane.

TABLE 2.1.1.1.25.3
Reported Henry's law constants of *n*-nonane at various temperatures and temperature dependence equations

$\ln K_{AW} = A - B/(T/K)$	(1)	$\log K_{AW} = A - B/(T/K)$ (1a)
$\ln (1/K_{AW}) = A - B/(T/K)$	(2)	$\log (1/K_{AW}) = A - B/(T/K)$ (2a)
$\ln (k_H/\text{atm}) = A - B/(T/K)$	(3)	$\ln [H/(\text{atm·m}^3/\text{mol})] = A - B/(T/K)$ (4a)
$\ln [H/(\text{Pa m}^3/\text{mol})] = A - B/(T/K)$	(4)	
$K_{AW} = A - B\cdot(T/K) + C\cdot(T/K)^2$	(5)	

Jönsson et al. 1982		Ashworth et al. 1988	
equilibrium cell-GC		EPICS-GC	
t/°C	H/(Pa m³/mol)	t/°C	H/(Pa m³/mol)
14.8	244287	10	40530
14.8	221668	15	50257
15.0	237196	20	33640
20.0	348179	25	41949
20.05	369343	30	47116
20.05	363831		
		eq. 4a	H/(atm m³/mol)
		A	–0.1847
		B	202.1

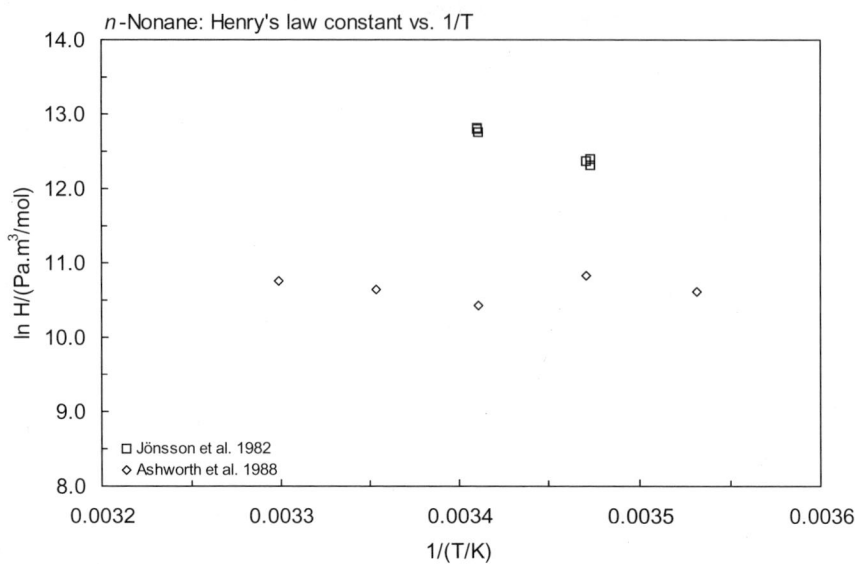

FIGURE 2.1.1.1.25.3 Logarithm of Henry's law constant versus reciprocal temperature for *n*-nonane.

Aliphatic and Cyclic Hydrocarbons

2.1.1.1.26 n-*Decane*

Common Name: *n*-Decane
Synonym: decane
Chemical Name: *n*-decane
CAS Registry No: 124-18-5
Molecular Formula: $C_{10}H_{22}$; $CH_3(CH_2)_8CH_3$
Molecular Weight: 142.282
Melting Point (°C):
    –29.6   (Lide 2003)
Boiling Point (°C):
    174.15   (Lide 2003)
Density (g/cm³ at 20°C):
    0.7301, 0.7273 (20°C, 25°C, Dreisbach 1959)
    0.7301, 0.7264 (20°C, 25°C, Riddick et al. 1986)
Molar Volume (cm³/mol):
    194.9, 195.9 (20°C, 25°C, calculated-density)
    229.4   (calculated-Le Bas method at normal boiling point)
Enthalpy of Vaporization, ΔH_V (kJ/mol):
    51.367, 31.279 (25°C, bp, Riddick et al. 1986)
    51.42   (209.15 K, recommended, Ruzicka & Majer 1994)
Enthalpy of Fusion, ΔH_{fus} (kJ/mol):
    28.72   (Dreisbach 1959)
    28.677   (Riddick et al. 1986)
    28.7   (Chickos et al. 1999)
Entropy of Fusion, ΔS_{fus} (J/mol K):
    117.99, 109.8 (exptl., calculated-group additivity method, Chickos et al. 1999)
Fugacity Ratio at 25°C, F: 1.0

Water Solubility (g/m³ or mg/L at 25°C or as indicated):
    0.016   (radiotracer, Baker 1958, 1959)
    0.0198   (shake flask-GC, Franks 1966)
    0.022   (Baker 1967)
    0.052   (shake flask-GC, McAuliffe 1969)
    0.0087   (shake flask-GC, Krasnoshchekova & Gubertritis 1973)
    0.182, 1.220 (shake flask-headspace-GC, Mackay et al. 1975)
    0.0029   (shake flask-refractometer, Becke & Quitzsch 1977)
    0.0524   (shake flask-GC, Coates et al. 1985)
    0.02, 0.015 (20°C, 25°C, tentative best values, IUPAC Solubility Data Series, Shaw 1989)
    0.0277, 0.0261 (20°C, 25°C, calculated-recommended liquid-liquid equilibrium LLE data, temp range 273.2–422.7 K, Mączyński et al. 2004)

Vapor Pressure (Pa at 25°C or as indicated and reported temperature dependence equations. Additional data at other temperatures designated * are compiled at the end of this section.):
    62.66*   (8.50°C, static-McLeod gauge, measured range –3.80 to 8.50°C, Linder 1931)
    7649*   (94.481°C, ebulliometry, measured range 94.481–175.121°C, Willingham et al. 1945)
    log (P/mmHg) = 6.95367 – 1501.268/(194.480 + t/°C); temp range 94.6–175.1°C (Antoine eq. from exptl. data, ebulliometry-manometer, Willingham et al. 1945)
    238*   (calculated-Antoine eq. regression, temp range 17.1–173°C, Stull 1947)
    182   (extrapolated-Antoine eq., Dreisbach 1959)
    log (P/mmHg) = 6.95367 – 1501.268/(194.480 + t/°C); temp range 75–210°C (Antoine eq. for liquid state, Dreisbach 1959)
    180*   (extrapolated-Antoine eq., temp range 57.7–202.9°C, Zwolinski & Wilhoit 1971)

log (P/mmHg) = 6.95367 − 1501.268/(194.480 + t/°C); temp range 57.7–202.9°C (Antoine eq., Zwolinski & Wilhoit 1971)

log (P/mmHg) = [−0.2185 × 10912.0/(T/K)] + 8.248089; temp range 17.1–173°C (Antoine eq., Weast 1972–73)

86.53*   (16.74°C, gas saturation, measured range −29.65 to 37.45°C, Carruth & Kobayashi 1973)

174   (extrapolated-Antoine eq., Boublik et al. 1984)

log (P/kPa) = 6.08321 − 1504.405/(194.831 + t/°C); temp range 94.48–175.1°C (Antoine eq. from reported exptl. data, Boublik et al. 1984)

173   (extrapolated-Antoine eq., temp range 58–203°C, Dean 1985, 1992)

log (P/mmHg) = 6.94365 − 1495.17/(193.86 + t/°C); temp range 58–203°C (Antoine eq., Dean 1985, 1992)

180*   (24.941°C, gas saturation, measured temp range 298.091–347.887 K, Allemand et al. 1986)

170   (lit. average, Riddick et al. 1986)

log (P/kPa) = 5.55216 − 1594.49/(126.36 + t/°C); temp range not specified (Antoine eq., Riddick et al. 1986)

171   (extrapolated-Antoine eq., Stephenson & Malanowski 1987)

log (P_L/kPa) = 6.80914 − 1900.343/(−47.319 + T/K); temp range 252–383 K (Antoine eq.-I, Stephenson & Malanowski 1987)

log (P_L/kPa) = 6.09206 − 1510.415/(−77.646 + T/K); temp range 373–443 K (Antoine eq.-II, Stephenson & Malanowski 1987)

log (P_L/kPa) = 6.04899 − 1482.502/(−80.635 + T/K); temp range 447–526 K (Antoine eq.-III, Stephenson & Malanowski 1987)

log (P_L/kPa) = 9.71412 − 6858.314/(454.63 + T/K); temp range 524–617 K (Antoine eq.IV, Stephenson & Malanowski 1987)

874.1*   (50.64°C, static-quartz pressure gauge, measured range 50.64–314.982°C, Morgan & Kobayashi 1994)

182*   (recommended, Ruzicka & Majer 1994)

ln [(P/kPa)/(P_o/kPa)] = [1 − (T_o/K)/(T/K)]·exp{2.96690 − 1.932579 × 10^{-3}·(T/K) + 1.644626 × 10^{-6}·(T/K)^2}; reference state at P_o = 101.325 kPa, T_o = 447.269 K (Cox equation, Ruzicka & Majer 1994)

log (P/mmHg) = 26.5125 − 3.3584 × 10^3/(T/K) − 6.1174·log (T/K) − 3.3225 × 10^{-10}·(T/K) + 4.8554 × 10^{-7}·(T/K)^2; temp range 243–618 K (vapor pressure eq., Yaws 1994)

127.6*   (20°C, ebulliometer and inclined piston gauge, measured temp range 268–490 K. Chirico et al. 1989)

324   (liquid P_L, GC-RT correlation; Donovan 1996)

520*   (41.6°C, ebulliometry, measured range 314.75–458.45 K, Dejoz et al. 1996)

ln (P/kPa) = 13.9735 − 3441.40/[(T/K) − 79.434]; temp range 314.75–458.45 K (ebulliometry, Dejoz et al. 1996)

Henry's Law Constant (Pa m^3/mol at 25°C):

326300   (calculated-P/C, Mackay & Shiu 1975)

499500   (calculated-P/C, Bobra et al. 1979; Mackay et al. 1979; selected, Mills et al. 1982)

489400   (calculated-P/C, Mackay 1981)

700000; 500000, 108000 (recommended; calculated-P/C, Mackay & Shiu 1981)

431100   (calculated-P/C, Eastcott et al. 1988)

477870   (calculated-vapor-liquid equilibrium (VLE) data, Yaws et al. 1991)

Octanol/Water Partition Coefficient, log K_{OW}:

5.67   (estimated-fragment const., Lyman 1982)

6.69, 5.98   (estimated-HPLC/MS, calculated-fragment const., Burkhard et al. 1985)

5.01   (estimated, Coates et al. 1985)

6.25 ± 0.70 (recommended, Sangster 1989)

Octanol/Air Partition Coefficient, log K_{OA}:

Bioconcentration Factor, log BCF:

Sorption Partition Coefficient, log K_{OC}:

Environmental Fate Rate Constants, k, and Half-Lives, $t_{1/2}$:

Aliphatic and Cyclic Hydrocarbons

Volatilization:

Photolysis:

Oxidation: rate constant k, for gas-phase second order rate constants, k_{OH} for reaction with OH radical, k_{NO_3} with NO_3 radical and k_{O_3} with O_3 or as indicated, *data at other temperatures see reference:

$k_{OH} = (11.4 \pm 0.6) \times 10^{-12}$ cm^3 molecule^{-1} s^{-1} at 299 ± 2 K (relative rate method, Atkinson et al. 1982a, 1984c)

$k_{OH} = 11.7 \times 10^{-12}$ cm^3 molecule^{-1} s^{-1} at 312 K in a smog chamber (Nolting et al. 1988)

$k_{OH} = 1.16 \times 10^{-12}$ cm^3 molecule^{-1} s^{-1} at 298 K (recommended, Atkinson 1989, 1990, 1991)

$k_{OH} = 11.6 \times 10^{-12}$ cm^3 molecule^{-1} s^{-1} at 298 K, atmospheric lifetime of 12 h during summer daylight (Altshuller 1991)

k_{NO_3}(exptl) = 2.59×10^{-16} cm^3 molecule^{-1} s^{-1}, k_{NO_3}(calc) = 2.47×10^{-16} cm^3 molecule^{-1} s^{-1} at 296 ± K (relative rate method, Aschmann & Atkinson 1995)

$k_{OH}^* = 11.2 \times 10^{-12}$ cm^3 molecule^{-1} s^{-1}, $k_{NO_3} = 2.8 \times 10^{-16}$ cm^3 molecule^{-1} s^{-1} at 298 K (recommended, Atkinson 1997)

Hydrolysis:

Biodegradation:

Biotransformation:

Bioconcentration, Uptake (k_1) and Elimination (k_2) Rate Constants or Half-Lives:

Half-Lives in the Environment:

Air: atmospheric $t_{1/2}$ ~ 2.4–24 h for C_4H_{10} and higher paraffins for the reaction with hydroxyl radical, based on the EPA Reactivity Classification of Organics (Darnall et al. 1976);

photooxidation reaction rate constant k = 1.16×10^{-11} cm^3 molecule^{-1} s^{-1} with hydroxyl radical with an estimated lifetime of 12 h during summer daylight (Altshuller 1991).

TABLE 2.1.1.1.26.1
Reported vapor pressures of *n*-decane at various temperatures and the coefficients for the vapor pressure equations

$\log P = A - B/(T/K)$ (1) $\quad\quad$ $\ln P = A - B/(T/K)$ (1a)

$\log P = A - B/(C + t/°C)$ (2) $\quad\quad$ $\ln P = A - B/(C + t/°C)$ (2a)

$\log P = A - B/(C + T/K)$ (3) $\quad\quad$ $\ln P = A - B/(C + T/K)$ (3a)

$\log P = A - B/(T/K) - C \cdot \log (T/K)$ (4)

$\ln (P/P_{ref}) = [1 - (T_{ref}/T)] \cdot \exp(a + bT + cT^2)$ (5)

1.

Linder 1931		Willingham et al. 1945		Stull 1947		Zwolinski & Wilhoit 1971	
static-McLeod gauge		ebulliometry		summary of literature data		selected values	
t/°C	P/Pa	t/°C	P/Pa	t/°C	P/Pa	t/°C	P/Pa
−3.80	22.0	94.481	7649	17.1	133.3	57.6	1333
0.20	28.0	98.352	8954	42.5	666.6	71.06	2666
0.50	30.66	101.859	10292	55.4	1333	79.61	4000
8.50	62.66	105.118	11686	69.1	2666	86.02	5333
		109.526	13812	84.6	5333	91.20	6666
		114.540	16609	94.6	7999	95.576	7999
		119.640	19913	108.0	13332	102.759	10666
		124.372	23451	127.8	26664	108.579	13332
		130.690	28951	149.9	53329	119.759	19998
		136.490	34892	173.0	101325	128.203	26664

(Continued)

TABLE 2.1.1.1.26.1 (*Continued*)

Linder 1931		Willingham et al. 1945		Stull 1947		Zwolinski & Wilhoit 1971	
static-McLeod gauge		ebulliometry		summary of literature data		selected values	
t/°C	P/Pa	t/°C	P/Pa	t/°C	P/Pa	t/°C	P/Pa
		143.495	43318			135.071	33331
		150.718	53654	mp/°C	−29.7	140.903	39997
		158.419	66757			150.535	53329
		166.772	83722			158.39	66661
		172.661	97609			165.088	79993
		173.295	99206			170.947	93326
		173.882	100701			172.039	95992
		174.538	102401			173.106	98659
		175.121	103921			174.152	101325
		bp/°C	174.123			eq. 2	P/mmHg
						A	6.95375
		eq. 2	P/mmHg			B	1508.75
		A	6.95367			C	193.374
		B	1501.268			bp/°C	174.152
		C	194.480			ΔH_V/(kJ mol^{-1})	
		temp range: 94.5–175.1°C				at 25°C	51.37
						at bp	39.28

2.

Carruth & Kobayashi 1973		Allemand et al. 1986		Chirico et al. 1989			
gas saturation		gas saturation		ebulliometry		inclined piston gauge	
t/°C	P/Pa	t/°C	P/Pa	t/°C	P/Pa	t/°C	P/Pa
−29.65	1.72	24.941	180.0	100.086	9596	−5.002	17.0
−21.84	3.20	30.022	253.3	103.317	10897	−0.002	26.4
−18.68	4.586	35.019	352.0	106.561	12348	9.994	59.9
−9.54	10.60	40.005	481.3	109.82	13961	19.999	127.6
4.35	33.20	45.002	650.6	113.09	15752	30.001	256.1
16.74	86.53	50.006	874.6	116.373	17737	34.999	355.0
31.45	186.7	55.001	1161	119.67	19933	40.0	486.3
37.45	205.3	59.996	1520	126.299	25023	44.0	657.8
		64.916	1969	132.983	31177	50.0	879.9
mp/°C	−29.71	70.049	2545	139.716	39565	60.006	1523.8
		74.737	3190	146.502	47375	80.002	2536.1
eq. 1a	P/mmHg			153.399	57817	75.001	3227.1
A	20.8865			160.227	70120		
B	6170.32			167.167	84533		
				174.157	101325	Cox eq.	
				181.199	120790	eq. 5a	P/kPa
				188.291	143250	P_{ref}/kPa	101.325
				195.43	169920	T_{ref}/K	447.307
				202.622	198490	a	2.96081
				209.859	232020	$10^3 b$/K^{-1}	−1.90111
				217.142	270020	$10^6 c$/K^{-2}	1.60359
						temp range: 268–490 K	

Aliphatic and Cyclic Hydrocarbons

TABLE 2.1.1.1.26.1 (*Continued*)

3.

Morgan & Kobayashi 1994		Ruzicka & Majer 1994		Dejoz et al. 1996			
static-quartz pressure gauge		recommended		vapor-liquid equilibrium			
t/°C	P/Pa	T/K	P/Pa	t/°C	P/Pa	t/°C	P/Pa
50.64	874.1	222.67	0.1	41.6	520	144.1	44090
59.933	1509	240.65	1.0	45.2	650	146.7	47640
79.938	4050	262.50	10	49.5	840	149.3	51420
99.970	9562	289.84	100	55.2	1170	151.2	54380
119.981	20150	325.64	1000	60.1	1520	153.4	57920
119.971	20154	374.25	10000	69.1	2410	156.5	63220
139.972	38892	447.27	101325	75.8	3330	159.3	69330
159.973	69695	298.15	182.0	84.2	4900	162.3	74190
179.974	117250			93.1	7200	165.5	80890
199.976	187250	Cox eq.		97.9	8760	168.4	87350
219.976	285770	eq. 5	P/kPa	100.7	9820	174.6	102460
239.979	.420380	P_{ref}/kPa	101.325	103.5	10970	178.9	114160
259.979	598790	T_{ref}/K	447.27	105.9	12020	182.7	125370
279.98	830150	a	2.96690	108.3	13180	185.3	133470
299.98	1125900	$10^3 b$/K^{-1}	−1.932579	110.	14250		
314.982	1395000	$10^6 c$/K^{-2}	1.644626	113.3	15870	eq. 3(a)	P/kPa
				117.1	18200	A	13.9735
data fitted to Wagner eq.				121.8	21470	B	3441.40
				126.6	25280	C	−79.434
				129.4	27750	temp range 315–458 K	
				133.4	31600		
				137.2	35660		
				141.9	41250		

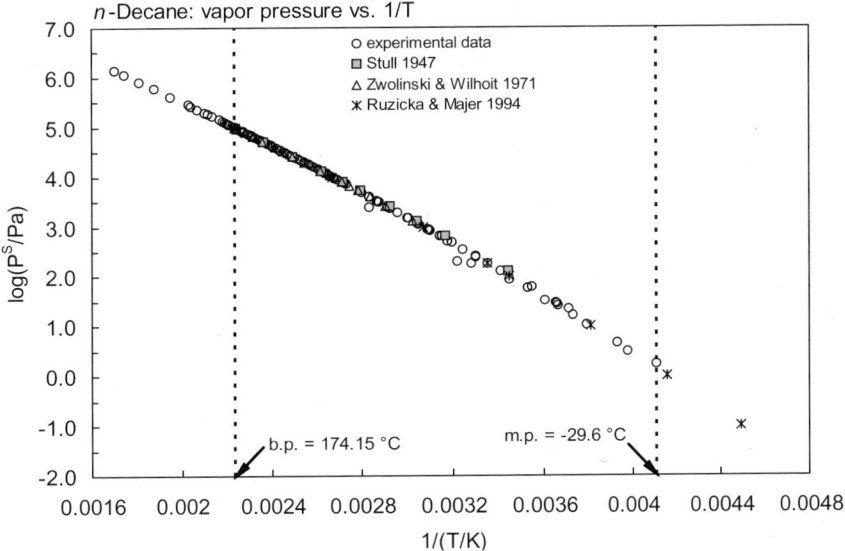

FIGURE 2.1.1.1.26.1 Logarithm of vapor pressure versus reciprocal temperature for *n*-decane.

2.1.1.1.27   n-Undecane

Common Name: *n*-Undecane
Synonym: undecane
Chemical Name: *n*-undecane
CAS Registry No: 1120-21-4
Molecular Formula: $C_{11}H_{24}$; $CH_3(CH_2)_9CH_3$
Molecular Weight: 156.309
Melting Point (°C):
    –25.5   (Lide 2003)
Boiling Point (°C):
    195.9   (Lide 2003)
Density (g/cm³ at 20°C):
    0.74024, 0.73652   (20°C, 25°C, Camin & Rossini 1955)
    0.7402, 0.7366   (20°C, 25°C, Dreisbach 1959)
Molar Volume (cm³/mol):
    211.2   (20°C, calculated-density)
    251.6   (calculated-Le Bas method at normal boiling point)
Enthalpy of Vaporization, ΔH_V (kJ/mol):
    56.5, 41.524 (25°C, bp, Dreisbach 1959)
    56.58   (298.15 K, recommended, Ruzicka & Majer 1994)
Enthalpy of Fusion, ΔH_{fus} (kJ/mol):
    6.86, 22.18, 29.03   (–36.55, –25.55°C, total phase change enthalpy, Chickos et al. 1999)
Entropy of Fusion, ΔS_{fus} (J/mol K):
    118.6, 119.1   (exptl., calculated-group additivity method, total phase change entropy, Chickos et al. 1999)
Fugacity Ratio at 25°C, F: 1.0

Water Solubility (g/m³ or mg/L at 25°C):
    0.0044   (shake flask-GC, McAuliffe 1969)
    0.0036   (shake flask-GC, Krashoshhchekova & Gubertritis 1973)
    0.0040   ("best" value, IUPAC Solubility Data Series, Shaw 1989)
    0.0042   (calculated-recommended liquid-liquid equilibrium LLE data, Mączyński et al. 2004)

Vapor Pressure (Pa at 25°C or as indicated and reported temperature dependence equations. Additional data at other temperatures designated * are compiled at the end of this section.):
    133.3*   (31.4°C, summary of literature data, temp range 31.4–194.5°C, Stull 1947)
    log (P/mmHg) = 6.97674 – 1572.477/(188.022 + t/°C); temp range 105.4–197.3°C (Antoine eq., ebulliometry-manometer measurement, Camin & Rossini 1955)
    57.18   (extrapolated-Antoine eq., Dreisbach 1959)
    log (P/mmHg) = 6.97674 – 1572.477/(188.022 + t/°C); temp range 98–258°C (Antoine eq. for liquid state, Dreisbach 1959)
    52.20*   (extrapolated-Antoine eq., temp range 75.1–225.8°C, Zwolinski & Wilhoit 1971)
    log (P/mmHg) = 6.97220 – 1569.57/(187.700 + t/°C); temp range 75.1–225.8°C (Antoine eq., Zwolinski & Wilhoit 1971)
    log (P/mmHg) = [–0.2185 × 11481.7/(T/K)] + 8.260477; temp range 31.4–194.5°C (Antoine eq., Weast 1972–73)
    52.5   (extrapolated-Antoine eq., Boublik et al. 1984)
    log (P/kPa) = 6.12013 – 1572.031/(188.062 + t/°C); temp range 104.5–197.3°C (Antoine eq. from reported exptl. data of Camin & Rossini 1955, Boublik et al. 1984)
    log (P/mmHg) = 6.97220 – 1569.57/(187.70 + t/°C); temp range 75–226°C (Antoine eq., Dean 1985, 1992)
    54.8   (interpolated-Antoine eq., Stephenson & Malanowski 1987)
    log (P_L/kPa) = 6.10154 – 1572.411/(–85.128 + T/K); temp range 278–470 K (Antoine eq., Stephenson & Malanowski 1987)
    56.89*   (recommended, Ruzicka & Majer 1994)

Aliphatic and Cyclic Hydrocarbons

$\ln[(P/kPa)/(P_o/kPa)] = [1 - (T_o/K)/(T/K)] \cdot \exp\{3.02771 - 2.045579 \times 10^{-3} \cdot (T/K) + 1.712658 \times 10^{-6} \cdot (T/K)^2\}$; reference state at $P_o = 101.325$ kPa, $T_o = 469.042$ K (Cox equation, Ruzicka & Majer 1994)

$\log(P/mmHg) = 82.9230 - 5.6085 \times 10^3/(T/K) - 23.7327 \cdot \log(T/K) + 1.0469 \times 10^{-2} \cdot (T/K) + 7.087 \times 10^{-13} \cdot (T/K)^2$; temp range 248–639 K (vapor pressure eq., Yaws 1994)

Henry's Law Constant (Pa m³/mol at 25°C):
- 185000 (calculated-P/C, Mackay & Shiu 1981)
- 185390 (calculated-vapor-liquid equilibrium (VLE) data, Yaws et al. 1991)

Octanol/Water Partition Coefficient, log K_{OW}:
- 6.94 (estimated-HPLC/MS, Burkhard et al. 1985)
- 6.51 (calculated-fragment const., Burkhard et al. 1985)

Octanol/Air Partition Coefficient, log K_{OA}:

Bioconcentration Factor, log BCF:

Sorption Partition Coefficient, log K_{OC}:

Environmental Fate Rate Constants, k, and Half-Lives, $t_{1/2}$:

Volatilization:

Photolysis:

Oxidation: rate constant k for gas-phase second order rate constants, k_{OH} for reaction with OH radical, k_{NO3} with NO_3 radical and k_{O3} with O_3 or as indicated, *data at other temperatures see reference:
- $k_{OH} = 13.7 \times 10^{-12}$ cm³ molecule⁻¹ s⁻¹ at 312 K in a smog chamber (Nolting et al. 1988)
- $k_{OH} = 1.32 \times 10^{-11}$ cm³ molecule⁻¹ s⁻¹ at 298 K (recommended, Atkinson 1989, 1990)
- $k_{OH} = 1.29 \times 10^{-11}$ cm³ molecule⁻¹ s⁻¹ at 298 K (recommended, Atkinson 1997)

Hydrolysis:

Biodegradation:

Biotransformation:

Bioconcentration, Uptake (k_1) and Elimination (k_2) Rate Constants or Half-Lives:

Half-Lives in the Environment:

TABLE 2.1.1.1.27.1
Reported vapor pressures of *n*-undecane at various temperatures and the coefficients for the vapor pressure equations

$\log P = A - B/(T/K)$ (1) $\ln P = A - B/(T/K)$ (1a)
$\log P = A - B/(C + t/°C)$ (2) $\ln P = A - B/(C + t/°C)$ (2a)
$\log P = A - B/(C + T/K)$ (3)
$\log P = A - B/(T/K) - C \cdot \log(T/K)$ (4)
$\ln(P/P_{ref}) = [1 - (T_{ref}/T)] \cdot \exp(a + bT + cT^2)$ (5) - Cox eq.

Stull 1947		Camin & Rossini 1955		Zwolinski & Wilhoit 1971		Ruzicka & Majer 1994	
summary of literature data		ebulliometry		selected values		recommended	
t/°C	P/Pa	t/°C	P/Pa	t/°C	P/Pa	T/K	P/Pa
31.4	133.3	104.58	5540	75.1	1333	236.04	0.1
58.4	666.6	110.952	8287	89.06	2666	254.71	1.0
72.2	1333	115.522	8337	97.93	4000	277.39	10
86.3	2666	118.963	9358	104.58	5333	305.75	100
103.1	5333	122.607	10954	109.95	6666	342.64	1000
113.9	7999	127.467	13104	114.486	7999	393.33	10000
126.8	13332	132.757	15837	121.934	10666	469.04	101325

(Continued)

TABLE 2.1.1.1.27.1 (*Continued*)

Stull 1947		Camin & Rossini 1955		Zwolinski & Wilhoit 1971		Ruzicka & Majer 1994	
summary of literature data		ebulliometry		selected values		recommended	
t/°C	P/Pa	t/°C	P/Pa	t/°C	P/Pa	T/K	P/Pa
148.0	26664	138.713	19448	127.969	13332	298.15	56.89
170.6	53329	143.585	22889	139.559	19998		
194.5	101325	150.437	28550	148.312	26664	Cox eq.	
		156.841	34826	155.431	33331	eq. 5	P/kPa
mp/°C		164.039	43031	161.475	39997	P_{ref}/kPa	101.325
		171.724	53780	171.458	53329	T_{ref}/K	469.64
		179.802	67077	179.603	66661	a	3.02711
		188.431	84053	186.537	79993	$10^3 b$/K^{-1}	−2.045579
		194.595	98132	192.608	93326	$10^6 c$/K^{-2}	1.712658
		195.242	101081	193.739	95992		
		196.511	102882	194.845	98659		
		197.272	104822	195.928	101325		
		bp/°C	195.890	bp/°C	195.928		
		eq. 2	P/mmHg	eq. 2	P/mmHg		
		A	6.97674	A	6.97220		
		B	1572.477	B	1569.57		
		C	188.022	C	187.700		
				ΔH_V/(kJ mol^{-1}) =			
				at 25°C	56.34		
				at bp	41.50		

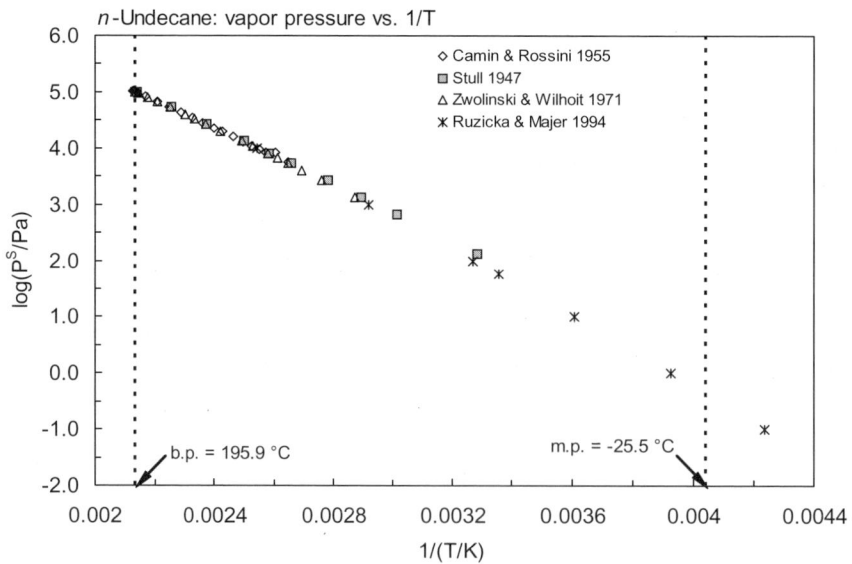

FIGURE 2.1.1.1.27.1 Logarithm of vapor pressure versus reciprocal temperature for *n*-undecane.

Aliphatic and Cyclic Hydrocarbons

2.1.1.1.28   n-Dodecane

Common Name: *n*-Dodecane
Synonym: dodecane
Chemical Name: *n*-dodecane
CAS Registry No: 112-40-3
Molecular Formula: $C_{12}H_{26}$; $CH_3(CH_2)_{10}CH_3$
Molecular Weight: 170.334
Melting Point (°C):
    –9.57   (Lide 2003)
Boiling Point (°C):
    216.32   (Lide 2003)
Density (g/cm³ at 20°C):
    0.7487, 0.7452   (20°C, 25° C, Dreisbach 1959; Riddick et al. 1986)
    0.74941   (20°C, densimeter, Dejoz et al. 1996)
Molar Volume (cm³/mol):
    227.5   (20°C, calculated-density, Stephenson & Malanowski 1987)
    273.8   (calculated-Le Bas method at normal boiling point)
Enthalpy of Vaporization, ΔH_V (kJ/mol):
    61.287, 43.64   (25°, bp, Riddick et al. 1986)
    61.52   (298.15 K, recommended, Ruzicka & Majer 1994)
Enthalpy of Fusion, ΔH_{fus} (kJ/mol):
    36.84   (Dreisbach 1959)
    35.86   (Riddick et al. 1986)
    36.82   (Chickos et al. 1999)
Entropy of Fusion, ΔS_{fus} (J/mol K):
    139.75, 128.5   (exptl., calculated-group additivity method, Chickos et al. 1999)
Fugacity Ratio at 25°C, F: 1.0

Water Solubility (g/m³ or mg/L at 25°C):
    0.0084   (shake flask-GC, Franks 1966)
    0.0034   (shake flask-GC, McAuliffe 1969)
    0.0037   (shake flask-GC, Sutton & Calder 1974)
    0.0037   (recommended, IUPAC Solubility Data Series, Shaw 1989)

Vapor Pressure (Pa at 25°C or as indicated and reported temperature dependence equations. Additional data at other temperatures designated * are compiled at the end of this section.):
    6365*   (126.31°C, ebulliometry, measured range 126.31–217.345°C, Willingham et al. 1945)
    log (P/mmHg) = 6.98059 – 1625.928/(180.311 + t/°C); temp range 126.4–217.3°C (Antoine eq. from exptl. data, ebulliometry-manometer, Willingham et al. 1945)
    133.3*   (47.7°C, summary of literature data, temp range 47.7–214.5°C, Stull 1947)
    17.60   (extrapolated-Antoine eq., Dreisbach 1959)
    log (P/mmHg) = 6.98059 – 1625.928/(180.311 + t/°C); temp range 150–280°C (Antoine eq. for liquid state, Dreisbach 1959)
    15.70*   (extrapolated-Antoine eq., temp range 91.47–247.08°C, Zwolinski & Wilhoit 1971)
    7.60   (derived from compiled data, Zwolinski & Wilhoit 1971)
    log (P/mmHg) = 6.99795 – 1639.27/(181.835 + t/°C); temp range 91.47–247.08°C (Antoine eq., Zwolinski & Wilhoit 1971)
    log (P/mmHg) = [–0.2185 × 11857.7/(T/K)] + 8.150997; temp range 47.7–345.8°C (Antoine eq., Weast 1972–73)
    15.5   (Antoine eq., Boublik et al. 1973, 1984)
    log (P/mmHg) = 6.9829 – 1627.714/(180.521 + t/°C); temp range 104.5–197.3°C (Antoine eq. from reported exptl. data of Willingham et al. 1945, Boublik et al. 1973)
    32.53   (calculated-bp, Mackay et al. 1982)

log (P/kPa) = 6.1074 − 1627.417/(180.489 + t/°C); temp range 104.5–197.3°C (Antoine eq. from reported exptl. data of Willingham et al. 1945, Boublik et al. 1984)
log (P/mmHg) = 6.99795 − 1639.27/(181.84 + t/°C); temp range 91–247°C (Antoine eq., Dean 1985, 1992)
17.3*   (24.931°C, gas saturation, measured temp range 298.081–389.66 K, Allemand et al. 1986)
16.0   (lit. average, Riddick et al. 1986)
log (P/kPa) = 6.12285 − 1639.27/(181.835 + t/°C); temp range not specified (Antoine eq., Riddick et al. 1986)
18.6   (interpolated-Antoine eq.-I, Stephenson & Malanowski 1987)
log (P_L/kPa) = 6.62064 − 1942.122/(−65.587 + T/K); temp range 278–400 K (Antoine eq.-I, Stephenson & Malanowski 1987)
log (P_L/kPa) = 6.12285 − 1639.27/(−91.315 + T/K); temp range 400–492 K (Antoine eq.-II, Stephenson & Malanowski 1987)
18.40, 18.67* (25.30, 25.35°C, electronic manometry, measured range −9.27 to 98.10°C, Sasse et al. 1988)
733.8*   (79.969°C, static-differential pressure, measured range 79.969–314.982°C, Morgan & Kobayashi 1994)
18.02   (recommended, Ruzicka & Majer 1994)
ln [(P/kPa)/(P_o/kPa)] = [1 − (T_o/K)/(T/K)]·exp{3.05854 − 2.018454 × 10^{-3}·(T/K) + 1.606849 × 10^{-6}·(T/K)^2}; reference state at P_o = 101.325 kPa, T_o = 489.438 K (Cox equation, Ruzicka & Majer 1994)
log (P/mmHg) = −5.6532 − 3.4698 × 10^3/(T/K) + 9.0272·log (T/K) − 2.3185 × 10^{-2}·(T/K) + 1.1235 × 10^{-5}·(T/K)^2; temp range 264–658 K (vapor pressure eq., Yaws 1994)
14.1; 17.8 (liquid P_L, GC-RT correlation; quoted lit., Donovan 1996)
440*   (71.6°C, ebulliometry, measured range 344.25–501.55 K, Dejoz et al. 1996)
ln (P/kPa) = 14.1090 − 3781.84/[(T/K) − 90.975]; temp range 344.25–501.55 K (ebulliometry, Dejoz et al. 1996)

Henry's Law Constant (Pa m^3/mol at 25°C):
    723000   (calculated-P/C, Bobra et al. 1979)
    723000, 786000, 317000; 750000   (calculated-P/C values; recommended, Mackay & Shiu 1981)
    721400   (selected, Mills et al. 1982)
    726900   (calculated-vapor-liquid equilibrium (VLE) data, Yaws et al. 1991)

Octanol/Water Partition Coefficient, log K_{OW}:
    7.24, 7.04   (estimated-HPLC/MS, calculated-fragment const., Burkhard et al. 1985)
    6.10   (Coates et al. 1985)
    6.80 ± 1.00 (recommended, Sangster 1989)

Octanol/Air Partition Coefficient, log K_{OA}:

Bioconcentration Factor, log BCF:

Sorption Partition Coefficient, log K_{OC}:

Environmental Fate Rate Constant, k, and Half-Lives, $t_{1/2}$:
    Volatilization: rate constants: k = 0.60 d^{-1}, $t_{1/2}$ = 1.1 d in spring at 8–16°C, k = 0.97 d^{-1}, $t_{1/2}$ = 0.7 d in summer at 20–22°C, k = 0.20 d^{-1}, $t_{1/2}$ = 3.6 d in winter at 3–7°C for the periods when volatilization appears to dominate, and k = 0.377 d^{-1}, $t_{1/2}$ = 1.8 d with HgCl$_2$, and k = 1.085 d^{-1}, $t_{1/2}$ = 0.64 d without HgCl$_2$ in September 9–15, in marine mesocosm experiments (Wakeham et al. 1983)
    Photolysis:
    Oxidation: rate constant k for gas-phase second order rate constants, k_{OH} for reaction with OH radical, k_{NO3} with NO$_3$ radical and k_{O3} with O$_3$ or as indicated, *data at other temperatures see reference:
        k_{OH} = 15.1 × 10^{-12} cm^3 molecule^{-1} s^{-1} at 312 K in a smog chamber (Nolting et al. 1988)
        k_{OH} = 14.2 × 10^{-12} cm^3 molecule^{-1} s^{-1} at 298 K (recommended, Atkinson 1989, 1990)
        k_{OH} = 13.9 × 10^{-12} cm^3 molecule^{-1} s^{-1} at 298 K (recommended, Atkinson 1997)
    Hydrolysis:
    Biodegradation:
    Biotransformation:
    Bioconcentration, Uptake (k_1) and Elimination (k_2) Rate Constants or Half-Lives:

Aliphatic and Cyclic Hydrocarbons

Half-Lives in the Environment:

    Air: Surface water: estimated $t_{1/2}$ = 0.5 d for surface waters in case of first order reduction process (Zoeteman et al. 1980)

        marine mesocosm $t_{1/2}$ = 1.1 d at 8–16°C in spring, $t_{1/2}$ = 0.7 d at 20–22°C in summer and $t_{1/2}$ = 3.6 h at 3–7°C in winter when volatilization dominates, and $t_{1/2}$ = 1.8 d with $HgCl_2$ as poison, and k = 1.085 d^{-1}, $t_{1/2}$ = 0.64 d without poison in mid-September (Wakeham et al. 1983)

TABLE 2.1.1.1.28.1
Reported vapor pressures of *n*-dodecane at various temperatures and the coefficients for the vapor pressure equations

$\log P = A - B/(T/K)$	(1)	$\ln P = A - B/(T/K)$ (1a)
$\log P = A - B/(C + t/°C)$	(2)	$\ln P = A - B/(C + t/°C)$ (2a)
$\log P = A - B/(C + T/K)$	(3)	$\ln P = A - B/(C + T/K)$ (3a)
$\log P = A - B/(T/K) - C \cdot \log(T/K)$	(4)	$\ln P = A - B/(T/K)$ (1a)
$\ln(P/P_{ref}) = [1 - (T_{ref}/T)] \cdot \exp(a + bT + cT^2)$	(5) - Cox eq.	

1.

Willingham et al. 1945		Stull 1947		Zwolinski & Wilhoit 1971		Allemand et al. 1986	
ebulliometry		summary of literature data		selected values		gas saturation	
t/°C	P/Pa	t/°C	P/Pa	t/°C	P/Pa	t/°C	P/Pa
126.31	6365	47.7	133.3	91.47	1333	24.931	17.33
131.108	7663	75.7	666.6	105.91	2666	28.95	24.66
135.223	8971	89.9	1333	115.09	4000	29.966	26.66
138.962	10306	104.3	2666	121.96	5333	35.043	40.00
142.444	11696	121.4	5333	127.521	6666	40.023	58.13
147.152	13816	131.7	7999	132.213	7999	40.29	60.26
152.529	16611	145.5	13332	139.915	10666	45.023	84.79
157.986	19917	165.8	26664	146.153	13332	49.978	120.5
163.030	23443	188.4	53329	158.131	19998	50.0	118.8
169.814	28948	214.5	101325	167.175	26664	55.32	170.7
176.039	34896			174.527	33331	59.936	232.0
183.537	43331	mp/°C	–9.6	180.769	39997	64.936	310.6
191.255	53654			191.075	53329	69.23	409.3
199.488	66750			199.481	66661	70.093	422.6
208.417	83701			206.636	79993	74.947	558.6
214.709	97594			212.898	93326	78.83	691.9
215.303	99189			214.065	95992	80.174	745.3
216.006	100671			215.206	98659	99.87	2013
216.712	102385			216.323	101325	116.5	4258
217.345	103910						
				bp/°C	216.323		
bp/°C	216.278						
eq. 2	P/mmHg			eq. 2	P/mmHg		
A	6.98059			A	6.97220		
B	1625.928			B	1639.27		
C	180.311			C	181.835		
				ΔH_V/(kJ mol^{-1}) =			
				at 25°C	61.59		
				at bp	43.64		

(Continued)

TABLE 2.1.1.1.28.1 (Continued)

2.

Sasse et al. 1988				Dejoz et al. 1996		Ruzicka & Majer 1994	
electronic manometry				vapor-liquid equilibrium		recommended	
t/°C	P/Pa	t/°C	P/Pa	t/°C	P/Pa	T/K	P/Pa
–9.27	0.587	92.17	1380	71.2	440	248.34	0.1
–3.94	1.069	95.15	1599	75.6	570	267.74	1.0
0.49	1.720	98.10	1847	81.6	800	291.28	10
5.48	2.986			86.7	1040	320.68	100
5.53	2.920	eq. 2	P/mmHg	91.1	1310	358.87	1000
10.47	4.800	A	1.17283	98.2	1860	411.27	10000
15.41	6.906	B	1766.802	102	2230	489.44	101325
15.46	7.706	C	194.662	106.8	2780	298.15	18.02
20.35	12.36			111.2	3380		
20.38	12.05			118.5	4430	Cox eq.	
25.30	18.40	Morgan & Kobayashi 1994		122.9	5540	eq. 5	P/kPa
25.35	18.67	static-differential pressure gauge		126.5	6380	P_{ref}/kPa	101.325
30.27	27.60	t/°C	P/Pa	130.6	7520	T_{ref}/K	489.438
30.30	28.13			135.2	8960	a	3.05854
35.27	40.80	79.969	733.8	143	11930	$10^3 b$/K^{-1}	–2.018454
35.30	41.33	99.970	2031	149.5	14980	$10^6 c$/K^{-2}	1.606849
40.25	59.33	119.921	4912	154.8	17930		
40.30	60.13	139.927	10690	160.8	21810		
45.15	84.93	159.973	21248	166.1	25800		
45.22	85.86	179.974	39125	171.0	29990		
50.04	119.3	199.975	67572	175.7	34530		
50.18	121.2	219.976	110500	180.8	40060		
55.02	164.9	239.976	172410	183.3	42990		
55.12	167.3	259.979	258710	186.0	46390		
60.08	227.0	279.980	375270	189.8	51530		
60.12	230.4	299.981	528820	193.2	56510		
65.11	313.2	314.982	671130	197.6	63480		
70.09	420.2			200.0	67570		
75.08	558.4	data fitted to Wagner eq.					
76.68	612.2		see ref.	eq. 3(a)	P/kPa		
80.12	735.8			A	14.1090		
82.10	820.0			B	3781.84		
85.11	949.1			C	–90.975		
87.26	1069			temp range 345–502 K			

Aliphatic and Cyclic Hydrocarbons

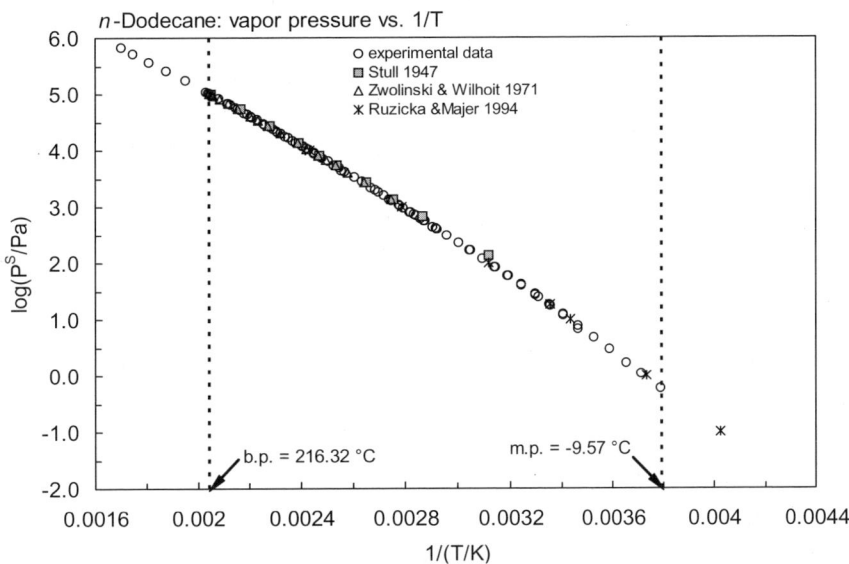

FIGURE 2.1.1.1.28.1 Logarithm of vapor pressure versus reciprocal temperature for *n*-dodecane.

2.1.1.1.29 n-Tridecane

Common Name: Tridecane
Synonym:
Chemical Name: *n*-tridecane
CAS Registry No: 629-50-5
Molecular Formula: $C_{13}H_{28}$; $CH_3(CH_2)_{11}CH_3$
Molecular Weight: 184.361
Melting Point (°C):
    –5.4   (Lide 2003)
Boiling Point (°C):
    235.47   (Lide 2003)
Density (g/cm³):
    0.76522, 0.75270   (20°C, 25°C, Camin & Rossini 1955)
    0.7564, 0.7528   (20°C, 25°C, Dreisbach 1959)
Molar Volume (cm³/mol):
    296.0   (calculated-Le Bas method at normal boiling point)
    243.7   (20°C, Stephenson & Malanowski 1987)
Enthalpy of Vaporization, ΔH_V (kJ/mol):
    66.23, 45.65 (25°C, bp, Riddick et al. 1986)
    66.68   (298.15 K, recommended, Ruzicka & Majer 1994)
Enthalpy of Fusion, ΔH_{fus} (kJ/mol):
    28.501   (Riddick et al. 1986)
    7.66, 28.49; 36.15   (–18.15, –5.35°C; total phase change enthalpy, Chickos et al. 1999)
Entropy of Fusion, ΔS_{fus} (J/mol K):
    136.31, 137.8   (exptl., calculated-group additivity method, total phase change entropy, Chickos et al. 1999)
Fugacity Ratio at 25°C, F: 1.0

Water Solubility (g/m³ or mg/L at 25°C):
    0.00104   (extrapolated from data of McAuliffe 1966, Coates et al. 1985)
    0.0047–0.0217   (estimated, Coates et al. 1985)
    0.06   (Riddick et al. 1986)
    0.33; 0.198   (measured; calculated molar volume correlation, Wang et al. 1992)

Vapor Pressure (Pa at 25°C or as indicated and reported temperature dependence equations. Additional data at other temperatures designated * are compiled at the end of this section.):
    133.3*   (59.4°C, summary of literature data, temp range 59.4–234.0°C, Stull 1947)
    5530*   (139.3°C, ebulliometry, measured range 139.3–236.065°C, Camin & Rossini 1955)
    log (P/mmHg) = 7.00339 – 1689.093/(174.284 + t/°C); temp range 145.1–236.1°C (Antoine eq., ebulliometry-manometer measurements, Camin & Rossini 1955)
    5.30   (extrapolated-Antoine eq., Dreisbach 1959; quoted, Riddick et al. 1986)
    log (P/mmHg) = 6.9887 – 1677.43/(172.90 + t/°C); temp range 131–302°C (Antoine eq. for liquid state, Dreisbach 1959)
    5.73*   (derived from compiled data, Zwolinski & Wilhoit 1971)
    log (P/mmHg) = 7.00756 – 1690.67/(107.20 + t/°C); temp range 107.2–267.04°C (Antoine eq., Zwolinski & Wilhoit 1971)
    log (P/mmHg) = [–0.2185 × 12991.3/(T/K)] + 8.481732; temp range 59.4–234°C (Antoine eq., Weast 197–73)
    4.55   (extrapolated-Antoine eq., Boublik et al. 1973, 1984)
    log (P/mmHg = 7.00925 – 1693.684/(t/°C + 174.815); temp range 139–236°C (Antoine eq. from reported exptl. data of Camin & Rossini 1955, Boublik et al. 1973)
    log (P/kPa) = 6.13542 – 1694.624/(t/°C + 174.916); temp. range 139–236°C (Antoine eq. from reported exptl data of Camin & Rossini 1955, Boublik et al. 1984)
    log (P/mmHg) = 7.00756 – 1690.67/(174.22 + t/°C); temp range 107–267°C (Antoine eq., Dean 1985, 1992)

Aliphatic and Cyclic Hydrocarbons

log (P/kPa) = 6.13246 – 1690.67/(174.220 + t/°C); temp range not specified (Antoine eq., Riddick et al. 1986)

log (P/mmHg) = 7.00756 – 1690.67/(t/°C + 174.22); temp range 107–267°C, (Antoine eq., Dean 1985; 1992)

log (P/kPa) = 6.13546 – 1690.67/(T/K – 98.93); temp range 417–511 K (Antoine eq., liquid, Stephenson & Malanowski 1987)

5.682*   (recommended, Ruzicka & Majer 1994)

ln [(P/kPa)/(P_o/kPa)] = [1 – (T_o/K)/(T/K)]·exp{3.10403 – 2.071819 × 10^{-3}·(T/K) + 1.61160 × 10^{-6}·(T/K)^2}; reference state at P_o = 101.325 kPa, T_o = 508.602 K (Cox equation, Ruzicka & Majer 1994)

log (P/mmHg) = 49.2391 – 4.9649 × 10^3/(T/K) – 13.769·log (T/K) –2.1146 × 10^{-9}·(T/K) + 2.5902 × 10^{-6}·(T/K)^2; temp range 268–676 K (vapor pressure eq., Yaws 1994)

Henry's Law Constant (Pa·m^3/mol at 25°C):

233351   (calculated-vapor-liquid equilibrium (VLE) data, Yaws et al. 1991)

Octanol/Water Partition Coefficient, log K_{OW}:

6.65   (HPLC-k' correlation, Coates et al. 1985)

6.50; 6.05 (calculated-fragment const.; calculated-molar volume, Wang et al. 1992)

Octanol/Air Partition Coefficient, log K_{OA}:

Bioconcentration Factor, log BCF or log K_B:

Sorption Partition Coefficient, log K_{OC}:

Environmental Fate Rate Constants, k, and Half-Lives, $t_{1/2}$:

Volatilization:

Photolysis:

Oxidation: rate constant k for gas-phase second order rate constants, k_{OH} for reaction with OH radical, k_{NO3} with NO_3 radical and k_{O3} with O_3 or as indicated, *data at other temperatures see reference:

k_{OH} = 17.5 × 10^{-12} cm^3 molecule^{-1} s^{-1} at 312 K in a smog chamber (Nolting et al. 1988)

k_{OH} = 15.4 × 10^{-12} cm^3 molecule^{-1} s^{-1} at 300 K, 17.4 × 10^{-12} cm^3 molecule^{-1} s^{-1} at 302 K (Atkinson 1989)

k_{OH} = 16.0 × 10^{-12} cm^3 molecule^{-1} s^{-1} at 298 K (recommended, Atkinson 1989, 1990, 1997)

Hydrolysis:

Biodegradation:

Biotransformation:

Bioconcentration and Uptake and Elimination Rate Constants (k_1 and k_2):

Half-Lives in the Environment:

TABLE 2.1.1.1.29.1
Reported vapor pressures of *n*-tridecane at various temperatures and the coefficients for the vapor pressure equations

log P = A – B/(T/K)	(1)	ln P = A – B/(T/K)	(1a)
log P = A – B/(C + t/°C)	(2)	ln P = A – B/(C + t/°C)	(2a)
log P = A – B/(C + T/K)	(3)		
log P = A – B/(T/K) – C·log (T/K)	(4)		
ln (P/P_{ref}) = [1 – (T_{ref}/T)]·exp(a + bT + cT^2)	(5) - Cox eq.		

Stull 1947		Camin & Rossini 1955		Zwolinski & Wilhoit 1971		Ruzicka & Majer 1994	
summary of literature data		ebulliometry		selected values		recommended	
t/°C	P/Pa	t/°C	P/Pa	t/°C	P/Pa	T/K	P/Pa
59.4	133.3	139.3	5530	107.2	1333	260.30	0.1
98.3	666.6	145.16	6931	122.05	2666	280.31	1.0

(Continued)

TABLE 2.1.1.1.29.1 (Continued)

| Stull 1947 | | Camin & Rossini 1955 | | Zwolinski & Wilhoit 1971 | | Ruzicka & Majer 1994 | |
| summary of literature data | | ebulliometry | | selected values | | recommended | |
t/°C	P/Pa	t/°C	P/Pa	t/°C	P/Pa	T/K	P/Pa
104.0	1333	150.011	8326	131.48	4000	304.58	10
120.2	2666	157.603	10934	138.55	5333	335.89	100
137.7	5333	162.749	13079	144.258	6666	374.24	1000
148.2	7999	174.699	19416	149.08	7999	428.21	10000
162.5	13332	187.176	28544	156.993	10666	508.60	101325
185.0	26664	201.634	43155	163.404	13332	298.15	5.682
209.4	53329	209.788	53768	175.709	19998		
234.0	101325	218.367	67061	184.998	26664	Cox eq.	
		227.524	84040	192.55	33331	eq. 5	P/kPa
mp/°C	–6.2	234.052	85041	198.96	39997	P_{ref}/kPa	101.325
		235.316	101054	209.543	53329	T_{ref}/K	469.64
		236.065	102829	218.175	66661	a	3.10403
				225.521	79993	$10^3 b$/K^{-1}	–2.071819
				231.95	93326	$10^6 c$/K^{-2}	1.712658
				233.148	95992		
				234.319	98659		
				235.466	101325		
				25.0	5.73		
				bp/°C	235.466		
				eq. 2	P/mmHg		
				A	7.00756		
				B	1690.67		
				C	174.220		
				ΔH_V/(kJ mol^{-1}) =			
				at 25°C	66.23		
				at bp	45.65		

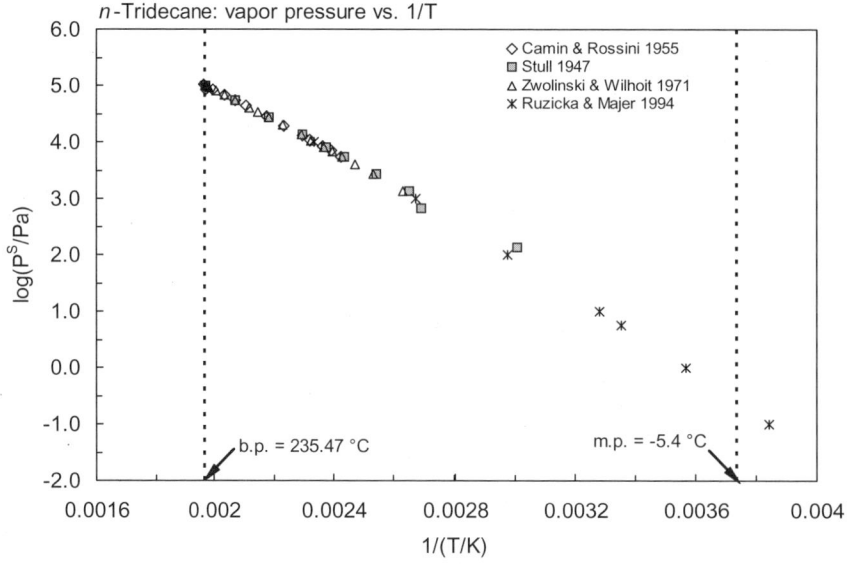

FIGURE 2.1.1.1.29.1 Logarithm of vapor pressure versus reciprocal temperature for *n*-tridecane.

Aliphatic and Cyclic Hydrocarbons

2.1.1.1.30   n-Tetradecane

Common Name: Tetradecane
Synonym:
Chemical Name: *n*-tetradecane
CAS Registry No: 629-59-4
Molecular Formula: $C_{14}H_{30}$; $CH_3(CH_2)_{12}CH_3$
Molecular Weight: 198.388
Melting Point (°C):
    5.82   (Lide 2003)
Boiling Point (°C):
    253.58   (Lide 2003)
Density (g/cm³):
    0.76275, 0.75917   (20°C, 25°C, Camin & Rossini 1955)
    0.7628, 0.7593   (20°C, 25°C, Dreisbach 1959)
Molar Volume (cm³/mol):
    260.1   (20°C, Stephenson & Malanowski 1987)
    318.2   (calculated-Le Bas method at normal boiling point, Eastcott et al. 1988)
Enthalpy of Vaporization, ΔH_V (kJ/mol):
    71.13, 47.73 (25°C, bp, Dreisbach 1961)
    71.73   (298.15 K, recommended, Ruzicka & Majer 1994)
Enthalpy of Fusion, ΔH_{fus} (kJ/mol):
    45.07   (Chickos et al. 1999)
Entropy of Fusion, ΔS_{fus} (J/mol K):
    161.54, 147.1   (exptl., calculated-group additivity method, Chickos et al. 1999)
Fugacity Ratio at 25°C, F: 1.0

Water Solubility (g/m³ or mg/L at 25°C):
    0.0069   (shake flask-GC, Frank 1966)
    0.00655   (extrapolated, McAuliffe 1966)
    0.0022; 0.0017 (shake flask-GC, distilled water; seawater, Sutton & Calder 1974)
    0.00033   (shake flask-GC, Coates et al. 1985)

Vapor Pressure (Pa at 25°C or as indicated and reported temperature dependence equations. Additional data at other temperatures designated * are compiled at the end of this section.):
    133.3*   (76.4°C, summary of literature data, temp range 76.4–252.5°C, Stull 1947)
    5532*   (154.860°C, ebulliometry, measured range 154.860–254.165°C, Camin & Rossini 1955)
    log (P/mmHg) = 7.01245 – 1739.623/(167.534 + t/°C); temp range 165.9–254.2°C (Antoine eq., ebulliometry-manometer measurements, Camin & Rossini 1955)
    1.56   (extrapolated-Antoine eq., Dreisbach 1959)
    log (P/mmHg) = 6.9957 – 1725.46/(165.75 + t/°C); temp range 147–325°C (Antoine eq. for liquid state, Dreisbach 1959)
    1.27*   (Antoine eq., temp range 121.80–286.0°C, Zwolinski & Wilhoit 1971)
    1.867   (derived from compiled data, Zwolinski & Wilhoit 1971)
    log (P/mmHg) = 7.01300 – 1740.88/(167.720 + t/°C); temp range 121.80–286.0°C (Antoine eq., Zwolinski & Wilhoit 1971)
    log (P/mmHg) = [–0.2185 × 13750.0/(T/K)] + 8.628699; temp range 76.4–252.5°C (Antoine eq., Weast 1972–73)
    1.30   (extrapolated-Antoine eq., Boublik et al. 1973, 1984)
    log (P/mmHg) = 7.02216 – 1747.452/(t/°C + 168.437); temp range 155–254°C (Antoine eq. from reported exptl. data of Camin & Rossini 1955, Boublik et al. 1973)
    log (P/kPa) = 6.14914 – 1749.052/(t/°C + 168.611); temp range 155–254°C (Antoine eq. from reported exptl. data of Camin & Rossini 1955, Boublik et al. 1984)

log (P/mmHg) = 7.01300 – 1740.88/(t/°C + 167.72); temp. range 112–286°C (Antoine eq., Dean 1985; 1992)

72.0*   (70.01°C, gas saturation, measured temp range 343.16–394.73 K, Allemand et al. 1986)

log (P_L/kPa) = 6.62828 – 2063.84/(T/K – 77.378); temp range 313–433 K (Antoine eq., liquid, Stephenson & Malanowski 1987)

log (P_L/kPa) = 6.1379 – 1740.88/(T/K – 105.43); temp. range 432–529 K (Antoine eq., liquid, Stephenson & Malanowski 1987)

442.9*   (99.926°C, static-differential pressure, measured range 99.926–314.982°C, Morgan & Kobayashi 1994)

1.804*   (recommended, Ruzicka & Majer 1994)

ln [(P/kPa)/(P_o/kPa)] = [1 – (T_o/K)/(T/K)]·exp{3.13624 – 2.063853 × 10^{-3}·(T/K) + 1.541507 × 10^{-6}·(T/K)^2}; reference state at P_o = 101.325 kPa, T_o = 526.691 K (Cox equation, Ruzicka & Majer 1994)

log (P/mmHg) = 106.1056 – 7.3461 × 10^3/(T/K) – 31.5195·log (T/K) + 1.2356 × 10^{-2}·(T/K) – 8.3955 × 10^{-13}·(T/K)^2; temp range 279–692 K (vapor pressure eq., Yaws 1994)

Henry's Law Constant (Pa·m³/mol at 25°C):

347000   (calculated-P/C, Mackay & Shiu 1981)

387000   (calculated-P/C, Eastcott et al. 1988)

114497   (calculated-vapor-liquid equilibrium (VLE) data, Yaws et al. 1991)

Octanol/Water Partition Coefficient, log K_{OW}:

7.20   (HPLC-k' correlation, Coates et al. 1985)

7.88, 8.10   (RP-HPLC-MS correlation, Burkhard et al. 1985)

7.00; 6.45   (calculated-fragment const.; calculated-molar volume, Wang et al. 1992)

8.0   (recommended, Sangster 1989, 1993)

6.49   (calculated-UNIFAC, Chen et al. 1993)

Octanol/Air Partition Coefficient, log K_{OA}:

Bioconcentration Factor, log BCF or log K_B:

Sorption Partition Coefficient, log K_{OC}:

Environmental Fate Rate Constants, k, and Half-Lives, $t_{1/2}$:

Volatilization:

Photolysis:

Oxidation: rate constant k for gas-phase second order rate constants, k_{OH} for reaction with OH radical, k_{NO3} with NO_3 radical and k_{O3} with O_3 or as indicated, *data at other temperatures see reference:

k_{OH} = 19.3 × 10^{-12} cm³ molecule^{-1} s^{-1} at 312 K in a smog chamber (Nolting et al. 1988)

k_{OH} = 19.2 × 10^{-12} cm³ molecule^{-1} s^{-1} at 312 K (Atkinson 1989)

k_{OH} = 19.0 × 10^{-12} cm³ molecule^{-1} s^{-1} at 298 K (Atkinson 1990)

k_{OH} = 18.0 × 10^{-12} cm³ molecule^{-1} s^{-1} at 298 K (recommended, Atkinson 1997)

Hydrolysis:

Biodegradation: microbial degradation $t_{1/2}$ < 15 d by *Pseudomonas sp.* (Setti et al. 1993)

Biotransformation:

Bioconcentration and Uptake and Elimination Rate Constants (k_1 and k_2):

Half-Lives in the Environment:

Aliphatic and Cyclic Hydrocarbons

TABLE 2.1.1.1.30.1
Reported vapor pressures of *n*-tetradecane at various temperatures and the coefficients for the vapor pressure equations

$$\log P = A - B/(T/K) \quad (1) \qquad \ln P = A - B/(T/K) \quad (1a)$$
$$\log P = A - B/(C + t/°C) \quad (2) \qquad \ln P = A - B/(C + t/°C) \quad (2a)$$
$$\log P = A - B/(C + T/K) \quad (3)$$
$$\log P = A - B/(T/K) - C \cdot \log(T/K) \quad (4)$$
$$\ln(P/P_{ref}) = [1 - (T_{ref}/T)] \cdot \exp(a + bT + cT^2) \quad (5) \text{ - Cox eq.}$$

1.

Stull 1947		Camin & Rossini 1955		Zwolinski & Wilhoit 1971		Allemand et al. 1986	
summary of literature data		ebulliometry		selected values		gas saturation	
t/°C	P/Pa	t/°C	P/Pa	t/°C	P/Pa	t/°C	P/Pa
76.4	133.3	154.860	5532	121.8	1333	70.01	71.99
106.0	666.6	165.911	6325	137.06	2666	70.45	74.53
120.7	1333	173.737	10928	146.75	4000	80.0	136.8
135.6	2666	191.234	19417	154.01	5333	91.0	262.7
152.7	5333	204.019	28546	159.881	6666	106.78	634.6
164.0	7999	218.84	43156	164.836	7999	121.58	1324
178.5	13332	236.013	67061	172.967	10666		
201.8	26664	245.408	84041	179.553	13332		
226.8	53329	252.104	98106	192.196	19998		
252.5	101325	253.401	101061	201.739	26664		
		254.165	102850	209.497	33331		
mp/°C	5.5			216.082	39997		
		bp/°C	253.516	226.953	53329		
				235.819	66661		
				243.364	79993		
				249.967	93326		
				251.197	95992		
				252.4	98659		
				253.577	101325		

(Continued)

TABLE 2.1.1.1.30.1 (*Continued*)

2.

Morgan & Kobayashi 1994		Ruzicka & Majer 1994	
differential pressure gauge		recommended	
t/°C	P/Pa	T/K	P/Pa
99.926	442.9	271.60	0.1
119.936	1227	292.22	1.0
139.946	3026	317.21	10
159.956	6693	348.38	100
179.965	13526	388.82	1000
199.976	25267	444.22	10000
219.976	44274	526.69	101325
229.964	57442	298.15	1.804
239.977	73530		
249.979	93059	Cox eq.	
259.979	116500	eq. 5	P/kPa
249.979	116520	P_{ref}/kPa	101.325
279.980	177570	T_{ref}/K	526.691
299.980	261560	a	3.13624
309.982	314050	$10^3 b/K^{-1}$	−2.063853
314.982	343450	$10^6 c/K^{-2}$	1.541507

data fitted to Wagner eq.

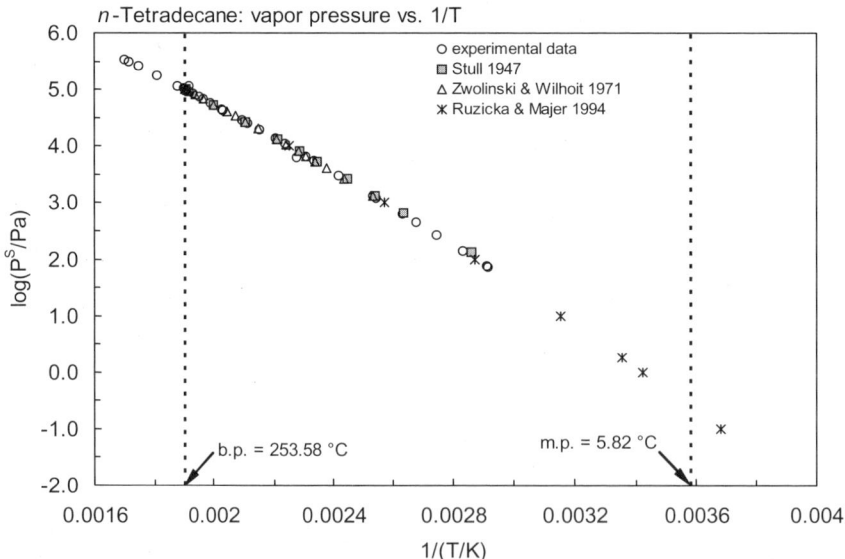

FIGURE 2.1.1.1.30.1 Logarithm of vapor pressure versus reciprocal temperature for *n*-tetradecane.

Aliphatic and Cyclic Hydrocarbons

2.1.1.1.31 n-Pentadecane

Common Name: Pentadecane
Synonym:
Chemical Name: *n*-pentadecane
CAS Registry No: 629-62-9
Molecular Formula: $C_{15}H_{32}$; $CH_3(CH_2)_{13}CH_3$
Molecular Weight: 212.415
Melting Point (°C):
    9.95   (Lide 2003)
Boiling Point (°C):
    270.6   (Camin & Rossini 1955; Dreisbach 1959; Stephenson & Malanowski 1987; Lide 2003)
Density (g/cm³):
    0.76830, 0.76488 (20°C, 25°C, Camin & Rossini 1955)
    0.7685, 0.7650 (20°C, 25°C, Dreisbach 1959)
Molar Volume (cm³/mol):
    276.4   (20°C, calculated-density)
    340.4   (calculated-Le Bas method at normal boiling point)
Enthalpy of Vaporization, ΔH_V (kJ/mol):
    76.16, 49.41 (25°C, bp, Dreisbach 1961)
    76.77   (298.15 K, recommended, Ruzicka & Majer 1994)
Enthalpy of Fusion, ΔH_{fus} (kJ/mol):
    9.17, 34.6; 43.77 (–2.25, 9.95; total phase change enthalpy, Chickos et al. 1999)
Entropy of Fusion, ΔS_{fus} (J/mol K):
    156.02, 156.5 (exptl., calculated-group additivity method, total phase change entropy, Chickos et al. 1999)
Fugacity Ratio at 25°C, F: 1.0

Water Solubility (g/m³ or mg/L at 25°C):
    7.6×10^{-5} (extrapolated from data of McAuliffe 1966, Coates et al. 1985)
    0.0612; 0.0613 (measured; calculated-molar volume correlation, Wang et al. 1992)

Vapor Pressure (Pa at 25°C or as indicated and reported temperature dependence equations. Additional data at other temperatures designated * are compiled at the end of this section.):
    133.3*   (91.6°C, summary of literature data, temp range 91.6–270.5°C, Stull 1947)
    5532*   (169.686°C, ebulliometry, measured range 169.686–270.449°C, Camin & Rossini 1955)
    log (P/mmHg) = 7.02445 – 1789.658/(161.291 + t/°C); temp range 169.6–270.6°C (Antoine eq., ebulliometry-manometer measurements, Camin & Rossini 1955)
    0.311   (extrapolated-Antoine eq., Dreisbach 1959)
    log (P/mmHg) = 7.0017 – 1768.82/(158.60 + t/°C); temp range 160–338°C (Antoine eq. for liquid state, Dreisbach 1959)
    1333*   (135.8°C, derived from compiled data, temp range 135.8–270.685°C, Zwolinski & Wilhoit 1971)
    log (P/mmHg) = 7.01359 – 1789.95/(161.380 + t/°C); temp range 135.8–303.8°C (Antoine eq., Zwolinski & Wilhoit 1971)
    log (P/mmHg) = [–0.2185 × 14635.9/(T/K)] + 8.822087; temp range 91.6–270.5°C (Antoine eq., Weast 1972–73)
    0.356   (extrapolated-Antoine eq., Boublik et al. 1973)
    log P/mmHg = 7.03121 – 1797.239/(t/°C + 164.128); temp range 170–270°C (Antoine eq. from exptl. data of Camin & Rossini 1955, Boublik et al. 1973)
    0.359   (extrapolated-Antoine eq., Boublik et al. 1984)
    log (P/kPa) = 6.15888 – 1797.239/(t/°C + 162.128); temp range 170–270.5°C (Antoine eq. derived from exptl. data of Camin & Rossini 1955, Boublik et al. 1984)
    log (P/mmHg) = 7.02359 – 1789.95/(161.38 + t/°C); temp range 136–304°C (Antoine eq., Dean 1985, 1992)
    15.87*   (60.0°C, gas saturation, measured temp range 60.0–136.0°C, Allemand et al. 1986)

log $(P_L/kPa) = 6.38149 - 1945.469/(T/K - 97.875)$; temp range 366–409 K (Antoine eq., liquid, Stephenson & Malanowski 1987)

log $(P_L/kPa) = 6.14849 - 1789.95/(T/K - 111.77)$; temp range 447–546 K (Antoine eq., liquid, Stephenson & Malanowski 1987)

0.576*   (recommended, Ruzicka & Majer 1994)

ln $[(P/kPa)/(P_o/kPa)] = [1 - (T_o/K)/(T/K)] \cdot \exp\{3.16144 - 2.062348 \times 10^{-3} \cdot (T/K) + 1.487263 \times 10^{-6} \cdot (T/K)^2\}$; reference state at $P_o = 101.325$ kPa, $T_o = 543.797$ K (Cox equation, Ruzicka & Majer 1994)

log $(P/mmHg) = 116.5157 - 8.041 \times 10^3/(T/K) - 38.799 \cdot \log(T/K) - 1.3398 \times 10^{-2} \cdot (T/K) - 4.4444 \times 10^{-6} \cdot (T/K)^2$; temp range 283–707 K (vapor pressure eq., Yaws 1994)

Henry's Law Constant (Pa·m^3/mol at 25°C):
48535   (calculated-vapor-liquid equilibrium (VLE) data, Yaws et al. 1991)

Octanol/Water Partition Coefficient, log K_{OW}:
7.72   (HPLC-k′ correlation, Coates et al. 1985)
7.50; 6.78   (calculated-fragment const.; calculated-molar volume correlation, Wang et al. 1992)

Octanol/Air Partition Coefficient, log K_{OA}:

Bioconcentration Factor, log BCF or log K_B:

Sorption Partition Coefficient, log K_{OC}:

Environmental Fate Rate Constant, k, and Half-Lives, $t_{1/2}$:

Volatilization: volatilization rate constant of k = 0.69 d^{-1} with a water column $t_{1/2}$ = 1.0 d at 6–18°C, in spring; k = 0.85 d^{-1} with a water column $t_{1/2}$ = 0.8 d at 20–22°C in summer; k = 0.16 d^{-1} with a water column $t_{1/2}$ = 4.3 d at 3–7°C in winter for mesocosm experiment in coastal marine environment when volatilization dominates, volatilization k = 0.343 d^{-1} with a water column $t_{1/2}$ = 2.0 d with HgCl$_2$ poisoned water tank and k = 1.241 d^{-1} with a water column $t_{1/2}$ = 0.56 d for non-poisoned water tank in late summer (Wakeham et al. 1983).

Photolysis:

Oxidation: rate constant k, for gas-phase second order rate constants, k_{OH} for reaction with OH radical, k_{NO3} with NO$_3$ radical and k_{O3} with O$_3$ or as indicated, *data at other temperatures see reference:

$k_{OH} = 22.3 \times 10^{-12}$ cm^3 molecule^{-1} s^{-1} at 312 K in a smog chamber (Nolting et al. 1988)

$k_{OH} = 22.2 \times 10^{-12}$ cm^3 molecule^{-1} s^{-1} at 312 K (Atkinson 1989)

$k_{OH} = 22.0 \times 10^{-12}$ cm^3 molecule^{-1} s^{-1} at 298 K (Atkinson 1990)

$k_{OH} = 22.2 \times 10^{-12}$ cm^3 molecule^{-1} s^{-1} (experimental); 17.87×10^{-12} cm^3 molecule^{-1} s^{-1} (Atmospheric Oxidation Program); and 7.47×10^{-12} cm^3 molecule^{-1} s^{-1} (Fate of Atmospheric Pollutants) for gas-phase reaction with OH radicals (Meylan & Howard 1993)

$k_{OH} = 21.0 \times 10^{-12}$ cm^3 molecule^{-1} s^{-1} at 298 K (recommended, Atkinson 1997)

Hydrolysis:

Biodegradation: microbial degradation $t_{1/2}$ < 15 d by *Pseudomonas sp.* (Setti et al. 1993)

Biotransformation:

Bioconcentration and Uptake and Elimination Rate Constants (k_1 and k_2):

Half-Lives in the Environment:

Surface water: water column half-lives, $t_{1/2}$ = 1.0 d, 6–18°C, in spring; $t_{1/2}$ = 0.8 d, 20–22°C, in summer; $t_{1/2}$ = 4.3 d, 3–7°C, in winter for mesocosm experiment in coastal marine mesocosm; $t_{1/2}$ = 2.0 d with HgCl poisoned water tank and $t_{1/2}$ = 0.56 d for non-poisoned water tank in late summer (Wakeham et al. 1983).

Aliphatic and Cyclic Hydrocarbons

TABLE 2.1.1.1.31.1
Reported vapor pressures of *n*-pentadecane at various temperatures and the coefficients for the vapor pressure equations

$\log P = A - B/(T/K)$ (1) $\quad\quad \ln P = A - B/(T/K)$ (1a)
$\log P = A - B/(C + t/°C)$ (2) $\quad\quad \ln P = A - B/(C + t/°C)$ (2a)
$\log P = A - B/(C + T/K)$ (3)
$\log P = A - B/(T/K) - C \cdot \log(T/K)$ (4)
$\ln(P/P_{ref}) = [1 - (T_{ref}/T)] \cdot \exp(a + bT + cT^2)$ (5) - Cox eq.

Stull 1947		Camin & Rossini 1955		Zwolinski & Wilhoit 1971		Allemand et al. 1986	
summary of literature data		ebulliometry		selected values		gas saturation	
t/°C	P/Pa	t/°C	P/Pa	t/°C	P/Pa	t/°C	P/Pa
91.6	133.3	169.686	5532	135.8	1333	60.0	15.87
121.0	666.6	180.919	8325	151.41	2666	72.84	35.60
135.4	1333	188.905	10927	161.45	4000	81.53	64.13
150.2	2666	206.886	19417	168.78	5333	94.31	148.0
167.7	5333	219.982	26546	174.785	6666	111.0	394.0
178.4	7999	235.150	43159	179.859	7999	136.0	1465
194.0	13332	151.703	67061	188.185	10666		
216.1	26664	262.310	84081	194.949	13332		
242.8	53329	269.164	98106	207.872	19998	**Ruzicka & Majer 1994**	
270.5	101325	270.449	101069	217.641	26664	recommended	
				225.582	33331	T/K	P/Pa
mp/°C	10	bp/°C	270.613	232.517	39997		
				243.446	53329	282.39	0.1
				252.517	66661	303.58	1.0
				260.237	79993	329.24	10
				266.992	93326	361.22	100
				268.25	95992	402.67	1000
				269.48	98659	459.40	10000
				270.685	101325	543.80	101325
						298.15	0.576
				bp/°C	270.685		
						Cox eq.	
				eq. 2	P/mmHg	eq. 5	P/kPa
				A	7.02359	P_{ref}/kPa	101.325
				B	1789.95	T_{ref}/K	543.64
				C	161.380	a	3.16774
						$10^3 b$/K^{-1}	−2.062348
				ΔH_V/(kJ mol^{-1}) =		$10^6 c$/K^{-2}	1.487263
				at 25°C	76.15		
				at bp	49.45		

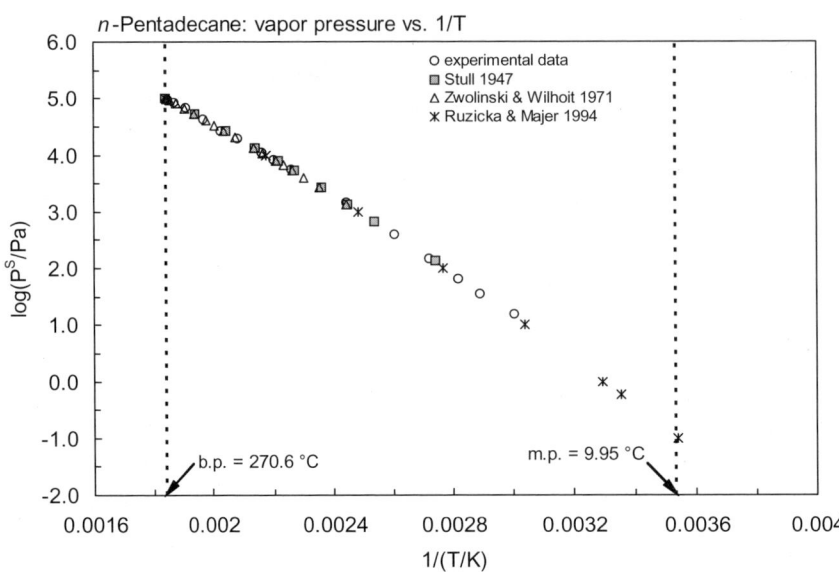

FIGURE 2.1.1.1.31.1 Logarithm of vapor pressure versus reciprocal temperature for *n*-pentadecane.

Aliphatic and Cyclic Hydrocarbons

2.1.1.1.32   n-*Hexadecane*

Common Name: Hexadecane
Synonym: cetane
Chemical Name: *n*-hexadecane
CAS Registry No: 544-76-3
Molecular Formula: $C_{16}H_{34}$; $CH_3(CH_2)_{14}CH_3$
Molecular Weight: 226.441
Melting Point (°C):
    18.12   (Lide 2003)
Boiling Point (°C):
    286.86   (Lide 2003)
Density (g/cm³):
    0.77344, 0.76996 (20°C, 25°C, Camin et al. 1954; Dreisbach 1959)
    0.7733   (20°C, Weast 1982–83)
Molar Volume (cm³/mol):
    292.8   (20°C, calculated-density, Stephenson & Malanowski 1987)
    362.2   (calculated-Le Bas molar volume at normal boiling point, Eastcott et al. 1988)
Enthalpy of Vaporization, ΔH_V (kJ/mol):
    81.35   (298.15 K, recommended, Ruzicka & Majer 1994)
Enthalpy of Fusion, ΔH_{fus} (kJ/mol):
    53.35, 51.46; 53.35 (18.15, 17.95°C; total phase change enthalpy, Chickos et al. 1999)
Entropy of Fusion, ΔS_{fus} (J/mol K):
    176.79, 165.8 (exptl., calculated-group additivity method, total phase change entropy, Chickos et al. 1999)
Fugacity Ratio at 25°C, F: 1.0

Water Solubility (g/m³ or mg/L at 25°C):

    6.28×10^{-3} (shake flask, Franks 1966)
    2.1×10^{-5} (extrapolated data of McAuliffe 1966, Coates et al. 1985)
    5.21×10^{-5} (extrapolated from data of McAuliffe 1996, Eastcott et al. 1988)
    9.0×10^{-4}, 4.0×10^{-4} (shake flask-GC, distilled water; seawater, Sutton & Calder 1974)
    2.33×10^{-6} (calculated-TSA, Lande et al. 1985)
    0.0272   (calculated-molar volume correlation, Wang et al. 1992)
    4.95×10^{-5} (calculated-molar volume and mp., Ruelle Kesselring 1997)

Vapor Pressure (Pa at 25°C or as indicated and reported temperature dependence equations. Additional data at other temperatures designated * are compiled at the end of this section.):

    133.3*   (105.3°C, summary of literature data, temp range 105.3–287.5°C, Stull 1947)
    6945*   (190.054°C, ebulliometry, measured range 190.054–286.704°C, Camin et al. 1954)
    13.33*   (81.0°C, static method-Hg manometer, measured range 81.0–286.0°C, Myers & Fenske 1955)
    log (P/mmHg) = 7.03044 – 1831.317/(154.528 + t/°C); temp range 190.0–286.8°C (Antoine eq., ebulliometry-manometer measurements, Camin et al. 1954)
    0.14   (71.87°C, calculated-Antoine eq., Dreisbach 1959)
    0.221*   (derived from compiled data, Zwolinski & Wilhoit 1971)
    log (P/mmHg) = 7.02867 – 1830.51/(154.450 + t/°C); temp range 149.18–320.7°C (Antoine eq., Zwolinski & Wilhoit 1971)
    log (P/mmHg) = [–0.2185 × 15405.5/(T/K)] + 8.956267; temp range 105.3–287.5°C (Antoine eq., Weast 1972–73)
    0.092   (extrapolated-Antoine eq., Boublik et al. 1973, 1984)
    log (P/mmHg) = 7.03519 – 1835.24/(t/°C + 154.968); temp range 188–285°C (Antoine eq. from exptl. data of Camin et al. 1945, Boublik et al. 1973)
    log (P/kPa) = 6.16189 – 1836.287/(t/°C + 155.125); temp range 190–287°C (Antoine eq. from exptl. data of Camin et al. 1954, Boublik et al. 1984)

log (P/mmHg) = 7.02867 − 1830.51/(154.45 + t/°C); temp range 149–321°C (Antoine eq., Dean 1985, 1992)
log (P$_L$/kPa) = 6.77064 − 2273.168/(T/K − 80.252); temp range 323–425 K (Antoine eq., liquid, Stephenson & Malanowski 1987)
log (P$_L$/kPa) = 6.15357 − 1830.57/(T/K − 118.7); temp range 467–563 K (Antoine eq., liquid, Stephenson & Malanowski 1987)
log (P/mmHg) = 7.02967 − 1830.51/(t/°C + 154.45); temp range 149–321°C (Antoine eq., Dean 1992)
log (P/mmHg) = 99.1091 − 7.5333 × 10^3/(T/K) − 32.251·log (T/K) + 1.0453 × 10^{-2}·(T/K) + 1.2328 × 10^{-12}·(T/K)^2; temp range 291–721 K (vapor pressure eq., Yaws 1994)

317.6*   (119.896°C, static-differential pressure, measured range 119.896–309.982°C, Morgan & Kobayashi 1994)
0.191*   (recommended, Ruzicka & Majer 1994)
ln [(P/kPa)/(P$_o$/kPa)] = [1 − (T$_o$/K)/(T/K)]·exp{3.18271 − 2.002545 × 10^{-3}·(T/K) + 1.384476 × 10^{-6}·(T/K)^2}; reference state at P$_o$ = 101.325 kPa, T$_o$ = 559.978 K (Cox equation, Ruzicka & Majer 1994)
0.190   (GC-retention time correlation, Chickos & Hanshaw 2004)

Henry's Law Constant (Pa·m^3/mol at 25°C):
389000   (calculated-P/C, Eastcott et al. 1988)
23072   (calculated-vapor-liquid equilibrium (VLE) data, Yaws et al. 1991)

Octanol/Water Partition Coefficient, log K_{OW}:
8.25   (HPLC-k′ correlation, Coates et al. 1985)
8.00; 7.26   (calculated-fragment const.; calculated-molar volume correlation, Wang et al. 1992)

Octanol/Air Partition Coefficient, log K_{OA}:

Bioconcentration Factor, log BCF or log K_B:

Sorption Partition Coefficient, log K_{OC}:

Environmental Fate Rate Constants, k, and Half-Lives, $t_{1/2}$:
Volatilization:
Photolysis:
Oxidation: rate constant k, for gas-phase second order rate constants, k_{OH} for reaction with OH radical, k_{NO3} with NO$_3$ radical and k_{O3} with O$_3$ or as indicated, *data at other temperatures see reference:
    k_{OH} = 25.0 × 10^{-12} cm^3 molecule^{-1} s^{-1} at 312 K (Atkinson 1989)
    k_{OH} = 25.0 × 10^{-12} cm^3 molecule^{-1} s^{-1} at 298 K (Atkinson 1990)
    k_{OH} = 23.0 × 10^{-12} cm^3 molecule^{-1} s^{-1} at 298 K (recommended, Atkinson 1997)
Hydrolysis:
Biodegradation: microbial degradation $t_{1/2}$ < 31 d by *Pseudomonas* sp. (Setti et al. 1993)
Biotransformation:
Bioconcentration and Uptake and Elimination Rate Constants (k$_1$ and k$_2$):

Half-Lives in the Environment:

Aliphatic and Cyclic Hydrocarbons

TABLE 2.1.1.1.32.1
Reported vapor pressures of *n*-hexadecane at various temperatures and the coefficients for the vapor pressure equations

$\log P = A - B/(T/K)$ (1) $\qquad \ln P = A - B/(T/K)$ (1a)

$\log P = A - B/(C + t/°C)$ (2) $\qquad \ln P = A - B/(C + t/°C)$ (2a)

$\log P = A - B/(C + T/K)$ (3)

$\log P = A - B/(T/K) - C \cdot \log (T/K)$ (4)

$\ln (P/P_{ref}) = [1 - (T_{ref}/T)] \cdot \exp(a + bT + cT^2)$ (5) - Cox eq.

1.

Stull 1947		Camin et al. 1954		Myers & Fenske 1955		Zwolinski & Wilhoit 1971	
summary of literature data		ebulliometry		static method-Hg manometer		selected values	
t/°C	P/Pa	t/°C	P/Pa	t/°C	P/Pa	t/°C	P/Pa
105.3	133.3	190.054	6945	81.0	13.33	149.18	1333
135.2	666.6	193.301	8330	94.2	66.66	165.14	2666
149.8	1333	199.273	9351	105.3	133.3	175.28	4000
164.7	2666	203.437	10951	117.2	266.6	182.87	5333
181.2	5333	208.962	13098	124.9	400.0	189.00	6666
193.2	7999	215.00	15824	130.7	533.3	194.184	7999
208.5	13332	221.78	19448	135.5	666.6	202.682	10666
231.7	26664	227.336	22886	138.2	799.9	209.565	13332
258.3	53329	235.145	28558	145.0	1067	222.774	19998
287.5	101325	242.432	34817	150.0	1333	232.743	26664
		250.605	43171	159.5	2000	240.846	33331
mp/°C	18.5	259.336	53777	166.0	2666	247.723	39997
		268.540	67076	176.0	4000	259.074	53329
		278.333	84065	183.6	5333	268.33	66661
		285.337	98168	189.7	6666	276.205	79993
		286.704	101325	194.3	7999	283.097	93326
				202.6	10666	284.38	95992
		bp/°C	286.792	209.8	13332	285.635	98659
				222.3	19998	286.864	101325
		eq. 2	P/mmHg	232.2	26664	25.0	0.221
		A	7.03044	245.8	39997	bp/°C	286.854
		B	1831.317	258.0	53329		
		C	154.528	267.0	66661	eq. 2	P/mmHg
				274.5	79993	A	7.02867
				282.2	93326	B	1830.51
				286.0	101325	C	154.450
						$\Delta H_V/(kJ\ mol^{-1}) =$	
						at 25°C	81.09
						at bp	51.21

(*Continued*)

TABLE 2.1.1.1.32.1 (*Continued*)

2.

Morgan & Kobayashi 1994		Ruzicka & Majer 1994	
static method pressure gauge		recommended	
t/°C	P/Pa	T/K	P/Pa
119.896	317.6	292.41	0.1
129.903	541.2	314.20	1.0
139.909	886.1	340.55	10
149.916	1408	373.36	100
159.929	2172	415.82	1000
169.929	3273	473.84	10000
179.935	4810	559.98	101325
189.943	6912	298.15	0.191
199.976	9774		
209.956	13530	Cox eq.	
219.976	18418	eq. 5	P/kPa
229.969	24691	P_{ref}/kPa	101.325
239.977	32600	T_{ref}/K	469.64
249.979	42464	a	3.18271
259.979	54642	$10^3 b$/K^{-1}	−2.002545
269.979	69438	$10^6 c$/K^{-2}	1.384476
279.980	87294		
289.980	108620		
299.980	133940		
309.982	163530		
data fitted to Wagner eq.			

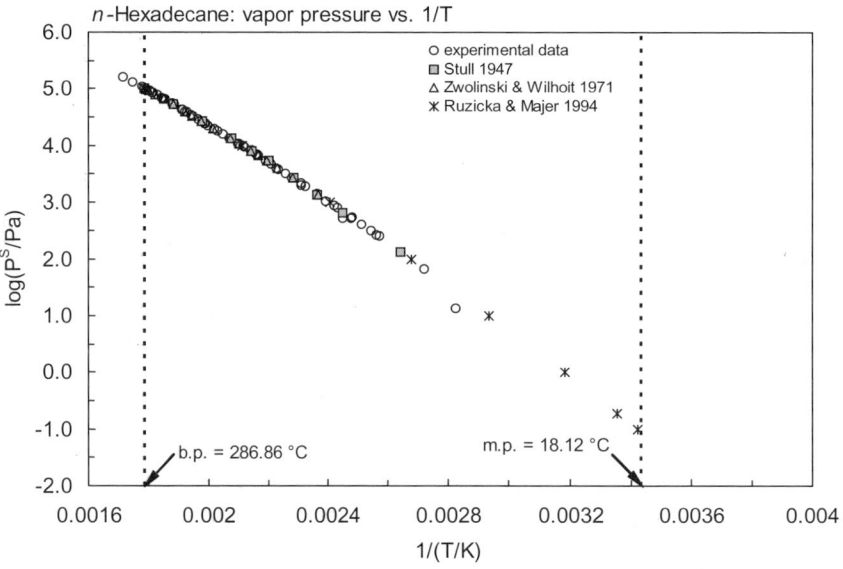

FIGURE 2.1.1.1.32.1 Logarithm of vapor pressure versus reciprocal temperature for *n*-hexadecane.

Aliphatic and Cyclic Hydrocarbons

2.1.1.1.33   n-Heptadecane

Common Name: Heptadecane
Synonym:
Chemical Name: *n*-heptadecane
CAS Registry No: 629-78-7
Molecular Formula: $C_{17}H_{36}$; $CH_3(CH_2)_{15}CH_3$
Molecular Weight: 240.468
Melting Point (°C):
   22   (Lide 2003)
Boiling Point (°C):
   302   (Lide 2003)
Density (g/cm³):
   0.7780, 0.7745   (20°C, 25°C, Dreisbach 1959)
   0.7780   (Weast 1982–83)
Molar Volume (cm³/mol):
   384.8   (calculated-Le Bas method at normal boiling point)
   309.1   (20°C, calculated-density)
Enthalpy of Vaporization, ΔH_V (kJ/mol):
   61.65, 53.12 (25°C, bp, Dreisbach 1961)
   86.47   (298.15 K, recommended, Ruzicka & Majer 1994)
Enthalpy of Fusion, ΔH_{fus} (kJ/mol):
   10.96, 40.17; 51.13   (1.15, 21.95°C; total phase change enthalpy, Chickos et al. 1999)
Entropy of Fusion, ΔS_{fus} (J/mol K):
   174.61, 175.1   (exptl., calculated-group additivity method, total phase change entropy, Chickos et al.1999)
Fugacity Ratio at 25°C, F: 1.0

Water Solubility (g/m³ or mg/L at 25°C):
   5.5×10^{-6} (extrapolated from data of McAuliffe 1966, Coates et al. 1985)
   0.0014   (reported as –log S (mol/L) = 7.24, calculated-molar volume, Wang et al. 1992)

Vapor Pressure (Pa at 25°C or as indicated and reported temperature dependence equations. Additional data at other temperatures designated * are compiled at the end of this section.):
   1467*   (163.5°C, temp range 163.5–303.0°C, Krafft 1882; quoted, Boublik et al. 1984)
   133.3*   (115.0°C, summary of literature data, temp range 115.0–303.0°C, Stull 1947)
   $\log (P/mmHg) = 7.0115 - 1847.82/(145.52 + t/°C)$; temp range 188–374°C (Antoine eq. for liquid state, Dreisbach 1959)
   0.0253*   (derived from compiled data, temp range 160.9–337°C, Zwolinski & Wilhoit 1971)
   $\log (P/mmHg) = 7.0143 - 1865.1/(149.20 + t/°C)$; temp range 160.9–337°C (Antoine eq., Zwolinski & Wilhoit 1971)
   $\log (P/mmHg) = [-0.2185 \times 15608.5/(T/K)] + 8.847487$; temp range 115–303°C (Antoine eq., Weast 1972–73)
   0.030   (extrapolated-Antoine eq., Boublik et al. 1973, 1984)
   $\log (P/mmHg) = 6.97509 - 1851.699/(t/°C + 149.263)$; temp range 164–303°C (Antoine eq., Boublik et al. 1973)
   $\log (P/kPa) = 6.09247 - 1845.726/(t/°C + 148.633)$; temp range 164–303°C (Antoine eq., Boublik et al. 1984)
   $\log (P/mmHg) = 7.10143 - 1865.1/(t/°C + 149.20)$; temp range 161–337°C (Antoine eq., Dean 1985, 1992)
   0.015   (interpolated-Antoine eq.-I, Stephenson & Malanowski 1987)
   $\log (P_L/kPa) = 11.1197 - 4757.087/(T/K)$; temp range 289–320 K (Antoine eq.-I, liquid, Stephenson & Malanowski 1987)
   $\log (P_L/kPa) = 6.1392 - 1865.1/(T/K - 123.95)$; temp range 488–577 K (Antoine eq.-II, liquid, Stephenson & Malanowski 1987)
   0.06148*   (recommended, Ruzicka & Majer 1994)
   $\ln [(P/kPa)/(P_o/kPa)] = [1 - (T_o/K)/(T/K)]\_\exp\{3.21826 - 2.036553 \times 10^{-3} \cdot (T/K) + 1.383899 \times 10^{-6} \cdot (T/K)^2\}$; reference state at $P_o = 101.325$ kPa, $T_o = 575.375$ K (Cox equation, Ruzicka & Majer 1994)

log (P/mmHg) = 173.4039 −1.0943 × 10⁴/(T/K) −59.212·log (T/K) + 2.0705 × 10⁻²·(T/K) −1.3433 × 10⁻¹²·(T/K)²;
   temp range 295–733 K (vapor pressure eq., Yaws 1994)
0.0627   (GC-retention time correlation, Chickos & Hanshaw 2004)

Henry's Law Constant (Pa m³/mol at 25°C):
5415   (calculated-vapor-liquid equilibrium (VLE) data, Yaws et al. 1991)

Octanol/Water Partition Coefficient, log K_{OW}:
8.92; 9.69   (estimated-RP-HPLC-MS; calculated-CLOGP, Burkhard et al. 1985)
8.79   (HPLC-k′ correlation, Coates et al. 1985)
8.50; 7.68   (calculated-fragment const.; calculated-molar volume correlation, Wang et al. 1992)

Octanol/Air Partition Coefficient, log K_{OA}:

Bioconcentration Factor, log BCF or log K_B:

Sorption Partition Coefficient, log K_{OC}:

Environmental Fate Rate Constant, k, and Half-Lives, $t_{1/2}$:
   Volatilization: volatilization k =1.23 d⁻¹ with a water column $t_{1/2}$ = 0.6 d at 6–18°C in spring; k = 0.79 d⁻¹ with a water column $t_{1/2}$ = 0.9 d at 20–22°C in summer; k = 0.14 d⁻¹ with a water column $t_{1/2}$ = 5.0 d at 3–7°C in winter for mesocosm experiments in coastal marine environment; volatilization k = 0.359 d⁻¹ with a water column $t_{1/2}$ = 1.9 d with HgCl poisoned water tank and k = 1.362 d⁻¹ with a water column $t_{1/2}$ = 0.51 d for non-poisoned water tank in late summer (Wakeham et al. 1983).
   Photolysis:
   Oxidation:
   Hydrolysis:
   Biodegradation: microbial degradation $t_{1/2}$ < 31 d by *Pseudomonas sp.* (Setti et al. 1993)
   Biotransformation:
   Bioconcentration and Uptake and Elimination Rate Constants (k_1 and k_2):

Half-Lives in the Environment:
   Surface water: water column $t_{1/2}$ = 0.6 d at 6–18°C, in spring; $t_{1/2}$ = 0.9 d at 20–22°C, in summer; $t_{1/2}$ = 5.0 d at 3–7°C, in winter for mesocosm experiment in coastal marine environment; $t_{1/2}$ = 1.9 d with HgCl poisoned water tank and $t_{1/2}$ = 0.51 d for non-poisoned water tank in late summer (Wakeham et al. 1983).

TABLE 2.1.1.1.33.1
Reported vapor pressures of *n*-heptadecane at various temperatures and the coefficients for the vapor pressure equations

log P = A − B/(T/K)   (1)   ln P = A − B/(T/K)   (1a)
log P = A − B/(C + t/°C)   (2)   ln P = A − B/(C + t/°C)   (2a)
log P = A − B/(C + T/K)   (3)
log P = A − B/(T/K) − C·log (T/K)   (4)
ln (P/P_{ref}) = [1 − (T_{ref}/T)]·exp(a + bT + cT²)   (5) - Cox eq.

Stull 1947		Krafft 1882		Zwolinski & Wilhoit 1971		Ruzicka & Majer 1994	
summary of literature data		in Boublik et al. 1984		selected values		recommended	
t/°C	P/Pa	t/°C	P/Pa	t/°C	P/Pa	T/K	P/Pa
115.0	133.3	163.5	1467	160.9	1333	302.38	0.1
145.2	666.6	170.0	2000	177.3	2666	324.64	1.0
160.0	1333	187.5	4000	187.6	4000	351.56	10
177.7	2666	201.5	6666	195.4	5333	385.06	100
195.8	5333	223.0	13332	201.7	6666	428.4	1000

TABLE 2.1.1.1.33.1 (*Continued*)

Stull 1947		Krafft 1882		Zwolinski & Wilhoit 1971		Ruzicka & Majer 1994	
summary of literature data		in Boublik et al. 1984		selected values		recommended	
t/°C	P/Pa	t/°C	P/Pa	t/°C	P/Pa	T/K	P/Pa
207.3	7999	303.0	101325	207	7999	487.59	10000
223.0	13332			215.7	10666	575.38	101325
247.8	26664	bp/°C	303.003	222.76	13332	298.15	0.06148
274.5	53329			236.30	19998		
303.0	101325			246.50	26664	Cox eq.	
				254.82	33331	eq. 5	P/kPa
mp/°C	22.5			261.87	39997	P_{ref}/kPa	101.325
				273.51	53329	T_{ref}/K	469.64
				283.0	66661	a	3.21826
				291.08	79993	$10^3 b/K^{-1}$	−2.036553
				298.14	93326	$10^6 c/K^{-2}$	1.383899
				299.47	95992		
				300.76	98659		
				302.02	101325		
				bp/°C	302.02		
				eq. 2	P/mmHg		
				A	7.0143		
				B	1865.10		
				C	149.20		
				ΔH_V/(kJ mol^{-1}) =			
				at 25°C	86.20		
				at bp	52.89		

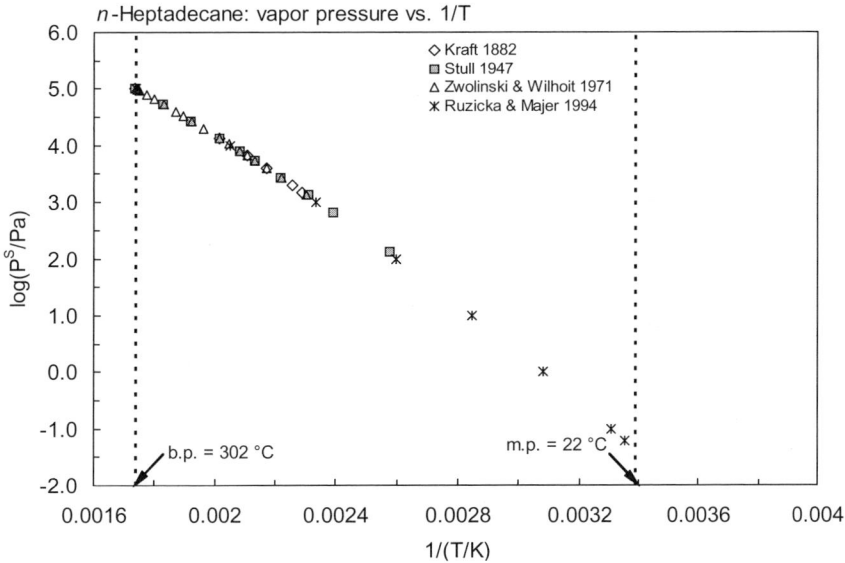

FIGURE 2.1.1.1.33.1 Logarithm of vapor pressure versus reciprocal temperature for *n*-heptadecane.

2.1.1.1.34   n-*Octadecane*

Common Name: Octadecane
Synonym:
Chemical Name: *n*-octadecane
CAS Registry No: 543-45-3
Molecular Formula: $C_{18}H_{38}$; $CH_3(CH_2)_{16}CH_3$
Molecular Weight: 254.495
Melting Point (°C):
    28.2   (Dreisbach 1959; Weast 1982–83; Stephenson & Malanowski 1987; Lide 2003)
Boiling Point (°C):
    316.3   (Lide 2003)
Density (g/cm^3):
    0.7819, 0.7785   (20°C, 25°C, Dreisbach 1959)
    0.7768   (Weast 1982–83)
Molar Volume (cm^3/mol):
    325.5   (20°C, calculated-density)
    407.0   (calculated-Le Bas method at normal boiling point, Eastcott et al. 1988)
Enthalpy of vaporization, ΔH_V (kJ/mol):
    90.824, 54.84   (25°C, bp, Dreisbach 1959)
    91.6   (25°C, Piacente et al. 1994)
    91.44   (298.15K, recommended, Ruzicka & Majer 1994)
Enthalpy of Fusion, ΔH_{fus} (kJ/mol):
    61.384   (Dreisbach 1959)
    61.13   (28°C, Piacente et al. 1994)
    61.5   (Chickos et al. 1999)
Entropy of Fusion, ΔS_{fus} (J/mol K):
    204.6, 184.5 (exptl., calculated-group additivity method, Chickos et al. 1999)
Fugacity Ratio at 25°C (assuming ΔS_{fus} = 56 J/mol K), F: 0.930 (mp at 28.2°C)

Water Solubility (g/m^3 or mg/L at 25°C):
    7.75×10^{-3} (Baker 1959)
    2.1×10^{-3}; 8.0×10^{-4} (shake flask-GC, distilled water; seawater, Sutton & Calder 1974)
    1.40×10^{-6} (extrapolated from data of McAuliffe 1966, Coates et al. 1988)
    4.05×10^{-6} (extrapolated from data of McAuliffe 1966, Eastcott et al. 1988)

Vapor Pressure (Pa at 25°C or as indicated and reported temperature dependence equations. Additional data at other temperatures designated * are compiled at the end of this section.):
    133.3*   (119.6°C, summary of literature data, temp range 119.6–317.0°C, Stull 1947)
    26.66*   (102.4°C, static method-Hg manometer, measured range 102.4–313°C, Myers & Fenske 1955)
    log (P/mmHg) = 7.0156 – 1883.73/(139.46 + t/°C); temp range 201–387°C (Antoine eq. for liquid state, Dreisbach 1959)
    1333*   (172.3°C, derived from compiled data, temp range 172.3–316.3°C, Zwolinski & Wilhoit 1971)
    log (P/mmHg) = 7.0022 – 1894.3/(143.3 + t/°C); temp range 172.3–352°C (Antoine eq., Zwolinski & Wilhoit 1971)
    log (P/mmHg) = [–0.2185 × 15447.0/(T/K)] + 8.619864; temp range 119.6–317°C (Antoine eq., Weast 1972–73)
    0.013   (extrapolated-Antoine eq., Boublik et al. 1973)
    log (P/mmHg) = 7.14067 – 2012.745/(t/°C + 155.492); temp range 174–317°C (Antoine eq., Boublik et al. 1973)
    0.0259   (liquid P_L, extrapolated-Antoine eq., Macknick & Prausnitz 1979)
    0.220*   (45.0°C, gas saturation, measured range 45.0–88.10°C, Macknick & Prausnitz 1979)
    ln (P_L/mmHg) = 25.548 – 10165/(T/K); temp range 45–88.1°C (Antoine eq. on exptl. data, gas saturation, liquid state, Macknick & Prausnitz 1979)

Aliphatic and Cyclic Hydrocarbons

0.0133 (extrapolated-Antoine eq., Boublik et al. 1984)

log (P/kPa) = 6.27065 − 2016.983/(t/°C + 155.924); temp range 174–317°C (Antoine eq., Boublik et al. 1984)

log (P/mmHg) = 7.0022 − 1894.3/(143.30 + t/°C); temp range 172–352°C (Antoine eq., Dean 1985, 1992)

1.15* (62.04°C, gas saturation, measured temp range 335.19–439.82 K, Allemand et al. 1986)

0.0261 (extrapolated-Antoine eq.-I, liquid, Stephenson & Malanowski 1987)

log (P_L/kPa) = 10.18833 − 4404.095/(T/K); temp range 310–361 K Antoine eq., liquid, Stephenson & Malanowski 1987)

log (P_L/kPa) = 6.1392 − 1894.3/(T/K − 129.85); temp range 501–550 K (Antoine eq., liquid, Stephenson & Malanowski 1987)

log (P/mmHg) = 7.0022 − 1894.3/(t/°C + 143.30); temp range 172–352°C (Antoine eq., Dean 1992)

256.6* (139.919°C, static-differential pressure, measured range 139.919–314.982°C, Morgan & Kobayashi 1994)

0.02007* (recommended, temp range 396–500 K, Ruzicka & Majer 1994)

ln $[(P/kPa)/(P_o/kPa)] = [1 − (T_o/K)/(T/K)] \cdot \exp\{3.24741 − 2.048039 \times 10^{-3} \cdot (T/K) + 1.362445 \times 10^{-6} \cdot (T/K)^2\}$; reference state at P_o = 101.325 kPa, T_o = 590.023 K (Cox equation, Ruzicka & Majer 1994)

log (P/mmHg) = $-15.0772 − 4.8702 \times 10^3/(T/K) + 14.501 \cdot \log(T/K) − 3.1625 \times 10^{-2} \cdot (T/K) + 1.3478 \times 10^{-5} \cdot (T/K)^2$; temp range 273–591 K (vapor pressure eq., Yaws 1994)

Henry's Law Constant (Pa·m³/mol at 25°C or as indicated):
1013 (15°C, Wakeham et al. 1986)
622200 (calculated-P/C, Eastcott et al. 1988)
893 (calculated-vapor-liquid equilibrium (VLE) data, Yaws et al. 1991)

Octanol/Water Partition Coefficient, log K_{OW}:
9.32 (HPLC-k′ correlation, Coates et al. 1985)
9.00; 8.13 (calculated-fragment const.; calculated-molar volume correlation, Wang et al. 1992)

Octanol/Air Partition Coefficient, log K_{OA}:

Bioconcentration Factor, log BCF or log K_B:

Sorption Partition Coefficient, log K_{OC}:
5.90 (K_{OM}, Wakeham et al. 1986)

Environmental Fate Rate Constants, k, and Half-Lives, $t_{1/2}$:
Volatilization: volatilization rate constant k_v = 0.03 d⁻¹, microcosm exptl. (Wakeham et al. 1986).
Photolysis:
Photooxidation:
Hydrolysis:
Biodegradation: degradation rate constant of about 0.66 d⁻¹ in a microcosm expt. (Wakeham et al. 1986); microbial degradation $t_{1/2}$ < 31 d by *Pseudomonas sp.* (Setti et al. 1993).
Biotransformation:
Bioconcentration and Uptake and Elimination Rate Constants (k_1 and k_2):

Half-Lives in the Environment:
Surface water: an estimated $t_{1/2}$ = 1.5 d in Rhine River for a first order reduction process in river water (Zoeteman et al. 1980)
$t_{1/2}$ ~ 23 d in a seawater microcosm experiment (Wakeham et al. 1986).

TABLE 2.1.1.1.34.1
Reported vapor pressures of *n*-octadecane at various temperatures and the coefficients for the vapor pressure equations

$\log P = A - B/(T/K)$ (1) $\ln P = A - B/(T/K)$ (1a)

$\log P = A - B/(C + t/°C)$ (2) $\ln P = A - B/(C + t/°C)$ (2a)

$\log P = A - B/(C + T/K)$ (3)

$\log P = A - B/(T/K) - C \cdot \log(T/K)$ (4)

$\ln(P/P_{ref}) = [1 - (T_{ref}/T)] \cdot \exp(a + bT + cT^2)$ (5) - Cox eq.

1.

Stull 1947		Myers & Fenske 1955		Zwolinski & Wilhoit 1971		Macknick & Prausnitz 1979	
summary of literature data		ebulliometry		selected values		gas saturation	
t/°C	P/Pa	t/°C	P/Pa	t/°C	P/Pa	t/°C	P/Pa
119.6	133.3	102.4	26.66	172.3	1333	45.0	0.220
152.1	666.6	115.9	66.66	189.0	2666	51.45	0.400
169.6	1333	127.0	133.3	199.6	4000	54.5	0.568
187.5	2666	139.6	266.6	207.5	5333	54.85	0.591
207.4	5333	147.0	400.0	213.9	6666	59.85	0.935
219.7	7999	153.3	533.3	219.3	7999	65.60	1.529
236.0	13332	157.7	666.6	228.2	10666	71.25	2.493
260.6	26664	161.8	799.9	234.39	13332	80.85	5.826
288.0	53329	167.7	1067	249.21	19998	84.10	7.148
317.0	101325	172.8	1333	259.64	26664	88.10	9.719
		182.8	2000	268.12	33331		
mp/°C	28	189.6	2666	275.32	39997	eq. 1a	P/mmHg
		199.9	4000	287.2	53329	A	25.548
		208.0	5333	296.9	66661	B	10165
		214.3	6666	305.2	79993		
		219.5	7999	312.4	93326		
		228.5	10666	313.7	95992		
		235.7	13332	315.0	98659		
		249.0	19998	316.3	101325		
		259.2	26664				
		273.8	39997	bp/°C	316.3		
		286.0	53329				
		294.7	66661	eq. 2	P/mmHg		
		302	79993	A	7.0022		
		308	93326	B	1894.3		
		313	101325	C	143.30		
				$\Delta H_V/(kJ\,mol^{-1}) =$			
				at 25°C	90.8		
				at bp	54.48		

TABLE 2.1.1.1.34.1 (*Continued*)

2.

Allemand et al. 1986		Morgan & Kobayashi 1994		Ruzicka & Majer 1994	
gas saturation		vapor-liquid equilibrium		recommended	
t/°C	P/Pa	t/°C	P/Pa	T/K	P/Pa
62.04	1.1506	139.919	256.6	311.88	0.1
71.39	2.680	149.924	443.6	334.61	1.0
84.0	6.599	159.93	729.4	362.08	10
88.0	9.479	169.936	1148	396.24	100
97.0	18.40	179.974	1761	440.41	1000
100.4	22.00	189.948	2637	500.70	10000
110.0	43.86	199.976	3870	590.02	101325
124.21	106.1	209.976	5566	298.15	0.02007
130.0	152.0	219.976	7834		
140.0	264.0	229.977	10835	Cox eq.	
166.67	995.9	239.970	14744	eq. 5	P/kPa
		254.979	22736	P_{ref}/kPa	101.325
		254.979	27300	T_{ref}/K	469.64
		264.979	34026	a	3.24741
		284.980	49559	$10^3 b/K^{-1}$	−2.048039
		299.98	70490	$10^6 c/K^{-2}$	1.362445
		314.982	98240		
		data fitted to Wagner eq.			

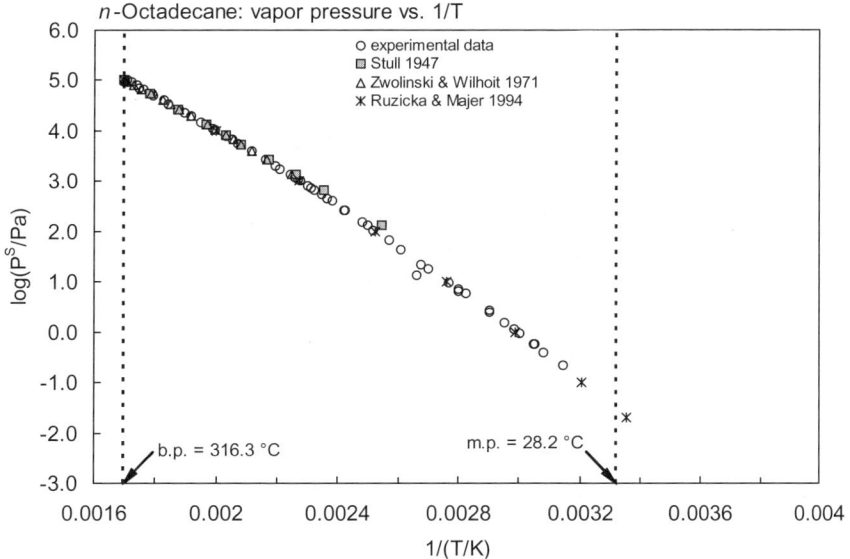

FIGURE 2.1.1.1.34.1 Logarithm of vapor pressure versus reciprocal temperature for *n*-octadecane.

2.1.1.1.35   n-*Eicosane*

Common Name: Eicosane
Synonym: didecyl
Chemical Name: *n*-eicosane
CAS Registry No: 112-95-8
Molecular Formula: $C_{20}H_{42}$; $CH_3(CH_2)_{18}CH_3$
Molecular Weight: 282.547
Melting Point (°C):
    36.6   (Lide 2003)
Boiling Point (°C):
    334.8   (Chirico et al. 1989)
Density (g/cm³):
    0.7887, 0.7853   (20°C, 25°C, Dreisbach 1959)
    0.7886   (Weast 1982–83)
Molar Volume (cm³/mol):
    451.4   (calculated-Le Bas molar volume at normal boiling point)
    358.2   (20°C, calculated-density)
    358   (Wang et al. 1992)
Enthalpy of vaporization, ΔH_V (kJ/mol):
    63.93; 64.35   (exptl., calculated, Macknick & Prausnitz 1979b)
    100.9; 110   (25, 94°C, Piacente et al. 1994)
    101.81   (298.15 K, recommended, Ruzicka & Majer 1994)
Enthalpy of Fusion, ΔH_{fus} (kJ/mol):
    69.875   (Dreisbach 1959)
    69.5   (Piacente et al. 1994)
    67.8   (Chickos et al. 1999)
Entropy of Fusion, ΔS_{fus} (J/mol K):
    219.6, 203.1 (exptl., calculated-group additivity method, Chickos et al. 1999)
Fugacity Ratio at 25°C (assuming ΔS_{fus} = 56 J/mol K), F: 0.769 (mp at 36.6°C)

Water Solubility (g/m³ or mg/L at 25°C):
    1.9×10^{-3}, 8.0×10^{-4} (shake flask-GC, distilled water; seawater, Sutton & Calder 1974)
    1.10×10^{-7} (extrapolated from data of McAuliffe 1966, Coates et al. 1985)
    3.11×10^{-7} (extrapolated from data of McAuliffe 1966, Eastcott et al. 1985)

Vapor Pressure (Pa at 25°C or as indicated and reported temperature dependence equations. Additional data at other temperatures designated * are compiled at the end of this section.):
    13.33*   (121.9°C, static method-Hg manometer, measured range 121.9–342.0°C, Myers & Fenske 1955)
    log (P/mmHg) = 7.0225 – 1948.7/(127.8 + t/°C), temp range: 224–417°C, (Antoine eq. for liquid state, Dreisbach 1959)
    1333*   (198.3°C, derived from compiled data, temp range 198.3–343.8°C, Zwolinski & Wilhoit 1971)
    log (P/mmHg) = 7.1522 – 2032.7/(132.1 + t/°C); temp range:198.3–379°C (Antoine eq., Zwolinski & Wilhoit 1971)
    0.41*   (71.5°C, P_L, gas saturation-IR, Macknick & Prausnitz 1979)
    ln (P_L/mmHg) = 26.849 – 11230/(T/K); temp range: 71.15–107.3°C, (Antoine eq., gas saturation, liquid state, Macknick & Prausnitz 1979)
    log (P_L/kPa) = 10.77373 – 4872.63/(T/K); temp range: 344–380 K (Antoine eq., liquid, Stephenson & Malanowski 1987)
    log (P_L/kPa) = 6.2771 – 2032.7/(T/K – 141.05); temp range 528–620 K (Antoine eq., liquid, Stephenson & Malanowski 1987)
    0.201*   (79.97°C, pressure gauge, measured range 90.07–194.21°C, Sasse et al. 1988)

Aliphatic and Cyclic Hydrocarbons

log (P/mmHg) = 7.95834 − 2665.762/(t/°C + 167.047); (Antoine eq. derived from exptl. data, liquid phase, pressure gauge measurement, Sasse et al. 1988)

15.6* (115°C, inclined piston measurement, temp range 115–215°C, Chirico et al. 1989)

ln (P/kPa) = 19.36 − 9083/(T/K); temp range 406–472 K (transpiration method, Piacente et al. 1991)

ln (P/kPa) = 22.53 − 10649/(T/K); temp range 345–393 K (torsion method, Piacente et al. 1991)

ln (P/kPa) = 18.10 − 7889/[(T/K) + 32]; temp range 315–472 K (transpiration, torsion and Knudsen methods, Antoine eq., Piacente et al. 1991)

log (P/mmHg) = 7.1522 − 2032.1/(t/°C + 132.10); temp range 198–379°C (Antoine eq., Dean 1992)

0.32*, 0.641 (74, 78°C, torsion-effusion method, measured range 74–115°C, Piacente et al. 1994)

log (P/kPa) = 13.37 − 5785/(T/K); temp range 351–384 K (torsion-effusion, Antoine eq., Piacente et al. 1994)

log (P/kPa) = 12.96 − 5709/(T/K); temp range 347–389 K (torsion-effusion, Antoine eq., Piacente et al. 1994)

log (P/kPa) = 13.16 − 5747/(T/K); temp range ~ 347–388 K, ΔH_V = 110 kJ mol^{-1} (selected Antoine eq. based on exptl. data, torsion-effusion method, Piacente et al. 1994)

0.002091* (recommended, Ruzicka & Majer 1994)

ln [(P/kPa)/(P$_o$/kPa)] = [1 − (T$_o$/K)/(T/K)]·exp{3.31181 − 2.102218 × 10^{-3}·(T/K) + 1.348780 × 10^{-6}·(T/K)^2}; reference state at P$_o$ = 101.325 kPa, T$_o$ = 617.415 K (Cox equation, Ruzicka & Majer 1994)

log (P/mmHg) = 19.4193 − 5.8699 × 10^3/(T/K) − 44.282·log (T/K) − 1.2606 × 10^{-2}·(T/K) + 5.2241 × 10^{-6}·(T/K)^2; temp range 310–767 K (vapor pressure eq., Yaws 1994)

0.00209* (GC-retention time correlation, Chickos & Hanshaw 2004)

Henry's Law Constant (Pa·m^3/mol):

198300 (calculated-P/C, Eastcott et al. 1988)

32.73 (calculated-vapor-liquid equilibrium (VLE) data, Yaws et al. 1991)

Octanol/Water Partition Coefficient, log K_{OW}:

10.39 (HPLC-k' correlation, Coates et al. 1985)

10.0; 8.92 (calculated-fragment const.; calculated-molar volume correlation, Wang et al. 1992)

Octanol/Air Partition Coefficient, log K_{OA}:

Bioconcentration Factor, log BCF or log K_B:

Sorption Partition Coefficient, log K_{OC}:

Environmental Fate Rate Constants, k, and Half-Lives, $t_{1/2}$:

Biodegradation: microbial degradation $t_{1/2}$ < 31 d by *Pseudomonas* sp. (Setti et al. 1993)

Half-Lives in the Environment:

TABLE 2.1.1.1.35.1
Reported vapor pressures of eicosane at various temperatures and the coefficients for the vapor pressure equations

$\log P = A - B/(T/K)$ (1) $\ln P = A - B/(T/K)$ (1a)
$\log P = A - B/(C + t/°C)$ (2) $\ln P = A - B/(C + t/°C)$ (2a)
$\log P = A - B/(C + T/K)$ (3) $\ln P = A - B/(C/T/K)$ (3a)
$\log P = A - B/(T/K) - C \cdot \log (T/K)$ (4)
$\ln (P/P_{ref}) = [1 - (T_{ref}/T)] \cdot \exp(a + bT + cT^2)$ (5)

1.

Myers & Fenske 1955		Zwolinski & Wilhoit 1971		Macknick & Prausnitz 1979		Sasse et al. 1988	
Hg manometer		selected values		gas saturation		electronic manometry	
t/°C	P/Pa	t/°C	P/Pa	t/°C	P/Pa	t/°C	P/Pa
121.9	13.33	198.3	1333	71.15	0.4106	90.07	2.333
136.5	66.6	215.3	2666	79.90	0.973	90.12	2.360
148.6	133.3	326.1	4000	86.70	1.693	100.17	5.526
162.0	266.6	234.1	5333	90.35	2.333	100.17	5.266
170.0	400.0	240.7	6666	94.50	3.266	110.16	11.49
176.7	533.3	246.1	7999	102.95	6.719	110.16	11.43
182.0	666.6	255.15	10666	107.30	9.133	120.06	22.53
185.7	799.9	262.41	13332			120.10	22.80
192.6	1067	276.39	19998			130.11	43.20
198.0	1333	286.91	26664	eq. 1a	P/mmHg	130.13	43.73
208.8	2000	295.45	33331	A	26.849	139.99	80.66
215.7	2666	302.70	39997	B	11230	140.03	79.86
226.4	4000	314.60	53329			149.88	141.61
234.7	5333	324.4	66661			149.97	144.9
241.3	6666	332.6	79993			159.73	243.7
246.5	7999	339.8	93326			159.83	245.7
255.7	10666	341.2	95992			169.61	404.2
262.5	13332	342.5	98659			169.68	410.6
276.1	19998	343.8	101325			179.49	657.3
286.4	26664					189.25	1025
300.2	39997	bp/°C	343.8			194.21	1266
312.0	53329						
320.0	66661	eq. 2	P/mmHg			eq. 2	P/mmHg
327.0	79993	A	7.1522			A	7.99897
333.0	93326	B	2032.7			B	2067.622
342.0	101325	C	132.1			C	177.32
		$\Delta H_V/(kJ\ mol^{-1}) =$					
		at 25°C	100.8				
		at bp	57.49				

Aliphatic and Cyclic Hydrocarbons

TABLE 2.1.1.1.35.1 (*Continued*)

2.

Chirico et al. 1989				Piacente et al. 1991			
ebulliometry		inclined piston gauge		transpiration method			
t/°C	P/Pa	t/°C	P/Pa	t/°C	P/Pa	t/°C	P/Pa
				1st set		2nd set data	
250.795	9596	115.0	15.6	150.85	114.8	124.85	23.99
254.896	10897	120.001	21.9	155.85	138.0	129.85	43.65
259.013	12348	125.001	30.7	160.35	182.0	134.35	37.15
263.144	13961	130.0	42.3	163.85	218.8	149.35	97.72
267.291	15752	135.0	58.3	166.85	275.4	152.85	123.0
271.454	17737	140.0	78.7	169.85	309.0	154.35	134.9
275.628	19933	150.0	139.9	171.85	371.5	156.35	147.9
284.032	25023	170.0	397.9	175.85	416.9	159.35	177.8
292.481	31177	180.0	641.9	177.85	537.0	165.85	218.8
300.985	38565	190.0	1007.7	180.35	537.0	169.85	281.8
309.540	46375	200.0	1539.6	182.85	537.0	172.35	275.4
318.146	57817	210.0	2300.3	185.85	645.7	174.35	354.8
326.761	70120	215.0	2789.0	188.35	831.8	176.35	426.6
335.416	84533			190.35	812.8	179.85	489.8
344.053	101325	Cox vapor pressure eq.		192.35	871.0	184.85	537.0
352.842	120790	eq. 5	P/kPa	193.85	912.0	188.85	645.7
		P_{ref}/kPa	101.325	196.85	1096.5	191.85	741.3
enthalpy of vaporization:		T_{ref}/K	617.456	198.85	1175	194.35	851.1
ΔH_V/(kJ mol^{-1}) =		a	3.31018			196.35	831.8
380 K	89.07	10^3b/K^{-1}	−2.09538				
400 K	86.133	10^6c/K^{-2}	1.34198				
420 K	83.448	temp range: 388–626 K					
440 K	80.78						
460 K	78.20						
480 K	75.68						
500 K	73.21						
520 K	70.75						
540 K	68.27						
560 K	65.74						
580 K	63.12						
600 K	60.37						
620 K	57.44						

(*Continued*)

TABLE 2.1.1.1.35.1 (Continued)

3.

Piacente et al. 1991 (cont'd)

transpiration (cont'd)		torsion method				Knudsen method	
t/°C	P/Pa	t/°C	P/Pa	t/°C	P/Pa	t/°C	P/Pa
3rd set data		cell B, #1		cell B, #3		cell AK	
131.35	64.57	71.85	0.575	66.35	0.3802	42	0.0288
135.35	67.61	77.85	1.148	73.85	0.7586	51	0.10
141.35	107.15	81.85	1.514	78.85	1.148	60	0.269
145.85	114.82	84.35	1.905	82.85	1.698	70	0.5888
150.35	158.49	87.85	2.455	84.85	2.089	cell AK	
157.35	194.98	90.85	3.020	87.85	2.630	70	0.741
159.85	229.09	96.85	4.169	89.85	3.388	77	1.318
162.35	257.04	97.85	5.888	91.85	3.981	83	1.738
163.85	295.12	101.85	6.761	95.85	4.786	93	4.467
166.35	281.84	104.35	8.710	97.85	6.026		
179.35	338.84	105.35	9.333	101.85	7.943		
174.35	436.52	108.35	11.48	110.35	15.14	Eq. for transpiration, torsion	
177.85	524.81	112.35	16.22	117.35	25.12	and Knudsen effusion	
183.35	707.95	112.85	17.38	117.85	25.70	eq. 3a	P/kPa
185.35	691.83	114.85	19.20	118.85	26.92	A	18.10 ± 0.23
187.85	812.83	cell B #2		cell B, #4		B	7889 ± 230
191.85	977.24	88.85	2.8184	71.35	0.5754	C	−32 ± 10
192.85	1023.3	89.85	3.388	78.35	1.148		
		90.85	3.388	82.85	1.698		
		91.85	3.802	86.85	2.291	bp /°C	343.65
Antoine eq. for transpiration		95.85	4.571	89.85	3.388	mp/°C	36.65
eq. 1a	P/kPa	96.85	4.898	104.85	10.0		
A	19.36 ± 0.46	97.35	5.129	107.85	11.48	$\Delta H_{fus}/(kJ\ mol^{-1}) = 69.882$	
B	9083 ± 207	1–2.35	7.943	109.35	13.18	at 298.15 K	
		104.35	8.710	110.85	15.49	$\Delta H_{subl}/(kJ\ mol^{-1}) = 152.3$	
		106.83	9.550	112.85	18.62	at 298.15 K	
		107.85	10.72	115.85	22.91		
		108.85	12.02	117.85	28.18		
		109.85	13.18	119.85	32.36		
		110.85	14.79				
		111.85	16.98	Antoine eq. for torsion			
		112.85	19.50	eq. 1a	P/kPa		
				A	22.53 ± 0.30		
				B	10649 ± 230		

TABLE 2.1.1.1.35.1 (Continued)

4.

Piacente et al. 1994				Morgan & Kobayashi 1994		Ruzicka & Majer 1994	
torsion-effusion				static pressure gauge		recommended	
t/°C	P/Pa	t/°C	P/Pa	t/°C	P/Pa	T/K	P/Pa
run a		run b		top-cut			
77.85	0.641	73.85	0.32	159.903	244	330.01	0.1
79.85	0.961	77.85	0.481	169.91	401.8	353.55	1.0
81.85	1.28	80.85	0.641	179.918	641.1	381.97	10
83.85	1.52	82.85	0.801	189.926	1006	417.30	100
85.85	1.84	85.85	1.28	199.934	1537	462.95	1000
88.85	2.40	88.85	1.60	199.976	1538	525.23	10000
91.85	3.36	91.85	2.40	209.976	2301	617.41	101325
93.85	4.17	94.85	3.04	219.947	3358	298.15	0.002091
94.85	4.97	97.85	4.17	229.954	4812		
98.85	6.57	100.85	4.65	239.961	6771	Cox eq.	
101.85	8.33	103.85	6.89	249.969	9359	eq. 5	P/kPa
103.85	10.40	106.85	8.65	259.979	12738	P_{ref}/kPa	101.325
107.85	15.1	112.85	14.3	269.979	17040	T_{ref}/K	617.415
110.85	19.4	114.85	17.1	284.980	25728	a	3.31181
				299.980	37745	$10^3 b$/K^{-1}	−2.102218
eq. for run a		eq. for run b		314.982	54210	$10^6 c$/K^{-2}	1.248780
eq. 1	P/kPa	eq. 1	P/kPa	mid-cut			
A	13.37	A	12.96	159.847	244.9		
B	5785	B	5709	169.853	401		
temp range: 351–384 K		temp range: 347–388 K		179.860	639		
				189.875	999.1		
ΔH_V/(kJ mol^{-1}) =		By weighing the slopes and		199.874	1541		
at 298.15 K	100.9	intercepts of above 2 eq.,		209.884	2302		
at 367 K	110 ± 2	selected vapor pressure eq.		219.894	3349		
		eq. 1	P/kPa	229.911	4813		
mp/°C	28	A	13.16	239.915	6773		
ΔH_{fus}/(kJ mol^{-1}) = 61.10		B	5747	249.931	9349		
				259.938	12736		
ΔH_{subl}/(kJ mol^{-1}) = 152.7		ΔH_V/(kJ mol^{-1}) = 110 ± 2		269.949	17029		
		at 367 K		279.957	22549		
				289.970	29252		
				299.970	37754		
				309.982	48080		
				data fitted to Wagner eq.			

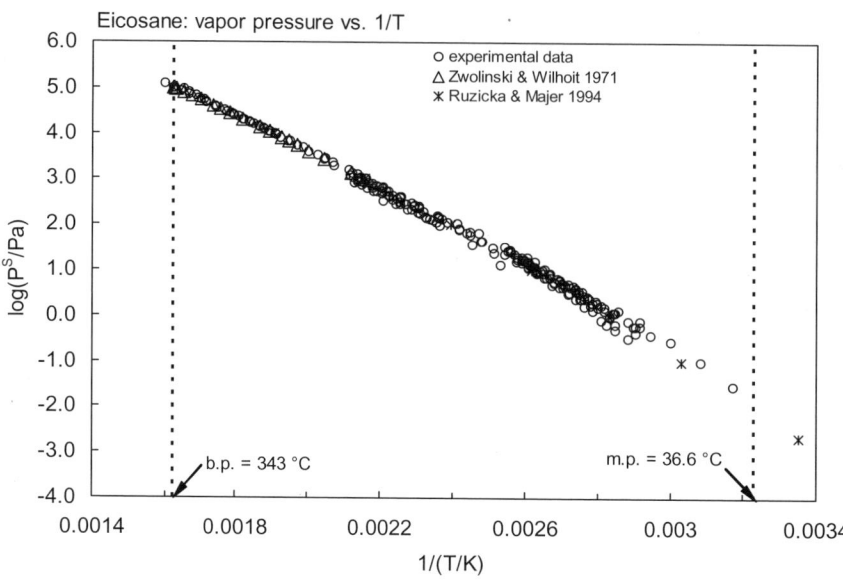

FIGURE 2.1.1.1.35.1 Logarithm of vapor pressure versus reciprocal temperature for *n*-eicosane.

Aliphatic and Cyclic Hydrocarbons

2.1.1.1.36   n-*Tetracosane*

Common Name: Tetracosane
Synonym:
Chemical Name: *n*-tetracosane
CAS Registry No: 646-31-1
Molecular Formula: $C_{24}H_{50}$; $CH_3(CH_2)_{22}CH_3$
Molecular Weight: 338.654
Melting Point (°C):
    50.4   (Lide 2003)
Boiling Point (°C):
    391.3   (Dreisbach 1959; Weast 1982–83; Stephenson & Malanowski 1987; Lide 2003)
Density (g/cm³):
    0.7991, 0.7958 (20°C, 25°C, Dreisbach 1959)
Molar Volume (cm³/mol):
    540.2   (calculated-Le Bas method at normal boiling point)
    423.8   (20°C, calculated-density)
Enthalpy of vaporization, ΔH_V (kJ/mol):
    126   (132°C, Piacente et al. 1994)
    121.9   (calculated, Chickos & Hanshaw 2004)
Enthalpy of Fusion, ΔH_{fus} (kJ/mol):
    54.9   (Dreisbach 1959; Piacente et al. 1994)
    31.3, 54.89; 86.19   (48.15, 50.95°C; total phase change enthalpy, Chickos et al. 1999)
Entropy of Fusion, ΔS_{fus} (J/mol K):
    266.79, 240.4   (exptl., calculated-group additivity method, total phase change entropy, Chickos et al. 1999)
Fugacity Ratio at 25°C (assuming ΔS_{fus} = 56 J/mol K), F: 0.563 (mp at 50.4°C)

Water Solubility (g/m³ or mg/L at 25°C):
    5.8×10^{-10} (extrapolated from data of McAuliffe 1966, Coates et al. 1985)

Vapor Pressure (Pa at 25°C or as indicated and reported temperature dependence equations. Additional data at other temperatures designated * are compiled at the end of this section.):
    133.3*   (183.8°C, summary of literature data, temp range 183.8–386.4°C, Stull 1947)
    log (P/mmHg) = 7.53923 – 2591.9/(165.1 + t/°C); temp range 260–500°C (Antoine eq. for liquid state, Dreisbach 1959)
    log (P/mmHg) = 7.0976 – 2112.0/(109.6 + t/°C) (Antoine eq., Kudchadker & Zwolinski 1966)
    66.66*   (175.9°C, derived from compiled data, temp range 175.9–391.3°C, Zwolinski & Wilhoit 1971)
    log (P/mmHg) = 7.0976 – 2112.0/(109.6 + t/°C); temp range 175.9–391.3°C (Antoine eq., Zwolinski & Wilhoit 1971)
    log (P/mmHg) = [–0.2185 × 19642.5/(T/K)] + 9.408166; temp range 183.8–386.4°C (Antoine eq., Weast 1972–73)
    log (P_L/kPa) = 6.44051 – 2289.02/(T/K – 147.92); temp range 498–573 K (liquid, Antoine eq., Stephenson & Malanowski 1987)
    0.253*   (100.26°C, pressure gauge, Sasse et al. 1988)
    log (P/mmHg) = 7.17666 – 2243.665/(t/°C + 126.236); temp range 100.26–149.23°C (Antoine eq. derived from exptl. data, liquid phase, Sasse et al. 1988)
    log (P/kPa) = (8.76 ± 0.50) – (4501 ± 250)/(T/K); temp range 451–497 K (Antoine eq. from exptl. data, transpiration method, Piacente & Scardala 1990; quoted, Pompili & Piacente 1990)
    ln (P/kPa) = 21.25 – 10946/(T/K); temp range 501–523 K (transpiration method, Piacente et al. 1991)
    ln (P/kPa) = 25.35 – 12399/(T/K); temp range 376–438 K (torsion method, Piacente et al. 1991)
    ln (P/kPa) = 18.38 – 8349/[(T/K) + 58]; temp range 343–523 K (transpiration, torsion and Knudsen methods, Antoine eq., Piacente et al. 1991)

90.1*   (179.916°C, static-differential pressure, measured range 179.916–314.820°C, Morgan & Kobayashi 1994)

0.721*, 0.801 (114, 115°C, torsion-effusion, measured range 386–425 K, Piacente et al. 1994)

log (P/kPa) = 13.57 – 6459/(T/K); temp range 388–413 K (torsion-effusion, Antoine eq., Piacente et al. 1994)

log (P/kPa) = 14.25 – 6726/(T/K); temp range 387–423 K (torsion-effusion, Antoine eq., Piacente et al. 1994)

log (P/kPa) = 13.92 – 6591/(T/K); temp range 386–425 K (torsion-effusion, Antoine eq., Piacente et al. 1994)

log (P/kPa) = 13.96 – 6608/(T/K); temp range ~ 386–425 K, ΔH_V = 126 kJ mol^{-1} (selected Antoine eq. based on exptl. data, torsion-effusion method, Piacente et al. 1994)

3.30×10^{-5} (quoted from Daubert & Danner 1997, Goss & Schwarzenbach 1999)

2.37×10^{-5} (GC-retention time correlation, Chickos & Hanshaw 2004)

Henry's Law Constant (Pa·m^3/mol):

Octanol/Water Partition Coefficient, log K_{OW}:

12.53   (HPLC-k′ correlation, Coates et al. 1985)

12.0; 10.5   (calculated-fragment const; calculated-molar volume correlation, Wang et al. 1992)

Octanol/Air Partition Coefficient, log K_{OA}:

Bioconcentration Factor, log BCF or log K_B:

Sorption Partition Coefficient, log K_{OC}:

Environmental Fate Rate Constants, k, and Half-Lives, $t_{\frac{1}{2}}$:

Biodegradation: microbial degradation $t_{\frac{1}{2}}$ < 31 d by *Pseudomonas* sp. (Setti et al. 1993)

Half-Lives in the Environment:

TABLE 2.1.1.1.36.1
Reported vapor pressures of tetracosane at various temperatures and the coefficients for the vapor pressure equations

log P = A – B/(T/K)	(1)	ln P = A – B/(T/K)	(1a)
log P = A – B/(C + t/°C)	(2)	ln P = A – B/(C + t/°C)	(2a)
log P = A – B/(C + T/K)	(3)	ln P = A – B/(C + t/K)	(3a)
log P = A – B/(T/K) – C·log (T/K)	(4)		

1.

Stull 1947		Zwolinski & Wilhoit 1971		Sasse et al. 1988		Piacente et al. 1990	
summary of literature data		selected values		electronic manometry		transpiration	
t/°C	P/Pa	t/°C	P/Pa	t/°C	P/Pa	t/°C	P/Pa
						run A	
183.8	133.3	175.9	66.66	100.26	0.255	183.85	85.11
219.6	666.6	188.0	133.3	110.25	0.613	185.85	100.0
237.6	1333	201.1	266.6	120.21	1.560	187.85	97.72
255.3	2666	220.5	666.6	130.15	3.506	189.85	107.2
276.3	5333	236.8	1333	139.93	7.386	193.85	134.9
288.4	7999	391.3	101325	140.09	7.506	195.85	151.4
305.2	13332			149.92	14.80	For run A:-	
330.3	66661	bp/°C	391.3	149.97	14.80	eq. 1	P/kPa
358.0	53329			159.77	28.53	A	8.17
386.4	101325	eq. 2	P/mmHg	159.62	52.13	B	4221
		A	7.0976	179.45	91.06	For temp range: 457–469 K	
mp/°C	51.1	B	2112.0	189.23	148.9	Run B	

TABLE 2.1.1.1.36.1 (Continued)

Stull 1947 summary of literature data		Zwolinski & Wilhoit 1971 selected values		Sasse et al. 1988 electronic manometry		Piacente et al. 1990 transpiration	
t/°C	P/Pa	t/°C	P/Pa	t/°C	P/Pa	t/°C	P/Pa
		C	109.6			187.85	104.7
				eq. 2	P/mmHg	192.85	138.0
				A	7.16777	197.85	182.0
				B	2243.665	206.85	288.4
				C	126.336	209.85	295.1
						212.35	316.2
						215.85	426.6
						216.85	436.5
						217.35	446.7
						217.88	457.1
						222.35	588.8
						223.35	616.6
						for run B:-	
						eq. 1	P/kPa
						A	9.53
						B	4843
						for temp range: 461–496 K	

2.

Piacente et al. 1990 (cont'd) transpiration method		Piacente et al. 1991 transpiration method		Piacente et al. 1991 torsion method			
t/°C	P/Pa	t/°C	P/Pa	t/°C	P/Pa	t/°C	P/Pa
run C				cell B		cont'd	
177.85	61.66	227.85	537.0	109.05	0.955	136.85	9.772
184.85	79.43	230.85	602.6	118.35	1.514	138.85	11.22
187.85	85.11	234.85	741.3	122.35	2.089	141.85	13.49
189.85	97.72	238.85	871.0	126.35	2.291	142.85	14.79
192.35	102.3	240.85	933.3	129.35	3.020	144.85	15.85
193.85	117.5	242.85	1096	131.35	3.388	147.85	19.05
194.85	141.25	245.85	1175	134.35	4.365	148.85	21.38
196.85	123.0	246.85	1259	136.35	4.898	149.85	23.44
201.35	141.25	247.85	1047	138.35	6.457	152.35	26.30
205.85	199.5	249.85	1148	141.35	7.943	154.85	29.51
209.85	245.5			143.35	10.23	158.85	39.81
212.85	288.4	For transpiration data:		145.35	11.75	160.35	44.67
213.85	302.0	eq. 1a	P/kPa	147.35	13.80	162.35	51.29
		A	21.25 ± 1.08	cell D		cell D	
		B	10946 ± 554	123.85	2.951	126.85	2.344
for transpiration data:				126.85	3.981	130.85	3.548
eq. 1	P/kPa			129.85	4.677	134.8513885	4.577
A	8.31 ± 1.08			133.85	5.888	139.85	5.888
B	4314 ± 211			135.85	7.079	140.85	6.457
for temp range: 451–487 K				137.85	8.128	142.85	9.333
				139.85	8.128	144.85	10.47
overall vapor pressure eq. by				142.85	9.333	147.85	12.88

(Continued)

TABLE 2.1.1.1.36.1 (Continued)

Piacente et al. 1990 (cont'd)		Piacente et al. 1991					
transpiration method		transpiration method		torsion method			
t/°C	P/Pa	t/°C	P/Pa	t/°C	P/Pa	t/°C	P/Pa
weighted slopes and intercepts				145.85	14.125	149.85	15.14
eq. 1	P/kPa			148.85	18.20	152.85	17.78
A	8.76 ± 0.50			151.85	36.31	153.85	18.62
B	4501 ± 250			cell A		156.85	22.39
				102.85	0.7413	162.85	33.11
ΔH_V/(kJ mol^{-1}) = 86 ± 5				104.85	0.8511	164.85	36.31
at 474 K				106.95	0.9772		
				109.85	1.2303	for torsion method data:-	
				114.85	1.5849	eq. 1a	P/kPa
				118.85	2.344	A	25.35 ± 0.60
				126.85	4.169	B	12399 ± 575
				127.85	4.898		
				129.85	5.623		
				131.85	6.918		
				133.85	8.128		

3.

Piacente et al. 1991 (cont'd)		Piacente et al. 1994					
Knudsen effusion		torsion-effusion					
t/°C	P/Pa	t/°C	P/Pa	t/°C	P/Pa	t/°C	P/Pa
		run a		run b		run c	
69.85	0.010	114.85	0.801	113.85	0.721	112.85	0.721
76.85	0.0282	120.85	1.60	117.85	1.12	113.85	0.801
83.85	0.0617	125.85	2.40	120.85	1.44	115.85	0.881
91.85	0.1995	128.85	3.20	123.85	2.08	117.85	1.12
119.35	1.995	132.35	4.33	127.85	2.88	119.85	1.44
		134.85	5.61	130.85	4.17	121.85	1.76
bp/K	664.3	138.35	7.37	133.85	5.29	124.85	2.32
mp/K	323.9	139.85	8.49	136.85	6.89	125.85	2.56
				138.85	8.49	126.85	2.80
eq. for transpiration, torsion		eq. 1	P/kPa	140.85	10.1	128.85	3.12
and Knudsen effusion		A	13.57 ± 0.26	143.85	12.8	130.85	4.17
eq. 3a	P/kPa	B	6459 ± 106	155.85	17.1	131.85	4.65
A	18.38 ± 0.34	temp range: 388–413 K		159.85	21.6	133.85	5.61
B	8349 ± 460					136.85	7.21
C	58 ± 15			eq. 1	P/kPa	139.85	9.29
ΔH_{fus}/(kJ mol^{-1}) = 58.893				A	14.25 ± 0.12	142.85	11.9
ΔH_{sub}/(kJ mol^{-1}) = 162.0				B	6726 ± 49	144.85	14.1
at 298.15 K				temp range: 387–423 K		146.85	17.0
						151.85	25.3
Morgan & Kobayashi 1994				selected vapor pressure eq.			
differential pressure gauge				eq. 1	P/kPa	eq. 1	P/kPa
t/°C	P/Pa			A	13.96 ± 0.30	A	13.92 ± 0.13
				B	6608 ± 100	B	6591 ± 0.51
179.916	90.1					temp range: 387–423 K	
189.925	154.7						
199.931	257			ΔH_V/(kJ mol^{-1}) = 126 ± 2			
				at 405 K			

TABLE 2.1.1.1.36.1 (Continued)

Piacente et al. 1991 (cont'd)		Piacente et al. 1994					
Knudsen effusion		torsion-effusion					
t/°C	P/Pa	t/°C	P/Pa	t/°C	P/Pa	t/°C	P/Pa
209.939	415.6						
219.946	652.9			mp/K	322		
229.950	1001			ΔH_{fus}/(kJ mol^{-1}) = 54.9			
239.960	1506			ΔH_{sub}/(kJ mol^{-1}) = 180.9 ± 2			
249.971	2209			at 298.15 K			
259.979	3176						
260.979	4489						
284.980	7303						
299.980	11470						
314.820	17460						

data fitted to Wagner eq.

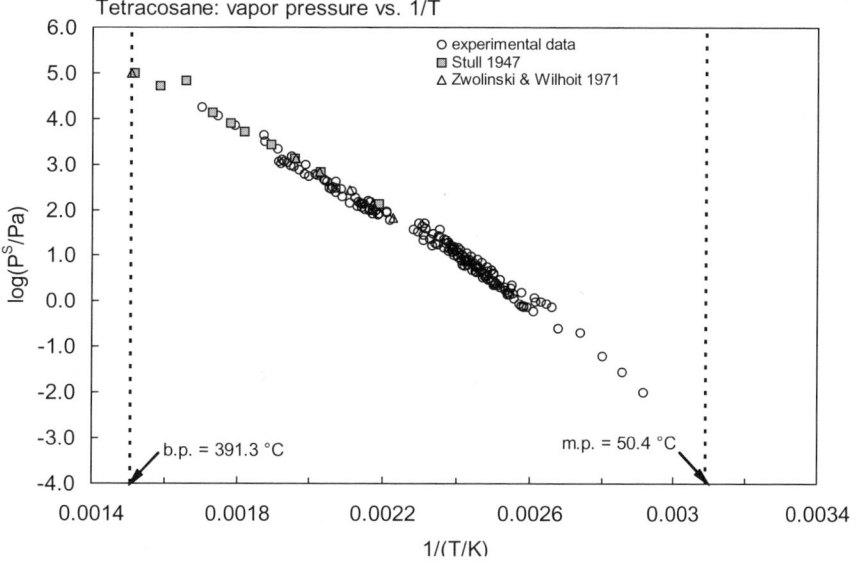

FIGURE 2.1.1.1.36.1 Logarithm of vapor pressure versus reciprocal temperature for *n*-tetracosane.

2.1.1.1.37   n-Hexacosane

Common Name: Hexacosane
Synonym: cerane
Chemical Name: n-hexacosane
CAS Registry No: 631-01-3
Molecular Formula: $C_{26}H_{54}$; $CH_3(CH_2)_{24}CH_3$
Molecular Weight: 366.707
Melting Point (°C):
> 56.1   (Lide 2003)

Boiling Point (°C):
> 412.2   (Dreisbach 1959; Weast 1982–83; Stephenson & Malanowski 1987; Lide 2003)

Density (g/cm³):
> 0.8032, 0.7998   (20°C, 25°C, Dreisbach 1959)
> 0.8032   (20°C, Weast 1982–83)

Molar Volume (cm³/mol):
> 584.6   (calculated-Le Bas method at normal boiling point, Eastcott et al. 1988)
> 456.6   (20°C, calculated-density)
> 457   (Wang et al. 1992)

Enthalpy of vaporization, ΔH_V (kJ/mol):
> 64.806   (bp, Dreisbach 1959)
> 131.2   (calculated, Chickos & Hanshaw 2004)

Enthalpy of Fusion, ΔH_{fus} (kJ/mol):
> 32.3, 59.5; 95.3   (53.35, 56.35°C, total phase change enthalpy, Chickos et al. 1999)

Entropy of Fusion, ΔS_{fus} (J/mol K):
> 289.03, 259.1   (exptl., calculated-group additivity method, total phase change entropy, Chickos et al. 1999)

Fugacity Ratio at 25°C (assuming ΔS_{fus} = 56 J/mol K): F: 0.495 (mp at 56.1°C)

Water Solubility (g/m³ or mg/L at 25°C):
> 1.7×10^{-3}; 1.0×10^{-4} (shake flask-GC, distilled water; seawater, Sutton & Calder 1974)
> 1.33×10^{-10} (extrapolated from data of McAuliffe 1966, Eastcott et al. 1988)

Vapor Pressure (Pa at 25°C or as indicated and reported temperature dependence equations. Additional data at other temperatures designated * are compiled at the end of this section.):
> 133.3*   (204.0°C, summary of literature data, temp range 204.0–399.8°C, Stull 1947)
> log (P/mmHg) = 7.57689 – 2692.73/(161.2 + t/°C); temp range 278–500°C (Antoine eq. for liquid state, Dreisbach 1959)
> log (P/mmHg) = 7.1096 – 2164.3/(99.6 + t/°C) (Antoine eq., Kudchadker & Zwolinski 1966)
> 66.66*   (192.5°C, derived from compiled data, temp range 192.5–412.2°C, Zwolinski & Wilhoit 1971)
> log (P/mmHg) = 7.1096 – 2164.3/(99.6 + t/°C); temp range 192.6–412.2°C (Antoine eq., Zwolinski & Wilhoit 1971)
> log (P/mmHg) = [–0.2185 × 21605.7/(T/K)] + 9.899820; temp range 204–399.8°C (Antoine eq., Weast 1972–73)
> log (P_L/kPa) = 6.2345 – 2164.3/(T/K – 173.55); temp range 466–685 K (Antoine eq., liquid, Stephenson & Malanowski 1987)
> log (P_L/kPa) = 9.44384 – 4935.969/(T/K); temp range 478–530 K (Antoine eq., liquid, Stephenson & Malanowski 1987)
> log (P/kPa) = (9.93 ± 0.50) – (5168 ± 200)/(T/K); temp range 455–519 K, (Antoine eq. from exptl. data, transpiration, Piacente & Scardala 1990; Pompili & Piacente 1990)
> ln (P/kPa) = 18.63 – 9892/(T/K); temp range 506–546 K (transpiration method, Piacente et al. 1991)
> ln (P/kPa) = 28.91 – 14285/(T/K); temp range 391–442 K (torsion method, Piacente et al. 1991)
> ln (P/kPa) = 17.76 – 8050/[(T/K) + 72]; temp range 356–546 K (transpiration, torsion and Knudsen methods, Antoine eq., Piacente et al. 1991)
> log (P/kPa) = 14.50 – 7084/(T/K); temp range 420–437 K (torsion-effusion, Antoine eq., Piacente et al. 1994)

Aliphatic and Cyclic Hydrocarbons

log (P/kPa) = 13.75 – 6765/(T/K); temp range 391–433 K (torsion-effusion, Antoine eq., Piacente et al. 1994)
log (P/kPa) = 13.65 – 6748/(T/K); temp range 392–431 K (torsion-effusion, Antoine eq., Piacente et al. 1994)
log (P/kPa) = 14.01 – 6682/(T/K); temp range ~ 392–437 K, ΔH_V = 132 kJ mol^{-1} (selected Antoine eq. based on exptl. data, torsion-effusion method, Piacente et al. 1994)
5.03 × 10^{-6} (quoted from Daubert & Danner 1997, Goss & Schwarzenbach 1999)
2.82 × 10^{-6} (GC-retention time correlation, Chickos & Hanshaw 2004)

Henry's Law Constant (Pa·m^3/mol):
21200 (calculated-P/C, Eastcott et al. 1988)

Octanol/Water Partition Coefficient, log K_{OW}:
13.0, 11.4 (calculated-fragment const., calculated-molar volume correlation, Wang et al. 1992)

Octanol/Air Partition Coefficient, log K_{OA}:

Bioconcentration Factor, log BCF or log K_B:

Sorption Partition Coefficient, log K_{OC}:

Environmental Fate Rate Constants, k, and Half-Lives, $t_{1/2}$:
Biodegradation: microbial degradation $t_{1/2}$ < 31 d by *Pseudomonas* sp. (Setti et al. 1993)

Half-Lives in the Environment:

TABLE 2.1.1.1.37.1
Reported vapor pressures of hexacosane at various temperatures and the coefficients for the vapor pressure equations

log P = A – B/(T/K)	(1)	ln P = A – B/(T/K)	(1a)
log P = A – B/(C + t/°C)	(2)	ln P = A – B/(C + t/°C)	(2a)
log P = A – B/(C + T/K)	(3)	ln P = A – B/(C + T/K)	(3a)
log P = A – B/(T/K) – C·log (T/K)	(4)		

1.

Stull 1947		Zwolinski & Wilhoit 1971		Piacente et al. 1990			
summary of literature data		selected values		transpiration		transpiration (continued)	
t/°C	P/Pa	t/°C	P/Pa	t/°C	P/Pa	t/°C	P/Pa
				run A		run C	
204.0	133.3	192.5	66.66	187.85	64.57	182.35	37.15
240.0	666.6	204.8	133.3	194.35	74.13	188.85	54.95
257.4	1333	218.3	266.6	197.35	91.20	193.35	70.79
275.8	2666	138.0	666.6	197.85	100.0	196.85	89.13
295.2	5333	254.6	527.75	198.35	97.72	199.85	104.7
307.8	7999	412.2	101325	199.85	109.65	202.35	123.0
323.2	13332			202.85	129.8	203.85	131.8
348.4	66661			205.85	131.8	204.85	112.2
374.5	53329	bp/°C	412.2	208.85	162.2	206.85	134.9
399.8	101325			211.85	251.2	210.85	229.1
		eq. 2	P/mmHg	214.85	263.0	215.35	288.4
mp/°C	56.6	A	7.1096	eq. for run A:-		218.85	263.0
		B	2164.3	eq. 1	P/kPa	221.85	309.0
		C	99.6	A	10.40 ± 0.42	223.35	346.7

(Continued)

TABLE 2.1.1.1.37.1 (Continued)

Stull 1947 summary of literature data		Zwolinski & Wilhoit 1971 selected values		Piacente et al. 1990 transpiration		Piacente et al. 1990 transpiration (continued)	
t/°C	P/Pa	t/°C	P/Pa	t/°C	P/Pa	t/°C	P/Pa
				B	5355 ± 198	243.35	691.8
				for temp range: 461–488 K		245.85	812.8
				run B		eq. for run C:-	
				202.35	100	eq. 1	P/kPa
				209.35	128.8	A	9.53 ± 0.21
				215.85	166.0	B	4979 ± 103
				220.85	269.2	temp range: 455.5–519 K	
				224.85	295.1		
				227.35	331.1	overall vapor pressure eq. by	
				229.35	407.4	weighted slopes and intercepts	
				234.85	457.1	eq. 1	P/kPa
						A	9.93 ± 0.50
				eq. 1	P/kPa	B	5168 ± 200
				A	10.09 ± 0.64		
				B	5289 ± 315	$\Delta H_V/(kJ\ mol^{-1})$ =	
				for temp range: 475.5–508 K		at 488 K	99 ± 4

2.

Piacente et al. 1991

transpiration method		torsion method		torsion (continued)		torsion (continued)	
t/°C	P/Pa	t/°C	P/Pa	t/°C	P/Pa	t/°C	P/Pa
		cell E		cell A		cell B	
232.85	354.8	131.35	1.202	141.35	1.479	138.85	3.020
238.35	616.6	132.35	1.380	143.35	1.950	142.85	4.571
242.85	616.6	134.35	1.585	145.35	2.692	144.85	4.169
242.85	616.6	136.35	1.778	146.35	3.162	144.85	4.898
246.35	575.4	137.35	1.995	146.85	3.715	147.35	6.166
247.85	660.7	139.35	2.570	147.35	4.169	150.35	6.457
249.35	758.6	140.85	3.020	147.85	4.898	153.85	9.550
251.35	871.0	142.35	3.388	148.35	5.623	154.35	9.120
252.35	933.3	146.35	5.129	148.85	6.310	154.35	9.550
254.35	933.3	148.35	6.166	149.35	7.586	154.85	9.550
255.85	1023	150.35	6.918	150.35	9.120	155.85	10.72
256.35	871.0	153.35	7.762	152.35	10.47	155.85	10.96
258.85	1047	155.35	9.120	154.35	11.22	158.85	13.80
259.35	1148	158.35	12.30	165.35	12.88	cell A	
261.85	1288	160.35	14.13	158.35	15.49	132.85	2.455
262.85	1288	163.35	17.38	160.35	18.62	135.35	2.951
266.35	1230	165.35	20.42	cell A		138.85	3.890
266.35	1259	167.35	23.99	117.85	0.490	142.35	5.129
268.35	1445	169.35	28.18	123.85	0.741	143.85	5.623
269.35	1413			126.85	0.977	145.35	6.310
271.85	1380			128.35	0.977	146.85	7.586
272.85	1585			129.35	1.230	147.35	7.413
272.85	1514			131.35	1.479	147.85	8.511
				133.35	1.698	148.85	9.550
eq. 1a	P/kPa			134.35	1.950	150.35	9.120

TABLE 2.1.1.1.37.1 (Continued)

Piacente et al. 1991

transpiration method		torsion method		torsion (continued)		torsion (continued)	
t/°C	P/Pa	t/°C	P/Pa	t/°C	P/Pa	t/°C	P/Pa
A	18.63 ± 1.03			135.35	2.188	150.85	10.00
B	9892 ± 552			136.35	2.455	151.85	10.47
				137.35	2.951	152.35	12.88
				138.35	3.467	152.85	13.18
				139.35	3.890	153.85	14.13
				141.35	4.677	155.85	15.85
				143.35	5.379		
						eq. 1a	P/kPa
						A	28.91 ± 0.40
						B	14285 ± 345

3.

Piacente et al. 1991 (cont'd) — Piacente et al. 1994

Knudsen effusion		torsion-effusion		torsion-effusion (continued)		torsion-effusion (continued)	
t/°C	P/Pa	t/°C	P/Pa	t/°C	P/Pa	t/°C	P/Pa
		run a		run b		run c	
83	0.01995	146.85	4.01	117.85	0.320	118.85	0.320
92	0.04677	147.85	4.49	120.85	0.401	124.85	0.481
102	0.1514	148.85	5.13	124.85	0.481	127.85	0.641
111	0.2818	149.85	5.77	128.85	0.801	130.85	0.801
119	0.5623	150.85	6.42	131.85	1.120	133.85	1.12
		151.85	7.05	134.85	1.52	136.85	1.60
bp/K	658.2	152.85	7.69	137.85	2.08	139.85	2.08
mp/K	329.4	153.85	8.33	141.85	2.88	142.85	2.72
		154.85	9.13	145.85	4.17	146.85	3.85
eq. for transpiration, torsion		155.85	9.61	149.85	5.77	151.85	6.09
and Knudsen effusion		156.85	10.9	156.85	10.7	154.85	7.85
eq. 3a	P/kPa	157.85	11.7	159.85	13.5	157.85	10.4
A	17.76 ± 0.46	159.85	13.3				
B	8050 ± 460	161.85	15.5				
C	72 ± 10	163.85	18.9				
				eq. 1	P/kPa	eq. 1	P/kPa
		eq. 1	P/kPa	A	13.75 ± 0.29	A	13.65 ± 0.29
		A	14.50 ± 0.36	B	6765 ± 119	B	6748 ± 120
ΔH_{fus}/(kJ mol^{-1}) = 60.70		B	7084 ± 153	temp range: 391–433 K		temp range: 392–431 K	
		temp range: 420–437 K					
ΔH_{sub}/(kJ mol^{-1}) = 177.2		overall vapor pressure eq.					
at 298.15 K				eq. 1	P/kPa		
				A	14.01 ± 0.30		
				B	6882 ± 50		
				ΔH_V/(kJ mol^{-1}) =			
				at 414 K	132 ± 1		
				mp/K	329		
				ΔH_{fus}/(kJ mol^{-1}) = 59.9			
				ΔH_{sub}/(kJ mol^{-1}) = 191.5 ± 1			
				at 298.15 K			

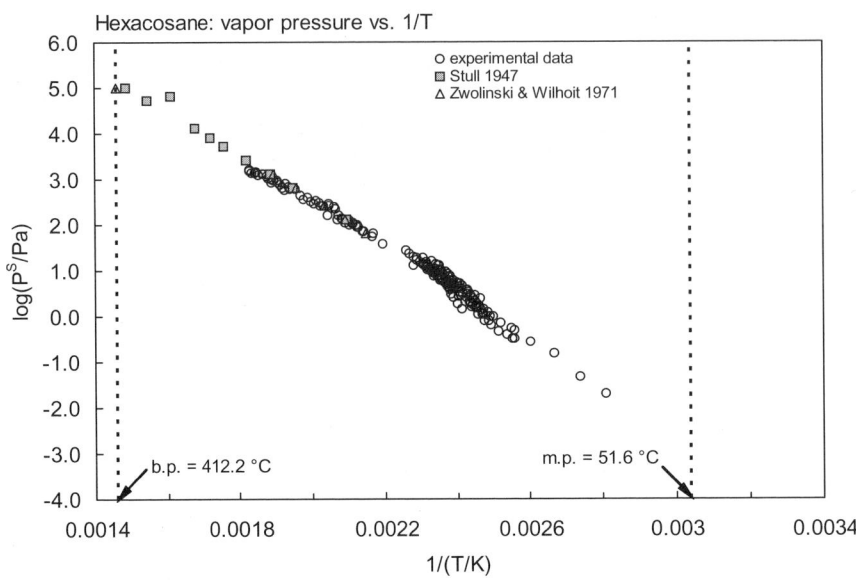

FIGURE 2.1.1.1.37.1 Logarithm of vapor pressure versus reciprocal temperature for *n*-hexacosane.

Aliphatic and Cyclic Hydrocarbons

2.1.1.2 Cycloalkanes

2.1.1.2.1 Cyclopentane

Common Name: Cyclopentane
Synonym: pentamethylene
Chemical Name: cyclopentane
CAS Registry No: 287-92-37
Molecular Formula: C_5H_{10}
Molecular Weight: 70.133
Melting Point (°C):
    –93.4   (Lide 2003)
Boiling Point (°C):
    49.3   (Lide 2003)
Density (g/cm³ at 20°C):
    0.7457   (Weast 1984)
    0.7454, 0.7440   (20°C, 25°C, Dreisbach 1959; Riddick et al. 1986)
Molar Volume (cm³/mol):
    94.10   (20°C, calculated-density; McAuliffe 1966)
    99.5   (calculated-Le Bas method at normal boiling point, Abernethy et al. 1988; Mackay & Shiu 1990)
Enthalpy of Vaporization, ΔH_V (kJ/mol):
    28.527, 27.296 (25°C, bp, Riddick et al. 1986)
Enthalpy of Fusion, ΔH_{fus} (kJ/mol):
    0.609   (Riddick et al. 1986)
    4.9, 0.34, 0.60; 5.84 (–151.15, –135.15, –93.45°C; total phase change enthalpy, Chickos et al. 1999)
Entropy of Fusion, ΔS_{fus} (J/mol K):
    45.96, 40.8 (exptl., calculated-group additivity method, total phase change entropy, Chickos et al. 1999)
Fugacity Ratio at 25°C, F: 1.0

Water Solubility (g/m³ or mg/L at 25°C. Additional data at other temperatures designated * are compiled at the end of this section.):
    156   (shake flask-GC, McAuliffe 1963, 1966)
    342*   (shake flask-GC, measured range 5.11–45.21°C, Pierotti & Liabastre 1972)
    160*   (shake flask-GC, measured range 25–153.1°C, Price 1976)
    160   (shake flask-GC, Krzyzanowska & Szeliga 1978)
    164   (shake flask-GC, Groves 1988)
    156*   (IUPAC "tentative" best, IUPAC Solubility Data Series, Shaw 1989)
    166*   (calculated-liquid-liquid equilibrium LLE data, temp range 273.2–426.3 K, Mączyński et al. 2004)

Vapor Pressure (Pa at 25°C or as indicated and reported temperature dependence equations. Additional data at other temperatures designated * are compiled at the end of this section.):
    27250*   (14.25°C, static method-manometry, measured range –47.24 to 14.25°C, Ashton et al. 1943)
    34890*   (20.2°C, ebulliometry-manometer, measured range 15.7–50°C, Willingham et al. 1945)
    log (P/mmHg) = 6.87798 – 1119.208/(230.738 + t/°C); temp range 15.7–50.0°C (Antoine eq. from exptl. data, ebulliometry-manometer, Willingham et al. 1945)
    43150*   (Antoine eq. regression, temp range –68 to 49.4°C, Stull 1947)
    42330   (calculated from exptl. determined data, Dreisbach 1955; quoted, Hine & Mookerjee 1975)
    log (P/mmHg) = 6,88676 – 1124.162/(231.361 + t/°C); temp range –25 to 110°C (Antoine eq. for liquid state, Dreisbach 1955)
    29036*   (selected exptl. data, temp range –39 to 230°C, Pasek & Thodos 1962)

42400 (interpolated-Antoine eq., Zwolinski & Wilhoit 1971; quoted, Mackay & Shiu 1981)
42330* (derived from compiled data, temp range –40.4 to 71.6°C, Zwolinski & Wilhoit 1971)
log (P/mmHg) = 6.88676 – 1124.162/(231.361 + t/°C); temp range –40.4 to 71.6°C (Antoine eq., Zwolinski & Wilhoit 1971)
log (P/mmHg) = [–0.2185 × 7411.1/(T/K)] + 7.940722; temp range –68 to 49.3°C (Antoine eq., Weast 1972–73)
log (P/atm) = [1– 322.386/(T/K)] × 10^{0.818603 – 7.52365 × 10^{-4}·(T/K) + 8.27395 × 10^{-7}·(T/K)^2}; temp range: 190.20–503.20 K (Cox eq., Chao et al. 1983)
42570, 42320 (calculated-Antoine equations, Boublik et al. 1984)
log (P/kPa) = 6.25832 – 1240.438/(242.957 + t/°C); temp range –47.25 to 14°C (Antoine eq. from reported exptl. data, Boublik et al. 1984)
log (P/kPa) = 6.82877 – 1133.199/(232.415 + t/°C), temp range 15.7–50.3°C (Antoine eq. from reported exptl. data of Willingham et al. 1945, Boublik et al. 1984)
42320 (interpolated-Antoine eq., Dean 1985, 1992)
log (P/mmHg) = 6.88676 – 1124.162/(231.36 + t/°C); temp range –40 to 72°C (Antoine eq., Dean 1985, 1992)
42400 (selected lit., Riddick et al. 1986)
log (P/kPa) = 6.04584 – 1142.30/(233.463 + t/°C); temp range not specified (Antoine eq., Riddick et al. 1986)
42340 (interpolated-Antoine eq., Stephenson & Malanowski 1987)
log (P_L/kPa) = 6.0080 – 1122.21/(–42.011 + T/K); temp range 280–311 K (Antoine eq.-I, Stephenson & Malanowski 1987)
log (P_L/kPa) = 6.08918 – 1174.132/(–34.864 + T/K); temp range 322–384 K (Antoine eq.-II, Stephenson & Malanowski 1987)
log (P_L/kPa) = 6.41769 – 1415.096/(–0.66 + T/K) temp range 381–455 K (Antoine eq.-III, Stephenson & Malanowski 1987)
log (P_L/kPa) = 6.77782 – 1749.65/(48.533 + T/K); temp range 452–511 K (Antoine eq.-III, Stephenson & Malanowski 1987)
log (P/mmHg) = 29.1547 – 2.3512 × 10^3/(T/K) – 7.6965·log (T/K) – 1.6212 × 10^{-10}·(T/K) + 3.125 × 10^{-6}·(T/K)^2; temp range 179–512 K (vapor pressure eq., Yaws 1994)
35000* (20.14°C, differential pressure gauge, measured range –68.59 to 70.06°C, Mokbel et al. 1995)

Henry's Law Constant (Pa m^3/mol at 25°C or as indicated and reported temperature dependence equations. Additional data at other temperatures designated * are compiled at the end of this section.):
18800 (calculated as 1/K_{AW}, C_W/C_A, reported as exptl., Hine & Mookerjee 1975)
13310; 18380 (calculated-group contribution, bond contribution, Hine & Mookerjee 1975)
18500; 19100, 18600 (recommended; calculated-P/C, Mackay & Shiu 1981)
17550 (calculated-MCI χ, Nirmalakhandan & Speece 1988)
19030 (calculated-vapor-liquid equilibrium (VLE) data, Yaws et al. 1991)
16617* (27.9°C, EPICS-GC, measured range 27.9–45°C, Hansen et al. 1993)
ln [H/(kPa·m^3/mol)] = –3351/(T/K) + 14.0; temp range 27.9–45°C (EPICS-GC, Hansen et al. 1993)
12796 (20°C, selected from reported experimental determined values, Staudinger & Roberts 1996, 2001)
log K_{AW} = 5.162 – 1302/(T/K) (summary of literature data, Staudinger & Roberts 2001)

Octanol/Water Partition Coefficient, log K_{OW}:
2.05 (calculated-π substituent constant, Hansch et al. 1968)
3.00 (shake flask-GC, Leo et al. 1975; Hansch & Leo 1979)
3.00 (recommended, Sangster 1989, 1993)
3.00 (recommended, Hansch et al. 1995)

Octanol/Air Partition Coefficient, log K_{OA}:

Bioconcentration Factor, log BCF:

Sorption Partition Coefficient, log K_{OC}:

Environmental Fate Rate Constants, k, and Half-Lives, $t_{1/2}$:
 Volatilization:
 Photolysis:

Oxidation: rate constant k, for gas-phase second order rate constants, k_{OH} for reaction with OH radical, k_{NO3} with NO_3 radical and k_{O3} with O_3 or as indicated, *data at other temperatures and/or the Arrhenius expression see reference:

$k_{O(3P)} = 1.30 \times 10^{-13}$ cm^3 molecule^{-1} s^{-1} for reaction with O(^{3}P) (Herron & Huie 1973)

$k_{OH} = (4.72 \pm 0.28) \times 10^{-12}$ cm^3 molecule^{-1} s^{-1}, k_{OH} (calc) = 5.80×10^{-12} cm^3 molecule^{-1} s^{-1} at 300 K (relative rate method, Darnall et al. 1978)

$k_{OH} = 5.40 \times 10^{-12}$ cm^3 molecule^{-1} s^{-1} (Atkinson et al. 1979)

$k_{OH} = 5.40 \times 10^{-12}$ cm^3 molecule^{-1} s^{-1}; $k_{O(3P)} = 1.3 \times 10^{-13}$ cm^3 molecule^{-1} s^{-1} for reaction with O(^{3}P) atoms at room temp. (Gaffney & Levine 1979)

$k_{OH} = 6.20 \times 10^{-12}$ cm^3 molecule^{-1} s^{-1} at 298 K and 5.18×10^{-12} cm^3 molecule^{-1} s^{-1} at 298 K and 5.24×10^{-12} cm^3·molecule^{-1} s^{-1} at 299 K and 4.43×10^{-12} cm^3·molecule^{-1} s^{-1} at 300 K (Atkinson 1985)

$k_{OH} = (3.12 \pm 0.23) \times 10^{12}$ cm^3 mol^{-1} s^{-1} at 298 ± 2 K (flash photolysis-resonance absorption technique, Jolly et al. 1985)

$k_{OH} = 5.02 \times 10^{-12}$ cm^3 molecule^{-1} s^{-1} at 295 K, measured range 295–491 K (Droege & Tully 1987)

$k_{OH}* = 5.16 \times 10^{-12}$ cm^3 molecule^{-1} s^{-1} at 298 K (recommended, Atkinson 1989, 1990)

$k_{OH} = 5.16 \times 10^{-12}$ cm^3 molecule^{-1} s^{-1} at 298 K with an estimated lifetime of 27 h in summer daylight (Altshuller 1991)

$k_{OH}* = 5.02 \times 10^{-12}$ cm^3 molecule^{-1} s^{-1} at 298 K (recommended, Atkinson 1997)

$k_{OH}* = 4.83 \times 10^{-12}$ cm^3 molecule^{-1} s^{-1} at 298 K, measured range 230–400 K (relative rate method, DeMore & Bayes 1999)

Hydrolysis:
Biodegradation:
Biotransformation:
Bioconcentration, Uptake (k_1) and Elimination (k_2) Rate Constants or Half-Lives:

Half-Lives in the Environment:

Air: atmospheric $t_{1/2}$ ~ 2.4–24 h for cycloparaffins, based on the EPA Reactivity Classification of Organics (Darnall et al. 1976);

photooxidation reaction rate constant k = 5.16×10^{-12} cm^3 molecule^{-1} s^{-1} with hydroxyl radical in air at 298 K (Atkinson 1990; Altshuller 1991) with an estimated lifetime of 27 h, based on reaction rate with OH radical in summer daylight (Altshuller 1991).

TABLE 2.1.1.2.1.1
Reported aqueous solubilities of cyclopentane at various temperatures

Pierotti & Liabastre 1972		Price 1976		Shaw 1989a		Mączyński et al. 2004	
shake flask-GC/FID		shake flask-GC/FID		IUPAC "tentative" values		calc-recommended LLE data	
t/°C	S/g·m^{-3}	t/°C	S/g·m^{-3}	t/°C	S/g·m^{-3}	t/°C	S/g·m^{-3}
5.11	338.6	25.0	160	5	339	25	168
15.21	341.7	40.1	163	15	342	40.1	175
25.11	341.9	55.7	180	25	156	55.7	195
35.21	368.5	99.1	296	30	160	99.1	343
45.21	341.5	118.0	372	40	350	118	468
		137.3	611	60	500	137.3	701
		153.1	792	80	750	153.1	974
ΔH_{sol}/(kJ mol^{-1}) = –2.50				100	1100		
25°C				120	1600		
		ΔH_{sol}/(kJ mol^{-1}) = –2.80		140	2600		
		25°C		160	3850		
				180	6700		
				200	14000		

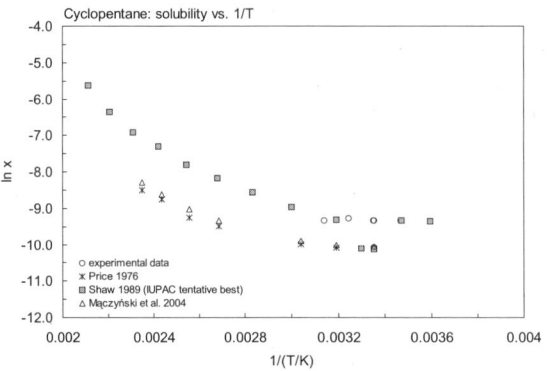

FIGURE 2.1.1.2.1.1 Logarithm of mole fraction solubility (ln x) versus reciprocal temperature for cyclopentane.

TABLE 2.1.1.2.1.2
Reported vapor pressures of cyclopentane at various temperatures and the coefficients for the vapor pressure equations

$$\log P = A - B/(T/K) \quad (1) \qquad \ln P = A - B/(T/K) \quad (1a)$$
$$\log P = A - B/(C + t/°C) \quad (2) \qquad \ln P = A - B/(C + t/°C) \quad (2a)$$
$$\log P = A - B/(C + T/K) \quad (3)$$
$$\log P = A - B/(T/K) - C·\log(T/K) \quad (4)$$
$$\log P = A + B/(T/K) + C·\log(T/K) + D[P/(T/K)^2] \quad (5)$$
$$\log P = A - B/T(T/K) - C·\log(T/K) + D \times 10^{-3}(T/K) - E \times 10^{-8}(T/K)^2 \quad (6)$$

1.

Ashton et al. 1943		Willingham et al. 1945		Stull 1947		Pasek & Thodos 1962	
static method-manometry		ebulliometry		summary of literature data		selected experimental data	
t/°C	P/Pa	t/°C	P/Pa	t/°C	P/Pa	t/°C	P/Pa
−47.24	830.6	15.707	28956	−68.0	133.3	−39.064	1571
−38.35	1570	20.196	34892	−49.6	666.6	9.335	21949
−31.63	2446	25.598	43322	−40.4	1333	15.722	29036
−24.7	3753	31.172	53656	−30.1	2666	48.146	97608
−18.6	5360	37.119	66760	−18.6	5333	57.885	133322
−12.91	7350	43.574	83722	−11.3	7999	71.612	1999836
−7.29	9875	48.131	97608	−1.30	13332	100.005	4197790
−1.93	12931	48.621	99205	13.8	26664	230.065	4023402
4.47	17569	49.073	100698	31.0	53329		
9.33	21949	49.587	102401	49.3	101325	bp/°C	49.307
14.25	27250	50.031	103921				
				mp/°C	−93.7	Frost-Kalkwarf equation:	
bp/°C	49.20					derived from exptl. data	
mp/°C	−93.62	bp/°C	49.262			eq 5	P/mmHg
						A	21.62180
eq. 6	P/mmHg	eq. 2	P/mmHg			B	−2131.85
A	30.957385	A	6.87798			C	−4.83947
B	2298.386	B	1119.208			D	1.41701
C	8.91170	C	230.738				
D	4.385677					Frost-Kalkwarf equation:	
E	1.054940					calculated from molecular	
						structures, normal bp and	

Aliphatic and Cyclic Hydrocarbons

TABLE 2.1.1.2.1.2 (*Continued*)

Ashton et al. 1943		Willingham et al. 1945		Stull 1947		Pasek & Thodos 1962	
static method-manometry		ebulliometry		summary of literature data		selected experimental data	
t/°C	P/Pa	t/°C	P/Pa	t/°C	P/Pa	t/°C	P/Pa
ΔH_V/(kJ mol^{-1}) = 29.21 at 25°C						deviations of resulting v.p.:-	
						eq 5	P/mmHg
						A	21.63300
						B	−2132.5
						C	−4.84313
						D	1.41701

2.

Zwolinski & Wilhoit 1971		Mokbel et al. 1995	
selected values		static method-manometry	
t/°C	P/Pa	t/°C	P/Pa
−40.4	1333	−68.59	138
−30.1	2666	−59.24	313
−23.6	4000	−49.42	692
−18.6	5333	−39.51	1446
−14.7	6666	−39.42	1456
−11.3	7999	−29.66	2775
−5.80	10666	−29.53	2781
−1.30	13332	−19.77	5034
7.28	19998	−19.64	5057
13.78	26664	−9.86	8679
19.08	33331	−9.73	8718
23.57	39997	0.15	14340
31.01	53329	0.22	14430
37.08	66661	10.11	22750
42.35	79993	10.17	22840
46.78	93326	20.08	34860
47.627	95992	20.14	35000
48.453	98659	30.27	51930
49.262	101325	40.23	74670
		50.14	104580
		60.12	143100
eq. 2	P/mmHg	65.10	166090
A	6.88676	70.06	191640
B	1124.162		
C	231.361	Wagner eq. given in ref.	
bp/°C	49.262		
ΔH_V/(kJ mol^{-1}) =			
at 25°C	28.53		
at bp	27.30		

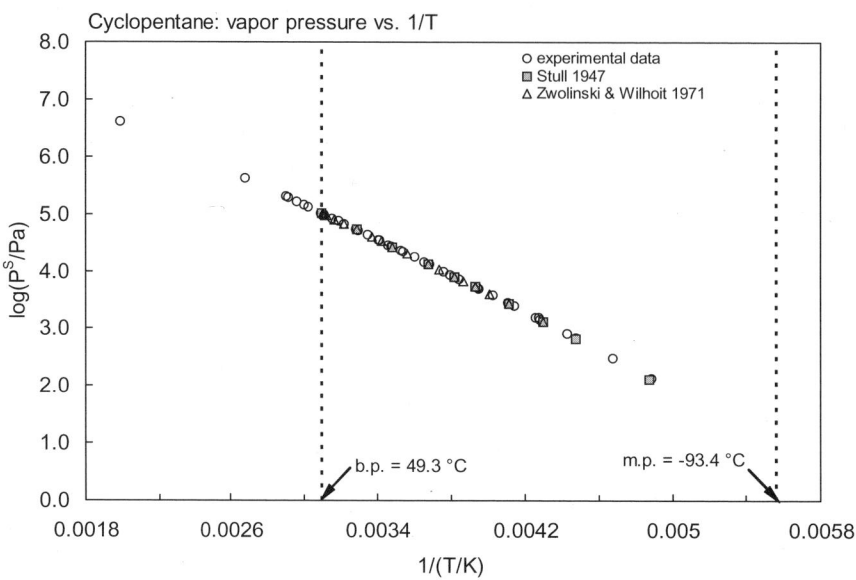

FIGURE 2.1.1.2.1.2 Logarithm of vapor pressure versus reciprocal temperature for cyclopentane.

TABLE 2.1.1.2.1.3
Reported Henry's law constants of cyclopentane at various temperatures

	Hansen et al. 1993
	EPICS-GC
t/°C	H/(kPa m^3/mol)
27.9	16.617
35.8	24.318
45.0	30.398

$\ln [H/(Pa\ m^3/mol)] = A - B/(T/K)$

	H/(kPa m^3/mol)
A	14 ± 2.03
B	3351 ± 633

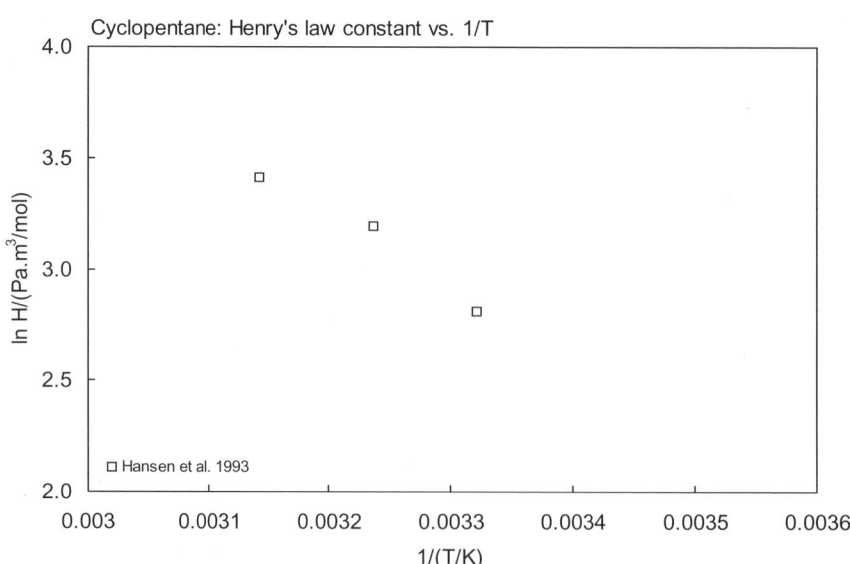

FIGURE 2.1.1.2.1.3 Logarithm of Henry's law constant versus reciprocal temperature for cyclopentane.

Aliphatic and Cyclic Hydrocarbons

2.1.1.2.2 Methylcyclopentane

Common Name: Methylcyclopentane
Synonym:
Chemical Name: methylcyclopentane
CAS Registry No: 96-37-3
Molecular Formula: C_6H_{12}
Molecular Weight: 84.159
Melting Point (°C):
    –142.42   (Lide 2003)
Boiling Point (°C):
    71.8   (Lide 2003)
Density (g/cm³ at 25°C):
    0.7487, 0.7439   (20°C, 25°C, Dreisbach 1955; Riddick et al. 1986)
Molar Volume (cm³/mol):
    112.4, 113.1   (20°C, 25°C, calculated from density)
    121.7   (calculated-Le Bas method at normal boiling point)
Enthalpy of Vaporization, ΔH_V (kJ/mol):
    41.59, 29.08   (25°C, bp, Riddick et al. 1986)
Enthalpy of Fusion, ΔH_{fus} (kJ/mol):
    6.929   (Riddick et al. 1986)
    6.93   (Chickos et al. 1999)
Entropy of Fusion, ΔS_{fus} (J/mol K):
    53.01, 43.9   (exptl., calculated-group additivity method, Chickos et al. 1999)
Fugacity Ratio at 25°C, F: 1.0

Water Solubility (g/m³ or mg/L at 25°C):
    42.6   (shake flask-GC, McAuliffe 1963)
    42.0   (shake flask-GC, McAuliffe 1966)
    41.8   (shake flask-GC, Price 1976)
    45.0   (partition coefficient-GC, Rudakov & Lutsyk 1979)
    43.0   (recommended, IUPAC Solubility Data Series, Shaw 1989)
    74.8   (calculated-recommended liquid-liquid equilibrium LLE data, Mączyński et al. 2004)

Vapor Pressure (Pa at 25°C or as indicated and reported temperature dependence equations):
    16680   (22.75°C, ebulliometry-manometer, measured range 15.0–22.6°C, Willingham et al. 1945)
    $\log (P/\text{mmHg}) = 6.86283 - 1186.059/(229.042 + t/°C)$; temp range 15.0–22.6°C (Antoine eq. from exptl. data, ebulliometry-manometer, Willingham et al. 1945)
    17870   (Antoine eq. regression, temp range –53.7 to 71.8°C, Stull 1947)
    18330   (calculated from exptl. determined data, Dreisbach 1955)
    $\log (P/\text{mmHg}) = 6.86283 - 1186.059/(226.042 + t/°C)$; temp range –5 to 125°C (Antoine eq. for liquid state, Dreisbach 1955)
    18302   (interpolated-Antoine eq., temp range –23.7 to 95.7°C, Zwolinski & Wilhoit 1971)
    $\log (P/\text{mmHg}) = 6.86283 - 1186.059/(226.042 + t/°C)$; temp range –23.7 to 95.7°C (Antoine eq., Zwolinski & Wilhoit 1971)
    17850   (interpolated-Antoine eq., temp range –53.7 to 71.8°C, Weast 1972–73)
    $\log (P/\text{mmHg}) = [-0.2185 \times 7490.0/(T/K)] + 7.945471$; temp range –53.7 to 71.8°C (Antoine eq., Weast 1972–73)
    $\log (P/\text{atm}) = [1 - 344.830/(T/K)] \times 10^{\{0.872156 - 9.88091 \times 10^{-4} \cdot (T/K) + 10.8367 \times 10^{-7} \cdot (T/K)^2\}}$; temp range: 183.15–513.15 K (Cox eq., Chao et al. 1983)

18330 (interpolated-Antoine eq., Boublik et al. 1984)
log (P/kPa) = 5.99178 − 1188.32/(226.307 + t/°C); temp range 15.035–72.6°C (Antoine eq. from reported exptl. data, Boublik et al. 1984)
18330 (interpolated-Antoine eq., Dean 1985)
log (P/mmHg) = 6.86283 − 1186.059/(226.04 + t/°C); temp range −24 to 96°C (Antoine eq., Dean 1985, 1992)
18400 (lit. average, Riddick et al. 1986)
log (P/kPa) = 5.98773 − 1186.059/(226.042 + t/°C); temp range not specified (Antoine eq., Riddick et al. 1986)
18340 (interpolated-Antoine eq., Stephenson & Malanowski 1987)
log (P_L/kPa) = 5.98551 − 1184.874/(−47.232 + T/K); temp range 288–346 K (Antoine eq., Stephenson & Malanowski 1987)
18400, 17066 (quoted, calculated-UNIFAC activity coeff., Banerjee et al. 1990)
log (P/mmHg) = 32.4766 − 2.6434 × 10^3/(T/K) − 8.733·log (T/K) + 2.0749 × 10^{-11}·(T/K) + 3.2158 × 10^{-6}·(T/K)^2; temp range 131–533 K (vapor pressure eq., Yaws 1994)

Henry's Law Constant (Pa m^3/mol at 25°C):
36664 (calculated as 1/K_{AW}, C_W/C_A reported as exptl., Hine & Mookerjee 1975)
22090, 27810 (calculated-group contribution, calculated-bond contribution, Hine & Mookerjee 1975)
36700; 36700, 36800 (recommended, calculated-P/C, Mackay & Shiu 1981)
25370 (calculated-MCI χ, Nirmalakhandan & Speece 1988)
36934 (calculated-P/C, Eastcott et al. 1988)
36180 (calculated-vapor-liquid equilibrium (VLE) data, Yaws et al. 1991)

Octanol/Water Partition Coefficient, log K_{OW}:
2.35 (calculated-π substituent constant, Hansch et al. 1968)
3.37 (shake flask, Log P Database, Hansch & Leo 1987)
3.37 (recommended, Sangster 1989)
3.37 (Hansch et al. 1995)

Octanol/Air Partition Coefficient, log K_{OA}:

Bioconcentration Factor, log BCF:

Sorption Partition Coefficient, log K_{OC}:

Environmental Fate Rate Constants, k, and Half-Lives, $t_{1/2}$:
Volatilization:
Photolysis:
Oxidation: rate constant k, for gas-phase second order rate constants, k_{OH} for reaction with OH radical, k_{NO3} with NO_3 radical and k_{O3} with O_3 or as indicated, *data at other temperatures see reference:
k_{OH} = 7.10 × 10^{-12} cm^3 molecule^{-1} s^{-1} (Atkinson 1990, 1991)
k_{OH} = 7.10 × 10^{-12} cm^3 molecule^{-1} s^{-1} with a estimated lifetime τ = 20 h in summer daylight (Altshuller 1991)
Hydrolysis:
Biodegradation:
Biotransformation:
Bioconcentration, Uptake (k_1) and Elimination (k_2) Rate Constants or Half-Lives:

Half-Lives in the Environment:
Air: atmospheric $t_{1/2}$ ~ 2.4–24 h for cycloparaffins, based on the EPA Reactivity Classification of Organics (Darnall et al. 1976);
rate constant k = 7.10 × 10^{-12} cm^3 molecule^{-1} s^{-1} for the reaction with hydroxyl radical in air (Atkinson 1990, 1991, Altshuller 1991); and an estimated reaction lifetime τ = 20 h in summer daylight (Altshuller 1991).

Aliphatic and Cyclic Hydrocarbons

2.1.1.2.3 1,1,3-Trimethylcyclopentane

Common Name: 1,1,3-Trimethylcyclopentane
Synonym:
Chemical Name: 1,1,3-trimethylcyclopentane
CAS Registry No: 4516-69-2
Molecular Formula: C_8H_{16}
Molecular Weight: 112.213
Melting Point (°C):
    –142.4   (Lide 2003)
Boiling Point (°C):
    104.9   (Lide 2003)
Density (g/cm³ at 20°C):
    0.7483, 0.7430 (20°C, 25°C, Dreisbach 1955)
Molar Volume (cm³/mol):
    146.0   (20°C, calculated-density, Wang et al. 1992)
    166.1   (calculated-Le Bas method at normal boiling point)
Enthalpy of Vaporization, ΔH_V (kJ/mol):
    36.23, 32.35 (25°C, bp, Dreisbach 1955)
Enthalpy of Fusion, ΔH_{fus} (kJ/mol):
Entropy of Fusion, ΔS_{fus} (J/mol K):
Fugacity Ratio at 25°C, F: 1.0

Water Solubility (g/m³ or mg/L at 25°C):
    3.73   (shake flask-GC, Price 1976)

Vapor Pressure (Pa at 25°C and reported temperature dependence equations):
    log (P/mmHg) = 6.80948 – 1275.998/(219.899 + t/°C); temp range 28.9–105.8°C (Antoine eq., ebulliometry-manometer measurements, Forziati et al. 1949)
    5300   (calculated from determined data, Dreisbach 1955)
    log (P/mmHg) = 6.80947 – 1275.988/(219.899 + t/°C); temp range 20–140°C (interpolated-Antoine eq. for liquid state, Dreisbach 1955)
    5300   (interpolated-Antoine eq., temp range –0.30 to 131.3°C, Zwolinski & Wilhoit 1971)
    log (P/mmHg) = 6.80947 – 1275.998/(219.899 + t/°C); temp range –0.30 to 131.3°C (interpolated-Antoine eq., Zwolinski & Wilhoit 1971)
    log (P/atm) = [1– 378.056/(T/K)] × 10^{0.848231 – 8.28174 × 10^{-4}·(T/K) + 8.81168 × 10^{-7}·(T/K)²}; temp range: 272.85–404.45 K (interpolated-Cox eq., Chao et al. 1983)
    5300   (interpolated-Antoine eq., Boublik et al. 1984)
    log (P/kPa) = 5.93423 – 1275.928/(219.893 + t/°C); temp range 28.944–105.8°C (extrapolated-Antoine eq. from reported exptl. data, Boublik et al. 1984)
    5070   (extrapolated-Antoine eq., temp range 29–106°C, Dean 1985. 1992)
    log (P/mmHg) = 6.80931 – 1275.92/(219.89 + t/°C); temp range 29–106°C (extrapolated-Antoine eq., Dean 1985, 1992)
    log (P$_L$/kPa) = 5.93036 – 1273.902/(–53.454 + T/K); temp range 301–379 K (extrapolated-Antoine eq., Stephenson & Malanowski 1987)

Henry's Law Constant (Pa m³/mol at 25°C):
    159000   (calculated-P/C, Mackay & Shiu 1981)

Octanol/Water Partition Coefficient, log K_{OW}:
    3.28   (calculated-regression eq. from Lyman et al. 1982, Wang et al. 1992)
    3.34   (calculated-molar volume V_M, Wang et al. 1992)

Octanol/Air Partition Coefficient, log K_{OA}:

Bioconcentration Factor, log BCF:

Sorption Partition Coefficient, log K_{OC}:

Environmental Fate Rate Constants, k and Half-Lives, $t_{1/2}$:

Half-Lives in the Environment:
    Air: atmospheric $t_{1/2}$ ~ 2.4–24 h for cycloparaffins, based on the EPA Reactivity Classification of Organics (Darnall et al. 1976).

2.1.1.2.4 n-Propylcyclopentane

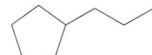

Common Name: *n*-Propylcyclopentane
Synonym:
Chemical Name: *n*-propylcyclopentane
CAS Registry No: 2040-96-2
Molecular Formula: C_8H_{16}
Molecular Weight: 112.213
Melting Point (°C):
    –117.3   (Dreisbach 1955; Lide 2003)
Boiling Point (°C):
    130.937   (Willingham et al. 1945)
    131   (Lide 2003)
Density (g/cm³ at 20°C):
    0.7763, 07723(20°C, 25°C, Dreisbach 1955)
Molar Volume (cm³/mol):
    144.6   (20°C, calculated-density)
    166.1   (calculated-Le Bas method at normal boiling point)
Enthalpy of Vaporization, ΔH_V (kJ/mol):
    41.197, 34.746(25°C, bp, Dreisbach 1955)
Enthalpy of Fusion, ΔH_{fus} (kJ/mol):
    10.04   (Chickos et al. 1999)
Entropy of Fusion, ΔS_{fus} (J/mol K):
    64.45, 57.9 (exptl., calculated-group additivity method, Chickos et al. 1999)
Fugacity Ratio at 25°C, F: 1.0

Water Solubility (g/m³ or mg/L at 25°C):
    2.04   (shake flask-GC, Price 1976)
    1.77   (shake flask-GC, Krzyzanowska & Szeliga 1978)

Vapor Pressure (Pa at 25°C and reported temperature dependence equations):
    log (P/mmHg) = 6.89887 – 1380.391/(212.610 + t/°C); temp range 51.7–131.9°C (Antoine eq. from exptl. data, ebulliometry-manometer, Willingham et al. 1945)
    log (P/mmHg) = 6.30392 – 1384.386/(213.159 + t/°C); temp range 51.9–131.9°C (Antoine eq., ebulliometry-manometer measurements, Forziati et al. 1949)
    1650   (calculated by formula, Dreisbach 1955)
    log (P/mmHg) = 6.90392 – 1384.386/(213.159 + t/°C); temp range 40–170°C (extrapolated-Antoine eq. for liquid state, Dreisbach 1955)
    1640   (interpolated-Antoine eq., temp range –21.3 to158.2°C, Zwolinski & Wilhoit 1971)
    log (P/mmHg) = 6.30392 – 1384.386/(213.159 + t/°C); temp range –21.3 to 158.2°C (Antoine eq., Zwolinski & Wilhoit 1971)
    1646   (extrapolated-Antoine eq., Boublik et al. 1973)
    log (P/mmHg) = 6.91061 – 1388.511/(213.615 + t/°C); temp range 51.88–131.97°C (Antoine eq. from reported exptl. data of Forziati et al. 1949, Boublik et al. 1973)
    1640, 5320 (quoted, calculated-bp, Mackay et al. 1982)
    log (P/atm) = [1– 427.713/(T/K)] × 10^{0.865420 – 7.04026 × 10^{-4}·(T/K) + 5.98562 × 10^{-7}·(T/K)²}; temp range: 313.35–458.95 K (Cox eq., Chao et al. 1983)
    1644   (extrapolated-Antoine eq., temp range 40–186°C, Dean 1985, 1992)
    log (P/mmHg) = 6.88646 – 1460.80/(207.94 + t/°C); temp range 40–186°C (Antoine eq., Dean 1985, 1992)
    log (P_L/kPa) = 6.04236 – 1393.284/(–58.949 + T/K); temp range 323–406 K (Antoine eq., Stephenson & Malanowski 1987)

log (P/mmHg) = 33.922 − 3.2097 × 10^3/(T/K) − 8.9914·log (T/K) − 3.2992 × 10$^{−11}$·(T/K) + 2.0684 × 10$^{−6}$·(T/K)^2; temp range 156–603 K (vapor pressure eq., Yaws 1994)

Henry's Law Constant (Pa m^3/mol at 25°C):
- 90200 (calculated-P/C, Mackay & Shiu 1981)
- 90430 (calculated-vapor-liquid equilibrium (VLE) data, Yaws et al. 1991)

Octanol/Water Partition Coefficient, log K_{OW}:
- 3.95 (calculated-regression eq. of Lyman et al. 1982, Wang et al. 1992)
- 2.65 (calculated-molar volume V_M, Wang et al. 1992)
- 4.37 (calculated-fragment const., Müller & Klein 1992)

Octanol/Air Partition Coefficient, log K_{OA}:

Bioconcentration Factor, log BCF:

Sorption Partition Coefficient, log K_{OC}:

Environmental Fate Rate Constants, k and Half-Lives, $t_{½}$:

Half-Lives in the Environment:

2.1.1.2.5 Pentylcyclopentane

Common Name: Pentylcyclopentane
Synonym: 1-cyclopentylpentane
Chemical Name: pentylcyclopentane
CAS Registry No: 3741-00-2
Molecular Formula: $C_{10}H_{20}$
Molecular Weight: 140.266
Melting Point (°C):
    –83   (Dreisbach 1955; Lide 2003)
Boiling Point (°C):
    180   (Dreisbach 1955; Lide 2003)
Density (g/cm^3 at 20°C):
    0.7912, 0.7874   (20°C, 25°C, Dreisbach 1959)
Molar Volume (cm^3/mol):
    177.3, 178.1 (20°C, 25°C, calculated-density)
    210.5   (calculated-Le Bas method at normal boiling point)
Enthalpy of Vaporization, ΔH_V (kJ/mol):
    51.12, 39.94 (25°, bp, Dreisbach 1955)
Enthalpy of Fusion, ΔH_{fus} (kJ/mol):
Entropy of Fusion, ΔS_{fus} (J/mol K):
Fugacity Ratio at 25°C, F: 1.0

Water Solubility (g/m^3 or mg/L at 25°C):
    0.115   (shake flask-GC, Price 1976)
    0.13   (calculated-recommended liquid-liquid equilibrium LLE data, Mączyński et al. 2004)

Vapor Pressure (Pa at 25°C and reported temperature dependence equations):
    159   (calculated by formula, Dreisbach 1959)
    log (P/mmHg) = 6.929 – 1526.0/(197.0 + t/°C); temp range 85–220°C (Antoine eq. for liquid state, Dreisbach 1955)
    152   (extrapolated-Antoine eq., temp range 60–210°C, Zwolinski & Wilhoit 1971)
    log (P/mmHg) = 6.9414 – 1540.6/(198.8 + t/°C); temp range 60–210°C (Antoine eq., Zwolinski & Wilhoit 1971)

Henry's Law Constant (Pa m^3/mol at 25°C):
    18500   (calculated-P/C, Mackay & Shiu 1981)
    18600   (calculated-vapor-liquid equilibrium (VLE) data, Yaws et al. 1991)

Octanol/Water Partition Coefficient, log K_{OW}:

Octanol/Air Partition Coefficient, log K_{OA}:

Bioconcentration Factor, log BCF:

Sorption Partition Coefficient, log K_{OC}:

Environmental Fate Rate Constants, k, and Half-Lives, $t_{1/2}$:

Half-Lives in the Environment:

2.1.1.2.6 Cyclohexane

Common Name: Cyclohexane
Synonym: hexahydrobenzene, hexamethylene
Chemical Name: cyclohexane
CAS Registry No: 110-82-7
Molecular Formula: C_6H_{12}
Molecular Weight: 84.159
Melting Point (°C):
    6.59   (Lide 2003)
Boiling Point (°C):
    80.73   (Lide 2003)
Density (g/cm³ at 20°C):
    0.7786, 0.7739   (20°C, 25°C, Dreisbach 1955; Riddick et al. 1986)
Molar Volume (cm³/mol):
    108.1   (20°C, calculated-density, McAuliffe 1966; Lande & Banerjee 1981)
    118.2   (calculated-Le Bas method at normal boiling point)
Enthalpy of Vaporization, ΔH_V (kJ/mol):
    32.89, 30.05   (25°C, bp, Riddick et al. 1986)
Enthalpy of Fusion, ΔH_{fus} (kJ/mol):
    2.677   (Riddick et al. 1986)
    5.84, 2.68; 9.41   (–87.05, 6.65°C; total phase change enthalpy, Chickos et al. 1999)
Entropy of Fusion, ΔS_{fus} (J/mol K):
    45.77, 44.55   (exptl., calculated-group additivity method, total phase change entropy, Chickos et al. 1999)
Fugacity Ratio at 25°C, F: 1.0

Water Solubility (g/m³ or mg/L at 25°C, additional data at other temperatures designated * are compiled at the end of this section.):
    80.0   (shake flask-gravitational method, McBain & Lissant 1951)
    55.0   (shake flask-GC, McAuliffe 1963, 1966)
    88.84*   (shake flask-GC/FID, measured range 5.11–45.21°C, Pierotti & Liabastre 1972)
    56.7   (shake flask-GC/FID, Leinonen & Mackay 1973)
    57.5   (shake flask-vapor extraction-GC/FID, Mackay & Shiu 1975)
    55.8, 50.2, 61.7 (shake flask-GC, Mackay et al. 1975)
    66.5   (shake flask-GC/FID, Price 1976)
    66.5   (shake flask-GC, Krzyzanowski & Szeliga 1978)
    52.0   (23.5°C, elution chromatography, Schwarz 1980)
    72.4   (calculated-HPLC-k′ correlation, converted from reported γ_W, Hafkenscheid & Tomlinson 1983)
    72.8*   (40°C, shake flask-GC, measured range 40–209.06°C, Tsonopoulos & Wilson 1983)
    58.4   (shake flask-GC, Groves 1988)
    58.0*   (IUPAC recommended best value, Shaw 1989a)
    $\ln x = -301.366 + 12924.45/(T/K) + 43.2980 \cdot \ln (T/K)$; temp range 290–400 K (eq. derived from literature calorimetric and solubility data, Tsonopoulos 2001)
    $\ln x = -219.863 + 6693.78/(T/K) + 31.3744 \cdot \ln (T/K)$; temp range 290–400 K (eq. derived from direct fit of solubility data, Tsonopoulos 2001)
    60.78   (calculated-liquid-liquid equilibrium LLE data, temp range 278.2–482.2 K, Mączyński et al. 2004)

Vapor Pressure (Pa at 25°C or as indicated and reported temperature dependence equations. Additional data at other temperatures designated * are compiled at the end of this section.):
    12972   (gas saturation/air-bubbling method, Washburn & Handorf 1935)
    16212*   (30°C, vapor-liquid equilibrium VLE data, measured range 30–80°C, Scatchard et al. 1939)

log (P/mmHg) = 6.65859 − 1040.641/(T/K) − 104.865/(T/K)^2; temp range 30–60°C (VLE data, Scatchard et al. 1939)

10910* (20.97°C, manometry, measured range 6.33–20.97 °C, Aston et al. 1943b)

11700* (20.96°C, ebulliometry-manometer, measured range 19.9–81.6°C, Willingham et al. 1945)

log (P/mmHg) = 6.84498 − 1203.526/(222.863 + t/°C); temp range 19.9–81.6°C (Antoine eq. from exptl. data, ebulliometry-manometer, Willingham et al. 1945)

12280* (calculated-Antoine eq. regression, temp range −45.3 to 80.7°C, Stull 1947)

10253*; 10375 (20.015°C, selected exptl., calculated-Frost-Kalkwarf vapor pressure eq., temp range 293.185–551.225 K, Pasek & Thodos 1962)

log (P/mmHg) = 23.14002 − 2411.8/(T/K) − 5.17900·log (T/K) + 1.84394·[(P/mmHg)/(T/K)^2]; temp range 293.185–551.225 K (Frost-Kalkwarf eq., Pasek & Thodos 1962)

13040* (static method, measured range 25–75°C, Cruickshank & Cutler 1967)

log (P/mmHg) = 6.85875 − 1212.014/(233.956 + t/°C); temp range 25–75°C (static method, Cruickshank & Cutler 1967)

13010 (calculated from determined data, Dreisbach 1955)

log (P/mmHg) = 6.84498 − 1203.526/(222.863 + t/°C); temp range −20 to 142°C (Antoine eq. for liquid state, Dreisbach 1955)

13159* (25.26°C, temp range 17.55–80.22°C, Boublik 1960; quoted, Boublik et al. 1984)

13014* (interpolated-Antoine eq., temp range 6.59–105.2°C, Zwolinski & Wilhoit 1971)

log (P/mmHg) = 6.84130 − 1201.531/(222.647 + t/°C); temp range 6.59–105.2°C (Antoine eq., Zwolinski & Wilhoit 1971)

11170 (interpolated-Antoine eq., temp range −45.3 to 257.5°C, Weast 1972–73)

log (P/mmHg) = [−0.2185 × 7830.9/(T/K)] + 7.662126; temp range −45.3 to 257.5°C (Antoine eq., Weast 1972–73)

log (P/atm) = [1 − 353.663/(T/K)] × 10^{0.881199 − 9.58655 × 10^{-4}·(T/K) + 9.72305 × 10^{-7}·(T/K)^2}; temp range 227.85–553.64 K (Cox eq., Chao et al. 1983)

13010, 13040 (interpolated-Antoine equations, Boublik et al. 1984)

log (P/kPa) = 5.97561 − 1206.731/(223.223 + t/°C), temp range 19.9–81.6°C (Antoine eq. from reported exptl. data of Willingham et al. 1945, Boublik et al. 1984)

log (P/kPa) = 6.98226 − 1211.248/(223.869 + t/°C); temp range 25–75°C (Antoine eq. from reported exptl. data, Boublik et al. 1984)

log (P/kPa) = 6.00569 − 1223.273/(225.089 + t/°C); temp range 17.55–80.22°C (Antoine eq. from reported exptl. data of Cruickshank & Cutler 1967, Boublik et al. 1984)

13020 (interpolated-Antoine eq., temp range 20–81°C, Dean 1985, 1992)

log (P/mmHg) = 6.84130 − 1201.53/(222.65 + t/°C); temp range 20–81°C (Antoine eq., Dean 1985, 1992)

13040 (lit. average, Riddick et al. 1986)

log (P/kPa) = 5.96407 − 1200.31/(222.504 + t/°C); temp range not specified (Antoine eq., Riddick et al. 1986)

13070 (interpolated-Antoine eq.-II, Stephenson & Malanowski 1987)

log (P_S/kPa) = 7.2778 − 1747.2/(26.84 + T/K); temp range 223–280 K (Antoine eq.-I, solid, Stephenson & Malanowski 1987)

log (P_L/kPa) = 5.9682 − 1201.531/(−50.503 + T/K); temp range 293–335 K (Antoine eq.-II, Stephenson & Malanowski 1987)

log (P_L/kPa) = 6.03245 − 1244.124/(−44.911 + T/K); temp range 353–414 K (Antoine eq.-III, Stephenson & Malanowski 1987)

log (P_L/kPa) = 6.36849 − 1519.732/(−4.032 + T/K); temp range 412–491 K (Antoine eq.-IV, Stephenson & Malanowski 1987)

log (P_L/kPa) = 7.37347 − 2683.075/(159.31 + T/K); temp range 489–533 K (Antoine eq.-V, Stephenson & Malanowski 1987)

12920, 1730 (quoted, calculated-UNIFAC activity coeff., Banerjee et al. 1990)

log (P/mmHg) = 48.5529 − 3.0874 × 10^3/(T/K) − 15.521·log (T/K) + 7.383 × 10^{-3}·(T/K) + 6.3563 × 10^{-12}·(T/K)^2; temp range 280–554 K (vapor pressure eq., Yaws 1994)

12068* (23.363°C, comparative ebulliometry, measured range 281–552 K, data fitted to Wagner eq., Ewing & Ochoa 2000)

16230, 36180 (30, 50°C, VLE equilibrium data, Carmona et al. 2000)

Henry's Law Constant (Pa m³/mol at 25°C or as indicated and reported temperature dependence equations. Additional data at other temperatures designated * are compiled at the end of this section.):

    19860   (calculated-P/C, Mackay & Shiu 1975)
    19690   (calculated as $1/K_{AW}$, C_W/C_A, reported as exptl., Hine & Mookerjee 1975)
    18380, 27810 (calculated-group contribution, calculated-bond contribution, Hine & Mookerjee 1975)
    18000   (recommended; Mackay & Shiu 1981)
    22092   (calculated-MCI χ, Nirmalakhandan & Speece 1988)
    5532, 17935, 19353, 1450 (bubble column non-equilibrium measurement, EPICS-GC, direct concentration. ratio, calculated-UNIFAC activity coeff., Ashworth et al. 1988)
    17935*   (EPICS-GC/FID, measured range 10–30°C, Ashworth et al. 1988)
    $\ln [H/(atm \cdot m^3/mol)] = 9.141 - 3238/(T/K)$; temp range: 10–30°C (EPICS measurements, Ashworth et al. 1988)
    19980   (calculated-vapor-liquid equilibrium (VLE) data, Yaws et al. 1991)
    37190*   (40°C, equilibrium headspace-GC, measured range 40–80°C, Kolb et al. 1992)
    $\ln (1/K_{AW}) = -12.90 + 3228/(T/K)$; temp range: 40–80°C (equilibrium headspace-GC measurements, Kolb et al. 1992)
    15062   (20°C, selected from reported experimental determined values, Staudinger & Roberts 1996, 2001)
    $\log K_{AW} = 5.154 - 1279/(T/K)$ (summary of literature data, Staudinger & Roberts 2001)

Octanol/Water Partition Coefficient, log K_{OW}:

    2.46   (calculated-π substituent constant, Hansch et al. 1968)
    3.44   (shake flask-GC, Leo et al. 1975; Hansch & Leo 1979)
    3.40   (Cramer 1977)
    3.18, 3.48, 3.48   (calculated-fragment const., Rekker 1977)
    3.69   (HPLC-k′ correlation, Hafkenscheid & Tomlinson 1983)
    3.70   (from activity coefficient measurement, Berti et al. 1986)
    3.44   (recommended, Sangster 1989, 1993)
    3.73   (from activity coefficient measurement, Tse & Sandler 1994)
    3.44   (recommended, Hansch et al. 1995)

Octanol/Air Partition Coefficient, log K_{OA} at 25°C or as indicated. Additional data at other temperatures designated* are compiled at the end of this section:

    2.83*   (20.29°C, from GC-determined γ^∞ in octanol, measured range 20.29–50.28°C, Gruber et al. 1997)
    2.71   (calculated-measured γ^∞ in pure octanol and vapor pressure P, Abraham et al. 2001)

Bioconcentration Factor, log BCF:

    2.38   (estimated, Howard 1990)

Sorption Partition Coefficient, log K_{OC}:

    2.68   (estimated-S, Howard 1990)

Environmental Fate Rate Constants, k, and Half-Lives, $t_{1/2}$:

    Volatilization: $t_{1/2}$ = 2.8 h from a model river 1 m deep with a 1 m/s current and a 3 m/s wind (Lyman et al. 1982, quoted, Howard 1990).
    Photolysis:
    Oxidation: rate constant k, for gas-phase second order rate constants, k_{OH} for reaction with OH radical, k_{NO3} with NO_3 radical and k_{O3} with O_3 or as indicated, *data at other temperatures and/or the Arrhenius expression see reference:
        k_{OH}*(exptl) = 5.38×10^{12} cm³ mol⁻¹ s⁻¹, k_{OH}(calc) = 4.79×10^{12} cm³ mol⁻¹ s⁻¹ at 295 K, measured range 295–497 K (flash photolysis-kinetic spectroscopy, Greiner 1970)
        k_{OH} (exptl) = 6.7×10^{-12} cm³ molecule⁻¹ s⁻¹ at 298 K; k_{OH} (calc) = 6.7×10^{-12} cm³·molecule⁻¹s⁻¹ at 300 K (Darnall et al. 1978)
        $k_{O(^3P)} = 1.40 \times 10^{-13}$ cm³·molecule⁻¹ s⁻¹ for the reaction with O(^{3}P) atoms at room temp. (Herron & Huie 1973)
        $k_{OH} = 7.0 \times 10^{-12}$ cm³ molecule⁻¹ s⁻¹ at room temp. (Atkinson et al. 1979)
        $k_{OH} = (7.57 \pm 0.05) \times 10^{-12}$ cm³ molecule⁻¹ s⁻¹ at room temp. (relative rate, Atkinson et al. 1984c)

k_{OH} = (6.20 ± 0.44) × 10^{-12} cm^3 $molecule^{-1}$ s^{-1} at (24.4 ± 0.4)°C with an atmospheric lifetime of 1.9 d for an average OH radical concentration of 1.0 × 10^6 molecules/cm^3 (Edney et al. 1986)

k_{OH} = 7.34 × 10^{-12} cm^3 $molecule^{-1}$ s^{-1} at 297 K (Atkinson 1986; quoted, Edney et al. 1986)

k_{OH} = 7.14 × 10^{-12} cm^3 $molecule^{-1}$ s^{-1} at 292 K, measured range 292–491 K (Droege & Tully 1987)

k_{OH} = 7.38 × 10^{-12} cm^3 $molecule^{-1}$ s^{-1} (Dilling et al. 1988)

k_{OH}* = 7.49 × 10^{-12} cm^3 $molecule^{-1}$ s^{-1} at 298 K (recommended, Atkinson 1989)

k_{OH}* = 7.49 × 10^{-12} cm^3 $molecule^{-1}$ s^{-1} at 298 K, k_{NO3} = 13.4 × 10^{-17} cm^3 $molecule^{-1}$ s^{-1} at 296 K (Atkinson 1990)

k_{NO3} = 1.35 × 10^{-16} cm^3 $molecule^{-1}$ s^{-1} at 296 K (Atkinson 1991)

k_{NO3}(exptl) = 1.35 × 10^{-16} cm^3 $molecule^{-1}$ s^{-1}, k_{NO3}(calc) = 2.04 × 10^{-16} cm^3 $molecule^{-1}$ s^{-1} at 296 ± 2 K (relative rate method, Aschmann & Atkinson 1995)

k_{OH}* = 7.21 × 10^{-12} cm^3 $molecule^{-1}$ s^{-1}, k_{NO3} = 1.4 × 10^{-16} cm^3 $molecule^{-1}$ s^{-1} at 298 K (recommended, Atkinson 1997)

k_{OH}* = 6.69 × 10^{-12} cm^3 $molecule^{-1}$ s^{-1} at 298 K, measured range 230–400 K (relative rate method, DeMore & Bayes 1999)

Hydrolysis:

Biodegradation: highly resistant to biodegradation (Howard 1990);

$t_{½}$(aq. aerobic) = 672 to 4032 h, based on unacclimated grab sample of aerobic soil and aerobic aqueous screening test data (Howard et al. 1991);

$t_{½}$(aq. anaerobic) = 2688 to 16280 h, based on estimated unacclimated aqueous aerobic biodegradation half-life (Howard et al. 1991).

Biotransformation:

Bioconcentration, Uptake (k_1) and Elimination (k_2) Rate Constants or Half-Lives:

Half-Lives in the Environment:

Air: atmospheric $t_{½}$ ~ 2.4–24 h for cycloparaffins, based on the EPA Reactivity Classification of Organics (Darnall et al. 1976);

atmospheric lifetime τ = 1.9 d for an average OH radical concentration of 1.0 × 10^6 molecules/cm^3 (Edney et al. 1986);

$t_{½}$ = 52 h, based on photooxidation rate constant (Howard 1990);

an atmospheric lifetime τ ~ 19 h in summer daylight, based on the photooxidation reaction rate constant of 7.49 × 10^{-12} cm^3 $molecule^{-1}$ s^{-1} for the reaction with OH radical in air during summer daylight (Altshuller 1991);

will degrade photochemically by hydroxyl radicals with $t_{½}$ = 52 h and much faster under photochemical smog conditions with $t_{½}$ = 6 h (Howard 1990);

$t_{½}$ = 8.7 - 87 h, based on reaction with OH radical half-life in air (Howard et al. 1991).

Surface water: volatilization $t_{½}$ = 2 h in a model river (Howard 1990);

$t_{½}$ = 672 - 4320 h, based on estimated unacclimated aqueous aerobic biodegradation half-life (Howard et al. 1991);

photooxidation $t_{½}$ = 1.4 × 10^9 to 6.9 × 10^{10} h (16000 to 780000 yr), based on measured rate data for alkylperoxyl radicals in aqueous solution (Howard et al. 1991).

Ground water: $t_{½}$ = 1344–8640 h, based on estimated unacclimated aqueous aerobic biodegradation half-life (Howard et al. 1991).

Sediment:

Soil: $t_{½}$ = 672–4320 h, based on unacclimated grab sample of aerobic soil and aerobic aqueous screening test data (Howard et al. 1991).

Biota:

TABLE 2.1.1.2.6.1
Reported aqueous solubilities of cyclohexane at various temperatures

Pierotti & Liabastre 1972		Tsonopoulos & Wilson 1983		Shaw 1989a	
shake flask-GC		shake flask-GC		IUPAC recommended	
t/°C	S/g·m⁻³	t/°C	S/g·m⁻³	t/°C	S/g·m⁻³
5.11	81.93	40	72.8	5	-
15.21	88.70	100	1770	15	-
25.11	88.84	146.89	4810	20	-
35.21	88.84	150	6070	25	58.0
45.21	91.32	200	18300	45	-
		209.06	23000	50	-
				56	-
				70	-
				71	-

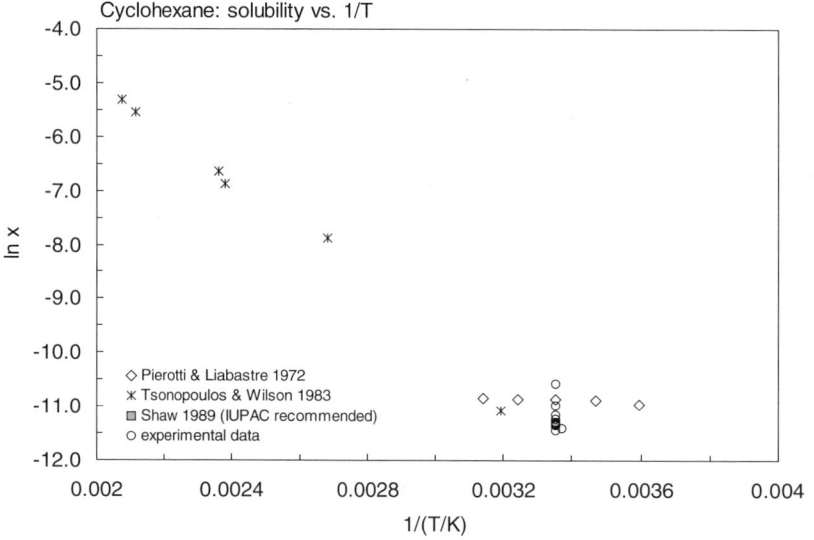

FIGURE 2.1.1.2.6.1 Logarithm of mole fraction solubility (ln x) versus reciprocal temperature for cyclohexane.

TABLE 2.1.1.2.6.2
Reported vapor pressures of cyclohexane at various temperatures and the coefficients for the vapor pressure equations:

$$\log P = A - B/(T/K) \quad (1)$$
$$\log P = A - B/(C + t/°C) \quad (2)$$
$$\log P = A - B/(C + T/K) \quad (3)$$
$$\log P = A - B/(T/K) - C \cdot \log(T/K) \quad (4)$$
$$\log P = A - B/(T/K) - C \cdot \log(T/K) + D[P/(T/K)^2] \quad (5)$$
$$\log P = A - B/T(T/K) - C/(T/K)^2 \quad (6)$$

$$\ln P = A - B/(T/K) \quad (1a)$$
$$\ln P = A - B/(C + t/°C) \quad (2a)$$

1.

Ashton et al. 1943b		Willingham et al. 1945		Stull 1947		Pasek & Thodos 1962	
manometry		ebulliometry		summary of literature data		selected exptl. data	
t/°C	P/Pa	t/°C	P/Pa	t/°C	P/Pa	t/°C	P/Pa
6.33	5388	19.915	10303	−45.3	133.3	20.015	10253
9.89	6421	22.657	11695	−25.4	666.6	52.693	39997
13.63	7697	26.347	13821	−15.9	1333	82.469	106658
17.65	9327	30.556	16610	−5.0	2666	105.215	199984
20.97	10910	34.821	19917	6.7	5333	160.015	670345
		38.789	23453	14.7	7999	219.905	1813184
mp/°C	6.69	44.108	28960	25.5	13332	240.395	3420467
bp/°C	80.8	48.991	43324	42.0	26664	278.075	3959408
		54.884	53659	60.8	53329		
ΔH_V/(kJ mol^{-1}) = 33.33		60.969	66762	80.7	101325	bp/°C	80.75
at 25°C		67.467	83723			Frost-Kalkwarf equation	
ΔH_{fus}/(kJ mol^{-1}) = 2.627		74.520	97609	mp/°C	6.60	derived from exptl. data	
		79.052	99209			eq. 5	P/mmHg
		80.037	100701			A	23.14002
Scatchard et al. 1939		80.534	102403			B	−2411.8
vapor-liquid equilibrium		81.093	103923			C	−5.27900
t/°C	P/Pa	81.582	103923			D	1.84394
30	16212	bp/°C	80.738			Frost-Kalkwarf equation	
40	24613	eq. 2	P/mmHg			calculated from molecular	
50	36237	A	6.84498			structures, normal bp and	
60	51901	B	1203.526			deviations of resulting v.p.	
70	72521	C	222.863			eq. 5	P/mmHg
80	88995					A	23.09550
eq. 6	P/mmHg					B	−2411.82
A	6.65859					C	−5.26151
B	1040.641					D	1.84394
C	116.197						

(Continued)

2.

Boublik 1960 in Boublik et al. 1984		Cruickshank & Cutler 1966 static method-manometry		Zwolinski & Wilhoit 1971 selected values		Ewing & Ochoa 2000 comparative ebulliometry	
t/°C	P/Pa	t/°C	P/Pa	t/°C	P/Pa	t/°C	P/Pa
17.55	9210	25	13040	6.69	5333	8.188	5777
20.16	10426	35	20104	11.01	6666	9.698	6243
22.98	11876	45	30008	14.66	7999	11.356	6791
25.26	13159	55	43538	20.666	10666	15.215	8225
28.77	15372	65	61551	25.536	13332	17.163	9039
32.77	18252	75	85012	34.904	19998	19.868	10284
37.10	21891			41.991	26664	22.608	11675
40.66	25264	mp/°C	6.55	47.763	33331	23.263	12086
45.77	30864	bp/°C	80.728	52.669	39997	28.346	15109
50.96	37543	density, d^{25}	0.77386	60.784	53329	44.379	29290
56.18	45356			67.415	66661	49.393	35456
62.46	56448	$\Delta H_V/(kJ\ mol^{-1})$ =		73.067	79993	55.302	44015
68.71	69467	at 20°C	33.23	78.021	93326	60.881	53541
75.42	86053	25°C	32.97	78.944	95992	66.269	64227
80.22	99818	54.1°C	31.45	79.848	98659	71.034	75023
		73.3°C	30.44	80.292	99992	75.094	86306
bp/°C	80.731	80.7°C	30.03	80.732	101325	78.498	94753
						80.80	101589
		eq. 2	P/mmHg	eq. 2	P/mmHg	83.489	110504
		A	6.85875	A	6.84130	87.149	122451
		B	1212.014	B	1201.531		
		C	233.956	C	222.647		
				bp/°C	80.732	temp range 281–552 K	
				$\Delta H_V/(kJ\ mol^{-1})$ =		data fitted to Wagner eq.	
				at 25°C	33.04		
				at bp	29.96		

Aliphatic and Cyclic Hydrocarbons

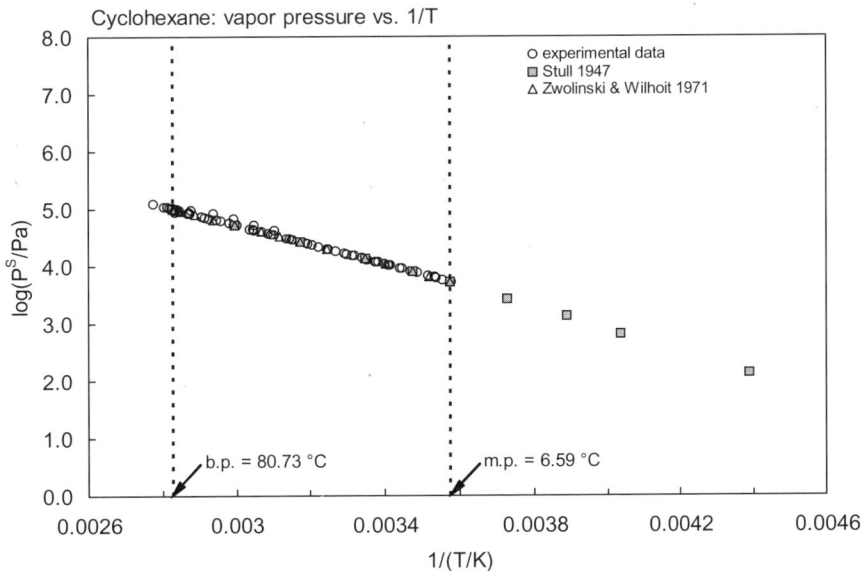

FIGURE 2.1.1.2.6.2 Logarithm of vapor pressure versus reciprocal temperature for cyclohexane.

TABLE 2.1.1.2.6.3
Reported Henry's law constants and octanol-air partition coefficients of cyclohexane at various temperatures and temperature dependence equations

$\ln K_{AW} = A - B/(T/K)$ (1) $\log K_{AW} = A - B/(T/K)$ (1a)

$\ln (1/K_{AW}) = A - B/(T/K)$ (2) $\log (1/K_{AW}) = A - B/(T/K)$ (2a)

$\ln (k_H/\text{atm}) = A - B/(T/K)$ (3)

$\ln [H/(\text{Pa m}^3/\text{mol})] = A - B/(T/K)$ (4) $\ln [H/(\text{atm·m}^3/\text{mol})] = A - B/(T/K)$ (4a)

$K_{AW} = A - B·(T/K) + C·(T/K)^2$ (5)

Henry's law constant				log K_{OA}	
Ashworth et al. 1988		Kolb et al. 1992		Gruber et al. 1997	
EPICS-GC		equilibrium headspace-GC		GC det'd activity coefficient	
t/°C	H/(Pa m³/mol)	t/°C	H/(Pa m³/mol)	t/°C	log K_{OA}
10	10436	40	37190	20.29	2.835
15	12767	60	55400	30.3	2.652
20	14189	70	95100	40.4	2.511
25	17935	80	146800	50.28	2.365
30	22595				
		eq. 2	1/K_{AW}		
		A	−12.90		
eq. 4a	H/(atm m³/mol)	B	−3228		
A	9.141				
B	3238				

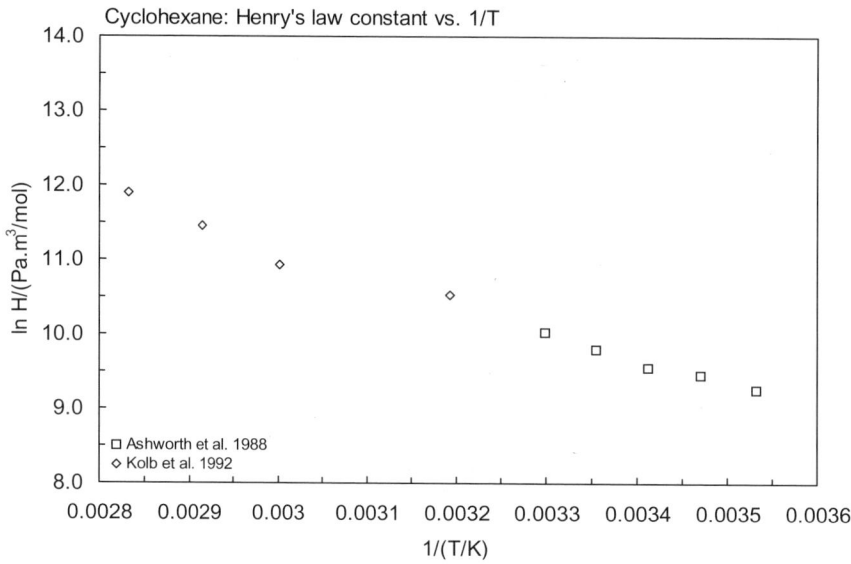

FIGURE 2.1.1.2.6.3 Logarithm of Henry's law constant versus reciprocal temperature for cyclohexane.

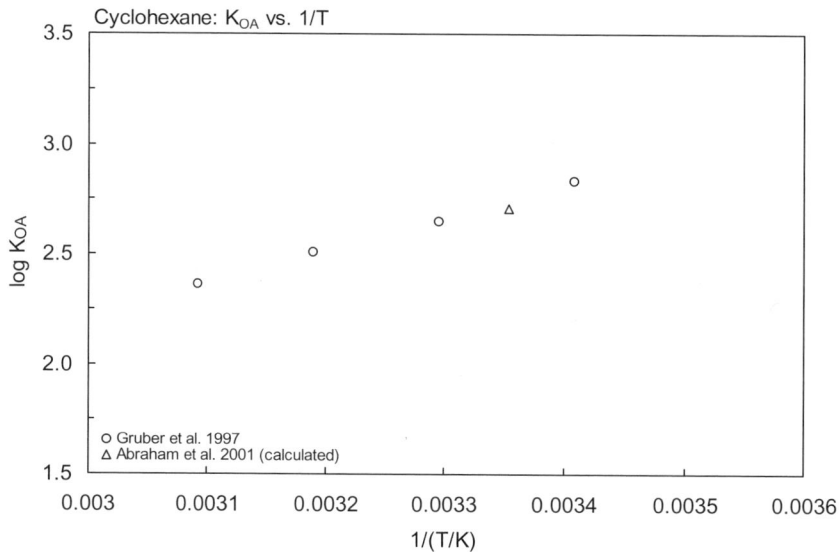

FIGURE 2.1.1.2.6.4 Logarithm of K_{OA} versus reciprocal temperature for cyclohexane.

2.1.1.2.7 Methylcyclohexane

Common Name: Methylcyclohexane
Synonym: hexahydrotoluene, cyclohexylmethane
Chemical Name: methylcyclohexane
CAS Registry No: 108-87-2
Molecular Formula: C_7H_{14}
Molecular Weight: 98.186
Melting Point (°C):
    –126.6   (Dreisbach 1959; Weast 1984; Riddick et al. 1986; Stephenson & Malanowski 1987; Lide 2003)
Boiling Point (°C):
    100.93   (Lide 2003)
Density (g/cm³ at 20°C):
    0.7694, 0.7651 (20°C, 25°C, Dreisbach 1955; Riddick et al. 1986)
Molar Volume (cm³/mol):
    127.6   (20°C, calculated-density, McAuliffe 1966; Stephenson & Malanowski 1987)
    140.4   (calculated-Le Bas method at normal boiling point)
Enthalpy of Vaporization, ΔH_V (kJ/mol):
    35.359, 31.13 (25°C, bp, Riddick et al. 1986)
Enthalpy of Fusion, ΔH_{fus} (kJ/mol):
    6.757   (Riddick et al. 1986)
    6.75   (Chickos et al. 1999)
Entropy of Fusion, ΔS_{fus} (J/mol K):
    46.1, 47.3 (exptl., calculated-group additivity method, Chickos et al. 1999)
Fugacity Ratio at 25°C, F: 1.0

Water Solubility (g/m³ or mg/L at 25°C or as indicated and reported temperature dependence equations. Additional data at other temperatures designated * are compiled at the end of this section.):
    14.0   (shake flask-GC, McAuliffe 1963, 1966)
    16.0*   (shake flask-GC, measured range 25–149.5°C, Price 1976; quoted, Eastcott et al. 1988)
    15.3   (partition coefficient-GC, Rudakov & Lutsyk 1979)
    15.2   (20°C, shake flask-GC, Burris & MacIntyre 1986)
    16.7   (shake flask-GC, Groves 1988)
    15.1*   (IUPAC recommended, temp range 25–150°C, Shaw 1989)
    15.82*   (calculated-liquid-liquid equilibrium LLE data, temp range 298.2–410.5 K, Mączyński et al. 2004)
    16.15*   (26.1°C, shake flask-GC, measured range 26.1–170.8°C, Marche et al. 2004)
    $\ln (S/\text{ppm}) = 13.091 - 7085.522/(T/K) + 1055594/(T/K)^2$ (Marche et al. 2004)

Vapor Pressure (Pa at 25°C or as indicated and reported temperature dependence equations. Additional data at other temperatures designated * are compiled at the end of this section.):
    6128   (interpolated-Antoine eq., Stuckey & Saylor 1940)
    $\log (P/\text{mmHg}) = 6.95423 - 1336.93/(T/K - 45.52)$; temp range 4–75°C (Antoine eq. based on exptl. data, Ramsay-Young method-Hg manometer, Stuckey & Saylor 1940)
    6354*   (25.59°C, ebulliometry-manometer, Willingham et al. 1945)
    $\log (P/\text{mmHg}) = 6.82689 - 1272.864/(221.630 + t/°C)$; temp range 25.6–101.8°C (Antoine eq. from exptl. data, ebulliometry-manometer, Willingham et al. 1945)
    5887*   (calculated-Antoine eq. regression, temp range –35.9 to 100.0°C, Stull 1947)
    6180   (calculated from determined data, Dreisbach 1955)
    $\log (P/\text{mmHg}) = 6.82689 - 1272.864/(221.630 + t/°C)$; temp range 10–155°C (Antoine eq. for liquid state, Dreisbach 1955)
    10207*   (35.901°C, temp range 35.901–95.946°C, Varushchenko et al. 1970)

6180*   (interpolated-Antoine eq., temp range –3.20 to 127.0°C, Zwolinski & Wilhoit 1971)
log (P/mmHg) = 6.82300 – 1272.763/(221.416 + t/°C); temp range –3.20 to 127.0°C (Antoine eq., Zwolinski & Wilhoit 1971)
5880   (interpolated-Antoine eq., temp range –35.9 to 100°C, Weast 1972–73)
log (P/mmHg) = [–0.2185 × 8549.2/(T/K)] + 7.909762; temp range –35.9 to 100°C (Antoine eq., Weast 1972–73)
5806   (calculated-bp, Mackay et al. 1982)
log (P/atm) = [1– 373.957/(T/K)] × 10^{0.862568 – 8.71426 × 10^{-4}·(T/K) + 8.69685 × 10^{-7}·(T/K)^2}; temp range 203.20–563.20 K (Cox eq., Chao et al. 1983)
5364, 6111, 6177 (Antoine equations, Boublik et al. 1984)
log (P/kPa) = 5.95366 – 1273.962/(221.755 + t/°C); temp range 25.6–101.8°C (Antoine eq. from reported exptl. data of Willingham et al. 1945, Boublik et al. 1984)
log (P/kPa) = 5.95497 – 1275.047/(221.678 + t/°C); temp range 35.9–95.9°C (Antoine eq. from reported exptl. data, Boublik et al. 1984)
log (P/kPa) = 5.92856 – 1253.199/(216.058 + t/°C); temp range 58.645–113.6°C (Antoine eq. from reported exptl. data of Meyer et al. 1976, Boublik et al. 1984)
6180   (interpolated-Antoine eq., temp range –3 to 127°C, Dean 1985, 1992)
log (P/mmHg) = 6.82300 – 1270.763/(221.42 + t/°C); temp range –3 to 127°C (Antoine eq., Dean 1985, 1992)
6100   (quoted lit., Riddick et al. 1986)
log (P/kPa) = 5.94790 – 1270.763/(221.416 + t/°C); temp range not specified (Antoine eq., Riddick et al. 1986)
6160   (interpolated-Antoine eq., Stephenson & Malanowski 1987)
log (P_L/kPa) = 5.9428 – 1266.954/(–52.282 + T/K); temp range 308–368 K (Antoine eq.-I, Stephenson & Malanowski 1987)
log (P_L/kPa) = 6.14677 – 1413.495/(–32.726 + T/K); temp range 373–511 K (Antoine eq.-II, Stephenson & Malanowski 1987)
log (P_L/kPa) = 7.29186 – 2700.205/(147.549 + T/K); temp range 501–573 K (Antoine eq.-III, Stephenson & Malanowski 1987)
log (P/mmHg) = 38.0955 – 3.0738 × 10^3/(T/K) –10.684·log (T/K) – 5.1766 × 10^{-11}·(T/K) + 3.5282 × 10^{-6}·(T/K)^2; temp range 147–572 K (vapor pressure eq., Yaws 1994)
5205*   (21.46°C, static method-manometry, measured range –76.96 to 41.44°C, Mokbel et al. 1995)
ln x = –328.666 + 14073.29/(T/K) + 47.1467·ln (T/K); temp range 290–400 K (eq. derived from literature calorimetric and solubility data, Tsonopoulos 2001)
ln x = –491.070 + 22132.10/(T/K) + 70.9150·ln (T/K); temp range 290–400 K (eq. derived from direct fit of solubility data, Tsonopoulos 2001)

Henry's Law Constant (Pa m^3/mol at 25°C or as indicated and reported temperature dependence equations. Additional data at other temperatures designated * are compiled at the end of this section.):
44080   (calculated as 1/K_{AW}, C_W/C_A, reported as exptl., Hine & Mookerjee 1975)
31030, 41340   (calculated-group contribution, calculated-bond contribution, Hine & Mookerjee 1975)
40000   (recommended; Mackay & Shiu 1981)
31934   (calculated-MCI χ, Nirmalakhandan & Speece 1988)
37930   (calculated-P/C, Eastcott et al. 1988)
43300   (calculated-vapor-liquid equilibrium (VLE) data, Yaws et al. 1991)
12666*   (27.3°C, EPICS-GC, measured range 27.3–45°C, Hansen et al. 1993)
ln [H/(kPa·m^3/mol)] = –9406/(T/K) + 34.0; temp range 27.3–45°C (EPICS-GC, Hansen et al. 1993)
6410   (20°C, selected from reported experimental determined values, Staudinger & Roberts 1996, 2001)
log K_{AW} = 13.507 – 3836/(T/K) (summary of literature data, Staudinger & Roberts 2001)

Octanol/Water Partition Coefficient, log K_{OW}:
2.76   (calculated-π substituent constant, Hansch et al. 1968)
2.96   (calculated- MCI π, Murray et al. 1975)
2.82   (Hansch & Leo 1979)
3.88   (recommended, Sangster 1989. 1993)
4.10   (calculated-fragment const. per Lyman 1982, Thoms & Lion 1992)
2.89   (calculated-molar volume V_M, Wang et al. 1992)

Aliphatic and Cyclic Hydrocarbons

    3.87   (calculated-fragment const., Müller & Klein 1992)
    3.61   (recommended, Hansch et al. 1995)

Octanol/Air Partition Coefficient, log K_{OA} at 25°C or as indicated. Additional data at other temperatures designated * are compiled at the end of this section:
    3.14*   (20.29°C, from GC-determined γ^∞ in octanol, measured range 20.29–50.28°C Gruber et al. 1997)
    3.05   (calculated-measured γ^∞ in pure octanol and vapor pressure P, Abraham et al. 2001)

Bioconcentration Factor, log BCF:

Sorption Partition Coefficient, log K_{OC}:

Environmental Fate Rate Constants, k and Half-Lives, $t_{1/2}$:
    Volatilization:
    Photolysis:
    Oxidation: rate constant k, for gas-phase second order rate constants, k_{OH} for reaction with OH radical, k_{NO3} with NO_3 radical and k_{O3} with O_3 or as indicated, *data at other temperatures see reference:
        $k_{OH} = (10.6 \pm 0.3) \times 10^{-12}$ cm³ molecule⁻¹ s⁻¹ at room temp. (relative rate, Atkinson et al. 1984c)
        $k_{OH} = 1.04 \times 10^{-11}$ cm³·molecule⁻¹ s⁻¹ at 298 K (recommended, Atkinson 1989, 1990)
        $k_{OH} = 1.04 \times 10^{-11}$ cm³·molecule⁻¹ s⁻¹ at 298 K with an estimated half-life of 13 h (Altshuller 1991)
        $k_{OH} = 1.0 \times 10^{-11}$ cm³ molecule⁻¹ s⁻¹ at 298 K (recommended, Atkinson 1997)
    Hydrolysis:
    Biotransformation:
    Biodegradation:
    Bioconcentration, Uptake (k_1) and Elimination (k_2) Rate Constants or Half-Lives:

Half-Lives in the Environment:
    Air: an atmospheric lifetime was estimated to be 13 h in summer daylight, based on the photooxidation rate constant of 1.04×10^{-11} cm³ molecule⁻¹ s⁻¹ with hydroxyl radicals in air (Altshuller 1991).

TABLE 2.1.1.2.7.1
Reported aqueous solubilities of methylcyclohexane at various temperatures

Price 1976		Shaw 1989a		Marche et al. 2004		Mączyński et al. 2004	
Shake flask-GC		IUPAC recommended		Shake flask-GC		Calc-recommended LLE data	
t/°C	S/g·m⁻³	t/°C	S/g·m⁻³	t/°C	S/g·m⁻³	t/°C	S/g·m⁻³
		25	15.1	26.1	16.15	25	15.82
25.0	16.0	30	17.0	70.5	27.49	40.1	16.91
40.1	18.0	40	18.0	100.5	54.88	55.7	19.64
55.7	18.9	50	19.0	131.0	133.1	99.1	44.18
99.1	33.8	70	22.0	151.4	230.7	120	76.37
120.0	79.5	90	29.0	170.8	386.2	137.3	125.5
137.3	139	110	52.0				
149.5	244	130	110	$S = A - B/(T/K) + C/(T/K)^2$			
		150	250	S	g/m³		
				A	13.091		
				B	7085.522		
				C	1055594		

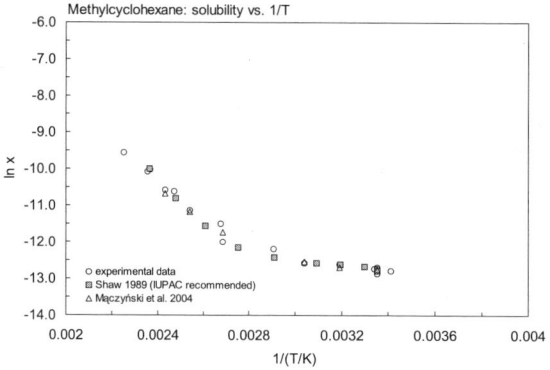

FIGURE 2.1.1.2.7.1 Logarithm of mole fraction solubility (ln x) versus reciprocal temperature for methylcyclohexane.

TABLE 2.1.1.2.7.2
Reported vapor pressures of methylcyclohexane at various temperatures and the coefficients for the vapor pressure equations

$$\log P = A - B/(T/K) \quad (1)$$
$$\log P = A - B/(C + t/°C) \quad (2)$$
$$\log P = A - B/(C + T/K) \quad (3)$$
$$\log P = A - B/(T/K) - C\cdot\log(T/K) \quad (4)$$
$$\log P = A - B/(T/K) - C\cdot\log(T/K) + D[P/(T/K)^2] \quad (5)$$

$$\ln P = A - B/(T/K) \quad (1a)$$
$$\ln P = A - B/(C + t/°C) \quad (2a)$$

1.

Willingham et al. 1945		Stull 1947		Varushchenko et al. 1970		Zwolinski & Wilhoit 1971	
ebulliometry		summary of literature data				selected values	
t/°C	P/Pa	t/°C	P/Pa	t/°C	P/Pa	t/°C	P/Pa
25.585	6354	–35.9	133.3	35.901	10207	–3.20	1333
29.533	7655	–14.0	666.6	45.809	15552	8.70	2666
32.976	8962	–3.2	1333	50.799	19018	16.3	4000
36.089	10303	8.7	2666	55.419	22754	21.98	5333
38.998	11696	22.0	5333	61.644	28719	26.585	6666
42.929	13820	35.0	7999	67.476	35390	30.477	7999
47.407	16919	42.1	13332	74.177	44530	36.874	10666
51.964	19918	59.6	26664	81.020	55721	42.063	13332
56.194	23453	79.6	53329	88.101	69542	52.048	19998
61.857	28958	100.0	101325	93.567	81944	59.604	26664
67.067	34896			94.159	93376	65.758	33331
73.349	43323	mp/°C	–126.4	94.766	84890	70.990	39997
79.842	53657			95.307	86243	79.646	53329
86.771	66760			95.946	87847	86.720	66661
94.299	83721					92.752	79993
99.614	97628					98.039	93326
100.185	99205					99.025	95992
100.715	100697					99.989	98659
101.312	102398					100.464	99992
101.832	103919					100.934	101325

Aliphatic and Cyclic Hydrocarbons

TABLE 2.1.1.2.7.2 *(Continued)*

Willingham et al. 1945		Stull 1947		Varushchenko et al. 1970		Zwolinski & Wilhoit 1971	
ebulliometry		summary of literature data				selected values	
t/°C	P/Pa	t/°C	P/Pa	t/°C	P/Pa	t/°C	P/Pa
100.715	100697					99.989	98659
101.312	102398					100.464	99992
101.832	103919					100.934	101325
bp/°C	100.934					eq. 2	P/mmHg
						A	6.82300
eq. 2	P/mmHg					B	1270.763
A	6.82689					C	221.416
B	1272.864					bp/°C	100.934
C	221.630					$\Delta H_V/(kJ\ mol^{-1}) =$	
						at 25°C	35.36
						at bp	31.13

2.

Mokbel et al. 1995					
static method-manometry					
t/°C	P/Pa	t/°C	P/Pa	t/°C	P/Pa
−76.96	1.68	−28.28	240	21.46	5205
−67.0	5.87	−18.38	497	31.45	8368
−58.55	15.8	−8.44	970	41.44	12950
−48.64	42.4	1.50	1785		
−38.15	108	11.47	3128	data fitted to Wagner eq.	

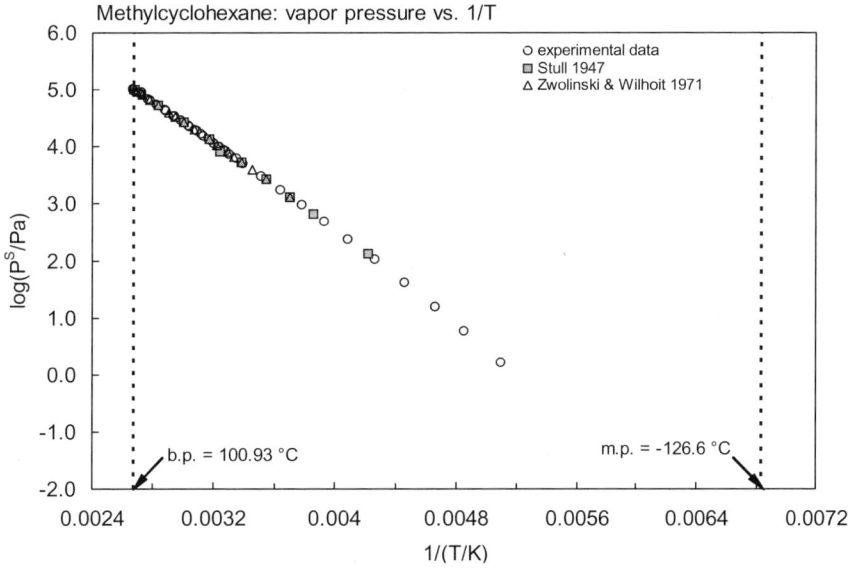

FIGURE 2.1.1.2.7.2 Logarithm of vapor pressure versus reciprocal temperature for methylcyclohexane.

TABLE 2.1.1.2.7.3
Reported Henry's law constants and octanol-air partition coefficients of methylcyclohexane at various temperatures and temperature dependence equations

$\ln K_{AW} = A - B/(T/K)$	(1)	$\log K_{AW} = A - B/(T/K)$	(1a)
$\ln (1/K_{AW}) = A - B/(T/K)$	(2)	$\log (1/K_{AW}) = A - B/(T/K)$	(2a)
$\ln (k_H/atm) = A - B/(T/K)$	(3)		
$\ln [H/(Pa\ m^3/mol)] = A - B/(T/K)$	(4)	$\ln [H/(atm \cdot m^3/mol)] = A - B/(T/K)$	(4a)
$K_{AW} = A - B \cdot (T/K) + C \cdot (T/K)^2$	(5)		

Henry's law constant		log K_{OA}	
Hansen et al. 1993		Gruber et al. 1997	
EPICS-GC		GC det'd activity coefficient	
t/°C	H/(kPa m³/mol)	t/°C	log K_{OA}
27.3	12.666	20.29	3.142
35.8	34.653	30.3	2.943
45.0	72.447	40.4	2.784
		50.28	2.632
eq. 4	H/(kPa m³/mol)		
A	34 ± 3.39		
B	9406 ± 1046		

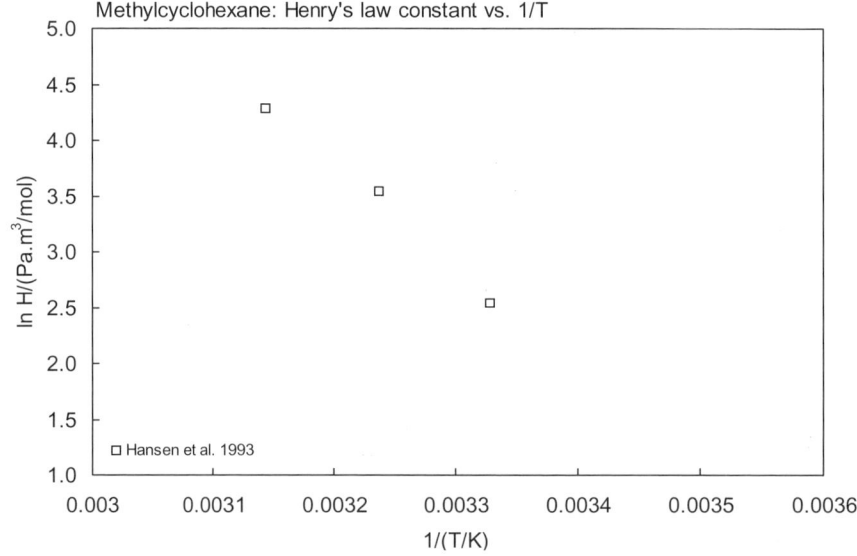

FIGURE 2.1.1.2.7.3 Logarithm of Henry's law constant versus reciprocal temperature for methylcyclohexane.

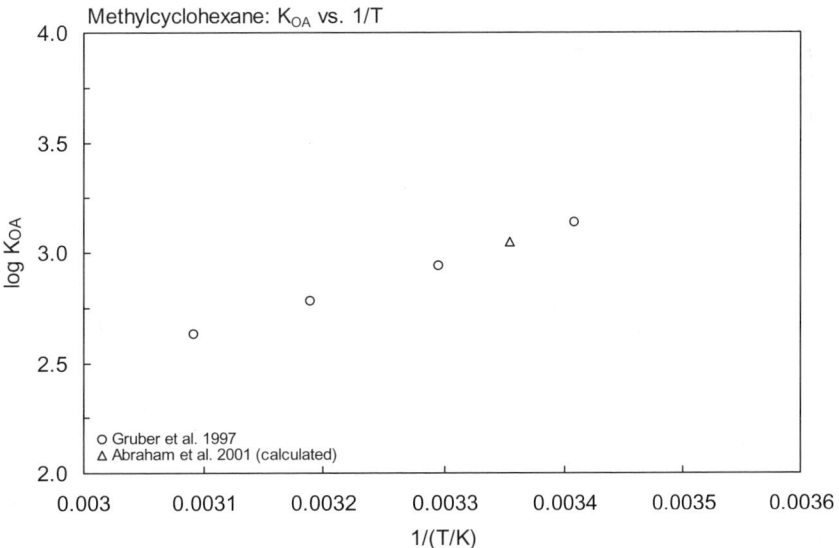

FIGURE 2.1.1.2.7.4 Logarithm of K_{OA} versus reciprocal temperature for methylcyclohexane.

2.1.1.2.8   1,2,-cis-Dimethylcyclohexane

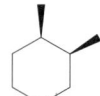

Common Name: 1,2-*cis*-Dimethylcyclohexane
Synonym: *cis*-1,2-dimethylcyclohexane
Chemical Name: 1,2-*cis*-dimethylcyclohexane
CAS Registry No: 2207-01-4
Molecular Formula: C_8H_{16}
Molecular Weight: 112.213
Melting Point (°C):
    –49.8   (Lide 2003)
Boiling Point (°C):
    129.8   (Lide 2003)
Density (g/cm³ at 20°C):
    0.7963, 0.7922   (20°C, 25°C, Dreisbach 1955; Riddick et al. 1986)
Molar Volume (cm³/mol):
    140.9   (20°C, calculated-density, McAuliffe 1966)
    141.6   (25°C, calculated-density; Ruelle & Kesselring 1997)
    162.6   (calculated-Le Bas method at normal boiling point)
Enthalpy of Vaporization, ΔH_V (kJ/mol):
    39.715, 34.196   (25°C, bp, Riddick et al. 1986)
Enthalpy of Fusion, ΔH_{fus} (kJ/mol):
    1.644   (Dreisbach 1955; Riddick et al. 1986)
    8.26, 1.64   (–100.65, –39.05°C, Chickos et al. 1999)
Entropy of Fusion, ΔS_{fus} (J/mol K):
    55.22, 50.2   (exptl., calculated-group additivity method, total phase change entropy, Chickos et al. 1999)
Fugacity Ratio at 25°C, F: 1.0

Water Solubility (g/m³ or mg/L at 25°C or as indicated and reported temperature dependence equations. Additional data at other temperatures designated * are compiled at the end of this section.):
    6.0   (shake flask-GC, McAuliffe 1966)
    5.01   (calculated-recommended liquid-liquid equilibrium LLE data, Mączyński et al. 2004)
    5.94*   (generator column-GC, measured range 273.15–313.15 K, Dohányosová et al. 2004)
    $\ln x = -59.7348 + 45.8700/\tau + 46.6282 \cdot \ln \tau$; $\tau = T/298.15$ K (empirical eq., generator column-GC, Dohányosová et al. 2004)
    $\ln (S/\text{ppm}) = 11.610 - 8455.943/(T/K) + 961943.2/(T/K)^2$, temp range 30.3–170°C (Marche et al. 2004)

Vapor Pressure (Pa at 25°C or as indicated and reported temperature dependence equations. Additional data at other temperatures designated * are compiled at the end of this section.):
    6352*   (49.185°C, ebulliometry, measured range 49.185–130.684°C, Willingham et al. 1945)
    1333*   (18.4°C, summary of literature data, temp range –15.9 to 129.7°C, Stull 1947)
    $\log (P/\text{mmHg}) = 6.84164 - 1369.525/(216.040 + t/°C)$; temp range 49.2–130.7°C (Antoine eq. from exptl. data, ebulliometry-manometer, Willingham et al. 1945)
    1929   (calculated by formula, Dreisbach 1955)
    $\log (P/\text{mmHg}) = 6.84164 - 1369.525/(216.040 + t/°C)$; temp range 40–170°C (Antoine eq. for liquid state, Dreisbach 1955)
    1933*   (interpolated-Antoine eq., temp range 18.4–157.6°C, Zwolinski & Wilhoit 1971)
    $\log (P/\text{mmHg}) = 6.83746 - 1367.311/(215.835 + t/°C)$; temp range 18.4–157.6°C (Antoine eq., Zwolinski & Wilhoit 1971)
    $\log (P/\text{mmHg}) = [-0.2185 \times 9364.9/(T/K)] + 8.001159$; temp range –15.9 to 129.7°C (Antoine eq., Weast 1972–73)

Aliphatic and Cyclic Hydrocarbons

$\log (P/\text{atm}) = [1 - 402.894/(T/K)] \times 10^{\wedge}\{0.841813 - 8.56119 \times 10^{-4} \cdot (T/K) + 5.01855 \times 10^{-7} \cdot (T/K)^2\}$; temp range: 257.25–430.75 K (Cox eq., Chao et al. 1983)

1927   (extrapolated-Antoine eq., Boublik et al. 1984)

$\log (P/\text{kPa}) = 5.96885 - 1370.962/(216.202 + t/°C)$; temp range 49.2–130.68°C (Antoine eq. from reported exptl. data of Willingham et al. 1945, Boublik et al. 1984)

1928   (interpolated-Antoine eq., temp range 18–158°C, Dean 1985, 1992)

$\log (P/\text{mmHg}) = 6.83746 - 1367.311/(215.84 + t/°C)$; temp range 18–158°C (Antoine eq., Dean 1985, 1992)

$\log (P/\text{kPa}) = 5.96654 - 1369.525/(216.040 + t/°C)$; temp range not specified (Antoine eq., Riddick et al. 1986)

$\log (P_L/\text{kPa}) = 5.96232 - 1367.306/(-57.314 + T/K)$; temp range 322–405 K (Antoine eq., Stephenson & Malanowski 1987)

$\log (P/\text{mmHg}) = 32.1535 - 3.0728 \times 10^3/(T/K) - 8.4344 \cdot \log (T/K) + 6.8943 \times 10^{-10} \cdot (T/K) + 1.9558 \times 10^{-6} \cdot (T/K)^2$; temp range 223–606 K (vapor pressure eq., Yaws 1994)

Henry's Law Constant (Pa m^3/mol at 25°C or as indicated. Additional data at other temperatures designated * are compiled at the end of this section.):

36000   (calculated-P/C, Mackay & Shiu 1981)
35830   (calculated-$1/K_{AW}$, C_W/C_A, reported as exptl., Hine & Mookerjee 1975)
53000, 62270   (calculated-group contribution, calculated-bond contribution, Hine & Mookerjee 1975)
44080   (calculated-MCI χ, Nirmalakhandan & Speece 1988)
36045   (calculated-vapor-liquid equilibrium (VLE) data, Yaws et al. 1991)
36180*   (derived from solute fugacity and mole fraction solubility, temp range 273.15–313.15 K, Dohányosová et al. 2004)

Octanol/Water Partition Coefficient, log K_{OW}:

3.06   (calculated-π substituent constant, Hansch et al. 1968)
3.33   (calculated-MCI χ, Murray et al. 1975)
3.21   (calculated-molar volume V_M, Wang et al. 1992)

Octanol/Air Partition Coefficient, log K_{OA}:

Bioconcentration Factor, log BCF:

Sorption Partition Coefficient, log K_{OC}:

Environmental Fate Rate Constants, k, and Half-Lives, $t_{1/2}$:

Half-Lives in the Environment:

Air: atmospheric $t_{1/2}$ ~ 2.4–24 h for cycloparaffins, based on the EPA Reactivity Classification of Organics (estimated, Darnall et al. 1976).

TABLE 2.1.1.2.8.1
Reported aqueous solubilities and Henry's law constants of 1,2-*cis*-dimethylcyclohexane at various temperatures

Aqueous solubility				Henry's law constant	
Doháyosová et al. 2004				Doháyosová et al. 2004	
generator column-GC/FID		smoothed raw exptl data		from solute fugacity f and x	
T/K	S/g·m^{-3}	T/K	S/g·m^{-3}	T/K	H/(Pa m^3/mol)
	raw data				
274.15	6.608	273.15	6.59	273.15	7218
274.15	6.670	278.15	6.27	278.15	10566
274.15	6.421	283.15	6.05	283.15	14994
278.15	6.184	288.15	5.93	288.15	20700
278.15	6.234	293.15	5.90	293.15	27720
278.15	6.078	298.15	5.94	298.15	36180
283.75	6.159	303.15	6.06	303.15	46260
283.75	6.041	308.15	6.25	308.15	57780
288.15	5.767	313.15	6.51	313.15	70740
288.15	6.234				
288.15	6.147	$\ln x = A + B/\tau + C \ln \tau$			
293.15	5.960	$\tau = T/298.15$			
293.15	5.679	A	−59.7348		
293.15	5.980	B	45.8700		
298.15	5.979	C	46.6282		
298.15	5.866				
298.15	6.003	ΔH_{sol}/(kJ mol^{-1})			
303.15	6.190	25°C			
308.15	6.483				
308.15	6.016				
313.15	6.546				
313.15	6.483				
313.15	6.795				

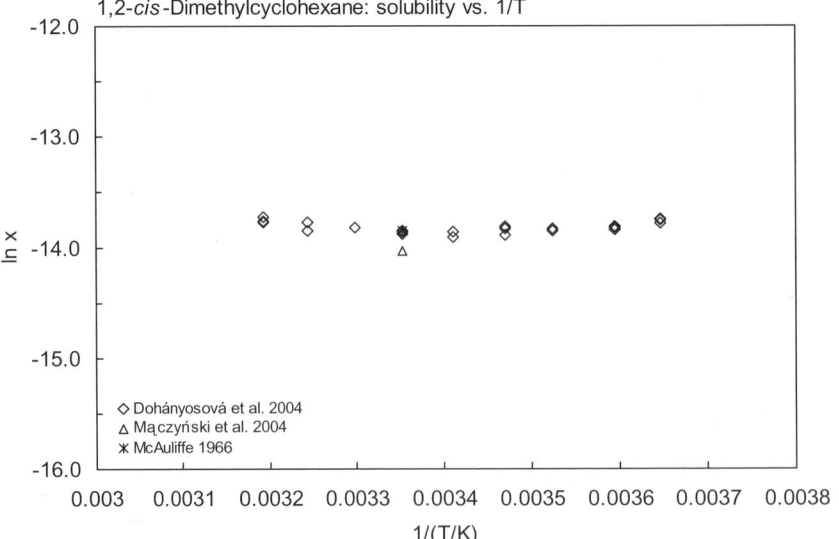

FIGURE 2.1.1.2.8.1 Logarithm of mole fraction solubility (ln x) versus reciprocal temperature for 1,2-*cis*-dimethylcyclohexane.

Aliphatic and Cyclic Hydrocarbons

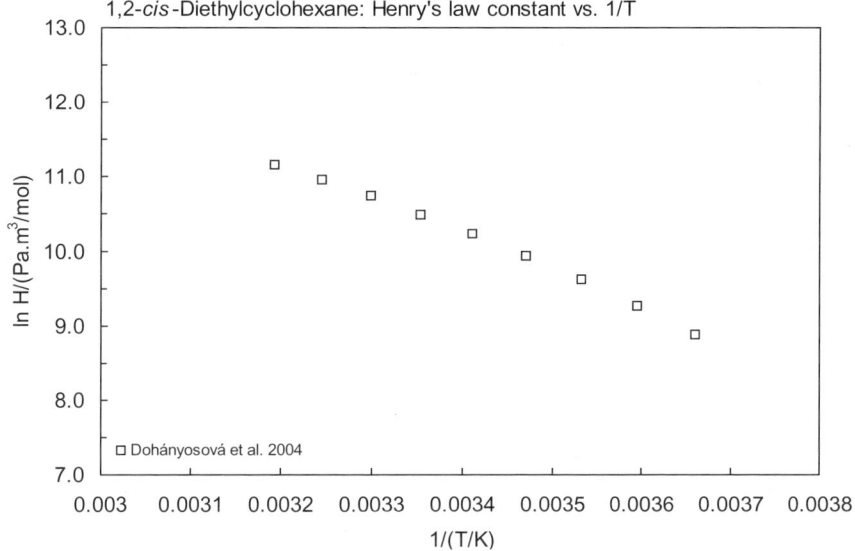

FIGURE 2.1.1.2.8.2 Logarithm of Henry's law constant versus reciprocal temperature for 1,2-cis-dimethylcyclohexane.

TABLE 2.1.1.2.8.2
Reported vapor pressures of 1,2-cis-dimethylcyclohexane at various temperatures and the coefficients for the vapor pressure equations

$$\log P = A - B/(T/K) \quad (1) \qquad \ln P = A - B/(T/K) \quad (1a)$$
$$\log P = A - B/(C + t/°C) \quad (2) \qquad \ln P = A - B/(C + t/°C) \quad (2a)$$
$$\log P = A - B/(C + T/K) \quad (3) \qquad \ln P = A - B/(C + T/K) \quad (3a)$$
$$\log P = A - B/(T/K) - C \cdot \log(T/K) \quad (4)$$
$$\ln (P/P_{ref}) = [1 - (T_{ref}/T)] \cdot \exp(a + bT + cT^2) \quad (5)$$

Willingham et al. 1945		Stull 1947		Zwolinski & Wilhoit 1971	
ebulliometry		summary of literature data		selected values	
t/°C	P/Pa	t/°C	P/Pa	t/°C	P/Pa
49.185	6352	−15.9	133.3	18.4	1333
53.413	7655	7.30	666.6	31.1	2666
57.094	8966	18.4	1333	39.24	4000
60.429	10304	31.1	2666	45.33	5333
63.543	11696	45.3	5333	50.257	6666
67.742	13820	54.4	7999	54.421	7999
72.553	16608	66.8	13332	61.264	10666
77.402	19924	85.6	26664	66.815	13332
81.921	23451	107.0	53329	77.493	19998
87.974	28955	129.7	101325	85.572	26664
93.548	34897			92.151	33331
100.258	43324	mp/°C	−50	97.744	39997
107.192	53656			106.944	53329
114.600	66759			114.554	66661
122.639	83718			120.998	79993
128.315	97607			126.647	93326
128.926	99203			127.699	95992
129.491	100694			128.729	98659
130.125	102389			129.738	101325

(Continued)

TABLE 2.1.1.2.8.2 *(Continued)*

Willingham et al. 1945		Stull 1947		Zwolinski & Wilhoit 1971	
ebulliometry		summary of literature data		selected values	
t/°C	P/Pa	t/°C	P/Pa	t/°C	P/Pa
130.684	103909			25.0	1933
bp/°C	124.450			eq. 2	P/mmHg
				A	6.83746
eq. 2	P/mmHg			B	1367.311
A	6.83866			C	215.835
B	1345.859			bp/°C	129.738
C	215.598			$\Delta H_V/(\text{kJ mol}^{-1})$ =	
temp range 45.2–125.4°C				a 25°C	39.71
pressure range 48–780 mmHg				a bp	33.64

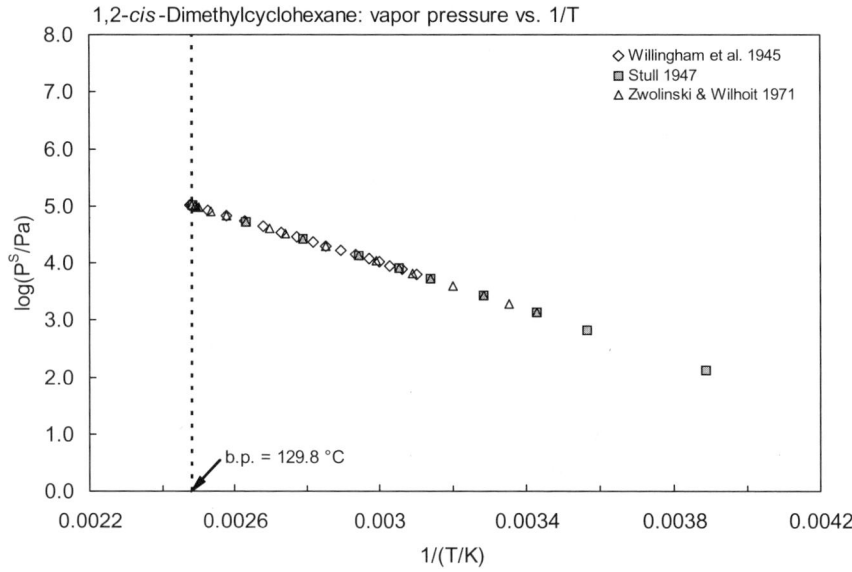

FIGURE 2.1.1.2.8.3 Logarithm of vapor pressure versus reciprocal temperature for 1,2-*cis*-dimethylcyclohexane.

2.1.1.2.9 1,4-trans-Dimethylcyclohexane

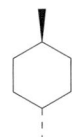

Common Name: 1,4-*trans*-Dimethylcyclohexane
Synonym: *trans*-1,4-dimethylcyclohexane
Chemical Name: 1,4-*trans*-dimethylcyclohexane
CAS Registry No: 2207-04-7
Molecular Formula: C_8H_{16}
Molecular Weight: 112.213
Melting Point (°C):
    –36.93   (Lide 2003)
Boiling Point (°C):
    119.4   (Lide 2003)
Density (g/cm³ at 20°C):
    0.7626, 0.7584   (20°C, 25°C, Dreisbach 1955)
Molar Volume (cm³/mol):
    147.2, 148   (20°C, 25°C, calculated-density)
    162.6   (calculated-Le Bas method at normal boiling point)
Enthalpy of Vaporization, ΔH_V (kJ/mol):
    38.14, 33.05   (25°, bp, Dreisbach 1955)
Enthalpy of Fusion, ΔH_{fus} (kJ/mol):
    11.422   (Dreisbach 1955)
    12.34   (Chickos et al. 1999)
Entropy of Fusion, ΔS_{fus} (J/mol K):
    52.26, 50.2   (exptl., calculated-group additivity method, total phase change entropy, Chickos et al. 1999)
Fugacity Ratio at 25°C, F: 1.0

Water Solubility (g/m³ or mg/L at 25°C):
    3.84   (shake flask-GC, Price 1976; quoted, Shaw 1989)

Vapor Pressure (Pa at 25°C and reported temperature dependence equations):
    $\log (P/mmHg) = 6.82180 - 1332.613/(218.791 + t/°C)$; temp range 40.3–120.3°C (Antoine eq. from exptl. data, ebulliometry-manometer, Willingham et al. 1945)
    3025   (calculated by formula, Dreisbach 1955)
    $\log (P/mmHg) = 6.82180 - 1332.613/(218.791 + t/°C)$; temp range 30–155°C (Antoine eq. for liquid state, Dreisbach 1955)
    3026   (interpolated-Antoine eq., temp range 10.1–146.8°C, Zwolinski & Wilhoit 1971)
    $\log (P/mmHg) = 6.81773 - 1330.437/(218.581 + t/°C)$; temp range 10.1–146.8°C (Antoine eq., Zwolinski & Wilhoit 1971)
    $\log (P/mmHg) = [-0.2185 \times 8951.2/(T/K)] + 7.898079$; temp range –24.3 to 119.5°C (Antoine eq., Weast 1972–73)
    $\log (P/atm) = [1 - 396.346/(T/K)] \times 10\wedge\{0.827486 - 6.12608 \times 10^{-4} \cdot (T/K) + 4.53086 \times 10^{-7} \cdot (T/K)^2\}$; temp range: 252.05–424.25 K (Cox eq., Chao et al. 1983)
    3024   (interpolated-Antoine eq., temp range 10–147°C, Dean 1985, 1992)
    $\log (P/mmHg) = 6.81773 - 1330.437/(218.58 + t/°C)$; temp range 10–147°C (Antoine eq., Dean 1985, 1992)
    $\log (P_L/kPa) = 5.94449 - 1331.612/(-54.43 + T/K)$; temp range 313–395 K (Antoine eq., Stephenson & Malanowski 1987)
    $\log (P/mmHg) = 32.5731 - 2.9872 \times 10^3/(T/K) - 8.6494 \cdot \log (T/K) - 2.1355 \times 10^{-9} \cdot (T/K) + 2.2946 \times 10^{-6} \cdot (T/K)^2$; temp range 236–590 K (vapor pressure eq., Yaws 1994)

Henry's Law Constant (Pa m^3/mol at 25°C):
    88200   (calculated-P/C, Mackay & Shiu 1981)
    88360   (calculated-vapor-liquid equilibrium (VLE) data, Yaws et al. 1991)

Octanol/Water Partition Coefficient, log K_{OW}:

Octanol/Air Partition Coefficient, log K_{OA}:

Bioconcentration Factor, log BCF:

Sorption Partition Coefficient, log K_{OC}:

Environmental Fate Rate Constants, k, and Half-Lives, $t_{1/2}$:

Half-Lives in the Environment:

    Air: atmospheric $t_{1/2}$ ~ 2.4–24 h for cycloparaffins, based on the EPA Reactivity Classification of Organics (estimated, Darnall et al. 1976).

Aliphatic and Cyclic Hydrocarbons

2.1.1.2.10   1,1,3-Trimethylcyclohexane

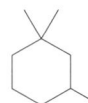

Common Name: 1,1,3-Trimethylcyclohexane
Synonym:
Chemical Name: 1,1,3-trimethylcyclohexane
CAS Registry No: 3073-66-3
Molecular Formula: C_9H_{18}
Molecular Weight: 126.239
Melting Point (°C):
    –65.7   (Lide 2003)
Boiling Point (°C):
    136.6   (Lide 2003)
Density (g/cm³ at 20°C): 0.7664
Molar Volume (cm³/mol):
    164.7   (20°C, calculated-density, Ruelle & Kesselring 1997)
    184.8   (calculated-Le Bas method at normal boiling point)
Enthalpy of Fusion, ΔH_{fus} (kJ/mol):
Entropy of Fusion, ΔS_{fus} (J/mol K):
Fugacity Ratio at 25°C, F: 1.0

Water Solubility (g/m³ or mg/L at 25°C):
    1.77   (shake flask-GC, Price 1976)

Vapor Pressure (Pa at 25°C and reported temperature dependence equations):
    log (P/mmHg) = 6.83705 – 1393.299/(215.551 + t/°C); temp range 54.7–137.6°C (Antoine eq., ebulliometry-manometer measurements, Forziati et al. 1949)
    log (P/atm) = [1– 409.802/(T/K)] × 10^{0.838270 – 6.63916 × 10^{-4}·(T/K) + 5.61172 × 10^{-7}·(T/K)²}; temp range: 327.82–410.80 K (Cox eq., Chao et al. 1983)
    1480   (extrapolated-Antoine eq., Boublik et al. 1984)
    log (P/kPa) = 5.96492 – 1395.206/(215.77 + t/°C); temp range 54.67–137.6°C (Antoine eq. from reported exptl. data of Forziati et al. 1949, Boublik et al. 1984)
    log (P/kPa) = 5.96816 – 1397.161/(215.961 + t/°C); temp range 54.69–137.65°C (Antoine eq. from reported exptl. data of Pasek & Thodos 1962, Boublik et al. 1984)
    1480   (interpolated-Antoine eq., temp range 55–137°C, Dean 1985, 1992)
    log (P/mmHg) = 6.83951 – 1394.88/(215.73 + t/°C); temp range 55–137°C (Antoine eq., Dean 1985, 1992)
    log (P_L/kPa) = 5.96449 – 1395.396/(–57.308 + T/K); temp range 348–411 K (Antoine eq., Stephenson & Malanowski 1987)

Henry's Law Constant (Pa m³/mol at 25°C):
    105600   (calculated-P/C from selected data)

Octanol/Water Partition Coefficient, log K_{OW}:

Octanol/Air Partition Coefficient, log K_{OA}:

Bioconcentration Factor, log BCF:

Sorption Partition Coefficient, log K_{OC}:

Environmental Fate Rate Constants, k, and Half-Lives, $t_{½}$:
  Volatilization:
  Photolysis:
  Oxidation: rate constant k for gas-phase second order rate constants, k_{OH} for reaction with OH radical, k_{NO3} with NO$_3$ radical and k_{O3} with O$_3$ or as indicated, *data at other temperatures see reference:
    k_{OH} = 8.73 × 10^{-12} cm^3 molecule^{-1} s^{-1} at 300 K (Atkinson 1989)
    k_{OH} = 8.7 × 10^{-12} cm^3 molecule^{-1} s^{-1} at 298 K (recommended, Atkinson 1997)
  Hydrolysis:
  Biodegradation:
  Biotransformation:
  Bioconcentration, Uptake (k_1) and Elimination (k_2) Rate Constants or Half-Lives:

Half-Lives in the Environment:

2.1.1.2.11 Ethylcyclohexane

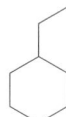

Common Name: Ethylcyclohexane
Synonym:
Chemical Name: ethylcyclohexane
CAS Registry No: 1678-91-7
Molecular Formula: C_8H_{16}, $C_2H_5C_6H_{11}$
Molecular Weight: 112.213
Melting Point (°C):
    –111.3   (Weast 1982–83; Lide 2003)
Boiling Point (°C):
    131.9   (Lide 2003)
Density (g/cm³):
    0.7880   (20°C, Weast 1982–82)
Molar Volume (cm³/mol):
    142.4   (20°C, calculated-density, Stephenson & Malanowski 1987)
    162.6   (calculated-Le Bas method at normal boiling point)
Enthalpy of Vaporization, ΔH_V (kJ/mol):
Enthalpy of Sublimation, ΔH_{subl} (kJ/mol):
Enthalpy of Fusion, ΔH_{fus} (kJ/mol):
Entropy of Fusion, ΔS_{fus} (J/mol K):
Fugacity Ratio at 25°C (assuming ΔS_{fus} = 56 J/mol K), F: 1.0

Water Solubility (g/m³ or mg/L at 25°C or as indicated and reported temperature dependence equations. Additional data at other temperatures designated * are compiled at the end of this section.):
    7.0*   (40°C, shake flask-solid phase extraction-GC, measured range 38.35–280°C, Heidman et al. 1985)
    $\ln x = -334.2468 + 14105.21/(T/K) + 47.93102 \cdot \ln (T/K)$; measured range 40–280°C (shake flask-solid phased extraction-GC, Heidman et al. 1985)
    7.0*   (40°C, IUPAC tentative value, temp range 40–280°C, Shaw 1989a)
    3.89*   (generator column-GC/FID, measured range 273.15–313.15 K, Dohányosová et al. 2004)
    $\ln x = -53.6687 + 39.4055/\tau + 41.1210 \cdot \ln \tau$, $\tau = [(T/K)/298.15]$, temp range 273.15–313.15 K (generator column-GC/FID, Dohányosová et al. 2004)
    4.36*   (30.3°C, shake flask-GC, measured range 30.3–170.8°C, Marche et al. 2004)
    $\ln x = -344.02468 + 14105.21/(T/K) + 47.93102 \cdot \ln (T/K)$; temp range 30.3–170.8°C (Marche et al. 2004)

Vapor Pressure (Pa at 25°C or as indicated and reported temperature dependence equations. Additional data at other temperatures designated * are compiled at the end of this section.):
    6351*   (51.412°C, ebulliometry, measured range 51.4–132.7°C, Willingham et al. 1945)
    1612   (extrapolated-Antoine eq., Willingham et al. 1945)
    $\log (P/mmHg) = 6.87041 - 1384.036/(214.128 + t/°C)$; temp range 51.4–132.7°C (Antoine eq., ebulliometry, Willingham et al. 1945)
    1333*   (20.6°C, summary of literature data, temp range –14.5 to 131.9°C, Stull 1947)
    1705*   (extrapolated-Antoine eq., temp range 20.6–159.5°C, Zwolinski & Wilhoit 1971)
    $\log (P/mmHg) = 6.86728 - 1382.466/(214.995 + t/°C)$; temp range 20.6–159.5°C (Antoine eq., Zwolinski & Wilhoit 1971)
    $\log (P_L/kPa) = 5.99043 - 1381.396/(-58.271 + T/K)$; temp range 323–407 K (Antoine eq., Stephenson & Malanowski 1987)

Henry's Law Constant (Pa m³/mol at 25°C Additional data at other temperatures designated * are compiled at the end of this section.):

ln (k_H/MPa) = 325.570 − 18496.5/(T/K) − 10.9666·(T/K)^2 − 44.7690·ln (T/K); maximum k_H = 1.186 × 10^4 MPa at 385.2 K (Heidman et al. 1985)

48960* (derived from solute fugacity and mole fraction solubility, temp range 273.15–323.15 K, Dohányosová et al. 2004)

Octanol/Water Partition Coefficient, log K_{OW}:

Octanol/Air Partition Coefficient, log K_{OA}:

Bioconcentration Factor, log BCF or log K_B:

Sorption Partition Coefficient, log K_{OC}:

Environmental Fate Rate Constants, k and Half-Lives, $t_{1/2}$:

Half-Lives in the Environment:

TABLE 2.1.1.2.11.1
Reported aqueous solubilities of ethylcyclohexane at various temperatures

ln x = A + B/(T/K) + C ln (T/K)   (1)

Heidman et al. 1985		Shaw 1989a		Dohányosová et al. 2004		Marche et al. 2004	
shake flask-GC		IUPAC "tentative" best		generator column-GC		shake flask-GC	
t/°C	S/g·m^{-3}	t/°C	S/g·m^{-3}	T/K	S/g·m^{-3}	t/°C	S/g·m^{-3}
38.35	6.8	40	7.0	274.15	3.803	30.3	4.358
94.45	15.0	50	7.0	274.15	4.108	70.4	8.790
150.25	120	60	7.0	278.15	3.996	100.5	28.18
206.35	750	70	10	278.15	3.678	131.0	67.95
263.15	7300	80	13	283.15	3.734	151.2	110.8
279.65	14600	90	16	288.15	3.759	170.8	183.3
		100	21	288.15	3.666		
eq. 1	x	120	55	293.15	3.766		
A	−334.2468	140	95	293.15	3.797		
B	14105.21	160	160	298.15	3.965		
C	47.93102	180	280	303.15	3.971		
		200	550	303.15	4.034		
		220	1600	313.15	4.358		
		240	3400	313.15	4.520		
		260	6500	313.15	4.364		
		280	15000				

empirical eq.
ln x = A + B/τ + C ln τ
τ = (T/K)/298.15
A   −53.6887
B   39.4055
C   41.1210

Aliphatic and Cyclic Hydrocarbons

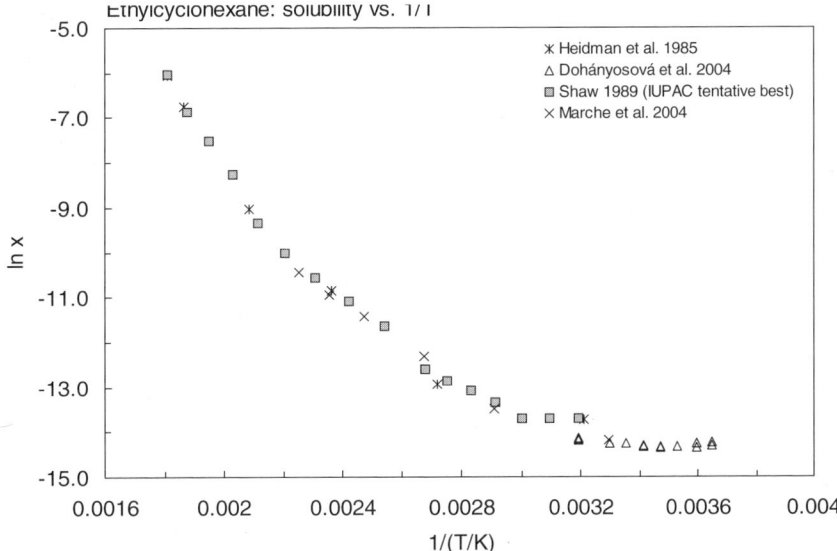

FIGURE 2.1.1.2.11.1 Logarithm of mole fraction solubility (ln x) versus reciprocal temperature for ethylcyclohexane.

TABLE 2.1.1.2.11.2
Reported vapor pressures and Henry's law constants of ethylcyclohexane at various temperatures and the coefficients for the vapor pressure equations

$$\log P = A - B/(T/K) \quad (1)$$
$$\log P = A - B/(C + t/°C) \quad (2)$$
$$\log P = A - B/(C + T/K) \quad (3)$$
$$\log P = A - B/(T/K) - C \cdot \log (T/K) \quad (4)$$

$$\ln P = A - B/(T/K) \quad (1a)$$
$$\ln P = A - B/(C + t/°C) \quad (2a)$$
$$\ln P = A - B/(C + T/K) \quad (3a)$$

Vapor pressure						Henry's law constant	
Willingham et al. 1945		Stull 1947		Zwolinski & Wilhoit 1971		Dohányosová et al. 2004	
Ebulliometry		Summary of literature data		Selected values		From solute fugacity f and x	
t/°C	P/Pa	t/°C	P/Pa	t/°C	P/Pa	t/°C	H/(Pa m³ mol⁻¹)
51.412	6351	−14.5	133.3	20.6	1333	0	10422
55.636	7653	9.2	666.6	33.4	2666	5	14994
59.315	8959	20.6	1333	41.48	4000	10	20880
62.655	10299	33.4	2666	47.57	5333	15	28440
65.755	11691	47.6	5333	51.494	6666	20	37800
69.948	13818	56.7	7999	56.656	7999	25	48960
74.738	16615	69.0	13332	63.493	10666	30	62100
79.587	19916	87.8	26664	69.093	13332	35	77040
84.115	23451	109.1	53329	79.699	19998	40	93780
90.158	28956	131.8	101325	87.762	26664		
95.716	34893			94.327	33331		
102.412	43320	mp/°C	−111.3	99.906	39997		
109.327	53657			109.13	53329		
116.709	66760			116.666	66661		
124.723	83722			123.088	79993		
130.379	97609			128.715	93326		

(Continued)

TABLE 2.1.1.2.11.2 (Continued)

Vapor pressure						Henry's law constant	
Willingham et al. 1945		Stull 1947		Zwolinski & Wilhoit 1971		Dohányosová et al. 2004	
Ebulliometry		Summary of literature data		Selected values		From solute fugacity f and x	
t/°C	P/Pa	t/°C	P/Pa	t/°C	P/Pa	t/°C	H/(Pa m³ mol⁻¹)
130.988	99208			129.764	95992		
131.551	100700			130.79	98659		
132.181	102404			131.795	101325		
132.742	103922			25.0	1707		
bp/°C	131.783			eq. 2	P/mmHg		
Antoine eq.				A	6.86728		
eq. 2	P/mmHg			B	1382.466		
A	6.87041			C	214.995		
B	1384.036			bp/°C	131.795		
C	214.128			$\Delta H_V/(kJ\ mol^{-1}) =$			
temp range 51.4–132.7°C				at 25°C	40.48		
pressure range 48–780 mmHg				at bp	34.31		

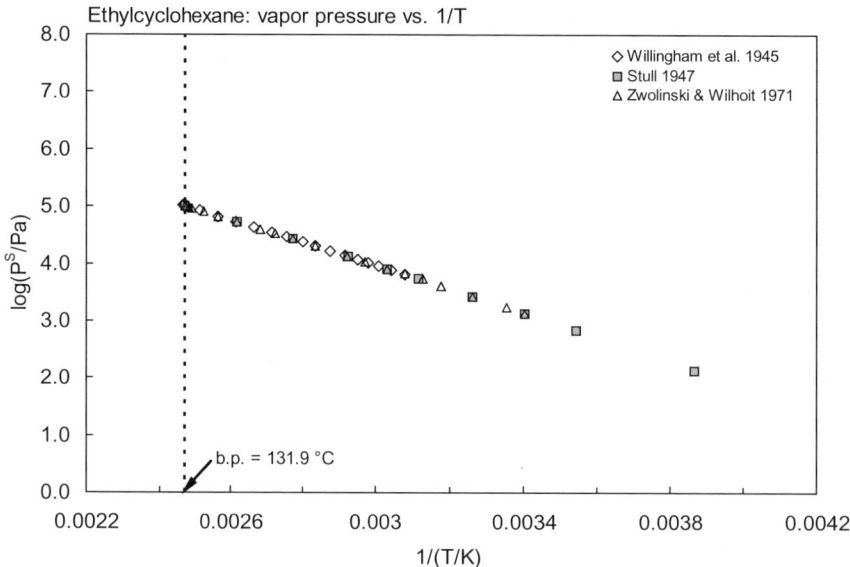

FIGURE 2.1.1.2.11.2 Logarithm of vapor pressure versus reciprocal temperature for ethylcyclohexane.

Aliphatic and Cyclic Hydrocarbons

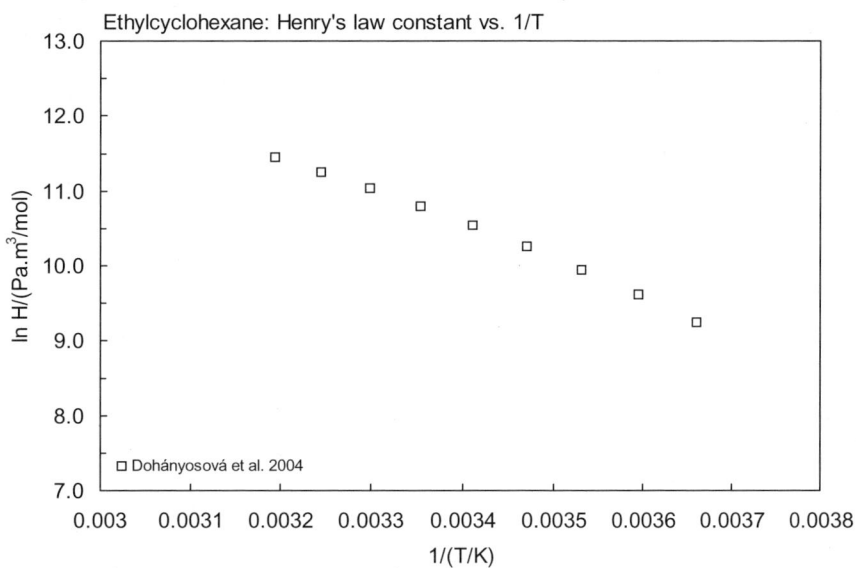

FIGURE 2.1.1.2.11.3 Logarithm of Henry's law constant versus reciprocal temperature for ethylcyclohexane.

2.1.1.2.12   Cycloheptane

Common Name: Cycloheptane
Synonym: suberane
Chemical Name: cycloheptane
CAS Registry No: 291-64-5
Molecular Formula: C_7H_{14}
Molecular Weight: 98.186
Melting Point (°C): –12
    –8.46   (Lide 2003)
Boiling Point (°C): 118.5
    118.4   (Lide 2003)
Density (g/cm³):
    0.8098   (20°C, Weast 1984)
    0.80656   (measured, Anand et al. 1975)
Molar Volume (cm³/mol):
    121.3   (20°C, calculated-density, Stephenson & Malanowski 1987)
    136.4   (calculated-Le Bas method at normal boiling point)
Enthalpy of Fusion, ΔH_{fus} (kJ/mol):
    4.98, 0.29, 0.45, 1.88   (–138.35, –74.95, –60.75, –8.05°C, Chickos et al. 1999)
Entropy of Fusion, ΔS_{fus} (J/mol K):
    47.6, 48.2   (exptl., calculated-group additivity method, total phase change entropy, Chickos et al. 1999)
Fugacity Ratio at 25°C, F: 1.0

Water Solubility (g/m³ or mg/L at 25°C or as indicated):
    30.0   (shake flask-GC, McAuliffe 1966)
    27.1   (30°C, shake flask-GC, Groves 1988)
    23.5, 23.5   (25, 30°C, calculated-liquid-liquid equilibrium LLE data, Mączyński et al. 2004)

Vapor Pressure (Pa at 25°C or as indicated and reported temperature dependence equations. Additional data at other temperatures designated * are compiled at the end of this section.):
    19920*   (68.2°C, ebulliometry, measured range 68.2–159°C, Finke et al. 1956)
    $\log (P/mmHg) = 6.85271 - 1330.742/(t/°C + 216.246)$; temp range 68.2–159°C (Antoine eq., ebulliometry, Finke et al. 1956)
    2924*   (static method-quartz spiral gauge, measured range 283.048–323.551 K, Anand et al. 1975)
    $\log (P/kPa) = 6.19317 - 1450.17/(T/K - 44.91)$; temp range 283.048–323.551 K (static method, vapor-liquid equilibria study, Anand et al. 1975)
    16312*   (63.03°C, comparative ebulliometry, measured range 60–121.7°C, Meyer & Hotz 1976)
    $\log (P/mmHg) = 5.85683 - 1333.780/(t/°C + 216.6438)$; temp range 60–121.7°C (Antoine eq., comparative ebulliometry, Meyer & Hotz 1976)
    $\log (P/atm) = [1 - 391.896/(T/K)] \times 10^{\{0.885524 - 8.19621 \times 10^{-4} \cdot (T/K) + 7.88065 \times 10^{-7} \cdot (T/K)^2\}}$; temp range: 284.35–432.17 K (Cox eq., Chao et al. 1983)
    2895, 2898   (extrapolated-Antoine eq., Boublik et al. 1984)
    $\log (P/kPa) = 5.97858 - 1331.383/(214.325 + t/°C)$; temp range 60.2–159°C (Antoine eq. from reported exptl. data, Boublik et al. 1984)
    $\log (P/kPa) = 5.98198 - 1333.899/(216.657 + t/°C)$; temp range 63.03–121.68°C (Antoine eq. from reported exptl. data, Boublik et al. 1984)
    2895   (extrapolated-Antoine eq., Dean 1985, 1992)
    $\log (P/mmHg) = 6.85395 - 1331.57/(216.35 + t/°C)$; temp range: 68–159°C (Antoine eq., Dean 1985, 1992)
    2930   (interpolated-Antoine eq., Stephenson & Malanowski 1987)

Aliphatic and Cyclic Hydrocarbons

log (P_L/kPa) = 5.98143 – 1333.833/(–56.458 + T/K); temp range 341–433 K (Antoine eq.-I, Stephenson & Malanowski 1987)

log (P_L/kPa) = 6.12682 – 1417.738/(–47.665 + T/K); temp range 282–333 K (Antoine eq.-II, Stephenson & Malanowski 1987)

log (P_L/kPa) = 5.97596 – 1329.98/(–56.968 + T/K); temp range 333–398 K (Antoine eq.-III, Stephenson & Malanowski 1987)

log (P_L/kPa) = 7.05325 – 2475.271/(108.392 + T/K); temp range 476–604 K (Antoine eq.-IV, Stephenson & Malanowski 1987)

log (P/mmHg) = 54.0858 – 3.6109 × 10^3/(T/K) – 17.331·log (T/K) + 7.5292 × 10^{-3}·(T/K) + 1.7553 × 10^{-6}·(T/K)^2; temp range 265–604 K (vapor pressure eq., Yaws 1994)

Henry's Law Constant (Pa m^3/mol at 25°C):

9977 (calculated-vapor-liquid equilibrium (VLE) data, Yaws et al. 1991)

Octanol/Water Partition Coefficient, log K_{OW}:

2.87 (calculated-π substituent constant, Hansch et al. 1968)
3.06 (calculated-MCI χ, Murray et al. 1975)
3.76 (calculated-fragment const., Yalkowsky & Morozowich 1980)
2.72 (calculated-molar volume V_M, Wang et al. 1992)
3.91 (calculated-fragment const., Müller & Klein 1992)
3.1648 (calculated-UNIFAC group contribution, Chen et al. 1993)
4.00 (recommended, Hansch et al. 1995)

Octanol/Air Partition Coefficient, log K_{OA}:

Bioconcentration Factor, log BCF:

Sorption Partition Coefficient, log K_{OC}:

Environmental Fate Rate Constants, k, and Half-Lives, $t_{1/2}$:

Volatilization:
Photolysis:
Oxidation: rate constant k, for gas-phase second order rate constants, k_{OH} for reaction with OH radical, k_{NO3} with NO_3 radical and k_{O3} with O_3 or as indicated, *data at other temperatures see reference:

k_{OH} = 13.1 × 10^{-12} cm^3 molecule^{-1} s^{-1} at 298 K (Atkinson 1985)
k_{OH} = (7.88 ± 1.38) × 10^{12} cm^3 mol^{-1} s^{-1} at 298 K (flash photolysis-resonance absorption, Jolly et al. 1985)
k_{OH} = (11.8 – 13.1) × 10^{-12} cm^3 molecule^{-1} s^{-1} at 298–300 K (review, Atkinson 1989)
k_{OH} = 1.25 × 10^{-11} cm^3 molecule^{-1} s^{-1} at 298 K (recommended, Atkinson 1990)
k_{OH} = 13.0 × 10^{-12} cm^3 molecule^{-1} s^{-1} at 298 K (recommended, Atkinson 1997)

Hydrolysis:
Biodegradation:
Biotransformation:
Bioconcentration, Uptake (k_1) and Elimination (k_2) Rate Constants or Half-Lives:

Half-Lives in the Environment:

Air: atmospheric $t_{1/2}$ ~ 2.4–24 h for cycloparaffins, based on the EPA Reactivity Classification of Organics (estimated, Darnall et al. 1976).

TABLE 2.1.1.2.12.1
Reported vapor pressures of cycloheptane at various temperatures and the coefficients for the vapor pressure equations

$$\log P = A - B/(T/K) \quad (1) \qquad \ln P = A - B/(T/K) \quad (1a)$$
$$\log P = A - B/(C + t/°C) \quad (2) \qquad \ln P = A - B/(C + t/°C) \quad (2a)$$
$$\log P = A - B/(C + T/K) \quad (3)$$
$$\log P = A - B/(T/K) - C \cdot \log(T/K) \quad (4)$$
$$\log P = A'[1 - (T_B/K)/(T/K)] \quad (5) \text{ where } A' = a + bT + cT^2$$

Finke et al. 1956		Anand et al. 1975		Meyer & Hotz 1976	
ebulliometric method		static-quartz spiral gauge		comparative ebulliometry	
t/°C	P/Pa	T/K	P/Pa	t/°C	P/Pa
68.204	19920	283.043	1272	63.03	16312
74.338	25007	288.237	1711	68.929	20748
80.529	31160	292.906	2218	77.747	28258
86.771	38547	298.150	2924	85.702	37172
93.068	47359	301.924	3550	96.324	52481
99.416	57803	309.193	5080	106.462	71394
105.820	70109	312.723	6006	117.547	97871
112.281	84525	317.813	7573	121.681	109500
118.793	101325	323.551	9741		
125.364	120798			bp/°C	118.813
131.985	143268	Antoine eq.			
138.665	169052	eq. 3	P/kPa	Antoine eq.	
145.387	198530	A	6.19317	eq. 2	P/cmHg
152.178	232087	B	1450.17	A	5.85683
159.022	270110	C	−44.91	B	1333.780
				C	216.6438
bp/°C	118.79			temp range: 63–122°C	
ΔH_V =	38.53 kJ/mol				
	at 25°C			Cox equation:	
Antoine eq.				eq 5	P/atm
eq. 2	P/mmHg			a	0.878453
A	6.85271			$-b \times 10^3$	0.916539
B	1330.742			$c \times 10^6$	0.965009
C	216.246				
Cox eq.					
eq. 5	P/atm				
T_B	391.953				
a	0.839608				
$-b \times 10^4$	6.9133				
$c \times 10^7$	6.4035				

Aliphatic and Cyclic Hydrocarbons

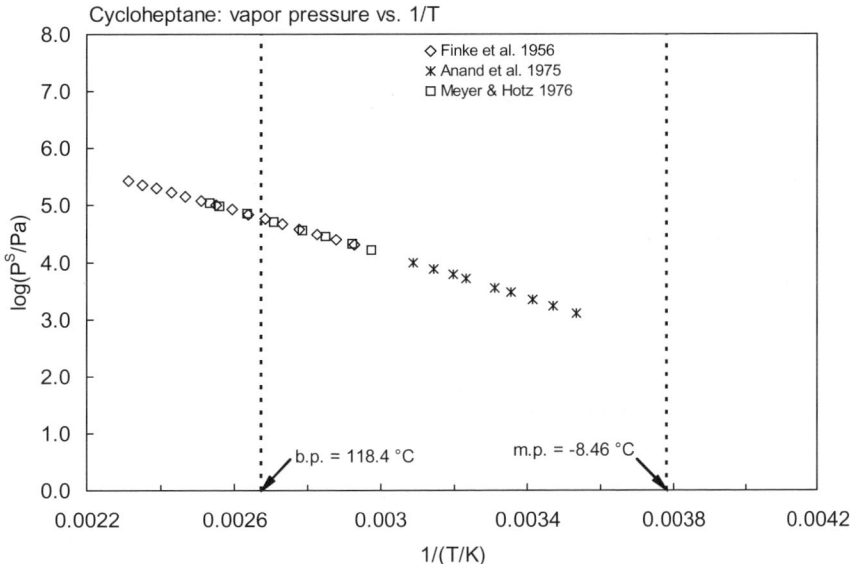

FIGURE 2.1.1.2.12.1 Logarithm of vapor pressure versus reciprocal temperature for cycloheptane.

2.1.1.2.13   Cyclooctane

Common Name: Cyclooctane
Synonym:
Chemical Name: cyclooctane
CAS Registry No: 292-64-8
Molecular Formula: C_8H_{16}
Molecular Weight: 112.213
Melting Point (°C):
    14.59   (Lide 2003)
Boiling Point (°C): 151
    149   (Lide 2003)
Density (g/cm³ at 20°C):
    0.8340   (Weast 1984)
Molar Volume (cm³/mol):
    134.4   (20°C, calculated-density, Stephenson & Malanowski 1987)
    154.1   (calculated-Le Bas method at normal boiling point)
Enthalpy of Fusion, ΔH_{fus} (kJ/mol):
    6.32, 0.48, 2.41 (−106.65, −89.35, 14.85°C, Chickos et al. 1999)
Entropy of Fusion, ΔS_{fus} (J/mol K):
    48.89, 51.9 (exptl., calculated-group additivity method, total phase change entropy, Chickos et al. 1999)
Fugacity Ratio at 25°C, F: 1.0

Water Solubility (g/m³ or mg/L at 25°C and the reported temperature dependence equations. Additional data at other temperatures designated * are compiled at the end of this section.):
    7.90   (shake flask-GC, McAuliffe 1966)
    7.48   (calculated-recommended liquid-liquid equilibrium LLE data, Mączyński et al. 2004)
    5.80*   (generator column-GC, measured range 27.15–313.15 K, DoháNyosová et al. 2004)
    $\ln x = -55.1375 + 41.2528/\tau + 43.2804 \cdot \ln \tau$; $\tau = T/298.15$ K (empirical eq., generator column-GC, DoháNyosová et al. 2004)

Vapor Pressure (Pa at 25°C or as indicated and reported temperature dependence equations. Additional data at other temperatures designated * are compiled at the end of this section.):
    19920*   (96.7°C, ebulliometry, measured range 96.7–194.4°C, Finke et al. 1956)
    log (P/mmHg) = 6.86173 − 1437.682/(t/°C + 210.003); temp range 96.7–194.4°C (Antoine eq., ebulliometry, Finke et al. 1956)
    748*   (static-quartz spiral gauge, measured range 290.961–323.326 K, Anand et al. 1975)
    log (P/kPa) = 5.97188 − 1447.45/(T/K − 60.67); temp range 291–323 K (static method, vapor-liquid equilibria VLE study, Anand et al. 1975)
    22454*   (100.133°C, comparative ebulliometry, measured range 100.1–161°C, Meyer & Hotz 1976)
    log (P/mmHg) = 5.861786 − 1438.455/(t/°C + 210.1844); temp range 100.1–161°C (Antoine eq., comparative ebulliometry, Meyer & Hotz 1976)
    767, 740   (extrapolated-Antoine eq., extrapolated, Boublik et al. 1984)
    log (P/kPa) = 5.98693 − 1437.751/(210.012 + t/°C); temp range 96.7–194.4°C (Antoine eq. from reported exptl. data, Boublik et al. 1984)
    log (P/kPa) = 6.06524 − 1492.101/(216.413 + t/°C); temp range 100.1–160.9°C (Antoine eq. from reported exptl. data, Boublik et al. 1984)
    740   (extrapolated-Antoine eq., Dean 1985, 1992)
    log (P/mmHg) = 6.86187 − 1437.79/(210.02 + t/°C); temp range 97–194°C (Antoine eq., Dean 1985, 1992)
    753   (interpolated-Antoine eq.-III, Stephenson & Malanowski 1987)

Aliphatic and Cyclic Hydrocarbons

log (P_L/kPa) = 5.98663 − 1437.682/(−63.147 + T/K); temp range 369–487 K (Antoine eq.-I, Stephenson & Malanowski 1987)

log (P_L/kPa) = 5.9899 − 1440.707/(−62.701 + T/K); temp range 369–468 K (Antoine eq.-II, Stephenson & Malanowski 1987)

log (P_L/kPa) = 6.20474 − 1564.985/(−50.842 + T/K); temp range 289–369 K (Antoine eq.-III, Stephenson & Malanowski 1987)

Henry's Law Constant (Pa m^3/mol at 25°C or as indicated. Additional data at other temperatures designated * are compiled at the end of this section.):

- 10485   (calculated-vapor-liquid equilibrium (VLE) data, Yaws et al. 1991)
- 14526*   (derived from solute fugacity and mole fraction solubility, temp range 273.15–313.15 K, Dohányosová et al. 2004)

Octanol/Water Partition Coefficient, log K_{OW}:

- 3.28   (calculated-π substituent constant, Hansch et al. 1968)
- 3.50   (calculated-MCI χ, Murray et al. 1975)
- 3.28   (Hutchinson et al. 1980; Sangster 1989)
- 3.28   (calculated-fragment const., Lyman 1982)
- 4.45   (recommended, Sangster 1989)
- 3.04   (calculated-molar volume V_M, Wang et al. 1992)
- 4.47   (calculated-fragment const., Müller & Klein 1992)
- 3.6117 (calculated-UNIFAC group contribution, Chen et al. 1993)

Octanol/Air Partition Coefficient, log K_{OA}:

Bioconcentration Factor, log BCF:

Sorption Partition Coefficient, log K_{OC}:

Environmental Fate Rate Constants, k, and Half-Lives, $t_{½}$:

Volatilization:

Photolysis:

Oxidation: rate constant k, for gas-phase second order rate constants, k_{OH} for reaction with OH radical, k_{NO3} with NO$_3$ radical and k_{O3} with O$_3$ or as indicated, *data at other temperatures see reference:

k_{OH} = 14.0 × 10^{-12} cm^3 molecule^{-1} s^{-1} at 298 K (recommended, Atkinson 1997)

Oxidation:

Hydrolysis:

Biodegradation:

Biotransformation:

Bioconcentration, Uptake (k_1) and Elimination (k_2) Rate Constants or Half-Lives:

Half-Lives in the Environment:

Air: atmospheric $t_{½}$ ~ 2.4–24 h for cycloparaffins, based on the EPA Reactivity Classification of Organics (estimated, Darnall et al. 1976).

TABLE 2.1.1.2.13.1
Reported aqueous solubilities and Henry's law constants of cyclooctane at various temperatures

Aqueous solubility				Henry's law constant	
Dohányosová et al. 2004				Dohányosová et al. 2004	
generator column-GC		smoothed exptl raw data		from solute fugacity f and x	
T/K	S/g·m^{-3}	T/K	S/g·m^{-3}	T/K	H/(Pa m^3/mol)
	raw data				
274.15	5.53	273.15	5.73	273.15	2826
274.15	5.885	278.15	5.59	278.15	4158
278.15	5.548	283.15	5.53	283.15	5922
278.15	5.442	288.15	5.55	288.15	8190
283.15	5.548	293.15	5.64	293.15	11052
288.15	5.754	298.15	5.80	298.15	14526
293.15	5.492	303.15	6.02	303.15	18720
298.15	5.848	308.15	6.32	308.15	23400
298.15	5.530	313.15	6.68	313.15	28800
303.15	5.922				
303.15	5.835	$\ln x = A + B/\tau + C \ln \tau$			
308.15	6.483	$\tau = T/298.15$			
308.15	6.421	A	−55.1375		
313.15	6.733	B	41.2528		
313.15	6.858	C	43.2804		

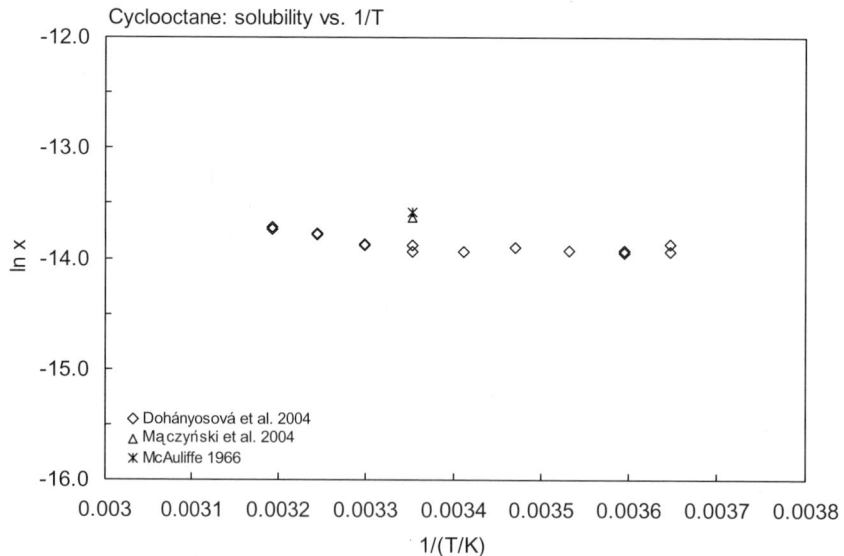

FIGURE 2.1.1.2.13.1 Logarithm of mole fraction solubility (ln x) versus reciprocal temperature for cyclooctane.

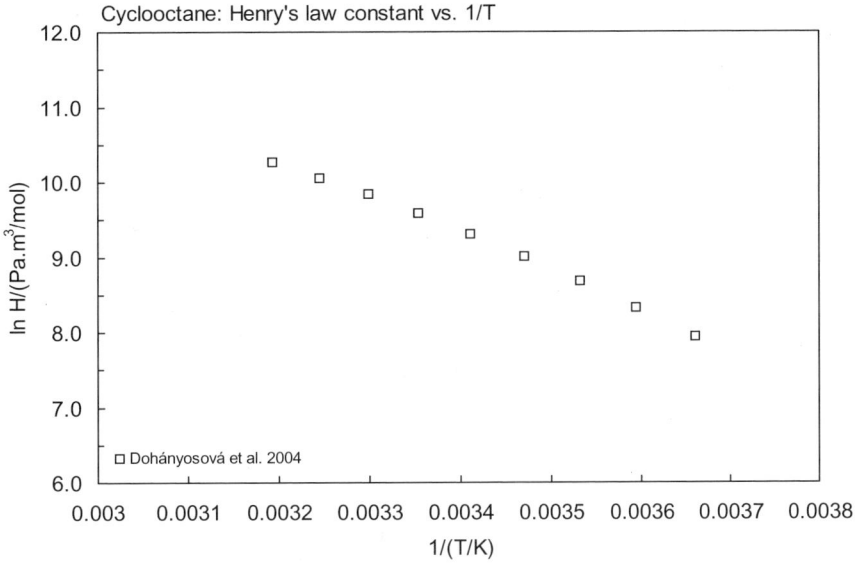

FIGURE 2.1.1.2.13.2 Logarithm of Henry's law constant versus reciprocal temperature for cyclooctane.

TABLE 2.1.1.2.13.2
Reported vapor pressures of cyclooctane at various temperatures and the coefficients for the vapor pressure equations

log P = A − B/(T/K)	(1)	ln P = A − B/(T/K)	(1a)
log P = A − B/(C + t/°C)	(2)	ln P = A − B/(C + t/°C)	(2a)
log P = A − B/(C + T/K)	(3)		
log P = A − B/(T/K) − C·log (T/K)	(4)		
log P = A′[1 − (T_B/K)/(T/K)]	(5) where A′ = a + bT + cT²		

Finke et al. 1956		Anand et al. 1975		Meyer & Hotz 1976		Doháynosová et al. 2004	
ebulliometry		static-quartz spiral gauge		comparative ebulliometry		from Anand et al. 1975	
t/°C	P/Pa	T/K	P/Pa	t/°C	P/Pa	T/K	P/Pa
96.711	19920	290.961	488	100.133	22454	273.15	207
103.318	25007	294.376	601	108.375	29598	278.15	292
109.977	31160	298.150	748	118.813	41155	283.15	406
116.694	38457	302.060	944	127.569	53410	288.15	557
123.472	47359	306.922	1244	139.435	74511	293.15	753
130.301	70109	311.678	1605	149.031	95919	303.15	1010
137.190	84525	316.472	2056	157.388	118293	308.15	1330
144.133	101325	323.326	2890	160.911	129263	313.15	1730
151.146	120798						
158.203	143268	Antoine eq.		bp/°C	151.148		
165.321	169052	eq. 3	P/kPa	Antoine eq.			
172.502	198530	A	6.91788	eq. 2	P/cmHg		
187.040	232087	B	1447.45	A	5.861786		
194.397	270110	C	−60.67	B	1438.455		
				C	210.1844		
bp/°C	151.14			temp range 100–161°C			
Antoine eq.							
eq. 2	P/mmHg			Cox equation:			

(*Continued*)

TABLE 2.1.1.2.13.2 (*Continued*)

Finke et al. 1956 ebulliometry		Anand et al. 1975 static-quartz spiral gauge		Meyer & Hotz 1976 comparative ebulliometry		Dohányosová et al. 2004 from Anand et al. 1975	
t/°C	P/Pa	T/K	P/Pa	t/°C	P/Pa	T/K	P/Pa
A	686173			eq 5	P/atm		
B	1437.682			a	0.869777		
C	210.003			$-b \times 10^3$	0.775348		
				$c \times 10^6$	0.716695		
Cox equation:							
eq 5	P/atm						
a	0.839609						
$-b \times 10^4$	6.2033						
$c \times 10^7$	5.177						
T_B	424.300						

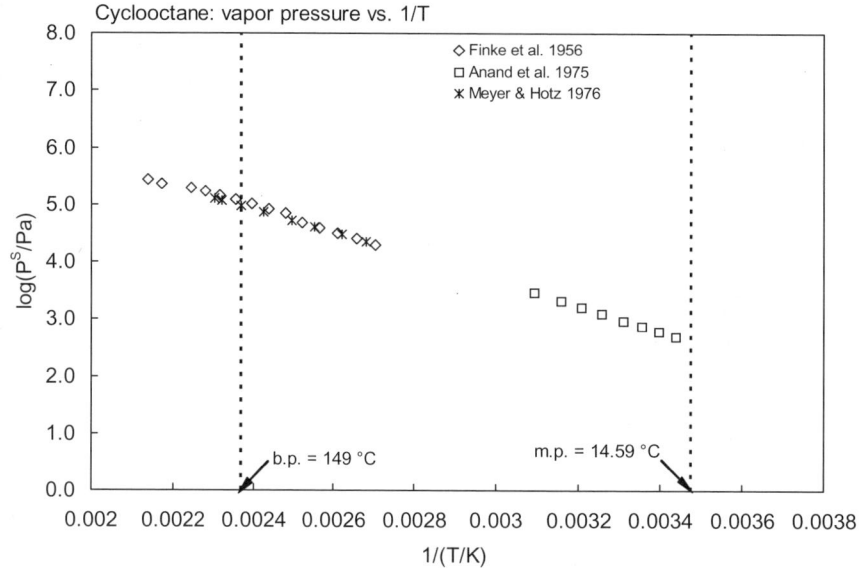

FIGURE 2.1.1.2.13.3 Logarithm of vapor pressure versus reciprocal temperature for cyclooctane.

2.1.1.2.14 Decalin

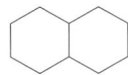

Common Name: Decalin
Synonym: bicyclo[4.4.0]decane, naphthalane, naphthane
Chemical Name: decahydronaphthalene (mixed isomers)
CAS Registry No: 91-17-8
Molecular Formula: $C_{10}H_{18}$
Molecular Weight: 138.250
Melting Point (°C):
    −42.9   (*cis*-decalin, Lide 2003)
    −30.4   (*trans*-decalin, Lide 2003)
Boiling Point (°C): 191.7
    195.774   (*cis*-decalin, Camin & Rossini 1955)
    187.273   (*trans*-decalin, Camin & Rossini 1955)
    191.7   (Riddick et al. 1986)
Density (g/cm³ at 20°C):
    0.8865, 0.8789   (20°C, 25°C, mixed isomers, Riddick et al. 1986)
Molar Volume (cm³/mol):
    154.8, 159.6 (*cis*-decalin, *trans*-decalin, calculated-density, Stephenson & Malanowski 1987)
    184.6   (calculated-Le Bas method at normal boiling point)
Enthalpy of Vaporization, ΔH_V (kJ/mol):
    41.09   (bp, mixed isomers, Riddick et al. 1986)
    51.342, 40.999   (*cis*-decalin, 25°C, bp, Riddick et al. 1986)
    49.87, 40.229   (*trans*-decalin, 25°C, bp, Riddick et al. 1986)
Enthalpy of Fusion, ΔH_{fus} (kJ/mol):
    14.414, 9.489   (*cis*-decalin, *trans*-decalin, Riddick et al. 1986)
    14.43   (*cis*-decalin, Chickos et al. 1999)
    2,13, 9.49   (*trans*-decalin, −57.05, 42.95°C, Chickos et al. 1999)
Entropy of Fusion, ΔS_{fus} (J/mol K):
    59.45, 52.1   (*cis*-decalin: exptl., calculated-group additivity method, Chickos et al. 1999)
    51.1, 52.1   (*trans*-decalin: exptl, calculated-group additivity method, total phase change entropy, Chickos et al. 1999)
Fugacity Ratio at 25°C, F: 1.0

Water Solubility (g/m³ or mg/L at 25°C):
    0.889   (shake flask-GC, Price 1976; quoted as more reliable value, Shaw 1989)
    6.21   (shake flask-GC, Hutchinson et al. 1980)
    1.99   (calculated-QSAR, Passino & Smith 1987)
    2.14   (calculated-molar volume, mp and mobile order thermodynamics, Ruelle & Kesselring 1997)

Vapor Pressure (Pa at 25°C or as indicated and reported temperature dependence equations. Additional data at other temperatures designated * are compiled at the end of this section.):
    133.3*   (23.3°C, isomer not specified, ebulliometry, measured range 23.3–150.0°C, Gardner & Brewer 1937)
    241*   (20°C, *cis*-decalin, manometry, measured range −29.5 to 194.7°C, Seyer & Mann 1945)
    381*   (13°C, *trans*-decalin, manometry, measured range −30.0 to 235.3°C, Seyer & Mann 1945)
    133.3*   (22.5°C, *cis*-decalin, summary of literature data, temp range 22.5–194.6°C, Stull 1947)
    666.6*   (30.6°C, *trans*-decalin, summary of literature data, temp range −0.80 to 186.7°C, Stull 1947)
    5529*   (99.883°C, *cis*-decalin, ebulliometry, measured range 99.883–196.376°C, Camin & Rossini 1955)
    5530*   (92.36°C, *trans*-decalin, ebulliometry, measured range 92.36–187.867°C, Camin & Rossini 1955)

log (P/mmHg) = 6.87529 − 1594.460/(203.392 + t/°C); temp range 99.8–196.4°C (*cis*-decalin, Antoine eq., ebulliometry-manometer measurement, Camin & Rossini 1955)

log (P/mmHg) = 6.85681 − 1564.683/(206.259 + t/°C); temp range 92.3–187.0°C (*trans*-decalin, Antoine eq., ebulliometry-manometer measurement, Camin & Rossini 1955)

104*, 164* (*cis*-decalin, *trans*-decalin, interpolated-Antoine eq., Zwolinski & Wilhoit 1971)

log (P/mmHg) = 6.87529 − 1594.460/(203.392 + t/°C); temp range 68.0–277.67°C (*cis*-decalin, Antoine eq., Zwolinski & Wilhoit 1971)

log (P/mmHg) = 6.85681 − 1564.683/(206.259 + t/°C); temp range 60.91–218.88°C (*trans*-decalin, Antoine eq., Zwolinski & Wilhoit 1971)

log (P/mmHg) = [−0.2185 × 10515.4/(T/K)] + 7.797540; temp range 22.5–194.6°C (*cis*-decalin, Antoine eq., Weast 1972–73)

log (P/mmHg) = [−0.2185 × 8749.1/(T/K)] + 6.973042; temp range −0.80 to 186.7°C (*trans*-decalin, Antoine eq., Weast 1972–73)

log (P/atm) = [1− 468.915/(T/K)] × 10^{0.683577 − 0.900942 × 10^{-4}·(T/K) + 2.28255 × 10^{-7}·(T/K)^2}; temp range 295.65–727.59 K (*cis*-decalin, Cox eq., Chao et al. 1983)

log (P/atm) = [1− 460.458/(T/K)] × 10^{0.880979 − 6.38749 × 10^{-4}·(T/K) + 4.59180 × 10^{-7}·(T/K)^2}; temp range: 365.51–461.02 K, (*trans*-decalin, Cox eq., Chao et al. 1983)

178, 104 (*cis*-decalin, calculated-Antoine eq., Boublik et al. 1984)

log (P/kPa) = 6.96043 − 2358.398/(280.79 + t/°C); temp range −29.5 to 194.7°C (*cis*-decalin, Antoine eq. from reported exptl. data, Boublik et al. 1984)

log (P/kPa) = 6.00042 − 1594.653/(203.415 + t/°C); temp range 99.88–196.4°C (*cis*-decalin, Antoine eq. from reported exptl. data of Camin & Rossini 1955, Boublik et al. 1984)

434, 165 (*trans*-decalin, calculated-Antoine eq., Boublik et al. 1984)

log (P/kPa) = 7.69594 − 3126.688/(363.012 + t/°C); temp range −30.0 to 253.3°C (*trans*-decalin, Antoine eq. from reported exptl. data, Boublik et al. 1984)

log (P/kPa) = 5.98704 − 1568.642/(206.726 + t/°C); temp range 92.36–187.9°C (*trans*-decalin, Antoine eq. from reported exptl. data of Camin & Rossini 1955, Boublik et al. 1984)

104, 168 (*cis*-, *trans*-decalin, Antoine eq., Dean 1985, 1992)

log (P/mmHg) = 6.87529 − 1594.81/(203.39 + t/°C), temp range 68–228°C (*cis*-decalin Antoine eq., Dean 1985, 1992)

log (P/mmHg) = 6.86581 − 1564.683/(206.26 + t/°C); temp range 61–219°C (*trans*-decalin Antoine eq., Dean 1985, 1992)

130 (mixed isomer, 23.3°C, lit. average, Riddick et al. 1986)

100, 164 (selected lit., *cis*-, *trans*-decalin, Riddick et al. 1986)

log (P/kPa) = 6.00019 − 1594.460/(203.392 + t/°C), temp range not specified (*cis*-decalin, Antoine eq., Riddick et al. 1986)

log (P/kPa) = 5.98171 − 1564.683/(206.259 + t/°C), temp range not specified (*trans*-decalin, Antoine eq., Riddick et al. 1986)

105, 168 (*cis*-decalin, *trans*-decalin, extrapolated-Antoine eq., Stephenson & Malanowski 1987)

log (P_L/kPa) = 6.00019 − 1595.176/(−69.622 + T/K); temp range 371–473 K (*cis*-decalin, Antoine eq., Stephenson & Malanowski 1987)

log (P_L/kPa) = 5.99363 − 1573.981/(−65.77 + T/K); temp range 363–461 K (*trans*-decalin, Antoine eq., Stephenson & Malanowski 1987)

log (P/mmHg) = 45.6345 − 4.21 × 10^3/(T/K) − 12.881·log (T/K) − 7.8083 × 10^{-11}·(T/K) + 2.8637 × 10^{-6}·(T/K)^2; temp range 230–702 K (*cis*-decalin, vapor pressure eq., Yaws 1994)

log (P/mmHg) = 76.1002 − 5.03 × 10^3/(T/K) − 25.078·log (T/K) + 9.7608 × 10^{-3}·(T/K) − 2.5814 × 10^{-6}·(T/K)^2; temp range 243–687 K (*trans*-decalin, vapor pressure eq., Yaws 1994)

128* (20.42°C, *trans*-decalin, differential pressure gauge, measured range −29.31 to 160.7°C, Mokbel et al. 1995)

Henry's Law Constant (Pa m^3/mol at 25°C or as indicated and reported temperature dependence equations. Additional data at other temperatures designated * are compiled at the end of this section.):

11855* (EPICS-GC/FID, measured range 10–30°C Ashworth et al. 1988)

ln [H/(atm·m^3/mol)] = 11.85 − 4125/(T/K); temp range 10–30°C (EPICS measurements, Ashworth et al. 1988)

Octanol/Water Partition Coefficient, log K_{OW}:
   4.79   (calculated-fragment const., Müller & Klein 1992)

Octanol/Air Partition Coefficient, log K_{OA}:

Bioconcentration Factor, log BCF:

Sorption Partition Coefficient, log K_{OC}:

Environmental Fate Rate Constants, k, and Half-Lives, $t_{½}$:
   Volatilization:
   Photolysis:
   Oxidation: rate constant k, for gas-phase second order rate constants, k_{OH} for reaction with OH radical, k_{NO3} with NO_3 radical and k_{O3} with O_3 or as indicated, *data at other temperatures see reference:
      k_{OH} = 1.96 × 10^{-11} cm^3 $molecule^{-1}$ s^{-1} of *cis*-decalin and 2.02 × 10^{-11} cm^3 $molecule^{-1}$ s^{-1} of *trans*-decalin at 299 K (Atkinson 1985)
      k_{OH} = 2.0 × 10^{-11} cm^3 $molecule^{-1}$ s^{-1} for the reaction at 298 K (Atkinson 1990)
   Hydrolysis:
   Biodegradation:
   Biotransformation:
   Bioconcentration, Uptake (k_1) and Elimination (k_2) Rate Constants or Half-Lives:

Half-Lives in the Environment:

TABLE 2.1.1.2.14.1
Reported vapor pressures and Henry's law constants of decalin (isomer not specified) at various temperatures and the coefficients for the vapor pressure equations

log P = A − B/(T/K)	(1)		ln P = A − B/(T/K)	(1a)
log P = A − B/(C + t/°C)	(2)		ln P = A − B/(C + t/°C)	(2a)
log P = A − B/(C + T/K)	(3)			
log P = A − B/(T/K) − C·log (T/K)	(4)			

Vapor pressure				Henry's law constant	
Gardner & Brewer 1937				Ashworth et al. 1988	
ebulliometry				EPICS-GC	
t/°C	P/Pa	t/°C	P/Pa	t/°C	H/(Pa m^3/mol)
23.3	133.3	60.0	987	10	7093
32.8	240	66.1	1293	15	8481
37.3	280	86.8	3520	20	10740
38.4	320	96.0	5186	25	11855
42.5	400	104.5	7293	30	20164
43.0	413	118.8	11652		
49.6	587	133.8	20278	ln H = A − B/(T/K)	
50.6	627	150.0	31997	eq. 4a	H/(atm m^3/mol)
54.3	747			A	11.85
54.6	800	bp/°C	193.8	B	4125
59.8	1027				

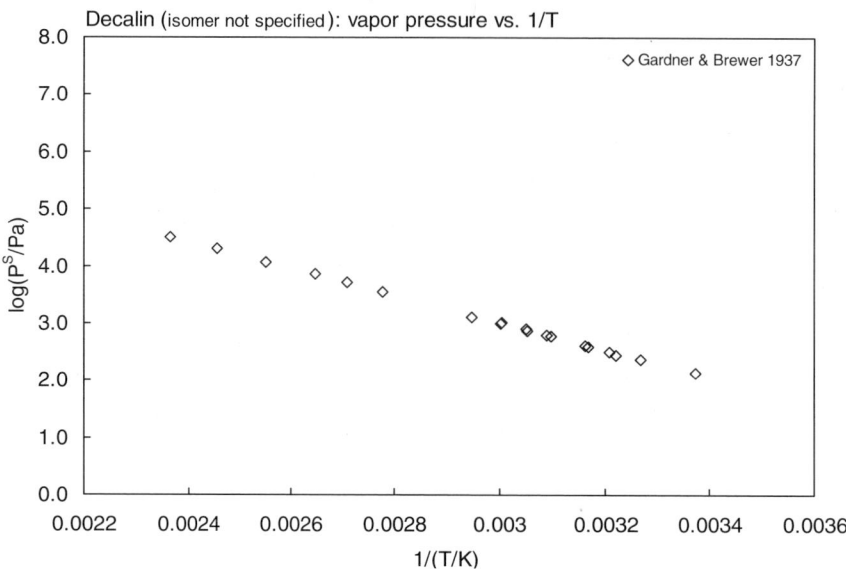

FIGURE 2.1.1.2.14.1 Logarithm of vapor pressure versus reciprocal temperature for decalin.

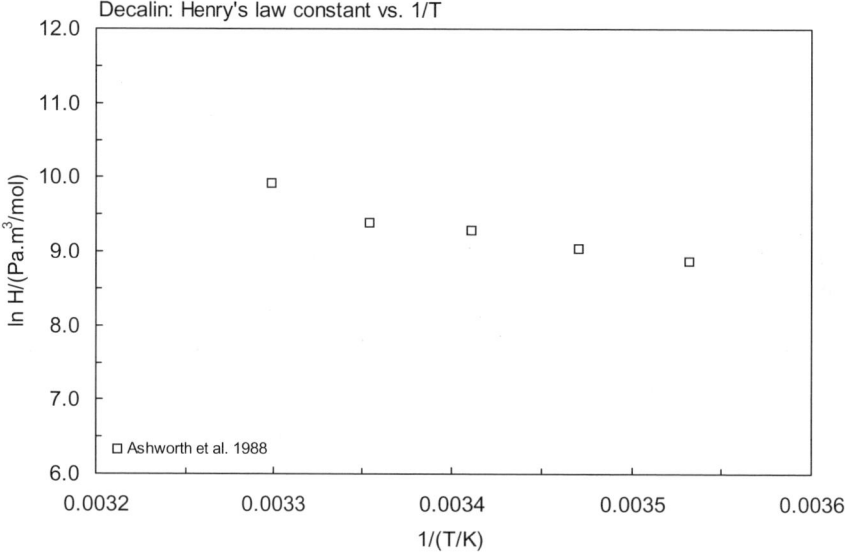

FIGURE 2.1.1.2.14.2 Logarithm of Henry's law constant versus reciprocal temperature for decalin.

Aliphatic and Cyclic Hydrocarbons

TABLE 2.1.1.2.14.2
Reported vapor pressures of *cis*-decalin at various temperatures and the coefficients for the vapor pressure equations

$\log P = A - B/(T/K)$ (1) $\ln P = A - B/(T/K)$ (1a)

$\log P = A - B/(C + t/°C)$ (2) $\ln P = A - B/(C + t/°C)$ (2a)

$\log P = A - B/(C + T/K)$ (3)

$\log P = A - B/(T/K) - C \cdot \log (T/K)$ (4) $\ln P = A - B/(T/K) - C \cdot \ln (T/K)$ (4a)

Seyer & Mann 1945		Stull 1947		Camin & Rossini 1955		Zwolinski & Wilhoit 1971	
manometry		summary of literature data		ebulliometry		selected values	
t/°C	P/Pa	t/°C	P/Pa	t/°C	P/Pa	t/°C	P/Pa
−29.5	51	22.5	133.3	99.883	5529	68.0	1333
−19.2	87	50.1	666.6	105.685	6941	82.64	2666
−10.0	115	64.2	1333	110.490	8331	91.97	4000
0.0	145	79.8	2666	114.152	9523	98.998	5333
12.0	196	97.2	5333	118.004	10943	104.609	6666
20.0	241	108.0	7999	123.132	13008	109.391	7999
38.0	457	123.2	3332	128.731	15827	117.280	10666
43.4	532	145.4	26664	135.021	19433	123.676	13332
49.9	704	169.9	53329	140.176	22881	135.934	19998
50.4	719	194.6	101325	147.456	28546	145.171	26664
52.9	805			154.245	34817	152.718	33331
56.9	971	mp/°C	−43.3	161.885	43164	159.127	39997
10.0	1095			170.056	53773	169.767	53329
70.0	1669			178.629	67068	178.349	66661
80.1	2590			187.823	83031	185.724	79993
92.4	4358			194.370	98080	192.251	93326
105.1	7239			195.055	99658	193.389	95992
109.8	8389			195.635	101001	194.622	98659
112.4	9121			196.376	102727	195.774	101325
124.6	13976						
148.7	29997			bp/°C	195.774		
172.7	57690			eq. 2	P/mmHg	eq. 2	P/mmHg
194.700	101093			A	6.87529	A	6.87529
				B	1594.460	B	1594.460
eq. 4a	P/cmHg			C	203.392	C	203.392
A	6.8139					bp/°C	195.774
B	1702.20					ΔH_V/(kJ mol⁻¹) = 39.30	
C	34.32						at bp
bp/°C	194.5						

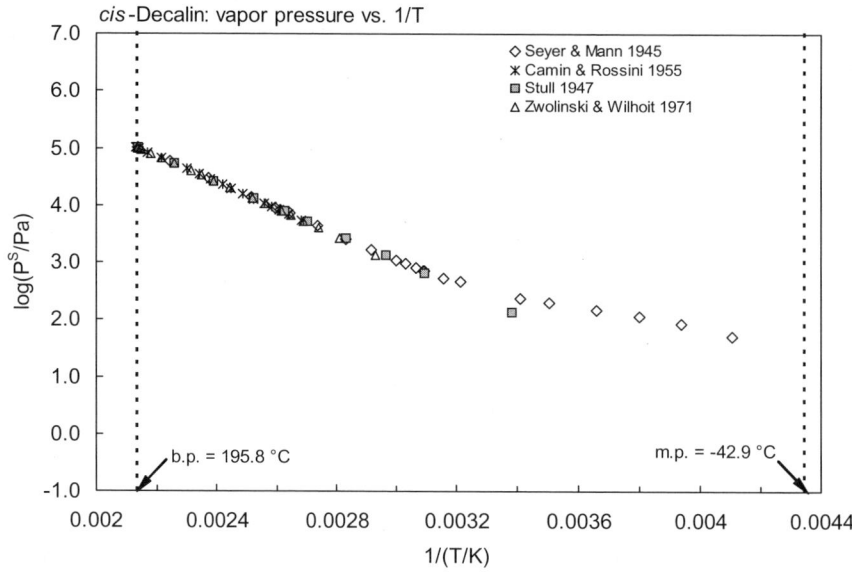

FIGURE 2.1.1.2.14.3 Logarithm of vapor pressure versus reciprocal temperature for *cis*-decalin.

TABLE 2.1.1.2.14.3
Reported vapor pressures of *trans*-decalin at various temperatures and the coefficients for the vapor pressure equations

$\log P = A - B/(T/K)$ (1) $\ln P = A - B/(T/K)$ (1a)

$\log P = A - B/(C + t/°C)$ (2) $\ln P = A - B/(C + t/°C)$ (2a)

$\log P = A - B/(C + T/K)$ (3)

$\log P = A - B/(T/K) - C \cdot \log(T/K)$ (4) $\ln P = A - B/(T/K) - C \cdot \ln(T/K)$ (4a)

1.

Seyer & Mann 1945		Stull 1947		Camin & Rossini 1955		Zwolinski & Wilhoit 1971	
manometry		summary of literature data		ebulliometry		selected values	
t/°C	P/Pa	t/°C	P/Pa	t/°C	P/Pa	t/°C	P/Pa
−30.0	59	−0.80	133.3	92.36	5530	60.91	1333
−24.1	80	30.6	666.6	98.129	6942	75.37	2666
−11.1	143	47.2	1333	102.891	8331	84.59	4000
−0.70	229	65.3	2666	106.500	9522	91.527	5333
0.0	241	85.7	5333	110.316	10946	97.072	6666
13.0	381	98.4	7999	115.356	13086	101.800	7999
30.9	801	114.6	13332	120.918	15828	109.599	10666
51.5	1597	136.2	26664	127.140	19434	115.922	13332
59.7	2221	160.1	53329	132.255	22434	128.045	19998
65.3	2657	186.7	101325	146.156	34817	137.182	26664
74.4	3562			153.719	43163	144.649	33331
83.5	5072	mp/°C	−30.7	161.801	53773	150.991	39997
95.5	7307			170.297	67069	161.522	53329
112.4	12635			179.395	84031	170.018	66661
119.4	16087			185.885	98081	177.320	79993
136.7	27478			186.563	99665	183.783	93326
152.3	42595			187.140	101005	184.911	95992
168.0	65024			187.867	102731	186.132	98659

TABLE 2.1.1.2.14.3 (Continued)

Seyer & Mann 1945		Stull 1947		Camin & Rossini 1955		Zwolinski & Wilhoit 1971	
manometry		summary of literature data		ebulliometry		selected values	
t/°C	P/Pa	t/°C	P/Pa	t/°C	P/Pa	t/°C	P/Pa
187.1	103009					187.274	101325
209.8	170060			bp/°C	187.273		
212.9	185496						
223.4	228787			eq. 2	P/mmHg	eq. 2	P/mmHg
235.3	284978			A	6.85681	A	6.83561
				B	1564.683	B	1564.683
bp/°C	185.8			C	206.259	C	206.269
eq. 4a	P/cmHg					bp/°C	187.274
A	6.8509					$\Delta H_V/(kJ\ mol^{-1}) = 38.50$	
B	2182.38					at bp	
C	32.64						

2.

Mokbel et al. 1995					
static method-manometry					
t/°C	P/Pa	t/°C	P/Pa	t/°C	P/Pa
−29.31	1.91	40.37	448	99.60	7347
−19.37	5.23	50.42	790	109.58	10670
−9.42	13.2	60.41	1329	119.69	15160
0.46	29.4	64.63	1636		
10.44	63.1	69.55	2050	data fitted to Wagner eq.	
20.42	128	79.48	3219		
30.35	245	89.54	4939		

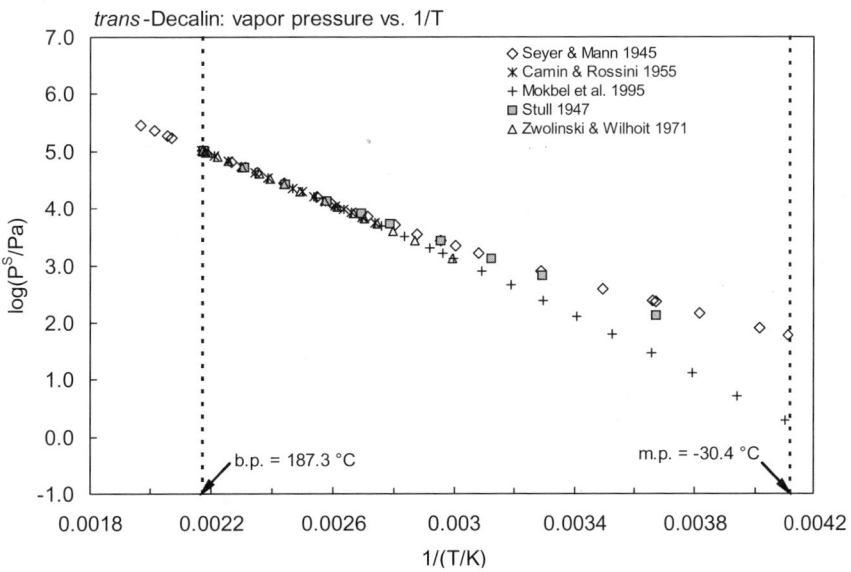

FIGURE 2.1.1.2.14.4 Logarithm of vapor pressure versus reciprocal temperature for *trans*-decalin.

2.1.2 UNSATURATED HYDROCARBONS

2.1.2.1 Alkenes

2.1.2.1.1 2-Methylpropene

Common Name: 2-Methylpropene
Synonym: isobutene, isobutylene
Chemical Name: 2-methylpropene
CAS Registry No: 115-11-7
Molecular Formula: C_4H_8; $CH_3C(CH_3)CH_2$
Molecular Weight: 56.107
Melting Point (°C):
    –140.7   (Lide 2003)
Boiling Point (°C):
    –6.9   (Dreisbach 1959; Lide 2003)
Density (g/cm³ at 20°C):
    0.5942, 0.5879   (20°C, 25°C, Dreisbach 1959)
Molar Volume (cm³/mol):
    94.4   (20°C, calculated-density, McAuliffe 1966)
    95.4   (25°C, calculated-density)
    88.8   (calculated-Le Bas method at normal boiling point)
Enthalpy of Fusion, ΔH_{fus} (kJ/mol):
    5.92   (Chickos et al. 1999)
Entropy of Fusion, ΔS_{fus} (J/mol K):
    44.72, 41.8   (exptl., calculated-group additivity method, total phase change entropy, Chickos et al. 1999)
Fugacity Ratio at 25°C, F: 1.0

Water Solubility (g/m³ or mg/L at 25°C):
    263   (shake flask-GC, of liquid at 1 atmospheric pressure, McAuliffe 1966)

Vapor Pressure (Pa at 25°C or as indicated and reported temperature dependence equations. Additional data at other temperatures designated * are compiled at the end of this section.):
    131695*, 131855   (0°C, static method-manometer, measured range –56.75 to 0°C, Lamb & Roper 1940)
    101325*   (–6.9°C, summary of literature data, temp range –105.1 to –6.9°C, Stull 1947)
    303700   (calculated from determined data, Dreisbach 1959)
    log (P/mmHg) = 6.84134 – 923.2/(240.0 + t/°C); temp range –68 to 39°C (Antoine eq. for liquid state, Dreisbach 1959)
    302642*   (extrapolated-Antoine eq., temp range –81.95 to 11.88°C, Zwolinski & Wilhoit 1971)
    log (P/mmHg) = 6.84134 – 923.200/(240.00 + t/°C); temp range –81.95 to 11.88°C (Antoine eq., Zwolinski & Wilhoit 1971)
    log (P/mmHg) = [–0.2185 × 5742.9/(T/K)] + 7.601563; temp range –105.1 to –6.90°C (Antoine eq., Weast 1972–73)
    log (P/mmHg) = 6.68466 – 886.25/(234.64 + t/°C); temp range –82 to 12°C (Antoine eq., Dean 1985, 1992)
    303255   (interpolated-Antoine eq.-II, Stephenson & Malanowski 1987)
    log (P_L/kPa) = 5.96624 – 923.2/(–33.15 + T/K); temp range 212–279 K (Antoine eq.-I, Stephenson & Malanowski 1987)
    log (P_L/kPa) = 5.93211 – 907.644/(–35.082 + T/K); temp range 266–313 K (Antoine eq.-II, Stephenson & Malanowski 1987)
    log (P_L/kPa) = 6.27428 – 1095.288/(–9.441 + T/K); temp range 310–376 K (Antoine eq.-III, Stephenson & Malanowski 1987)

Aliphatic and Cyclic Hydrocarbons

log (P_L/kPa) = 7.64267 – 2336.466/(160.311 + T/K); temp range 371–418 K (Antoine eq.-IV, Stephenson & Malanowski 1987)

log (P/mmHg) = 39.2295 – 2.1094 × 10^3/(T/K) – 12.567·log (T/K) + 7.7304 × 10^{-3}·(T/K) – 1.3659 × 10^{-6}·(T/K)^2; temp range 133–418 K (vapor pressure eq., Yaws 1994)

607940   (45.44°C, vapor-liquid equilibrium VLE data, Pasanen et al. 2004)

Henry's Law Constant (Pa m^3/mol at 25°C):

21600   (calculated-1/K_{AW}, C_W/C_A, reported as exptl., Hine & Mookerjee 1975)
23100, 10800   (calculated-group contribution, calculated-bond contribution, Hine & Mookerjee 1975)
14800   (calculated-P/C, Mackay & Shiu 1981)
35800   (calculated-MCI χ, Nirmalakhandan & Speece 1988)
20994   (calculated-vapor-liquid equilibrium (VLE) data, Yaws et al. 1991)

Octanol/Water Partition Coefficient, log K_{OW}:

0.64, 1.32   (quoted, calculated-molar volume V_M, Wang et al. 1992)

Bioconcentration Factor, log BCF:

Sorption Partition Coefficient, log K_{OC}:

Environmental Fate Rate Constants, k, and Half-Lives, $t_{1/2}$:

Volatilization:
Photolysis:
Oxidation: rate constant k, for gas-phase second order rate constants, k_{OH} for reaction with OH radical, k_{NO3} with NO_3 radical and k_{O3} with O_3 or as indicated, *data at other temperatures and/or the Arrhenius expression see reference:

k_{O3} = 6.2 × 10^{-18} cm^3 molecule^{-1} s^{-1} at room temp. (Hanst et al. 1958)
k_{O3} = 2.32 × 10^{-17} cm^3 molecule^{-1} s^{-1} at 30°C (flow system, Bufalini & Altshuller 1965)
k_{OH} = 6.46 × 10^{-11} cm^3 molecule^{-1} s^{-1} at 298 K (discharge flow system-MS, Morris & Niki 1971)
k_{O3} = 13.6 × 10^{-18} cm^3 molecule^{-1}·s^{-1} at room temp (Japar et al. 1974)
k_{O3} = 11.7 × 10^{-18} cm^3 molecule^{-1} s^{-1} (Huie & Herron 1975)
k_{OH} = (3.05 ± 0.31) × 10^{10} cm^3 M^{-1} s^{-1} at 305 ± 2 K (relative rate method, Winer et al. 1976)
$k_{O(3P)}$ = 1.60 × 10^{-11} cm^3 molecule^{-1} s^{-1} for the reaction with O(^{3}P) atoms (Singleton & Cvetanovic 1976; Atkinson & Pitts Jr. 1977; quoted, Gaffney & Levine 1979)
k_{OH}* = 5.07 × 10^{-11} cm^3 molecule^{-1} s^{-1} at 297.2 K, measured range 297–425 K (flash photolysis-resonance fluorescence, Atkinson & Pitts 1977)
k_{O3} = 11.7 × 10^{-18} cm^3 molecule^{-1} s^{-1} at 294 ± 2 K (chemiluminescence, Adeniji et al. 1981)
k_{OH} = 5.13 × 10^{-11} cm^3 molecule^{-1} s^{-1} at 295 K (relative rate method, Atkinson & Aschmann 1984)
k_{OH} = (6.46 ± 0.13) × 10^{-11} cm^3 molecule^{-1} s^{-1} at 298 ± 2 K (relative rate method, Ohta 1984)
k_{O3} = 5.14 × 10^{-11} cm^3 molecule^{-1} s^{-1} at 298 K (recommended, Atkinson 1989)
k_{OH} = 5.14 × 10^{-11} cm^3 molecule^{-1} s^{-1}; k_{O3} = 1.21 × 10^{-17} cm^3 molecule^{-1} s^{-1} (review, Atkinson 1990)
k_{OH} = 5.14 × 10^{-11} cm^3 molecule^{-1} s^{-1}, k_{NO3} = 31.3 × 10^{-12} cm^3 molecule^{-1} s^{-1} at 298 K (Sabljic & Güsten 1990)
k_{NO3} = (3.15 – 3.38) × 10^{-12} cm^3 molecule^{-1} s^{-1} at 298 K (Atkinson 1990)
k_{OH}* = 5.14 × 10^{-11} cm^3 molecule^{-1} s^{-1}, k_{NO3} = 3.32 × 10^{-13} cm^3 molecule^{-1} s^{-1}, k_{O3}* = 11.3 × 10^{-18} cm^3 molecule^{-1} s^{-1} and $k_{O(3P)}$ = 1.69 × 10^{-11} cm^3 molecule^{-1} s^{-1} for reaction with O(^{3}P) atom at 298 K (recommended, Atkinson 1997)

Hydrolysis:
Biodegradation:
Biotransformation:
Bioconcentration, Uptake (k_1) and Elimination (k_2) Rate Constants or Half-Lives:

Half-Lives in the Environment:

Air: atmospheric lifetime was estimated to be 5.3 h, based on photooxidation rate constant k = 5.14 × 10^{-11} cm^3·molecule^{-1} s^{-1} with OH radicals in air during summer daylight (Altshuller 1991).

Surface water: $t_{1/2}$ ~ 320 h and 9×10^4 d for oxidation by OH and RO_2 radicals for olefins and $t_{1/2}$ = 8.0 d for substituted olefins, based on rate constant k = 1×10^6 M^{-1} s^{-1} for oxidation by singlet oxygen in aquatic system (Mill & Mabey 1985).

TABLE 2.1.2.1.1.1
Reported vapor pressures of 2-methylpropene (isobutene) at various temperatures and the coefficients for the vapor pressure equations

$$\log P = A - B/(T/K) \quad (1)$$
$$\log P = A - B/(C + t/°C) \quad (2)$$
$$\log P = A - B/(C + T/K) \quad (3)$$
$$\log P = A - B/(T/K) - C \cdot \log(T/K) \quad (4)$$
$$\log P = A - B/(T/K) - C \cdot (T/K) \quad (5)$$
$$\ln P = A - B/(T/K) \quad (1a)$$
$$\ln P = A - B/(C + t/°C) \quad (2a)$$

Lamb & Roper 1940		Stull 1947		Zwolinski & Wilhoit 1971			
static method-manometer		summary of literature data		selected values			
t/°C	P/Pa	t/°C	P/Pa	t/°C	P/Pa	t/°C	P/Pa
–56.75	8725	–105.1	133.3	–81.95	1333	–8.983	93326
–25.30	47703	–96.5	666.6	–73.37	2666	–8.274	95992
–20.88	57462	–81.9	1333	–67.90	4000	–7.580	98659
–7.47	99791	–73.4	2666	–63.79	5333	–6.900	101325
–0.67	128789	–63.8	5333	–60.472	6666	25.0	302642
0.0	131695	–57.7	7999	–57.664	7999		
0.0	131855	–49.3	13332	–53.051	10666		
		–36.7	26664	–49.309	13332	eq. 2	P/mmHg
bp/°C	–7.12	–22.2	53329	–42.111	19998	A	6.84134
eq. 5	P/mmHg	–6.9	101325	–36.666	26664	B	923.200
A	9.77465			–32.231	33331	C	240.000
B	1503.866	mp/°C	–140.3	–28.462	39997	bp/°C	–6.90
C	0.0046649			–22.227	53329	ΔH_V/(kJ mol^{-1}) =	
				–17.133	66661	at 25°C	20.59
				–12.789	79993	at bp	22.12

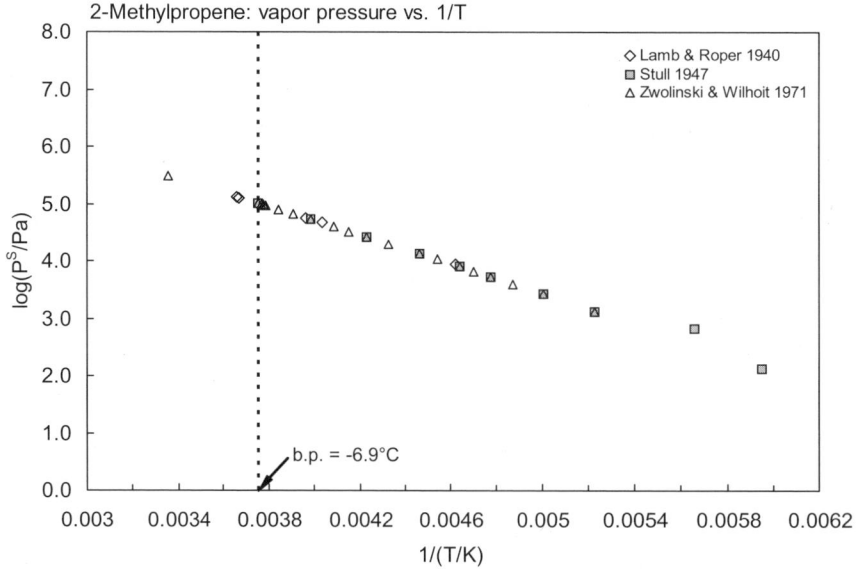

FIGURE 2.1.2.1.1.1 Logarithm of vapor pressure versus reciprocal temperature for 2-methylpropene.

Aliphatic and Cyclic Hydrocarbons

2.1.2.1.2   1-Butene

Common Name: 1-Butene
Synonym: butylene
Chemical Name: 1-butene
CAS Registry No: 106-98-9
Molecular Formula: C_4H_8; $CH_3CH_2CHCH_2$
Molecular Weight: 56.107
Melting Point (°C):
    –185.34 (Lide 2003)
Boiling Point (°C):
    –6.26   (Dreisbach 1959; Stephenson & Malanowski 1987; Lide 2003)
Density (g/cm³ at 20°C):
    0.5951, 0.5888 (20°C, 25°C, at saturation pressure, Dreisbach 1959)
Molar Volume (cm³/mol):
    94.3   (20°C, calculated-density, McAuliffe 1966; Ruelle & Kesselring 1997)
    95.3   (25°C, calculated-density)
    88.8   (calculated-Le Bas method at normal boiling point)
Enthalpy of Fusion, ΔH_{fus} (kJ/mol):
    3.85   (Chickos et al. 1999)
Entropy of Fusion, ΔS_{fus} (J/mol K):
    43.84, 47.3 (exptl., calculated-group additivity method, total phase change entropy, Chickos et al. 1999)
Fugacity Ratio at 25°C, F: 1.0

Water Solubility (g/m³ or mg/L at 25°C):
    222   (shake flask-GC, liquid at 1 atmospheric pressure, McAuliffe 1966)

Vapor Pressure (Pa at 25°C or as indicated and reported temperature dependence equations. Additional data at other temperatures designated * are compiled at the end of this section.):
    128536*   (0°C, static method-manometer, measured range –56.75 to 0°C, Lamb & Roper 1940)
    $\log (P/mmHg) = -1330.977/(T/K) - 0.0017607 \cdot (T/K) + 8.33816$; temp range 195–274 K (static method, Lamb & Roper 1940)
    361100*   (Antoine eq. regression, temp range –104.8 to –6.3°C, Stull 1947)
    296000   (calculated-Antoine eq., Dreisbach 1959)
    $\log (P/mmHg) = 6.84290 - 926.1/(240.0 + t/°C)$; temp range –67 to 40°C (Antoine eq. for liquid state, Dreisbach 1959)
    297309*   (derived from compiled data, Zwolinski & Wilhoit 1971)
    $\log (P/mmHg) = 6.84290 - 926.10/(240.00 + t/°C)$; temp range –81.5 to 12.6°C (Antoine eq., Zwolinski & Wilhoit 1971)
    $\log (P/mmHg) = [-0.2185 \times 5996.7/(T/K)] + 7.826754$; temp range –104.8 to –6.3°C (Antoine eq., Weast 1972–73)
    $\log (P/mmHg) = 6.531 - 810.261/(228.066 + t/°C)$ (Antoine eq. from reported exptl. data, Boublik et al. 1984)
    $\log (P/mmHg) = 6.79290 - 908.80/(238.54 + t/°C)$; temp range –82 to 13°C (Antoine eq., Dean 1985, 1992)
    297020, 295800 (interpolated-Antoine eq.-III, V, Stephenson & Malanowski 1987)
    $\log (P_L/kPa) = 5.9678 - 926.1/(-33.15 + T/K)$; temp range 200–274 K (Antoine eq.-I, Stephenson & Malanowski 1987)
    $\log (P_L/kPa) = 8.1706 - 1601.52/(7.059 + T/K)$; temp range 126–192 K (Antoine eq.-II, Stephenson & Malanowski 1987)
    $\log (P_L/kPa) = 6.05416 - 970.771/(-27.089 + T/K)$; temp range 267–345 K (Antoine eq.-III, Stephenson & Malanowski 1987)
    $\log (P_L/kPa) = 6.77294 - 1482/801/(48.073 + T/K)$; temp range 342–411 K (Antoine eq.-IV, Stephenson & Malanowski 1987)
    $\log (P_L/kPa) = 6.27411 - 1097.171/(-9.657 + T/K)$; temp range 267–411 K (Antoine eq.-V, Stephenson & Malanowski 1987)

log (P/mmHg) = 27.3116 −1.9235 × 10³/(T/K) −70.2064·log (T/K) + 7.4852 × 10⁻¹²·(T/K) + 3.6481 × 10⁻⁶·(T/K)²; temp range 88–420 K (vapor pressure eq., Yaws 1994)

596140 (50.12°C, vapor-liquid equilibrium VLE data, Pasanen et al., 2004)

Henry's Law Constant (Pa m³/mol at 25°C):
25610 (calculated-P/C, Mackay & Shiu 1981)
25370 (calculated-1/K_{AW}, C_W/C_A, reported as exptl., Hine & Mookerjee 1975)
26560, 15280 (calculated-group contribution, calculated-bond contribution, Hine & Mookerjee 1975)
29800 (calculated-MCI χ, Nirmalakhandan & Speece 1988)
24800 (calculated-vapor-liquid equilibrium (VLE) data, Yaws et al. 1991)

Octanol/Water Partition Coefficient, log K_{OW}:
2.40; 2.17, 2.26, 2.43 (quoted; calculated-f const., Rekker 1977)
1.59, 1.32 (quoted, calculated-molar volume V_M, Wang et al. 1992)

Octanol/Air Partition Coefficient, log K_{OA}:

Bioconcentration Factor, log BCF:

Sorption Partition Coefficient, log K_{OC}:

Environmental Fate Rate Constants, k, and Half-Lives, $t_{½}$:
Volatilization:
Photolysis:
Oxidation: rate constant k, for gas-phase second order rate constants, k_{OH} for reaction with OH radical, k_{NO3} with NO_3 radical and k_{O3} with O_3 or as indicated, *data at other temperatures and/or the Arrhenius expression see reference:

k_{O3} = 1.03 × 10⁻¹⁷ cm³ molecule⁻¹ s⁻¹ at 30°C (flow system, Bufalini & Altshuller 1965)
k_{OH} = 4.08 × 10⁻¹¹ cm³ molecule⁻¹ s⁻¹ at 298 K (discharge flow system-MS, Morris & Niki 1971)
k_{O3} = 1.23 × 10⁻¹⁷ cm³ molecule⁻¹ s⁻¹ at room temp (Japar et al. 1974)
k_{O3} = 1.03 × 10⁻¹⁷ cm³ molecule⁻¹ s⁻¹ at room temp (Huie & Herron 1975)
k_{OH} = (2.94, 2.96) × 10⁻¹² cm³ molecule⁻¹ s⁻¹ at 298 K (flash photolysis-resonance fluorescence, Ravishankara et al. 1978)
k_{O3} = 1.26 × 10⁻¹⁷ cm³ molecule⁻¹ s⁻¹ at 294 ± 2 K (static system-chemiluminescence, Adeniji et al. 1981)
$k_{O(3P)}$ = 4.20 × 10⁻¹² cm³ molecule⁻¹ s⁻¹ for the reaction with O(³P) atom (Singleton & Cvetanovic 1976; Atkinson & Pitts Jr. 1977; quoted, Gaffney & Levine 1979)
k_{OH}* = 3.53 × 10⁻¹¹ cm³ molecule⁻¹ s⁻¹ at 297.2 K, measured range 297–425 K (flash photolysis-resonance fluorescence, Atkinson & Pitts, Jr. 1977)
k_{OH} = 3.13 × 10⁻¹¹ cm³ molecule⁻¹ s⁻¹ at 295 K (relative rate method, Atkinson & Aschmann 1984)
k_{OH} = 3.14 × 10⁻¹¹ cm³ molecule⁻¹ s⁻¹ at 298 K (recommended, Atkinson 1989)
k_{OH} = 3.14 × 10⁻¹¹ cm³ molecule⁻¹ s⁻¹, k_{NO3} = 1.10 × 10⁻¹⁷ cm³ molecule⁻¹ s⁻¹ at 298 K (Atkinson 1990)
k_{OH} = 3.14 × 10⁻¹¹ cm³·molecule⁻¹ s⁻¹, k_{NO3} = 1.23 × 10⁻¹⁷ cm³ molecule⁻¹ s⁻¹ at 298 K (Sabljic & Güsten 1990)
k_{OH} = 3.14 × 10⁻¹¹ cm³ molecule⁻¹ s⁻¹ with an estimated lifetime of 5.5 h during summer daylight hours (Altshuller 1991)
k_{OH}* = 3.14 × 10⁻¹¹ cm³ molecule⁻¹ s⁻¹, k_{NO3}* = 1.35 × 10⁻¹⁴ cm³ molecule⁻¹ s⁻¹, and k_{O3}* = 9.64 × 10⁻¹⁸ cm³ molecule⁻¹ s⁻¹, and $k_{O(3P)}$ = 4.15 × 10⁻¹² cm³ molecule⁻¹ s⁻¹ for reaction with O(³P) atom at 298 K (recommended, Atkinson 1997)

Hydrolysis:
Biodegradation:
Biotransformation:
Bioconcentration, Uptake (k_1) and Elimination (k_2) Rate Constants or Half-Lives:

Half-Lives in the Environment:
Air: atmospheric lifetime was estimated to be 5.5 h, based on the reaction rate constant k = 3.14 × 10⁻¹¹ cm³ molecule⁻¹ s⁻¹ with OH radicals during summer daylight in the gas phase (Altshuller 1991).

Aliphatic and Cyclic Hydrocarbons

Surface water: $t_{1/2}$ = 320 h and 9 × 10^4 d for reaction with OH and RO$_2$ radicals of olefins in aquatic system, and $t_{1/2}$ = 7.3 d, based on oxidation reaction rate constant k = 3 × 10^3 M^{-1} s^{-1} with singlet O$_2$ for unsubstituted olefins in aquatic system (Mill & Mabey 1985).

TABLE 2.1.2.1.2.1
Reported vapor pressures of 1-butene at various temperatures and the coefficients for the vapor pressure equations

$\log P = A - B/(T/K)$ (1) $\ln P = A - B/(T/K)$ (1a)
$\log P = A - B/(C + t/°C)$ (2) $\ln P = A - B/(C + t/°C)$ (2a)
$\log P = A - B/(C + T/K)$ (3)
$\log P = A - B/(T/K) - C \cdot \log (T/K)$ (4)
$\log P = A - B/(T/K) - C \cdot (T/K)$ (5)

Lamb & Roper 1940		Stull 1947		Zwolinski & Wilhoit 1971			
static method-manometer		summary of literature data		selected values			
t/°C	P/Pa	t/°C	P/Pa	t/°C	P/Pa	t/°C	P/Pa
−56.75	8549	−104.8	133.3	−81.5	1333	−8.35	93326
−25.3	45423	−89.4	666.6	−72.89	2666	−7.64	95992
−22.91	50436	−81.6	1333	−67.41	4000	−6.94	98659
−10.60	85513	−73.0	2666	−63.29	5333	−6.26	101325
−7.47	96992	−63.4	5333	−59.96	6666	25.0	297309
−0.67	125549	−57.2	7999	−57.15	7999		
0.0	128536	−48.9	13332	−52.52	10666	eq. 2	P/mmHg
		−36.2	26664	−48.77	13332	A	6.84290
bp/°C	−6.30	−21.7	53329	−41.56	19998	B	926.10
eq. 5	P/mmHg	−6.3	101325	−36.10	26664	C	240.00
A	6.33816			−31.65	33331	bp/°C	−6.26
B	1330.977	mp/°C	−130	−27.87	39997		
C	0.0017607			−21.62	53329	ΔH$_V$/(kJ mol^{-1}) =	
				−16.52	66661	at 25°C	20.13
				−12.16	79993	at bp	21.92

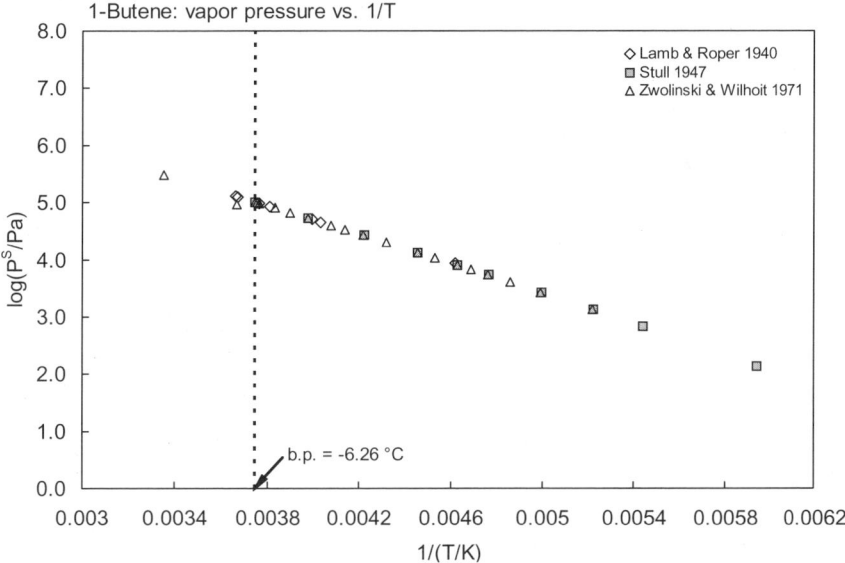

FIGURE 2.1.2.1.2.1 Logarithm of vapor pressure versus reciprocal temperature for 1-butene.

2.1.2.1.3   2-Methyl-1-butene

Common Name: 2-Methyl-1-butene
Synonym:
Chemical Name: 2-methyl-1-butene
CAS Registry No: 563-46-2
Molecular Formula: C_5H_{10}, $CH_3CH_2C(CH_3)=CH_2$
Molecular Weight: 70.133
Melting Point (°C):
    –137.53   (Lide 2003)
Boiling Point (°C):
    31.2   (Lide 2003)
Density (g/cm³ at 20°C):
    0.6504, 0.6451   (20°C, 25°C, at saturation pressure, Dreisbach 1959; Dean 1985)
Molar Volume (cm³/mol):
    107.8, 108.7   (20°C, 25°C, calculated-density)
    111.0   (calculated-Le Bas method at normal boiling point)
Enthalpy of Fusion, ΔH_{fus} (kJ/mol):
    7.91   (Chickos et al. 1999)
Entropy of Fusion, ΔS_{fus} (J/mol K):
    58.34, 48.9   (exptl., calculated-group additivity method, total phase change entropy, Chickos et al. 1999)
Fugacity Ratio at 25°C, F: 1.0

Water Solubility (g/m³ or mg/L at 25°C or as indicated. Additional data at other temperatures designated * are compiled at the end of this section.):

    215*   (20°C, shake flask-GC, measured range 20–60°C, Pavlova et al. 1966)
    155   (estimated-nomograph of Kabadi & Danner 1979; Brookman et al. 1985)
    130   (misquoted from 3-methyl-1-butene, Wakita et al. 1986)
    168   (calculated-fragment solubility constants, Wakita et al. 1986)
    260   (calculated-regression eq. of Lyman et al. 1982, Wang et al. 1992)
    137   (calculated-molar volume V_M, Wang et al. 1992)
    128, 198   (calculated-molar volume, mp and mobile order thermodynamics, Ruelle & Kesselring 1997)

Vapor Pressure (Pa at 25°C or as indicated and reported temperature dependence equations. Additional data at other temperatures designated * are compiled at the end of this section.):

    135540*   (interpolated-Antoine eq. regression, temp range –89.1 to 20.2°C, Stull 1947)
    70242*   (20.996°C, ebulliometry, measured range 1.155–62.675°C, Scott et al. 1949)
    $\log (P/\text{mmHg}) = 6.87314 - 1053.780/(232.768 + t/°C)$; temp range 1.155 to 62.675°C (Antoine eq., ebulliometry, Scott et al. 1949)
    81320   (calculated from determined data, Dreisbach 1959)
    $\log (P/\text{mmHg}) = 6.87314 - 1053.8/(233.0 + t/°C)$; temp range –38 to 75°C (Antoine eq. for liquid state, Dreisbach 1959)
    81327*, 81330   (derived from compiled data, interpolated-Antoine eq., temp range –53.4 to 52.24°C Zwolinski & Wilhoit 1971)
    $\log (P/\text{mmHg}) = 6.87314 - 1053.780/(232.788 + t/°C)$; temp range –53.4 to 52.24°C (Antoine eq., Zwolinski & Wilhoit 1971)
    $\log (P/\text{mmHg}) = [-0.2185 \times 6474.6/(T/K)] + 7.751419$; temp range –89.1 to 20.2°C (Antoine eq., Weast 1972–73)
    82830   (interpolated-Antoine eq., Boublik et al. 1984)
    $\log (P/\text{kPa}) = 5.98834 - 1046.771/(232.181 + t/°C)$ temp range 1.115–63.68°C (Antoine eq. from reported exptl. data, Boublik et al. 1984)
    99500   (interpolated-Antoine eq., temp range –53 to 52°C, Dean 1985, 1992)
    $\log (P/\text{mmHg}) = 6.84637 - 1039.69/(236.65 + t/°C)$; temp range –53 to 52°C (Antoine eq., Dean 1985, 1992)
    81360   (interpolated-Antoine eq., Stephenson & Malanowski 1987)

Aliphatic and Cyclic Hydrocarbons

log (P_L/kPa) = 5.99292 − 1050.937/(−40.727 + T/K); temp range 240–336 K (Antoine eq., Stephenson & Malanowski 1987)

log (P/mmHg) = 30.2418 −2.2723 × 10^3/(T/K) − 8.1482·log (T/K) + 5.2331 × 10^{-11}·(T/K) + 3.6802 × 10^{-6}·(T/K)^2; temp range 136–465 K (vapor pressure eq., Yaws 1994)

Henry's Law Constant (Pa m^3/mol at 25°C):

  43080   (calculated-MCI χ, Nirmalakhandan & Speece 1988)

Octanol/Water Partition Coefficient, log K_{OW}:

  2.07   (calculated-regression of Lyman et al. 1982, Wang et al. 1992)
  1.89   (calculated-molar volume V_M, Wang et al. 1992)

Octanol/Air Partition Coefficient, log K_{OA}:

Bioconcentration Factor, log BCF:

Sorption Partition Coefficient, log K_{OC}:

Environmental Fate Rate Constants, k, and Half-Lives, $t_{1/2}$:

  Volatilization:
  Photolysis:
  Oxidation: rate constant k, for gas-phase second order rate constants, k_{OH} for reaction with OH radical, k_{NO3} with NO$_3$ radical and k_{O3} with O$_3$ or as indicated, *data at other temperatures and/or the Arrhenius expression see reference:

  k_{OH} = 9.01 × 10^{-11} cm^3 molecule^{-1} s^{-1} at 298 K (discharge flow system-MS, Morris & Niki 1971)
  k_{OH} = (6.37 ± 0.16) × 10^{-11} cm^3 molecule^{-1} s^{-1} at 298 ± 2 K (relative rate method, Ohta 1984)
  k_{OH} = 9.01 × 10^{-11} cm^3 molecule^{-1} s^{-1} to 6.07 × 10^{-11} cm^3 molecule^{-1} s^{-1} at 298 K (review, Atkinson 1985)
  k_{OH} = (60.7 − 90.1) × 10^{-12} cm^3 molecule^{-1} s^{-1} at 298 K (review, Atkinson 1989)
  k_{OH} = 6.10 × 10^{-11} cm^3 molecule^{-1} s^{-1} with an estimated atmospheric lifetime of 2.3 h during summer daylight hours (Altshuller 1991)
  k_{OH} = 6.1 × 10^{-11} cm^3 molecule^{-1} s^{-1}, and k_{O3} = 16.0 × 10^{-18} cm^3 molecule^{-1} s^{-1} at 298 K (recommended, Atkinson 1997)

  Hydrolysis:
  Biodegradation:
  Biotransformation:
  Bioconcentration, Uptake (k_1) and Elimination (k_2) Rate Constants or Half-Lives:

Half-Lives in the Environment:

  Air: atmospheric lifetime τ ~ 2.3 h, based on the photooxidation rate constant k = 6.10 × 10^{-11} cm^3 molecule^{-1} s^{-1} with hydroxyl radical in air during summer daylight (Altshuller 1991).
  Surface water: $t_{1/2}$ = 320 h and 9 × 10^4 d for reaction with OH and RO$_2$ radicals for olefins in aquatic system, and $t_{1/2}$ = 7.3 d, based on oxidation reaction rate constant of 3 × 10^3 M^{-1} s^{-1} with singlet oxygen for unsubstituted olefins in aquatic system (Mill & Mabey 1985).

TABLE 2.1.2.1.3.1
Reported aqueous solubilities of 2-methyl-1-butene at various temperatures

Pavlova et al. 1966	
Shake flask-GC	
t/°C	S/g·m^{-3}
20	215
40	326
50	250
60	267

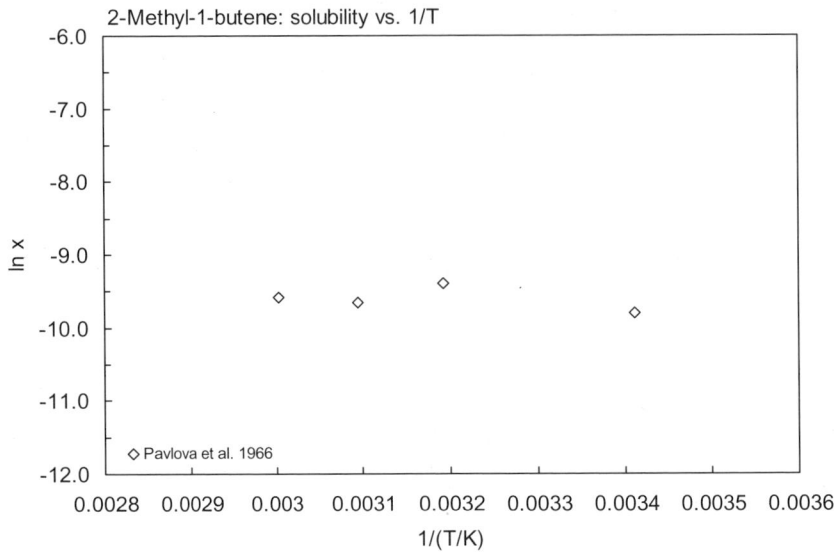

FIGURE 2.1.2.1.3.1 Logarithm of mole fraction solubility (ln x) versus reciprocal temperature for 2-methyl-1-butene.

TABLE 2.1.2.1.3.2
Reported vapor pressures of 2-methyl-1-butene at various temperatures and the coefficients for the vapor pressure equations

$\log P = A - B/(T/K)$ (1) $\ln P = A - B/(T/K)$ (1a)
$\log P = A - B/(C + t/°C)$ (2) $\ln P = A - B/(C + t/°C)$ (2a)
$\log P = A - B/(C + T/K)$ (3)
$\log P = A - B/(T/K) - C \cdot \log (T/K)$ (4)

Stull 1947		Scott et al. 1949		Zwolinski & Wilhoit 1971			
summary of literature data		ebulliometry		selected values			
t/°C	P/Pa	t/°C	P/Pa	t/°C	P/Pa	t/°C	P/Pa
–89.1	133.3	1.155	31163	–53.4	1333	eq. 2	P/mmHg
–72.8	666.6	6.054	38547	–43.7	2666	A	6.87314
–64.3	1333	10.993	47357	–37.5	4000	B	1053.780
–54.8	2666	15.973	57800	–32.87	5333	C	232.788
–44.1	5333	20.996	70242	–29.13	6666	bp/°C	31.163
–37.3	7999	26.062	84534	–25.96	7999	$\Delta H_V/(kJ\ mol^{-1})$ =	
–28.0	13332	31.162	101319	–20.76	10666	at 25°C	25.86
–13.8	26664	36.308	120792	–16.54	13332	at bp	25.50
2.5	53329	41.500	143281	–8.44	19998		
20.2	101325	46.728	169952	–2.308	26664		
		52.005	198556	2.683	33331		
mp/°C	–135	57.320	232100	6.924	39997		
		62.675	270083	13.936	53329		
				19.865	66661		
		bp/°C	31.16	24.546	79993		
		Antoine eq.		28.623	93326		
		eq. 2	P/mmHg	29.620	95992		
		A	6.87314	30.400	98659		
		B	1052.780	31.163	101325		
		C	232.788	25.0	81327		

Aliphatic and Cyclic Hydrocarbons

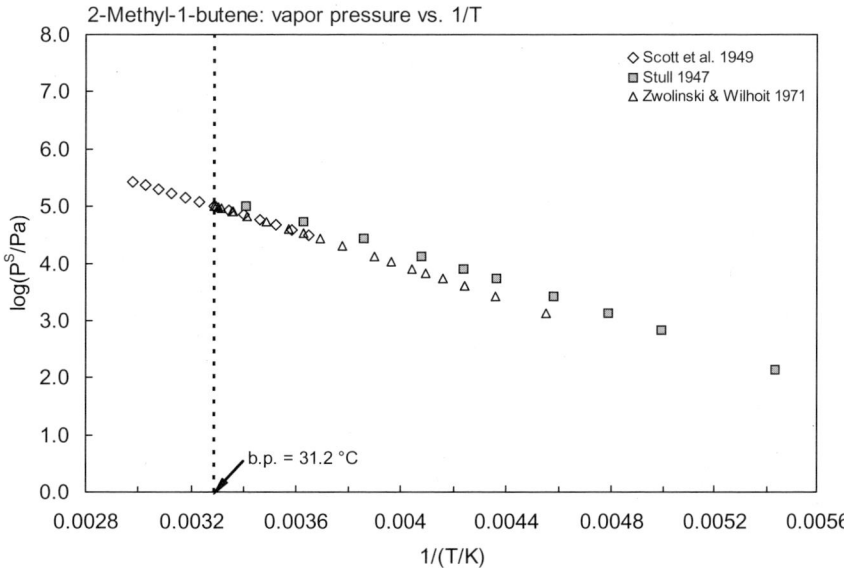

FIGURE 2.1.2.1.3.2 Logarithm of vapor pressure versus reciprocal temperature for 2-methyl-1-butene.

2.1.2.1.4 3-Methyl-1-butene

Common Name: 3-Methyl-1-butene
Synonym:
Chemical Name: 3-methyl-1-butene
CAS Registry No: 563-45-1
Molecular Formula: C_5H_{10}, $(CH_3)_2CCHCH=CH_2$
Molecular Weight: 70.133
Melting Point (°C):
    –168.43   (Lide 2003)
Boiling Point (°C):
    20.1   (Lide 2003)
Density (g/cm³ at 20°C):
    0.6272, 0.6219   (20°C, 25°C, at saturation pressure, Dreisbach 1959)
Molar Volume (cm³/mol):
    111.8   (20°C, calculated-density, McAuliffe 1966; Ruelle & Kesselring 1997)
    111.0   (calculated-Le Bas method at normal boiling point)
Enthalpy of Fusion, ΔH_{fus} (kJ/mol):
    5.36   (Chickos et al. 1999)
Entropy of Fusion, ΔS_{fus} (J/mol K):
    51.19, 41.4   (exptl., calculated-group additivity method, total phase change entropy, Chickos et al. 1999)
Fugacity Ratio at 25°C, F: 1.0

Water Solubility (g/m³ or mg/L at 25°C):
    130   (shake flask-GC, McAuliffe 1966)

Vapor Pressure (Pa at 25°C or as indicated and reported temperature dependence equations. Additional data at other temperatures designated * are compiled at the end of this section.):
    101325   (20.2°C, summary of literature data, temp range –89.1 to 20.2°C, Stull 1947)
    120790*   (25.128°C, ebulliometry, measured range 0.218–51.139°C, Scott & Waddington 1950)
    log (P/mmHg) = 6.82618 – 1013.474/(236.816 + t/°C); temp range 0.219–51.139°C (Antoine eq., ebulliometric method, Scott & Waddington 1950)
    120260   (calculated-Antoine eq., Dreisbach 1959)
    log (P/mmHg) = 6.82618 – 1013.474/(237.0 + t/°C); temp range –47 to 60°C (Antoine eq. for liquid state, Dreisbach 1959)
    120000   (interpolated-Antoine eq., Zwolinski & Wilhoit 1971)
    120270*   (derived from compiled data, temp range –62.9 to 40.84°C, Zwolinski & Wilhoit 1971)
    log (P/mmHg) = 6.82618 – 1013.474/(236.816 + t/°C); temp range –62.9 to 40.84°C (Antoine eq., Zwolinski & Wilhoit 1971)
    120180   (interpolated-Antoine eq., temp range –63 to 41°C, Dean 1985, 1992)
    log (P/mmHg) = 6.82455 – 1012.37/(236.65 + t/°C); temp range –63 to 41°C (Antoine eq., Dean 1985, 1992)
    120300   (interpolated-Antoine eq., Stephenson & Malanowski 1987)
    log (P_L/kPa) = 5.94656 – 1010.866/(–36.694 + T/K); temp range 237–324 K (Antoine eq., Stephenson & Malanowski 1987)
    log (P/mmHg) = 31.1486 – 2.1764 × 10³/(T/K) –8.6146·log (T/K) + 5.9672 × 10⁻¹¹·(T/K) + 4.7555 × 10⁻⁶·(T/K)²; temp range 105–450 K (vapor pressure eq., Yaws 1994)

Henry's Law Constant (Pa m³/mol at 25°C):
    54230   (calculated-1/K_{AW}, C_W/C_A, reported as exptl., Hine & Mookerjee 1975)
    63715, 22610   (calculated-group contribution, calculated-bond contribution, Hine & Mookerjee 1975)
    54700   (calculated-P/C, Mackay & Shiu 1981)
    43080   (calculated-MCI χ, Nirmalakhandan & Speece 1988)

52940   (calculated-vapor-liquid equilibrium (VLE) data, Yaws et al. 1991)

Octanol/Water Partition Coefficient, log K_{OW}:
   2.07   (calculated-regression of Lyman et al. 1982, Wang et al. 1992)
   2.05   (calculated-molar volume V_M, Wang et al. 1992)

Octanol/Air Partition Coefficient, log K_{OA}:

Bioconcentration Factor, log BCF:

Sorption Partition Coefficient, log K_{OC}:

Environmental Fate Rate Constants, k, and Half-Lives, $t_{½}$:
   Volatilization:
   Photolysis:
   Oxidation: rate constant k, for gas-phase second order rate constants, k_{OH} for reaction with OH radical, k_{NO3} with NO_3 radical and k_{O3} with O_3 or as indicated, *data at other temperatures and/or the Arrhenius expression see reference:
      k_{OH} = (3.10 ± 0.31) × 10^{-11} cm^3 $molecule^{-1}$ s^{-1} at 299.2 K, measured range 299–433 K (flash photolysis-resonance fluorescence, Atkinson et al. 1977)
      $k_{O(3P)}$ = 4.30 × 10^{-12} cm^3 $molecule^{-1}$ s^{-1} for the reaction with $O(^3P)$ atom (Singleton & Cvetanovic 1976; quoted, Gaffney & Levine 1979)
      k_{OH} = 3.18 × 10^{-11} cm^3 $molecule^{-1}$ s^{-1} at 295 K (relative rate method, Atkinson & Aschmann 1984)
      k_{OH} = 9.01 × 10^{-11} cm^3 $molecule^{-1}$ s^{-1} and 6.07 × 10^{-11} cm^3 $molecule^{-1}$ s^{-1} at 298 K (Atkinson 1985)
      k_{OH}* = 3.18 × 10^{-11} cm^3 $molecule^{-1}$ s^{-1} at 298 K (recommended, Atkinson 1989, 1990)
      k_{NO3} = 9.8 × 10^{-17} cm^3 $molecule^{-1}$ s^{-1} at 296 K (Atkinson 1991)
      k_{OH}* = 3.18 × 10^{-11} cm^3 $molecule^{-1}$ s^{-1}, k_{O3} = 11.0 × 10^{-18} cm^3 $molecule^{-1}$ s^{-1}, and $k_{O(3P)}$ = 4.15 × 10^{-12} cm^3 $molecule^{-1}$ s^{-1} for reaction with $O(^3P)$ atom at 298 K (recommended, Atkinson 1997)
   Hydrolysis:
   Biodegradation:
   Biotransformation:
   Bioconcentration, Uptake (k_1) and Elimination (k_2) Rate Constants or Half-Lives:

Half-Lives in the Environment:
   Surface water: $t_{½}$ = 320 h and 9 × 10^4 d for reaction with OH and RO_2 radicals for olefins in aquatic system, and $t_{½}$ = 7.3 d, based on oxidation reaction rate constant k = 3 × 10^3 M^{-1} s^{-1} with singlet oxygen for (unsubstituted olefins in aquatic system (Mill & Mabey 1985).

TABLE 2.1.2.1.4.1
Reported vapor pressures of 3-methyl-1-butene at various temperatures and the coefficients for the vapor pressure equations

$$\log P = A - B/(T/K) \quad (1) \qquad \ln P = A - B/(T/K) \quad (1a)$$
$$\log P = A - B/(C + t/°C) \quad (2) \qquad \ln P = A - B/(C + t/°C) \quad (2a)$$
$$\log P = A - B/(C + T/K) \quad (3)$$
$$\log P = A - B/(T/K) - C \cdot \log (T/K) \quad (4)$$

Scott & Waddington 1950				Zwolinski & Wilhoit 1971			
ebulliometry				selected values			
t/°C	P/Pa	t/°C	P/Pa	t/°C	P/Pa	t/°C	P/Pa
0.210	47363	bp/°C	20.06	−62.9	1333	17.757	93326
5.112	57799	Antoine eq.		−53.4	2666	18.541	95992
10.053	70109	eq. 2	P/mmHg	−47.3	4000	19.309	98659
15.033	84158	A	6.82618	−42.6	5333	20.061	101325
20.061	101325	B	1913.474	−39.2	6666	25.0	120270
25.128	120790	C	236.816	−36.05	7999		
30.245	143268			−30.96	10666	eq. 2	P/mmHg
35.402	169066			−26.82	13332	A	6.82618
40.602	198543			−18.87	19998	B	1013.474
45.847	232073			−12.85	26664	C	236.816
51.139	280097			−7.95	33331	bp/°C	20.061
				−3.783	39997	$\Delta H_V/(kJ\ mol^{-1}) =$	
				3.109	53329	at 25°C	23.85
				8.743	66661	at bp	24.06
				13.546	79993		

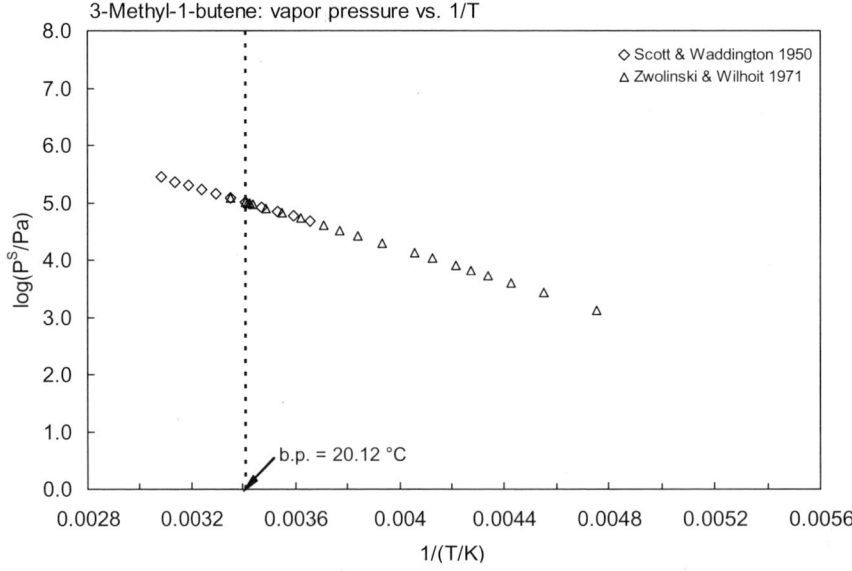

FIGURE 2.1.2.1.4.1 Logarithm of vapor pressure versus reciprocal temperature for 3-methyl-1-butene.

Aliphatic and Cyclic Hydrocarbons

2.1.2.1.5   2-Methyl-2-butene

Common Name: 2-Methyl-2-butene
Synonym:
Chemical Name: 2-methyl-2-butene
CAS Registry No: 513-35-9
Molecular Formula: C_5H_{10}, $CH_3CH=C(CH_3)CH_3$
Molecular Weight: 70.133
Melting Point (°C):
  –133.72   (Lide 2003)
Boiling Point (°C):
  38.56   (Lide 2003)
Density (g/cm³ at 20°C):
  0.6623, 0.6570   (20°C, 25°C, Dreisbach 1959)
Molar Volume (cm³/mol):
  105.9, 106.8 (20°C, 25°C, calculated-density)
  111.0   (calculated-Le Bas method at normal boiling point)
Enthalpy of Fusion, ΔH_{fus} (kJ/mol):
  7.60   (Chickos et al. 1999)
Entropy of Fusion, ΔS_{fus} (J/mol K):
  54.47, 59.4 (exptl., calculated-group additivity method, total phase change entropy, Chickos et al. 1999)
Fugacity Ratio at 25°C, F: 1.0

Water Solubility (g/m³ or mg/L at 25°C or as indicated. Additional data at other temperatures designated * are compiled at the end of this section.):
  215*   (20°C, shake flask-GC, measured range 20–60°C, Pavlova et al. 1966)
  325*   (calculated-liquid-liquid equilibrium LLE data, temp range 288–333.2 K, Góral et al. 2004)

Vapor Pressure (Pa at 25°C or as indicated and reported temperature dependence equations. Additional data at other temperatures designated * are compiled at the end of this section.):
  47876*   (18.07°C, static method, measured range –78.85 to 18.07°C, Lamb & Roper 1940)
  log (P/mmHg) = 9.86840 – 1773.506/(T/K) + 0.0035747·log (T/K); temp range –78.85 to 18.07°C (static method, Lamb & Roper 1940)
  53329*   (21.6, summary of literature data, temp range –75.4 to 38.5°C, Stull 1947)
  57798*   (ebulliometry, measured range 3.042 to 70.59°C, Scott et al. 1949)
  log (P/mmHg) = 6.91562 – 1095.088/(232.842 + t/°C); temp range 3.042 to 70.59°C (Antoine eq., ebulliometry, Scott et al. 1949)
  62140   (calculated-Antoine eq., Dreisbach 1959)
  log (P/mmHg) = 6.91562 – 1095.088/(233.0 + t/°C); temp range –31 to 85°C (Antoine eq. for liquid state, Dreisbach 1959)
  62142*, 62140   (derived from compiled data, interpolated-Antoine eq., Zwolinski & Wilhoit 1971)
  log (P/mmHg) = 6.91562 – 1095.088/(232.842 + t/°C); temp range –47.7 to 60.0°C (Antoine eq., Zwolinski & Wilhoit 1971)
  62890, 62140   (calculated-Antoine eq., Boublik et al. 1984)
  log (P/kPa) = 6.15017 – 1146.28/(238.416 + t/°C); temp range –78.85 to 18.07°C (Antoine eq. from reported exptl. data, Boublik et al. 1984)
  log (P/kPa) = 6.04808 – 1099.054/(233.314 + t/°C); temp range 3.04–70.59°C (Antoine eq. from reported exptl. data, Boublik et al. 1984)
  62240   (interpolated-Antoine eq., Dean 1985, 1992)
  log (P/mmHg) = 6.96659 – 1124.33/(236.63 + t/°C); temp range –48 to 60°C (Antoine eq., Dean 1985, 1992)
  62170   (interpolated-Antoine eq., Stephenson & Malanowski 1987)

log (P_L/kPa) = 6.04475 − 1097.501/(−39.985 + T/K); temp range 271–343 K (Antoine eq., Stephenson & Malanowski 1987)

log (P/mmHg) = 33.7539 − 2.426 × 10^3/(T/K) − 9.4429·log (T/K) + 9.8488 × 10^{-11}·(T/K) + 4.7156 × 10^{-6}·(T/K)^2; temp range 139–471 K (vapor pressure eq., Yaws 1994)

Henry's Law Constant (Pa m^3/mol at 25°C):
  24650   (calculated-P/C from selected data)

Octanol/Water Partition Coefficient, log K_{OW}:

Octanol/Air Partition Coefficient, log K_{OA}:

Bioconcentration Factor, log BCF:

Sorption Partition Coefficient, log K_{OC}:

Environmental Fate Rate Constants, k, and Half-Lives, $t_{1/2}$:
  Volatilization:
  Photolysis:
  Oxidation: rate constant k; for gas-phase second-order rate constants, k_{OH} for reaction with OH radical, k_{NO_3} with NO_3 radical and k_{O_3} with O_3 or as indicated, *data at other temperatures and/or the Arrhenius expression see reference:
    k_{O_3} = 7.47 × 10^{-16} cm^3 molecule^{-1} s^{-1} at 30°C (flow system, Bufalini & Altshuller 1965)
    k_{OH} = 11.9 × 10^{-11} cm^3 molecule^{-1} s^{-1} at 298 K (discharge flow system-MS, Morris & Niki 1971)
    $k_{O(^3P)}$ = 5.4 × 10^{-11} cm^3 molecule^{-1} s^{-1} for reaction with O(^{3}P) (Herron & Huie 1973; Furuyama et al. 1974; Atkinson & Pitts Jr. 1978; quoted, Gaffney & Levine 1979)
    k_{O_3} = 4.93 × 10^{-16} cm^3 molecule^{-1} s^{-1} at 299 K (Japar et al. 1974)
    k_{OH} = 4.8 × 10^9 L mol^{-1} s^{-1} with atmospheric $t_{1/2}$ < 0.24 h (Darnall et al.1976, Lloyd et al. 1976)
    k_{OH} = 8.4 × 10^{-11} cm^3 molecule^{-1} s^{-1} (Atkinson et al. 1979; quoted, Gaffney & Levine 1979)
    k_{OH} = 8.4 × 10^{-11} cm^3 molecule^{-1} s^{-1}; $k_{O(^3P)}$ = 54 × 10^{-12} cm^3 molecule^{-1} s^{-1} at room temp. (LFE correlation, Gaffney & Levine 1979)
    k_{OH} = (87.1 ± 2.6) × 10^{-11} cm^3 molecule^{-1} s^{-1} at 297 ± 2 K (relative rate method, Ohta 1984)
    k_{NO_3} = (5.5 ± 1.2) × 10^{-12} cm^3 molecule^{-1} s^{-1} 295 K (relative rate method, Atkinson et al. 1984a)
    k_{O_3} = 4.2 × 10^{-16} cm^3 molecule^{-1} s^{-1} with calculated lifetimes τ = 55 min and 17 min in clean and moderately polluted atmosphere respectively, k_{OH} = 8.7 × 10^{-11} cm^3 molecule^{-1} s^{-1} with calculated lifetimes τ = 3.2 h and 1.6 h in clean and moderately polluted atmosphere, respectively; k_{NO_3} = 5.5 × 10^{-12} cm^3 molecule^{-1} s^{-1} with calculated lifetimes τ = 13 min and 1.3 min in clean and moderately polluted atmosphere, respectively (Atkinson et al. 1984a)
    k_{NO_2} < 0.5 × 10^{-20} cm^3 molecule^{-1} s^{-1} at 295 K (Atkinson et al. 1984b)
    k_{O_3} = 4.5 × 10^{-16} cm^3 molecule^{-1} s^{-1}; k_{OH} = 8.6 × 10^{-11} cm^3 molecule^{-1}s^{-1}; k_{NO_3} = 5.5 × 10^{-12} cm^3 molecule^{-1} s^{-1}; and $k_{O(^3P)}$ = 47.6 × 10^{-12} cm^3 molecule^{-1} s^{-1} with O(^{3}P) atom at room temp. (Atkinson et al. 1984b)
    k_{OH} = 8.68 × 10^{-11} cm^3 molecule^{-1} s^{-1} at 295 ± 1 K (relative rate method, Atkinson & Aschmann 1984)
    k_{O_3} = (6.79–7.97) × 10^{-16} cm^3 molecule^{-1} s^{-1}, 296–299 K (Atkinson & Carter 1984)
    k_{O_3} = 4.2 × 10^{-16} cm^3 molecule^{-1} s^{-1} with a loss rate of 25 d^{-1}; k_{OH} = 8.7 × 10^{-11} cm^3 molecule^{-1}s^{-1} with a loss rate of 8 d^{-1}, and k_{NO_3} = 9.9 × 10^{-11} cm^3 molecule^{-1} s^{-1} with a loss rate of 205 d^{-1} (Atkinson & Carter 1984)
    k_{OH} = 7.7 × 10^{-11} to 1.19 × 10^{-10} cm^3 molecule^{-1} s^{-1} between 297.7–299.5 K (Atkinson 1985)
    k_{O_3} = 4.2 × 10^{-16} cm^3 molecule^{-1} s^{-1} with a loss rate of 25 d^{-1}; k_{OH} = 8.7 × 10^{-11} cm^3 molecule^{-1}s^{-1} with a loss rate of 3.8 d^{-1}, and k_{NO_3} = 9.9 × 10^{-11} cm^3 molecule^{-1} s^{-1} with a loss rate of 205 d^{-1} at room temp (Atkinson 1985)
    k_{NO_3} = (9.33 ± 1.18) × 10^{-12} cm^3 molecule^{-1} s^{-1} 296 ± 2 K (relative rate method, Atkinson 1988)
    k_{OH} = 8.69 × 10^{-11} cm^3 molecule^{-1} s^{-1} at 298 K (recommended, Atkinson 1989)
    k_{OH} = 6.89 × 10^{-11} cm^3 molecule^{-1} s^{-1}, k_{O_3} = 4.23 × 10^{-16} cm^3 molecule^{-1} s^{-1} at 198 K (Atkinson 1990)
    k_{OH} = 8.69 × 10^{-11} cm^3 molecule^{-1} s^{-1}, k_{NO_3} = 9.33 × 10^{-12} cm^3 molecule^{-1} s^{-1} at 298 K (Sabljic & Güsten 1990)

k_{NO3} = 9.37 × 10⁻¹³ cm³ molecule⁻¹ s⁻¹ at 298 K (recommended, Atkinson 1991)
k_{OH}* = 8.69 × 10⁻¹¹ cm³ molecule⁻¹ s⁻¹, k_{NO3} = 9.37 × 10⁻¹² cm³ molecule⁻¹ s⁻¹, k_{O3}* = 4.03 × 10⁻¹⁶ cm³ molecule⁻¹ s⁻¹, and $k_{O(^3P)}$ = 5.1 × 10⁻¹¹ cm³ molecule⁻¹ s⁻¹ for reaction with O(³P) atom at 298 K (recommended, Atkinson 1997)

Hydrolysis:

Biodegradation:

Biotransformation:

Bioconcentration, Uptake (k_1) and Elimination (k_2) Rate Constants or Half-Lives:

Half-Lives in the Environment:

Air: atmospheric $t_{1/2}$ < 0.24 h, based on the photooxidation rate constant k = 4.8 × 10⁹ L mol⁻¹ s⁻¹ with hydroxyl radical in air (Darnall et al. 1976; Lloyd et al. 1976);
atmospheric lifetimes, τ(calc) = 0.95 h for the reaction with O₃, τ = 3.2 h with OH radical and τ = 0.12 h with NO₃ radical, based on the rate constants and environmental concentrations of OH, O₃ and NO₃ in the gas phase (Atkinson & Carter 1984);
calculated lifetimes: τ = 55 min due to reaction with O₃ in 24-h period, τ = 3.2 h with OH radical during daytime, and τ = 13 min for NO₃ radical during nighttime for "clean" atmosphere; τ = 17 min for reaction with O₃ in 24-h period, τ = 1.6 h with OH radical during daytime, and τ = 1.3 min with NO₃ radical during nighttime in "moderately" polluted atmosphere (Atkinson et al. 1984a);
atmospheric lifetimes τ(calc) = 6.38 h for the reaction with OH radical, τ(calc) = 0.92 h with O₃ and τ(calc) = 0.12 h with NO₃ radical in the gas phase (Atkinson 1985).

Surface water: $t_{1/2}$ ~ 320 h and 9 × 10⁴ d for reaction with OH and RO₂ radicals respectively in aquatic system, and $t_{1/2}$ = 8.0 d, based on rate constant of 10⁶ M⁻¹ s⁻¹ for the reaction with singlet oxygen in aquatic system (Mill & Mabey 1985).

TABLE 2.1.2.1.5.1
Reported aqueous solubilities of 2-methyl-2-butene at various temperatures

Pavlova et al. 1966		Góral et al. 2004	
shake flask-GC		calc-recommended LLE data	
t/°C	S/g·m⁻³	t/°C	S/g·m⁻³
20	215	15	343
40	236	20	334
50	250	25	325
60	267	40	325
		50	338
		60	361

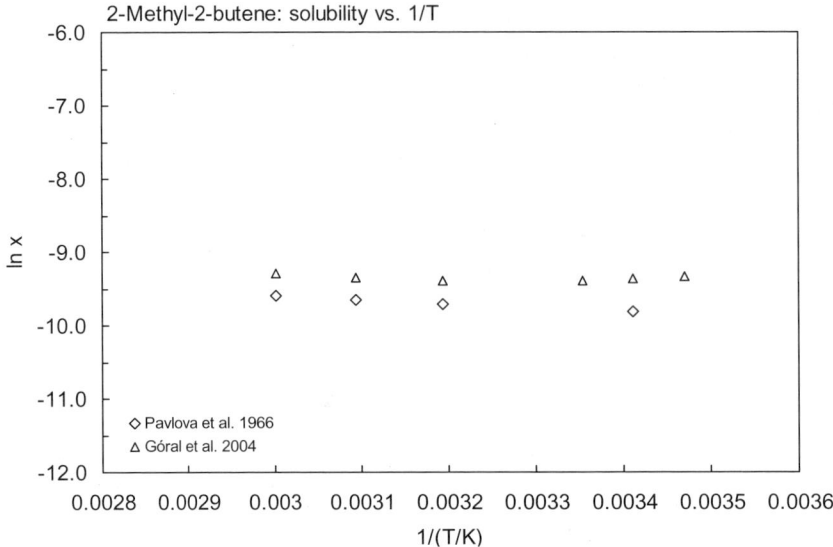

FIGURE 2.1.2.1.5.1 Logarithm of mole fraction solubility (ln x) versus reciprocal temperature for 2-methyl-2-butene.

TABLE 2.1.2.1.5.2
Reported vapor pressures of 2-methyl-2-butene at various temperatures and the coefficients for the vapor pressure equations

$\log P = A - B/(T/K)$ (1) $\ln P = A - B/(T/K)$ (1a)
$\log P = A - B/(C + t/°C)$ (2) $\ln P = A - B/(C + t/°C)$ (2a)
$\log P = A - B/(C + T/K)$ (3)
$\log P = A - B/(T/K) - C \cdot \log (T/K)$ (4)
$\log P = A - B/(T/K) - C \cdot (T/K)$ (5)

Lamb & Roper 1940		Stull 1947		Scott et al. 1949		Zwolinski & Wilhoit 1971	
Static method-manometer		Summary of literature data		Ebulliometry		Selected values	
t/°C	P/Pa	t/°C	P/Pa	t/°C	P/Pa	t/°C	P/Pa
–78.85	97	–75.4	133.3	3.042	25007	–47.7	1333
–54.78	811	–57.0	666.6	8.008	31172	–37.8	2666
–37.98	2696	–47.0	1333	12.987	38534	–31.46	4000
–29.84	4517	–37.9	2666	18.033	47364	–26.75	5333
–21.06	7537	–26.7	5333	23.103	57798	–22.92	6666
–12.14	11767	–19.4	7999	28.220	70110	–19.68	7999
–3.98	18238	–9.9	13332	33.373	84522	–14.37	10666
0.0	21918	4.0	26664	38.567	101319	–10.06	13332
0.0	22025	21.6	53329	43.806	120810	–1.786	19998
2.44	24625	38.5	101325	49.078	143268	4.486	26664
4.51	27024			54.399	169079	9.559	33331
16.31	44783	mp/°C	–133	59.753	198543	13.883	39997
18.07	47876			65.151	232087	21.029	53329
				70.590	272230	26.864	66661
bp/°C	38.43					31.834	79993
eq. 5	P/mmHg					36.187	93326
A	9.68640					36.998	95992
B	1773.506					37.791	98659
C	0.0035747					38.568	101325

Aliphatic and Cyclic Hydrocarbons

TABLE 2.1.2.1.5.2 *(Continued)*

Lamb & Roper 1940		Stull 1947		Scott et al. 1949		Zwolinski & Wilhoit 1971	
Static method-manometer		Summary of literature data		Ebulliometry		Selected values	
t/°C	P/Pa	t/°C	P/Pa	t/°C	P/Pa	t/°C	P/Pa
						25.0	62142
						eq. 2	P/mmHg
						A	6.91562
						B	1095.088
						C	232.842
						bp/°C	38.568
						ΔH_v/(kJ mol^{-1}) =	
						at 25°C	27.06
						at bp	26.30

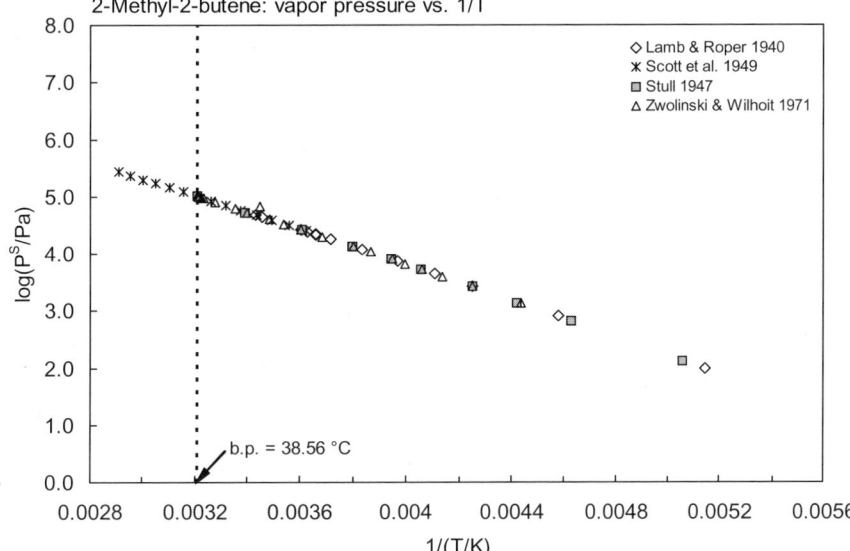

FIGURE 2.1.2.1.5.2 Logarithm of vapor pressure versus reciprocal temperature for 2-methyl-2-butene.

2.1.2.1.6   1-Pentene

Common Name: 1-Pentene
Synonym: amylene, α-*n*-amylene, propylethylene
Chemical Name: 1-pentene
CAS Registry No: 109-67-1
Molecular Formula: C_5H_{10}
Molecular Weight: 70.133
Melting Point (°C):
    –165.12   (Lide 2003)
Boiling Point (°C):
    29.96   (Lide 2003)
Density (g/cm^3):
    0.6405, 0.63533   (20°C, 25°C, Forziati et al. 1950, Dreisbach 1959)
    0.6353   (20°C, Riddick et al. 1986)
Molar Volume (cm^3/mol):
    109.5   (20°C, calculated-density, Stephenson & Malanowski 1987)
    111.0   (calculated-Le Bas method at normal boiling point)
Enthalpy of Vaporization, ΔH_V (kJ/mol):
    25.47, 25.20 (25 °, bp, Riddick et al. 1986)
Enthalpy of Fusion, ΔH_{fus} (kJ/mol):
    5.998   (Riddick et al. 1986)
    5.81   (Chickos et al. 1999)
Entropy of Fusion, ΔS_{fus} (J/mol K):
    53.82, 54.4 (exptl., calculated-group additivity method, total phase change entropy, Chickos et al. 1999)
Fugacity Ratio at 25°C, F: 1.0

Water Solubility (g/m^3 or mg/L at 25°C):
    148   (shake flask-GC, McAuliffe 1966)
    191   (calculated-recommended liquid-liquid equilibrium LLE data, Góral et al. 2004)

Vapor Pressure (Pa at 25°C or as indicated and reported temperature dependence equations. Additional data at other temperatures designated * are compiled at the end of this section.):
    91420*   (interpolated-Antoine eq. regression, temp range –80.4 to 30.1°C, Stull 1947)
    70834   (20°C, static method, measured range 0–200°C, Day et al. 1948)
    log (P/mmHg) = 7.40607 – 1372.194/(T/K); temp range 0–30°C (static method, Day et al. 1948)
    log (P/mmHg) = 7.31561 – 1342.407/(T/K); temp range 40–95°C (static method, Day et al. 1948)
    84508*   (ebulliometry, measured range –0.159 to 61.64°C, Scott et al. 1949)
    log (P/mmHg) = 6.85487 – 1049.00/(233.994 + t/°C); temp range –0.159 to 61.64°C (Antoine eq., ebulliometry, Scott et al. 1949)
    83750*   (24.6°C, ebulliometry-manometer, measured range 12.8–30.7°C, Forziati et al. 1950)
    log (P/mmHg) = 6.78568 – 1014.293/(229.783 + t/°C); temp range 12.8–30.7°C (Antoine eq., ebulliometry measurements, Forziati et al. 1950)
    86500   (calculated from determined data, Dreisbach 1959)
    log (P/mmHg) = 6.84650 – 1044.9/(234.0 + t/°C); temp range –39 to 73°C (Antoine eq. for liquid state, Dreisbach 1959)
    85000*   (interpolated-Antoine eq., temp range –63 to 41°C, Zwolinski & Wilhoit 1971)
    log (P/mmHg) = 6.84650 – 1044.895/(233.516 + t/°C); temp range –63 to 41°C (Antoine eq., Zwolinski & Wilhoit 1971)
    log (P/mmHg) = [–0.2185 × 6931.2/(T/K)] + 7.914969; temp range –80.4 to 30.1°C (Antoine eq., Weast 1972–73)
    85020   (interpolated-Antoine eq., Boublik et al. 1984)

log (P/kPa) = 5.9716 − 1045.212/(233.598 + t/°C); temp range 12.84–30.7°C (Antoine eq. from reported exptl. data of Forziati et al. 1950, Boublik et al. 1984)

85200   (interpolated-Antoine eq., temp range −55 to 51°C, Dean 1985, 1992)

log (P/mmHg) = 6.84424 − 1044.015/(233.50 + t/°C); temp range −55 to 51°C (Antoine eq., Dean 1985, 1992)

85100   (literature average, Riddick et al. 1986)

log (P/kPa) = 5.96914 − 1044.01/(233.49 + t/°C); temp range not specified (Antoine eq., Riddick et al. 1986)

85040   (interpolated-Antoine eq., Stephenson & Malanowski 1987)

log (P_L/kPa) = 5.96999 − 1043.962/(−39.767 + T/K); temp range 218–311 K (Antoine eq., Stephenson & Malanowski 1987)

log (P/mmHg) = 36.2741 − 2.4452 × 10^3/(T/K) − 10.405·log (T/K) − 7.4629 × 10^{-11}·(T/K) + 5.4070 × 10^{-6}·(T/K)^2; temp range 110–465 K (vapor pressure eq., Yaws 1994)

Henry's Law Constant (Pa m^3/mol at 25°C):

40330   (calculated-P/C, Mackay & Shiu 1975; selected, Mills et al. 1982)

41140   (calculated as 1/K_{AW}, C_W/C_A, reported as exptl., Hine & Mookerjee 1975)

37520, 22610 (calculated-group contribution, calculated-bond contribution, Hine & Mookerjee 1975)

40300   (calculated-P/C, Mackay & Shiu 1981)

37520   (calculated-MCI χ, Nirmalakhandan & Speece 1988)

40405   (calculated-P/C, Eastcott et al. 1988)

40280   (calculated-vapor-liquid equilibrium (VLE) data, Yaws et al. 1991)

Octanol/Water Partition Coefficient, log K_{OW}:

2.20   (calculated-π substituent constant, Hansch et al. 1968)

2.69   (calculated-f const., Yalkowsky & Morozowich 1980)

2.20   (calculated-MCI χ, Murray et al. 1975)

2.80   (selected, Müller & Klein 1992)

2.3970   (calculated-UNIFAC group contribution, Chen et al. 1993)

Octanol/Air Partition Coefficient, log K_{OA} at 25°C or as indicated. Additional data at other temperatures designated * are compiled at the end of this section:

2.0*   (20.29°C, from GC determined γ^∞ in octanol, measured range 20.29–50.2°C, Gruber et al. 1997)

1.93   (calculated-measured γ^∞ in pure octanol and vapor pressure P, Abraham et al. 2001)

Bioconcentration Factor, log BCF:

Sorption Partition Coefficient, log K_{OC}:

Environmental Fate Rate Constants, k, and Half-Lives, $t_{1/2}$:

    Volatilization:

    Photolysis:

    Oxidation: rate constant k for gas-phase second order rate constants, k_{OH} for reaction with OH radical, k_{NO3} with NO$_3$ radical and k_{O3} with O$_3$ or as indicated, *data at other temperatures see reference:

        k_{OH} = 4.25 × 10^{-11} cm^3 molecule^{-1} s^{-1} at 298 K (discharge flow system-MS, Morris & Niki 1971)

        k_{O3} = 1.07 × 10^{-17} cm^3 molecule^{-1} s^{-1} (Japar et al. 1974)

        k_{O3} = 5.3 × 10^{-18} cm^3 molecule^{-1} s^{-1}, 7.4 × 10^{-18} cm^3 molecule^{-1} s^{-1}, and 1.07 × 10^{-17} cm^3 molecule^{-1} s^{-1} (review, Atkinson & Carter 1984)

        k_{OH} = 3.13 × 10^{-11} cm^3 molecule^{-1} s^{-1} at 295 K (relative rate method, Atkinson & Aschmann 1984)

        k_{OH} = 3.14 × 10^{-11} cm^3 molecule^{-1} s^{-1} at 298 K (recommended, Atkinson 1989)

        k_{OH} = 3.14 × 10^{-11} cm^3 molecule^{-1} s^{-1}, k_{O3} = 1.10 × 10^{-17} cm^3 molecule^{-1} s^{-1} at 298 K (Atkinson 1990)

        k_{O3} = 1.00 × 10^{-17} cm^3 molecule^{-1} s^{-1} at 298 K (recommended, Atkinson 1994)

        k_{OH} = 3.14 × 10^{-11} cm^3 molecule^{-1} s^{-1}, k_{O3}* = 10.0 × 10^{-18} cm^3 molecule^{-1} s^{-1}, and $k_{O(3P)}$ = 4.65 × 10^{-12} cm^3 molecule^{-1} s^{-1} for reaction with O(^{3}P) atom at 298 K (recommended, Atkinson 1997)

    Hydrolysis:

    Biodegradation:

Biotransformation:
Bioconcentration, Uptake (k_1) and Elimination (k_2) Rate Constants or Half-Lives:

Half-Lives in the Environment:

Surface water: $t_{1/2}$ ~ 320 h and 9×10^4 d for olefins in aquatic system by oxidation with OH and RO_2 radicals; while $t_{1/2}$ = 7.3 d based on rate constant k = 3×10^5 M^{-1} s^{-1} for the oxidation of unsubstituted olefins with singlet oxygen in aquatic system (Mill & Mabey 1985).

TABLE 2.1.2.1.6.1
Reported vapor pressures of 1-pentene at various temperatures and the coefficients for the vapor pressure equations

$\log P = A - B/(T/K)$ (1) $\ln P = A - B/(T/K)$ (1a)
$\log P = A - B/(C + t/°C)$ (2) $\ln P = A - B/(C + t/°C)$ (2a)
$\log P = A - B/(C + T/K)$ (3)
$\log P = A - B/(T/K) - C \cdot \log (T/K)$ (4)

Stull 1947		Scott et al. 1949		Forziati et al. 1950		Zwolinski & Wilhoit 1971	
summary of literature data		ebulliometry		ebulliometry		selected values	
t/°C	P/Pa	t/°C	P/Pa	t/°C	P/Pa	t/°C	P/Pa
–80.4	133.3	–0.159	31168	12.834	53703	–54.8	1333
–63.3	666.6	4.751	38545	18.468	66797	–45.1	2666
–54.5	1333	9.706	47348	24.584	63754	–38.91	4000
–46.0	2666	14.706	57789	28.900	97645	–34.28	5333
–34.1	5333	19.750	70094	29.362	99227	–30.53	6666
–27.1	7999	24.834	84508	29.796	100727	–27.36	7999
–17.7	13332	29.967	101322	30.289	102453	–22.14	10666
–3.40	26664	35.142	120813	30.723	103988	–17.92	13332
12.8	53329	40.359	143295			–9.789	19998
30.1	101325	45.614	169079	bp/°C	29.968	–3.640	26664
		50.914	198557			1.368	33331
mp/°C		56.253	232061	eq. 2	P/mmHg	5.624	39997
		61.641	270071	A	6.78568	12.664	53329
				B	1014.294	18.416	66661
		bp/°C	29.97	C	229.783	23.319	79993
		Antoine eq.				27.616	93326
		eq. 2	P/mmHg			28.417	95992
		A	6.85487			29.201	98659
		B	1049.00			29.968	101325
		C	233.994			25.0	85020
						eq. 2	P/mmHg
						A	6.84650
						B	1044.895
						C	233.516
						bp/°C	29.968
						ΔH_V/(kJ mol^{-1}) =	
						at 25°C	25.47
						at bp	25.20

Aliphatic and Cyclic Hydrocarbons

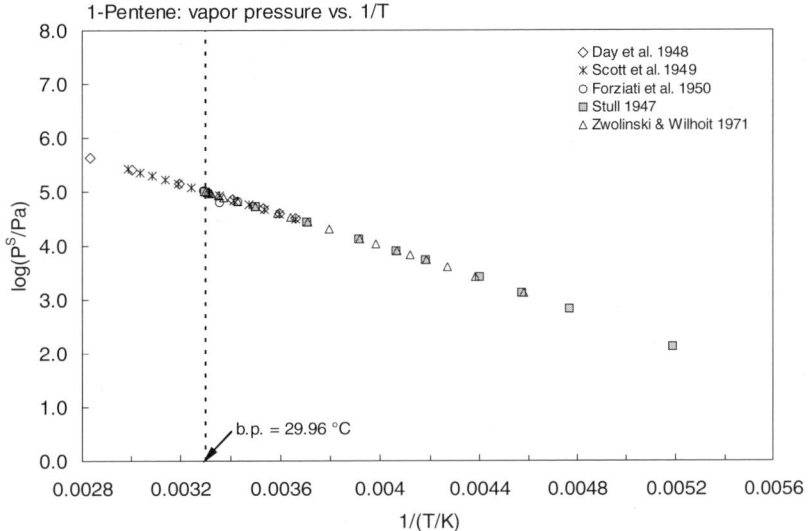

FIGURE 2.1.2.1.6.1 Logarithm of vapor pressure versus reciprocal temperature for 1-pentene.

TABLE 2.1.2.1.6.2
Reported octanol-air partition coefficients of 1-pentene at various temperatures and temperature dependence equations

Gruber et al. 1997	
GC det'd activity coefficient	
t/°C	log K_{OA}
20.29	1.995
30.3	1.852
40.4	1.740
50.28	1.630

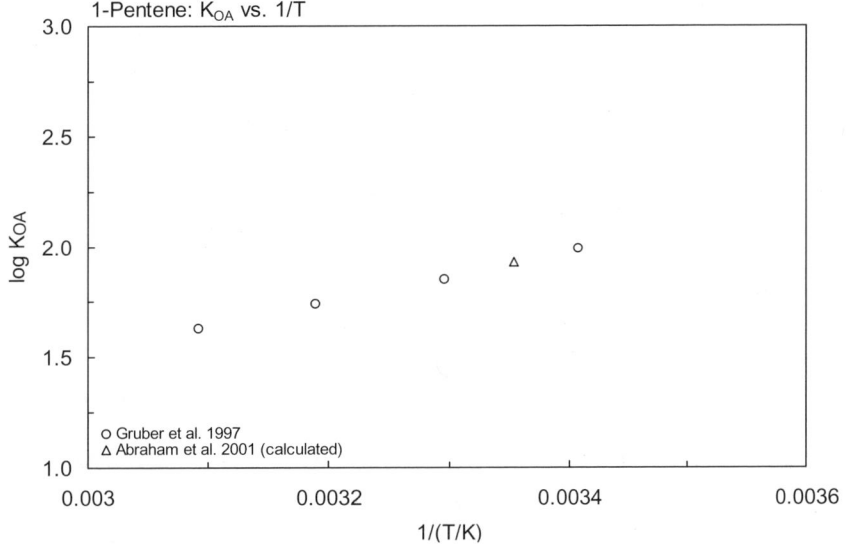

FIGURE 2.1.2.1.6.2 Logarithm of K_{OA} versus reciprocal temperature for 1-pentene.

2.1.2.1.7   cis-2-Pentene

Common Name: *cis*-2-Pentene
Synonym: (Z)-2-pentene
Chemical Name: *cis*-2-pentene
CAS Registry No: 627-20-3
Molecular Formula: C_5H_{10}
Molecular Weight: 70.133
Melting Point (°C):
    –151.36   (Lide 2003)
Boiling Point (°C):
    36.93   (Lide 2003)
Density (g/cm³ at 20°C):
    0.6556, 0.6504 (20°C, 25°C, Dreisbach 1959)
Molar Volume (cm³/mol):
    107.0, 107.8 (20°C, 25°C, calculated-density)
    111.0   (calculated-Le Bas method at normal boiling point)
Enthalpy of Fusion, ΔH_{fus} (kJ/mol):
    7.112   (Riddick et al. 1986)
    7.11   (Chickos et al. 1999)
Entropy of Fusion, ΔS_{fus} (J/mol K):
    58.39, 52.8 (exptl., calculated-group additivity method, total phase change entropy, Chickos et al. 1999)
Fugacity Ratio at 25°C, F: 1.0

Water Solubility (g/m³ or mg/L at 25°C):
    203   (shake flask-GC, *cis-trans* form not specified, McAuliffe 1966)

Vapor Pressure (Pa at 25°C or as indicated and reported temperature dependence equations. Additional data at other temperatures designated * are compiled at the end of this section.):
    57795*   (21.541°C, ebulliometry, measured range 1.595–68.842°C, Scott & Waddington 1950)
    $\log (P/\text{mmHg}) = 6.87540 - 1069.460/(240.786 + t/°C)$; temp range 1.595–68.842°C (Antoine eq., ebulliometric method, Scott & Waddington 1950)
    65940   (calculated-Antoine eq., Dreisbach 1959)
    $\log (P/\text{mmHg}) = 6.87274 - 1068.0/(231.0 + t/°C)$; temp range –33 to 82°C (Antoine eq. for liquid state, Dreisbach 1959)
    66000, 65941 (interpolated-Antoine eq., derived from compiled data, Zwolinski & Wilhoit 1971)
    $\log (P/\text{mmHg}) = 6.87274 - 1067.951/(230.585 + t/°C)$; temp range –48.7 to 58.31°C (Antoine eq., Zwolinski & Wilhoit 1971)
    65950   (interpolated-Antoine eq., Boublik et al. 1984)
    $\log (P/\text{kPa}) = 6.99984 - 1069.227/(230.757 + t/°C)$; temp range 1.595–68.88°C (Antoine eq. from reported exptl. data, Boublik et al. 1984)
    66640   (interpolated-Antoine eq., Dean 1985, 1992)
    66000   (quoted lit., Riddick et al. 1986)
    $\log (P/\text{kPa}) = 5.96798 - 1052.44/(228.693 + t/°C)$; temp range not specified (Antoine eq., Riddick et al. 1986)
    $\log (P/\text{mmHg}) = 6.84308 - 1052.44/(228.69 + t/°C)$; temp range –49 to 58°C (Antoine eq., Dean 1985, 1992)
    65970   (interpolated-Antoine eq., Stephenson & Malanowski 1987)
    $\log (P_L/\text{kPa}) = 5.99069 - 1064.178/(-43.035 + T/K)$; temp range 234–318 K (Antoine eq., Stephenson & Malanowski 1987)
    $\log (P/\text{mmHg}) = 34.0427 - 2.4524 \times 10^3/(T/K) - 9.5014 \cdot \log (T/K) - 5.0816 \times 10^{-11} \cdot (T/K) + 4.3638 \times 10^{-6} \cdot (T/K)^2$; temp range 122–476 K (vapor pressure eq., Yaws 1994)

Aliphatic and Cyclic Hydrocarbons

Henry's Law Constant (Pa m^3/mol at 25°C):
    22800   (calculated-P/C, Mackay & Shiu 1981)
    22770   (calculated-vapor-liquid equilibrium (VLE) data, Yaws et al. 1991)

Octanol/Water Partition Coefficient, log K_{OW}:
    2.20   (calculated-π substituent constant, Hansch et al. 1968)
    2.20   (calculated-MCI χ, *cis-trans* form not specified, Murray et al. 1975)
    2.3772   (calculated-UNIFAC group contribution, Chen et al. 1993)

Octanol/Air Partition Coefficient, log K_{OA}:

Bioconcentration Factor, log BCF:

Sorption Partition Coefficient, log K_{OC}:

Environmental Fate Rate Constants, k, and Half-Lives, $t_{1/2}$:
    Volatilization:
    Photolysis:
    Oxidation: rate constant k, for gas-phase second order rate constants, k_{OH} for reaction with OH radical, k_{NO3} with NO$_3$ radical and k_{O3} with O$_3$ or as indicated, *data at other temperatures see reference:
        $k_{O(3P)} = 1.80 \times 10^{-11}$ cm^3 molecule^{-1} s^{-1} for the reaction with O(^{3}P) atom (Herron & Huie 1973; quoted, Gaffney & Levine 1979)
        $k_{OH} = 6.20 \times 10^{-11}$ cm^3 molecule^{-1} s^{-1} (Atkinson et al. 1979; quoted, Gaffney & Levine 1979)
        $k_{OH} = (6.23 \pm 0.1) \times 10^{-11}$ cm^3 molecule^{-1} s^{-1} at 298 ± 2 K (relative rate method, Ohta 1984)
        $k_{OH} = (65.4 - 65.9) \times 10^{-12}$ cm^3 molecule^{-1} s^{-1} at 298–303 K (Atkinson 1989)
        $k_{OH} = 6.50 \times 10^{-11}$ cm^3 molecule^{-1} s^{-1} at 298 K (Atkinson 1990)
        $k_{OH} = 6.50 \times 10^{-11}$ cm^3 molecule^{-1} s^{-1} with an estimated atmospheric lifetime of 2.29 h in summer daylight (Altshuller 1991)
        $k_{OH} = 6.5 \times 10^{-11}$ cm^3 molecule^{-1} s^{-1}, and $k_{O(3P)} = 1.7 \times 10^{-11}$ cm^3 molecule^{-1} s^{-1} for reaction with O(^{3}P) atom at 298 K (recommended, Atkinson 1997)
    Hydrolysis:
    Biodegradation:
    Biotransformation:
    Bioconcentration, Uptake (k_1) and Elimination (k_2) Rate Constants or Half-Lives:

Half-Lives in the Environment:
    Air: photooxidation reaction rate constant k = 6.50×10^{-11} cm^3 molecule^{-1} s^{-1} with hydroxyl radical in air (Atkinson 1990, Altshuller 1991) with an estimated atmospheric lifetime of 2.29 h in summer daylight (Altshuller 1991).
    Surface water: $t_{1/2} \sim$ 320 h for oxidation by OH radicals, $t_{1/2} = 9 \times 10^4$ d for olefins in aquatic system, and $t_{1/2}$ = 7.3 d based on rate constant k = 3×10^3 M^{-1} s^{-1} for oxidation of unsubstituted olefins by singlet oxygen in aquatic system (Mill & Mabey 1985).

TABLE 2.1.2.1.7.1
Reported vapor pressures of *cis*-2-pentene at various temperatures

Scott & Waddington 1950			
ebulliometric method			
t/°C	P/Pa	t/°C	P/Pa
1.595	25009	58.070	198556
6.522	31163	63.456	232087
11.486	38546	68.882	270057
16.494	47357		
21.541	57795	$\log P = A - B/(C + t/°C)$	
26.633	70102	bp/°C	36.94
31.766	84518		P/mmHg
36.944	101329	A	6.87540
42.161	120804	B	1069.460
47.423	143281	C	240.786
52.724	169066		

Aliphatic and Cyclic Hydrocarbons

2.1.2.1.8   2-Methyl-1-pentene

Common Name: 2-Methyl-1-pentene
Synonym:
Chemical Name: 2-methyl-1-pentene
CAS Registry No: 763-29-1
Molecular Formula: C_6H_{12}; $CH_3(CH_2)_2C(CH_3)CH_2$
Molecular Weight: 84.159
Melting Point (°C):
    –135.7   (Dreisbach 1959; Lide 2003)
Boiling Point (°C):
    60.7   (Dreisbach 1959)
    62.1   (Lide 2003)
Density (g/cm³ at 20°C):
    0.6799, 0.6751   (20°C, 25°C, Dreisbach 1959)
Molar Volume (cm³/mol):
    123.8, 124.7 (20°C, 25°C, calculated-density)
    133.2   (calculated-Le Bas method at normal boiling point)
Enthalpy of Fusion, ΔH_{fus} (kJ/mol):
Entropy of Fusion, ΔS_{fus} (J/mol K):
Fugacity Ratio at 25°C, F: 1.0

Water Solubility (g/m³ or mg/L at 25°C):
    78.0   (shake flask-GC, McAuliffe 1966)
    98.2   (calculated-recommended liquid-liquid equilibrium LLE data, Góral et al. 2004)

Vapor Pressure (Pa at 25°C and reported temperature dependence equations):
    27464   (calculated-Antoine eq., Dreisbach 1959)
    log (P/mmHg) = 6.88772 – 1154.7/(227.0 + t/°C); temp range –14 to 100°C (Antoine eq. for liquid state, Dreisbach 1959)
    26000, 26051 (interpolated-Antoine eq., derived from compiled data, Zwolinski & Wilhoit 1971)
    log (P/mmHg) = 6.85030 – 1138.516/(224.764 + t/°C); temp range –30.1 to 85.16°C (Antoine eq., Zwolinski & Wilhoit 1971)
    log (P/mmHg) = 6.85030 – 1138.516/(224.70 + t/°C); temp range –30 to 85°C (Antoine eq., Dean 1985, 1992)
    26060   (interpolated-Antoine eq., Stephenson & Malanowski 1987)
    log (P_L/kPa) = 5.89056 – 1091.679/(–46.306 + T/K); temp range 265–333 K (Antoine eq.-I, Stephenson & Malanowski 1987)
    log (P_L/kPa) = 5.99434 – 1148.616/(–49.853 + T/K); temp range: 275–344 K (Antoine eq.-II, Stephenson & Malanowski 1987)
    log (P/mmHg) = 32.9509 – 2.8171 × 10³/(T/K) – 8.9572·log (T/K) – 8.7635 × 10⁻¹¹·(T/K) + 3.1710 × 10⁻⁶·(T/K)²; temp range 137–507 K (vapor pressure eq., Yaws 1994)

Henry's Law Constant (Pa m³/mol at 25°C):
    28100   (calculated-P/C, Mackay & Shiu 1981)
    28093   (calculated-vapor-liquid equilibrium (VLE) data, Yaws et al. 1991)

Octanol/Water Partition Coefficient, log K_{OW}:

Octanol/Air Partition Coefficient, log K_{OA}:

Bioconcentration Factor, log BCF:

Sorption Partition Coefficient, log K_{OC}:

Environmental Fate Rate Constants, k, and Half-Lives, $t_{1/2}$:
  Volatilization:
  Photolysis:
  Oxidation: rate constant k, for gas-phase second order rate constants, k_{OH} for reaction with OH radical, k_{NO3} with NO_3 radical and k_{O3} with O_3 or as indicated, *data at other temperatures see reference:
    k_{O3} = 1.05 × 10^{-17} cm^3 molecule^{-1} s^{-1} under atmospheric conditions (Atkinson & Carter 1984)
    k_{OH} = (8.76 ± 0.14) × 10^{-11} cm^3 molecule^{-1} s^{-1} at 298 ± 2 K (relative rate method, Ohta 1984)
    k_{OH} = 62.6 × 10^{-12} cm^3 molecule^{-1} s^{-1} at 298 K (recommended, Atkinson 1985, Atkinson 1989)
    k_{OH} = 6.3 × 10^{-11} cm^3 molecule^{-1} s^{-1}, k_{O3} = 15.0 × 10^{-18} cm^3 molecule^{-1} s^{-1} at 298 K (recommended, Atkinson 1997)
  Hydrolysis:
  Biodegradation:
  Biotransformation:
  Bioconcentration, Uptake (k_1) and Elimination (k_2) Rate Constants or Half-Lives:

Half-Lives in the Environment:
  Surface water: $t_{1/2}$ ~ 320 h and 9 × 10^4 d for oxidation by OH and RO_2 radicals for olefins in aquatic system, and $t_{1/2}$ = 8.0 d, based on rate constant k = 1.0 × 10^6 M^{-1} s^{-1} for oxidation of substituted olefins with singlet oxygen in aquatic system (Mill & Mabey 1985).

Aliphatic and Cyclic Hydrocarbons

2.1.2.1.9 4-Methyl-1-pentene

Common Name: 4-Methyl-1-pentene
Synonym:
Chemical Name: 4-methyl-1-pentene
CAS Registry No: 691-37-2
Molecular Formula: C_6H_{12}
Molecular Weight: 84.159
Melting Point (°C):
    –153.6   (Dreisbach 1959; Lide 2003)
Boiling Point (°C):
    53.88   (Dreisbach 1959)
    53.9   (Lide 2003)
Density (g/cm³ at 20°C):
    0.6642, 0.6594 (20°C, 25°C, Dreisbach 1059)
Molar Volume (cm³/mol):
    126.7   (20°C, calculated-density, McAuliffe 1966, Ruelle & Kesselring 1997)
    127.6   (25°C, calculated-density)
    133.2   (calculated-Le Bas method at normal boiling point)
Enthalpy of Fusion, ΔH_{fus} (kJ/mol):
Entropy of Fusion, ΔS_{fus} (J/mol K):
Fugacity Ratio at 25°C, F: 1.0

Water Solubility (g/m³ or mg/L at 25°C):
    48.0   (shake flask-GC, McAuliffe 1966)

Vapor Pressure (Pa at 25°C and reported temperature dependence equations):
    35600   (calculated-Antoine eq., Dreisbach 1959)
    $\log (P/\text{mmHg}) = 6.87757 - 1130.0/(229.0 + t/°C)$; temp range –20 to 91°C (Antoine eq. for liquid state, Dreisbach 1959)
    36104, 36100 (derived from compiled data, interpolated-Antoine eq., Zwolinski & Wilhoit 1971)
    $\log (P/\text{mmHg}) = 6.83529 - 1121.302/(229.687 + t/°C)$; temp range –37.5 to 76.75°C (Antoine eq., Zwolinski & Wilhoit 1971)
    $\log (P/\text{mmHg}) = 6.83529 - 1121.302/(229.687 + t/°C)$; temp range –38 to 77°C (Antoine eq., Dean 1985, 1992)
    36110   (interpolated-Antoine eq., Stephenson & Malanowski 1987)
    $\log (P_L/\text{kPa}) = 5.94694 - 1114.082/(-44.332 + T/K)$; temp range 265–333 K (Antoine eq., Stephenson & Malanowski 1987)
    $\log (P/\text{mmHg}) = 44.7746 - 2.7364 \times 10^3/(T/K) - 14.283 \cdot \log (T/K) + 7.31 \times 10^{-3} \cdot (T/K) + 4.8402 \times 10^{-14} \cdot (T/K)^2$; temp range 120–496 K (vapor pressure eq., Yaws 1994)

Henry's Law Constant (Pa m³/mol at 25°C):
    62270   (calculated as $1/K_{AW}$, C_W/C_A, reported as exptl., Hine & Mookerjee 1975)
    65200, 34220 (calculated-group contribution, calculated-bond contribution, Hine & Mookerjee 1975)
    63200   (calculated-P/C, Mackay & Shiu 1981)
    63270   (calculated-vapor-liquid equilibrium (VLE) data, Yaws et al. 1991)

Octanol/Water Partition Coefficient, log K_{OW}:
    2.50   (calculated-π constant, Hansch et al. 1968)
    2.51   (calculated-MCI χ, Murray et al. 1975)

Octanol/Air Partition Coefficient, log K_{OA}:

Bioconcentration Factor, log BCF:

Sorption Partition Coefficient, log K_{OC}:

Environmental Fate Rate Constants, k, and Half-Lives, $t_{1/2}$:
- Volatilization:
- Photolysis:
- Oxidation: rate constant k, for gas-phase second order rate constants, k_{OH} for reaction with OH radical, k_{NO3} with NO_3 radical and k_{O3} with O_3 or as indicated, *data at other temperatures see reference:
  $k_{O3} = 1.06 \times 10^{-17}$ cm^3 molecule^{-1} s^{-1} at room temp. (Atkinson & Carter 1984)
  $k_{O3} = 9.2 \times 10^{-18}$ cm^3 molecule^{-1} s^{-1} at 298 K (recommended, Atkinson 1997)
- Hydrolysis:
- Biodegradation:
- Biotransformation:
- Bioconcentration, Uptake (k_1) and Elimination (k_2) Rate Constants or Half-Lives:

Half-Lives in the Environment:
- Surface water: $t_{1/2}$ ~ 320 h and 9×10^4 d for oxidation by OH and RO_2 radicals for olefins in aquatic system, and $t_{1/2}$ = 8.0 d, based on rate constant k = 1.0×10^6 M^{-1} s^{-1} for oxidation of substituted olefins with singlet oxygen in aquatic system (Mill & Mabey 1985).

Aliphatic and Cyclic Hydrocarbons

2.1.2.1.10 1-Hexene

Common Name: 1-Hexene
Synonym: α-hexene
Chemical Name: 1-hexene
CAS Registry No: 646-04-8
Molecular Formula: C_6H_{12}; $CH_3(CH_2)_3CHCH_2$
Molecular Weight: 84.159
Melting Point (°C):
    –139.76   (Lide 2003)
Boiling Point (°C):
    63.48   (Lide 2003)
Density (g/cm³ at 20°C):
    0.6732, 0.6685   (20°C, 25°C, Forziati et al. 1950; Dreisbach 1959; Riddick et al. 1986)
Molar Volume (cm³/mol):
    125.0   (20°C, calculated-density, McAuliffe 1966; Wang et al. 1992; Ruelle & Kesselring 1997)
    125.9   (25°C, calculated-density)
    133.2   (calculated-Le Bas method at normal boiling point, Eastcott et al. 1988)
Enthalpy of Fusion, ΔH_{fus} (kJ/mol):
    9.347   (Riddick et al. 1986)
    9.35   (Chickos et al. 1999)
Entropy of Fusion, ΔS_{fus} (J/mol K):
    70.1, 61.6   (exptl., calculated-group additivity method, total phase change entropy, Chickos et al. 1999)
Fugacity Ratio at 25°C, F: 1.0

Water Solubility (g/m³ or mg/L at 25°C or as indicated. Additional data at other temperatures designated * are compiled at the end of this section.):
    50.0   (shake flask-GC, McAuliffe 1966)
    65.5   (shake flask-titration, Natarajan & Venkatachalam 1972)
    65.5, 54.12, 42.16 (25, 30, 35°C, shake flask-titration, in 0.001M HNO_3 solution, Natarajan & Venkatachalam 1972)
    55.4   (shake flask-GC, Leinonen & Mackay 1973)
    60.0   (20°C, shake flask-GC, Budantseva et al. 1976)
    69.7   (generate column-GC, Tewari et al. 1982a)
    100, 53   (20°C, 25°C, "best" values, IUPAC Solubility Data Series, Shaw 1989a)
    $\ln x = -268.791 + 11353.70/(T/K) + 38.4871 \cdot \ln (T/K)$; temp range 290–400 K (eq. derived from literature calorimetric and solubility data, Tsonopoulos 2001)
    $\ln x = -276.423 + 11833.54/(T/K) + 39.5126 \cdot \ln (T/K)$; temp range 290–400 K (eq. derived from direct fit of solubility data, Tsonopoulos 2001)
    51.43*   (calculated-liquid-liquid equilibrium LLE data, temp range 293.2–494.3 K, Góral et al. 2004)

Vapor Pressure (Pa at 25°C or as indicated and reported temperature dependence equations. Additional data at other temperatures designated * are compiled at the end of this section.):
    22300*   (Antoine eq. regression, temp range –57.5 to 66°C, Stull 1947)
    23500*   (23.7°C, ebulliometry-manometer, measured range 15.9–64.3°C, Forziati et al. 1950)
    $\log (P/mmHg) = 6.86573 - 1152.971/(225.849 + t/°C)$; temp range 15.9–64.3°C (Antoine eq., ebulliometry-manometer measurements, Forziati et al. 1950)
    25000   (calculated-Antoine eq., Dreisbach 1959)
    $\log (P/mmHg) = 6.86572 - 1152.971/(226.0 + t/°C)$; temp range –12 to 79°C (Antoine eq. for liquid state, Dreisbach 1959)
    24800*, 24798 (interpolated-Antoine eq., derived from compiled data, Zwolinski & Wilhoit 1971)
    $\log (P/mmHg) = 6.86573 - 1152.971/(225.849 + t/°C)$; temp range –29.3 to 86.64°C (Antoine eq., Zwolinski & Wilhoit 1971)

log (P/mmHg) = [–0.2185 × 7787.6/(T/K)] + 7.930324; temp range –57.5 to 66°C (Antoine eq., Weast 1972–73)
24800 (interpolated-Antoine eq., Boublik et al. 1984)
log (P/kPa) = 5.99426 – 1154.952/(226.002 + t/°C); temp range 15.89–64.311°C (Antoine eq. from reported exptl. data of Forziati et al. 1950, Boublik et al. 1984)
24800 (interpolated-Antoine eq., temp range –16 to 64°C, Dean 1985, 1992)
log (P/mmHg) = 6.85770 – 1148.62/(225.25 + t/°C); temp range –16 to 64°C (Antoine eq., Dean 1985, 1992)
24800 (selected lit., Riddick et al. 1986)
log (P/kPa) = 5.98260 – 1148.62/(225.346 + t/°C); temp range not specified (Antoine eq., Riddick et al. 1986)
24800 (interpolated-Antoine eq., Stephenson & Malanowski 1987)
log (P_L/kPa) = 5.98336 – 1149.029/(–47.755 + T/K); temp range 273–343 K (Antoine eq., Stephenson & Malanowski 1987)
log (P/mmHg) = 33.4486 – 2.6221 × 10^3/(T/K) – 9.1784·log (T/K) + 3.093 × 10^{-12}·(T/K) + 3.678 × 10^{-6}·(T/K)^2; temp range 133–504 K (vapor pressure eq., Yaws 1994)

Henry's Law Constant (Pa m^3/mol at 25°C):
41750 (calculated-P/C, Mackay & Shiu 1975)
44080 (calculated-1/K_{AW}, C_W/C_A, reported as exptl., Hine & Mookerjee 1975)
51790, 34220 (calculated-group contribution, calculated-bond contribution, Hine & Mookerjee 1975)
41800 (calculated-P/C, Mackay & Shiu 1981)
47230 (calculated- χ, Nirmalakhandan & Speece 1988)
41640 (calculated-P/C, Eastcott et al. 1988)
29940 (calculated-vapor–liquid equilibrium (VLE) data, Yaws et al. 1991)

Octanol/Water Partition Coefficient, log K_{OW}:
3.39 (generator column-GC, Tewari et al. 1982a,b)
3.47 (calculated-activity coeff. γ, Wasik et al. 1981)
3.48 (calculated-activity coeff. γ, Wasik et al. 1982)
3.39, 3.40 (generator column-GC, calculated-activity coeff. γ, Schantz & Martire 1987)
3.40 (recommended, Sangster 1989)
3.39 (recommended, Hansch et al. 1995)

Octanol/Air Partition Coefficient, log K_{OA} at 25°C or as indicated. Additional data at other temperatures designated * are compiled at the end of this section:
2.50* (20.29°C, from GC-determined γ$^\infty$ in octanol, measured range 20.29–50.28°C, Gruber et al. 1997)
2.41 (calculated-measured γ$^\infty$ in pure octanol and vapor pressure P, Abraham et al. 2001)

Bioconcentration Factor, log BCF:

Sorption Partition Coefficient, log K_{OC}:

Environmental Fate Rate Constants, k, and Half-Lives, $t_{1/2}$:
Volatilization:
Photolysis:
Oxidation: rate constant k, for gas-phase second order rate constants, k_{OH} for reaction with OH radical, k_{NO3} with NO_3 radical and k_{O3} with O_3 or as indicated, *data at other temperatures see reference:
k_{O3} = 0.90 × 10^{-17} cm^3 molecule^{-1} s^{-1} at room temp. (Cadle & Schadt 1952)
k_{O3} = 1.00 × 10^{-17} cm^3 molecule^{-1} s^{-1} at room temp. (Hanst et al. 1958)
k_{O3} = 1.10 × 10^{-17} cm^3 molecule^{-1} s^{-1} at 30°C (flow system, Bufalini & Altshuller 1965)
k_{O3} = 1.40 × 10^{-17} cm^3 molecule^{-1} s^{-1} at 296 K (static system-chemiluminescence, Cox & Penkett 1972)
k_{O3} = 1.10 × 10^{-17} cm^3 molecule^{-1} s^{-1} at 299 K (static system-chemiluminescence, Stedman et al. 1973)
k_{O3} = 1.11 × 10^{-17} cm^3 molecule^{-1} s^{-1} (static system-chemiluminescence, Japar et al. 1974)
k_{O3} = 1.08 × 10^{-17} cm^3 molecule^{-1} s^{-1} at 294 ± 2 K (static system-chemiluminescence, Adeniji et al. 1981)
k_{O3} = 1.21 × 10^{-17} cm^3 molecule^{-1} s^{-1} at 296 K (static system-chemiluminescence, Atkinson et al. 1982)
k_{O3} = (0.91 to 1.36) × 10^{-17} cm^3 molecule^{-1} s^{-1} at 294–303 K (Atkinson & Carter 1984)
k_{OH} = 3.68 × 10^{-11} cm^3 molecule^{-1} s^{-1} at 295 K (relative rate method, Atkinson & Aschmann 1984)

k_{OH} = (3.75 – 3.25) × 10^{-11} cm³ molecule⁻¹ s⁻¹ at 295–303 K (Atkinson 1985)
k_{OH} = (32.9 – 37.5) × 10^{-12} cm³ molecule⁻¹ s⁻¹ at 295–303 K (Atkinson 1989)
k_{OH} = 3.18 × 10^{-11} cm³ molecule⁻¹ s⁻¹, k_{O3} = 1.17 × 10^{-17} cm³ molecule⁻¹ s⁻¹ at 298 K (Atkinson 1990)
k_{O3} = 1.10 × 10^{-17} cm³ molecule⁻¹ s⁻¹ at 298 K (recommended, Atkinson 1997)
k_{OH} = 3.7 × 10^{-11} cm³ molecule⁻¹ s⁻¹, k_{O3}* = 11.0 × 10^{-18} cm³ molecule⁻¹ s⁻¹, and $k_{O(3P)}$ = 4.65 × 10^{-12} cm³ molecule⁻¹ s⁻¹ for reaction with O(³P) atom at 298 K (recommended, Atkinson 1997)

Hydrolysis:

Biodegradation:

Biotransformation:

Bioconcentration, Uptake (k_1) and Elimination (k_2) Rate Constants or Half-Lives:

Half-Lives in the Environment:

Surface water: $t_{½}$ ~ 320 h and 9 × 10^4 d for oxidation by OH and RO_2 radicals for olefins in aquatic system, and $t_{½}$ = 7.3 d, based on rate constant k = 3 × 10^3 M⁻¹ s⁻¹ for oxidation of unsubstituted olefins with singlet oxygen in aquatic system (Mill & Mabey 1985).

TABLE 2.1.2.1.10.1
Reported aqueous solubilities of 1-hexene at various temperatures

Góral et al. 2004

Calc-recommended LLE data

t/°C	S/g·m⁻³
20	56.11
25	51.43
36.8	51.43
93.3	102.9
148.8	392.8
204.4	2151
221.1	3740

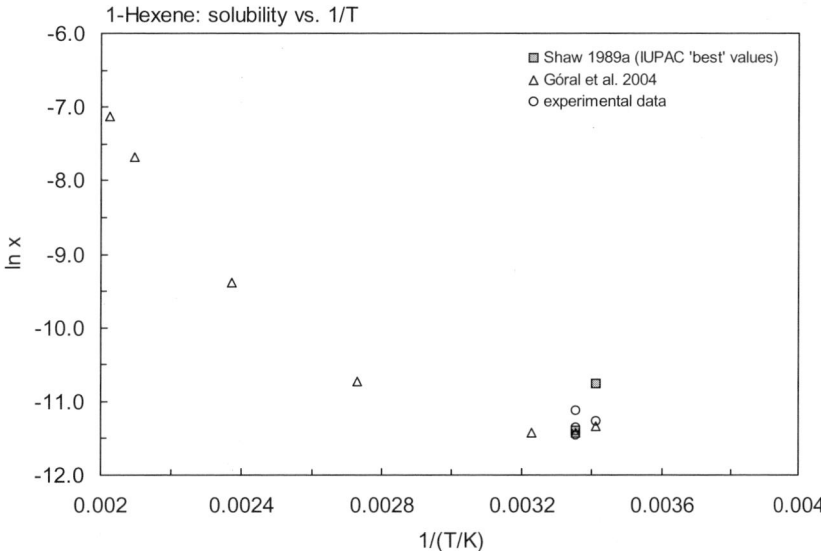

FIGURE 2.1.2.1.10.1 Logarithm of mole fraction solubility (ln x) versus reciprocal temperature for 1-hexene.

TABLE 2.1.2.1.10.2
Reported vapor pressures and octanol-air partition coefficients of 1-hexene at various temperatures and the coefficients for the vapor pressure equations

$\log P = A - B/(T/K)$ (1) $\quad\quad \ln P = A - B/(T/K)$ (1a)
$\log P = A - B/(C + t/°C)$ (2) $\quad\quad \ln P = A - B/(C + t/°C)$ (2a)
$\log P = A - B/(C + T/K)$ (3)
$\log P = A - B/(T/K) - C \cdot \log (T/K)$ (4)

Vapor pressure						log K_{OA}	
Stull 1947		Forziati et al. 1950		Zwolinski & Wilhoit 1971		Gruber et al. 1997	
Summary of literature data		Ebulliometry		Selected values		GC det'd activity coefficient	
t/°C	P/Pa	t/°C	P/Pa	t/°C	P/Pa	t/°C	log K_{OA}
−57.5	133.3	15.890	16645	−29.30	1333	20.29	2.503
−38.0	666.6	19.950	19946	−18.65	2666	30.3	2.331
−28.1	1333	23.720	23485	−11.88	4000	40.4	2.20
−17.2	2666	28.762	28990	−6.81	5333	50.28	2.068
−5.0	5333	33.399	34936	−2.70	6666		
2.8	7999	38.993	43366	0.776	7999		
13.0	13332	44.763	53705	8.482	10666		
29.0	26664	50.914	66798	11.109	13332		
46.8	53329	62.323	97648	20.006	19998		
66.0	101325	62.827	99230	26.736	26664		
		63.299	100730	32.215	33331		
mp/°C		62.837	102457	38.871	39997		
		64.311	103995	44.569	53329		
				50.859	66661		
		eq. 2	P/mmHg	56.219	79993		
		A	6.86573	60.915	93326		
		B	1152.971	61.790	95992		
		C	225.849	62.647	98659		
		bp/°C	63.485	63.585	101325		
				25.0	24798		
				eq. 2	P/mmHg		
				A	6.86572		
				B	1152.971		
				C	225.849		
				bp/°C	63.485		
				ΔH_V/(kJ mol^{-1}) =			
				at 25°C	28.28		
				at bp	30.63		

Aliphatic and Cyclic Hydrocarbons

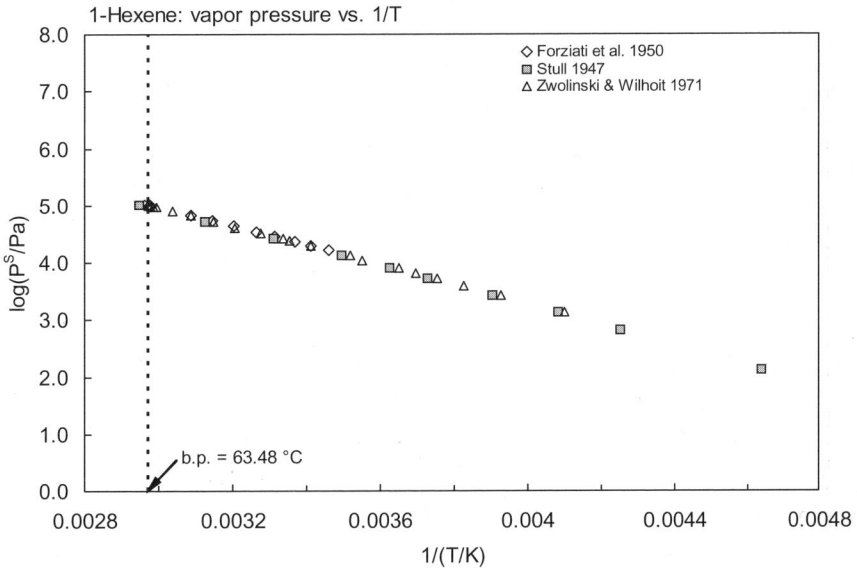

FIGURE 2.1.2.1.10.2 Logarithm of vapor pressure versus reciprocal temperature for 1-hexene.

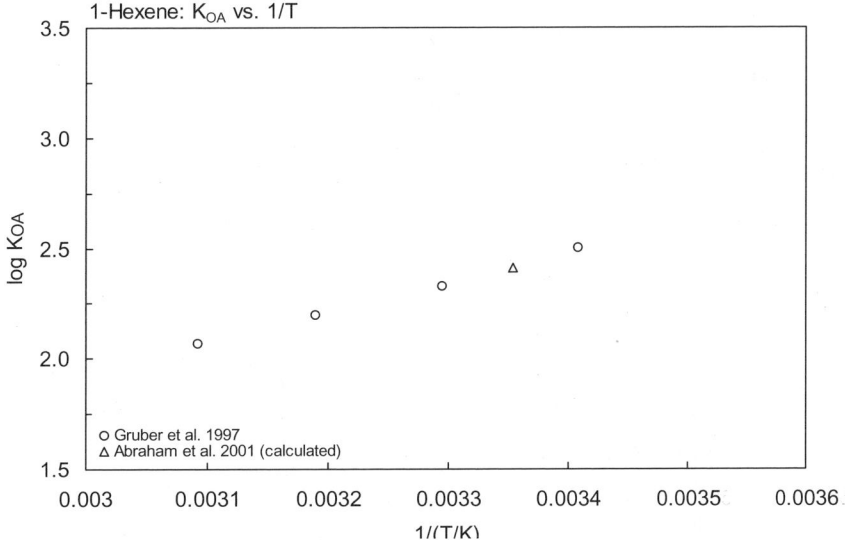

FIGURE 2.1.2.1.10.3 Logarithm of K_{OA} versus reciprocal temperature for 1-hexene.

2.1.2.1.11   1-Heptene

Common Name: 1-Heptene
Synonym: 1-heptylene, α-heptene
Chemical Name: 1-heptene
CAS Registry No: 592-76-7
Molecular Formula: C_7H_{14}, $CH_3(CH_2)_4CH=CH_2$
Molecular Weight: 98.186
Melting Point (°C):
    –118.9   (Lide 2003)
Boiling Point (°C):
    93.64   (Lide 2003)
Density (g/cm³ at 20°C):
    0.6970, 0.6927   (20°C, 25°C, Forziati et al. 1950; Dreisbach 1959)
Molar Volume (cm³/mol):
    140.9   (20°C, calculated-density, Stephenson & Malanowski 1987; Ruelle & Kesselring 1997)
    141.8   (25°C, calculated-density)
    155.4   (calculated-Le Bas method at normal boiling point)
Enthalpy of Vaporization, ΔH_V (kJ/mol):
    35.65, 31.09 (25°C, bp, Riddick et al. 1986)
Enthalpy of Fusion, ΔH_{fus} (kJ/mol):
    12.401   (Riddick et al. 1986)
    12.66   (Chickos et al. 1999)
Entropy of Fusion, ΔS_{fus} (J/mol K):
    82.5, 77.5   (exptl., calculated-group additivity method, total phase change entropy, Chickos et al. 1999)
    Fugacity Ratio at 25°C, F: 1.0

Water Solubility (g/m³ or mg/L at 25°C or as indicated):
    31.03, 27.6, 24.06 (20°C, 25°C, 30°C, shake flask-titration, in 0.001M HNO_3 solution, Natarajan & Venkatachalam 1972)
    18.16   (generator column-GC, Tewari et al. 1982a)
    23.6   (calculated-activity coeff. γ and K_{OW}, Tewari et al. 1982b)
    19, 25, 32, 38 (10, 20, 24, 30°C, "best values", IUPAC Solubility Data Series, Shaw 1989a)
    15.27, 13.6, 13.1 (10, 20, 30°C, calculated-recommended liquid-liquid equilibrium LLE data, Góral et al. 2004)

Vapor Pressure (Pa at 25°C or as indicated and reported temperature dependence equations. Additional data at other temperatures designated * are compiled at the end of this section.):
    7690*   (25.5°C, ebulliometry-manometer, measured range 21.6–94.5°C, Forziati et al. 1950)
    log (P/mmHg) = 6.90069 – 1257.505/(219.179 + t/°C); temp range 21.6–94.5°C (Antoine eq., ebulliometry-manometer measurements, Forziati et al. 1950)
    7510   (calculated-Antoine eq., Dreisbach 1959)
    log (P/mmHg) = 6.90069 – 1257.505/(219.18 + t/°C); temp range 10–128°C (Antoine eq. for liquid state, Dreisbach 1959)
    26663* (53.17°C, temp range 53.17–93.61°C, Eisen & Orav 1970; quoted, Boublik et al. 1984)
    7506*, 7510 (derived from compiled data, interpolated-Antoine eq., Zwolinski & Wilhoit 1971)
    log (P/mmHg) = 6.90069 – 1257.505/(219.179 + t/°C); temp range –6.07 to 118.44°C (Antoine eq., Zwolinski & Wilhoit 1971)
    log (P/mmHg) = [–0.2185 × 8643.2/(T/K)] + 7.991519; temp range –35.8 to 98.5°C (Antoine eq., Weast 1972–73)
    7520, 7530 (interpolated-Antoine equations., Boublik et al. 1984)
    log (P/kPa) = 6.04107 – 1266.473/(220.202 + t/°C); temp range 21.6–94.53°C (Antoine eq. from reported exptl. data of Forziati et al. 1950, Boublik et al. 1984)

Aliphatic and Cyclic Hydrocarbons

log (P/kPa) = 6.03512 – 1263.343/(219.922 + t/°C); temp range 54.17–93.61°C (Antoine eq. from reported exptl. data, Boublik et al. 1984)

7515 (interpolated-Antoine eq., Dean 1985, 1992)

log (P/mmHg) = 6.91087 – 1258.345/(219.30 + t/°C); temp range –6 to 118°C (Antoine eq., Dean 1985, 1992)

7500 (quoted lit., Riddick et al. 1986)

log (P/kPa) = 6.02677 – 1258.34/(219.299 + t/°C); temp range not specified (Antoine eq., Riddick et al. 1986)

7500 (extrapolated-Antoine eq., Stephenson & Malanowski 1987)

log (P_L/kPa) = 5.99079 – 1237.44/(–56.26 + T/K); temp range 311–368 K (Antoine eq., Stephenson & Malanowski 1987)

log (P/mmHg) = 38.1255 – 3.064 × 10^3/(T/K) – 10.679·log (T/K) + 1.2244 × 10^{-10}·(T/K) + 3.668 × 10^{-6}·(T/K)^2; temp range 154–537 K (vapor pressure eq., Yaws 1994)

Henry's Law Constant (Pa m^3/mol at 25°C):

40580 (calculated-vapor-liquid equilibrium (VLE) data, Yaws et al. 1991)

Octanol/Water Partition Coefficient, log K_{OW}:

3.99 (generator column-concn. ratio-GC, Tewari et al. 1982a,b)
4.09 (calculated-activity coeff. γ, Wasik et al. 1982)
4.06 (generator column-GC, Schantz & Martire 1987)
3.99 (recommended, Sangster 1989, 1993)
3.99 (recommended, Hansch et al. 1995)

Octanol/Air Partition Coefficient, log K_{OA}:

Bioconcentration Factor, log BCF:

Sorption Partition Coefficient, log K_{OC}:

Environmental Fate Rate Constants, k, and Half-Lives, $t_{½}$:

Volatilization:

Photolysis:

Oxidation: rate constant k, for gas-phase second order rate constants, k_{OH} for reaction with OH radical, k_{NO3} with NO$_3$ radical and k_{O3} with O$_3$ or as indicated, *data at other temperatures see reference:

k_{O3} = 8.1 × 10^{-18} cm^3 molecule^{-1} s^{-1} at room temp. (Cadle & Schadt 1952)

k_{O3} = 1.73 × 10^{-17} cm^3 molecule^{-1} s^{-1} at 296 K (static system-chemiluminescence, Atkinson et al. 1982)

k_{OH} = 3.97 × 10^{-11} cm^3 molecule^{-1} s^{-1} at 295 K (relative rate method, Atkinson & Aschmann 1984)

k_{OH} = 40.5 × 10^{-12} cm^3 molecule^{-1} s^{-1} and 36.1 × 10^{-12} cm^3 molecule^{-1} s^{-1} at 298 K and 305 K respectively (Atkinson 1989)

k_{OH} = 4.0 × 10^{-11} cm^3 molecule^{-1} s^{-1}, k_{O3} = 1.73 × 10^{-17} cm^3 molecule^{-1} s^{-1} at 298 K (Atkinson 1990)

k_{OH} = 4.0 × 10^{-11} cm^3 molecule^{-1} s^{-1}, k_{O3} = 12.0 × 10^{-18} cm^3 molecule^{-1} s^{-1} at 298 K (recommended, Atkinson 1997)

Hydrolysis:

Biodegradation:

Biotransformation:

Bioconcentration, Uptake (k_1) and Elimination (k_2) Rate Constants or Half-Lives:

Half-Lives in the Environment:

Surface water: $t_{½}$ ~ 320 h and 9 × 10^4 d for oxidation by OH and RO$_2$ radicals for olefins in aquatic system, and $t_{½}$ = 7.3 d, based on rate constant k = 3 × 10^3 M^{-1} s^{-1} for oxidation of unsubstituted olefins with singlet oxygen in aquatic system (Mill & Mabey 1985).

TABLE 2.1.2.1.11.1
Reported vapor pressures of 1-heptene at various temperatures and the coefficients for the vapor pressure equations

$\log P = A - B/(T/K)$ (1) $\ln P = A - B/(T/K)$ (1a)

$\log P = A - B/(C + t/°C)$ (2) $\ln P = A - B/(C + t/°C)$ (2a)

$\log P = A - B/(C + T/K)$ (3)

$\log P = A - B/(T/K) - C \cdot \log(T/K)$ (4)

Forziati et al. 1950		Eisen & Orav 1970		Zwolinski & Wilhoit 1971			
Ebulliometry		in Boublik et al. 1984		Selected values			
t/°C	P/Pa	t/°C	P/Pa	t/°C	P/Pa	t/°C	P/Pa
21.609	6485	53.17	26663	−6.07	1333	eq. 2	P/mmHg
25.492	7691	65.06	39997	5.39	2666	A	6.90069
28.768	8991	73.32	53329	12.68	4000	B	1257.505
34.525	11700	80.07	66661	18.15	5333	C	219.179
38.281	13845	85.82	79993	22.569	6666	bp/°C	93.643
42.564	16644	90.86	93325	26.306	7999		
46.923	19945	91.33	94659	32.443	10666	$\Delta H_V/(kJ\ mol^{-1})$ =	
50.970	23482	91.79	95992	37.418	13332	at 25°C	31.09
56.384	28988	92.25	97325	46.982	19998	at bp	35.65
67.366	43366	92.71	98658	54.212	26664		
73.563	53705	93.16	99991	60.096	33331		
80.179	66801	93.61	101325	65.095	39997		
92.391	97650			73.357	53329		
92.941	99233			80.104	66661		
93.444	100733			84.853	79993		
94.022	102462			90.888	93326		
94.531	104002			91.826	95992		
				92.744	98659		
eq. 2	P/mmHg			93.643	101325		
A	6.90069			25.0	7506		
B	1257.505						
C	219.179						
bp/°C	93.643						

Aliphatic and Cyclic Hydrocarbons

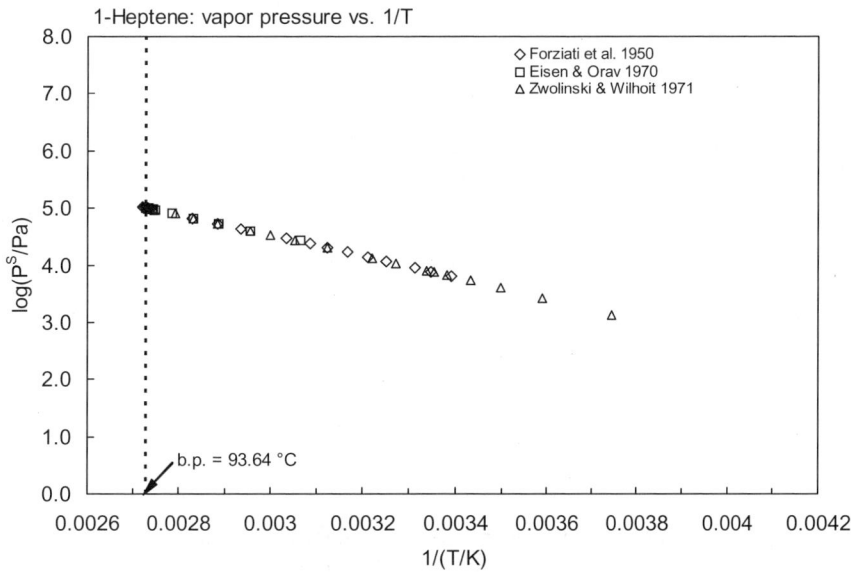

FIGURE 2.1.2.1.11.1 Logarithm of vapor pressure versus reciprocal temperature for 1-heptene.

2.1.2.1.12   1-Octene

Common Name: 1-Octene
Synonym: α-octene, caprylene, α-octylene
Chemical Name: 1-octene
CAS Registry No: 111-66-0
Molecular Formula: C_8H_{16}; $CH_3(CH_2)_5CH=CH_2$
Molecular Weight: 112.213
Melting Point (°C):
    −101.7   (Lide 2003)
Boiling Point (°C):
    121.29   (Lide 2003)
Density (g/cm³ at 20°C):
    0.7149, 0.7109   (20°C, 25°C, Forziati et al. 1950; Dreisbach 1959; Riddick et al. 1986)
Molar Volume (cm³/mol):
    154.9   (20°C, calculated-density, Stephenson & Malanowski 1987)
    177.6   (calculated-Le Bas method at normal boiling point)
Enthalpy of Vaporization, ΔH_V (kJ/mol):
    40.35, 33.95 (25°C, bp, Riddick et al. 1986)
Enthalpy of Fusion, ΔH_{fus} (kJ/mol):
    15.569   (Riddick et al. 1986)
    15.31   (Chickos et al. 1999)
Entropy of Fusion, ΔS_{fus} (J/mol K):
    89.29, 86.8   (exptl., calculated-group additivity method, total phase change entropy, Chickos et al. 1999)
Fugacity Ratio at 25°C, F: 1.0

Water Solubility (g/m³ or mg/L at 25°C or as indicated):
    2.70   (shake flask-GC, McAuliffe 1966)
    4.10   (generator column-GC, Tewari et al. 1982a)
    6.82   (calculated-activity coeff. γ and K_{OW}, Tewari et al. 1982b)
    2.70, 22.2   (quoted, IUPAC Solubility Data Series, Shaw 1989)
    2.93, 2.93   (25, 37.78°C, calculated-recommended liquid-liquid equilibrium LLE data, Góral et al. 2004)

Vapor Pressure (Pa at 25°C or as indicated and reported temperature dependence equations. Additional data at other temperatures designated * are compiled at the end of this section.):
    6382*   (44.893°C, ebulliometry, measured range 44.893–122.223°C, Forziati et al. 1950)
    log (P/mmHg) = 6.93262 − 1353.486/(212.765 + t/°C); temp range 44.8–122.2°C (Antoine eq., ebulliometry-manometer measurements, Forziati et al. 1950)
    2317   (calculated-Antoine eq., Dreisbach 1959)
    log (P/mmHg) = 6.93263 − 1253.5/(212.764 + t/°C); temp range 0–151°C (Antoine eq. for liquid state, Dreisbach 1959)
    2320*   (interpolated-Antoine eq., temp range 15.38–147.54°C, Zwolinski & Wilhoit 1971)
    log (P/mmHg) = 6.93263 − 1253.486/(212.764 + t/°C); temp range 15.38–147.54°C (Antoine eq., Zwolinski & Wilhoit 1971)
    2320   (extrapolated-Antoine eq., Boublik et al. 1984)
    log (P/kPa) = 6.06421 − 1356.472/(213.099 + t/°C); temp range 44.89–122.2°C (Antoine eq. from reported exptl. data Forziati et al. 1950, Boublik et al. 1984)
    2320   (interpolated-Antoine eq., temp range 15–147°C, Dean 1985, 1992)
    log (P/mmHg) = 6.93495 − 1355.46/(213.05 + t/°C); temp range 15–147°C (Antoine eq., Dean 1985, 1992)
    2300   (quoted lit., Riddick et al. 1986)
    log (P/kPa) = 6.05985 − 1355.46/(213.054 + t/°C); temp range not specified (Antoine eq., Riddick et al. 1986)
    2320   (extrapolated-Antoine eq., Stephenson & Malanowski 1987)

log $(P_L/kPa) = 6.05178 - 1350.245/(-60.716 + T/K)$; temp range 317–400 K (Antoine eq., Stephenson & Malanowski 1987)

log $(P/mmHg) = 56.1183 - 3.7657 \times 10^3/(T/K) - 10.006 \cdot \log(T/K) + 7.7387 \times 10^{-3} \cdot (T/K) - 1.3036 \times 10^{-6} \cdot (T/K)^2$; temp range 171–567 K (vapor pressure eq., Yaws 1994)

Henry's Law Constant (Pa m^3/mol at 25°C):

91700	(calculated-P/C, Mackay & Shiu 1975; selected, Mills et al. 1982)
96440	(calculated-$1/K_{AW}$, C_W/C_A, reported as exptl., Hine & Mookerjee 1975)
101000, 75000	(calculated-group contribution, calculated-bond contribution, Hine & Mookerjee 1975)
96400	(calculated-P/C, Mackay et al. 1981; Eastcott et al. 1988)
74860	(calculated-MCI χ, Nirmalakhandan & Speece 1988)
63500	(calculated-vapor-liquid equilibrium (VLE) data, Yaws et al. 1991)

Octanol/Water Partition Coefficient, log K_{OW}:

4.57	(generator column-GC, Tewari et al. 1982a)
4.76	(calculated-activity coeff. γ, Wasik et al. 1981, 1982)
4.56, 4.72	(generator column-GC, calculated-activity coeff. γ, Schantz & Martire 1987)
4.57	(recommended, Sangster 1989)
4.57	(recommended, Hansch et al. 1995)

Octanol/Air Partition Coefficient, log K_{OA}:

3.53	(calculated-measured γ^∞ in pure octanol and vapor pressure P, Abraham et al. 2001)

Bioconcentration Factor, log BCF:

Sorption Partition Coefficient, log K_{OC}:

Environmental Fate Rate Constants, k, and Half-Lives, $t_{1/2}$:

Volatilization:

Photolysis:

Oxidation: rate constant k; for gas-phase second-order rate constants, k_{OH} for reaction with OH radical, k_{NO3} with NO$_3$ radical and k_{O3} with O$_3$ or as indicated, *data at other temperatures see reference:

$k_{O3} = 8.1 \times 10^{-18}$ cm^3 molecule^{-1} s^{-1} for the reaction with ozone in air (Atkinson & Carter 1984)

$k_{OH} = 4.0 \times 10^{-11}$ cm^3 molecule^{-1} s^{-1}, $k_{O3} = 1.70 \times 10^{-17}$ cm^3 molecule^{-1} s^{-1}, and $k_{O(3P)} = 1.10 \times 10^{-11}$ cm^3 molecule^{-1} s^{-1} for the reaction with O(^{3}P) atom in gas phase (Paulson & Seinfeld 1992)

$k_{O3} = 14.0 \times 10^{-18}$ cm^3 molecule^{-1} s^{-1} at 298 K (recommended, Atkinson 1997)

Hydrolysis:

Biodegradation:

Biotransformation:

Bioconcentration, Uptake (k_1) and Elimination (k_2) Rate Constants or Half-Lives:

Half-Lives in the Environment:

Surface water: $t_{1/2} \sim 320$ h and 9×10^4 d for oxidation by OH and RO$_2$ radicals for olefins in aquatic system, and $t_{1/2} = 7.3$ d, based on rate constant $k = 3 \times 10^3$ M^{-1} s^{-1} for oxidation of unsubstituted olefins with singlet oxygen in aquatic system (Mill & Mabey 1985).

TABLE 2.1.2.1.12.1
Reported vapor pressures of 1-octene at various temperatures and the coefficients for the vapor pressure equations

$\log P = A - B/(T/K)$ (1) $\qquad$ $\ln P = A - B/(T/K)$ (1a)

$\log P = A - B/(C + t/°C)$ (2) $\qquad$ $\ln P = A - B/(C + t/°C)$ (2a)

$\log P = A - B/(C + T/K)$ (3)

$\log P = A - B/(T/K) - C \cdot \log (T/K)$ (4)

Forziati et al. 1950				Zwolinski & Wilhoit 1971			
ebulliometry				selected values			
t/°C	P/Pa	t/°C	P/Pa	t/°C	P/Pa	t/°C	P/Pa
44.893	6382	121.075	100743	15.38	1333	118.362	93326
48.975	7690	121.685	102474	27.57	2666	119.355	95992
52.140	8994	122.223	102019	35.33	4000	120.328	98659
55.581	10330			41.15	5333	121.2880	101325
58.557	11720	eq. 2	P/mmHg	45.848	6666	25.0	2320
62.557	13844	A	6.93262	49.820	7999		
67.096	16644	B	1353.486	56.343	10666	eq. 2	P/mmHg
71.736	19945	C	212.764	61.630	13332	A	6.93263
76.022	23482	bp/°C	121.280	71.789	19998	B	1353.486
81.779	2890			79.645	26664	C	212.764
87.053	34934			85.710	33331	bp/°C	121.280
93.428	43366			91.014	39997	$\Delta H_V/(kJ\ mol^{-1}) =$	
106.997	66806			99.778	53329	at 25°C	37.95
119.967	97658			106.932	66661	at bp	33.76
120.539	99242			113.026	79993		

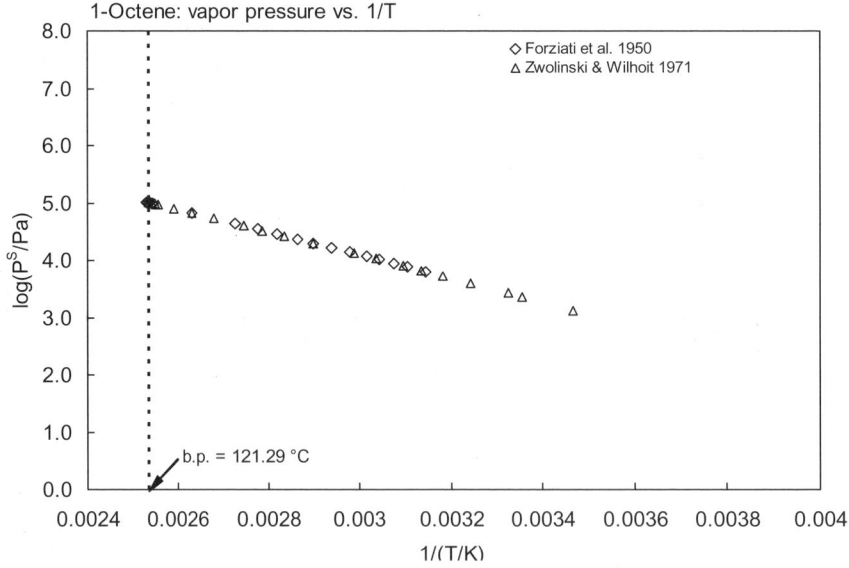

FIGURE 2.1.2.1.12.1 Logarithm of vapor pressure versus reciprocal temperature for 1-octene.

Aliphatic and Cyclic Hydrocarbons

2.1.2.1.13   1-Nonene

Common Name: 1-Nonene
Synonym: α-nonene, *n*-heptylethylene, 1-nonylene
Chemical Name: 1-nonene
CAS Registry No: 124-11-8
Molecular Formula: C_9H_{18}; $CH_3(CH_2)_6CH=CH_2$
Molecular Weight: 126.239
Melting Point (°C):
  –81.3   (Lide 2003)
Boiling Point (°C):
  146.9   (Lide 2003)
Density (g/cm³ at 20°C):
  0.7292, 0.7235 (20°C, 25°C, Forziati et al. 1950; Dreisbach 1959; Riddick et al. 1986)
Molar Volume (cm³/mol):
  173.1, 174.1 (20°C, 25°C, calculated-density)
  199.8   (calculated-Le Bas method at normal boiling point)
Enthalpy of Vaporization, ΔH_V (kJ/mol):
  45.52, 36.31 (25°C, bp, Riddick et al. 1986)
Enthalpy of Fusion, ΔH_{fus} (kJ/mol):
  19.075   (Riddick et al. 1986)
  19.37   (Chickos et al. 1999)
Entropy of Fusion, ΔS_{fus} (J/mol K):
  104.23, 96.1 (exptl., calculated-group additivity method, total phase change entropy, Chickos et al. 1999)
Fugacity Ratio at 25°C, F: 1.0

Water Solubility (g/m³ or mg/L at 25°C):
  0.63   (estimated-nomograph, Brookman et al. 1986)
  1.12   (generator column-GC, Tewari et al. 1982a)
  2.09   (calculated-activity coeff. γ and K_{OW}, Tewari et al. 1982b)

Vapor Pressure (Pa at 25°C or as indicated and reported temperature dependence equations. Additional data at other temperatures designated * are compiled at the end of this section.):
  6385*   (66.607°C, ebulliometry, measured range 66.607–147.860°C, Forziati et al. 1950)
  log (P/mmHg) = 6.95389 – 1435.359/(205.535 + t/°C); temp range 66.6–147.9°C (Antoine eq., ebulliometry-manometer measurements, Forziati et al. 1950)
  712   (calculated-Antoine eq., Dreisbach 1959)
  log (P/mmHg) = 6.95387 – 1435.295/(205.535 + t/°C); temp range 25–173°C (Antoine eq. for liquid state, Dreisbach 1959)
  707*, 712 (derived from compiled data, extrapolated-Antoine eq., Zwolinski & Wilhoit 1971)
  log (P/mmHg) = 6.95430 – 1436.20/(205.69 + t/°C); temp range 35–175°C (Antoine eq., Dean 1985, 1992)
  710   (quoted lit., Riddick et al. 1986)
  log (P/kPa) = 6.07920 – 1436.20/(205.690 + t/°C); temp range not specified (Antoine eq., Riddick et al. 1986)
  712   (extrapolated-Antoine eq., Stephenson & Malanowski 1987)
  log (P_L/kPa) = 6.07341 – 1432,435/(–67.884 + T/K); temp range 339–423 K (Antoine eq., Stephenson & Malanowski 1987)
  log (P/mmHg) = 60.6089 – 4.2023 × 10³/(T/K) – 19.446·log (T/K) + 7.8308 × 10⁻³·(T/K) + 1.591 × 10⁻¹³·(T/K)²; temp range 192–593 K (vapor pressure eq., Yaws 1994)
  650.4   (23.25°C, transpiration method, Verevkin et al. 2000)
  ln (P/Pa) = 24.60 – 5379/(T/K); temp range 278.5–318.3 K (transpiration method, Verevkin et al. 2000)

Henry's Law Constant (Pa m^3/mol at 25°C):
    80450   (calculated-vapor-liquid equilibrium (VLE) data, Yaws et al. 1991)

Octanol/Water Partition Coefficient, log K_{OW}:
    5.15   (generator column-GC, Tewari et al. 1982a,b)
    5.34   (calculated-activity coeff. γ, Wasik et al. 1981, 1982)
    5.31   (calculated-activity coeff. γ, Schantz & Martire 1987)
    5.15   (recommended, Sangster 1989)
    5.15   (recommended, Hansch et al. 1995)

Octanol/Air Partition Coefficient, log K_{OA}:
    3.83   (calculated-measured $γ^∞$ in pure octanol and vapor pressure P, Abraham et al. 2001)

Bioconcentration Factor, log BCF:

Sorption Partition Coefficient, log K_{OC}:

Environmental Fate Rate Constants, k and Half-Lives, $t_{½}$:

Half-Lives in the Environment:

    Surface water: $t_{½}$ ~ 320 h and $9 × 10^4$ d for oxidation by OH and RO_2 radicals for olefins in aquatic system, and $t_{½}$ = 7.3 d, based on rate constant k = $3 × 10^3$ M^{-1} s^{-1} for oxidation of unsubstituted olefins with singlet oxygen in aquatic system (Mill & Mabey 1985).

TABLE 2.1.2.1.13.1
Reported vapor pressures of 1-nonene at various temperatures and the coefficients for the vapor pressure equations

$$\log P = A - B/(T/K) \quad (1)$$
$$\log P = A - B/(C + t/°C) \quad (2)$$
$$\log P = A - B/(C + T/K) \quad (3)$$
$$\log P = A - B/(T/K) - C·\log (T/K) \quad (4)$$
$$\ln P = A - B/(T/K) \quad (1a)$$
$$\ln P = A - B/(C + t/°C) \quad (2a)$$

Forziati et al. 1950				Zwolinski & Wilhoit 1971			
ebulliometry				selected values			
t/°C	P/Pa	t/°C	P/Pa	t/°C	P/Pa	t/°C	P/Pa
66.607	6385	146.091	99242	35.55	1333	143.805	93326
70.874	7691	146.653	100742	48.38	2666	144.848	95992
74.517	8994	147.289	102474	56.55	4000	145.869	98659
77.861	10331	147.860	104020	62.67	5333	146.868	101325
81.001	11722			67.612	6666	25.0	706.6
85.202	13805	bp/°C	146.868	71.790	7999		
89.942	16644			78.651	10666	eq. 2	P/mmHg
94.829	19945	eq. 2	P/mmHg	84.210	13332	A	6.95387
99.341	23482	A	6.95389	94.889	19998	B	1435.359
110.935	34934	B	1435.359	102.956	26664	C	205.535
117.622	43364	C	146.868	109.518	33331	bp/°C	146.868
124.521	53707			115.090	39997	$ΔH_V$/(kJ mol^{-1}) =	
131.881	66806			124.295	53329	at 25°C	36.32
139.859	83770			131.808	66661	at bp	45.52
145.488	97658			138.204	79993		

Aliphatic and Cyclic Hydrocarbons

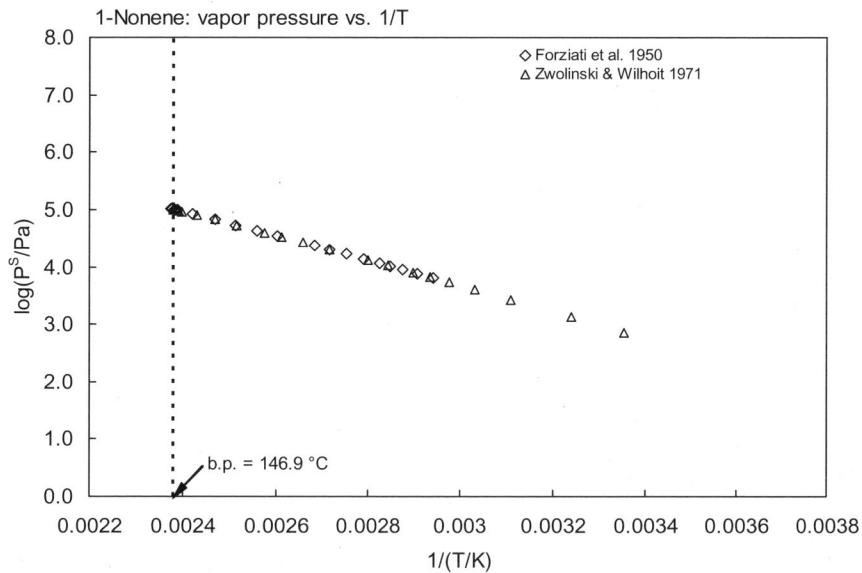

FIGURE 2.1.2.1.13.1 Logarithm of vapor pressure versus reciprocal temperature for 1-nonene.

2.1.2.1.14 1-Decene

Common Name: 1-Decene
Synonym: α-decene
Chemical Name: 1-decene
CAS Registry No: 872-05-9
Molecular Formula: $C_{10}H_{20}$
Molecular Weight: 140.266
Melting Point (°C):
    –66.3   (Lide 2003)
Boiling Point (°C):
    170.5   (Lide 2003)
Density (g/cm³ at 20°C):
    0.7408, 0.7369 (20°C, 25°C, Forziati et al. 1950; Dreisbach 1959)
Molar Volume (cm³/mol):
    189.3   (20°C, calculated-density, Stephenson & Malanowski 1987; Ruelle & Kesselring 1997)
    190.3   (25°C, calculated-density)
    222.0   (calculated-Le Bas method at normal boiling point)
Enthalpy of Vaporization, ΔH_V (kJ/mol):
    50.43, 38.66 (25°, bp, Riddick et al. 1986)
Enthalpy of Fusion, ΔH_{fus} (kJ/mol):
    21.75   (Chickos et al. 1999)
Entropy of Fusion, ΔS_{fus} (J/mol K):
    106.8, 105.5 (exptl., calculated-group additivity method, Chickos et al. 1999)
Fugacity Ratio at 25°C, F: 1.0

Water Solubility (g/m³ or mg/L at 25°C or as indicated):
    5.70   (shake flask-titration with bromine, Natarajan & Venkatachalam 1972; quoted Shaw 1989)
    11.0, 8.50, 5.70 (15, 20, 25°C, shake flask-titration, in 0.001M HNO_3 solution, Natarajan & Venkatachalam 1972)
    0.161   (calculated-K_{OW}, Wang et al. 1992)
    0.433   (calculated-molar volume V_M, Wang et al. 1992)
    0.222, 0.344 (calculated-molar volume, mp and mobile order thermodynamics, Ruelle & Kesselring 1997)
    0.429, 4.29, 70.13 (101, 151.5, 202°C, calculated-recommended liquid-liquid equilibrium LLE data, Góral et al. 2004)

Vapor Pressure (Pa at 25°C or as indicated and reported temperature dependence equations. Additional data at other temperatures designated * are compiled at the end of this section.):
    133.3*   (14.7°C, summary of literature data, temp range 14.7–192.0°C, Stull 1947)
    6397*   (86.774°C, ebulliometry, measured range 86.774–171.605°C, Forziati et al. 1950)
    $\log (P/\text{mmHg}) = 6.96036 - 1501.812/(197.578 + t/°C)$; temp range 86.7–171.6°C (Antoine eq., ebulliometry-manometer measurements, Forziati et al. 1950)
    218   (calculated-Antoine eq., Dreisbach 1959)
    $\log (P/\text{mmHg}) = 6.96034 - 1501.872/(197.58 + t/°C)$; temp range 25–253°C (Antoine eq. for liquid state, Dreisbach 1959)
    218*   (extrapolated-Antoine eq., temp range 54.4–199.3°C, Zwolinski & Wilhoit 1971)
    $\log (P/\text{mmHg}) = 6.96034 - 1501.872/(197.578 + t/°C)$; temp range 54.4–199.3°C (Antoine eq., Zwolinski & Wilhoit 1971)
    $\log (P/\text{kPa}) = 6.07985 - 1497.943/(197.102 + t/°C)$; temp range 86.77–171.6°C (Antoine eq. from reported exptl. data of Forziati et al. 1950, Boublik et al. 1984)
    215   (extrapolated-Antoine eq., temp range: 54–199°C, Dean 1985, 1992)
    $\log (P/\text{mmHg}) = 6.93477 - 1484.98/(195.707 + t/°C)$, temp range: 54–199°C (Antoine eq., Dean 1985, 1992)
    210   (quoted lit., Riddick et al. 1986)

log (P/kPa) = 6.05967 − 1484.98/(195.707 + t/°C); temp range not specified (Antoine eq., Riddick et al. 1986)
223 (extrapolated-Antoine eq., Stephenson & Malanowski 1987)
log (P_L/kPa) = 6.12458 − 1528.811/(−72.566 + T/K); temp range 383–445 K (Antoine eq., Stephenson & Malanowski 1987)
log (P/mmHg) = 2.2678 − 3.1244 × 10^3/(T/K) + 5.432·log (T/K) − 2.0137 × 10^{-2}·(T/K) + 1.1221 × 10^{-5}·(T/K)^2; temp range 207–617 K (vapor pressure eq., Yaws 1994)

Henry's Law Constant (Pa m^3/mol at 25°C):

Octanol/Water Partition Coefficient, log K_{OW}:
4.78 (calculated-regression eq. of Lyman et al. 1982, Wang et al. 1992)
5.18 (calculated-molar volume V_M, Wang et al. 1992)
4.7037 (calculated-UNIFAC group contribution, Chen et al. 1993)

Octanol/Air Partition Coefficient, log K_{OA}:

Bioconcentration Factor, log BCF:

Sorption Partition Coefficient, log K_{OC}:

Environmental Fate Rate Constants, k, and Half-Lives, $t_{1/2}$:
Volatilization:
Photolysis:
Oxidation: rate constant k, for gas-phase second order rate constants, k_{OH} for reaction with OH radical, k_{NO3} with NO_3 radical and k_{O3} with O_3 or as indicated, *data at other temperatures see reference:
k_{O3} = 1.08 × 10^{-17} cm^3 molecule^{-1} s^{-1} for reaction with ozone in the gas phase (Atkinson & Carter 1984)
k_{O3} = 9.3 × 10^{-18} cm^3 molecule^{-1} s^{-1} at 298 K (recommended, Atkinson 1997)
Hydrolysis:
Biodegradation:
Biotransformation:
Bioconcentration, Uptake (k_1) and Elimination (k_2) Rate Constants or Half-Lives:

Half-Lives in the Environment:
Surface water: $t_{1/2}$ ~ 320 h and 9 × 10^4 d for oxidation by OH and RO_2 radicals for olefins in aquatic system, and $t_{1/2}$ = 7.3 d, based on rate constant k = 3 × 10^3 M^{-1} s^{-1} for the oxidation of unsubstituted olefins with singlet oxygen in aquatic system (Mill & Mabey 1985).

TABLE 2.1.2.1.14.1
Reported vapor pressures of 1-decene at various temperatures and the coefficients for the vapor pressure equations

log P = A − B/(T/K)	(1)	ln P = A − B/(T/K)	(1a)
log P = A − B/(C + t/°C)	(2)	ln P = A − B/(C + t/°C)	(2a)
log P = A − B/(C + T/K)	(3)		
log P = A − B/(T/K) − C·log (T/K)	(4)		

Stull 1947		Forziati et al. 1950		Zwolinski & Wilhoit 1971	
summary of literature data		ebulliometry		selected values	
t/°C	P/Pa	t/°C	P/Pa	t/°C	P/Pa
14.7	133.3	86.774	6397	54.40	1333
40.3	666.6	91.308	7694	67.80	2666
53.7	1333	95.134	8997	76.33	4000
67.8	2666	98.604	10334	82.71	5333
83.3	5333	101.844	11723	87.875	6666

(Continued)

TABLE 2.1.2.1.14.1 (*Continued*)

Stull 1947		Forziati et al. 1950		Zwolinski & Wilhoit 1971	
summary of literature data		ebulliometry		selected values	
t/°C	P/Pa	t/°C	P/Pa	t/°C	P/Pa
93.5	7999	106.223	13848	92.236	7999
14.7	133.3	86.774	6397	54.40	1333
106.5	13332	111.213	16647	99.396	10666
126.7	26664	116.283	19950	105.198	13332
149.2	53329	120.995	23463	116.342	19998
192.0	101325	127.265	28990	124.760	26664
		140.063	43759	131.607	33331
mp/°C		147.265	53710	137.421	39997
		154.939	66810	147.024	53329
		169.134	97662	154.861	66661
		169.762	99247	161.533	79993
		170.345	100747	163.376	93326
		171.012	102487	168.464	95992
		171.605	104026	170.052	98659
				170.570	101325
		eq. 2	P/mmHg		
		A	6.96036	eq. 2	P/mmHg
		B	1501.872	A	6.96034
		C	197.578	B	1501.872
		bp/°C	170.570	C	197.578
				bp/°C	170.570
				$\Delta H_V/(\text{kJ mol}^{-1})$ =	
				at 25°C	38.66
				at bp	50.46

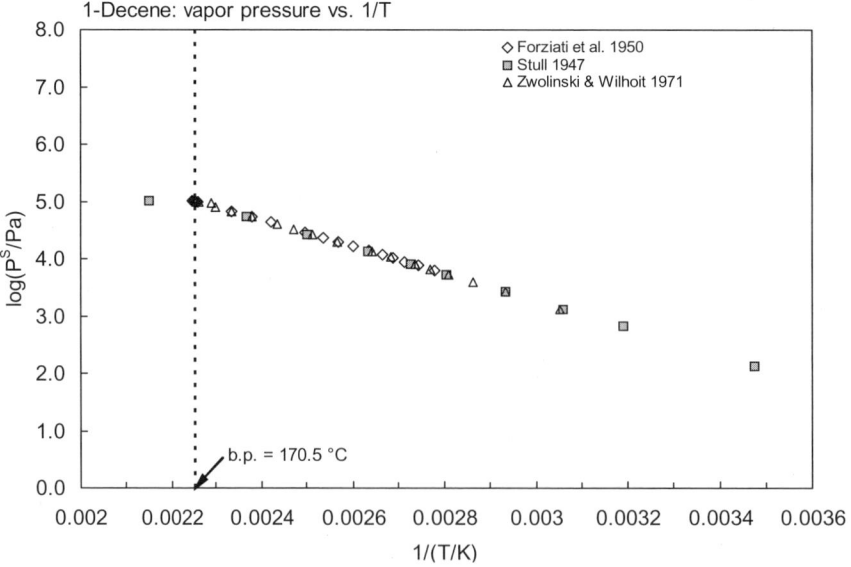

FIGURE 2.1.2.1.14.1 Logarithm of vapor pressure versus reciprocal temperature for 1-decene.

Aliphatic and Cyclic Hydrocarbons

2.1.2.2 Dienes

2.1.2.2.1   1,3-Butadiene

Common Name: 1,3-Butadiene
Synonym: α,γ-butadiene, bivinyl, divinyl, erythrene, vinylethylene, biethylene, pyrrolylene
Chemical Name: 1,3-butadiene
CAS Registry No: 106-99-0
Molecular Formula: C_4H_6; $CH_2=CHCH=CH_2$
Molecular Weight: 54.091
Melting Point (°C):
- –108.91   (Lide 2003)

Boiling Point (°C):
- –4.41   (Lide 2003)

Density (g/cm^3 at 20°C):
- 0.6211, 0.6149 (20°C, 25°C, at saturation pressure, Dreisbach 1959)

Molar Volume (cm^3/mol):
- 87.1   (20°C, calculated-density, McAuliffe 1966; Stephenson & Malanowski 1987; Wang et al. 1992 Ruelle & Kesselring 1997)
- 81.4   (calculated-Le Bas method at normal boiling point)

Enthalpy of Fusion, ΔH_{fus} (kJ/mol):
- 7.98   (Chickos et al. 1999)

Entropy of Fusion, ΔS_{fus} (J/mol K):
- 48.62, 45.2 (exptl., calculated-group additivity method, Chickos et al. 1999)

Fugacity Ratio at 25°C, F: 1.0

Water Solubility (g/m^3 or mg/L at 25°C):
- 735   (shake flask-GC, at 1 atmospheric pressure, McAuliffe 1966)

Vapor Pressure (Pa at 25°C or as indicated and reported temperature dependence equations. Additional data at other temperatures designated * are compiled at the end of this section.):
- 113857*   (–1.50°C, static method-manometer, measured range –75.5 to –1.50°C, Heisig 1933)
- 336200*   (calculated-Antoine eq. regression, temp range –102.8 to 4.6°C, Stull 1947)
- 280600   (calculated-Antoine eq., Dreisbach 1959)
- log (P/mmHg) = 6.85941 – 935.53/(239.55 + t/°C); temp range –66 to 46°C (Antoine eq. for liquid state, Dreisbach 1959)
- 281000*   (extrapolated-Antoine eq., temp range –58.201 to 14.43°C, Zwolinski & Wilhoit 1971)
- 280644   (derived from compiled data, Zwolinski & Wilhoit 1971)
- log (P/mmHg) = 6.84999 – 930.546/(238.854 + t/°C); temp range –58.201 to 14.43°C (Antoine eq., Zwolinski & Wilhoit 1971)
- 61295*   (20.211°C, temp range –16.204 to 33.257°C, Boublikova 1972; quoted, Boublik et al. 1984)
- 247700   (extrapolated-Antoine eq., temp range –82.5 to 9.7°C, Weast 1972–73)
- log (P/mmHg) = [–0.2185 × 7761.0/(T/K)] + 8.997505; temp range –82.5 to 9.7°C (Antoine eq., Weast 1972–73)
- 281650, 281510* (static method quartz manometer, measured range 15–55°C, Flebbe et al. 1982)
- 281230, 310400 (extrapolated-Antoine equations, Boublik et al. 1984)
- log (P/kPa) = 6.86369 – 1313.687/(275.492 + t/°C); temp range –81 to –24°C (Antoine eq. from reported exptl. data, Boublik et al. 1984)
- log (P/kPa) = 5.97484 – 931.996/(239.329 + t/°C); temp range –75 to –1.5°C (Antoine eq. from reported exptl. data, Boublik et al. 1984)
- log (P/mmHg) = 7.03555 – 998.106/(245.233 + t/°C); temp range –87 to –62°C (Antoine eq., Dean 1985, 1992)
- log (P/mmHg) = 6.84999 – 930.546/(238.854 + t/°C); temp range –58 to 15°C (Antoine eq., Dean 1985, 1992)
- 281000   (interpolated-Antoine eq.-III, Stephenson & Malanowski 1987)

log (P_L/kPa) = 6.16045 – 998.106/(–27.916 + T/K); temp range 193–213 K (Antoine eq.-I, Stephenson & Malanowski 1987)

log (P_L/kPa) = 5.97489 – 930.546/(–34.306 + T/K); temp range 213–276 K (Antoine eq.-II, Stephenson & Malanowski 1987)

log (P_L/kPa) = 5.99667 – 940.687/(–33.017 + T/K); temp range 270–318 K (Antoine eq.-III, Stephenson & Malanowski 1987)

log (P_L/kPa) = 6.31615 – 1130.927/(–5.606 + T/K); temp range 315–382 K (Antoine eq.-IV, Stephenson & Malanowski 1987)

log (P_L/kPa) = 8.86984 – 3877.451/(315.612 + T/K); temp range 380–425 K (Antoine eq.-V, Stephenson & Malanowski 1987)

log (P/mmHg) = 30.0572 – 1.9891 × 10^3/(T/K) – 8.2922·log (T/K) + 2.5664 × 10^{-10}·(T/K) + 5.1334 × 10^{-6}·(T/K)^2; temp range 164–425 K (vapor pressure eq., Yaws 1994)

ln (P/atm) = 9.16107 – 2154.139/(T/K – 33.596); temp range 207–319 K (Antoine eq., Oliveira & Uller 1996)

Henry's Law Constant (Pa m^3/mol at 25°C):
- 7460 (calculated-P/C, Mackay & Shiu 1981)
- 6370 (calculated-1/K_{AW}, C_W/C_A, reported as exptl., Hine & Mookerjee 1975; quoted, Howard 1989)
- 7150, 6230 (calculated-group contribution, calculated-bond contribution, Hine & Mookerjee 1975)
- 10820 (calculated-MCI χ, Nirmalakhandan & Speece 1988)
- 7720 (calculated-vapor-liquid equilibrium (VLE) data, Yaws et al. 1991)

Octanol/Water Partition Coefficient, log K_{OW}:
- 1.99 (shake flask-GC, Leo et al. 1975; Hansch & Leo 1979)
- 1.87, 1.68, 1.90 (calculated-fragment const., Rekker 1977)
- 2.22 (calculated-UNIFAC, Banerjee & Howard 1988)
- 1.99 (recommended, Sangster 1989)
- 1.56 (calculated-V_M, Wang et al. 1992)

Octanol/Air Partition Coefficient, log K_{OA}:

Bioconcentration Factor, log BCF:
- 1.28 (calculated-K_{OW}, Lyman et al. 1982; quoted, Howard 1989)

Sorption Partition Coefficient, log K_{OC}:
- 1.86–2.36 (soils and sediments, calculated-K_{OW} and S, Lyman et al. 1982; quoted, Howard 1989)

Environmental Fate Rate Constants, k, and Half-Lives, $t_{1/2}$:

Volatilization: volatilizes rapidly from water and land (Howard 1989).

Photolysis:

Oxidation: rate constant k, for gas-phase second order rate constants, k_{OH} for reaction with OH radical, k_{NO3} with NO_3 radical and k_{O3} with O_3 or as indicated, *data at other temperatures and/or the Arrhenius expression see reference:

k_{O3} = 8.4 × 10^{-18} cm^3 molecule^{-1} s^{-1} at 299 K (Japer et al. 1974)

k_{OH} = (46.4 ± 9.3) × 10^9 L mol^{-1} s^{-1} at 305 ± 2 K (relative rate method, Lloyd et al. 1976)

k_{OH} = 4.64 × 10^{10} L mol^{-1} s^{-1}, $t_{1/2}$ = 0.25 h for reaction with OH radical only, $t_{1/2}$ = 0.24 h with an average concn of 0.1 ppm of O_3 at 300 K (Darnall et al. 1976)

$k_{O(^3P)}$ = 1.9 × 10^{-11} cm^3 molecule^{-1} s^{-1} for reaction with O(^{3}P) atom at room temp. (Atkinson & Pitts, Jr. 1977)

k_{OH}* = (6.85 ± 0.69) × 10^{-11} cm^3 molecule^{-1} s^{-1} at 299.5 K, measured range 299.9–424 K (flash photolysis-resonance fluorescence, Atkinson et al. 1977)

k_{OH} = 6.9 × 10^{-11} cm^3 molecule^{-1} s^{-1} at room temp. (Atkinson et al. 1979; quoted, Gaffney & Levine 1979)

k_{O3}* = (1.17 ± 0.19) × 10^{-17} cm^3 molecule^{-1} s^{-1} at 296 K, measured range 276–324 K with atmospheric lifetime τ ~ 24 h due to reaction with O_3 and τ ~ 4 h with OH radical (Atkinson et al. 1982)

$k_{OH} = 6.85 \times 10^{-11}$ cm^3 molecule^{-1} s^{-1} at 297 ± K, k_{OH}(calc) = 6.22×10^{-11} cm^3 molecule^{-1} s^{-1} (relative rate method, Ohta 1983)

$k_{O3} = (6.1 - 8.4) \times 10^{-18}$ cm^3 molecule^{-1} s^{-1} at room temp. to 299 K (Atkinson & Carter 1984)

$k_{NO3} = (5.34 \pm 0.62) \times 10^{-14}$ cm^3 molecule^{-1} s^{-1} at 295 K (relative rate method, Atkinson et al. 1984a)

$k_{NO2} = (3.1 \pm 0.3) \times 10^{-20}$ cm^3 molecule^{-1} s^{-1} with NO$_2$ at 295 ± 2 K (relative rate method, Atkinson et al. 1984b)

$k_{OH} = 6.8 \times 10^{-11}$ cm^3 molecule^{-1} s^{-1}, $k_{NO3} = 0.053 \times 10^{-12}$ cm^3 molecule^{-1} s^{-1}, $k_{O3} = 8.4 \times 10^{-17}$ cm^3 molecule^{-1} s^{-1}; $k_{O(3P)} = 6.4 \times 10^{-18}$ cm^3 molecule^{-1} s^{-1}, and $k_{NO2} = 3.1 \times 10^{-20}$ cm^3 molecule^{-1} s^{-1} with NO$_2$, at room temp. (Atkinson et al. 1984b)

$k_{OH} = 6.65 \times 10^{-11}$ cm^3 molecule^{-1} s^{-1} at 295 K (relative rate method, Atkinson & Aschmann 1984)

$k_{NO3} = 2.1 \times 10^{-13}$ cm^3 molecule^{-1} s^{-1} at 298 K (fast flow system/MS, Benter & Schindler 1988)

$k_{OH}^* = 6.66 \times 10^{-11}$ cm^3 molecule^{-1} s^{-1} at 298 K (recommended, Atkinson 1989)

$k_{NO3} = (4.4 \pm 0.8) \times 10^{-14}$ cm^3 molecule^{-1} s^{-1} at 296 ± 1 K (FTIR, Andersson & Ljungström 1989)

$k_{OH} = 66.6 \times 10^{-12}$ cm^3 molecule^{-1} s^{-1}, $k_{NO3} = 7.5 \times 10^{-18}$ cm^3 molecule^{-1} s^{-1} (Atkinson 1990)

$k_{NO3} = 1.0 \times 10^{-13}$ cm^3 molecule^{-1} s^{-1} at 298 K (recommended, Atkinson 1991)

$k_{OH} = 6.7 \times 10^{-11}$ cm^3 molecule^{-1} s^{-1}, $k_{NO3} = 7.5 \times 10^{-18}$ cm^3 molecule^{-1} s^{-1} (Sabljic & Gusten 1990)

$k_{OH} = 2.6 \times 10^{-11}$ cm^3 molecule^{-1} s^{-1}, $k_{NO3} = 1.0 \times 10^{-13}$ cm^3 molecule^{-1} s^{-1}, $k_{O3}^* = 6.3 \times 10^{-18}$ cm^3 molecule^{-1} s^{-1}, and $k_{O(3P)} = 1.98 \times 10^{-11}$ cm^3 molecule^{-1} s^{-1} for reaction with O(^{3}P) atom at 298 K (recommended, Atkinson 1997)

Hydrolysis: will hydrolyze appreciably (Howard 1989).
Biodegradation: $t_{1/2}$(aerobic) = 7 d, $t_{1/2}$(anaerobic) = 28 d in natural waters (Capel & Larson 1995)
Biotransformation:
Bioconcentration, Uptake (k_1) and Elimination (k_2) Rate Constants or Half-Lives:

Half-Lives in the Environment:
Air: estimated photooxidation $t_{1/2}$ = 0.24–24 h (Darnall et al 1976) for the reaction with hydroxyl radical; photooxidation with OH radicals with an estimated $t_{1/2}$ ~ 3.1 h (Lyman et al. 1982; quoted, Howard 1989); completely degraded within 6 h in a smog chamber irradiated by sunlight (Kopcynski et al 1972; quoted, Howard 1989);
$t_{1/2}$ = 15 h in air for the reaction with nitrate radical (Atkinson et al. 1984a; quoted, Howard 1989);
$t_{1/2}$ = 0.76–7.8 h, based on measured photooxidation rate constants in air (Howard et al. 1991).
Surface water: $t_{1/2}$ = 1200 to 48000 h, based on measured photooxidation rate constants with OH radicals in water (Güsten et al. 1981; quoted, Howard et al. 1991);
estimated $t_{1/2}$ = 3.8 h for evaporation from a model river 1 m deep with a 1 m/s current and a 3 m/s wind (Lyman et al. 1982; quoted, Howard 1989);
$t_{1/2}$ ~ 320 h and $t_{1/2}$ = 9 × 10^4 for oxidation by OH and RO$_2$ radicals and $t_{1/2}$ = 9 × 10^4 d for olefins in aquatic system, and $t_{1/2}$ = 19 h, based on rate constant of 1.0 × 10^7 M^{-1} s^{-1} for oxidation of dienes with singlet oxygen in aquatic system (Mill & Mabey 1985);
volatilizes rapidly with a half-life estimated to be several hours (Howard 1989);
$t_{1/2}$ = 168–672 h, based on estimated aqueous aerobic biodegradation half-lives (Howard et al. 1991)
$t_{1/2}$(aerobic) = 7 d, $t_{1/2}$(anaerobic) = 28 d in natural waters (Capel & Larson 1995).
Ground water: $t_{1/2}$ = 336–1344 h, based on estimated aqueous aerobic biodegradation half-lives (Howard et al. 1991).
Sediment:
Soil: $t_{1/2}$ = 168–672 h, based on estimated aqueous aerobic biodegradation half-lives (Howard et al. 1991).
Biota:

TABLE 2.1.2.2.1.1
Reported vapor pressures of 1,3-butadiene at various temperatures and the coefficients for the vapor pressure equations

$\log P = A - B/(T/K)$ (1) $\ln P = A - B/(T/K)$ (1a)
$\log P = A - B/(C + t/°C)$ (2) $\ln P = A - B/(C + t/°C)$ (2a)
$\log P = A - B/(C + T/K)$ (3)
$\log P = A - B/(T/K) - C \cdot \log (T/K)$ (4)
$\log P = A[1 - \Phi /(T/K)]$ (5) where $\log A = a + bT + cT^2$

Heisig 1933		Stull 1947		Zwolinski & Wilhoit 1971		Flebbe et al. 1982	
static method-manometer		summary of literature data		selected values		static-quartz manometer	
t/°C	P/Pa	t/°C	P/Pa	t/°C	P/Pa	t/°C	P/Pa
−75.5	1946	−102.8	133.3	−79.86	1333	15	144720
−63.4	4720	−87.6	666.6	−71.18	2666	15	144700
−51.6	10279	−79.7	1333	−65.676	4000	25	281650
−39.4	20452	−71.0	2666	−61.44	5333	25	281510
−38.6	21558	−61.3	5333	−58.201	6666	35	499680
−32.7	29197	−55.1	7999	−55.381	7999	35	499250
−26.1	40183	−46.8	13332	−50.747	10666	55	824250
−19.9	53395	−33.9	26664	−46.989	13332	55	823810
−15.5	64554	−19.3	53329	−39.760	19998		
−10.4	79980	−4.60	101325	−34.292	26664	vapor pressure eq. given	
−5.60	97285			−29.839	33331	in reference	
−1.50	113857	mp/°C	−108.0	−26.054	39997		
				−19.796	53329		
				−14.681	66661		
				−10.322	79993		
				−6.502	93326		
				−5.979	95992		
				−5.093	98659		
				−4.411	101325		
				25.0	280644		
				eq. 2	P/mmHg		
				A	6.84999		
				B	930.546		
				C	238.854		
				temp range −58 to 14.4°C			
				bp/°C	−4.411		
				ΔH_V/(kJ mol^{-1}) =			
				at 25°C	21.05		
				at bp	22.68		

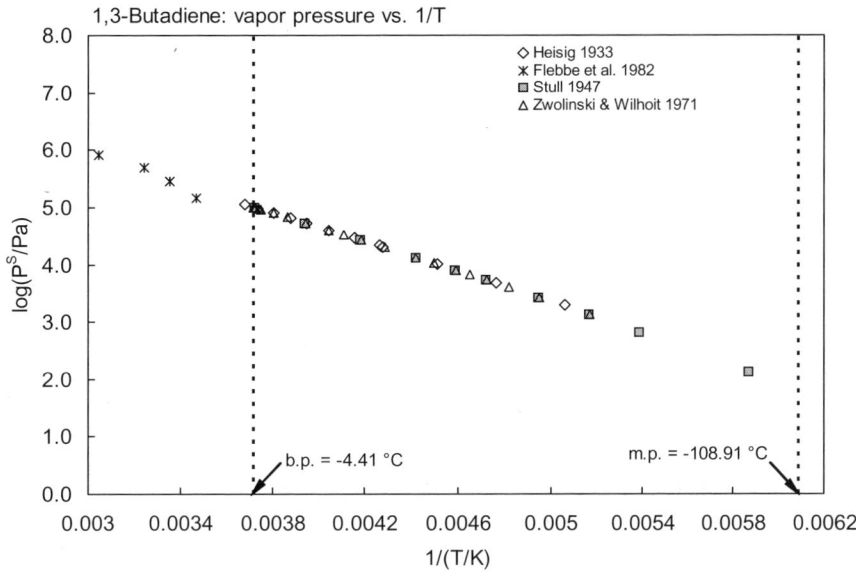

FIGURE 2.1.2.2.1.1 Logarithm of vapor pressure versus reciprocal temperature for 1,3-butadiene.

2.1.2.2.2 2-Methyl-1,3-butadiene (Isoprene)

Common Name: 2-Methyl-1,3-butadiene
Synonym: isoprene
Chemical Name: 2-methyl-1,3-butadiene
CAS Registry No: 78-79-5
Molecular Formula: C_5H_8; $CH_2=C(CH_3)CH=CH_2$
Molecular Weight: 68.118
Melting Point (°C):
    –145.9   (Lide 2003)
Boiling Point (°C):
    34   (Lide 2003)
Density (g/cm^3 at 20°C):
    0.6809, 0.6759 (20°C, 25°C, Dreisbach 1959)
Molar Volume (cm^3/mol):
    100.4   (20°C, calculated-density)
    103.6   (calculated-Le Bas method at normal boiling point)
Enthalpy of Fusion, ΔH_{fus} (kJ/mol):
    4.92   (Chickos et al. 1999)
Entropy of Fusion, ΔS_{fus} (J/mol K):
    38.68, 34.7 (exptl., calculated-group additivity method, Chickos et al. 1999)
Fugacity Ratio at 25°C, F: 1.0

Water Solubility (g/m^3 or mg/L at 25°C or as indicated. Additional data at other temperatures designated * are compiled at the end of this section.):
    642   (shake flask-GC, McAuliffe 1966)
    545*   (20°C, shake flask-GC, measured range 20–60°C, Pavlova et al. 1966)
    610*   (recommended best value, IUPAC Solubility Data Series, temp range 20–60°C, Shaw 1989a)
    530*   (calculated-liquid-liquid equilibrium LLE data, temp range 293.2–313.2 K, Góral et al. 2004)

Vapor Pressure (Pa at 25°C or as indicated and reported temperature dependence equations. Additional data at other temperatures designated * are compiled at the end of this section.):
    53329*   (15.4°C, summary of literature data, temp range –79.8 to 32.6°C, Stull 1947)
    66816*   (22.5°C, ebulliometry, measured range 16.8–34.8°C, Forziati et al. 1950)
    log (P/mmHg) = 6.90335 – 1080.996/(234.668 + t/°C); temp range 16.8–34.8°C (Antoine eq., ebulliometry measurements, Forziati et al. 1950)
    3349*   (–38.227°C, inclined-piston manometer, measured range –57.598 to –38.227°C, Osborn & Douslin 1969)
    73330   (calculated-Antoine eq., Dreisbach 1959; quoted, Hine & Mookerjee 1975)
    log (P/mmHg) = 6.90334 – 1080.996/(234.67 + t/°C); temp range –35 to 84°C (Antoine eq. for liquid state, Dreisbach 1959)
    73300   (interpolated-Antoine eq., temp range –18.477 to 55.36°C, Zwolinski & Wilhoit 1971)
    73340*   (derived from compiled data, Zwolinski & Wilhoit 1971)
    log (P/mmHg) = 6.88564 – 1071.578/(233.513 + t/°C); temp range –18.477 to 55.36°C (Antoine eq., Zwolinski & Wilhoit 1971)
    73000, 78380 (calculated-Antoine eq., Boublik et al. 1984)
    log (P/kPa) = 6.05468 – 1095.41/(236.322 + t/°C); temp range –58 to –38.2°C (Antoine eq. from reported exptl. data, Boublik et al. 1984)
    log (P/kPa) = 6.05329 – 1092.997/(236.002 + t/°C); temp range –16 to 33°C (Antoine eq. from reported exptl. data, Boublik et al. 1984)

Aliphatic and Cyclic Hydrocarbons

73350 (interpolated-Antoine eq., temp range –52 to –24°C, Dean 1985, 1992)
log (P/mmHg) = 7.01187 – 1126.159/(238.88 + t/°C); temp range –52 to –24°C (Antoine eq., Dean 1985, 1992)
log (P/mmHg) = 6.88564 – 1071.518/(233.51 + t/°C); temp range –19 to 55°C (Antoine eq., Dean 1985, 1992)
73350 (interpolated-Antoine eq., Stephenson & Malanowski 1987)
log (P_L/kPa) = 6.13677 – 1126,159/(–34.266 + T/K); temp range 221–254 K (Antoine eq.-I, Stephenson & Malanowski 1987)
log (P_L/kPa) = 6.01–54 – 1071.578/(–39.637 + T/K); temp range 254–316 K (Antoine eq.-II, Stephenson & Malanowski 1987)
73330, 10770 (quoted, calculated-UNIFAC activity coeff., Banerjee et al. 1990)

Henry's Law Constant (Pa m^3/mol at 25°C):
7840 (calculated-1/K_{AW}, C_W/C_A, reported as exptl., Hine & Mookerjee 1975)
6230, 6520 (calculated-group contribution, calculated-bond contribution, Hine & Mookerjee 1975)
7780 (calculated-P/C, Mackay & Shiu 1981)
14940 (calculated-MCI χ, Nirmalakhandan & Speece 1988)
7780 (calculated-vapor-liquid equilibrium (VLE) data, Yaws et al. 1991)

Octanol/Water Partition Coefficient, log K_{OW}:
2.05 (calculated-regression eq. from Lyman et al. 1982, Wang et al. 1992)
1.91 (calculated-molar volume V_M, Wang et al. 1992)

Octanol/Air Partition Coefficient, log K_{OA}:
2.06 (calculated-measured γ^∞ in pure octanol and vapor pressure P, Abraham et al. 2001)

Bioconcentration Factor, log BCF:

Sorption Partition Coefficient, log K_{OC}:

Environmental Fate Rate Constant and Half-Lives:
Volatilization:
Photolysis:
Oxidation: rate constant k, for gas-phase second order rate constants, k_{OH} for reaction with OH radical, k_{NO3} with NO_3 radical and k_{O3} with O_3 or as indicated, *data at other temperatures and/or the Arrhenius expression see reference:
k_{O3} = 16.5 × 10^{-18} cm^3 molecule^{-1} s^{-1} at 294 ± 2 K (chemiluminescence, Adeniji et al. 1981)
k_{O3}* = 1.25 × 10^{-17} cm^3 molecule^{-1} s^{-1} for the reaction with ozone in air at 296 K, measured range 278–323 K (static system-chemiluminescence, Atkinson et al. 1982)
k_{OH}* = (9.26 ± 1.5) × 10^{-11} cm^3 molecule^{-1} s^{-1} at 299 K, measured range 299–422 K (flash photolysis-resonance fluorescence, Kleindienst et al. 1982)
k_{OH} = 10.1 × 10^{-11} cm^3 molecule^{-1} s^{-1} at 297 ± K, k_{OH}(calc) = 9.40 × 10^{-11} cm^3 molecule^{-1} s^{-1} (relative rate method, Ohta 1983)
k_{O3}* = (0.58 – 1.25) × 10^{-17} cm^3 molecule^{-1} s^{-1} between 260–296 K (Atkinson & Carter 1984)
k_{O3} = 1.4 × 10^{-17} cm^3 molecule^{-1} s^{-1} with a loss rate of 0.8 d^{-1}, k_{OH} = 9.6 × 10^{-11} cm^3 molecule^{-1} s^{-1} with a loss rate of 8 d^{-1}, and k_{NO3} = 5.8 × 10^{-13} cm^3 molecule^{-1} s^{-1} with a loss rate of 12 d^{-1} (Atkinson & Carter 1984)
k_{NO3} = (3.23 ± 0.38) × 10^{-13} cm^3 molecule^{-1} s^{-1} at 295 ± 1 K (relative rate method, Atkinson et al. 1984a)
k_{NO2} = (10.3 ± 0.3) × 10^{-20} cm^3 molecule^{-1} s^{-1} with NO_2 at 295 ± 2 K (relative rate method, Atkinson et al. 1984b)
k_{O3} = 1.2 × 10^{-17} cm^3 molecule^{-1} s^{-1}, k_{OH} = 9.6 × 10^{-11} cm^3 molecule^{-1} s^{-1}, k_{NO3} = 3.2 × 10^{-13} cm^3 molecule^{-1} s^{-1} at room temp. (Atkinson et al. 1984b)

k_{OH} = 10.2 × 10⁻¹¹ cm³ molecule⁻¹ s⁻¹ at 295 K (relative rate method, Atkinson & Aschmann 1984)
k_{OH} = 9.98 × 10⁻¹¹ cm³ molecule⁻¹ s⁻¹, 9.26 × 10⁻¹¹ cm³ molecule⁻¹ s⁻¹ at 299 K (Atkinson 1985)
k_{OH} = (101 ± 2) × 10⁻¹² cm³ molecule⁻¹ s⁻¹ at 23.7 ± 0.5°C (Edney et al. 1986)
k_{NO3} = 1.3 × 10⁻¹² cm³ molecule⁻¹ s⁻¹ at 298 K (Benter & Schindler 1988)
k_{OH}* = 1.01 × 10⁻¹⁰ cm³ molecule⁻¹ s⁻¹ at 298 K (recommended, Atkinson 1989)
k_{OH} = 1.01 × 10⁻¹⁰ cm³ molecule⁻¹ s⁻¹, k_{O3} = 1.43 × 10⁻¹⁷ cm³ molecule⁻¹ s⁻¹ at 298 K (Atkinson 1990)
k_{OH} = 5.91 × 10⁻¹³ cm³ molecule⁻¹ s⁻¹, k_{NO3} = 1.01 × 10⁻¹⁰ cm³ molecule⁻¹ s⁻¹ at 298 K (Sabljic & Güsten 1990)
k_{NO3}* = 6.78 × 10⁻¹³ cm³ molecule⁻¹ s⁻¹ at 298 K (recommended, Atkinson 1991)
k_{OH}* = 1.01 × 10⁻¹⁰ cm³ molecule⁻¹ s⁻¹, k_{NO3}* = 6.78 × 10⁻¹³ cm³ molecule⁻¹ s⁻¹, k_{O3}* = 12.8 × 10⁻¹⁸ cm³ molecule⁻¹ s⁻¹, and $k_{O(3P)}$ = 3.5 × 10⁻¹¹ cm³ molecule⁻¹ s⁻¹ for the reaction with O(³P) atom at 298 K (recommended, Atkinson 1997)

Hydrolysis:

Biodegradation:

Biotransformation:

Bioconcentration, Uptake (k_1) and Elimination (k_2) Rate Constants or Half-Lives:

Half-Lives in the Environment:

Air: calculated atmospheric lifetimes: 32 h due to reaction with O_3 in 24-h period, 2.9 h with OH radical during daytime, and 3.6 h for NO_3 radical during nighttime for "clean atmosphere"; 10 h for reaction with O_3 in 24-h period, 1.4 h with OH radical during daytime, and 22 min with NO_3 radical during nighttime in "moderately polluted atmosphere" (Atkinson et al.1984a, Winer et al. 1984);

atmospheric lifetimes are calculated to be 28.3 h for the reaction with O_3, 2.9 h with OH radicals and 0.083 h with NO_3 radicals, all based on the reaction rate constants with O_3, OH and NO_3 radicals in the gas phase (Atkinson & Carter 1984)

Surface water: $t_{1/2}$ ~ 320 h and 9 × 10⁴ d for oxidation by OH and RO_2 radicals for olefins in aquatic system, and $t_{1/2}$ = 19 h, based on rate constant of 1.0 × 10⁷ M⁻¹ s⁻¹ for oxidation of dienes with singlet oxygen in aquatic system (Mill & Mabey 1985).

TABLE 2.1.2.2.2.1
Reported aqueous solubilities of 2-methyl-1,3-butadiene (isoprene) at various temperatures

Pavlova et al. 1966		Shaw 1989a		Góral et al. 2004	
shake flask-GC		IUPAC tentative values		calc-recommended LLE data	
t/°C	S/g·m⁻³	t/°C	S/g·m⁻³	t/°C	S/g·m⁻³
20	544.8	20	540	20	530
40	664.6	25	610	25	530
50	760.9	40	660	40	530
60	867.1	50	760		
		60	870		

Aliphatic and Cyclic Hydrocarbons

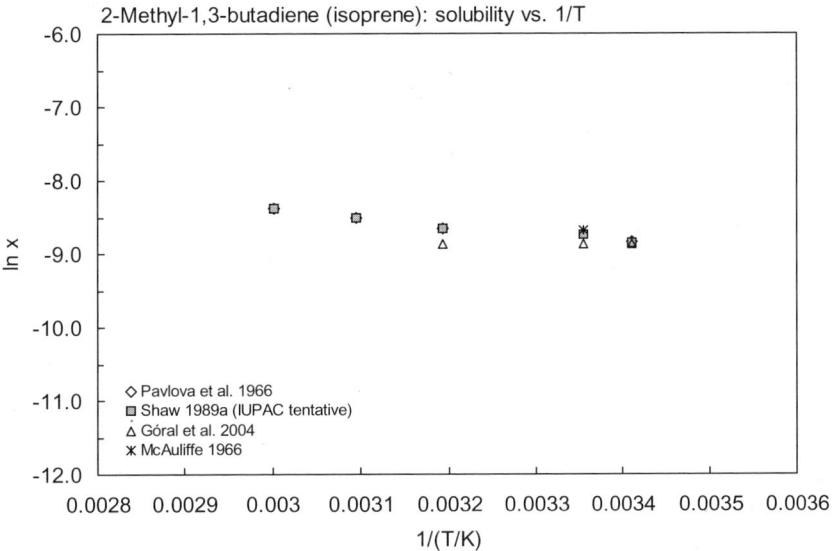

FIGURE 2.1.2.2.2.1 Logarithm of mole fraction solubility (ln x) versus reciprocal temperature for 2-methyl-1,3-butadiene.

TABLE 2.1.2.2.2.2
Reported vapor pressures of 2-methyl-1,3-butadiene (isoprene) at various temperatures and the coefficients for the vapor pressure equations

$\log P = A - B/(T/K)$ (1) $\ln P = A - B/(T/K)$ (1a)

$\log P = A - B/(C + t/°C)$ (2) $\ln P = A - B/(C + t/°C)$ (2a)

$\log P = A - B/(C + T/K)$ (3)

$\log P = A - B/(T/K) - C \cdot \log(T/K)$ (4)

$\log P = A[1 - \Phi/(T/K)]$ (5) where $\log A = a + bT + cT^2$

1.

Stull 1947 summary of literature data		Forziati et al. 1950 ebulliometry		Osborn & Douslin 1969 inclined-piston manometer		Zwolinski & Wilhoit 1971 selected values	
t/°C	P/Pa	t/°C	P/Pa	t/°C	P/Pa	t/°C	P/Pa
							liquid
−79.8	133.3	16.836	53718	−57.598	842.86	−51.561	1333
−62.3	666.6	22.506	66816	−55.186	1017	−41.587	2666
−53.3	1333	28.061	83780	−52.77	1221	−35.413	4000
−43.5	2666	33.006	97665	−50.35	1460	−30.714	5333
−32.6	5333	33.469	99260	−47.93	1738	−26.917	6666
−25.4	7999	33.903	100757	−45.507	2060	−23.710	7999
−16.0	13332	34.399	102484	−43.083	2432		liquid
−1.2	26664	34.834	104030	−40.656	2860	−18.447	10666
15.4	53329			−38.227	3349	−14.181	13332
32.6	101325	Antoine eq.				−5.98	19998
		eq. 2	P/mmHg	Cox eq.		0.221	26664
mp/°C	−146.7	A	6.90335	eq. 5	P/atm	5.268	33331
		B	1080.996	Φ	307.217	9.557	39997
		C	234.668	a	0.820543	16.646	53329

(*Continued*)

TABLE 2.1.2.2.2.2 (*Continued*)

Stull 1947 summary of literature data		Forziati et al. 1950 ebulliometry		Osborn & Douslin 1969 inclined-piston manometer		Zwolinski & Wilhoit 1971 selected values	
t/°C	P/Pa	t/°C	P/Pa	t/°C	P/Pa	t/°C	P/Pa
				$-b \times 10^{-4}$	8.31178	22.437	66661
		bp/°C	34.067	$c \times 10^{-7}$	10.32622	27.371	79993
						31.693	93326
						32.499	95992
						33.287	98659
						34.059	101325
						bp/°C	34.059
						eq. 2	P/mmHg
						A	6.88654
						B	1071.578
						C	233.513
						for temp range -18–$55°C$	
						$\Delta H_V/(kJ\ mol^{-1}) =$	
						at 25°C	26.44
						at bp	26.07

2.

Boublikova 1972 (thesis)

in Boublik et al. 1984

t/°C	P/Pa
-16.204	12046
-12.508	14549
-8.291	17921
-4.649	21314
-0.541	25780
2.793	29923
6.725	35493
10.416	41470
14.852	49704
20.211	61295
26.339	77125
33.257	98578
bp/°C	34.036
eq. 2	P/kPa
A	6.05329
B	1092.997
C	236.002

Aliphatic and Cyclic Hydrocarbons

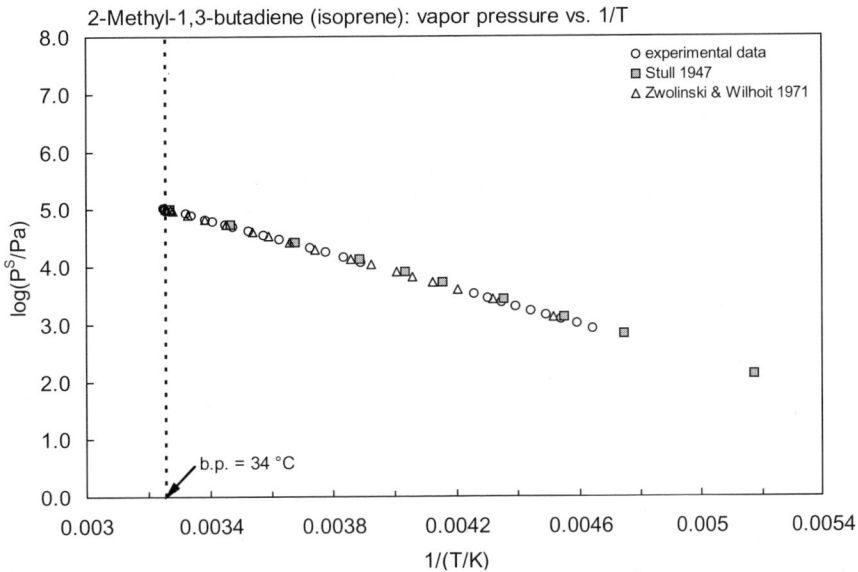

FIGURE 2.1.2.2.2.2 Logarithm of vapor pressure versus reciprocal temperature for 2-methyl-1,3-butadiene.

2.1.2.2.3 2,3-Dimethyl-1,3-butadiene

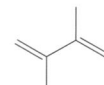

Common Name: 2,3-Dimethyl-1,3-butadiene
Synonym:
Chemical Name: 2,3-dimethyl-1,3-butadiene
CAS Registry No: 513-81-5
Molecular Formula: C_6H_{10}; $CH_2 = (CH_3)C(CH_3)=CH_2$
Molecular Weight: 82.143
Melting Point (°C):
    –76   (Lide 2003)
Boiling Point (°C):
    68.8   (Lide 2003)
Density (g/cm³ at 20°C):
    0.7267, 0.7222   (20°C, 25°C, Dreisbach 1959)
Molar Volume (cm³/mol):
    113.0, 113.7 (20°C, 25°C, calculated-density)
    125.8   (calculated-Le Bas method at normal boiling point)
Enthalpy of Fusion, ΔH_{fus} (kJ/mol):
Entropy of Fusion, ΔS_{fus} (J/mol K):
Fugacity Ratio at 25°C, F: 1.0

Water Solubility (g/m³ or mg/L at 25°C):
    327   (quoted, Hine & Mookerjee 1975)
    226   (calculated-fragment solubility constants, Wakita et al. 1986)
    94.3, 226   (calculated-molar volume, mp and mobile order thermodynamics, Ruelle & Kesselring 1997)

Vapor Pressure (Pa at 25°C and reported temperature dependence equations):
    19200   (calculated-Antoine eq., Dreisbach 1959)
    $\log (P/\text{mmHg}) = 7.02388 - 1220.88/(225.9 + t/°C)$; temp range –6 to 116°C (Antoine eq. for liquid state, Dreisbach 1959)
    20160   (interpolated-Antoine eq., Boublik et al. 1984)
    $\log (P/\text{kPa}) = 6.25005 - 1302.766/(238.42 + t/°C)$; temp range 0.04–68.6°C (Antoine eq. from reported exptl. data, Boublik et al. 1984)
    20160   (interpolated-Antoine eq., Dean 1985, 1992)
    $\log (P/\text{mmHg}) = 7.1197 - 1299.69/(238.09 + t/°C)$; temp range 0–68.5°C (Antoine eq., Dean 1985, 1992)
    20150   (interpolated-Stephenson & Malanowski 1987)
    $\log (P_L/\text{kPa}) = 6.3266 - 1346.0/(-30.15 + T/K)$; temp range 273–342 K (Antoine eq., Stephenson & Malanowski 1987)
    $\log (P/\text{mmHg}) = 29.9755 - 2.5677 \times 10^3/(T/K) - 7.8544 \cdot \log (T/K) + 2.2361 \times 10^{-10} \cdot (T/K) + 2.4591 \times 10^{-6} \cdot (T/K)^2$; temp range 197–526 K (vapor pressure eq., Yaws 1994)

Henry's Law Constant (Pa m³/mol at 25°C):
    4830   (calculated-1/K_{AW}, C_W/C_A, reported as exptl., Hine & Mookerjee 1975)
    5420, 6990   (calculated-group contribution, calculated-bond contribution, Hine & Mookerjee 1975)
    21100   (calculated-MCI χ, Nirmalakhandan & Speece 1988)

Octanol/Water Partition Coefficient, log K_{OW}:

Octanol/Air Partition Coefficient, log K_{OA}:

Bioconcentration Factor, log BCF:

Aliphatic and Cyclic Hydrocarbons

Sorption Partition Coefficient, log K_{OC}:

Environmental Fate Rate Constant and Half-Lives:

Volatilization:

Photolysis:

Oxidation: rate constant k, for gas-phase second order rate constants, k_{OH} for reaction with OH radical, k_{NO3} with NO_3 radical and k_{O3} with O_3 or as indicated, *data at other temperatures see reference:

k_{OH} = (1.25 ± 0.05) × 10^{-10} cm^3 $molecule^{-1}$ s^{-1} at 297 ± 2 K; k_{OH}(calc) = 1.26 × 10^{-10} cm^3 $molecule^{-1}$ s^{-1} (relative rate method, Ohta 1983)

k_{OH} = 1.22 × 10^{-10} cm^3 $molecule^{-1}$ s^{-1} at 297 K (Atkinson 1985, 1989)

k_{NO3} = 2.3 × 10^{-12} cm^3 $molecule^{-1}$ s^{-1} at 298 K (Benter & Shindler 1988)

k_{OH} = 1.22 × 10^{-10} cm^3 $molecule^{-1}$ s^{-1}, k_{NO3} = 1.052 × 10^{-12} cm^3 $molecule^{-1}$ s^{-1} at 298 K (Sabljic & Güsten 1990)

k_{NO3} = 2.1 × 10^{-12} cm^3 $molecule^{-1}$ s^{-1} at 298 K (recommended, Atkinson 1991)

k_{OH} = 1.22 × 10^{-10} cm^3 $molecule^{-1}$ s^{-1}, k_{NO3} = 2.1 × 10^{-13} cm^3 $molecule^{-1}$ s^{-1}, and k_{O3}* = 26.5 × 10^{-18} cm^3 $molecule^{-1}$ s^{-1} at 298 K (recommended, Atkinson 1997)

Hydrolysis:

Biodegradation:

Biotransformation:

Bioconcentration, Uptake (k_1) and Elimination (k_2) Rate Constants or Half-Lives:

Half-Lives in the Environment:

Surface water: $t_{½}$ ~ 320 h and 9 × 10^4 d for oxidation by OH and RO_2 radicals for olefins in aquatic system, and $t_{½}$ = 19 h, based on rate constant k = 1.0 × 10^7 M^{-1} s^{-1} for oxidation of dienes by singlet oxygen in aquatic system (Mill & Mabey 1985).

2.1.2.2.4   1,4-Pentadiene

Common Name: 1,4-Pentadiene
Synonym:
Chemical Name: 1,4-pentadiene
CAS Registry No: 591-93-5
Molecular Formula: C_5H_8; $CH_2=CHCH_2CH=CH_2$
Molecular Weight: 68.118
Melting Point (°C):
    –148.2   (Lide 2003)
Boiling Point (°C):
    26   (Lide 2003)
Density (g/cm³ at 20°C):
    0.6608, 0.6557   (20°C, 25°C, Dreisbach 1959)
Molar Volume (cm³/mol):
    103.1, 103.9 (20°C, 25°C, calculated-density)
    103.6   (calculated-Le Bas method at normal boiling point)
Enthalpy of Fusion, ΔH_{fus} (kJ/mol):
    6.14   (Chickos et al. 1999)
Entropy of Fusion, ΔS_{fus} (J/mol K):
    49.41, 52/3   (exptl., calculated-group additivity method, Chickos et al. 1999)
Fugacity Ratio at 25°C, F: 1.0

Water Solubility (g/m³ or mg/L at 25°C):
    558   (shake flask-GC, McAuliffe 1966)

Vapor Pressure (Pa at 25°C or as indicated and reported temperature dependence equations. Additional data at other temperatures designated * are compiled at the end of this section.):
    105100*   (interpolated-Antoine eq. regression, temp range –83.5 to 26.1°C, Stull 1947)
    63774*   (20.669°C, ebulliometry, measured range 14.7–26.7°C (Forziati et al. 1950)
    log (P/mmHg) = 6.84880 – 1025.016/(232.354 + t/°C); temp range 14.7–26.7°C (Antoine eq., ebulliometry measurements, Forziati et al. 1950)
    97900   (calculated-Antoine eq., Dreisbach 1959; quoted, Hine & Mookerjee 1975)
    log (P/mmHg) = 6.84880 – 1025.016/(232.354 + t/°C); temp range –41 to 72°C (Antoine eq. for liquid state, Dreisbach 1959)
    3611*   (–43.083°C, inclined-piston manometer, measured range –60.01 to –43.083°C, Osborn & Douslin 1969)
    98000   (interpolated-Antoine eq., Zwolinski & Wilhoit 1971; quoted, Mackay & Shiu 1981)
    97940*   (derived from compiled data, temp range –57.16 to 26°C, Zwolinski & Wilhoit 1971)
    log (P/mmHg) = 6.83543 – 1017.995/(231.461 + t/°C); temp range –33.271 to 46.73°C (Antoine eq., Zwolinski & Wilhoit 1971)
    log (P/mmHg) = [–0.2185 × 6826.6/(T/K)] + 7.899113; temp range –83.5 to 26.1°C (Antoine eq., Weast 1972–73)
    98300, 99400 (extrapolated-Antoine equations, Boublik et al. 1984)
    log (P/kPa) = 6.06018 – 1063.485/(236.447 + t/°C); temp range –78.84 to –18.08°C (Antoine eq. from reported exptl. data, Boublik et al. 1984)
    log (P/kPa) = 6.34694 – 1239.949/(238.278 + t/°C); temp range –60.0 to –26.05°C (Antoine eq. from reported exptl. data, Boublik et al. 1984)
    97930   (interpolated-Antoine eq., Dean 1985, 1992)
    log (P/mmHg) = 7.17401 – 1155.378/(244.30 + t/°C); temp range –57 to –37°C (Antoine eq., Dean 1985, 1992)
    log (P/mmHg) = 6.83543 – 1017.995/(231.46 + t/°C); temp range –33 to 47°C (Antoine eq., Dean 1985, 1992)
    97900   (interpolated-Antoine eq., Stephenson & Malanowski 1987)

Aliphatic and Cyclic Hydrocarbons

log (P_L/kPa) = 6.29891 – 1155.378/(–28.852 + T/K); temp range 216–234 K (Antoine eq.-I, Stephenson & Malanowski 1987)

log (P_L/kPa) = 5.96033 – 1017.995/(–41.698 + T/K); temp range 236–307 K (Antoine eq.-II, Stephenson & Malanowski 1987)

log $(P/mmHg)$ = 23.7408 – 2.0505 × 10^3/(T/K) – 5.679·log (T/K) – 5.9671 × 10^{-11}·(T/K) + 1.1242 × 10^{-6}·(T/K)^2; temp range 125–479 K (vapor pressure eq., Yaws 1994)

Henry's Law Constant (Pa m^3/mol at 25°C):

12140	(calculated-1/K_{AW}, C_W/C_A, reported as exptl., Hine & Mookerjee 1975)
15640	(calculated-bond contribution, Hine & Mookerjee 1975)
12000	(calculated-P/C, Mackay & Shiu 1981)
13620	(calculated-MCI χ, Nirmalakhandan & Speece 1988)
11946	(calculated-vapor-liquid equilibrium (VLE) data, Yaws et al. 1991)

Octanol/Water Partition Coefficient, log K_{OW}:

2.48	(shake flask, Log P Database, Hansch & Leo 1987)
2.48	(recommended, Sangster 1989)

Octanol/Air Partition Coefficient, log K_{OA}:

Bioconcentration Factor, log BCF:

Sorption Partition Coefficient, log K_{OC}:

Environmental Fate Rate Constant and Half-Lives:
  Volatilization:
  Photolysis:
  Oxidation: rate constant k, for gas-phase second order rate constants, k_{OH} for reaction with OH radical, k_{NO3} with NO$_3$ radical and k_{O3} with O$_3$ or as indicated, *data at other temperatures see reference:
    k_{OH} = (50.6 ± 1.3) × 10^{-12} cm^3 molecule^{-1} s^{-1} at 297 ± 2 K; k_{OH}(calc) = 50.2 × 10^{-10} cm^3 molecule^{-1} s^{-1} (relative rate method, Ohta 1983)
    k_{OH} = 5.33 × 10^{-11} cm^3 molecule^{-1} s^{-1} at 297 K (Atkinson 1985, Atkinson 1989)
    k_{OH} = 5.1 × 10^{-11} cm^3 molecule^{-1} s^{-1} at room temp (Atkinson et al. 1984b)
    k_{NO3} = 7.8 × 10^{-13} cm^3 molecule^{-1} s^{-1} at 298 K (fast flow system, Benter & Shindler 1988)
    k_{OH} = 5.3 × 10^{-11} cm^3 molecule^{-1} s^{-1} at 298 K (recommended, Atkinson 1997)
  Hydrolysis:
  Biodegradation:
  Biotransformation:
  Bioconcentration, Uptake (k_1) and Elimination (k_2) Rate Constants or Half-Lives:

Half-Lives in the Environment:

  Surface water: $t_{½}$ ~ 320 h and 9 × 10^4 d for oxidation by OH and RO$_2$ radicals for olefins in aquatic system, and $t_{½}$ = 19 h, based on rate constant k = 1.0 × 10^7 M^{-1} s^{-1} for oxidation of dienes with singlet oxygen in aquatic system (Mill & Mabey 1985).

TABLE 2.1.2.2.4.1
Reported vapor pressures of 1,4-pentadiene at various temperatures and the coefficients for the vapor pressure equations

$$\log P = A - B/(T/K) \quad (1) \qquad \ln P = A - B/(T/K) \quad (1a)$$
$$\log P = A - B/(C + t/°C) \quad (2) \qquad \ln P = A - B/(C + t/°C) \quad (2a)$$
$$\log P = A - B/(C + T/K) \quad (3)$$
$$\log P = A - B/(T/K) - C \cdot \log (T/K) \quad (4)$$
$$\log P = A[1 - \Phi /(T/K)] \quad (5) \text{ where } \log A = a + bT + cT^2$$

Stull 1947		Forziati et al. 1950		Osborn & Douslin 1969		Zwolinski & Wilhoit 1971	
summary of literature data		ebulliometry		inclined-piston manometer		selected values	
t/°C	P/Pa	t/°C	P/Pa	t/°C	P/Pa	t/°C	P/Pa
−83.5	133.3	14.706	66811	−60.01	1071.4	−57.162	1333
−66.2	666.6	20.699	83774	−57.598	1291	−47.57	2666
−57.1	1333	24.931	97660	−55.186	1550	−41.49	4000
−47.7	2666	25.384	99253	−52.77	1851	−36.492	5333
−37.0	5333	25.806	100752	−50.35	2201	−33.271	6666
−30.0	7999	26.287	102480	−47.93	2605	−30.168	7999
−20.0	13332	26.714	104026	−45.507	3072	−25.069	10666
−6.70	26664			−43.083	3611	−20.933	13332
8.30	53329	bp/°C	25.967			−12.976	19998
26.1	101325					−6.956	26664
		Antoine eq.		Cox eq.		−2.053	33331
mp/°C		eq. 2	P/mmHg	eq. 5	P/atm	2.115	39997
		A	6.84880	Φ	299.117	9.009	53329
		B	1025.016	a	0.812446	14.642	66661
		C	232.354	−b × 10⁻⁴	7.52279	19.445	79993
				c × 10⁻⁷	8.34048	23.654	93326
						24.439	95992
						25.206	98659
						25.958	101325
						eq. for temp−33−46.7°C:	
						eq. 2	P/mmHg
						A	6.83543
						B	1017.995
						C	231.461
						bp/°C	25.958
						ΔH_V/(kJ mol⁻¹) =	
						at 25°C	25.15
						at bp	25.15

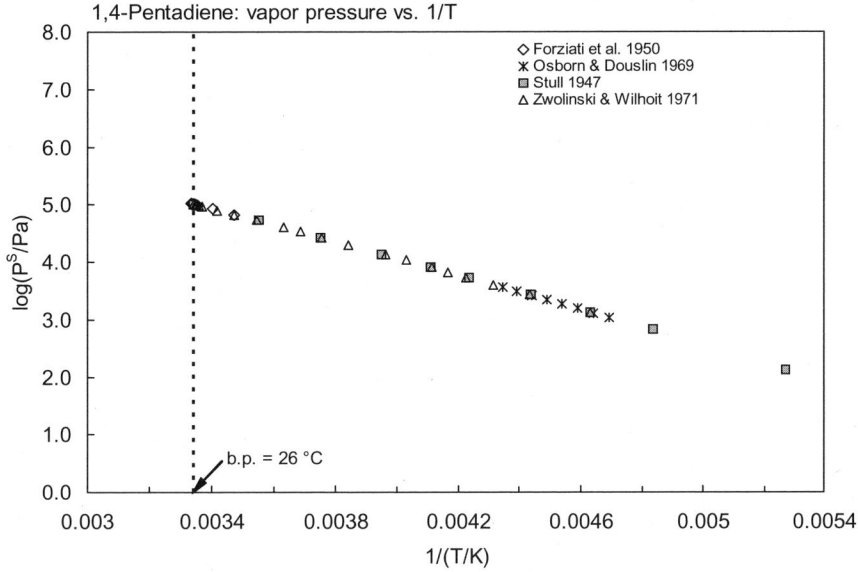

FIGURE 2.1.2.2.4.1 Logarithm of vapor pressure versus reciprocal temperature for 1,4-pentadiene.

2.1.2.2.5 1,5-Hexadiene

Common Name: 1,5-Hexadiene
Synonym:
Chemical Name: 1,5-Hexadiene
CAS Registry No: 592-42-7
Molecular Formula: C_6H_{10}; $CH_2=CH(CH_2)_2CH=CH_2$
Molecular Weight: 82.143
Melting Point (°C):
    −140.7   (Lide 2003)
Boiling Point (°C):
    59.4   (Lide 2003)
Density (g/cm³ at 20°C):
    0.6920   (Weast 1984)
    0.6923, 0.6878   (20°C, 25°C, Dreisbach 1959)
Molar Volume (cm³/mol):
    119.0   (20°C, calculated-density, Wang et al. 1992)
    125.8   (calculated-Le Bas method at normal boiling point)
Enthalpy of Fusion, ΔH_{fus} (kJ/mol):
Entropy of Fusion, ΔS_{fus} (J/mol K):
Fugacity Ratio at 25°C, F: 1.0

Water Solubility (g/m³ or mg/L at 25°C or as indicated):
    169   (shake flask-GC, McAuliffe 1966)
    320   (14°C, calculated-recommended liquid-liquid equilibrium LLE data, Góral et al. 2004)

Vapor Pressure (Pa at 25°C or as indicated and reported temperature dependence equations. Additional data at other temperatures designated * are compiled at the end of this section.):
    27730   (calculated-Antoine eq., Dreisbach 1959)
    $\log (P/\text{mmHg}) = 7.00740 - 1184.99/(227.7 + t/°C)$; temp range −13 to 102°C (Antoine eq. for liquid state, Dreisbach 1959)
    32077*   (26.95°C, static method-Hg manometer, measured range 26.95–46.13°C, Letcher & Marsicano 1974)
    29690   (interpolated-Antoine eq., Dean 1985, 1992)
    $\log (P/\text{mmHg}) = 6.5741 - 1013.5/(214.8 + t/°C)$; temp range 0–59°C (Antoine eq., Dean 1985, 1992)
    29670, 29690   (interpolated-Antoine eq.-I, extrapolated-Antoine eq.-II, Stephenson & Malanowski 1987)
    $\log (P_L/\text{kPa}) = 5.7368 - 1032.0/(-56.15 + T/K)$; temp range 273–333 K (Antoine eq.-I, Stephenson & Malanowski 1987)
    $\log (P_L/\text{kPa}) = 5.98314 - 1159.908/(-40.998 + T/K)$; temp range 299–333 K (Antoine eq.-II, Stephenson & Malanowski 1987)
    $\log (P/\text{mmHg}) = 10.5886 - 2.0106 \times 10^3/(T/K) + 0.28813 \cdot \log (T/K) - 9.562 \times 10^{-3} \cdot (T/K) + 7.164 \times 10^{-6} \cdot (T/K)^2$; temp range 132–507 K (vapor pressure eq., Yaws 1994)

Henry's Law Constant (Pa m³/mol at 25°C):
    13620   (calculated-1/K_{AW}, C_W/C_A, reported as exptl., Hine & Mookerjee 1975)
    17550, 23130   (calculated-group contribution, calculated-bond contribution, Hine & Mookerjee 1975)
    17150   (calculated-MCI χ, Nirmalakhandan & Speece 1988)

Octanol/Water Partition Coefficient, log K_{OW}:
    2.40   (calculated-π substituent constant, Hansch et al. 1968)
    2.29   (calculated-MCI χ, Murray et al. 1975)
    2.45   (calculated-fragment const., Yalkowsky & Morozowich 1980)
    2.68   (calculated-hydrophobicity const., Iwase et al. 1985)

Aliphatic and Cyclic Hydrocarbons

2.80 (recommended, Sangster 1989)
2.8208 (calculated-UNIFAC group contribution, Chen et al. 1993)
2.78 (calculated-f const., Müller & Klein 1992)
2.43 (calculated-molar volume V_M, Wang et al. 1992)
2.75 (recommended, Sangster 1993)
2.73 (selected, Hansch et al. 1995)

Octanol/Air Partition Coefficient, log K_{OA}:

Bioconcentration Factor, log BCF:

Sorption Partition Coefficient, log K_{OC}:

Environmental Fate Rate Constant and Half-Lives:
Volatilization:
Photolysis:
Oxidation: rate constant k, for gas-phase second order rate constants, k_{OH} for reaction with OH radical, k_{NO3} with NO_3 radical and k_{O3} with O_3 or as indicated, *data at other temperatures see reference:
$k_{OH} = 6.16 \times 10^{-11}$ cm^3 molecule^{-1} s^{-1} at 297 ± 2 K (relative rate method, Ohta 1983)
$k_{OH} = 6.35 \times 10^{-11}$ cm^3 molecule^{-1} s^{-1}, 5.85×10^{-11} cm^3 molecule^{-1} s^{-1} with 1,3-butadiene and propylene as standard substances respectively at 297 ± 2 K (relative rate method, Ohta 1983)
$k_{OH} = 6.2 \times 10^{-11}$ cm^3 molecule^{-1} s^{-1} at 298 K (recommended, Atkinson 1997)
Hydrolysis:
Biodegradation:
Biotransformation:
Bioconcentration, Uptake (k_1) and Elimination (k_2) Rate Constants or Half-Lives:

Half-Lives in the Environment:
Surface water: $t_{½}$ ~ 320 h and 9×10^4 d for oxidation by OH and RO_2 radicals for olefins in aquatic system, and $t_{½}$ = 19 h, based on rate constant of 1.0×10^7 M^{-1} s^{-1} for oxidation of dienes by singlet oxygen in aquatic system (Mill & Mabey 1985).

TABLE 2.1.2.2.5.1
Reported vapor pressures of 1,5-hexadiene at various temperatures

Letcher & Marsicano 1974

static method-Hg manometer

t/°C	P/Pa
26.95	32077
28.93	34690
31.62	38530
34.54	43036
36.94	47169
39.96	52622
41.77	56022
43.04	58622
46.13	65221

Antoine eq.
log P = A − B/(C + T/K)

	P/mmHg
A	6.6228
B	1037.35
C	−55.52

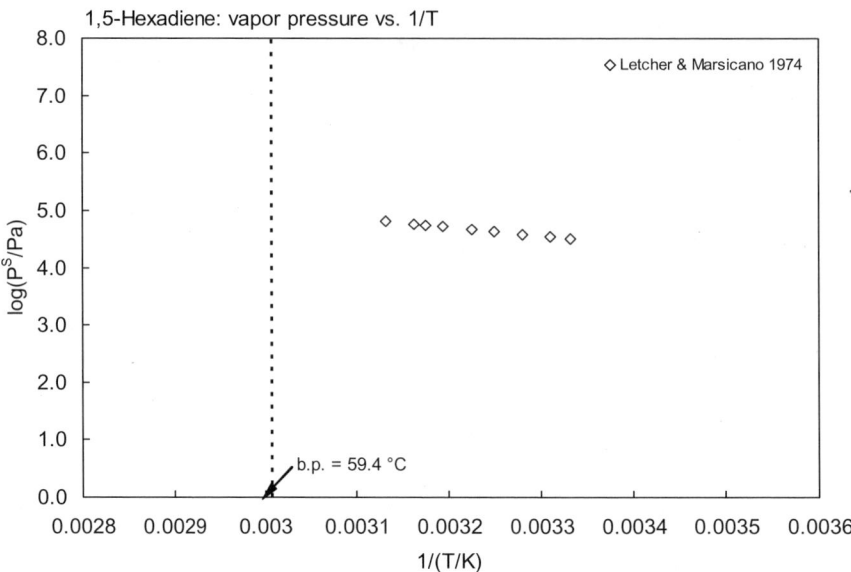

FIGURE 2.1.2.2.5.1 Logarithm of vapor pressure versus reciprocal temperature for 1,5-hexadiene.

Aliphatic and Cyclic Hydrocarbons

2.1.2.2.6   1,6-Heptadiene

Common Name: 1,6-Heptadiene
Synonym:
Chemical Name: 1,6-heptadiene
CAS Registry No: 3070-53-9
Molecular Formula: C_7C_{12}
Molecular Weight: 96.170
Melting Point (°C):
    –129   (Lide 2003)
Boiling Point (°C):
    90   (Lide 2003)
Density (g/cm³ at 20°C): 0.714
Molar Volume (cm³/mol):
    134.0   (20°C, calculated-density, McAuliffe 1966; Wang et al. 1992)
    148.0   (calculated-Le Bas method at normal boiling point)
Enthalpy of Fusion, ΔH_{fus} (kJ/mol):
Entropy of Fusion, ΔS_{fus} (J/mol K):
Fugacity Ratio at 25°C, F: 1.0

Water Solubility (g/m³ or mg/L at 25°C):
    44.0   (shake flask-GC, McAuliffe 1966)

Vapor Pressure (Pa at 25°C):

Henry's Law Constant (Pa m³/mol):

Octanol/Water Partition Coefficient, log K_{OW}:
    2.90   (calculated-π substituent constants, Hansch et al. 1968)
    2.73   (calculated-MCI χ, Murray et al. 1975)
    3.31   (calculated-fragment const., Müller & Klein 1992)
    2.85   (calculated-molar volume V_M, Wang et al. 1992)
    3.2189   (calculated-UNIFAC group contribution, Chen et al. 1993)

Octanol/Air Partition Coefficient, log K_{OA}:

Bioconcentration Factor, log BCF:

Sorption Partition Coefficient, log K_{OC}:

Environmental Fate Rate Constant and Half-Lives:

Half-Lives in the Environment:
    Surface water: $t_{1/2}$ ~ 320 h and 9×10^4 d for oxidation by OH and RO_2 radicals for olefins in aquatic system, and $t_{1/2}$ = 19 h, based on rate constant of 1.0×10^7 M^{-1} s^{-1} for oxidation of dienes by singlet oxygen in aquatic system (Mill & Mabey 1985).

2.1.2.3 Alkynes

2.1.2.3.1 1-Butyne

Common Name: 1-Butyne
Synonym: ethyl acetylene, but-1-yne
Chemical Name: 1-butyne
CAS Registry No: 107-00-6
Molecular Formula: C_4H_6, $CH_3CH_2C{\equiv}CH$
Molecular Weight: 54.091
Melting Point (°C):
    –125.7   (Lide 2003)
Boiling Point (°C):
    8.08   (Lide 2003)
Density (g/cm³ at 20°C):
    0.650. 0.65   (20°C, 25°C, at saturation pressure, Dreisbach 1959)
Molar Volume (cm³/mol):
    83.22   (20°C, calculated-density)
    81.4   (calculated-Le Bas method at normal boiling point)
Enthalpy of Fusion, ΔH_{fus} (kJ/mol):
    6.03   (Chickos et al. 1999)
Entropy of Fusion, ΔS_{fus} (J/mol K):
    40.9, 36.9   (exptl., calculated-group additivity method, Chickos et al. 1999)
Fugacity Ratio at 25°C, F: 1.0

Water Solubility (g/m³ or mg/L at 25°C):
    2870   (shake flask-GC, at 1 atmospheric pressure, McAuliffe 1966)

Vapor Pressure (Pa at 25°C and reported temperature dependence equations):
    $\log (P/\text{mmHg}) = 6.97497 - 986.46/(232.85 + t/°C)$; temp range –67 to 43°C (Antoine eq. for liquid state, Dreisbach 1959)
    188251   (derived from compiled data, Zwolinski & Wilhoit 1971)
    $\log (P/\text{mmHg}) = 7.07338 - 1101.71/(235.81 + t/°C)$; temp range –30.8 to 26.8°C (Antoine eq., Zwolinski & Wilhoit 1971)
    $\log (P/\text{mmHg}) = [-0.2185 \times 6596.9/(T/K)] + 8.032581$; temp range –92.5 to 8.7°C (Antoine eq., Weast 1972–73)
    $\log (P/\text{mmHg}) = 6.98198 - 988.75/(233.01 + t/°C)$; temp range –68 to 27°C (Antoine eq., Dean 1985, 1992)
    188220   (extrapolated-Antoine eq., Stephenson & Malanowski 1987)
    $\log (P_L/\text{kPa}) = 6.10688 - 988.75/(-40.14 + T/K)$; temp range 205–289 K (Antoine eq., Stephenson & Malanowski 1987)
    $\log (P/\text{mmHg}) = 43.8278 - 2.4255 \times 10^3/(T/K) - 14.141 \cdot \log (T/K) + 8.2138 \times 10^{-3} \cdot (T/K) + 7.4889 \times 10^{-14} \cdot (T/K)^2$; temp range 147–443 K (vapor pressure eq., Yaws 1994)

Henry's Law Constant (Pa m³/mol at 25°C):
    1880   (calculated-$1/K_{AW}$, C_W/C_A, reported as exptl., Hine & Mookerjee 1975)
    2210, 1800   (calculated-group contribution, calculated-bond contribution, Hine & Mookerjee 1975)
    1910   (calculated-P/C, Mackay & Shiu 1981)
    2820   (calculated-MCI χ, Nirmalakhandan & Speece 1988)
    1846   (calculated-vapor-liquid equilibrium (VLE) data, Yaws et al. 1991)

Octanol/Water Partition Coefficient, log K_{OW}:
    1.44, 1.48   (quoted, calculated-molar volume V_M, Wang et al. 1992)

Aliphatic and Cyclic Hydrocarbons

Octanol/Air Partition Coefficient, log K_{OA}:

Bioconcentration Factor, log BCF:

Sorption Partition Coefficient, log K_{OC}:

Environmental Fate Rate Constants, k, and Half-Lives, $t_{½}$:

    Volatilization:

    Photolysis:

    Oxidation: rate constant k, for gas-phase second order rate constants, k_{OH} for reaction with OH radical, k_{NO3} with NO_3 radical and k_{O3} with O_3 or as indicated, *data at other temperatures see reference:

        k_{O3} = 1.79 × 10^{-18} cm^3 molecule^{-1} s^{-1} at 298 K (static system-IR, Dillemuth et al. 1963)

        k_{O3} = (33 ± 5) × 10^{-21} cm^3 molecule^{-1} s^{-1} at 294 ± 1 K (static system-UV, DeMore 1971)

        k_{O3} = (19.7 ± 2.6) × 10^{-21} cm^3 molecule^{-1} s^{-1} at 294 ± 2 K (relative rate method, Atkinson et al.1984b)

        k_{OH}* = (10.42 ± 1.38) × 10^{-12} cm^3 molecule^{-1} s^{-1} at 300 K, measured range 253–343 K (discharge flow-resonance fluorescence, Boodaghians et al. 1987)

        k_{OH} = 8.0 × 10^{-12} cm^3 molecule^{-1} s^{-1} at 298 K (recommended, Atkinson 1989)

        k_{OH} = 8.0 × 10^{-12} cm^3 molecule^{-1} s^{-1} cm^3 molecule^{-1} s^{-1}, k_{O3} = 2.0 × 10^{-20} cm^3 molecule^{-1} s^{-1} at 298 K (Atkinson 1990)

        k_{O3} = 1.79 × 10^{-18} cm^3 molecule^{-1} s^{-1} at 298 K; and 1.79 × 10^{-20} cm^3 molecule^{-1} s^{-1}, 4.0 × 10^{-21} cm^3 molecule^{-1} s^{-1} at 294 K (literature review, Atkinson 1991)

    Hydrolysis:

    Biodegradation:

    Biotransformation:

    Bioconcentration, Uptake (k_1) and Elimination (k_2) Rate Constants or Half-Lives:

Half-Lives in the Environment:

2.1.2.3.2   1-Pentyne

Common Name: 1-Pentyne
Synonym: pent-1-yne
Chemical Name: 1-pentyne
CAS Registry No: 627-19-0
Molecular Formula: C_5H_8; $CH\equiv C(CH_2)_2CH_3$
Molecular Weight: 68.118
Melting Point (°C):
    –90   (Lide 2003)
Boiling Point (°C):
    40.1   (Lide 2003)
Density (g/cm³ at 20°C):
    0.6901, 0.6849 (20°C, 25°C, Dreisbach 1959)
Molar Volume (cm³/mol):
    98.71, 99.46 (20°C, 25°C, calculated-density)
    103.6   (calculated-Le Bas method at normal boiling point)
Enthalpy of Fusion, ΔH_{fus} (kJ/mol):
Entropy of Fusion, ΔS_{fus} (J/mol K):
Fugacity Ratio at 25°C, F: 1.0

Water Solubility (g/m³ or mg/L at 25°C):
    1570   (shake flask-GC, McAuliffe 1966)
    1049   (generator column-GC, Tewari et al. 1982a,b)
    1363   (calculated-recommended liquid-liquid equilibrium LLE data, Góral et al. 2004)

Vapor Pressure (Pa at 25°C and reported temperature dependence equations):
    57520   (calculated from determined data, Dreisbach 1959)
    $\log (P/mmHg) = 6.97263 - 1095.42/(227.53 + t/°C)$; temp range –50 to 70°C (Antoine eq. for liquid state, Dreisbach 1959)
    57600   (interpolated-Antoine eq., temp range –33 to 61°C Zwolinski & Wilhoit 1971)
    57462   (derived from compiled data, Zwolinski & Wilhoit 1971)
    $\log (P/mmHg) = 7.04614 - 1092.52/(227.18 + t/°C)$; temp range –33 to 61°C (Antoine eq., Zwolinski & Wilhoit 1971)
    57540   (interpolated-Antoine eq., temp range –44 to 61°C, Dean 1985)
    $\log (P/mmHg) = 6.96734 - 1092.52/(227.18 + t/°C)$; temp range –44 to 61°C (Antoine eq., Dean 1985, 1992)
    57540   (interpolated-Antoine eq., Stephenson & Malanowski 1987)
    $\log (P_L/kPa) = 6.09224 - 1092.52/(-45.97 + T/K)$; temp range 229–315 K (Antoine eq., Stephenson & Malanowski 1987)
    $\log (P/mmHg) = 33.8369 - 2.4684 \times 10^3/(T/K) - 9.4301 \cdot \log (T/K) + 6.1345 \times 10^{-10} \cdot (T/K) + 4.676 \times 10^{-6} \cdot (T/K)^2$; temp range 167–481 K (vapor pressure eq., Yaws 1994)

Henry's Law Constant (Pa m³/mol at 25°C):
    2536   (calculated-$1/K_{AW}$, C_W/C_A, reported as exptl., Hine & Mookerjee 1975)
    2980, 2660 (calculated-group contribution, calculated-bond contribution, Hine & Mookerjee 1975)
    2500   (calculated-P/C, Mackay & Shiu 1981)
    3422   (calculated-MCI χ, Nirmalakhandan & Speece 1988)
    4983   (calculated-vapor-liquid equilibrium (VLE) data, Yaws et al. 1991)

Octanol/Water Partition Coefficient, log K_{OW}:
    1.98   (shake flask-UV, Hansch et al. 1968, Hansch & Anderson 1967)
    2.12   (generator column-GC, Tewari et al. 1982a,b)

Aliphatic and Cyclic Hydrocarbons

    1.98   (recommended, Sangster 1989, 1993)
    1.98   (recommended, Hansch et al. 1995)

Octanol/Air Partition Coefficient, log K_{OA}:

Bioconcentration Factor, log BCF:

Sorption Partition Coefficient, log K_{OC}:

Environmental Fate Rate Constants, k, and Half-Lives, $t_{½}$:

    Volatilization:

    Photolysis:

    Oxidation: rate constant k, for gas-phase second order rate constants, k_{OH} for reaction with OH radical, k_{NO3} with NO_3 radical and k_{O3} with O_3 or as indicated, *data at other temperatures and/or the Arrhenius expression see reference:

        k_{OH}* = $(11.17 \pm 0.8) \times 10^{-12}$ cm^3 molecule^{-1} s^{-1} at 298 K, measured range 253–343 K (discharge flow-resonance fluorescence, Boodaghians et al. 1987, quoted, Atkinson 1989)

        k_{NO3} = 7.54×10^{-16} cm^3 molecule^{-1} s^{-1} at 295 K (Atkinson 1991)

    Hydrolysis:

    Biodegradation:

    Biotransformation:

    Bioconcentration, Uptake (k_1) and Elimination (k_2) Rate Constants or Half-Lives:

2.1.2.3.3   1-Hexyne

Common Name: 1-Hexyne
Synonym: hex-1-yne
Chemical Name: 1-hexyne
CAS Registry No: 693-02-7
Molecular Formula: C_6H_{10}; $C_4H_9C\equiv CH$
Molecular Weight: 82.143
Melting Point (°C):
    –131.9   (Dreisbach 1959; Lide 2003)
Boiling Point (°C):
    71.33   (Dreisbach 1959)
    71.3   (Lide 2003)
Density (g/cm³ at 20°C):
    0.7155, 0.7155   (20°C, 25°C, Dreisbach 1959)
Molar Volume (cm³/mol):
    114.8   (20°C, calculated-density, McAuliffe 1966; Stephenson & Malanowski 1987)
    125.8   (calculated-Le Bas method at normal boiling point)
Enthalpy of Fusion, ΔH_{fus} (kJ/mol):
Entropy of Fusion, ΔS_{fus} (J/mol K):
Fugacity Ratio at 25°C, F: 1.0

Water Solubility (g/m³ or mg/L at 25°C):
    360   (shake flask-GC, McAuliffe 1966)
    686   (generator column-GC, Tewari et al. 1982a,b)
    688   (generator column-GC, Miller et al. 1985)
    392   (calculated-recommended liquid–liquid equilibrium LLE data, Góral et al. 2004)

Vapor Pressure (Pa at 25°C):
    18140   (calculated-Antoine eq., Dreisbach 1959)
    $\log (P/mmHg) = 6.91212 - 1194.6/(225.0 + t/°C)$; temp range –8 to 118°C (Antoine eq. for liquid state, Dreisbach 1959)
    18145   (interpolated-Antoine eq., Stephenson & Malanowski 1987)
    $\log (P_L/kPa) = 6.03702 - 1194.6/(-48.15 + T/K)$; temp range 265–391 K (Antoine eq., Stephenson & Malanowski 1987)
    $\log (P/mmHg) = 55.7231 - 3.2541 \times 10^3/(T/K) - 18.405 \cdot \log (T/K) + 9.5814 \times 10^{-3} \cdot (T/K) + 9.2278 \times 10^{-14} \cdot (T/K)^2$; temp range 141–516 K (vapor pressure eq., Yaws 1994)

Henry's Law Constant (Pa m³/mol):
    4020   (calculated-$1/K_{AW}$, C_W/C_A, reported as exptl., Hine & Mookerjee 1975)
    4210, 4020   (calculated-group contribution, calculated-bond contribution, Hine & Mookerjee 1975)
    4310   (calculated-MCI χ, Nirmalakhandan & Speece 1988)
    2166   (calculated-vapor-liquid equilibrium (VLE) data, Yaws et al. 1991)

Octanol/Water Partition Coefficient, log K_{OW}:
    2.48   (calculated-π substituent constants, Hansch et al. 1968)
    2.73   (generator column-GC, Tewari et al. 1982a,b)
    2.73   (recommended, Sangster 1989, 1993)
    2.73   (selected, Hansch et al. 1995)

Aliphatic and Cyclic Hydrocarbons

Octanol/Air Partition Coefficient, log K_{OA}:

Bioconcentration Factor, log BCF:

Sorption Partition Coefficient, log K_{OC}:

Environmental Fate Rate Constants, k, and Half-Lives, $t_{1/2}$:
    Volatilization:
    Photolysis:
    Oxidation: rate constant k, for gas-phase second order rate constants, k_{OH} for reaction with OH radical, k_{NO3} with NO_3 radical and k_{O3} with O_3 or as indicated, *data at other temperatures and/or the Arrhenius expression see reference:
        $k_{OH}^* = (12.6 \pm 0.04) \times 10^{-12}$ cm^3 molecule^{-1} s^{-1} at 298 K, measured range 253–343 K (discharge flow-resonance fluorescence, Boodaghians et al. 1987; quoted, Atkinson 1989)
        $k_{NO3} = 1.60 \times 10^{-15}$ cm^3 molecule^{-1} s^{-1} at 295 K (Atkinson 1991)
    Hydrolysis:
    Biodegradation:
    Biotransformation:
    Bioconcentration, Uptake (k_1) and Elimination (k_2) Rate Constants or Half-Lives:

Half-Lives in the Environment:

2.1.2.3.4   1-Heptyne

Common Name: 1-Heptyne
Synonym:
Chemical Name: 1-heptyne
CAS Registry No: 628-71-7
Molecular Formula: C_7H_{12}
Molecular Weight: 96.170
Melting Point (°C):
    –81   (Lide 2003)
Boiling Point (°C):
    99.7   (Lide 2003)
Density (g/cm³ at 20°C):
    0.7330   (Weast 1984)
    0.7328, 0.7283   (20°C, 25°C, Dreisbach 1959)
Molar Volume (cm³/mol):
    131.2   (20°C, calculated-density, McAuliffe 1966; Stephenson & Malanowski 1987)
    148.0   (calculated-Le Bas method at normal boiling point)
Enthalpy of Fusion, ΔH_{fus} (kJ/mol):
Entropy of Fusion, ΔS_{fus} (J/mol K):
Fugacity Ratio at 25°C, F: 1.0

Water Solubility (g/m³ or mg/L at 25°C):
    94.0   (shake flask-GC, McAuliffe 1966)
    107   (calculated-recommended liquid-liquid equilibrium LLE data, Góral et al. 2004)

Vapor Pressure (Pa at 25°C and reported temperature dependence equations):
    7000   (Antoine eq., Dreisbach 1959)
    4298   (extrapolated-Antoine eq., Boublik et al. 1984)
    $\log (P/kPa) = 6.27249 - 1314.492/(208.097 + t/°C)$; temp range 63.1–99.98°C (Antoine eq. from reported exptl. data, Boublik et al. 1984)
    7500   (extrapolated-Antoine eq., Stephenson & Malanowski 1987)
    $\log (P_L/kPa) = 6.4039 - 1392.4/(-56.55 + T/K)$; temp range 336–373 K (Antoine eq., Stephenson & Malanowski 1987)
    $\log (P/mmHg) = 38.1255 - 3.064 \times 10^3/(T/K) - 10.679 \cdot \log (T/K) + 1.2244 \times 10^{-10} \cdot (T/K) + 3.668 \times 10^{-6} \cdot (T/K)^2$; temp range 154–537 K (vapor pressure eq., Yaws 1994)

Henry's Law Constant (Pa m³/mol at 25°C):
    6830   (calculated-$1/K_{AW}$, C_W/C_A, reported as exptl., Hine & Mookerjee 1975)
    5950, 6090   (calculated-group contribution, calculated-bond contribution, Hine & Mookerjee 1975)
    5420   (calculated-MCI χ, Nirmalakhandan & Speece 1988)
    7160   (calculated-vapor-liquid equilibrium (VLE) data, Yaws et al. 1991)

Octanol/Water Partition Coefficient, log K_{OW}:
    2.98   (calculated-π substituent constants, Hansch et al. 1968)
    2.93   (calculated-MCI χ, Murray et al. 1975)
    2.98   (calculated-molar volume V_M, Wang et al. 1992)
    2.99   (calculated-fragment const., Müller & Klein 1992)
    3.18   (calculated-UNIFAC group contribution, Chen et al. 1993)

Aliphatic and Cyclic Hydrocarbons

Octanol/Air Partition Coefficient, log K_{OA}:

Bioconcentration Factor, log BCF:

Sorption Partition Coefficient, log K_{OC}:

Environmental Fate Rate Constants, k, and Half-Lives, $t_{1/2}$:

Half-Lives in the Environment:

2.1.2.3.5   1-Octyne

Common Name: 1-Octyne
Synonym:
Chemical Name: 1-octyne
CAS Registry No: 629-05-0
Molecular Formula: C_8H_{14}; $C_6H_{13}C{\equiv}CH$
Molecular Weight: 110.197
Melting Point (°C):
    –79.3   (Dreisbach 1959; Lide 2003)
Boiling Point (°C): 127–128
    126.3   (Lide 2003)
Density (g/cm³ at 20°C):
    0.7461, 0.7419   (20°C, 25°C, Dreisbach 1959)
Molar Volume (cm³/mol):
    147.7   (20°C, calculated-density, McAuliffe 1966; Stephenson & Malanowski 1987)
    170.2   (calculated-Le Bas method at normal boiling point)
Enthalpy of Fusion, ΔH_{fus} (kJ/mol):
Entropy of Fusion, ΔS_{fus} (J/mol K):
Fugacity Ratio at 25°C, F: 1.0

Water Solubility (g/m³ or mg/L at 25°C):
    24.0   (shake flask-GC, McAuliffe 1966)
    25.4   (calculated-recommended liquid-liquid equilibrium LLE data, Góral et al. 2004)

Vapor Pressure (Pa at 25°C and reported temperature dependence equations):
    1813   (calculated-Antoine eq., Dreisbach 1959)
    $\log(P/mmHg) = 7.02447 - 1413.8/(215.0 + t/°C)$; temp range 25–170°C (Antoine eq. for liquid state, Dreisbach 1959)
    1723   (extrapolated-Antoine eq., Boublik et al. 1984)
    $\log(P/kPa) = 6.36895 - 1997.558/(202.608 + t/°C)$; temp range 69.2–271.1°C (Antoine eq. from reported exptl. data, Boublik et al. 1984)
    1715   (extrapolated-Antoine eq., Stephenson & Malanowski 1987)
    $\log(P_L/kPa) = 6.19321 - 1427.434/(-214.625 + T/K)$; temp range 84.8–126.26 K (Antoine eq., Stephenson & Malanowski 1987)

Henry's Law Constant (Pa m³/mol at 25°C):
    8208   (calculated-$1/K_{AW}$, C_W/C_A, reported as exptl., Hine & Mookerjee 1975)
    8208, 9000   (calculated-group contribution, calculated-bond contribution, Hine & Mookerjee 1975)
    6827   (calculated-MCI χ, Nirmalakhandan & Speece 1988)
    8325   (calculated-vapor-liquid equilibrium (VLE) data, Yaws et al. 1991)

Octanol/Water Partition Coefficient, log K_{OW}:
    3.48   (calculated-π substituent constants, Hansch et al. 1968)
    3.37   (calculated-MCI χ, Murray et al. 1975)
    3.49   (calculated-molar volume V_M, Wang et al. 1992)
    3.52   (calculated-f const., Müller & Klein 1992)
    3.84   (calculated-UNIFAC group contribution, Chen et al. 1993)

Aliphatic and Cyclic Hydrocarbons

Octanol/Air Partition Coefficient, log K_{OA}:

Bioconcentration Factor, log BCF:

Sorption Partition Coefficient, log K_{OC}:

Environmental Fate Rate Constants, k, and Half-Lives, $t_{1/2}$:

Half-Lives in the Environment:

2.1.2.3.6   1-Nonyne

Common Name: 1-Nonyne
Synonym:
Chemical Name: 1-nonyne
CAS Registry No: 3452-09-3
Molecular Formula: C_9H_{16}; $C_7H_{15}C{\equiv}CH$
Molecular Weight: 124.223
Melting Point (°C):
    –50   (Dreisbach 1959; Lide 2003)
Boiling Point (°C): 150–151
    150.8   (Dreisbach 1959; Lide 2003)
Density (g/cm^3 at 20°C):
    0.7570   (Weast 1984)
    0.7568, 0.7527   (20°C, 25°C, Dreisbach 1959)
Molar Volume (cm^3/mol):
    164.1   (20°C, calculated-density, McAuliffe 1966; Ruelle & Kesselring 1997)
    192.4   (calculated-Le Bas method at normal boiling point)
Enthalpy of Fusion, ΔH_{fus} (kJ/mol):
Entropy of Fusion, ΔS_{fus} (J/mol K):
Fugacity Ratio at 25°C, F: 1.0

Water Solubility (g/m^3 or mg/L at 25°C):
    7.20   (shake flask-GC, McAuliffe 1966)
    6.30   (calculated-recommended liquid-liquid equilibrium LLE data, Góral et al. 2004)

Vapor Pressure (Pa at 25°C):
    835   (calculated-Antoine eq., Dreisbach 1959; quoted, Hine & Mookerjee 1975)
    log (P/mmHg) = 6.77410 – 1404.7/(210.0 + t/°C); temp range 50–223°C (Antoine eq. for liquid state, Dreisbach 1959)

Henry's Law Constant (Pa m^3/mol):
    14600   (calculated-1/K_{AW}, C_W/C_A, reported as exptl., Hine & Mookerjee 1975)
    11600, 13010 (calculated-group contribution, calculated-bond contribution, Hine & Mookerjee 1975)
    8700   (calculated-MCI χ, Nirmalakhandan & Speece 1988)
    14400   (calculated-vapor-liquid equilibrium (VLE) data, Yaws et al. 1991)

Octanol/Water Partition Coefficient, log K_{OW}:
    3.98   (calculated-π substituent constants, Hansch et al. 1968)
    3.81   (calculated-MCI χ, Murray et al. 1975)
    3.98   (calculated-molar volume V_M, Wang et al. 1992)
    4.05   (calculated-fragment const., Müller & Klein 1992)
    4.50   (calculated-UNIFAC group contribution, Chen et al. 1993)

Octanol/Air Partition Coefficient, log K_{OA}:

Bioconcentration Factor, log BCF:

Sorption Partition Coefficient, log K_{OC}:

Environmental Fate Rate Constants, k, and Half-Lives, $t_{1/2}$:

Half-Lives in the Environment:

Aliphatic and Cyclic Hydrocarbons

2.1.2.4 Cycloalkenes

2.1.2.4.1 Cyclopentene

Common Name: Cyclopentene
Synonym:
Chemical Name: cyclopentene
CAS Registry No: 142-29-0
Molecular Formula: C_5H_8
Molecular Weight: 68.118
Melting Point (°C):
    –135.0   (Lide 2003)
Boiling Point (°C):
    44.2   (Lide 2003)
Density (g/cm³ at 20°C):
    0.7720, 0.7665 (20°C, 25°C, Dreisbach 1955)
Molar Volume (cm³/mol):
    88.2   (20°C, calculated-density, McAuliffe 1966; Stephenson & Malanowski 1987)
    92.1   (calculated-Le Bas method at normal boiling point)
Enthalpy of Vaporization, ΔH_V (kJ/mol):
    27.92, 26.96   (25°C, bp, Dreisbach 1955)
Enthalpy of Fusion, ΔH_{fus} (kJ/mol):
    0.48, 3.36; 3.84(–186.08, –135.05°C; total phase change enthalpy, Chickos et al. 1999)
Entropy of Fusion, ΔS_{fus} (J/mol K):
    29.83, 37.6 (exptl., calculated-group additivity method, total phase change entropy, Chickos et al. 1999)
Fugacity Ratio at 25°C, F: 1.0

Water Solubility (g/m³ or mg/L at 25°C):
    535   (shake flask-GC, McAuliffe 1966)
    611   (shake flask-titration with bromine, Natarajan & Venkatachalam 1972)
    1645   (shake flask-GC, Pierotti & Liabastre 1972)
    540   (suggested "tentative" value, IUPAC Solubility Data Series, Shaw 1989a)
    719   (calculated-liquid-recommended liquid equilibrium LLE data, Góral et al. 2004)

Vapor Pressure (Pa at 25°C or as indicated and reported temperature dependence equations. Additional data at other temperatures designated * are compiled at the end of this section.):
    39890*   (19.77°C, static method-quartz spiral gauge, measured range –42.75 to 19.77°C, Lister 1941)
    43375*   (21.028°C, ebulliometry, measured range 11.3–45.024°C, Forziati et al. 1950)
    $\log (P/\text{mmHg}) = 6.92066 - 1121.818/(233.446 + t/°C)$; temp range 11.3–45.0°C (Antoine eq., ebulliometry, Forziati et al. 1950)
    50690   (calculated by formula, Dreisbach 1955)
    $\log (P/\text{mmHg}) = 6.92066 - 1121.818/(233.446 + t/°C)$; temp range –30 to 105°C (Antoine eq. for liquid state, Dreisbach 1955)
    $\log (P/\text{atm}) = [1 - 317.520/(T/K)] \times 10^{\{0.814441 - 7.42372 \times 10^{-4} \cdot (T/K) + 8.49035 \times 10^{-7} \cdot (T/K)^2\}}$; temp range: 223.2–393.2 K (Cox eq., Chao et al. 1983)
    $\log (P/\text{mmHg}) = 6.92066 - 1121.818/(223.45 + t/°C)$; temp range not specified (Antoine eq., Dean 1985, 1992)
    50710   (interpolated-Antoine eq., Stephenson & Malanowski 1987)
    $\log (P_L/\text{kPa}) = 6.01617 - 1105.926/(-41.615 + T/K)$; temp range 249–318 K (Antoine eq., Stephenson & Malanowski 1987)
    $\log (P/\text{mmHg}) = 30.1132 - 2.3537 \times 10^3/(T/K) - 8.0609 \cdot \log (T/K) - 5.7786 \times 10^{-11} \cdot (T/K) + 3.4591 \times 10^{-6} \cdot (T/K)^2$; temp range 138–507 K (vapor pressure eq., Yaws 1994)

Henry's Law Constant (Pa m³/mol at 25°C or as indicated. Additional data at other temperatures designated * are compiled at the end of this section.):

6370 (calculated-$1/K_{AW}$, C_W/C_A, reported as exptl., Hine & Mookerjee 1975)
3580, 9650 (calculated-group contribution, calculated-bond contribution, Hine & Mookerjee 1975)
6460 (calculated-vapor-liquid equilibrium (VLE) data, Yaws et al. 1991)
2957, 3408, 3863, 4372* (10, 15, 20, 25°C, headspace-GC, measured range 10–25°C, Bakierowska & Trzeszczyñski 2003)
ln $(1/K_{AW})$ = 6.989 – 1915/(T/K); temp range 10–25°C, headspace-GC, Bakierowska & Trzeszczyñski 2003)

Octanol/Water Partition Coefficient, log K_{OW}:
1.75 (calculated-π substituent const., Hansch et al. 1968)
1.76 (calculated-molar volume V_M, Wang et al. 1992)
2.25 (calculated-fragment const., Müller & Klein 1992)
2.1791 (calculated-UNIFAC group contribution, Chen et al. 1993)
2.80 (calculated-UNIFAC activity coeff., Dallos et al. 1993)

Octanol/Air Partition Coefficient, log K_{OA}:

Bioconcentration Factor, log BCF:

Sorption Partition Coefficient, log K_{OC}:

Environmental Fate Rate Constants, k, and Half-Lives, $t_{1/2}$:
Volatilization:
Photolysis:
Oxidation: rate constant k, for gas-phase second order rate constants, k_{OH} for reaction with OH radical, k_{NO3} with NO_3 radical and k_{O3} with O_3 or as indicated, *data at other temperatures see reference:
k_{O3} = 8.13 × 10^{-16} cm³ molecule⁻¹ s⁻¹ at 298 K (Japar et al. 1974; quoted, Adeniji et al. 1981)
k_{O3} = 9.69 × 10^{-16} cm³ molecule⁻¹ s⁻¹ at 294 ± 2 K (chemiluminescence, Adeniji et al. 1981)
k_{O3} = (2.75 ± 0.33) × 10^{-16} cm³ molecule⁻¹ s⁻¹ at 297 ± 1 K (Atkinson et al. 1983a)
k_{OH} = (6.39 ± 0.23) × 10^{-11} cm³ molecule⁻¹ s⁻¹ at 298 ± 2 K (relative rate method, Atkinson et al. 1983b)
k_{O3} = 4.97 × 10^{-16} cm³ molecule⁻¹ s⁻¹ at 291.5 K in synthetic air (Bennett et al. 1987)
k_{O3} = (62.4 ± 3.5) × 10^{-17} cm³ molecule⁻¹ s⁻¹ at 297 ± 2 K in a smog chamber (Nolting et al. 1988)
k_{OH} = 4.99 × 10^{-11} cm³ molecule⁻¹ s⁻¹ at 298 ± 3 K, and k_{OH} = 5.7 × 10^{-11} cm³ molecule⁻¹ s⁻¹ relative to propene (relative rate method, Rogers 1989)
k_{OH} = (5.02 – 6.73) × 10^{-11} cm³ molecule⁻¹ s⁻¹ at 298 K (literature review, Atkinson 1989)
k_{OH} = 4.0 × 10^{-11} cm³ molecule⁻¹ s⁻¹ at 298 K, k_{NO3} = 4.6 × 10^{-13} cm³ molecule⁻¹ s⁻¹ at 298 K (Atkinson 1990)
k_{OH} = 6.7 × 10^{-11} cm³ molecule⁻¹ s⁻¹, k_{NO3} = 5.81 × 10^{-13} cm³ molecule⁻¹ s⁻¹ at 298 K (Sabljic & Güsten 1990)
k_{OH} = 6.7 × 10^{-11} cm³ molecule⁻¹ s⁻¹, k_{NO3} = 5.3 × 10^{-13} cm³ molecule⁻¹ s⁻¹, k_{O3}* = 5.7 × 10^{-16} cm³ molecule⁻¹ s⁻¹, and $k_{O(^3P)}$ = 2.1 × 10^{-11} cm³ molecule⁻¹ s⁻¹ for the reaction with O(^{3}P) atom, at 298 K (recommended, Atkinson 1997)
Hydrolysis:
Biodegradation:
Biotransformation:
Bioconcentration, Uptake (k_1) and Elimination (k_2) Rate Constants or Half-Lives:

Half-Lives in the Environment:
Air: photooxidation rate constant of 4.97 × 10^{-16} cm³ molecule⁻¹ s⁻¹ for the reaction with O_3 in synthetic air was determined at atmospheric pressure at 291.5 K (Bennett et al. 1987);
rate constant k = 4.99 × 10^{-11} cm³ molecule⁻¹ s⁻¹ for the reaction with OH radicals in air at 298 K (Rogers 1989).
Surface water: $t_{1/2}$ ~ 320 h and 9 × 10^4 d for oxidation by OH and RO_2 radicals in aquatic system, and $t_{1/2}$ = 40 d, based on rate constant k = 2 × 10^5 M⁻¹ s⁻¹ for the oxidation of cyclic olefins with singlet oxygen in aquatic system (Mill & Mabey 1985).

TABLE 2.1.2.4.1.1
Reported vapor pressures and Henry's law constants of cyclopentene at various temperatures and the coefficients for the vapor pressure equations

$\log P = A - B/(T/K)$ (1) $\quad\quad \ln P = A - B/(T/K)$ (1a)
$\log P = A - B/(C + t/°C)$ (2) $\quad\quad \ln P = A - B/(C + t/°C)$ (2a)
$\log P = A - B/(C + T/K)$ (3)
$\log P = A - B/(T/K) - C \cdot \log (T/K)$ (4)
$\log P = A - B/(T/K) - C \cdot \log (T/K) + D \cdot [P/(T/K)^2]$ (5)
$\log (P/\text{atm}) = A'[1- (T_B/T)]$ (6) where $\log A' = a + bT + cT^2$

Vapor pressure				Henry's law constant	
Lister 1941		Forziati et al. 1950		Bakierowska & Trzeszczyński 2003	
static-quartz spiral gauge		ebulliometry		headspace-GC	
t/°C	P/Pa	t/°C	P/Pa	t/°C	Pa m³ mol⁻¹
−42.75	1447	11.325	28994	10	2957
−24.18	4742	15.718	34940	15	3408
0	17292	21.028	43375	20	3863
19.77	39890	26.506	53710	25	4372
		32.34	66807		
		39.678	93760	$\ln (1/K_{AW}) = 6.989 - 1915/(T/K)$	
		43.146	97653		
		43.624	99247		
		44.071	100744		
		44.576	102469		
		45.024	104009		
		bp/°C	44.242		
		Antoine eq.			
		eq. 2	P/mmHg		
		A	6.92066		
		B	1121.818		
		C	233.446		

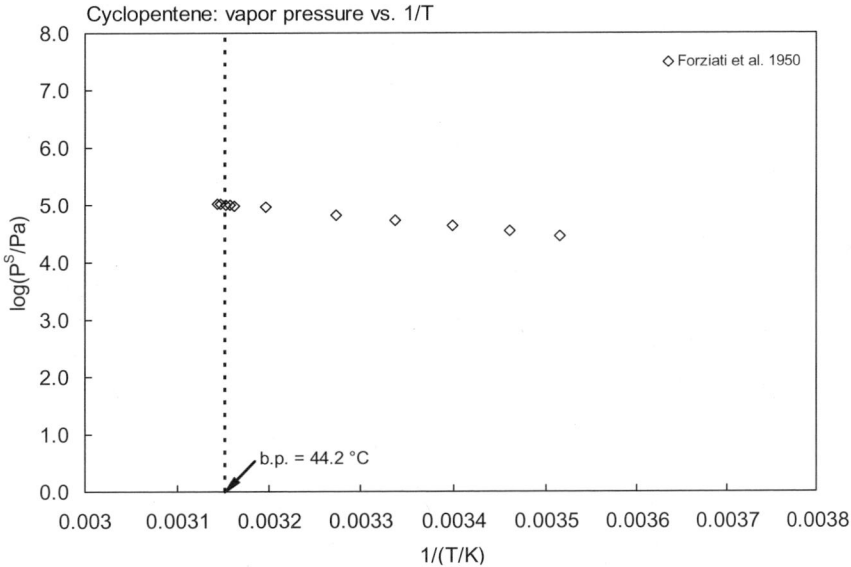

FIGURE 2.1.2.4.1.1 Logarithm of vapor pressure versus reciprocal temperature for cyclopentene.

2.1.2.4.2 Cyclohexene

Common Name: Cyclohexene
Synonym: 1,2,3,4-tetrahydrobenzene, tetrahydrobenzene
Chemical Name: cyclohexene
CAS Registry No: 110-83-8
Molecular Formula: C_6H_{10}
Molecular Weight: 82.143
Melting Point (°C):
    –103.5   (Weast 1982–83; Lide 2003)
Boiling Point (°C):
    82.98   (Lide 2003)
Density (g/cm³ at 20°C):
    0.8110, 0.8061 (20°C, 25°C, Dreisbach 1959)
    0.8102   (20°C, Weast 1982–83)
Molar Volume (cm³/mol):
    101.4   (calculated-density, Stephenson & Malanowski 1987; Ruelle & Kesselring 1997)
    110.8   (calculated-Le Bas method at normal boiling point)
Enthalpy of Vaporization, ΔH_V (kJ/mol):
    33.142, 30.485 (25°C, bp, Riddick et al. 1986)
Enthalpy of Vaporization, ΔH_V (kJ/mol):
    29.125, 27.276 (25°C, bp, Riddick et al. 1986)
Enthalpy of Fusion, ΔH_{fus} (kJ/mol):
    3.28   (Riddick et al. 1986)
    4.23, 3.28; 7.51(–134.45, –104.45°C; total phase change enthalpy, Chickos et al. 1999)
Entropy of Fusion, ΔS_{fus} (J/mol K):
    49.85, 41.3 (exptl., calculated-group additivity method, total phase change entropy, Chickos et al. 1999)
Fugacity Ratio at 25°C, F: 1.0

Water Solubility (g/m³ or mg/L at 25°C or as indicated and reported temperature dependence equations. Additional data at other temperatures designated * are compiled at the end of this section.):
    130   (shake flask-cloud point, McBain & Lissant 1951)
    160   (Farkas 1964)
    213   (shake flask-GC, McAuliffe 1966)
    299*   (25.11°C, shake flask-GC, measured 5.11–45.21°C, Pierotti & Liabastre 1972)
    281, 286   (23.5°C, elution chromatography, Schwarz 1980)
    160   (recommended best value, IUPAC Solubility Data Series, Shaw 1989a)
    246*   (calculated-liquid-liquid equilibrium LLE data, temp range 278.2–318.2 K, Góral et al. 2004)

Vapor Pressure (Pa at 25°C or as indicated and reported temperature dependence equations. Additional data at other temperatures designated * are compiled at the end of this section.):
    160, 752, 3345, 8723* (–44.42, –24.85, 0. 18.45°C, static method-quartz spiral gauge, Lister 1941)
    11734*   (24.794°C, ebulliometry, measured range 12.2–83.9°C (Forziati et al. 1950)
    $\log (P/mmHg) = 6.888617 - 1229.973/(224.104 + t/°C)$; temp range 12.2–83.9°C (Antoine eq., ebulliometry measurements, Forziati et al. 1950)
    11840   (calculated by formula, Dreisbach 1955)
    $\log (P/mmHg) = 6.88617 - 1229.973/(224.104 + t/°C)$; temp range 3.0–146°C (Antoine eq. for liquid state, Dreisbach 1955)
    19885*   (36.875°C, comparative ebulliometry, measured range 36.875–91.378°C, Meyer & Hotz 1973)
    $\log (P/mmHg) = [1 - 356.172/(T/K)] \times 10^{\wedge}\{0.873674 - 9.73841 \times 10^{-4} \cdot (T/K) + 10.9078 \times 10^{-7} \cdot (T/K)^2\}$; temp range 213.2–364.53 K (Cox eq., Chao et al. 1983)

Aliphatic and Cyclic Hydrocarbons

11800 (selected lit., Riddick et al. 1986)
log (P/kPa) = 7.109 − 2289.0/(T/K); temp range: not specified (Antoine eq., Riddick et al. 1986)
11850 (extrapolated-Antoine eq., Stephenson & Malanowski 1987)
log (P_L/kPa) = 5.997323 − 1221,899/(−49.978 + T/K); temp range 309–385 K (Antoine eq., Stephenson & Malanowski 1987)
log (P/mmHg) = 52.1749 − 3.238 × 10^3/(T/K) − 16.878·log (T/K) + 8.0388 × 10^{-3}·(T/K) + 1.3259 × 10^{-13}·(T/K)^2; temp range 170–560 K (vapor pressure eq., Yaws 1994)

Henry's Law Constant (Pa m^3/mol at 25°C or as indicated):
4020 (calculated-1/K_{AW}, C_W/C_A, reported as exptl., Hine & Mookerjee 1975)
4946, 13310 (calculated-group contribution, calculated-bond contribution, Hine & Mookerjee 1975)
4568 (calculated-vapor-liquid equilibrium (VLE) data, Yaws et al. 1991)
3960 (23°C, batch air stripping-IR, Nielsen et al. 1994)
2069, 2467, 2618, 2965 (10, 15, 20, 25°C, headspace-GC, Bakierowska & Trzeszczyński 2003)
ln (1/K_{AW}) = 5.860 − 1691/(T/K); temp range 10–25°C, headspace-GC, Bakierowska & Trzeszczyński 2003)

Octanol/Water Partition Coefficient, log K_{OW}:
2.16 (calculated-π substituent constants, Hansch et al. 1968)
2.86 (shake flask-GC, Leo et al. 1975)
1.90 (shake flask-GC, Canton & Wegman 1983)
2.86 (recommended, Sangster 1989, 1993)
2.86 (recommended, Hansch et al. 1995)

Octanol/Air Partition Coefficient, log K_{OA} at 25°C or as indicated. Additional data at other temperatures designated * are compiled at the end of this section:
2.92* (20.29°C, from GC-determined γ^∞ in octanol, measured range 20.29–50.28°C, Gruber et al. 1997)
2.83 (calculated-measured γ^∞ in pure octanol and vapor pressure P, Abraham et al. 2001)

Bioconcentration Factor, log BCF:

Sorption Partition Coefficient, log K_{OC}:

Environmental Fate Rate Constants, k, and Half-Lives, $t_{1/2}$:
  Volatilization:
  Photolysis:
  Oxidation: rate constant k, for gas-phase second order rate constants, k_{OH} for reaction with OH radical, k_{NO3} with NO_3 radical and k_{O3} with O_3 or as indicated, *data at other temperatures see reference:
   $k_{O(3P)}$ = 2.20 × 10^{-11} cm^3 molecule^{-1} s^{-1} for the reaction with O(^{3}P) atom (Herron & Huie 1973)
   k_{OH} = 6.77 × 10^{-11} cm^3 molecule^{-1} s^{-1} at 298 K (relative rate method, Atkinson et al. 1979)
   k_{O3} = 1.69 × 10^{-16} cm^3 molecule^{-1} at 298 K (Japar et al. 1974)
   k_{O3} = 2.04 × 10^{-16} cm^3 molecule^{-1} s^{-1} at 294 ± 2 K (chemiluminescence, Adeniji et al. 1981)
   k_{O3} = (1.04 ± 0.14) × 10^{-16} cm^3 molecule^{-1} s^{-1} at 297 ± 1 K (Atkinson et al. 1983a)
   k_{OH} = (6.43 ± 0.17) × 10^{-11} cm^3 molecule^{-1} s^{-1} at 298 ± 2 K (relative rate method, Atkinson et al. 1983b)
   k_{OH} = (64.1 ± 2.5) × 10^{-11} cm^3 molecule^{-1} s^{-1} at 297 ± 2 K (relative rate method, Ohta 1983)
   k_{O3} = 1.04 × 10^{-16} cm^3 molecule^{-1} s^{-1} at 297 K (Atkinson & Carter 1984)
   k_{O3} = 1.04 × 10^{-16} cm^3 molecule^{-1} s^{-1}, k_{OH} = 6.4 × 10^{-11} cm^3 molecule^{-1} s^{-1}, k_{NO3} = 0.29 × 10^{-12} cm^3 molecule^{-1} s^{-1}, $k_{O(3P)}$ = 21 × 10^{-12} cm^3 molecule^{-1} s^{-1} with O(^{3}P) atom and k_{NO2} < 0.2 × 10^{-20} cm^3 molecule^{-1} s^{-1} with NO_2 (Atkinson & Aschmann 1984)
   k_{O3} = 1.51 × 10^{-16} cm^3 molecule^{-1} s^{-1} in synthetic air at 295 K (Bennett et al. 1987)
   k_{O3} = (7.8 ± 0.5) × 10^{-17} cm^3 molecule^{-1} s^{-1} at 297 ± 2 K in a smog chamber (Nolting et al. 1988)
   k_{OH} = 5.40 × 10^{-11} cm^3 molecule^{-1} s^{-1} at 298 K and k_{OH} = 6.1 × 10^{-11} cm^3 molecule^{-1} s^{-1} relative to propene (relative rate method, Rogers 1989)
   k_{OH} = 6.77 × 10^{-11} cm^3 molecule^{-1} s^{-1} at 298 K (recommended, Atkinson 1989)
   k_{OH} = 6.77 × 10^{-11} cm^3 molecule^{-1} s^{-1}, k_{NO3} = 5.3 × 10^{-13} cm^3 molecule^{-1} s^{-1} at 298 K (Atkinson 1990)
   k_{OH} = 6.75 × 10^{-11} cm^3 molecule^{-1} s^{-1}, k_{NO3} = 5.3 × 10^{-13} cm^3 molecule^{-1} s^{-1} at 298 K (Sabljic & Güsten 1990)

k_{NO3} = 5.28 × 10^{-13} cm³ molecule^{-1} s^{-1} at 295 K (quoted, Atkinson 1991)

k_{OH} = 6.77 × 10^{-11} cm³ molecule^{-1} s^{-1}, k_{NO3}* = 5.9 × 10^{-13} cm³ molecule^{-1} s^{-1}, and k_{O3}* = 81.4 × 10^{-18} cm³ molecule^{-1} s^{-1}, and $k_{O(3P)}$ = 2.0 × 10^{-11} cm³ molecule^{-1} s^{-1} for the reaction with O(^{3}P) atom, at 298 K (recommended, Atkinson 1997)

Hydrolysis:

Biodegradation:

Biotransformation:

Bioconcentration, Uptake (k_1) and Elimination (k_2) Rate Constants or Half-Lives:

Half-Lives in the Environment:

Surface water: $t_{½}$ ~ 320 h and 9 × 10^4 d for oxidation by OH and RO$_2$ radicals in aquatic system, and $t_{½}$ = 40 d, based on rate constant k = 2 × 10^5 M^{-1} s^{-1} for oxidation of cyclic olefins by singlet oxygen in aquatic system (Mill & Mabey 1985).

TABLE 2.1.2.4.2.1
Reported aqueous solubilities of cyclohexene at various temperatures

Pierotti & Liabastre 1972		Shaw 1989a		Góral et al. 2004	
shake flask-GC/FID		IUPAC "tentative" values		calc-recommended LLE data	
t/°C	S/g·m^{-3}	t/°C	S/g·m^{-3}	t/°C	S/g·m^{-3}
5.11	280	5	–	5.1	265
15.21	298.5	15	–	15.2	251
25.11	299	20	–	20	246
35.21	302.5	25	160	23.5	246
45.21	310.5	35	–	25	246
		45	–	25.1	246
				35.2	251
				45.2	265

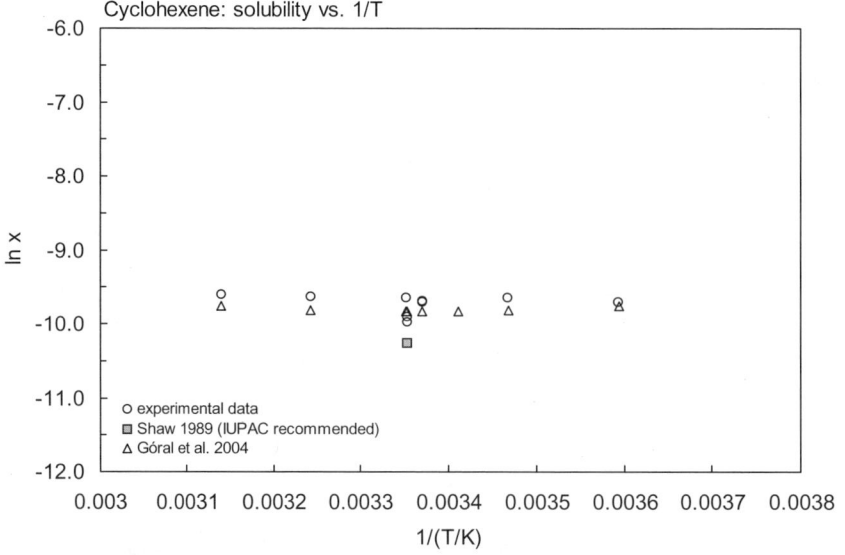

FIGURE 2.1.2.4.2.1 Logarithm of mole fraction solubility (ln x) versus reciprocal temperature for cyclohexene.

Aliphatic and Cyclic Hydrocarbons

TABLE 2.1.2.4.2.2
Reported vapor pressures and octanol-air partition coefficients of cyclohexene at various temperatures and the coefficients for the vapor pressure equations

$\log P = A - B/(T/K)$ (1) $\quad\quad \ln P = A - B/(T/K)$ (1a)

$\log P = A - B/(C + t/°C)$ (2) $\quad\quad \ln P = A - B/(C + t/°C)$ (2a)

$\log P = A - B/(C + T/K)$ (3)

$\log P = A - B/(T/K) - C \cdot \log (T/K)$ (4)

$\log P = A - B/(T/K) - C \cdot \log (T/K) + D \cdot [P/(T/K)^2]$ (5)

$\log (P/atm) = A'[1 - (T_B/T)]$ (6) where $\log A' = a + bT + cT^2$

Vapor pressure							log K_{OA}	
Lister 1941		Forziati et al. 1950		Meyer & Hotz 1973			Gruber et al. 1997	
static-quartz spiral gauge		ebulliometry		comparative ebulliometry			GC det'd activity coeff.	
t/°C	P/Pa	t/°C	P/Pa	t/°C	P/Pa		t/°C	P/Pa
−44.42	160	12.236	6417	36.875	19885		20.29	2.926
−24.85	752	15.920	7709	43.560	26084		30.3	2.762
0	3345	19.137	9006	49.795	33189		40.4	2.594
18.45	8723	22.063	10344	55.785	41404		50.28	2.447
		24.794	11734	62.537	52551				
$\Delta H_V/(kJ\ mol^{-1}) = 32.59$		28.490	13858	68.815	64943				
at 27°C		32.702	16651	75.354	80213				
		36.996	19958	81.075	95764				
		40.976	23358	86.112	111325				
		46.302	28995	91.378	129633				
		51.191	35060						
		57.107	43376	bp/°C	82.945				
		63.200	53712						
		69.708	66808	Antoine eq.					
		76.766	83761	eq. 3	P/cmHg				
		81.757	97648	A	5.872420				
		82.292	99248	B	1221.899				
		82.791	100746	C	223.1720				
		83.353	102470						
		83.852	104010	Cox eq.					
				eq. 6	P/atm				
		eq. 2	P/mmHg	a	0.833958				
		A	6.88617	−b × 10³	0.742586				
		B	1229.973	c × 10⁶	0.767278				
		C	224.104	T_B/K	356.0954				
		bp/°C	82.979						

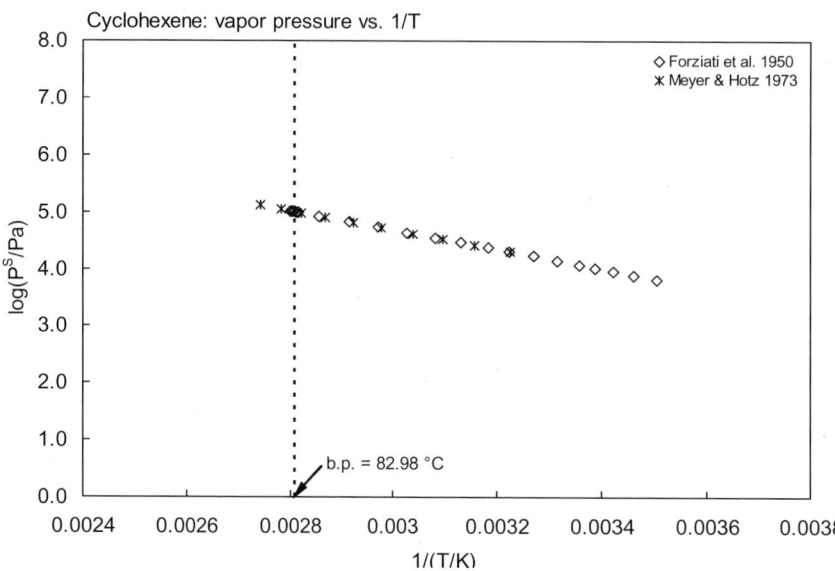

FIGURE 2.1.2.4.2.2 Logarithm of vapor pressure versus reciprocal temperature for cyclohexene.

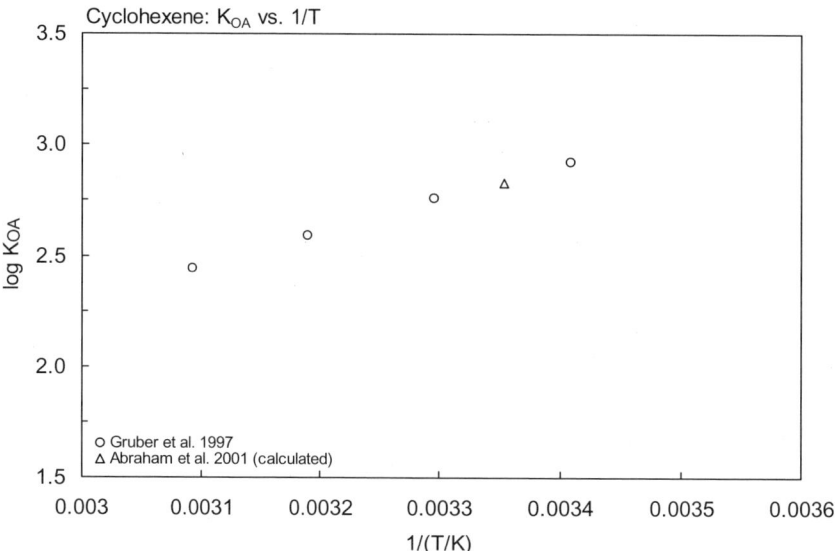

FIGURE 2.1.2.4.2.3 Logarithm of K_{OA} versus reciprocal temperature for cyclohexene.

Aliphatic and Cyclic Hydrocarbons

2.1.2.4.3   1-Methylcyclohexene

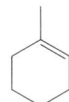

Common Name: 1-Methylcyclohexene
Synonym:
Chemical Name: 1-methylcyclohexene
CAS Registry No: 591-49-1
Molecular Formula: C_7H_{12}
Molecular Weight: 96.170
Melting Point (°C):
    –120.4   (Lide 2003)
Boiling Point (°C):
    110.3   (Lide 2003)
Density (g/cm³ at 20°C):
    0.8102, 0.8058 (20°C, 25°C, Dreisbach 1959; Weast 1982–83)
Molar Volume (cm³/mol):
    118.7   (20°C, calculated-density, McAuliffe 1966; Stephenson & Malanowski 1987)
    133.0   (calculated-Le Bas method at normal boiling point)
Enthalpy of Vaporization, ΔH_V (kJ/mol):
    37.75, 32.70 (25°, bp, Dreisbach 1955)
Enthalpy of Fusion, ΔH_{fus} (kJ/mol):
    6.63   (Chickos et al. 1999)
Entropy of Fusion, ΔS_{fus} (J/mol K):
    43.16, 44.1 (exptl., calculated-group additivity method, total phase change entropy, Chickos et al. 1999)
Fugacity Ratio at 25°C, F: 1.0

Water Solubility (g/m³ or mg/L at 25°C):
    52.0   (shake flask-GC, McAuliffe 1966)
    64.1   (calculated-recommended liquid-liquid equilibrium LLE data, Góral et al. 2004)

Vapor Pressure (Pa at 25°C or as indicated and reported temperature dependence equations):
    4080   (calculated by formula, Dreisbach 1955)
    log (P/mmHg) = 6.86861 – 1308.0/(218.0 + t/°C); temp range 25–165°C, (Antoine eq. for liquid state, Dreisbach 1955)
    3933   (extrapolated-Antoine eq., Stephenson & Malanowski 1987)
    log (P_L/kPa) = 6.0101 – 1311.087/(–56.045 + T/K); temp range 333–384 K (Antoine eq., Stephenson & Malanowski 1987)
    4858   (25.25°C, transpiration method, Verevkin et al. 2000)
    ln (P/Pa) = 23.65 – 4531/(T/K); temp range 275.4–313.4 K (transpiration method, Verevkin et al. 2000)

Henry's Law Constant (Pa m³/mol at 25°C):
    7660   (calculated-1/K_{AW}, C_W/C_A, reported as exptl., Hine & Mookerjee 1975)
    7485, 14260 (calculated-group contribution, calculated-bond contribution, Hine & Mookerjee 1975)
    11070   (calculated-MCI χ, Nirmalakhandan & Speece 1988)

Octanol/Water Partition Coefficient, log K_{OW}:
    1.05   (calculated-regression of Lyman et al. 1982, Wang et al. 1992)
    2.20   (calculated-molar volume V_M, Wang et al. 1992)

Octanol/Air Partition Coefficient, log K_{OA}:

Bioconcentration Factor, log BCF:

Sorption Partition Coefficient, log K_{OC}:

Environmental Fate Rate Constants, k, and Half-Lives., $t_{1/2}$:

    Volatilization:

    Photolysis:

    Oxidation: rate constant k, for gas-phase second order rate constants, k_{OH} for reaction with OH radical, k_{NO3} with NO_3 radical and k_{O3} with O_3 or as indicated, *data at other temperatures see reference:

        k_{OH} = 9.44 × 10^{-11} cm^3 molecule^{-1} s^{-1} at 305 K (Darnall et al. 1976, Atkinson 1989)

        k_{NO3} = (2.87 ± 0.34) × 10^{-13} cm^3 molecule^{-1} s^{-1} at 295 K (relative rate method, Atkinson et al. 1984a)

        k_{NO2} < 0.20 × 10^{-20} cm^3 molecule^{-1} s^{-1}; k_{O3} = 1.4 × 10^{-16} cm^3 molecule^{-1} s^{-1}; k_{OH} = 6.4 × 10^{-11} cm^3 molecule^{-1} s^{-1}; k_{NO3} = 2.9 × 10^{-13} cm^3 molecule^{-1} s^{-1}; and $k_{O(3P)}$ = 2.1 × 10^{-11} cm^3 molecule^{-1} s^{-1} with O(^{3}P) atom at room temp. (relative rate method, Atkinson et al. 1984b)

        k_{OH} = 9.45 × 10^{-11} cm^3 molecule^{-1} s^{-1} at 294 K (Atkinson 1985)

        k_{OH} = 9.4 × 10^{-11} cm^3 molecule^{-1} s^{-1} at 298 K (recommended, Atkinson 1990)

        k_{OH} = 9.4 × 10^{-11} cm^3 molecule^{-1} s^{-1}, k_{NO3}* = 1.7 × 10^{-11} cm^3 molecule^{-1} s^{-1}, k_{O3} = 1.65 × 10^{-16} cm^3 molecule^{-1} s^{-1}, and $k_{O(3P)}$ = 9.0 × 10^{-11} cm^3 molecule^{-1} s^{-1} for the reaction with O(^{3}P) atom, at 298 K (recommended, Atkinson 1997)

    Hydrolysis:

    Biodegradation:

    Biotransformation:

    Bioconcentration, Uptake (k_1) and Elimination (k_2) Rate Constants or Half-Lives:

Half-Lives in the Environment:

    Surface water: $t_{1/2}$ ~ 320 h and 9 × 10^4 d for oxidation by OH and RO_2 radicals in aquatic system, and $t_{1/2}$ = 40 d, based on rate constant k = 2 × 10^5 M^{-1} s^{-1} for the oxidation of cyclic olefins by singlet oxygen in aquatic system (Mill & Mabey 1985).

Aliphatic and Cyclic Hydrocarbons

2.1.2.4.4 Cycloheptene

Common Name: Cycloheptene
Synonym: suberene
Chemical Name: cycloheptene
CAS Registry No: 628-92-2
Molecular Formula: C_7H_{12}
Molecular Weight: 96.170
Melting Point (°C):
    –56   (Weast 1982–83; Stephenson & Malanowski 1987; Lide 2003)
Boiling Point (°C):
    115   (Weast 1982–83; Lide 2003)
Density (g/cm³ at 20°C):
    0.8228   (Weast 1982–83)
Molar Volume (cm³/mol):
    116.9   (20°C, calculated-density, McAuliffe 1966; Lande & Banerjee 1981; Wang et al. 1992)
    129.0   (calculated-Le Bas method at normal boiling point)
Enthalpy of Vaporization, ΔH_V (kJ/mol):
    36.73   (27°C, Lister 1941)
Enthalpy of Fusion, ΔH_{fus} (kJ/mol):
    5.28, 0.71, 0.97; 6.96 (–119.15, –63.15, –56.15°C; total phase change enthalpy, Chickos et al. 1999)
Entropy of Fusion, ΔS_{fus} (J/mol K):
    42.14, 45.0 (exptl., calculated-group additivity method, total phase change entropy, Chickos et al. 1999)
Fugacity Ratio at 25°C, F: 1.0

Water Solubility (g/m³ or mg/L at 25°C):
    66.0   (shake flask-GC, McAuliffe 1966)

Vapor Pressure (Pa at 25°C or as indicated and reported temperature dependence equations):
    188, 821, 2636, 6547 (–21.44, 0, 19.82, 39.06°C, static method-quartz spiral gauge, Lister 1941)
    2670   (interpolated-Antoine eq., Stephenson & Malanowski 1987)
    $\log (P_L/kPa) = 7.27243 - 2011.9/(T/K)$; temp range 251–313 K (Antoine eq., Stephenson & Malanowski 1987)

Henry's Law Constant (Pa m³/mol):

Octanol/Water Partition Coefficient, log K_{OW}:
    2.57   (calculated-π substituent constants, Hansch et al. 1968)
    2.75   (calculated-MCI χ, Murray et al. 1975)
    3.073   (calculated-UNIFAC group contribution, Chen et al. 1993)
    2.58   (calculated-molar volume V_M, Wang et al. 1992)
    3.37   (calculated-fragment const., Müller & Klein 1992)

Octanol/Air Partition Coefficient, log K_{OA}:

Bioconcentration Factor, log BCF:

Sorption Partition Coefficient, log K_{OC}:

Environmental Fate Rate Constants, k, and Half-Lives, $t_{½}$:
    Volatilization:
    Photolysis:
    Oxidation: rate constant k, for gas-phase second order rate constants, k_{OH} for reaction with OH radical, k_{NO3} with NO_3 radical and k_{O3} with O_3 or as indicated, *data at other temperatures see reference:

k_{O3} = (3.19 ± 0.36) × 10^{-16} cm^3 molecule^{-1} s^{-1} at 297 ± 1 K (Atkinson et al. 1983a; quoted, Atkinson & Carter 1984)

k_{OH} = (7.08 ± 0.11) × 10^{-11} cm^3 molecule^{-1} s^{-1} at 298 K (relative rate method, Atkinson et al. 1983b)

k_{OH} = 7.44 × 10^{-11} cm^3 molecule^{-1} s^{-1} at 298 K (Atkinson et al. 1983b, Atkinson 1989)

k_{O3} = (28.3 ± 1.5) × 10^{-17} cm^3 molecule^{-1} s^{-1} at 297 ± 2 K in a smog chamber (Nolting et al. 1988)

k_{NO3} = 2.80 × 10^{-13} cm^3 molecule^{-1} s^{-1}, k_{OH} = 7.13 × 10^{-11} cm^3 molecule^{-1} s^{-1} at 298 K (Sabljic & Güsten 1990)

k_{NO3} = 4.84 × 10^{-13} cm^3 molecule^{-1} s^{-1} at 298 K (Atkinson 1991)

k_{OH} = 7.4 × 10^{-11} cm^3 molecule^{-1} s^{-1}, k_{NO3} = 4.8 × 10^{-13} cm^3 molecule^{-1} s^{-1}, and k_{O3}* = 2.45 × 10^{-16} cm^3 molecule^{-1} s^{-1} at 298 K (recommended, Atkinson 1997)

Hydrolysis:

Biodegradation:

Biotransformation:

Bioconcentration, Uptake (k_1) and Elimination (k_2) Rate Constants or Half-Lives:

Half-Lives in the Environment:

Surface water: $t_{½}$ ~ 320 h and 9 × 10^4 d for oxidation by OH and RO$_2$ radicals in aquatic system, and $t_{½}$ = 40 d, based on rate constant k = 2 × 10^5 M^{-1} s^{-1} for oxidation of cyclic olefins by singlet oxygen in aquatic system (Mill & Mabey 1985).

2.1.2.4.5 Cyclooctene

Common Name: Cyclooctene
Synonym:
Chemical Name: cyclooctene
CAS Registry No: 931-87-3 (*cis*-octene), 931-89-5 (*trans*-octene)
Molecular Formula: C_8H_{14}
Molecular Weight: 110.197
Melting Point (°C):
    –12, –59   (*cis*-, *trans*-cyclooctene, Weast 1982–83; Lide 2003)
    –14.5 to –15.5   (*cis*-cyclooctene, Stephenson & Malanowski 1987)
Boiling Point (°C):
    138, 143   (*cis*-, *trans*-cyclooctene, Weast 1982–83; Lide 2003)
Density (g/cm³):
    0.8472, 0.8483   (20°C, *cis*-, *trans*-cyclooctene, Weast 1982–83)
Molar Volume (cm³/mol):
    130.1   (*cis*-, 20°C, calculated-density, McAuliffe 1966; Lande & Banerjee 1981; Wang et al. 1992)
    129.9   (*trans*-, 20°C, calculated-density, McAuliffe 1966; Lande & Banerjee 1981; Wang et al. 1992)
    146.7   (calculated-Le Bas method at normal boiling point)
Enthalpy of Vaporization, ΔH_V (kJ/mol):
    41.57   (27°C, Lister 1941)
Enthalpy of Sublimation, ΔH_{subl} (kJ/mol):
Enthalpy of Fusion, ΔH_{fus} (kJ/mol):
Entropy of Fusion, ΔS_{fus} (J/mol K):
Fugacity Ratio at 25°C, F: 1.0

Water Solubility (g/m³ or mg/L at 25°C or as indicated and reported temperature dependence equations. Additional data at other temperatures designated * are compiled at the end of this section.):
    22.9*   (generator column-GC/FID, measured range 273.15–313.15 K, Dohányosová et al. 2004)
    $\ln x = -33.3561 + 20.8640/\tau + 43.2804 \cdot \ln \tau$, $\tau = [(T/K)/298.15]$, temp range 273.15–313.15 K (generator column-GC/FID, Dohányosová et al. 2004)

Vapor Pressure (Pa at 25°C or as indicated and reported temperature dependence equations):
    209.3, 774.6, 2333, 5948   (0, 19.83, 40.36, 60.22°C, static method-quartz spiral manometer, Lister 1941)
    $\log (P_L/\text{kPa}) = 7.3641 - 2194.3/(T/K)$, temp range 273–441 K (*cis*-cyclooctene, Antoine eq., Stephenson & Malanowski 1987)
    1010   (interpolated from data of Lister 1941, temp range 273.15–313.15 K, Dohányosová et al. 2004)

Henry's Law Constant (Pa m³/mol at 25°C. Additional data at other temperatures designated * are compiled at the end of this section.):
    4842*   (derived from measured mole fraction solubility and solute fugacity, temp range 273.15–313.15 K, Dohányosová et al. 2004)

Octanol/Water Partition Coefficient, log K_{OW}:

Octanol/Air Partition Coefficient, log K_{OA}:

Bioconcentration Factor, log BCF or log K_B:

Sorption Partition Coefficient, log K_{OC}:

Environmental Fate Rate Constants, k and Half-Lives, $t_{1/2}$:
  Volatilization:
  Photolysis:
  Oxidation: rate constant k, for gas-phase second order rate constants, k_{OH} for reaction with OH radical, k_{NO3} with NO_3 radical and k_{O3} with O_3 or as indicated, *data at other temperatures see reference:
    k_{O3}* = 3.75 × 10^{-16} cm^3 molecule^{-1} s^{-1} at 298 K (*cis*-cyclooctene, recommended, Atkinson 1997)
  Hydrolysis:
  Biodegradation:
  Biotransformation:
  Bioconcentration and Uptake and Elimination Rate Constants (k_1 and k_2):

Half-Lives in the Environment:

TABLE 2.1.2.4.5.1
Reorted aqueous solubilities and Henry's law constants of cyclooctene at various temperatures

Aqueous solubility				Henry's law constant	
Doháryosová et al. 2004				Doháryosová et al. 2004	
generator column-GC/FID		smoothed raw exptl data		from solute fugacity f and x	
T/K	S/g·m^{-3}	T/K	S/g·m^{-3}	T/K	H/(Pa m^3/mol)
	raw data				
274.15	20.39	273.15	19.9	273.15	1184.4
274.15	20.14	278.15	20.2	278.15	1625.4
278.15	20.39	283.15	20.7	283.15	2196
278.15	19.59	288.15	21.3	288.15	2898
283.15	21.24	293.15	22.0	293.15	3780
283.15	20.45	298.15	22.9	298.15	4842
288.15	20.94	303.15	23.9	303.15	6120
288.15	20.63	308.15	25.1	308.15	7614
293.15	21.92	313.15	26.5	313.15	9378
298.15	23.63				
298.15	24.61	ln x = A + B/τ + C ln τ			
303.15	22.84	τ = T/298.15			
303.15	23.14	A	−33.3561		
308.15	27.24	B	20.8640		
308.15	24.73	C	23.4396		
308.15	26.88				
313.15	27.31				
313.15	24.80				

Aliphatic and Cyclic Hydrocarbons

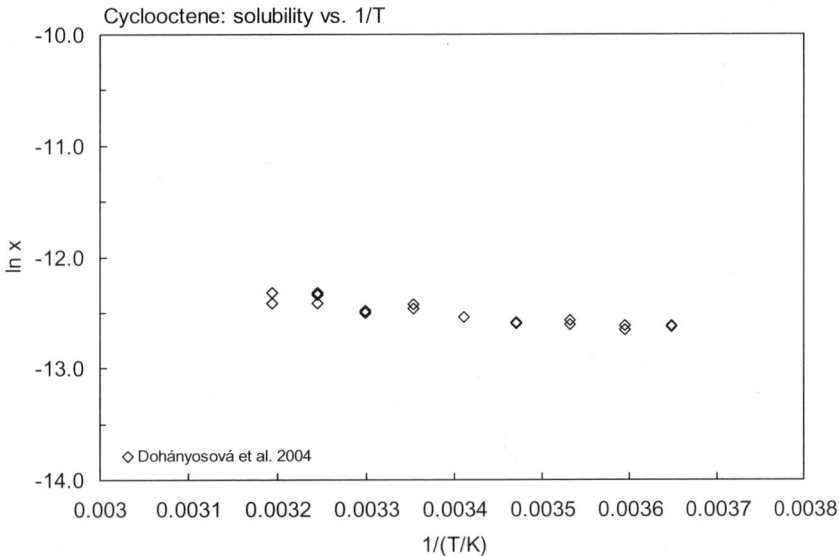

FIGURE 2.1.2.4.5.1 Logarithm of mole fraction solubility (ln x) versus reciprocal temperature for cyclooctene.

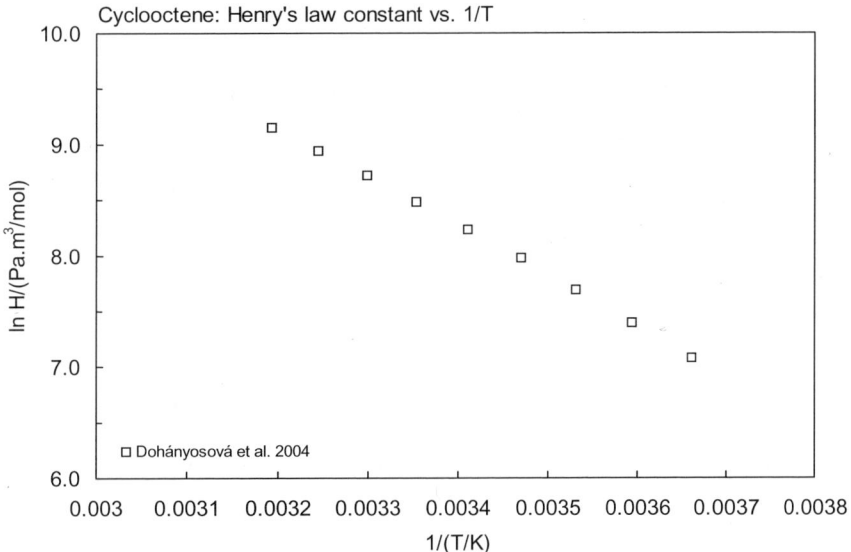

FIGURE 2.1.2.4.5.2 Logarithm of Henry's law constant versus reciprocal temperature for cyclooctene.

2.1.2.4.6   1,4-Cyclohexadiene

Common Name: 1,4-Cyclohexadiene
Synonym: 1,4-dihyrobenzene
Chemical Name: 1,4-cyclohexadiene
CAS Registry No: 628-41-1
Molecular Formula: C_6H_8
Molecular Weight: 80.128
Melting Point (°C):
    –49.2   (Weast 1983–83; Lide 2003)
Boiling Point (°C):
    85.5   (Lide 2003)
Density (g/cm³ at 20°C):
    0.8471   (Weast 1982–83)
Molar Volume (cm³/mol):
    93.60   (20°C, calculated-density, McAuliffe 1966; Stephenson & Malanowski 1987)
    103.4   (calculated-Le Bas method at normal boiling point)
Enthalpy of Fusion, ΔH_{fus} (kJ/mol):
    0.82, 5.72; 6.53 (–81.15, –49.15°C, total phase change enthalpy, Chickos et al. 1999)
Entropy of Fusion, ΔS_{fus} (J/mol K):
    29.16, 38.0 (exptl., calculated-group additivity method, total phase change entropy, Chickos et al. 1999)
Fugacity Ratio at 25°C, F: 1.0

Water Solubility (g/m³ or mg/L at 25°C or as indicated. Additional data at other temperatures designated * are compiled at the end of this section.):
    700   (shake flask-GC, McAuliffe 1966)
    930*   (shake flask-GC, measured range 5.11–45.21°C, Pierotti & Liabstre 1972)
    800*   (recommended, temp range 5–45°C, IUPAC Solubility Data Series, Shaw 1989a)
    979*   (calculated-liquid-liquid equilibrium LLE data, temp range 278.3–318.4 K, Góral et al. 2004)

Vapor Pressure (Pa at 25°C or as indicated and reported temperature dependence equations. Additional data at other temperatures designated * are compiled at the end of this section.):
    11892*   (31.1°C, static method-Hg manometer, measured range 304.25–322.23 K, Letcher & Marsicano 1974)
    $\log (P/\text{mmHg}) = [1 - 368.566/(T/K)] \times 10^{\wedge}\{0.916704 - 6.81678 \times 10^{-4} \cdot (T/K) - 7.02362 \times 10^{-7} \cdot (T/K)^2\}$; temp range 304.25–322.23 K (Cox eq., Chao et al. 1983)
    8973   (extrapolated-Antoine eq., Boublik et al. 1984)
    $\log (P/\text{kPa}) = 5.86553 - 1176.707/(214.528 + t/°C)$; temp range 31.1–49.08°C (Antoine eq. from reported exptl. data, Boublik et al. 1984)
    9009   (extrapolated-Antoine eq., Stephenson & Malanowski 1987)
    $\log (P_L/\text{kPa}) = 6.41736 - 1475.149/(-26.108 + T/K)$; temp range: 304–360 K (Antoine eq., Stephenson & Malanowski 1987)

Henry's Law Constant (Pa m³/mol):

Octanol/Water Partition Coefficient, log K_{OW}
    2.30   (shake flask, Log P Database, Hansch & Leo 1987)
    2.48   (calculated-UNIFAC activity coeff., Banerjee & Howard 1988)
    2.30   (recommended, Sangster 1989, 1993)
    2.30   (recommended, Hansch et al. 1995)

Octanol/Air Partition Coefficient, log K_{OA}:

Aliphatic and Cyclic Hydrocarbons

Bioconcentration Factor, log BCF:

Sorption Partition Coefficient, log K_{OC}:

Environmental Fate Rate Constants, k, and Half-Lives, $t_{½}$:
  Volatilization:
  Photolysis:
  Oxidation: rate constant k, for gas-phase second order rate constants, k_{OH} for reaction with OH radical, k_{NO3} with NO_3 radical and k_{O3} with O_3 or as indicated, *data at other temperatures see reference:
    k_{O3} = (0.639 ± 0.074) × 10^{-16} cm^3 $molecule^{-1}$ s^{-1} at 297 ± 1 K (Atkinson et al. 1983a)
    k_{OH} = (9.48 ± 0.39) × 10^{-11} cm^3 $molecule^{-1}$ s^{-1} at 298 ± 2 K (relative rate method, Atkinson et al. 1983b)
    k_{NO3} = (2.89 ± 0.035) × 10^{-13} cm^3 $molecule^{-1}$ s^{-1} at 295 K (Atkinson et al. 1984a)
    k_{O3} = 63.9 × 10^{-18} cm^3 $molecule^{-1}$ s^{-1}, k_{OH} = 9.5 × 10^{-11} cm^3 $molecule^{-1}$ s^{-1}, k_{NO3} = 0.29 × 10^{-12} cm^3 $molecule^{-1}$ s^{-1}, and k_{NO2} < 0.4 × 10^{-20} cm^3 $molecule^{-1}$ s^{-1} with NO_2 (Atkinson et al. 1984b)
    k_{OH}(exptl) = 9.90 × 10^{-11} cm^3 $molecule^{-1}$ s^{-1}, k_{OH}(calc) = 1.03 × 10^{-10} cm^3 $molecule^{-1}$ s^{-1} (Atkinson 1985)
    k_{NO3} = 7.8 × 10^{-13} cm^3 $molecule^{-1}$ s^{-1} at 298 K (fast flow system, Benter & Schindler 1988)
    k_{OH} = (99.2 – 99.8) × 10^{-12} cm^3 $molecule^{-1}$ s^{-1} at 298 K (review, Atkinson 1989)
    k_{OH} = 9.91 × 10^{-11} cm^3 $molecule^{-1}$ s^{-1} and k_{NO3} = 5.30 × 10^{-13} cm^3 $molecule^{-1}$ s^{-1} for the reaction with NO_3 radical at nights (Sabljic & Güsten 1990)
    k_{NO3} = 6.6 × 10^{-13} cm^3 $molecule^{-1}$ s^{-1} at 298 K (recommended, Atkinson 1991)
    k_{OH} = 9.95 × 10^{-11} cm^3 $molecule^{-1}$ s^{-1}, k_{NO3} = 6.6 × 10^{-13} cm^3 $molecule^{-1}$ s^{-1}, and k_{O3}* = 4.6 × 10^{-17} cm^3 $molecule^{-1}$ s^{-1} at 298 K (recommended, Atkinson 1997)
  Hydrolysis:
  Biodegradation:
  Biotransformation:
  Bioconcentration, Uptake (k_1) and Elimination (k_2) Rate Constants or Half-Lives:

Half-Lives in the Environment:
  Surface water: $t_{½}$ ~ 320 h and 9 × 10^4 d for oxidation by OH and RO_2 radicals for olefins in aquatic system, and $t_{½}$ = 19 h, based on rate constant k = 1.0 × 10^7 M^{-1} s^{-1} for oxidation of dienes by singlet oxygen in aquatic system (Mill & Mabey 1985).

TABLE 2.1.2.4.6.1
Reported aqueous solubilities of 1,4-cyclohexadiene at various temperatures

Aqueous solubility						Vapor pressure	
Pierotti & Liabastre 1972		Shaw 1989a		Góral et al. 2004		Letcher & Marsicano 1974	
shake flask-GC/FID		IUPAC "tentative" values		calc-recommended LLE data		static method-Hg manometer	
t/°C	S/g·m⁻³	t/°C	S/g·m⁻³	t/°C	S/g·m⁻³	T/K	S/g·m⁻³
5.11	851.9	5	900	5.1	1068	304.25	11892
15.21	958.5	15	900	15.2	979	307.67	13826
25.11	936.2	25	800	25.0	979	310.76	15812
35.21	963.4	35	1000	25.1	979	312.92	42655
45.21	1010	45	1000	35.2	979	316.67	20238
				45.2	1068	319.94	22998
						322.23	25251
						Antoine eq.,	P/mmHg
						log P = A – B/(C + T/K)	
						A	7.2687
						B	1461.75
						C	–29.4

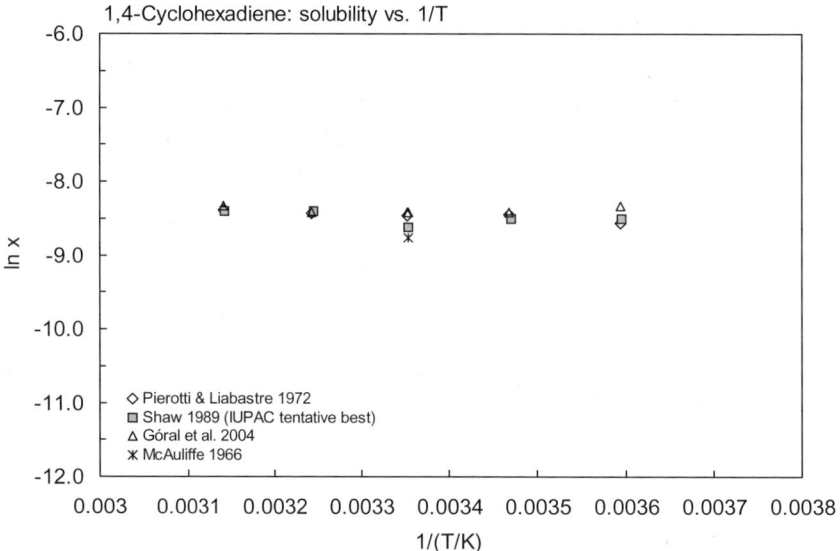

FIGURE 2.1.2.4.6.1 Logarithm of mole fraction solubility (ln x) versus reciprocal temperature for 1,4-cyclohexadiene.

Aliphatic and Cyclic Hydrocarbons

2.1.2.4.7 Cycloheptatriene

Common Name: Cycloheptatriene
Synonym: tropilidene
Chemical Name: 1,3,5-cycloheptatriene
CAS Registry No: 544-25-2
Molecular Formula: C_7H_8
Molecular Weight: 92.139
Melting Point (°C):
    −79.5   (Weast 1982–83; Lide 2003)
Boiling Point (°C):
    117   (Weast 1982–83; Lide 2003)
Density (g/cm³ at 20°C):
    0.8875   (Weast 1982–83)
Molar Volume (cm³/mol):
    103.0   (20°C, calculated-density, McAuliffe 1966; Lande & Banerjee 1981; Wang et al. 1992)
    114.2   (calculated-Le Bas method at normal boiling point)
Enthalpy of Fusion, ΔH_{fus} (kJ/mol):
    2.35, 1.16; 3.51   (−93.15, −75.15°C, Chickos et al. 1999)
Entropy of Fusion, ΔS_{fus} (J/mol K):
    21.11, 38.5 (exptl., calculated-group additivity method, total phase change entropy, Chickos et al. 1999)
Fugacity Ratio at 25°C, F: 1.0

Water Solubility (g/m³ or mg/L at 25°C or as indicated. Additional data at other temperatures designated * are compiled at the end of this section.):
    620   (shake flask-GC, McAuliffe 1966)
    669*   (25.11°C, shake flask-GC, measured range 5.11–40.21°C, Pierotti & Liabastre 1972)
    640*   (recommended best value, temp range 5–45°C, IUPAC Solubility Data Series, Shaw 1989)
    563*   (calculated-liquid-liquid equilibrium LLE data, temp range 278.3–318.4 K, Góral et al. 2004)

Vapor Pressure (Pa at 25°C or as indicated and reported temperature dependence equations. Additional data at other temperatures designated * are compiled at the end of this section.):
    3136*   (ebulliometry, measured range 0–65°C, Finke et al. 1956)
    log (P/mmHg) = 6.97032 − 1374.065/(t/°C + 220.538); temp range 0–65°C (Antoine eq., ebulliometry, Finke et al. 1956)
    2825, 3138   (interpolated-Antoine eq-I, II, Stephenson & Malanowski 1987)
    log (P_L/kPa) = 6.09522 − 1374.656/(−52.612 + T/K); temp range 273–390 K (Antoine eq.-I, Stephenson & Malanowski 1987)
    log (P_L/kPa) = 6.12574 − 1390.771/(−53.069 + T/K); temp range 273–390 K (Antoine eq.-II, Stephenson & Malanowski 1987)

Henry's Law Constant (Pa m³/mol at 25°C):
    432   (calculated-P/C from selected data)
    466   (calculated-vapor-liquid equilibrium (VLE) data, Yaws et al. 1991)

Octanol/Water Partition Coefficient, log K_{OW}:
    2.63   (shake flask, Eadsforth & Moser 1983)
    3.03   (HPLC-RT correlation, Eadsforth & Moser 1983)
    2.63   (recommended, Sangster 1989)
    2.63   (recommended, Hansch et al. 1995)

Octanol/Air Partition Coefficient, log K_{OA}:

Bioconcentration Factor, log BCF:

Sorption Partition Coefficient, log K_{OC}:

Environmental Fate Rate Constants, k, and Half-Lives, $t_{½}$:
- Volatilization:
- Photolysis:
- Oxidation: rate constant k, for gas-phase second order rate constants, k_{OH} for reaction with OH radical, k_{NO3} with NO_3 radical and k_{O3} with O_3 or as indicated, *data at other temperatures see reference:
  - k_{O3} = (5.39 ± 0.078) × 10^{-17} cm^3 $molecule^{-1}$ s^{-1}; k_{OH} = (9.12 ± 0.23) × 10^{-11} cm^3 $molecule^{-1}$ s^{-1} for at 294 ± 2 K (Atkinson et al. 1984b)
  - k_{O3} = 5.39 × 10^{-17} cm^3 $molecule^{-1}$ s^{-1} for at 294 K (Atkinson & Carter 1984)
  - k_{OH} = 9.74 × 10^{-11} cm^3 $molecule^{-1}$ s^{-1} at 294 K (Atkinson 1985)
  - k_{NO3} = 1.18 × 10^{-12} cm^3 $molecule^{-1}$ s^{-1} and k_{OH} = 9.44 × 10^{-11} cm^3 $molecule^{-1}$ s^{-1} at 298 K (Sabljic & Güsten 1990)
  - k_{OH} = 96.9 × 10^{-12} cm^3 $molecule^{-1}$ s^{-1} at 294 K (Atkinson 1989)
  - k_{NO3} = 1.19 × 10^{-12} cm^3 $molecule^{-1}$ s^{-1} at 298 K (quoted, Atkinson 1991)
  - k_{OH} = 9.7 × 10^{-11} cm^3 $molecule^{-1}$ s^{-1}, k_{NO3} = 1.2 × 10^{-12} cm^3 $molecule^{-1}$ s^{-1}, and k_{O3} = 5.4 × 10^{-17} cm^3 $molecule^{-1}$ s^{-1} at 298 K (recommended, Atkinson 1997)
- Hydrolysis:
- Biodegradation:
- Biotransformation:
- Bioconcentration, Uptake (k_1) and Elimination (k_2) Rate Constants or Half-Lives:

Half-Lives in the Environment:

TABLE 2.1.2.4.7.1
Reported aqueous solubilities and vapor pressures of cycloheptatriene at various temperatures

log P = A – B/(T/K)	(1)	ln P = A – B/(T/K)	(1a)
log P = A – B/(C + t/°C)	(2)	ln P = A – B/(C + t/°C)	(2a)
log P = A – B/(C + T/K)	(3)		
log P = A – B/(T/K) – C·log (T/K)	(4)		

Aqueous solubility						Vapor pressure	
Pierotti & Liabastre 1972		Shaw 1989a		Góral et al. 2004		Finke et al. 1956	
shake flask-GC		IUPAC recommended		calc-recommended LLE data		ebulliometry	
t/°C	S/g·m^{-3}	t/°C	S/g·m^{-3}	t/°C	S/g·m^{-3}	t/°C	P/Pa
5.11	580.9	5	580	5.1	614	0	733.3
10.21	664.5	15	660	15.2	614	15	1815
25.11	669.4	25	640	25.0	563	20	2400
30.21	741.8	35	740	25.1	563	25	3136
40.21	764.8	45	760	35.2	614	30	4058
				45.2	614	35	5198
						40	6591
						45	8286
						50	10327
						55	12774
						60	15672
						65	19094

TABLE 2.1.2.4.7.1 (*Continued*)

Aqueous solubility						Vapor pressure	
Pierotti & Liabastre 1972		Shaw 1989a		Góral et al. 2004		Finke et al. 1956	
shake flask-GC		IUPAC recommended		calc-recommended LLE data		ebulliometry	
t/°C	S/g·m^{-3}	t/°C	S/g·m^{-3}	t/°C	S/g·m^{-3}	t/°C	P/Pa
						bp/°C	115.60
						ΔH_V/(kJ mol^{-1}) = 38.70	
							at 25°C
						eq. 2	P/mmHg
						A	6.97032
						B	1374.656
						C	220.538

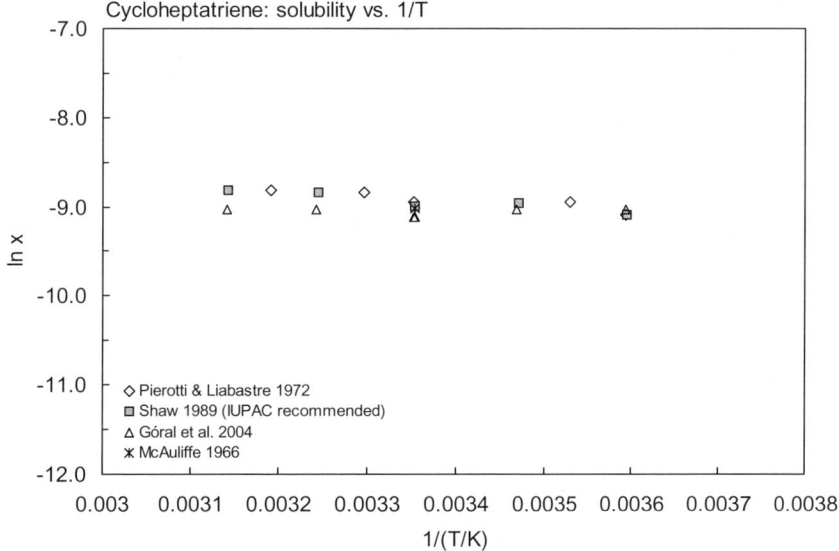

FIGURE 2.1.2.4.7.1 Logarithm of mole fraction solubility (ln x) versus reciprocal temperature for cyclohepta-triene.

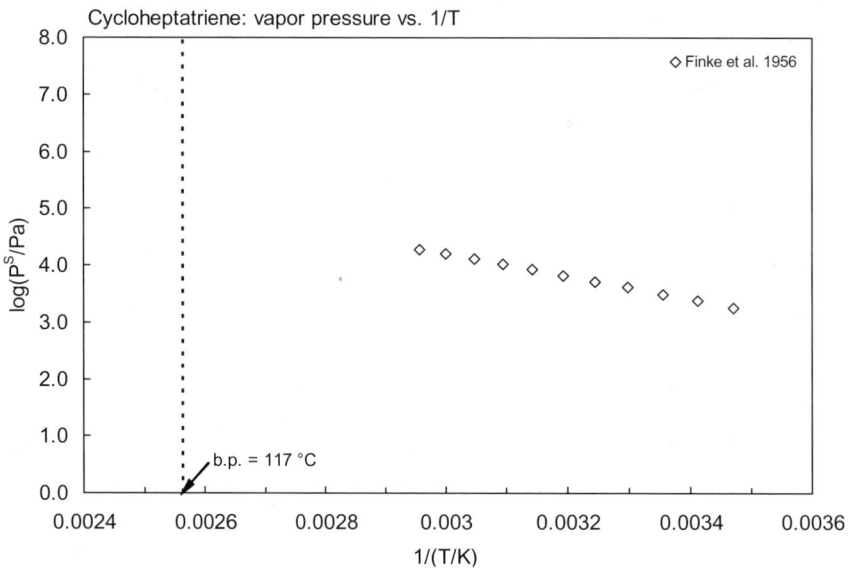

FIGURE 2.1.2.4.7.2 Logarithm of vapor pressure versus reciprocal temperature for cycloheptatriene.

2.1.2.4.8 dextro-Limonene [(R)-(+)-Limonene]

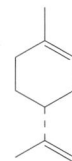

Common Name: *d*-Limonene
Synonym: *d-p*-mentha-1,8,-diene, (R)-(+)-*p*-mentha-1,8-diene, (+)-1-methyl-4-(1-methylethenyl)cyclohexene, *p*-mentha-1,8-diene, carvene, cinene, citrene, cajeputene, kautschin
Chemical Name: *dextro*-limonene, (R)-(+)-limonene
CAS Registry No: 5989-27-5
Molecular Formula: $C_{10}H_{16}$
Molecular Weight: 136.234
Melting Point (°C):
    –74   (Lide 2003)
Boiling Point (°C):
    178   (Weast 1982–83; Lide 2003)
Density (g/cm³ at 25°C):
    0.8403, 0.8383 (20°C, 25°C, Riddick et al. 1986)
Molar Volume (cm³/mol):
    162.1   (20°C, calculated-density, Stephenson & Malanowski 1987)
    192.2   (calculated-Le Bas method at normal boiling point)
Enthalpy of Vaporization, ΔH_V (kJ/mol):
    45.1   (25°C, Riddick et al. 1986)
Enthalpy of Fusion, ΔH_{fus} (kJ/mol):
Entropy of Fusion, ΔS_{fus} (J/mol K):
Fugacity Ratio at 25°C, F: 1.0

Water Solubility (g/m³ or mg/L at 25°C):
    13.49   (shake flask-GC, Massaldi & King 1973)
    13.8   (selected lit., Riddick et al. 1986)
    20.44   (shake flask-GC/FID, Fichan et al. 1999)

Vapor Pressure (Pa at 25°C or as indicated and reported temperature dependence equations):
    275.64   (calculated-Antoine eq. regression, Stull 1947)
    275.5   (interpolated-Antoine eq., Weast 1972–73)
    log (P/mmHg) = [–0.2185 × 10508.4/(T/K)] + 8.016262; temp range 14.0–175°C (Antoine eq., Weast 1972–73)
    2670   (68.2°C, Riddick et al. 1986)
    278, 202 (calculated-Antoine eq.-I, II, Stephenson & Malanowski 1987)
    log (P_L/kPa) = 6.81591 – 2075.62/(–16.65 + T/K); temp range 287–448 K (Antoine eq.-I, Stephenson & Malanowski 1987)
    log (P_L/kPa) = 7.67098 – 2494.342/(T/K); temp range 288–323 K (Antoine eq.-II, Stephenson & Malanowski 1987)
    log (P/mmHg) = 9.3771 – 2.8246 × 10³/(T/K) + 1.0584·log (T/K) – 8.9107 × 10⁻³·(T/K) + 4.8462 × 10⁻⁶·(T/K)²; temp range 199–660 K (vapor pressure eq., Yaws 1994)
    213   (activity coefficient-GC, Fichan et al. 1999)

Henry's Law Constant (Pa m³/mol at 25°C):
    2725   (calculated-P/C from selected data)

Octanol/Water Partition Coefficient, log K_{OW}:
    4.38   (RP-HPLC-RT correlation, Griffin et al. 1999)

Octanol/Air Partition Coefficient, log K_{OA}:

Bioconcentration Factor, log BCF:

Sorption Partition Coefficient, log K_{OC}:

Environmental Fate Rate Constants, k, and Half-Lives, $t_{1/2}$:
- Volatilization:
- Photolysis:
- Oxidation: rate constant k, for gas-phase second order rate constants, k_{OH} for reaction with OH radical, k_{NO3} with NO_3 radical and k_{O3} with O_3 or as indicated *data at other temperatures and/or the Arrhenius expression see reference:

  $k_{OH} = 9 \times 10^{10}$ cm^3 mol^{-1} s^{-1}, with $t_{1/2} < 0.24$ h (Darnall et al. 1976)

  $k_{OH} = (9.0 \pm 1.35) \times 10^{10}$ M^{-1} s^{-1} at 1 atm and 305 ± 2 K (relative rate method, Winer et al. 1976)

  $k_{OH} = 9.0 \times 10^{10}$ M^{-1} s^{-1}, $k_{O3} = 3.9 \times 10^5$ M^{-1} s^{-1}, and $k_{O(3P)} = (6.50 \pm 0.52) \times 10^{10}$ cm^3 M^{-1} s^{-1} for reaction with O(^{3}P) atom at room temp. (Winer et al. 1976)

  k_{OH}(calc) = 13.8×10^{-11} cm^3 molecule^{-1} s^{-1}, k_{OH}(obs.) = $(14.0, 14.2) \times 10^{-11}$ cm^3 molecule^{-1} s^{-1} (Atkinson et al. 1983b)

  $k_{NO3} = (7.7 \pm 1.7) \times 10^{-12}$ cm^3 molecule^{-1} s^{-1} at 295 K (relative rate technique, Atkinson et al. 1984a)

  $k_{O3} = 6.4 \times 10^{-16}$ cm^3 molecule^{-1} s^{-1} with calculated lifetimes $\tau = 6$ min and 11 min in 24-h in clean and moderately polluted atmosphere, respectively; $k_{OH} = 1.42 \times 10^{-10}$ cm^3 molecule^{-1} s^{-1} with calculated $\tau = 2.0$ h and 1.0 h during daytime in clean and moderately polluted atmosphere respectively, $k_{NO3} = 7.7 \times 10^{-12}$ cm^3 molecule^{-1} s^{-1} with calculated $\tau = 9$ min and 0.9 min during nighttime in clean and moderately polluted atmosphere, respectively, at room temp. (Atkinson et al. 1984a)

  $k_{NO2} < 3.5 \times 10^{-20}$ cm^3 molecule^{-1} s^{-1} for gas phase reaction with NO_2 at 295 K (Atkinson et al. 1984b)

  $k_{O3} = 6.4 \times 10^{-16}$ cm^3 molecule^{-1} s^{-1}; $k_{OH} = 1.42 \times 10^{-10}$ cm^3 molecule^{-1} s^{-1}, $k_{NO3} = 7.7 \times 10^{-12}$ cm^3 molecule^{-1} s^{-1} and $k_{O(3P)} = 1.29 \times 10^{-10}$ cm^3 molecule^{-1} s^{-1} with O(^{3}P) atom at room temp. (Atkinson et al. 1984b)

  $k_{O3} = 6.0 \times 10^{-16}$ cm^3 molecule^{-1} s^{-1} with a loss rate of 36 d^{-1}; $k_{OH} = 1.4 \times 10^{-10}$ cm^3 molecule^{-1} s^{-1} with a loss rate of 12 d^{-1}, and $k_{NO3} = 1.4 \times 10^{-11}$ cm^3 molecule^{-1} s^{-1} with a loss rate of 290 d^{-1} (Atkinson & Carter 1984)

  $k_{O3} = 6.0 \times 10^{-16}$ cm^3 molecule^{-1} s^{-1} with a loss rate of 36 d^{-1}; $k_{OH} = 1.7 \times 10^{-10}$ cm^3 molecule^{-1} s^{-1} with a loss rate of 7.3 d^{-1}, and $k_{NO3} = 1.4 \times 10^{-11}$ cm^3 molecule^{-1} s^{-1} with a loss rate of 290 d^{-1} at room temp. (Atkinson 1985)

  $k_{OH} = (16.9 \pm 0.5) \times 10^{-11}$ cm^3 molecule^{-1} s^{-1} at 294 ± 1 K (relative rate method, Atkinson et al.1986)

  $k_{O3} = 6.4 \times 10^{-16}$ cm^3 molecule^{-1} s^{-1} with calculated $\tau = 36$ min; $k_{OH} = 1.7 \times 10^{-10}$ cm^3 molecule^{-1} s^{-1} with τ(calc) = 1.6 h, $k_{NO3} = 1.4 \times 10^{-12}$ cm^3 molecule^{-1} s^{-1} with τ(calc) = 5.0 min for clean tropospheric conditions at room temp. (Atkinson et al. 1986)

  $k_{OH} = (146–171) \times 10^{-12}$ cm^3 molecule^{-1} s^{-1} at 294–305 K (review, Atkinson 1989)

  $k_{OH} = 1.71 \times 10^{-10}$ cm^3 molecule^{-1} s^{-1}, $k_{NO3} = 1.22 \times 10^{-11}$ cm^3 molecule^{-1} s^{-1}, $k_{O3} = 2.03 \times 10^{-16}$ cm^3 molecule^{-1} s^{-1}, and $k_{O(3P)} = 7.2 \times 10^{-11}$ cm^3 molecule^{-1} s^{-1} for the reaction with O(^{3}P) atom, at 298 K (recommended, Atkinson 1997)
- Hydrolysis:
- Biodegradation:
- Biotransformation:
- Bioconcentration, Uptake (k_1) and Elimination (k_2) Rate Constants or Half-Lives:

Half-Lives in the Environment:
- Air: $t_{1/2} < .24$ h in air based on its photooxidation rate constant of 9×10^{13} cm^3 mol^{-1} s^{-1} for the gas phase reaction with hydroxyl radical (Darnall et al. 1976; Lloyd et al. 1976);

  calculated lifetimes: $\tau = 36$ min due to reaction with O_3 in 24-h period, $\tau = 2.0$ h with OH radical during daytime, and $\tau = 9$ min for NO_3 radical during nighttime for "clean" atmosphere; $\tau = 11$ min for reaction with O_3 in 24-h period, $\tau = 1.0$ h with OH radical during daytime, and $\tau = 0.9$ min for NO_3 radical during nighttime in moderately polluted atmosphere (Atkinson et al. 1984a);

  calculated atmospheric lifetimes, $\tau = 36$ min, 1.6 h and 5.0 min for reaction with O_3, OH and NO_3 radicals respectively for clean tropospheric conditions at room temp. (Atkinson et al. 1986);

  calculated tropospheric lifetimes $\tau = 1.1$ h, 1.9 h and 53 min due to reactions with OH radical, O_3 and NO_3 radical, respectively, at room temp. (Corchnoy & Atkinson 1990).

2.1.2.4.9 α-Pinene

Common Name: α-Pinene
Synonym: *dl*-pinene, 2-pinene
Chemical Name: 2,6,6-trimethylbicyclo[3,1,1]hept-2-ene
CAS Registry No: 7785-70-8
     α-pinene *d*-Form 80-56-8
Molecular Formula: $C_{10}H_{16}$
Molecular Weight: 136.234
Melting Point (°C):
    –55   (*dl*-Form, Weast 1982–83)
    –64   (*d*-Form, Riddick et al. 1986; Lide 2003)
    –50   (*d*-Form, Stephenson & Malanowski 1987)
Boiling Point (°C):
    156.2   (*dl*-Form, Weast 1982–83; Lide 2003)
    156   (*d*-Form, Stephenson & Malanowski 1987)
    155–156   (*d*-, *l*-Form, Budavari 1989)
Density (g/cm³):
    0.8582   (20°C, *dl*-Form, Weast 1982–83)
    0.8582, 0.8539   (*d*-Form, Riddick et al. 1986)
    0.8592, 0.8591, 0.8590 (20°C, *dl*-, *d*-, *l*-Form, Budavari 1989)
Molar Volume (cm³/mol):
    157.4   (*d*-Form, Stephenson & Malanowski 1987)
    183.7   (calculated-Le Bas method at normal boiling point)
Enthalpy of Vaporization, ΔH_V (kJ/mol):
    46.61, 39.673   (25°C, bp, Riddick et al. 1986)
Enthalpy of Fusion, ΔH_{fus} (kJ/mol):
Entropy of Fusion, ΔS_{fus} (J/mol K):
Fugacity Ratio at 25°C, F: 1.0

Water Solubility (g/m³ or mg/L at 25°C):
    21.8, 3.42; 5.04   (quoted lit. values; shake flask-GC/FID, Fichan et al. 1999)

Vapor Pressure (Pa at 25°C or as indicated and reported temperature dependence equations. Additional data at other temperatures designated * are compiled at the end of this section.):
    640*, 800   (22.2, 22.5°C, measured range 21.1–148°C, Pickett & Peterson 1929)
    237.3*   (13.25°C, Hg manometer, measured range –6.0 to 13.25°C, Linder 1931)
    666.6*   (24.6°C, summary of literature data, temp range –1.0 to 155.0°C, Stull 1947)
    457*, 655   (21.2, 27.2°C, measured range 19.4–155.75°C, Hawkins & Armstrong 1954)
    678   (interpolated-Antoine eq., Weast 1972–73)
    log (P/mmHg) = [–0.2185 × 9813.6/(T/K)] + 7.898207; temp range –1.0 to 155°C (Antoine eq., Weast 1972–73)
    667   (Verschueren 1983)
    605, 587   (interpolated-Antoine equations, Boublik et al. 1984)
    log (P/kPa) = 6.37971 – 1692.803/(231.558 + t/°C); temp range 21.1–148°C (Antoine eq. from reported exptl. data of Pickett & Peterson 1929, Boublik et al. 1984)
    log (P/kPa) = 6.95174 – 1430.936/(206.42 + t/°C); temp range 19.44–155.75°C (Antoine eq. from reported exptl. data of Hawkins & Armstrong 1954, Boublik et al. 1984)
    655   (selected, Riddick et al. 1986)
    log (P/kPa) = 25.52644 – 3134.525/(T/K) – 6.16045·log (T/K) (Riddick et al. 1986)

582 (α-pinene *d*-Form, interpolated-Antoine eq., temp range 292–433 K, Stephenson & Malanowski 1987)

log (P/kPa) = 5.92666 – 1414.16/(T/K) (Antoine eq., liquid, temp range 292–433 K (α-pinene *d*-Form, Stephenson & Malanowski 1987)

588 (interpolated-Antoine eq., temp range 19–156°C, Dean 1992)

log (P/mmHg) = 6.8525 – 1446.4/(t/°C + 208.0); temp range 19–156°C (Antoine eq., Dean 1992)

log (P/mbar) = 7.076588 – 1511.961/[(T/K) – 57.730]; temp range 365–430 K (vapor-liquid equilibrium (VLE)-Fischer still, Reich & Sanhueza 1993)

log (P/mmHg) = 21.4735 – 2.7156 × 10³/(T/K) – 5.0076·log (T/K) + 2.8146 × 10⁻³·(T/K) – 1.5389 × 10⁻⁶·(T/K)²; temp range 209–632 K (vapor pressure eq., Yaws 1994)

613, 581, 465; 529 (quoted lit. values; deduced from exptl. determined activity coeff. at infinite dilution, Fichan et al. 1999)

Henry's Law Constant (Pa m³/mol at 25°C or indicated. Additional data at other temperatures designated * are compiled at the end of this section.):

0.194* (20°C, calculated from measured liquid-phase diffusion coefficients, measured range –10 to 20°C, Zhang et al. 2003)

ln [H'/(M/atm)] = –6.590 + 3800/(T/K); temp range 263–293 K (Zhang et al. 2003)

Octanol/Water Partition Coefficient, log K_{OW}:

4.44 ((+)-α-pinene, RP-HPLC-RT correlation, Griffin et al. 1999)

4.48 ((–)-α-pinene, RP-HPLC-RT correlation, Griffin et al. 1999)

Octanol/Air Partition Coefficient, log K_{OA}:

Bioconcentration Factor, log BCF or log K_B:

Sorption Partition Coefficient, log K_{OC}:

Environmental Fate Rate Constants, k, and Half-Lives, $t_{1/2}$:

Volatilization:

Photolysis:

Oxidation: rate constant k, for gas-phase second order rate constants k_{O3} for reaction with O_3, k_{OH} with OH radical and k_{NO3} with NO_3 radical at 25°C or as indicated, *data at other temperatures and/or the Arrhenius expression see reference:

k_{OH} = (3.48 ± 0.52) × 10¹⁰ cm³ M⁻¹ s⁻¹ at 305 ± 2 K (relative rate method, Winer et al. 1976)

k_{OH} = 3.5 × 10¹⁰ M⁻¹ s⁻¹, k_{O3} = 2.0 × 10⁵ M⁻¹ s⁻¹, and $k_{O(3P)}$ = (1.60 ± 0.06) × 10¹⁰ M⁻¹ s⁻¹ for reaction with O(³P) atom at room temp. (Winer et al. 1976)

k_{OH} = 32 × 10⁻¹² cm³ molecule⁻¹ s⁻¹; k_{O3} = 67 × 10⁻¹² cm³ molecule⁻¹ s⁻¹ at room temp. (Gaffney & Levine 1979)

k_{O3}* = (8.4 ± 1.9) × 10⁻¹⁷ cm³ molecule⁻¹ s⁻¹ at 296 K, measured range 276–324 K, atmospheric lifetime τ = 3–4 h due to reaction with O_3, and τ ~ 4 h due to reaction with OH radical (Atkinson et al. 1982)

k_{OH}* = (6.01 ± 0.82) × 10⁻¹¹ cm³ molecule⁻¹ s⁻¹ at 298 K, measured range 298–422 K (flash photolysis-resonance fluorescence, Kleindlenst et al. 1982)

k_{OH}(calc) = 8.7 × 10⁻¹¹ cm³ molecule⁻¹ s⁻¹, k_{OH}(obs) = (7.6, 5.5, 6.01) × 10⁻¹¹ cm³ molecule⁻¹ s⁻¹ (Atkinson et al. 1983b)

k_{OH}* = 60.1 × 10⁻¹² cm³ molecule⁻¹ s⁻¹ at 298 K, measured range 298–422K (Flash photolysis-resonance fluorescence, Kleindienst et al. 1982; Atkinson 1985)

k_{NO3} = (3.4 ± 0.8) × 10⁻¹² cm³ molecule⁻¹ s⁻¹ at 295 ± 1 K (relative rate method, Atkinson et al. 1984a)

k_{O3} = 8.4 × 10⁻¹⁷ cm³ molecule⁻¹ s⁻¹ with calculated lifetimes of 4.6 h and 1.4 h in 24-h in clean and moderately polluted atmosphere, respectively; k_{OH} = 6.0 × 10⁻¹¹ cm³ molecule⁻¹ s⁻¹ with calculated lifetimes of 4.6 h and 2.3 h during daytime in clean and moderately polluted atmosphere, respectively, k_{NO3} = 3.4 × 10⁻¹² cm³ molecule⁻¹ s⁻¹ with calculated lifetimes of 20 min and 2 min during nighttime in clean and moderately polluted atmosphere respectively at room temp. (Atkinson et al. 1984a)

k_{NO2} < 2.1 × 10⁻²⁰ cm³ molecule⁻¹ s⁻¹ for gas phase reaction with NO_2 at 295 K (Atkinson et al. 1984b)

$k_{O3} = 8.4 \times 10^{-17}$ cm^3 molecule^{-1} s^{-1}; $k_{OH} = 6.0 \times 10^{-11}$ cm^3 molecule^{-1} s^{-1}, $k_{NO3} = 3.4 \times 10^{-12}$ cm^3 molecule^{-1} s^{-1} and $k_{O(3P)} = 640 \times 10^{-18}$ cm^3 molecule^{-1} s^{-1} with O(^{3}P) at room temp. (Atkinson et al. 1984b)

$k_{O3} = 8 \times 10^{-17}$ cm^3 molecule^{-1} s^{-1} with a loss rate of 5 d^{-1}; $k_{OH} = 6.0 \times 10^{-12}$ cm^3 molecule^{-1}s^{-1} with a loss rate of 5 d^{-1}, and $k_{NO3} = 6.1 \times 10^{-12}$ cm^3 molecule^{-1} s^{-1} with a loss rate of 130 d^{-1} (Atkinson & Carter 1984)

$k_{OH} = (5.45 \pm 0.32) \times 10^{-11}$ cm^3 molecule^{-1} s^{-1} at 294 $\pm$ 1 K (relative rate method, Atkinson et al.1986)

$k_{O3} = 8.4 \times 10^{-17}$ cm^3 molecule^{-1} s^{-1} with calculated τ = 5.6 h; $k_{OH} = 5.5 \times 10^{-11}$ cm^3 molecule^{-1} s^{-1} with τ(calc) = 5.1 h, $k_{NO3} = 6.1 \times 10^{-12}$ cm^3 molecule^{-1} s^{-1} with τ(calc) = 11 min for clean tropospheric conditions at room temp. (Atkinson et al. 1986)

$k_{O3} = (8.6 \pm 1.3) \times 10^{-17}$ cm^3 molecule^{-1} s^{-1} at 297 $\pm$ 2 K in a smog chamber (Nolting et al. 1988)

$k_{OH}* = 5.37 \times 10^{-11}$ cm^3 molecule^{-1} s^{-1} at 298 K (recommended, Atkinson 1989)

$k_{OH}* = 5.37 \times 10^{-11}$ cm^3 molecule^{-1} s^{-1}, $k_{NO3} = 5.79 \times 10^{-12}$ cm^3 molecule^{-1} s^{-1}, $k_{O3} = (8.4, 8.6) \times 10^{-12}$ cm^3 molecule^{-1} s^{-1} at 296 K (Atkinson et al. 1990)

$k_{OH} = 9.12 \times 10^{-11}$ cm^3 molecule^{-1} s^{-1} and $k_{NO3} = 5.75 \times 10^{-12}$ cm^3 molecule^{-1} s^{-1} (Müller & Klein 1991)

$k_{OH} = 5.32 \times 10^{-11}$ cm^3 molecule^{-1} s^{-1}, $k_{NO3} = 5.79 \times 10^{-12}$ cm^3 molecule^{-1} s^{-1} at 298 K (Sabljic & Güsten 1990)

$k_{NO3}* = 6.16 \times 10^{-12}$ cm^3 molecule^{-1} s^{-1} at 298 K (recommended, Atkinson 1991)

$k_{OH}* = 5.37 \times 10^{-11}$ cm^3 molecule^{-1} s^{-1}, $k_{NO3}* = 6.16 \times 10^{-12}$ cm^3 molecule^{-1} s^{-1}, $k_{O3}* = 86.6 \times 10^{-18}$ cm^3 molecule^{-1} s^{-1}, and $k_{O(3P)} = 3.2 \times 10^{-11}$ cm^3 molecule^{-1} s^{-1} for the reaction with O(^{3}P) atom, at 298 K (recommended, Atkinson 1997)

k_{OH}(lit.) = 5.45 $\times$ 10^{-11} cm^3 molecule^{-1} s^{-1}; k_{OH}(calc) = 8.5 $\times$ 10^{-11} cm^3 molecule^{-1} s^{-1} (quoted; calculated-QSAR, Peeters et al. 1999)

Hydrolysis:
Biodegradation:
Biotransformation:
Bioconcentration and Uptake and Elimination Rate Constants (k_1 and k_2):

Half-Lives in the Environment:

Air: calculated lifetimes: τ = 4.6 h due to reaction with O$_3$ in 24-h period, τ = 4.6 h with OH radical during daytime, and τ = 20 min for NO$_3$ radical during nighttime for clean atmosphere; τ = 1.4 h for reaction with O$_3$ in 24-h period, τ = 2.3 h with OH radical during daytime, and τ = 2 min with NO$_3$ radical during nighttime in moderately polluted atmosphere (Atkinson et al.1984a, Winer et al. 1984);

calculated atmospheric lifetimes of 5.6 h, 5.1 h and 11 min for reaction with O$_3$, OH and NO$_3$ radicals respectively for clean tropospheric conditions at room temp. (Atkinson et al. 1986);

calculated tropospheric lifetimes of 3.4 h, 4.6 h and 2.0 h due to reactions with OH radical, O$_3$ and NO$_3$ radical respectively at room temp. (Corchnoy & Atkinson 1990).

TABLE 2.1.2.4.9.1
Reported vapor pressures of α-pinene at various temperatures and the coefficients for the vapor pressure equations

log P = A – B/(T/K)	(1)	ln P = A – B/(T/K)	(1a)
log P = A – B/(C + t/°C)	(2)	ln P = A – B/(C + t/°C)	(2a)
log P = A – B/(C + T/K)	(3)		
log P = A – B/(T/K) – C·log (T/K)	(4)		

1.

Pickett & Peterson 1929				Linder 1931		Stull 1947	
Ramsay & Young method				Hg manometer		summary of literature data	
t/°C	P/Pa	t/°C	P/Pa	t/°C	P/Pa	t/°C	P/Pa
21.2	533	134.9	57128	–6.0	54.66	–1.0	133
22.0	573	135.4	57608	1.0	96.0	24.6	666.6
22.2	640	138.5	63728	13.25	237.3	37.3	1333
22.5	800	138.8	64261			51.4	2666

(Continued)

TABLE 2.1.2.4.9.1 (*Continued*)

Pickett & Peterson 1929				Linder 1931		Stull 1947	
Ramsay & Young method				Hg manometer		summary of literature data	
t/°C	P/Pa	t/°C	P/Pa	t/°C	P/Pa	t/°C	P/Pa
54.7	3040	139.8	65861			66.8	5333
55.3	3200	140.1	66794			76.8	7999
55.5	3306	140.5	67461			90.1	13332
77.4	7839	141.2	68794			110.2	26664
77.5	7906	141.3	69327			132.3	53329
77.7	7959	141.9	70394			155.0	101325
77.9	8026	147.0	80526				
78.6	8253	148.0	83060			mp/°C	−55
79.3	8519						
79.6	8639	bp/°C	155–158				
101.1	19625						
102.8	20758						
103.2	21025						
103.5	21238						
103.7	21371						
103.8	21451						
131.2	52196						
132.3	53462						
132.4	53862						
132.5	53902						
132.7	54089						
132.9	54275						
133.0	54755						
133.1	54862						
133.2	54995						

2.

Hawkins & Armstrong 1954							
Hg manometer							
t/°C	P/Pa	t/°C	P/Pa	t/°C	P/Pa	t/°C	P/Pa
19.45	408.0	65.63	4858	95.27	16164	155.77	100601
21.3	457.3	66.54	5121	102.25	20717	155.76	100793
27.25	654.6	68.27	5554	105.9	23430		
29.72	790.6	70.45	6073	106.76	24217	bp/°C	155.9
37.07	1201	75.1	7430	110.53	27438		
46.92	2015	76.04	7698	113.5	30135	eq. 4	P/mmHg
53.69	2832	77.12	8033	115.07	31795	A	26.40174
54.18	2877	79.75	8905	122.11	39478	B	3134.525
56.57	3234	84.71	10859	125.02	42963	C	6.16045
57.09	3309	86.23	11536	129.87	49509	155.77	100601
57.3	3393	88.42	12512	135.72	58678		
62.28	4221	92.01	14299	147.48	80731		

Aliphatic and Cyclic Hydrocarbons

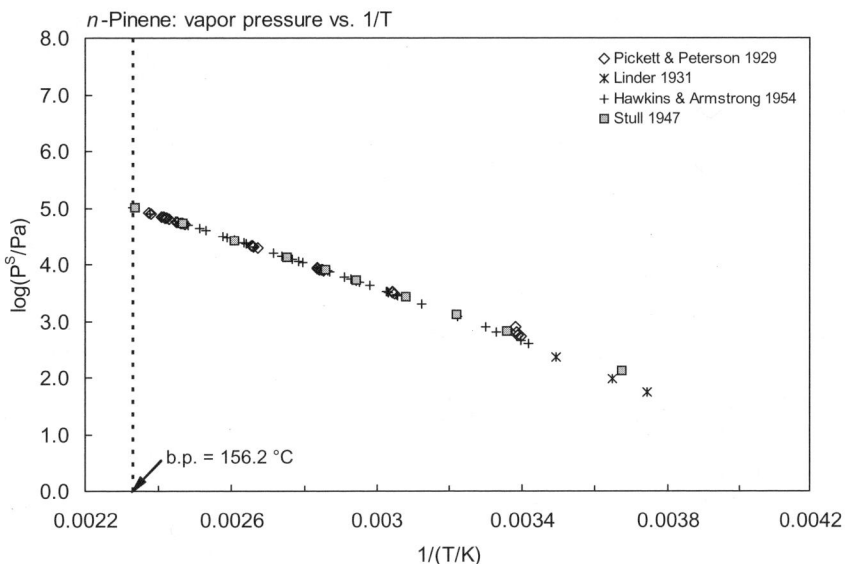

FIGURE 2.1.2.4.9.1 Logarithm of vapor pressure versus reciprocal temperature for α-pinene.

TABLE 2.1.2.4.9.2
Reported Henry's law constants of α-pinene at various temperatures

Zhang et al. 2003

liquid-phase diffusion coefficient

t/°C	H/(Pa m³/mol)
−10	0.0397
0	0.0618
10	0.113
20	0.194

ln (kH/atm) = A − B/(T/K)
eq. 3   H′/(M atm⁻¹)
A   −6.59
B   3800

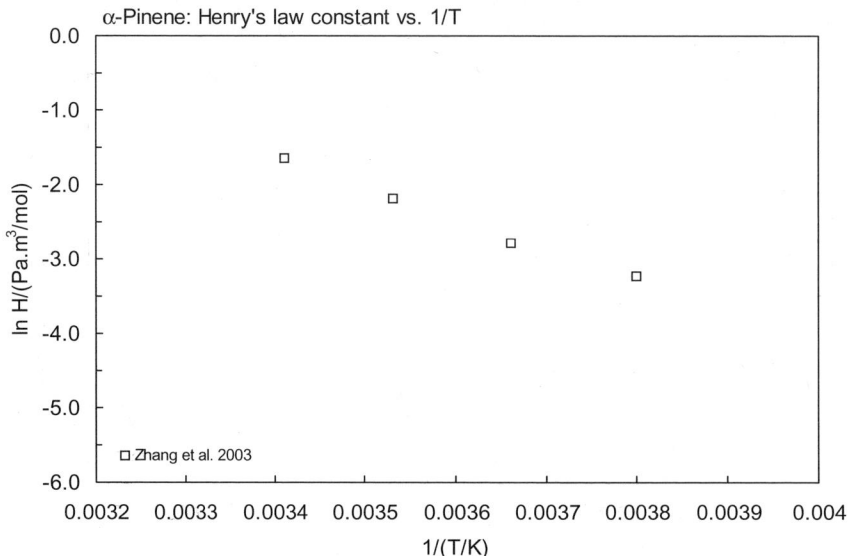

FIGURE 2.1.2.4.9.2 Logarithm of Henry's law constant versus reciprocal temperature for α-pinene.

2.1.2.4.10 β-Pinene

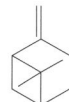

Common Name: β-Pinene
Synonym: β-Pinene *d*, or nopinene; β-Pinene *l*, or 2(10)-pinene
Chemical Name: 6,6-dimethyl-2-methylene bicyclo[3,1,1]heptane
CAS Registry No: 19172-67-3
      β-pinene *l*-Form 127-91-3
Molecular Formula: $C_{10}H_{16}$
Molecular Weight: 136.234
Melting Point (°C):
   –61.54   (*l*-Form, Riddick et al. 1986)
   –50   (*l*-Form, Stephenson & Malanowski 1987)
   –61.5   (l-Form, Lide 2003)
Boiling Point (°C):
   164–166, 162.4   (*d*-Form, *l*-Form, Weast 1982–83)
   166   (*l*-Form, Riddick et al. 1986; Lide 2003)
   163   (*l*-Form, Stephenson & Malanowski 1987)
   165–166, 164–166, 162–163 (*dl*-, *d*-, *l*-Form, Budavari 1989)
Density (g/cm³):
   0.8654, 0.8694 (20°C, *d*-, *l*-Form, Weast 1982–83)
   0.8667   (25°C, *l*-Form, Riddick et al. 1986)
Molar Volume (cm³/mol):
   157.4   (*l*-Form, Stephenson & Malanowski 1987)
   183.7   (calculated-Le Bas method at normal boiling point)
Enthalpy of Vaporization, ΔH_V (kJ/mol):
   43.471, 40.208 (25°C, bp, Riddick et al. 1986)
Enthalpy of Fusion, ΔH_{fus} (kJ/mol):
Entropy of Fusion, ΔS_{fus} (J/mol K):
Fugacity Ratio at 25°C, F: 1.0

Water Solubility (g/m³ or mg/L at 25°C):
   32.7, 6.27; 11.04 (quoted lit. values; shake flask-GC, Fichan et al. 1999)

Vapor Pressure (Pa at 25°C or as indicated and reported temperature dependence equations. Additional data at other temperatures designated * are compiled at the end of this section.):
   613*   (Ramsay & Young method, measured range 24.3–158.1°C, Pickett & Peterson 1929)
   666.6*   (30.0°C, summary of literature data, temp range 4.2–158.3°C, Stull 1947)
   345*, 447   (23.06, 26.8°C, measured range 18.71–165.91°C, Hawkins & Armstrong 1954)
   501   (interpolated-Antoine eq., Weast 1972–73)
   log (P/mmHg) = [–0.2185 × 10235.8/(T/K)] + 8.633424; temp range 4.2–158.3°C (Antoine eq., Weast 1972–73)
   667   (Verschueren 1983)
   395   (interpolated-Antoine eq., Boublik et al. 1984)
   log (P/kPa) = 6.02052 – 1509.944/(210.05 + t/°C); temp range 18.71–165.9°C (Antoine eq. from reported exptl. data of Hawkins & Armstrong 1954, Boublik et al. 1984)
   610   (selected, Riddick et al. 1986)
   log (P/kPa) = 27.90258 – 3318.845/(T/K) – 6.94263·log (T/K) (Riddick et al. 1986)
   391   (*l*-Form, interpolated- Antoine eq., temp range 291–441K, Stephenson & Malanowski 1987)
   log (P/kPa) = 6.04993 – 1520.15/(T/K + 62.75); temp range 291–441 K (*l*-Form, Antoine eq., liquid, Stephenson & Malanowski 1987)

394   (interpolated-Antoine eq., Dean 1992)

log (P/mmHg) = 6.8984 − 1511.7/(t/°C + 210.2); temp range 19–156°C (Antoine eq., Dean 1992)

log (P/mbar) = 7.067997 − 1539.348/[(T/K) − 59.937]; temp range 364–439 K (vapor-liquid equilibrium (VLE)-Fischer still, Reich & Sanhueza 1993)

log (P/mmHg) = 46.3728 − 3.9789 × 10³/(T/K) − 13.284·log (T/K) − 1.3113 × 10⁻¹⁰·(T/K) + 3.4783 × 10⁻⁶·(T/K)²; temp range 120–651 K (vapor pressure eq., Yaws 1994)

Henry's Law Constant (Pa m³/mol):

Octanol/Water Partition Coefficient, log K_{OW}:
4.16   (RP-HPLC-RT correlation, Griffin et al. 1999)

Octanol/Air Partition Coefficient, log K_{OA}:

Bioconcentration Factor, log BCF or log K_B:

Sorption Partition Coefficient, log K_{OC}:

Environmental Fate Rate Constants, k, and Half-Lives, $t_{1/2}$:

Volatilization:

Photolysis:

Oxidation: rate constant k, for second order gas-phase rate constants k_{OH}, k_{O3} and k_{NO3} for reactions with OH radicals, O_3 and NO_3 radicals or as indicated, *data at other temperatures and/or the Arrhenius expression see reference:

k_{OH} = 42 × 10⁹ L mol⁻¹ s⁻¹, with $t_{1/2}$ = 0.24–2.4 h (Darnall et al. 1976)

k_{OH} = (4.06 ± 0.61) × 10¹⁰ M⁻¹ s⁻¹ at 1 atm and 305 ± 2 K (relative rate method, Winer et al. 1976)

k_{OH} = 4.1 × 10¹⁰ M⁻¹ s⁻¹; k_{O3} = 2.2 × 10⁴ M⁻¹ s⁻¹; and $k_{O(3P)}$ = (1.51 ± 0.06) × 10¹⁰ M⁻¹ s⁻¹ for reaction with O(³P) atom (Winer et al. 1976)

k_{O3} = 30 × 10⁻¹² cm³ molecule⁻¹ s⁻¹; k_{OH} = 65 × 10⁻¹² cm³ molecule⁻¹ s⁻¹ at room temp. (Gaffney & Levine 1979)

k_{O3} = (2.1 ± 0.5) × 10⁻¹⁷ cm³ molecule⁻¹ s⁻¹ at 296 ± 2 K with atmospheric lifetime τ ∼ 13 h due to reaction with O_3 and τ ∼ 4 h due to reaction with OH radical (Atkinson et al. 1982)

k_{OH}* = (7.76 ± 1.1) × 10⁻¹¹ cm³ molecule⁻¹ s⁻¹ at 298 K, measured range 297–423 K (flash photolysis-resonance fluorescence, Kleindienst et al. 1982; Atkinson 1985)

k_{OH}(calc) = 5.1 × 10⁻¹¹ cm³ molecule⁻¹ s⁻¹, k_{OH}(obs) = (6.4, 7.76) × 10⁻¹¹ cm³ molecule⁻¹ s⁻¹ (Atkinson et al. 1983b)

k_{NO3} = (1.4 ± 0.3) × 10⁻¹² cm³ molecule⁻¹ s⁻¹ at 295 ± 1 K (relative rate method, Atkinson et al. 1984a)

k_{O3} = 2.1 × 10⁻¹⁷ cm³ molecule⁻¹ s⁻¹ with calculated lifetimes τ = 18 h and 5.5 h in 24-h in clean and moderately polluted atmosphere respectively; k_{OH} = 7.8 × 10⁻¹¹ cm³ molecule⁻¹ s⁻¹ with calculated τ = 3.6 h and 1.8 h during daytime in clean and moderately polluted atmosphere, respectively; k_{NO3} = 1.4 × 10⁻¹² cm³ molecule⁻¹ s⁻¹ with calculated τ = 50 min and 5 min during nightime in clean and moderately polluted atmosphere, respectively, at room temp. (Atkinson et al. 1984a)

k_{NO2} = < 2.4 × 10⁻²⁰ cm³ molecule⁻¹ s⁻¹ for gas phase reaction with NO_2 at 295 K (Atkinson et al. 1984b)

k_{O3} = 2.1 × 10⁻¹⁷ cm³ molecule⁻¹ s⁻¹; k_{OH} = 7.8 × 10⁻¹¹ cm³ molecule⁻¹ s⁻¹, k_{NO3} = 1.4 × 10⁻¹² cm³ molecule⁻¹ s⁻¹; and $k_{O(3P)}$ = 640 × 10⁻¹⁸ cm³ molecule⁻¹ s⁻¹ for the reaction with O(³P) atom, at room temp. (Atkinson et al. 1984b)

k_{OH} = (7.95 ± 0.52) × 10⁻¹¹ cm³ molecule⁻¹ s⁻¹ at 294 ± 1 K (relative rate method, Atkinson et al.1986)

k_{O3} = 2.1 × 10⁻¹⁷ cm³ molecule⁻¹ s⁻¹ with calculated τ = 18 h; k_{OH} = 8.0 × 10⁻¹¹ cm³ molecule⁻¹ s⁻¹ with τ(calc) = 3.5 h, k_{NO3} = 2.5 × 10⁻¹² cm³ molecule⁻¹ s⁻¹ with τ(calc) = 28 min for clean tropospheric conditions at room temp. (Atkinson et al. 1986)

k_{NO3} = 2.36 × 10⁻¹² cm³ molecule⁻¹ s⁻¹ at 296 ± 2 K (relative rate method, Atkinson et al. 1988)

k_{O3} = (1.4 ± 0.2) × 10⁻¹⁷ cm³ molecule⁻¹ s⁻¹ at 297 ± 2 K in a smog chamber (Nolting et al. 1988)

k_{OH}* = 7.89 × 10⁻¹¹ cm³ molecule⁻¹ s⁻¹ at 298 K (recommended, Atkinson 1989)

k_{OH}* = 7.89 × 10⁻¹¹ cm³ molecule⁻¹ s⁻¹, k_{NO3} = 2.36 × 10⁻¹² cm³ molecule⁻¹ s⁻¹, and k_{O3} = (2.1; 1.4) × 10⁻¹⁷ cm³ molecule⁻¹ s⁻¹ at 296 K (Atkinson et al. 1990)

$k_{OH} = 5.62 \times 10^{-11}$ cm^3 molecule^{-1} s^{-1}; $k_{NO3} = 2.34 \times 10^{-12}$ cm^3 molecule^{-1} s^{-1} (Müller & Klein 1991)

$k_{OH} = 7.82 \times 10^{-11}$ cm^3 molecule^{-1} s^{-1}, $k_{NO3} = 2.36 \times 10^{-12}$ cm^3 molecule^{-1} s^{-1}, at 298K (Sabljic & Güsten 1990)

$k_{NO3} = 2.51 \times 10^{-12}$ cm^3 molecule^{-1} s^{-1} at 298 K (recommended, Atkinson 1991)

$k_{O3} = 12.2 \pm 1.3 \times 10^{-18}$ cm^3 molecule^{-1} s^{-1} at 22 ± 1°C (Grosjean et al. 1993)

$k_{OH}* = 7.89 \times 10^{-11}$ cm^3 molecule^{-1} s^{-1}, $k_{NO3} = 2.51 \times 10^{-12}$ cm^3 molecule^{-1} s^{-1}, and $k_{O3}* = 1.5 \times 10^{-17}$ cm^3 molecule^{-1} s^{-1}, and $k_{O(3P)} = 2.7 \times 10^{-11}$ cm^3 molecule^{-1} s^{-1} for the reaction with O(^{3}P) atom, at 298 K (recommended, Atkinson 1997)

$k_{OH, lit} = 7.95 \times 10^{-11}$ cm^3 molecule^{-1} s^{-1}; k_{OH}(calc) = 6.0×10^{-11} cm^3 molecule^{-1} s^{-1} (quoted; calculated-QSAR, Peters et al. 1999)

Hydrolysis:

Biodegradation:

Biotransformation:

Bioconcentration and Uptake and Elimination Rate Constants (k_1 and k_2):

Half-Lives in the Environment:

Air: calculated lifetimes: $\tau = 18$ h due to reaction with O$_3$ in 24-h period, $\tau = 3.6$ h with OH radical during daytime, and $\tau = 50$ min for NO$_3$ radical during nighttime for "clean" atmosphere; $\tau = 5.5$ h for reaction with O$_3$ in 24-h period, $\tau = 1.8$ h with OH radical during daytime, and $\tau = 5$ min with NO$_3$ radical during nighttime hours in "moderately polluted" atmosphere (Atkinson et al.1984a, Winer et al. 1984);

calculated atmospheric lifetimes of 18 h, 3.5 h and 28 min for reaction with O$_3$, OH and NO$_3$ radicals respectively for clean tropospheric conditions at room temp. (Atkinson et al. 1986);

calculated tropospheric lifetimes of 2.3 h, 1.1 d and 4.9 h due to reactions with OH radical, O$_3$ and NO$_3$ radical respectively at room temp. (Corchnoy & Atkinson 1990).

TABLE 2.1.2.4.10.1
Reported vapor pressures of β-pinene at various temperatures and the coefficients for the vapor pressure equations

log P = A – B/(T/K)	(1)	ln P = A – B/(T/K)	(1a)
log P = A – B/(C + t/°C)	(2)	ln P = A – B/(C + t/°C)	(2a)
log P = A – B/(C + T/K)	(3)		
log P = A – B/(T/K) – C·log (T/K)	(4)		

1.

Pickett & Peterson 1929						Stull 1947	
Ramsay & Young method						summary of literature data	
t/°C	P/Pa	t/°C	P/Pa	t/°C	P/Pa	t/°C	P/Pa
24.3	453.3	94.7	13839	131.9	47809	4.2	133.3
24.5	480	95.1	14225	132.1	48329	30.0	666.6
24.7	520	95.7	14585	132.6	48889	42.3	1333
24.8	400	96.0	14705	132.9	49583	58.1	2666
25.0	613	96.2	14905	133.6	50623	71.5	5333
25.4	653	96.5	14905	133.9	50796	81.2	7999
53.5	2486	105.8	20825	154.7	91792	94.0	13332
54.0	2440	106.0	20998	154.9	92126	114.1	26664
54.4	2466	106.4	21305	155.2	92459	136.1	53329
54.6	2520	106.9	21665	156.2	94859	158.3	101325
55.0	2586	107.1	21812	158.0	100258		
68.7	4680	108.2	21932	158.1	100926	mp/°C	–
68.9	4720	131.2	47529				
69.1	4866	131.6	47929	bp/°C	160.2–163.8		
70.9	5240	131.7	47929				

(Continued)

TABLE 2.1.2.4.10.1 (*Continued*)

2.

Hawkins & Armstrong 1954							
Hg manometer							
t/°C	P/Pa	t/°C	P/Pa	t/°C	P/Pa	t/°C	P/Pa
18.72	252.0	50.11	1644	81.23	6949	161.22	89634
20.05	284.0	52.32	1812	85.96	8266	165.8	100788
23.09	345.3	56.57	2268	85.86	9983	157.82	101194
26.82	446.6	56.94	2341	90.80	11346		
29.45	517.3	58.86	2550	94.12	12828	bp/°C	166.0
31.68	592.0	59.48	3056	97.33	15948		
32.03	592.0	62.62	3084	103.41	19288	eq. 4	P/mmHg
36.78	793.3	66.52	3534	108.81	24117	A	28.77768
37.09	803.9	68.75	4014	115.50	27972	B	3318.845
39.41	929.3	72.23	4677	125.33	33079	C	6.94243
41.34	1037	75.43	5384	131.39	39762		
45.52	1289	78.59	6141	136.02	45106		
49.41	1580	80.88	6810	149.55	65069		

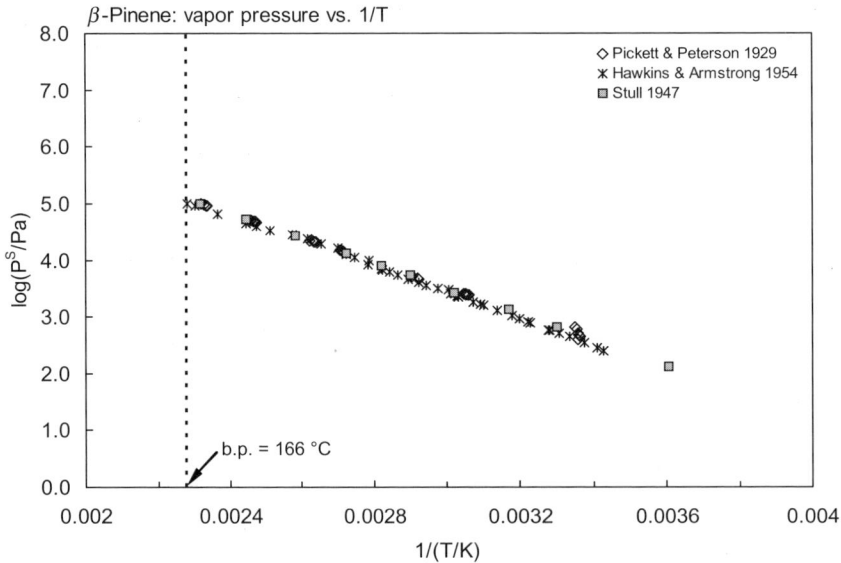

FIGURE 2.1.2.4.10.1 Logarithm of vapor pressure versus reciprocal temperature for β-pinene.

2.2 SUMMARY TABLES AND QSPR PLOTS

TABLE 2.2.1
Summary of physical properties of aliphatic and cyclic hydrocarbons

Compound	CAS no.	Molecular formula	Molecular weight, MW g/mol	m.p. °C	b.p. °C	Fugacity ratio, F at 25°C*	Density, ρ g/cm^3 at 20°C	Molar volume, V_M cm^3/mol MW/ρ at 20°C	Le Bas
Alkanes:									
Isobutane (2-Methylpropane)	75-28-5	C$_4$H$_{10}$	58.122	−159.4	−11.73	1	0.5571	104.33	96.2
2,2-Dimethylpropane (Neopentane)	463-82-1	C$_5$H$_{12}$	72.149	−16.4	9.48	1	0.5910	122.08	118.4
n-Butane	106-97-8	C$_4$H$_{10}$	58.122	−138.3	−0.5	1	0.5786	100.45	96.2
2-Methylbutane (Isopentane)	78-78-4	C$_5$H$_{12}$	72.149	−159.77	27.88	1	0.6193	116.50	118.4
2,2-Dimethylbutane	75-83-2	C$_6$H$_{14}$	86.175	−98.8	49.73	1	0.6492	132.74	140.6
2,3-Dimethylbutane	79-29-8	C$_6$H$_{14}$	86.175	−128.10	57.93	1	0.6616	130.25	140.6
2,2,3-Trimethylbutane	464-06-2	C$_7$H$_{16}$	100.202	−24.6	80.86	1	0.6901	145.20	162.8
n-Pentane	109-66-0	C$_5$H$_{12}$	72.149	−129.67	36.06	1	0.6262	115.22	118.4
2-Methylpentane (Isohexane)	107-83-5	C$_6$H$_{14}$	86.175	−153.6	60.26	1	0.6322	136.31	140.6
3-Methylpentane	96-14-0	C$_6$H$_{14}$	86.175	−162.90	63.27	1	0.66431	129.72	140.6
2,2-Dimethylpentane	590-35-2	C$_7$H$_{16}$	100.202	−123.7	79.2	1	0.6739	148.69	162.8
2,4-Dimethylpentane	108-08-7	C$_7$H$_{16}$	100.202	−119.5	80.49	1	0.6727	148.95	162.8
3,3-Dimethylpentane	562-49-2	C$_7$H$_{16}$	100.202	−134.4	86.06	1	0.6933	144.53	162.8
2,2,4-Trimethylpentane (Isooctane)	540-84-1	C$_8$H$_{18}$	114.229	−107.3	99.22	1	0.6919	165.09	185.0
2,3,4-Trimethylpentane	565-75-3	C$_8$H$_{18}$	114.229	−109.2	113.5	1	0.7191	158.85	185.0
n-Hexane	110-54-3	C$_6$H$_{14}$	86.175	−95.35	68.73	1	0.6593	130.71	140.6
2-Methylhexane (Isoheptane)	591-76-4	C$_7$H$_{16}$	100.202	−118.2	90.04	1	0.6786	147.66	162.8
3-Methylhexane	589-34-4	C$_7$H$_{16}$	100.202	−119.4	92	1	0.6871	145.83	162.8
2,2,5-Trimethylhexane	3522-94-9	C$_9$H$_{20}$	128.255	−105.7	124.09	1	0.7072	181.36	207.2
n-Heptane	142-82-5	C$_7$H$_{16}$	100.202	−90.55	98.4	1	0.6837	146.56	162.8
2-Methylheptane	592-27-8	C$_8$H$_{18}$	114.229	−109.02	117.66	1	0.698	163.65	185.0
3-Methylheptane	589-81-1	C$_8$H$_{18}$	114.229	−120.48	118.9	1	0.7075	161.45	185.0
n-Octane	111-65-9	C$_8$H$_{18}$	114.229	−56.82	125.67	1	0.70256	162.59	185.0
4-Methyloctane	2216-34-4	C$_9$H$_{20}$	128.255	−113.3	142.4	1	0.7199	178.16	207.2
n-Nonane	111-84-2	C$_9$H$_{20}$	128.255	−53.46	150.82	1	0.7177	178.70	207.2
n-Decane	124-18-5	C$_{10}$H$_{22}$	142.282	−29.6	174.15	1	0.7301	194.88	229.4
n-Undecane	1120-21-4	C$_{11}$H$_{24}$	156.309	−25.5	195.9	1	0.7402	211.17	251.6
n-Dodecane	112-40-3	C$_{12}$H$_{26}$	170.334	−9.57	216.32	1	0.7487	227.51	273.8
n-Tridecane	629-50-5	C$_{13}$H$_{28}$	184.361	−5.4	235.47	1	0.7564	243.73	296.0

(Continued)

TABLE 2.2.1 (Continued)

Compound	CAS no.	Molecular formula	Molecular weight, MW g/mol	m.p. °C	b.p. °C	Fugacity ratio, F at 25°C*	Density, ρ g/cm³ at 20°C	Molar volume, V_M cm³/mol MW/ρ at 20°C	Le Bas
n-Tetradecane	629-59-4	$C_{14}H_{30}$	198.388	5.82	253.58	1	0.7628	260.08	318.2
n-Pentadecane	629-62-9	$C_{15}H_{32}$	212.415	9.95	270.6	1	0.7685	276.40	340.4
n-Hexadecane	544-76-3	$C_{16}H_{34}$	226.441	18.12	286.86	1	0.77344	292.77	362.6
n-Heptadecane	629-78-7	$C_{17}H_{36}$	240.468	22.0	302.0	1	0.7780	309.08	384.8
n-Octadecane	593-45-3	$C_{18}H_{38}$	254.495	28.2	316.3	0.930	0.7819	325.48	407.0
n-Eicosane	112-95-8	$C_{20}H_{42}$	282.547	36.6	343	0.769	0.7887	358.24	451.4
n-Tetracosane	646-31-1	$C_{24}H_{50}$	338.654	50.4	391.3	0.563	0.7991	423.79	540.2
n-Hexacosane	630-01-3	$C_{26}H_{54}$	366.707	56.1	412.2	0.495	0.8032	456.56	584.6
Cycloalkanes:									
Cyclopentane	287-92-3	C_5H_{10}	70.133	–93.4	49.3	1	0.7454	94.09	99.5
Methylcyclopentane	96-37-7	C_6H_{12}	84.159	–142.42	71.8	1	0.7487	112.41	121.7
1,1,3-Trimethylcyclopentane	4516-69-2	C_8H_{16}	112.213	–142.4	104.9	1	0.7483	149.96	166.1
Propylcyclopentane	2040-96-2	C_8H_{16}	112.213	–117.3	131	1	0.7763	144.55	166.1
Pentylcyclopentane	3741-00-2	$C_{10}H_{20}$	140.266	–83	180	1	0.7912	177.28	210.5
Cyclohexane	110-82-7	C_6H_{12}	84.159	6.59	80.73	1	0.7786	108.09	118.2
Methylcyclohexane	108-87-2	C_7H_{14}	98.186	–126.6	100.93	1	0.7694	127.61	140.4
1,2-cis-Dimethylcyclohexane	2207-01-4	C_8H_{16}	112.213	–49.8	129.8	1	0.7963	140.92	162.6
1,4-trans-Dimethylcyclohexane	2207-04-7	C_8H_{16}	112.213	–36.93	119.4	1	0.7626	147.15	162.6
1,1,3-Trimethylcyclohexane	3073-66-3	C_9H_{18}	126.239	–65.7	136.6	1	0.7664	165.72	184.8
Ethylcyclohexane	1678-91-7	C_8H_{16}	112.213	–111.3	131.9	1	0.7880	142.40	162.6
Cycloheptane	291-64-5	C_7H_{14}	98.186	–8.46	118.4	1	0.8098	121.25	136.4
Cyclooctane	292-64-8	C_8H_{16}	112.213	14.59	149	1	0.8340	134.55	154.1
cis-Decalin	493-01-6	$C_{10}H_{18}$	138.250	–42.9	195.8	1	0.8931	154.80	184.6
trans-Decalin	493-02-7	$C_{10}H_{18}$	138.250	–30.4	187.3	1	0.8662	159.60	184.6
Alkenes:									
2-Methylpropene	115-11-7	C_4H_8	56.107	–140.7	–6.9	1	0.5942	94.42	88.8
1-Butene	106-98-9	C_4H_8	56.107	–185.34	–6.26	1	0.5951	94.28	88.8
2-Methyl-1-butene	563-46-2	C_5H_{10}	70.133	–137.53	31.2	1	0.6504	107.83	111.0
3-Methyl-1-butene	563-45-1	C_5H_{10}	70.133	–168.43	20.1	1	0.6272	111.82	111.0
2-Methyl-2-butene	513-35-9	C_5H_{10}	70.133	–133.72	38.56	1	0.6623	105.89	111.0
1-Pentene	109-67-1	C_5H_{10}	70.133	–165.12	29.96	1	0.6405	109.50	111.0
cis-2-Pentene	627-20-3	C_5H_{10}	70.133	–151.36	36.93	1	0.6556	106.98	111.0
2-Methyl-1-pentene	763-29-1	C_6H_{12}	84.159	–135.7	62.1	1	0.6799	123.78	133.2
4-Methyl-1-pentene	691-37-2	C_6H_{12}	84.159	–153.6	53.9	1	0.6642	126.71	133.2

Aliphatic and Cyclic Hydrocarbons

Name	CAS	Formula	MW	mp	bp		density		
1-Hexene	592-41-6	C₆H₁₂	84.159	−139.76	63.48	1	0.6732	125.01	133.2
1-Heptene	592-76-7	C₇H₁₄	98.186	−118.9	93.64	1	0.6970	140.87	155.4
1-Octene	111-66-0	C₈H₁₆	112.213	−101.7	121.29	1	0.7149	156.96	177.6
1-Nonene	124-11-8	C₉H₁₈	126.239	−81.3	146.9	1	0.7292	173.12	199.8
1-Decene	872-05-9	C₁₀H₂₀	140.266	−66.3	170.5	1	0.7408	189.34	222.0
Dienes:									
1,3-Butadiene	106-99-0	C₄H₆	54.091	−108.91	−4.41	1	0.6211	87.09	81.4
2-Methyl-1,3-butadiene (Isoprene)	78-79-5	C₅H₈	68.118	−145.9	34.0	1	0.6809	100.04	103.6
2,3-Dimethyl-1,3-butadiene	513-81-5	C₆H₁₀	82.143	−76	68.8	1	0.7267	113.04	125.8
1,4-Pentadiene	591-93-5	C₅H₈	68.118	−148.2	26	1	0.6608	103.08	103.6
1,5-Hexadiene	592-42-7	C₆H₁₀	82.143	−140.7	59.4	1	0.6923	118.65	125.8
1,6-Heptadiene	3070-53-9	C₇H₁₂	96.170	−129	90	1	0.714	134.69	148.0
Alkynes:									
1-Butyne	107-00-6	C₄H₆	54.091	−125.7	8.08	1	0.650	83.22	81.4
1-Pentyne	627-19-0	C₅H₈	68.118	−90	40.1	1	0.6901	98.71	103.6
1-Hexyne	693-02-7	C₆H₁₀	82.143	−131.9	71.3	1	0.7155	114.81	125.8
1-Heptyne	628-71-7	C₇H₁₂	96.170	−81	99.7	1	0.7328	131.24	148.0
1-Octyne	629-05-0	C₈H₁₄	110.197	−79.3	126.3	1	0.7461	147.70	170.2
1-Nonyne	3452-09-3	C₉H₁₆	124.223	−50	150.8	1	0.7568	164.14	192.4
Cycloalkenes:									
Cyclopentene	142-29-0	C₅H₈	68.118	−135.0	44.2	1	0.772	88.24	92.1
Cyclohexene	110-83-8	C₆H₁₀	82.143	−103.5	82.98	1	0.8110	101.29	110.8
1-Methylcyclohexene	591-49-1	C₇H₁₂	96.170	−120.4	110.3	1	0.8102	118.70	133.0
Cycloheptene	628-92-2	C₇H₁₂	96.170	−56	115	1	0.8228	116.88	129.0
cis-Cyclooctene	931-87-3	C₈H₁₄	110.197	−12	138	1	0.8472	130.07	146.7
trans-Cyclooctene	931-89-5	C₈H₁₄	110.197	−59	143	1	0.8483	129.90	146.7
1,4-Cyclohexadiene	628-41-1	C₆H₈	80.128	−49.2	85.5	1	0.8471	94.59	103.4
1,3,5-Cycloheptatriene	544-25-2	C₇H₈	92.139	−79.5	117	1	0.8875	103.82	114.2
d-Limonene	5989-27-5	C₁₀H₁₆	136.234	−74.0	178	1	0.8403	162.13	192.2
α-Pinene	80-56-8	C₁₀H₁₆	136.234	−64	156.2	1	0.8582	158.74	183.7
β-Pinene	127-91-3	C₁₀H₁₆	136.234	−61.5	166	1	0.8694	156.70	183.7

* Assuming $\Delta S_{fus} = 56$ J/mol K.

TABLE 2.2.2
Summary of selected physical-chemical properties of aliphatic and cyclic hydrocarbons at 25°C

Selected properties

Compound	Vapor pressure		Solubility			log K_{ow}	Henry's law constant H/(Pa·m³/mol) calculated P/C
	P^S/Pa	P_L/Pa	S/(g/m³)	C^S/(mol/m³)	C_L/(mol/m³)		
Alkanes:							
Isobutane (2-Methylpropane)	357000	357000	48.9	0.8413	0.8413	2.76	120435*
2,2-Dimethylpropane	172000	172000	33.2	0.4602	0.4602	3.11	220195*
n-Butane	243000	243000	61.4	1.0564	1.0564	2.90	95915*
2-Methylbutane (Isopentane)	91640	91640	47.8	0.6625	0.6625	2.30	138320
2,2-Dimethylbutane	42600	42600v	18.4	0.2135	0.2135	3.82	199515
2,3-Dimethylbutane	32010	32010	19.1	0.2216	0.2216	3.85	144422
2,2,3-Trimethylbutane	13652	13652	4.38	0.0437	0.0437		312320
n-Pentane	68400	68400	38.5	0.5336	0.5336	3.45	128180
2-Methylpentane (Isohexane)	28200	28200	13.8	0.1601	0.1601	2.80	176097
3-Methylpentane	25300	25300	12.8	0.1485	0.1485	3.60	170330
2,2-Dimethylpentane	14000	14000	4.4	0.0439	0.0439	3.10	318825
2,4-Dimethylpentane	13100	13100	4.06	0.0405	0.0405	3.10	323312
3,3-Dimethylpentane	10940	10940	5.94	0.0593	0.0593		184550
2,2,4-Trimethylpentane (Isooctane)	6560	6560	2.44	0.0214	0.0214		307110
2,3,4-Trimethylpentane	3600	3600	2.0	0.0119	0.0119		205614
n-Hexane	20200	20200	9.5	0.1102	0.1102	4.11	183235
2-Methylhexane (Isoheptane)	8780	8780	2.54	0.0253	0.0253		346370
3-Methylhexane	8210	8210	3.3	0.0329	0.0329		249290
2,2,5-Trimethylhexane	2210	2210	1.15	0.0090	0.0090	4.50	246472
n-Heptane	6110	6110	2.93	0.0292	0.0292	5.00	208955
2-Methylheptane	2600	2600	0.85	0.00744	0.0074		349410
n-Octane	1800	1800	0.66	0.005778	0.0058	5.15	311536
n-Nonane	571	571	0.22	0.001715	0.0017	5.65	332880
n-Decane	175	175	0.052	0.000365	0.00037	6.25	478840
n-Undecane	52.2	52.2	0.004	0.000026	0.000026		2039835
n-Dodecane	18.02	18.02	0.0037	0.000022	0.000022	6.80	829570
n-Tridecane	6.682	6.682					
n-Tetradecane	1.804	1.804				8.00	
n-Hexadecane	0.191	0.191					
n-Heptadecane	0.0615	0.0615					
n-Octadecane	0.02	0.019					

Aliphatic and Cyclic Hydrocarbons

Compound					
n-Eicosane	0.00209	0.00161			
Cycloalkanes:					
Cyclopentane	42400	42400	166	2.3669	2.3669	3.00	17915
Methylcyclopentane	18300	18300	43	0.5109	0.5109	3.37	35815
1,1,3-Trimethylcyclopentane	5300	5300	3.73	0.0332	0.0332		159440
Propylcyclopentane	1640	1640	2.04	0.0182	0.0182		90210
Pentylcyclopentane	152	152	0.115	0.0008	0.0008		185395
Cyclohexane	13014	13014	58	0.6892	0.6892	3.44	18885
Methylcyclohexane	6180	6180	15.1	0.1538	0.1538	3.88	40185
1,2-cis-Dimethylcyclohexane	1930	1930	6	0.0535	0.0535		36095
1,4-trans-Dimethylcyclohexane	3020	3020	3.84	0.0342	0.0342		88250
1,1,3-Trimethylcyclohexane	1480	1480	1.77	0.0140	0.0140		105560
Cycloheptane	2924	2924	23.5	0.2393	0.2393		12220
Cyclooctane	748	748	5.80	0.0517	0.0517	4.45	14470
Alkenes:							
2-Methylpropene	304000	304000	263	4.6875	4.6875		21620*
1-Butene	297000	297000	222	3.9567	3.9567		25610*
2-Methyl-1-butene	81330	81330					
3-Methyl-1-butene	120000	120000	130	1.8536	1.8536		54670*
2-Methyl-2-butene	62410	62410	325	4.634	4.634		13470
1-Pentene	85000	85000	148	2.1103	2.1103	2.20	40280
cis-2-Pentene	66000	66000				2.20	
2-Methyl-1-pentene	26000	26000	78	0.9268	0.9268	2.50	28050
4-Methyl-1-pentene	36100	36100	48	0.5703	0.5703		63295
1-Hexene	24800	24800	50	0.5941	0.5941	3.39	41743
1-Heptene	7510	7510	18.3	0.1864	0.1864	3.99	40295
1-Octene	2320	2320	2.7	0.0241	0.0241	4.57	96420
1-Nonene	712	712	1.12	0.0089	0.0089	5.15	80250
1-Decene	218	218					
Dienes:							
1,3-Butadiene	281000	281000	735	13.588	13.588	1.99	7458*
2-Methyl-1,3-butadiene (Isoprene)	73300	73300	642	9.4248	9.4248		7780
2,3-Dimethyl-1,3-butadiene	20160	20160	327	3.9809	3.9809		5065
1,4-Pentadiene	98000	98000	558	8.1917	8.1917	2.48	11965
1,5-Hexadiene	29690	29690	169	2.0574	2.0574	2.75	14430
1,6-Heptadiene			44	0.4575	0.4575		

(Continued)

TABLE 2.2.2 (Continued)

Compound	Vapor pressure		Solubility			log K_{ow}	Henry's law constant $H/(Pa \cdot m^3/mol)$
	P^S/Pa	P_L/Pa	$S/(g/m^3)$	$C^S/(mol/m^3)$	$C_L/(mol/m^3)$		calculated P/C
Alkynes:							
1-Butyne	188000	188000	2870	53.059	53.059		1910*
1-Pentyne	57600	57600	1570	23.048	23.048	1.98	2500
1-Hexyne	18140	18140	360	4.3830	4.3830	2.73	4140
1-Heptyne	7500	7000	94	0.9774	0.9774	2.98	7675
1-Octyne	1715	1715	24	0.2178	0.2178		7875
1-Nonyne			7.2	0.0580	0.0580		
Cycloalkenes:							
Cyclopentene	50710	50706	535	7.8540	7.8540	2.48	6455
Cyclohexene	11850	11850	213	2.5928	2.5928	2.86	4570
1-Methylcyclohexene	4689	4689	52	0.5407	0.5407		8670
Cycloheptene	2670	2670	66	0.6863	0.6863		3890
Cyclooctene	1010	1010	22.9	0.2078	0.2078	2.47	4860
1,4-Cyclohexadiene	9009	9009	800	9.9840	9.9840		902
1,3,5-Cycloheptatriene	3140	3140	620	7.7376	7.7376	2.63	467
d-Limonene	270	270	13.8	0.1013	0.1013		2665
α-Pinene	582	582					
β-Pinene	395	395					

* Vapor pressure exceeds atmospheric pressure, Henry's law constant $H/(Pa \cdot m^3/mol) = 101325$ Pa/C^S mol/m^3.

Aliphatic and Cyclic Hydrocarbons

TABLE 2.2.3
Suggested half-life classes of hydrocarbons in various environmental compartments at 25°C

Compound	Air class	Water class	Soil class	Sediment class
Alkanes:				
n-Pentane	2	5	6	7
n-Hexane	2	5	6	7
n-Octane	2	5	6	7
n-Decane	2	5	6	7
n-Dodecane	2	5	6	7
Cycloalkanes:				
Cyclopentane	2	5	5	6
Methylcyclopentane	2	5	5	6
Cyclohexane	2	5	5	6
Methylcyclohexane	2	5	5	6
Cyclooctane	2	5	5	6
Alkenes:				
1-Pentene	1	4	5	6
1-Octene	1	4	5	6
1,3-Butadiene	1	4	5	6
1,4-Pentadiene	1	4	5	6
Alkynes:				
1-Hexyne	1	4	5	6
Cycloalkenes:				
Cyclopentene	1	4	5	6
Cyclohexene	1	4	5	6

where,

Class	Mean half-life (hours)	Range (hours)
1	5	< 10
2	17 (~ 1 day)	10–30
3	55 (~ 2 days)	30–100
4	170 (~ 1 week)	100–300
5	550 (~ 3 weeks)	300–1,000
6	1700 (~ 2 months)	1,000–3,000
7	5500 (~ 8 months)	3,000–10,000
8	17000 (~ 2 years)	10,000–30,000
9	55000 (~ 6 years)	> 30,000

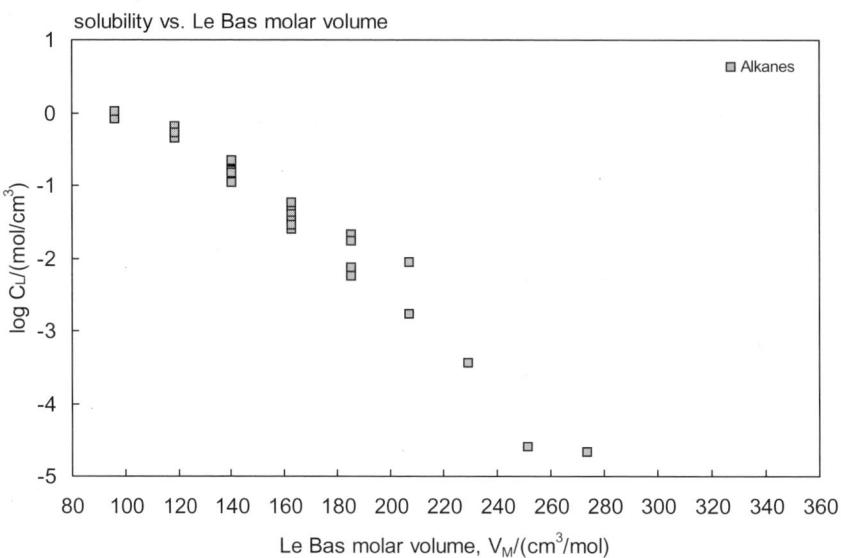

FIGURE 2.2.1 Molar solubility (liquid or supercooled liquid) versus Le Bas molar volume for alkanes.

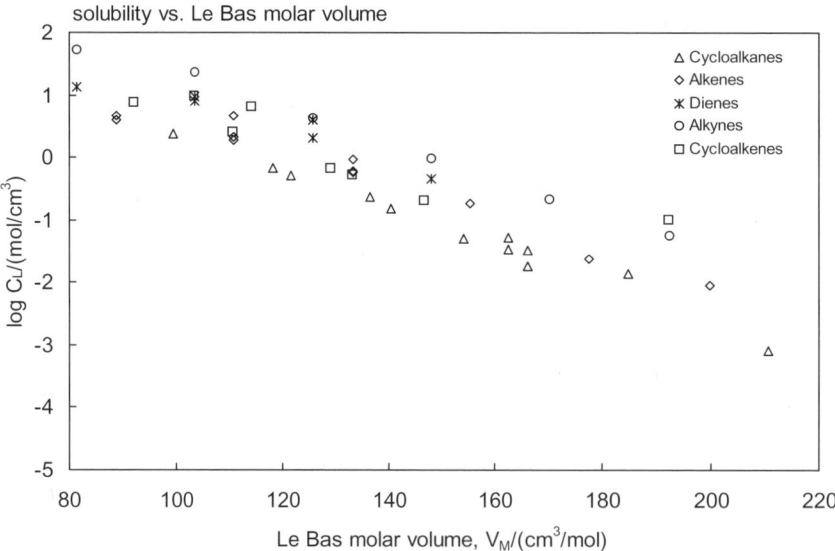

FIGURE 2.2.2 Molar solubility (liquid or supercooled liquid) versus Le Bas molar volume for aliphatic and cyclic hydrocarbons.

Aliphatic and Cyclic Hydrocarbons

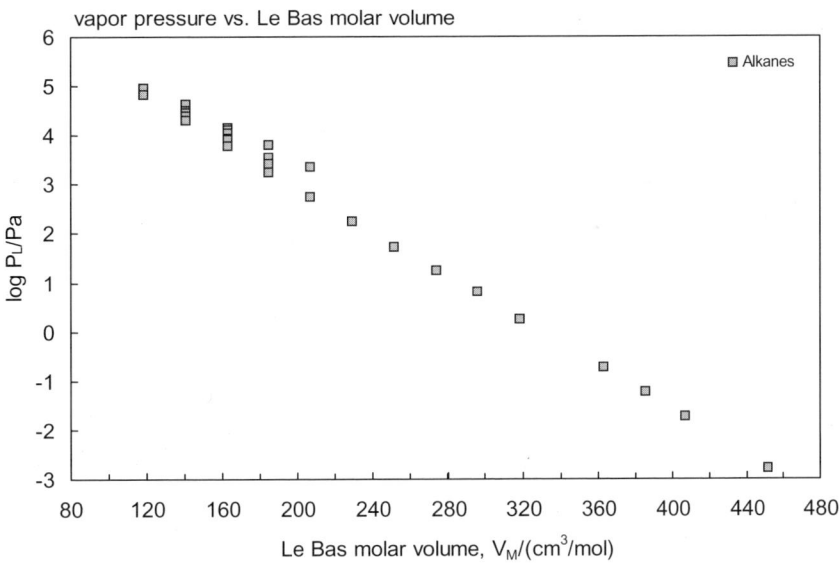

FIGURE 2.2.3 Vapor pressure (liquid or supercooled liquid) versus Le Bas molar volume for alkanes.

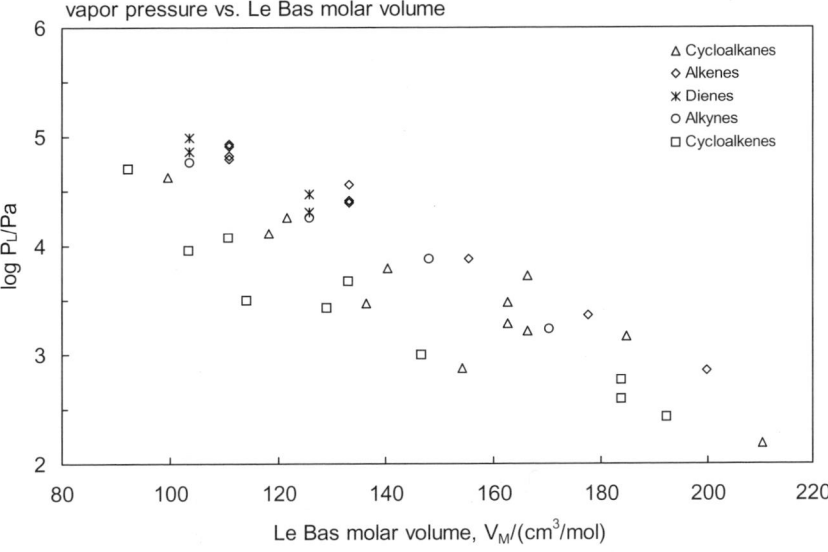

FIGURE 2.2.4 Vapor pressure (liquid or supercooled liquid) versus Le Bas molar volume for aliphatic and cyclic hydrocarbons.

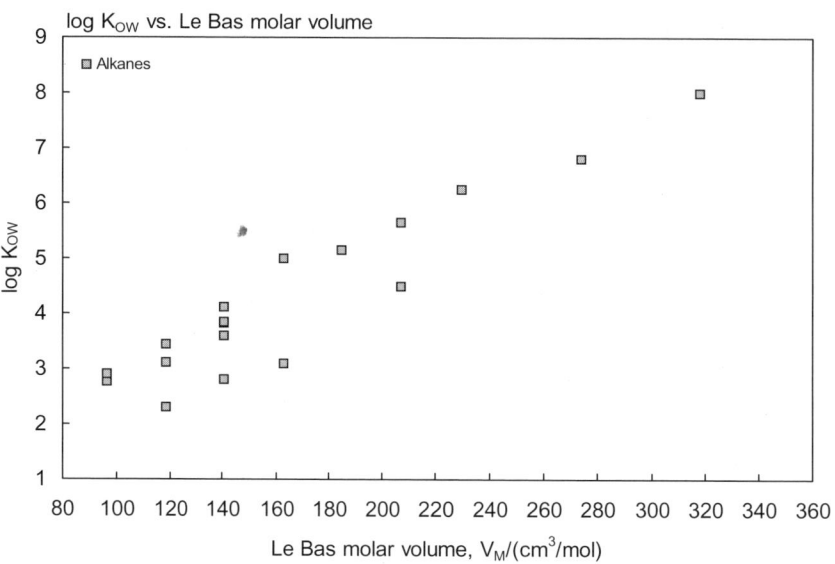

FIGURE 2.2.5 Octanol-water partition coefficient versus Le Bas molar volume for alkanes.

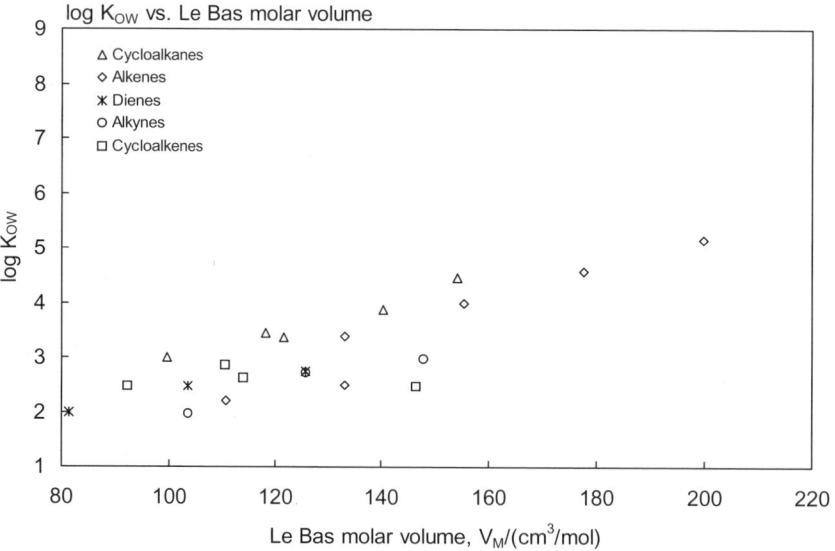

FIGURE 2.2.6 Octanol-water partition coefficient versus Le Bas molar volume for aliphatic and cyclic hydrocarbons.

Aliphatic and Cyclic Hydrocarbons

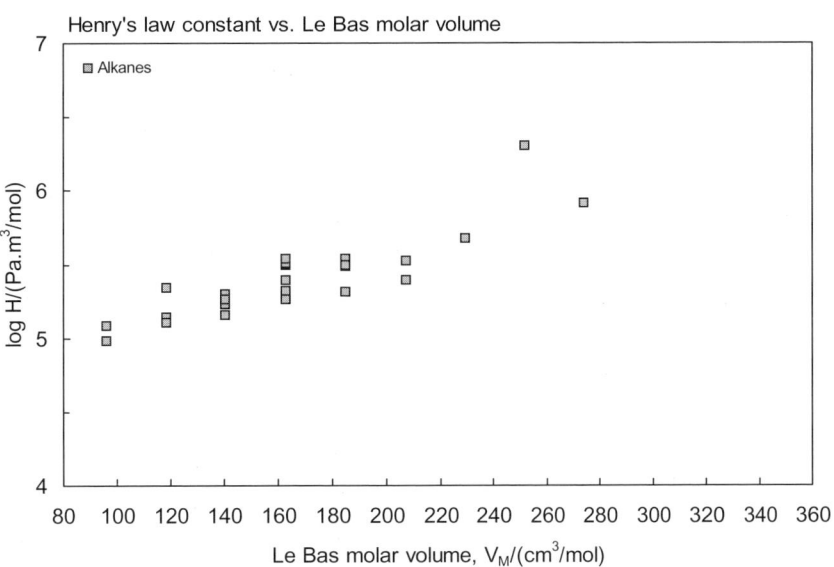

FIGURE 2.2.7 Henry's law constant versus Le Bas molar volume for alkanes.

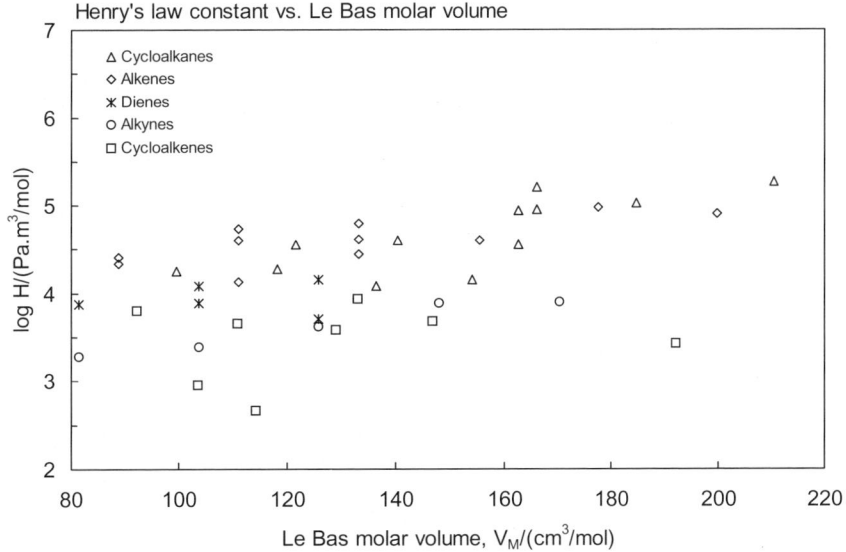

FIGURE 2.2.8 Henry's law constant versus Le Bas molar volume for aliphatic and cyclic hydrocarbons.

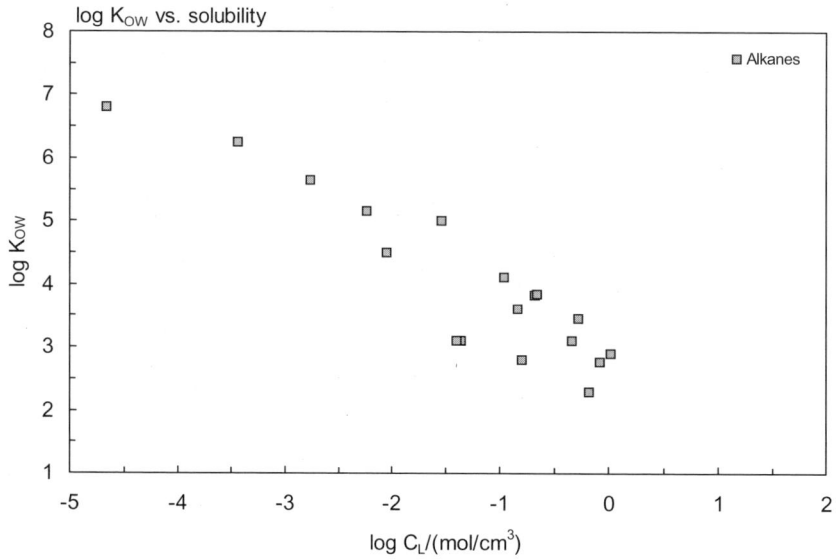

FIGURE 2.2.9 Octanol-water partition coefficient versus molar solubility (liquid or supercooled liquid) for alkanes.

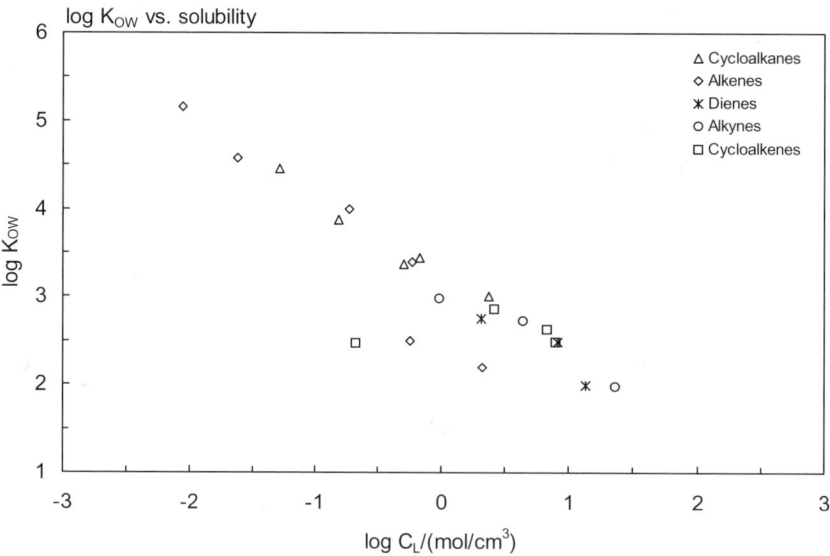

FIGURE 2.2.10 Octanol-water partition coefficient versus molar solubility (liquid or supercooled liquid) for aliphatic and cyclic hydrocarbons.

2.3 REFERENCES

Abernethy, S., Mackay, D., McCarty, L.S. (1988) "Volume fraction" correlation for narcosis in aquatic organisms: the key role of partitioning. *Environ. Toxicol. Chem.* 7, 469–481.
Abraham, M.H., Le J., Acree, Jr., W.E., Carr, P.W., Dallas, A.J. (2001) The solubility of gases and vapours in dry octan-1-ol at 298 K. *Chemosphere* 44, 855–863.
Adeniji, S.A., Kerr, J.A., Williams, M.R. (1981) Rate constants for ozone-alkene reactions under atmospheric conditions. *Int. J. Chem. Kinet.* 13, 209–217.
Allemand, N., Jose, J., Merlin, J.C. (1986) Mesure des pressions de vapeur d'hydrocarbures C_{10} A C_{18} n-alcanes et n-alkylbenzenes dans le domaine 3-1000 pascal. *Thermochim. Acta* 105, 79–90.
Altshuller, A.P. (1991) Chemical reactions and transport of alkanes and their products in the troposphere. *J. Atmos. Chem.* 12, 19–61.
Altshuller, A.P., Bufalini, J.J. (1971) Photochemical aspects of air pollution: A review. *Environ. Sci. Technol.* 5(1), 39–64.
Anand, S.C., Grolier, J.-P.E., Kiyohara, O., Halpin, C.J., Benson, G.C. (1975) Thermodynamic properties of some cycloalkane-cycloalkanol systems at 298.15 K. III. *J. Chem. Eng. Data* 20, 184–189.
Andersson, Y., Ljungström, E. (1989) Gas phase reaction of the NO_3 radical with organic compound in the dark. *Atmos. Environ.* 23, 1153–1155.
Aquan-Yuen, M., Mackay, D., Shiu, W.Y. (1979) Solubility of hexane, phenanthrene, chlorobenzene, and p-dichlorobenzene in aqueous electrolyte solutions. *J. Chem. Eng. Data* 24, 30–34.
Aschmann, S.M., Atkinson, R. (1995) Rate constants for the reactions of the NO_3 radical with alkanes at 296 ± 2 K. *Atmos. Environ.* 29, 2311–2316.
Ashton, J.G., Fink, H.L., Schumann, S.C. (1943a) The heat capacity, heats of transition, fusion and vaporization and the vapor pressures of cyclopentane. Evidence for a non-planar structure. *J. Am. Chem. Soc.* 65, 341–346.
Ashton, J.G., Kennedy, R.M., Schumann, S.C. (1940) The heat capacity and entropy, heats of fusion and vaporization and the vapor pressure of isobutane. *J. Am. Chem. Soc.* 62, 2059–2063.
Ashton, J.G., Szasz, G.J., Fink, H.L. (1943b) The heat capacity, heats of transition, fusion and vaporization and the vapor pressures of cyclohexane. The vibrational frequencies of alicyclic ring systems. *J. Am. Chem. Soc.* 65, 1135–1139.
Ashworth, R.A., Howe, G.B., Mullins, M.E., Rogers, T.N. (1988) Air-water partitioning coefficients of organics in dilute aqueous solutions. *J. Hazard. Materials* 18, 25–36.
Atkinson, R. (1985) Kinetics and mechanisms of the gas phase reaction of hydroxyl radicals with organic compounds under atmospheric conditions. *Chem. Rev.* 85, 69–201.
Atkinson, R. (1987) Structure-activity relationship for the estimation of the rate constants for the gas phase reactions of OH radicals with organic compounds. *Int. J. Chem. Kinetics* 19, 799–828.
Atkinson, R. (1989) Kinetics and Mechanisms of the gas-phase reactions of the hydroxyl radical with organic compounds. *J. Phys. Chem. Data* Monograph No.1.
Atkinson, R. (1990) Gas-phase tropospheric chemistry of organic compounds, a review. *Atmos. Environ.* 24A, 1–41.
Atkinson, R. (1991) Kinetics and mechanisms of the gas-phase reactions of the NO_3 radical with organic compounds. *J. Phys. Chem. Data* 20, 450–507.
Atkinson, R. (1997) Gas-phase tropospheric chemistry of volatile organic compounds:1. Alkanes and alkenes. *J. Phys. Chem. Ref. Data* 26, 215–289.
Atkinson, R. (2000) Atmospheric chemistry of VOCs and No_x. *Atmos. Environ.* 34, 2063–2101.
Atkinson, R., Aschmann, S.M. (1984) Rate constants for the reactions of O_3 and OH radicals with a series of alkynes. *Int. J. Chem. Kinet.* 16, 259–268.
Atkinson, R., Aschmann, S.M. (1995) Rate constants for the reactions of the NO_3 radical with alkanes at 296 ± 2 K. *Atmos. Environ.* 29, 2311–2316.
Atkinson, R., Carter, W.L. (1984) Kinetics and mechanisms of the gas-phase reactions of ozone with organic compounds under atmospheric conditions. *Chem. Rev.* 84, 437–470.
Atkinson, R., Aschmann, S.M., Carter, W.L., Pitts, Jr., J.N. (1983) Effects of ring strain on gas-phase rate constants. 1. Ozone reactions with cycloalkenes. *Int. J. Chem. Kinet.* 15, 721–731.
Atkinson, R., Aschmann, S.M., Carter, W.P.L., Winer, A.M., Pitts, Jr., J.N. (1982a) Kinetics of the reactions of OH radicals with n-alkanes at 299 ± 2 K. *Int. J. Chem. Kinet.* 14, 781–788.
Atkinson, R., Aschmann, S.M., Pitts, J.N., Jr.. (1986) Rate constants for the gas-phase reactions of the OH radical with a series of monoterpenes at 294 ± 1 K. *Int. J. Chem. Kinet.* 18, 287–299.
Atkinson, R., Aschmann, S.M., Pitts, J.N. Jr. (1988) Rate constants for the gas-phase reactions of the NO_3 radicals with a series of organic compounds at 296 ± 2 K. *J. Phys. Chem.* 92, 3454–3457.
Atkinson, R., Aschmann, S.M., Winer, A.M., Pitts, Jr., J.N. (1984a) Kinetics of the gas-phase reactions of NO_3 radicals with a series of dialkenes, cycloalkenes, and monoterpenes at 295 ± 1 K. *Environ. Sci. Technol.* 18, 370–375.
Atkinson, R., Aschmann, S.M., Winer, A.M., Pitts, Jr., J.N. (1984b) Gas phase reaction of NO_2 with alkenes and dialkenes. *Int. J. Chem. Kinet.* 16, 697–706.
Atkinson, R., Carter, W.P.L., Aschmann, S.M., Carter, W.P.L. (1984d) Kinetics of the reaction of O_3 and OH radicals with a series of dialkenes and trialkenes at 294 ± 2 K. *Int. J. Chem. Kinet.* 16, 967–976.

Atkinson, R., Carter, W.P.L., Aschmann, S.M., Winer, A.M., Pitts, Jr., J.N. (1984c) Kinetics of the reaction of OH radicals with a series of branched alkanes at 297 ± 2 K. *Int. J. Chem. Kinet.* 16, 469–481.

Atkinson, R., Darnall, K.R., Lloyd, A.C., Winer, A.M., Pitts, Jr., J.N. (1979) Kinetics and mechanisms of reaction of hydroxyl radicals with organic compounds in the gas phase. *Adv. Photochem.* 11, 375–488.

Atkinson, R., Pitts, J.N., Jr. (1977a) Absolute rate constants for the reaction of oxygen (^{3}P) atoms with a series of olefins over temperature range 293–439 K. *J. Phys. Chem.* 67, 38–43.

Atkinson, R., Pitts, J.N., Jr. (1977b) Absolute rate constants for the reaction of oxygen (^{3}P) atoms with allene, 1,3-butadiene and vinyl methyl ether over temperature range 297–439 K. *J. Phys. Chem.* 67, 2492–2495.

Atkinson, R., Pitts, J.N., Jr. (1978) Kinetics for the reaction of oxygen (^{3}P) atoms and hydroxyl radicals with 2-methyl-2-butene. *J. Phys. Chem.* 68, 2992–2994.

Bagley, J.A., Canosa-Mas, C., Little, M.R., Parr, A.D., Smith, S.J., Waygood, S.J., Wayne, R.P. (1990) Temperature dependence of reactions of the nitrate radical with alkanes. *J. Chem. Soc. Farad. Trans.* 86(12), 2109–2114.

Baker, E.G. (1958) American Chemical Society, Division of Petroleum Chemistry, Preprints 3, No. 4, C61–68.

Baker, E.G. (1959) Origin and migration of oil. *Science* 129, 871–874.

Baker, E.G. (1967) A geochemical evaluation of petroleum migration and accumulation. In: *Fundamental Aspects of Petroleum Geochemistry.* Nagy, B., Colombo, V. Eds., pp. 299–330, Elsevier, New York, New York.

Bakierowska, A.-M., Trzeszczyñski, J. (2003) Graphical method for the determination of water/gas partition coefficients of volatile organic compounds by a headspace gas chromatography technique. *Fluid Phase Equil.* 213, 139–146.

Banerjee, S., Howard, P.H. (1988) Improved estimation of solubility and partitioning through correction of UNIFAC-derived activity coefficients. *Environ. Sci. Technol.* 22, 839–841.

Banerjee, S., Howard, P.H., Lande, S.S. (1990) General structure vapor pressure relationship for organics. *Chemosphere* 21, 1173–1180.

Barone, G., Crescenzi, V., Pispisa, B., Quadrifoglio, B. (1966) Hydrophobic interactions in polyelectrolytes solutions. II. Solubility of some C_3-C_6 alkanes in poly(methacrylic acid) aqueous solutions. *J. Macromol. Chem.* 1, 761–771.

Becke, A., Quitzsch, G. (1977) Das phasengleichgewichtsverhalten ternärer systeme der art C_4-alkohol-wasser-kohlenwasserstoff. *Chem. Tech.* 29, 49–51.

Bennet, P.J., Harris, S.J., Kerr, J. A. (1987) A reinvestigation of the rate constants for the reactions of ozone with cyclopentene and cyclohexene under atmospheric conditions. *Int. J. Chem. Kinet.* 19, 609–614.

Benter, Th., Schindler, R.N. (1988) Absolute rate coefficients for the reaction of NO_3 radicals with simple dienes. *Chem. Phys. Lett.* 145, 67–70.

Berti, P., Cabani, S. Conti, G., Mollica, V. (1986) The thermodynamic study of organic compounds in octan-1-ol. *J. Chem. Soc., Faraday, Trans.* 1, 82, 2547.

Bissell, T.G., Williamson, A.G. (1975) Vapour pressures and excess Gibbs energies of n-hexane and of n-heptane + carbon tetrachloride and + chloroform at 298.15 K. *J. Chem. Thermodyn.* 7, 131–136.

Bittrich, H.J., Gedan, H., Feix, G. (1979) Zur löslichkeitsbeeinflussung von kohlenwasserstoffen in wasser. *Z. Phys. Chem.* (Leibzig) 260, 1009–1013.

Black, C., Joris, G.G., Taylor, H.S. (1948) The solubility of water in hydrocarbons. *J. Chem. Phys.* 16, 537–548.

Bobra, A.M., Shiu, W.Y., Mackay, D. (1979) Distribution of hydrocarbons among oil, water and vapor phases during oil dispersant toxicity tests. *Bull. Environ. Contam. Toxicol.* 4, 297–305.

Bodor, N., Gabanyi, Z., Wong, C.-K. (1989) A new method for the estimation of partition coefficient. *J. Am. Chem. Soc.* 111, 3783–3786.

Boodaghians, R.B., Hall, I.W., Toby, F.S., Wayne, R.P. (1987) Absolute determinations of the kinetics and temperature dependences of the reactions of OH with a series of alkynes. *J. Chem. Soc. Farad. Trans.* 2, 83(11), 2073–2080.

Booth, H.S., Everson, H.E. (1948) Hydrotropic solubilities: solubilities in 40 percent sodium xylenesulfonate. *Ind. Eng. Chem.* 40(8), 1491–1493.

Boublik, T. (1960) Thesis, Utzcht, Prague.—reference from Boublik et al. 1984.

Boublik, T., Fried, V., Hala, E. (1973) *The Vapour Pressure of Pure Substances.* Elsevier, Amsterdam.

Boublik, T., Fried, V., Hala, E. (1984) *The Vapour Pressures of Pure Substances.* (second revised edition), Elsevier, Amsterdam.

Boublikova, L. (1972) Thesis, Utzcht, Prague.—reference from Boublik et al. 1984.

Brookman, G.T., Flanagan, M., Kebe, J.O. (1985) *Literature Survey: Hydrocarbon Solubilities and Attenuation Mechanisms,* prepared for Environmental Affairs Dept. of American Petroleum Institute. *API Publication* No. 4414, August, 1985, Washington D.C.

Budavari, S., Ed. (1989) *The Merck Index. An Encyclopedia of Chemicals, Drugs and Biologicals.* 11th edition, Merck & Co. Inc., Rahway, New Jersey.

Budantseva, L.S., Lesteva, T.M., Nemtsov, M.S. (1976) *Zh. Fiz. Khim.* 50, 1344.

Bufalini, J.J., Altshuller, A.P. (1965) Kinetics of vapor-phase hydrocarbon-ozone reactions. *Can. J. Chem.* 43, 2243–2250.

Burkhard, L.P., Kuehl, D.W., Veith, G.D. (1985) Evaluation of reverse phase liquid chromatograph/mass spectrometry for estimation of *n*-octanol-water partition coefficients. *Chemosphere* 14(10), 1551–1560.

Burris, D.R., MacIntyre, W.G. (1986) A thermodynamic study of solutions of liquid hydrocarbon mixtures in water. *Geochim. Cosmochim. Acta* 50, 1545–1549.

Cadle, R.D., Schadt, C. (1952) Kinetics of the gas-phase reaction of olefins with ozone. *J. Am. Chem. Soc.* 74, 6002–6004.

Camin, D.L., Forziati, A.F., Rossini, F. (1954) Physical properties of *n*-hexadecane, *n*-decylcyclopentane, *n*-decylcyclohexane, 1-hexadecene and *n*-decylbenzene. *J. Phys. Chem.* 58, 440–442.

Camin, D.L., Rossini, F. (1955) Physical properties of 14 American Petroleum Institute research hydrocarbons, C_9 to C_{15}. *J. Phys. Chem.* 59, 1173–1179.

Campbell, A.N., Kartzmark, E.M., Anand. S.C., Cheng, Y., Dzikowski, H.P., Skrynyk, S.M. (1968) Partially miscible liquid systems: the density, change of volume on mixing, vapor pressure, surface tension, and viscosity in the system: aniline-hexane. *Can. J. Chem.* 46, 2399–2407.

Canton, J.H., Wegman, R.C.C. (1983) Studies on the toxicity of tribromomethane, cyclohexene, and bromocyclohexane to different fresh water organisms. *Water Res.* 17, 743–747.

Capel, P.D., Larson, S.J. (1995) A chemodynamic approach for estimating losses of target organic chemicals form water during sample holding time. *Chemosphere* 30, 1097–1107.

Carmona, F.J., Gonzalez, J.A., Carcia de la Fuente, I., Cobos, J.C., Bhethanabotla, V.R., Campbell, S.W. (2000) Thermodynamic properties of *n*-alkoxyethanols + organic solvent mixtures. XI. Total vapor pressure measurements for *n*-hexane, cyclohexane or *n*-heptane + 2-ethoxyethanol at 303.15 and 323.15 K. *J. Chem. Eng. Data* 45, 699–703.

Carruth, G.F., Kobayashi, R. (1973) Vapor pressure of normal paraffins ethane through *n*-decane from their triple points to about 10 mm mercury. *J. Chem. Eng. Data* 18(2), 115–126.

Chao, J., Lin, C.T., Chung, T.H. (1983) Vapor pressure of coal chemicals. *J. Phys. Chem. Ref. Data* 12, 1033–1063.

Chen, F., Holten-Andersen, J., Tyle, P. (1993) New developments of the UNIFAC model for environmental application. *Chemosphere* 26, 1225–1354.

Chickos, J.S., Acree, Jr., W.E., Liebman, J.F. (1999) Estimating solid-liquid phase change enthalpies and entropies. *J. Phys. Chem. Ref. Data* 28, 1535–1673.

Chickos, J.S., Hanshaw, W. (2004) Vapor pressures and vaporization enthalpies of the *n*-alkanes form C_{21} to C_{30} at T = 298.15 K by correlation gas chromatography. *J. Chem. Eng. Data* 49, 77–85.

Chirico, R.D, Nguyen, A., Steele, W.V., Strube, M.M. (1989) Vapor pressure of n-alkanes revisited. New high-precision vapor pressure data on *n*-decane, *n*-eicosane, and *n*-octacosane. *J. Chem. Eng. Data* 34, 149–156.

Claussen, W.F., Polglase, M.F. (1952) Solubilities and structure in aqueous aliphatic hydrocarbons solution. *J. Am. Chem. Soc.* 74, 4817–4819.

Coates, M., Connell, D.W., Barron, D.M. (1985) Aqueous solubility and octan-1-ol to water partition coefficients of aliphatic hydrocarbons. *Environ. Sci. Technol.* 19, 628–632.

Connolly, J.F. (1966) Solubility of hydrocarbons in water near the critical solution temperatures. *J. Chem. Eng. Data* 11, 13–16.

Corchnoy, S.B., Atkinson, R. (1990) Kinetics of the gas-phase reactions of OH and NO_3 radicals with 2-carene, 1,8-cineole, *p*-cymene, and terpinolene. *Environ. Sci. Technol* 24, 1497–1502.

Coutant, R.W., Keigley, G.W. (1988) An alternative method for gas chromatographic determination of volatile organic compounds in water. *Anal. Chem.* 60, 2436–2537.

Cox, R.A., Penkett, S.A. (1972) Aerosol formation from sulfur dioxide in the presence of ozone and olefinic hydrocarbons. *J. Chem. Soc. Farad. Trans.* 1, 68, 1735–1753.

Cramer III, R.D. (1977) "Hydrophobic interaction" and solvation energies: Discrepancies between theory and experimental data. *J. Am. Chem. Soc.* 99, 5408–5412.

Cruickshank, A.J.B., Cutler, A.J.B. (1967) Vapor pressure of cyclohexane, 25–75°C. *J. Chem. Eng. Data* 12, 326–329.

Dallos, A., Wienke, G., Ilchmann, A., Gmehling, J. (1993) Vorausberechnung von octanol/wasser-verteilungskoeffizienten mit hilfe der UNIFAC-methode. *Chem.-Ing.-Tech.* 65, 291–303.

D'Amboise, M., Hanai, T. (1982) Hydrophobicity and retention in reverse phase liquid chromatography. *J. Liq. Chromatogr.* 5, 229–244.

Darnall, K.R., Atkinson, R., Pitts, J.N., Jr. (1978) Rate constants for the reaction of the OH radical with selected alkanes at 300 K. *J. Phys. Chem.* 82, 1581–1584.

Darnall, K.R., Lloyd, A.C., Winer, A.M., Pitts, J.N. (1976) Reactivity scale for atmospheric hydrocarbons based on reaction with hydroxyl radicals. *Environ. Sci. Technol.* 10, 692–696.

Daubert, T.E., Danner, R.P. (1997) *Physical and Thermodynamic Properties of Pure Chemicals*. Taylor and Francis, Philadelphia, PA.

Day, H.O., Nicholson, D.E., Felsing, W.A. (1948) The vapor pressures and some related quantities of pentene-1 from 0–200°C. *J. Am. Chem. Soc.* 70, 1784–1785.

Dean, J.D., Ed. (1985) *Lange's Handbook of Chemistry*. 13th ed. McGraw-Hill, Inc., New York.

Dean, J.D., Ed. (1992) *Lange's Handbook of Chemistry*. 14th ed. McGraw-Hill, Inc., New York.

Dejoz, A., Gonzáles-Alfaro, V., Miguel, P.J., Vázquez, M.I. (1996) Isobaric vapor-liquid equilibria for binary systems composed of octane, decane, and dodecane at 20 kPa. *J. Chem. Eng. Data* 41, 93–96.

DeMore, W.B. (1971) Rates and mechanism of alkyne ozonation. *Int. J. Chem. Kinet.* 3, 161–173.

DeMore, W.B., Bayes, K.D.(1999) Rate constants for the reactions of hydroxyl radical with several alkanes, cycloalkanes, and dimethyl ether. *J. Phys. Chem. A*, 103, 2649–2654.

Deno, N.C., Berkheimer, H.E. (1960) Phase equilibria molecular transport thermodynamics: activity coefficients as a function of structure and media. *J. Chem. Eng. Data* 5, 1–5.

Dillemuth, F.J., Schubert, C.C., Skidmore, D.R. (1963) The reaction of O_3 with acetylenic hydrocarbons. *Combust. Flame* 6(3), 211–212.

Dilling, W.L., Gonsior, S.J., Boggs, G.U., Mendoza, C.G. (1988) Organic photochemistry. 20. A method for estimating gas-phase rate for reactions of hydroxyl radicals with organic compounds from their relative rates of reaction with hydrogen peroxide under photolysis in 1,1,2-trichlorotrifluoroethane solution. *Environ. Sci. Technol.* 22, 1447–1453.

Doháňyosová, P., Sarraute, S., Dohnal, V., Majer, V., Costa Gomes M. (2004) Aqueous solubility and related thermodynamic functions of nonaromatic hydrocarbons as a function of molecular structure. *Ind. Eng. Chem. Res.* 43, 2805–2815.

Donovan, S.F. (1996) New method for estimating vapor pressure by the use of gas chromatography. *J. Chromatogr.* A, 749, 123–129.

Doyle, G.J., Lloyd, A.C., Darnall, K.R., Winer, A.M., Pitts Jr., J.N. (1975) Gas phase kinetic study of relative rates of reaction of selected aromatic compounds with hydroxy radicals in environmental chamber. *Environ. Sci. Technol.* 9, 237–241.

Dreisbach, R.R. (1955) *Physical Properties of Chemical Compounds. Adv. Chem. Ser.* 15, American Chemical Society, Washington DC.

Dreisbach, R.R. (1959) *Physical Properties of Chemical Compounds II. Adv. Chem. Ser.* 22, American Chemical Society, Washington DC.

Dreisbach, R.R. (1961) *Physical Properties of Chemical Compounds—III*. Advances in Chemistry Series, American Chemical Society Applied Publications. American Chemical Society.

Droege, A.T., Tully, F.P. (1987) Hydrogen-atom abstraction from alkanes by OH. 6. Cyclopentane and cyclohexane. *J. Phys. Chem.* 91, 1222–1225.

Durand, R. (1948) Investigations on hydrotropy. The solubility of benzene, hexane and cyclohexane in aqueous solutions of fatty acid salts. *Compt. Rend.* 226, 409–410.

Eadsforth, C.V., Moser, P. (1983) Assessments of reversed phase chromatographic methods for determining partition coefficients. *Chemosphere* 12, 1459–1475.

Eastcott, L., Shiu, W.Y., Mackay, D. (1988) Environmentally relevant physical-chemical properties of hydrocarbons: A review of data and development of simple correlations. *Oil Chem. Pollut.* 4, 191–216.

Edney, E.O., Kleindienst, T.E., Corse, E.W. (1986) Room temperature rate constants for the reaction of OH with selected chlorinated and oxygenated hydrocarbons. *Int J. Chem. Kinet.* 18, 1355–1371.

Eisen, O., Drav, A.I. (1970) *Eesti. Nsv. Tead. Alad. Tdim. Keem. Geol.* 19, 202. - ref. see Boublik et al. 1984

Ewing, M.B., Sanchez Ochoa, J.C. (2000) The vapour pressure of cyclohexane over the whole fluid range determined using comparative ebulliometry. *J. Chem. Thermodyn.* 32, 1157–1167.

Ewing, M.B., Sanchez Ochoa, J.C. (2003) The vapour pressures of *n*-octane determined using comparative ebulliometry. *Fluid Phase Equil.* 210, 277–285.

Farkas, E.J. (1965) New method for determining hydrocarbon-in-water solubilities. *Anal. Chem.* 37, 1173–1175.

Fichan, I., Larroche, C., Gros, J.B. (1999) Water solubility, vapor pressure, and activity coefficients of terpenes and terpenoids. *J. Chem. Eng. Data* 44, 56–62.

Finke, H.L., Scott, D.W., Gross, M.E., Messerly, J.F., Waddington, G. (1956) Cycloheptane, cycloöctane and 1,3,5-cycloheptatriene. Low temperature thermal properties, vapor pressure and derived chemical thermodynamic properties. *J. Am. Chem. Soc.* 78, 5469–5476.

Flebbe, J.L., Barclay, D.A., Manley, D.B. (1982) Vapor pressures of some C_4 hydrocarbons and their mixtures. *J. Chem. Eng. Data* 27, 405–412.

Forziati, A.F., Norris, W.R., Rossini, F.D. (1949) Vapor pressures and boiling points of sixty API-NBS hydrocarbons. *J. Res. Natl. Bur. Std.* 43, 555–563.

Forziati, A.F., Camin, D.L., Rossini, F.D. (1950) Density, refractive index, boiling point, and vapor pressure of eight monoolefin (1-alkene), six pentadiene, and two cyclomonoolefin hydrocarbons. *J. Res. Natl. Bur. Std.* 45, 406–410.

Franks, F. (1966) Solute-water interactions and the solubility behaviour of long-chain paraffin hydrocarbons. *Nature* 210, 87–88.

Führer, H. (1924) Die wasserlöslichkeit in homologen reihen. *Chem. Ber.* 57, 510–515.

Furuyama, S., Atkinson, R., Colussi, A.J., Cvetanovic, R.J. (1974) Determination by the phase shift method of the absolute rate constants of reactions of oxygen (^{3}P) atoms with olefins at 25°C. *Int'l. J. Chem. Kinet.* 6, 741.

Gaffney, J.S., Levine, S.Z. (1979) Predicting gas phase organic molecule reaction rates using linear free-energy correlations. I. O(^{3}P) and OH addition and abstraction reactions. *Int. J. Chem. Kinet.* 11, 1197–1209.

Gardner, G. S., Brewer, J.E. (1937) Vapor pressure of commercial high-boiling organic solvents, *Ind. Eng. Chem.* 29, 179–181.

Góral, M., Mączyński, A., Wiœniewska-Goclowska, B. (2004) Recommended liquid-liquid equilibrium data. Part 2. Unsaturated hydrocarbon-water systems. *J. Phys. Chem. Ref. Data* 33, 579–591.

Goss, R.-U., Schwarzenbach, R.P. (1999) Empirical prediction of vaporization and heats of adsorption of organic compounds. *Environ. Sci. Technol.* 33, 3390–3393.

Gracia, M., Sánchez, Pérez, P., Valero, J., Getiérrez Losa, C. (1992) Vapour pressures of (butan-1-ol + hexane) at temperatures between 283.10 K and 323.12 K. *J. Chem. Thermodyn.* 24, 463–471.

Greiner, N.R. (1967) Hydroxyl-radical kinetics by kinetic spectroscopy. II. Reactions with C_2H_6, C_3H_8, and *iso*-C_4H_{10} at 300 K. *J. Chem. Phys.* 46, 3389–3392.

Greiner, N.R. (1970) Hydroxyl radical kinetics by kinetic spectroscopy. VI. Reactions with alkanes in the range 300–500 K. *J. Chem. Phys.* 53, 1070–1076.

Griffin, S., Grant Willie, S., Markham, J. (1999) Determination of octanol-water partition coefficient for terpenoids using reversed-phase high performance liquid chromatography. *J. Chromatog.* A, 864, 221–228.

Grosjean, D., Grosjean, E., Williams, II, E.L. (1994) Atmospheric chemistry of olefins: A product study of the ozone — alkene reaction with cyclohexane added to scavenge OH. *Environ. Sci. Technol.* 26, 186–196.

Groves, Jr, F.R. (1988) Solubility of cycloparaffins in distilled water and salt water. *J. Chem. Eng. Data* 33, 136–138.

Gruber, D., Langenheim, D., Gmehling, J. (1997) Measurement of activity coefficients at infinite dilution using gas-liquid chromatography. 6. Results for systems exhibiting gas-liquid interface adsorption with 1-octanol. *J. Chem. Eng. Data* 42, 882–885.

Güesten, H., Filby, W.G., Schoop, S. (1981) Prediction of hydroxyl radical reaction rates with organic compounds in the gas phase. *Atmos. Environ.* 15, 1763–1765.

Gutsche, B., Knapp, H. (1982) Isothermal measurements of vapor-liquid equilibria for three *n*-alkane-chloroalkane mixtures. *Fluid Phase Equil.* 8, 285–300.

Haag, W.R., Yao, C.C.D. (1992) Rate constants for reaction of hydroxyl radicals with several drinking water contaminants. *Environ. Sci. Technol.* 26, 1005–1013.

Hafkenscheid, T.L., Tomlinson, E. (1983) Correlations between alkane/water and octan-1-ol/water distribution coefficients and isocratic reversed-phase liquid chromatographic capacity factors of acids, bases and neutrals. *Int'l. J. Pharmaceu.* 16, 225–239.

Hansch C., Anderson, S. (1967) The effect of intramolecular hydrophobic bonding on partition coefficients. *J. Org. Chem.* 32, 2583–2586.

Hansch, C., Leo, A. (1979) *Substituent Constants for Correlation Analysis in Chemistry and Biology.* Wiley, New York.

Hansch, C., Leo, A. (1987) Medchem Project, Pomona College, Claremont, CA.

Hansch. C., Leo, A.J., Hoekman, D. (1995) *Exploring QSAR, Hydrophobic, Electronic, and Steric Constants.* ACS Professional Reference Book, American Chemical Society, Washington, DC.

Hansch, C., Quinlan, J.E., Lawrance, G.L. (1968) The linear free-energy relationship between partition coefficients and the aqueous solubility of organic liquids. *J. Org Chem.* 33, 345–350.

Hansen, K.C., Zhou, Z., Yaws, C.L., Aminabhavi, T.M.(1993) Determination of Henry's law constants of organics in dilute aqueous solutions. *J. Chem. Eng. Data* 38, 546–550.

Hanst, P.L., Stephens, E.R., Scott, W.E., Doerr, R.C. (1958) *Atmospheric Ozone-Olefin Reactions.* Franklin Institute, Philadelphia, Pa.

Harris, K.R., Dunlop, P.J. (1970) Vapor pressures and excess Gibbs energies of mixtures of benzene with chlorobenzene, *n*-hexane and *n*-heptane at 25°C. *J. Chem. Thermodyn.* 2, 801–811.

Harris, S.J., Kerr, J.A. (1988) Relative rate measurements of some reactions of hydroxyl radicals with alkanes studied under atmospheric conditions. *Int. J. Chem. Kinet.* 20, 939–955.

Hawkins, J.E., Armstrong, G.T. (1054) Physical and thermodynamic properties of terpenes. III. Vapor pressures of α-pinene and β-pinene. *J. Am. Chem. Soc.* 76, 3756–3758.

Heidman, J.L., Tsonopoulos, C., Brady, C.J., Wilson, G.M. (1985) High-temperature mutual solubilities of hydrocarbons and water. *AIChE J.* 31, 376–384.

Heisig, G.B. (1933) Action of radon on some unsaturated hydrocarbons. III. Vinylacetylene and butadiene. *J. Am. Chem. Soc.* 55, 2304–2311.

Hermann, R.B. (1972) Theory of hydrophobic bonding. II. The correlation of hydrocarbon solubility in water with solvent cavity surface area. *J. Phys. Chem.* 76, 2754–2758.

Herron, J.T., Huie, R.E. (1973) Rate constants for the reactions of atomic oxygen (^{3}P) with organic compounds in the gas phase. *J. Phys. Chem. Ref. Data* 2, 467–518.

Hine, J., Mookerjee, P.K. (1975) The intrinsic hydrophilic character of organic compounds. Correlations in terms of structural contributions. *J. Org. Chem.* 40(3), 292–298.

Horstmann, S., Wilken, M., Fischer, K., Gmehling, J. (2004) Isothermal vapor-liquid equilibrium and excess enthalpy data for the binary systems propylene oxide + 2-methylpentane and difluoromethane (R32) + pentafluoroethane (R125). *J. Chem. Eng. Data* 49,1504–1507.

Howard, P.H., Ed. (1989) *Handbook of Fate and Exposure Data for Organic Chemicals.* Vol. I - *Large Production and Priority Pollutants.* Lewis Publishers, Chelsea, Michigan.

Howard, P.H., Ed., (1990) *Handbook of Fate and Exposure Data for Organic Chemicals.* Vol. II - *Solvents.* Lewis Publishers, Inc., Chelsea, Michigan.

Howard, P.H., Boethling, R.S., Jarvis, W.F., Meylan, W.M., Michalenko, E.M. (1991) *Handbook of Environmental Degradation Rates.* Lewis Publishers, Chelsea, MI.

Huie, R.E., Herron, J.T. (1975) Temperature dependence of the rate constants for reaction of ozone with some olefins. *Int'l. J. Chem. Kinet.* S1, 165.

Hussam, A., Carr, P.W. (1985) Rapid and precise method for the measurement of vapor/liquid equilibria by headspace gas chromatography. *Anal. Chem.* 57, 793–801.

Hutchinson, T.C., Hellebust, J.A., Tam, D., Mackay, D., Mascarenhas, R.A., Shiu, W.Y. (1980) The correlation of the toxicity to algae of hydrocarbons and halogenated hydrocarbons with their physical-chemical properties. In: *Hydrocarbons and Halogenated Hydrocarbons in the Aquatic Environment.* Afghan, B.K., Mackay, D., Eds., pp. 577–586. Plenum Press, New York.

IUPAC Solubility Data Series (1989) Vol. 37: *Hydrocarbons (C_5-C_7) with Water and Seawater.* Shaw, D.G., Ed., Pergamon Press, Oxford, England.

Iwase, K., Komatsu, K., Hirono, S., Nakagawa, S., Moriguchi, I. (1985) Estimation of hydrophobicity based on the solvent-accessible surface area of molecules. *Chem. Pharm. Bull.* 33, 2114–2121.

Japar, S.M., Wu,. C.H., Niki, H. (1974) Rate constants for the reaction of ozone with olefins in the gas phase. *J. Phys. Chem.* 23, 2318–2320.

Jolly, G.S., Paraskevopoulos, G., Singleton, D.L. (1985) Rate of OH radical reactions. XII. The reactions of OH with c-C_3H_6, c-C_5H_{10}, and c-C_7H_{14}. Correlation of hydroxyl rate constants with bond dissociation energies. *Int. J. Chem. Kinet.* 17, 1–10.

Jönsson, J.Å., Vejrosta, J., Novak, J. (1982) Air/water partition coefficients for normal alkanes (n-pentane to n-nonane) *Fluid Phase Equil.* 9, 279–286.

Kabadi, V.N., Danner, R.P. (1979) Nomograph solves for solubilities of hydrocarbons in water. *Hydrocarbon Processing* 58, 245–246.

Kleindienst, T.E., Harris, G.W., Pitts, Jr., J.N. (1982) Rates and temperature dependences of the reaction of hydroxyl radical with isoprene, its oxidation products, and selected terpenes. *Environ. Sci. Technol.* 16, 844–846.

Kolb, B., Welter, C., Bichler, C. (1992) Determination of partition coefficients by automatic equilibrium headspace gas chromatography by vapor phase calibration. *Chromatographia* 34, 235–240.

Kopcynski, S.L., Lonneman. W.A., Sutterfield, F.D., Darley, P.E. (1972) Photochemistry of atmospheric samples in Los Angeles. *Environ. Sci. Technol.* 6, 342–347.

Krafft, F. (1982) Ber. 15, 1687. — reference from Boublik et al. 1984.

Krasnoshchekova, R.Ya., Gubertritis, M.Ya. (1973) Solubility of paraffin hydrocarbons in fresh and saltwater. *Neftekhimiya* 13, 885–887.

Krasnoshchekova, R.Ya., Gubertritis, M.Ya. (1975) Solubility of alkylbenzenes in fresh and salt waters. *Vodnye. Resursy.* 2, 170–173.

Kretschmer, C.B., Nowakowska, J., Wiebe, R. (1948) Densities and liquid-vapor equilibria of the system ethanol-isooctane (2,2,4-trimethylpentane) between 0 and 50°. *J. Am. Chem. Soc.* 70, 1785–1790.

Krzyzanowska, T., Szeliga, J. (1978) A method for determining the solubility of individual hydrocarbons. *Nafta* (Katowice) 28, 414–417.

Kudchadker, A.P., Zwolinski, B.J. (1966) Vapor pressures and boiling points of normal alkanes, C_{21} to C_{100}. *J. Chem. Eng. Data* 11, 253–255.

Kühne, R., Ebert, R.-U., Kleint, F., Schmidt, G., Schüürmann, G. (1995) Group contribution methods to estimate water solubility of organic chemicals. *Chemosphere* 30, 2061–2077.

Lamb, A.B., Roper, E.E. (1940) The vapor pressures of certain unsaturated hydrocarbons. *J. Am. Chem. Soc.* 62, 806–814.

Lande, S.S., Banerjee, S. (1981) Predicting aqueous solubility of organic nonelectrolytes from molar volume. *Chemosphere* 10, 751–759.

Lande, S.S., Hagen, D.F., Seaver, A.E. (1985) Computation of total molecular surface area from gas phase ion mobility data and its correlation with aqueous solubilities of hydrocarbons. *Environ. Toxicol. Chem.* 4, 325–334.

Leahy, D.E. (1986) Intrinsic molecular volume as a measure of the cavity term in linear solvation energy relationships: octanol-water partition coefficients and aqueous solubilities. *J. Pharm. Sci.* 75, 629–636.

Leinonen, P.J., Mackay, D. (1973) The multicomponent solubility of hydrocarbons in water. *Can. J. Chem. Eng.* 51, 230–233.

Leo, A., Hansch, C., Elkins, D. (1971) Partition coefficients and their uses. *Chem. Rev.* 71, 525–616.

Leo, A., Jow, P.Y.C., Silipo, C., Hansch, C. (1975) Calculation of hydrophobic constant (Log P) from π and f constants. *J. Med. Chem.* 18(9), 865–868.

Letcher, T.M., Marsicano, F. (1974) Vapour pressures and densities of some unsaturated C_6 acyclic and cyclic hydrocarbons between 300 and 320 K. *J. Chem. Thermodyn.* 6, 509–514.

Lide, D.R., Editor (2003) *Handbook of Chemistry and Physics*. 84th ed., CRC Press, LLC. Boca Raton, Florida.

Linder, E.G. (1931) Vapor pressures of some hydrocarbons. *J. Phys. Chem.* 35, 531–535.

Lister, M.W. (1941) Heats of organic reactions. X. Heats of bromination of cyclic olefins. *J. Am. Chem. Sco.* 63, 143–149.

Lloyd, A.C., Darnall, K.R., Winer, A.M., Pitts, Jr., J.N. (1976) Relative rate constants for reaction of the hydroxyl radical with a series of alkanes, alkenes, and aromatic hydrocarbons. *J. Phys. Chem.* 80, 189–794.

Lyman, W.J. (1982) Adsorption coefficients for soil and sediments. Chapter 4, In: *Handbook of Chemical Property Estimation Methods*, W.J. Lyman, W.F. Reehl, D.H. Rosenblatt, Eds., McGraw-Hill, New York.

Lyman, W.J., Reehl, W.F., Rosenblatt, D.H. (1982) *Handbook of Chemical Property Estimation Methods*, McGraw-Hill, New York.

Mackay, D. (1981) Environmental and laboratory rates of volatilization of toxic chemicals from water. In: *Hazardous Assessment of Chemicals, Current Development*. Volume 1, Academic Press.

Mackay, D. (1982) Correlation of bioconcentration factors. *Environ. Sci. Technol.* 16, 274–278.

Mackay, D., Bobra, A.M., Chan, D.W., Shiu, W.Y. (1982) Vapor pressure correlation for low-volatility environmental chemicals. *Environ. Sci. Technol.* 16, 645–649.

Mackay, D., Leinonen, P.J. (1975) Rate of evaporation of low-solubility contaminants from water bodies to atmosphere. *Environ. Sci. Technol.* 7, 1178–1180.

Mackay, D., Shiu, W.Y. (1975) The aqueous solubility and air-water exchange characteristics of hydrocarbons under environmental conditions. In: *Chemistry and Physics of Aqueous Gas Solutions*. Adams, W.A., Greer, G., Desnoyers, J.E., Atkinson, G., Kell, K.B., Oldham, K.B., Walkey, J., Eds., pp. 93–110, Electrochem. Soc. , Inc., Princeton, N.J.

Mackay, D., Shiu, W.Y. (1981) A critical review of Henry's law constants for chemicals of environmental interest. *J. Phys. Chem. Ref. Data* 10, 1175–1199.

Mackay, D., Shiu, W.Y. (1990) Physical-chemical properties and fate of volatile organic compounds: an application of the fugacity approach. In: *Significance and Treatment of Volatile Organic Compounds in Water Supplies*. Ram, N.M., Christman, R.F., Cantor, K.P., Eds., pp. 183–203, Lewis Publishers, Chelsea, Michigan.

Mackay, D., Shiu, W.Y., Sutherland, R.P. (1979) Determination of air-water Henry's law constants for hydrophobic pollutants. *Environ. Sci. Technol.* 13, 333–337.

Mackay, D., Shiu, W.Y., Wolkoff, A.W. (1975) Gas chromatographic determination of low concentrations of hydrocarbons in water by vapor phase extraction. *ASTM STP* 573, pp. 251–258, Am. Soc. Testing and Materials, Philadelphia, Pennsylvania.

Macknick, A.B., Prausnitz, J.M. (1979) Vapor pressures of high-molecular-weight hydrocarbons. *J. Chem. Eng. Data* 24, 175–178.

Mączyński, A., Wiœniewska-Goclowska, B., Góral, M. (2004) Recommended liquid-liquid equilibrium data. Part 1. Binary alkane-water systems. *J. Phys. Chem. Ref. Data* 33, 549–577.

Marche, C., Ferronato, C., Jose, J. (2003) Solubilities of n-alkanes (C_6 to C_8) in water from 30°C to 180°C. *J. Chem. Eng. Data* 48, 967–971.

Marche, C., Ferronato, C., Jose, J. (2004) Solubilities of alkylcyclohexanes in water from 30°C to 180°C. *J. Chem. Eng. Data* 49, 937–940.

Massaldi, H.A., King, C.J. (1973) Simple technique to determine solubilities of sparingly soluble organics: solubility and activity coefficients of d-limonene, butylbenzene, and n-hexyl acetate in water and sucrose solutions. *J. Chem. Eng. Data* 18, 393–397.

McAuliffe, C. (1963) Solubility in water of C_1 - C_9 hydrocarbons. *Nature* (London) 200, 1092–1093.

McAuliffe, C. (1966) Solubility in water of paraffin, cycloparaffin, olefin, acetylene, cycloolefin and aromatic hydrocarbons. *J. Phys. Chem.* 76, 1267–1275.

McAuliffe, C. (1969) Solubility in water of normal C_9 and C_{10} alkane hydrocarbons. *Science* 163, 478–479.

McBain, J.W., Lissant, K.J. (1951) The solubilization of four typical hydrocarbons in aqueous solution by three typical detergents. *J. Phys. Colloid Chem.* 55, 655–662.

Messerly, G.H., Kennedy, R.M. (1940) The heat capacity and entropy, heats of fusion and vaporization and the vapor pressure of n-pentane. *J. Am. Chem. Soc.* 62, 2988–2991.

Meyer, E.F., Hotz, R.D. (1973) High-precision vapor-pressure data for eight organic compounds. *J. Chem. Eng. Data* 18, 359–362.

Meyer, E.F., Hotz, R.D. (1976) Cohesive energies in polar organic liquids. 3. Cyclic ketones. *J. Chem. Eng. Data* 21, 274–279.

Meylan, W.M., Howard, P.H., Boethling, R.S. (1992) Molecular topograph/fragment contribution method for predicting soil sorption coefficients. *Environ. Sci. Technol.* 26, 1560–1567.

Mill, T., Mabey, W. (1985) Photochemical transformations. In: *Environmental Exposure from Chemicals*. Vol. I, Neely, W.B., Blau, G.E., Eds., Chap. 8, pp. 175–216. CRC Press, Boca Raton, Florida.

Miller, M.M., Wasik, S.P., Huang, G.L., Shiu, W.Y., Mackay, D. (1985) Relationships between octanol-water partition coefficient and aqueous solubility. *Environ. Sci. Technol.* 19, 522–529.

Mills, W.B., Dean, J.D., Porcella, D.B., Gherini, S.A., Hudson, R.J.M., Frick, W.E., Rupp, G.L., Bowie, G.L. (1982) Water Quality Assessment: A Screening Procedure for Toxic and Conventional Pollutants. Part 1, U.S. EPA, EPA-600/6–82–004a.

Mokbel, I., Rauzy, E., Loiseleur, H., Berro, C., Jose, J. (1995) Vapor pressures of 12 alkylcyclohexanes, cyclopentane, butylcyclopentane and *trans*-decahydronaphthalene down to 0.5 Pa. Experimental results, correlation and prediction by an equation of state. *Fluid Phase Equil.* 108, 103–120.

Morgan, D.L., Kobayashi, R. (1994) Direct vapor pressure measurements of 10 n-alkanes in the C_{10} - C_{28} range. *Fluid Phase Equil.* 97, 211–242.

Morris, Jr., E.D., Niki, H. (1971) Reactivity of hydroxyl radicals with olefins. *J. Phys. Chem.* 75, 3640–3641.

Morrison, T.J., Bilett, F. (1952) The salting out of non-electrolytes. Part II. The effect of variation in non-electrolyte. *J. Chem. Soc.* 3819–3822.

Müller, M., Klein, W. (1991) Estimating atmosphere degradation processes by SARs. *Sci. Total Environ.* 109/110, 261–273.

Müller, M., Klein, W. (1992) Comparative evaluation of methods predicting water solubility for organic compounds. *Chemosphere* 25, 769–782.

Murray, W.J., Hall, L.H., Kier, LB. (1975) Molecular connectivity III: Relationship to partition coefficients. *J. Pharm. Sci.* 64, 1978–1981.

Myers, H.S., Fenske, M.R. (1955) Measurement and correlation of vapor pressure data for high boiling hydrocarbons. *Ind. Eng. Chem.* 47, 1652–1658.

Myrdal, P., Ward, G.H., Dannenfelser, R.-M., Mishra, D., Yalkowsky, S.H. (1992) AQUAFAC 1: Aqueous functional group activity coefficients; application to hydrocarbons. *Chemosphere* 24, 1047–1061.

Natarajan, G.S., Venkatachalam, K.A. (1972) Solubilities of some olefins in aqueous solutions. *J. Chem. Eng. Data* 17, 328–329.

Nelson, H.D., De Ligny, C.L. (1968) The determination of the solubilities of some n-alkanes in water at different temperatures by means of gas chromatography. *Rec. Trav. Chim. Payus-Bae* (Recueil) 87, 528–544.

Nicolini, E., Laffitte, P. (1949) Vapor pressure of some pure organic liquids. *Comept. Rend.* 229, 757–759.

Nirmalakhandan, N.N., Speece, R.E. (1988a) Prediction of aqueous solubility of organic chemicals based on molecular structure. *Environ. Sci. Technol.* 22, 328–338.

Nirmalakhandan, N.N., Speece, R.E. (1988b) QSAR model for predicting Henry's law constant. *Environ. Sci. Technol.* 22, 1349–1357.

Nolting, F., Behnke, W., Zetzsch C. (1988) A smog chamber for studies of the reactions of terpenes and alkanes with ozone and OH. *J. Atmos. Chem.* 6, 47–59.

Ohta, T. (1983) Rate constants for the reactions of OH radicals with alkyl substituted olefins. *Int. J. Chem. Kinet.* 16, 879–886.
Ohta, T. (1984) Rate constants for the reactions of diolefins with OH radicals in the gas phase. Estimate of the rate constants from those for monoolefins. *J. Phys. Chem.* 87, 1209–1213.
Oliveira, J.V., Uller, A.M.C. (1996) Solubility of pure 1,3-butadiene and methyl propene and their mextures ain pure *n*-methyl-2-pyrrolidone and in its aqueous solutions. *Fluid Phase Equil.* 118, 133–141.
Osborn, A.G., Douslin, D.R. (1969) Vapor pressure relations for the seven pentadienes. *J. Chem. Eng. Data* 14, 208–209.
Osborn, A.G., Douslin, D.R. (1974) Vapor pressure relations for 15 hydrocarbons. *J. Chem. Eng. Data* 19, 114–117.
Parks, G.S., Huffman, H.M. (1931) Some fusion and transition data for hydrocarbons. *Ind. Eng. Chem.* 23, 1138–1139.
Pasanen, M., Uusi-Kyyny, P., Pokki, J.-P., Pakkanen, M., Aittamaa, J. (2004) Vapor-liquid equilibrium for 1-propanol + 1-butene, + *cis*-2-butene, + 2-methyl-1-propene, + *trans*-2-butene, + *n*-butane, and + 2-methyl-propane. *J. Chem. Eng. Data* 49, 1628–1634.
Pasek, G.J., Thodos, G. (1962) Vapor pressures of naphthenic hydrocarbons. *J. Chem. Eng. Data* 7, 21–26.
Passino, D.R.M., Smith, S.B. (1987) Quantitative structure-activity relationships (QSAR) and toxicity data in hazard assessment. In: *QSAR in Environmental Toxicology-II.* Kaiser, K.L.E., Editor, D. Reidel Publishing Co., Dordrecht, Holland. pp. 261–270.
Paulson, S.E., Seinfeld, J.H. (1992) Atmospheric photochemical oxidation of 1-octene: OH, O_3, $O(^3P)$ reactions. *Environ. Sci. Technol.* 26, 1165–1173.
Pavlova, S.P., Pavlov, S.Yu., Serafimov, L.A., Kofman, L.S. (1966) Mutual solubility of C_5 hydrocarbons and water. *Promyshlennost. Sinteticheskogo Kouchuka* 3, 18–20.
Peeters, J., Vandenberk, S., Piessens, E, Pultan, V. (1999) H-atom abstraction in reactions of cyclic polyalkenes with OH. *Chemosphere* 38, 1189–1193.
Piacente, V., Fontana, D., Scardala, P. (1994) Enthalpies of vaporization of a homologous series of *n*-alkanes determined from vapor pressure measurements. *J. Chem. Eng. Data* 39, 231–237.
Piacente, V., Pompili, T., Scardala, P., Ferro, D. (1991) Temperature dependence of the vaporization enthalpies of *n*-alkanes from vapour-pressure measurements. *J. Chem. Thermodyn.* 23, 379–396.
Piacente, V., Scardala, P. (1990) Vaporization enthalpies and entropies of some *n*-alkanes. *Thermochim. Acta* 159, 193–200.
Pickett, O.A., Peterson, J.M. (1929) Terpenes and terpene alcohols. I.- Vapor pressure-temperature relationship. *Ind. Eng. Chem.* 21, 325–326.
Pierotti, R.A., Liabastre, A.A. (1972) Structure and Properties of Water Solutions. U.S. Natl. Tech. Inform. Ser., PB rep. No. 21163, 113 pp.
Platford, R.F. (1979) Glyceryl trioleate-water partition coefficients for three simple organic compounds. *Bull. Environ. Contam. Toxicol.* 21, 68.
Platford, R.F. (1983) The octanol-water partitioning of some hydrophobic and hydrophilic compounds. *Chemosphere* 12(7/8), 1107–1111.
Polak, J., Lu, B.C.Y. (1973) Mutual solubilities of hydrocarbons and water at 0 and 25°C. *Can. J. Chem.* 51, 4018–4023.
Pompili, T., Piacente, V. (1990) Enthalpy of vaporization of n-heptacosane and n-nonacosane from their vapour pressure determinations. *Thermochim. Acta* 170, 289–291.
Price, L.C. (1976) Aqueous solubility of petroleum as applied to its origin and primary migration. *Am. Assoc. Petrol. Geol. Bull.* 60, 213–244.
Ravishankara, A.R., Wagner, S., Fischer, S., Smith, G., Schiff, R., Watson, R.T., Tesi, G., Davis, D.D. (1978) A kinetics study of the reactions of OH with several aromatic and olefinic compounds. *Int. J. Chem. Kinet.* Vol. X, 783–804.
Reich, R., Sanhueza, V. (1993) Vapor-liquid equilibria for α-pinene or β-pinene with anisole. *J. Chem. Eng. Data* 38, 341–343.
Rekker, R.F. (1977) *The Hydrophobic Fragmental Constants. Its Derivation and Application, A Means of Characterizing Membrane Systems.* Elsevier Sci. Publ. Co., Oxford, England.
Rhodes, J.M., Bhethanabotla, V.R., Campbell, S.C. (1997) Total vapor pressure measurements for heptane + 1-pentanol, + 2-pentanol, + 3-pentanol, + 2-methyl-1-butanol, + 2-methyl-2-butanol, + 3-methyl-1-butanol, and + 3-methyl-2-butanol at 313.15 K. *J. Chem. Eng. Data* 42, 731–734.
Riddick, J.A., Bunger, W.B., Sakano, T.K. (1986) *Organic Solvents.* Wiley Interscience, New York.
Rogers, J.D. (1989) Rate constant measurements for the reaction of the hydroxyl radical with cyclohexene, cyclopentene, and glutaraldehyde. *Environ. Sci. Technol.* 23, 177–181.
Rudakov, E.S., Lutsyk, A.I. (1979) Solubility of saturated hydrocarbons in aqueous sulfuric acid. *Zh. Fiz. Khim.* 53, 1298–1300.
Ruelle, P., Kesselring, U.W. (1997) Aqueous solubility prediction of environmentally important chemicals from the mobile order thermodynamics. *Chemosphere* 34, 273–298.
Ruzicka, K., Majer, V. (1994) Simultaneous treatment of vapor pressures and related thermal data between the triple and normal boiling temperatures for *n*-alkanes C_5 - C_{20}. *J. Phys. Chem. Ref. Data* 23, 1–39.
Ryu, S.-A., Park, S.-J. (1999) A rapid determination method of the air/water partition coefficient and its application. *Fluid Phase Equil.* 161, 295–304.
Sabljic, A., Güsten, H. (1990) Predicting the night-time NO_3 radical reactivity in the troposphere. *Atmos. Environ.* 24A, 73–78.
Sangster, J. (1989) Octanol-water partition coefficients of simple organic compounds. *J. Phys. Chem. Ref. Data* 18, 1111–1230.
Sangster, J. (1993) LOGKOW A Databank of Evaluated Octanol-Water Partition Coefficients. 1st ed., Montreal, Quebec, Canada.

Sarraute, S., Delepine, H., Costa Gomes, M.F., Majer, V.(2004) Aqueous solubility, Henry's law constants and air/water partition coefficients of *n*-octane and two halogenated octanes. *Chemosphere* 57, 1543–1551.

Sasse, K., Jose, J., Merlin, J.-C. (1988) A static apparatus for measurement of low vapor pressures. Experimental results on high molecular-weight hydrocarbons. *Fluid Phase Equil.* 42, 287–304.

Scatchard, G., Wood, S.E., Mochel, J.M. (1939) Vapor-liquid equilibrium. IV. Carbon tetrachloride-cyclohexane mixtures. *J. Am. Chem. Soc.* 61, 3206–3210.

Schantz, M.M., Martire, D.E. (1987) Determination of hydrocarbon-water partition coefficients from chromatographic data and based on solution thermodynamics and theory. *J. Chromatogr.* 391, 35–51.

Schubert, C.C., Pease, R.N. (1956) The oxidation of lower paraffin hydrocarbons. I. Room temperature reaction of methane, propane, n-butane and isobutane with ozonized oxygen. *J. Am. Chem. Soc.* 78, 2044–2048.

Schumann, S.C., Aston, J.S., Sagenkahn, M. (1942) The heat capacity and entropy, heats of fusion and vaporization and the vapor pressures of isopentane. *J. Am. Chem. Soc.* 64, 1039–1043.

Schwarz, F.P. (1980) Measurement of the solubilities of slightly soluble organic liquids in water by elution chromatography. *Anal. Chem.* 52, 10–15.

Scott, D.E., Waddington, G. (1950) Vapor pressure of *cis*-pentene, *trans*-2-pentene and 3-methyl-1-butene. *J. Am. Chem. Soc.* 72, 4310–4311.

Scott, D.W., Waddington, G., Smith, J.C., Huffman, H.M. (1949) Thermodynamic properties of three isomeric pentenes. *J. Am. Chem. Soc.* 71, 2767–2773.

Setti, L., Lanzarini, G., Pifferi, P.G., Spagna, G. (1993) Further research into the aerobic degradation of n-alkanes in a heavy oil by a pure culture of a *Pseudomonas* sp. *Chemosphere* 26(6), 1151–1157.

Seyer, Wm.F., Mann, C.W. (1945) The vapor pressures of *cis*- and *trans*-decahydronaphthalene. *J. Am. Chem. Soc.* 67, 328–329.

Shaw, D.G., Ed. (1989a) IUPAC Solubility Data Series Vol. 37: *Hydrocarbons (C_5-C_7) with Water and Seawater*. Pergamon Press, Oxford, England.

Shaw, D.G., Ed. (1989b) IUPAC Solubility Data Series Vol. 38: *Hydrocarbons (C_8-C_{36}) with Water and Seawater*. Pergamon Press, Oxford, England.

Singleton, D.L., Cvetanovic, R.J. (1976) Temperature dependence of oxygen atoms with olefins. *J. Am. Chem. Soc.* 98, 6812–6819.

Smyth, C.P., Engel, E.W. (1929) Molecular orientation and the partial vapor pressures of binary mixtures. I. Systems compounds of normal liquids. *J. Am. Chem. Soc.* 51, 2646–2660.

Staudinger, J., Roberts, P.V. (1996) A critical review of Henry's law constants for environmental applications. *Crit. Rev. Environ. Sci. Technol.* 26, 205–297.

Staudinger, J., Roberts, P.V. (2001) A critical compilation of Henry's law constant temperature dependence relations for organic compounds in dilute aqueous solutions. *Chemosphere* 44, 561–576.

Stedman, D.H., Wu, C.H., Niki, H. (1973) Kinetics of gas-phase reactions of ozone with some olefins. *J. Phys. Chem.* 77, 2511–2514.

Steele, K., Poling, B.E., Manley, D.B. (1976) Vapor pressures for the system 1-butene, isobutane, and 1,3-butadiene. *J. Chem. Eng. Data* 21, 399–403.

Stephenson, R.M., Malanowski, S. (1987) *Handbook of the Thermodynamic of Organic Compounds*. Elsevier Science New York, N.Y.

Stuckey, J.M., Saylor, J.H. (1940) The vapor pressures of some organic compounds. I. *J. Am. Chem. Soc.* 62, 2922–2925.

Stull, D.R. (1947) Vapor pressure of pure substances organic compounds. *Ind. Eng. Chem.* 39(4), 517–560.

Sutton, C., Calder, J.A. (1974) Solubility of higher-molecular-weight-paraffins in distilled water and seawater. *Environ. Sci. Technol.* 8, 654–657.

Tewari, Y.B., Martire, D.E., Wasik, S.P., Miller, M.M. (1982a) Aqueous solubilities and octanol-water partition coefficients of binary liquid mixtures of organic compounds at 25°C. *J. Solution Chem.* 11, 435–445.

Tewari, Y.B., Miller, M.M., Wasik, S.P. (1982b) Calculation of aqueous solubilities of organic compounds. *NBS J. Res.* 87, 155–158.

Tewari, Y.B., Miller, M.M., Wasik, S.P., Martire, D.E. (1982c) Aqueous solubility and octanol/water partition coefficient of organic compounds at 25.0°C. *J. Chem. Eng. Data* 27, 451–454.

Thoms, S.R., Lion, L.W. (1992) Vapor-phase partitioning of volatile organic compounds: a regression approach. *Environ. Toxicol. Chem.* 11, 1377–1388.

Tse, G., Sandler, S.I. (1994) Determination of infinite dilution activity coefficients and 1-octanol/water partition coefficients of volatile organic pollutants. *J. Chem. Eng. Data* 39, 354–357.

Tsonopoulos, C. (1999) Thermodynamic analysis of the mutual solubilities of normal alkanes and water. *Fluid Phase Equil.* 156, 21–33.

Tsonopoulos, C. (2001) Thermodynamic analysis of the mutual solubilities of hydrocarbons and water. *Fluid Phase Equil.* 186, 185–206.

Tsonopoulos, C., Prausnitz, J.M. (1971) Activity coefficients of aromatic solutes in dilute aqueous solutions. *I & EC Fundam.* 593–600.

Tsonopoulos, C., Wilson, G.M. (1983) High-temperature mutual solubilities of hydrocarbons and water. Part I: Benzene, cyclohexane and *n*-hexane. *AIChE Journal* 29, 990–999.

Valsaraj, K.T. (1988) On the physico-chemical aspects of partitioning of non-polar hydrophobic organics at the air-water interface. *Chemosphere* 17, 875–887.

Valvani, S.C., Yalkowsky, S.H., Roseman, T.J. (1981) Solubility and partitioning IV. Aqueous solubility and octanol-water partition coefficient of liquid electrolytes. *J. Pharm. Sci.* 70, 502–507.

Van der Linden, A.C. (1978) Degradation of oil in the marine environment. *Dev. Biodegrad. Hydrocarbons* 1, 165–200.

Varushchenko et al. (1970) *Zh. Phys. Khim.* 40, 3022. — reference from Boublik et al. 1984.

Vejrosta, J., Novák, J., Jönsson, J.Å. (1982) A method for measuring infinite-dilution partition coefficients of volatile compounds between the gas and liquid phases of aqueous systems. *Fluid Phase Equil.* 8, 25–35.

Verevkin, S.P., Wandschneider, D., Heintz, A. (2000) Determination of vaporization enthalpies of selected linear and branched C_7, C_8, C_9, C_{11} and C_{12} monoolefin hydrocarbons from transpiration and correlation gas-chromatography methods. *J. Chem. Eng. Data* 45, 618–625.

Verschueren, K. (1983) *Handbook of Environmental Data on Organic Chemicals.* 2nd ed., Van Nostrand Reinhold, New York.

Wakeham, S.G., Davis, A.C., Karas, J.L. (1983) Microcosm experiments to determine the fate and persistence of volatile organic compounds in coastal seawater. *Environ. Sci. Technol.* 17, 611–617.

Wakita, K., Yoshimoto, M., Miyamoto, S., Watanabe, H. (1986) A method for calculation of the aqueous solubility of organic compounds by using new fragment solubility constants. *Chem. Pharm. Bull.* 34, 4663–4681.

Wang, L., Zhao, Y., Hong, G. (1992) Predicting aqueous solubility and octanol/water partition coefficients of organic chemicals from molar volume. *Environ. Chem.* 11, 55–70.

Washburn, E.R., Handorf, B.H. (1935) The vapor pressure of binary solutions of ethyl alcohol and cyclohexane at 25°C. *J. Am. Chem. Soc.* 57, 441–443.

Wasik, S.P., Tewari, Y.B., Miller, M.M., Martire, D.E. (1981) *Octanol/Water Partition Coefficients and Aqueous Solubilities of Organic Compounds.* NBSIR 81–2406, report prepared for Office of Toxic Substances, Environmental Protection Agency, Washington, DC.

Wasik, S.P., Miller, M.M., Tewari, Y.B., May, W.E., Sonnefeld, W.J., DeVoe, H., Zoller, W.H. (1983) Determination of the vapor pressure, aqueous solubility, and octanol/water partition coefficient of hydrophobic substances by coupled generator column/liquid chromatographic methods. *Residue Rev.* 85, 29–42.

Wasik, S.P., Tewari, Y.B., Miller, M.M. (1982) Measurements of octanol/water partition coefficient by chromatographic method. *J. Res. Natl. Bur. Std.* 87, 311–315.

Weast, R.C., Ed. (1972–73) *Handbook of Chemistry and Physics,* 53th ed. CRC Press, Cleveland.

Weast, R.C. (1983–84) *Handbook of Chemistry and Physics*, 64th ed., CRC Press, Florida.

Weber, L.A. (2000) Vapor pressure of heptane from the triple point to the critical point. *J. Chem. Eng. Data* 45, 173–176.

Willingham, C.B., Taylor, W.J., Pignocco, J.M., Rossini, F.D. (1945) Vapor pressure and boiling points of some paraffin, alkylcyclopentane, alkylcyclohexane, and alkylbenzene hydrocarbons. *J. Res. Natl. Bur. Std.* 34, 219–244.

Winer, A.M., Atkinson, R., Pitts, J.N., Jr. (1984) Gaseous nitrate radical: possible nighttime atmospheric sink for biogenic organic compounds. *Science* 224, 156–159.

Winer, A.M., Darnall, K.R., Atkinson, R. Pitts, Jr., J.N. (1979) Smog chamber study of the correlation of hydroxyl radical rate constants with ozone formation. *Environ. Sci. Technol.* 7, 622–626.

Winer, A.M., Lloyd, A.C., Darnall, KR., Pitts, Jr., J.N. (1976) Relative rate constants for the reaction of the hydroxyl radical with selected ketones, chloroethenes, and monoterpene hydrocarbons. *J. Phys. Chem.* 80, 1635-1639.

Yalkowsky, S.H., Morozowich, W. (1980) A physical chemical basis for the design of orally active prodrugs. In: *Drug Design*, Vol IX, pp. 121–185, Academic Press, New York.

Yaws, C.L. (1994) *Handbook of Vapor Pressure.* Volume 1: C_1 to C_4 Compounds, Volume 2: C_5 to C_7 Compounds. Volume 3: C_5 to C_{28} Compounds. Gulf Publishing Co., Houston, Texas.

Zafonte, L., Bonamassa, F. (1977) Relative photochemical reactivity of propane and *n*-butane. *Environ. Sci. Technol.* 11, 1015–1017.

Zhang, H.Z., Li, Y.Q., Xia, J.R., Davidovits, Williams, L.R., Jayne, J.T., Kolb, C.E., Worsnop, D.R. (2003) Uptake of gas-phase species by 1-octanol. 1. Uptake of α-pinene, γ-terpinene, *p*-cymene, and 2-methyl-2-hexanol as a function of relative humidity and temperature. *J. Phys. Chem. A* 107, 6388–6397.

Zoeteman, B.C.J., Harmsen, K., Linders, J.B.H. (1980) Persistent organic pollutants in river water and ground water of the Netherlands. *Chemosphere* 9, 231–249.

Zwolinski, B.J., Wilhoit, R.C. (1971) *Handbook of Vapor Pressures and Heats of Vaporization of Hydrocarbons and Related Compounds.* API-44 TRC Publication No. 101, Texas A. & M. University, Evans Press, Fort Worth, Texas.

3 Mononuclear Aromatic Hydrocarbons

CONTENTS

3.1 List of Chemicals and Data Compilations 407
    3.1.1 Mononuclear aromatic hydrocarbons 407
        3.1.1.1 Benzene 407
        3.1.1.2 Toluene 425
        3.1.1.3 Ethylbenzene 439
        3.1.1.4 *o*-Xylene 450
        3.1.1.5 *m*-Xylene 459
        3.1.1.6 *p*-Xylene 467
        3.1.1.7 1,2,3-Trimethylbenzene 476
        3.1.1.8 1,2,4-Trimethylbenzene 481
        3.1.1.9 1,3,5-Trimethylbenzene 486
        3.1.1.10 *n*-Propylbenzene 493
        3.1.1.11 Isopropylbenzene 500
        3.1.1.12 1-Ethyl-2-methylbenzene (*o*-Ethyltoluene) 505
        3.1.1.13 1-Ethyl-3-methylbenzene (*m*-Ethyltoluene) 508
        3.1.1.14 1-Ethyl-4-methylbenzene (*p*-Ethyltoluene) 512
        3.1.1.15 1-Isopropyl-4-methylbenzene (*p*-Cymene) 516
        3.1.1.16 *n*-Butylbenzene 520
        3.1.1.17 Isobutylbenzene 525
        3.1.1.18 *sec*-Butylbenzene 528
        3.1.1.19 *tert*-Butylbenzene 532
        3.1.1.20 1,2,3,4-Tetramethylbenzene 536
        3.1.1.21 1,2,3,5-Tetramethylbenzene 539
        3.1.1.22 1,2,4,5-Tetramethylbenzene 542
        3.1.1.23 Pentamethylbenzene 545
        3.1.1.24 Pentylbenzene 547
        3.1.1.25 Hexamethylbenzene 550
        3.1.1.26 *n*-Hexylbenzene 553
        3.1.1.27 Heptylbenzene 557
        3.1.1.28 *n*-Octylbenzene 559
        3.1.1.29 Nonylbenzene 562
        3.1.1.30 Decylbenzene 564
        3.1.1.31 Undecylbenzene 567
        3.1.1.32 Dodecylbenzene 569
        3.1.1.33 Tridecylbenzene 572
        3.1.1.34 Tetradecylbenzene 574
        3.1.1.35 Styrene 576
        3.1.1.36 α-Methylstyrene 582
        3.1.1.37 β-Methylstyrene 584
        3.1.1.38 *o*-Methylstyrene 586
        3.1.1.39 *m*-Methylstyrene 588

        3.1.1.40   *p*-Methylstyrene 591
        3.1.1.41   Tetralin 594
3.2   Summary Tables and QSPR Plots 598
3.3   References 605

3.1 LIST OF CHEMICALS AND DATA COMPILATIONS

3.1.1 MONONUCLEAR AROMATIC HYDROCARBONS

3.1.1.1 Benzene

Common Name: Benzene
Synonym: benzol, cyclohexatriene
Chemical Name: benzene
CAS Registry No: 71-43-2
Molecular Formula: C_6H_6
Molecular Weight: 78.112
Melting Point (°C):
    5.49   (Lide 2003)
Boiling Point (°C):
    80.09   (Lide 2003)
Density (g/cm³ at 20°C):
    0.8765   (Weast 1982–1983)
Molar Volume (cm³/mol):
    89.1   (20°C, calculated from density)
    96.0   (calculated-Le Bas method at normal boiling point)
Enthalpy of Vaporization, ΔH_V (kJ/mol):
    33.843, 30.726   (25°C, bp, Riddick et al. 1986)
Enthalpy of Fusion ΔH_{fus} (kJ/mol):
    9.916   (Tsonopoulos & Prausnitz 1971)
    9.866   (Riddick et al. 1986)
    9.87   (exptl., Chickos et al. 1999)
Entropy of Fusion ΔS_{fus} (J/mol K):
    35.564   (Tsonopoulos & Prausnitz 1971)
    35.4, 44.5   (exptl., calculated-group additivity method, Chickos et al. 1999)
Fugacity Ratio at 25°C, F: 1.0

Water Solubility (g/m³ or mg/L at 25°C or as indicated and reported temperature dependence equations. Additional data at other temperatures designated * are compiled at the end of this section):
    1850   (30°C, shake flask-interferometer, Gross & Saylor 1931)
    1786   (shake flask-turbidimetric method, Stearns et al. 1947)
    1402   (residue-volume method, Booth & Everson 1948)
    1740   (shake flask-UV spec., Andrews & Keefer 1949)
    1860   (shake flask-UV, Klevens 1950)
    1790*   (shake flask-UV, Bohon & Claussen 1951)
    1755   (shake flask-UV, McDevit & Long 1952)
    1718   (shake flask-UV, Morrison & Billett 1952)
    1796   (Hayashi & Sasaki 1956; quoted, Keeley et al. 1988)
    1780, 1823   (selected, calculated-molar volume, Lindenburg 1956; quoted, Horvath 1982)
    1760   (Brady & Huff 1958)
    1740*   (shake flask-UV, measured range 0.4–45°C, Arnold et al. 1958)
    $S/(wt.\%) = 0.1806 - 0.001095 \cdot (t/°C) + 3.170 \times 10^{-5} \cdot (t/°C)^2$; temp range 5–45°C (shake flask-UV, Arnold et al. 1958); or
    $S/(wt.\%) = 0.1784 - 0.0007436 \cdot (t/°C) + 1.1906 \times 10^{-5} \cdot (t/°C)^2 + 1.217 \times 10^{-7} \cdot (t/°C)^3$; temp range 5–45°C (shake flask-UV, Arnold et al. 1958)
    1800*   (24°C, shake flask-UV, measured range 0.8–64.5°C, Alexander 1959)
    1890   (35°C, shake flask-UV spectrophotometry, Hine et al. 1962)
    1742*   (shake flask-UV, measured range 17–63°C, Franks et al. 1963)

1780   (shake flask-GC, McAuliffe 1963, 1966)
2100*   (20°C, polythermic method, measured range 20–79.5°C, Udovenko & Aleksandrova 1963)
1778   (calculated-group contribution, Irmann 1965; quoted, Horvath 1982)
2167   (vapor saturation-UV, Worley 1967)
1740   (21°C, extraction by nonpolar resins/elution, Chey & Calder 1972)
1765*   (shake flask-GC, measured range 4–25°C, Leinonen 1972)
1830*   (shake flask-UV spectroscopy, measured range 25–55°C, Bradley et al. 1973)
1755   (shake flask-GC, Polak & Lu 1973)
1755*   (shake flask-GC, measured range 25–84.7°C, Price 1973)
1765   (shake flask-GC, Leinonen & Mackay 1973)
1760*   (shake flask-UV, measured range 4.5–20.1°C, Brown & Wasik 1974)
1906   (shake flask-UV, Vesala 1974)
1769   (shake flask-GC, Mackay et al. 1975)
1780   (shake flask-GC, Mackay & Shiu 1975)
1740   (shake flask-GC, Price 1976)
1791*   (generator column-HPLC/UV, May et al. 1978; May 1980)

$S/(\mu g/kg) = [1833 + 0.3166 \cdot (t/°C)^2 - 0.6838 \cdot (t/°C)^3] \times 10^3$; temp range 0.2–25.8°C (generator column-HPLC/UV, May et al. 1978, 1980)

$\log x = 424.544/(T/K)^2 - 2955.82/(T/K) + 1.6606$; temp range 0–55°C (Ueda et al. 1978)

1769   (shake flask-fluorescence spectrophotometry, Aquan-Yuen et al. 1979)
1734*   (20°C, shake flask-UV, Ben-Naim & Wilf 1979)
1790*   (20°C, shake flask-GC, Bittrich et al. 1979)
1820–1930   (elution chromatography-UV, Schwarz 1980)
1750   (shake flask-LSC, Banerjee et al. 1980)
1610*   (vapor saturation-UV spec., measured range 5–45°C, Sanemasa et al. 1981)
1787   (shake flask-GC, Chiou et al. 1982; 1983)
1620*   (vapor saturation-UV spec., measured range 5–45°C, Sanemasa et al. 1982)
1792*   (generator column-HPLC/UV, May et al. 1983)
1789   (generator column-HPLC, Wasik et al. 1983)
1809   (HPLC-k' correlation, converted from reported γ_w, Hafkenscheid & Tomlinson 1983a)
1617   (vapor saturation-UV spec., Sanemasa et al. 1984)
1810   (shake flask-radiometric method, Lo et al. 1986)
1695   (shake flask-GC, Keeley et al. 1988)
1650   (shake flask-GC, Coutant & Keigley 1988)
1770*   (IUPAC recommended, temp range 0–70°C, Shaw 1989a)

$S/(g/100\ g\ soln) = 5.5773 - 4.6067 \times 10^{-2} \cdot (T/K) + 1.2504 \times 10^{-4} \cdot (T/K)^2 - 1.0489 \times 10^{-7} \cdot (T/K)^3$; temp range 0–70°C (summary of literature data, Shaw 1989a)

1732*   (20°C, activity coefficient-GC, Cooling et al. 1992)
1840*   (30°C, equilibrium flow cell-GC, measured range 30–100°C, Chen & Wagner 1994a)

$\ln(1/x) = -6.191 + 14.03 \cdot [(T/K)/562.2]^{-1} - 3.511 \cdot [(T/K)/562.2]^{-2}$; temp range 303.15–373.15 K (equilibrium flow cell-GC, Chen & Wagner 1994a)

$\ln x = 6.191 - 14.03 \cdot (T_r/K)^{-1} + 3.511 \cdot (T_r/K)^{-2}$, $T_r = T/T_c$, the reduced temp, system temp T divided by critical temp T_c (Chen & Wagner 1994c)

1760   (dialysis tubing equilibration-GC. Etzweiler et al. 1995)

$\ln x = -15.544647 - 1442.4276/(T/K) - 3.283 \times 10^{-5} \cdot (T/K)^2$, temp range 5–50°C (regression eq. of literature data, Shiu & Ma 2000)

$\ln x = -180.368 + 7524.83/(T/K) + 25.8585 \cdot \ln(T/K)$; temp range 290–400 K (eq. derived from literature calorimetric and solubility data, Tsonopoulos 1999)

Vapor Pressure (Pa at 25°C or as indicated and reported temperature dependence equations. Additional data at other temperatures designated * are compiled at the end of this section):

12654   (Hg manometer, Hovorka & Dreisbach 1934)
9960*   (20°C, manometer, measured range 0–50°C, Stuckey & Saylor 1940)

$\log(P/mmHg) = 7.12491 - 1323.06/(T/K - 41.23)$; temp range 0–75°C (manometer, Stuckey & Saylor 1940)

Mononuclear Aromatic Hydrocarbons

11700* (23.7°C, ebulliometry-manometer, measured range 14.5–80.9°C, Willingham et al. 1945)

log (P/mmHg) = 6.89324 − 1203.835/(219.924 + t/°C); temp range 14.5–80.9°C (ebulliometry-manometer, Antoine eq. from exptl. data, Willingham et al. 1945)

13332* (26.1°C, summary of literature data, Stull 1947)

11720* (23.27°C, ebulliometry, measured range 10.9–80.9°C, Forziati et al. 1949)

log (P/mmHg) = 6.91210 − 1214.645/(221.205 + t/°C); temp range 10.9–80.9°C (ebulliometry-manometer, Antoine eq. from exptl. data, Forziati et al. 1949)

12690 (interpolated-Antoine eq., Dreisbach 1955)

log (P/mmHg) = 6.90565 − 1211.033/(220.79 + t/°C); temp range 0–160°C (Antoine eq. for liquid state, Dreisbach 1955)

23450* (39.093°C, summary of literature data, temp range 7.565–260°C, Bond & Thodos 1960)

545800* (146.85°C, ebulliometry, measured range 146.85–286.85°C, Ambrose et al. 1967)

32045* (46.85°C, summary of literature data, temp range 46.85–286.85°C, Ambrose et al.1970)

12700* (extrapolated-Antoine eq., Zwolinski & Wilhoit, 1971)

log (P/mmHg) = 6.90565 − 1211.033/(220.790 + t/°C); temp range −11.6 to 103.92°C (Antoine eq., Zwolinski & Wilhoit 1971)

12680 (extrapolated, Antoine eq., Boublik et al. 1973; 1984)

log (P/mmHg) = [−0.2185 × 10254.2/(T/K)] + 9.5560; temp range −58 to −30°C (Antoine eq., Weast 1972–73)

log (P/mmHg) = [−0.2185 × 8146.5/(T/K)] + 7.833714; temp range −36.7 to 290.3°C (Antoine eq., Weast 1972–73)

12339* (24.396°C, ebulliometry, measured range 19.071–32.467°C, Osborn & Scott 1978)

19933* (32.182°C, ebulliometry, measured range 32.182–115.697°C, Scott & Osborn 1979)

12640* (average, ebulliometry-bubble cap boilers, measured range 290–378 K, Ambrose 1981)

12100 (gas saturation-GC, Politzki et al. 1982)

log (P/atm) = (1 − 353.214/T) × 10^(0.832632 − 6.72598 × 10^4·T + 6.38324 × 10^7·T^2); T in K, temp range 280.0–562.6 K (Cox vapor pressure eq., Chao et al. 1983)

12690, 12680 (interpolated-Antoine equations, Boublik et al. 1984)

log (P/kPa) = 6.01905 − 1204.637/(220.069 + t/°C); temp range 21.2–105°C (Antoine eq. from reported exptl. data of Ambrose 1981, Boublik et al. 1984)

log (P/kPa) = 6.01788 − 1203.677/(219.904 + t/°C); temp range 14.5–80.9°C (Antoine eq. from reported exptl. data of Willingham et al. 1945, Boublik et al. 1984)

12690 (extrapolated, Antoine eq., Dean 1985, 1992)

log (P/mmHg) = 9.1064 − 1885.9/(244.2 + t/°C); temp range −12 to 3°C (Antoine eq., Dean 1985, 1992)

log (P/mmHg) = 6.90565 − 1211.033/(220.79 + t/°C); temp range 8–103°C (Antoine eq., Dean 1985, 1992)

12716 (headspace-GC, Hussam & Carr 1985)

log (P/kPa) = 6.02232 − 1206.33/(220.91 + t/°C); temp range not specified (Antoine eq., Riddick et al. 1986)

12700 (interpolated-Antoine eq.-III, Stephenson & Malanowski 1987)

log (P_S/kPa) = 10.0091 − 2836/(25.31 + T/K); temp range 223–279 K (solid, Antoine eq.-I, Stephenson & Malanowski 1987)

log (P_S/kPa) = 8.45261 − 1986.69/(−23.089C + T/K); temp range 218–279 K (solid, Antoine eq.-II, Stephenson & Malanowski 1987)

log (P_L/kPa) = 6.01907 − 1204.682/(−53,072 + T/K); temp range 279–377 K (liquid, Antoine eq.-III, Stephenson & Malanowski 1987)

log (P_L/kPa) = 6.06832 − 1236.034/(−48.99 + T/K); temp range 353–422 K (Antoine eq.-IV, Stephenson & Malanowski 1987)

log (P_L/kPa) = 6.3607 − 1466.083/(−15.44 + T/K); temp range 420–502 K (Antoine eq.-V, Stephenson & Malanowski 1987)

log (P_L/kPa) = 7.51922 − 2809.514/(171.489 + T/K); temp range 501–562 K (Antoine eq.-VI, Stephenson & Malanowski 1987)

13100* (gas saturation, measured range −15.4 to 40°C, Liu & Dickhut 1994)

log (P/mmHg) = 31.7718 − 2.7254 × 10^3/(T/K) − 8.4442·log (T/K) − 5.3534 × 10^{-9}·(T/K) + 2.7187 × 10^{-6}·(T/K)^2, temp range 279–562 K (vapor pressure eq., Yaws 1994)

log (P/kPa) = 6.02994 − 1211.033/[(T/K) − 52.36]; temp range 5–50°C (regression eq. from literature data, Shiu & Ma 2000)

Henry's Law Constant (Pa m³/mol at 25°C or as indicated and reported temperature dependence equations. Additional data at other temperatures designated * are compiled at the end of this section):

653	(30°C, concn ratio-UV, Saylor et al. 1938)
576	(Taha et al. 1966)
442*	(20.06°C, headspace-GC, Brown & Wasik 1974)

$\ln [H/(Pa\ m^3/mol)] = 21.26071 - 4445.58/(T/K)$; temp range 4.5–20°C (regression eq. of exptl. data of Brown & Wasik 1974, Shiu & Ma 2000)

555, 530	(calculated as $1/K_{AW}$, calculated-bond contribution, Hine & Mookerjee 1975)
551	(headspace-GC, Vitenberg et al. 1975)
562, 556	(batch air stripping-GC, calculated-P/C, Mackay et al. 1979)
552*	(shake flask-concn. ratio-UV, measured range 10–30°C, Green & Frank 1979)

$\log (H/atm) = 8.58 - 1852.308/(T/K)$; temp range 10–30°C (shake flask-concn-UV, Green & Frank 1979)

$\ln (H/atm) = 8.58 - 1852.038/(T/K)$ (Kavanaugh & Trussell 1980)

554*	(equilibrium cell-concentration ratio-GC/FID, Leighton & Calo 1981)

$\ln (k_H/atm) = 19.02 - 3964/(T/K)$; temp range 1.0–27.2°C (equilibrium cell-concn ratio, Leighton & Calo 1981)

610*	(vapor-liquid equilibrium-GC, measured range 15–45°C, Sanemasa et al. 1981)
608*	(vapor-liquid equilibrium-GC, measured range 5–45°C, Sanemasa et al. 1982)
562	(gas stripping-GC, Warner et al. 1987)
740; 441	(20°C, EPICS-GC, calculated-P/C, Yurteri et al. 1987)
535; 588; 557; 554; 555	(EPICS-GC/FID; batch air stripping-GC; calculated P/C; direct concentration ratio; calculated-UNIFAC, Ashworth et al. 1988)
535*	(EPICS-GC/FID, measured range 10–30°C, Ashworth et al. 1988)

$\ln [H/(atm\ m^3/mol)] = 5.534 - 3194/(T/K)$; temp range 10–30°C (EPICS measurements, Ashworth et al. 1988)

586	(concentration ratio, Keeley et al. 1988)
555	(infinite activity coeff. γ^∞ from solubility measurement, Abraham et al. 1990)
564	(calculated-vapor-liquid equilibrium (VLE) data, Yaws et al. 1991)
570*	(extrapolated from equilibrium headspace-GC data, measured range 40–80°C, Ettre et al. 1993)

$\log (1/K_{AW}) = -2.1678537 + 836.2228/(T/K)$; temp range: 45–80°C (equilibrium headspace-GC measurements, Ettre et al. 1993)

569	(infinite activity coeff. γ^∞ in water determined by inert gas stripping-GC, Li et al. 1993)
604*	(equilibrium headspace-GC, measured range 10–30°C, Perlinger et al. 1993)
535*	(static headspace-GC, measured range 25–50°C, Robbins et al. 1993)
644	(headspace solid-phase microextraction (SPME)-GC, Zhang & Pawliszyn 1993)
488	(23°C, gas stripping-IR, Nielsen et al. 1994)
481*	(EPICS-GC/FID, measured range 2–25°C, Dewulf et al. 1995)
267, 612	(6.0, 25°C, EPICS-GC/FID, natural seawater with salinity of 35‰, Dewulf et al. 1995)

$\ln K_{AW} = -3640/(T/K) + 0.00786 \cdot Z + 10.577$; with Z salinity 0–35.5‰, temp range: 2–35°C (EPICS-GC/FID, Dewulf et al. 1995)

552*	(25.4°C, gas stripping-HPLC/UV/fluorescence, Alaee et al. 1996)

$\ln [H/(Pa\ m^3/mol)] = 21.87689 - 4672.28/(T/K)$; temp range: 4–34.9°C and enthalpy of volatilization $\Delta H_{vol} = 32.2$ kJ/mol at 20°C; (gas stripping-HPLC/UV measurements, Alaee et al. 1996)

485	(20°C, selected from literature experimentally measured data, Staudinger & Roberts 1996)
640*	(vapor-liquid equilibrium-GC, measured range 10–35°C, Turner et al. 1996)

$K_{AW} = 0.0763 + 0.00211 \cdot (T/K) + 0.000162 \cdot (T/K)^2$; temp range 0–50°C (vapor-liquid equilibrium-GC measurements with additional lit. data, Turner et al. 1996)

538*	(headspace equilibrium-GC, Peng & Wan 1997)

$\ln K_{AW} = 7.15 - 1397/(T/K)$; temp range 15–45°C (headspace equilibrium-GC, Peng & Wan 1997)

272	(gas stripping-GC, Altschuh et al. 1999)
439	(20°C, headspace equilibrium-GC, Peng & Wan 1998)

$\ln K_{AW} = 7.44 - 1448/(T/K)$; temp range 0–45°C (seawater with salinity of 36‰, headspace-GC, Peng & Wan 1998)

466	(21°C, headspace equilibrium-GC, de Wolf & Lieder 1998)
558	(exponential saturator EXPSAT technique, Dohnal & Hovorka 1999)
580.6	(modified EPICS method-GC, Ryu & Park 1999)
556	(EPICS-static headspace method-GC/FID, Miller & Stuart 2000)
466	(20°C, selected from literature experimentally measured data, Staudinger & Roberts 2001)

Mononuclear Aromatic Hydrocarbons

log K_{AW} = 5.053 − 1693/(T/K) (van't Hoff eq. derived from lit. data, Staudinger & Roberts 2001)
538*   (solid-phase microextraction-GC, measured range 15–40°C, Bierwagen & Keller 2001)
ln K_{AW} = 8.1648 − 2889.4/(T/K); temp range 15–40°C (SPME-GC, Bierwagen & Keller 2001)
573.4*   (EPICS-SPME, measured range 2–60°C, Görgényi et al. 2002)
ln K_{AW} = 10.01 − 3430.4/(T/K); temp range 2–60°C (EPICS-SPME method, Görgényi et al. 2002)
514–606   (27°C, headspace equilibrium-GC, at different solute concn: 0.48–19.1 mg/L, measured temp range 300–315 K, Cheng et al. 2003)
558*   (headspace-GC, measured range 10–25°C, Bakierowska & Trzeszczyński 2003)
ln (1/K_{AW}) = 11.663 − 3920/(T/K); temp range 10–25°C, headspace-GC, Bakierowska & Trzeszczyński 2003)

Octanol/Water Partition Coefficient, log K_{OW} at 25°C or as indicated. Additional data at other temperatures designated * are compiled at the end of this section:

2.13   (shake flask-UV, Fujita et al. 1964)
1.56, 1.65 (shake flask-UV, calculated-M.O. indices, Rogers & Cammarata 1969)
2.13   (calculated-fragment const., Rekker 1977)
2.13, 1.56, 2.15, 2.03, 2.04   (Hansch & Leo 1979)
2.39   (HPLC-RT correlation, Veith et al. 1979a)
2.12   (shake flask-LSC, Banerjee et al. 1980)
2.28   (HPLC-k′ correlation, Hanai et al. 1981)
2.11   (HPLC-RT correlation, McDuffie 1981)
2.43   (HPLC-k′ correlation, McDuffie 1981)
2.16   (HPLC-k′ correlation, D'Amboise & Hanai 1982)
2.13   (shake flask-GC, Watarai et al. 1982)
2.20   (shake flask-HPLC, Hammers et al. 1982)
2.18   (HPLC-k′ correlation, Miyake & Terada 1982)
2.02   (shake flask method, Eadsforth & Moser 1983)
2.38   (HPLC method, Eadsforth & Moser 1983)
2.10   (shake flask-GC, Platford 1983)
2.48   (HPLC-RT correlation, Swann et al. 1983)
2.10   (HPLC-k′ correlation, Hafkenscheid & Tomlinson 1983b)
2.04   (HPLC-RV correlation, Garst 1984)
2.25   (RP-HPLC-k′ correlation, Rapaport & Eisenreich 1984)
2.13   (generator column-GC/ECD, Miller et al. 1984)
2.26   (HPLC-k′ correlation, De Kock & Lord 1987)
2.01   (generator column-reversed phase-LC, Schantz & Martire 1987)
2.16   (RP-HPLC-capacity factor correlation, Sherblom & Eganhouse 1988)
1.91   (RP-HPLC-RT correlation, ODS column with masking agent, Bechalany et al. 1989)
2.13   (recommended, Sangster 1989, 1993)
2.186   (shake flask/slow stirring-GC, De Bruijn et al. 1989)
2.21   (normal phase-HPLC-k′ correlation, Govers & Evers 1992)
2.13   (recommended, Hansch et al. 1995)
1.97*   (24.8°C, EPICS-GC, measured range 2.2–24.8°C, Dewulf et al. 1999)

Octanol/Air Partition Coefficient, log K_{OA} at 25°C or as indicated. Additional data at other temperatures designated * are compiled at the end of this section:

2.90*   (20.29°C, from GC-determined γ^∞ in octanol, measured range 20.290–50.28°C, Gruber et al. 1997)
2.80   (head-space GC-FID both phases, Abraham et al. 2001)

Bioconcentration Factor, log BCF:

0.64   (pacific herring, Korn et al. 1977)
0.54   (eels, Ogata & Miyake 1978; Ogata et al. 1984)
1.10   (fathead minnow, Veith et al. 1980)
1.48, 1.0   (algae, fish, Freitag et al. 1984)
1.48   (algae, Geyer et al. 1984)
0.63   (gold fish, Ogata et al. 1984)

< 1.0, 3.23 (fish, activated sludge, Freitag et al. 1985)
0.54, 0.64, 0.63; 1.38 (selected: eels, pacific herring, gold fish; calculated, Howard 1990)
1.63 (*S. capricornutum*, Herman et al. 1991)

Sorption Partition Coefficient, log K_{OC} at 25°C or as indicated:
1.92 (sediment, sorption isotherms by batch equilibrium-UV spec., Karickhoff et al. 1979)
1.63; 1.82 (Hastings soil pH 5.6; Overton soil pH 7.8, batch equilibrium, Rogers et al. 1980)
1.78 (average of 17 sediments and soils, sorption isotherms by batch equilibrium, Karickhoff 1981)
1.58; 1.73; 1.64 (forest soil pH 5.6; forest soil pH 4.2; agricultural soil pH 7.4, Seip et al. 1986)
1.42 (sediment 4.02% OC from Tamar estuary, batch equilibrium-GC, Vowles & Mantoura 1987)
1.34 (untreated Marlette soil A horizon, organic carbon OC 2.59%, batch equilibrium-adsorption isotherm, Lee et al. 1989)
2.08, 2.04 (organic cations treated Marlette soil A horizon: HDTMA treated OC 6.48%; DDTMA treated, OC 4.37%, batch equilibrium-adsorption isotherm, Lee et al. 1989)
2.65, 2.59, 2.25 (organic cations treated Marlette soil B_t horizon: HDTMA treated OC 3.72%, DDTMA treated OC 1.98%, NTMA treated, OC 1.18%, batch equilibrium-adsorption isotherm, Lee et al. 1989)
2.69, 2.66 (organic cations HDTMA treated soils: St. Clair soil B_t horizon OC 3.25%; Oshtemo soil B_t horizon OC 0.83%, batch equilibrium-adsorption isotherm, Lee et al. 1989)
1.89 (aquifer material with f_{OC} of 0.006 and measured partition coeff. K_P = 0.47 mL/g., Abdul et al. 1990)
1.58, 1.49 (Riddles soil top layer, pH 5.0; below top layer pH 5.3, batch equilibrium, Boyd et al. 1990)
1.82, 1.87 (RP-HPLC-k' correlation, humic acid-silica column, Szabo et al. 1990a,b)
1.74; 1.81 (Captina silt loam pH 4.97; McLaurin sandy loam pH 4.43, batch equilibrium, Walton et al. 1992)
1.75 (average of 5 soils, sorption isotherms by batch equilibrium method-GC, Xing et al. 1994)
1.96 (soil, calculated-molecular connectivity indices, Sabljic et al. 1995)
1.57, 1.62, 1.74 (RP-HPLC-k' correlation on 3 different stationary phases, Szabo et al. 1995)
1.82, 1.84 (RP-HPLC-k' correlation including MCI related to non-dispersive intermolecular interactions, hydrogen-bonding indicator variable, Hong et al. 1996)
1.84, 1.86, 1.87, 1.88, 1.90, 1.87, 1.90 (2.3, 3.8, 6.2, 8.0, 13.5, 18.6, 25°C, natural sediment from River Leie, organic carbon f_{OC} = 4.12%, EPICS-GC/FID, Dewulf et al. 1999)
2.76, 2.41 (natural zeolite modified with a cation surfactant HDTMA with surface coverage of 100, 200 mmol/kg at pH 7, batch equilibrium-sorption isotherm, Li et al. 2000)
1.64, 1.58, 1.78 (soils: organic carbon OC ≥ 0.1%, OC ≥ 0.5%, 0.1 ≤ OC < 0.5%, average, Delle Site 2001)

Sorption Partition Coefficient, log K_{OM}:
1.26 (Woodburn silt loam soil, 1.9% organic matter, equilibrium isotherm-GC, Chiou et al. 1983)
1.04 (untreated Marlette soil A horizon, organic matter OM 5.18%, batch equilibrium-adsorption isotherm, Lee et al. 1989)
1.89, 1.81 (organic cations treated Marlette soil A horizon: HDTMA treated, organic matter 10.03%; DDTMA treated, OM 5.18%, batch equilibrium-adsorption isotherm, Lee et al. 1989)
2.53, 2.46, 2.08 (organic cations treated Marlette soil B_t horizon: HDTMA treated OM 4.85%, DDTMA treated OM 2.73%, NTMA treated, OM 1.74%, batch equilibrium-adsorption isotherm, Lee et al. 1989)
2.56, 2.53 (organic cations HDTMA treated soils: St. Clair soil B_t horizon OM 4.38%; Oshtemo soil B_t horizon OM 1.12%, batch equilibrium-adsorption isotherm, Lee et al. 1989)
1.34; 1.14 (high-organic-content soils: Florida peat - 57.1% C; Michigan muck - 53.7% C, equilibrium isotherm-GC, Rutherford & Chiou 1992)

Environmental Fate Rate Constants, k, or Half-Lives, $t_{1/2}$:
Volatilization: $t_{1/2}$ = 4.81 h from water depth of 1 m (calculated, Mackay & Leinonen 1975; Haque et al. 1980); k = 0.03 d^{-1} with $t_{1/2}$ = 23 d in spring at 8–16°C, k = 0.22 d^{-1} with $t_{1/2}$ = 31 d in summer at 20–22°C, k = 0.054 d^{-1} with $t_{1/2}$ = 13 d in winter at 3–7°C during the periods when volatilization appears to dominate, and k = 0.101 d^{-1} with $t_{1/2}$ = 6.9 d with $HgCl_2$ in September 9–15, 1980 in marine mesocosm experiments (Wakeham et al. 1983);
$t_{1/2}$ ~ 27 h from a river of 1 m depth with wind speed 3 m/s and water current of 1 m/s is 2.7 h at 20°C (Lyman et al. 1982).

Photolysis: atmospheric photolysis $t_{1/2}$ = 2808–16152 h, based on measured photolysis half-lives in deionized water (Hustert et al. 1981; Howard et al. 1991);
    aqueous photolysis $t_{1/2}$ = 2808–16152 h, based on measured photolysis half-lives in deionized water (Hustert et al. 1981; Howard et al. 1991);
    reaction rate constants, k = 8.64×10^{-4} h^{-1} in air, and k = 1.8×10^{-4} h^{-1} in water (Mackay et al. 1985).
Oxidation: rate constant k; and gas-phase second order rate constants, k_{OH} for reaction with OH radical, k_{NO3} with NO$_3$ radical and k_{O3} with O$_3$ or as indicated. Data at other temperatures and/or the Arrhenius expression are designated *, see reference:
    k_{OH} = 1.24×10^{-12} cm^3 molecule^{-1} s^{-1}; $k_{O(3P)}$ = 0.24×10^{-13} cm^3 molecule^{-1} s^{-1} for the reaction of O(^{3}P) atom at room temp. (flash photolysis-resonance fluorescence, Hansen et al. 1975)
    k_{OH} ≤ 2.3×10^{9} L mol^{-1} s^{-1} with $t_{1/2}$ ≥ 5.1 h; $k_{O(3P)}$ = $(0.144 \pm 0.2) \times 10^{8}$ L mol^{-1} s^{-1} with O(^{3}P) atom at room temp. (relative rate method, Doyle et al. 1975; Lloyd et al. 1976)
    k_{OH} = 0.85×10^{9} L mol^{-1} s^{-1}, with $t_{1/2}$ = 2.4–24 h (Darnall et al. 1976)
    k_{OH}* = $(1.20 \pm 0.15) \times 10^{-12}$ cm^3 molecule^{-1} s^{-1} at room temp., measured over temp range 296–473 K (flash photolysis-resonance fluorescence, Perry et al. 1977)
    photooxidation $t_{1/2}$ = 8.021×10^{3} – 3.21×10^{5} h in water, based on measured rate constant for reaction with OH radical in water (Güesten et al. 1981)
    k_{OH} = 1.4×10^{-12} cm^3 molecule^{-1} s^{-1} and residence time of 8.3 d, loss of 11.4% in one day or 12 sunlit hours at 300 K in urban environments (Singh et al. 1981)
    k_{OH} = 28 cm^3 mol^{-1} s^{-1} at 300 K (Lyman et al. 1982)
    k_{OH} = 0.82×10^{-9} M^{-1} s^{-1} with $t_{1/2}$ = 6.8 d in the atmosphere (Mill 1982)
    k = (2.0 ± 0.4) M^{-1} s^{-1} for the reaction with ozone in water using 50–1000 mM t-BuOH as scavenger at pH 2.0 and 20–23°C (Hoigné & Bader 1983)
    k_{OH} = $(8.8 \pm 0.4)1.45 \times 10^{-13}$ cm^3 molecule^{-1} s^{-1} at 295 K (flash photolysis-resonance fluorescence, Wahner & Zetzsch 1983)
    k_{NO3} < 2.3×10^{-17} cm^3 molecule^{-1} s^{-1} at 296 K (Atkinson et al. 1984)
    k_{OH} = 1.19×10^{-12} cm^3 molecule^{-1} s^{-1} at room temp. (Atkinson et al. 1985)
    k_{OH}* = 1.28×10^{-12} cm^3 molecule^{-1} s^{-1} at 298 K (recommended, Atkinson 1985)
    k_{OH}(calc) = 2.3×10^{-12} cm^3 molecule^{-1} s^{-1}, k_{OH}(obs.) = 1.19×10^{-12} cm^3 molecule^{-1} s^{-1} at room temp. (SAR structure-activity relationship, Atkinson 1985)
    k_{OH} = 1.45×10^{-12} cm^3 molecule^{-1} s^{-1} at 298 K. (relative rate method, Ohta & Ohyama 1985)
    k_{OH} = 1.26×10^{-12} cm^3 molecule^{-1} s^{-1} at 23.8°C, with an atmospheric lifetime of 9.1 d (Edney et al. 1986)
    k_{OH}* = 1.14×10^{-12} cm^3 molecule^{-1} s^{-1} at room temp., measured range 239–354 K (flash photolysis-resonance fluorescence, Witte et al. 1986)
    k_{OH}(calc) = 2.0×10^{-12} cm^3 molecule^{-1} s^{-1}, k_{OH}(obs.) = 1.28×10^{-12} cm^3 molecule^{-1} s^{-1} at room temp. (SAR [structure-activity relationship], Atkinson 1987)
    k_{OH}* = $(1.29 \pm 1.4) \times 10^{-12}$ cm^3 molecule^{-1} s^{-1} at 296 K, measured range 234–438 K (flash photolysis-resonance fluorescence, Wallington et al. 1987)
    k_{O3} < 0.01×10^{-18} cm^3 molecule^{-1} s^{-1}; k_{OH} = 1.28×10^{-12} cm^3·molecule^{-1} s^{-1}, and k_{NO3} < 3.2×10^{-17} cm^3 molecule^{-1} s^{-1} at room temp. (Atkinson & Aschmann 1988)
    k_{OH}* = 1.40×10^{-12} cm^3 molecule^{-1} s^{-1} at 298 K (recommended, Atkinson 1989)
    k_{OH} = 1.23×10^{-12} cm^3 molecule^{-1} s^{-1} at 298 K (Atkinson 1990)
    k_{OH}(calc) = 1.51×10^{-12} cm^3 molecule^{-1} s^{-1} (molecular orbital calculations, Klamt 1993)
Hydrolysis: no hydrolyzable functional groups (Mabey et al. 1982).
Biodegradation:
    $t_{1/2}$ = 6 d in estuarine water (estimated, Lee & Ryan 1976)
    $t_{1/2}$(aq. aerobic) = 120–384 h, based on seawater dieaway test data (Van der Linden 1978) and river dieaway data (Vaishnav & Babeu, 1987; Howard et al. 1991)
    k = 4.58×10^{-3} h^{-1} in water (Lee & Ryan 1979; Mackay et al. 1985)
    k = 0.2 yr^{-1} with $t_{1/2}$ = 110 d (Zoeteman et al 1981; Olsen & Davis 1990)
    k = 0.5 d^{-1} significant degradation in favourable aerobic environment (Tabak et al. 1981; Mills et al. 1982)
    $t_{1/2}$(aq. anaerobic) = 2688–17280 h, based on unacclimated aqueous anaerobic biodegradation screening test data (Horowitz et al. 1982; Howard et al. 1991)
    k = 0.12 d^{-1} in river water (estimated, Bartholomew & Pfaender 1983; quoted, Battersby 1990)
    $t_{1/2}$ = 8.6 d in activated sludge (estimated, Freitag et al. 1985, quoted, Anderson et al. 1991)

$k = 0.025$ d^{-1} with $t_{1/2} = 28$ d in groundwater, $k = 0.044$ d^{-1} with $t_{1/2} = 16$ d in Lester River with nutrient and microbial addition, and $k = 0.082$ d^{-1} with $t_{1/2} = 8$ d in Superior harbor waters (Vaishnav & Babeu 1987)

$t_{1/2}$(aerobic) = 5 d, $t_{1/2}$(anaerobic) = 110 d in natural waters (Capel & Larson 1995)

$k = 0.58$ d^{-1} associated with microbial population growth initially followed by a slower second phase with $k = 0.12$ d^{-1} degradation by *P. aeruginosa* is a two-stage process (Kim et al. 2003).

Bioconcentration, Uptake (k_1) and Elimination (k_2) Constants or Half-Lives:

$t_{1/2} = 0.5$ d for elimination from eels, 0.5 d (Ogata & Miyake 1978).

Half-Lives in the Environment:

Air: $t_{1/2} \geq 5.1$ h, based on a determined rate of disappearance in ambient LA basin air for reaction with OH radical at 300 K (Doyle et al. 1975);

$t_{1/2} = 2.4–24$ h, based on rate of disappearance for the reaction with OH radical (Darnall et al. 1976);

residence time of 8.3 d, loss of 11.4% in one day or 12 sunlit hours at 300 K in urban environments (Singh et al. 1981);

$t_{1/2} = 50.1–501$ h, based on photooxidation half-life in air (Atkinson 1985; Howard et al. 1991);

calculated lifetime of 9.1 d due to reaction with OH radical (Edney et al. 1986);

summer daylight lifetime of 115 h due to reaction with OH radical (Altshuller 1991);

calculated lifetimes of 9.4 d, > 4 yr and > 4.5 yr for reactions with OH radical, NO$_3$ radical and O$_3$, respectively (Atkinson 2000).

Surface Water:

$t_{1/2} = 4.81$ h, based on evaporation loss at 25°C and 1 m depth of water (Mackay & Leinonen 1975)

biodegradation $t_{1/2} \sim 6$ d in estuarine water (Lee & Ryan 1976)

$t_{1/2} = 120–384$ h, based on unacclimated aerobic biodegradation half-life (Van der Linden 1978; Vaishnav & Babeu 1987; Howard et al. 1991);

$t_{1/2} = 23$ d at 8–16°C in the spring, $t_{1/2} = 3.1$ d at 20–22°C in the summer and $t_{1/2} = 13$ d at 3–7°C in the winter, and $t_{1/2} = 6.9$ d with HgCl$_2$ in September 9–15 from mesocosm experiments (Wakeham et al. 1983).

Ground water: $t_{1/2} \sim 1$ yr from persistence observed in the groundwater of Netherlands (Zoeteman et al. 1981),

$t_{1/2} = 240–17280$ h, based on unacclimated aqueous aerobic biodegradation half-life (Van der Linden 1978; Vaishnav & Babeu 1987; Howard et al. 1991).

Soil: $t_{1/2} = 120–384$ h, based on unacclimated aqueous aerobic biodegradation half-life (Van der Linden 1978; Vaishnav & Babeu 1987; Howard et al. 1991);

$t_{1/2} < 10$ d (Ryan et al. 1988);

$t_{1/2} = 365$ d, assumed first-order biological/chemical degradation in the soil (Jury et al. 1990);

disappearance $t_{1/2} < 2$ d for test soils (Anderson et al. 1991).

Biota:

TABLE 3.1.1.1.1
Reported aqueous solubilities of benzene at various temperatures and reported temperature dependence equations

$$R \cdot \ln x = -[\Delta H_{fus}/(T/K)] + (0.000408)[(T/K) - 291.15]^2 - c + b \cdot (T/K) \quad (1)$$
$$\text{'S}/(\mu g/kg) = a \cdot t^3 + b \cdot t_{½} + c \cdot t + d \quad (2)$$
$$\ln x = A - B/T(K) \quad (3)$$
$$\ln x = A + B/\tau + C \ln \tau, \text{ where } \tau = T/T_o, T_o = 298.15 \text{ K} \quad (4)$$

1.

Bohon & Claussen 1951		Arnold et al. 1958		Alexander 1959		Franks 1963	
shake flask-UV		shake flask-UV		shake flask-UV		shake flask-UV	
t/°C	S/g·m⁻³	t/°C	S/g·m⁻³	t/°C	S/g·m⁻³	t/°C	S/g·m⁻³
0.4	1741	4.5	1720	0.8	1840	17	1714
5.2	1810	4.9	1770	9.4	1790	22	1723
10	1800	5.0	1740	16.8	1770	25	1742
14.9	1770	6.7	1740	24	1800	29	1745
21	1790	9.0	1730	31	1830	32	1788
25.6	1790	12.5	1720	38	1920	35	1823
30.2	1843	15	1730	44.7	2030	40.5	1905
34.9	1877	20	1710	51.5	2140	42	1910
42.8	1998	20.6	1720	65.4	2340	44	1931
		24.8	1710			46	1983
		24.9	1740			51	2075
ΔH_{sol}/(kJ mol⁻¹) =		27.3	1745			56	2183
25	2.42	30	1775			61	2305
12	−2.30	30.9	1884			63	2352
17	−0.25	45	1975				
18	0	49.8	2044				
22	1.34	54.5	2152				
27	3.01	59.8	2265				
32	4.435	64.8	2313				
37	5.86						
		ΔH_{sol}/(kJ mol⁻¹) = 2.27 at 25°C					

2.

Leinonen 1972		Udovenko & Aleksandrova 1963		Price 1973		Bradley et al. 1973	
shake flask-GC		polythermic method		shake flask-GC		shake flask-UV	
t/°C	S/g·m⁻³	t/°C	S/g·m⁻³	t/°C	S/g·m⁻³	t/°C	S/g·m⁻³
4	1710	20	2100	25	1755	25	1830
5	1737	40	2270	55.3	3980	45	2160
5.4	1746	40.5	2480	84.7	6468	55	2380
6.1	1735	44.5	2590				
7.0	1781	56.5	2880				
10.3	1748	60	3000				
13	1741	65	3190				
16	1730	79.5	3730				
19.1	1721						
22.1	1739						

(Continued)

TABLE 3.1.1.1.1 *(Continued)*

3.

Brown & Wasik 1974		May et al. 1980, 1983		Ben-Naim & Wiff 1979		Bittrich et al. 1979	
shake flask-UV		generator column-HPLC/UV		shake flask-UV		shake flask-GC	
t/°C	S/g·m^{-3}	t/°C	S/g·m^{-3}	t/°C	S/g·m^{-3}	t/°C	S/g·m^{-3}
4.5	1840	0.2	1836	10	1625	20	1790
6.3	1850	6.2	1804	20	1734	40	2025
7.1	1810	11	1799			69	2442
9.0	1810	13	1770				
11.8	1770	16.9	1762				
12.1	1770	18.6	1767				
15.1	1790	25.0	1790				
17.9	1790	25.8	1819				
20.1	1760						

temp dependence eq. 2 given in May et al. 1978b and May 1980

S	mg/kg
a	0.0247
b	–0.6838
c	0.3166
d	1833

4.

Sanemasa et al. 1981		Shaw 1989a (IUPAC)		Cooling et al. 1992		Chen & Wagner 1994a	
vapor saturation-UV		recommended values		activity coefficient-GC		equilibrium flow cell-GC	
t/°C	S/g·m^{-3}	t/°C	S/g·m^{-3}	t/°C	S/g·m^{-3}	t/°C	S/g·m^{-3}
15	1540	0	1690	20	1732	30	1840
25	1610	5	1800	30	1688	40	2014
35	1770	10	1780	40	1712	50	2213
45	1870	15	1760	50	1760	60	2452
		20	1760			70	2713
		25	1770			80	3033
		30	1810			90	3472
		35	1860			100	4123
Sanemasa et al. 1982		40	1930				
vapor saturation-UV		45	1990			ΔH_{sol}/(kJ mol^{-1}) = 3.69	
t/°C	S/g·m^{-3}	50	2080			25°C	
5	1620	55	2190				
15	1580	60	2310				
25	1620	65	2410				
35	1710	70	2670				
45	1800						

ΔH_{sol}/(kJ mol^{-1}) = 2.07 at 25°C

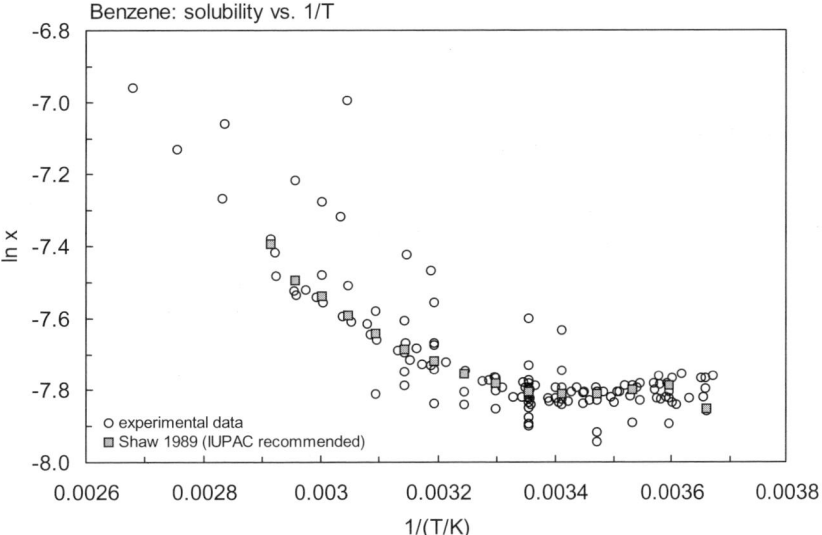

FIGURE 3.1.1.1.1 Logarithm of mole fraction solubility (ln x) versus reciprocal temperature for benzene.

TABLE 3.1.1.1.2
Reported vapor pressures of benzene at various temperatures and the coefficients for the vapor pressure equations

$\log P = A - B/(T/K)$ (1) $\ln P = A - B/(T/K)$ (1a)

$\log P = A - B/(C + t/°C)$ (2) $\ln P = A - B/(C + t/°C)$ (2a)

$\log P = A - B/(C + T/K)$ (3)

$\log P = A - B/(T/K) - C \cdot \log (T/K)$ (4)

$\log P = A - B/(T/K) - C \cdot \log (T/K) + D \cdot P/(T/K)^2$ (5)

1.

Stuckey & Saylor 1940		Willingham et al. 1945		Stull 1947		Forziati et al. 1949	
mercury manometer		ebulliometry		summary of lit. data		ebulliometry	
t/°C	P/Pa	t/°C	P/Pa	t/°C	P/Pa	t/°C	P/Pa
0	3509	14.548	7654	–36.7	133.3*	10.983	6397
12.5	6040	17.720	8962	–19.6	266.6*	14.575	7690
25	9960	20.504	10303	–11.5	1333*	17.697	8989
37.5	15800	23.270	11699	–2.60	2666*	20.028	11700
40	24240	26.886	13818	7.60	5333	23.271	11720
50	36050	31.004	16621	15.4	7999	26.908	13840
		35.191	19922	26.1	13332	31.013	16631
eq. 4	P/mmHg	30.078	23450	42.2	26664	35.207	19942
A	7.12491	44.284	28952	60.5	53323	39.095	23465
B	1323.06	49.066	34897	80.1	101325	44.294	28976
C	41.23	54.832	43320			49.084	34924
		60.784	53652		*solid	54.852	43358
bp/°C	80.06	67.135	66753	mp/°C	5.5	60.803	53697
		74.028	83717			67.148	66795
		78.891	97601			74.035	83746
		79.413	99197			78.903	97643
		79.898	100689			79.424	99230

(Continued)

TABLE 3.1.1.1.2 *(Continued)*

1.

Stuckey & Saylor 1940		Willingham et al. 1945		Stull 1947		Forziati et al. 1949	
mercury manometer		ebulliometry		summary of lit. data		ebulliometry	
t/°C	P/Pa	t/°C	P/Pa	t/°C	P/Pa	t/°C	P/Pa
		80.442	102384			79.090	100722
		80.922	106570			80.461	102460
						80.948	104000
		eq. 2	P/mmHg				
		A	6.89324			eq. 2	P/mmHg
		B	1203.835			A	6.9210
		C	219.924			B	1214.645
						B	221.205
		bp/°C	80.103				
						bp/°C	80.099

2.

Bond & Thodos 1960		Ambrose et al. 1967		Ambrose et al. 1970		Zwolinski & Wilhoit 1971	
compiled data		ebulliometry		compiled data		selected values	
t/°C	P/Pa	t/°C	P/Pa	t/°C	P/Pa	t/°C	P/Pa
7.565	5333	146.85	545800	46.85	32045	−11.6	1333*
39.093	23450	166.85	809200	66.85	66116	−2.60	2666*
67.15	66753	186.85	1156700	86.85	124180	3.0	4000*
95.713	159987	206.85	1603700	106.85	215960	7.55	5333
110.015	231848	226.85	2166700	126.85	352460	11.80	6666
150.015	577419	246.85	2864400	146.85	545760	15.39	7999
180.015	1017250	266.85	3719300	166.85	809050	21.293	10666
240.015	2589787	286.85	4771700	186.85	1156600	26.075	13332
260.015	3376922			206.85	1603800	35.266	19998
		eq. 5	P/mmHg	226.85	2166900	42.214	26664
eq. 5	P/mmHg	A	20.87440	246.85	2864000	47.868	33331
A	23.36128	B	2472.77	266.85	3718800	52.672	39997
B	2457.12	C	5.44671	286.85	4772600	60.611	53329
C	5.28840	D	1238			67.093	66661
D	1.56738					72.616	79993
						77.454	93326
bp/°C	80.115					78.354	95991
						79.236	98659
						80.100	101325
						25.0	12690
						*solid	
						eq. 2	P/mmHg
						A	9.1064
						B	1885.9
						C	244.2
						for liquid	
						eq. 2	P/mmHg
						A	6.90565
						B	1211.033
						C	220.790

Mononuclear Aromatic Hydrocarbons

TABLE 3.1.1.1.2 *(Continued)*

Bond & Thodos 1960		Ambrose et al. 1967		Ambrose et al. 1970		Zwolinski & Wilhoit 1971	
compiled data		ebulliometry		compiled data		selected values	
t/°C	P/Pa	t/°C	P/Pa	t/°C	P/Pa	t/°C	P/Pa
						bp/°C	80.100
						ΔH_V/(kJ mol^{-1})	
						at 25°C	33.85
						at bp	30.76

3.

Osborn & Scott 1978		Scott & Osborn 1979		Ambrose 1981			
ebulliometry		ebulliometry		ebulliometry		ebulliometry	
t/°C	P/Pa	t/°C	P/Pa	T/K	P/Pa	T/K	P/Pa
				set 1		set 2	
19.071	9585	32.182	19933	290.076	8634	294.165	10527
21.728	10887	40.637	25023	302.392	15388	297.699	12431
24.396	12339	46.139	31177	311.186	22484	303.293	16017
27.0755	13955	51.684	38565	318.694	30464	306.060	18080
29.765	15748	57.276	47375	325.097	38953	314.942	26227
32.467	17735	62.991	57817	330.437	47571	318.971	30791
		68.591	70120	334.886	55511	323.246	36326
		74.319	84532	338.935	63815	328.325	43899
		80.092	101325	342.946	72985	334.125	54050
		85.911	120791	346.244	81275	338.779	63474
		91.777	143240	349.910	91346	348.400	87089
		97.689	169030	353.469	102043	352.356	98297
		103.645	198490	356.187	110854	353.212	101226
		109.648	232020	358.873	120137	353.802	103094
		115.697	270030	362.286	132779	357.611	115693
				365.234	144528	363.086	135887
				367.897	155797	367.789	155309
				370.527	167605	372.897	178812
				373.151	180059	378.152	205747
				375.844	193529		
				378.523	207939		
				381.325	223436		

4.

Ambrose 1981 (continued)				Liu & Dickhut 1994	
ebulliometry		ebulliometry		gas saturation-GC	
T/K	P/Pa	T/K	P/Pa	t/°C	P/Pa
set 3		set 4			
285.957	7014	297.769	12471	−15.4	880
292.893	9903	302.633	15553	−5.0	1750
298.684	13007	307.159	18957	10	6540
302.619	15545	308.384	19979	25	13100
304.302	16745	314.503	25716	40	26400
310.167	21539	319.907	31940		
314.406	25665	324.512	38019		
319.512	31444	329.473	45774		

(Continued)

TABLE 3.1.1.1.2 *(Continued)*

4.

Ambrose 1981 (continued)				Liu & Dickhut 1994	
ebulliometry		ebulliometry		gas saturation-GC	
T/K	P/Pa	T/K	P/Pa	t/°C	P/Pa
323.921	37264	333.842	53510		
329.536	45876	338.144	62115		
334.406	54580	342.687	77356		
339.063	64081	347.828	85511		
343.751	74940	352.112	97852		
348.357	86959	352.617	99394		
352.594	99320	352.955	100439		
353.078	100816	358.109	117426		
353.660	102633	362.258	132666		
358.727	119597	367.239	152938		
363.062	135783	372.175	175319		
367.554	154281	377.584	202669		
372.630	177510				
377.875	204219				
383.175	234260				

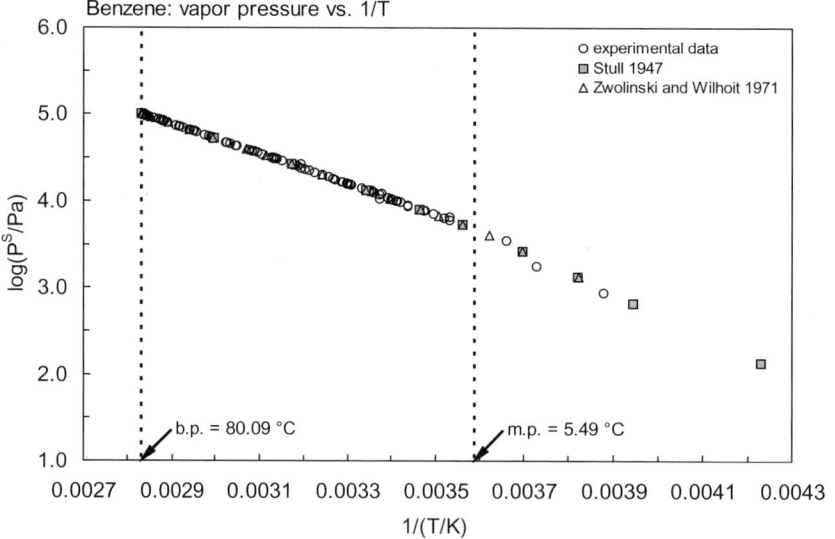

FIGURE 3.1.1.1.2 Logarithm of vapor pressure versus reciprocal temperature for benzene.

TABLE 3.1.1.1.3
Reported Henry's law constants of benzene at various temperatures and reported temperature dependence equations

$$\ln K_{AW} = A - B/(T/K) \quad (1)$$
$$\ln (1/K_{AW}) = A - B/(T/K) \quad (2)$$
$$\ln (k_H/atm) = A - B/(T/K) \quad (3)$$
$$\ln [H/(Pa\ m^3/mol)] = A - B/(T/K) \quad (4)$$
$$K_{AW} = A - B\cdot(T/K) + C\cdot(T/K)^2 \quad (5)$$

$$\log K_{AW} = A - B/(T/K) \quad (1a)$$
$$\log (1/K_{AW}) = A - B/(T/K) \quad (2a)$$

$$\ln [H/(atm\cdot m^3/mol)] = A - B/(T/K) \quad (4a)$$

1.

Brown & Wasik 1974		Green & Frank 1979		Leighton & Calo 1981		Sanemasa et al. 1981	
headspace-GC		concentration ratio-UV		equilibrium cell-GC		vapor-liquid equilibrium	
t/°C	H/(Pa m³/mol)	t/°C	H/(Pa m³/mol)	t/°C	H/(Pa m³/mol)	t/°C	H/(Pa m³/mol)
4.5	187.2	10	262	1.0	178	15	396
6.33	209.3	15	332	1.3	174	25	610
7.06	222.9	20	430	11	280	35	877
8.96	246.4	25	552	13	330	45	1267
11.75	289.9	30	688	21	470		
12.1	295.3			33	482		
15.1	346.8	eq. 3	k_H/atm	27.2	597	Sanemasa et al. 1982	
17.93	391.6	A	8.58	25	554	vapor liquid-equilibrium	
20.06	442.4	B	1852.308			t/°C	H/(Pa m³/mol)
				eq. 3	k_H/atm	5	225
				A	19.02	15	387
				B	3964	25	608
						35	905
						45	1321

2.

Ashworth et al. 1988		Robbins et al. 1993		Perlinger et al. 1993		Ettre et al. 1993	
EPICS-GC		static headspace-GC		headspace-GC		equil. headspace-GC	
t/°C	H/(Pa m³/mol)	t/°C	H/(Pa m³/mol)	t/°C	H/(Pa m³/mol)	t/°C	H/(Pa m³/mol)
10	334	25	535	10	290	25	570
15	391	30	679	15	380	45	912
20	458	40	890	20	460	60	1220
25	535	45	1236	25	604	70	1668
30	730	50	1450	30	741	80	1767
eq. 4	H/(atm·m³/mol)					eq. 2	1/K_AW
A	5.534					A	−2.1678537
B	3194					B	836.2228

3.

Dewulf et al. 1995		Alaee et al. 1996		Turner et al. 1996		Peng & Wan 1997	
EPICS-GC		gas stripping-GC		vapor phase-equilibrium		headspace-GC	
t/°C	H/(Pa m³/mol)	t/°C	H/(Pa m³/mol)	t/°C	H/(Pa m³/mol)	t/°C	H/(Pa m³/mol)
2.0	162	4	169	10	287	15	366
6.0	208	10	228	15	390	20	436
10	228	15	326	25	640	25	538

(Continued)

TABLE 3.1.1.1.3 *(Continued)*

3.

Dewulf et al. 1995		Alaee et al. 1996		Turner et al. 1996		Peng & Wan 1997	
EPICS-GC		gas stripping-GC		vapor phase-equilibrium		headspace-GC	
t/°C	H/(Pa m^3/mol)	t/°C	H/(Pa m^3/mol)	t/°C	H/(Pa m^3/mol)	t/°C	H/(Pa m^3/mol)
18.2	366	20.6	441	35	986	30	675
25	481	25.4	552			35	766
		30.1	744			40	947
eq. 1	K_{AW}	34.9	874	eq. 5	K_{AW}	45	1053
A	10.577			A	0.0763		
B	3640			B	0.00211		
		enthalpy of volatilization:		C	0.000162	eq.1	K_{AW}
		ΔH_{vol}/(kJ·mol^{-1}) = 32.2				A	7.15
		at 20°C				B	1397
		eq. 1	K_{AW}				
		A	11.5352				
		B	3873				

4.

Bierwagen & Keller 2001		Görgényi et al. 2002		Bakierowska & T. 2003	
SPME-GC		EPICS-SPME method		headspace-GC	
t/°C	H/(Pa m^3/mol)	t/°C	H/(Pa m^3/mol)	t/°C	H/(Pa m^3/mol)
15	376	2	172.2	10	294
25	538	6	235.6	15	353.5
30	675	10	279.2	20	440
40	893	18	441.3	25	558
		25	573.4		
eq. 1	K_{AW}	30	740.0	Eq. 2	$1/K_{AW}$
A	8.1648	40	1033	A	11.663
B	2889.4	50	1429	B	3920
		60	1844		
		eq. 1	K_{AW}		
		A	10.01		
		B	3430.4		

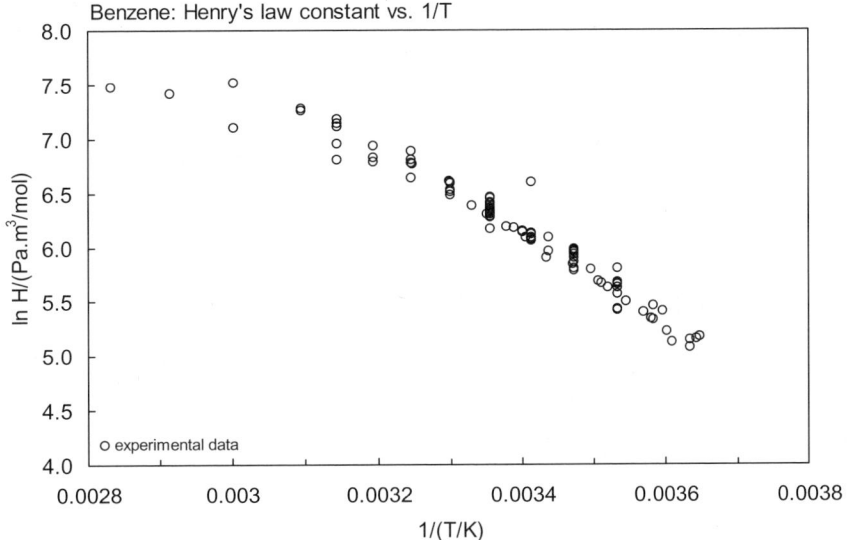

FIGURE 3.1.1.1.3 Logarithm of Henry's law constant versus reciprocal temperature for benzene.

TABLE 3.1.1.1.4
Reported octanol-water partition coefficients and octanol-air partition coefficients of benzene at various temperatures

log K_{OW}		log K_{OA}	
Dewulf et al. 1999		Gruber et al. 1997	
EPICS-GC, both phases		activity coefficient-GC	
t/°C	log K_{OW}	t/°C	log K_{OA}
2.2	1.973	20.29	2.9
6	1.961	30.3	2.71
10	2.053	40.4	2.56
14.1	2.01	50.28	2.42
18.7	2.04		
24.8	1.974		
change in enthalpy: ΔH_{OW}/(kJ mol^{-1}) = 1.7 (−8.3 to 11.8) enthalpy of transfer ΔH_{oct}/(kJ mol^{-1}) = 8.1 (−1.9 to 18.2)			

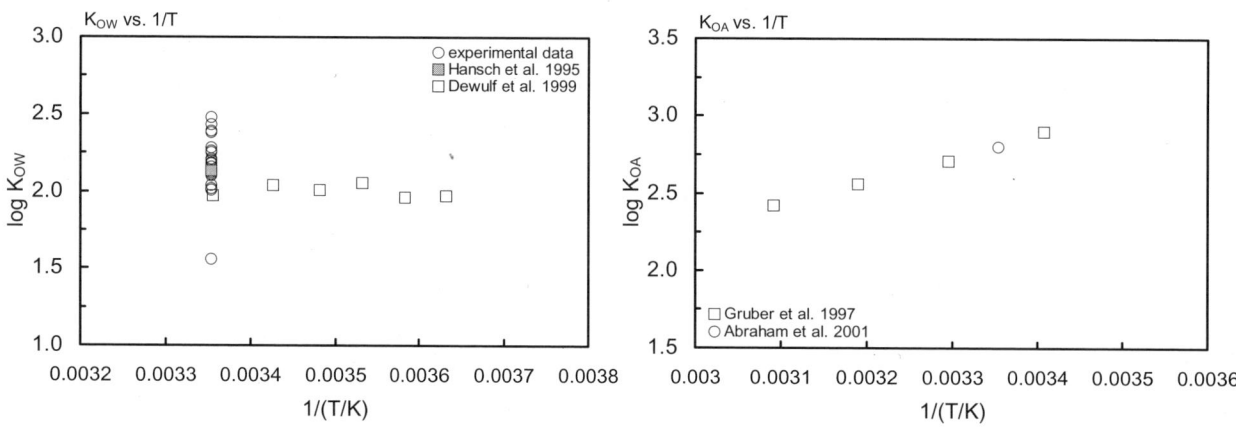

FIGURE 3.1.1.1.4 Logarithm of K_{OW} and K_{OA} versus reciprocal temperature for benzene.

3.1.1.2 Toluene

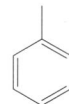

Common Name: Toluene
Synonym: methyl benzene, phenylmethane, toluol, methylbenzol, methacide
Chemical Name: toluene
CAS Registry No: 108-88-3
Molecular Formula: C_7H_8, $C_6H_5CH_3$
Molecular Weight: 92.139
Melting Point (°C):
    –94.95   (Lide 2003)
Boiling Point (°C):
    110.63   (Lide 2003)
Density (g/cm³ at 20°C):
    0.8669   (Weast 1982–83)
Molar Volume (cm³/mol):
    106.3   (20°C, calculated from density)
    118.2   (calculated-Le Bas method at normal boiling point)
Enthalpy of Vaporization, ΔH_V (kJ/mol):
    37.99, 33.183   (25°C, bp, Riddick et al. 1986)
Enthalpy of Fusion ΔH_{fus} (kJ/mol):
    6.636   (Riddick et al. 1986)
    6.62   (exptl., Chickos et al. 1999)
Entropy of Fusion ΔS_{fus} (J/mol K):
    37.15, 45.0   (exptl., calculated-group additivity method, Chickos et al. 1999)
Fugacity Ratio at 25°C, F: 1.0

Water Solubility (g/m³ or mg/L at 25°C or indicated. Additional data at other temperatures designated * are compiled at the end of this section):

470	(16°C, shake flask, Fühner, 1924)
570	(30°C, shake flask-interferometer, Gross & Saylor 1931)
347	(residue-volume method, Booth & Everson 1948)
530	(shake flask-UV, Andrews & Keffer 1949)
500	(flask flask-UV, Klevens 1950)
627*	(shake flask-UV, measured range 0.4–45.3°C, Bohon & Claussen 1951)
546	(shake flask-UV, Morrison & Billett 1952)
550	(Dreisbach 1955)
595	(quoted, Deno & Berkheimer 1960)
538	(shake flask-GC, McAuliffe 1963)
515	(shake flask-GC, McAuliffe 1966)
479	(21°C, shake flask-GC, Chey & Calder 1972)
530*	(shake flask-GC, measured range 5–45°C, Pierotti & Liabastre 1972)
547*	(shake flask-UV, measured 25–55°C, Bradley et al. 1973)
573*	(shake flask-GC, Polak & Lu 1973)
517	(shake flask-GC, Mackay & Wolkoff 1973)
573*	(headspace-GC, measured range 4.5–20.1°C, Brown & Wasik 1974)
627	(shake flask-UV, Vesala 1974)
520	(shake flask-GC, Mackay & Shiu 1975)
534.8	(shake flask-GC, Sutton & Calder 1975)
554	(shake flask-GC, Price 1976)
488; 563	(shake flask-titration, shake flask-cloud point, Sada et al. 1975)

534   (shake flask-fluorescence, Schwarz 1977)
554   (shake flask-GC, Krzyzanowska & Szeliga 1978)
log x = 626.526/(T/K)^2 − 4300.59/(T/K) + 3.3585, temp range 0–50°C (Ueda et al. 1978)
572, 587   (10, 20°C, shake flask-UV, Ben-Naim & Wiff 1979)
660   (elution chromatography, Schwarz 1980)
732*, 739* (20°C, exptl.-elution chromatography, shake flask-UV, Schwarz & Miller 1980)
155   (shake flask-LSC, Banerjee et al. 1980)
507   (shake flask-GC, Rossi & Thomas 1981)
557*   (vapor saturation-UV spec., measured range 15–45°C, Sanemasa et al. 1981)
526*   (vapor saturation-UV spec., measured range 15–45°C, Sanemasa et al. 1982)
585   (generator column-HPLC/UV, Tewari et al. 1982b)
578   (generator column-HPLC/UV, Tewari et al. 1982c)
580   (generator column-HPLC/UV, Wasik et al. 1983)
524   (shake flask-HPLC/UV, Banerjee 1984)
521   (vapor saturation-UV spec., Sanemasa et al. 1984)
ln x = −185.1695 + 7348.55/(T/K) + 26.34525·ln (T/K); temp range 310–560 K (Heidman et al. 1985)
520   (shake flask-radiometry, Lo et al. 1986)
580   (shake flask-GC, Keeley et al. 1988)
538   (shake flask-GC, Coutant & Keigley 1988)
530*   (IUPAC recommended value, temp range 5–55°C, Shaw 1989a)
599*   (30°C, equilibrium flow cell-GC; measured range 30–100°C Chen & Wagner 1994b)
ln (1/x) = −12.21 + 21.39·[(T/K)/591.8]$^{-1}$ − 3.572·[(T/K)/591.8]$^{-2}$; temp range 303.15–373.15 K (equilibrium flow cell-GC, Chen & Wagner 1994b)
ln x = 12.21 − 21.39·(T$_r$/K)$^{-1}$ + 5.372·(T$_r$/K)$^{-2}$, T$_r$ = T/T$_c$, the reduced temp, system temp T divided by critical temp T$_c$ (Chen & Wagner 1994c)
562.9   (shake flask-UV spectrophotometry, Poulson et al. 1999)
ln x = −46.05 − 7268.85/(T/K) − 1.411 × 10^{-4}·(T/K)^2; temp range 5–50°C (regression eq. of literature data, Shiu & Ma 2000)
519*   (shake flask-GC/FID, measured range 5–45°C, Ma et al. 2001)
556*   (vapor absorption technique-HPLC/UV, measured range 0.5–55°C, Dohányosová et al. 2001)
558*   (shake flask-UV, measured range 0–50°C, Sawamura et al. 2001)
ln x = −221.739 + 9274.79/(T/K) + 31.8721·ln (T/K); temp range 290–400 K (eq. derived from literature calorimetric and solubility data, Tsonopoulos 1999)

Vapor Pressure (Pa at 25°C and reported temperature dependence equation. Additional data at other temperatures designated * are compiled at the end of this section):
920*   (0°C, mercury manometer, measured range −9.70 to 0°C, Linder 1931)
3786*   (Hg manometer measurements, Pitzer & Scott 1943)
log (P/mmHg) = −2866.53/(T/K) − 6.7 log (T/K) + 27.6470; temp range: 0–50°C (manometer, three-constant vapor pressure eq. from exptl. data, Pitzer & Scott 1943)
6357*   (35.366°C, ebulliometry, measured range 35.366–111.509°C, Willingham et al. 1945)
log (P/mmHg) = 6.95337 − 1343.943/(219.377 + t/°C); temp range 35.4–111.5°C (manometer, Antoine eq. from exptl. data, Willingham et al. 1945)
2666*   (18.4°C, summary of literature data, temp range −16.7 to 110.6°C, Stull 1947)
6386*   (35.504°C, ebulliometry, measured range 35.504–111.545°C, Forziati et al. 1949)
log (P/mmHg) = 6.95508 − 1345.087/(219.516 + t/°C); temp range 35.5–111.5°C (manometer, Antoine eq. from exptl. data, Forziati et al. 1949)
3792   (calculated from det. data, Dreisbach 1955)
log (P/mmHg) = 6.95334 − 1343.943/(219.377 + t/°C); temp range 20–200°C (Antoine eq. for liquid state, Dreisbach 1955)
1333*   (6.375°C, compiled data, temp range 6.375–136.435°C, Bond & Thodos 1960)
256200*   (146.85°C, ebulliometry, measured range 146.85–306.85°C, Ambrose et al. 1967)
48898*   (86.85°C, compiled data, temp range 86.85–306.85°C, Ambrose et al. 1970)
3792*   (interpolated, Antoine eq., Zwolinski & Wilhoit 1971)

Mononuclear Aromatic Hydrocarbons

log (P/mmHg) = 6.95464 – 1344.80/(219.482 + t/°C); temp range; 6.36–136.42°C (Antoine eq., Zwolinski & Wilhoit 1971)

log (P/mmHg) = [–0.2185 × 9368.5/(T/K)] + 8.3300; temp range –92 to –15°C (Antoine eq., Weast 1972–73)

log (P/mmHg) = [–0.2185 × 8586.5/(T/K)] + 8.719392; temp range –26.7 to 319°C (Antoine eq., Weast 1972–73)

2904* (19.99°C, differential capacitance gauge, measured range 0–49.26°C, Munday et al. 1980)

log (P/mmHg) = –5541.623/(T/K) + 25.08047 – 0.01055321(T/K); temp range 0–49.26°C (differential capacitance gauge, Munday et al. 1980)

3560 (gas saturation-GC, Politzki et al. 1982)

log (P/atm) = (1 – 383.737/T) × 10^(0.837122 – 6.48791 × $10^4 \cdot T$ + 5.91293 × $10^{7} \cdot T^2$); T in K, temp range 245.0–590.0 K (Cox vapor pressure eq., Chao et al. 1983)

log (P/kPa) = 6.08436 – 1347.62/(219.787 + t/°C); temp range 35.37–111.5°C (Antoine eq. from reported exptl. data, Boublik et al. 1984)

log (P/kPa) = 6.37988 – 1575.007/(249.372 + t/°C); temp range 86.85–306.8°C (Antoine eq. from reported exptl. data, Boublik et al. 1984)

3786 (Daubert & Danner 1985)

3790 (interpolated-Antoine eq., Dean 1985, 1992)

log (P/mmHg) = 6.95464 – 1344.80/(219.48 + t/°C); temp range 6–137°C (Antoine eq., Dean 1985, 1992)

3790 (headspace-GC, Hussam & Carr 1985)

log (P/kPa) = 6.08540 – 1348.77/(219.976 + t/°C); temp range not specified (Antoine eq., Riddick et al. 1986)

3800 (extrapolated-Antoine eq., Stephenson & Malanowski 1987)

log (P_L/kPa) = 6.08627 – 1349.122/(–53.154 + T/K); temp range 308–386 K (liquid, Antoine eq.-I, Stephenson & Malanowski 1987)

log (P_L/kPa) = 6.1258 – 1376.61/(–51.1 + T/K); temp range 210–219 K (liquid, Antoine eq.-II, Stephenson & Malanowski 1987)

log (P_L/kPa) = 6.12012 – 1374.901/(–49.657 + T/K); temp range 383–445 K (liquid, Antoine eq.-III, Stephenson & Malanowski 1987)

log (P_L/kPa) = 6.40815 – 1615.834/(–15.897 + T/K); temp range 440–531 K (liquid, Antoine eq.-IV, Stephenson & Malanowski 1987)

log (P_L/kPa) = 7.65383 – 3153.235/(188.566 + T/K); temp range 530–592 K (liquid, Antoine eq.-V, Stephenson & Malanowski 1987)

log (P_L/kPa) = 6.16273 – 1391.005/(–48.974 + T/K); temp range 273–295 K (liquid, Antoine eq.-VI, Stephenson & Malanowski 1987, selected, Shiu & Ma 2000)

3090* (20.98°C, static method, measured range 199.22–402.21 K, Mokbel et al. 1998)

Henry's Law Constant (Pa m^3/mol at 25°C and reported temperature dependence equations. Additional data at other temperatures designated * are compiled at the end of this section):

474* (headspace-GC, Brown & Wasik 1974)

527 (headspace-GC, Vitenberg et al. 1975)

673 (batch air stripping-GC, Mackay et al. 1979)

625* (23.0°C, equilibrium cell-concn ratio-GC/FID, Leighton & Calo 1981)

ln (k_H/atm) = 18.46 – 3751/(T/K); temp range:1.0–23.0°C (equilibrium cell-concn ratio, Leighton & Calo 1981)

628* (vapor-liquid equilibrium-GC., Sanemasa et al. 1981)

664* (vapor-liquid equilibrium, Sanemasa et al. 1982)

647 (EPICS-GC, Garbarnini & Lion 1985)

634* (20°C, headspace-GC, measured range 20–46°C, Schoene & Steinhanses 1985)

601 (gas stripping-GC, Warner et al. 1987)

594 (20°C, EPICS, Yurteri et al. 1987)

651* (EPICS-GC, Ashworth et al. 1988)

ln [H/(atm m^3/mol)] = 5.133 – 3024/(T/K); temp range 10–30°C (EPICS measurements, Ashworth et al. 1988)

652 (infinite activity coeff. γ^∞ from solubility measurement, Abraham et al. 1990)

644 (calculated-vapor-liquid equilibrium (VLE) data, Yaws et al. 1991)

933* (40°C, static headspace-GC, measured range 40–80°C, Kolb et al. 1992)

ln (1/K_{AW}) = –6.03 + 2198/(T/K); temp range 40–80°C (equilibrium headspace-GC measurements, Kolb et al. 1992)

1116* (45°C, headspace-GC, measured range 45–80°C, Ettre et al. 1993)

log $(1/K_{AW})$ = –2.5323790 + 928.3536/(T/K); temp range 45–80°C (equilibrium headspace-GC measurements, Ettre et al. 1993)
631   (infinite activity coeff. γ^∞ in water determined by inert gas stripping-GC, Li et al. 1993)
660*   (equilibrium headspace-GC, Perlinger et al. 1993)
652*   (static headspace-GC, Robbins et al. 1993)
644   (headspace solid-phase microextraction (SPME)-GC, Zhang & Pawliszyn 1993)
676   (23°C, gas stripping-IR, Nielsen et al. 1994)
555*   (EPICS-GC/FID, Dewulf et al. 1995)
699   (EPICS-GC/FID, natural seawater with salinity of 35‰ Dewulf et al. 1995)
ln K_{AW} = –4064/(T/K) + 0.00834·Z + 12.150; with Z salinity 0–35.5l, temp range 2–35°C, (EPICS-GC/FID, Dewulf et al. 1995)
541   (20°C, selected from literature experimentally measured data, Staudinger & Roberts 1996)
684*   (vapor-liquid equilibrium.-GC, Turner et al. 1996)
595   (gas stripping-GC, Altschuh et al. 1999)
605*   (headspace equilibrium-GC, Peng & Wan 1997)
ln K_{AW} = 7.94 – 1621/(T/K); temp range 15–45°C (headspace equilibrium-GC, Peng & Wan 1997)
478   (headspace-GC, Peng & Wan 1998)
ln K_{AW} = 7.89 – 1565/(T/K); temp range 0–45°C (seawater with salinity of 36%, headspace-GC, Peng & Wan 1998)
637.2   (exponential saturator EXPSAT technique, Dohnal & Hovorka 1999)
674.8   (modified EPICS method-GC, Ryu & Park 1999)
652   (EPICS-GC, David et al. 2000)
644   (EPICS-static headspace method-GC/FID, Miller & Stuart 2000)
959   (EPICS-GC, Ayuttaya et al. 2001)
509   (20°C, selected from literature experimentally measured data, Staudinger & Roberts 2001)
log K_{AW} = 5.271 – 1745/(T/K) (van = t Hoff eq. derived from lit. data, Staudinger & Roberts 2001)
647.5*   (EPICS-SPME, measured range 2–60°C, Görgényi et al. 2002)
ln K_{AW} = 11.25 – 3770.4/(T/K); temp range 2–60°C (EPICS-SPME method, Görgényi et al. 2002)
556–574 (27°C, solid-phase microextraction-GC, solute concn 0.47–19.21 mg/L, measured range 15–40°C, Cheng et al. 2003)
612*   (headspace-GC, measured range 10–25°C, Bakierowska & Trzeszczyński 2003)
ln $(1/K_{AW})$ = 11.926 – 3977/(T/K); temp range 10–25°C, headspace-GC, Bakierowska & Trzeszczyński 2003)

Octanol/Water Partition Coefficient, log K_{OW} at 25°C or as indicated. Additional data at other temperatures designated * are compiled at the end of this section:
2.69   (shake flask-UV, Fujita et al. 1964; quoted, Hansch et al. 1968; Hansch et al. 1972)
2.11   (shake flask-UV, Rogers & Cammarata 1969)
2.69, 2.73, 2.11, 2.80   (Leo et al. 1971; Hansch & Leo 1979)
2.21   (shake flask-LSC, Banerjee et al. 1980)
2.68   (shake flask-HPLC, Nahum & Horvath 1980)
2.59   (HPLC-k' correlation, Hanai et al. 1981)
2.97   (HPLC-k' correlation, McDuffie 1981)
2.78   (HPLC-k' correlation, Hammers et al. 1982)
2.59   (HPLC-k' correlation, D'Amboise & Hanai 1982)
2.65   (generator column-HPLC/UV, Tewari et al. 1982b,c)
2.62   (HPLC-k' correlation, Miyake & Terada 1982)
2.65   (generator column-HPLC/UV, Wasik et al. 1983)
2.74   (HPLC-k' correlation, Hafkanscheid & Tomlinson 1983b)
2.11–2.80, 2.65   (range, mean; shake flask method, Eadsforth & Moser 1983)
2.51–3.06, 2.88   (range, mean; HPLC method, Eadsforth & Moser 1983)
2.10   (shake flask, Platford 1979, 1983)
2.72   (HPLC-RV correlation, Garst & Wilson 1984)
2.89   (HPLC-RT correlation, Rapaport & Eisenreich 1984)
2.78   (HPLC/MS correlation, Burkhard et al. 1985)
3.00   (HPLC-k' correlation, De Kock & Lord 1987)
2.65   (generator column-RP-LC, Schantz & Martire 1987)

2.62 (RP-HPLC-RT correlation, ODS column with masking agent, Bechalany et al. 1989)
2.73 (recommended, Sangster 1989, 1993)
2.66, 2.69 (RP-HPLC capacity factor correlations, Sherblom & Eganhouse 1988)
2.786 (shake flask/slow stirring-GC, De Brujin et al. 1989)
2.63 ± 0.05, 2.786 ± 0.005 (shake flask/slow stirring, interlaboratory studies, Brooke et al. 1990)
2.76 (normal phase HPLC-k' correlation, Govers & Evers 1992)
2.73 (recommended, Hansch et al. 1995)
2.77 ± 0.02 (HPLC-k' correlation, Poulson et al. 1997)
2.32* (24.8°C, EPICS-GC, measured range 2.2–24.8°C, Dewulf et al. 1999)

Octanol/Air Partition Coefficient, log K_{OA} at 25°C or as indicated. Additional data at other temperatures designated * are compiled at the end of this section:

3.42* (20.29°C, from GC-determined γ^∞ in octanol, measured range 20.29–50.28°C, Gruber et al. 1997)
3.31 (head-space GC/FID both phases, Abraham et al. 2001)

Bioconcentration Factor, log BCF:

1.12 (eels, Ogata & Miyake 1978)
0.22 (Manila clam, Nunes & Benville 1979)
0.62 (mussels, Geyer et al. 1982)
0.92 (goldfish, Ogata et al. 1984),
3.28, 2.58, 1.95 (activated sludge, algae, fish, Freitag et al. 1985)
1.99 (*S. capricornutum*, Herman et al. 1991)

Sorption Partition Coefficient, log K_{OC} at 25°C or as indicated:

2.39 (average 5 soils and 3 sediments, sorption isotherms by batch equilibrium and column experiments, Schwarzenbach & Westall 1981)
2.27, 1.89 (ICN humic acid, ICN HA coated Al_2O_3, headspace equilibrium, Garbarnini & Lion 1985)
2.28, 1.89 (Offutt AFB soil, Whitean AFB soil, headspace equilibrium, Garbarnini & Lion 1985)
1.91, 1.13, 1.19, 2.18, 2.09, −1.30, 1.94 (Sapsucker Woods humic acid, Sapsucker Woods fulvic acid, tannic acid, lignin, zein, cellulose, Aldrich humic acid, headspace equilibrium, Garbarnini & Lion 1986)
2.18, 2.21, 2.43, 2.54 (Sapsucker Woods S.W. soil, S.W. ethyl ether extracted soil, humin, oxidized humin, headspace equilibrium, Garbarnini & Lion 1986)
1.74. 2.13, 1.98 (forest soil pH 5.6, forest soil pH 4.2, agricultural soil pH 7.4, Seip et al. 1986)
2.0 (sediment 4.02% OC from Tamar estuary, batch equilibrium-GC, Vowles & Mantoura 1987)
1.70 (untreated Marlette soil A horizon, organic carbon OC 2.59%, batch equilibrium-adsorption isotherm, Lee et al. 1989)
2.50, 2.39 (organic cations treated Marlette soil A horizon: HDTMA treated OC 6.48%; DDTMA treated, OC 4.37%, batch equilibrium-adsorption isotherm, Lee et al. 1989)
2.86, 2.86, 2.43 (organic cations treated Marlette soil B_t horizon: HDTMA treated OC 3.72%, DDTMA treated OC 1.98%, NTMA treated, OC 1.18%, batch equilibrium-adsorption isotherm, Lee et al. 1989)
1.59 (untreated St Clair soil B_t horizon, OC 0.44%, batch equilibrium, Lee et al. 1989)
3.03, 2.90 (organic cations HDTMA treated soils: St. Clair soil B_t horizon OC 3.25%; Oshtemo soil B_t horizon OC 0.83%, batch equilibrium-adsorption isotherm, Lee et al. 1989)
2.01 (aquifer material with f_{OC} = 0.006 and measured partition coeff. K_P = 0.61 mL/g., Abdul et al. 1990)
2.10, 2.26 (HPLC-k' correlation, humic acid-silica column, Szabo et al. 1990a,b)
2.22, 2.16 (Captina silt loam, Mclaurin sandy loam, batch equilibrium, Walton et al. 1992)
2.10 (average of 5 soils, sorption isotherms by batch equilibrium method-GC, Xing et al. 1994)
2.21, 2.31, 2.21 (RP-HPLC-k' correlation on 3 different stationary phases, Szabo et al. 1995)
2.17, 2.18 (RP-HPLC-k' correlation including MCI related to non-dispersive intermolecular interactions, hydrogen-bonding indicator variable, Hong et al. 1996)
2.12 (HPLC-screening method, Müller & Kördel 1996)
2.23, 2.31, 2.33, 234, 2.40, 2.31, 2.34 (2.3, 3.8, 6.2, 8.0, 13.5, 18.6, 25°C, natural sediment from River Leie, organic carbon f_{OC} = 4.12%, EPICS-GC/FID, Dewulf et al. 1999)
1.89, 2.00, 1.79 (soils: organic carbon OC ≥ 0.1%, OC ≥ 0.5%, 0.1 ≤ OC < 0.5%, average, Delle Site 2001)

Sorption Partition Coefficient, log K_{OM}:

1.39 (untreated Marlette soil A horizon, organic matter OM 5.18%, batch equilibrium-adsorption isotherm, Lee et al. 1989)

2.30, 2.16 (organic cations treated Marlette soil A horizon: HDTMA treated, organic matter 10.03%; DDTMA treated, OM 5.18%, batch equilibrium-adsorption isotherm, Lee et al. 1989)

2.74, 2.72, 2.28 (organic cations treated Marlette soil B_t horizon: HDTMA treated OM 4.85%, DDTMA treated OM 2.73%, NTMA treated, OM 1.74%, batch equilibrium-adsorption isotherm, Lee et al. 1989)

1.29 (untreated St Clair soil B_t horizon, OM 0.88%, batch equilibrium, Lee et al. 1989)

2.89; 2.74 (organic cations HDTMA treated soils: St. Clair soil B_t horizon OM 4.38%; Oshtemo soil B_t horizon OM 1.12%, batch equilibrium-adsorption isotherm, Lee et al. 1989)

Environmental Fate Rate Constants, k, or Half-Lives, $t_{1/2}$:

Volatilization: $t_{1/2}$ = 5.18 h from water depth of 1-m (Mackay & Leinonen 1975; Haque et al. 1980);

k = 0.043 d^{-1} with $t_{1/2}$ = 16 d in spring at 8–16°C, k = 0.0463 d^{-1} with $t_{1/2}$ = 1.5 d in summer at 20–22°C, k = 0.053 d^{-1} with $t_{1/2}$ = 13 d in winter at 3–7°C for the periods when volatilization appears to dominate, and k = 0.088 d^{-1} with $t_{1/2}$ = 7.6 d with $HgCl_2$, in September 9–15, in marine mesocosm experiments (Wakeham et al. 1983);

evaporation $t_{1/2}$ ~2.9 d from a river of 1-m depth with wind speed of 3 m/s and water current of 1 m/s at 20°C (Lyman et al. 1982);

estimated $t_{1/2}$ = 1 and 4 d for evaporation from a river and lake, respectively (Howard 1990).

Photolysis: not environmentally significant or relevant (Mabey et al. 1982);

k = 3.39 × 10^{-3} h^{-1} with H_2O_2 under photolysis at 25°C in F-113 solution and with HO· in the gas (Dilling et al. 1988); $t_{1/2}$ < 0.25 h on silica gel under indoor artificial UV-light "continuous" condition (Söderström et al. 2004).

Oxidation: rate constant k; for gas-phase second order rate constants, k_{OH} for reaction with OH radical, k_{NO3} with NO_3 radical and k_{O3} with O_3 or as indicated. Data at other temperatures and/or the Arrhenius expression are designated *, see reference:

k_{OH} = (2.5 ± 0.9) × 10^9 L mol^{-1} s^{-1} with $t_{1/2}$ = 4.6 h; $k_{O(^3P)}$ = (0.450 ± 0.045) × 10^8 L mol^{-1} s^{-1} with $O(^3P)$ atom at room temp. (relative rate method, Doyle et al. 1975; Lloyd et al. 1976)

k_{OH} = 5.78 × 10^{-12} cm^3 $molecule^{-1}$ s^{-1}; k_{O3} = 0.75 × 10^{-13} cm^3 $molecule^{-1}$ s^{-1} for the reaction of $O(^3P)$ atom at room temp. (flash photolysis-resonance fluorescence, Hansen et al. 1975)

k_{OH} = 3.6 × 10^9 L mol^{-1} s^{-1}, with $t_{1/2}$ = 2.4–24 h at room temp. (Darnall et al. 1976)

k_{OH}* = (6.40 ± 0.64) × 10^{-12} cm^3 $molecule^{-1}$ s^{-1} at room temp., measured over temp range 296–473 K (flash photolysis-resonance fluorescence, Perry et al. 1977)

k_{OH} = 6 × 10^{-12} cm^3 $molecule^{-1}$ s^{-1}, k_{NO3} ≤ 3 × 10^{-14} cm^3 $molecule^{-1}$ s^{-1} at 300 ± 1 K (Carter et al. 1981)

k_{OH} = 6.0 × 10^{-12} cm^3 $molecule^{-1}$ s^{-1} and residence time of 1.9 d, loss of 40.9% in one day or 12 sunlit hours at 300 K in urban environments (Singh et al. 1981)

k << 360 M^{-1} h^{-1} for singlet oxygen and k = 144 M^{-1} h^{-1} for RO_2 radical (Mabey et al. 1982)

k_{O3} = 160 cm^3 mol^{-1} s^{-1} at 300 K (Lyman et al. 1982)

k_{OH} = 3.5 × 10^9 M^{-1} s^{-1} with $t_{1/2}$ = 1.6 d in the atmosphere (Mill 1982)

k = (14 ± 3) M^{-1} s^{-1} for the reaction with ozone in water at pH 1.7 and 20–23°C (Hoigné & Bader 1983)

k_{O3} < 1.0 × 10^{-20} cm^3 $molecule^{-1}$ s^{-1} with a loss rate of < 0.0006 d^{-1}; k_{OH} = 6.4 × 10^{-12} cm^3 $molecule^{-1}$ s^{-1} with a loss rate of 0.6 d^{-1} and k_{NO3} = 3.7 × 10^{-17} cm^3 $molecule^{-1}$ s^{-1} with a loss rate of 0.0007 d^{-1} at room temp. (review, Atkinson & Carter 1984)

k_{NO3} = 1.8 × 10^{-17} cm^3 $molecule^{-1}$ s^{-1} at 296 K (Atkinson et al. 1984)

k_{OH}* = 6.19 × 10^{-12} cm^3 $molecule^{-1}$ s^{-1} at 298 K (recommended, Atkinson 1985)

k_{OH}(calc) = 5.9 × 10^{-12} cm^3 $molecule^{-1}$ s^{-1}, k_{OH}(obs.) = 5.7 × 10^{-12} cm^3 $molecule^{-1}$ s^{-1} at room temp. (SAR [structure-activity relationship], Atkinson 1985)

k_{OH} = 5.7 × 10^{-12} cm^3 $molecule^{-1}$ s^{-1} with $t_{1/2}$ = 10–104 h (Atkinson 1985; Howard 1991)

k_{O3} < 1.0 × 10^{-20} cm^3 $molecule^{-1}$ s^{-1} with a loss rate of < 0.0006 d^{-1}; k_{OH} = 6.2 × 10^{-12} cm^3 $molecule^{-1}$ s^{-1} with a loss rate of 0.27 d^{-1} and k_{NO3} = 3.6 × 10^{-17} cm^3 $molecule^{-1}$ s^{-1} with a loss rate of 0.0007 d^{-1} at room temp. (review, Atkinson 1985)

k_{OH} = 6.03 × 10^{-12} cm^3 $molecule^{-1}$ s^{-1} at 298 K (relative rate method, Ohta & Ohyama 1985)

k_{OH} = 5.35 × 10^{-12} cm^3 $molecule^{-1}$ s^{-1} at 24.2°C, with a calculated atmospheric lifetime τ = 2.2 d (Edney et al. 1986)

k_{OH}(calc) = 5.5 × 10^{-12} cm^3 molecule^{-1} s^{-1}, k_{OH}(obs.) = 6.19 × 10^{-12} cm^3 molecule^{-1} s^{-1} at room temp. (SAR structure-activity relationship, Atkinson 1987)

k_{NO3} = 6.46 × 10^{-17} cm^3 molecule^{-1} s^{-1}, k_{OH} = 6.19 × 10^{-12} cm^3 molecule^{-1} s^{-1} at 298 K (Atkinson et al. 1988; quoted, Sabljic & Güsten 1990; Müller & Klein 1992)

k_{O3} < 0.01 × 10^{-18} cm^3 molecule^{-1} s^{-1}; k_{OH} = 6.19 × 10^{-12} cm^3 molecule^{-1} s^{-1}, and k_{NO3} = (7.8 ± 1.5) × 10^{-17} cm^3 molecule^{-1} s^{-1} at room temp. (relative rate method, Atkinson & Aschmann 1988)

k_{OH}* = 5.96 × 10^{-12} cm^3 molecule^{-1} s^{-1} at 298 K (recommended, Atkinson 1989, 1990)

photooxidation $t_{1/2}$ = 10–104 h, based on measured rate data for the vapor phase reaction with OH radicals in air; $t_{1/2}$(aq.) = 321–1284 h, based on measured rate data for hydroxyl radicals in aqueous solution (Howard et al. 1991)

k_{OH}(calc) = 5.50 × 10^{-12} cm^3 molecule^{-1} s^{-1} at 298 K (estimated by SARs, Müller & Klein 1992)

k_{OH}(calc) = 4.79 × 10^{-12} cm^3 molecule^{-1} s^{-1} (based on molecular orbital calculations, Klamt 1993)

Hydrolysis: not aquatically significant (Callahan et al. 1979); no hydrolyzable functional groups (Mabey et al. 1982).

Biodegradation:

100% biodegraded after 192 h at 13°C with an initial concn of 2.22 × 10^{-6} L/L (Jamison et al. 1976)

$t_{1/2}$ ≈ 90 d in uncontaminated estuarine water; and $t_{1/2}$ ≈ 30 d in oil polluted water (Lee 1977)

k = 0.5 d^{-1}, significant degradation in aerobic environment (Tabak et al. 1981; Mills et al. 1982)

k = 0.07 yr^{-1} with $t_{1/2}$ = 39 d (Zoeteman et al. 1981; Olsen & Davis 1990)

$t_{1/2}$(aq. anaerobic) = 1344–5040 h, based on anaerobic screening test data and anaerobic sediment grab sample data (Horowitz et al. 1982; Howard et al. 1991)

$t_{1/2}$ = 9.5 d in activated sludge (estimated, Freitag et al. 1985; quoted, Anderson et al. 1991)

$t_{1/2}$(aq. aerobic) =: 96–528 h, based on an acclimated seawater dieaway test (Howard et al. 1991)

$t_{1/2}$(aerobic) = 4 d, $t_{1/2}$(anaerobic) = 56 d in natural waters (Capel & Larson 1995)

$t_{1/2}$ = 31–220 h for toluene concn range from 0.5–200 µg/g in sandy loam soil and degradation rate k = 1.76 × 10^{-2} and 0.42 µg g^{-1} h^{-1} of soil for 0.5 and 5.0 -g/g, respectively; $t_{1/2}$ = 172 and 165 h in sand and degradation rate k = 1.05 × 10^{-2} and 0.22 µg g^{-1} h^{-1} of soil for toluene concn 0.5 and 5 µg/g respectively in sand (Davis & Madsen 1996)

Biotransformation: 1.0 × 10^{-7} mL cell^{-1} h^{-1} (Mabey et al. 1982).

Bioconcentration, Uptake (k_1) and Elimination (k_2) Rate Constants or Half-Lives:

$t_{1/2}$ ~ 1.4 d elimination from eels in seawater (Ogata & Miyake 1978).

Half-Lives in the Environment:

Air: $t_{1/2}$ = 4.6 h in ambient air based on reaction with OH radical at ~300K (Doyle et al. 1975);

$t_{1/2}$ = 2.4–24 h based on rate of disappearance for the reaction with hydroxyl radicals (Darnall et al. 1976)

photodecomposition $t_{1/2}$ = 6.8 h under simulated atmospheric conditions, with NO (Dilling et al. 1976)

residence time of 1.9 d, loss of 40.9% in one day or 12 sunlit hours at 300 K in urban environments (Singh et al. 1981)

calculated lifetime τ = 2.2 d due to reaction with OH radical (Edney et al. 1986)

summer daylight lifetime of 23 h due to reaction with OH radical (Altshuller 1991)

$t_{1/2}$ = 10–104 h, based on photooxidation in air (Howard et al. 1991)

calculated lifetimes, τ = 1.9 d, 1.9 yr and >4.5 yr for reactions with OH radical, NO$_3$ radical and O$_3$ respectively (Atkinson 2000).

Surface Water: $t_{1/2}$ = 5.55 h, based on evaporative loss at 25°C and 1-m depth of water (calculated, Mackay & Leinonen 1975; Haque et al. 1980)

photooxidation $t_{1/2}$ = 321–1284 h in water, based on measured rate data for hydroxyl radical in aqueous solution (Dorfman & Adams 1973; Howard et al. 1991)

$t_{1/2}$ = 16 d in spring at 6–16°C, $t_{1/2}$ = 1.5 d in summer at 20–22°C, $t_{1/2}$ = 13 d in winter at 3–7°C when volatilization dominates and $t_{1/2}$ = 7.9 d with HgCl$_2$ in marine mesocosm experiments (Wakeham et al. 1983)

$t_{1/2}$ = 96–528 h, based on estimated aqueous aerobic biodegradation half-life (Howard et al. 1991).

Ground water: $t_{1/2}$ ≈ 0.3 yr from observed persistence in groundwater of Netherlands (Zoeteman et al. 1981);

$t_{1/2}$ = 168–672 h, based on unacclimated grab sample data of aerobic soil from groundwater aquifers (Wilson et al. 1983; Swindoll et al. 1987; Howard et al. 1991).

Soil: $t_{1/2}$ < 10 d (Ryan et al. 1988)

$t_{1/2}$ = 5 d assumed first-order biological/chemical degradation in soil (Jury et al. 1990);

reported lit. $t_{½}$ = 0.1 – 1.7 d and 7 d in soil, measured disappearance $t_{½}$ < 2.0 d from test soils (Anderson et al. 1991)

$t_{½}$ = 96–528 h, based on estimated aqueous aerobic biodegradation half-life (Howard et al. 1991)

$t_{½}$ = 31, 57, 96 and 220 h with toluene concn range from 0.5, 5, 50 and 200 µg/g in sandy loam soil, $t_{½}$ = 172 and 165 h with toluene concn 0.5 and 5 µg/g, respectively, in sand (Davis & Madsen 1996).

Biota: $t_{½}$ = 10 h clearance from fish (Neely 1980).

TABLE 3.1.1.2.1
Reported aqueous solubilities of toluene at various temperatures and reported enthalpy of solution

1.

Bohon & Claussen 1951		Pierotti & Liabastre 1972		Bradley et al. 1973		Brown & Wasik 1974	
shake flask-UV		shake flask-GC		shake flask-UV		headspace-GC	
t/°C	S/g·m⁻³	t/°C	S/g·m⁻³	t/°C	S/g·m⁻³	t/°C	S/g·m⁻³
0.4	658	5	634	25	547	25	573
3.6	646	10	632	45	722	4.5	612
10	628	20	661	55	860	6.3	601
11.2	624	25	630			7.1	586
14.9	623	35	672			9	587
15.9	621	45	672	**Polak & Lu 1973**		11.8	573
25	627			shake flask-GC		12.1	575
25.6	625	$\Delta_{sol}H/(kJ\ mol^{-1})$ = 1.50		t/°C	S/g·m⁻³	15.1	569
30	640	at 25°C				17.9	577
30.2	642			0	724	20.1	566
35.2	657			25	573		
42.8	701					$\Delta_{sol}H/(kJ\ mol^{-1})$ = 4.70	
45.3	717					at 25°C	
$\Delta_{sol}H/(kJ\ mol^{-1})$							
25°C	2.3						
2°C	–3.93						
7°C	–3.01						
12°C	–1.38						
17°C	0						
22°C	0.962						
27°C	3.22						
32°C	4.435						
37°C	5.73						
42°C	7.15						

2.

Ben-Naim & Wiff 1979		Schwarz & Miller 1980		Sanemasa et al. 1981		Shaw 1989a (IUPAC)	
shake flask-UV		shake flask-UV		vapor saturation-UV		recommended values	
t/°C	S/g·m⁻³	t/°C	S/g·m⁻³	t/°C	S/g·m⁻³	t/°C	S/g·m⁻³
10	572	10	777	15	533	0	690
20	587	20	739	25	557	5	630
		30	754	35	587	10	590
				45	635	20	570
						25	530
		Elution chromatography		$\Delta H_{sol}/(kJ\ mol^{-1})$ = 3.70		30	590
		t/°C	S/g·m⁻³	at 25°C		40	640
		10	758			45	660

Mononuclear Aromatic Hydrocarbons

TABLE 3.1.1.2.1 *(Continued)*

Ben-Naim & Wiff 1979		Schwarz & Miller 1980		Sanemasa et al. 1981		Shaw 1989 (IUPAC)	
shake flask-UV		shake flask-UV		vapor saturation-UV		recommended values	
t/°C	S/g·m⁻³	t/°C	S/g·m⁻³	t/°C	S/g·m⁻³	t/°C	S/g·m⁻³
		20	732	\multicolumn{2}{c}{Sanemasa et al. 1982}	55	860
		30	920	\multicolumn{2}{c}{vapor saturation-UV}		
				15	514	\multicolumn{2}{c}{$\Delta H_{sol}/(kJ\,mol^{-1}) = 1.70$ at 25°C}
				25	526		
				35	545		
				45	584		
				\multicolumn{2}{c}{$\Delta H_{sol}/(kJ\,mol^{-1}) = 2.10$ at 25°C}		

3.

Chen & Wagner 1994b		Ma et al. 2001		Doháyosová et al. 2001		Sawamura et al. 2001	
equilibrium flow cell-GC		shake flask-GC/FID		vapor absorption-HPLC/UV		shake flask-UV	
t/°C	S/g·m⁻³	t/°C	S/g·m⁻³	t/°C	S/g·m⁻³	t/°C	S/g·m⁻³
30	599	5	540	0.5	563	0	583.5
40	609	15	516	5	552	5	670
50	650	25	519	15	542	10	558
60	737	35	555	25	556	15	553
70	875	45	632	35	590	20	553
80	1013			45	632	25	558
90	1187			55	704	30	573.8
100	1371					35	589.1
\multicolumn{2}{c}{$\Delta H_{sol}/(kJ\,mol^{-1}) = 0.37$ at 25°C}			\multicolumn{2}{c}{$\Delta H_{sol}/(kJ\,mol^{-1}) = 3.0$ at 25°C}		40	614.7
						45	640.3
						50	676.6

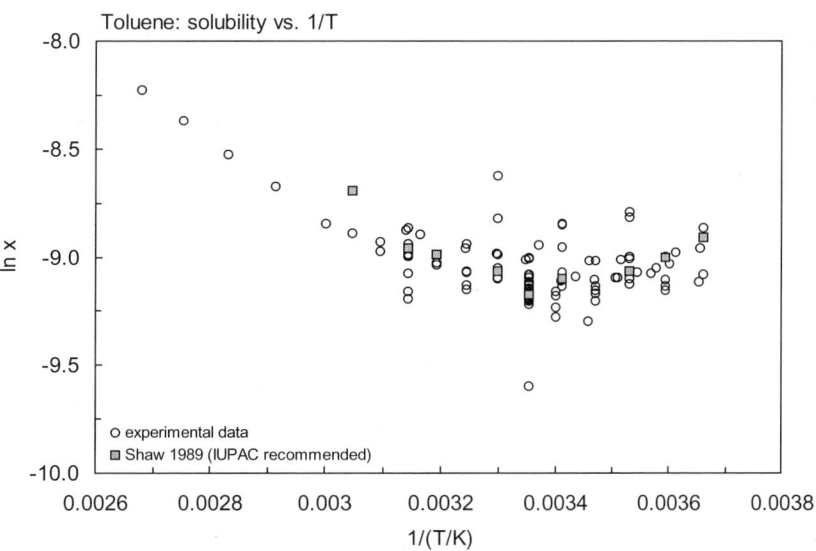

FIGURE 3.1.1.2.1 Logarithm of mole fraction solubility (ln x) versus reciprocal temperature for toluene.

TABLE 3.1.1.2.2
Reported vapor pressures of toluene at various temperatures and the coefficients for the vapor pressure equations

$$\log P = A - B/(T/K) \quad (1)$$
$$\log P = A - B/(C + t/°C) \quad (2)$$
$$\log P = A - B/(C + T/K) \quad (3)$$
$$\log P = A - B/(T/K) - C \cdot \log (T/K) \quad (4)$$
$$\log P = A - B/(T/K) - C \cdot \log (T/K) + D \cdot P/(T/K)^2 \quad (5)$$
$$\log P = A - B/(T/K) - C \cdot (T/K) + D \cdot (T/K)^2 \quad (6)$$
$$\ln P = A - B/(T/K) \quad (1a)$$
$$\ln P = A - B/(C + t/°C) \quad (2a)$$

1.

Pitzer & Scott 1943		Stull 1947		Willingham et al. 1945		Forziati et al. 1949	
mercury manometer		summary of literature data		ebulliometry		ebulliometry	
t/°C	P/Pa	t/°C	P/Pa	t/°C	P/Pa	t/°C	P/Pa
0	900	−16.7	133.3	35.366	6357	35.504	6386
12.5	1920	−4.40	666.6	39.343	7654	39.437	7690
25	3786	6.40	1333	42.810	8962	45.997	10328
37.5	7026	18.4	2666	45.948	10303	48.894	11719
50	12266	31.8	5333	48.867	11700	52.848	13840
		40.3	7999	52.802	13818	57.315	16632
eq. 4	P/mmHg	51.9	13332	57.293	16621	61.869	19942
A	27.6470	69.5	26664	61.851	19929	66.107	23473
B	2866.53	89.5	53329	66.079	23450	71.758	28976
C	6.70	110.6	101325	71.738	28952	76.965	34924
				76.942	34897	83.230	43359
$\Delta H_V/(kJ\ mol^{-1}) = 38.137$		mp/°C	−95	83.202	43320	89.659	53697
at 298 K				89.667	53600	96.580	66795
				96.559	66672	104.052	83748
				104.037	83717	109.328	97644
Linder 1931				109.312	97603	109.894	99232
mercury manometer				109.879	99200	110.420	100724
t°/C	P/Pa			110.403	100690	111.018	102460
−9.70	470.6			110.991	102385	111.545	104000
−8.70	502.6			111.509	103930		
−7.20	556						
−4.40	664			eq. 2	P/mmHg	eq. 2	P/mmHg
−4.35	666.6			A	6.95337	A	6.95508
−3.70	706.6			B	1343.943	B	1345.087
−3.50	714.6			C	219.377	C	219.516
−2.75	742.6						
0	920			bp/°C	110.623	bp/°C	110.626

2.

Bond & Thodos 1960		Ambrose et al. 1970		Zwolinski & Wilhoit 1971		Munday et al. 1980	
compiled data		compiled data		selected values		differential capacitance gauge	
t/°C	P/Pa	t/°C	P/Pa	t/°C	P/Pa	t/°C	P/Pa
6.375	1333	86.85	48898	6.36	1333	0.0	901.3
46.748	10666	106.85	90907	18.38	2666	6.06	1309
90.682	53643	126.85	157180	26.03	4000	10.2	1763
120.585	133322	146.85	255940	31.76	5333	19.99	3706
136.435	199984	166.85	396580	36.394	6666	29.89	4864
		186.85	589400	40.308	7999	39.60	7743

TABLE 3.1.1.2.2 (Continued)

Bond & Thodos 1960 compiled data		Ambrose et al. 1970 compiled data		Zwolinski & Wilhoit 1971 selected values		Munday et al. 1980 differential capacitance gauge	
t/°C	P/Pa	t/°C	P/Pa	t/°C	P/Pa	t/°C	P/Pa
eq. 5	P/mmHg	206.85	845520	46.733	10666	49.26	11870
A	24.27652	226.85	1176900	51.940	13332		
B	2690.80	246.85	1597500	61.942	19998	eq. 6	P/mmHg
C	5.45371	266.85	2122500	69.498	26664	A	25.08047
D	2.04294	286.85	2768000	75.644	33331	B	5541.623
		306.85	3556300	80.863	39997	C	0.01055321
bp/°C	110.64			89.484	53329	D	0
				96.512	66661		
				102.511	79993	Mokbel et al. 1998 static method*	
Ambrose et al. 1967 ebulliometry				107.575	93326	t/°C	P/Pa
t/°C	P/Pa			108.733	95992	199.22	1.051×10^{-3}
146.85	256200			109.689	98659	215.91	7.660×10^{-3}
166.85	396600			110.625	101325	234.88	5.001×10^{-2}
186.85	589400			25.0	3792	254.64	0.252
206.85	845400					274.01	0.956
226.85	1177100			bp/°C	110.625	278.99	1.305
246.85	1597700					284.17	1.764
266.85	2122200			eq. 2	P/mmHg	294.13	3.090
286.85	2767600			A	6.95464	304.06	5.145
306.85	3556800			B	1344.80	314.03	8.257
				C	219.482	324.06	12.836
						334.04	19.275
				$\Delta H_v/(\text{kJ mol}^{-1}) =$		344.02	28.185
eq. 5	P/bar			at 25°C	37.99	353.98	40.016
A	20.17980			at bp	33.18	363.95	55.744
B	2640.04					372.03	71.857
C	5.14885					382.08	96.602
D	1434					392.10	127.860
1 bar = 1×10^5 Pa						402.21	166.405

*complete list see ref.

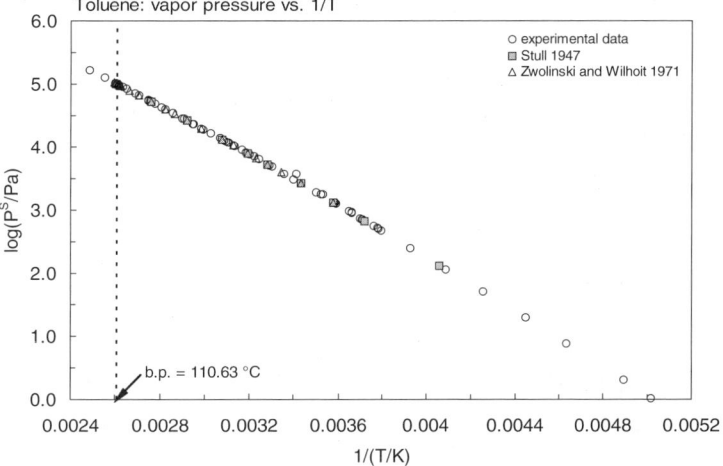

FIGURE 3.1.1.2.2 Logarithm of vapor pressure versus reciprocal temperature for toluene.

TABLE 3.1.1.2.3
Reported Henry's law constants of toluene at various temperatures and temperature dependence equations

$\ln K_{AW} = A - B/(T/K)$ (1) $\qquad$ $\log K_{AW} = A - B/(T/K)$ (1a)
$\ln (1/K_{AW}) = A - B/(T/K)$ (2) $\qquad$ $\log (1/K_{AW}) = A - B/(T/K)$ (2a)
$\ln (k_H/\text{atm}) = A - B/(T/K)$ (3)
$\ln [H/(\text{Pa m}^3/\text{mol})] = A - B/(T/K)$ (4) $\qquad$ $\ln [H/(\text{atm·m}^3/\text{mol})] = A - B/(T/K)$ (4a)
$K_{AW} = A - B·(T/K) + C·(T/K)^2$ (5)

1.

Brown & Wasik 1974		Leighton & Calo 1981		Sanemasa et al. 1981		Schoene & S. 1985	
head space-GC		equilibrium cell-GC		vapor liquid-equilibrium		headspace-GC	
t/°C	H/(Pa m³/mol)	t/°C	H/(Pa m³/mol)	t/°C	H/(Pa m³/mol)	t/°C	H/(Pa m³/mol)
4.5	179	1.0	222	15	382	20	634
6.33	204	1.3	236	25	628	36.9	670
7.06	218	12.4	373	35	979	41.5	680
8.96	244	12.5	361	45	1404	46	690
11.75	294	17.9	459				
12.1	299	19.1	525			eq. 1	K_{AW}
15.1	259	22.7	565	Sanemasa et al. 1982		A	6.90
17.93	414	23	625	vapor-liquid equilibrium		B	2194
20.06	474			t/°C	H/Pa m³/mol		
		eq. 3	k_H/atm				
		A	18.46	15	396		
		B	3751	25	664		
				35	1060		
				45	1571		

2.

Ashworth et al. 1988		Kolb et al. 1992		Ettre et al. 1993		Perlinger et al. 1993	
EPICS-GC		static headspace-GC		headspace-GC		equil. headspace-GC	
t/°C	H/(Pa m³/mol)	t/°C	H/(Pa m³/mol)	t/°C	H/(Pa m³/mol)	t/°C	H/(Pa m³/mol)
10	386	40	933	45	1116	10	293
15	499	60	1565	60	1489	15	390
20	562	70	1915	70	1877	20	499
25	651	80	2312	80	2427	25	660
30	819					30	838
		eq. 2	$1/K_{AW}$	eq. 2	$1/K_{AW}$		
eq 4a	H/(atm m³/mol)	A	7.61	A	−2.532379		
A	5.133	B	2647	B	928.3536		
B	3024						

3.

Robbins et al. 1993		Dewulf et al. 1995		Turner et al. 1996		Peng & Wan 1997	
static headspace-GC		EPICS-GC		vapor phase-equilibrium		headspace-GC	
t/°C	H/(Pa m³/mol)	t/°C	H/(Pa m³/mol)	t/°C	H/(Pa m³/mol)	t/°C	H/(Pa m³/mol)
25	652	2.0	175	11	376	15	391
30	835	6.0	203	15	460	20	475
40	1086	10	250	25	684	25	605

TABLE 3.1.1.2.3 (Continued)

Robbins et al. 1993		Dewulf et al. 1995		Turner et al. 1996		Peng & Wan 1997	
static headspace-GC		EPICS-GC		vapor phase-equilibrium		headspace-GC	
t/°C	H/(Pa m³/mol)	t/°C	H/(Pa m³/mol)	t/°C	H/(Pa m³/mol)	t/°C	H/(Pa m³/mol)
45	1351	18.2	424	35	1202	30	774
50	1450	25	555			35	984
				eq. 5	K_{AW}	40	1104
eq. 4a	H/(atm m³/mol)	eq 1	K_{AW}	A	0.115	45	1309
A	7.14	A	12.40	B	0.00474		
B	3689	B	4064	C	0.000466	eq. 1	K_{AW}
						A	7.94
						B	1621

4.

Görgényi et al. 2002		Bakierowska & T. 2003	
EPICS-SPME method		headspace-GC	
t/°C	H/(Pa m³/mol)	t/°C	H/(Pa m³/mol)
2.0	169.3	10	288
6.0	238.2	15	358
10	304.9	20	467
18	476.9	25	612
25	647.5		
30	822.4	eq. 2	$1/K_{AW}$
40	1214	A	11.926
50	1758	B	3977
60	2286		
eq. 1	K_{AW}		
A	11.25		
B	3770.4		

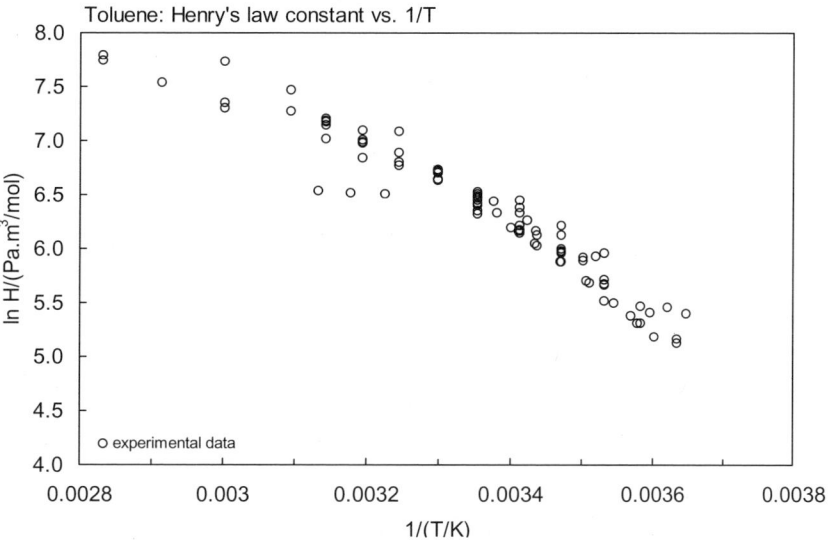

FIGURE 3.1.1.2.3 Logarithm of Henry's law constant versus reciprocal temperature for toluene.

TABLE 3.1.1.2.4
Reported octanol/water partition coefficients and octanol-air partition coefficients of toluene at various temperatures

log K_{OW}		log K_{OA}	
Dewulf et al. 1999		Gruber et al. 1997	
EPICS-GC, both phases		GC det'd activity coeff.	
t/°C	log K_{OW}	t/°C	log K_{OA}
2.2	2.316	20.29	3.42
6	2.405	30.3	3.2
10	2.464	40.4	3.03
14.1	2.38	50.28	2.86
18.7	2.41		
24.8	2.32		
enthalpy change $\Delta H/(kJ\ mol^{-1}) = -1.0$ (−16 to 13.9)			

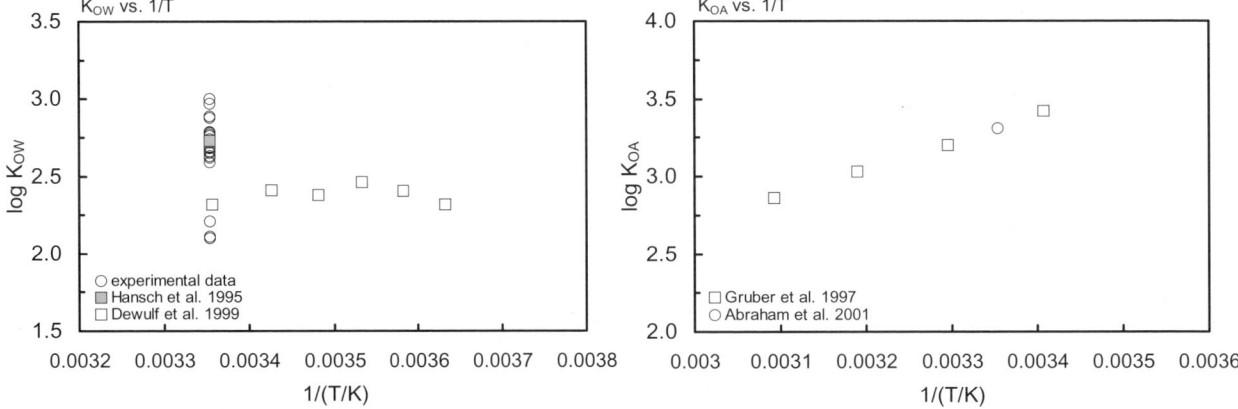

FIGURE 3.1.1.2.4 Logarithm of K_{OW} and K_{OA} versus reciprocal temperature for toluene.

3.1.1.3 Ethylbenzene

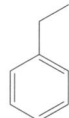

Common Name: Ethylbenzene
Synonym: phenylethane, ethylbenzol
Chemical Name: ethylbenzene
CAS Registry No: 100-41-4
Molecular Formula: C_8H_{10}, $C_2H_5C_6H_5$
Molecular Weight: 106.165
Melting Point (°C):
    –94.96   (Lide 2003)
Boiling Point (°C):
    136.2   (Weast 1982–83; Lide 2003)
Density (g/cm³ at 20°C):
    0.867   (Weast 1982–83)
Molar Volume (cm³/mol):
    122.4   (20°C, calculated-density, McAuliffe 1966; Stephenson & Malanowski 1987)
    140.4   (calculated-Le Bas method at normal boiling point)
Enthalpy of Vaporization, ΔH_V (kJ/mol):
    42.25, 35.2   (25°C, bp, Riddick et al. 1986)
Enthalpy of Fusion ΔH_{fus} (kJ/mol):
    9.184   (Riddick et al. 1986)
    9.16   (Chickos et al. 1999)
Entropy of Fusion ΔS_{fus} (J/mol K):
    51.43, 52.2   (exptl., calculated-group additivity method, Chickos et al. 1999)
Fugacity Ratio at 25°C, F: 1.0

Water Solubility (g/m³ or mg/L at 25°C or as indicated. Additional data at other temperatures designated * are compiled at the end of this section):
    140   (15°C, shake flask, Fühner 1924)
    168   (shake flask-UV, Andrews & Keefer 1950)
    175   (shake flask-UV, Klevens 1950)
    208*   (shake flask-UV, measured range 0.4–42.8°C, Bohon & Claussen 1951)
    165   (shake flask-UV, Morrison & Billett 1952)
    159   (shake flask-GC, McAuliffe 1963)
    152   (shake flask-GC, McAuliffe 1966)
    177*   (shake flask-GC, Polak & Lu 1973)
    180*   (shake flask-GC, measured range 4.5–20.1°C, Brown & Wasik 1974)
    203   (shake flask-UV, Vesala 1974)
    161   (shake flask-GC, Sutton & Calder 1975)
    131   (shake flask-GC, Price 1976)
    131   (shake flask-GC, Krzyzanowska & Szeliga 1978)
    203, 212   (10, 20°C, shake flask-UV, Ben-Naim & Wiff 1979)
    208*, 184*   (20°C, elution chromatography, shake flask-UV, measured range 10–30°C, Schwarz & Miller 1980)
    181*   (vapor saturation-UV, measured range 15–45°C, Sanemasa et al. 1981)
    169*   (vapor saturation-UV, measured range 15–45°C, Sanemasa et al. 1982)
    172   (generator column-HPLC/UV, Tewari et al. 1982a)
    187   (generator column-HPLC/UV, Tewari et al. 1982c)
    166   (HPLC-k′ correlation, converted from γ_W, Hafkenscheid & Tomlinson 1983a)
    187   (generator column-HPLC/UV, Wasik et al. 1983)
    172   (vapor saturation-UV, Sanemasa et al. 1984)

192*			(generator column-HPLC/UV, measured range 10–45°C, Owens et al. 1986)
172			(shake flask-purge and trap-GC, Coutant & Keigley 1988)
169*			(IUPAC recommended value, temp range 0–100°C, Shaw1989b)
170*			(30°C, equilibrium flow cell-GC, measured range 30–100°C, Chen & Wagner 1994c)
ln x = 11.59 – 20.52·$(T_r/K)^{-1}$ + 4.750·$(T_r/K)^{-2}$; T_r = T/T_c, the reduced temp, system temp T divided by critical temp T_c, temp range 303.15–373.15 K (equilibrium flow cell-GC, Chen & Wagner 1994c)
ln x = –30.799 + 3986.26/(T/K) + 7.9095 × 10^{-5}·$(T/K)^2$; temp range 5–50°C (regression eq. of literature data, Shiu & Ma 2000)
170*			(vapor absorption technique-HPLC/UV, measured range 0.5–55°C, Doháryosová et al. 2001)
170*			(shake flask-UV, measured range 0–50°C, Sawamura et al. 2001)
ln x = –263.220 + 11024.75/(T/K) + 37.8858·ln (T/K); temp range 290–400 K (eq. derived from literature calorimetric and solubility data, Tsonopoulos 1999)

Vapor Pressure (Pa at 25°C or as indicated and reported temperature dependence equations. Additional data at other temperatures designated * are compiled at the end of this section):

209.3*			(0°C, mercury manometer, measured range –11.6 to 0°C, Linder 1931)
1546*			(30°C, Hg manometer, measured range 10–50°C, Rintelen 1937)
log (P/mmHg) = 22.90283 – 2847.75/(T/K) – 5·log (T/K); temp range 4–75°C (vapor pressure eq. from manometer measurements, Stuckey & Saylor 1940)
6277*			(56.589°C, ebulliometry, measured range 56.589–137.124°C, Willingham et al. 1945)
log (P/mmHg) = 6.94998 – 1419.315/(212.611 + t/°C); temp range 56.6–137.1°C (manometer, Antoine eq. from exptl. data, Willingham et al. 1945)
1333*			(25.9°C, summary of literature data, Stull 1947)
8399*			(63.3°C, static-Hg manometer, measured range 63.3–135.9°C, Buck et al. 1949)
6398*			(56.689°C, ebulliometry, measured range 56.689–137.16°C, Forziati et al. 1949)
log (P/mmHg) = 6.95904 – 1425.464/(213.345 + t/°C); temp range 56.7–137.2°C (manometer, Antoine eq. from exptl. data, Forziati et al. 1949)
log (P/mmHg) = –3225/(T/K) – 7.553·log (T/K) + 30.49; temp range 80–120°C (vapor pressure eq. from Hg manometer measurements, Buck et al. 1949)
1276(extrapolated-Antoine eq., Dreisbach 1955)
log (P/mmHg) = 6.95719 – 1424.255/(213.206 + t/°C), temp range: 45–190°C (Antoine eq. for liquid state, Dreisbach 1955)
133800*		(146.85°C, ebulliometry, measured range 146.85–326.85°C, Ambrose et al. 1967)
1270*			(interpolated-Antoine eq., Zwolinski & Wilhoit 1971)
log (P/mmHg) = 6.95719 – 1424.255/(213.206 + t/°C); temp range 25.88–163.47°C (Antoine eq., Zwolinski & Wilhoit 1971)
log (P/mmHg) = [–0.2185 × 9303.3/(T/K)] + 7.809470; temp range –9.8 to 326.5°C, (Antoine eq., Weast 1972–73)
9585*			(66.031°C, comparative ebulliometry, measured range 66.031–176.953°C, Osborn & Scott 1980)
log (P/atm) = (1 – 409.229/T) × 10^(0.859833 – 6.85948 × 10^4·T + 5.94439 × 10^7·T^2); T in K, temp range 243.2–615.0 K (Cox vapor pressure eq., Chao et al. 1983)
1270, 1265	(extrapolated-Antoine eq., Boublik et al. 1984; quoted, Howard 1989)
log (P/kPa) = 6.08206 – 1425.305/(213.415 + t/°C), temp range: 25.88–92.7°C (Antoine eq. from reported exptl. data, Boublik et al. 1984)
log (P/kPa) = 6.0785 – 1421.653/(212.816 + t/°C); temp range 56.6–137.12°C (Antoine eq. from reported exptl. data of Willingham et al. 1945, Boublik et al. 1984)
1268			(interpolated-Antoine eq., Dean 1985, 1992)
log (P/mmHg) = 6.95719 – 1424.255/(213.21 + t/°C); temp range 26–164°C (Antoine eq., Dean 1985, 1992)
1300			(selected value., Riddick et al. 1986)
log (P/kPa) = 6.09280 – 1431.71/(214.099 + t/°C); temp range not specified (Antoine eq., Riddick et al. 1986)
1266			(extrapolated-Antoine eq., Stephenson & Malanowski 1987)
log (P_L/kPa) = 6.06991 – 1416.922/(–69.716 + T/K); temp range 298–420 K (liquid, Antoine eq.-I, Stephenson & Malanowski 1987)
log (P_L/kPa) = 6.10898 – 1445.262/(–57.128 + T/K); temp range 409–459 K (liquid, Antoine eq.-II, Stephenson & Malanowski 1987)

Mononuclear Aromatic Hydrocarbons

log (P_L/kPa) = 6.36656 − 1665.991/(−26.716 + T/K); temp range 457–554 K (liquid, Antoine eq.-III, Stephenson & Malanowski 1987)

log (P_L/kPa) = 7.49119 − 3056.747/(159.496 + T/K); temp range 549–617 K (liquid, Antoine eq.-IV, Stephenson & Malanowski 1987)

1280, 283 (quoted, calculated-UNIFAC activity coeff., Banerjee et al. 1990)

log (P/mmHg) = 36.1998 − 3.3402 × 10^3/(T/K) − 9.7970·log (T/K) − 1.1467 × 10^{-11}·(T/K) + 2.5758 × 10^{-6}·(T/K)^2, temp range 178–617 K (vapor pressure eq., Yaws 1994)

Henry's Law Constant (Pa m^3/mol at 25°C or as indicated and reported temperature dependence equations. Additional data at other temperatures designated * are compiled at the end of this section):

559.1*	(20.06°C, headspace-GC, Brown & Wasik 1974)
757	(calculated-bond contribution, Hine & Mookerjee 1975)
879	(calculated as 1/K_{AW}, C_W/C_A, reported as exptl., Hine & Mookerjee 1975)
854	(batch stripping-GC, Mackay et al. 1979; quoted, Howard 1989)
734*	(vapor-liquid equilibrium-GC, Sanemasa et al. 1981)
797*	(vapor-liquid equilibrium.-GC, Sanemasa et al. 1982)
653	(gas stripping-GC, Warner et al. 1987)
798*	(EPICS-GC/FID, Ashworth et al. 1988)

ln [H/(atm·m^3/mol)] = 11.92 − 4994/(T/K), temp range 10–30°C, EPICS measurements, Ashworth et al. 1988)

815	(calculated-vapor-liquid equilibrium (VLE) data, Yaws et al. 1991)
756	(infinite activity coeff. γ^∞ in water determined by inert gas stripping-GC, Li et al. 1993)
660*	(equilibrium headspace-GC, Perlinger et al. 1993)
788*	(static headspace-GC, Robbins et al. 1993)
397	(headspace solid-phase microextraction (SPME)-GC, Zhang & Pawliszyn 1993)
669*	(EPICS-GC/FID, measured range 2–25°C, Dewulf et al. 1995)
302, 838	(6.0, 25°C, EPICS-GC/FID, natural seawater with salinity of 35‰, Dewulf et al. 1995)

ln K_{AW} = −4567/(T/K) + 0.01047·Z + 14.001; with Z salinity 0–35.5‰, temp range: 2–35°C (EPICS-GC/FID, Dewulf et al. 1995)

602	(20°C, selected from literature experimentally measured data, Staudinger & Roberts 1996)
895	(vapor-liquid equilibrium.-GC, Turner et al. 1996)
629	(EPICS-static headspace method-GC/FID, Miller & Stuart 2000)
943.2	(modified EPICS method-GC, Ryu & Park 1999)
583	(20°C, selected from literature experimentally measured data, Staudinger & Roberts 2001)

log K_{AW} = 6.541 − 2100/(T/K) (van't Hoff eq. derived from lit. data, Staudinger & Roberts 2001)

1173–1273 (27°C, equilibrium headspace-GC, solute concn 0.43–18.66 mg/L, measured range 300–315 K, Cheng et al. 2003)

Octanol/Water Partition Coefficient, log K_{OW} at 25°C:

3.15	(shake flask-UV, Hansch et al. 1968; Hansch & Leo 1979; Hansch & Leo 1985)
3.13	(calculated-fragment const., Rekker 1977)
3.12	(HPLC-k' correlation, Hanai et al. 1981)
3.12	(HPLC-k' correlation, D'Amboise & Hanai 1982)
3.26	(HPLC-k' correlation, Hammers et al. 1982)
3.15	(generator column-HPLC/UV, Tewari et al. 1982a)
3.13	(generator column-HPLC/UV, Tewari et al. 1982c)
3.16	(HPLC-k' correlation, Miyake & Terada 1982)
3.24	(HPLC-k' correlation, Hafkenscheid & Tomlinson 1983b)
3.13	(generator column-HPLC/UV, Wasik et al. 1983)
3.13	(generator column-RP-LC, Schantz & Martire 1987)
3.13, 3.21	(RP-HPLC capacity factor correlations, Sherblom & Eganhouse 1988)
3.15	(recommended, Sangster 1989, 1993)
3.15	(recommended, Hansch et al. 1995)
3.32, 3.32, 3.53, 3.51	(HPLC-k' correlation, different combinations of stationary and mobile phases under isocratic conditions, Makovsakya et al. 1995)
3.05	(RP-HPLC-RT correlation, short ODP column, Donovan & Pescatore 2002)

Octanol/Air Partition Coefficient, log K_{OA} at 25°C or as indicated
- 3.85, 3.62, 3.41, 3.24 (20.29, 30.3, 40.4, 50.28°C, from GC-determined γ^∞ in octanol, Gruber et al. 1997)
- 3.72, 3.698 (interpolated value from exptl value of Gruber et al. 1997, calculated using measured γ^∞ in pure octanol by Tewari et al. 1982, Abraham et al. 2001)

Bioconcentration Factor, log BCF:
- 0.67 (clams, exposed to water-soluble fraction of crude oil, Nunes & Benville 1979; selected, Howard 1989)
- 2.16 (fish, calculated, Lyman et al. 1982; quoted, Howard 1989)
- 2.67 (microorganisms-water, calculated-K_{OW}, Mabey et al., 1982)
- 1.19 (goldfish, Ogata et al. 1984)
- 1.20, 1.19 (fish: calculated, correlated, Sabljic 1987b)
- 2.31 (*S. capricornutum*, Herman et al. 1991)

Sorption Partition Coefficient, log K_{OC} at 25°C or as indicated:
- 1.98 (soil, sorption isotherm, Chiou et al. 1983)
- 2.41 (sediment 4.02% OC from Tamar estuary, batch equilibrium-GC, Vowles & Mantoura 1987)
- 2.30 (RP-HPLC-k' correlation, cyanopropyl column, Hodson & Williams 1988)
- 2.03 (untreated Marlette soil A horizon, organic carbon OC 2.59%, batch equilibrium-adsorption isotherm, Lee et al. 1989)
- 2.83, 2.61 (organic cations treated Marlette soil A horizon: HDTMA treated OC 10.03%; DDTMA treated, OC 4.37%, batch equilibrium-adsorption isotherm, Lee et al. 1989)
- 2.13 (untreated Marlette soil B_t horizon, organic carbon OC 0.30%, batch equilibrium-adsorption isotherm, Lee et al. 1989)
- 3.23, 3.12, 2.58 (organic cations treated Marlette soil B_t horizon: HDTMA treated OC 3.72%, DDTMA treated OC 1.98%, NTMA treated, OC 1.18%, batch equilibrium-adsorption isotherm, Lee et al. 1989)
- 2.26, 2.38 (untreated soils: St. Clair soil B_t horizon OC 0.44%; Oshtemo soil B_t horizon OC 0.11%, batch equilibrium-adsorption isotherm, Lee et al. 1989)
- 3.37, 3.19 (organic cations HDTMA treated soils: St. Clair soil B_t horizon OC 3.25%; Oshtemo soil B_t horizon OC 0.83%, batch equilibrium-adsorption isotherm, Lee et al. 1989)
- 2.27, 2.05 (Riddles soil: top layer pH 5.0, below top layer pH 5.3, batch equilibrium, Boyd et al. 1990)
- 2.52, 2.47 (RP-HPLC-k' correlation, humic acid-silica column, Szabo et al. 1990a,b)
- 2.35, 2.40, 2.42 (RP-HPLC-k' correlation on 3 different stationary phases, Szabo et al.1995)
- 2.51, 2.51 (RP-HPLC-k' correlation including MCI related to non-dispersive intermolecular interactions, hydrogen-bonding indicator variable, Hong et al. 1996)
- 2.32 (HPLC-screening method, Müller & Kördel 1996)
- 2.49, 2.73, 2.65, 2.73, 2.77, 2.73 2.74 (2.3, 3.8, 6.2, 8.0, 13.5, 18.6, 25°C, natural sediment from River Leie, organic carbon f_{OC} = 4.12%, EPICS-GC/FID, Dewulf et al. 1999)
- 2.04, 2.18, 1.90 (soils: organic carbon OC $\geq$ 0.1%, OC $\geq$ 0.5%, 0.1 $\leq$ OC < 0.5%, average, Delle Site 2001)

Sorption Partition Coefficient, log K_{OM}:
- 1.98 (Woodburn silt loam soil, 1.9% organic matter, equilibrium isotherm-GC, Chiou et al. 1983)
- 1.04 (untreated Marlette soil A horizon, organic matter OM 5.18%, batch equilibrium-adsorption isotherm, Lee et al. 1989)
- 1.89, 1.81 (organic cations treated Marlette soil A horizon: HDTMA treated, organic matter 10.03%; DDTMA treated, OM 5.18%, batch equilibrium-adsorption isotherm, Lee et al. 1989)
- 2.53, 2.46, 2.08 (organic cations treated Marlette soil B_t horizon: HDTMA treated OM 4.85%, DDTMA treated OM 2.73%, NTMA treated OM 1.74%, batch equilibrium-adsorption isotherm, Lee et al. 1989)
- 2.56, 2.53 (organic cations HDTMA treated soils: St. Clair soil B_t horizon OM 4.38%; Oshtemo soil B_t horizon OM 1.12%, batch equilibrium-adsorption isotherm, Lee et al. 1989)

Environmental Fate Rate Constants, k, or Half-lives, $t_{1/2}$:

Volatilization: rate constants: k = 0.035 d^{-1} with $t_{1/2}$ = 20 d in spring at 8–16°C, k = 0.331 d^{-1} with $t_{1/2}$ = 2.1 d in summer at 20–22°C, k = 0.054 d^{-1} with $t_{1/2}$ = 13 d in winter at 3–7°C for the periods when volatilization appears to dominate, and k = 0.097 d^{-1} with t$_{½}$ = 7.1 d with HgCl$_2$ in September 9–15, in marine mesocosm experiments (Wakeham et al. 1983);

t$_{½}$ ~ 3.1 h of evaporation from a river of 1 m depth with wind speed 3 m/s and water current of 1 m/s at 20°C (Lyman et al. 1982; quoted, Howard 1989).

Photolysis: not environmentally significant or relevant (Mabey et al. 1982).

Oxidation: rate constant k; for gas-phase second order rate constants, k$_{OH}$ for reaction with OH radical, k$_{NO3}$ with NO$_3$ radical and k$_{O3}$ with O$_3$ or as indicated. Data at other temperatures and/or the Arrhenius expression are designated *, see reference:

k$_{OH}$ = (4.8 ± 1.0) × 10^9 L mol^{-1} s^{-1} at 305 ± 2 K (relative rate method, Lloyd et al. 1976)

k$_{OH}$ = 4.8 × 10^9 L mol^{-1} s^{-1} with t$_{½}$ = 0.24–24 h (Darnall et al. 1976)

t$_{½}$ ≈ 15 h in water, probably not important as aquatic fate (Callahan et al. 1979)

k$_{OH}$ = 8.0 × 10^{-12} cm^3 molecule^{-1} s^{-1} and residence time of 1.4 d, loss of 51% in one day or 12 sunlit hours at 300 K in urban environments (Singh et al. 1981)

k << 360 M^{-1} h^{-1} for singlet oxygen and 720 M^{-1} h^{-1} for RO$_2$ radical (Mabey et al. 1982)

k$_{OH}$ = 4.4 × 10^9 M^{-1} s^{-1} with t$_{½}$ = 1.3 d in the atmosphere (Mill 1982)

k$_{O3}$ = 340 cm^3 mol^{-1} s^{-1} at 300 K (Lyman et al. 1982)

k = (14 ± 4) M^{-1} s^{-1} for the reaction with ozone in water using 100 mM t-BuOH as scavenger at pH 2.0 and 20–23°C (Hoigné & Bader 1983)

k$_{OH}$ = 7.5 × 10^{-12} cm^3 molecule^{-1} s^{-1} with t$_{½}$ = 8.56–85.6 h (Atkinson 1985)

k$_{OH}$ = 6.47 × 10^{-12} cm^3 molecule^{-1} s^{-1} at 298 K (relative rate method, Ohta & Ohyama 1985)

k$_{OH}$(calc) = 6.1 × 10^{-12} cm^3 molecule^{-1} s^{-1}, k$_{OH}$(obs.) = 7.5 × 10^{-12} cm^3 molecule^{-1} s^{-1} at room temp. (SAR [structure-activity relationship], Atkinson 1987)

k$_{OH}$ = 7.1 × 10^{-12} cm^3 molecule^{-1} s^{-1} at ~298 K (recommended, Atkinson 1989, 1990)

k$_{OH}$(calc) = 6.06 × 10^{-12} cm^3 molecule^{-1} s^{-1} (molecular orbital calculations, Klamt 1993)

Hydrolysis: not aquatically significant (Callahan et al. 1979);
no hydrolyzable functional groups (Mabey et al. 1982)

Biodegradation:
100% biodegraded after 192 h at 13°C with an initial concn of 1.36 × 10^{-6} L/L (Jamison et al. 1976)

t$_{½}$(aq. aerobic) = 72–240 h, based on unacclimated aqueous aerobic biodegradation half-life and seawater dieaway test data (Van der Linden 1978; Howard et al. 1991);

t$_{½}$(aq. anaerobic) = 4224–5472 h, based on anaerobic groundwater die-away test data (Wilson et al. 1986; Howard et al. 1991)

k = 0.5 d^{-1}, significant degradation under favourable conditions in an aerobic environment (Tabak et al. 1981; Mills et al. 1982)

t$_{½}$ ~ 2 d degradation by established microorganisms depending on body of water and its temperature (Howard 1989)

k = 0.07 yr^{-1} with t$_{½}$ = 37 d (Olsen & Davis 1990).

Biotransformation: 3 × 10^{-9} mL cell^{-1} h^{-1} (Mabey et al. 1982).

Half-Lives in the Environment:

Air: t$_{½}$ = 0.24–24 h, based on rate of disappearance for the reaction with hydroxyl radical (Darnall et al. 1976); photodecomposition t$_{½}$ = 5.0 h under simulated atmospheric conditions, with NO (Dilling et al. 1976); residence time of 1.4 d, loss of 51% in one day or 12 sunlit hours at 300 K in urban environments (Singh et al. 1981)

summer daylight lifetime τ = 20 h due to reaction with OH radical (Altshuller 1991);

t$_{½}$ = 8.56–85.6 h, based on photooxidation half-life in air (Atkinson 1985; Howard et al. 1991).

Surface Water: t$_{½}$ = 5–6 h (Callahan et al. 1979), based on the estimated evaporative loss of toluene at 25°C and 1 m depth of water (Mackay & Leinonen 1975);

t$_{½}$ = 20 d in spring at 6–16°C, t$_{½}$ = 2.1 d in summer at 20–22°C, t$_{½}$ = 13 d in winter at 3–7°C and t$_{½}$ = 7.1 d with HgCl$_2$ in marine mesocosm experiments (Wakeham et al. 1983);

t$_{½}$ = 72–240 h, based on unacclimated aqueous aerobic biodegradation half-life (Van der Linden 1978; Howard et al. 1991).

Ground water: t$_{½}$ = 144–5472 h, based on unacclimated aqueous aerobic biodegradation half-life and seawater dieaway test data (Van der Linden 1978; Howard et al. 1991);

t$_{½}$ ~ 0.3 yr from observed persistence in groundwater of the Netherlands (Zoeteman et al. 1981).

Soil: $t_{1/2}$ = 72–240 h, based on unacclimated aqueous aerobic biodegradation half-life (Van der Linden 1978; Howard et al. 1991);
$t_{1/2}$ < 10 d (Ryan et al. 1988).

Biota:

TABLE 3.1.1.3.1
Reported aqueous solubilities of ethylbenzene at various temperatures

1.

Bohon & Claussen 1951		Polak & Lu 1973		Brown & Wasik 1974		Ben-Naim & Wiffb 1979	
shake flask-UV		shake flask-GC		headspace-GC		shake flask-UV	
t/°C	S/g·m^{-3}	t/°C	S/g·m^{-3}	t/°C	S/g·m^{-3}	t/°C	S/g·m^{-3}
0.4	219	0	197	4.5	196	10	203
5.2	213	25	177	6.3	192	20	212
20.7	207			7.1	186		
21.2	207			9	187		
25	208			11.8	181		
25.6	209			12.1	183		
30.2	211			15.1	180		
34.9	221			17.9	184		
42.8	231			20.1	180		
ΔH_{sol}/(kJ mol^{-1}) =				ΔH_{sol}/(kJ mol^{-1}) = 11.9			
25°C	1.6			at 25°C			
2°C	–3.98						
7°C	–2.74						
12°C	–1.63						
17°C	–0.343						
22°C	–0.167						
27°C	1.97						
32°C	3.615						
37°C	7.36						

2.

Schwarz & Miller 1980		Sanemasa et al. 1981		Owens et al. 1986		Shaw 1989b (IUPAC)	
shake flask-UV		vapor saturation-UV		generator column-HPLC/UV		recommended values	
t/°C	S/g·m^{-3}	t/°C	S/g·m^{-3}	t/°C	S/g·m^{-3}	t/°C	S/g·m^{-3}
10	180	15	176	10	197	0	200
20	184	25	181	14	192	10	180
		35	194	17	189	20	181
elution chromatography		45	215	18	183	25	169
10	211			19	178	30	190
20	208	ΔH_{sol}/(kJ mol^{-1}) = 3.60		20	188	40	200
		at 25°C		21	183	50	220
				22	182	60	250
		Sanemasa et al. 1982		23.5	186	70	280
		vapor saturation-UV		25	192	80	330
		t/°C	S/g·m^{-3}	25.8	186	90	390
				28	185.5	100	460
		15	160	30	188.7		
		25	169	35	193	ΔH_{sol}/(kJ mol^{-1}) = 2.1	

TABLE 3.1.1.3.1 (Continued)

Schwarz & Miller 1980		Sanemasa et al. 1981		Owens et al. 1986		Shaw 1989b (IUPAC)	
shake flask-UV		vapor saturation-UV		generator column-HPLC/UV		recommended values	
t/°C	S/g·m^{-3}	t/°C	S/g·m^{-3}	t/°C	S/g·m^{-3}	t/°C	S/g·m^{-3}
		35	176	40	205	at 25°C	
		45	196	45	211.4		
		ΔH_{sol}/(kJ mol^{-1}) = 3.90 at 25°C		ΔH_{sol}/(kJ mol^{-1}) = 1.30 25°C			

3.

Chen & Wagner 1994c		Dohányosová et al. 2001		Sawamura et al. 2001	
equil. flow cell-GC		vapor abs-HPLC/UV		shake flask-UV	
t/°C	S/g·m^{-3}	t/°C	S/g·m^{-3}	t/°C	S/g·m^{-3}
30	170	0.5	169	0	190.8
40	172	5	167	5	182
50	198	15	167	10	175
60	240	25	170	15	170
70	291	35	177	20	169
80	353	45	188	25	170
90	410	55	213.5	30	173.8
100	504			35	179.8
		ΔH_{sol}/(kJ mol^{-1}) = 3.0 at 25°C		40	188.1
ΔH_{sol}/(kJ mol^{-1}) = 1.30 at 25°C				45	196.9
				50	208

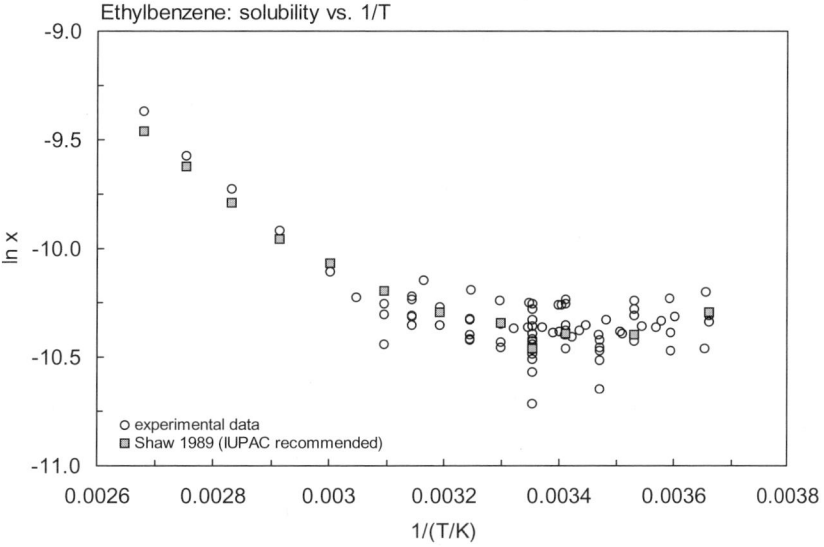

FIGURE 3.1.1.3.1 Logarithm of mole fraction solubility (ln x) versus reciprocal temperature for ethylbenzene.

TABLE 3.1.1.3.2
Reported vapor pressures of ethylbenzene at various temperatures and the coefficients for the vapor pressure equations

$$\log P = A - B/(T/K) \quad (1) \qquad \ln P = A - B/(T/K) \quad (1a)$$
$$\log P = A - B/(C + t/°C) \quad (2) \qquad \ln P = A - B/(C + t/°C) \quad (2a)$$
$$\log P = A - B/(C + T/K) \quad (3)$$
$$\log P = A - B/(T/K) - C \cdot \log(T/K) \quad (4)$$

1.

Linder 1931		Willingham et al. 1945		Stull 1947		Buck et al. 1949	
mercury manometer		ebulliometry		summary of literature data		static-Hg manometer	
t/°C	P/Pa	t/°C	P/Pa	t/°C	P/Pa	t/°C	P/Pa
–11.6	74.7	56.589	6277	–9.8	133.3	63.3	8399
–1.20	184	60.796	7654	13.9	266.6	67.7	10133
0	209.3	64.463	8962	25.9	1333	75.2	13999
		67.775	10303	38.6	2666	82.6	18665
		70.862	11699	52.8	5333	91.6	25998
		75.027	13919	61.8	7999	99	33864
		79.777	16608	74.1	13332	108.1	45596
Rintelin et al. 1937		84.599	19924	92.7	26664	112.7	52662
mercury manometer		89.071	23450	113.8	53323	117.6	60928
t/°C	P/Pa	95.056	28954	136.2	101325	120.5	66128
10	387	100.561	34897			125.6	75460
30	1546	107.183	43320	mp/°C	–94.9	129.3	85060
50	4613	114.02	53654			135.9	100792
		121.312	66755				
		129.221	83717			eq. 4	P/mmHg
		134.8	97603			A	3225
		135.399	99199			B	7.553
Stuckey & Saylor 1940		135.954	100690			C	30.49
static-Hg manometer		136.574	102385			temp range:	
t/°C	P/Pa	137.124	103903			80–120°C	
measured 4–75°C		eq. 2	P/mmHg				
eq. 4	P/mmHg	A	6.94998				
A	22.9028	B	1419.315				
B	2857.75	C	212.611				
C	5						
		bp/°C	136.187				
bp/°C	136.32						

2.

Forziati et al. 1949		Ambrose et al. 1967		Zwolinski & Wilhoit 1971		Osborn & Scott 1980	
ebulliometry		ebulliometry		selected values		comparative ebulliometry	
t/°C	P/Pa	t/°C	P/Pa	t/°C	P/Pa	t/°C	P/Pa
56.689	6398	146.85	133800	25.88	1333	66.031	9585
60.887	7690	166.85	216100	38.6	2666	69.091	10887
64.51	8990	186.85	332500	46.69	4000	72.167	12339
67.827	10330	206.85	492000	52.75	5333	75.258	13956
70.891	11720	226.85	703300	57.657	6666	78.254	15748
75.054	13840	246.85	976500	61.789	7999	81.465	17734

TABLE 3.1.1.3.2 *(Continued)*

Forziati et al. 1949		Ambrose et al. 1967		Zwolinski & Wilhoit 1971		Osborn & Scott 1980	
ebulliometry		ebulliometry		selected values		comparative ebulliometry	
t/°C	P/Pa	t/°C	P/Pa	t/°C	P/Pa	t/°C	P/Pa
79.791	16633	266.85	1321500	68.596	10666	84.587	19933
83.619	19944	286.85	1751100	74.105	13332	90.869	25023
89.09	23474	306.85	2279100	84.687	19998	97.199	31177
95.074	28978	326.85	2924900	92.68	26664	103.575	38565
100.675	34925			99.182	33331	110.001	47375
107.21	43360	eq. 5	P/bar	104.703	39997	116.474	57817
114.046	53698	A	21.956	113.823	53329	122.998	70120
121.331	66796	B	2923.84	121.266	66661	129.571	84532
129.234	83749	C	5.67301	127.603	79993	136.193	101325
134.815	97645	D	1772	133.152	93326	142.863	120790
135.413	99235			134.185	95991	149.587	143240
135.969	100725			135.196	98659	156.356	169020
136.602	102462			135.164	101325	163.174	198490
137.16	103999			25	1266.6	170.038	232020
						176.953	270030
eq. 2	P/mmHg			eq. 2	P/mmHg		
A	6.959			A	6.95719	data fitted to 4-constant	
B	1425.46			B	1424.255	vapor pressure eq.	
C	213.345			C	213.206		
bp/°C	136.186			bp/°C	136.186		
				$\Delta H_V/(kJ\ mol^{-1})$			
				at 25°C	42.25		
				at bp	35.56		

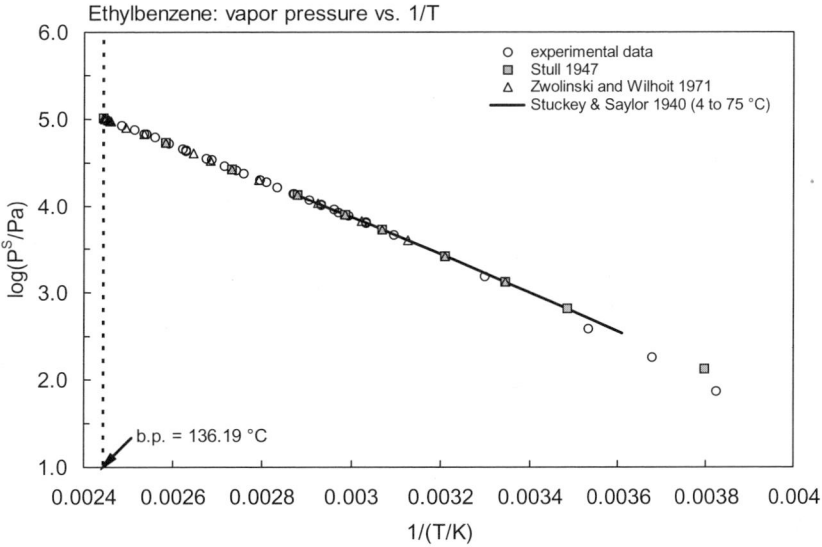

FIGURE 3.1.1.3.2 Logarithm of vapor pressure versus reciprocal temperature for ethylbenzene.

TABLE 3.1.1.3.3
Reported Henry's law constants of ethylbenzene at various temperatures and temperature dependence equations

$\ln K_{AW} = A - B/(T/K)$	(1)	$\log K_{AW} = A - B/(T/K)$	(1a)
$\ln (1/K_{AW}) = A - B/(T/K)$	(2)	$\log (1/K_{AW}) = A - B/(T/K)$	(2a)
$\ln (k_H/atm) = A - B/(T/K)$	(3)		
$\ln [H/(Pa\ m^3/mol)] = A - B/(T/K)$	(4)	$\ln [H/(atm \cdot m^3/mol)] = A - B/(T/K)$	(4a)
$K_{AW} = A - B \cdot (T/K) + C \cdot (T/K)^2$	(5)		

1.

Brown & Wasik 1974		Sanemasa et al. 1981		Sanemasa et al. 1982		Ashworth et al. 1988	
head space-GC		vapor-liquid equilibrium		vapor-liquid equilibrium		EPICS-GC	
t/°C	H/(Pa m³/mol)	t/°C	H/(Pa m³/mol)	t/°C	H/(Pa m³/mol)	t/°C	H/(Pa m³/mol)
4.5	187.5	15	459	15	418	10	330
6.33	217.2	25	797	25	734	15	457
7.06	235.1	35	1339	35	1211	20	609
8.96	265	45	1436	45	1436	25	798
11.75	329					30	1064
12.1	333.6						
15.1	409.7					eq. 4a	H/(atm m³/mol)
17.93	480.2					A	11.92
20.06	559.1					B	4994

2.

Robbins et al. 1993		Perlinger et al. 1993		Dewulf et al. 1995	
static headspace-GC		equilibrium headspace-GC		EPICS-GC	
t/°C	H/(Pa m³/mol)	t/°C	H/(Pa m³/mol)	t/°C	H/(Pa m³/mol)
25	788	10	306	2	180
30	1034	15	428	6	194
40	1662	20	583	10	257
		25	660	18.2	497
eq. 4	H/(Pa m³/mol)	30	1044	25	669
A	7.14				
B	3689			eq 1	K_{AW}
				A	14.001
				B	4567

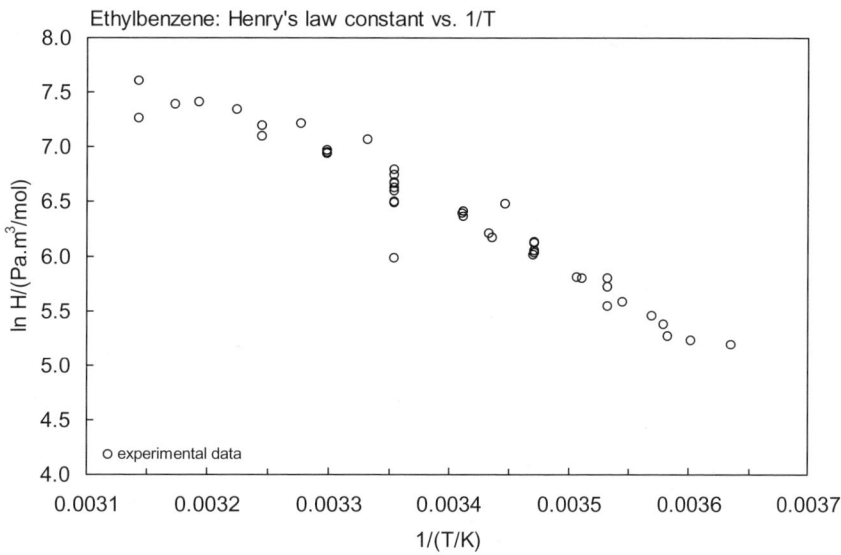

FIGURE 3.1.1.3.3 Logarithm of Henry's law constant versus reciprocal temperature for ethylbenzene.

3.1.1.4   o-Xylene

Common Name: o-Xylene
Synonym: 1,2-dimethylbenzene, o-xylol, 2-methyltoluene
Chemical Name: o-xylene
CAS Registry No: 95-47-6
Molecular Formula: C_8H_{10}, $C_6H_4(CH_3)_2$
Molecular Weight: 106.165
Melting Point (°C):
    –25.2   (Weast 1982–83; Lide 2003)
Boiling Point (°C):
    144.5   (Lide 2003)
Density (g/cm³ at 20°C):
    0.8802   (Weast 1982–83)
Molar Volume (cm³/mol):
    120.6   (20°C, calculated-density)
    140.4   (calculated-Le Bas method at normal boiling point)
Enthalpy of Vaporization, ΔH_V (kJ/mol):
    43.434, 36.82   (25°C, bp, Riddick et al. 1986)
Enthalpy of Fusion ΔH_{fus} (kJ/mol):
    13.6   (Chickos et al. 1999)
Entropy of Fusion ΔS_{fus} (J/mol K)
    55.23   (Yalkowsky & Valvani 1980)
    54.9, 45.5   (exptl., calculated-group additivity method, Chickos et al. 1999)
Fugacity Ratio at 25°C, F: 1.0

Water Solubility (g/m³ or mg/L at 25°C or as indicated. Additional data at other temperatures designated *, are compiled at the end of this section):
    204   (shake flask-UV, Andrews & Keefer 1949)
    175   (shake flask-GC, McAuliffe 1963)
    175   (shake flask-GC, McAuliffe 1966)
    176   (shake flask-GC, Hermann 1972)
    213*   (shake flask-GC, Polak & Lu 1973)
    170.5   (shake flask-GC, Sutton & Calder 1975)
    167   (shake flask-GC, Price 1976)
    167   (shake flask-GC, Krzyzanowska & Szeliga 1978)
    240*   (20°C, shake flask-UV, Ben-Naim & Wiff 1979)
    179*   (vapor saturation-UV spec., measured range 15–45°C, Sanemasa et al. 1982)
    221   (generator column-HPLC/UV, Tewari et al. 1982c)
    221   (generator column-HPLC/UV, Wasik et al. 1983)
    176   (shake flask-purge and trap-GC, Coutant & Keigley 1988)
    173*   (IUPAC recommended value, temp range 0–45°C, Shaw 1989b)

Vapor Pressure (Pa at 25°C or as indicated and reported temperature dependence equations. Additional data at other temperatures designated * are compiled at the end of this section):
    146.7*   (0.60°C, mercury manometer, measured range –17 to 0.60°C, Linder 1931)
    767*   (20°C, Hg manometer, Kassel 1936)
    log (P/mmHg) = –2830.0/(T/K) – 5·log (T/K) + 22.7480; temp range 0–80°C (vapor pressure eq. from Hg manometer measurements, Kassel 1936)
    987*   (30°C, Hg manometer, measured range 10–50°C, Rintelen et al. 1937)

log (P/mmHg) = –2908.07/(T/K) – 5·log (T/K) + 22.95279; temp range 4–75°C (vapor pressure eq. from manometer measurements, Stuckey & Saylor 1940)

880*   (Hg manometer measurements, Pitzer & Scott 1943)

log (P/mmHg) = –3327.16/(T/K) – 8.0 log (T/K) + 31.7771; temp range 0–60°C (manometer, three-constant vapor pressure eq. from exptl. data, Pitzer & Scott 1943)

6354*   (63.460°C, ebulliometry, measured range 63.460–145.367°C, Willingham et al. 1945)

6401*   (63.608°C, ebulliometry, measured range 63.608–145.400°C, Forziati et al. 1949)

log (P/mmHg) = 6.99937 – 1474.969/(213.714 + t/°C); temp range 63.5–145.4°C (manometer, Antoine eq. from exptl. data, Willingham et al. 1945)

266.6*   (20.2°C, summary of literature data, Stull 1947)

log (P/mmHg) = 6.99891 – 1474.679/(213.686 + t/°C); temp range 63.6–145.4°C (manometer, Antoine eq. from exptl. data, Forziati et al. 1949)

892   (extrapolated-Antoine eq., Dreisbach 1955)

log (P/mmHg) = 6.99891 – 1474.679/(213.686 + t/°C); temp range 50–200°C (Antoine eq. for liquid state, Dreisbach 1955)

1333*   (32.155°C, compiled data, temp range 32.155–172.095°C, Bond & Thodos 1960)

108000*   (146.85°C, ebulliometry, measured range 146.85–346.85°C, Ambrose et al. 1967)

882*   (interpolated-Antoine eq., Zwolinski & Wilhoit 1971)

log (P/mmHg) = 6.99891 – 1474.679/(213.686 + t/°C); temp range 32.14–172.07°C (Antoine eq., Zwolinski & Wilhoit 1971)

log (P/mmHg) = [–0.2185 × 9998.5/(T/K)] + 8.147551; temp range –3.8 to 144.4°C (Antoine eq., Weast 1972–73)

log (P/atm) = $(1 - 417.496/T) \times 10^{\wedge}(0.855257 - 6.48662 \times 10^4 \cdot T + 5.53883 \times 10^7 \cdot T^2)$; T in K, temp range: 253.2–631.64 K (Cox vapor pressure eq., Chao et al. 1983)

882, 885   (extrapolated-Antoine eq., Boublik et al. 1984)

log (P/kPa) = 6.12699 – 1476.753/(213.911 + t/°C); temp range 63.46–145.4°C (Antoine eq. from reported exptl. data of Willingham et al. 1945, Boublik et al. 1984)

log (P/kPa) = 5.9422 – 1387.336/(206.409 + t/°C); temp range 0–50°C (Antoine eq. from reported exptl. data of Pitzer & Scott 1943, Boublik et al. 1984)

882   (extrapolated-Antoine eq., Dean 1985, 1992)

log (P/mmHg) = 6.99891 – 1474.679/(213.69 + t/°C); temp range 32–172°C (Antoine eq., Dean 1985, 1992)

880   (Riddick et al. 1986)

log (P/kPa) = 6.13072 – 1479.82/(214.315 + t/°C); temp range not specified (Antoine eq., Riddick et al. 1986)

885   (extrapolated-Antoine eq., Stephenson & Malanowski 1987)

log (P_L/kPa) = 6.13132 – 1480.155/(–58.804 + T/K); temp range 333–419 K (liquid, Antoine eq.-I, Stephenson & Malanowski 1987)

log (P_L/kPa) = 6.15921 – 1502.949/(–55.725 + T/K); temp range 416–473 K (liquid, Antoine eq.-II, Stephenson & Malanowski 1987)

log (P_L/kPa) = 6.46119 – 1772.963/(–18.84 + T/K); temp range 471–571 K (liquid, Antoine eq.-III, Stephenson & Malanowski 1987)

log (P_L/kPa) = 7.91427 – 3735.582/(229.953 + T/K); temp range 567–630 K (liquid, Antoine eq.-IV, Stephenson & Malanowski 1987)

log (P/mmHg) = $37.2413 - 3.4573 \times 10^3/(T/K) - 10.126 \cdot \log (T/K) + 9.0676 \times 10^{-11} \cdot (T/K) + 2.6123 \times 10^{-6} \cdot (T/K)^2$, temp range 248–630 K (vapor pressure eq., Yaws 1994)

Henry's Law Constant (Pa m³/mol at 25°C or as indicated and reported temperature dependence equations. Additional data at other temperatures designated * are compiled at the end of this section):

542, 506   (calculated as 1/K_{AW}, calculated-bond contribution, Hine & Mookerjee 1975)

647   (vapor liquid equilibrium-concentration ratio, Leighton & Calo 1981)

526*   (vapor-liquid equilibrium, Sanemasa et al. 1982)

594   (20°C, EPICS-GC, Yurteri et al. 1987)

493*   (EPICS-GC/FID, Ashworth et al. 1988)

ln [H/(atm·m³/mol)] = 5.541 – 3220/(T/K); temp range 10–30°C (EPICS measurements, Ashworth et al. 1988)

424   (calculated-vapor-liquid equilibrium (VLE) data, Yaws et al. 1991)

592   (concentration ratio, Anderson 1992)

1067*   (40°C, equilibrium headspace-GC, Kolb et al. 1992)

ln (1/K_{AW}) = –7.61 + 2647/(T/K), temp range 40–80°C (equilibrium headspace-GC measurements, Kolb et al. 1992)
- 485 (infinite activity coeff. γ^∞ in water determined by inert gas stripping-GC, Li et al. 1993)
- 506* (static headspace-GC, Robbins et al. 1993)
- 372 (headspace solid-phase microextraction (SPME)-GC, Zhang & Pawliszyn 1993)
- 429* (EPICS-GC/FID, measured range 2–25°C, Dewulf et al. 1995)
- 189, 496 (6.0, 25°C, EPICS-GC/FID, natural seawater with salinity of 35‰ Dewulf et al. 1995)

ln K_{AW} = –4232/(T/K) + 0.01115·Z + 12.400; with Z salinity 0–35.5%, temp range: 2–35°C (EPICS-GC/FID, Dewulf et al. 1995)
- 412 (20°C, selected from literature experimentally measured data, Staudinger & Roberts 1996)
- 731 (vapor-liquid equilibrium-GC, Turner et al. 1996)
- 464.4 (exponential saturator EXPSAT technique, Dohnal & Hovorka 1999)
- 390 (20°C, selected from literature experimentally measured data, Staudinger & Roberts 2001)

log K_{AW} = 5.064 – 1719/(T/K) (van't Hoff eq. derived from lit. data, Staudinger & Roberts 2001)

Octanol/Water Partition Coefficient, log K_{OW} at 25°C:
- 3.15 (calculated-π substituent constant, Hansch et al. 1968)
- 2.73 (shake flask-LSC, Banerjee et al. 1980)
- 3.19 (HPLC-k' correlation, Hammers et al. 1982)
- 3.13 (generator column-HPLC/UV, Tewari et al. 1982b,c)
- 3.13 (generator column-HPLC/UV, Wasik et al. 1983)
- 3.13; 3.14, 3.14, 3.06, 3.16, 3.42 (quoted exptl.; calculated-π const., f const., MW, MCI χ, TSA, Doucette & Andren 1988)
- 3.25, 3.35 (RP-HPLC-k' capacity factor correlations, Sherblom & Eganhouse 1988)
- 3.12 (recommended, Sangster 1989)
- 3.18 (normal phase HPLC-k' correlation, Govers & Evers 1992)
- 3.12 (recommended, Hansch et al. 1995)

Octanol/Air Partition Coefficient, log K_{OA} at 25°C or as indicated. Additional data at other temperatures designated * are compiled at the end of this section:
- 3.80* (30.3°C, from GC-determined γ^∞ in octanol, measured range 30.3–50.28°C, Gruber et al. 1997)
- 3.72 (calculated-measured γ^∞ in pure octanol of Tewari et al. 1982, Abraham et al. 2001)

Bioconcentration Factor, log BCF:
- 1.33 (eels, Ogata & Miyake 1978)
- 0.79 (clams, Nunes & Benville 1979)
- 1.15 (goldfish, Ogata et al. 1984)
- 2.34 (*S. capricornutum*, Herman et al. 1991)

Sorption Partition Coefficient, log K_{OC} at 25°C:
- 1.68–1.83 (Nathwani & Philip 1977)
- 2.35 (sediment 4.02% OC from Tamar estuary, batch equilibrium-GC, Vowles & Mantoura 1987)
- 2.73 (HPLC-k' correlation, cyanopropyl column, Hodson & Williams 1988)
- 2.37, 2.40 (RP-HPLC-k' correlation, humic acid-silica column, Szabo et al. 1990a,b)
- 3.13 (average of 5 soils, sorption isotherms by batch equilibrium method, Xing et al. 1994)
- 2.36, 2.65, 2.65 (RP-HPLC-k' correlation on 3 different stationary phases, Szabo et al. 1995)
- 2.45, 2.45 (RP-HPLC-k' correlation including MCI related to non-dispersive intermolecular interactions, hydrogen-bonding indicator variable, Hong et al. 1996)
- 2.40, 2.70, 2.58, 2.68, 2.73, 2.69, 2.68 (2.3, 3.8, 6.2, 8.0, 13.5, 18.6, 25°C, natural sediment from River Leie, organic carbon f_{OC} = 4.12%, EPICS-GC/FID, Dewulf et al. 1999)

Environmental Fate Rate Constants, k, or Half-Lives, $t_{1/2}$:

Volatilization: $t_{1/2}$ = 5.61 h from water depth of 1 m (Mackay & Leinonen 1975; Haque et al. 1980);
$t_{1/2}$ ~ 3.2 h of evaporation from water of 1 m depth with wind speed of 3 m/s and water current of 1 m/s (Lyman et al. 1982);

$t_{\frac{1}{2}} \sim 31\text{--}125$ h of evaporation from a typical river or pond (Howard 1990).

Photolysis: $k = 7.46 \times 10^{-3}$ h^{-1} with H$_2$O$_2$ under photolysis at 25°C in F-113 solution and with HO- in the gas (Dilling et al. 1988).

Oxidation: rate constant k; for gas-phase second order rate constants, k_{OH} for reaction with OH radical, k_{NO3} with NO$_3$ radical and k_{O3} with O$_3$ or as indicated. Data at other temperatures and/or the Arrhenius expression are designated *, see reference:

$k_{OH} = (7.7 \pm 2.3) \times 10^9$ L mol^{-1} s^{-1}; $k_{O(3P)} = (1.05 \pm 0.11) \times 10^8$ L mol^{-1} s^{-1} with O(^{3}P) atom at room temp. (relative rate method, Doyle et al. 1975; Lloyd et al. 1976)

$k_{OH} = (15.3 \pm 1.5) \times 10^{-12}$ cm^3 molecule^{-1} s^{-1}, $k_{O(3P)} = (1.74 \pm 0.18) \times 10^{-13}$ cm^3 molecule^{-1} s^{-1} for the reaction of O(^{3}P) atom at room temp. (flash photolysis-resonance fluorescence, Hansen et al. 1975)

$k_{OH} = 8.4 \times 10^9$ L mol^{-1} s^{-1} with $t_{\frac{1}{2}} = 0.24\text{--}24$ h (Darnall et al. 1976)

$k_{OH}* = (14.3 \pm 1.5) \times 10^{-12}$ cm^3 molecule^{-1} s^{-1} at room temp., measured over temp range 296–473 K (flash photolysis-resonance fluorescence, Perry et al. 1977)

$k_{OH} = (12.9, 13.0, 12.4) \times 10^{-12}$ cm^3 molecule^{-1} s^{-1} with different dilute gas, Ar or He at 298 K (flash photolysis-resonance fluorescence, Ravishankara et al. 1978)

$k_{OH} = 13.9 \times 10^{-12}$ cm^3 molecule^{-1} s^{-1} and residence time of 0.8 d, loss of 71.3% in one day or 12 sunlit hours at 300 K in urban environments (Singh et al. 1981)

$k_{O3} < 0.01 \times 10^{-18}$ cm^3 molecule^{-1} s^{-1} at 296 ± 2 K with a calculated lifetime $\tau > 2300$ d, and a calculated lifetime of 0.8 d due to reaction with OH radical at room temp. (Atkinson et al. 1982)

$k_{O3} = 950$ cm^3 mol^{-1} s^{-1} at 300 K (Lyman et al. 1982)

$k_{OH} = (5.9\text{--}12) \times 10^9$ M^{-1} s^{-1} with $t_{\frac{1}{2}} = 0.47\text{--}1.0$ d (Mill 1982)

$k = (90 \pm 20)$ M^{-1} s^{-1} for the reaction with ozone in water at pH 1.7–5.0 and 20–23°C (Hoigné & Bader 1983)

$k_{NO3} = 1.1 \times 10^{-16}$ cm^3 molecule^{-1} s^{-1} at 296 K (Atkinson et al. 1984)

$k_{OH} = 13.4 \times 10^{-12}$ cm^3 molecule^{-1} s^{-1} with $t_{\frac{1}{2}} = 4.4\text{--}44$ h (Atkinson 1985)

$k_{OH} = 12.5 \times 10^{-12}$ cm^3 molecule^{-1} s^{-1} at 298 K (relative rate method, Ohta & Ohyama 1985)

$k_{OH} = 12.4 \times 10^{-12}$ cm^3 molecule^{-1} s^{-1} at 24.2°C, with a calculated atmospheric lifetime of 0.93 d (Edney et al. 1986)

k_{OH}(calc) = 6.9×10^{-12} cm^3 molecule^{-1} s^{-1}, k_{OH}(obs.) = 14.7×10^{-12} cm^3 molecule^{-1} s^{-1} at room temp. (SAR [structure-activity relationship] Atkinson 1987)

$k_{OH} = 1.47 \times 10^{-11}$ cm^3 molecule^{-1} s^{-1}, $k_{NO3} = 3.74 \times 10^{-16}$ cm^3 molecule^{-1} s^{-1} at 298 K (Atkinson et al. 1988; quoted, Sabljic & Güsten 1990; Müller & Klein 1992)

$k_{O3} < 0.01 \times 10^{-18}$ cm^3 molecule^{-1} s^{-1}, $k_{OH} = 1.47 \times 10^{-11}$ cm^3 molecule^{-1} s^{-1}, $k_{NO3} = 3.7 \times 10^{-16}$ cm^3 molecule^{-1} s^{-1} at room temp. (Atkinson & Aschmann 1988)

$k_{OH}* = 1.37 \times 10^{-11}$ cm^3 molecule^{-1} s^{-1} independent over 296–320 K (recommended, Atkinson 1989)

$k_{OH} = 13.7 \times 10^{-12}$ cm^3 molecule^{-1} s^{-1} at 298 K (Atkinson 1990)

k_{OH}(calc) = 6.92×10^{-11} cm^3 molecule^{-1} s^{-1} at 298 K (estimated by SARs, Müller & Klein 1992)

k_{OH}(calc) = 14.75×10^{-12} cm^3 molecule^{-1} s^{-1} (molecular orbital calculations, Klamt 1993)

k_{OH}(calc) = 6.51×10^{-12} cm^3 molecule^{-1} s^{-1}, k_{OH}(exptl) = 13.7×10^{-12} cm^3 molecule^{-1} s^{-1} at 298 K (SAR structure-activity relationship, Kwok & Atkinson 1995)

Hydrolysis: no hydrolyzable functional groups (Mabey et al. 1982).

Biodegradation:

100% biodegraded after 192 h at 13°C with an initial concn of 1.62×10^{-6} L/L (Jamison et al. 1976);

$t_{\frac{1}{2}}$(aq. Aerobic) = 168–672 h, estimated based on aqueous screening test data (Bridie et al. 1979; Howard et al. 1991) and soil column study simulating an aerobic river/ground-water infiltration system (Kuhn et al. 1985; Howard et al. 1991)

$t_{\frac{1}{2}}$(aq. anaerobic) = 4320–8640 h, estimated based on acclimated grab sample data for anaerobic soil from a groundwater aquifer receiving landfill leachate (Wilson et al. 1986) and a soil column study simulating an anaerobic river/groundwater infiltration system (Kuhn et al. 1985; Howard et al. 1991)

$k = 0.06$ yr^{-1} with $t_{\frac{1}{2}} = 32$ d (Olsen & Davis 1990)

$t_{\frac{1}{2}}$(aerobic) = 7 d, $t_{\frac{1}{2}}$(anaerobic) = 180 d in natural waters (Capel & Larson 1995)

Bioconcentration, Uptake (k_1) and Elimination (k_2) Rate Constants or Half-Lives:

Half-Lives in the Environment:

Air: $t_{\frac{1}{2}} = 0.24\text{--}2.4$ h, based on rate of disappearance for the reaction with hydroxyl radical (Darnall et al. 1976);

residence time of 0.8 d, loss of 71.3% in one day or 12 sunlit hours at 300 K in urban environments (Singh et al. 1981)

calculated lifetimes τ > 2300 d and 0.8 d due to reactions with O_3 and OH radical respectively at room temp. (Atkinson et al. 1982);

$t_{1/2}$ = 4.4–44 h, based on photooxidation half-life in air (Atkinson 1985; Howard et al. 1991);

summer daylight lifetime τ = 10 h due to reaction with OH radical (Altshuller 1991);

calculated lifetime τ = 0.93 d due to reaction with OH radical (Edney et al. 1986).

Surface Water: photooxidation $t_{1/2}$ = 3.9×10^5–2.7×10^8 h in water, based on estimated rate data for alkoxyl radical in aqueous solution (Hendry et al. 1974);

$t_{1/2}$ = 5.18 h, based on evaporative loss at 25°C and 1 m depth of water (Mackay & Leinonen 1975; Haque et al. 1980);

$t_{1/2}$ = 168–672 h, based on estimated aqueous aerobic biodegradation half-life (Bridie et al. 1979; Kuhn et al. 1985; Howard et al. 1991);

$t_{1/2}$ = 1–5 d, volatilization to be the dominant removal process (Howard 1990).

$t_{1/2}$(aerobic) = 7 d, $t_{1/2}$(anaerobic) = 180 d in natural waters (Capel & Larson 1995)

Ground water: $t_{1/2}$ = 336–8640 h, based on estimated aqueous aerobic and anaerobic biodegradation half-life (Bridie et al. 1979; Kuhn et al. 1985; Wilson et al. 1986; Howard et al. 1991);

$t_{1/2}$ ~0.3 yr from observed persistence in groundwater of the Netherlands (Zoeteman et al. 1981).

Soil: $t_{1/2}$ = 168–672 h, based on estimated aqueous aerobic biodegradation half-life (Bridie et al. 1979; Kuhn et al. 1985; Howard et al. 1991).

Biota: $t_{1/2}$ = 2 d to eliminate from eels in seawater (Ogata & Miyake 1978).

TABLE 3.1.1.4.1
Reported aqueous solubilities of o-xylene at various temperatures

Polak & Lu 1973		Ben-Naim & Wiff 1979		Sanemasa 1982		Shaw 1989b (IUPAC)	
shake flask-GC		shake flask-UV		vapor saturation-UV		recommended values	
t/°C	S/g·m^{-3}	t/°C	S/g·m^{-3}	t/°C	S/g·m^{-3}	t/°C	S/g·m^{-3}
0	142	10	212.4	15	168	0	140
25	213	20	240	25	179	15	170
				35	198	25	173
				45	214	35	200
						45	210

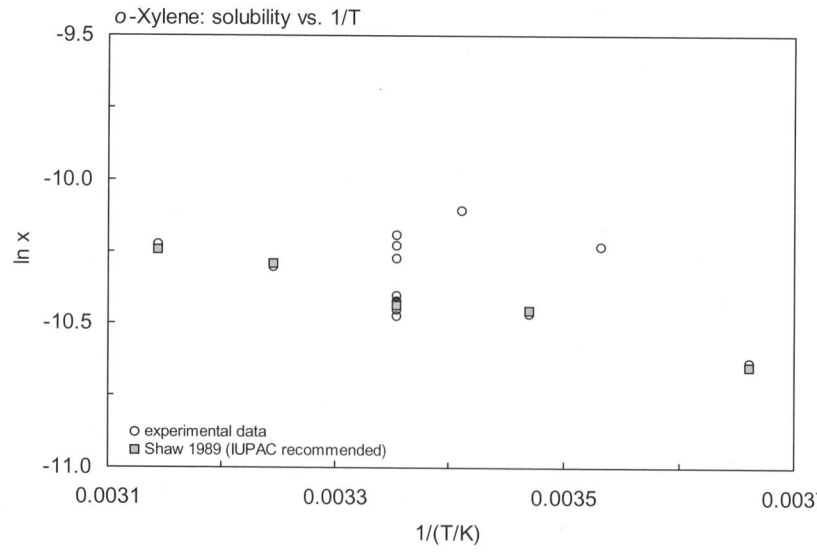

FIGURE 3.1.1.4.1 Logarithm of mole fraction solubility (ln x) versus reciprocal temperature for o-xylene.

Mononuclear Aromatic Hydrocarbons

TABLE 3.1.1.4.2
Reported vapor pressures of *o*-xylene at various temperatures and the coefficients for the vapor pressure equations

$$\log P = A - B/(T/K) \quad (1) \qquad \ln P = A - B/(T/K) \quad (1a)$$
$$\log P = A - B/(C + t/°C) \quad (2) \qquad \ln P = A - B/(C + t/°C) \quad (2a)$$
$$\log P = A - B/(C + T/K) \quad (3)$$
$$\log P = A - B/(T/K) - C \cdot \log (T/K) \quad (4)$$
$$\log P = A - B/(T/K) - C \cdot \log (T/K) + D \cdot P/(T/K)^2 \quad (5)$$

1.

Kassel 1936		Rintelen et al. 1937		Stuckey & Saylor 1940		Pitzer & Scott 1943	
mercury manometer		mercury manometer		mercury manometer		mercury manometer	
t/°C	P/Pa	t/°C	P/Pa	t/°C	P/Pa	t/°C	P/Pa
0	215	10	213	measured 4–75°C		0	173
10	417	30	987	eq. 4	P/mmHg	12.5	400
20	767	50	3346	A	22.95279	25.0	880
30	1347			B	908.07	37.5	1800
40	2280	bp 144.0–144.1°C		C	5.0	50	3400
50	3680					60	5413
60	5826	**Linder 1931**		bp/°C	144.39–144.41		
70	8892	mercury manometer				eq. 4	P/mmHg
80	13186	t/°C	P/Pa			A	31.7771
		–17	26.66			B	3327.16
eq. 4	P/mmHg	–10.7	56.0			C	8.0
A	23.7480	0	141.3				
B	2830.06	0.60	146.7			$\Delta H_V/(kJ\,mol^{-1}) = 43.806$	
C	5.0					at 25°C	

2.

Willingham et al. 1945		Stull 1947		Forziati et al. 1949		Bond & Thodos 1960	
ebulliometry		summary of literature data		ebulliometry		compiled data	
t/°C	P/Pa	t/°C	P/Pa	t/°C	P/Pa	t/°C	P/Pa
63.460	6354	–3.80	133.3	63.608	6401	32.155	1333
67.746	7654	20.2	266.6	67.852	7693	63.48	6354
71.481	8963	32.1	1333	71.548	8991	74.87	10303
74.857	10303	45.1	2666	74.916	10331	112.46	39997
77.993	11699	59.5	5333	78.048	11722	141.35	93326
82.242	13819	68,8	7999	82.285	13843	172.095	199984
87.081	16621	81.3	13332	87.101	16635		
91.987	19924	100.2	26664	92.015	19945	eq. 5	P/mmHg
96.541	23450	121.7	53329	96.568	23475	A	25.82849
102.632	28954	144.4	101325	102.657	28979	B	3040.72
108.227	34897			198.250	34926	C	5.94175
114.965	43322	mp/°C	–25.2	114.988	43362	D	2.61456
121.909	53654			121.935	53700		
129.318	66756			129.333	66797	bp/°C	144.426
137.346	83717			137.356	83750		
143.007	97604			143.619	97647		
143.614	99200			143.626	99236		
144.176	100692			144.190	87395		
144.809	102385			144.832	90465		

(Continued)

TABLE 3.1.1.4.2 *(Continued)*

Willingham et al. 1945		Stull 1947		Forziati et al. 1949		Bond & Thodos 1960	
ebulliometry		summary of literature data		ebulliometry		compiled data	
t/°C	P/Pa	t/°C	P/Pa	t/°C	P/Pa	t/°C	P/Pa
145.367	103905			145.400	104000		
eq. 2	P/mmHg			eq. 2	P/mmHg		
A	6.99937			A	6.99891		
B	1474.967			B	1474.679		
C	144.414			C	213.686		
bp/°C	144.414			bp/°C	144.411		

3.

Ambrose et al. 1967		Zwolinski & Wilhoit 1971			
ebulliometry		selected values			
t/°C	P/Pa	t/°C	P/Pa	t/°C	P/Pa
146.85	108000	32.14	1333	145.395	103991
166.85	177200	45.13	2666	146.359	106658
186.85	276700	53.38	4000	150.912	119990
206.85	413900	59.56	5333	155.08	13322
226.85	598300	64.558	6666	162.53	159987
246.85	839100	68.778	7999	172.07	199984
266.85	1146500	75.704	10666	25.0	879.9
286.85	1532100	81.314	13332		
306.85	2009000	92.085	19998	eq. 2	P/mmHg
326.85	2592500	100.217	26664	A	6.99891
346.85	3304100	106.829	33331	B	1474.679
		112.441	39997	C	213.686
eq. 5	P/bar	121.708	53329		
A	20.79970	129.267	66661	bp/°C	144.411
B	2921.11	135.700	79993		
C	5.26888	141.332	93326	ΔH_V/(kJ mol^{-1})	
D	1672	142.380	95991	at 25°C	43.43
		143.407	98659	at bp	36.82
1 bar = 1×10^5 Pa		144.411	101325		

Mononuclear Aromatic Hydrocarbons

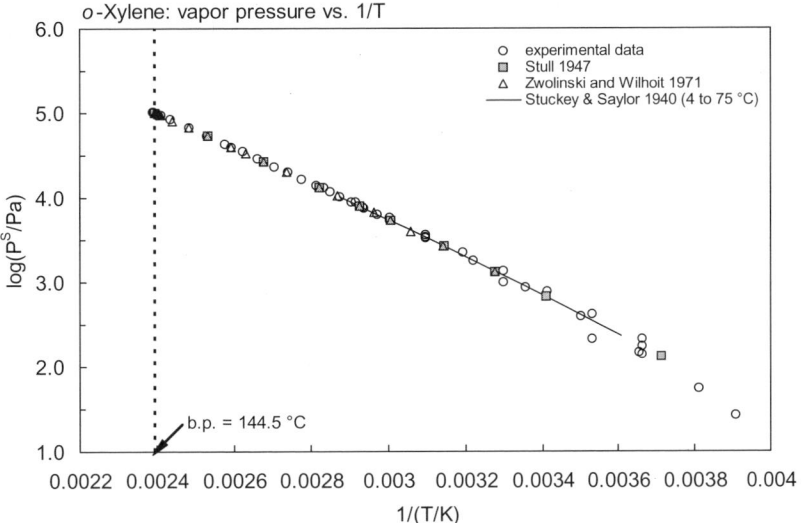

FIGURE 3.1.1.4.2 Logarithm of vapor pressure versus reciprocal temperature for o-xylene.

TABLE 3.1.1.4.3
Reported Henry's law constants of o-xylene at various temperatures and temperature dependence equations

$\ln K_{AW} = A - B/(T/K)$	(1)	$\log K_{AW} = A - B/(T/K)$	(1a)
$\ln (1/K_{AW}) = A - B/(T/K)$	(2)	$\log (1/K_{AW}) = A - B/(T/K)$	(2a)
$\ln (k_H/atm) = A - B/(T/K)$	(3)		
$\ln [H/(Pa\ m^3/mol)] = A - B/(T/K)$	(4)	$\ln [H/(atm \cdot m^3/mol)] = A - B/(T/K)$	(4a)
$K_{AW} = A - B(T/K) + C(T/K)^2$	(5)		

Sanemasa et al. 1982		Ashworth et al. 1988		Kolb et al. 1992		Robbins et al. 1993		Dewulf et al. 1995	
vapor-liquid equilibrium		EPICS-GC		headspace-GC		static headspace-GC		EPICS-GC	
t/°C	H/(Pa m³/mol)	t/°C	H/(Pa m³/mol)	t/°C	H/(Pa m³/mol)	t/°C	H/(Pa m³/mol)	t/°C	H/(Pa m³/mol)
15	299	10	289	40	1067	25	506	2	133
25	526	15	366	60	2114	30	637	6	118
35	844	20	480	70	2825	40	1104	10	155
45	1323	25	493	80	2966	45	1074	18	325
		30	634			50	1175	25	429
		eq. 4a		eq. 2	K_{AW}	eq. 4		eq 1	K_{AW}
		H/(atm m³/mol)		A	761	H/(Pa m³/mol)		A	12.4
		A	5.541	B	2647/R	A	17.67818	B	4243
		B	3220			B	3397.97		

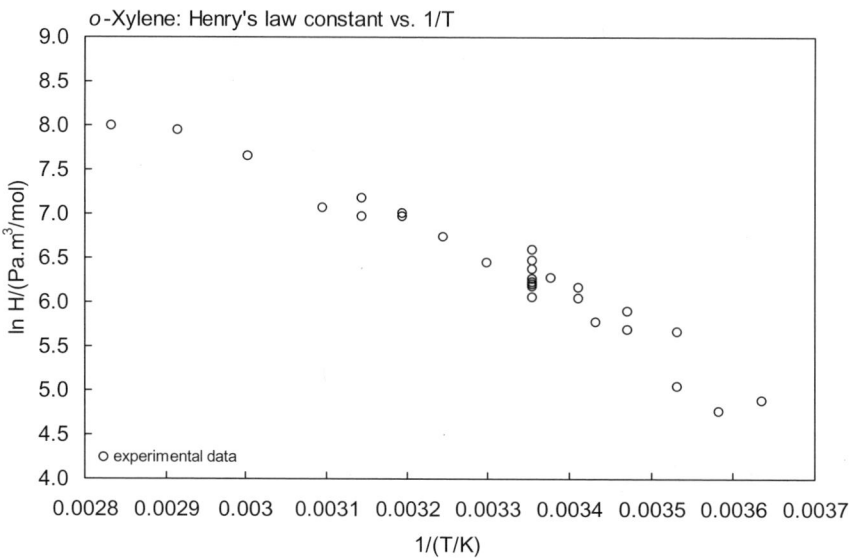

FIGURE 3.1.1.4.3 Logarithm of Henry's law constant versus reciprocal temperature for o-xylene.

TABLE 3.1.1.4.4
Reported octanol-air partition coefficients of o-xylene at various temperatures

Gruber et al. 1997	
GC det'd activity coefficient	
t/°C	log K_{OA}
20.29	–
30.3	3.80
40.4	3.59
50.28	3.38

3.1.1.5 *m*-Xylene

Common Name: *m*-Xylene
Synonym: 1,3-dimethylbenzene, *m*-xylol, 3-methyltoluene
Chemical Name: *m*-xylene
CAS Registry No: 108-38-3
Molecular Formula: C_8H_{10}, $C_6H_4(CH_3)_2$
Molecular Weight: 106.165
Melting Point (°C)
    –47.8   (Lide 2003)
Boiling Point (°C):
    139.12   (Lide 2003)
Density (g/cm³ at 20°C):
    0.8842   (Weast 1982–83)
Molar Volume (cm³/mol):
    120.1   (20°C, calculated-density)
    140.4   (calculated-Le Bas method at normal boiling point)
Enthalpy of Vaporization, ΔH_V (kJ/mol):
    42.656, 36.36   (25°C, bp, Riddick et al. 1986)
Enthalpy of Fusion ΔH_{fus} (kJ/mol):
    11.57   (Chickos et al. 1999)
Entropy of Fusion ΔS_{fus} (J/mol K):
    51.88   (Yalkowsky & Valvani 1980)
    51.4, 45.6   (exptl., calculated-group additivity method, Chickos et al. 1999)
Fugacity Ratio at 25°C, F: 1.0

Water Solubility (g/m³ or mg/L at 25°C, or as indicated and reported temperature dependence equations. Additional data at other temperatures designated * are compiled at the end of this section):
    173   (shake flask-UV, Andrews & Keefer 1949)
    196*   (shake flask-UV, measured range 0.4–39.6°C, Bohon & Claussen 1951)
    157   (shake flask-GC, Hermann 1972)
    162*   (shake flask-GC, Polak & Lu 1973)
    206   (shake flask-UV, Vesala 1974)
    146   (shake flask-GC, Sutton & Calder 1975)
    160*   (synthetic method-GC, measured range 20–70°C, Chernoglazova & Simulin 1976)
    134   (shake flask-GC, Price 1976)
    134   (shake flask-GC, Krzyzanowska & Szeliga 1978)
    162*   (vapor saturation-UV spec., measured range 15–45°C, Sanemasa et al. 1982)
    159   (generator column-HPLC/UV, Tewari et al. 1982c)
    160   (generator column-HPLC/UV, Wasik et al. 1983)
    160*   (IUPAC recommended value, temp range 0–70°C, Shaw 1989b)

Vapor Pressure (Pa at 25°C or as indicated and reported temperature dependence equations. Additional data at other temperatures designated * are compiled at the end of this section):
    1812.7*   (–2.80°C, mercury manometer, measured range –8.40 to –2.80°C, Linder 1931)
    833*   (20°C, Hg manometer, Kassel 1936)
    log (P/mmHg) = –2876.3/(T/K) – 5·log (T/K) + 22.9425; temp range 0–80°C (vapor pressure eq. from Hg manometer measurements, Kassel 1936)
    1213*   (30°C, Hg manometer, measured range 10–50°C, Rintelen et al. 1937)

log (P/mmHg) = –2870.38/(T/K) – 5·log (T/K) + 22.92341; temp range 4–75°C (vapor pressure eq. from manometer measurements, Stuckey & Saylor 1940)
1113*   (Hg manometer, measured range 0–60°C, Pitzer & Scott 1943)
log (P/mmHg) = –2871.66/(T/K) – 5.0 log (T/K) + 22.9270; temp range 0–60°C (manometer, three-constant vapor pressure eq. from exptl. data, Pitzer & Scott 1943)
6355*   (59.203°C, ebulliometry, measured range 59.203–140.041°C, Willingham et al. 1945)
log (P/mmHg) = 7.00343 – 1458.214/(214.609 + t/°C); temp range 59.2–140.0°C (manometer, Antoine eq. from exptl. data, Willingham et al. 1945)
266.6*   (16.8°C, summary of literature data, temp range –6.9 to 139.1°C, Stull 1947)
6400*   (59.335°C, ebulliometry, measured range 59.335–140.078°C, Forziati et al. 1949)
log (P/mmHg) = 8.00849 – 1461.925/(215.073 + t/°C); temp range 59.3–140.1°C (manometer, Antoine eq. from exptl. data, Forziati et al. 1949)
1115   (extrapolated-Antoine eq., Dreisbach 1955)
log (P/mmHg) = 7.00908 – 1462.266/(215.105 + t/°C); temp range 45–195°C (Antoine eq. for liquid state, Dreisbach 1955)
124200*   (146.85°C, ebulliometry, measured range 146.85–316.85°C, Ambrose et al. 1967)
1100*   (extrapolated-Antoine eq., Zwolinski & Wilhoit 1971)
log (P/mmHg) = 7.00908 – 1462.266/(215.105 + t/°C); temp range 28.24–166.39°C (Antoine eq., Zwolinski & Wilhoit 1971)
log (P/mmHg) = [–0.2185 × 9904.2/(T/K)] + 8.167049; temp range –6.9 to 139.1°C (Antoine eq., Weast 1972–73)
log (P/atm) = (1 – 3412.335/T) × 10^(0.859841 – 6.73249 × 10^4·T + 5.87438 × 10^7·T^2); T in K, temp range 243.2–619.2 K (Cox vapor pressure eq., Chao et al. 1983)
1104, 1142   (extrapolated-Antoine eq., interpolated-Antoine eq., Boublik et al. 1984)
log (P/kPa) = 6.13232 – 1460.805/(214.895 + t/°C); temp range 59.2–140.4°C (Antoine eq. from reported exptl. data of Willingham et al. 1945, Boublik et al. 1984)
log (P/kPa) = 6.4729 – 1641.628/(230.899 + t/°C); temp range: 0–60°C (Antoine eq. from reported exptl. data of Pitzer & Scott 1943, Boublik et al. 1984)
1106   (extrapolated-Antoine eq., Dean 1985, 1992)
log (P/mmHg) = 7.00908 – 1462.266/(215.11 + t/°C); temp range: 28–166°C (Antoine eq., Dean 1985, 1992)
1100   (selected lit. value., Riddick et al. 1986)
log (P/kPa) = 6.13785 – 1465.39/(215.512 + t/°C); temp range not specified (Antoine eq., Riddick et al. 1986)
1110   (extrapolated-Antoine eq., Stephenson & Malanowski 1987)
log (P_L/kPa) = 6.14083 – 1457.244/(–57.442 + T/K); temp range 331–414 K (Antoine eq.-I, Stephenson & Malanowski 1987)
log (P_L/kPa) = 5.76037 – 1292.22/(–72.052 + T/K); temp range 267–301 K (Antoine eq.-II, Stephenson & Malanowski 1987)
log (P_L/kPa) = 6.17035 – 1490.184/(–54.184 + T/K); temp range 412–462 K (Antoine eq.-III, Stephenson & Malanowski 1987)
log (P_L/kPa) = 6.42535 – 1710.901/(–24.591 + T/K); temp range 461–554 K (Antoine eq.-IV, Stephenson & Malanowski 1987)
log (P_L/kPa) = 7.59221 – 3163.74/(165.278 + T/K); temp range 550–617 K (Antoine eq.-V, Stephenson & Malanowski 1987)
log (P/mmHg) = 34.6803 – 3.2981 × 10^3/(T/K) – 9.2570·log (T/K) – 4.3563 × 10^{-10}·(T/K) + 2.4103 × 10^{-6}·$(T/K)^2$, temp range 226–617 K (vapor pressure eq., Yaws 1994)

Henry's Law Constant (Pa m^3/mol at 25°C or as indicated and reported temperature dependence equations. Additional data at other temperatures designated *, are compiled at the end of this section):
731*   (vapor-liquid equilibrium, Sanemasa et al. 1982)
754*   (EPICS-GC/FID, Ashworth et al. 1988)
ln [H/(atm·m^3/mol)] = 6.280 – 3337/(T/K); temp range: 10–30°C (EPICS measurements, Ashworth et al. 1988)
675   (calculated-vapor-liquid equilibrium (VLE) data, Yaws et al. 1991)
665   (infinite activity coeff. γ^∞ in water determined by inert gas stripping-GC, Li et al. 1993)
739*   (static headspace-GC, same as p-xylene, Robbins et al. 1993)
615*   (EPICS-GC/FID, measured range 2–25°C, Dewulf et al. 1995)

297, 771 (6.0, 25°C, EPICS-GC/FID, natural seawater with salinity of 35‰, Dewulf et al. 1995)
ln K_{AW} = –4026/(T/K) + 0.00846·Z + 12.123; with Z salinity 0–35.5‰, temp range: 2–35°C (EPICS-GC/FID, Dewulf et al. 1995)
590 (20°C, selected from literature experimentally measured data, Staudinger & Roberts 1996)
658.5 (exponential saturator EXPSAT technique, Dohnal & Hovorka 1999)
561 (20°C, selected from literature experimentally measured data, Staudinger & Roberts 2001)
log K_{AW} = 5.204 – 1713/(T/K) (van't Hoff eq. derived from lit. data, Staudinger & Roberts 2001)

Octanol/Water Partition Coefficient, log K_{OW} at 25°C:
3.20 (Hansch et al. 1968; Leo et al. 1971; Hansch & Leo 1979; Hansch & Leo 1985)
3.18 (generator column-HPLC/UV, Wasik et al. 1981)
3.29 (HPLC-k' correlation, Hammers et al. 1982)
3.13 (generator column-HPLC/UV, Tewari et al. 1982b,c)
3.20 (generator column-HPLC/UV, Wasik et al. 1983)
3.28 (HPLC-RV retention volume correlation, Garst & Wilson 1984)
3.37 (HPLC-k' correlation, Haky & Young 1984)
3.33, 3.45 (RP-HPLC-k' capacity factor correlations, Sherblom & Eganhouse 1988)
3.20 (recommended, Sangster 1989, 1993)
3.31 (normal phase HPLC-k' correlation, Govers & Evers 1992)
3.20 (recommended, Hansch et al. 1995)

Octanol/Air Partition Coefficient, log K_{OA} at 25°C or as indicated. Additional data at other temperatures designated * are compiled at the end of this section:
3.69* (30.3°C, from GC-determined γ^∞ in octanol, measured range 30.3–50.28°C, Gruber et al. 1997)
3.79 (calculated-measured γ^∞ in pure octanol of Tewari et al. 1982, Abraham et al. 2001)

Bioconcentration Factor, log BCF:
1.37 (eels, Ogata & Miyake 1978)
0.78 (clams, Nunes & Benville 1979)
1.17 (goldfish, Ogata et al. 1984)
2.40 (*S. capricornutum*, Herman et al. 1991)

Sorption Partition Coefficient, log K_{OC} at 25°C:
2.11, 2.48, 2.20 (forest soil pH 5.6, forest soil pH 4.2, agricultural soil pH 7.4, Seip et al. 1986)
2.37, 2.40 (RP-HPLC-k' correlation, Szabo et al. 1990a,b)
2.62, 2.63 (RP-HPLC-k' correlation including MCI related to non-dispersive intermolecular interactions, hydrogen-bonding indicator variable, Hong et al. 1996)
2.06, 2.33 (soils: organic carbon OC ≥ 0.1%, OC ≥ 0.5%, average, Delle Site 2001)

Environmental Fate Rate Constants, k, or Half-Lives, $t_{1/2}$:
Volatilization: $t_{1/2}$ ~ 3.1 h for evaporation from water of 1 m depth with wind speed of 3 m/s and water current of 1 m/s (Lyman et al. 1982);
$t_{1/2}$ ~ 27–135 h for evaporation from a typical river or pond (Howard 1990).
Photolysis:
Oxidation: rate constant k; for gas-phase second-order rate constants, k_{OH} for reaction with OH radical, k_{NO3} with NO_3 radical and k_{O3} with O_3 or as indicated. Data at other temperatures and/or the Arrhenius expression are designated *, see reference:
photooxidation $t_{1/2}$ = $4.8 \times 10^6 - 2.4 \times 10^8$ h in water, based on estimated rate data for alkoxy radical in aqueous solution (Hendry et al. 1974)
k_{OH} = $(14 \pm 1) \times 10^9$ L mol^{-1} s^{-1} with $t_{1/2}$ = 0.83 h; $k_{O(3P)}$ = $(2.12 \pm 0.21) \times 10^8$ L mol^{-1} s^{-1} with $O(^3P)$ atom at room temp. (relative rate method, Doyle et al. 1975)
k_{OH} = 1.4×10^{10} L mol^{-1} s^{-1} with an initial concn of 2.0×10^{-10} mol L^{-1} at 300 K (Doyle et al. 1975)
k_{OH} = $(23.6 \pm 2.4) \times 10^{-12}$ cm^3 molecule^{-1} s^{-1} and $k_{O(3P)}$ = $(3.52 \pm 0.35) \times 10^{-13}$ cm^3 molecule^{-1} s^{-1} for the reaction of $O(^3P)$ atom at room temp. (flash photolysis-resonance fluorescence, Hansen et al. 1975)
k_{OH} = $(12.9 \pm 2.6) \times 10^9$ L mol^{-1} s^{-1} at 305 ± 2 K (relative rate method, Lloyd et al. 1976)

$k_{OH} = 14.1 \times 10^9$ L mol^{-1} s^{-1} with $t_{1/2} = 0.24$–2.4 h (Darnall et al. 1976)

$k_{OH}* = (24.0 \pm 2.5) \times 10^{-12}$ cm^3·molecule^{-1} s^{-1} at room temp., measured over temp range 296–473 K (flash photolysis-resonance fluorescence, Perry et al. 1977)

$k_{OH} = (15.6 - 21.4) \times 10^{-12}$ cm^3 molecule^{-1} s^{-1} with different dilute gas, Ar or He at 298 K (flash photolysis-resonance fluorescence, Ravishanakara et al. 1978)

$k_{OH} = 23.4 \times 10^{-12}$ cm^3 molecule^{-1} s^{-1} and residence time of 0.5 d, loss of 86.5% in one day or 12 sunlit hours at 300 K in urban environments (Singh et al. 1981)

$k_{O3} < 0.005 \times 10^{-18}$ cm^3 molecule^{-1} s^{-1} with a calculated lifetime $\tau > 1150$ d and a lifetime $\tau = 0.5$ d due to reaction with OH radical at room temp. (Atkinson et al. 1982)

$k_{O3} = 780$ cm^3 mol^{-1} s^{-1} for the reaction with ozone at 300 K (Lyman et al. 1982)

$k_{OH} = (5.9$–$12) \times 10^9$ M^{-1} s^{-1} with $t_{1/2} = 0.47$–1.0 d for xylenes (Mill 1982)

$k = (94 \pm 20)$ M^{-1} s^{-1} for the reaction with ozone in water using 1 mM t-BuOH as scavenger at pH 2.0 and 20–23°C (Hoigné & Bader 1983)

$k_{NO3} = 7.1 \times 10^{-17}$ cm^3 molecule^{-1} s^{-1} at 296 K (Atkinson et al. 1984)

$k_{OH} = 23.5 \times 10^{-12}$ cm^3 molecule^{-1} s^{-1} with $t_{1/2} = 2.6$–26 h in air (Atkinson 1985; Howard et al. 1991)

$k_{OH} = 22.2 \times 10^{-12}$ cm^3 molecule^{-1} s^{-1} at 298 K (Ohta & Ohyama 1985)

$k_{OH} = 22.7 \times 10^{-12}$ cm^3 molecule^{-1} s^{-1} 23.5°C, with a calculated atmospheric lifetime $\tau = 0.51$ d (Edney et al. 1986)

k_{OH}(calc) $= 14 \times 10^{-12}$ cm^3 molecule^{-1} s^{-1}, k_{OH}(obs.) $= 24.5 \times 10^{-12}$ cm^3 molecule^{-1} s^{-1} at room temp. (SAR [structure-activity relationship], Atkinson 1987)

$k_{NO3} = 2.49 \times 10^{-16}$ cm^3 molecule^{-1} s^{-1}, $k_{OH} = 2.45 \times 10^{-11}$ cm^3 molecule^{-1} s^{-1} at 298 K (Atkinson et al. 1988; quoted, Sabljic & Güsten 1990; Müller & Klein 1992)

$k_{OH}* = 2.36 \times 10^{-12}$ cm^3 molecule^{-1} s^{-1} at 298 K (recommended, Atkinson 1989, 1990)

k_{OH}(calc) $= 1.45 \times 10^{-11}$ cm^3 molecule^{-1} s^{-1} at 298 K (estimated by SARs, Müller & Klein 1992)

k_{OH}(calc) $= 17.47 \times 10^{-12}$ cm^3 molecule^{-1} s^{-1} (molecular orbital calculations, Klamt 1993)

$k_{OH} = (1.81 \pm 0.40) \times 10^{-11}$ cm^3 molecule^{-1} s^{-1} and $(2.03 \pm 0.10) \times 10^{-11}$ cm^3 molecule^{-1} s^{-1} at 298 ± 2 K (relative rate method, Phousongphouang & Arey 2002)

Hydrolysis: no hydrolyzable functional groups (Mabey et al. 1982).

Biodegradation:

100% biodegraded after 192 h at 13°C with an initial concn of 3.28×10^{-6} L/L (Jamison et al. 1976)

$t_{1/2}$(aq. aerobic) = 168–672 h, based on aqueous screening test data (Bridie et al. 1979; Howard et al. 1991) and soil column study simulating an aerobic river/groundwater infiltration system (Kuhn et al. 1985; Howard et al. 1991)

$t_{1/2}$(aq. anaerobic) = 672–2688 h, based on unacclimated aqueous aerobic biodegradation half-life (Bridie et al. 1979; Kuhn et al. 1985; Howard et al. 1991)

$t_{1/2} = 0.03$ d (Olsen & Davis 1990)

Bioconcentration, Uptake (k_1) and Elimination (k_2) Rate Constants or Half-Lives:

Half-Lives in the Environment:

Air: $t_{1/2} = 0.83$ h, based on rate of disappearance for the reaction with OH radical in ambient LA basin air at 300 K (Doyle et al. 1975);

photodecomposition $t_{1/2} = 2.9$ h under simulated atmospheric conditions, with NO (Dilling et al. 1976);

estimated lifetime $\tau = 1.5$ h under photochemical smog conditions in S.E. England (Brice & Derwent 1978) and (Perry et al. 1977);

residence time of 0.5 d, loss of 86.5% in one day or 12 sunlit hours at 300 K in urban environments (Singh et al. 1981)

calculated lifetimes $\tau > 1150$ d and 0.5 d due to reactions with O_3 and OH radical, respectively, at room temp. (Atkinson et al. 1982);

$t_{1/2} = 2.6$–26 h, based on photooxidation half-life in air (Atkinson 1985; Howard et al. 1991);

calculated atmospheric lifetime $\tau = 0.51$ d due to reaction with OH radical (Edney et al. 1986);

summer daylight lifetime $\tau = 5.9$ h due to reaction with OH radical (Altshuller 1991);

calculated lifetimes of 5.9 h, 200 d and > 4.5 yr for reactions with OH radical, NO_3 radical and O_3, respectively (Atkinson 2000).

Surface Water: $t_{1/2} = 168$–672 h, based on estimated aqueous aerobic biodegradation half-life (Bridie et al. 1979; Kuhn et al. 1985; Howard et al. 1991);

volatilization appears to be dominant removal process with $t_{1/2} = 1$–5.5 d (Howard 1990).

Ground water: $t_{1/2}$ = 336–8640 h, based on estimated aqueous aerobic and anaerobic biodegradation half-lives (Bridie et al. 1979; Kuhn et al. 1985; Howard et al. 1991);
 $t_{1/2}$ ~ 0.3 yr from observed persistence in groundwater of the Netherlands (Zoeteman et al. 1981);
 abiotic hydrolysis or dehydro-halogenation $t_{1/2}$ = 377 months (Olsen & Davis 1990).
Soil: $t_{1/2}$ = 168–672 h, based on estimated aqueous aerobic biodegradation half-life (Bridie et al. 1979; Kuhn et al. 1985; Howard et al. 1991).
Biota: $t_{1/2}$ = 2 d, half-life to eliminate from eels in seawater (Ogata & Miyake 1978).

TABLE 3.1.1.5.1
Reported aqueous solubilities of *m*-xylene at various temperatures

Bohon & Claussen 1951		Polak & Lu 1973		Sanemasa et al. 1982		Shaw 1989b (IUPAC)	
shake flask-UV		shake flask-GC		vapor saturation-UV		recommended values	
t/°C	S/g·m^{-3}	t/°C	S/g·m^{-3}	t/°C	S/g·m^{-3}	t/°C	S/g·m^{-3}
0.4	209	0	196	15	158	0	203
5.2	201	25	162	25	162	10	200
14.9	192			35	168	20	170
21	196			45	186	25	160
25	196					30	180
25.6	196					40	220
30.3	198			ΔH_{sol}/(kJ mol^{-1}) = 2.60		50	230
34.9	203			at 25°C		60	320
39.6	218					70	350
ΔH_{sol}/(kJ mol^{-1}) =		Chernoglazova & Simulin 1976				ΔH_{sol}/(kJ mol^{-1}) = 2.9	
25°C	2.8	synthetic method-GC				(calc from Van't Hoff eq.)	
2°C	−5.506	t/°C	S/g·m^{-3}				
7°C	−3.828						
12°C	−1.59	20	160				
17°C	0	40	220				
22°C	1.22	70	380				
27°C	1.99						
32°C	3.92	ΔH_{sol}/(kJ mol^{-1}) = 11.9					
37°C	8.87	between 127–239°C					

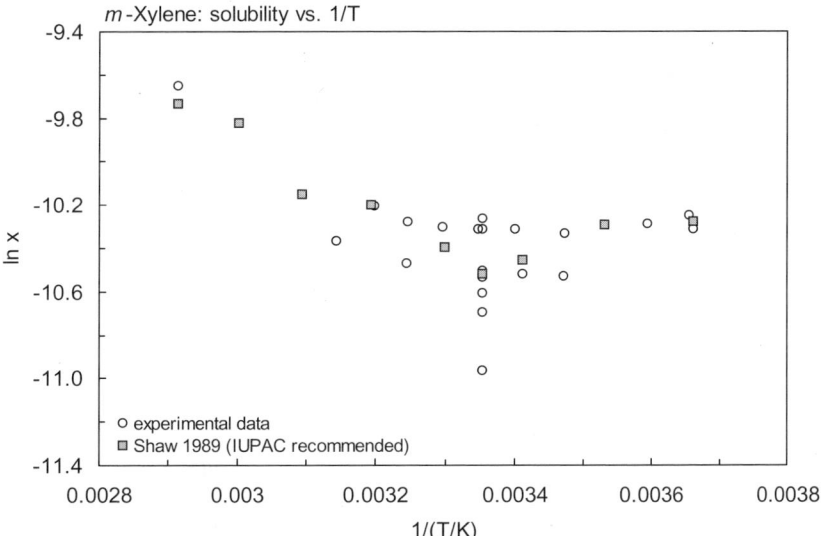

FIGURE 3.1.1.5.1 Logarithm of mole fraction solubility (ln x) versus reciprocal temperature for *m*-xylene.

TABLE 3.1.1.5.2
Reported vapor pressures of *m*-xylene at various temperatures and the coefficients for the vapor pressure equations

$\log P = A - B/(T/K)$ (1)	$\ln P = A - B/(T/K)$ (1a)
$\log P = A - B/(C + t/°C)$ (2)	$\ln P = A - B/(C + t/°C)$ (2a)
$\log P = A - B/(C + T/K)$ (3)	
$\log P = A - B/(T/K) - C \cdot \log (T/K)$ (4)	
$\log P = A - B/(T/K) - C \cdot \log (T/K) + D(T/K)^2$ (5)	

1.

Kassel 1936		Rintelen et al. 1937		Pitzer & Scott 1943		Willingham et al. 1945	
mercury manometer		mercury manometer		mercury manometer		ebulliometry	
t/°C	P/Pa	t/°C	P/Pa	t/°C	P/Pa	t/°C	P/Pa
0	228	10	293	0	233	59.203	6355
10	433	30	1213	12.5	520	68.436	7654
20	833	50	3906	25	1113	67.123	8963
30	1480			37.7	2220	70.458	9272
40	2533			50	4160	73.588	11698
50	4173	bp/°C	138.99	60	6600	77.747	13819
60	6626		139.15			82.522	16621
70	10212					87.367	19924
80	15280					91.86	23450
				eq. 4	P/mmHg	97.87	28954
eq. 4	P/mmHg			A	22.927	103.396	34897
A	22.9425			B	2871.66	110.041	43322
B	2876.3			C	5	116.896	53654
C	5					124.205	66755
				$\Delta H_V/(kJ\ mol^{-1})$		132.128	83717
				298 K	42.51	137.713	97604
						138.314	99200
Linder 1931		Stuckey & Saylor 1940				138.869	100692
mercury manometer		mercury manometer				139.493	102385
t/°C	P/Pa	t/°C	P/Pa			140.041	103906
−8.40	120	measured 4–75°C				eq. 2	P/mmHg
−6.75	137.3	eq. 4	P/mmHg			A	7.0034
−2.80	1812.7	A	22.92341			B	1458.21
		B	2708.38			C	214.609
		C	5.0				
		bp/°C	139.2			bp/°C	139.2

2.

Stull 1947		Forziati et al. 1949		Ambrose et al. 1967		Zwolinski & Wilhoit 1971	
summary of literature data		ebulliometry		ebulliometry		selected values	
t/°C	P/Pa	t/°C	P/Pa	t/°C	P/Pa	t/°C	P/Pa
−6.9	133.3	59.335	6400	146.85	124200	28.24	1333
16.8	266.6	63.518	7692	166.85	202100	41.07	2666
28.3	1333	67.157	8991	186.85	313700	49.23	4000
41.1	2666	70.506	10330	206.85	467000	56.33	5333

TABLE 3.1.1.5.2 *(Continued)*

Stull 1947		Forziati et al. 1949		Ambrose et al. 1967		Zwolinski & Wilhoit 1971	
summary of literature data		ebulliometry		ebulliometry		selected values	
t/°C	P/Pa	t/°C	P/Pa	t/°C	P/Pa	t/°C	P/Pa
55.3	5333	73.601	11720	226.85	671500	60.269	6666
64.4	7999	77.778	13842	246.85	937400	64.437	7999
76.8	13332	81.527	16633	266.85	1276300	71.277	10666
95.5	26664	87.387	19945	268.86	1700900	76.818	13332
116.7	53329	91.874	23474	306.85	2226600	87.454	19998
139.1	101325	97.887	28979	316.85	2872100	95.483	26664
		103.412	34925			102.01	33331
mp/°C	−47.9	110.067	43360	eq. 5	P/bar	107.55	39997
		116.923	53700	A	21.9924	116.69	53329
		124.226	66796	B	2957.79	124.15	66661
		132.144	83749	C	5.66789	130.58	79993
		137.731	97647	D	1776	136.06	93326
		138.329	99235			137.1	95991
		138.887	100726			138.112	98659
		129.52	102464			138.1	101325
		140.078	104000			25	1106.6
		eq. 2	P/mmHg			eq. 2	p/mmHg
		A	7.00849			A	7.00908
		B	1461.925			B	1462.266
		C	215.073			C	215.105
		bp/°C	139.104			bp/°C	139.103
						ΔH_v/(kJ mol^{-1}) =	
						at 25°C	42.66
						at bp	36.36

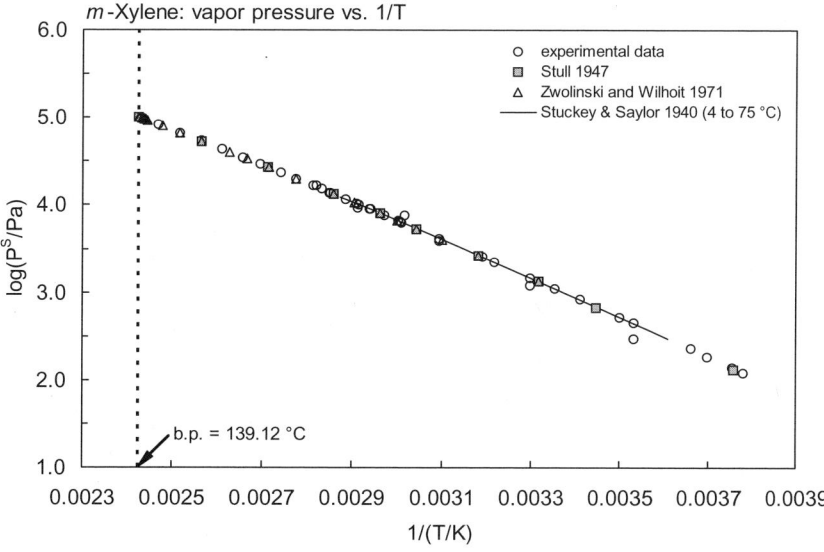

FIGURE 3.1.1.5.2 Logarithm of vapor pressure versus reciprocal temperature for *m*-xylene.

TABLE 3.1.1.5.3
Reported Henry's law constants of *m*-xylene at various temperatures and temperature dependence equations

$\ln K_{AW} = A - B/(T/K)$	(1)	$\log K_{AW} = A - B/(T/K)$	(1a)
$\ln (1/K_{AW}) = A - B/(T/K)$	(2)	$\log (1/K_{AW}) = A - B/(T/K)$	(2a)
$\ln (k_H/atm) = A - B/(T/K)$	(3)		
$\ln [H/(Pa\ m^3/mol)] = A - B/(T/K)$	(4)	$\ln [H/(atm \cdot m^3/mol)] = A - B/(T/K)$	(4a)
$K_{AW} = A - B \cdot (T/K) + C \cdot (T/K)^2$	(5)		

Sanemasa et al. 1982		Ashworth et al. 1988		Robbins et al. 1993		Dewulf et al. 1995	
vapor-liquid equilibrium		EPICS-GC		static headspace-GC		EPICS-GC	
t/°C	H/(Pa m³/mol)	t/°C	H/(Pa m³/mol)	t/°C	H/(Pa m³/mol)	t/°C	H/(Pa m³/mol)
15	405	10	416	25	739	2	209
25	731	15	503	30	900	6	204
35	1329	20	606	40	1489	10	264
45	1872	25	754	45	1591	18.2	472
		30	899	50	1652	25	615
		eq. 4a	H/(atm m³/mol)	eq. 4	H/(Pa m³/mol)	eq 1	K_{AW}
		A	6.28	A	17.83472	A	12.13
		B	3337	B	3337.45	B	4026

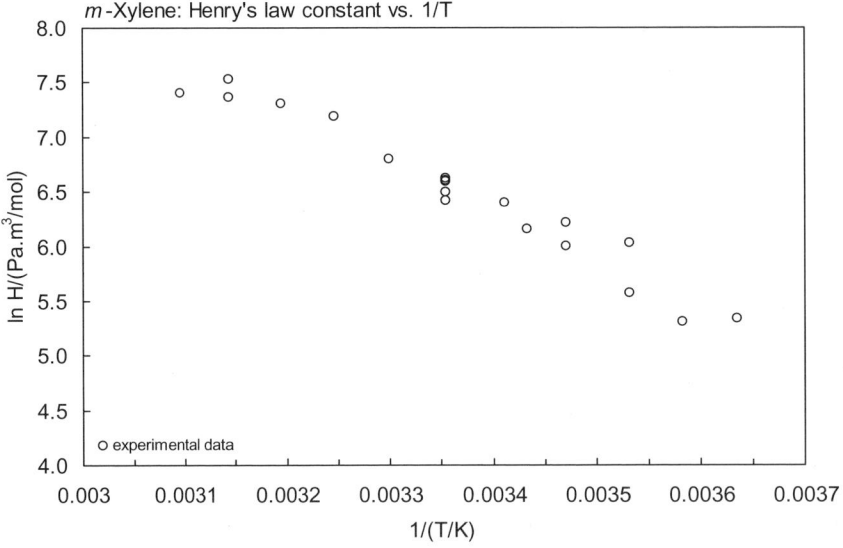

FIGURE 3.1.1.5.3 Logarithm of Henry's law constant versus reciprocal temperature for *m*-xylene.

TABLE 3.1.1.5.4
Reported octanol-air partition coefficients of *m*-xylene at various temperatures

Gruber et al. 1997	
GC det'd activity coefficient	
t/°C	log K_{OA}
20.29	-
30.3	3.69
40.4	3.51
50.28	3.30

3.1.1.6 *p*-Xylene

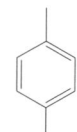

Common Name: *p*-Xylene
Synonym: 1,4-dimethylbenzene, *p*-xylol, 4-methyltoluene
Chemical Name: *p*-xylene
CAS Registry No: 106-42-3
Molecular Formula: C_8H_{10}, $C_6H_4(CH_3)_2$
Molecular Weight: 106.165
Melting Point (°C):
    13.25   (Lide 2003)
Boiling Point (°C):
    138.37   (Lide 2003)
Density (g/cm³ at 20°C):
    0.8611   (Weast 1982–83)
Molar Volume (cm³/mol):
    123.3   (20°C, calculated-density)
    140.4   (calculated-Le Bas method at normal boiling point)
Enthalpy of Vaporization, ΔH_V (kJ/mol):
    42.376, 35.98   (25°C, bp, Riddick et al. 1986)
Enthalpy of Fusion ΔH_{fus} (kJ/mol):
    17.113   (Tsonopoulos & Prausnitz 1971; Riddick et al. 1986)
    17.11   (Chickos et al. 1999)
Entropy of Fusion ΔS_{fus} (J/mol K):
    59.413   (Tsonopoulos & Prausnitz 1971; Yalkowsky & Valvani 1980)
    59.77, 45.6 (exptl., calculated-group additivity method, Chickos et al. 1999)
Fugacity Ratio at 25°C, F: 1.0

Water Solubility (g/m³ or mg/L at 25°C or as indicated and reported temperature dependence equations. Additional data at other temperatures designated * are compiled at the end of this section):
    200   (shake flask-UV, Andrews & Keefer 1949)
    198*   (shake flask-UV, measured range 0.4–85°C, Bohon & Claussen 1951)
    185*   (shake flask-GC, Polak & Lu 1973)
    156   (shake flask-GC, Sutton & Calder 1975)
    163   (shake flask-GC, Hermann 1972)
    157   (shake flask-GC, Price 1976)
    157   (shake flask-GC, Krzyzanowska & Szeliga 1978)
    191*   (20°C, shake flask-UV, Ben-Naim & Wiff 1979)
    163*   (vapor saturation-UV spec., measured range 15–45°C, Sanemasa et al. 1982)
    214.5   (generator column-HPLC/UV, GC/ECD, Tewari et al. 1982c)
    182   (HPLC-k′ correlation, converted from reported γ_W, Hafkenscheid & Tomlinson 1983a)
    214   (generator column-HPLC, Wasik et al. 1983)
    215   (generator column-GC/ECD, Miller et al. 1985)
    190   (shake flask-radiometric, Lo et al. 1986)
    180*   (IUPAC recommended value, temp range 0–90°C, Shaw 1989b)
    169*   (30°C, equilibrium flow cell-GC, measured range 30–100°C, Chen & Wagner 1994c)
    $\ln x = 11.79 - 20.89\cdot(T_r/K)^{-1} + 4.892\cdot(T_r/K)^{-2}$; $T_r = T/T_c$, the reduced temp, system temp T divided by critical temp T_c, temp range 303.15–373.15 K (equilibrium flow cell-GC, Chen & Wagner 1994c)
    $\ln x = -27.937 + 3230.3/(T/K) + 7.595 \times 10^{-5}\cdot(T/K)^2$; temp range 5–50°C (regression eq. of literature data, Shiu & Ma 2000)

Vapor Pressure (Pa at 25°C or as indicated and reported temperature dependence equations. Additional data at other temperatures designated * are compiled at the end of this section):

154.7*   (0.2°C, mercury manometer, Linder 1931)
787*   (20°C, Hg manometer, Kassel 1936)
log (P/mmHg) = –2930.0/(T/K) – 5·log (T/K) + 23.1000; temp range 0–80°C (vapor pressure eq. from Hg manometer measurements, Kassel 1936)
1437*   (30°C, Hg manometer, measured range 10–50°C, Rintelen et al. 1937)
log (P/mmHg) = –2851.90/(T/K) – 5·log (T/K) + 22.88436; temp range 4–75°C (vapor pressure eq. from manometer measurement, Stuckey & Saylor 1940)
1187*   (Hg manometer, Pitzer & Scott 1943)
log (P_S/mmHg) = –3141.33/(T/K) + 11.6092; temp range 0–13.23°C (manometer, solid, two-constant vapor pressure eq. from exptl. data, Pitzer & Scott 1943)
log (P_L/mmHg) = –3080.31/(T/K) – 6.7 log (T/K) + 27.8581; temp range 13.23–60°C (manometer, liquid, three-constant eq. from exptl. data, Pitzer & Scott 1943)
6354*   (58.288°C, ebulliometry, measured range 58.288–139.289°C, Willingham et al. 1945)
log (P/mmHg) = 6.98648 – 1491.548/(207.171 + t/°C); temp range 58.3–139.8°C (manometer, Antoine eq. from exptl. data, Willingham et al. 1945)
1333*   (27.3°C, summary of literature data, Stull 1947)
6398*   (58.419°C, ebulliometry, measured range 58.419–139.329°C, Forziati et al. 1949)
log (P/mmHg) = 6.99184 – 1454.328/(215.411 + t/°C); temp range 58.4–139.3°C (manometer, Antoine eq. from exptl. data, Forziati et al. 1949)
1175   (extrapolated-Antoine eq., Dreisbach 1955)
log (P/mmHg) = 6.99052 – 1453.430/(215.307 + t/°C); temp range 45–190°C (Antoine eq. for liquid state, Dreisbach 1955)
126500*   (146.85°C, ebulliometry, measured range 146.85–316.85°C, Ambrose et al. 1967)
1170*   (extrapolated-Antoine eq., Zwolinski & Wilhoit 1971)
log (P/mmHg) = 6.9052 – 1453,430/(215.307 + t/°C); temp range 27.32–165.73°C (Antoine eq., Zwolinski & Wilhoit 1971)
log (P/mmHg) = [–0.2185 × 9809.9/(T/K)] + 8.124805; temp range –8.1 to138.3°C (Antoine eq., Weast 1972–73)
880.3*   (20.015 °C, inclined-piston gauge, measured range –26.043 to 20.015, Osborn & Douslin 1974)
log (P/atm) = (1 – 411.503/T) × 10^(0.847730 – 6.39489 × 10^4·T + 5.59094 × 10^7·T^2); T in K, temp range 290.0–618.2 K (Cox vapor pressure eq., Chao et al. 1983)
1170   (extrapolated-Antoine eq., Boublik et al. 1984)
log (P/kPa) = 6.11376 – 1452.215/(215.518 + t/°C), temp range 58.3–139.3°C (Antoine eq. from reported exptl. data of Willingham et al. 1945, Boublik et al. 1984)
log (P/kPa) = 6.11513 – 1453.812/(215.242 + t/°C); temp range 99.17–179.23°C (Antoine eq. from reported exptl. data of Osborn & Douslin 1974, Boublik et al. 1984)
1167   (extrapolated-Antoine eq., Dean 1985, 1992)
log (P/mmHg) = 6.90052 – 1453.430/(215.31 + t/°C); temp range 27–166°C (Antoine eq., Dean 1985, 1992)
1200; 1160 (quoted lit.; calculated-Antoine eq., Riddick et al. 1986; quoted, Howard 1990)
log (P/kPa) = 6.11140 – 1451.39/(215.148 + t/°C); temp range not specified (Antoine eq., Riddick et al. 1986)
1180   (extrapolated-Antoine eq., Stephenson & Malanowski 1987)
log (P_S/kPa) = 15.50091 – 6327.014/(115.724 + T/K); temp range 247–286 K (solid, Antoine eq.-I, Stephenson & Malanowski 1987)
log (P_L/kPa) = 6.14779 – 1475.767/(–55.241 + T/K); temp range 286–453 K (liquid, Antoine eq.-II, Stephenson & Malanowski 1987)
log (P_L/kPa) = 6.14049 – 1472.773/(–55.342 + T/K); temp range 411–463 K (liquid, Antoine eq.-III, Stephenson & Malanowski 1987)
log (P_L/kPa) = 6.44333 – 1735.196/(–19.846 + T/K); temp range 460–553 K (liquid, Antoine eq.-IV, Stephenson & Malanowski 1987)
log (P_L/kPa) = 7.84182 – 3543.356/(208.522 + T/K); temp range 551–616 K (liquid, Antoine eq.-V, Stephenson & Malanowski 1987)
1165*   (McLeod gauge, measured range 20.0–50.07, Smith 1990)
log (P/mmHg) = 60.0531 – 4.1059 × 10^3/(T/K) – 19.441·log (T/K) + 8.2881 × 10^{-3}·(T/K) – 2.3647 × 10^{-12}·$(T/K)^2$, temp range 286–616 K (vapor pressure eq., Yaws 1994)

Henry's Law Constant (Pa m³/mol at 25°C or as indicated and reported temperature dependence equations. Additional data at other temperatures designated * are compiled at the end of this section):

762* (vapor-liquid equilibrium, measured range 15–45°C, Sanemasa et al. 1982)
754*; 752 (EPICS-GC/FID; batch air stripping-GC, Ashworth et al. 1988)
ln [H/(atm·m³/mol)] = 6.931 − 3520/(T/K); temp range 10–30°C (EPICS measurements, Ashworth et al. 1988)
614 (calculated-vapor-liquid equilibrium (VLE) data, Yaws et al. 1991)
856, 1189, 1576 (27, 35.8, 46°C, EPICS-GC, Hansen et al. 1993)
ln [H/(kPa·m³/mol)] = −3072/(T/K) + 10.0; temp range 27–46°C (EPICS measurements, Hansen et al. 1993)
696 (infinite activity coeff. γ^∞ in water determined by inert gas stripping-GC, Li et al. 1993)
739* (static headspace-GC, same as m-xylene, Robbins et al. 1993)
595 (headspace solid-phase microextraction (SPME)-GC, Zhang & Pawliszyn 1993)
575* (EPICS-GC/FID, measured range 2–25°C, Dewulf et al. 1995)
318, 763 (6.0, 25°C, EPICS-GC/FID, natural seawater with salinity of 35‰, Dewulf et al. 1995)
ln K_{AW} = −4479/(T/K) + 0.01196·Z + 13.597; with Z salinity 0–35.5‰, temp range 2–25°C (EPICS-GC/FID, Dewulf et al. 1995)
641 (20°C, selected from literature experimentally measured data, Staudinger & Roberts 1996)
678.6 (exponential saturator EXPSAT technique, Dohnal & Hovorka 1999)
669.1 (modified EPICS method-GC, Ryu & Park 1999)
604 (20°C, selected from literature experimentally measured data, Staudinger & Roberts 2001)
log K_{AW} = 4.900 − 1615/(T/K) (van't Hoff eq. derived from lit. data, Staudinger & Roberts 2001)

Octanol/Water Partition Coefficient, log K_{OW} at 25°C:

3.15 (Leo et al. 1971; Hansch & Leo 1985)
3.20 (Hansch & Leo 1979)
3.10 (HPLC-k' correlation, Hanai et al. 1981)
3.28 (HPLC-k' correlation, Hammers et al. 1982)
3.18 (generator column-HPLC/UV, GC/ECD, Tewari et al. 1982b,c)
3.29 (HPLC-k' correlation, Hafkenscheid & Tomlinson 1983b)
3.18 (generator column-HPLC/UV, Wasik et al. 1983)
3.29 (HPLC-RV correlation, Garst 1984)
3.36, 3.48 (RP-HPLC-k' capacity factor correlations; Sherblom & Eganhouse 1988)
3.15 (recommended, Sangster 1989, 1993)
3.35 (normal phase HPLC-k' correlation, Govers & Evers 1992)
3.15 (recommended, Hansch et al. 1995)

Octanol/Air Partition Coefficient, log K_{OA} at 25°C or as indicated. Additional data at other temperatures designated * are compiled at the end of this section:

3.68* (30.3°C, from GC-determined γ^∞ in octanol, measured range 30.3–50.28°C, Gruber et al. 1997)
3.79 (calculated-measured γ^∞ in pure octanol of Gruber et al. 1997, Abraham et al. 2001)

Bioconcentration Factor, log BCF:

1.37 (eels, Ogata & Miyake 1978)
1.17 (goldfish, Ogata et al. 1984)
2.41 ($S.\ capricornutum$, Herman et al. 1991)

Sorption Partition Coefficient, log K_{OC} at 25°C or as indicated:

2.52 (average 5 soils and 3 sediments, sorption isotherms by batch equilibrium and column experiments, Schwarzenbach & Westall 1981)
2.42 (sediment 4.02% OC from Tamar estuary, batch equilibrium-GC, Vowles & Mantoura 1987)
2.24 (aquifer material with f_{OC} of 0.006 and measured partition coeff. K_P = 1.04 mL/g., Abdul et al. 1990)
1.87, 2.22 (Webster soil, Webster soil HP, batch equilibrium, Pennell et al. 1992)
3.53, 2.63 (sorbent: Silica gel, kaolinite, batch equilibrium, Pennell et al. 1992)
2.72, 2.17 (Captina silt loam pH 4.97, McLaurin sandy loam pH 4.43, Walton et al. 1992)
2.43, 2.44 (RP-HPLC-k' correlation including MCI related to non-dispersive intermolecular interactions, hydrogen-bonding indicator variable, Hong et al. 1996)

2.37   (HPLC-screening method, Müller & Kördel 1996)
2.49, 2.75, 2.65, 2.76, 2.79, 2.76, 2.78   (2.3, 3.8, 6.2, 8.0, 13.5, 18.6, 25°C, natural sediment from River Leie, organic carbon f_{OC} = 4.12%, EPICS-GC/FID, Dewulf et al. 1999)
2.27, 2.31, 2.21   (soils: organic carbon OC ≥ 0.1%, OC ≥ 0.5%, 0.1 ≤ OC < 0.5%, average, Delle Site 2001)

Environmental Fate Rate Constants, k, or Half-Lives, $t_{½}$:

Volatilization: $t_{½}$ ~ 3.1 h for evaporation from water of 1 m depth with wind speed of 3 m/s and water current of 1 m/s h (Lyman et al. 1982);
  $t_{½}$ ~ 27–135 h for evaporation from a typical river or pond (Howard 1990).
Photolysis:
Oxidation: rate constant k for gas-phase second order rate constants, k_{OH} for reaction with OH radical, k_{NO3} with NO_3 radical and k_{O3} with O_3 or as indicated. Additional data at other temperatures designated * are compiled at the end of this section:
  photooxidation $t_{½}$ = 2.8 × 10⁶–1.4 × 10⁸ h, based on estimated rate data for alkoxy radical in aqueous solution (Hendry 1974)
  k_{OH} = (7.4 ± 1.5) × 10⁹ L mol⁻¹ s⁻¹; $k_{O(3P)}$ = (1.09 ± 0.11) × 10⁸ L mol⁻¹ s⁻¹ with O(³P) atom at room temp. (relative rate method, Doyle et al. 1975; Lloyd et al. 1976)
  k_{OH} = (12.2 ± 1.2) × 10⁻¹² cm³ molecule⁻¹ s⁻¹; $k_{O(3P)}$ = (1.81 ± 0.18) × 10⁻¹³ cm³ molecule⁻¹ s⁻¹ for the reaction of O(³P) atom at room temp. (flash photolysis-resonance fluorescence, Hansen et al. 1975)
  k_{OH} = 7.45 × 10⁹ L mol⁻¹ s⁻¹ with $t_{½}$ = 0.24–2.4 h (Darnall et al. 1976)
  k_{OH}* = (15.3 ± 1.7) × 10⁻¹² cm³ molecule⁻¹ s⁻¹ at room temp., measured range 296–473 K (flash photolysis-resonance fluorescence, Perry et al. 1977)
  k_{OH} = (8.8–10.5) × 10⁻¹² cm³ molecule⁻¹ s⁻¹ with different dilute gas, Ar or He at 298 K (flash photolysis-resonance fluorescence, Ravishanakara et al. 1978)
  k_{OH} = 12.3 × 10⁻¹² cm³ molecule⁻¹ s⁻¹ and residence time of 0.9 d, loss of 67% in one day or 12 sunlit hours at 300 K in urban environments (Singh et al. 1981)
  k_{OH} = 950 cm³ mol⁻¹ s⁻¹ at 300 K (Lyman et al. 1982)
  k_{OH} = (5.9–12) × 10⁹ M⁻¹ s⁻¹ with $t_{½}$ = 0.47–1.0 d for xylenes (Mill 1982)
  k = (140 ± 30) M⁻¹ s⁻¹ for the reaction with ozone in water using 1 mM t-BuOH as scavenger at pH 2.0 and 20–23°C (Hoigné & Bader 1983)
  k_{NO3} = 1.4 × 10⁻¹⁶ cm³ molecule⁻¹ s⁻¹ at 296 K (Atkinson et al. 1984)
  k_{OH} = 14.1 × 10⁻¹² cm³ molecule⁻¹ s⁻¹ with half-life of 4.2–42 h (Atkinson 1985; Howard et al. 1991)
  k_{OH} = 12.9 × 10⁻¹² cm³ molecule⁻¹ s⁻¹ at 25°C (Ohta & Ohyama 1985)
  k_{OH} = 13.6 × 10⁻¹² cm³ molecule⁻¹ s⁻¹ at 22.8°C, with a calculated atmospheric lifetime τ = 0.84 d (Edney et al. 1986)
  k_{OH}(calc) = 6.9 × 10⁻¹² cm³ molecule⁻¹ s⁻¹, k_{OH}(obs.) = 15.2 × 10⁻¹² cm³ molecule⁻¹ s⁻¹ at room temp. (SAR [structure-activity relationship], Atkinson 1987)
  k_{NO3} = 4.50 × 10⁻¹⁶ cm³ molecule⁻¹ s⁻¹, k_{OH} = 1.52 × 10⁻¹¹ cm³·molecule⁻¹ s⁻¹ at 298 K (Atkinson et al. 1988; quoted, Sabljic & Güsten 1990; Müller & Klein 1992)
  k_{OH}* = 1.43 × 10⁻¹² cm³ molecule⁻¹ s⁻¹ over temp range at 296–335 K (recommended, Atkinson 1989)
  k_{OH} = 14.3 × 10⁻¹² cm³ molecule⁻¹ s⁻¹ at 298 K (Atkinson 1990)
  k_{OH}(calc) = 6.92 × 10⁻¹¹ cm³ molecule⁻¹ s⁻¹ at 298 K (estimated by SARs, Müller & Klein 1992)
  k_{OH}(calc) = 17.40 × 10⁻¹² cm³ molecule⁻¹ s⁻¹ (molecular orbital calculations, Klamt 1993)
  k_{OH}(calc) = 6.51 × 10⁻¹² cm³ molecule⁻¹ s⁻¹, k_{OH}(exptl) = 14.3 × 10⁻¹² cm³ molecule⁻¹ s⁻¹ at 298 K (SAR structure-activity relationship, Kwok & Atkinson 1995)
Hydrolysis: no hydrolyzable functional groups (Mabey et al. 1982)
  $t_{½}$ = 1150 months, abiotic hydrolysis or dehydro-halogenation half-life (Olsen & Davis 1990).
Biodegradation:
  100% biodegraded after 192 h at 13°C with an initial concn of 1.03 × 10⁻⁶ L/L (Jamison et al. 1976)
  $t_{½}$(aq. aqueous) = 168–672 h, based on aqueous screening test data (Bridie et al. 1979) and soil column study simulating an aerobic river/groundwater infiltration system (Kuhn et al. 1985; Howard et al. 1991)
  $t_{½}$(aq. anaerobic) = 672–2688 h, based on unacclimated aqueous aerobic biodegradation half-life (Bridie et al. 1979; Kuhn et al. 1985; Howard et al. 1991)
  $t_{½}$ = 0.03 d (Olsen & Davis 1990).
Bioconcentration, Uptake (k_1) and Elimination (k_2) Rate Constants or Half-Lives:

Half-Lives in the Environment:
  Air: $t_{1/2}$ = 0.24–2.4 h, based on rate of disappearance for the reaction with hydroxyl radical (Darnall et al. 1976)
    residence time of 0.9 d, loss of 67% in one day or 12 sunlit hours at 300 K in urban environments (Singh et al. 1981)
    photodecomposition $t_{1/2}$ = 3.1 h under simulated atmospheric conditions, with NO (Dilling et al. 1976)
    $t_{1/2}$ = 4.2–42 h, based on photooxidation half-life (Howard et al. 1991);
    calculated atmospheric lifetime τ = 0.84 d due to reaction with OH radical (Edney et al. 1986);
    summer daylight lifetime τ = 10 h due to reaction with OH radical (Altshuller 1991).
  Surface Water: $t_{1/2}$ = 168–672 h, based on estimated aqueous aerobic biodegradation half-life (Howard et al. 1991).
  Ground water: $t_{1/2}$ = 336–8640 h, based on estimated aqueous aerobic and anaerobic biodegradation half-lives (Howard et al. 1991);
    $t_{1/2}$ ~ 0.3 yr from observed persistence in groundwater of the Netherlands (Zoeteman et al. 1981);
    abiotic hydrolysis or dehydro-halogenation $t_{1/2}$ = 1150 months (Olsen & Davis 1990).
  Soil: $t_{1/2}$ = 168–672 h, based on estimated aqueous aerobic biodegradation half-life (Bridie et al. 1979; Kuhn et al. 1985; Howard et al. 1991);
    disappearance $t_{1/2}$ = 2.2 d from test soils (Anderson et al. 1991).
  Biota: $t_{1/2}$ = 2.6 d, half-life to eliminate from eels in seawater (Ogata & Miyake 1978).

TABLE 3.1.1.6.1
Reported aqueous solubilities of *p*-xylene at various temperatures

1.

Bohon & Claussen 1951		Polak & Lu 1973		Shaw 1989b (IUPAC)		Chen & Wagner 1994c	
shake flask-UV		shake flask-GC		recommended values		equilibrium flow cell-GC	
t/°C	S/g·m^{-3}	t/°C	S/g·m^{-3}	t/°C	S/g·m^{-3}	t/°C	S/g·m^{-3}
0.4	156	0	164	0	160	30	169
10	188	25	185	10	200	30	188
10	197			20	180	50	102
21	195			25	180	60	230
25	198			30	190	70	289
25.6	199	Ben-Naim & Wiff 1979		40	220	80	337
30.2	201	shake flask-UV		50	280	90	395
30.3	204	t/°C	S/g·m^{-3}	60	320	100	516
34.9	207	10	189	70	360		
35.2	207	20	191	80	420	ΔH_{sol}/(kJ mol^{-1}) = 3.48	
42.8	222			90	480	at 25°C	
42.8	256						
54.4	310			ΔH_{sol}/(kJ mol^{-1}) = 6.90			
61.7	340	Sanemasa et al. 1982		at 25°C			
73.9	387	vapor saturation-UV					
85	459	t/°C	S/g·m^{-3}				
ΔH_{sol}/(kJ mol^{-1}) =		15	157				
25°C	3.9	25	163				
7°C	−0.422	35	176				
12°C	−1.34	45	178				
17°C	0.515						
22°C	1.36						
27°C	2.29	ΔH_{sol}/(kJ mol^{-1}) = 3.20					
32°C	4.23	at 25°C					
37°C	7.36						

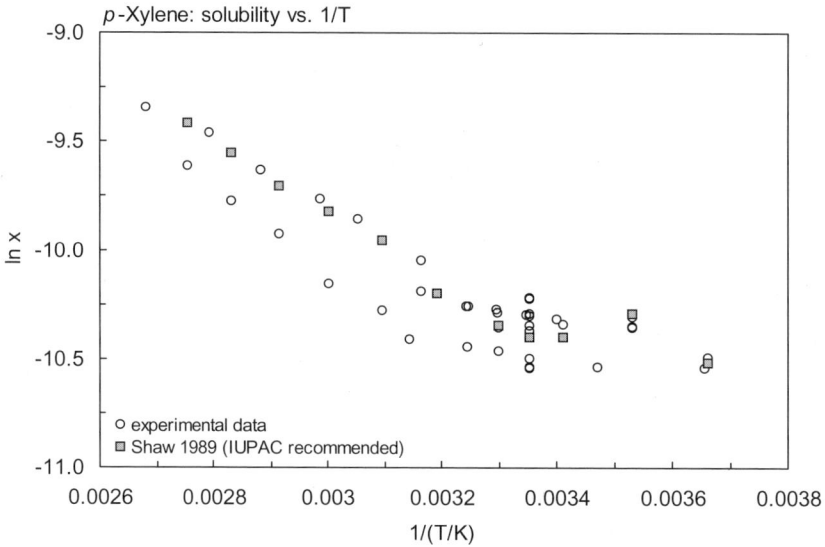

FIGURE 3.1.1.6.1 Logarithm of mole fraction solubility (ln x) versus reciprocal temperature for p-xylene.

TABLE 3.1.1.6.2
Reported vapor pressures of p-xylene at various temperatures and the coefficients for the vapor pressure equations

$\log P = A - B/(T/K)$ (1) $\qquad \ln P = A - B/(T/K)$ (1a)
$\log P = A - B/(C + t/°C)$ (2) $\qquad \ln P = A - B/(C + t/°C)$ (2a)
$\log P = A - B/(C + T/K)$ (3)
$\log P = A - B/(T/K) - C \cdot \log(T/K)$ (4)
$\log P = A - B/(T/K) - C \cdot \log(T/K) + D(T/K)^2$ (5)

1.

Kassel 1936		Rintelen et al. 1937		Pitzer & Scott 1943		Willingham et al. 1945	
mercury manometer		mercury manometer		mercury manometer		ebulliometry	
t/°C	P/Pa	t/°C	P/Pa	t/°C	P/Pa	t/°C	P/Pa
0	208	10	307	0	178	58.288	6354
10	415	30	1427	12.5	540	62.523	7657
20	787	50	4306	25	1187	66.216	8966
30	1427			37.7	2333	69.549	10306
40	2453	bp 138.27–138.37°C		50	4346	72.657	11696
50	4093			60	6846	76.832	13822
60	6573					81.636	16623
70	10226					86.488	18725
80	15452			0–13.23°C solid		90.990	23454
				eq. 1	P/mmHg	97.013	28956
eq. 4	P/mmHg			A	11.6092	102.546	34901
A	23.100			B	3141.33	109.211	43326
B	2930.0					116.083	53656
C	5.0			13.23–60°C liquid		123.049	66756
				eq. 4	P/mmHg	131.355	83717
				A	27.8581	136.956	97605
				B	3080.31	137.558	99201

TABLE 3.1.1.6.2 (Continued)

Kassel 1936		Rintelen et al. 1937		Pitzer & Scott 1943		Willingham et al. 1945	
mercury manometer		mercury manometer		mercury manometer		ebulliometry	
t/°C	P/Pa	t/°C	P/Pa	t/°C	P/Pa	t/°C	P/Pa
Linder 1931		Stuckey & Saylor 1940		C	6.7	138.114	100694
mercury manometer		mercury manometer				138.742	102392
t/°C	P/Pa	t/°C	P/Pa	$\Delta H_V/(kJ\ mol^{-1}) = 42.30$		139.289	103906
				at 298 K			
−9.5	44.0	measured 4–75°C				eq. 2	P/mmHg
−2.5	116	eq. 4	P/mmHg			A	6.98648
0	126.7	A	22.88436			B	1450.688
0.2	154.7	B	2851.90			C	214.990
		C	5.0				
						bp/°C	138.348
		bp/°C 138.33–138.38					
		mp/°C 13.20–12.95					

2.

Stull 1947		Forziati et al. 1949		Ambrose et al. 1967		Zwolinski & Wilhoit 1971	
summary of literature data		ebulliometry		ebulliometry		selected values	
t/°C	P/Pa	t/°C	P/Pa	t/°C	P/Pa	t/°C	P/Pa
−8.10	133.3	58.419	6398	146.85	126500	27.32	1333
15.5	266.6	62.419	7690	166.85	205200	40.15	2666
27.3	1333	66.280	8990	186.85	317100	48.31	4000
40.1	2666	69.605	10330	206.85	470900	54.42	5333
54.4	5333	72.684	11720	226.85	675800	59.363	6666
63.5	7999	76.885	13842	246.85	941900	63.535	7999
75.9	13332	81.658	16633	266.85	1280800	70.383	10666
94.6	26664	86.506	19944	286.85	1705700	75.931	13332
115.9	53323	91.017	23474	306.85	2232400	86.583	19998
138.3	101325	97.032	28978	316.85	2880400	94.626	26664
		102.573	34925			101.167	33331
mp/°C	13.3	109.240	43360	eq. 5	P/bar	106.719	39997
		123.431	66796	A	21.14250	115.887	53329
		131.371	83749	B	2892.27	123.366	66661
		137.574	99235	C	5.40051	129.372	79993
		138.132	100726	D	1759	135.304	93326
		138.768	102462			136.341	95991
		139.329	104000			137.347	98659
						138.351	101325
		eq. 2	P/mmHg			25.0	1173.2
		A	6.99184				
		B	1454.328			eq. 2	P/mmHg
		C	215.411			A	6.99052
						B	1453.430
		bp/°C	138.351			C	215.307
						bp/°C	138.351
						$\Delta H_V/(kJ\ mol^{-1}) =$	
						at 25°C	42.38
						at bp	35.98

(Continued)

TABLE 3.1.1.6.2 *(Continued)*

3.

Osborn & Douslin 1974				Smith 1990	
inclined-piston gauge		ebulliometry		McLeod gauge	
t/°C	P/Pa	t/°C	P/Pa	t/°C	P/Pa
−26.043	11.60	67.795	9590	20.0	865.3
−21.164	19.60	70.881	10892	25.0	1165
−16.277	33.20	73.978	12344	30.0	1541
−10.403	60.79	77.086	13960	35.24	2075
−4.995	103.86	80.206	15752	40.0	2645
−0.006	170.5	83.339	17737	44.27	3306
4.912	272.4	86.483	19933	50.07	4374
9.838	429.2	92.804	25023		
12.483	544.4				
13.285	581.7			eq. 2a	P/mmHg
15.004	648.9			A	16.19136
17.474	755.4			B	3371.18
20.015	880.3			C	215.367
				$\Delta H_V/(kJ\ mol^{-1}) = 42.98$	
				at 25°C	
				$\Delta S_V/(J\ mol^{-1}\ K^{-1}) = 107.0$	
				at 25°C	

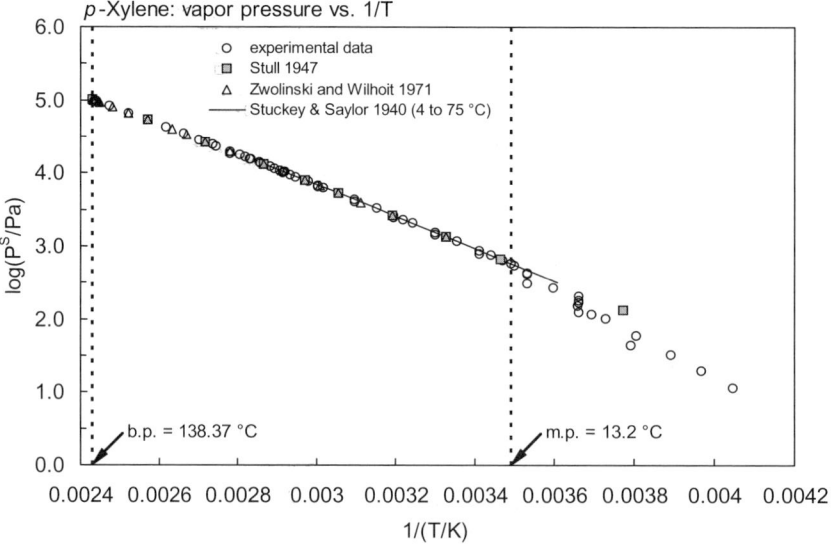

FIGURE 3.1.1.6.2 Logarithm of vapor pressure versus reciprocal temperature for *p*-xylene.

TABLE 3.1.1.6.3
Reported Henry's law constants of *p*-xylene at various temperatures and temperature dependence equations

$\ln K_{AW} = A - B/(T/K)$ (1)	$\log K_{AW} = A - B/(T/K)$ (1a)
$\ln (1/K_{AW}) = A - B/(T/K)$ (2)	$\log (1/K_{AW}) = A - B/(T/K)$ (2a)
$\ln (k_H/\text{atm}) = A - B/(T/K)$ (3)	
$\ln [H/(\text{Pa m}^3/\text{mol})] = A - B/(T/K)$ (4)	$\ln [H/(\text{atm} \cdot \text{m}^3/\text{mol})] = A - B/(T/K)$ (4a)
$K_{AW} = A - B \cdot (T/K) + C \cdot (T/K)^2$ (5)	

Sanemasa et al. 1982		Ashworth et al. 1988		Robbins et al. 1993		Hansen et al. 1993		Dewulf et al. 1995	
vapor-liquid equil		EPICS-GC		static headspace-GC		EPICS-GC		EPICS-GC	
t/°C	H/Pa m³/mol	t/°C	H/Pa m³/mol	t/°C	H/Pa m³/mol	t/°C	H/Pa m³/mol	t/°C	H/Pa m³/mol
15	430	10	426	25	739	27	856	2	176
25	762	15	489	30	900	35.8	1189	6	158
35	1265	20	654	40	1489	46	1576	10	252
45	2052	25	754	45	1591			18.2	468
		30	958	50	1652			25	575
		eq. 4a		eq. 4		eq. 4		eq 1	K_{AW}
		H/(atm m³/mol)		H/(Pa m³/mol)		H/(kPa m³/mol)		A	13.597
		A	6.931	A	17.83472	A	10.1	B	4479
		B	3520	B	3337.45	B	3072		

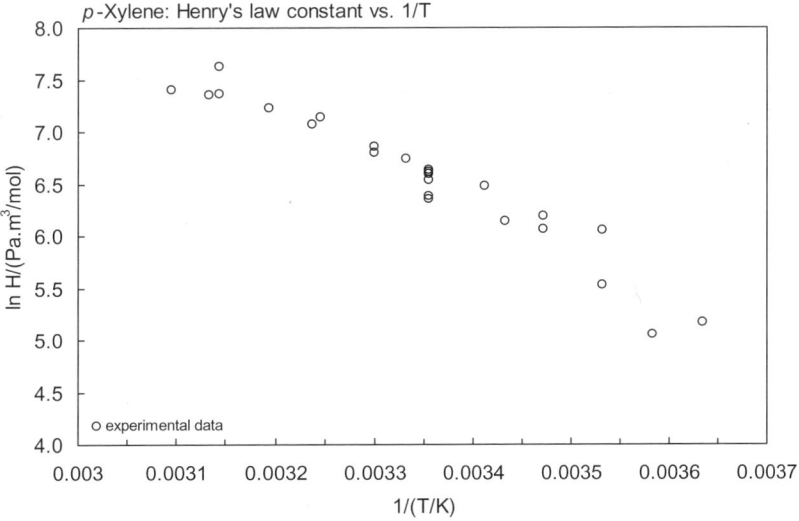

FIGURE 3.1.1.6.3 Logarithm of Henry's law constant versus reciprocal temperature for *p*-xylene.

TABLE 3.1.1.6.4
Reported octanol-air partition coefficients of *p*-xylene at various temperatures

Gruber et al. 1997	
GC det'd activity coefficient	
t/°C	log K_{OA}
20.29	–
30.3	3.68
40.4	3.48
50.28	3.29

3.1.1.7 1,2,3-Trimethylbenzene

Common Name: 1,2,3-Trimethylbenzene
Synonym: hemimellitene
Chemical Name: 1,2,3-trimethylbenzene
CAS Registry No: 526-73-8
Molecular Formula: C_9H_{12}, $C_6H_3(CH_3)_3$
Molecular Weight: 120.191
Melting Point (°C):
    –25.4   (Weast 1982–83; Lide 2003)
Boiling Point (°C):
    176.1   (Weast 1982–83; Lide 2003)
Density (g/cm³ at 20°C):
    0.8944   (Weast 1982–83)
Molar Volume (cm³/mol):
    134.4   (20°C, calculated-density)
    162.6   (calculated-Le Bas method at normal boiling point)
Enthalpy of Fusion ΔH_{fus} (kJ/mol):
    0.66, 1.33, 8.18; 10.17   (–54.45, –42.85, –25.35°C; total phase change enthalpy, Chickos et al. 1999)
Entropy of Fusion ΔS_{fus} (J/mol K):
    41.81, 46.2   (exptl., calculated-group additivity method, total phase change entropy, Chickos et al. 1999)
Fugacity Ratio at 25°C, F: 1.0

Water Solubility (g/m³ or mg/L at 25°C or as indicated and reported temperature dependence equations. Additional data at other temperatures designated * are compiled at the end of this section):
    75.2   (shake flask-GC, Sutton & Calder 1975)
    62.7*   (vapor saturation-UV spec., measured range 15–45°C, Sanemasa et al. 1982)
    65.5   (generator column-HPLC/UV, Tewari et al. 1982c)
    69*   (IUPAC recommended value, temp range 15–45°C, Shaw 1989b)
    $\ln x = -39.5173 + 5289.13/(T/K) + 1.149 \times 10^{-4} \cdot (T/K)^2$; temp range 5–50°C (regression eq. of literature data, Shiu & Ma 2000)

Vapor Pressure (Pa at 25°C or as indicated and reported temperature dependence equations. Additional data at other temperatures designated * are compiled at the end of this section):
    133.3*   (16.8°C, summary of literature data, Stull 1947)
    6417*   (90.332°C, ebulliometry, measured range 90.332–177.126°C, Forziati et al. 1949)
    $\log (P/\text{mmHg}) = 7.04082 - 1593.958/(207.078 + t/°C)$; temp range 90.3–177.1°C (manometer, Antoine eq. from exptl. data, Forziati et al. 1949)
    206   (extrapolated-Antoine eq., Dreisbach 1955)
    $\log (P/\text{mmHg}) = 7.04082 - 1593.958/(207.078C + t/°C)$; temp range 75–230°C (Antoine eq. for liquid state, Dreisbach 1955)
    198*   (extrapolated-Antoine eq., Zwolinski & Wilhoit 1971)
    $\log (P/\text{mmHg}) = 7.04082 - 1593.958/(207.078 + t/°C)$; temp range 56.79–205.36°C (Antoine eq., Zwolinski & Wilhoit 1971)
    $\log (P/\text{mmHg}) = [-0.2185 \times 10781.9/(T/K)] + 8.154069$; temp range 16.8–176°C (Antoine eq., Weast 1972–73)
    217   (calculated-bp, Mackay et al. 1982; Eastcott et al. 1988)
    $\log (P/\text{atm}) = (1 - 449.175/T) \times 10 \wedge (0.869047 - 6.33423 \times 10^4 \cdot T + 5.14963 \times 10^7 \cdot T^2)$; T in K, temp range 290.0–660.0 K (Cox vapor pressure eq., Chao et al. 1983)
    157   (extrapolated-Antoine eq., Boublik et al. 1984)

Mononuclear Aromatic Hydrocarbons

log (P/kPa) = 6.16365 − 1592.422/(206.905 + t/°C); temp range 90.33–177.1°C (Antoine eq. from reported exptl. data of Forziati et al. 1949, Boublik et al. 1984)

198.4   (extrapolated-Antoine eq., Dean 1985, 1992)

log (P/mmHg) = 7.04082 − 1593.958/(207.08 + t/°C); temp range 57–205°C (Antoine eq., Dean 1985, 1992)

199   (extrapolated, Antoine eq., Stephenson & Malanowski 1987)

log (P_L/kPa) = 6.16477 − 1593.776/(−66.032 + T/K); temp range 363–456 K (Antoine eq., Stephenson & Malanowski 1987)

log (P/mmHg) = 2.7492 − 2.6428 × 10^3/(T/K) + 3.6120·log (T/K) − 1.0213 × 10^{-2}·(T/K) + 5.0553 × 10^{-6}·(T/K)^2, temp range 248–665 K (vapor pressure eq., Yaws 1994)

log (P/kPa) = 6.17303 − 1593.958/[(T/K) −66.072]; temp range 5–50°C (regression eq. from literature data, Shiu & Ma 2000)

Henry's Law Constant (Pa m^3/mol at 25°C or as indicated and reported temperature dependence equations. Additional data at other temperatures designated * are compiled at the end of this section):

441*   (vapor-liquid equilibrium-GC, measured range 15–45°C, Sanemasa et al. 1982)

364   (calculated-vapor-liquid equilibrium (VLE) data, Yaws et al. 1991)

Octanol/Water Partition Coefficient, log K_{OW}:

3.66   (Hansch & Leo 1979)

3.66   (HPLC-k' correlation, Hammers et al. 1982)

3.55   (generator column-HPLC/UV, Tewari et al. 1982b, 1982c)

3.70, 3.86   (RP-HPLC-k' correlation, Sherblom & Eganhouse 1988)

3.63   (recommended, Sangster 1989, 1993)

3.55, 3.59, 3.66   (quoted lit. values, Hansch et al. 1995)

Octanol/Air Partition Coefficient, log K_{OA}:

Bioconcentration Factor, log BCF:

Sorption Partition Coefficient, log K_{OC}:

2.80   (average 5 soils and 3 sediments, sorption isotherms by batch equilibrium and column experiments, Schwarzenbach & Westall 1981)

Environmental Fate Rate Constants, k, or Half-Lives, $t_{1/2}$:

Volatilization:

Photolysis:

Oxidation: rate constant k; for gas-phase second-order rate constants, k_{OH} for reaction with OH radical, k_{NO3} with NO_3 radical and k_{O3} with O_3 or as indicated, *data at other temperatures and/or the Arrhenius expression see reference:

k_{OH} = (14 ± 3) × 10^9 L mol^{-1} s^{-1}; $k_{O(3P)}$ = (6.9 ± 0.7) × 10^8 L mol^{-1} s^{-1} with O(^{3}P) atom at room temp. (relative rate method, Doyle et al. 1975; Lloyd et al. 1976)

k_{OH} = (26.4 ± 2.6) × 10^{-12} cm^3 molecule^{-1} s^{-1}, $k_{O(3P)}$ = 11.5 × 10^{-13} cm^3 molecule^{-1} s^{-1} for the reaction of O(^{3}P) atom at room temp (flash photolysis-resonance fluorescence, Hansen et al. 1975)

k_{OH} = 14.9 × 10^9 L mol^{-1} s^{-1} with $t_{1/2}$ = 0.24–2.4 h (Darnall et al. 1976)

k_{OH}* = (33.3 ± 4.5) × 10^{-12} cm^3·molecule^{-1} s^{-1} at room temp., measured range 296–473 K (flash photolysis-resonance fluorescence Perry et al. 1977)

k_{OH} = (15–30) × 10^9 M^{-1} s^{-1} with $t_{1/2}$ = 0.2–0.4 d for trimethylbenzenes (Mill 1982)

k = (400 ± 100) M^{-1} s^{-1} for the reaction with ozone in water at pH 1.7 and 20–23°C (Hoigné & Bader 1983)

k_{NO3} = 5.6 × 10^{-16} cm^3 molecule^{-1} s^{-1} at 296 K (Atkinson et al. 1984)

k_{OH} = 3.16 × 10^{-11} cm^3 molecule^{-1} s^{-1} at room temp. (Atkinson 1985)

k_{OH} = 2.96 × 10^{-11} cm^3 molecule^{-1} s^{-1} at 298 K (Ohta & Ohyama 1985)

k_{OH}(calc) = 1.8 × 10^{-11} cm^3 molecule^{-1} s^{-1}, k_{OH}(obs.) = 3.33 × 10^{-11} cm^3 molecule^{-1} s^{-1} at room temp. (SAR [structure-activity relationship], Atkinson 1987)

k_{NO3} = 1.85 × 10^{-15} cm^3 molecule^{-1} s^{-1}, k_{OH} = 3.16 × 10^{-11} cm^3 molecule^{-1} s^{-1} at 298 K (Atkinson et al. 1988; quoted, Sabljic & Güsten 1990; Müller & Klein 1992)

k_{OH} = 3.27 × 10^{-11} cm^3 molecule^{-1} s^{-1} at 298 K (recommended, Atkinson 1989, 1990)
k_{OH}(calc) = 1.78 × 10^{-11} cm^3 molecule^{-1} s^{-1} at 298 K (estimated by SARs, Müller & Klein 1992
k_{OH}(calc) = 3.99 × 10^{-11} cm^3 molecule^{-1} s^{-1} (molecular orbital calculations, Klamt 1993)

Hydrolysis:

Biotransformation:

Biodegradation:

Bioconcentration:

Half-Lives in the Environment:

Air: $t_{1/2}$ = 0.24–2.4 h, based on rate of disappearance for the reaction with hydroxyl radical (Darnall et al. 1976; Howard et al. 1991);
estimated lifetime τ = 1.5 h under photochemical smog conditions in S.E. England (Brice & Derwent 1978) and (Perry et al. 1977);
summer daylight lifetime τ = 4.2 h due to reaction with OH radical (Altshuller 1991)

TABLE 3.1.1.7.1
Reported aqueous solubilities and Henry's law constants of 1,2,3-trimethylbenzene at various temperatures

Aqueous solubility				Henry's law constant	
Sanemasa et al. 1982		Shaw 1989b (IUPAC)		Sanemasa et al. 1982	
vapor saturation-UV		recommended values		vapor-liquid equilibrium	
t/°C	S/g·m^{-3}	t/°C	S/g·m^{-3}	t/°C	H/(Pa m^3/mol)
15	75.2	15	60	15	241.2
25	65.5	25	69	25	441
35	72.2	35	72	35	706
45	85.2	45	85	45	1058

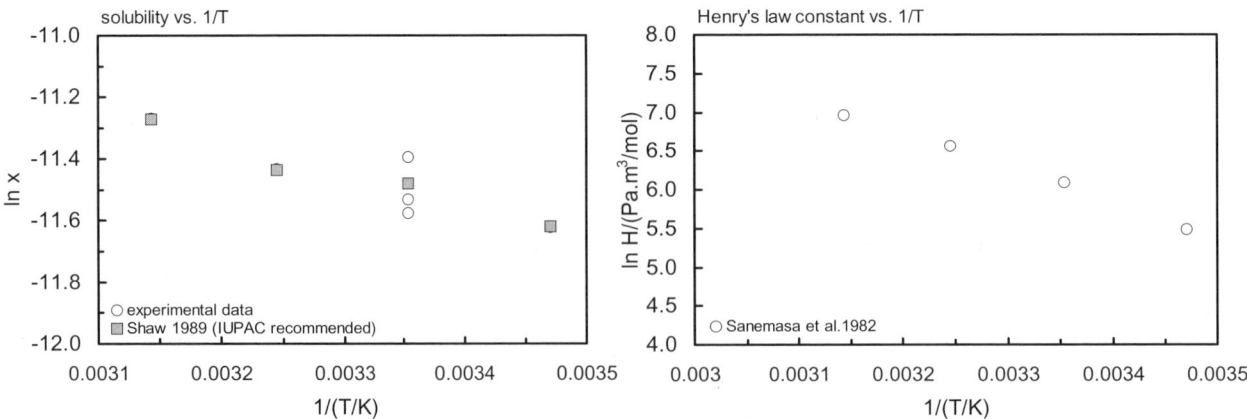

FIGURE 3.1.1.7.1 Logarithm of mole fraction solubility and Henry's law constant versus reciprocal temperature for 1,2,3-trimethylbenzene.

TABLE 3.1.1.7.2
Reported vapor pressures of 1,2,3-trimethylbenzene at various temperatures and the coefficients for the vapor pressure equations

$$\log P = A - B/(T/K) \quad (1)$$
$$\log P = A - B/(C + t/°C) \quad (2)$$
$$\log P = A - B/(C + T/K) \quad (3)$$
$$\log P = A - B/(T/K) - C \cdot \log(T/K) \quad (4)$$
$$\ln P = A - B/(T/K) \quad (1a)$$
$$\ln P = A - B/(C + t/°C) \quad (2a)$$

Stull 1947		Forziati et al. 1949		Zwolinski & Wilhoit 1971	
summary of literature data		ebulliometry		selected values	
t/°C	P/Pa	t/°C	P/Pa	t/°C	P/Pa
16.8	133.3	90.332	6417	56.79	1333
42.9	666.6	94.826	7697	70.62	2666
55.9	1333	98.77	8993	79.41	4000
69.9	2666	102.336	10328	86	5333
85.4	5333	105.663	11720	91.313	6666
95.3	7999	110.157	13840	95.802	7999
1088	13332	115.287	16640	103.168	10666
129	26664	120.504	19944	109.132	13332
152	53329	125.333	23474	120.578	19998
176.1	101325	131.8	28978	129.215	26664
		137.737	34918	136.214	33331
mp/°C	−25.5	144.882	43351	142.191	39997
		152.26	53692	152.022	53329
		160.106	66792	160.037	66661
		168.614	83750	166.856	79993
		174.606	97644	172.823	93326
		175.252	99237	173.934	95992
		175.852	100732	175.02	98659
		176.527	102453	176.084	101325
		177.126	103985		
				eq. 2	P/mmHg
				A	7.04082
		eq. 2	P/mmHg	B	1593.958
		A	7.04082	C	207.078
		B	1593.958	bp/°C	176.084
		C	207.078	$\Delta H_V/(kJ\,mol^{-1}) =$	
				at 25°C	49.06
		bp/°C	176.084	at bp	40.04

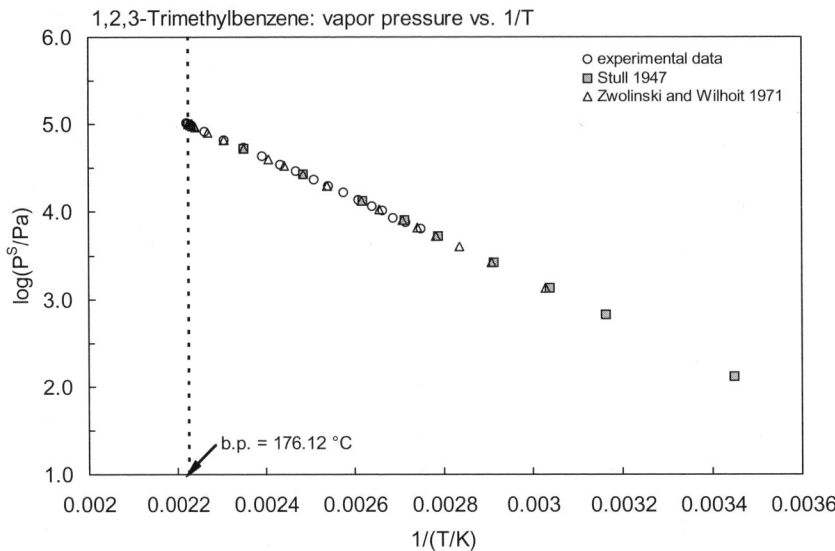

FIGURE 3.1.1.7.2 Logarithm of vapor pressure versus reciprocal temperature for 1,2,3-trimethylbenzene.

3.1.1.8 1,2,4-Trimethylbenzene

Common Name: 1,2,4-Trimethylbenzene
Synonym: pseudocumene
Chemical Name: 1,2,4-trimethylbenzene
CAS Registry No: 95-63-6
Molecular Formula: C_9H_{12}, $C_6H_3(CH_3)_3$
Molecular Weight: 120.191
Melting Point (°C):
    –43.77   (Lide 2003)
Boiling Point (°C):
    169.38   (Lide 2003)
Density (g/cm³ at 20°C):
    0.8758   (Weast 1982–83)
Molar Volume (cm³/mol):
    137.2   (20°C, calculated-density)
    162.6   (calculated-Le Bas method at normal boiling point)
Enthalpy of Fusion ΔH_{fus} (kJ/mol):
    13.19   (Chickos et al. 1999)
Entropy of Fusion ΔS_{fus} (J/mol K):
    57.53, 46.2   (exptl., calculated-group additivity method, Chickos et al. 1999)
Fugacity Ratio at 25°C, F: 1.0

Water Solubility (g/m³ or mg/L at 25°C or as indicated and reported temperature dependence equations. Additional data at other temperatures designated * are compiled at the end of this section):
    57   (shake flask-GC, McAuliffe 1966)
    59.0   (shake flask-GC, Sutton & Calder 1975)
    51.9   (shake flask-GC, Price 1976)
    51.9   (shake flask-GC, Krzyzanowska & Szeliga 1978)
    56.5*   (vapor saturation-UV spec., measured range 15–45°C, Sanemasa et al. 1982)
    56*   (IUPAC recommended, temp range 15–45°C, Shaw 1989b)
    $\ln x = -8.760 - 868.70/(T/K)$; temp range 5–50°C (regression eq. of literature data, Shiu & Ma 2000)

Vapor Pressure (Pa at 25°C or as indicated and reported temperature dependence equations. Additional data at other temperatures designated * are compiled at the end of this section):
    133.3*   (13.6°C, summary of literature data, Stull 1947)
    6417*   (84.804°C, ebulliometry, measured range 84.804–170.377°C, Forziati et al. 1949)
    $\log (P/mmHg) = 7.04393 - 1573.267/(208.564 + t/°C)$; temp range 84.8–170.4°C (manometer, Antoine eq. from exptl. data, Forziati et al. 1949)
    280   (extrapolated-Antoine eq., Dreisbach 1955)
    $\log (P/mmHg) = 7.04383 - 1573.267/(208.564 + t/°C)$; temp range 70–220°C (Antoine eq. for liquid state, Dreisbach 1955)
    2666*   (65.405°C, compiled data, temp range 65.405–198.215°C, Bond & Thodos 1960)
    $\log (P/mmHg) = 23.2393 - 3301.19/(T/K) - 6.21412 \cdot \log (T/K) + 3.15835[P(mmHg)/(T/K)^2]$, temp range 65.4–198°C (Bond & Thodos 1960)
    271*   (extrapolated-Antoine eq., Zwolinski & Wilhoit 1971; quoted, Mackay & Shiu 1981; Eastcott et al. 1988)
    $\log (P/mmHg) = 7.04383 - 1573.267/(208.564 + t/°C)$; temp range 51.75–198.2°C (Antoine eq., Zwolinski & Wilhoit 1971)

log (P/mmHg) = [–0.2185 × 10710.2/(T/K)] + 8.209013; temp range 13.6–169.2°C (Antoine eq., Weast 1972–73)

log (P/atm) = (1 – 442.537/T) × 10^(0.846724 – 5.41424 × 10^4·T + 4.22211 × 10^7·T^2); T in K, temp range 253.0–645.0 K (Cox vapor pressure eq., Chao et al. 1983)

270 (extrapolated-Antoine eq., Boublik et al. 1984)

log (P/kPa) = 6.16282 – 1569.06/(208.089 + t/°C); temp range 84.8–170.4°C (Antoine eq. from reported exptl. data of Forziati et al. 1949, Boublik et al. 1984)

log (P/mmHg) = 7.04383 – 1573.83/(208.56 + t/°C); temp range 52–198°C (Antoine eq., Dean 1985, 1992)

log (P_L/kPa) = 6.16695 – 1572.687/(–64.593 + T/K); temp range 357–450 K (liquid, Antoine eq., Stephenson & Malanowski 1987)

log (P/mmHg) = 2.1667 – 2.6318 × 10^3/(T/K) + 4.0350·log (T/K) – 1.1776 × 10^{-2}·(T/K) + 6.0956 × 10^{-6}·$(T/K)^2$, temp range 229–649 K (vapor pressure eq., Yaws 1994)

log (P/kPa) = 6.16866 – 1573.267/[(T/K) – 64.586]; temp range 5–50°C (regression eq. from literature data, Shiu & Ma 2000)

Henry's Law Constant (Pa m^3/mol at 25°C or as indicated and reported temperature dependence equations. Additional data at other temperatures designated * are compiled at the end of this section):

619* (vapor-liquid equilibrium, measured range 15–45°C, Sanemasa et al. 1982)
475 (20°C, EPICS-GC, Yurteri et al. 1987)
571 (calculated-vapor-liquid equilibrium (VLE) data, Yaws et al. 1991)
704, 1135, 1591 (27, 35, 45°C, EPICS-GC, Hansen et al. 1993)
ln [H/(kPa·m^3/mol)] = –4298/(T/K) + 14.0; temp range 27–45°C (EPICS-GC, Hansen et al. 1993)
529 (20°C, selected from literature experimentally measured data, Staudinger & Roberts, 1996, 2001)
log K_{AW} = 5.125 – 1697/(T/K) (van't Hoff eq. derived from lit. data, Staudinger & Roberts 2001)

Octanol/Water Partition Coefficient, log K_{OW}:

3.65 (calculated-π substituent constant, Hansch et al. 1968)
3.63 (shake flask-HPLC/UV both phases, Wasik et al. 1981)
3.78 (HPLC-k′ correlation, Hammers et al. 1982)
3.82, 4.00 (RP-HPLC-k′ correlations, Sherblom & Eganhouse 1988)
3.63 (recommended value, Sangster 1989)
3.78 (normal phase HPLC-k′ correlation, Govers & Evers 1992)
3.70 (recommended, Sangster 1993)
3.83, 3.78 (quoted lit., Hansch et al. 1995)

Octanol/Air Partition Coefficient, log K_{OA}:

Bioconcentration Factor, log BCF:

Sorption Partition Coefficient, log K_{OC}:

3.28 (computed-K_{OW}, Kollig 1995)

Environmental Fate Rate Constants, k, or Half-Lives, $t_{1/2}$:

Photolysis: rate constant k = 2.686 × 10^{-2} h^{-1} with H_2O_2 under photolysis at 25°C in F-113 solution and with HO- in the gas (Dilling et al. 1988).
no photolyzable functional groups (Howard et al. 1991).

Oxidation: rate constant k, for gas-phase second order rate constants, k_{OH} for reaction with OH radical, k_{NO3} with NO_3 radical and k_{O3} with O_3 or as indicated, *data at other temperatures and/or the Arrhenius expression see reference:

k_{OH} = (2.0 ± 0.3) × 10^{10} L mol^{-1} s^{-1} with $t_{1/2}$ = 0.58 h; $k_{O(^3P)}$ = (6.0 ± 0.6) × 10^8 L mol^{-1} s^{-1} with $O(^3P)$ atom at room temp. (relative rate method, Doyle et al. 1975; Lloyd et al. 1976)

k_{OH} = (33.5 ± 3.4) × 10^{-12} cm^3 $molecule^{-1}$ s^{-1}, and k_{NO3} = 10.0 × 10^{-13} cm^3 $molecule^{-1}$ s^{-1} for the reaction of $O(^3P)$ atom at room temp. (flash photolysis-resonance fluorescence, Hansen et al. 1975)

k_{OH} = 20 × 10^9 L mol^{-1} s^{-1} with $t_{1/2}$ = 0.24–2.4 h (Darnall et al. 1976)

k_{OH}* = (40.0 ± 4.5) × 10^{-12} cm^3·$molecule^{-1}$ s^{-1} at room temp., measured range 296–473 K (flash photolysis-resonance fluorescence Perry et al. 1977) photooxidation $t_{1/2}$ = 1056–43000 h in water, based on measured rate data with hydroxy radical in aqueous solution (Güesten et al. 1981)

k_{OH} = 33.2 × 10^{-12} cm^3 molecule^{-1} s^{-1} and residence time of 0.3 d, loss of 96.4% in one day or 12 sunlit hours at 300 K in urban environments (Singh et al. 1981)

k_{OH} = (1.5–30) × 10^9 M^{-1} s^{-1} with $t_{1/2}$ = 0.2–0.4 d for trimethylbenzenes (Mill 1982)

k_{NO3} = 5.4 × 10^{-16} cm^3 molecule^{-1} s^{-1} at 296 K (Atkinson et al. 1984)

k_{OH} = 38.4 × 10^{-12} cm^3 molecule^{-1} s^{-1} with $t_{1/2}$ = 1.6–16 h (Atkinson 1985)

k_{OH} = 31.5 × 10^{-12} cm^3 molecule^{-1} s^{-1} at 25°C (Ohta & Ohyama 1985)

k_{OH}(calc) = 18 × 10^{-12} cm^3 molecule^{-1} s^{-1}, k_{OH}(obs.) = 40 × 10^{-12} cm^3 molecule^{-1} s^{-1} at room temp. (SAR [structure-activity relationship], Atkinson 1987)

k_{NO3} = 1.80 × 10^{-15} cm^3 molecule^{-1} s^{-1}, and k_{OH} = 3.84 × 10^{-11} cm^3 molecule^{-1} s^{-1} at 298 K (Atkinson et al. 1988; quoted, Sabljic & Güsten 1990; Müller & Klein 1992)

k_{OH} ≈ 32.5 × 10^{-12} cm^3 molecule^{-1} s^{-1} at 298 K (recommended, Atkinson 1990)

k_{OH}(calc) = 1.78 × 10^{-11} cm^3 molecule^{-1} s^{-1} at 298 K (estimated by SARs, Müller & Klein 1992)

k_{OH}(calc) = 39.72 × 10^{-12} cm^3 molecule^{-1} s^{-1} (molecular orbital calculations, Klamt 1993)

Hydrolysis: no hydrolyzable functional groups (Mabey et al. 1982).

Biodegradation: aqueous aerobic biodegradation $t_{1/2}$ =168–672 h, based on aqueous screening studies (Marion & Malaney 1964; Kitano 1978; Van der Linden 1978; Tester & Harker 1981; Trzilova & Horska 1988; Howard et al. 1991);

anaerobic aqueous biodegradation $t_{1/2}$ = 672–2688 h, based on estimated aqueous aerobic biodegradation half-lives (Howard et al. 1991).

Half-Lives in the Environment:

Air: $t_{1/2}$ = 0.58 h estimated from the rate of disappearance for the reaction with OH radical (Doyle et al. 1975) $t_{1/2}$ = 0.24–2.4 h (Darnall et al. 1976);

residence time of 0.3 d, loss of 96.4% in one day or 12 sunlit hours at 300 K in urban environments (Singh et al. 1981)

$t_{1/2}$ = 1.6–16 h, based on photooxidation half-life in air (Atkinson 1985; Howard et al. 1991);

summer daylight lifetime τ = 4.3 h due to reaction with OH radical (Altshuller 1991);

calculated lifetimes of 4.3 h, 26 d and > 4.5 yr for reactions with OH radical, NO$_3$ radical and O$_3$, respectively (Atkinson 2000).

Surface Water: $t_{1/2}$ = 168–672 h, based on estimated aqueous aerobic biodegradation half-lives (Kitano 1978; Van der Linden 1978; Tester & Harker 1981; Trzilova & Horska 1988; Marion & Melaney 1964; Howard et al. 1991);

photooxidation $t_{1/2}$ = 1056–43000 h, based on measured rate data with OH radical in aqueous solution (Güesten et al. 1981; Howard et al. 1991).

Ground water: $t_{1/2}$ = 336–1344 h, based on estimated aqueous aerobic biodegradation half-lives (Howard et al. 1991).

Soil: $t_{1/2}$ = 168–672 h, based on estimated aqueous aerobic biodegradation half-lives (Howard et al. 1991).

Biota:

TABLE 3.1.1.8.1
Reported aqueous solubilities and Henry's law constants of 1,2,4-trimethylbenzene at various temperatures

Aqueous solubility				Henry's law constant			
Sanemasa et al. 1982		Shaw 1989b (IUPAC)		Sanemasa et al. 1982		Hansen et al. 1993	
vapor saturation-UV		recommended values		vapor-liquid equilibrium		EPICS-GC	
t/°C	S/g·m⁻³	t/°C	S/g·m⁻³	t/°C	H/(Pa m³/mol)	t/°C	H/(Pa m³/mol)
15	52.3	15	52	15	377	27	704
25	56.5	25	56	25	619	35	1135
35	62.1	35	62	35	1042	45	1591
45	69.3	45	69	45	1663		
						ln [H/(kPa m³/mol)] = A − B/(T/K)	
						A	14.0
						B	4298

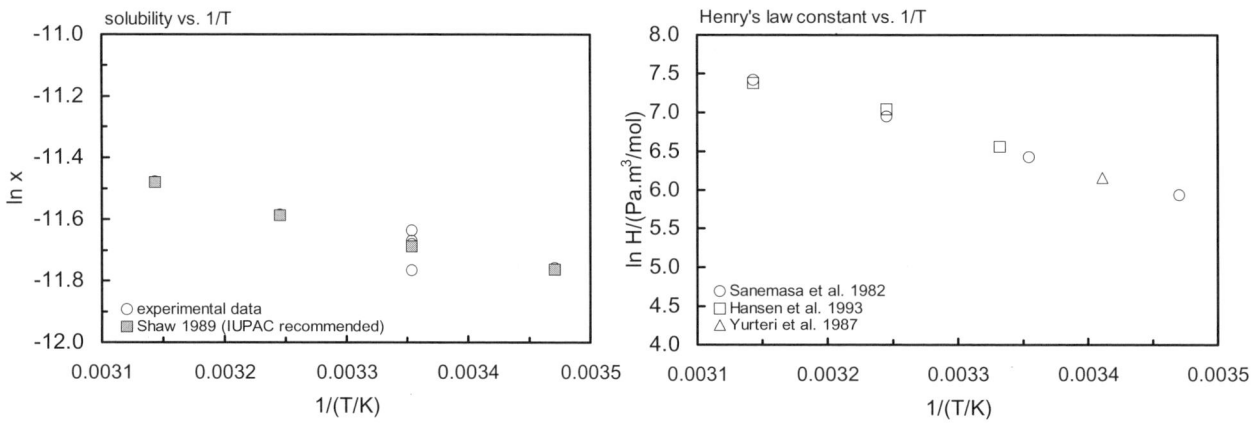

FIGURE 3.1.1.8.1 Logarithm of mole fraction solubility and Henry's law constant versus reciprocal temperature for 1,2,4-trimethylbenzene.

TABLE 3.1.1.8.2
Reported vapor pressures of 1,2,4-trimethylbenzene at various temperatures and the coefficients for the vapor pressure equations

$$\log P = A - B/(T/K) \quad (1) \qquad \ln P = A - B/(T/K) \quad (1a)$$
$$\log P = A - B/(C + t/°C) \quad (2) \qquad \ln P = A - B/(C + t/°C) \quad (2a)$$
$$\log P = A - B/(C + T/K) \quad (3)$$
$$\log P = A - B/(T/K) - C \cdot \log (T/K) \quad (4)$$
$$\log P = A - B/(T/K) - C \cdot \log (T/K) + D \cdot P/(T/K)^2 \quad (5)$$

Stull 1947		Forziati et al. 1949		Bond & Thodos 1960		Zwolinski & Wilhoit 1971	
summary of literature data		ebulliometry		summary of literature data		selected values	
t/°C	P/Pa	t/°C	P/Pa	t/°C	P/Pa	t/°C	P/Pa
13.6	133.3	84.804	6417	65.405	2666	51.75	1333
38.4	666.6	89.259	7697	97.49	10666	65.39	2666
50.7	1333	93.155	8991	153.55	66661	74.05	4000
64.5	2666	96.65	10328	180.505	133322	80.54	5333

TABLE 3.1.1.8.2 *(Continued)*

Stull 1947 summary of literature data		Forziati et al. 1949 ebulliometry		Bond & Thodos 1960 summary of literature data		Zwolinski & Wilhoit 1971 selected values	
t/°C	P/Pa	t/°C	P/Pa	t/°C	P/Pa	t/°C	P/Pa
79.8	5333	99.94	11720	198.215	199984	85.787	6666
89.5	7999	104.369	13840			90.214	7999
102.8	13332	109.418	16640	bp/°C	169.366	97.475	10666
122.7	26664	114.572	19944			103.355	13332
145.4	53329	119.328	23474	eq. 5	P/mmHg	114.639	19998
169.2	101325	125.694	28978	A	23.2393	123.153	26664
		131.556	34918	B	3301.19	130.072	33331
mp/°C	−44.1	138.599	43351	C	6.21412	135.944	39997
		145.867	53692	D	3.15835	145.634	53329
		153.604	66792			153.534	66661
		161.991	83750			160.256	79993
		167.896	97644			166.137	93326
		168.534	99237			167.231	95992
		169.121	100732			168.302	98659
		169.788	102453			169.351	101325
		170.377	103985				
						eq. 2	P/mmHg
		eq. 2	P/mmHg			A	7.04383
		A	7.04383			B	1573.267
		B	1573.267			C	208.564
		C	208.564			bp/°C	169.351
						$\Delta H_V/(kJ\ mol^{-1})=$	
		bp/°C	169.351			at 25°C	47.94
						at bp	39.25

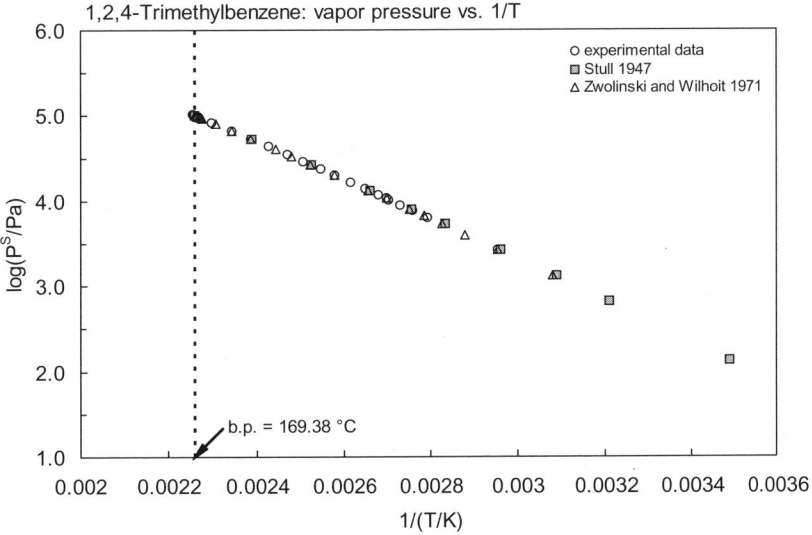

FIGURE 3.1.1.8.2 Logarithm of vapor pressure versus reciprocal temperature for 1,2,4-trimethylbenzene.

3.1.1.9 1,3,5-Trimethylbenzene

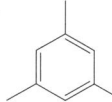

Common Name: 1,3,5-Trimethylbenzene
Synonym: mesitylene
Chemical Name: 1,3,5-trimethylbenzene
CAS Registry No: 108-67-8
Molecular Formula: C_9H_{12}, $C_6H_3(CH_3)_3$
Molecular Weight: 120.191
Melting Point (°C):
    –44.7   (Weast 1982–83; Lide 2003)
Boiling Point (°C):
    164.74   (Lide 2003)
Density (g/cm³ at 20°C):
    0.880   (Weast 1982–83)
Molar Volume (cm³/mol):
    136.6   (20°C, calculated-density)
    162.6   (calculated-Le Bas method at normal boiling point)
Enthalpy of Vaporization, ΔH_V (kJ/mol):
    47.48, 39.04   (25°C, bp, Riddick et al. 1986)
    51.85   (calculated-bp, Govers & Evers 1992)
Enthalpy of Fusion ΔH_{fus} (kJ/mol):
    9.514   (Riddick et al. 1986)
    9.51   (Chickos et al. 1999)
Entropy of Fusion ΔS_{fus} (J/mol K):
    41.66, 46.2   (exptl., calculated-group additivity method, Chickos et al. 1999)
Fugacity Ratio at 25°C, F: 1.0

Water Solubility (g/m³ or mg/L at 25°C or as indicated and reported temperature dependence equations. Additional data at other temperatures designated * are compiled at the end of this section):
    173   (residue-volume method, Booth & Everson 1948)
    97   (shake flask-UV, Andrews & Keffer 1950)
    39.4   (shake flask-UV, Vesala 1974)
    48.2   (shake flask-GC, Sutton & Calder 1975)
    49.5*   (vapor saturation-UV, temp range 15–45°C, Sanemasa et al. 1981)
    50*   (vapor saturation-UV, temp range 15–45°C, Sanemasa et al. 1982)
    49.5   (HPLC-k′ correlation, converted from reported γ_W, Hafkenscheid & Tomlinson 1983)
    48.9*   (recommended, temp range 15–45°C, Shaw 1989b)
    64*   (30°C, equilibrium flow cell-GC, measured range 30–100°C, Chen & Wagner 1994c)
    $\ln x = 26.26 - 35.26 \cdot (T_r/K)^{-1} + 7.905 \cdot (T_r/K)^{-2}$; $T_r = T/T_c$, the reduced temp, system temp T divided by critical temp T_c, temp range 303.15–373.15 K (equilibrium flow cell-GC, Chen & Wagner 1994c)
    $\ln x = -9.533 - 678.83/(T/K)$; temp range 5–50°C (regression eq. of literature data, Shiu & Ma 2000)

Vapor Pressure (Pa at 25°C or as indicated and reported temperature dependence equations. Additional data at other temperatures designated * are compiled at the end of this section):
    121.3*   (10.6°C, mercury manometer, Linder 1931)
    248*   (20°C, mercury manometer, Kassel 1936)
    $\log (P/\text{mmHg}) = -3104.5/(T/K) - 5 \cdot \log (T/K) + 23.1929$; temp range 0–80°C (vapor pressure eq. from Hg manometer measurements, Kassel 1936)
    507*   (30°C, mercury manometer, measured range 10–50°C, Rintelen et al. 1937)

Mononuclear Aromatic Hydrocarbons

log (P/mmHg) = –3122.45/(T/K) – 5·log (T/K) + 22.23680; temp range 4–75°C (manometer, vapor pressure eq. from exptl. data, Stuckey & Saylor 1940)

133.3*   (9.6°C, summary of literature data, Stull 1947)

6415*   (81.488°C, ebulliometry, measured range 81.488–165.725°C, Forziati et al. 1949)

log (P/mmHg) = 7.07437 – 1569.622/(209.578 + t/°C); temp range 81.5–165.7°C (manometer, Antoine eq. from exptl. data, Forziati et al. 1949)

331   (extrapolated-Antoine eq., Dreisbach 1955)

log (P/mmHg) = 7.07436 – 1569.622/(209.578 + t/°C); temp range 70–210°C (Antoine eq. for liquid state, Dreisbach 1955)

328*   (extrapolated-Antoine eq., Zwolinski & Wilhoit 1971; quoted, Mackay & Shiu 1981; Eastcott et al. 1988)

log (P/mmHg) = 7.07435 – 1569.622/(209.578 + t/°C); temp range 48.82–193.07°C (Antoine eq., Zwolinski & Wilhoit 1971)

log (P/mmHg) = [–0.2185 × 10516.8/(T/K)] + 8.161663; temp range 9.6–164.7°C (Antoine eq., Weast 1972–73)

366   (calculated-bp, Mackay et al. 1982)

log (P/atm) = (1 – 437.769/T) × $10^{\wedge}(0.872945 – 6.55508 \times 10^{4} \cdot T + 5.47586 \times 10^{7} \cdot T^{2})$; T in K, temp range 253.0–635.0 K (Cox vapor pressure eq., Chao et al. 1983)

323   (extrapolated-Antoine eq., Boublik et al. 1984)

log (P/kPa) = 6.20212 – 1571.575/(209.79 + t/°C); temp range 81.488–165.7°C (Antoine eq. from reported exptl. data, Boublik et al. 1984)

322   (extrapolated-Antoine eq., Dean 1985)

330   (selected lit., Riddick et al. 1986)

log (P/kPa) = 6.21017 – 1577.80/(210.526 + t/°C); temp range not specified (Antoine eq., Riddick et al. 1986)

log (P/mmHg) = 7.07436 – 1569.622/(209.58 + t/°C); temp range 49–193°C (Antoine eq., Dean 1985, 1992)

330   (interpolated-Antoine eq.-II, Stephenson & Malanowski 1987)

log (P_L/kPa) = 6.19762 – 1569.749/(–63.565 + T/K); temp range 354–445 K (liquid, Antoine eq.-I, Stephenson & Malanowski 1987)

log (P_L/kPa) = 6.62312 – 1810.653/(–43.307 + T/K); temp range 249–356 K (liquid, Antoine eq.-II, Stephenson & Malanowski 1987)

360   (computed-expert system SPARC, Kollig 1995)

log (P/kPa) = 6.18965 – 1569.622/[(T/K) ± 63.572]; temp range 5–50°C (regression eq. from literature data, Shiu & Ma 2000)

Henry's Law Constant (Pa m^3/mol at 25°C or as indicated and reported temperature dependence equations. Additional data at other temperatures designated * are compiled at the end of this section):

929*   (vapor-liquid equilibrium, measured range 15–45°C, Sanemasa et al. 1981)

887*   (vapor-liquid equilibrium, measured range 15–45°C, Sanemasa et al. 1982)

682; 849   (EPICS; batch stripping, Ashworth et al. 1988)

682*   (EPICS-GC/FID, measured range 10–30°C, Ashworth et al. 1988)

ln [H/(atm m^3/mol)] = 7.241 – 3628/(T/K); temp range 10–30°C (EPICS measurements, Ashworth et al. 1988)

803   (calculated-vapor-liquid equilibrium (VLE) data, Yaws et al. 1991)

704   (infinite activity coeff. γ^{∞} in water determined by inert gas stripping-GC, Li et al. 1993)

597   (20°C, selected from literature experimentally measured data, Staudinger & Roberts 1996, 2001)

log K_{AW} = 4.329 – 1448/(T/K) (van't Hoff eq. derived from lit. data, Staudinger & Roberts 2001)

Octanol/Water Partition Coefficient, log K_{OW}:

3.42   (Leo et al. 1971; Hansch & Leo 1979)

3.78   (HPLC-k' correlation, Hammers et al. 1982)

3.82   (HPLC-k' correlation, Hafkenscheid & Tomlinson 1983)

3.42   (HPLC-RV correlation, Garst 1984)

3.89   (normal phase HPLC-k' correlation, Govers & Evers 1992)

3.42   (recommended value, Sangster 1993)

3.42   (recommended, Hansch et al. 1995)

Octanol/Air Partition Coefficient, log K_{OA}:

Bioconcentration Factor, log BCF:

Sorption Partition Coefficient, log K_{OC}:
- 2.82   (average 5 soils and 3 sediments, sorption isotherms by batch equilibrium and column experiments, Schwarzenbach & Westall 1981)
- 2.77   (soil, calculated-MCI χ, Sabljic 1987a)
- 2.75   (soil, calculated-MCI χ, Sabljic 1987b)
- 2.85   (soil, calculated-MCI χ, Bahnick & Doucette 1988)
- 2.82   (soil, calculated-QSAR-χ, Sabljic et al. 1995)
- 3.37   (computed-K_{OW}, Kollig 1995)

Environmental Fate Rate Constants, k, or Half-Lives, $t_{\frac{1}{2}}$:

Volatilization:

Photolysis: rate constant $k = 1.606 \times 10^{-2}$ h^{-1} with H_2O_2 under photolysis at 25°C in F-113 solution and with HO· in the gas (Dilling et al. 1988).

Oxidation: rate constant k, for gas-phase second order rate constants, k_{OH} for reaction with OH radical, k_{NO3} with NO_3 radical and k_{O3} with O_3 or as indicated, *data at other temperatures and/or the Arrhenius expression see reference:

$k_{O3} < 60$ L mol^{-1} s^{-1} for vapor phase reaction with ozone at 30°C (Bufalini & Altshuller 1965)

$k_{OH} = (31 \pm 4) \times 10^9$ L mol^{-1} s^{-1}; $k_{O(^3P)} = (16.8 \pm 2.0) \times 10^8$ L mol^{-1} s^{-1} with O(^{3}P) atom at room temp. (relative rate method, Doyle et al. 1975; Lloyd et al. 1976)

$k_{OH} = (47.2 \pm 4.8) \times 10^{-12}$ cm^3 molecule^{-1} s^{-1}, and $k_{O(^3P)} = (27.9 \pm 3.3) \times 10^{-13}$ cm^3 molecule^{-1} s^{-1} for the reaction of O(^{3}P) atom at room temp. (flash photolysis-resonance fluorescence, Hansen et al. 1975)

$k_{OH} = 29.7 \times 10^9$ L mol^{-1} s^{-1} with $t_{\frac{1}{2}} = 0.24$–2.4 h (Darnall et al. 1976)

$k_{OH}* = (62.4 \pm 7.5) \times 10^{-12}$ cm^3 molecule^{-1} s^{-1} at room temp., measured range 296–473 K (flash photolysis-resonance fluorescence, Perry et al. 1977)

photooxidation in water, $t_{\frac{1}{2}} = 3208$–1.28×10^5 h, based on measured rate constant for reaction with hydroxyl radical in water (Mill et al. 1980)

$k_{OH} = 49.3 \times 10^{-12}$ cm^3 molecule^{-1} s^{-1} and residence time of 0.2 d, loss of 99.3% in one day or 12 sunlit hour at 300 K in urban environments (Singh et al. 1981)

$k_{O3} = 4200$ cm^3 mol^{-1} s^{-1} for the reaction with ozone at 300 K (Lyman et al. 1982)

$k_{OH} = (15–30) \times 10^9$ M^{-1} s^{-1} with $t_{\frac{1}{2}} = 0.2$–0.4 d for trimethylbenzenes (Mill 1982)

$k = (700 \pm 200)$ M^{-1} s^{-1} for the reaction with ozone in water at pH 1.7 and 20–23°C (Hoigné & Bader 1983)

$k_{NO3} = 2.4 \times 10^{-16}$ cm^3 molecule^{-1} s^{-1} at 296 K (Atkinson et al. 1984)

$k_{OH} = 60.5 \times 10^{-12}$ cm^3 molecule^{-1} s^{-1} at room temp. (Atkinson 1985)

$k_{OH} = 38.7 \times 10^{-12}$ cm^3 molecule^{-1} s^{-1} at room temp. (Ohta & Ohyama 1985)

$k_{NO3} = 7.91 \times 10^{-16}$ cm^3 molecule^{-1} s^{-1}, and $k_{OH} = 6.05 \times 10^{-11}$ cm^3 molecule^{-1} s^{-1} at 298 K (Atkinson 1985; Atkinson et al. 1988; quoted, Sabljic & Güsten 1990; Müller & Klein 1992)

k_{OH}(calc) = 38×10^{-12} cm^3 molecule^{-1} s^{-1}, k_{OH}(obs.) = 62.4×10^{-12} cm^3 molecule^{-1} s^{-1} at room temp. (SAR [structure-activity relationship], Atkinson 1987)

$k_{OH}* = 57.5 \times 10^{-12}$ cm^3 molecule^{-1} s^{-1} at 298 K (recommended, Atkinson 1989, 1990)

k_{OH}(calc) = 3.72×10^{-11} cm^3 molecule^{-1} s^{-1} at 298 K (estimated by SARs, Müller & Klein 1992)

k_{OH}(calc) = 54.16×10^{-12} cm^3 molecule^{-1} s^{-1} (molecular orbital calculations, Klamt 1993)

Hydrolysis:

Biodegradation: unacclimated aerobic aqueous biodegradation $t_{\frac{1}{2}} = 48$–192 h, based on a soil column study in which aerobic groundwater was continuously percolated through quartz sand (Kappeler & Wuhrmann 1978; Howard et al. 1991); $t_{\frac{1}{2}}$(aq. anaerobic) = 192–768 h, based on unacclimated aqueous aerobic biodegradation half-life (Howard et al. 1991).

Bioconcentration:

Half-Lives in the Environment:

Air: 0.24–2.4 h, based on rate of disappearance for the reaction with hydroxyl radical (Darnall et al. 1976; Howard et al. 1991);

estimated lifetime under photochemical smog conditions in S.E. England is 0.7 h (Brice & Derwent 1978; Perry et al. 1977 and Darnall et al. 1976);

residence time of 0.2 d, loss of 99.3% in one day or 12 sunlit hours at 300 K in urban environments (Singh et al. 1981)

$t_{1/2}$ = 9.72–97.2 h, based on estimated photooxidation half-life in air (Atkinson 1987).

Surface Water: $t_{1/2}$ = 48–192 h, based on a soil column study in which aerobic groundwater was continuously percolated through quartz sand (Kappeler & Wuhrmann 1978; Howard et al. 1991);

$t_{1/2}$ = 1 d in surface waters in case a first order reduction process may be assumed (estimated, Zoeteman et al. 1980).

Ground water: $t_{1/2}$ = 96–384 h, based on a soil column study in which aerobic ground water was continuously percolated through quartz sand (Kappeler & Wuhrmann 1978; Howard et al. 1991).

Soil: $t_{1/2}$ = 48–192 h, based on a soil column study in which aerobic groundwater was continuously percolated through quartz sand (Kappeler & Wuhrmann 1978; Howard et al. 1991).

Biota:

TABLE 3.1.1.9.1
Reported aqueous solubilities of 1,3,5-trimethylbenzene at various temperatures

Sanemasa et al. 1981		Sanemasa et al. 1982		Shaw 1989b (IUPAC)		Chen & Wagner 1994c	
vapor saturation-UV		vapor saturation-UV		recommended values		shake flask-GC	
t/°C	S/g·m⁻³	t/°C	S/g·m⁻³	t/°C	S/g·m⁻³	t/°C	S/g·m⁻³
15	45.6	15	46	15	46	30	64
25	49.5	25	50	25	48.9	40	67.8
35	54.2	35	54.9	35	54	50	74.12
45	56.5	45	58.9	45	57	60	90.82
						70	111
						80	140
						90	164
						100	194
						ΔH_{sol}/(kJ mol⁻¹) = 4.49 at 25°C	

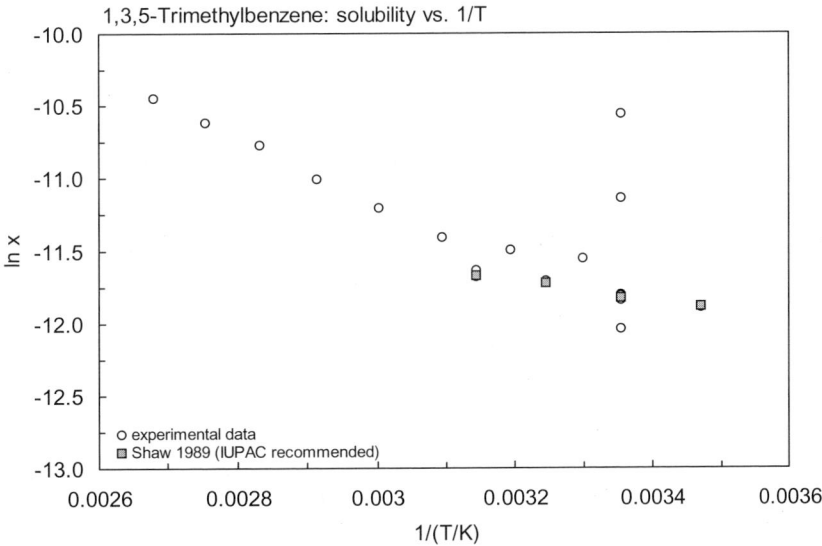

FIGURE 3.1.1.9.1 Logarithm of mole fraction solubility (ln x) versus reciprocal temperature for 1,3,5-trimethylbenzene.

TABLE 3.1.1.9.2
Reported vapor pressures of 1,3,5-trimethylbenzene at various temperatures and the coefficients for the vapor pressure equations

$\log P = A - B/(T/K)$ (1) $\qquad$ $\ln P = A - B/(T/K)$ (1a)

$\log P = A - B/(C + t/°C)$ (2) $\qquad$ $\ln P = A - B/(C + t/°C)$ (2a)

$\log P = A - B/(C + T/K)$ (3)

$\log P = A - B/(T/K) - C \cdot \log (T/K)$ (4)

1.

Linder 1931		Kassel 1936		Rintelen et al. 1937		Stull 1947	
Hg manometer		mercury manometer		mercury manometer		summary of literature data	
t/°C	P/Pa	t/°C	P/Pa	t/°C	P/Pa	t/°C	P/Pa
−1.7	45.3	0	58.7	10	80	9.6	133.3
0	50	10	124	30	507	34.7	266.6
2.3	66.7	20	248	50	1523	47.4	1333
3.2	70.7	30	521	Stuckey & Saylor 1940		61	2666
−2.75	38.66	40	844	mercury manometer		76.1	5333
−1.20	44	50	1467	t/°C	P/Pa	85.8	7999
1.5	57.3	60	2440	measured 4–75°C		98.9	13332
10.6	121.3	70	3933	eq. 4	P/mmHg	118.6	26664
−4.20	37.33	80	6133	A	23.2367	141	53323
2.7	66.66	90	9319	B	3122.45	164.7	101325
10.2	121.32	100	13786	C	5	mp/°C	44.8
		eq. 4	P/mmHg				
		A	22.1929	bp/°C	164.54		
		B	3104.5				
		C	5.0				

2.

Forziati et al. 1949				Zwolinski & Wilhoit 1971			
ebulliometry				selected values			
t/°C	P/Pa	t/°C	P/Pa	t/°C	P/Pa	t/°C	P/Pa
81.488	6415	163.911	99244	48.82	1333	161.556	93326
85.857	7697	164.489	100733	62.3	2666	162.632	95991
89.662	8991	165.146	102454	70.85	4000	163.686	98659
93.131	10328	165.725	103987	77.25	5333	164.716	101325
96.386	11722			82.424	6666		
100.747	13840	eq. 2	P/mmHg	86.789	7999	eq. 2	P/mmHg
105.716	16641	A	7.07437	93.949	10666	A	7.07436
110.789	19944	B	1573.267	99.746	13332	B	1569.622
115.498	23474	C	208.564	110.866	19998	C	209.578
121.765	28979			119.254	26664		
134.464	43352	bp/°C	164.716	126.068	33331	bp/°C	164.716
141.618	53692			131.849	39997	$\Delta H_V/(kJ\ mol^{-1}) =$	
149.238	66792			141.387	53329	at 25°C	47.48
157.477	83752			149.161	66661	at bp	39.04
163.289	97644			155.772	79993		

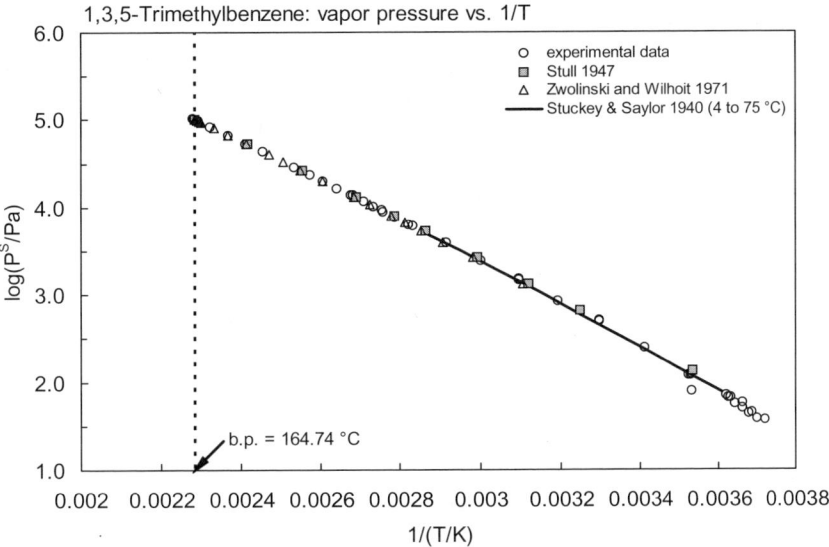

FIGURE 3.1.1.9.2 Logarithm of vapor pressure versus reciprocal temperature for 1,3,5-trimethylbenzene.

TABLE 3.1.1.9.3
Reported Henry's law constants of 1,3,5-trimethylbenzene at various temperatures and temperature dependence equations

$\ln K_{AW} = A - B/(T/K)$ (1) $\log K_{AW} = A - B/(T/K)$ (1a)

$\ln (1/K_{AW}) = A - B/(T/K)$ (2) $\log (1/K_{AW}) = A - B/(T/K)$ (2a)

$\ln (k_H/\text{atm}) = A - B/(T/K)$ (3)

$\ln [H/(\text{Pa m}^3/\text{mol})] = A - B/(T/K)$ (4) $\ln [H/(\text{atm·m}^3/\text{mol})] = A - B/(T/K)$ (4a)

$K_{AW} = A - B\cdot(T/K) + C\cdot(T/K)^2$ (5)

Sanemasa et al. 1981		Sanemasa et al. 1982		Ashworth et al. 1988	
vapor-liquid equilibrium		vapor-liquid equilibrium		EPICS-GC	
t/°C	H/(Pa m³/mol)	t/°C	H/(Pa m³/mol)	t/°C	H/(Pa m³/mol)
15	547	15	511	10	408
25	929	25	887	15	466
35	1501	35	1365	20	477
45	2466	45	2394	25	682
				30	976
				eq. 4a	
				H/(atm m³/mol)	
				A	7.241
				B	3628

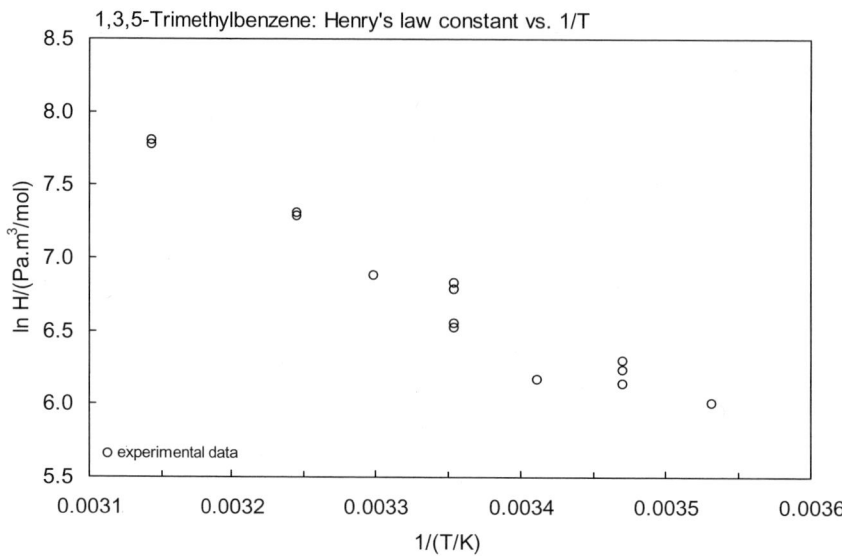

FIGURE 3.1.1.9.3 Logarithm of Henry's law constant versus reciprocal temperature for 1,3,5-trimethylbenzene.

3.1.1.10 *n*-Propylbenzene

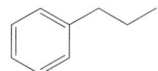

Common Name: *n*-Propylbenzene
Synonym: 1-phenylpropane, propylbenzene
Chemical Name: *n*-propylbenzene
CAS Registry No: 103-65-1
Molecular Formula: C_9H_{12}, $C_6H_5(CH_2)_2CH_3$
Molecular Weight: 120.191
Melting Point (°C):
    –99.6   (Lide 2003)
Boiling Point (°C):
    159.24   (Lide 2003)
Density (g/cm³ at 20°C):
    0.862   (Weast 1982–83)
Molar Volume (cm³/mol):
    139.4   (20°C, calculated-density)
    170.0   (calculated-Le Bas method at normal boiling point)
Enthalpy of Fusion ΔH_{fus} (kJ/mol):
    9.27   (Chickos et al. 1999)
Entropy of Fusion ΔS_{fus} (J/mol K):
    53.39, 59.3 (exptl., calculated-group additivity method, Chickos et al. 1999)
Fugacity Ratio at 25°C, F: 1.0

Water Solubility (g/m³ or mg/L at 25°C or as indicated. Additional data at other temperatures designated * are compiled at the end of this section):

    60   (15°C, volumetric, Führner 1924; quoted, Chiou et al. 1982; Chiou 1985)
    120   (shake flask-turbidimetric, Stearns et al. 1947)
    55   (shake flask-UV, Andrews & Keffer 1950)
    120   (shake flask-UV, Klevens 1951)
    60   (shake flask-GC, Hermann 1972)
    70   (shake flask-GC, Krasnoshchekova & Gubergrits 1975)
    66.4   (shake flask-UV spec., Ben-Naim & Wiff 1979)
    51.9*   (generator column-HPLC/UV, 15–30°C, DeVoe et al. 1981)
    51.0*   (vapor saturation-UV spec., measured range 15–45°C, Sanemasa et al. 1982)
    47.1   (generator column.-HPLC/UV, Tewari et al. 1982a)
    52.2   (generator column-HPLC/UV, GC/ECD, Tewari et al. 1982c)
    59.5   (HPLC-k′ correlation, converted from reported γ_W, Hafkenscheid & Tomlinson 1983a)
    52.1   (generator column-HPLC/UV, Wasik et al. 1983)
    45.2   (vapor saturation-UV spec., Sanemasa et al. 1984)
    51.7*   (generator column-HPLC/UV, measured range 10–45°C, Owens et al. 1986)
    55.0*   (IUPAC recommended, temp range 15–45°C, Shaw 1989b)
    48.2*   (vapor absorption technique-HPLC/UV, measured range 0.5–55°C, Dohányosová et al. 2001)
    55.0*   (shake flask-UV, measured range 0–50°C, Sawamura et al. 2001)
    ln x = –304.679 + 12774.71/(T/K) + 43.8994·ln (T/K); temp range 290–400 K (eq. derived from literature calorimetric and solubility data, Tsonopoulos 1999)

Vapor Pressure (Pa at 25°C or as indicated and reported temperature dependence equations; *data at other temperatures are tabulated at end of section):
    260*   (13.9°C, mercury manometer, Linder 1931)
    6353*   (75.646°C, ebulliometry, measured range 75.646–160.202°C, Willingham et al. 1945)

log (P/mmHg) = 6.95178 − 1649.548/(207.171 + t/°C); temp range 75.6–160.2°C (manometer, Antoine eq. from exptl. data, Willingham et al. 1945)

667* (31.3°C, summary of literature data, Stull 1947)

6402* (75.818°C, ebulliometry, measured range 75.818–160.239°C, Forziati et al. 1949)

log (P/mmHg) = 6.95094 − 1490.963/(207.100 + t/°C); temp range 75.8–160.2°C (manometer, Antoine eq. from exptl. data, Forziati et al. 1949)

458 (extrapolated-Antoine eq., Dreisbach 1955)

log (P/mmHg) = 6.95142 − 1491.297/(207.140 + t/°C); temp range 65–205°C (Antoine eq. for liquid state, Dreisbach 1955)

449* (extrapolated-Antoine eq., Zwolinski & Wilhoit 1971)

log (P/mmHg) = 6.95142 − 1491.297/(207.140 + t/°C); temp range 43.33–187.87°C (Antoine eq., Zwolinski & Wilhoit 1971)

log (P/mmHg) = [−0.2185 × 10424.1/(T/K)] + 8.185880; temp range 6.3–159.2°C (Antoine eq., Weast 1972–73)

log (P/atm) = (1 − 432.321/T) × 10^(0.891023 − 6.89092 × 10^4·T + 5.79948 × 10^7·T^2); T in K, temp range 280.0–635.0 K (Cox vapor pressure eq., Chao et al. 1983)

450 (extrapolated-Antoine eq., Boublik et al. 1984)

log (P/kPa) = 6.08028 − 1493.914/(207.427 + t/°C); temp range 75.2–160.24°C (Antoine eq. from reported exptl. data of Forziati et al. 1949, Boublik et al. 1984)

log (P/mmHg) = 6.95142 − 1491.297/(207.14 + t/°C); temp range 43–188°C (Antoine eq., Dean 1985, 1992)

449 (extrapolated-Antoine eq., Stephenson & Malanowski 1987)

log (P_L/kPa) = 6.07438 − 1490.61/(−66.029 + T/K); temp range 348–433 K (liquid, Antoine eq., Stephenson & Malanowski 1987)

log (P/mmHg) = 39.8219 − 3.6978 × 10^3/(T/K) − 10.962·log (T/K) + 8.7429 × 10^{-11}·(T/K) + 2.6959E × 10^{-6}·$(T/K)^2$; temp range 174–638 K (Yaws 1994)

log (P/kPa) = 6.07625 − 1490.903/[(T/K) − 66.05]; temp range 5–50°C (regression eq. from literature data, Shiu & Ma 2000)

Henry's Law Constant (Pa m³/mol at 25°C or as indicated and reported temperature dependence equations; *data at other temperatures are tabulated at end of section):

1062* (vapor-liquid equilibrium, measured range 15–45°C, Sanemasa et al. 1982)

1094* (EPICS-GC/FID, measured range 10–30°C, Ashworth et al. 1988)

ln [H/(atm·m3/mol)] = 7.835 − 3681/(T/K), temp range 10–30°C (EPICS measurements, Ashworth et al. 1988)

1034 (calculated-vapor-liquid equilibrium (VLE) data, Yaws et al. 1991)

1102 (infinite activity coeff. γ^∞ in water determined by inert gas stripping-GC, Li et al. 1993)

1175* (equilibrium headspace-GC, measured range 10–30°C, Perlinger et al. 1993)

902 (20°C, selected from literature experimentally measured data, Staudinger & Roberts 1996, 2001)

log K_{AW} = 4.587 − 1471/(T/K) (van't Hoff eq. derived from lit. data, Staudinger & Roberts 2001)

Octanol/Water Partition Coefficient, log K_{OW}:

3.68 (shake flask-UV, Iwasa et al. 1965; Hansch et al. 1968; 1972)

3.57, 3.68 (Leo et al. 1971; Hansch & Leo 1979)

3.66 (calculated-fragment const., Rekker 1977)

3.44 (shake flask-HPLC, Nahum & Horvath 1980)

3.691* (3.701, 3.72-HPLC/UV, DeVoe et al. 1981)

3.71 (generator column-HPLC/UV both phases, Tewari et al. 1982a)

3.63 (HPLC-k' correlation, Hammers et al. 1982)

3.69 (generator column-HPLC/GC, Tewari et al. 1982b,c; Wasik et al. 1983)

3.89 (HPLC-k' correlation, Hafkenscheid & Tomlinson 1983)

3.69 (generator column-HPLC/UV, Wasik et al. 1983)

3.69 (generator column-RP-HPLC, Schantz & Martire 1987)

3.71, 3.88 (RP-HPLC-k' correlations, Sherblom & Eganhouse 1988)

3.69 (recommended, Sangster 1989, 1993)

3.69 (HPLC-RT correlation, Jenke et al. 1990)

3.72 (recommended, Hansch et al. 1995)

Octanol/Air Partition Coefficient, log K_{OA} at 25°C:
    4.09   (calculated-measured γ^∞ in pure octanol of Tewari et al. 1982, Abraham et al. 2001)

Bioconcentration Factor, log BCF:

Sorption Partition Coefficient, log K_{OC}:
    2.87   (sediment 4.02% OC from Tamar estuary, batch equilibrium-GC, Vowles & Mantoura 1987)
    2.83, 2.98   (RP-HPLC-k' correlation, humic acid-silica column, Szabo et al. 1990a,b)
    2.81, 2.84, 2.87   (RP-HPLC-k' correlation on 3 different stationary phases, Szabo et al. 1995)

Environmental Fate Rate Constants, k, or Half-Lives, $t_{1/2}$:
    Volatilization: rate constants: k = 0.037 d^{-1}, $t_{1/2}$ = 19 d in spring at 8–16°C, k = 0.539 d^{-1}, $t_{1/2}$ = 1.3 d in summer at 20–22°C, k = 0.065 d^{-1}, $t_{1/2}$ = 11 d in winter at 3–7°C for the periods when volatilization appears to dominate, and k = 0.086 d^{-1}, $t_{1/2}$ = 8.1d with $HgCl_2$ in September 9–15, in marine mesocosm experiments (Wakeham et al. 1983)
    Photolysis: rate constant k = 6.96 × 10^{-3} h^{-1} with H_2O_2 under photolysis at 25°C in F-113 solution and with HO- in the gas (Dilling et al. 1988).
    Oxidation: rate constant k, for gas-phase second order rate constants, k_{OH} for reaction with OH radical, k_{NO3} with NO_3 radical and k_{O3} with O_3 or as indicated, *data at other temperatures see reference:
        k_{OH} = 3.7 × 10^9 L mol^{-1} s^{-1} with $t_{1/2}$ = 2.4–24 h (Darnall et al. 1976)
        k_{OH} = (3.7 ± 0.8) × 10^9 L mol^{-1} s^{-1} at 305 ± 2 K (relative rate method, Lloyd et al. 1976)
        k_{OH} = (6.40, 5.86) × 10^{-12} cm^3 molecule^{-1} s^{-1} with different dilute gas, Ar or He at 298 K (flash photolysis-resonance fluorescence, Ravishanakara et al. 1978)
        k_{OH} = 3.5 × 10^{-9} M^{-1} s^{-1} with $t_{1/2}$ = 1.6 d (Mill 1982)
        k_{OH} = 5.7 × 10^{-12} cm^3 molecule^{-1} s^{-1} at room temp. (Atkinson 1985)
        k_{OH} = 6.58 × 10^{-12} cm^3 molecule^{-1} s^{-1} at room temp. (Ohta & Ohyama 1985)
        k_{OH}(calc) = 7.5 × 10^{-12} cm^3 molecule^{-1} s^{-1}, k_{OH}(obs.) = 5.7 × 10^{-12} cm^3 molecule^{-1} s^{-1} at room temp. (SAR [structure-activity relationship], Atkinson 1987)
        k_{OH} - 6.0 × 10^{-12} cm^3 molecule^{-1} s^{-1} at 298 K (recommended, Atkinson 1989, 1990)
        k_{OH}(calc) = 5.99 × 10^{-12} cm^3 molecule^{-1} s^{-1} (molecular orbital calculations, Klamt 1993)
    Hydrolysis:
    Biodegradation:
    Bioconcentration

Half-Lives in the Environment:
    Air: $t_{1/2}$ = 2.4–24 h, based on rate of disappearance for the reaction with hydroxyl radical (Danrall et al. 1976; Howard et al. 1991);
        estimated lifetime τ = 6 h under photochemical smog conditions in S.E. England (Brice & Derwent 1978; Darnall et al. 1976).
    Surface water: $t_{1/2}$ = 19 d in spring at 6–16°C, $t_{1/2}$ = 1.3 d in summer at 20–22°C, $t_{1/2}$ = 11 d in winter at 3–7°C when volatilization dominates and $t_{1/2}$ = 8.1 d with $HgCl_2$ in marine mesocosm experiments (Wakeham et al. 1983)

TABLE 3.1.1.10.1
Reported aqueous solubilities of n-propylbenzene at various temperatures

1.

Sanemasa et al. 1982		DeVoe et al. 1981		Owens et al. 1986		Shaw 1989b (IUPAC)	
vapor saturation-UV		generator column-HPLC		generator column-HPLC		recommended values	
t/°C	S/g·m⁻³	t/°C	S/g·m⁻³	t/°C	S/g·m⁻³	t/°C	S/g·m⁻³
15	46.1	23	51.32	10	53.73	15	47
25	51	15	51.21	15	52.29	25	55
35	55	20	51.09	20	54.33	35	55
45	64.1	25	51.93	25	52.25	45	64
		30	53.49	30	52.53		
				35	56.62		
				40	63.95		
				45	66.59		
				$\Delta H_{sol}/(kJ\,mol^{-1}) = 3.70$ at 25°C			

2.

Dohányosová et al. 2001		Sawamura et al. 2001	
vapor absorption-HPLC/UV		shake flask-UV	
t/°C	S/g·m⁻³	t/°C	S/g·m⁻³
0.5	46.5	0	60.2
5	45.8	5	57.8
15	44.8	10	55.96
25	48.2	15	54.96
35	52.2	20	54.5
45	62.4	25	54.96
55	74.4	30	56.2
		35	58.4
$\Delta_{sol}H/(kJ\,mol^{-1}) = 6.20$ at 25°C		40	61.3
		45	65.6
		50	90.6

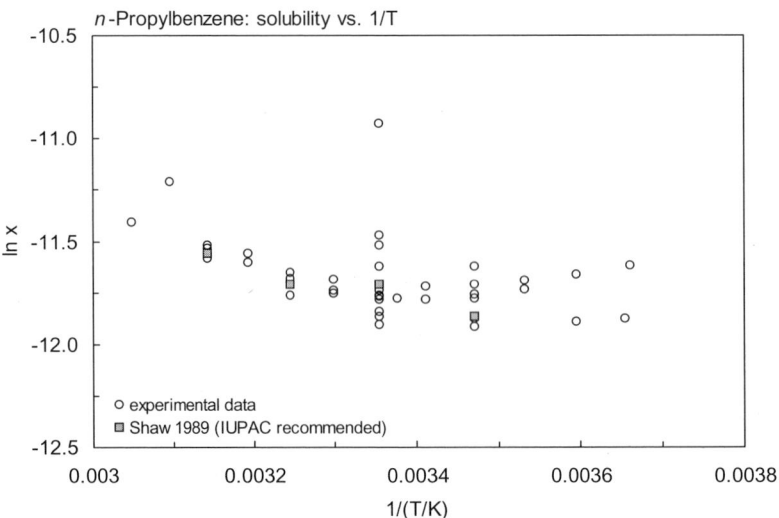

FIGURE 3.1.1.10.1 Logarithm of mole fraction solubility (ln x) versus reciprocal temperature for n-propylbenzene.

TABLE 3.1.1.10.2
Reported vapor pressures of *n*-propylbenzene at various temperatures and the coefficients for the vapor pressure equations

$\log P = A - B/(T/K)$ (1) $\quad\quad \ln P = A - B/(T/K)$ (1a)
$\log P = A - B/(C + t/°C)$ (2) $\quad\quad \ln P = A - B/(C + t/°C)$ (2a)
$\log P = A - B/(C + T/K)$ (3)
$\log P = A - B/(T/K) - C \cdot \log(T/K)$ (4)

Linder 1931		Willingham et al. 1945		Stull 1947		Forziati et al. 1949		Zwolinski & Wilhoit 71	
mercury manometer		ebulliometry		summary of lit. data		ebulliometry		selected values	
t/°C	P/Pa	t/°C	P/Pa	t/°C	P/Pa	t/°C	P/Pa	t/°C	P/Pa
−6.8	46.7	75.646	6353	6.3	133.3	75.818	6402	43.44	1333
−0.7	77.3	80.064	7654	31.3	666.6	80.181	7694	56.79	2666
3.6	113.2	83.909	8965	43.4	1333	83.993	8993	65.28	4000
13.9	260	87.383	10304	56.8	2666	87.457	10332	71.64	5333
		90.622	11696	71.6	5333	90.688	11723	76.784	6666
		94.993	13820	81.1	7999	95.049	13844	81.13	7999
		99.986	16621	94	13332	100.02	16636	88.264	10666
		105.046	19924	113.5	66664	105.085	19946	94.046	13332
		109.744	23450	137.7	53329	109.781	23557	105.142	19998
		116.032	28955	159.2	101325	116.06	23982	113.542	26664
		121.807	34898			128.794	43364	120.367	33331
		128.764	43323	mp/°C	−99.5	135.972	53702	126.163	39997
		135.942	53654			143.625	66799	135.737	53329
		143.598	66757			151.921	83753	143.551	66661
		151.908	83718			157.779	97649	150.205	79993
		157.76	97607			158.408	99239	156.031	93326
		158.389	99203			158.991	100730	157.116	95992
		158.972	100694			159.654	102465	158.178	98659
		159.625	102386			160.239	104003	159.217	101325
		160.202	103910						
						eq. 2	P/mmHg	eq. 2	P/mmHg
		eq. 2	P/mmHg			A	6.95094	A	6.95142
		A	6.95178			B	1490.963	B	1491.297
		B	1491.548			C	207.1	C	207.14
		C	207.171						
						bp/°C	159.218	bp/°C	159.217
		bp/°C	159.216					$\Delta H_V/(kJ\,mol^{-1}) =$	
								at 25°C	46.23
								at bp	38.24

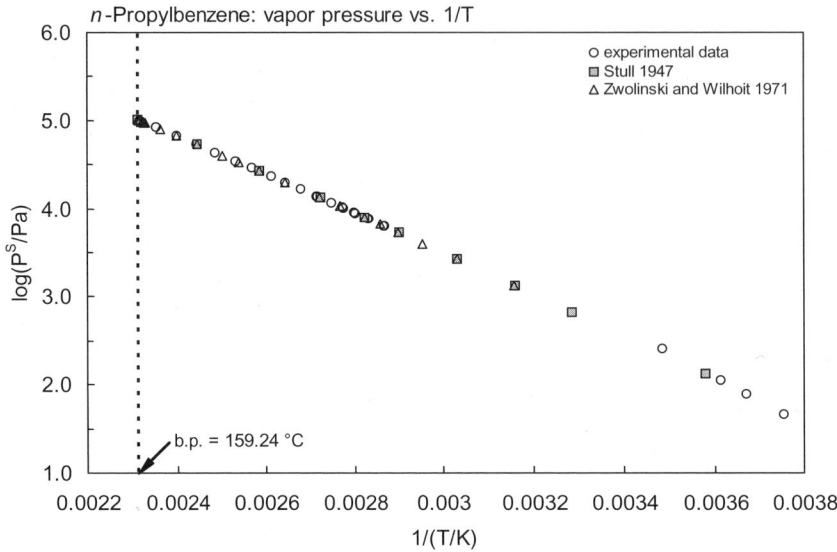

FIGURE 3.1.1.10.2 Logarithm of vapor pressure versus reciprocal temperature for *n*-propylbenzene.

TABLE 3.1.1.10.3
Reported Henry's law constants and octanol-water partition coefficients of *n*-propylbenzene at various temperatures

Henry's law constant						log K_{OW}	
Sanemasa et al. 1982		Ashworth et al. 1988		Perlinger et al. 1993		DeVoe et al. 1981	
vapor-liquid equilibrium		EPICS-GC		equilibrium headspace-GC		generator column-GC	
t/°C	H/(Pa m³/mol)	t/°C	H/(Pa m³/mol)	t/°C	H/(Pa m³/mol)	t/°C	log K_{OW}
15	594	10	576	10	441	25	3.691
25	1062	15	741	15	629	25	3.701
35	1818	20	893	20	848	10	3.705
45	2754	25	1094	25	1175	20	3.735
		30	1388	30	1550	25	3.72
						30	3.715
		ln H = A − B/(T/K)				35	3.682
		H/(atm m³/mol)					
		A	7.835			shake flask-GC	
		B	3681			ambient	3.734
						ambient	3.718
						ambient	3.711

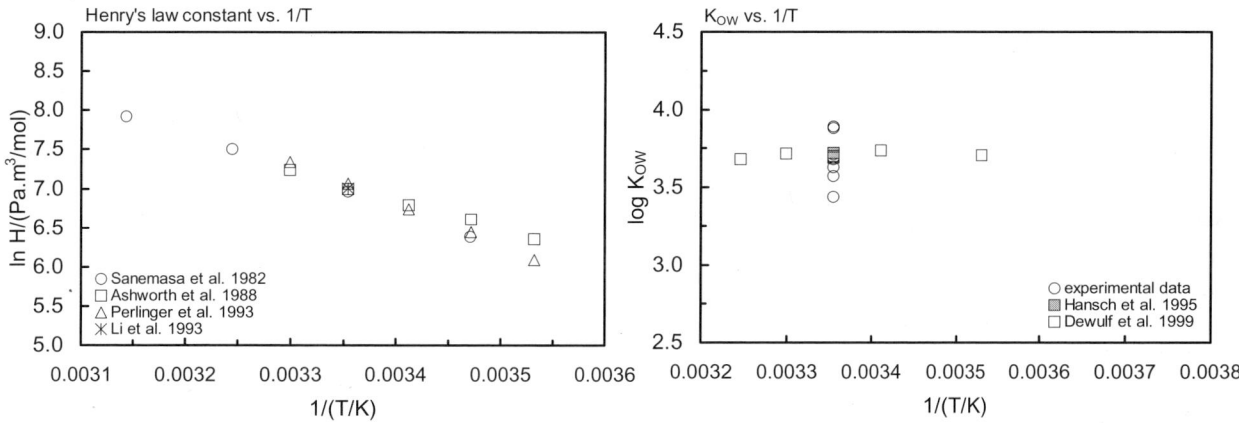

FIGURE 3.1.1.10.3 Logarithm of Henry's law constant and K_{OA} versus reciprocal temperature for *n*-propylbenzene.

3.1.1.11 Isopropylbenzene

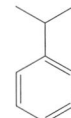

Common Name: Isopropylbenzene
Synonym: cumene, 2-phenylpropane, (1-methylethyl)benzene, cumol, *i*-propylbenzene
Chemical Name: isopropylbenzene
CAS Registry No: 98-82-8
Molecular Formula: C_9H_{12}, $C_6H_5CH(CH)_2$
Molecular Weight: 120.191
Melting Point (°C):
    –96.02   (Lide 2003)
Boiling Point (°C):
    152.41   (Lide 2003)
Density (g/cm³ at 20°C):
    0.8618   (Weast 1982–83)
Molar Volume (cm³/mol):
    139.5   (20°C, calculated-density)
    162.6   (calculated-Le Bas method at normal boiling point)
Enthalpy of Vaporization, ΔH_V (kJ/mol):
    45.141, 37.53   (25°C, bp, Riddick et al. 1986)
Enthalpy of Fusion ΔH_{fus} (kJ/mol):
    7.786   (Riddick et al. 1986)
    7.32   (Chickos et al. 1999)
Entropy of Fusion ΔS_{fus} (J/mol K):
    41.34, 46.3   (exptl., calculated-group additivity method, Chickos et al. 1999)
Fugacity Ratio at 25°C, F: 1.0

Water Solubility (g/m³ or mg/L at 25°C; or as indicated and reported temperature dependence equations. Additional data at other temperatures designated * are compiled at the end of this section):
    170   (shake flask-turbidimetric, Stearns et al. 1947)
    73   (shake flask-UV, Andrews & Keffer, 1950)
    80.5*   (shake flask-UV, measured range 25–80°C, Glew & Robertson 1956)
    53   (shake flask-GC, McAuliffe 1963)
    50   (shake flask-GC, McAuliffe 1966)
    50   (shake flask-GC, Hermann 1972)
    65.3   (shake flask-GC, Sutton & Calder 1975)
    48.3   (shake flask-GC, Price 1976)
    48.3   (shake flask-GC, Krzyzanowska & Szeliga 1978)
    61.5*   (vapor saturation-UV, measured range 15–45°C, Sanemasa et al. 1982)
    56*   (IUPAC recommended, temp range 15–80°C, Shaw 1989b)

Vapor Pressure (Pa at 25°C or as indicated and reported temperature dependence equations. Additional data at other temperatures designated * are compiled at the end of this section):
    300*   (13.7°C, mercury manometer, measured range –8.2–13.7°C, Linder 1931)
    log (P/mmHg) = –2175/(T/K) + 7.991 (isoteniscope method, temp range not specified, Kobe et al. 1941)
    6353*   (70.02°C, ebulliometry, measured range 70.02–153.367°C, Willingham et al. 1945)
    log (P/mmHg) = 6.92929 – 1455.811/(207.202 + t/°C); temp range 70.0–153.4°C (manometer, Antoine eq. from exptl. data, Willingham et al. 1945)
    666.6*   (31.3°C, summary of literature data, temp range 6.3–159.2°C, Stull 1947)
    6401*   (70.16°C, ebulliometry, measured range 70.16–153.4°C, Forziati et al. 1949)
    log (P/mmHg) = 6.93958 – 1462.717/(207.993 + t/°C); temp range 70.2–153.4°C (manometer, Antoine eq. from exptl. data, Forziati et al. 1949)

621   (extrapolated-Antoine eq., Dreisbach 1955; quoted, Hine & Mookerjee 1975)

log (P/mmHg) = 6.93666 − 1460.793/(207.777 + t/°C); temp range 60–200°C (Antoine eq. for liquid state, Dreisbach 1955)

609   (interpolated, Glew & Robertson 1956)

611*   (extrapolated-Antoine eq., Zwolinski & Wilhoit 1971)

log (P/mmHg) = 6.93666 − 1460.793/(207.777 + t/°C); temp range 38.29–180.67°C (Antoine eq., Zwolinski & Wilhoit 1971)

log (P/mmHg) = [−0.2185 × 10335.3/(T/K)] + 8.231760; temp range 2.9–152.4°C (Antoine eq., Weast 1972–73)

log (P/atm) = (1 − 425.438/T) × 10^(0.877964 − 7.34971 × 10^4·T + 6.06942 × 10^7·T^2); T in K, temp range 264.95–630.0 K (Cox vapor pressure eq., Chao et al. 1983)

610   (extrapolated-Antoine eq., Boublik et al. 1984)

log (P/kPa) = 6.0571 − 1457.715/(207.415 + t/°C); temp range 70.02–153.4°C (Antoine eq. from reported exptl. data of Forziati et al. 1949, Boublik et al. 1984)

log (P/kPa) = 6.06528 − 1464.366/(208.235 + t/°C); temp range 56.39–151.69°C (Antoine eq. from reported exptl. data of Dreyer et al. 1955, Boublik et al. 1984)

610   (selected lit., Riddick et al. 1986)

log (P/kPa) = 6.06588 − 1464.17/(208.207 + t/°C); temp range not specified (Antoine eq., Riddick et al. 1986)

log (P/mmHg) = 6.93666 − 1460.793/(207.78 + t/°C); temp range 39–181°C (Antoine eq., Dean 1985, 1992)

605   (extrapolated-Antoine eq., Stephenson & Malanowski 1987)

log (P_L/kPa) = 6.05949 − 1459.975/(−65.942 + T/K); temp range 339–433 K (liquid, Antoine eq., Stephenson & Malanowski 1987)

log (P/mmHg) = −0.9234 − 2.9558 × 10^3/(T/K) + 7.1685·log (T/K) − 2.5369 × 10^{-2}·(T/K) + 1.4858 × 10^{-6}·(T/K)^2; temp range 177–631 K (Yaws 1994)

log P/kPa = 6.06149 − 1460.793/[(T/K) − 65.373]; temp range 5–50°C (regression eq. from literature data, Shiu & Ma 2000)

Henry's Law Constant (Pa m^3/mol at 25°C and reported temperature dependence equations. Additional data at other temperatures designated * are compiled at the end of this section):

1469   (calculated-vapor-liquid equilibrium (VLE) data, Yaws et al. 1991)

1323*   (28°C, EPICS-GC, measured range 28–46.1°C, Hansen et al. 1993)

ln [H/(kPa·m^3/mol)] = −3269/(T/K) + 11.0; temp range 28–46.1°C (EPICS-GC, Hansen et al. 1993)

1126   (infinite activity coeff. γ^∞ in water determined by inert gas stripping-GC, Li et al. 1993)

960   (20°C, selected from literature experimentally measured data, Staudinger & Roberts 1996)

902   (20°C, selected from literature experimentally measured data, Staudinger & Roberts 2001)

log K_{AW} = 3.774 − 1256/(T/K) (van't Hoff eq. derived from literature data, Staudinger & Roberts 2001)

Octanol/Water Partition Coefficient, log K_{OW}:

3.43   (calculated-π substituent constant, Hansch et al. 1968)
3.66   (Leo et al. 1971; Hansch & Leo 1979)
3.63   (shake flask-GC, Chiou et al. 1977, 1982)
3.51   (headspace GC, Hutchinson et al. 1980)
3.52   (HPLC-k′ correlation, Hanai et al. 1981)
3.52   (HPLC-k′ correlation, D'Amboise & Hanai 1982)
3.40   (HPLC-k′ correlation, Miyake & Terada 1982)
3.89, 4.07   (RP-HPLC-k′ correlations, Sherblom & Eganhouse 1988)
3.66   (recommended, Sangster 1989, 1993)
3.82   (from measured activity coeff., Tse et al. 1994)
3.66   (recommended, Hansch et al. 1995)

Octanol/Air Partition Coefficient, log K_{OA} at 25°C:

3.98   (calculated-measured γ^∞ in pure octanol of Tewari et al. 1982, Abraham et al. 2001)

Bioconcentration Factor, log BCF:

1.55   (goldfish, Ogata et al. 1984)

Sorption Partition Coefficient, log K_{OC}:

Environmental Fate Rate Constants, k, or Half-Lives, $t_{1/2}$:

Volatilization: $t_{1/2}$ = 5.7 h from water depth of 1 m (calculated, Mackay & Leinonen 1975).

Photolysis:

Oxidation: rate constant k, for gas-phase second order rate constants, k_{OH} for reaction with OH radical, k_{NO3} with NO_3 radical and k_{O3} with O_3 or as indicated, *data at other temperatures see reference:

k_{OH} = (3.7 ± 0.8) × 10^9 L mol^{-1} s^{-1} at 305 ± 2 K (relative rate method, Lloyd et al. 1976)

k_{OH} = 3.7 × 10^9 L mol^{-1} s^{-1} with $t_{1/2}$ = 2.4–24 h (Darnall et al. 1976)

k_{OH} = (7.79 ± 0.40) × 10^{-12} cm^3 $molecule^{-1}$ s^{-1} at 200 torr He and 298 K (flash photolysis-resonance fluorescence, Ravishanakara et al. 1978)

k_{OH} = 4.6 × 10^{-1} M^{-1} s^{-1} with $t_{1/2}$ = 1.2 d (Mill 1982)

k = (11 ± 3) M^{-1} s^{-1} for the reaction with ozone in water using 100 mM *t*-BuOH as scavenger at pH 2.0 and 20–23°C (Hoigné & Bader 1983)

k_{OH} = 6.6 × 10^{-12} cm^3 $molecule^{-1}$ s^{-1} at room temp. (Atkinson 1985)

k_{OH} = 6.25 × 10^{-12} cm^3 $molecule^{-1}$ s^{-1} at room temp. (relative rate method, Ohta & Ohyama 1985)

k_{OH}(calc) = 7.1 × 10^{-12} cm^3 $molecule^{-1}$ s^{-1}, k_{OH}(obs.) = 6.6 × 10^{-12} cm^3 $molecule^{-1}$ s^{-1} at room temp. (SAR [structure-activity relationship], Atkinson 1987)

k_{OH} = 6.5 × 10^{-12} cm^3 $molecule^{-1}$ s^{-1} at 298 K (recommended, Atkinson 1989, 1990)

k_{OH}(calc) = 4.69 × 10^{-12} cm^3 $molecule^{-1}$ s^{-1} (molecular orbital calculations, Klamt 1993)

Hydrolysis:

Biodegradation:

Bioconcentration

Half-Lives in the Environment:

Air: $t_{1/2}$ = 2.4–24 h, based on rate of disappearance for the reaction with hydroxyl radical (Darnall et al. 1976; Howard et al. 1991);

estimated lifetime τ = 6 h under photochemical smog conditions in S.E. England (Brice & Derwent 1978) and (Darnall et al. 1976).

Surface Water: $t_{1/2}$ = 5.79 h, calculated half-life based on evaporative loss at 25°C and 1 m depth of water (Mackay & Leinonen 1975).

TABLE 3.1.1.11.1
Reported aqueous solubilities and Henry's law constants of isopropylbenzene at various temperatures

Aqueous solubility						Henry's law constant	
Glew & Robertson 1956		Sanemasa et al. 1982		Shaw 1989b (IUPAC)		Hansen et al. 1993	
shake flask-UV spec.		vapor equilibrium-UV spec.		recommended values		EPICS-GC	
t/°C	S/g·m^{-3}	t/°C	S/g·m^{-3}	t/°C	S/g·m^{-3}	t/°C	H/(Pa m^3/mol)
24.936	80.47	15	59.5	15	60	28	1323
29.984	82.91	25	61.5	25	56	36	1547
34.918	85.64	35	68.7	30	74	46.1	2422
39.958	89.79	45	77.5	40	82		
44.905	94.57			50	90	ln H = A – B/(T/K)	
49.902	100.4	ΔH_{sol}/(kJ mol^{-1}) = 5.10		60	120	H/(kPa m^3/mol)	
54.916	106.9	at 25°C		70	140	A	11
59.983	115			80	160	B	3299
65.165	124.4						
70.32	135.6						
75.097	147.3						
80.209	161.7						
ΔH_{sol}/(kJ mol^{-1}) = 3.57 at 25°C							

Mononuclear Aromatic Hydrocarbons

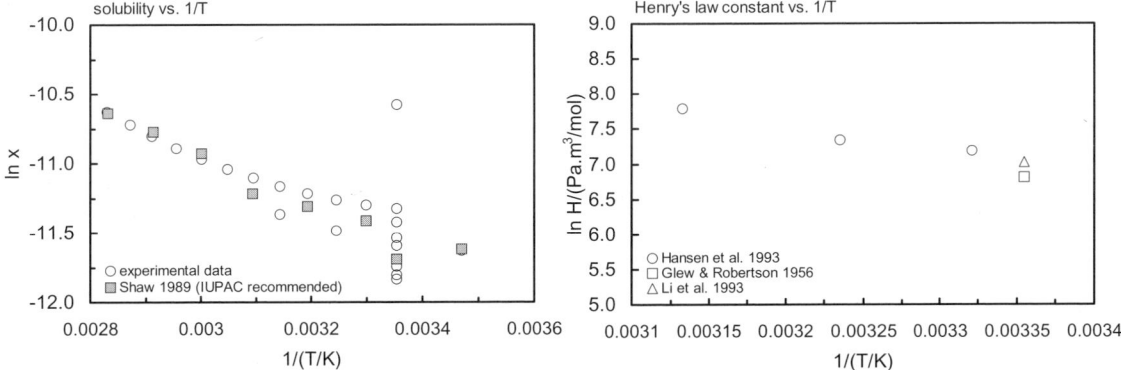

FIGURE 3.1.1.11.1 Logarithm of mole fraction solubility and Henry's law constant versus reciprocal temperature for isopropylbenzene.

TABLE 3.1.1.11.2
Reported vapor pressures of isopropylbenzene at various temperatures and the coefficients for the vapor pressure equations

$\log P = A - B/(T/K)$ (1) $\qquad$ $\ln P = A - B/(T/K)$ (1a)
$\log P = A - B/(C + t/°C)$ (2) $\qquad$ $\ln P = A - B/(C + t/°C)$ (2a)
$\log P = A - B/(C + T/K)$ (3)
$\log P = A - B/(T/K) - C \cdot \log (T/K)$ (4)

Linder 1931		Willingham et al. 1945		Forziati et al. 1949		Zwolinski & Wilhoit 1971	
Hg manometer		ebulliometry		ebulliometry		selected values	
t/°C	P/Pa	t/°C	P/Pa	t/°C	P/Pa	t/°C	P/Pa
−8.2	57.3	70.02	6353	70.16	6401	36.29	1333
1.3	124	74.365	7654	74.47	7693	51.43	2666
13.7	300	78.155	8965	78.23	8993	59.79	4000
		81.579	10304	81.64	10331	66.06	5333
		84.768	11696	89.11	11722	71.123	6666
		89.077	13820	94.01	13843	75.407	7999
		93.991	16621	99	16636	82.433	10666
		98.975	19924	103.64	19946	88.13	13332
Stull 1947		103.604	23450	109.82	23477	99.076	19998
summary of lit. data		109.802	28995	115.52	28980	107.346	26664
t/°C	P/Pa	115.495	34898	12.38	34928	114.076	33331
6.3	133	122.353	43323	129.46	43363	119.789	39997
31.3	666.6	129.433	53654	137.01	53701	129.23	53329
43.4	1333	136.983	66757	145.19	66799	136.937	66661
56.8	2666	145.176	83718	150.97	83752	143.501	79993
71.6	5333	150.956	97607	151.59	97649	149.249	93326
81.1	7999	151.576	99203	152.17	100730	150.319	95992
94	13332	152.152	100694	152.82	102465	151.367	98659
113.5	26664	152.798	102386	153.4	104003	152.392	101325
137.7	53329	153.367	103910				
159.2	101325			eq. 2	P/mmHg	eq. 2	P/mmHg
		eq. 2	P/mmHg	A	6.93958	A	6.93666
mp/°C	−99.5	A	6.92929	B	1462.717	B	1460.793
		B	1455.81	C	207.993	C	207.777
		C	207.202				
				bp/°C	152.392	bp/°C	152.392
		bp/°C	152.393			ΔH_V/(kJ mol^{-1}) =	
						at 25°C	45.14
						at bp	37.53

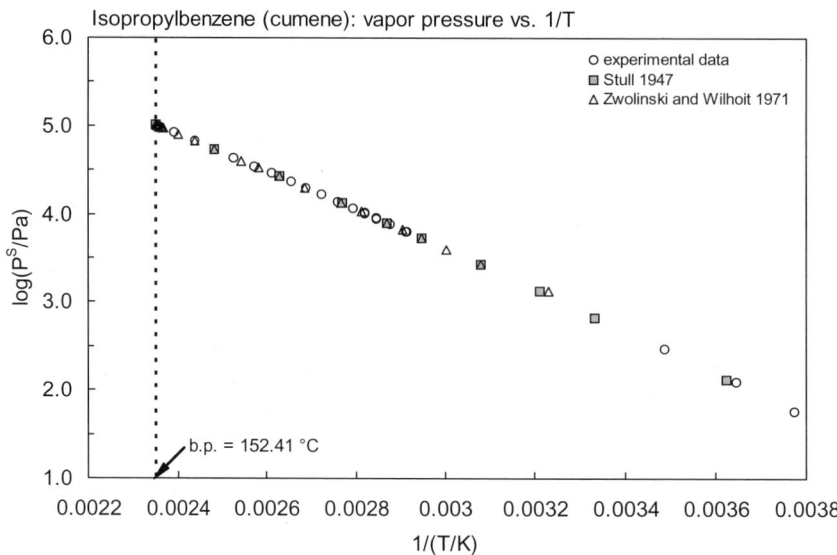

FIGURE 3.1.1.11.2 Logarithm of vapor pressure versus reciprocal temperature for isopropylbenzene.

3.1.1.12 1-Ethyl-2-methylbenzene (o-Ethyltoluene)

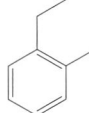

Common Name: 1-Ethyl-2-methylbenzene
Synonym: 2-ethyltoluene, o-ethyltoluene
Chemical Name: 1-ethyl-2-methylbenzene, 1-methyl-2-ethylbenzene
CAS Registry No: 611-14-3
Molecular Formula: C_9H_{12}, $C_6H_4CH_3C_2H_5$
Molecular Weight: 120.191
Melting Point (°C):
    –79.83   (Lide 2003)
Boiling Point (°C):
    165.2   (Weast 1982–83; Lide 2003)
Density (g/cm³ at 20°C):
    0.8807   (Weast 1982–83)
Molar Volume (cm³/mol):
    136.5   (20°C, calculated-density)
    162.6   (calculated-Le Bas method at normal boiling point)
Enthalpy of Fusion ΔH_{fus} (kJ/mol):
Entropy of Fusion ΔS_{fus} (J/mol K):
Fugacity Ratio at 25°C, F: 1.0

Water Solubility (g/m³ or mg/L at 25°C):
    40.0   (estimated from nomograph, Kabadi & Danner 1979)
    93.05   (shake flask-GC, Mackay & Shiu 1981)
    74.6   (generator column-HPLC/UV, Tewari et al. 1982c)

Vapor Pressure (Pa at 25°C or as indicated and reported temperature dependence equations. Additional data at other temperatures designated * are compiled at the end of this section):
    133.3*   (9.4°C, summary of literature data, Stull 1947)
    6417*   (81.146°C, ebulliometry, measured range 81.146–157.825°C, Forziati et al. 1949)
    log (P/mmHg) = 7.00314 – 1535.374/(207.300 + t/°C); temp range 81.1–166.2°C (manometer, Antoine eq. from exptl. data, Forziati et al. 1949)
    336   (extrapolated-Antoine eq., Dreisbach 1955)
    log (P/mmHg) = 7.00314 – 1535.374/(207.3 + t/°C); temp range 70–215°C (Antoine eq. for liquid state, Dreisbach 1955)
    330*   (extrapolated-Antoine eq., Zwolinski & Wilhoit 1971)
    log (P/mmHg) = 7.00314 – 1535.374/(207.300 + t/°C); temp range 48.46–193.89°C (Antoine eq., Zwolinski & Wilhoit 1971)
    log (P/mmHg) = [–0.2185 × 10448.8/(T/K)] + 8.141032; temp range 9.4–165.2°C (Antoine eq., Weast 1972–73)
    log (P/atm) = (1 – 438.357/T) × 10^(0.863837 – 6.34917 × 10⁻⁴·T + 5.19164 × 10⁻⁷·T²); T in K, temp range 285.0–645.0 K (Cox vapor pressure eq., Chao et al. 1983)
    328   (extrapolated-Antoine eq., Boublik et al. 1984)
    log (P/kPa) = 6.11997 – 1529.684/(206.648 + t/°C); temp range 81.146–166.2°C (Antoine eq. from reported exptl. data Forziati et al. 1949, Boublik et al. 1984)
    330   (extrapolated-Antoine eq, Dean 1985, 1992)
    log (P/mmHg) = 7.00314 – 1535.374/(207.30 + t/°C); temp range 48–194°C (Antoine eq., Dean 1985, 1992)
    log (P_L/kPa) = 6.1129 – 1532.449/(–66.123 + T/K); temp range 353–443 K (liquid, Antoine eq., Stephenson & Malanowski 1987)
    log (P/mmHg) = 15.1142 – 2.9821 × 10³/(T/K) – 1.2619·log (T/K) – 6.3248 × 10⁻³·(T/K) + 3.5155 × 10⁻⁶·(T/K)²; temp range 192–651 K (Yaws 1994)

Henry's Law Constant (Pa m³/mol at 25°C or as indicated and reported temperature dependence equations. Additional data at other temperatures designated * are compiled at the end of this section):

    565   (EPICS-GC/FID, Ashworth et al. 1988)

    ln [H/(atm m³/mol)] = 5.557 – 3179/(T/K); temp range 10–30°C (EPICS measurements, Ashworth et al. 1988)

    426   (calculated-vapor-liquid equilibrium (VLE) data, Yaws et al. 1991)

    512   (20°C, selected from literature experimentally measured data, Staudinger & Roberts 1996)

Octanol/Water Partition Coefficient, log K_{OW}:

    3.63   (headspace GC, Hutchinson et al. 1980)

    3.53   (generator column-HPLC/UV, DeVoe et al.1981; Tewari et al. 1982a)

    3.78, 3.95   (RP-HPLC-k′ correlations, Sherblom & Eganhouse 1988)

    3.53   (recommended, Sangster 1989, 1993)

    3.53   (recommended, Hansch et al. 1995)

Bioconcentration Factor, log BCF:

Sorption Partition Coefficient, log K_{OC}:

Environmental Fate Rate Constants, k, or Half-Lives, $t_{1/2}$:

    Volatilization:

    Photolysis:

    Oxidation: rate constant k, for gas-phase second order rate constants, k_{OH} for reaction with OH radical, k_{NO3} with NO_3 radical and k_{O3} with O_3 or as indicated, *data at other temperatures see reference:

        k_{OH} = (8.2 ± 1.6) × 10⁹ L mol⁻¹ s⁻¹ at 305 ± 2 K (relative rate method, Lloyd et al. 1976)

        k_{OH} = 8.2 × 10⁹ L mol⁻¹ s⁻¹ with half-life of 0.24–2.4 h (Darnall et al. 1976)

        k_{OH} = 12.0 × 10⁻¹² cm³ molecule⁻¹ s⁻¹ at room temp. (Atkinson 1985)

        k_{OH} = 12.4 × 10⁻¹² cm³ molecule⁻¹ s⁻¹ at 298 K (relative rate method, Ohta & Ohyama 1985)

        k_{OH} = 12.3 × 10⁻¹² cm³ molecule⁻¹ s⁻¹ at 298 K (recommended, Atkinson 1989, 1990)

        k_{OH}(calc) = 17.7 × 10⁻¹² cm³ molecule⁻¹ s⁻¹ (molecular orbital calculations, Klamt 1993)

    Hydrolysis:

    Biodegradation:

    Biotransformation:

    Bioconcentration

Half-Lives in the Environment:

    Air: $t_{1/2}$ = 0.24–2.4 h, based on rate of disappearance for the reaction with hydroxyl radical (Darnall et al. 1976; Howard et al. 1991);

        summer daylight lifetime τ = 11 h due to reaction with OH radical (Altshuller 1991).

    Surface water: $t_{1/2}$ = 0.5 d in surface water in case of a first order reduction process may be assumed (estimated, Zoeteman et al. 1980).

TABLE 3.1.1.12.1
Reported vapor pressures of *o*-ethyltoluene at various temperatures and the coefficients for the vapor pressure equations

$$\log P = A - B/(T/K) \quad (1)$$
$$\log P = A - B/(C + t/°C) \quad (2)$$
$$\log P = A - B/(C + T/K) \quad (3)$$
$$\log P = A - B/(T/K) - C \cdot \log(T/K) \quad (4)$$
$$\ln P = A - B/(T/K) \quad (1a)$$
$$\ln P = A - B/(C + t/°C) \quad (2a)$$

Stull 1947		Forziati et al. 1949		Zwolinski & Wilhoit 1971			
summary of literature data		ebulliometry		selected values			
t/°C	P/Pa	t/°C	P/Pa	t/°C	P/Pa		
9.4	133.3	81.146	6417	48.46	1333	eq. 2	P/mmHg
34.8	666.6	85.618	7697	61.96	2666	A	7.00314
47.6	1333	89.448	8993	70.54	4000	B	1535.374
61.2	2666	92.949	10323	76.97	5333	C	207.3
76.4	5333	96.2	11720	82.165	6666	bp/°C	165.153
86	7999	100.584	13840	86.552	7999	$\Delta H_V/(kJ\ mol^{-1})$ =	
99	13332	105.598	16641	93.751	10666	at 25°C	47.7
119	26664	110.711	19944	99.582	13332	at bp	38.87
141.4	53329	115.436	23473	110.777	19998		
165.1	101325	121.762	28978	119.29	26664		
		127.574	34920	126.1	33331		
mp/°C	–104.7	134.57	43351	131.933	39997		
		141.792	53693	141.563	53329		
		149.582	66792	149.418	66661		
		157.825	83749	156.103	79993		
				161.954	93326		
		eq. 2	P/mmHg	164.044	95992		
		A	7.00314	164.11	98659		
		B	1535.374	165.534	101325		
		C	207.3				
		bp/°C	165.153				

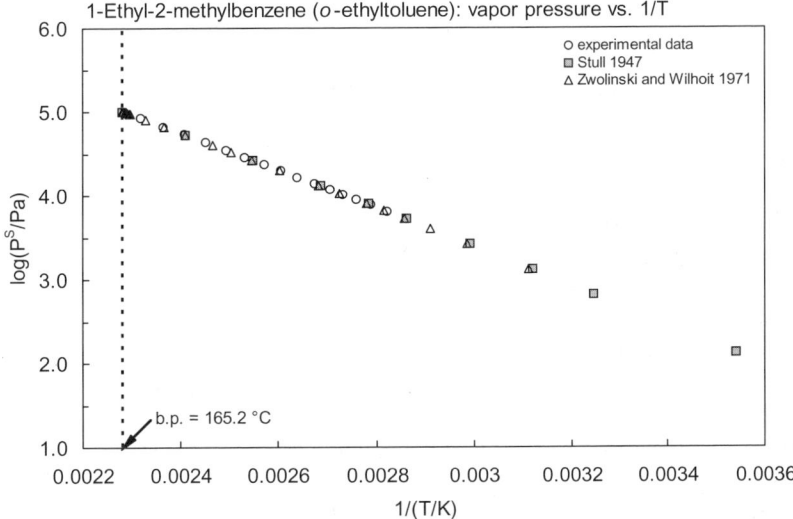

FIGURE 3.1.1.12.1 Logarithm of vapor pressure versus reciprocal temperature for 1-ethyl-2-methylbenzene (*o*-ethyltoluene).

3.1.1.13 1-Ethyl-3-methylbenzene (*m*-Ethyltoluene)

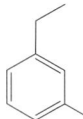

Common Name: 1-Ethyl-3-methylbenzene
Synonym: 3-ethyltoluene, *m*-ethyl toluene
Chemical Name: 1-ethyl-3-methylbenzene, 1-methyl-3-ethylbenzene
CAS Registry No: 620-14-4
Molecular Formula: C_9H_{12}, $C_6H_4CH_3C_2H_5$
Molecular Weight: 120.191
Melting Point (°C):
   – 95.6   (Lide 2003)
Boiling Point (°C):
   161.3   (Forziati et al. 1949, Weast 1982–83; Lide 2003)
Density (g/cm³):
   0.8645   (20°C, Weast 1982–83)
Molar Volume (cm³/mol):
   139.0   (20°C, calculated-density)
   162.6   (calculated-Le Bas method at normal boiling point)
Enthalpy of Fusion, ΔH_{fus} (kJ/mol):
Entropy of Fusion ΔS_{fus} (J/mol K):
Fugacity Ratio at 25°C, F: 1.0

Water Solubility (g/m³ or mg/L at 25°C):

Vapor Pressure (Pa at 25°C or as indicated and reported temperature dependence equations. Additional data at other temperatures designated * are compiled at the end of this section):
   666.6*   (32.3°C, summary of literature data, temp range 7.2–161.3°C, Stull 1947)
   3066*   (62.1°C, mercury manometer, measured range 62.1–160.3°C, Buck et al. 1949)
   6417*   (78.105°C, ebulliometry, measured range 78.105–154.053°C, Forziati et al. 1949)
   $\log (P/mmHg) = 7.01582 - 1529.784/(208.509 + t/°C)$; temp range 78.1–162.3°C (manometer, Antoine eq. from exptl. data, Forziati et al. 1949)
   399   (extrapolated-Antoine eq., Dreisbach 1955)
   $\log (P/mmHg) = 7.01582 - 1529.184/(208.509 + t/°C)$; temp range 65–210°C (Antoine eq. for liquid state, Dreisbach 1955)
   391*   (extrapolated-Antoine eq., Zwolinski & Wilhoit 1971)
   $\log (P/mmHg) = 7.01582 - 1529.184/(208.509 + t/°C)$; temp range 45.68–189.74°C (Antoine eq., Zwolinski & Wilhoit 1971)
   $\log (P/mmHg) = [-0.2185 \times 10416.6/(T/K)] + 8.152199$; temp range 7.2–161.3°C (Antoine eq., Weast 1972–73)
   $\log (P/atm) = (1 - 434.538/T) \times 10^{\wedge}(0.861399 - 6.30303 \times 10^4 \cdot T + 5.19848 \times 10^7 \cdot T^2)$; T in K, temp range 280.0–635.0 K (Cox vapor pressure eq., Chao et al. 1983)
   394   (extrapolated-Antoine eq., Boublik et al. 1984)
   $\log (P/kPa) = 6.12947 - 1531.584/(209.417 + t/°C)$; temp range 78.3–163°C (Antoine eq. from reported exptl. data, Boublik et al. 1984)
   391   (extrapolated-Antoine eq., Dean 1985, 1992)
   $\log (P/mmHg) = 7.01582 - 1529.184/(208.51 + t/°C)$; temp range 46–190°C (Antoine eq., Dean 1985, 1992)
   391   (extrapolated-Antoine eq., Stephenson & Malanowski 1987)
   $\log (P_L/kPa) = 6.13801 - 1527.983/(-64.715 + T/K)$; temp range 348–438 K (Antoine eq., Stephenson & Malanowski 1987)
   $\log (P/mmHg) = 39.8909 - 3.6042 \times 10^3/(T/K) - 11.466 \cdot \log (T/K) + 3.5274 \times 10^{-2} \cdot (T/K) + 7.3492 \times 10^{-14} \cdot (T/K)^2$, temp range 178–637 K (Yaws 1994)

Henry's Law Constant (Pa·m³/mol):

Octanol/Water Partition Coefficient, log K_{OW}:
- 3.88, 4.07   (RP-HPLC-k' correlations, Sherblom & Eganhouse 1988)
- 3.98   (lit. average value, Sangster 1993)
- 3.88   (quoted from Sherblom & Eganhouse 1988; Hansch et al. 1995)

Octanol/Air Partition Coefficient, log K_{OA}:

Bioconcentration Factor, log BCF or log K_B:

Sorption Partition Coefficient, log K_{OC}:
- 2.42   (aquifer material with f_{OC} of 0.006 and measured partition coeff. K_P = 1.58 mL/g., Abdul et al. 1990)

Environmental Fate Rate Constants, k, and Half-Lives, $t_{½}$:
  Volatilization:
  Photolysis:
  Oxidation: rate constant k, for gas-phase second order rate constants, k_{OH} for reaction with OH radical, k_{NO3} with NO_3 radical and k_{O3} with O_3 or as indicated, *data at other temperatures see reference:
  k_{OH} = (11.7 ± 2.3) × 10⁹ L mol⁻¹ s⁻¹ at 305 ± K (relative rate method, Lloyd et al. 1976)
  k_{OH} = 11.7 × 10⁹ L mol⁻¹ s⁻¹ with estimated $t_{½}$ ~ 0.24–2.4 h (Darnall et al. 1976)
  k_{OH} = (21.3 ± 1.1) × 10⁻¹² cm³ molecule⁻¹ s⁻¹ at room temp. (relative rate method, Ohta & Ohyama 1985; Atkinson 1989)
  k_{OH} = 1.92 × 10⁻¹¹ cm³ molecule⁻¹ s⁻¹ at 298 K (recommended, Atkinson 1989, 1990)
  k_{OH}(calc) = 20.93 × 10⁻¹² cm³ molecule⁻¹ s⁻¹ (molecular orbital calculations, Klamt 1993)
  Hydrolysis:
  Biodegradation:
  Biotransformation:
  Bioconcentration and Uptake and Elimination Rate Constants (k_1 and k_2):

Half-Lives in the Environment:
  Air: $t_{½}$ = 0.24–2.4 h, based on rate of disappearance for the reaction with hydroxyl radical (Darnall et al 1976; Howard et al. 1991);
  summer daylight lifetime τ = 7.2 h due to reaction with OH radical (Altshuller 1991).

TABLE 3.1.1.13.1
Reported vapor pressures of *m*-ethyltoluene at various temperatures and the coefficients for the vapor pressure equations

$\log P = A - B/(T/K)$ (1) $\ln P = A - B/(T/K)$ (1a)

$\log P = A - B/(C + t/°C)$ (2) $\ln P = A - B/(C + t/°C)$ (2a)

$\log P = A - B/(C + T/K)$ (3)

$\log P = A - B/(T/K) - C \cdot \log (T/K)$ (4)

Stull 1947		Forziati et al. 1949		Buck et al. 1949		Zwolinski & Wilhoit 1971	
summary of literature data		ebulliometry		mercury manometer		selected values	
t/°C	P/Pa	t/°C	P/Pa	t/°C	P/Pa	t/°C	P/Pa
7.2	133.3	78.105	6417	62.1	3066	45.68	1333
32.3	666.6	82.525	7697	74	5386	59.07	2666
44.7	1333	86.293	8993	77.9	6293	67.58	4000
58.2	2666	89.793	10328	85.4	8693	73.95	5333
73.3	5333	93.022	11720	91.3	10879	79.102	6666
82.9	7999	97.368	13840	95.1	12772	83.45	7999
95.9	13332	102.326	16641	102.1	16772	90.584	10666
115.5	26664	107.383	19944	115.3	26398	96.363	13332
137.8	53329	112.074	23473	122.5	34264	107.456	19998
161.3	101325	118.338	28978	129.8	42397	115.829	26664
		124.082	34920	140.7	58262	122.635	33331
mp/°C	–104.7	131.027	43351	148.5	72527	128.412	39997
		138.178	53693	160.3	99592	137.949	53329
		145.795	66792			145.727	66661
		154.053	83749	eq. 4	P/mmHg	152.346	79993
				A	25.08	158.318	93326
		eq. 2	P/mmHg	B	3155	159.217	95992
		A	7.01582	C	5.663	160.272	98659
		B	1529.184			161.305	101325
		C	208.509				
						eq. 2	P/mmHg
		bp/°C	161.305			A	7.01582
						B	1529.184
						C	208.509
						bp/°C	161.305
						ΔH_V/(kJ mol^{-1}) =	
						at 25°C	46.9
						at bp	38.53

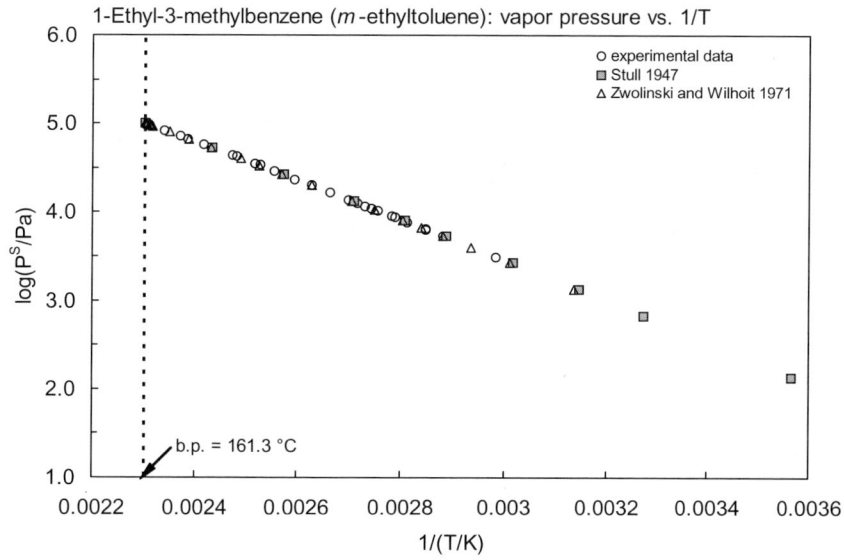

FIGURE 3.1.1.13.1 Logarithm of vapor pressure versus reciprocal temperature for 1-ethyl-3-methylbenzene (*m*-ethyltoluene).

3.1.1.14 1-Ethyl-4-methylbenzene (*p*-Ethyltoluene)

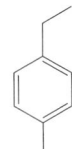

Common Name: 1-Ethyl-4-methylbenzene
Synonym: 4-ethyltoluene, *p*-ethyl toluene
Chemical Name: 1-ethyl-4-methylbenzene, 1-methyl-4-ethylbenzene
CAS Registry No: 622-96-8
Molecular Formula: C_9H_{12}, $C_6H_4CH_3C_2H_5$
Molecular Weight: 120.191
Melting Point (°C):
    –62.35   (Lide 2003)
Boiling Point (°C):
    162   (Weast 1982–83; Lide 2003)
Density (g/cm³ at 20°C):
    0.8614   (Weast 1982–83)
Molar Volume (cm³/mol):
    139.5   (20°C, calculated -density, Stephenson & Malanowski 1987)
    162.6   (calculated-Le Bas method at normal boiling point)
Enthalpy of Fusion ΔH_{fus} (kJ/mol):
Entropy of Fusion ΔS_{fus} (J/mol K):
Fugacity Ratio at 25°C, F: 1.0

Water Solubility (g/m³ or mg/L at 25°C):
    40.0   (estimated from nomograph, Kabadi & Danner 1979)
    94.85   (shake flask-GC, Mackay & Shiu 1981)

Vapor Pressure (Pa at 25°C or as indicated and reported temperature dependence equations. Additional data at other temperatures designated * are compiled at the end of this section):
    666.6*   (32.7°C, summary of literature data, temp range 7.6–162°C, Stull 1947)
    3840*   (66.8°C, mercury manometer, measured range 66.8–161.1°C, Buck et al. 1949)
    6417*   (78.396°C, ebulliometry, measured range 78.396–154.684°C, Forziati et al. 1949)
    $\log (P/\text{mmHg}) = 6.99801 - 1527.113/(208.921 + t/°C)$; temp range 78.3–163.0°C (manometer, Antoine eq. from exptl. data, Forziati et al. 1949)
    402   (extrapolated-Antoine eq., Dreisbach 1955)
    $\log (P/\text{mmHg}) = 6.99802 - 1527.113/(208.921 + t/°C)$; temp range 65–210°C (Antoine eq. for liquid state, Dreisbach 1955)
    393*   (extrapolated-Antoine eq., Zwolinski & Wilhoit 1971)
    $\log (P/\text{mmHg}) = 6.99802 - 1527.113/(208.921 + t/°C)$; temp range 45.68–190.64°C (Antoine eq., Zwolinski & Wilhoit 1971)
    $\log (P/\text{mmHg}) = [-0.2185 \times 10461.1/(T/K)] + 8.175267$; temp range 7.6–162°C (Antoine eq., Weast 1972–73)
    $\log (P/\text{atm}) = (1 - 345.228/T) \times 10^{\wedge}(0.856105 - 6.18307 \times 10^4 \cdot T + 5.08568 \times 10^7 \cdot T^2)$; T in K, temp range 280.0–635.0 K (Cox vapor pressure eq., Chao et al. 1983)
    393   (extrapolated-Antoine eq., Dean 1985, 1992)
    $\log (P/\text{mmHg}) = 6.99802 - 1527.113/(208.92 + t/°C)$; temp range 46–191°C (Antoine eq., Dean 1985, 1992)
    $\log (P_L/\text{kPa}) = 6.11098 - 1519.486/(-65.035 + T/K)$; temp range 349–442 K (liquid, Antoine eq., Stephenson & Malanowski 1987)
    $\log (P/\text{mmHg}) = 46.9026 - 3{,}8382 \times 10^3/(T/K) - 14.154 \cdot \log (T/K) + 4.9305 \times 10^{-3} \cdot (T/K) - 1.3901 \times 10^{-13} \cdot (T/K)^2$; temp range 211–640 K (Yaws 1994)

Henry's Law Constant (Pa m^3/mol at 25°C):
- 498 (calculated-P/C, Mackay & Shiu 1981, Eastcott et al. 1988)
- 498 (calculated-vapor-liquid equilibrium (VLE) data, Yaws et al. 1991)

Octanol/Water Partition Coefficient, log K_{OW}:
- 3.63 (headspace GC, Hutchinson et al. 1980)
- 3.90, 4.09 (RP-HPLC-k' correlations, Sherblom & Eganhouse 1988)
- 3.63 (recommended, Sangster 1989, 1993)
- 3.90 (quoted from Sherblom & Eganhouse 1988, Hansch et al. 1995)

Bioconcentration Factor, log BCF:

Sorption Partition Coefficient, log K_{OC}:

Environmental Fate Rate Constants, k, or Half-Lives, $t_{1/2}$:
- Volatilization:
- Photolysis:
- Oxidation: rate constant k, for gas-phase second order rate constants, k_{OH} for reaction with OH radical, k_{NO3} with NO$_3$ radical and k_{O3} with O$_3$ or as indicated, *data at other temperatures see reference:
  - $k_{OH} = (7.8 \pm 1.6) \times 10^9$ L mol^{-1} s^{-1} at 305 ± 2 K (relative rate method, Lloyd et al. 1976)
  - $k_{OH} = 7.8 \times 10^9$ L mol^{-1} s^{-1} with $t_{1/2}$ = 0.24–2.4 h (Darnall et al. 1976)
  - $k_{OH} = 12.9 \times 10^{-12}$ cm^3 molecule^{-1} s^{-1} and residence time of 0.9 d, loss of 67% in one day or 12 sunlit hours at 300 K in urban environments (Singh et al. 1981)
  - $k_{OH} = 11.3 \times 10^{-12}$ cm^3 molecule^{-1} s^{-1} at room temp. (Atkinson 1985)
  - $k_{OH} = 12.8 \times 10^{-12}$ cm^3 molecule^{-1} s^{-1} at 298 K (relative rate method, Ohta & Ohyama 1985)
  - $k_{OH} = 12.1 \times 10^{-12}$ cm^3 molecule^{-1} s^{-1} at 298 K (recommended, Atkinson 1989, 1990)
  - k_{OH}(calc) = 21.6 × 10^{-12} cm^3 molecule^{-1} s^{-1} (molecular orbital calculations, Klamt 1993)

Half-Lives in the Environment:
- Air: $t_{1/2}$ = 0.24–2.4 h, based on rate of disappearance for the reaction with hydroxyl radical (Darnall et al. 1976) residence time of 0.9 d, loss of 67% in 1 d or 12 sunlit hours at 300 K in urban environments (Singh et al. 1981)
summer daylight lifetime τ = 11 h due to reaction with OH radical (Altshuller 1991).

TABLE 3.1.1.14.1
Reported vapor pressures of *p*-ethyltoluene at various temperatures and the coefficients for the vapor pressure equations

$$\log P = A - B/(T/K) \quad (1)$$
$$\log P = A - B/(C + t/°C) \quad (2)$$
$$\log P = A - B/(C + T/K) \quad (3)$$
$$\log P = A - B/(T/K) - C \cdot \log(T/K) \quad (4)$$
$$\ln P = A - B/(T/K) \quad (1a)$$
$$\ln P = A - B/(C + t/°C) \quad (2a)$$

Stull 1947		Forziati et al. 1949		Buck et al. 1949		Zwolinski & Wilhoit 1971	
summary of literature data		ebulliometry		mercury manometer		selected values	
t/°C	P/Pa	t/°C	P/Pa	t/°C	P/Pa	t/°C	P/Pa
7.6	133.3	78.396	6417	66.8	3840	45.68	1333
32.7	666.6	82.701	7657	77.4	6106	59.14	2666
44.9	1333	86.523	8993	79.4	6586	67.68	4000
58.5	2666	89.988	10328	83.5	7999	74.09	5333
73.6	5333	93.252	11720	87.9	9466	79.265	6666
83.2	7999	97.63	13840	90.2	10399	83.637	7999
96.3	13332	102.619	16641	94	11999	90.811	10666
116.1	26664	107.71	19944	105.8	18638	96.623	13332
136.4	53329	112.422	23474	122.7	33197	107.781	19998
162	101325	118.727	28978	131.7	44130	116.205	26664
		131.499	43351	136.9	51329	123.054	33331
mp/°C	–	139.701	53693	144.2	63595	128.869	39997
		146.368	66792	152.5	79727	138.469	53329
		154.684	83750	161.1	100792	146.3	66661
						152.965	79993
		eq. 2	P/mmHg	eq. 4	P/mmHg	158.799	93326
		A	6.99801	A	21.27	159.885	95992
		B	1527.113	B	2939	160.948	98659
		C	208.921	C	4.406	161.989	101325
		bp/°C	161.989			eq. 2	P/mmHg
						A	6.99802
						B	1527.113
						C	208.921
						bp/°C	161.989
						$\Delta H_V/(kJ\ mol^{-1}) =$	
						at 25°C	46.61
						at bp	38.41

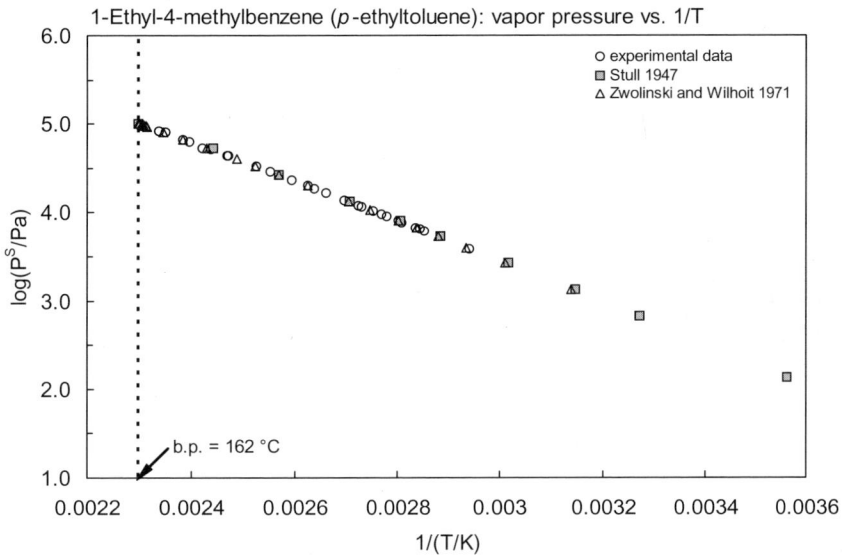

FIGURE 3.1.1.14.1 Logarithm of vapor pressure versus reciprocal temperature for 1-ethyl-4-methylbenzene (*p*-ethyltoluene).

3.1.1.15 1-Isopropyl-4-methylbenzene (*p*-Cymene)

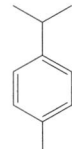

Common Name: 1-Isopropyl-4-methylbenzene
Synonym: *p*-cymene, *p*-isopropyltoluene, 1-methyl-4-isopropylbenzene
Chemical Name: 1-isopropyl-4-methylbenzene
CAS Registry No: 99-87-6
Molecular Formula: $C_{10}H_{14}$, $CH_3C_6H_4CH(CH_3)_2$
Molecular Weight: 134.218
Melting Point (°C):
    –67.94   (Lide 2003)
Boiling Point (°C):
    177.1   (Weast 1982–83; Lide 2003)
Density (g/cm^3 at 20°C):
    0.8573   (Weast 1982–83)
Molar Volume (cm^3/mol):
    156.6   (20°C, calculated-density)
    184.8   (calculated-Le Bas method at normal boiling point)
Enthalpy of Fusion ΔH_{fus} (kJ/mol):
    9.66   (Dreisbach 1955)
    9.67   (Chickos et al. 1999)
Entropy of Fusion ΔS_{fus} (J/mol K):
    47.33, 46.8   (exptl., calculated-group additivity method, Chickos et al. 1999)
Fugacity Ratio at 25°C, F: 1.0

Water Solubility (g/m^3 or mg/L at 25°C):
    34.15   (residue volume method, Booth & Everson 1948)
    23.35   (shake flask-LSC, Banerjee et al. 1980)
    50.7 ± 2.3   (shake flask-HPLC/UV, Lun et al. 1997)

Vapor Pressure (Pa at 25°C or as indicated and reported temperature dependence equations. Additional data at other temperatures designated * are compiled at the end of this section):
    90.7*   (13.3°C, mercury manometer, Linder 1931)
    log (P/mmHg) = 8.063 – 10670/(T/K) (isoteniscope method, measured range not specified, Kobe et al. 1941)
    133.3*   (19°C, summary of literature data, Stull 1947)
    212   (extrapolated-Antoine eq., Dreisbach 1955)
    log (P/mmHg) = 6.9260 – 1538.00/(203.10 + t/°C); temp range 80–215°C (Antoine eq. for liquid state, Dreisbach 1955)
    12026*   (107.04°C, ebulliometry, measured range 107.04–178.42°C, McDonald et al. 1959)
    204*   (extrapolated-Antoine eq., Zwolinski & Wilhoit 1971)
    log (P/mmHg) = 6.9237 – 1537.06/(203.05 + t/°C); temp range 56.4–207.1°C (Antoine eq., Zwolinski & Wilhoit 1971)
    log (P/atm) = (1 – 450.311/T) × 10^(0.875129 – 6.86627 × 10^4·T + 5.61507 × 10^7·T^2); T in K, temp range 290.0–650.0 K (Cox vapor pressure eq., Chao et al. 1983)
    log (P/mmHg) = 7.05074 – 1608.91/(208.72 + t/°C); temp range 107–178°C (Antoine eq., Dean 1985, 1992)
    194   (extrapolated-Antoine eq., Stephenson & Malanowski 1987)
    log (P_L/kPa) = 6.16214 – 1599.29/(–65.492 + T/K); temp range 380–452 K (liquid, Antoine eq., Stephenson & Malanowski 1987)
    log (P/mmHg) = –5.5137 – 3.0256 × 10^3/(T/K) + 8.9840·log (T/K) – 2.5597 × 10^{-2}·(T/K) + 1.3823 × 10^{-5}·(T/K)^2, temp range 205–653 K (Yaws 1994)

Mononuclear Aromatic Hydrocarbons

Henry's Law Constant (Pa m^3/mol at 25°C):
- 800 (calculated-P/C, Mackay & Shiu 1981)
- 942 (computed-expert system SPARC, Kollig 1995)

Octanol/Water Partition Coefficient, log K_{OW}:
- 4.10 (shake flask-LSC, Banerjee et al. 1980)
- 4.14 (calculated-UNIFAC activity coeff., Arbuckle 1983)
- 3.45 (calculated-UNIFAC activity coeff., Banerjee & Howard 1988)
- 4.10 (recommended, Sangster 1989)
- 4.10 (recommended, Hansch et al. 1995)
- 4.0 (computed-expert system SPARC, Kollig 1995)

Bioconcentration Factor, log BCF:

Sorption Partition Coefficient, log K_{OC}:
- 3.70 (computed-K_{OW}, Kollig 1995)

Environmental Fate Rate Constants, k, or Half-Lives, $t_{1/2}$:
  Volatilization:
  Photolysis: rate constant of 1.68×10^{-2} h^{-1} with H_2O_2 under photolysis at 25°C in F-113 solution and with HO· in the gas (Dilling et al. 1988).
  Oxidation: rate constant k, for gas-phase second order rate constants, k_{OH} for reaction with OH radical, k_{NO3} with NO$_3$ radical and k_{O3} with O$_3$ or as indicated, *data at other temperatures see reference:
  $k_{OH} = 0.92 \times 10^{11}$ M^{-1} s^{-1} at room temp. (estimated from structurally similar p-ethyltoluene, Winer et al. 1976)
  k_{OH}(calc) = 1.50×10^{-11} cm^3 molecule^{-1} s^{-1} at room temp. (SAR, Atkinson 1987)
  $k_{OH} = 1.53 \times 10^{-11}$ cm^3 molecule^{-1} s^{-1} at room temp. (Dilling et al. 1988)
  $k_{OH} = (1.51 \pm 0.41) \times 10^{-11}$ cm^3 molecule^{-1} s^{-1} with a tropospheric lifetime τ = 1.0–1.4 d; k_{NO3} = $(9.9 \pm 1.6) \times 10^{-16}$ cm^3 molecule^{-1} s^{-1} with a tropospheric lifetime τ = 1.3 yr and a calculated tropospheric lifetime τ > 330 d due to reaction with O$_3$ at 295 ± 2 K (relative rate method, Corchnoy & Atkinson 1990)
  Hydrolysis:
  Biodegradation:
  Biotransformation:
  Bioconcentration

Half-Lives in the Environment:
  Air: calculated tropospheric lifetimes of 1.0 d, > 330 d and 1.3 yr due to reactions with OH radical, O$_3$ and NO$_3$ radical, respectively (Corchnoy & Atkinson 1990)

TABLE 3.1.1.15.1
Reported vapor pressures of *p*-cymene at various temperatures and the coefficients for the vapor pressure equations

$$\log P = A - B/(T/K) \quad (1)$$
$$\log P = A - B/(C + t/°C) \quad (2)$$
$$\log P = A - B/(C + T/K) \quad (3)$$
$$\log P = A - B/(T/K) - C \cdot \log (T/K) \quad (4)$$
$$\ln P = A - B/(T/K) \quad (1a)$$
$$\ln P = A - B/(C + t/°C) \quad (2a)$$

Linder 1931		Stull 1947		McDonald et al. 1959		Zwolinski & Wilhoit 1971	
mercury manometer		summary of literature data		ebulliometry		selected values	
t/°C	P/Pa	t/°C	P/Pa	t/°C	P/Pa	t/°C	P/Pa
–3.5	18.67	19	133.3	107.04	12026	56.4	1333
–5.3	18.67	44.6	666.6	128.24	25189	70.3	2666
0	30.66	57.6	1333	151.9	51745	79.2	4000
0.8	33.33	71.5	2666	175.35	96965	85.8	5333
13.3	90.66	87	5333	176.46	99792	91.1	6666
		96.8	7999	177.36	101949	95.7	7999
		110.1	13332	178.42	104553	103.1	10666
		130	26664			109.3	13332
		151.8	53329	eq. 2	P/mmHg	120.7	19998
		175	101325	A	7.03724	129.5	26664
				B	1599.29	136.5	33331
		mp/°C		C	207.659	142.6	39997
						152.6	53329
				mp/°C	–67.98	108	66661
						167.7	79993
Kobe et al. 1941						173.8	93326
isoteniscope method						174.94	95992
t/°C	P/Pa					176.05	98659
data presented by						177.13	101325
eq. 1	P/mmHg						
A	8.063					eq. 2	P/mmHg
B	2332					A	6.9237
						B	1537.06
bp/°C	176.8					C	203.05
$\Delta H_V/(kJ\ mol^{-1})$ = 44.64							
						bp/°C	177.13
						$\Delta H_V/(kJ\ mol^{-1})$ =	
						at 25°C	50.29
						at bp	38.16

Mononuclear Aromatic Hydrocarbons

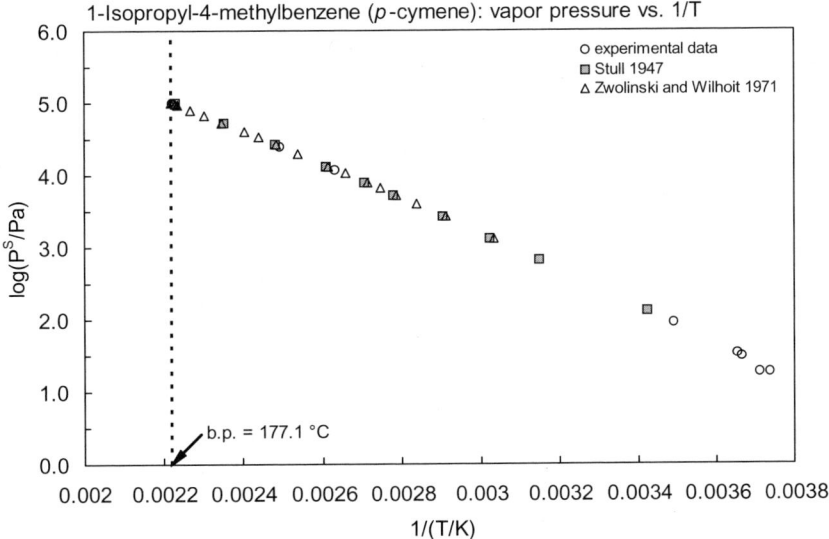

FIGURE 3.1.1.15.1 Logarithm of vapor pressure versus reciprocal temperature for 1-isopropyl-4-methylbenzene (*p*-cymene).

3.1.1.16 *n*-Butylbenzene

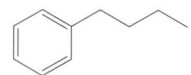

Common Name: *n*-Butylbenzene
Synonym: butylbenzene
Chemical Name: *n*-butylbenzene
CAS Registry No: 104-51-8
Molecular Formula: $C_{10}H_{14}$, $C_6H_5(CH_2)_3CH_3$
Molecular Weight: 134.218
Melting Point (°C):
    –87.85   (Lide 2003)
Boiling Point (°C):
    183.31   (Lide 2003)
Density (g/cm³ at 20°C):
    0.8601   (Weast 1982–83)
Molar Volume (cm³/mol):
    156.1   (20°C, calculated-density)
    184.8   (calculated-Le Bas method at normal boiling point)
Enthalpy of Fusion ΔH_{fus} (kJ/mol):
    10.98   (Dreisbach 1955)
    11.22   (Chickos et al. 1999)
Entropy of Fusion ΔS_{fus} (J/mol K):
    60.56, 66.5   (exptl., calculated-group additivity method, Chickos et al. 1999)
Fugacity Ratio at 25°C, F: 1.0

Water Solubility (g/m³ or mg/L at 25°C; *data at other temperatures are tabulated at end of section):
    12.6   (shake flask-UV, Andrews & Keefer 1950)
    50.5   (shake flask-UV, Klevens 1950)
    15.4   (estimated, Deno & Berkheimer 1960)
    17.7   (shake flask-GC/ECD, Massaldi & King 1973)
    11.8   (shake flask-GC, Sutton & Calder 1975)
    12.6   (shake flask-GC, Mackay & Shiu 1981)
    13.83   (generator column-HPLC/UV, GC/ECD, Tewari et al. 1982)
    13.8   (generator column-HPLC/UV, Wasik et al. 1983)
    13.76*   (generator column-HPLC/UV, measured range 7–45°C, Owens et al. 1986)
    15.0   (IUPAC recommended, Shaw 1989b)
    16.7*   (30°C, equilibrium flow cell-GC, measured range 30–100°C, Chen & Wagner 1994c)
    $\ln x = -43.2390 - 5720.35/(T/K) - 1.221 \times 10^{-4} \cdot (T/K)^2$; temp range 5–50°C (regression eq. of literature data, Shiu & Ma 2000)
    12.25*   (vapor absorption technique-HPLC/UV, measured range 0.5–55°C, Dohányosová et al. 2001)
    $\ln x = -346.295 + 14524.83/(T/K) + 49.9130 \cdot \ln(T/K)$; temp range 290–400 K (eq. derived from literature calorimetric and solubility data, Tsonopoulos 1999)

Vapor Pressure (Pa at 25°C or as indicated and reported temperature dependence equations; *data at other temperatures are tabulated at end of section):
    64*   (12.2°C, mercury manometer, Linder 1931)
    133.3*   (22.7°C, summary of literature data, Stull 1947)
    6415*   (96.233°C, ebulliometry, measured range 96.233–184.329°C, Forziati et al. 1949)
    $\log(P/\mathrm{mmHg}) = 6.98318 - 1577.965/(201.378 + t/°C)$; temp range 96.2–184.2°C (manometer, Antoine eq. from exptl. data, Forziati et al. 1949)
    145   (extrapolated-Antoine eq., Dreisbach 1955)
    $\log(P/\mathrm{mmHg}) = 6.98317 - 1577.965/(201.378 + t/°C)$; temp range 85–220°C (Antoine eq. for liquid state, Dreisbach 1955)

Mononuclear Aromatic Hydrocarbons

137*   (extrapolated-Antoine eq., Zwolinski & Wilhoit 1971)
log (P/mmHg) = 6.98317 − 1577.965/(201.378 + t/°C); temp range 62.36–213.1°C (Antoine eq., Zwolinski & Wilhoit 1971)
log (P/mmHg) = [−0.2185 × 11052.1/(T/K)] + 8.194170; temp range 22.7–183.1°C (Antoine eq., Weast 1972–73)
log (P/atm) = (1 − 456.368/T) × 10^(0.889482 − 7.01171 × 10^4·T + 5.65027 × 10^7·T^2); T in K, temp range 295.0–660.0 K (Cox vapor pressure eq., Chao et al. 1983)
138, 147   (extrapolated-Antoine equations, Boublik et al. 1984)
log (P/kPa) = 6.11624 − 1583.708/(202.013 + t/°C); temp range 96.2–184.3°C (Antoine eq. from reported exptl. data, Boublik et al. 1984)
log (P/kPa) = 6.22353 − 1660.274/(210.314 + t/°C); temp range 101.3–181.8°C (Antoine eq. from reported exptl. data, Boublik et al. 1984)
log (P/mmHg) = 6.98317 − 1577.965/(201.378 + t/°C); temp range 62–213°C (Antoine eq., Dean 1985, 1992)
log (P_L/kPa) = 6.09809 − 1571.648/(−72.413 + T/K); temp range: 369–463 K (Antoine eq., Stephenson & Malanowski 1987)
110   (20.51°C, static method, measured range 243.8–403.14 K, Kasehgari et al. 1993)
log (P/kPa) = 6.41845 − 1779.018/(220.982 + t/°C); temp range 243.8–403.14 K (static method, Kasehgari et al. 1993)
log (P/mmHg) = 49.9687 − 4.3981 × 10^3/(T/K) − 14.352·log (T/K) + 4.2054 × 10^{-11}·(T/K) + 3.4379 × 10^{-6}·(T/K)^2, temp range 185–661 K (Yaws 1994)
107   (20.16°C, static method, measured range 253.76–418.04 K, Mokbel et al. 1998)

Henry's Law Constant (Pa m^3/mol at 25°C; *data at other temperatures are tabulated at end of section):
1300   (calculated-P/C, Mackay & Shiu 1981)
1332   (calculated-vapor-liquid equilibrium (VLE) data, Yaws et al. 1991)
1502   (infinite activity coeff. γ^∞ in water determined by inert gas stripping-GC, Li et al. 1993)
1692*   (equilibrium headspace-GC, Perlinger et al. 1993)
1357.8   (modified EPICS method-GC, Ryu & Park 1999)

Octanol/Water Partition Coefficient, log K_{OW}:
4.26   (Hansch & Leo 1979)
4.19   (calculated-fragment const., Rekker 1977)
3.86   (headspace GC, Hutchinson et al. 1980)
4.26   (HPLC-k' correlation, Hammers et al. 1982)
4.28   (generator column-HPLC/UV, Tewari et al. 1982c)
4.21   (HPLC methods, Harnisch et al. 1983)
4.28   (generator column-HPLC/UV, Wasik et al. 1983)
4.29   (generator column-RP-LC, Schantz & Martire 1987)
4.26, 4.50   (RP-HPLC-k' correlations, Sherblom & Eganhouse 1988)
4.26   (recommended, Sangster 1989, 1993)
4.377   (shake flask/slow stirring-GC, De Brujin et al. 1989)
4.38   (recommended, Hansch et al. 1995)

Bioconcentration Factor, log BCF:

Sorption Partition Coefficient, log K_{OC}:
3.39   (average 5 soils and 3 sediments, sorption isotherms by batch equilibrium and column experiments, Schwarzenbach & Westall 1981)
3.40   (sediment 4.02% OC from Tamar estuary, batch equilibrium-GC, Vowles & Mantoura 1987)
3.52   (RP-HPLC-k' correlation, cyanopropyl column, Hodson & Williams 1988)
3.15, 3.32   (RP-HPLC-k' correlation, Szabo et al. 1990a,b)
3.35, 3.38, 3.39   (RP-HPLC-k' correlation on different stationary phases, Szabo et al. 1995)

Environmental Fate Rate Constants or Half-Lives:

Half-Lives in the Environment:

TABLE 3.1.1.16.1
Reported aqueous solubilities and Henry's law constants of n-butylbenzene at various temperatures

Aqueous solubility						Henry's law constant	
Owens et al. 1986		Chen & Wagner 1994c		Doháyosová et al. 2001		Perlinger et al. 1993	
generator column-HPLC		equilibrium flow cell-GC		vapor abs.-HPLC/UV		equil. headspace-GC	
t/°C	S/g·m^{-3}	t/°C	S/g·m^{-3}	t/°C	S/g·m^{-3}	t/°C	H/(Pa m^3/mol)
7	13.34	30	16.7	0.5	11.5	10	543
10	12.98	40	17.82	5	11.5	15	828
12.5	13.14	50	20.3	15	11.3	20	1115
15	12.97	60	26.9	25	12.25	25	1692
17.5	13.14	70	32.06	35	14.5	30	2168
20	13.66	80	47.72	45	17.6		
25	13.76	90	61.14	55	21.2		
30	14.58	100	83.5				
35	15.4			ΔH_{sol}/(kJ mol^{-1}) = 1.29			
40	1763	ΔH_{sol}/(kJ mol^{-1}) = 1.29					
45	20.21	at 25°C					
ΔH_{sol}/(kJ mol^{-1}) = 6.60 at 25°C							

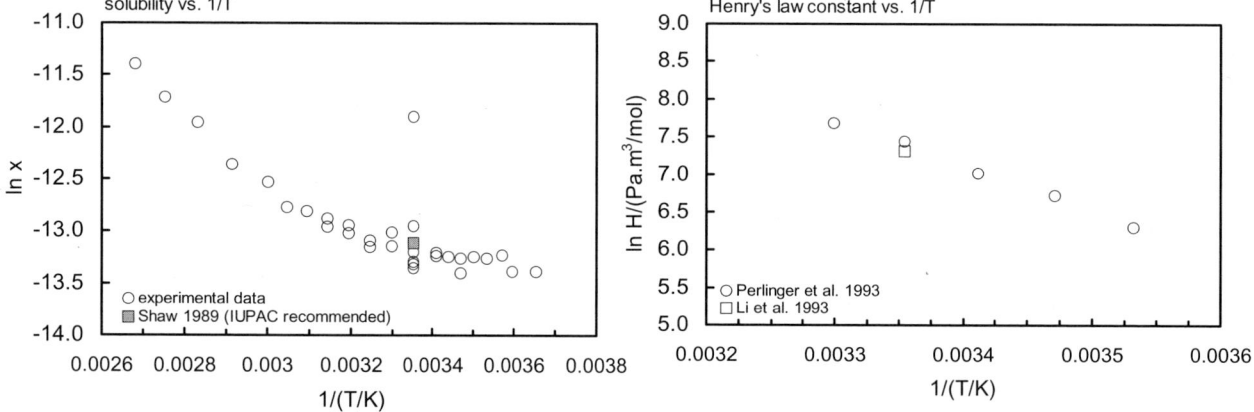

FIGURE 3.1.1.16.1 Logarithm of mole fraction solubility and Henry's law constant versus reciprocal temperature for *n*-butylbenzene.

TABLE 3.1.1.16.2
Reported vapor pressures of *n*-butylbenzene at various temperatures and the coefficients for the vapor pressure equations

$\log P = A - B/(T/K)$ (1) $\quad\quad\quad$ $\ln P = A - B/(T/K)$ (1a)

$\log P = A - B/(C + t/°C)$ (2) $\quad\quad\quad$ $\ln P = A - B/(C + t/°C)$ (2a)

$\log P = A - B/(C + T/K)$ (3)

$\log P = A - B/(T/K) - C \cdot \log (T/K)$ (4)

Linder 1931		Stull 1947		Forziati et al. 1949		Zwolinski & Wilhoit 1971	
mercury manometer		summary of literature data		ebulliometry		selected values	
t/°C	P/Pa	t/°C	P/Pa	t/°C	P/Pa	t/°C	P/Pa
−4.7	13.3	22.7	133.3	96.233	6415	62.35	1333
5.5	33.3	48.8	666.6	100.814	7697	76.32	2666
12.2	64	63	1333	104.778	8991	85.21	4000
		76.3	2666	108.403	10328	91.86	5333
		92.4	5333	111.762	11722	97.241	6666
		102.6	7999	116.322	13840	101.785	7999
		116.2	13332	121.506	16640	109.243	10666
		136.9	26664	126.797	19945	115.286	13332
		159.2	53329	138.3	28979	126.89	19998
		183.1	101325	151.541	43352	135.853	26664
				159.032	53693	142.779	33331
		mp/°C	−82.1	167.011	66793	149.829	39997
				175.666	83753	158.82	53329
				181.767	97645	166.971	66661
				182.429	99245	173.91	79993
				183.636	100734	179.985	93326
				183.725	102456	181.116	95992
				184.329	103989	182.223	98659
						183.306	101325
				eq. 2	P/mmHg		
				A	6.98318	eq. 2	P/mmHg
				B	1577.965	A	6.9808
				C	201.378	B	1577.008
						C	201.331
				bp/°C	183.27		
						bp/°C	183.306
						ΔH_V/(kJ mol^{-1})	
						at 25°C	50.12
						at bp	37.75

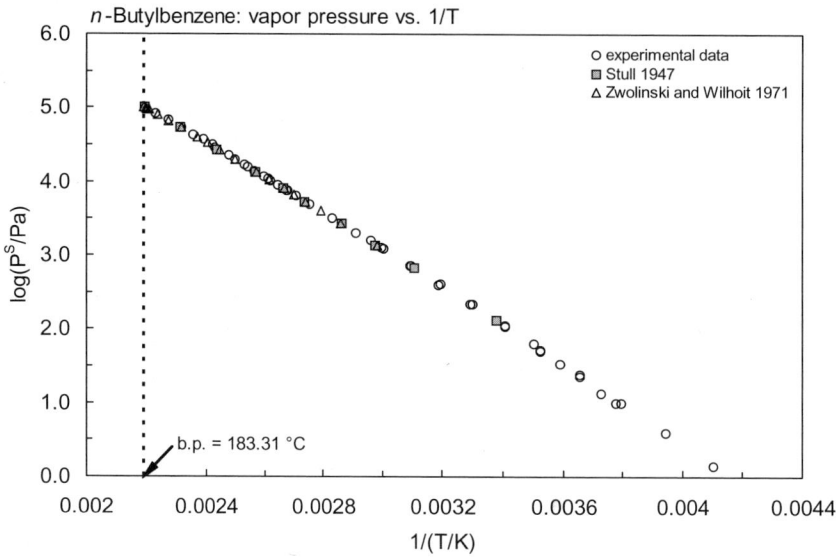

FIGURE 3.1.1.16.2 Logarithm of vapor pressure versus reciprocal temperature for *n*-butylbenzene.

3.1.1.17 Isobutylbenzene

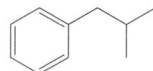

Common Name: Isobutylbenzene
Synonym: i-butylbenzene, 2-methylpropylbenzene, methyl-1-phenylpropane
Chemical Name: isobutylbenzene
CAS Registry No: 538-93-2
Molecular Formula: $C_{10}H_{14}$, $C_6H_5CH_2CH(CH_3)_2$
Molecular Weight: 134.218
Melting Point (°C):
    –51.4   (Lide 2003)
Boiling Point (°C):
    170.5   (Stephenson & Malanowski 1987)
Density (g/cm³ at 20°C):
    0.8532   (Weast 1982–83)
Molar Volume (cm³/mol):
    157.3   (20°C, calculated-density)
    184.8   (calculated-Le Bas method at normal boiling point)
Enthalpy of Fusion ΔH_{fus} (kJ/mol):
    12.51   (Dreisbach 1955)
Entropy of Fusion ΔS_{fus} (J/mol K):
Fugacity Ratio at 25°C, F: 1.0

Water Solubility (g/m³ or mg/L at 25°C):
    10.1   (shake flask-GC, Price 1976)

Vapor Pressure (Pa at 25°C or as indicated and reported temperature dependence equations. Additional data at other temperatures designated * are compiled at the end of this section):
    667*   (21.1°C, summary of literature data, Stull 1947)
    6415*   (86.637°C, ebulliometry, measured range 86.637–173.814°C, Forziati et al. 1949)
    log (P/mmHg) = 6.93033 – 1526.384/(204.171 + t/°C); temp range 86.6–173.8°C (manometer, Antoine eq. from exptl. data, Forziati et al. 1949)
    257   (extrapolated-Antoine eq., Dreisbach 1955)
    log (P/mmHg) = 6.93033 – 1526.384/(204.171 + t/°C); temp range 75–210°C (Antoine eq. for liquid state, Dreisbach 1955)
    248*   (extrapolated-Antoine eq., Zwolinski & Wilhoit 1971)
    log (P/mmHg) = 6.92804 – 1525.446/(204.122 + t/°C); temp range 53.21–202.45°C (liquid, Antoine eq., Zwolinski & Wilhoit 1971)
    log (P/mmHg) = [–0.2185 × 8567.8/(T/K)] + 7.048112; temp range: –9.8 to 170.5°C (Antoine eq., Weast 1972–73)
    log (P/atm) = (1 – 445.940/T) × 10^(0.870338 – 6.75481 × 10⁻⁴·T + 5.59009 × 10⁻⁷·T²); T in K, temp range 285.0–645.0 K (Cox vapor pressure eq., Chao et al. 1983)
    249   (extrapolated-Antoine eq., Boublik et al. 1984)
    log (P/kPa) = 6.06156 – 1530.811/(204.675 + t/°C); temp range 86.64–173.8°C (Antoine eq. from reported exptl. data, Boublik et al. 1984)
    log (P_L/kPa) = 6.06898 – 1536.514/(–67.788 + T/K); temp range 373–447 K (Antoine eq., Stephenson & Malanowski 1987)
    log (P/mmHg) = –7.0438 – 2.6892 × 10³/(T/K) + 8.7843·log (T/K) – 2.1426 × 10⁻²·(T/K) + 1.1248 × 10⁻⁵·(T/K)²; temp range 222–650 K (Yaws 1994)

Henry's Law Constant (Pa m³/mol at 25°C):
    1160, 1714   (calculated-C_W/C_A, calculated-bond contribution, Hine & Mookerjee 1975)
    3300   (calculated-P/C, Mackay & Shiu 1981)
    1393   (calculated-QSAR, Nirmalakhandan & Speece 1988b)

Octanol/Water Partition Coefficient, log K_{OW}:

4.01	(headspace GC, Hutchinson et al. 1980)
4.54, 4.82	(RP-HPLC-k' correlations, Sherblom & Eganhouse 1988)
4.68	(average lit. value, Sangster 1993)
4.54	(Hansch et al. 1995)

Bioconcentration Factor, log BCF:

Sorption Partition Coefficient, log K_{OC}:

Environmental Fate Rate Constants or Half-Lives:

Volatilization: estimated $t_{1/2} \sim 3.2$ h, evaporation from a river of 1 m depth with wind speed 3 m/s and water current of 1 m/s at 20°C (Lyman et al. 1982).

Half-Lives in the Environment:

TABLE 3.1.1.17.1
Reported vapor pressures of isobutylbenzene at various temperatures and the coefficients for vapor pressure equations

$\log P = A - B/(T/K)$ (1) $\ln P = A - B/(T/K)$ (1a)

$\log P = A - B/(C + t/°C)$ (2) $\ln P = A - B/(C + t/°C)$ (2a)

$\log P = A - B/(C + T/K)$ (3)

$\log P = A - B/(T/K) - C \cdot \log (T/K)$ (4)

Stull 1947		Forziati et al. 1949		Zwolinski & Wilhoit 1971	
summary of literature data		ebulliometry		selected values	
t/°C	P/Pa	t/°C	P/Pa	t/°C	P/Pa
–9.8	133.3	86.637	6415	53.21	1333
21.1	666.6	91.118	7697	66.97	2666
37.3	1333	95.026	8991	75.73	4000
54.7	2666	98.62	10328	82.29	5333
73.2	5333	101.946	11722	87.602	6666
84.1	7999	106.45	13840	92.087	7999
99	13332	111.582	16640	99.452	10666
120.7	26664	116.808	19945	105.422	13332
145.2	53329	121.659	23474	116.893	19998
170.2	101325	128.149	28979	125.561	26664
		134.112	34918	132.614	33331
mp/°C		141.301	43352	138.694	39997
		148.724	53693	149.502	53329
		156.632	66793	156.583	66661
		165.217	83754	163.465	79993
		171.27	97647	169.492	93326
		171.92	99247	170.615	95992
		172.526	100737	171.714	98659
		173.209	102457	172.789	101325
		173.814	103990		
				eq. 2	P/mmHg
		eq. 2	P/mmHg	A	6.92804
		A	6.93033	B	1525.446
		B	1526.384	C	204.122
		C	204.171	bp/°C	172.789
				$\Delta H_V/(kJ\ mol^{-1})$ =	
		bp/°C	172.759	at 25°C	49.45
				at bp	37.82

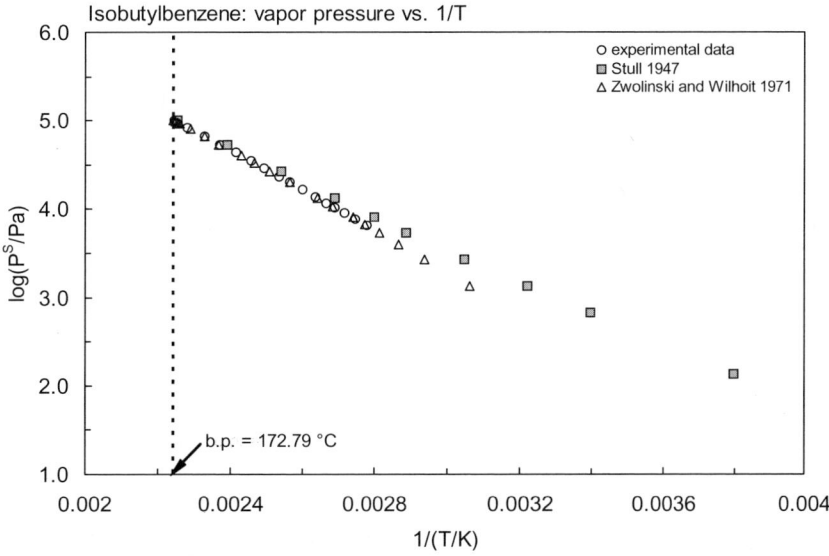

FIGURE 3.1.1.17.1 Logarithm of vapor pressure versus reciprocal temperature for isobutylbenzene.

3.1.1.18 sec-Butylbenzene

Common Name: *sec*-Butylbenzene
Synonym: 2-phenylbutane, (1-methylpropyl)benzene, *s*-butylbenzene
Chemical Name: *sec*-butylbenzene
CAS Registry No: 135-98-8
Molecular Formula: $C_{10}H_{14}$, $C_6H_5CH(CH_3)C_2H_5$
Molecular Weight: 134.218
Melting Point (°C):
    –82.7   (Lide 2003)
Boiling Point (°C):
    173.3   (Lide 2003)
Density (g/cm³ at 20°C):
    0.8621   (Weast 1982–83)
Molar Volume (cm³/mol):
    155.7   (20°C, calculated from density)
    184.8   (calculated-Le Bas method at normal boiling point)
Enthalpy of Fusion ΔH_{fus} (kJ/mol):
    9.83   (Dreisbach 1955)
Entropy of Fusion ΔS_{fus} (J/mol K):
Fugacity Ratio at 25°C, F: 1.0

Water Solubility (g/m³ or mg/L at 25°C):
    30.9   (shake flask-UV, Andrews & Keefer 1950)
    17.6   (shake flask-GC, Sutton & Calder 1975)

Vapor Pressure (Pa at 25°C or as indicated and reported temperature dependence equations. Additional data at other temperatures designated * are compiled at the end of this section):
    70.7*   (9.8°C, mercury manometer, measured range –8.6 to 9.8°C, Linder 1931)
    133.3*   (18.6°C, summary of literature data, temp range 18.6–173.5°C, Stull 1947)
    6415*   (87.118°C, ebulliometry, measured range 87.118–174.358°C, Forziati et al. 1949)
    log (P/mmHg) = 6.95097 – 1540.174/(205.101 + t/°C); temp range 87.1–174.4°C, (manometer, Antoine eq. from exptl. data, Forziati et al. 1949)
    250   (extrapolated, Antoine eq., Dreisbach 1955)
    log (P/mmHg) = 6.95097 – 1540.174/(205.101 + t/°C); temp range 75–210°C (Antoine eq. for liquid state, Dreisbach 1955)
    241*   (extrapolated, Antoine eq., Zwolinski & Wilhoit 1971)
    log (P/mmHg) = 6.94866 – 1539.233/(205.052 + t/°C); temp range 53.7–202.95°C (Antoine eq., Zwolinski & Wilhoit 1971)
    log (P/mmHg) = [–0.2185 × 11609.3/(T/K)] + 8.318014; temp range 18.6–173.5°C (Antoine eq., Weast 1972–73)
    log (P/atm) = (1 – 446.499/T) × 10^(0.870844 – 6.72060 × 10⁴·T + 5.52698 × 10⁷·T²); T in K, temp range 285.0–645.0 K (Cox vapor pressure eq., Chao et al. 1983)
    240   (extrapolated, Antoine eq., Boublik et al. 1984)
    log (P/kPa) = 6.05072 – 1533.897/(204.382 + t/°C); temp range 87.12–174.4°C (Antoine eq. from reported exptl. data, Boublik et al. 1984)
    log (P/mmHg) = 6.94219 – 1533.95/(204.39 + t/°C); temp range 87–174°C (Antoine eq., Dean 1985, 1992)
    log (P_L/kPa) = 6.10298 – 1559.452/(C + T/K); temp range 384–448 K (Antoine eq., Stephenson & Malanowski 1987)
    188*   (20.33°C, static method, measured range 243.92–373.39 K, Kasehgari et al. 1993)
    log (P/kPa) = 6.47915 – 1781.723/(208.35 + t/°C); temp range 243.92–373.39 K (static method, Kasehgari et al. 1993)

log (P/mmHg) = 61.5904 − 4.5093 × 10³/(T/K) − 19.522·log (T/K) + 6.9865 × 10⁻³·(T/K) + 7.8205 × 10⁻¹⁴·(T/K)²,
temp range 198–665 K (Yaws 1994)
186*   (20.23°C, static method, measured range 263.52–393.39 K, Mokbel et al. 1998)

Henry's Law Constant (Pa m³/mol at 25°C):

Octanol/Water Partition Coefficient, log K_{OW}:
- 4.44, 4.70   (RP-HPLC-k' correlations, Sherblom & Eganhouse 1988)
- 4.57   (average lit. value, Sangster 1993)
- 4.44   (Hansch et al. 1995)
- 3.90   (computed-expert system SPARC, Kollig 1995)

Octanol/Air Partition Coefficient, log K_{OA}:

Bioconcentration Factor, log BCF:

Sorption Partition Coefficient, log K_{OC}:
- 2.71   (aquifer material with f_{OC} of 0.006 and measured partition coeff. K_P = 3.06 mL/g., Abdul et al. 1990)
- 3.60   (computed-K_{OW}, Kollig 1995)

Environmental Fate Rate Constants or Half-Lives:

Half-Lives in the Environment:

TABLE 3.1.1.18.1
Reported vapor pressures of sec-butylbenzene at various temperatures and the coefficients for the vapor pressure equations

log P = A − B/(T/K)   (1)   ln P = A − B/(T/K)   (1a)
log P = A − B/(C + t/°C)   (2)   ln P = A − B/(C + t/°C)   (2a)
log P = A − B/(C + T/K)   (3)
log P = A − B/(T/K) − C·log (T/K)   (4)

1.

Linder 1931		Stull 1947		Forziati et al. 1949		Zwolinski & Wilhoit 1971	
mercury manometer		summary of literature data		ebulliometry		selected values	
t/°C	P/Pa	t/°C	P/Pa	t/°C	P/Pa	t/°C	P/Pa
−8.6	13.33	18.6	133.3	87.118	6415	53.7	1333
−3.0	24	44.2	666.6	91.684	7695	67.49	2666
2.6	42.66	57	1333	95.62	8990	76.26	4000
9.8	70.66	70.6	2666	99.179	10328	82.84	5333
		86.2	5333	102.523	11722	88.15	6666
		96	7999	107.009	13840	92.614	7999
		109.5	13332	112.151	16641	100.012	10666
		128.8	26664	117.387	19945	105.986	13332
		150.3	53329	122.232	23475	117.462	19998
		173.5	101325	128.715	28979	126.132	26664
				134.683	43352	133.185	33331
		mp/°C	−82.87	141.867	53693	139.175	39997
				149.288	66793	149.069	53329
				157.194	83754	157.133	66661
				165.768	97647	164.021	79993

(Continued)

TABLE 3.1.1.18.1 *(Continued)*

1.

Linder 1931		Stull 1947		Forziati et al. 1949		Zwolinski & Wilhoit 1971	
mercury manometer		summary of literature data		ebulliometry		selected values	
t/°C	P/Pa	t/°C	P/Pa	t/°C	P/Pa	t/°C	P/Pa
				171.82	99245	170.042	93326
				173.068	100737	171.164	95992
				173.754	102457	172.261	98659
				174.358	103990	173.864	101325
				eq. 2	P/mmHg	eq. 2	P/mmHg
				A	6.95097	A	6.94866
				B	1540.174	B	1539.223
				C	205.101	C	205.052
				bp/°C	173.035	bp/°C	173.335
						$\Delta H_V/(kJ\ mol^{-1})$ =	
						at 25°C	49.5
						at bp	37.95

2.

Kasehgari et al. 1993		Mokbel et al. 1998	
static method		static method	
T/K	P/Pa	T/K	P/Pa
243.92	2.92	263.52	16.31
253.79	7.99	273.42	43.34
263.69	19.3	283.40	92.39
273.55	44.9	293.38	186.0
283.50	93.3	303.42	357.0
293.48	188	313.44	652.0
303.49	358	323.55	1145
313.61	654	333.55	1943
313.61	655	343.57	3088
323.60	1136	353.62	4862
323.60	1139	363.43	7329
328.60	1478	373.39	10789
333.57	1904		
343.57	3088	data fitted to Wagner eq.	
363.43	4862		
373.39	7329		
Antoine eq.			
eq. 2	P/kPa		
A	6.47915		
B	1781.723		
C	226.989		

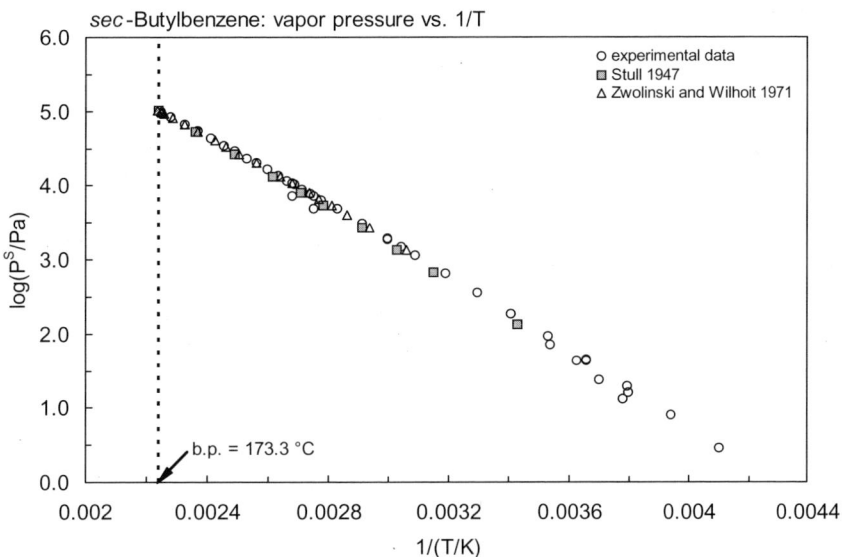

FIGURE 3.1.1.18.1 Logarithm of vapor pressure versus reciprocal temperature for *sec*-butylbenzene.

3.1.1.19 *tert*-Butylbenzene

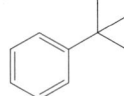

Common Name: *tert*-Butylbenzene
Synonym: (1,1-dimethylethyl)benzene, 2-methyl-2-phenylpropane, trimethylphenylmethane, pseudobutylbenzene, *t*-butylbenzene
Chemical Name: *tert*-butylbenzene
CAS Registry No: 98-06-6
Molecular Formula: $C_{10}H_{14}$, $C_6H_5C(CH_3)_3$
Molecular Weight: 134.218
Melting Point (°C):
    –58   (Stephenson & Malanowski 1987)
Boiling Point (°C):
    169.1   (Lide 2003)
Density (g/cm³ at 20°C):
    0.8665   (Weast 1982–83)
Molar Volume (cm³/mol):
    154.9   (20°C, calculated from density)
    184.8   (calculated-Le Bas method at normal boiling point)
Enthalpy of Fusion ΔH_{fus} (kJ/mol):
    8.38   (Dreisbach 1955)
    8.40   (Chickos et al. 1999)
Entropy of Fusion ΔS_{fus} (J/mol K):
    39.1, 45.4   (exptl., calculated-group additivity method, Chickos et al. 1999)
Fugacity Ratio at 25°C, F: 1.0

Water Solubility (g/m³ or mg/L at 25°C):
    34.0   (shake flask-UV, Andrews & Keefer 1950)
    29.5   (shake flask-GC, Sutton & Calder 1975)

Vapor Pressure (Pa at 25°C or as indicated and reported temperature dependence equations. Additional data at other temperatures designated * are compiled at the end of this section):
    144*   (13.7°C, mercury manometer, measured range –2 to 13.7°C, Linder 1931)
    133.3*   (13°C, summary of literature data, temp range 13–168.5°C, Stull 1947)
    6426*   (83.887°C, ebulliometry, measured range 83.887–170.165°C, Forziati et al. 1949)
    $\log (P/mmHg) = 9.2050 - 1504.572/(203.328 + t/°C)$; temp range 83.9–170.2°C (manometer, Antoine eq. from exptl. data, Forziati et al. 1949)
    295   (extrapolated-Antoine eq., Dreisbach 1955)
    $\log (P/mmHg) = 6.92050 - 1504.572/(203.328 + t/°C)$; temp range 70–205°C (Antoine eq. for liquid state, Dreisbach 1955)
    286*   (extrapolated-Antoine eq., Zwolinski & Wilhoit 1971)
    $\log (P/mmHg) = 6.91829 - 1503.651/(203.280 + t/°C)$; temp range 50.79–198.54°C (Antoine eq., Zwolinski & Wilhoit 1971)
    $\log (P/mmHg) = [-0.2185 \times 10705.5/(T/K)] + 8.195269$; temp range 13–168°C (Antoine eq., Weast 1972–73)
    $\log (P/atm) = (1 - 442.319/T) \times 10^{\wedge}(0.881530 - 7.21114 \times 10^4 \cdot T + 6.01764 \times 10^7 \cdot T^2)$; T in K, temp range 285.0–635.0 K (Cox vapor pressure eq., Chao et al. 1983)
    285   (extrapolated-Antoine eq., Boublik et al. 1984)
    $\log (P/kPa) = 6.03861 - 1499.886/(202.792 + t/°C)$; temp range 83.88–170.2°C (Antoine eq. from reported exptl. data of Forziati et al. 1949, Boublik et al. 1984)
    286   (extrapolated-Antoine eq., Dean 1985, 1992)
    $\log (P/mmHg) = 6.92255 - 1505.987/(203.49 + t/°C)$; temp range 84–170°C (Antoine eq., Dean 1985, 1992)

log (P_L/kPa) = 6.06067 − 1515.51/(−68.551 + T/K); temp range 368–444 K (Antoine eq., Stephenson & Malanowski 1987)

log (P/mmHg) = 41.4522 − 3.9027 × 10^3/(T/K) − 11.410·log (T/K) + 2.4320 × 10^{-10}·(T/K) + 2.2517 × 10^{-6}·(T/K)^2, temp range 215–660 K (Yaws 1994)

Henry's Law Constant (Pa m^3/mol at 25°C):
- 1200 (calculated-P/C, Mackay & Shiu 1981)
- 1300 (calculated-C_A/C_W, Eastcott et al. 1988)

Octanol/Water Partition Coefficient, log K_{OW}:
- 4.11 (Leo et al. 1971; Hansch & Leo 1979)
- 4.07 (shake flask-HPLC, Nahum & Horvath 1980)
- 4.25, 4.49 (RP-HPLC-k′ correlations, Sherblom & Eganhouse 1988)
- 4.11 (recommended, Sangster 1989, 1993)
- 4.11 (recommended, Hansch et al. 1995)
- 3.73 (RP-HPLC-RT correlation, short ODP column, Donovan & Pescatore 2002)

Octanol/Air Partition Coefficient, log K_{OA}:

Bioconcentration Factor, log BCF:

Sorption Partition Coefficient, log K_{OC}:

Environmental Fate Rate Constants, k, or Half-Lives, $t_{½}$:

Volatilization/Evaporation:

Photolysis:

Oxidation: rate constant k = 4.58 × 10^{-12} cm^3 molecule^{-1} s^{-1} for the gas phase reaction with OH radical at room temp. (Ohta & Ohyama 1985; Atkinson 1989);
  rate constant k = 4.6 × 10^{-12} cm^3 molecule^{-1} s^{-1} for the gas-phase reaction with OH radical at 298 K (Atkinson 1990).

Oxidation:

Hydrolysis:

Biodegradation:

Bioconcentration

Half-Lives in the Environment:

TABLE 3.1.1.19.1
Reported vapor pressures of *tert*-butylbenzene at various temperatures and the coefficients for the vapor pressure equations

$$\log P = A - B/(T/K) \quad (1)$$
$$\log P = A - B/(C + t/°C) \quad (2)$$
$$\log P = A - B/(C + T/K) \quad (3)$$
$$\log P = A - B/(T/K) - C \cdot \log(T/K) \quad (4)$$
$$\ln P = A - B/(T/K) \quad (1a)$$
$$\ln P = A - B/(C + t/°C) \quad (2a)$$

Linder 1931		Stull 1947		Forziati et al. 1949		Zwolinski & Wilhoit 1971	
mercury manometer		summary of literature data		ebulliometry		selected values	
t/°C	P/Pa	t/°C	P/Pa	t/°C	P/Pa	t/°C	P/Pa
–2.0	36	13	133.3	83.887	6426	50.79	1333
2.3	57.3	39	666.6	88.312	7695	64.41	2666
10.8	140	51.7	1333	92.194	8990	73.07	4000
13.7	144	65.6	2666	95.715	10332	79.56	5333
		80.8	5333	99.017	11722	84.816	6666
		90.6	7999	103.471	13840	89.254	7999
		103.8	13332	108.546	16641	96.542	10666
		123.7	26664	113.72	19945	102.449	13332
		145.8	53329	118.524	23475	113.802	19998
		168.5	101325	124.936	28979	122.449	26664
				137.968	43354	129.364	33331
		mp/°C	–58.0	145.315	53692	135.295	39997
				153.149	66781	145.095	53329
				161.649	83754	153.097	66661
				167.646	97647	159.913	79993
				168.287	99247	165.882	93326
				168.886	100737	166.994	95992
				169.565	102458	168.083	98659
				170.165	103991	169.018	101325
				eq. 2	P/mmHg	eq. 2	P/mmHg
				A	6.9205	A	6.91829
				B	1504.572	B	1503.651
				C	203.328	C	203.28
				bp/°C	169.119	bp/°C	169.148
						$\Delta H_v/(kJ\ mol^{-1})$ =	
						at 25°C	49.8
						at bp	37.61

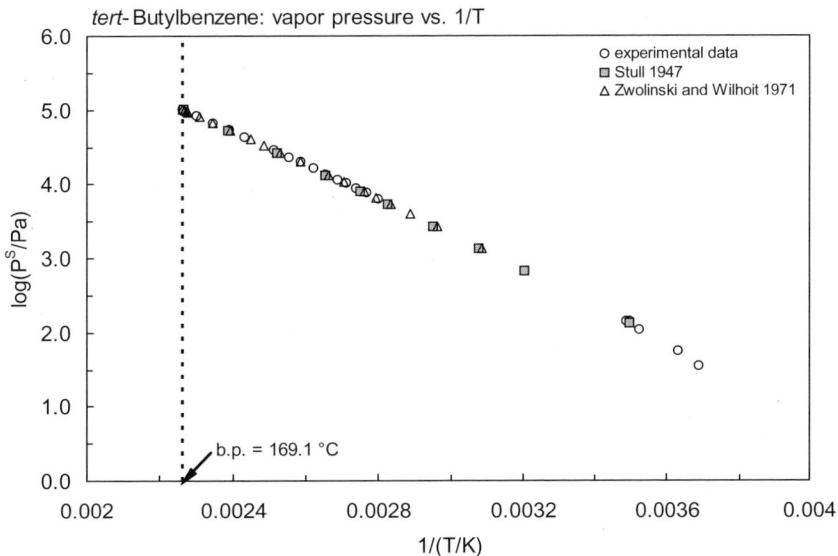

FIGURE 3.1.1.19.1 Logarithm of vapor pressure versus reciprocal temperature for *tert*-butylbenzene.

3.1.1.20 1,2,3,4-Tetramethylbenzene

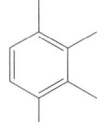

Common Name: 1,2,3,4-Tetramethylbenzene
Synonym: perhintene, prebnitene
Chemical Name: 1,2,3,4-tetramethylbenzene
CAS Registry No: 488-23-3
Molecular Formula: $C_{10}H_{14}$, $C_6H_2(CH_3)_4$
Molecular Weight: 134.218
Melting Point (°C):
    –6.2   (Weast 1982–83; Lide 2003)
Boiling Point (°C):
    205   (Weast 1982–83; Lide 2003)
Density (g/cm³ at 20°C):
    0.9052   (Weast 1982–83)
Molar Volume (cm³/mol):
    148.3   (20°C, calculated-density)
    184.8   (calculated-Le Bas method at normal boiling point)
Enthalpy of Fusion ΔH_{fus} (kJ/mol):
    11.21   (Dreisbach 1955)
    11.23   (Chickos et al. 1999)
Entropy of Fusion ΔS_{fus} (J/mol K):
    42.31, 45.7   (exptl., calculated-group additivity method, Chickos et al. 1999)
Fugacity Ratio at 25°C, F: 1.0

Water Solubility (g/m³ or mg/L at 25°C):

Vapor Pressure (Pa at 25°C or as indicated and reported temperature dependence equations. Additional data at other temperatures designated * are compiled at the end of this section):
    133.3*   (42.6°C, summary of literature data, temp range 42.6–204.4°C, Stull 1947)
    49.20   (extrapolated-Antoine eq., Dreisbach 1955)
    log (P/mmHg) = 7.0584 – 1689.10/(199.28 + t/°C); temp range 100–250°C (Antoine eq. for liquid state, Dreisbach 1955)
    1333*   (79.515°C, compiled data, temp range 79.515–235.815°C, Bond & Thodos 1960)
    45.01*   (extrapolated-Antoine eq., Zwolinski & Wilhoit 1971)
    log (P/mmHg) = 7.0594 – 1690.54/(199.48 + t/°C); temp range 79.5–235.9°C (liquid, Antoine eq., Zwolinski & Wilhoit 1971)
    log (P/mmHg) = [–0.2185 × 12258.0/(T/K)] + 8.534237; temp range 42.6–204.4°C (Antoine eq., Weast 1972–73)
    log (P/atm) = (1 – 478.255/T) × 10^(0.889494 – 6.47585 × 10⁴·T + 4.96841 × 10⁷·T²); T in K, temp range 310.0–690.0 K (Cox vapor pressure eq., Chao et al. 1983)
    45.01   (extrapolated-Antoine eq., Dean 1985, 1992)
    log (P/mmHg) = 7.0594 – 1690.54/(199.48 + t/°C); temp range 80–217°C (Antoine eq., Dean 1985, 1992)
    45.02   (extrapolated-Antoine eq., Stephenson & Malanowski 1987)
    log (P_L/kPa) = 6.1843 – 1690.54/(–73.67 + T/K); temp range 352–509 K (liquid, Antoine eq., Stephenson & Malanowski 1987)

Henry's Law Constant (Pa m³/mol):

Octanol/Water Partition Coefficient, log K_{OW}:
    3.84   (generator column-HPLC, Wasik et al. 1982)
    4.11   (HPLC-k' correlation, Hammers et al. 1982)

4.30, 4.53	(RP-HPLC-k' correlations, Sherblom & Eganhouse 1988)
4.00	(recommended, Sangster 1989, 1993)
4.09	(normal phase HPLC-k' correlation, Govers & Evers 1992)
3.98	(recommended, Hansch et al. 1995)

Octanol/Air Partition Coefficient, log K_{OA}:

Bioconcentration Factor, log BCF:

Sorption Partition Coefficient, log K_{OC}:

Environmental Fate Rate Constants, k, or Half-Lives, $t_{½}$:

Half-Lives in the Environment:

TABLE 3.1.1.20.1
Reported vapor pressures of 1,2,3,4-tetramethylbenzene at various temperatures and the coefficients for the vapor pressure equations

$$\log P = A - B/(T/K) \quad (1)$$
$$\log P = A - B/(C + t/°C) \quad (2)$$
$$\log P = A - B/(C + T/K) \quad (3)$$
$$\log P = A - B/(T/K) - C \cdot \log(T/K) \quad (4)$$
$$\log P = A - B/(T/K) - C \cdot \log(T/K) + D \cdot P/(T/K)^2 \quad (5)$$

$$\ln P = A - B/(T/K) \quad (1a)$$
$$\ln P = A - B/(C + t/°C) \quad (2a)$$

Stull 1947		Bond & Thodos 1960		Zwolinski & Wilhoit 1971			
summary of literature data		compiled data		selected values			
t/°C	P/Pa	t/°C	P/Pa	t/°C	P/Pa	t/°C	P/Pa
42.6	133.3	79.515	1333	79.5	1333	201.6	93326
68.7	666.6	128.415	10666	94.1	2666	202.83	95992
81.8	1333	188.185	66661	103.4	4000	204.54	98659
95.8	2666	216.925	133322	110.3	5333	205.09	101325
111.5	5333	235.815	199984	115.9	6666		
121.8	7999			120.6	7999	eq. 2	P/mmHg
135.7	13332	bp/°C	205.055	128.4	10666	A	7.0594
1155.7	26664			134.7	13332	B	1690.54
180	53329	eq. 5	P/mmHg	146.7	19998	C	199.48
204.4	101325	A	27.4323	155.8	26664	bp/°C	205.9
		B	3713.52	163.2	33331	ΔH_V/(kJ mol^{-1}) =	
mp/°C	–4.0	C	6.77416	169.2	39997	at 25°C	57.15
		D	3.81118	179.8	53329	at bp	45.02
				188.22	66661		
				195.99	79993		

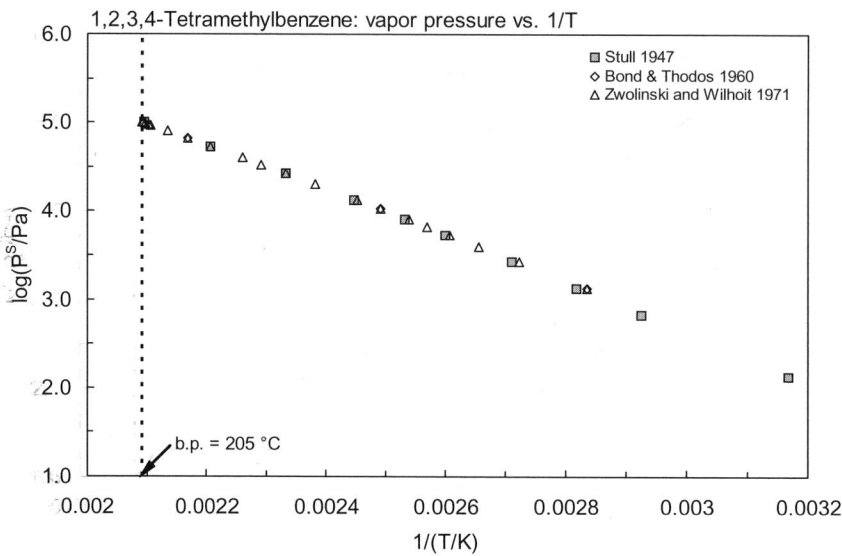

FIGURE 3.1.1.20.1 Logarithm of vapor pressure versus reciprocal temperature for 1,2,3,4-tetramethylbenzene.

Mononuclear Aromatic Hydrocarbons

3.1.1.21 1,2,3,5-Tetramethylbenzene

Common Name: 1,2,3,5-Tetramethylbenzene
Synonym: isodurene
Chemical Name: 1,2,3,5-tetramethylbenzene
CAS Registry No: 527-53-7
Molecular Formula: $C_{10}H_{14}$, $C_6H_2(CH_3)_4$
Molecular Weight: 134.218
Melting Point (°C):
    –24.1   (Stephenson & Malanowski 1987)
Boiling Point (°C):
    198   (Weast 1982–83; Stephenson & Malanowski 1987; Lide 2003)
Density (g/cm³ at 20°C):
    0.8903   (Weast 1982–83)
Molar Volume (cm³/mol):
    150.8   (20°C, calculated-density)
    184.8   (calculated-Le Bas method at normal boiling point)
Enthalpy of Fusion ΔH_{fus} (kJ/mol):
    12.93   (Chickos et al. 1999)
Entropy of Fusion ΔS_{fus} (J/mol K):
    52.01, 46.7 (exptl., calculated-group additivity method, Chickos et al. 1999)
Fugacity Ratio at 25°C, F: 1.0

Water Solubility (g/m³ or mg/L at 25°C):
    16.2, 16.4, 19.4   (15, 25, 35°C, estimated- RP-HPLC-k′ correlation, Finizio & Di Guardo 2001)

Vapor Pressure (Pa at 25°C or as indicated and reported temperature dependence equations. Additional data at other temperatures designated * are compiled at the end of this section):
    133.3*   (40.6°C, summary of literature data, temp range 40.6–197.9°C, Stull 1947)
    67.10   (extrapolated-Antoine eq., Dreisbach 1955)
    $\log (P/mmHg) = 7.0769 - 1674.00/(200.94 + t/°C)$; temp range 95–240°C (Antoine eq. for liquid state, Dreisbach 1955)
    62.22*   (extrapolated-Antoine eq., Zwolinski & Wilhoit 1971)
    $\log (P/mmHg) = 7.70779 - 1675.43/(201.14 + t/°C)$; temp range 74.5–228.3°C (Antoine eq., Zwolinski & Wilhoit 1971)
    $\log (P/mmHg) = [-0.2185 \times 12358.4/(T/K)] + 8.680246$; temp range 40.6–197.9°C (Antoine eq., Weast 1972–73)
    $\log (P/atm) = (1 - 471.208/T) \times 10^{\wedge}(0.891876 - 6.64575 \times 10^4 \cdot T + 5.21861 \times 10^7 \cdot T^2)$; T in K, temp range 305.0–675.0 K (Cox vapor pressure eq., Chao et al. 1983)
    62.22   (extrapolated, Antoine eq., Dean 1985)
    $\log (P/mmHg) = 7.0779 - 1675.43/(201.14 + t/°C)$; temp range 75–228°C (Antoine eq., Dean 1985, 1992)
    62.23   (extrapolated-Antoine eq., Stephenson & Malanowski 1987)
    $\log (P_L/kPa) = 6.2028 - 1675.43/(-72.01 + T/K)$; temp range 348–502 K (Antoine eq., Stephenson & Malanowski 1987)
    $\log (P/mmHg) = -3.9778 - 2.960 \times 10^3/(T/K) + 7.3226 \cdot \log (T/K) - 1.7725 \times 10^{-2} \cdot (T/K) + 8.6365 \times 10^{-6} \cdot (T/K)^2$, temp range 249–679 K (Yaws 1994)

Henry's Law Constant (Pa m³/mol):

Octanol/Water Partition Coefficient, log K_{OW}:
    4.04   (generator column-HPLC/UV, Wasik et al. 1982)
    4.17   (HPLC-k′ correlation, Hammers et al. 1982)

4.30 (average value, RP-HPLC-k′ correlation, Sherblom & Eganhouse 1988)
4.10 (recommended, Sangster 1989, 1993)
4.23 (normal phase HPLC-k′ correlation, Govers & Evers 1992)
4.04 (recommended, Hansch et al. 1995)

Octanol/Air Partition Coefficient, log K_{OA}:

Bioconcentration Factor, log BCF:

Sorption Partition Coefficient, log K_{OC}:

Environmental Fate Rate Constants, k, or Half-Lives, $t_{1/2}$:

Half-Lives in the Environment:

TABLE 3.1.1.21.1
Reported vapor pressures of 1,2,3,5-tetramethylbenzene at various temperatures and the coefficients for the vapor pressure equations

$$\log P = A - B/(T/K) \quad (1)$$
$$\log P = A - B/(C + t/°C) \quad (2)$$
$$\log P = A - B/(C + T/K) \quad (3)$$
$$\log P = A - B/(T/K) - C \cdot \log (T/K) \quad (4)$$
$$\ln P = A - B/(T/K) \quad (1a)$$
$$\ln P = A - B/(C + t/°C) \quad (2a)$$

Stull 1947		Zwolinski & Wilhoit 1971			
summary of literature data		selected values			
t/°C	P/Pa	t/°C	P/Pa	t/°C	P/Pa
40.6	133.3	74.5	1333	194.68	93326
65.8	666.6	88.9	2666	195.83	95992
77.8	1333	98	4000	196.95	98659
91	2666	104.8	5333	198.05	101325
105.8	5333	110.3	6666		
115.4	7999	115	7999	eq. 2	P/mmHg
128.3	13332	122.6	10666	A	7.0779
149.9	26664	128.8	13332	B	1675.43
173.7	53329	140.7	19998	C	201.14
197.9	101325	149.6	26664		
		156.9	33331	bp/°C	198.05
mp/°C	−24.0	163	39997	ΔH_V/(kJ mol^{-1}) =	
		173.2	53329	at 25°C	55.82
		181.5	66661	at bp	43.81
		188.5	79993		

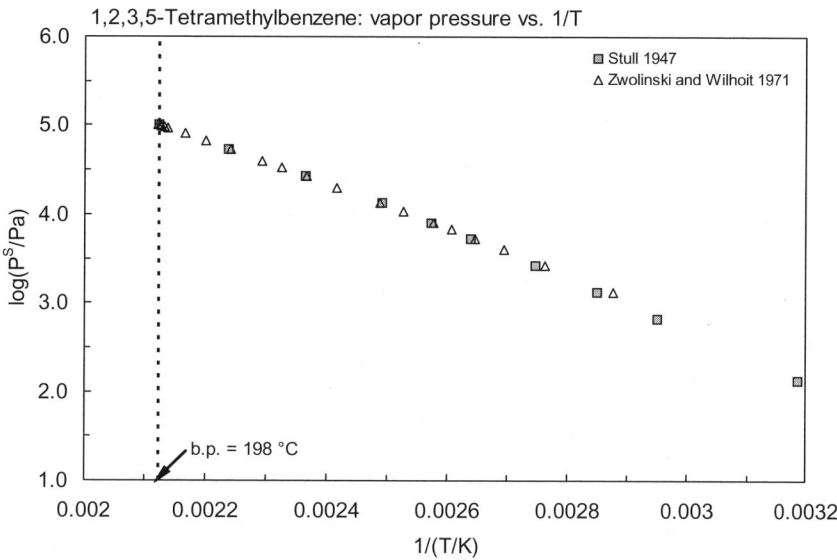

FIGURE 3.1.1.21.1 Logarithm of vapor pressure versus reciprocal temperature for 1,2,3,5-tetramethylbenzene.

3.1.1.22 1,2,4,5-Tetramethylbenzene

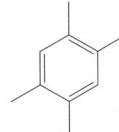

Common Name: 1,2,4,5-Tetramethylbenzene
Synonym: durene
Chemical Name: 1,2,4,5-tetramethylbenzene
CAS Registry No: 95-93-2
Molecular Formula: $C_{10}H_{14}$, $C_6H_2(CH_3)_4$
Molecular Weight: 134.218
Melting Point (°C):
    79.3   (Lide 2003)
Boiling Point (°C):
    196.8   (Weast 1982–83; Stephenson & Malanowski 1987; Lide 2003)
Density (g/cm³ at 20°C): 0.838
Molar Volume (cm³/mol):
    160.1   (20°C, calculated-density)
    184.8   (calculated-Le Bas method at normal boiling point)
Enthalpy of Fusion ΔH_{fus} (kJ/mol):
    21.0   (Dreisbach 1955; Tsonopoulos & Prausnitz 1971)
    20.88   (Chickos et al. 1999)
Entropy of Fusion ΔS_{fus} (J/mol K):
    59.83   (Tsonopoulos & Prausnitz 1971)
    59.25, 46.7   (exptl., calculated-group additivity method, Chickos et al. 1999)
Fugacity Ratio at 25°C (assuming ΔS_{fus} = 56 J/mol K), F: 0.293 (mp at 79.3°C)

Water Solubility (g/m³ or mg/L at 25°C or as indicated):
    19.5   (Deno & Berkheimer 1960)
    3.48   (shake flask-GC, Price 1976)
    3.48   (shake flask-GC, Krzyzanowska & Szeliga 1978)
    13.9   (HPLC-k′ correlation, converted from reported activity coeff γ_W, Hafkenscheid & Tomlinson 1983)
    17.2, 18.6, 28.2   (15, 25, 35°C, RP-HPLC-k′ correlation, Finizio & Di Guardo 2001)

Vapor Pressure (Pa at 25°C or as indicated and reported temperature dependence equations. Additional data at other temperatures designated * are compiled at the end of this section):
    4.4*   (1.75°C, mercury manometer, measured range –1.7–1.75 °C, Linder 1931)
    133.3*   (45°C, summary of literature data, temp range 45–195.9°C, Stull 1947)
    70.9   (extrapolated-Antoine eq., Dreisbach 1955)
    $\log (P/\text{mmHg}) = 7.0790 - 1671.0/(201.23 + t/°C)$; temp range 95–240°C (Antoine eq. for liquid state, Dreisbach 1955)
    65.9*   (extrapolated-Antoine eq., Zwolinski & Wilhoit 1971)
    $\log (P/\text{mmHg}) = 7.0800 - 1672.43/(201.43 + t/°C)$; temp range 73.6–227°C (Antoine eq., Zwolinski & Wilhoit 1971)
    $\log (P/\text{mmHg}) = [-0.2185 \times 12582.6/(T/K)] + 8.822113$; temp range 45–195.9°C (Antoine eq., Weast 1972–73)
    $\log (P/\text{atm}) = (1 - 470.032/T) \times 10^{\wedge}(0.884259 - 6.36677 \times 10^4 \cdot T + 4.97646 \times 10^7 \cdot T^2)$; T in K, temp range 346.75–675.0 K (Cox vapor pressure eq., Chao et al. 1983)
    65.9   (extrapolated-Antoine eq., Dean 1985, 1992)
    $\log (P/\text{mmHg}) = 7.0800 - 1672.43/(201.43 + t/°C)$; temp range 74–277°C (Antoine eq., Dean 1985, 1992)
    65.9   (extrapolated-Antoine eq., Stephenson & Malanowski 1987)
    $\log (P_L/\text{kPa}) = 6.2049 - 1672.43/(-71.72 + T/K)$; temp range 353–500 K (Antoine eq., Stephenson & Malanowski 1987)

$\log~(P/\text{mmHg}) = -51.3593 - 1.6523 \times 10^3/(T/K) + 26.656 \cdot \log~(T/K) - 3.5721 \times 10^{-2} \cdot (T/K) + 1.5018 \times 10^{-5} \cdot (T/K)^2$, temp range 352–Y675K (Yaws 1994)

Henry's Law Constant (Pa m^3/mol at 25°C):
- 2541 (calculated-vapor-liquid equilibrium (VLE) data, Yaws et al. 1991)

Octanol/Water Partition Coefficient, log K_{OW}:
- 4.00 (Hansch & Leo 1979)
- 4.24 (HPLC-RV correlation, Garst 1984)
- 4.13, 4.34 (RP-HPLC-k' correlations, Sherblom & Eganhouse 1988)
- 4.10 (recommended, Sangster 1989, 1993)
- 4.27 (normal phase HPLC-k' correlation, Govers & Evers 1992)
- 4.00 (recommended, Hansch et al. 1995)

Octanol/Air Partition Coefficient, log K_{OA}:

Bioconcentration Factor, log BCF:

Sorption Partition Coefficient, log K_{OC}:
- 3.12 (average 5 soils and 3 sediments, sorption isotherms by batch equilibrium and column experiments, Schwarzenbach & Westall 1981)
- 2.99 (soil, calculated-MCI χ, Sabljic 1987a,b)
- 2.76 (aquifer material with f_{OC} of 0.006 and measured K_P = 3.42 mL/g., Abdul et al. 1990)

Environmental Fate Rate Constants, k, or Half-Lives, $t_{1/2}$:
- Volatilization:
- Photolysis:
- Oxidation: rate constant k = 1.1×10^4 cm^3 mol^{-1} s^{-1} for reaction with ozone at 300 K (estimated, Lyman 1982).
- Hydrolysis:
- Biodegradation:
- Bioconcentration:

Half-Lives in the Environment:

TABLE 3.1.1.22.1
Reported vapor pressures of 1,2,4,5-tetramethylbenzene at various temperatures and the coefficients for the vapor pressure equations

$$\log P = A - B/(T/K) \quad (1)$$
$$\log P = A - B/(C + t/°C) \quad (2)$$
$$\log P = A - B/(C + T/K) \quad (3)$$
$$\log P = A - B/(T/K) - C \cdot \log(T/K) \quad (4)$$

$$\ln P = A - B/(T/K) \quad (1a)$$
$$\ln P = A - B/(C + t/°C) \quad (2a)$$

Linder 1931		Stull 1947		Zwolinski & Wilhoit 1971			
Hg manometer		summary of literature data		selected values			
t/°C	P/Pa	t/°C	P/Pa	t/°C	P/Pa	t/°C	P/Pa
–1.7	1.733	45	133.3	73.6	1333	193.49	93326
–1.3	21.33	63	666.6	88	2666	194.63	95992
1.3	4.4	74.6	1333	97.1	4000	195.75	98659
1.75	4.4	88	2666	103.9	5333	196.84	101325
		104.2	5333	109.4	6666		
		114.8	7999	114	7999	eq. 2	P/mmHg
		128.1	13332	121.6	10666	A	7.08
		149.5	26664	127.8	13332	B	1672.43
		172.1	53329	139.6	19998	C	201.43
		195.9	101325	148.5	26664		
				155.8	33331	bp/°C	196.84
		mp/°C	79.5	161.9	39997	$\Delta H_V/(kJ\ mol^{-1}) =$	
				172.1	53329	at 25°C	75
				180.3	66661	at bp	44.52
				187.3	79993		

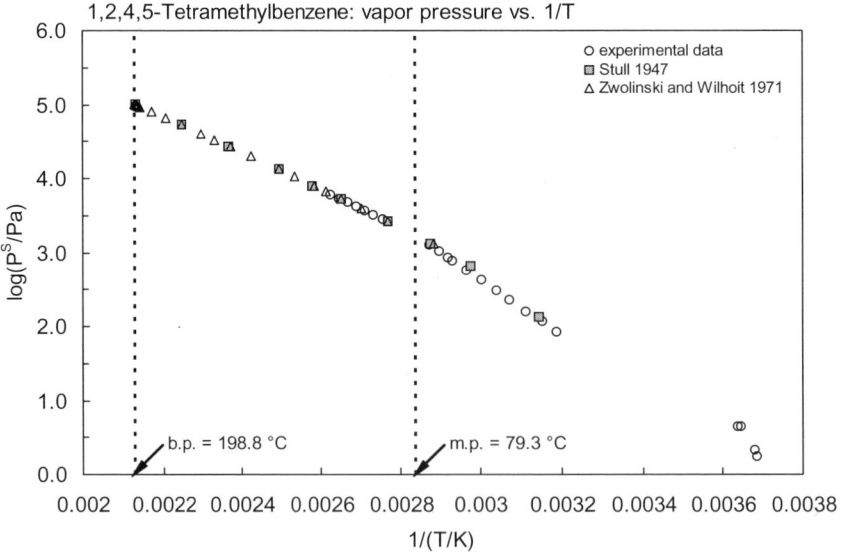

FIGURE 3.1.1.22.1 Logarithm of vapor pressure versus reciprocal temperature for 1,2,4,5-tetramethylbenzene.

3.1.1.23 Pentamethylbenzene

Common Name: Pentamethylbenzene
Synonym:
Chemical Name: pentamethylbenzene
CAS Registry No: 700-12-9
Molecular Formula: $C_{11}H_{16}$, $C_6H(CH_3)_5$
Molecular Weight: 148.245
Melting Point (°C):
    54.5   (Weast 1982–83; Lide 2003)
Boiling Point (°C):
    232   (Weast 1982–83; Stephenson & Malanowski 1987; Lide 2003)
Density (g/cm^3 at 20°C):
    0.917, 0.913   (20°C, 25°C, Dreisbach 1955)
    0.917   (Weast 1982–83)
Molar Volume (cm^3/mol):
    161.7   (20°C, calculated from density)
    207.0   (calculated-Le Bas method at normal boiling point)
Enthalpy of Fusion ΔH_{fus} (kJ/mol):
    12.34   (Tsonopoulos & Prausnitz 1971)
    1.98, 10.67; 12.65   (23.65, 55.05°C; total phase change enthalpy, Chickos et al. 1999)
Entropy of Fusion ΔS_{fus} (J/mol K):
    37.70   (Tsonopoulos & Prausnitz 1971)
    39.33, 47.3   (exptl., calculated-group additivity method, total phase change entropy, Chickos et al. 1999)
Fugacity Ratio at 25°C (assuming ΔS_{fus} = 56 J.mol K.), F: 0.514 (mp at 54.5°C)

Water Solubility (g/m^3 or mg/L at 25°C):
    15.6   (Deno & Berkheimer 1960)
    15.52   (calculated-K_{OW}, Yalkowsky et al. 1983)

Vapor Pressure (Pa at 25°C and reported temperature dependence equations):
    13.84   (extrapolated-Antoine eq., Dreisbach 1955)
    log (P/mmHg) = 7.13756 − 1833.8/(199.0 + t/°C); temp range 125–280°C (Antoine eq. for liquid state, Dreisbach 1955)
    9.52   (extrapolated-Antoine eq., Stephenson & Malanowski 1987)
    log (P_L/kPa) = 6.3509 − 1867/(−75.15 + T/K); temp range 338–503 K (Antoine eq., Stephenson & Malanowski 1987)

Henry's Law Constant (Pa m^3/mol):

Octanol/Water Partition Coefficient, log K_{OW}:
    4.56   (HPLC-k' correlation, Hammers et al. 1982)
    4.57   (HPLC-RV correlation, Garst 1984)
    4.56   (recommended, Sangster 1989, 1993)
    4.59   (normal phase HPLC-k' correlation, Govers & Evers 1992)
    4.56   (recommended, Hansch et al. 1995)

Octanol/Air Partition Coefficient, log K_{OA}:
Bioconcentration Factor, log BCF:
Sorption Partition Coefficient, log K_{OC}:

Environmental Fate Rate Constants, k, or Half-Lives, $t_{1/2}$:
Half-Lives in the Environment:

3.1.1.24 Pentylbenzene

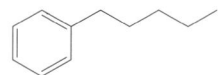

Common Name: *n*-Pentylbenzene
Synonym: phenylpentane
Chemical Name: pentylbenzene
CAS Registry No: 538-68-1
Molecular Formula: $C_{11}H_{16}$, $C_6H_5(CH_2)_4CH_3$
Molecular Weight: 148.245
Melting Point (°C):
    –75   (Dreisbach 1955; Weast 1982–83; Lide 2003)
Boiling Point (°C):
    205.4   (Weast 1982–83; Lide 2003)
Density (g/cm³ at 20°C):
    0.8585   (Weast 1982–83)
Molar Volume (cm³/mol):
    172.7   (20°C, calculated from density)
    207.0   (calculated-Le Bas method at normal boiling point)
Enthalpy of Fusion ΔH_{fus} (kJ/mol):
Entropy of Fusion ΔS_{fus} (J/mol K):
Fugacity Ratio at 25°C, F: 1.0

Water Solubility (g/m³ or mg/L at 25°C or as indicated and reported temperature dependence equations. Additional data at other temperatures designated * are compiled at the end of this section):
    10.5   (shake flask-UV, Andrews & Keefer 1950)
    3.84   (generator column-HPLC/UV, Tewari et al. 1982c)
    3.37*   (generator column-HPLC/UV, measured range 7–45°C, Owens et al. 1986)
    $\ln x = -387.920 + 16274.64/(T/K) + 55.9266 \cdot \ln (T/K)$; temp range 290–400 K (eq. derived from literature calorimetric and solubility data, Tsonopoulos 1999)

Vapor Pressure (Pa at 25°C or as indicated and reported temperature dependence equations. Additional data at other temperatures designated * are compiled at the end of this section):
    43.7   (extrapolated-Antoine eq., Dreisbach 1955)
    $\log (P/mmHg) = 7.04709 - 1670.68/(195.6 + t/°C)$; temp range 105–270°C (Antoine eq. for liquid state, Dreisbach 1955)
    43.7*   (extrapolated-Antoine eq., Zwolinski & Wilhoit 1971)
    $\log (P/mmHg) = 6.97833 - 1639.91/(194.76 + t/°C)$; temp range 80–237°C (Antoine eq., Zwolinski & Wilhoit 1971)
    $\log (P/mmHg) = 34.2755 - 3.6829 \times 10^3/(T/K) - 9.3387 \cdot \log (T/K) + 2.7727 \times 10^{-3} \cdot (T/K) - 8.8315 \times 10^{-15} \cdot (T/K)^2$, temp range 198–680 K (Yaws 1994)

Henry's Law Constant (Pa m³/mol at 25°C):
    600   (calculated-P/C, Mackay & Shiu 1981)
    617   (calculated-C_A/C_W, Eastcott et al. 1988)
    1689   (calculated-vapor-liquid equilibrium (VLE) data, Yaws et al. 1991)
    1628.2   (modified EPICS method-GC, Ryu & Park 1999)

Octanol/Water Partition Coefficient, log K_{OW}:
    4.56   (HPLC-k' correlation, Hammers et al. 1982)
    4.90   (generator column-HPLC/UV, Tewari et al. 1982c)
    4.90   (recommended, Sangster 1989, 1993)
    4.90   (recommended, Hansch et al. 1995)

Octanol/Air Partition Coefficient, log K_{OA}:

Bioconcentration Factor, log BCF:

Sorption Partition Coefficient, log K_{OC}:

Environment Fate Rate Constants, k, or Half-Lives, $t_{1/2}$:
    Volatilization:
    Photolysis:
    Oxidation: rate constant k = 5.0×10^4 cm^3 mol^{-1} s^{-1} for the reaction with ozone at 300 K (Lyman 1982).
    Hydrolysis:
    Biodegradation:
    Bioconcentration:

Half-Lives in the Environment:

TABLE 3.1.1.24.1
Reported aqueous solubilities and vapor pressures of *n*-pentylbenzene at various temperatures

Aqueous solubility		Vapor pressure			
Owens et al. 1986		Zwolinski & Wilhoit 1971			
generator column-HPLC		selected values			
t/°C	S/g·m^{-3}	t/°C	P/Pa	t/°C	P/Pa
7	3.48	80	1333	283.18	95992
10	3.18	94	2666	203.16	97325
12.5	3.44	103.3	4000	204.33	98659
15	3.19	110.3	5333	204.9	99992
17.5	3.43	115.87	6666	205.46	101325
20	3.18	120.6	7999		
25	3.37	128.36	10666	log P = A − B/(C + t/°C)	
30	3.61	134.65	13332	Antoine eq.	P/mmHg
35	3.92	146.73	19998	A	6.97833
40	4.25	155.85	26664	B	1639.91
45	4.69	163.27	33331	C	194.76
		169.57	39997		
ΔH_{sol}/(kJ mol^{-1}) = 6.50		179.97	53329	bp/°C	205.46
at 25°C		188.45	66661	ΔH_V/(kJ mol^{-1})	
		196.68	79993	at 25°C	55.06
		202	93326	at 25°C	41.21
		202.59	94659		

Mononuclear Aromatic Hydrocarbons

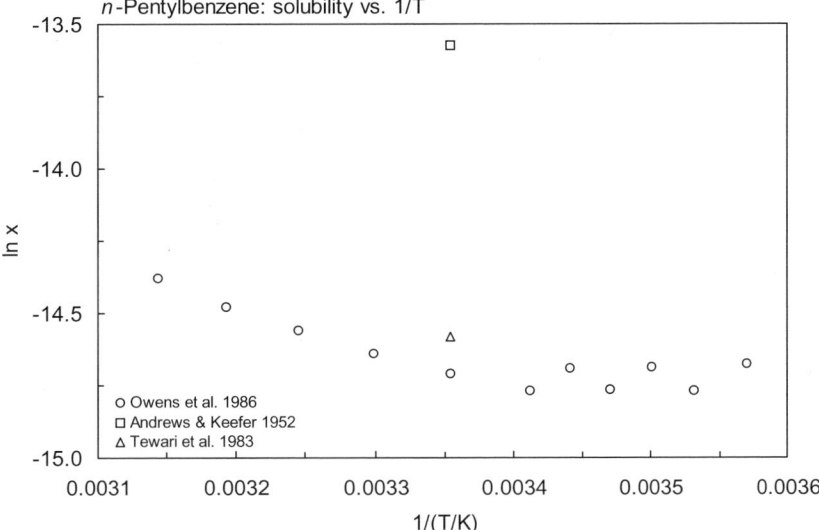

FIGURE 3.1.1.24.1 Logarithm of mole fraction solubility (ln x) versus reciprocal temperature for n-pentylbenzene.

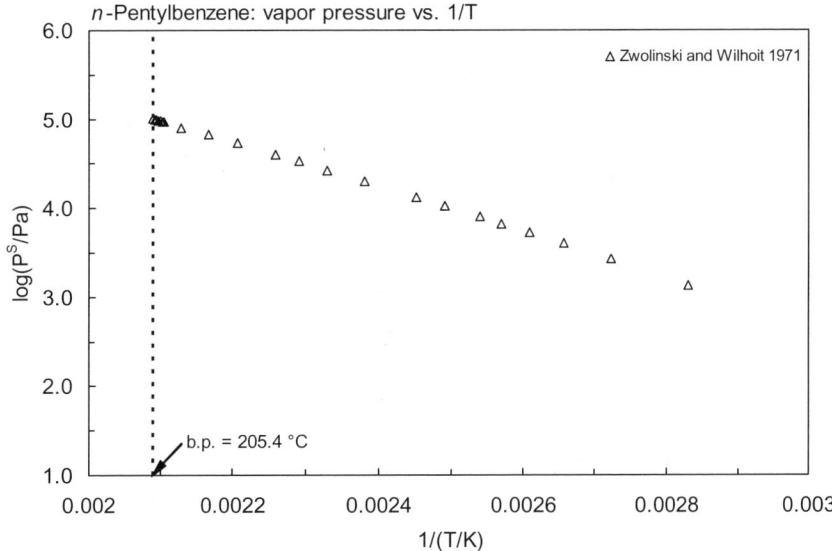

FIGURE 3.1.1.24.2 Logarithm of vapor pressure versus reciprocal temperature for *n*-pentylbenzene.

3.1.1.25 Hexamethylbenzene

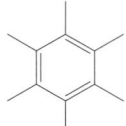

Common Name: *n*-Hexamethylbenzene
Synonym: mellitene
Chemical Name: *n*-hexamethylbenzene
CAS Registry No: 87-95-4
Molecular Formula: $C_{12}H_{18}$, $C_6(CH_3)_6$
Molecular Weight: 162.271
Melting Point (°C):
    165.5   (Lide 2003)
Boiling Point (°C):
    263.4   (Weast 1982–83)
Density (g/cm³ at 25°C):
    1.063   (Weast 1982–83)
Molar Volume (cm³/mol):
    152.6   (25°C, calculated-density)
    229.2   (calculated-Le Bas method at normal boiling point)
Enthalpy of Fusion ΔH_{fus} (kJ/mol):
    20.46   (Tsonopoulos & Prausnitz 1971)
    1.76, 20.63; 22.38   (110.5, 165.55°C; total phase change enthalpy, Chickos et al. 1999)
Entropy of Fusion ΔS_{fus} (J/mol K):
    46.44   (Tsonopoulos & Prausnitz 1971)
    51.6, 47.9   (exptl., calculated-group additivity method, total phase change entropy, Chickos et al. 1999)
Fugacity Ratio (assuming ΔS_{fus} = 56 J/mol K), F: 0.042 (mp = 166.5°C)

Water Solubility (g/m³ or mg/L at 25°C):
    0.235   (generator column-GC, Doucette & Andren 1988)

Vapor Pressure (Pa at 25°C or as indicated and reported temperature dependence equations. Additional data at other temperatures designated * are compiled at the end of this section):
    1.004*   (41.07°C, transpiration method, measured range 41–90.5°C, Overberger et al. 1969)
    0.28*   (30°C, diaphragm pressure gauge, measured range 30–70°C, Ambrose et al. 1976)
    0.160   (calculated-vapor pressure eq., Ambrose et al. 1976)
    log (P/Pa) = 13.134 – 3855/[(T/K) – 21]; temp range 202–343 K (diaphragm pressure gauge measurements, Antoine eq., Ambrose et al. 1976)
    log (P/atm) = (1 – 571.163/T) × 10^(1.00973 – 5.04725 × 10⁴·T – 6.310130 × 10⁷·T²); T in K, temp range 303.1–343.02 K (Cox vapor pressure eq., Chao et al. 1983)
    0.155   (extrapolated, Antoine eq., Stephenson & Malanowski 1987)
    log (P_S/kPa) = 8.6223 – 2965.633/(–59.583 + T/K); temp range 303–343 K (solid, Antoine eq.-I, Stephenson & Malanowski 1987)
    log (P_L/kPa) = 5.89588 – 1629.9/(–118.46 + T/K); temp range 443–537 K (liquid, Antoine eq.-II, Stephenson & Malanowski 1987)

Henry's Law Constant (Pa m³/mol):

Octanol/Water Partition Coefficient, log K_{OW}:
    4.61   (generator column-HPLC, Wasik et al. 1982)
    5.11   (HPLC-k′ correlation, Hammers et al. 1982)
    4.31   (HPLC-RV correlation, Garst & Wilson 1984)
    4.60   (shake flask/slow stirring-GC, Brooke et al. 1986)

Mononuclear Aromatic Hydrocarbons

4.60, 4.88	(reversed phase HPLC-k′ correlations, Sherblom & Eganhouse 1988)
4.75	(recommended, Sangster 1989)
4.95	(normal phase HPLC-k′ correlation, Govers & Evers 1992)
5.11	(recommended, Sangster 1993)
4.61	(recommended, Hansch et al. 1995)

Octanol/Air Partition Coefficient, log K_{OA}:

6.31	(calculated-S_{oct} and vapor pressure P, Abraham et al. 2001)

Bioconcentration Factor, log BCF:

Sorption Partition Coefficient, log K_{OC}:

Environmental Fate Rate Constants, k, or Half-Lives, $t_{½}$:
  Volatilization:
  Photolysis:
  Oxidation: rate constant $k = 2.4 \times 10^5$ cm^3 mol^{-1} s^{-1} for the reaction with ozone at 300K (Lyman et al. 1982)
  Hydrolysis:
  Biodegradation:
  Bioconcentration

Half-Lives in the Environment:

TABLE 3.1.1.25.1
Reported vapor pressures of hexamethylbenzene at various temperatures and the coefficients for the vapor pressure equations

$\log P = A - B/(T/K)$	(1)	$\ln P = A - B/(T/K)$	(1a)
$\log P = A - B/(C + t/°C)$	(2)	$\ln P = A - B/(C + t/°C)$	(2a)
$\log P = A - B/(C + T/K)$	(3)		
$\log P = A - B/(T/K) - C \cdot \log (T/K)$	(4)		

Overberger et al. 1969		Ambrose et al. 1976	
transpiration method		diaphragm gauge	
t/°C	P/Pa	t/°C	P/Pa
41.07	1.004	29.95	0.28
49.11	2.306	35.07	0.49
54.68	3.912	40.02	0.85
62.62	8.015	44.96	1.41
68.71	13.62	49.95	2.34
68.72	13.58	55.04	3.81
75.21	23.71	59.95	6.05
75.33	23.79	64.91	9.39
79.75	33.69	69.87	14.4
79.75	33.5		
84.2	48.3	mp/°C	165.55
89.46	72		
89.48	72.16	eq. 3	P/Pa
90.54	77.91	A	13.134
		B	3855
		C	−21.0

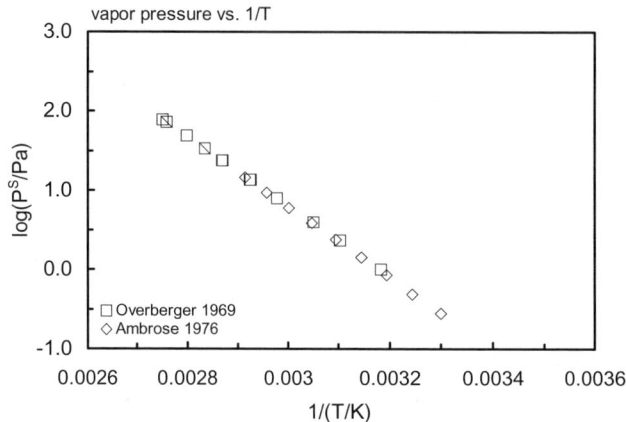

FIGURE 3.1.1.25.1 Logarithm of vapor pressure versus reciprocal temperature for *n*-hexamethylbenzene.

3.1.1.26 n-Hexylbenzene

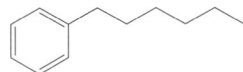

Common Name: n-Hexylbenzene
Synonym: 1-phenylhexane, hexylbenzene
Chemical Name: n-hexylbenzene
CAS Registry No: 1077-16-3
Molecular Formula: $C_{12}H_{18}$, $C_6H_5(CH_2)_5CH_3$
Molecular Weight: 162.271
Melting Point (°C):
    –61.0   (Dreisbach 1955; Lide 2003)
Boiling Point (°C):
    226.1   (Lide 2003)
Density (g/cm³ at 20°C):
    0.8613   (Weast 1982–83)
Molar Volume (cm³/mol):
    188.4   (20°C, calculated from density)
    229.2   (calculated-Le Bas method at normal boiling point)
Enthalpy of Fusion, ΔH_{fus} (kJ/mol):
Entropy of Fusion, ΔS_{fus} (J/mol K):
Fugacity Ratio at 25°C, F: 1.0

Water Solubility (g/m³ or mg/L at 25°C or as indicated and reported temperature dependence equations. Additional data at other temperatures designated * are compiled at the end of this section):
    1.02   (generator column-HPLC/UV, Tewari et al. 1982c)
    0.995*   (generator column-HPLC, measured range 5–29°C, May et al. 1983)
    0.902*   (generator column-HPLC/UV, measured range 7–45°C, Owens et al. 1986)
    $\ln x = -429.463 + 18024.83/(T/K) + 61.9402 \cdot \ln (T/K)$; temp range 290–400 K (eq. derived from literature calorimetric and solubility data, Tsonopoulos 1999)

Vapor Pressure (Pa at 25°C or as indicated and reported temperature dependence equations. Additional data at other temperatures designated * are compiled at the end of this section):
    14.01   (extrapolated-Antoine eq., Dreisbach 1955)
    $\log (P/\text{mmHg}) = 7.18284 - 1813.74/(195.5 + t/°C)$; temp range 120–290°C (Antoine eq. for liquid state, Dreisbach 1955)
    13.61*   (extrapolated, Antoine eq., Zwolinski & Wilhoit 1971)
    $\log (P/\text{mmHg}) = 6.9853 - 1700.5/(188.2 + t/°C)$; temp range 96–258°C (Antoine eq., Zwolinski & Wilhoit 1971)
    10.52*   (20.51°C, static method, measured range 273.73–462.97 K, Kasehgari et al. 1993)
    $\log (P/\text{kPa}) = 6.50020 - 1946.435/(208.935 + t/°C)$; temp range 273.73–462.97 K (static method, Kasehgari et al. 1993)
    $\log (P/\text{mmHg}) = 6.7694 - 3.6050 \times 10^3/(T/K) + 3.3416 \cdot \log (T/K) - 1.5306 \times 10^{-2} \cdot (T/K) + 7.8479 \times 10^{-6} \cdot (T/K)^2$, temp range 212–698 K (Yaws 1994)
    10.52*   (20.51°C, measured range 263.88–462.97 K, Mokbel et al. 1998)

Henry's Law Constant (Pa m³/mol):
    1977   (calculated-C_A/C_W, Eastcott et al. 1988)
    2172   (calculated-vapor-liquid equilibrium (VLE) data, Yaws et al. 1991)

Octanol/Water Partition Coefficient, log K_{OW}:
    5.25   (calculated-fragment const., Rekker 1977)
    5.52   (generator column-HPLC/UV, Tewari et al. 1982c)
    5.24   (TLC-RT correlation, Bruggeman et al. 1982)

5.45, 5.25   (quoted of HPLC methods, Harnisch et al. 1983)
5.26, 5.62   (RP-HPLC-k' correlations, Sherblom & Eganhouse 1988)
5.52   (recommended, Sangster 1989, 1993)
5.52   (recommended, Hansch et al. 1995)

Bioconcentration Factor, log BCF:

Sorption Partition Coefficient, log K_{OC}:

Environmental Fate Rate Constants, k, or Half-Lives, $t_{1/2}$:

Half-Lives in the Environment:

TABLE 3.1.1.26.1
Reported aqueous solubilities of *n*-hexylbenzene at various temperatures

	May et al. 1983			Owens et al. 1986	
	generator column-HPLC			generator column-HPLC	
t/°C	S/g·m⁻³	t/°C	S/g·m⁻³	t/°C	S/g·m⁻³
5	0.921	20	0.043	7	0.834
6	0.921	21	0.949	10	0.836
8	0.92	22	0.956	15	0.826
9	0.996	23	0.953	20	0.951
11	0.904	24	0.983	25	0.902
12	0.928	25	0.995	30	0.996
13	0.93	26	0.999	35	1.069
14	0.92	27	0.999	40	1.069
15	0.925	29	1.012	45	1.298
16.01	0.908	29	1.016		
17	0.919			ΔH_{sol}/(kJ mol⁻¹) = 8.0	
18	0.91	ΔH_{sol}/(kJ mol⁻¹) = 7.60		at 25°C	
19	0.921	at 25°C			

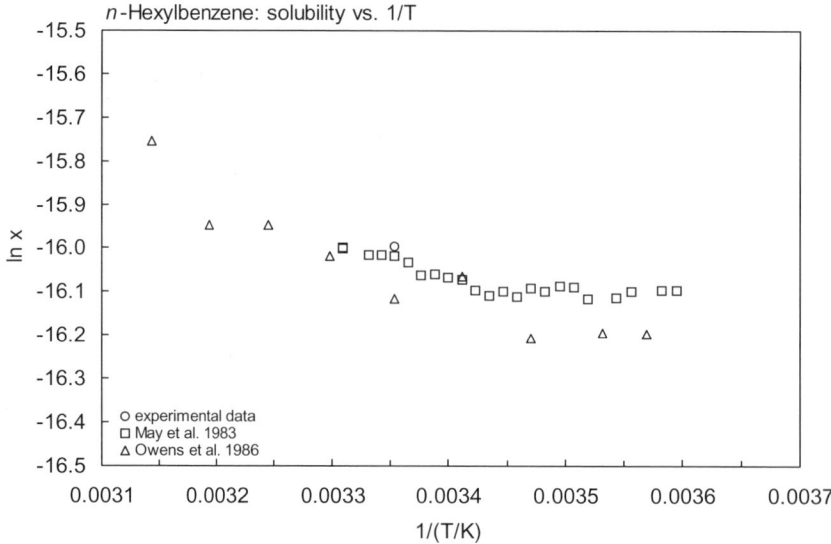

FIGURE 3.1.1.26.1 Logarithm of mole fraction solubility (ln *x*) versus reciprocal temperature for *n*-hexylbenzene.

TABLE 3.1.1.26.2
Reported vapor pressures of *n*-hexylbenzene at various temperatures and the coefficients for the vapor pressure equations

$$\log P = A - B/(T/K) \quad (1)$$
$$\log P = A - B/(C + t/°C) \quad (2)$$
$$\log P = A - B/(C + T/K) \quad (3)$$
$$\log P = A - B/(T/K) - C \cdot \log (T/K) \quad (4)$$
$$\ln P = A - B/(T/K) \quad (1a)$$
$$\ln P = A - B/(C + t/°C) \quad (2a)$$

Zwolinski & Wilhoit 1971		Kasehgari et al. 1993		Mokbel et al. 1998	
selected values		static method		static method	
t/°C	P/Pa	T/K	P/Pa	T/K	P/Pa
96	1333	273.73	1.653	263.88	0.609
111	2666	283.63	4.239	273.73	1.648
121	4000	293.66	10.52	283.63	4.236
128	5333	303.64	23.20	293.66	10.52
133.5	6666	313.62	48.66	303.64	23.20
138.4	7999	323.59	97.32	313.62	48.72
146.4	10666	333.58	184.9	323.59	97.36
152.9	13332	343.57	336.63	333.58	185.0
165.4	19998	353.58	589.1	343.57	336.0
174.8	26664	363.58	997.8	353.58	589.0
182.5	33331	372.84	1606	363.58	998.0
189.0	39997	382.77	2519	372.84	1606
199.8	53329	392.70	3841	382.77	2520
208.5	66661	402.74	5722	392.70	3841
216.0	79993	412.76	8320	402.74	5722
222.5	93326	422.76	11828	412.76	8321
223.1	94659	432.75	16516	422.76	11829
223.7	95992	442.89	22640	432.75	16517
224.3	97325	452.83	30575	442.80	22641
224.9	98659	462.97	40693	452.83	30576
225.5	99992			462.79	40694
226.1	101325	eq. 2	P/kPa		
		A	6.50020	data fitted to Wagner eq.	
Antoine eq.	P/mmHg	B	1946.435		
A	6.9853	C	208.935		
B	1700.5				
C	1882.2				
bp/°C	226.1				
$\Delta H_V/(kJ\ mol^{-1}) =$					
at 25°C	60				
at bp	43.1				

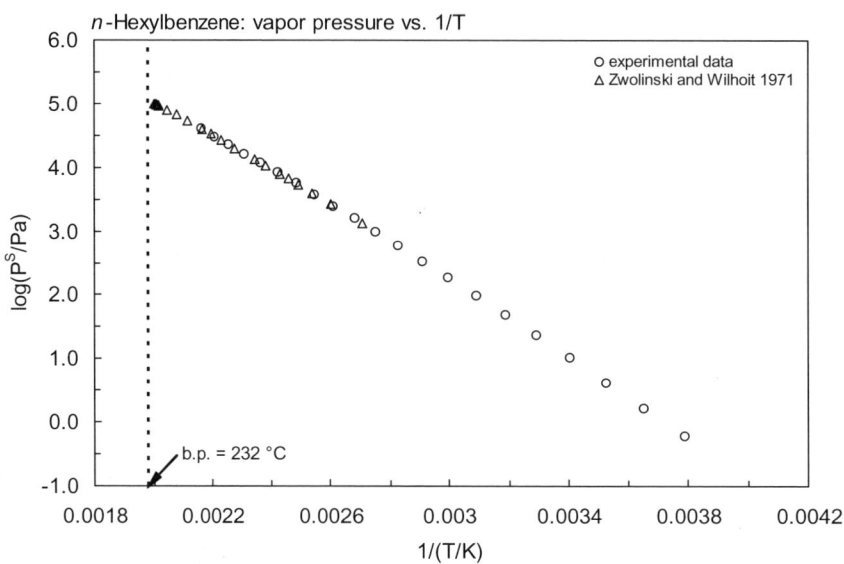

FIGURE 3.1.1.26.2 Logarithm of vapor pressure versus reciprocal temperature for *n*-hexylbenzene.

Mononuclear Aromatic Hydrocarbons

3.1.1.27 Heptylbenzene

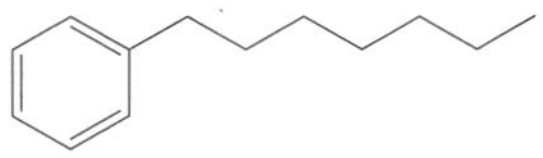

Common Name: Heptylbenzene
Synonym: 1-phenylheptane
Chemical Name: *n*-heptylbenzene
CAS Registry No: 1078-71-3
Molecular Formula: $C_{13}H_{20}$, $C_6H_5(CH_2)_6CH_3$
Molecular Weight: 176.298
Melting Point (°C):
    –48   (Weast 1982–83; Stephenson & Malanowski 1987; Lide 2003)
Boiling Point (°C):
    240   (Lide 2003)
Density (g/cm^3):
    0.8567   (20°C, Weast 1982–83)
Molar Volume (cm^3/mol):
    205.8   (20°C, calculated-density)
    251.4   (calculated-Le Bas method at normal boiling point))
Enthalpy of Fusion, ΔH_{fus} (kJ/mol):
Entropy of Fusion, ΔS_{fus} (J/mol K):
Fugacity Ratio at 25°C, F: 1.0

Water Solubility (g/m^3 or mg/L at 25°C):
    0.686; 0.925   (calculated-regression eq., calculated-molar volume correlation, Wang et al. 1992)

Vapor Pressure (Pa at 25°C and reported temperature dependence equations. Additional data at other temperatures designated * are compiled at the end of this section):
    133.3*   (66.2°C, summary of literature data, temp range 66.2–233°C, Stull 1947)
    1333*   (112°C, derived from compiled data, temp range 112–246°C, Zwolinski & Wilhoit 1971)
    log (P/mmHg) = 7.0006 – 1761.2/(181.5 + t/°C); temp range 112–279°C (Antoine eq., Zwolinski & Wilhoit 1971)
    log (P/kPa) = 6.1255 – [1761.2/(T – 91.65)]; temp range 423–527 K (liquid, Antoine equation, Stephenson & Malanowski 1987)
    log (P/mmHg) = 89.2811 – 6.4093 × 10^3/(T/K) – 29.248·log (T/K) + 1.0328 × 10^{-2}·(T/K) + 6.2451 × 10^{-14}·(T/K)^2, temp range 225–714 K (Yaws 1994)

Henry's Law Constant (Pa·m^3/mol at 25°C):

Octanol/Water Partition Coefficient, log K_{OW}:
    5.768   (HPLC-k' correlation, Hanai & Hubert 1981)
    5.37   (HPLC-k' correlation, Ritter et al. 1995)

Octanol/Air Partition Coefficient, log K_{OA}:

Bioconcentration Factor, log BCF or log K_B:

Sorption Partition Coefficient, log K_{OC}:

Environmental Fate Rate Constants, k, and Half-Lives, $t_{1/2}$:

Half-Lives in the Environment:

TABLE 3.1.1.27.1
Reported vapor pressures of heptylbenzene at various temperatures and the coefficients for the vapor pressure equations

$\log P = A - B/(T/K)$ (1) $\ln P = A - B/(T/K)$ (1a)

$\log P = A - B/(C + t/°C)$ (2) $\ln P = A - B/(C + t/°C)$ (2a)

$\log P = A - B/(C + T/K)$ (3)

$\log P = A - B/(T/K) - C \cdot \log (T/K)$ (4)

Stull 1947		Zwolinski & Wilhoit 1971			
summary of literature data		selected values			
t/°C	P/Pa	t/°C	P/Pa	t/°C	P/Pa
66.2	133.3	112	1333	243	94659
94.8	666.6	128	2666	243.6	95992
109	1333	137	4000	244.2	97325
124.2	2666	145	5333	244.8	98659
141.6	5333	150.7	6666	245.4	99992
151.5	7999	155.7	7999	246	101325
165.7	13332	164	10666		
186.6	26664	170.7	13332	eq. 2	P/mmHg
210	53329	183.6	19998	A	6.97833
233	101325	193.3	26664	B	1639.91
		201.1	33331	C	194.76
mp/°C	–	207.8	39997		
		218.9	53329	bp/°C	205.46
		227.9	66661	ΔH_V/(kJ mol^{-1}) =	
		235.6	79993	at 25°C	55.06
		242.3	93326	at bp	41.21

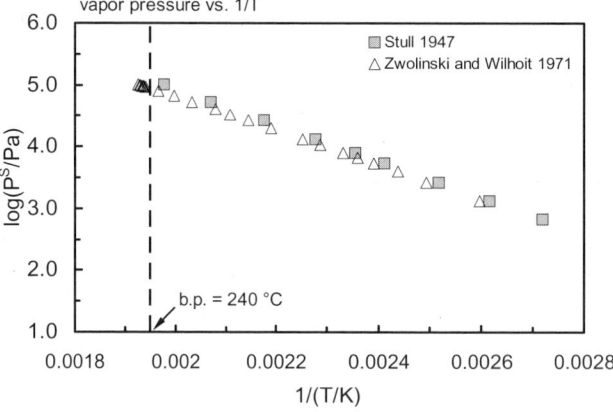

FIGURE 3.1.1.27.1 Logarithm of vapor pressure versus reciprocal temperature for heptylbenzene.

Mononuclear Aromatic Hydrocarbons

3.1.1.28 n-Octylbenzene

Common Name: *n*-Octylbenzene
Synonym: 1-phenyloctane, octylbenzene
Chemical Name: *n*-octylbenzene
CAS Registry No: 2189-60-8
Molecular Formula: $C_{14}H_{22}$, $C_6H_5(CH_2)_7CH_3$
Molecular Weight: 190.325
Melting Point (°C):
    −36   (Lide 2003)
Boiling Point (°C):
    264   (Lide 2003)
Density (g/cm³):
    0.8582   (20°C, Weast 1982–83)
Molar Volume (cm³/mol):
    222.2   (20°C, calculated-density)
    273.6   (calculated-Le Bas method at normal boiling point)
Enthalpy of Fusion, ΔH_{fus} (kJ/mol):
    29.96   (Chickos et al. 1999)
Entropy of Fusion, ΔS_{fus} (J/mol K):
    127.91, 110.4   (exptl., calculated-group additivity method, total phase change entropy, Chickos et al. 1999)
Fugacity Ratio at 25°C, F: 1.0

Water Solubility (g/m³ or mg/L at 25°C):
    0.219; 0.204   (calculated-regression eq.; calculated-molar volume correlation, Wang et al. 1992)

Vapor Pressure (Pa at 25°C or as indicated and reported temperature dependence equations. Additional data at other temperatures designated * are compiled at the end of this section):
    1333*   (127°C, derived from compiled data, temp range 127–264.4°C, Zwolinski & Wilhoit 1971)
    log (P/mmHg) = 7.0086 − 1812.2/(174.6 + t/°C); temp range 127–298°C (Antoine eq., Zwolinski & Wilhoit 1971)
    7.666*   (43°C, gas saturation, measured range 43–125.6°C, Allemand et al. 1986)
    log (P/kPa) = 8.35571 − 3293.744/(T/K); temp range 368–400 K (liquid, Antoine equation, Stephenson & Malanowski 1987)
    1.15*   (20.01°C, static method, measured range 293.16–462.87 K, Kasehgari et al. 1993)
    log (P/kPa) = 6.50210 − 2183.874/(207.887 + t/°C); temp range 293.16–462.87 K (static method, Kasehgari et al. 1993)
    log (P/mmHg) = 1.8919 − 4.1324 × 10³/(T/K) + 6.1473·log (T/K) − 2.0294 × 10⁻²·(T/K) + 9.6879 × 10⁻⁶·(T/K)², temp range 237–729 K (Yaws 1994)
    1.105*   (20.34°C, static method, measured range 293.49–462.87 K, Mokbel et al. 1998)

Henry's Law Constant (Pa·m³/mol at 25°C):

Octanol/Water Partition Coefficient, log K_{OW}:
    6.30   (RP-TLC-RT correlation, Bruggeman et al. 1982)
    6.52, 6.29   (RP-HPLC-k′ correlation, Harnisch et al. 1982)
    6.297   (HPLC-k′ correlation, Hanai & Hubert 1984)
    6.35, 6.85   (RP-HPLC-k′ correlations, Sherblom & Eganhouse 1988)
    6.30   (recommended, Sangster 1993)
    5.89   (HPLC-k′ correlation, Ritter et al. 1995)

Octanol/Air Partition Coefficient, log K_{OA}:

Bioconcentration Factor, log BCF or log K_B:

Sorption Partition Coefficient, log K_{OC}:

Environmental Fate Rate Constants, k, and Half-Lives, $t_{1/2}$:

Half-Lives in the Environment:

TABLE 3.1.1.28.1
Reported vapor pressures of octylbenzene at various temperatures and the coefficients for the vapor pressure equations

$\log P = A - B/(T/K)$ (1) $\qquad$ $\ln P = A - B/(T/K)$ (1a)

$\log P = A - B/(C + t/°C)$ (2) $\qquad$ $\ln P = A - B/(C + t/°C)$ (2a)

$\log P = A - B/(C + T/K)$ (3)

$\log P = A - B/(T/K) - C \cdot \log (T/K)$ (4)

Zwolinski & Wilhoit 1971		Allemand et al. 1986		Kasehgari et al. 1993		Mokbel et al. 1998	
selected values		gas saturation		static method		static method	
t/°C	P/Pa	t/°C	P/Pa	T/K	P/Pa	T/K	P/Pa
127	1333	43.0	7.666	293.16	1.15	293.49	1.105
143	2666	52.43	14.8	303.13	3.01	313.49	2.690
153	4000	72.75	66.8	313.01	6.87	303.48	6.316
161	5333	84.2	134.7	323.01	15.3	23.47	13.92
166.7	6666	95.39	262.7	333.15	31.5	333.48	28.76
171.9	7999	106.0	437.3	343.15	60.8	343.45	57.64
180.4	10666	125.64	1205	353.13	114	353.45	109.0
187.2	13332			363.14	207	363.42	198.0
200.3	19998			373.22	362	373.45	350.0
210.4	26664			383.25	609	403.00	1513
218.4	33331			393.32	994	412.96	2329
225.3	39997			403.33	1574	422.92	3504
236.7	53329			412.96	2329	432.94	5143
235.9	66661			422.92	3504	442.96	7395
253.8	79993			432.94	5143	452.90	10366
260.7	93326			442.96	7395	462.87	14266
261.9	95992			452.90	10370		
263.2	98659			462.87	14270	data fitted to Wagner eq.	
264.4	101325						
				Antoine eq.			
eq. 2	P/mmHg			eq. 2	P/kPa		
A	7.0086			A	6.50210		
B	1812.2			B	2183.874		
C	174.6			C	207.887		
bp/°C	264.4						
ΔH_V/(kJ mol^{-1}) =							
at 25°C	69.96						
at bp	46.9						

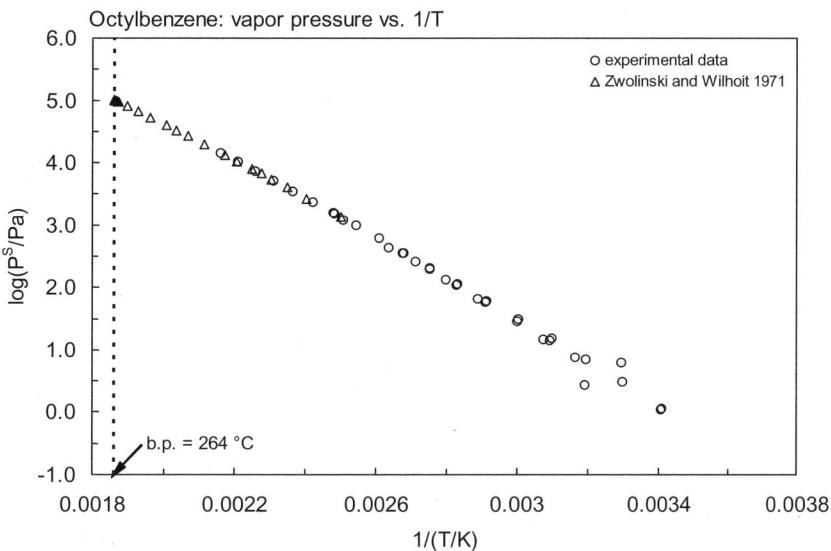

FIGURE 3.1.1.28.1 Logarithm of vapor pressure versus reciprocal temperature for octylbenzene.

3.1.1.29 Nonylbenzene

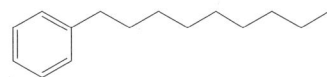

Common Name: Nonylbenzene
Synonym: 1-phenylnonane
Chemical Name: n-nonylbenzene
CAS Registry No: 1081-77-2
Molecular Formula: $C_{15}H_{24}$, $C_6H_5(CH_2)_8CH_3$
Molecular Weight: 204.352
Melting Point (°C):
    –24   (Dreisbach 1955; Lide 2003)
Boiling Point (°C):
    280.5   (Lide 2003)
Density (g/cm³):
    0.8558, 0.8522   (20°C, 25°C, Dreisbach 1955)
    0.8584   (20°C, Weast 1982–83)
Molar Volume (cm³/mol):
    238.1   (20°C, calculated-density)
    295.8   (calculated-Le Bas method at normal boiling point)
Enthalpy of Fusion, ΔH_{fus} (kJ/mol):
Entropy of Fusion, ΔS_{fus} (J/mol K):
Fugacity Ratio at 25°C, F: 1.0

Water Solubility (g/m³ or mg/L at 25°C):
    0.0725; 0.112   (calculated-regression eq., calculated-molar volume correlation, Wang et al. 1992)

Vapor Pressure (Pa at 25°C and reported temperature dependence equations. Additional data at other temperatures designated * are compiled at the end of this section):
    0.573   (extrapolated-Antoine eq., Dreisbach 1955)
    log (P/mmHg) = 7.19041 – 1991.0/(180.0 + t/°C); temp range 165–330°C (Antoine eq. for liquid state, Dreisbach 1955)
    1333*   (142°C, derived from compiled data, temp range 142–282°C, Zwolinski & Wilhoit 1971)
    log (P/mmHg) = 7.0245 – 1862.6/(167.5 + t/°C); temp range 142–316°C (Antoine eq., Zwolinski & Wilhoit 1971)
    2.906*   (43°C, gas saturation, measured range 43–142°C, Allemand et al. 1986)
    0.338   (GC-RT correlation, Sherblom et al. 1992)
    log (P/mmHg) = –0.9235 – 4.2232 × 10³/(T/K) + 7.3073·log (T/K) – 2.0964 × 10⁻²·(T/K) + 9.7152 × 10⁻⁶·(T/K)², temp range 249–741 K (Yaws 1994)
    0.8747*   (30.52°C, static method, measured range 313.67–466.46 K, Mokbel et al. 1998)

Henry's Law Constant (Pa·m³/mol at 25°C):

Octanol/Water Partition Coefficient, log K_{OW}:
    6.828   (HPLC-k′ correlation, Hanai & Hubert 1981)
    6.83, 7.40   (RP-HPLC-k′ correlations, Sherblom & Eganhouse 1988)
    7.11   (recommended, Sangster 1993)
    6.41   (HPLC-k′ correlation, Ritter et al. 1995)
    6.83   (quoted from Sherblom & Eganhouse 1988, Hansch et al. 1995)

Octanol/Air Partition Coefficient, log K_{OA}:
Bioconcentration Factor, log BCF or log K_B:
Sorption Partition Coefficient, log K_{OC}:

Environmental Fate Rate Constants, k, and Half-Lives, $t_{1/2}$:
Half-Lives in the Environment:

TABLE 3.1.1.29.1
Reported vapor pressures of nonylbenzene at various temperatures and the coefficients for the vapor pressure equations

$$\log P = A - B/(T/K) \quad (1) \qquad \ln P = A - B/(T/K) \quad (1a)$$
$$\log P = A - B/(C + t/°C) \quad (2) \qquad \ln P = A - B/(C + t/°C) \quad (2a)$$
$$\log P = A - B/(C + T/K) \quad (3)$$
$$\log P = A - B/(T/K) - C \cdot \log (T/K) \quad (4)$$

Zwolinski & Wilhoit 1971				Allemand et al. 1986		Mokbel et al. 1998	
selected values				gas saturation		static method	
t/°C	P/Pa			t/°C	P/Pa	T/K	P/Pa
142	1333	eq. 2	P/mmHg	43	2.906	303.67	0.8474
158	2666	A	7.0245	50	5.186	313.65	2.271
168	4000	B	1862.6	59	10.36	323.67	5.369
176	5333	C	167.5	64	15.07	333.61	12.07
182.3	6666			74.32	31.73	343.53	25.20
187.5	7999	bp/°C	282.0	85.0	64.26	351.83	44.97
196.2	10666	ΔH_V/(kJ mol^{-1})		91.3	95.33	361.84	85.44
203.2	13332	at 25°C	74.81	98.28	146.7	371.75	155
216.7	19998	at bp	49.0	108.57	260.0	381.65	270
226.8	26664			120.0	480.0	391.62	459
235.1	33331			130.73	780.0	401.56	752
247.1	39997			142.0	1352	411.48	1207
253.7	53329					411.60	1203
263.1	66661					421.45	1881
271.1	79993					431.44	2835
278.2	93326					441.49	4201
279.5	95992					451.48	6053
280.7	98659					461.48	8591
282	101325					466.46	10099

data fitted to Wagner eq.

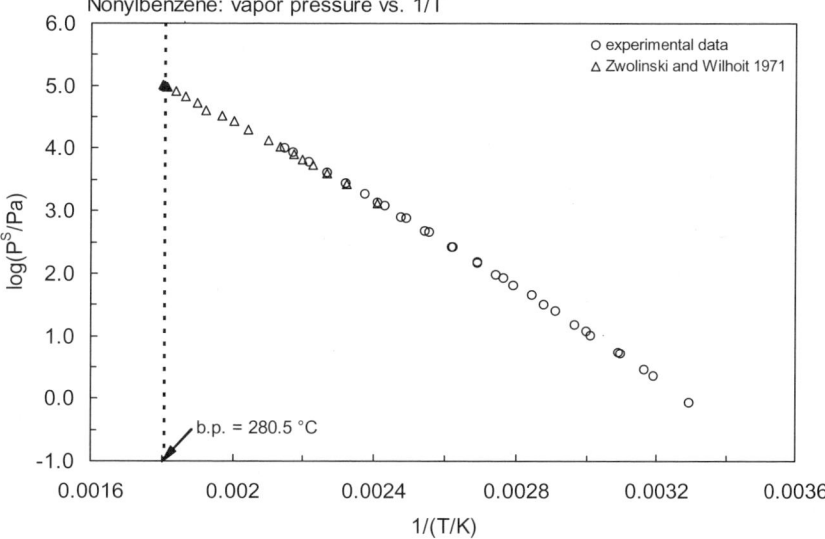

FIGURE 3.1.1.29.1 Logarithm of vapor pressure versus reciprocal temperature for nonylbenzene.

3.1.1.30 Decylbenzene

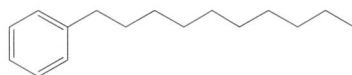

Common Name: Decylbenzene
Synonym: 1-phenyldecane
Chemical Name: *n*-decylbenzene
CAS Registry No: 104-72-3
Molecular Formula: $C_{16}H_{26}$, $C_6H_5(CH_2)_9CH_3$
Molecular Weight: 218.337
Melting Point (°C):
    −14.4   (Dreisbach 1955, Stephenson & Malanowski 1987: Lide 2003)
Boiling Point (°C):
    293   (Lide 2003)
Density (g/cm^3):
    0.85553, 0.85189   (20°C, 25°C, Camin et al. 1954; Dreisbach 1955)
Molar Volume (cm^3/mol):
    255.3   (20°C, calculated-density)
    318.0   (calculated-Le Bas method at normal boiling point)
Enthalpy of Fusion, ΔH_{fus} (kJ/mol):
Entropy of Fusion, ΔS_{fus} (J/mol K):
Fugacity Ratio at 25°C, F: 1.0

Water Solubility (g/m^3 or mg/L at 25°C):
    2.50   (vapor-phase saturation, shake flask-GC, Sherblom et al. 1992)
    5.59, 8.47, 10.55   (calculated-V_M, K_{OW}, TSA, Sherblom et al. 1992)
    0.023; 0.0188   (calculated-regression eq., calculated-molar volume correlation, Wang et al. 1992)

Vapor Pressure (Pa at 25°C and reported temperature dependence equations. Additional data at other temperatures designated * are compiled at the end of this section):
    8329*   (202.987°C, ebulliometry, measured range 202.987–297.799°C, Camin et al. 1954)
    log (P/mmHg) = 7.03642 − 1904.132/(160.318 + t/°C); temp range 202.9–297.9°C (Antoine eq., manometer measurements, Camin et al. 1954)
    0.20   (extrapolated-Antoine eq., Dreisbach 1961)
    log (P/mmHg) = 7.27177 − 2107.7/(180.0 + t/°C); temp range 185–345°C (Antoine eq. for liquid state, Dreisbach 1955)
    1333*   (155.1°C, derived from compiled data, temp range 155.1–297.89°C, Zwolinski & Wilhoit 1971)
    log (P/mmHg) = 7.03642 − 1904.132/(160.318 + t/°C); temp range 155.1–322.9°C (Antoine eq., Zwolinski & Wilhoit 1971)
    log (P/kPa) = 6.16274 − 1905.56/(160.503 + t/°C); temp range 203–297.8°C (Antoine eq. from reported exptl. data of Camin et al. 1954, Boublik et al. 1984)
    log (P/kPa) = 4.03653 − 876.208/(T/K − 203.15); temp range 317–427 K (liquid, Antoine equation, Stephenson & Malanowski 1987)
    log (P/kPa) = 6.15658 − [1900.916/(T/K − 113.16)]; temp range 475–571 K (liquid, Antoine equation, Stephenson & Malanowski 1987)
    0.133, 0.127 (P_L, GC-RT correlation, Sherblom et al. 1992)
    0.707*   (39.85°C, static method, measured range 313.0–433.23 K, Kasehgari et al. 1993)
    log (P/kPa) = 6.37655 − 2098.329/(180.620 + t/°C); temp range 313.0–433.23 K (static method, Kasehgari et al. 1993)
    log (P/mmHg) = −4.4754 − 4.4669 × 10^3/(T/K) + 9.1965·log (T/K) − 2.4010 × 10^{-2}·(T/K) + 1.0848 × 10^{-5}·(T/K)^2, temp range 259–753 K (Yaws 1994)

Mononuclear Aromatic Hydrocarbons

Henry's Law Constant (Pa·m^3/mol at 25°C):

Octanol/Water Partition Coefficient, log K_{OW}:
- 7.35 (RP-TLC retention time correlation, Bruggeman et al. 1982)
- 7.60, 7.33 (RP-HPLC-k' correlations, Harnisch et al. 1982)
- 7.358 (HPLC-k' correlation, Hanai & Hubert 1984)
- 7.37, 8.01 (RP-HPLC-k' correlation, Sherblom & Eganhouse 1988)
- 7.38 (HPLC-k' correlation, Sherblom et al. 1992)
- 7.35 (recommended, Sangster 1993)
- 6.94 (HPLC-k' correlation, Ritter et al. 1995)
- 7.37 (quoted from Sherblom & Eganhouse 1988, Hansch et al. 1995)

Octanol/Air Partition Coefficient, log K_{OA}:

Bioconcentration Factor, log BCF or log K_B:

Sorption Partition Coefficient, log K_{OC}:

Environmental Fate Rate Constants, k, and Half-Lives, $t_{\frac{1}{2}}$:

Half-Lives in the Environment:

TABLE 3.1.1.30.1
Reported vapor pressures of decylbenzene at various temperatures and the coefficients for the vapor pressure equations

$\log P = A - B/(T/K)$ (1) $\ln P = A - B/(T/K)$ (1a)
$\log P = A - B/(C + t/°C)$ (2) $\ln P = A - B/(C + t/°C)$ (2a)
$\log P = A - B/(C + T/K)$ (3)
$\log P = A - B/(T/K) - C \cdot \log(T/K)$ (4)

Camin et al. 1954		Zwolinski & Wilhoit 1971				Kasehgari et al. 1993	
ebulliometry		selected values				static method	
t/°C	P/Pa	t/°C	P/Pa			T/K	P/Pa
202.987	8329	155.1	1333	eq. 2	P/mmHg	313.0	0.797
211.392	10938	171.7	2666	A	7.03642	323.13	1.92
217.156	13088	182.19	4000	B	1904.132	343.15	10.3
230.476	19442	190.07	5333	C	160.318	353.11	21.6
244.331	28555	196.431	6666			363.12	41.9
260.372	43172	201.803	7999	bp/°C	297.89	373.25	78.5
278.950	67077	210.617	10666	ΔH_V/(kJ mol^{-1}) =		383.28	142
296.370	98146	217.755	13332	at 25°C	79.75	393.34	248
297.799	101113	231.452	19998	at bp	50.6	403.37	417
		241.789	26664			413.33	683
bp/°C	297.083	250.19	33331			413.33	684
		257.18	39997			423.21	1081
		269.86	53329			433.23	1687
		278.68	66661				
		286.843	79993			Antoine eq	
		293.986	93326			eq. 2	P/kPa
		295.969	95992			A	6.37655
		296.616	98659			B	2098.329
		297.89	101325			C	180.620

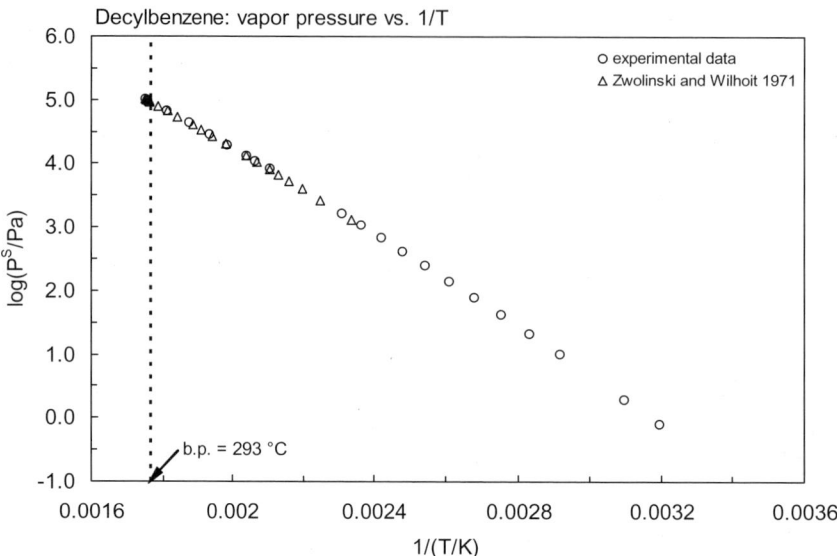

FIGURE 3.1.1.30.1 Logarithm of vapor pressure versus reciprocal temperature for decylbenzene.

3.1.1.31 Undecylbenzene

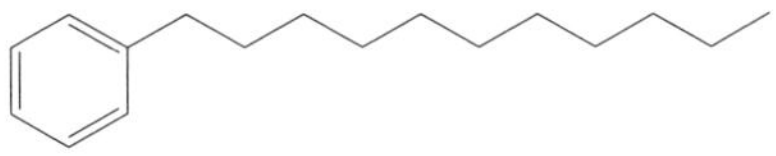

Common Name: Undecylbenzene
Synonym: 1-phenylundecane
Chemical Name: *n*-undecylbenzene
CAS Registry No: 6742-54-7
Molecular formula: $C_{17}H_{28}$, $C_6H_5(CH_2)_{10}CH_3$
Molecular Weight: 232.404
Melting Point (°C):
    –5.0   (Dreisbach 1955; Lide 2003)
Boiling Point (°C):
    316   (Dreisbach 1955; Lide 2003)
Density (g/cm³):
    0.8553, 0.8517   (20°C, 25°C, Dreisbach 1961)
Molar Volume (cm³/mol):
    271.7   (20°C, calculated-density)
    340.2   (calculated-Le Bas method at normal boiling point)
Enthalpy of Fusion, ΔH_{fus} (kJ/mol):
Entropy of Fusion, ΔS_{fus} (J/mol K):
Fugacity Ratio at 25°C, F: 1.0

Water Solubility (g/m³ or mg/L at 25°C):
    0.00702; 0.00377   (calculated-regression eq., calculated-molar volume correlation, Wang et al. 1992)

Vapor Pressure (Pa at 25°C and reported temperature dependence equations. Additional data at other temperatures designated * are compiled at the end of this section):
    0.080   (calculated-Antoine eq., Dreisbach 1955)
    log (P/mmHg) = 7.34672 – 2215.1/(180.0 + t/°C); temp range 195–375°C (Antoine eq. for liquid state, Dreisbach 1955)
    1333*   (168°C, derived from compiled data, temp range 168–313.2°C, Zwolinski & Wilhoit 1971)
    log (P/mmHg) = 7.0509 – 1944.1/(153.0 + t/°C); temp range 168–349°C (Antoine eq., Zwolinski & Wilhoit 1971)
    0.050, 0.047   (P_L, GC-RT correlation, Sherblom et al. 1992)
    log (P/mmHg) = 124.1549 – 8.8970 × 10³/(T/K) – 41.223·log (T/K) + 1.3662 × 10⁻²·(T/K) – 8.1321 × 10⁻¹⁴·(T/K)², temp range 268–764 K (Yaws 1994)
    0.7156*   (50.42°C, static method, measured range 323.57–467.33°C, data fitted to Wagner eq., Mokbel et al. 1998)

Henry's Law Constant (Pa·m³/mol at 25°C):

Octanol/Water Partition Coefficient, log K_{OW}:
    8.14   (RP-HPLC-RT correlation, Sherblom et al. 1992, quoted, Sangster 1993)

Octanol/Air Partition Coefficient, log K_{OA}:

Bioconcentration Factor, log BCF or log K_B:

Sorption Partition Coefficient, log K_{OC}:

Environmental Fate Rate Constants, k, and Half-Lives, $t_{½}$:

Half-Lives in the Environment:

TABLE 3.1.1.31.1
Reported vapor pressures of undecylbenzene at various temperatures and the coefficients for the vapor pressure equations

$\log P = A - B/(T/K)$ (1) $\ln P = A - B/(T/K)$ (1a)

$\log P = A - B/(C + t/°C)$ (2) $\ln P = A - B/(C + t/°C)$ (2a)

$\log P = A - B/(C + T/K)$ (3)

$\log P = A - B/(T/K) - C \cdot \log(T/K)$

Zwolinski & Wilhoit 1971				Mokbel et al. 1998			
selected values				static method			
t/°C	P/Pa	t/°C	P/Pa	T/K	P/Pa	T/K	P/Pa
168	1333	309.2	93326	323.57	0.7156	462.28	3088
185	2666	310.6	95992	333.59	1.820	467.33	37726
196	4000	311.9	98659	343.71	4.304		
204	5333	313.2	101325	352.69	9.405	data fitted to Wagner eq.	
210.3	6666			363.78	19.39		
215.7	7999	eq. 2	P/mmHg	372.07	34.36		
224.7	10666	A	7.0590	382.06	64.41		
231.9	13332	B	1944.1	392.06	117		
245.8	19998	C	153.0	402.11	204		
256.3	26664			412.11	344		
264.8	33331	bp/°C	313.2	422.09	564		
272.1	39997	$\Delta H_V/(kJ\, mol^{-1})$ =		422.19	568		
284	53329	at 25°C	84.68	432.25	908		
293.7	66661	at bp	52.3	442.27	1369		
302	79993			452.29	2080		

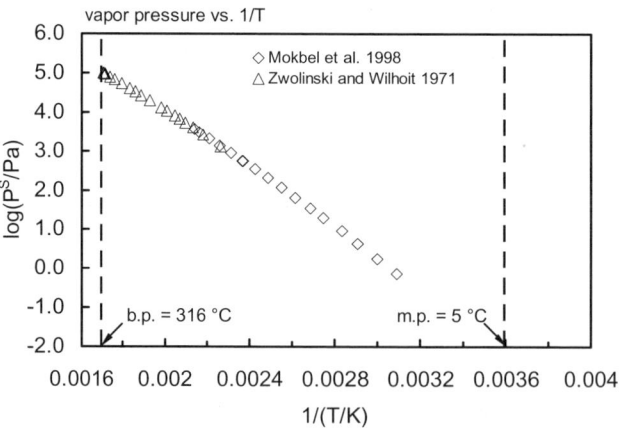

FIGURE 3.1.1.31.1 Logarithm of vapor pressure versus reciprocal temperature for undecylbenzene.

3.1.1.32 Dodecylbenzene

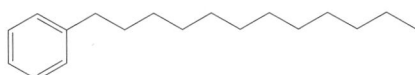

Common Name: Dodecylbenzene
Synonym: 1-phenyldodecane
Chemical Name: n-dodecylbenzene
CAS Registry No: 123-01-3
Molecular Formula: $C_{18}H_{30}$, $C_6H_5(CH_2)_{11}CH_3$
Molecular Weight: 246.431
Melting Point (°C):
    3.0   (Dreisbach 1955; Lide 2003)
Boiling Point (°C):
    328   (Lide 2003)
Density (g/cm³):
    0.8551, 0.8516   (20°C, 25°C, Dreisbach 1955)
Molar Volume (cm³/mol):
    288.2   (20°C, calculated-density)
    362.4   (calculated-Le Bas method at normal boiling point)
Enthalpy of Fusion, ΔH_{fus} (kJ/mol):
Entropy of Fusion, ΔS_{fus} (J/mol K):
Fugacity Ratio at 25°C, F: 1.0

Water Solubility (g/m³ or mg/L at 25°C):

Vapor Pressure (Pa at 25°C and reported temperature dependence equations. Additional data at other temperatures designated * are compiled at the end of this section):
    26.66*   (78.4°C, ebulliometry-McLeod gauge, measured range 78.4–288.6°C, Myers & Fenske 1955)
    0.032   (extrapolated-Antoine eq., Dreisbach 1955)
    $\log (P/\text{mmHg}) = 7.41934 - 2319.2/(180.0 + t/°C)$; temp range 210–385°C (Antoine eq. for liquid state, Dreisbach 1955)
    1333*   (181°C, derived from compiled data, temp range 181–327.6°C, Zwolinski & Wilhoit 1971)
    $\log (P/\text{mmHg}) = 7.0693 - 1981.6/(145.5 + t/°C)$; temp range 181–363°C (Antoine eq., Zwolinski & Wilhoit 1971)
    1.08*   (53.7°C, gas saturation, measured range 53.7–192.7°C, Allemand et al. 1986)
    0.019, 0.017   (P_L, GC-RT correlation, Sherblom et al. 1992)
    0.727*   (69.771°C, static method, measured range 332.92–453.26 K, Kasehgari et al. 1993)
    $\log (P/\text{kPa}) = 6.66087 - 2371.902/(182.311 + t/°C)$; temp range 332.92–453.26 K (static method, Kasehgari et al. 1993)
    $\log (P/\text{mmHg}) = 145.6916 - 1.0165 \times 10^3/(T/K) - 48.761 \cdot \log (T/K) + 1.5985 \times 10^{-2} \cdot (T/K) + 4.881 \times 10^{-13} \cdot (T/K)^2$, temp range 276–774 K (Yaws 1994)

Henry's Law Constant (Pa·m³/mol at 25°C):

Octanol/Water Partition Coefficient, log K_{OW}:
    8.65   (RP-HPLC-k′ correlation, Sherblom et al. 1992)

Octanol/Air Partition Coefficient, log K_{OA}:

Bioconcentration Factor, log BCF or log K_B:

Sorption Partition Coefficient, log K_{OC}:

Environmental Fate Rate Constants, k, and Half-Lives, $t_{½}$:

Half-Lives in the Environment:

TABLE 3.1.1.32.1
Reported vapor pressures of dodecylbenzene at various temperatures and the coefficients for the vapor pressure equations

$\log P = A - B/(T/K)$ (1) $\ln P = A - B/(T/K)$ (1a)
$\log P = A - B/(C + t/°C)$ (2) $\ln P = A - B/(C + t/°C)$ (2a)
$\log P = A - B/(C + T/K)$ (3)
$\log P = A - B/(T/K) - C \cdot \log (T/K)$ (4)

Myers & Fenske 1955		Zwolinski & Wilhoit 1971		Allemand et al. 1986		Kasehgari et al. 1993	
ebulliometry-McLeod gauge		selected values		gas saturation		static method	
t/°C	P/Pa	t/°C	P/Pa	t/°C	P/Pa	T/K	P/Pa
78.4	26.66	181	1333	53.7	1.080	343.21	0.640
91.2	66.66	198	2666	86.19	6.733	353.26	1.61
102.3	133.3	209	4000	93.0	11.15	363.23	3.72
114.2	266.6	217	5333	103.0	21.60	363.23	3.73
122.4	400.0	223.5	6666	111.0	36.40	373.21	8.11
128.2	533.3	229.0	7999	124.02	79.86	383.22	16.9
133.4	666.6	238.1	10666	136.47	158.7	393.22	32.8
137.0	800	245.4	13332	160.44	540	403.27	59.5
143.5	1067	259.5	19998	192.71	1464	413.28	106
148.8	1333	270.1	26664			423.29	182
158.5	2000	278.7	33331			433.37	306
165.6	2666	286	39997			443.39	493
176.4	4000	298.1	53329			453.33	760
184.2	5333	307.9	66661			463.38	1167
191.0	6666	316.3	79993				
196.0	7999	323.6	93326			Antoine eq	
205.5	10666	325	95992			eq. 2	P/kPa
212.0	13332	326.3	98659			A	6.66087
225.4	19998	327.6	101325			B	2371.902
235.8	26664					C	182.311
250.3	39997	eq. 2	P/mmHg				
262.0	53329	A	7.0693				
270.4	66661	B	1981.6				
278.2	79992	C	145.5				
285.5	93326						
288.6	101325	bp/°C	327.6				
		ΔH_V/(kJ mol^{-1})					
		at 25°C	89.62				
		at bp	54.4				

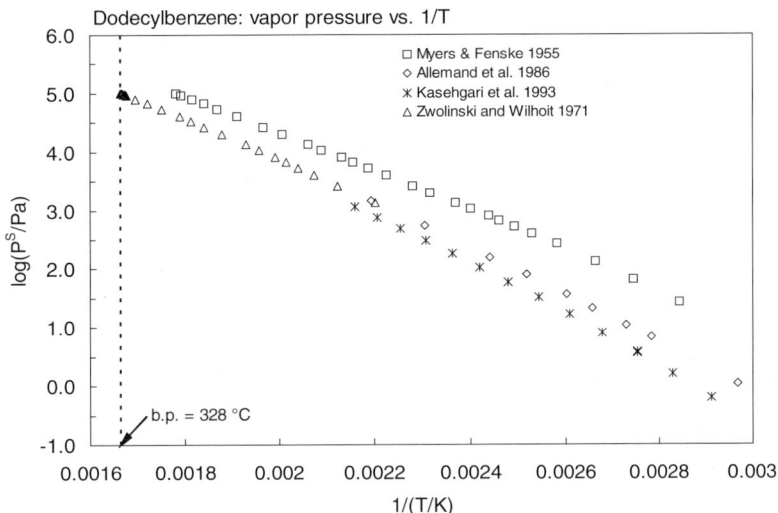

FIGURE 3.1.1.32.1 Logarithm of vapor pressure versus reciprocal temperature for dodecylbenzene.

3.1.1.33 Tridecylbenzene

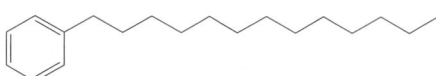

Common Name: Tridecylbenzene
Synonym: 1-phenyltridecane
Chemical Name: *n*-tridecylbenzene
CAS Registry No: 123-02-4
Molecular Formula: $C_{19}H_{32}$, $C_6H_5(CH_2)_{12}CH_3$
Molecular Weight: 260.457
Melting Point (°C):
    10   (Dreisbach 1955; Lide 2003)
Boiling Point (°C):
    346   (Dreisbach 1955; Lide 2003)
Density (g/cm³):
    0.8550, 0.8515   (20°C, 25°C, Dreisbach 1955)
Molar Volume (cm³/mol):
    304.6   (20°C, calculated-density)
    384.6   (calculated-Le Bas method at normal boiling point)
Enthalpy of Fusion, ΔH_{fus} (kJ/mol):
Entropy of Fusion, ΔS_{fus} (J/mol K):
Fugacity Ratio at 25°C, F: 1.0

Water Solubility (g/m³ or mg/L at 25°C):
    0.233   (vapor saturation-shake flask-GC, Sherblom et al. 1992)
    0.233, 0.0885, 0.181   (calculated-V_m, K_{OW}, TSA, Sherblom et al. 1992)
    0.00067; 0.00137   (calculated-regression eq., calculated-molar volume correlation, Wang et al. 1992)

Vapor Pressure (Pa at 25°C and reported temperature dependence equations. Additional data at other temperatures designated * are compiled at the end of this section):
    0.0125   (extrapolated-Antoine eq., Dreisbach 1955)
    log (P/mmHg) = 7.49437 – 2626.7/(180.0 + t/°C); temp range 226–405°C (Antoine eq. for liquid state, Dreisbach 1955)
    1333*   (193°C, derived from compiled data, temp range 193–341.2°C, Zwolinski & Wilhoit 1971)
    log (P/mmHg) = 7.0843 – 2013.9/(137.9 + t/°C); temp range 193–376°C (Antoine eq., Zwolinski & Wilhoit 1971)
    0.008, 0.07   (P_L, GC-RT correlation, Sherblom et al. 1992)
    0.64*   (70.06°C, static method, measured range 343.21–463.38 K, Kasehgari et al. 1993)
    log (P/kPa) = 6.13410 – 2087.968/(153.790 + t/°C); temp range 343.21–463.38 K (static method, Kasehgari et al. 1993)
    log (P/mmHg) = 160.3924 – 1.1093 × 10⁴/(T/K) – 53.875·log (T/K) + 1.7532 × 10⁻²·(T/K) + 3.727 × 10⁻¹³·(T/K)², temp range 283–783 K (Yaws 1994)

Henry's Law Constant (Pa·m³/mol at 25°C):

Octanol/Water Partition Coefficient, log K_{OW}:
    9.36   (RP-HPLC-k′ correlation, Sherblom et al. 1992)
    8.97; 8.54   (calculated-fragment const., calculated-molar volume correlation, Wang et al. 1992)

Octanol/Air Partition Coefficient, log K_{OA}:

Bioconcentration Factor, log BCF or log K_B:

Sorption Partition Coefficient, log K_{OC}:

Environmental Fate Rate Constants, k, and Half-Lives, $t_{1/2}$:

Half-Lives in the Environment:

TABLE 3.1.1.33.1
Reported vapor pressures of tridecylbenzene at various temperatures

Zwolinski & Wilhoit 1971				Kasehgari et al. 1993	
selected values				static method	
t/°C	P/Pa	t/°C	P/Pa	T/K	P/Pa
193	1333	338.5	95992	343.20	6.40×10^{-4}
210	2666	339.9	98659	353.26	1.61×10^{-3}
221	4000	341.2	101325	363.23	3.73×10^{-3}
229	5333			373.21	8.11×10^{-3}
236	6666	log P = A − B/(C + t/°C)		383.22	0.0169
242	7999		P/mmHg	393.22	0.0328
251	10666	A	7.0843	403.27	0.0595
258.2	13332	B	2013.9	413.28	0.106
271.6	19998	C	137.9	423.29	0.182
283.1	26664			433.37	0.306
291.8	33331	bp/°C	341.2	443.39	0.493
299.2	39997	ΔH_V/(kJ mol^{-1})		453.33	0.760
311.4	53329	at 25°C	94.6	463.38	1.167
321.3	66661	at bp	56.1		
329.8	79993				
337.2	93326				

3.1.1.34 Tetradecylbenzene

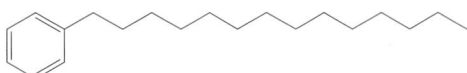

Common Name: Tetradecylbenzene
Synonym: 1-phenyltetradecane
Chemical Name: *n*-tetradecylbenzene
CAS Registry No: 1459-10-5
Molecular Formula: $C_{20}H_{34}$, $C_6H_5(CH_2)_{13}CH_3$
Molecular Weight: 274.484
Melting Point (°C):
    16   (Dreisbach 1955; Lide 2003)
Boiling Point (°C):
    359   (Dreisbach 1955; Lide 2003)
Density (g/cm³):
    0.8549, 0.8514   (20°C, 25°C, Dreisbach 1955)
Molar Volume (cm³/mol):
    321.7   (20°C, calculated-density)
    406.8   (calculated-Le Bas method at normal boiling point)
Enthalpy of Fusion, ΔH_{fus} (kJ/mol):
Entropy of Fusion, ΔS_{fus} (J/mol K):
Fugacity Ratio at 25°C, F: 1.0

Water Solubility (g/m³ or mg/L at 25°C):

Vapor Pressure (Pa at 25°C and reported temperature dependence equations. Additional data at other temperatures designated * are compiled at the end of this section):
    1333*   (205°C, derived from compiled data, temp range 205–354°C, Zwolinski & Wilhoit 1971)
    0.0055   (extrapolated-Antoine eq., Dreisbach 1955; quoted, Sherblom et al. 1992)
    log (P/mmHg) = 7.56143 − 2522.8/(180.0 + t/°C); temp range 235–410°C (Antoine eq. for liquid state, Dreisbach 1955)
    log (P/mmHg) = 7.101 − 2042/(130 + t/°C); temp range 205–300°C (Antoine eq., Zwolinski & Wilhoit 1971)
    0.002   (P_L, GC-RT correlation, Sherblom et al. 1992)

Henry's Law Constant (Pa·m³/mol at 25°C):

Octanol/Water Partition Coefficient, log K_{OW}:
    9.95   (RP-HPLC-k′ correlation, Sherblom et al. 1992)

Octanol/Air Partition Coefficient, log K_{OA}:

Bioconcentration Factor, log BCF or log K_B:

Sorption Partition Coefficient, log K_{OC}:

Environmental Fate Rate Constants, k, and Half-Lives, $t_{½}$:

Half-Lives in the Environment:

TABLE 3.1.1.34.1
Reported vapor pressures of tetradecylbenzene at various temperatures

Zwolinski & Wilhoit 1971

selected values			
t/°C	P/Pa	t/°C	P/Pa
205	1333	351	95992
222	2666	353	98659
233	4000	354	101325
241	5333		
248	6666	$\log P = A - B/(C + t/°C)$	
254	7999	eq. 2	P/mmHg
263	10666	A	7.010
270	13332	B	2042.0
285	19998	C	130.0
295	26664		
304	33331	bp/°C	354.0
312	39997	ΔH_V/(kJ mol^{-1})	
324	53329	at 25°C	99.6
334	66661	at bp	57.7
342	79993		
350	93326		

3.1.1.35 Styrene

Common Name: Styrene
Synonym: phenylethene, styrol, styrolene cinnamene, cinnamol, phenylethylene, vinylbenzene, ethenylbenzene
Chemical Name: styrene
CAS Registry No: 100-42-5
Molecular Formula: C_8H_8, $C_6H_5CH=CH_2$
Molecular Weight: 104.150
Melting Point (°C):
    −30.65   (Lide 2003)
Boiling Point (°C):
    145   (Lide 2003)
Density (g/cm³ at 20°C):
    0.9060, 0.9012   (20°C, 25°C, Dreisbach & Martin 1949; Dreisbach 1955; Riddick et al. 1986)
    0.906   (Weast 1982–83)
Molar Volume (cm³/mol):
    115.0   (20°C, calculated-density)
    133.0   (calculated-Le Bas method at normal boiling point)
Enthalpy of Vaporization, ΔH_V (kJ/mol):
    43.932, 38.7 (25°C, bp, Riddick et al. 1986)
Enthalpy of Fusion, ΔH_{fus} (kJ/mol):
    10.95   (Riddick et al. 1986; Chickos et al. 1999)
Entropy of Fusion, ΔS_{fus} (J/mol K):
    45.15, 52.2   (exptl., calculated-group additivity method, Chickos et al. 1999)
Fugacity Ratio at 25°C, F: 1.0

Water Solubility (g/m³ or mg/L at 25°C or as indicated and reported temperature dependence equations. Additional data at other temperatures designated * are compiled at the end of this section):
    330*   (24°C, shake flask-Karl Fischer titration, measured range 7–51°C, Lane 1946)
    310*   (cloud point method, measured range 10–60°C, Lane 1946)
    220   (shake flask-method not specified, Frilette & Hohenstein 1948)
    300   (shake flask-UV, Andrews & Keefer 1950)
    160   (shake flask-HPLC/UV, Banerjee et al. 1980)
    250*   (recommended best value, temp range 10–60°C, Shaw 1989b)
    $\ln x = -19.471 - 1655.9/(T/K) - 4.6244 \times 10^{-5} \cdot (T/K)^2$; temp range 5–50°C (regression eq. of literature data, Shiu & Ma 2000)

Vapor Pressure (Pa at 25°C or as indicated and reported temperature dependence equations. Additional data at other temperatures designated * are compiled at the end of this section):
    288*   (8.2°C, mercury manometer, measured range −7.7 to 8.2°C, Linder 1931)
    $\log (P/\text{mmHg}) = 7.929 - 2103/(T/K)$; temp range 33.5–116.3°C (isoteniscope method, Burchfield 1942)
    841.3*   (static-Hg manometer, measured range 12.5–60°C, Pitzer et al. 1946)
    969.4*   (calculated-Antoine eq. regression, Stull 1947)
    $\log (P/\text{mmHg}) = 7.22302 - 1629.2/(230 + t/°C)$ (Antoine eq., Dreisbach & Martin 1949)
    6799*   (66.7°C, mercury manometer, Buck et al. 1949)
    $\log (P/\text{mmHg}) = -3151/(T/K) - 6.294 \cdot \log (T/K) + 26.92$; temp range 80–120°C (vapor pressure eq. from Hg manometer measurements, Buck et al. 1949)
    807   (extrapolated by formula, Dreisbach 1955)
    $\log (P/\text{mmHg}) = 6.92409 - 1430.0/(206.0 + t/°C)$; temp range 55–205°C (Antoine eq. for liquid state, Dreisbach 1955)
    1093*   (29.92°C, measured range 29.92–110.06°C, Dreyer et al. 1955)
    1333*   (32.24°C, measured range 32.4–82.9°C, Chaiyavech & van Winkle 1959)

Mononuclear Aromatic Hydrocarbons

log (P/mmHg) = 8.2696 − 2221.21/(T/K); temp range 32.4–82.9°C (Chaiyavech & van Winkle 1959)

log (P/mmHg) = [−0.2185 × 9634.7/(T/K)] + 7.922049; temp range −7.0 to 145.2°C (Antoine eq., Weast 1972–73)

879, 812 (extrapolated-Antoine eq., Boublik et al. 1973)

log (P/mmHg) = 7.14016 − 1574.511/(224.087 + t/°C); temp range 32.4–62.19°C (Antoine eq. from reported exptl. data, Boublik et al. 1973)

log (P/mmHg) = 7.06623 − 1507.434/(214.985 + t/°C); temp range 29.92–144.77°C (Antoine eq. from reported exptl. data, Boublik et al. 1973)

log (P/atm) = (1 − 418.675/T) × 10^(0.886470 − 8.14267 × 10^4·T + 7.57896 × 10^7·T^2); T in K, temp range 281.35–417.92 K (Cox vapor pressure eq., Chao et al. 1983)

878, 811 (extrapolated-Antoine eq., Boublik et al. 1984)

log (P_L/kPa) = 6.235 − 1557.406/(222.538 + T/K); temp range 32.4–62.19°C (Antoine eq., Boublik et al. 1984)

log (P/kPa) = 6.18301 − 1501.162/(214.42 + t/°C); temp range 29.92–144.77°C (Antoine eq. from reported exptl. data, Boublik et al. 1984)

879 (extrapolated, Antoine eq., Dean 1985, 1992)

log (P/mmHg) = 7.14016 − 1774.51/(224.09 + t/°C); temp range 32–82°C (Antoine eq., Dean, 1985, 1992)

841 (lit. average, Riddick et al. 1986)

log (P/kPa) = 6.34792 − 1629.20/(230.0 + t/°C), temp range not specified (Antoine eq., Riddick et al. 1986)

880 (interpolated-Antoine eq., Stephenson & Malanowski 1987)

log (P_L/kPa) = 7.3945 − 2221.3/(T/K); temp range 245–334 K (Antoine eq., Stephenson & Malanowski 1987)

log (P_L/kPa) = 6.08201 − 1445.58/(−63.72 + T/K); temp range 334–419 K (Antoine eq., Stephenson & Malanowski 1987)

log (P/mmHg) = 55.8621 − 4.024 × 10^3/(T/K) − 17.609·log (T/K) + 6.6842 × 10^{-3}·(T/K) + 1.9438 × 10^{-13}·(T/K)^2, temp range 243–648 K (Yaws 1994)

log (P/kPa) = 7.3945 − 2221.3/(T/K); temp range 5–50°C (regression eq. from literature data, Shiu & Ma 2000)

Henry's Law Constant (Pa m^3/mol at 25°C):

233 (calculated-P/C, Mackay & Shiu 1990)

285, 527 (quoted, Howard et al. 1989)

267 (calculated-vapor-liquid equilibrium (VLE) data, Yaws et al. 1991)

297 (exponential saturator EXPSAT technique, Dohnal & Hovorka 1999)

261 (20°C, selected from reported experimental determined values, Staudinger & Roberts 1996, 2001)

log K_{AW} = 5.628 − 1935/(T/K) (summary of literature data, Staudinger & Roberts 2001)

Octanol/Water Partition Coefficient, log K_{OW}:

3.14 (calculated-fragment const., Rekker 1977)

2.95 (shake flask, Hansch & Leo 1979)

3.16 (shake flask-HPLC, Banerjee et al. 1980)

2.76 (HPLC-RT correlation Fujisawa & Masuhara 1981)

2.90 (HPLC-RT correlation, Wang et al. 1986)

3.05 (recommended, Sangster 1989, 1993)

2.95 (recommended, Hansch et al. 1995)

Octanol/Air Partition Coefficient, log K_{OA}:

Bioconcentration Factor, log BCF:

1.13 (goldfish, Ogata et al. 1984)

Sorption Partition Coefficient, log K_{OC}:

3.42–2.74 (Swann et al. 1983)

2.96, 2.71 (quoted exptl., calculated-MCI χ, Meylan et al. 1992)

Environmental Fate Rate Constants, k, and Half-Lives, $t_{1/2}$:

Volatilization: Volatilization and biodegradation may be dominant transport and transformation processes for styrene in water; calculated volatilization $t_{1/2}$ = 3 h from a river 1-m deep with a current speed of 1.0 m/s and wind velocity of 3 m/s (Howard 1989);

volatilized rapidly from shallow layers of lake water with $t_½$ = 1 to 3 h, but much slower from soil (Fu & Alexander 1992).

Photolysis:

Oxidation: rate constant k, for gas-phase second order rate constants, k_{OH} for reaction with OH radical, k_{NO3} with NO_3 radical and k_{O3} with O_3 or as indicated, *data at other temperatures see reference:

k_{O3} = 1.8 × 10⁴ L mol⁻¹ s⁻¹ for the reaction with ozone at 30°C (Bufalini & Altshuller 1965)

k_{O3} = (22.6 ± 4.6) × 10⁻¹⁸ cm³ molecule⁻¹ s⁻¹ at 296 ± 2 K (relative rate method, Atkinson et al. 1982)

k_{OH} = 5.25 × 10⁻¹¹ cm³ molecule⁻¹ s⁻¹ (Atkinson 1985)

k_{O3} = 22 × 10⁻¹⁸ cm³ molecule⁻¹ s⁻¹, k_{OH} = (5.87 ± 0.15) × 10⁻¹¹ cm³ molecule⁻¹ s⁻¹, k_{NO3} = (1.51 ± 0.2) × 10⁻¹³ cm³ molecule⁻¹ s⁻¹ at room temp. (Atkinson & Aschmann 1988)

k_{OH} = 5.8 × 10⁻¹¹ cm³ molecule⁻¹ s⁻¹ at 298 K (recommended, Atkinson 1989)

k_{O3} = 1.71 × 10⁻¹⁷ cm³ molecule⁻¹ s⁻¹ at 296 K (Tuazon et al. 1993)

Hydrolysis: no hydrolyzable groups (Howard et al. 1991).

Biodegradation: $t_½$(aerobic) = 336–672 h, based on unacclimated grab samples of aerobic soil and a subsurface sample; $t_½$(anaerobic) = 1344–2688 h, based on estimated unacclimated aqueous aerobic biodegradation half-life (Howard et al. 1991);

styrene will be rapidly destroyed by biodegradation in most aerobic environments, and the rate may be slow at low concentrations in aquifers and lake waters and in environments at low pH (Fu & Alexander 1992)

$t_½$(aerobic) = 14 d, $t_½$ = 56 d in natural waters (Capel & Larson 1995)

Biotransformation:

Bioconcentration, Uptake (k_1) and Elimination (k_2) Rate Constants or Half-Lives:

Half-Lives in the Environment:

Air: atmospheric $t_½$ ~ 2.4–24 h, based on the EPA Reactivity Classification of Organics (Darnall et al. 1976); will react rapidly with both hydroxyl radical and ozone in air with a combined $t_½$(calc) = 2.5 h, the reaction $t_½$ = 3.5 h with OH radicals and $t_½$ = 9 h with ozone (Howard 1989);

photooxidation $t_½$ = 0.9 to 7.3 h, based on measured rate data for the reaction with OH radical and O_3 in air (Howard et al. 1991);

calculated lifetimes of styrene due to reaction with OH radicals, NO_3 radicals and O_3 are ~3 h, ~4 h and 1 d, respectively, for ambient atmospheric conditions (Tuazon et al. 1993);

calculated lifetimes of 1.4 h, 3.7 h and 1.0 d for reactions with OH radical, NO_3 radical and O_3 respectively (Atkinson 2000).

Surface water: $t_½$ = 0.6 d in surface waters in case a first order reduction process may be assumed (Zoeteman et al. 1980)

$t_½$ = 336–672 h, based on estimated unacclimated aqueous aerobic biodegradation half-life (Howard et al. 1991);

$t_½$(calc) = 0.75–51 d for styrene loss from surface waters (Fu & Alexander 1992).

Ground water: $t_½$ = 672–5040 h, based on estimated unacclimated aqueous aerobic biodegradation half-life and acclimated aqueous screening test data (Howard et al. 1991).

Sediment:

Soil: $t_½$ = 336–672 h, based on unacclimated grab samples of aerobic soil and acclimated aqueous screening test data (Howard et al. 1991).

Biota:

TABLE 3.1.1.35.1
Reported aqueous solubilities of styrene at various temperatures

Lane 1946				Shaw 1989b (IUPAC)	
shake flask-titration		shake flask-cloud pt.		recommended values	
t/°C	S/g·m^{-3}	t/°C	S/g·m^{-3}	t/°C	S/g·m^{-3}
7	290	15	250	10	290
24	330	25	310	20	300
32	360	44	400	25	340
40	400	49	450	30	340
51	450	65	500	40	400
				50	460
				60	530

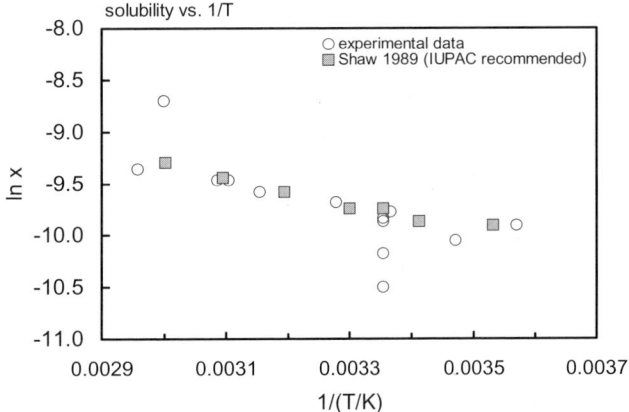

FIGURE 3.1.1.35.1 Logarithm of mole fraction solubility (ln x) versus reciprocal temperature for styrene.

TABLE 3.1.1.35.2
Reported vapor pressures of styrene at various temperatures and the coefficients for the vapor pressure equations

$$\log P = A - B/(T/K) \quad (1)$$
$$\log P = A - B/(C + t/°C) \quad (2)$$
$$\log P = A - B/(C + T/K) \quad (3)$$
$$\log P = A - B/(T/K) - C \cdot \log(T/K) \quad (4)$$

$$\ln P = A - B/(T/K) \quad (1a)$$
$$\ln P = A - B/(C + t/°C) \quad (2a)$$

1.

Linder 1931		Pitzer et al. 1946		Stull 1947		Buck et al. 1949	
Hg manometer		static-Hg manometer		summary of literature data		Hg manometer	
t/°C	P/Pa	t/°C	P/Pa	t/°C	P/Pa	t/°C	P/Pa
−7.7	86.66	12.5	384	−7.0	133.3	66.7	6799
−0.5	105.7	25	841.3	18	666.6	75.4	9866
8.2	288	37.5	1692	30.8	1333	82.4	13332
		50	3230	44.6	2666	88	16665
		60	5177	59.8	5333	91.7	18932
Burchfield 1942				69.5	7999	95.1	21732
isoteniscope method		mp/°C	−30.68	82	13332	98.2	24665
t/°C	P/Pa			101.3	26664	119.1	48129
		eq. 4	P/mmHg	122.5	53329	132.6	71461
data presented by		A	28.8631	145.2	101325	145	100658
Clausius-Clapeyron eq.		B	3203				
		C	7	mp/°C	−30.6		
eq. 1	P/mmHg					eq. 4	P/mmHg
A	7.929	ΔH_{fus}(kJ mol^{-1}) = 10.95				A	26.92
B	2103					B	3151
measured temp range:		ΔH_V(kJ mol^{-1}) = 43.93				C	6.294
33.5–116.3°C		at 25°C					

2.

Dreyer et al. 1955		Chaiyavech & Van Winkle 1959	
t/°C	P/Pa	t/°C	P/Pa
29.92	1093	32.4	1333
39.21	1827	35.6	2666
60.04	5106	53.86	4000
74.42	9639	60.05	5333
85.53	14999	65.45	6666
99.51	25105	76.6	10666
110.06	56877	82.9	13332
		eq. 1	P/mmHg
		A	8.2696
		B	2221.3

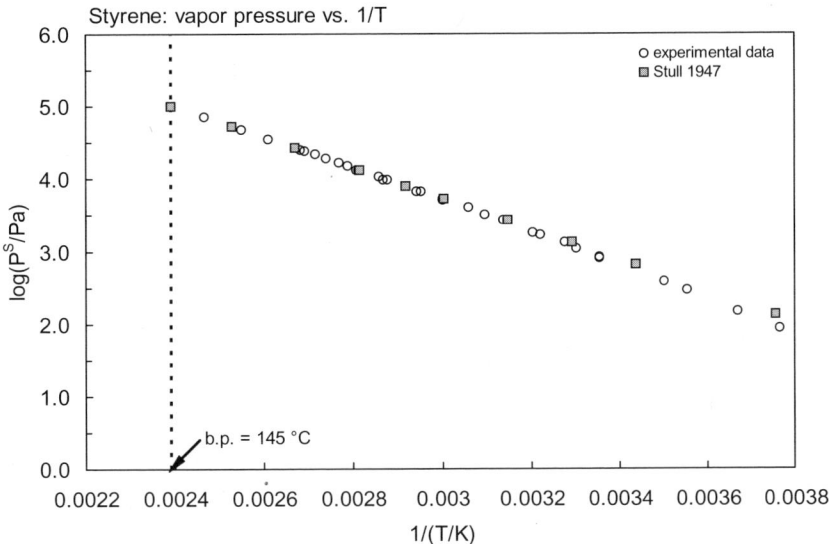

FIGURE 3.1.1.35.2 Logarithm of vapor pressure versus reciprocal temperature for styrene.

3.1.1.36 α-Methylstyrene

Common Name: α-Methylstyrene
Synonym: isopropenylbenzene
Chemical Name: α-Methylstyrene
CAS Registry No: 98-83-9
Molecular Formula: C_9H_{10}, $C_6H_5C(CH_3)=CH_2$
Molecular Weight: 118.175
Melting Point (°C):
    −23.2   (Lide 2003)
Boiling Point (°C):
    165.4   (Lide 2003)
Density (g/cm³):
    0.9106, 0.9062   (20°C, 25°C, Dreisbach 1955)
    0.9082   (20°C, Weast 1982–83)
Molar Volume (cm³/mol):
    129.8   (20°C, calculated-density)
    155.2   (calculated-Le Bas method at normal boiling point)
Enthalpy of Fusion, ΔH_{fus} (kJ/mol):
    11.92   (Chickos et al. 1999)
Entropy of Fusion, ΔS_{fus} (J/mol K):
    47.55, 53.8   (exptl., calculated-group additivity method, Chickos et al. 1999)
Fugacity Ratio at 25°C, F: 1.0

Water Solubility (g/m³ or mg/L at 25°C):

Vapor Pressure (Pa at 25°C or as indicated and reported temperature dependence equations. Additional data at other temperatures designated * are compiled at the end of this section):
    133.3*   (7.4°C, summary of literature data, temp range 7.4–165.4°C, Stull 1947)
    log (P/mmHg) = 7.28240 − 1740.3/(230 + t/°C) (Antoine eq., Dreisbach & Martin 1949)
    333   (extrapolated by formula., Dreisbach 1855)
    log (P/mmHg) = 6.92366 − 1486.88/(202.4 + t/°C); temp range 70–220°C (Antoine eq. for liquid state, Dreisbach 1955)
    log (P/mmHg) = [−0.2185 × 10214.6/(T/K)] + 7.959753; temp range 7.4–165.4°C (Antoine eq., Weast 1972–73)
    log (P/mmHg) = 6.92366 − 1486.88/(202.4 + t/°C); temp range not specified (Antoine eq., Dean 1985, 1992)
    log (P_L/kPa) = 6.04856 − 1486.88/(−70.75 + T/K); temp range 343–493 K (liquid, Antoine eq.-I, Stephenson & Malanowski 1987)
    log (P_L/kPa) = 6.294 − 1599.88/(−63.72 + T/K); temp range 353–413 K (liquid, Antoine eq.-II, Stephenson & Malanowski 1987)
    log (P/mmHg) = −0.8626 − 2.5638 × 10³/(T/K) + 5.3807·log (T/K) − 1.3516 × 10⁻²·(T/K) + 6.7181 × 10⁻⁶·(T/K)², temp range 250–654 K (Yaws 1994)

Henry's Law Constant (Pa·m³/mol at 25°C):

Octanol/Water Partition Coefficient, log K_{OW}:
    3.48   (recommended, Hansch et al. 1995)

Octanol/Air Partition Coefficient, log K_{OA}:

Bioconcentration Factor, log BCF or log K_B:

Mononuclear Aromatic Hydrocarbons

Sorption Partition Coefficient, log K_{OC}:

Environmental Fate Rate Constants, k, and Half-Lives, $t_{1/2}$:
    Volatilization:
    Photolysis:
    Photooxidation: k = (52 ± 6) × 10^{-12} cm^3 $molecule^{-1}$ s^{-1} for the gas-phase reactions with OH radical at 298 ± 2 K (Atkinson 1989).
    Hydrolysis:
    Biodegradation:
    Biotransformation:
    Bioconcentration and Uptake and Elimination Rate Constants (k_1 and k_2):

Half-Lives in the Environment:

TABLE 3.1.1.36.1
Reported vapor pressures of α-methylstyrene at various temperatures and the coefficients for the vapor pressure equations

log P = A – B/(T/K)	(1)	ln P = A – B/(T/K)	(1a)
log P = A – B/(C + t/°C)	(2)	ln P = A – B/(C + t/°C)	(2a)
log P = A – B/(C + T/K)	(3)		
log P = A – B/(T/K) – C·log (T/K)	(4)		

Stull 1947

summary of literature data

t/°C	P/Pa
7.4	133.3
34.0	666.6
47.1	1333
61.8	2666
77.8	5333
88.3	7999
102.2	13332
121.8	26664
143	53329
165.4	101325
mp/°C	–23.2

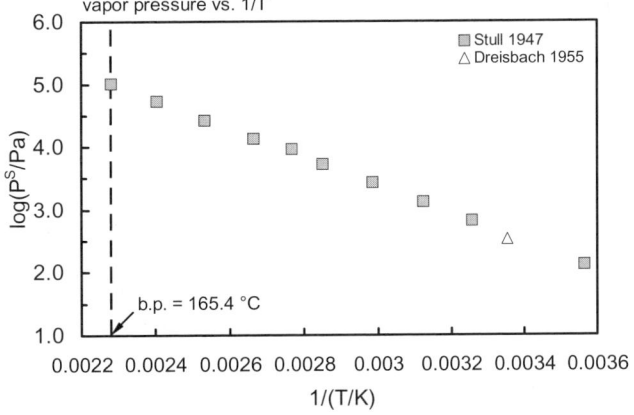

FIGURE 3.1.1.36.1 Logarithm of vapor pressure versus reciprocal temperature for α-methylstyrene.

3.1.1.37 β-Methylstyrene

Common Name: β-Methylstyrene
Synonym: propenylbenzene
Chemical Name: β-Methylstyrene
CAS Registry No: 766-90-5 (*cis*-); 873-66-5 (*trans*-)
Molecular Formula: C_9H_{10}, $C_6H_5CH=CHCH_3$
Molecular Weight: 118.175
Melting Point (°C):
    –61.6   (*cis*-, Stephenson & Malanowski 1987; Lide 2003)
    –29.3   (*trans*-, Stephenson & Malanowski 1987; Lide 2003)
Boiling Point (°C):
    174   (*cis*-, Stephenson & Malanowski 1987)
    175–176 (*trans*-, Stephenson & Malanowski 1987)
    167.5   (*cis*-, Lide 2003)
    178.3   (*trans*-, Lide 2003)
Density (g/cm³):
    0.9088   (20°C, *cis*-, Lide 2003)
    0.9023   (25°C, *trans*-, Lide 2003)
Molar Volume (cm³/mol):
    129.7   (*cis*-, Stephenson & Malanowski 1987)
    131.0   (*trans*-, Stephenson & Malanowski 1987)
    155.2   (calculated-Le Bas method at normal boiling point)
Enthalpy of Fusion, ΔH_{fus} (kJ/mol):
Entropy of Fusion, ΔS_{fus} (J/mol K):
Fugacity Ratio at 25°C, F: 1.0

Water Solubility (g/m³ or mg/L at 25°C):

Vapor Pressure (Pa at 25°C or as indicated and reported temperature dependence equations. Additional data at other temperatures designated * are compiled at the end of this section):
    133.3*   (17.5°C, summary of literature data, measured range 32.1–112°C, temp range 17.5–179°C, Stull 1947)
    267   (calculated by formula, Dreisbach 1955)
    $\log(P/\text{mmHg}) = 6.92339 - 1499.80/(201.0 + t/°C)$; temp range 75–200°C (Antoine eq. for liquid state, Dreisbach 1955)
    $\log(P/\text{mmHg}) = [-0.2185 \times 10701.3/(T/K)] + 8.071487$; temp range 17.5–179°C (Antoine eq., Weast 1972–73)
    $\log(P/\text{mmHg}) = 6.92339 - 1499.80/(201.0 + t/°C)$; temp range not specified (Antoine eq., Dean 1985, 1992)
    $\log(P_L/\text{kPa}) = 6.04829 - 1499.8/(-72.15 + T/K)$; temp range 348–498 K (*cis*-, liquid, Antoine eq., Stephenson & Malanowski 1987)
    $\log(P_L/\text{kPa}) = 6.58873 - 1915.94/(-33.996 + T/K)$; temp range 291–452 K (*trans*-, liquid, Antoine eq., Stephenson & Malanowski 1987)

Henry's Law Constant (Pa·m³/mol):

Octanol/Water Partition Coefficient, log K_{OW}:

Octanol/Air Partition Coefficient, log K_{OA}:

Bioconcentration Factor, log BCF or log K_B:

Sorption Partition Coefficient, log K_{OC}:

Mononuclear Aromatic Hydrocarbons

Environmental Fate Rate Constant and Half-Lives:
- Volatilization:
- Photolysis:
- Photooxidation: rate constant k = (59 ± 6) × 10^{-12} cm^3 $molecule^{-1}$ s^{-1} for the gas-phase reactions with OH radical at 298 ± 2 K (Atkinson 1989)
- Hydrolysis:
- Biodegradation:
- Biotransformation:
- Bioconcentration and Uptake and Elimination Rate Constants (k_1 and k_2):

Half-Lives in the Environment:

TABLE 3.1.1.37.1
Reported vapor pressures of ß-methylstyrene at various temperatures and the coefficients for the vapor pressure equations

log P = A – B/(T/K) (1)	ln P = A – B/(T/K) (1a)
log P = A – B/(C + t/°C) (2)	ln P = A – B/(C + t/°C) (2a)
log P = A – B/(C + T/K) (3)	
log P = A – B/(T/K) – C·log (T/K) (4)	

Stull 1947

summary of literature data

t/°C	P/Pa
17.5	133.3
43.8	666.6
57	1333
71.5	2666
87.7	5333
97.8	7999
111.7	13332
132	26664
154.7	53329
179.0	101325
mp/°C	–30.1

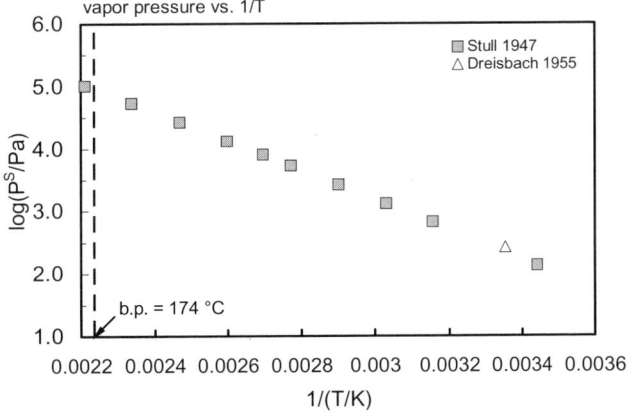

FIGURE 3.1.1.37.1 Logarithm of vapor pressure versus reciprocal temperature for β-methylstyrene.

3.1.1.38 *o*-Methylstyrene

Common Name: *o*-Methylstyrene
Synonym: 2-methylstyrene, 2-vinyl toluene, *o*-methylvinylbenzene
Chemical Name: 2-methylstyrene
CAS Registry No: 611-15-4
Molecular Formula: C_9H_{10}, 2-$CH_3C_6H_4CH=CH_2$
Molecular Weight: 118.175
Melting Point (°C):
    –68.5   (Lide 2003)
Boiling Point (°C):
    169.8   (Lide 2003)
Density (g/cm³):
    0.9106   (20°C, Weast 1982–83)
Molar Volume (cm³/mol):
    129.8   (20°C, calculated-density)
    155.2   (calculated-Le Bas method at normal boiling point)
Enthalpy of Fusion, ΔH_{fus} (kJ/mol):
Entropy of Fusion, ΔS_{fus} (J/mol K):
Fugacity Ratio at 25°C, F: 1.0

Water Solubility (g/m³ or mg/L at 25°C):

Vapor Pressure (Pa at 25°C or as indicated and reported temperature dependence equations. Additional data at other temperatures designated * are compiled at the end of this section):
    387*   (32.1°C, differential manometer, measured range 32.1–112°C, Clements et al. 1953)
    log (P/mmHg) = 7.15212 – 1628.405/(211.386 + t/°C); temp range 32.1–112°C (Antoine eq. from differential manometer measurements, Clements et al. 1953)
    240.8   (extrapolated-Antoine eq., Dreisbach 1955)
    log (P/mmHg) = 7.09235 – 1582.7/(206.0 + t/°C); temp range 75–200°C (Antoine eq. for liquid state, Dreisbach 1955)
    298.9   (extrapolated-Antoine eq., Boublik et al. 1973)
    log (P/mmHg) = 7.21287 – 1644.083/(214.585 + t/°C); temp range 32–112.4°C (Antoine eq. from reported exptl. data, Boublik et al. 1973)
    log (P/atm) = (1 – 443.504/(T/K) × 10^[0.890379 – 7.17666 × 10^4·(T/K) + 5.97058 × 10^7·(T/K)²]; temp range 305.16–385.5 K (Cox vapor pressure eq., Chao et al. 1983)
    246.4   (extrapolated-Antoine eq., Boublik et al. 1984)
    log (P/kPa) = 6.33107 – 1660.041/(214.219 + t/°C); temp range 32–112.4°C (Antoine eq. from reported exptl. data, Boublik et al. 1984)
    log (P/mmHg) = 7.2129 – 1644.08/(214.59 + t/°C); temp range 32–112°C (Antoine eq., Dean 1985, 1992)
    log (P/mmHg) = 6.88461 – 1485.41/(200.0 + t/°C); temp range 75–255°C (Antoine eq., Dean 1985, 1992)
    log (P_L/kPa) = 6.27762 – 1628.405/(–61.764 + T/K); temp range 305–385 K (liquid, Antoine eq., Stephenson & Malanowski 1987)
    log (P/mmHg) = 36.8413 – 3.7269 × 10^3/(T/K) – 9.7997·log (T/K) + 1.4115 × 10^{-10}·(T/K) + 1.9658 × 10^{-6}·(T/K)², temp range 205–659 K (Yaws 1994)

Henry's Law Constant (Pa·m³/mol):

Octanol/Water Partition Coefficient, log K_{OW}:

Octanol/Air Partition Coefficient, log K_{OA}:

Bioconcentration Factor, log BCF or log K_B:

Sorption Partition Coefficient, log K_{OC}:

Environmental Fate Rate Constants, k, and Half-Lives, $t_{\frac{1}{2}}$:

Half-Lives in the Environment:

TABLE 3.1.1.38.1
Reported vapor pressures of o-methylstyrene at various temperatures

Clements et al. 1953	
differential manometer	
t/°C	P/Pa
32.01	387
40.85	663
58.51	1753
58.54	1765
70.2	3113
82.8	5529
82.83	5534
100.35	11310
112.35	17698
mp/°C	−68.57
bp/°C	169.8
	ΔH_V = 35.54 kJ/mol
	log P = A − B/(C + t/°C)
eq. 2	mmHg
A	7.15272
B	1628.405
C	211.386

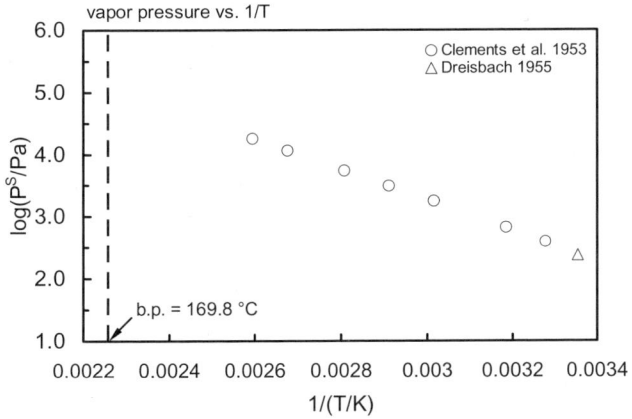

FIGURE 3.1.1.38.1 Logarithm of vapor pressure versus reciprocal temperature for o-methylstyrene.

3.1.1.39 *m*-Methylstyrene

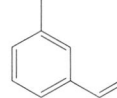

Common Name: *m*-Methylstyrene
Synonym: methylvinylbenzene, 3-vinyl toluene, *m*-methylvinylbenzene
Chemical Name: 3-methylstyrene
CAS Registry No: 100-42-1
Molecular Formula: C_9H_{10}, 3-$CH_3C_6H_4CH=CH_2$
Molecular Weight: 118.175
Melting Point (°C):
    –86.3   (Lide 2003)
Boiling Point (°C):
    164   (Lide 2003)
Density (g/cm³ at 20°C):
    0.9028   (20°C, Weast 1982–83)
Molar Volume (cm³/mol):
    130.9   ((20°C, calculated-density)
    155.2   (calculated-Le Bas method at normal boiling point)
Enthalpy of Fusion, ΔH_{fus} (kJ/mol):
Entropy of Fusion, ΔS_{fus} (J/mol K):
Fugacity Ratio at 25°C, F: 1.0

Water Solubility (g/m³ or mg/L at 25°C):
    89.0   (*m*- and *p*-methylstyrene commercial product, Dreisbach 1955)
    89.0, 100   (quoted, *m*- and *p*-methylstyrene, calculated-group contribution, Irmann 1965)

Vapor Pressure (Pa at 25°C or as indicated and reported temperature dependence equations. Additional data at other temperatures designated * are compiled at the end of this section):
    2693*   (67.6°C, mercury manometer, measured range 67.6–169.1°C, Buck et al. 1949)
    log (P/mmHg) = –3563/(T/K) – 7.553·log (T/K) + 30.90; temp range 80–120°C (vapor pressure eq. from Hg manometer measurements, Buck et al. 1949)
    687*   (41.48°C, differential manometer, Clements et al. 1953)
    log (P/mmHg) = 6.95079 – 1520.412/(210.967 + t/°C); temp range 41.48–111.8°C (Antoine eq. from differential manometer measurements, measured range 41.48–111.8°C, Clements et al. 1953)
    257   (calculated by formula, Dreisbach 1955)
    log (P/mmHg) = 6.99468 – 1553.4/(206.0 + t/°C); temp range 75–225°C (Antoine eq. for liquid state, Dreisbach 1955)
    228, 244   (extrapolated-Antoine eq., Boublik et al. 1973)
    log (P/mmHg) = 7.06423 – 1564.74/(204.083 + t/°C); temp range 67.6–169°C (Antoine eq. from reported exptl. data, Boublik et al. 1973)
    log (P/mmHg) = 7.11224 – 1615.091/(210.809 + t/°C); temp range 41.7–111.8°C (Antoine eq. from reported exptl. data, Boublik et al. 1973)
    log (P/atm) = (1 – 442.985/T) × 10^(0.885861 – 7.19653 × 10^4·T + 6.75359 × 10^7·T^2); T in K, temp range 314.93–442.15 K (Cox vapor pressure eq., Chao et al. 1983)
    227, 244   (extrapolated-Antoine eq., Boublik et al. 1984)
    log (P/kPa) = 6.17253 – 1553.744/(202.922 + t/°C); temp range 67.6–169.1°C (Antoine eq. from reported exptl. data, Boublik et al. 1984)
    log (P/kPa) = 6.22823 – 1609.825/(210.331 + t/°C); temp range 41.7–111.8°C (Antoine eq. from reported exptl. data, Boublik et al. 1984)
    log (P/mmHg) = 7.27534 – 1695.4/(220.0 + t/°C); temp range 10–72°C (Antoine eq., Dean 1985, 1992)
    log (P/mmHg) = 6.87928 – 1471.28/(200.0 + t/°C); temp range 72–250°C (Antoine eq., Dean 1985, 1992)

Mononuclear Aromatic Hydrocarbons

245   (Antoine eq., Stephenson & Malanowski 1987)

log (P_L/kPa) = 6.07569 − 1520.412/(−71.183 + T/K); temp range 314–385 K (Antoine eq., Stephenson & Malanowski 1987)

log (P/mmHg) = 11.6959 − 2.9912 × 10^3/(T/K) + 0.33334·log (T/K) − 8.8935 × 10^{-3}·(T/K) + 4.9793 × 10^{-6}·(T/K)^2, temp range 187–657 K (Yaws 1994)

Henry's Law Constant (Pa m^3/mol at 25°C):

387   (calculated-vapor-liquid equilibrium (VLE) data, Yaws et al. 1991)

Octanol/Water Partition Coefficient, log K_{OW}:

3.35   (Leo et al. 1971)

Octanol/Air Partition Coefficient, log K_{OA}:

Bioconcentration Factor, log BCF:

1.50   (gold fish, flow-through method, Ogata et al. 1984)
1.50, 1.63   (gold fish, quoted, calculated-MCI χ, Ogata et al. 1984)

Sorption Partition Coefficient, log K_{OC}:

Environmental Fate Rate Constants, k, and Half-Lives, $t_{1/2}$:

Half-Lives in the Environment:

TABLE 3.1.1.39.1
Reported vapor pressures of *m*-methylstyrene at various temperatures and the coefficients for the vapor pressure equations

log P = A − B/(T/K)   (1)   ln P = A − B/(T/K)   (1a)
log P = A − B/(C + t/°C)   (2)   ln P = A − B/(C + t/°C)   (2a)
log P = A − B/(C + T/K)   (3)
log P = A − B/(T/K) − C·log (T/K) (4)

Buck et al. 1949				Clements et al. 1953			
Hg manometer				differential manometer			
t/°C	P/Pa	t/°C	P/Pa	t/°C	P/Pa	t/°C	P/Pa
67.6	2693	eq. 4 t/°C	P/mmHg	41.48	687	99.33	10696
72.3	3440	A	30.9	43.22	744	111.8	17020
80.7	4933	B	3563	49.49	1104		
87	6533	C	7.553	52.12	1261	mp/°C	−86.34
93.8	8373			55.73	1499	bp/°C	171.6
96.6	9733			57.8	1673		
106.9	14399			65.2	2445	ΔH_V = 38.79 kJ/mol	
121	12998			70.95	3192		
128.3	30398			71.01	3202	eq. 2	mmHg
138.5	41597			71.02	3204	A	6.95079
147.6	54795			71.1	3248	B	1520.412
159.2	75727			77.12	4234	C	201.967
169.1	98392						

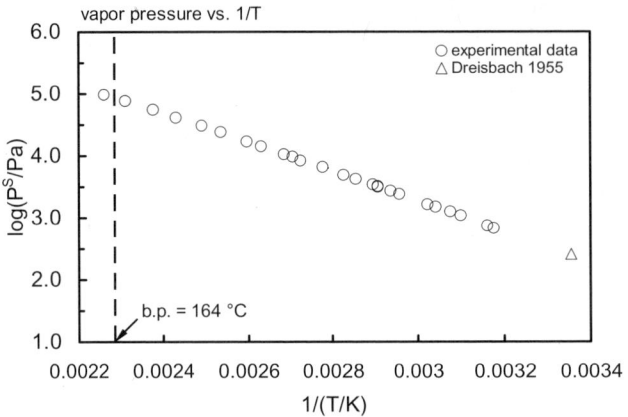

FIGURE 3.1.1.39.1 Logarithm of vapor pressure versus reciprocal temperature for *m*-methylstyrene.

3.1.1.40 p-Methylstyrene

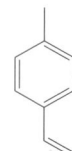

Common Name: p-Methylstyrene
Synonym: methylvinylbenzene, 4-vinyl toluene, p-methylvinylbenzene
Chemical Name: 4-methylstyrene
CAS Registry No: 622-97-9
Molecular Formula: C_9H_{10}, 4-$CH_3C_6H_4CH=CH_2$
Molecular Weight: 118.17
Melting Point (°C):
    –34.1   (Lide 2003)
Boiling Point (°C):
    172.8   (Lide 2003)
Density (g/cm³ at 20°C):
    0.9016, 0.9060   (20°C, 25°C, Dreisbach 1955)
    0.8760   (20°C, Weast 1982–83)
Molar Volume (cm³/mol):
    129.8   (20°C, calculated-density)
    155.2   (calculated-Le Bas method at normal boiling point)
Enthalpy of Fusion, ΔH_{fus} (kJ/mol):
Entropy of Fusion, ΔS_{fus} (J/mol K):
Fugacity Ratio at 25°C, F: 1.0

Water Solubility (g/m³ or mg/L at 25°C):
    89.0   (quoted, m- and p-methylstyrene commercial product, Dreisbach 1955)
    89.0, 100   (quoted, m- and p-methylstyrene, calculated-group contribution, Irmann 1965)

Vapor Pressure (Pa at 25°C or as indicated and reported temperature dependence equations. Additional data at other temperatures designated * are compiled at the end of this section):
    133.3*   (16°C, summary of literature data, temp range 16–175°C, Stull 1947)
    2773*   (68.6°C, mercury manometer, measured range 68.6–170°C, Buck et al. 1949)
    log (P/mmHg) = –3476/(T/K) – 6.923·log (T/K) + 29.03; temp range 80–120°C (vapor pressure eq. from Hg manometer measurements, Buck et al. 1949)
    log (P/mmHg) = 7.34046 – 1791.0/(230 + t/°C) (Antoine eq., Dreisbach & Martin 1949)
    4954*   (90.98°C, ebulliometry, measured range 90.98–171.06°C, Dreisbach & Shrader 1949)
    376*   (31.82°C, differential manometer, measured range 31.8–96.9°C, Clements et al. 1953)
    log (P/mmHg) = 7.0483 – 1594.747/(209.889 + t/°C); temp range 31.8–96.9°C (Antoine eq. from differential manometer measurements, Clements et al. 1953)
    241.5   (calculated by formula, Dreisbach 1955)
    log (P/mmHg) = 7.35420 – 1765.6/(223.8 + t/°C); temp range 75–205°C (Antoine eq. for liquid state, Dreisbach 1955)
    log (P/mmHg) = [–0.2185 × 10724.2/(T/K)] + 8.130903; temp range 16.0–175°C (Antoine eq., Weast 1972–73)
    216.6, 241   (extrapolated-Antoine eq., Boublik et al. 1973)
    log (P/mmHg) = 7.01119 – 1535.073/(200.732 + t/°C); temp range 68.6–170°C (Antoine eq. from reported exptl. data, Boublik et al. 1973)
    log (P/mmHg) = 6.11531 – 1591.082/(209.441 + t/°C); temp range 31.8–96.93°C (Antoine eq. from reported exptl. data, Boublik et al. 1973)
    log (P/atm) = (1 – 443.748/(T/K) × 10^[0.875061 – 7.08160 × 10⁴·(T/K) + 7.33467 × 10⁷·(T/K)²]; temp range: 289.15–443.15 K, (Cox vapor pressure eq., Chao et al. 1983)
    215, 241   (extrapolated-Antoine eq., Boublik et al. 1984)

log (P/kPa) = 6.11531 − 1521.514/(199.299 + t/°C); temp range 68.6–170°C (Antoine eq. from reported exptl. data, Boublik et al. 1984)

log (P/kPa) = 6.16144 − 1586.596/(209.046 + t/°C); temp range 31.8–96.93°C (Antoine eq. from reported exptl. data, Boublik et al. 1984)

log (P/mmHg) = 7.0112 − 1535.1/(200.7 + t/°C); temp range 68–170°C (Antoine eq., Dean 1985, 1992)

242 (extrapolated-Antoine eq., Stephenson & Malanowski 1987)

log (P_L/kPa) = 6.1732 − 1594.147/(−63.261 + T/K); temp range 304–370 K (liquid, Antoine eq., Stephenson & Malanowski 1987)

log (P/mmHg) = 50.6506 − 4.0628 × 10^3/(T/K) − 15.524·log (T/K) + 5.5381 × 10^{-3}·(T/K) − 1.1313 × 10^{-13}·(T/K)^2, temp range 239–665 K (Yaws 1994)

Henry's Law Constant (Pa m^3/mol at 25°C):

287 (calculated-vapor-liquid equilibrium (VLE) data, Yaws et al. 1991)

Octanol/Water Partition Coefficient, log K_{OW}:

3.35 (Leo et al. 1971; quoted, Ogata et al. 1984)

Octanol/Air Partition Coefficient, log K_{OA}:

Bioconcentration Factor, log BCF:

1.55 (gold fish, flow-through method, Ogata et al. 1984)

1.55, 1.62 (gold fish, quoted, calculated-MCI χ, Ogata et al. 1984)

Sorption Partition Coefficient, log K_{OC}:

Environmental Fate Rate Constants, k, and Half-Lives, $t_{1/2}$:

Half-Lives in the Environment:

TABLE 3.1.1.40.1

Reported vapor pressures of *p*-methylstyrene at various temperatures and the coefficients for the vapor pressure equations

log P = A − B/(T/K)	(1)	ln P = A − B/(T/K)	(1a)
log P = A − B/(C + t/°C)	(2)	ln P = A − B/(C + t/°C)	(2a)
log P = A − B/(C + T/K)	(3)		
log P = A − B/(T/K) − C·log (T/K)	(4)		

Stull 1947		Buck et al. 1949		Dreisbach & Shrader 1949		Clements et al. 1953	
summary of literature data		Hg manometer		ebulliometry		differential manometer	
t/°C	P/Pa	t/°C	P/Pa	t/°C	P/Pa	t/°C	P/Pa
16	133.3	68.6	2773	90.98	4954	31.82	376
42	666.6	70.6	2986	98.13	10351	41.76	689
55.1	1333	75.1	3733	116.06	16959	41.83	693
6.2	2666	78.8	4400	171.06	101325	52.17	1207
85	5333	80.6	4720			53.94	1331
95	7999	82.7	5200	mp/°C	−70.3	59.56	1797
108.6	13332	84.2	5573	bp/°C	171.06	59.91	1855
128.7	26664	87	6559			66.62	2538
151.2	53329	94.1	8559			75.4	3858
175	101325	104.4	12826			76.19	3974
		115.1	18958			80.2	4718

Mononuclear Aromatic Hydrocarbons

TABLE 3.1.1.40.1 *(Continued)*

Stull 1947 summary of literature data		Buck et al. 1949 Hg manometer		Dreisbach & Shrader 1949 ebulliometry		Clements et al. 1953 differential manometer	
t/°C	P/Pa	t/°C	P/Pa	t/°C	P/Pa	t/°C	P/Pa
mp/°C	–	131.9	33197			82.44	5208
		144	47996			90.9	7439
		154.7	65728			96.93	9466
		163.7	83193				
		170	99725			mp/°C	–34.15
						bp/°C	172.78
		mp/°C	–37.8			ΔH_v = 38.95 kJ/mol	
		eq. 4	P/mmHg			eq. 2	P/mmHg
		A	29.03			A	7.0483
		B	3476			B	1594.747
		C	6.923			C	209.889

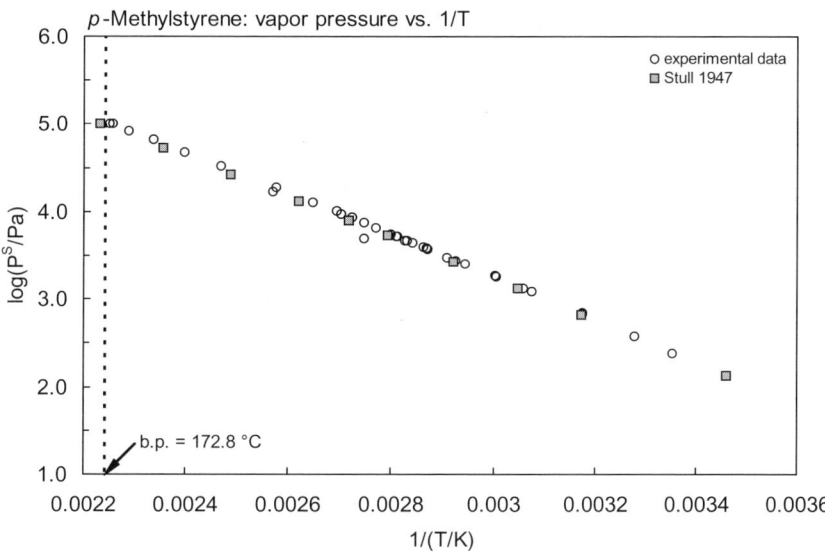

FIGURE 3.1.1.40.1 Logarithm of vapor pressure versus reciprocal temperature for *p*-methylstyrene.

3.1.1.41 Tetralin

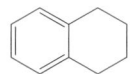

Common Name: Tetralin
Synonym: naphthalene-1,2,3,4-tetrahydride
Chemical Name: 1,2,3,4-tetrahydronaphthalene
CAS Registry No: 119-64-2
Molecular Formula: $C_{10}H_{12}$
Molecular Weight: 132.202
Melting Point (°C):
    –35.7   (Lide 2003)
Boiling Point (°C):
    207.6   (Weast 1982–83; Lide 2003)
Density (g/cm³ at 20°C):
    0.9702   (20°C, Weast 1982–83; Dean 1985)
    0.9695, 0.9660   (20°C, 25°C, Riddick et al. 1986)
Molar Volume (cm³/mol):
    136.4   (20°C, calculated- density)
    177.4   (calculated-Le Bas method at normal boiling point)
Enthalpy of Fusion, ΔH_{fus} (kJ/mol):
    12.477   (Riddick et al. 1986)
    12.45   (Chickos et al. 1999)
Entropy of Fusion, ΔS_{fus} (J/mol K):
    52.44, 49.6 (exptl., calculated-group additivity method, Chickos et al. 1999)
Fugacity Ratio at 25°C, F: 1.0

Water Solubility (g/m³ or mg/L at 25°C):
    14.94   (calculated-QSAR Data base, Passino & Smith 1987)

Vapor Pressure (Pa at 25°C or as indicated and reported temperature dependence equations. Additional data at other temperatures designated * are compiled at the end of this section):
    2666*   (93.8°C, temp range 93.8–171.06°C, Herz & Schuftan 1922)
    53.3*   (mercury manometer, measured range –2.4–65°C, Linder 1931)
    66.66*   (39.3°C, ebulliometry, measured range 39.3–148.6°C, Gardner & Brewer 1937)
    133.3*   (38°C, summary of literature data, temp range 38–207°C, Stull 1947)
    $\log (P/\text{mmHg}) = [-0.2185 \times 11613.0/(T/K)] + 8.194951$; temp range 38–207.2°C (Antoine eq., Weast 1972–73)
    1737*   (82.3°C, diaphragm gauge, measured range 82.3–276.21°C, Nasir et al. 1980)
    $\log (P/\text{mmHg}) = [1- 480.364/(T/K)] \times 10^{\{0.85916 - 5.75417 \times 10^{-4}\cdot(T/K) + 4.41971 \times 10^{-7}\cdot(T/K)^2\}}$; temp range 311.15–710.93 K, (Cox eq., Chao et al. 1983)
    53.75   (extrapolated-Antoine eq., Dean 1985, 1992)
    $\log (P/\text{mmHg}) = 7.07055 - 1741.30/(208.26 + t/°C)$; temp range 94–206°C (Antoine eq., Dean 1985, 1992)
    53.0   (selected lit. average, Riddick et al. 1986)
    $\log (P/\text{kPa}) = 11.079 - 2797.90/(T/K) + 1.187 \cdot \log (T/K)$, temp range not specified (Antoine eq., Riddick et al. 1986)
    56.7   (Antoine eq., extrapolated, Stephenson & Malanowski 1987)
    $\log (P_L/\text{kPa}) = 6.35719 - 1854.82/(-54.237 + T/K)$; temp range 311–481 K (Antoine eq., Stephenson & Malanowski 1987)
    $\log (P/\text{mmHg}) = 39.9174 - 4.132 \times 10^3/(T/K) - 10.78 \cdot \log (T/K) + 1.9691 \times 10^{-10} \cdot (T/K) + 2.0405 \times 10^{-6} \cdot (T/K)^2$, temp range 237–720 K (Yaws 1994)
    38.35*   (20.26°C, static method, measured range 253.8–442.8 K, Mokbel et al. 1998)

Henry's Law Constant (Pa m³/mol at 25°C or as indicated and reported temperature dependence equations. Additional data at other temperatures designated * are compiled at the end of this section):

    76.0, 106.4, 137.8, 189.5, 271.6   (0, 15, 20, 25. 30°C, EPICS-GC, Ashworth et al. 1988)

    $\ln [H/(\text{atm m}^3 \text{mol}^{-1})] = 11.83 - 5392/(T/K)$; temp range 10–30°C (EPICS-GC, Ashworth et al. 1988)

    142.8   (20°C, selected from literature experimentally measured data, Staudinger & Roberts 1996, 2001)

    $\log K_{AW} = 6.332 - 2215/(T/K)$ (van't Hoff eq. derived from literature data, Staudinger & Roberts 2001)

Octanol/Water Partition Coefficient, $\log K_{OW}$:

    3.83   (calculated-fragment const., Rekker 1977)

Octanol/Air Partition Coefficient, $\log K_{OA}$:

Bioconcentration Factor, $\log$ BCF:

Sorption Partition Coefficient, $\log K_{OC}$:

Environmental Fate Rate Constants, k, and Half-Lives, $t_{\frac{1}{2}}$:

    Volatilization:

    Photolysis:

    Oxidation: rate constant k, for gas-phase second order rate constants, k_{OH} for reaction with OH radical, k_{NO3} with NO_3 radical and k_{O3} with O_3 or as indicated, *data at other temperatures see reference:

        $k_{OH} = (3.43 \pm 0.06) \times 10^{-11}$ cm³ molecule⁻¹ s⁻¹, $k_{NO3} = (8.6 \pm 1.3) \times 10^{-15}$ cm³ molecule⁻¹ s⁻¹ at room temp. (relative rate method, Atkinson & Aschmann 1988)

        $k_{OH} = 3.43 \times 10^{-12}$ cm³ molecule⁻¹ s⁻¹ at 298 K (recommended, Atkinson 1990)

        $k_{OH}(\text{calc}) = 11.4 \times 10^{-12}$ cm³ molecule⁻¹ s⁻¹, $k_{OH}(\text{exptl}) = 34.3 \times 10^{-12}$ cm³ molecule⁻¹ s⁻¹ at 298 K (SAR [structure-activity relationship], Kwok & Atkinson 1995)

    Hydrolysis:

    Biodegradation:

    Biotransformation:

    Bioconcentration, Uptake (k_1) and Elimination (k_2) Rate Constants or Half-Lives:

Half-Lives in the Environment:

TABLE 3.1.1.41.1
Reported vapor pressures of tetralin at various temperatures and the coefficients for the vapor pressure equations

$\log P = A - B/(T/K)$ (1) $\ln P = A - B/(T/K)$ (1a)
$\log P = A - B/(C + t/°C)$ (2) $\ln P = A - B/(C + t/°C)$ (2a)
$\log P = A - B/(C + T/K)$ (3)
$\log P = A - B/(T/K) - C \cdot \log(T/K)$ (4)

Herz & Schuftan 1922		Stull 1947		Gardner & Brewer 1937		Mokbel et al. 1998	
		summary of literature data		ebulliometry		static method	
t/°C	P/Pa	t/°C	P/Pa	t/°C	P/Pa	T/K	P/Pa
93.8	2666	38	133	39.3	66.66	253.8	1.279
103	4000	65.3	666.6	40.1	79.99	264.01	3.349
140	15732	79	1333	40.8	93.33	273.65	7.984
150	21598	93.8	2666	41.8	93.33	283.54	17.94
167.5	36397	110	5333	46.6	226.7	293.51	38.35
206.2	98658	121.3	7999	48.9	253.3	303.52	78.52
		135.3	13332	49.4	253.3	313.51	153.0
		157.2	26664	54	293.3	323.55	284.0
Linder 1931		181.8	53329	74.4	946.6	333.57	506.0
mercury manometer		207.2	101325	75.1	959.9	343.62	868.0
t/°C	P/Pa			101.1	3040	353.66	1443
		mp/°C	−31.0	126.6	9040	358.64	1801
−2.4	2			148.6	19625	363.62	2254
−1.2	3.133					373.66	3508
25	53.33			bp/°C	210.5	383.00	5179
65	110.7					392.95	7617
						402.89	10948
				Nasir et al. 1980		412.86	15444
				diaphragm gauge		422.82	21388
				t/°C	P/Pa	432.84	29254
						442.80	39128
				82.3	1737		
				125.81	10500	data fitted to Wagner eq.	
				150.15	22419		
				200.11	84002		
				250.66	257307		
				276.21	348091		

Mononuclear Aromatic Hydrocarbons

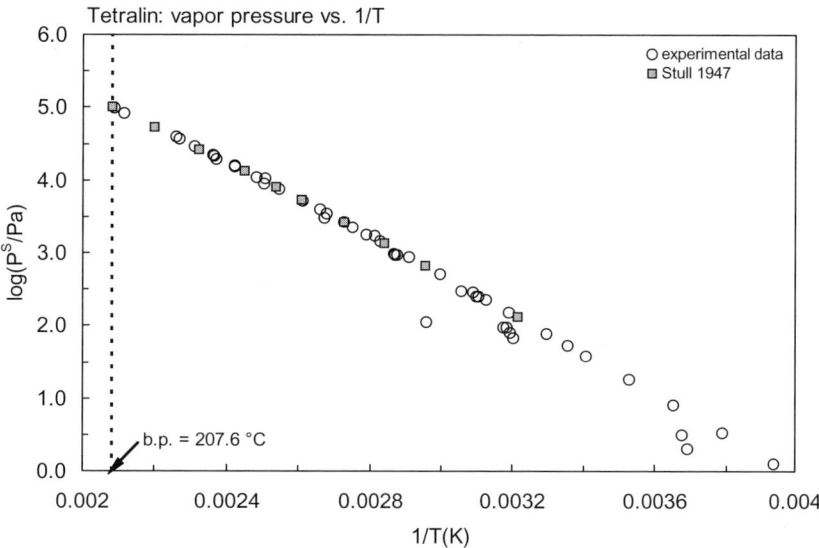

FIGURE 3.1.1.41.1 Logarithm of vapor pressure versus reciprocal temperature for tetralin.

3.2 SUMMARY TABLES AND QSPR PLOTS

TABLE 3.2.1
Summary of the physical properties of mononuclear aromatic hydrocarbons

Compound	CAS no.	Molecular formula	Molecular weight, MW g/mol	mp °C	bp °C	Fugacity ratio, F at 25 °C*	Density, ρ g/cm³ at 20°C	Molar volume, V_M cm³/mol MW/ρ at 20°C	Le Bas
Benzene	71-43-2	C_6H_6	78.112	5.49	80.09	1	0.8765	89.12	96.0
Toluene	108-88-3	C_7H_8	92.139	−94.95	110.63	1	0.8668	106.30	118.2
Ethylbenzene	100-41-4	C_8H_{10}	106.165	−94.96	136.19	1	0.867	122.45	140.4
o-Xylene	95-47-6	C_8H_{10}	106.165	−25.2	144.5	1	0.8802	120.61	140.4
m-Xylene	108-38-3	C_8H_{10}	106.165	−47.8	139.12	1	0.8842	120.07	140.4
p-Xylene	106-42-3	C_8H_{10}	106.165	13.25	138.37	1	0.8611	123.29	140.4
1,2,3-Trimethylbenzene	526-73-8	C_9H_{12}	120.191	−25.4	176.12	1	0.8944	134.38	162.6
1,2,4-Trimethylbenzene	95-63-6	C_9H_{12}	120.191	−43.77	169.38	1	0.8758	137.24	162.6
1,3,5-Trimethylbenzene	108-67-8	C_9H_{12}	120.191	−44.72	164.74	1	0.8800	136.58	162.6
n-Propylbenzene	103-65-1	C_9H_{12}	120.191	−99.6	159.24	1	0.862	139.43	162.6
Isopropylbenzene	98-82-8	C_9H_{12}	120.191	−96.02	152.41	1	0.8618	139.47	162.6
1-Ethyl-2-methylbenzene (o-Ethyltoluene)	611-14-3	C_9H_{12}	120.191	−79.83	165.2	1	0.8807	136.47	162.6
1-Ethyl-3-methylbenzene (m-Ethyltoluene)	620-14-4	C_9H_{12}	120.191	−95.6	161.3	1	0.8645	139.03	162.6
1-Ethyl-4-methylbenzene (p-Ethyltoluene)	622-96-8	C_9H_{12}	120.191	−62.35	162	1	0.8614	139.53	162.6
Isopropyl-4-methylbenzene (p-Cymene)	99-87-6	$C_{10}H_{14}$	134.218	−67.94	177.1	1	0.8573	156.56	184.8
n-Butylbenzene	104-51-8	$C_{10}H_{14}$	134.218	−87.85	183.31	1	0.8601	156.05	184.8
Isobutylbenzene	538-93-2	$C_{10}H_{14}$	134.218	−51.4	172.79	1	0.8532	157.31	184.8
sec-Butylbenzene	135-98-8	$C_{10}H_{14}$	134.218	−82.7	173.3	1	0.8621	155.69	184.4
tert-Butylbenzene	98-06-6	$C_{10}H_{14}$	134.218	−57.8	169.1	1	0.8665	154.90	184.8
1,2,3,4-Tetramethylbenzene	488-23-3	$C_{10}H_{14}$	134.218	−6.2	205	1	0.9052	148.27	184.8
1,2,3,5-Tetramethylbenzene	527-53-7	$C_{10}H_{14}$	134.218	−23.7	198	1	0.8903	150.76	184.8
1,2,4,5-Tetramethylbenzene	95-93-2	$C_{10}H_{14}$	134.218	79.3	196.8	0.293	0.8380	160.16	184.8
n-Pentylbenzene	538-68-1	$C_{11}H_{16}$	148.245	−75	205.4	1	0.8585	172.68	207.0

Mononuclear Aromatic Hydrocarbons

Name	CAS	Formula	MW	mp	bp		density		
Pentamethylbenzene	700-12-9	C₁₁H₁₆	148.245	54.5	232	0.514	0.917	161.66	207.0
n-Hexylbenzene	1077-16-3	C₁₂H₁₈	162.271	−61	226.1	1	0.8613	188.40	229.2
Hexamethylbenzene	87-85-4	C₁₂H₁₈	162.271	165.5	263.4	0.0418	1.063	152.65#	229.2
Heptylbenzene	1078-71-3	C₁₃H₂₀	176.298	−48	240	1	0.8567	205.79	251.4
Octylbenzene	2189-60-8	C₁₄H₂₂	190.325	−36	264	1	0.8562	222.29	273.6
Nonylbenzene	1081-77-2	C₁₅H₂₄	204.352	−24	280.5	1	0.8584	238.06	295.8
Decylbenzene	104-72-3	C₁₆H₂₆	218.377	−14.4	293	1	0.8555	255.26	318.0
Undecylbenzene	6742-54-7	C₁₇H₂₈	232.404	−5	316	1	0.8553	271.72	340.2
Dodecylbenzene	123-01-3	C₁₈H₃₀	246.431	3	328	1	0.8551	288.19	362.4
Tridecylbenzene	123-02-4	C₁₉H₃₂	260.457	10	346	1	0.8550	304.63	384.6
Tetradecylbenzene	1459-10-5	C₂₀H₃₄	274.484	16	359	1	0.8549	321.07	406.8
Styrene	100-42-5	C₈H₈	104.150	−30.65	145	1	0.9060	114.96	133.0
o-Methylstyrene	611-15-4	C₉H₁₀	118.175	−68.5	169.8	1	0.9106	129.78	155.2
m-Methylstyrene	100-80-1	C₉H₁₀	118.175	−86.3	164	1	0.9028	130.90	155.2
p-Methylstyrene	622-97-9	C₉H₁₀	118.175	−34.1	172.8	1	0.9016	131.07	155.2
α-Methylstyrene	98-83-9	C₉H₁₀	118.175	−23.2	165.4	1	0.9106	129.78	155.2
β-Methylstyrene, cis-	766-90-5	C₉H₁₀	118.175	−61.6	167.5	1	0.9088	130.03	155.2
β-Methylstyrene, trans-	873-66-5	C₉H₁₀	118.175	−29.3	178.3	1	0.9023	130.97#	155.2
Tetralin	119-64-2	C₁₀H₁₂	132.202	−35.7	207.6	1	0.9695	136.36	177.4

* Assuming ΔS_{fus} = 56 J/mol K.
# at 25°C.

TABLE 3.2.2
Summary of selected physical-chemical properties of mononuclear aromatic hydrocarbons at 25°C

Compound	Vapor pressure		Selected properties: Aqueous solubility			log K_{OW}	Henry's law constant
	P^S/Pa	P_L/Pa	S/(g/m^3)	C^S/(mol/m^3)	C_L/(mol/m^3)		H/(Pa·m^3/mol) calculated P/C
Benzene	12700	12700	1780	22.788	22.788	2.13	557
Toluene	3800	3800	515	5.590	5.590	2.69	680
Ethylbenzene	1270	1270	152	1.431	1.431	3.13	887
o-Xylene	1170	1170	220	2.072	2.072	3.15	565
m-Xylene	1100	1100	160	1.507	1.507	3.20	730
p-Xylene	1170	1170	215	2.024	2.024	3.18	578
1,2,3-Trimethylbenzene	200	200	70	0.582	0.582	3.55	343
1,2,4-Trimethylbenzene	270	270	57	0.474	0.474	3.60	569
1,3,5-Trimethylbenzene	325	325	50	0.416	0.416	3.58	781
n-Propylbenzene	450	450	52	0.433	0.433	3.69	1040
Isopropylbenzene	610	610	50	0.416	0.416	3.63	1466
1-Ethyl-2-methylbenzene	330	330	75	0.624	0.624	3.63	529
1-Ethyl-3-methylbenzene	391	391					
1-Ethyl-4-methylbenzene	395	395	95	0.790	0.790	3.63	500
Isopropyl-4-methylbenzene	204	204	34	0.253	0.253	4.10	805
n-Butylbenzene	137	137	13.8	0.103	0.103	4.26	1332
Isobutylbenzene	250	250	10.1	0.075	0.075	4.01	3322
sec-Butylbenzene	240	240	17	0.127	0.127		1890
tert-Butylbenzene	286	286	30	0.224	0.224	4.11	1280
1,2,3,4-Tetramethylbenzene	45	45				3.90	
1,2,3,5-Tetramethylbenzene	62	62				4.04	
1,2,4,5-Tetramethylbenzene	66	66	3.48	0.026	0.026	4.10	2546
n-Pentylbenzene	44	44	3.85	0.026	0.026	4.90	1694
Pentamethylbenzene	9.52	18.63	15.5	0.105	0.205	5.52	
n-Hexylbenzene	13.61	13.61	1.02	0.006	0.006	4.61	2165
Hexamethylbenzene	0.155	3.90	0.235	0.001	0.036	5.37	
Heptylbenzene	3.95	3.95				6.30	
Octylbenzene	2.03	2.03				6.82	
Nonylbenzene	0.30	0.30				7.37	
Decylbenzene	0.077	0.077					

Mononuclear Aromatic Hydrocarbons

Undecylbenzene	0.045				8.14
Dodecylbenzene	0.017				8.65
Tridecylbenzene	0.00070				9.36
Tetradecylbenzene	0.000113				9.95
Styrene	880	250	2.40		2.95
o-Methylstyrene	245				3.35
m-Methylstyrene	250				3.35
p-Methylstyrene	242				3.35
α-Methylstyrene	324				3.48
β-Methylstyrene, cis-	258				3.35
β-Methylstyrene, trans-	216.6			2.40	
Tetralin	54				3.83

TABLE 3.2.3
Suggested half-life classes of mononuclear aromatic hydrocarbons in various environmental compartments at 25°C

Compound	Air class	Water class	Soil class	Sediment class
Benzene	3	4	5	6
Toluene	2	5	6	7
Ethyl benzene	2	5	6	7
o-Xylene	2	5	6	7
m-Xylene	2	5	6	7
p-Xylene	2	5	6	7
1,2,3-Trimethylbenzene	2	5	6	7
1,2,4-Trimethylbenzene	2	5	6	7
1,3,5-Trimethylbenzene	2	5	6	7
n-Propylbenzene	2	5	6	7
Isopropylbenzene	2	5	6	7
1-Isopropyl-4-methylbenzene (p-Cymene)	2	5	6	7
1,2,4,5-Tetramethylbenzene	2	5	6	7

where,

Class	Mean half-life (hours)	Range (hours)
1	5	< 10
2	17 (~ 1 day)	10–30
3	55 (~ 2 days)	30–100
4	170 (~ 1 week)	100–300
5	550 (~ 3 weeks)	300–1,000
6	1700 (~ 2 months)	1,000–3,000
7	5500 (~ 8 months)	3,000–10,000
8	17000 (~ 2 years)	10,000–30,000
9	55000 (~ 6 years)	> 30,000

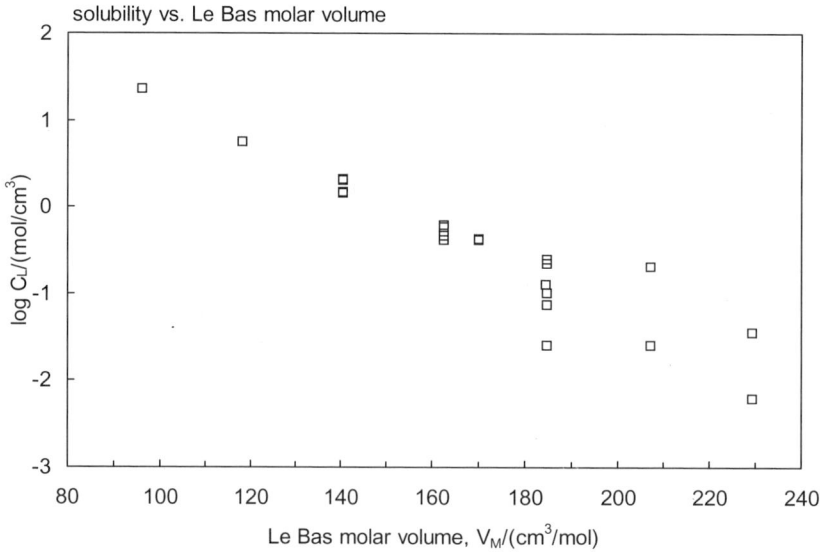

FIGURE 3.2.1 Molar solubility (liquid or supercooled liquid) versus Le Bas molar volume for mononuclear aromatic hydrocarbons.

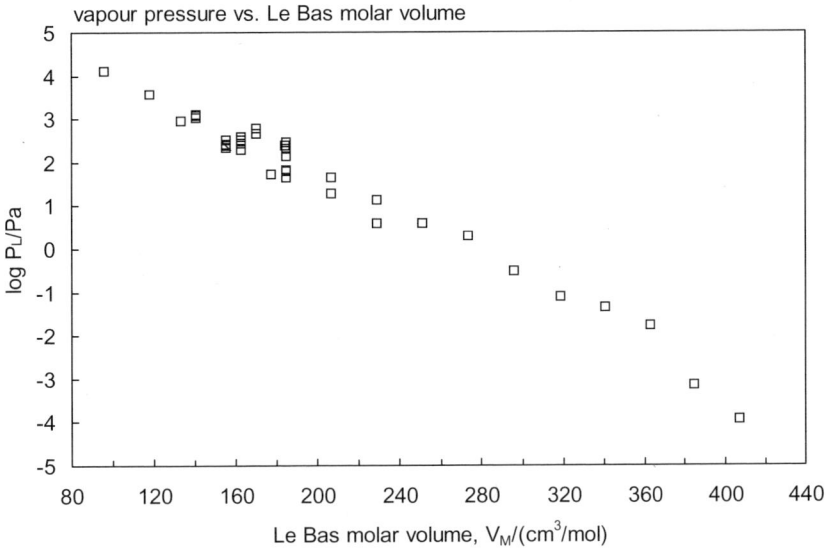

FIGURE 3.2.2 Vapor pressure (liquid or supercooled liquid) versus Le Bas molar volume for mononuclear aromatic hydrocarbons.

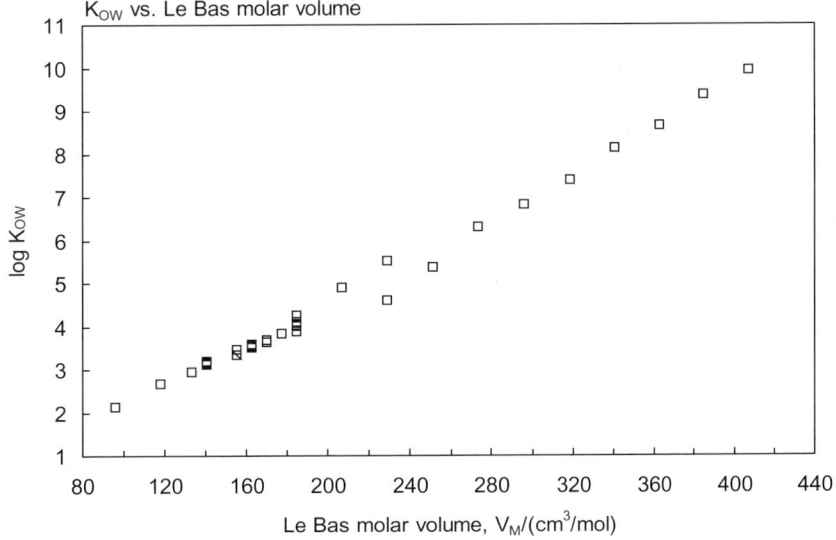

FIGURE 3.2.3 Octanol-water partition coefficient versus Le Bas molar volume for mononuclear aromatic hydrocarbons.

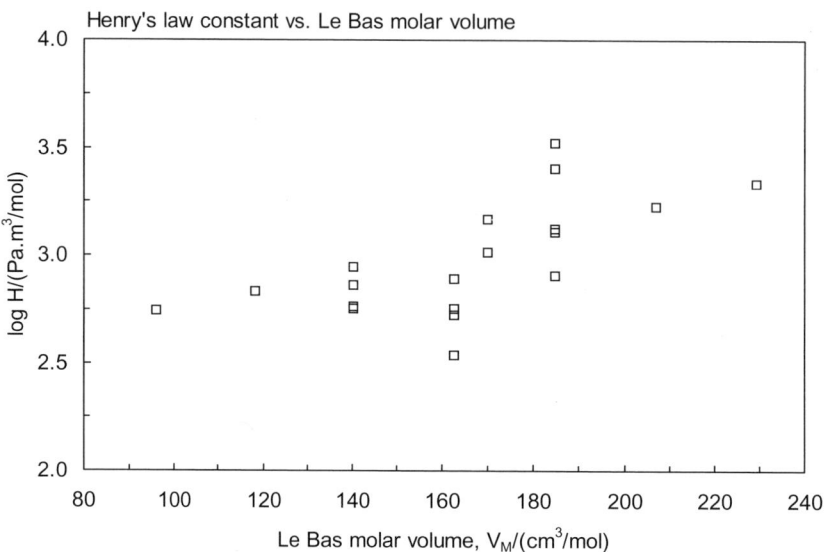

FIGURE 3.2.4 Henry's law constant versus Le Bas molar volume for mononuclear aromatic hydrocarbons.

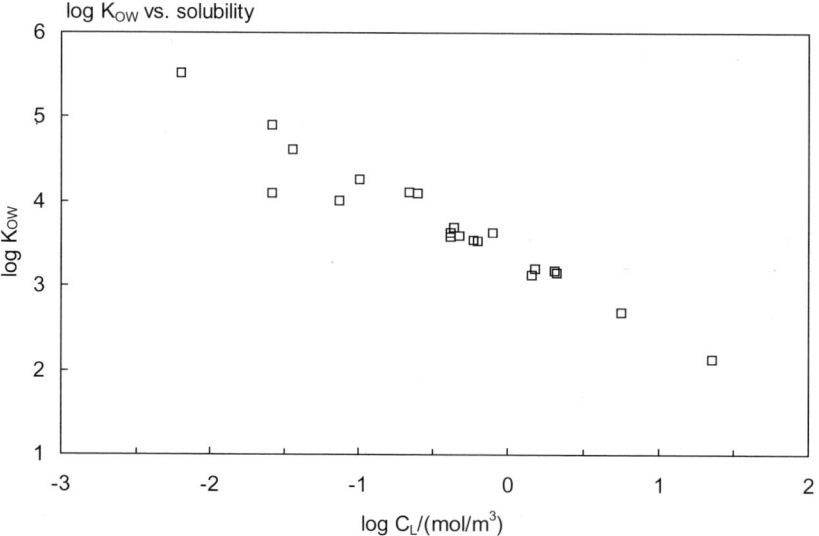

FIGURE 3.2.5 Octanol-water partition coefficient versus molar solubility (liquid or supercooled liquid) for mononuclear aromatic hydrocarbons.

3.3 REFERENCES

Abdul, A.S., Gibson, T.L., Rai, D.N. (1990) Use of humic acid solution to remove organic contaminants from hydrogeologic systems. *Environ. Sci. Technol.* 24(3), 328–333.

Abraham, M.H., Fuchs, R., Whiting, G.S., Chambers, E.C. (1990) *J. Chem. Soc. Parkin Trans* 2, 291.

Abraham, M.H., Le J., Acree, Jr., W.E., Carr, P.W., Dallas, A.J. (2001) The solubility of gases and vapours in dry octan-1-ol at 298 K. *Chemosphere* 44, 855–863.

Alaee, M., Whittal, R.M., Strachan, W.M.J. (1996) The effect of water temperature and composition on Henry's law constant for various PAH's. *Chemosphere* 32, 1153–1164.

Alexander, D.M. (1959) The solubility of benzene in water. *J. Phys. Chem.* 63, 1021–1022.

Allemand, N., Jose, J., Merlin, J.C. (1986) Mesure des pressions de vapeur d'hydrocarbures C_{10} a C_{18} *n*-alcanes et *n*-alkylbenzenes dans le domaine 301000 pascal. *Thermochim. Acta* 105, 79–90.

Altshuller, A.P. (1991) Chemical reactions and transport of alkanes and their products in the troposphere. *J. Atmos. Chem.* 12, 19–61.

Altschuh, J., Brüggemann, Santl, H., Eichinger, G., Piringer, O.G.(1999) Henry's law constants for a diverse set of organic chemicals: Experimental determination and comparison of estimation methods. *Chemosphere* 39, 1871–1887.

Ambrose, D. (1981) Reference value of vapour pressure. The vapour pressures of benzene and hexafluorobenzene. *J. Chem. Thermodyn.* 13, 1161–1167.

Ambrose, D., Broderick, B.E., Townsend, R. (1967) The vapour pressures above the normal boiling point and the critical pressures of some aromatic hydrocarbons. *J. Chem. Soc.* A, 633–641.

Ambrose, D., Counsell, J.F., Davenport, A.J. (1970) The use of Chebyshev polynomials for the representation of vapour pressures between the triple point and the critical point. *J. Chem. Thermodyn.* 2, 283–294.

Ambrose, D., Lawrenson, I.J., Sprake, C.H.S. (1976) The vapor pressure of hexamethylbenzene. *J. Chem. Thermodyn.* 8, 503–504.

Anderson, M.A. (1992) Influence of surfactants on vapor-liquid partitioning. *Environ. Sci. Technol.* 28, 2186–2191.

Anderson, T.A., Beauchamp, J.J., Walton, B.T. (1991) Organic chemicals in the environment. *J. Environ. Qual.* 20, 420–424.

Andrews L.J., Keefer, R.M. (1949) Cation complexes of compounds containing carbon-carbon double bonds. IV. The argentation of aromatic hydrocarbons. *J. Am. Chem. Soc.* 71, 3644–3647.

Andrews, L.J., Keefer, R.M. (1950) Cation complexes of compounds containing carbon-carbon double bonds. VII. Further studies on the argentation of substituted benzenes. *J. Am. Chem. Soc.* 72, 5034–5037.

Aquan-Yuen, M., Mackay, D., Shiu, W.Y. (1979) Solubility of hexane, phenanthrene, chlorobenzene, and *p*-dichlorobenzene in aqueous electrolyte solutions. *J. Chem. Eng. Data* 24, 30–34.

Arbuckle, W.B. (1983) Estimating activity coefficients for use in calculating environmental parameters. *Environ. Sci. Technol.* 17, 537–542.

Arnold, D.S., Plank, C.A., Erickson, E.E., Pike, F.P. (1958) Solubility of benzene in water. *Chem. Eng. Data* Ser. 3, 253–256.

Ashworth, R.A., Howe, G.B., Mullins, M.E., Rogers, T.N. (1988) Air-water partitioning coefficients of organics in dilute aqueous solutions. *J. Hazard. Materials* 18, 25–36.

Atkinson, R. (1985) Kinetics and mechanisms of the gas phase reaction of hydroxyl radicals with organic compounds under atmospheric conditions. *Chem. Rev.* 85, 69–201.

Atkinson, R. (1987) Structure-activity relationship for the estimation of the rate constants for the gas phase reactions of OH radicals with organic compounds. *Int. J. Chem. Kinetics* 19, 799–828.

Atkinson, R. (1989) Kinetics and Mechanisms of the gas-phase reactions of the hydroxyl radical with organic compounds. *J. Phys. Chem. Data* Monograph No. 1.

Atkinson, R. (1990) Gas-phase tropospheric chemistry of organic compounds: a review. *Atmos. Environ.* 24A, 1–41.

Atkinson, R. (2000) Atmospheric chemistry of VOCs and No_x. *Atmos. Environ.* 34, 2063–2101.

Atkinson, R., Aschmann, S.M. (1988) Kinetics of the reactions of acenaphthene and acenaphthylene and structurally-related aromatic compounds with OH and NO_3 radicals, N_2O_5 and O_3 at 296 ± 2 K. *Int. J. Chem. Kinet.* 20, 513–539.

Atkinson, R., Aschmann, S.M., Fitz, D.R., Winer, A.M., Pitts Jr., J.N. (1982) Rate constants for the gas-phase reactions of O_3 with selected organics at 296 K. *Int. J. Chem. Kinet.* 14, 13–18.

Atkinson, R., Aschmann, S.M., Pitts, J.N. Jr. (1988) Rate constants for the gas-phase reactions of the NO_3 radicals with a series of organic compounds at 296 ± 2 K. *J. Phys. Chem.* 92, 3454–3457.

Atkinson, R., Aschmann, S.M., Winer, A.M., Pitts, Jr., J.N. (1985) Atmospheric gas phase loss processes for chlorobenzene, benzotrifluoride and 4-chlorobenzotrifluoride, and generalization of predictive techniques for atmospheric lifetimes of aromatic compounds. *Environ. Contam. Toxicol.* 14, 417–425.

Atkinson, R., Carter, W.L. (1984) Kinetics and mechanisms of the gas-phase reactions of ozone with organic compounds under atmospheric conditions. *Chem. Rev.* 84, 437–470.

Atkinson, R., Carter, W.L., Plum, C.N., Winer, A.M., Pitts, Jr., J.N. (1984) Kinetics of the gas-phase reactions of NO_3 radicals with a series of aromatics at 296 ± 2 K. *Int. J. Chem. Kinet.* 16, 887–898.

Ayuttaya, P.C.N., Rogers, T.N., Mullins, M.E., Kline, A.A. (2001) Henry's law constants derived from equilibrium static cell measurements for dilute organic-water mixtures. *Fluid Phase Equil.* 185, 359–377.

Bahnick, D.A., Doucette, W.J. (1988) Use of molecular indices to estimate soil sorption coefficient for organic chemicals. *Chemosphere* 17, 1703–1715.

Bakierowska, A.-M., Trzeszczyński, J. (2003) Graphical method for the determination of water/gas partition coefficients of volatile organic compounds by a headspace gas chromatography technique. *Fluid Phase Equil.* 213, 139–146.

Banerjee, S. (1984) Solubility of organic mixture in water. *Environ. Sci. Technol.* 18, 587–591.

Banerjee, S., Howard, P.H. (1988) Improved estimation of solubility and partitioning through correction of UNIFAC-derived activity coefficients. *Environ. Sci. Technol.* 22, 839–841.

Banerjee, S., Howard, P.H., Lande, S.S. (1990) General structure vapor pressure relationship for organics. *Chemosphere* 21, 1173–1180.

Banerjee, S., Yalkowsky, S.H., Valvani, S.C. (1980) Water solubility and octanol/water partition coefficient of organics. Limitations of solubility-partition coefficient correlation. *Environ. Sci. Technol.* 14, 1227–1229.

Bartholomew, G.W., Pfaender, F.K. (1983) Influence of spatial and temporal variations on organic pollutant biodegradation rates in an estuarine environment. *Appl. Environ. Microbiol.* 45, 103–109.

Battersby, N.S. (1990) A review of biodegradation kinetics in the aquatic environment. *Chemosphere* 21, 1243–1284.

Bechalany, A. Röthlisberger, T., El Tayar, N., Testa, B. (1989) Comparison of various non-polar stationary phases used for assessing lipophilicity. *J. Chromatogr.* 473, 115–124.

Ben-Naim, A., Wiff, J. (1979) A direct measurement of intramolecular hydrophobic interactions. *J. Chem. Phys.* 70, 771–777.

Bierwagen, B.G., Keller, A.A. (2001) Measurement of Henry's law constant for methyl *tert*-butyl ether using solid-phase microextraction. *Environ. Toxicol. Chem.* 20, 1625–1629.

Bittrich, H.J., Gedan, H., Feix, G. (1979) Zur löslichkeitsbeeinflussung von kohlenwasserstoffen in wasser. *Z. Phys. Chem.* (Leibzig) 260, 1009–1013.

Bohon, R.L., Claussen, W.F. (1951) The solubility of aromatic hydrocarbons in water. *J. Am. Chem. Soc.* 72, 1571–1576.

Bond, D.L., Thodos, G. (1960) Vapor pressures of alkyl aromatic hydrocarbons. *J. Chem. Eng. Data* 5, 289–292.

Booth, H.S., Everson, H.E. (1948) Hydrotropic solubilities: solubilities in 40 per cent sodium xylenesulfonate. *Ind. Eng. Chem.* 40(8), 1491–1493.

Boublik, T., Fried, V., Hala, E. (1973) *The Vapour Pressure of Pure Substances.* Elsevier, Amsterdam.

Boublik, T., Fried, V., Hala, E. (1984) *The Vapour Pressures of Pure Substances* (revised second edition), Elsevier, Amsterdam.

Boyd, S.A., Jin, X., Lee, J.-F. (1990) Sorption of nonionic organic compounds by corn residues from a no-tillage field. *J. Environ. Qual.* 19, 734–938.

Brady, A.P., Huff, H. (1958) Vapor pressure of benzene over aqueous detergent solutions. *J. Phys. Chem.* 62, 644–649.

Bradley, R.S., Dew, M.J., Munro, D.C. (1973) Solubility of benzene and toluene in water and aqueous salt solutions under pressure. *High Temp. High Press.* 5, 169–176.

Brice, K.A., Derwent, R.G. (1978) Emissions inventory for hydrocarbons in United Kingdom. *Atom. Environ.* 12, 2045–2054.

Bridie, A.L., Wolff, C.J.M., Winter, M. (1979) BOD and COD of some petrochemicals. *Water Res.* 13, 627–630.

Brooke, D.N., Dobbs, A.J., Williams, N. (1986) Octanol/water partition coefficients P: measurement, estimation, and interpretation particularly for chemicals with P > 10^6. *Ecotoxicol. Environ. Saf.* 11, 251–260.

Brown, R.L., Wasik, S.P. (1974) A method of measuring the solubilities of hydrocarbons in aqueous solutions. *J. Res. Nat. Bur. Std.* 78A, 453–460.

Bruggeman, W.A, Van Der Steen, J., Hutzinger, O. (1982). Reversed-phase thin-layer chromatography of polynuclear aromatic hydrocarbons and chlorinated biphenyls. Relationship with hydrophobicity as measured by aqueous solubility and octanol-water partition coefficient. *J. Chromatogr.* 238, 335–346.

Buck, F.R., Coles, K.F., Kennedy, G.T., Morton, F. (1949) Some nuclear-methylated styrenes and related compounds. Part I. *J. Chem. Soc.* 2377–2383.

Bufalini, J.J., Altshuller, A.P. (1965) Kinetics of vapor-phase hydrocarbon-ozone reaction. *Can. J. Chem.* 43, 2243–2250.

Burchfield, P.E. (1942) Vapor pressures of indene, styrene and dicyclopentadiene. *J. Am. Chem. Soc.* 64, 2501.

Callahan, M.A., Slimak, M.W., Gabel, N.W., May, I.P., Fowler, C.F., Freed, J.R., Jennings, P., Durfee, R.L., Whitmore, F.C., Maestri, B., Mabey, W.R., Holt, B.R., Gould, C. (1979) *Water-Related Environmental Fate of 129 Priority Pollutants,* Vol. I, EPA Report No. 440/4–79–029a. Versar, Inc., Springfield, VA.

Camin, D.L., Forziati, A.F., Rossini, F. (1954) Physical properties of *n*-hexadecane, *n*-decylclopentane, *n*-decylcyclohexane, 1-hexadecene and *n*-decylbenzene. *J. Phys. Chem.* 58, 440–442.

Capel, P.D., Larson, S.J. (1995) A chemodynamic approach for estimating losses of target organic chemicals form water during sample holding time. *Chemosphere* 30, 1097–1107.

Carter, W.P.L., Winer, A.M., Pitts, Jr., J.N. (1981) Major atmospheric sink for phenol and the cresols. Reaction with the nitrate radical. *Environ. Sci. Technol.* 15, 829–831.

Chaiyavech, P., Van Winkle, M. (1959) Styrene-ethylbenzene vapor-liquid equilibria at reduced pressures. *J. Chem. Eng. Data* 4, 53–56.

Chao, J., Lin, C.T., Chung, T.H. (1983) Vapor pressure of coal chemicals. *J. Phys. Chem. Ref. Data* 12, 1033–1063.

Chen, H., Wagner, J. (1994a) An apparatus and procedure for measuring mutual solubilities of hydrocarbons + water: Benzene + water from 303 to 373 K. *J. Chem. Eng. Data* 39, 470–474.

Chen, H., Wagner, J. (1994b) An efficient and reliable gas chromatographic method for measuring liquid-liquid mutual solubilities in alkylbenzene + water mixtures: Toluene + water from 303 to 373 K. *J. Chem. Eng. Data* 39, 474–479.

Chen, H., Wagner, J. (1994c) Mutual solubilities of alkylbenzene + water systems at temperatures from 303 to 373 K: Ethylbenzene, *p*-xylene, 1,3,5-trimethylbenzene, and *n*-butylbenezene. *J. Chem. Eng. Data* 39, 679–684.

Cheng, W.-H., Chu, F.-S., Liou, J.-J. (2003) Air-water interface equilibrium partitioning coefficients of aromatic hydrocarbons. *Atmos. Environ.* 37, 4807–4815.

Chernoglazova, F.S., Simulin, Yu. N. (1976) Solubility tribenzyamine in benzene and isopropyl alcohol. *Zh. Fiz. Khim.* 50.809.

Chey, W., Calder, G.V. (1972) Method for determining solubility of slightly soluble organic compounds. *J. Chem. Eng. Data* 17(2), 199–200.

Chickos, J.S., Acree, Jr., W.E., Liebman, J.F. (1999) Estimating solid-liquid phase change enthalpies and entropies. *J. Phys. Chem. Ref. Data* 28, 1535–1673.

Chiou, C.T. (1985) Partition coefficients of organic compounds in lipid-water system and correlations with fish bioconcentration factors. *Environ. Sci. Technol.* 19, 57–62.

Chiou, C., Freed, D., Schmedding, D., Kohnert, R. (1977) Partition coefficient and bioaccumulation of selected organic chemicals. *Environ. Sci. Technol.* 11(5),475–478.

Chiou, C.T., Porter, P.E., Schmedding, D.W. (1983) Partition equilibria of nonionic organic compounds between soil organic matter and water. *Environ. Sci. Technol.* 17, 227–231.

Chiou, C.T., Schmedding, D.W., Manes, M. (1982) Partitioning of organic compounds in octanol-water systems. *Environ. Sci. Technol.* 16, 4–10.

Clements, H.E., Wise, K.V., Johnsen, S.E. (1953) Physical properties of *o*-, *m*-, and *p*-methylstyrene. *J. Am. Chem. Soc.* 75, 1593–1595.

Cooling, M.R., Khalfaoui, B., Newsham, D.M.T. (1992) Phase equilibria in very dilute mixtures of water and unsaturated chlorinated hydrocarbons and of water and benzene. *Fluid Phase Equil.* 81, 217–229.

Corchnoy, S.B., Atkinson, R. (1990) Kinetics of the gas-phase reactions of OH and NO_3 radicals with 2-carene, 1,8-cineole, *p*-cymene, and terpinolene. *Environ. Sci. Technol.* 24, 1497–1502.

Coutant, R.W., Keigley, G.W. (1988) An alternative method for gas chromatographic determination of volatile organic compounds in water. *Anal. Chem.* 60, 2436–2537.

D'Amboise, M., Hanai, T. (1982) Hydrophobicity and retention in reverse phase liquid chromatography. *J. Liq. Chromatogr.* 5, 229–244.

Darnall, K.R., Lloyd, A.C., Winer, A.M., Pitts, J.N. (1976) Reactivity scale for atmospheric hydrocarbons based on reaction with hydroxyl radicals. *Environ. Sci. Technol.* 10, 692–696.

Daubert, T.E., Danner, R.P. (1985) *Data Compilation Tables of Properties of Pure Compounds*. Am. Institute of Chem. Engineers. pp. 450.

David, M.O., Fendinger, N.J., Hand, V.C. (2000) Determination of Henry's law constants for organosilicones in actual and simulated wastewater. *Environ. Sci. Technol.* 34, 4554–4559.

Davis, J.W., Madsen, S. (1996) Factors affecting the biodegradation of toluene in soil. *Chemosphere* 33, 107–130.

Dean, J.D., Ed. (1985) *Lange's Handbook of Chemistry*. 13th ed. McGraw-Hill, New York.

Dean, J.D., Ed. (1992) *Lange's Handbook of Chemistry*. 14th ed. McGraw-Hill, New York.

De Bruijn, J., Busser, F., Seinen, W., Hermens, J. (1989) Determination of octanol/water partition coefficient for hydrophobic organic chemicals with the "slow-stirring" method. *Environ. Toxicol. Chem.* 8, 499–512.

De Kock, A.C., Lord, D.A. (1987) A simple procedure for determining octanol-water partition coefficients using reversed phase high performance liquid chromatography (RPHPLC). *Chemosphere* 16(1), 133–142.

Delle Site, A. (2001) Factors affecting sorption of organic compounds in natural sorbent/water systems and sorption coefficients for selected pollutants. A review. *J. Phys. Chem. Ref. Data* 30, 187–439.

Deno, N.C., Berkheimer, H.E. (1960) Phase equilibria molecular transport thermodynamics: activity coefficients as a function of structure and media. *J. Chem. Eng. Data* 5, 1–5.

DeVoe, H., Miller, M.M., Wasik, S.P. (1981) Generator columns and high pressure liquid chromatography for determining aqueous solubilities and octanol-water partition coefficients of hydrophobic substances. *J. Res. Natl. Bur. Std.* 86, 361.

de Wolf, W., Lieder, P.H. (1998) A novel method to determine uptake and elimination kinetics of volatile chemicals in fish. *Chemosphere* 36, 1713–1724.

Dewulf, J., Drijvers, D., van Langenhove, H. (1995) Measurement of Henry's law constant as function of temperature and salinity for low temperature range. *Environ. Toxicol. Chem.* 29, 323–331.

Dewulf, J., van Langenhove, H., Graré, S. (1999) Sediment/water and octanol/water equilibrium partitioning of volatile organic compounds: temperature dependence in the 2–25°C range. *Water Res.* 33, 2424–2436.

Dilling, W.L., Bredweg, C.J., Tefertiller, N.B. (1976) Organic photochemistry. Simulated atmospheric photodecomposition rate of methylene chloride, 1,1,1-trichloroethane, trichloroethylene, tetrachloroethylene, and other compounds. *Environ. Sci. Technol.* 10, 351–356.

Dilling, W.L., Gonsior, S.J., Boggs, G.U., Mendoza, C.G. (1988) Organic photochemistry. 20. A method for estimating gas-phase rate constants for reactions of hydroxyl radicals with organic compounds from their relative rates of reaction with hydrogen peroxide under photolysis in 1,1,2-tirchlorotrifluoroethane solution. *Environ. Sci. Technol.* 22, 1447–1553.

Dohányosová, P., Fenclová, D., Vrbka, P., Dohnal, V. (2001) Measurement of aqueous solubility of hydrophobic volatile organic compounds by solute vapor absorption technique: toluene, ethylbenzene, propylbenzene, and butylbenzene at temperatures from 273 K to 328 K. *J. Chem. Eng. Data* 46, 1533–1539.

Dohnal, V., Hovorka, Š. (1999) Exponential saturator: A novel gas-liquid partitioning technique for measurement of large limiting activity coefficients. *Ind. Eng. Chem. Res.* 38, 2036–2043.

Donovan, S.F., Pescatore, M.C. (2002) Method for measuring the logarithm of the octanol-water partition coefficient by using short octadecyl-poly(vinyl alcohol) high-performance liquid chromatography columns. *J. Chromatogr. A*, 952, 47–61.

Dorfman, L.M., Adams, G.E. (1973) *Reactivity of the hydroxyl radical in aqueous solution*. NSRD-NDB-46. NTIS COM-73-50623. National Bureau Standards, pp. 51, Washington, D.C.

Doucette, W.J., Andren, A.W. (1988) Estimation of octanol/water partition coefficients: Evaluation of six methods for highly hydrophobic aromatic hydrocarbons. *Chemosphere* 17, 345–359.

Doyle, G.J., Lloyd, A.C., Darnall, K.R., Winer, A.M., Pitts, J.N. Jr. (1975) Gas phase kinetic study of relative rates of reaction of selected aromatic compounds with hydroxyl radicals in an environmental chamber. *Environ. Sci. Technol.* 9, 237–241.

Dreisbach, R.R. (1955) *Physical Properties of Chemical Compounds*. Adv. Chem. Ser. 15, 134.

Dreisbach, R.,R., Martin, A.A.I. (1949) Physical data on some organic compounds. *Ind. Eng. Chem.* 41, 2875–2878.

Dreisbach, R.R., Shrader, A.A.I. (1949) Vapor pressure-temperature data on some organic compounds. *Ind. Eng. Chem.* 41, 2879–2880.

Dreyer, R., Martin, W., Von Webber, U. (1955) *J. Prakt. Chem.* 273, 324.

Eadsforth, C.V., Moser, P. (1983) Assessments of reversed phase chromatographic methods for determining partition coefficients. *Chemosphere* 12, 1459–1475.

Eastcott, L., Shiu, W.Y., Mackay, D. (1988) Environmentally relevant physical-chemical properties of hydrocarbons: A review of data and development of simple correlations. *Oil & Chem. Pollut.* 4, 191–216.

Edney, E.L., Kleindienst, T.E., Corse, E.W. (1986) Room temperature rate constants for the reaction of OH with selected chlorinated and oxygenated hydrocarbons. *Int. J. Chem. Kinet.* 18, 1355–1371.

Ettre, L.S., Welter, C., Kolb, B. (1993) Determination of gas-liquid partition coefficients by automatic equilibrium headspace-gas chromatography utilizing the phase ratio variation method. *Chromatographia* 35, 73–84.

Etzweiler, F., Senn, E., Schmidt, H.W.H. (1995) Method for measuring aqueous solubilities of organic compounds. *Anal. Chem.* 67, 655–658.

Finizio, A., Di Guardo, A. (2001) Estimating temperature dependence of solubility and octanol-water partition coefficient for organic compounds using RP-HPLC. *Chemosphere* 45, 1063–1070.

Forziati, A.F., Norris, W.R., Rossini, F.D. (1949) Vapor pressures and boiling points of sixty API-NBS hydrocarbons. *J. Res. Natl. Bur. Std.* 43, 555–563.

Franks, F., Gent, M., Johnson, H.H. (1963) The solubility of benzene in water. *J. Chem. Soc.* 2716–2723.

Freitag, D., Lay, J.P. Korte, F. (1984) Environmental hazard profile-test results as related to structures and translation into the environment. In: *QSAR in Environmental Toxicology*, Kaiser, K.L.E., Ed., D. Reidel Publ. Co., Dordrecht, Netherlands.

Freitag, D., Ballhorn, L., Geyer, H., Korte, F. (1985) Environmental hazard profile of organic chemicals. An experimental method for the assessment of the behaviour of organic chemicals in the ecosphere by means of simple laboratory tests with ^{14}C labelled chemicals. *Chemosphere* 14, 1589–1616.

Frilette, V.J., Hohenstein, W.P. (1948) Polymerization of styrene in soap solutions. *J. Polymer Sci.* 3, 22–31.

Fu, M.H., Alexander, M. (1992) Biodegradation of styrene in samples of natural environments. *Environ. Sci. Technol.* 26, 1540–1546.

Fühner, H. (1924) Die Wasserlöslichkeit in homologen Reihen. *Chem. Ber.* 57, 510–515.

Fujisawa, S., Masuhara, E. (1981) Determination of partition coefficients of acrylated, methacrylated, and vinyl monomers using high performance liquid chromatography (HPLC). *J. Biomed. Materials Res.* 15, 787–793.

Fujita, T., Iwasa, Hansch, C. (1964). A new substituent constant, "pi" derived from partition coefficients. *J. Am. Chem. Soc.* 86, 5175–5180.

Garbarnini, D.R., Lion, L.W. (1985) Evaluation of sorptive partitioning of nonionic pollutants in closed systems by headspace analysis. *Environ. Sci. Technol.* 19, 1122–1128.

Gardner, G.S, Brewer, J.E. (1937) Vapor pressure of commercial high-boiling organic solvents. *Ind. Eng. Chem.* 29, 179–181.

Garst, J.E. (1984) Accurate, wide range, automated, high performance liquid chromatographic method for the estimation of octanol/water partition coefficients. II: Equilibration in partition coefficient measurements, additivity of substituent-constants and correlation of biological data. *J. Pharm. Sci.* 73(11), 1623–1629.

Garst, J.E., Wilson, W.C. (1984) Accurate, wide range, automated, high performance liquid chromatographic method for the estimation of octanol/water partition coefficients. II: Effects of chromatographic method and procedure variables on accuracy and reproducibility of the method. *J. Pharm. Sci.* 73(11), 1616–1623.

Geyer, H., Sheenhan, P., Kotzias, D., Freitag, D., Korte, F. (1982) Prediction of ecotoxicological behaviour of chemicals: relationship between physico-chemical properties and bioaccumulation of organic chemicals in the mussel *mytilus edulis*. *Chemosphere* 11, 1121–1134.

Geyer, H.J., Politzki, G., Freitag, D. (1984) Prediction of ecotoxicological behaviour of chemicals: relationship between n-octanol/water partition coefficient and bioaccumulation of organic chemicals by *alga chlorella*. *Chemosphere* 13(2), 269–284.

Glew, D.N., Roberson, R.E. (1956) The spectrophotometric determination of the solubility of cumene in water by a kinetic method. *J. Phys. Chem.* 60, 332–337.

Görgényi, M., Dewulf, J., Van Langenhove, H. (2002) Temperature dependence of Henry's law constant in an extended temperature range. *Chemosphere* 48, 757–762.

Govers, H.A.J., Evers, E.H.G. (1992) Prediction of distribution properties by solubility parameters: description of the method and application to methylbenzenes. *Chemosphere* 24, 453–464.

Green, W.J., Frank, H.S. (1979) The state of dissolved benzene in aqueous solution. *J. Soln. Chem.* 8(3), 187–196.

Gross, P.M., Saylor, J.H. (1931) The solubilities of certain slightly soluble organic compounds in water. *J. Am. Chem. Soc.* 53, 1744–1751.

Gruber, D., Langenheim, D., Gmehling, J., Moollan, W. (1997) Measurement of activity coefficients at infinite dilution using gas-liquid chromatography. 6. Results for systems exhibiting gas-liquid interface adsorption with 1-octanol. *J. Chem. Eng.* 42, 882–885.

Güesten, H., Filby, W.G., Schoop, S. (1981) Prediction of hydroxyl radical reaction rates with organic compounds in the gas phase. *Atmos. Environ.* 15, 1763–1765.

Haky, J.E., Young, A.M. (1984) Evaluation of a simple HPLC correlation method for the estimation of the octanol-water partition coefficients of organic compounds. *J. Liq. Chromatogr.* 7, 675–689.

Hafkenscheid, T.L., Tomlinson, E. (1983a) Isocratic chromatographic retention data for estimating aqueous solubilities of acidic, basic and neutral drugs. *Intl. J. Pharm.* 16, 1–21.

Hafkenscheid, T.L., Tomlinson, E. (1983b) Correlations between alkane/water and octan-1-ol/water distribution coefficients and isocratic reversed-phase liquid chromatographic capacity factor of acids, bases and neutrals. *Intl. J. Pharm.* 16, 225–240.

Hammers, W.E., Meurs, G.J., De Ligny, C.L. (1982) Correlations between liquid chromatographic capacity ratio data on Lichrosorb RP-18 and partition coefficients in the octanol-water system. *J. Chromatogr.* 247, 1–13.

Hanai, T., Hubert, J. (1984) Retention versus van der Waals volume and δ energy in liquid chromatography. *J. Chromatogr.* 290, 197–206.

Hanai, T., Tran, C., Hubert, J. (1981) An approach to the prediction of retention times in liquid chromatography. *J. HRC & CC* 4, 454–460.

Hansch, C., Leo, A. (1979) *Substituent Constants for Correlation Analysis in Chemistry and Biology.* Wiley, New York.

Hansch, C., Leo, A. (1985) Medichem Project. Pomona College, Claremont, CA.

Hansch, C., Leo, A., Nickaitani, D. (1972) On the additive - constitutive character of partition coefficients. *J. Org. Chem.* 37, 3090–3092.

Hansch. C., Leo, A.J., Hoekman, D. (1995) *Exploring QSAR, Hydrophobic, Electronic, and Steric Constants.* ACS Professional Reference Book, American Chemical Society, Washington, DC.

Hansch, C., Quinlan, J.E., Lawrence, G.L. (1968) The linear free-energy relationship between partition coefficients and the aqueous solubility of organic liquids. *J. Org. Chem.* 33, 347–350.

Hansen, D.A., Atkinson, R., Pitts, J.N. Jr. (1975) Rate constants for the reaction of OH radicals with a series of aromatic hydrocarbons. *J. Phys. Chem.* 79, 1763–1766.

Hansen, K.C., Zhou, Z., Yaws, C.L., Aminabhavi, T.J. (1993) Determination of Henry's law constants of organics in dilute aqueous solutions. *J. Chem. Eng. Data* 38, 546–550.

Haque, R., Falco, J., Cohen, S., Riordan, C. (1980) Role of transport and fate studies in the exposure, assessment and screening of toxic chemicals. In: *Dynamics, Exposure and Hazard Assessment of Toxic Chemicals.* Haque, R., Ed., pp. 47–67, Ann Arbor Sci. Publ., Ann Arbor, MI.

Harnisch, M., Möckel, H.J., Schulze, G. (1983) Relationship between log P_{OW} shake-flask values and capacity factors derived from reversed-phase HPLC for n-alkylbenzenes and some OECD reference substances. *J. Chromatogr.* 282, 315–332.

Hayashi, M., Sasaki, T. (1956) Measurements of solubilities of sparingly soluble liquids in water and aqueous detergent solutions using nonionic surfactant. *Bull. Chem. Soc. Jpn.* 29, 857.

Heidman, J.L., Tsonopoulos, C., Brady, C.J., Wilson, G.M. (1985) High-temperature mutual solubilities of hydrocarbons and water. *AIChE J.* 31, 376–384.

Hendry, D.G., Mill, T., Piszkiewicz, L., Howard, J.A., Eigenman, H.K. (1974) A critical review of H-atom transfer in the liquid phase: chlorine atom, alkyl, trichloromethyl, alkoxy and alkylperoxy radicals. *J. Phys. Chem. Ref. Data* 3, 944–978.

Herman, D.C., Mayfield, C.I., Innis, W.E. (1991) The relationship between toxicity and bioconcentration of volatile aromatic hydrocarbons by the *alga selenastrum capricornutum*. *Chemosphere* 22(7), 665–676.

Hermann, R.B. (1972) Theory of hydrophobic bonding. II. The correlation of hydrocarbon solubility in water with solvent cavity surface area. *J. Phys. Chem.* 76, 2754–2758.

Herz, W., Schuftan, B. (1922) *Z. Phys. Chem.* 101, 269.—reference from Boublik et al. 1984.

Hine, J., Haworth, H.W., Ramsay, O.B. (1963) Polar effects on rates and equilibria. VI. The effect of solvent on the transmission of polar effects, *J. Am. Chem. Soc.* 85, 1473–1476.

Hine, J., Mookerjee, P.K. (1975) The intrinsic hydrophilic character of organic compounds. Correlations in terms of structural contributions. *J. Org. Chem.* 40(3), 292–298.

Hodson, J., Williams, N.A. (1988) The estimation of the adsorption coefficient (K_{OC}) for soils by high performance liquid chromatography. *Chemosphere* 17, 67–77.

Hoigné, J., Bader, H. (1983) Rate constants of reactions of ozone with organic and inorganic compounds in water. - I. Non-dissociating organic compounds. *Water Res.* 17, 173–183.

Hong, H., Wang, L., Han, S., Zou, G. (1996) Prediction adsorption coefficients (K_{OC}) for aromatic compounds by HPLC retention factors (k'). *Chemosphere* 32, 343–351.

Horowitz, A., Shelton, D.R., Cornell, C.P., Tiedje, J.M. (1982) Anaerobic degradation of aromatic compounds in sediment and digested sludge. *Dev. Ind. Microbiol.* 23, 435–444.

Horvath, A.L. (1982) *Halogenated Hydrocarbons, Solubility-Miscibility with Water.* Marcel Dekker, New York, N.Y.

Hovorka, F., Dreisbach, D. (1934) Vapor pressure of binary systems. I. Benzene and acetic acid. *J. Am. Chem. Soc.* 56, 1664–1666.

Howard, P.H., Ed. (1989) *Handbook of Fate and Exposure Data for Organic Chemicals. Vol. I - Large Production and Priority Pollutants.* Lewis Publishers, Chelsea, MI.

Howard, P.H., Ed., (1990) *Handbook of Fate and Exposure Data for Organic Chemicals. Vol. II - Solvents.* Lewis Publishers, Chelsea, MI.

Howard, P.H., Boethling, R.S., Jarvis, W.F., Meylan, W.M., Michalenko, E.M., Eds. (1991) *Handbook of Environmental Degradation Rates.* Lewis Publishers, Chelsea, MI.

Hussam, A., Carr, P.W. (1985) A study of a rapid and precise methodology for the measurement of vapor liquid equilibria by headspace gas chromatography. *Anal. Chem.* 57, 793–801.

Hustert, K., Mansour, M., Korte, F. (1981) The EPA Test- a method to determine the photochemical degradation of organic compounds in aqueous systems. *Chemosphere* 10, 995–998.

Hutchinson, T.C., Hellebust, J.A., Tam, D., Mackay, D., Mascarenhas, R.A., Shiu, W.Y. (1980) The correlation of the toxicity to algae of hydrocarbons and halogenated hydrocarbons with their physical-chemical properties. In: *Hydrocarbons and Halogenated Hydrocarbons in the Aquatic Environment.* Afghan, B.K., Mackay, D., Eds., pp. 577–586. Plenum Press, New York.

Irmann, F. (1965) Eine einfache korrelation zwischen wasserlöslichkeit und struktur vor kohlenwasserstoffen und hologen kohlen wasserstoffen. *Chem. Eng. Tech.* 37(8), 789–798.

Iwasa, J., Fujita, T., Hansch, C. (1965) Substituent constants for aliphatic functions obtained from partition coefficients. *J. Med. Chem.* 8, 150–153.

Jamison, V.W., Raymond, R.L., Hudson, J.O. (1976) *Biodegradation of high octane gasoline. Proceedings of The Third International Biodegradation Symposium.* J.M. Sharpley, A.M. Kaplan Eds., pp. 187–196.

Jenke, D.R., Hayward, D.S., Kenley, R.A. (1990) Liquid chromatographic measurement of solute/solvent partition coefficients: application to solute/container interactions. *J. Chromatogr. Sci.* 20, 609–6612.

Jury, W.A., Russo, D., Streile, G., El Abd, H. (1990) Evaluation of volatilization by organic chemicals residing below the soil surface. *Water Resources Res.* 26, 13–26.

Kabadi, V.N., Danner, R.P. (1979) Nomograph solves for solubilities of hydrocarbons in water. *Hydrocarbon Processing* 58, 245–246.

Kappeler, T., Wuhrmann, K. (1978) Microbial degradation of water soluble fraction of gas oil. *Water Res.* 12, 327–333.

Karickhoff, S.W. (1981) Semiempirical estimation of sorption of hydrophobic pollutants on natural sediments and soils. *Chemosphere* 10, 833–846.

Karickhoff, S.W., Brown, D.S., Scott, T.A. (1979) Sorption of hydrophobic pollutants on natural water sediments. *Water Res.* 13, 241–248.

Kasehgari, H., Mokbel, I., Viton, C., Jose, J. (1993) Vapor pressure of 11 alkylbenzenes in the range 10^{-3} - 280 torr. Correlation by equation of state. *Fluid Phase Equil.* 87, 133–152.

Kassel, L.S. (1936) Vapor pressures of the xylenes and mesitylene. *J. Am. Chem. Soc.* 58, 670–671.

Kavanaugh, M.C., Trussell, R.R. (1980) Design of aeration towers to strip volatile contaminants from drinking water. *J. Am. Water Works Assoc.* 72, 684–692.

Keeley, D.F., Hoffpauir, M.A., Meriwether, J.R. (1988) Solubility of aromatic hydrocarbons in water and sodium chloride solutions of different ionic strengths: Benzene and Toluene. *J. Chem. Eng. Data* 33, 87–89.

Kitano, M. (1978) Biodegradation and bioaccumulation of chemical substances. OECD Tokyo Meeting. Reference Book TSU-No. 3.

Klamt, A. (1993) Estimation of gas-phase hydroxyl radical rate constants of organic compounds from molecular orbital calculations. *Chemosphere* 26, 1273–1289.

Klevens, H.B. (1950) Solubilization of polycyclic hydrocarbons. *J. Phys. Colloid Chem.* 54, 283–298.

Kobe, K.A., Okabe, T. S., Ramstad, M.T., Huemmer, P.M. (1941) *p*-Cymene studies. VI. Vapor pressure p-cymene, some of its derivatives and related compounds. *J. Am. Chem. Soc.* 63, 2151–3252.

Kolb, B., Welter, C., Bichler, C. (1992) Determination of partition coefficients by automatic equilibrium headspace gas chromatography by vapor phase calibration. *Chromatographia* 34, 235–240.

Kollig, H.P. (1995) *Environmental Fate Constants for Additional 27 Organic Chemicals under Consideration for EPA's Hazardous Water Identification Projects.* EPA/600/R-95/039. Environmental Research Laboratory, Office of Research and Development, U.S. EPA, Athens, GA.

Korn, S. et al. (1977) The uptake, distribution and depuration of carbon-14 labelled benzene and carbon-14 labelled toluene in pacific herring. *Fish Bull. Natl. Marine Fish Ser.* 75, 633–636.

Krasnoshchekova, R.Ya., Gubergrits, M. (1975) Solubility of *n*-alkylbenzenes in fresh and salt waters. *Vodnye. Resursy.* 2, 170–173.

Krzyzanowska, T., Szeliga, J. (1978) A method for the solubility of individual hydrocarbons determining in water. *Nafta (Katowice)* 34(12), 413–417.

Kuhn, E.P., Coldberg, P.J., Schnoor, J.L., Waner, O., Zehnder, A.J.B. Schwarzenbach, R.P. (1985) Microbial transformation of substituted benzenes during infiltration of river water to ground water: laboratory column studies. *Environ. Sci. Technol.* 19, 961–968.

Kühne, R., Ebert, R.-U., Kleint, F., Schmidt, G., Schüürmann, G. (1995) Group contribution methods to estimate water solubility of organic chemicals. *Chemosphere* 30, 2061–2077.

Kwok, W.S.C., Atkinson, R. (1995) Estimation of hydroxyl radical reaction rate constants for gas-phase organic compounds using a structure reactivity relationship: an update. *Atmos. Environ.* 29, 1685–1695.

Lane, W.H. (1946) Determination of the solubility of styrene in water and of water in styrene. *Ind. Eng. Chem. Anal. Ed.* 18, 295–296.

Lee, R.F. (1977) Oil Spill Conference, Am. Petroleum Institute pp. 611–616.

Lee, J.F., Crum, J.R., Boyd, S.A. (1989) Enhanced retention of organic contaminants by soils exchanged with organic cations. *Environ. Sci. Technol.* 23, 1365–1372.

Lee, R.F., Ryan, C. (1976) Biodegradation of petroleum hydrocarbons by marine microbes. In: *Proceedings of the third International Biodegradation Symposium,* pp. 119–125.

Lee, R.F., Ryan, C. (1979) Microbial degradation of organochlorine compounds in estuarine waters and sediments. In: *Proceedings of the Workshop of Microbial Degradation of Pollutants in Marine Environments.* EPA-600/9-79-012. Washington, D.C.

Leighton, D.T., Calo, J.M. (1981) Distribution coefficients of chlorinated hydrocarbons in dilute air-water systems for groundwater contamination applications. *J. Chem. Eng. Data* 26, 381–385.

Leinonen, P.J. (1972) The Solubility of Hydrocarbons in Water. M.A.Sc. Thesis, University of Toronto, Toronto, Canada.

Leinonen, P.J., Mackay, D. (1973) The multicomponent solubility of hydrocarbons in water. *Can. J. Chem. Eng.* 51, 230–233.

Li, J., Dallas, A.J., Eikens, D.I., Carr, P.W., Bergmann, D.L., Hait, M.J., Eckert, C.A. (1993) Measurement of large infinite dilution activity coefficients of nonelectrolytes in water by inert gas stripping and gas chromatography. *Anal. Chem.* 65, 3212–3218.

Lide, D.R., Editor (2003) *Handbook of Chemistry and Physics*. 84th CRC Press, Boca Raton, FL.

Lindberg, A.B. (1956) Physicochime des solutions. - Sur une relation simple entre le volume moléculaire et la solubilité dans l'eau des hydrocarbures et dérivé halogénés. *C.R. Acad. Sci.* 243, 2057–2060.

Linder, E.G. (1931) Vapor pressures of some hydrocarbons. *J. Phys. Chem.* 35, 531–535.

Liu, K., Dickhut, R.M. (1994) Saturation vapor pressures and thermodynamic properties of benzene and selected chlorinated benzenes at environmental temperatures. *Chemosphere* 29, 581–589.

Lloyd, A.C., Darnall, K.R., Winer, A.M., Pitts, Jr., J.N. (1976) Relative rate constants for reaction of the hydroxyl radical with a series of alkanes, alkenes, and aromatic hydrocarbons. *J. Phys. Chem.* 80, 789–794.

Lo, J.M., Tseng, C.L., Yang, J.Y. (1986) Radiometric method for determining solubility of organic solvents in water. *Anal. Chem.* 58, 1896–1897.

Lun, R., Varhanickova, D., Shiu, W.-Y., Mackay, D. (1997) Aqueous solubilities and octanol-water partition coefficients of cymenes and chlorocymenes. *J. Chem. Eng. Data* 42, 951–953.

Lyman, W.J. (1982) Adsorption coefficients for soil and sediments. chapter 4, In: *Handbook of Chemical Property Estimation Methods,* W.J. Lyman, W.F. Reehl, D.H. Rosenblatt, Eds., McGraw-Hill, New York.

Lyman, W.J., Reehl, W.F., Rosenblatt, D.H. (1982) *Handbook of Chemical Property Estimation Methods,* McGraw-Hill, New York.

Ma, J.H.Y., Hung, H., Shiu, W.Y., Mackay, D. (2001) Temperature dependence of the aqueous solubility of selected chlorobenzenes and chlorotoluenes. *J. Chem. Eng. Data* 46, 619–622.

Mabey, W., Smith, , J.H., Podoll, R.T., Johnson, H.L., Mill, T., Chou, T.W., Gates, J., Waight-Partridge, I., Vanderberg, D. (1982) *Aquatic Fate Process for Organic Priority Pollutants.* EPA Report No. 440/4-81-014.

Mackay, D., Bobra, A.M., Chan, D.W., Shiu, W.Y. (1982) Vapor pressure correlation for low-volatility environmental chemicals. *Environ. Sci. Technol.* 16, 645–649.

Mackay, D., Leinonen, P.J. (1975) Rate of evaporation of low-solubility contaminants from water to atmosphere. *Environ. Sci. Technol.* 7, 1178–1180.

Mackay, D., Paterson, S., Chung, B., Neely, W.B. (1985) Evaluation of the environmental behavior of chemicals with a level III fugacity model. *Chemosphere* 14, 335–374.

Mackay, D., Shiu, W.Y. (1975) The aqueous solubility and air-water exchange characteristics of hydrocarbons under environmental conditions. In: *Chemistry and Physics of Aqueous Gas Solutions.* Adams, W.A., Greer, G., Desnoyers, J.E., Atkinson, G., Kell, K.B., Oldham, K.B., Walkey, J., Eds., pp. 93–110, Electrochem. Soc., Princeton, NJ.

Mackay, D., Shiu, W.Y. (1981) A critical review of Henry's law constants for chemicals of environmental interest. *J. Phys. Chem. Ref. Data* 10, 1175–1199.

Mackay, D., Shiu, W.Y. (1990) Physical-chemical properties and fate of volatile organic compounds: an application of the fugacity approach. In: *Significance and Treatment of Volatile Organic Compounds in Water Supplies.* Ram, N.M., Christman, R.F., Cantor, K.P., Eds., pp. 183–203, Lewis Publishers, Chelsea, MI.

Mackay, D., Shiu, W.Y., Sutherland, R.P. (1979) Determination of air-water Henry's law constants for hydrophobic pollutants. *Environ. Sci. Technol.* 13, 333–337.

Mackay, D., Shiu, W.Y., Wolkoff, A.W. (1975) Gas chromatographic determination of low concentrations of hydrocarbons in water by vapor phase extraction. *Water Quality Parameters. ASTM STP* 573, pp. 251–258, Am. Soc. Testing and Materials, Philadelphia.

Mackay, D., Wolkoff, A.W. (1973) Rate of evaporation of low-solubility contaminants from water bodies to atmosphere. *Environ. Sci. Technol.* 7, 611–614.

Makovskaya, V., Dean, J.R., Tomlinson, W.R., Comber, M. (1995) Determination of octanol-water partition coefficients using gradient liquid chromatography. *Anal. Chim. Acta* 315, 183–192.

Marion, C.V., Malaney, G.W. (1964) Ability of activated sludge microorganisms to oxidize aromatic organic compounds. In: *Proc. Ind. Waste Conf., Eng. Bull.,* Purdue Univ., Eng. Ext. Ser., pp. 297–308.

Massaldi, H.A., King, C.J. (1973) Simple technique to determine solubilities of sparingly soluble organics: solubility and activity coefficients of *d*-limonene, butylbenzene, and *n*-hexyl acetate in water and sucrose solutions. *J. Chem. Eng. Data* 18, 393–397.

May, W.E. (1980) The solubility behaviour of polycyclic aromatic hydrocarbons in aqueous systems. *Petroleum Mar. Environ.; Adv. Chem. Ser.* 185, Chapter 7, Am. Chem. Soc., Washington DC.

May, W.E., Wasik, S.P., Freeman, D.H. (1978) Determining of the solubility behavior of some polycyclic aromatic hydrocarbons in water. *Anal. Chem.* 50, 997–1000.

May, W.E., Wasik, S.P., Miller, M.M., Tewari Y.B., Brown-Thomas, J.M., Goldberg, R.N. (1983) Solution thermodynamics of some slightly soluble hydrocarbons in water. *J. Chem. Eng. Data* 28, 197–200.

McAuliffe, C. (1963) Solubility in water of C_1 - C_9 hydrocarbons. *Nature* (London) 200, 1092–1093.

McAuliffe, C. (1966) Solubility in water of paraffin, cycloparaffin, olefin, acetylene, cycloolefin and aromatic hydrocarbons. *J. Phys. Chem.* 76, 1267–1275.

McDevit, W.F., Long, F.A. (1952) The activity coefficient of benzene in aqueous salt solutions. *J. Am. Chem. Soc.* 74, 1773–1777.

McDonald, R.A., Shrader, S.A., Stull, D.R. (1959) Vapor pressures and freezing points of 30 organics. *J. Chem. Eng. Data* 4, 311–313.

McDuffie, B. (1981) Estimation of octanol/water partition coefficients for organic pollutants using reversed phase HPLC. *Chemosphere* 10, 73–83.

Meylan, W., Howard, P.H, Boethling R.S. (1992) Molecular topology/fragment contribution method for predicting soil sorption coefficients. *Environ. Sci. Technol.* 26, 1560–1567.

Mill, T. (1982) Hydrolysis and oxidation processes in the environment. *Environ. Toxicol. Chem.* 1, 135–141.

Mill, T., Hendry, D.G., Richardson, H. (1980) Free radical oxidants in natural waters. *Science* 207, 886–887.

Miller, M.E., Stuart, J.D. (2000) Measurement of aqueous Henry's law constants for oxygenates and aromatics found in gasolines by the static headspace method. *Anal. Chem.* 72, 622–625.

Miller, M.M., Ghodbane, S., Wasik, S.P., Tewari, Y.B., Martire, D.E. (1984) Aqueous solubilities, octanol/water partition coefficients and entropies of melting of chlorinated benzenes and biphenyls. *J. Chem. Eng. Data* 29, 184–190.

Miller, M.M., Wasik, S.P., Huang, G.L., Shiu, W.Y., Mackay, D. (1985) Relationships between octanol-water partition coefficient and aqueous solubility. *Environ. Sci. Technol.* 19, 522–529.

Mills, W.B., Dean, J.D., Porcella, D.B., Gherini, S.A., Hudson, R.J.M., Frick, W.E., Rupp, G.L., Bowie, G.L. (1982) *Water Quality Assessment: A Screening Procedure for Toxic and Conventional Pollutants.* Part 1, U.S. EPA, EPA-600/6-82-004a.

Miyake, K., Terada, H. (1982) Determination of partition coefficients of very hydrophobic compounds by high performance liquid chromatography on glyceryl-coated controlled-pore glass. *J. Chromatogr.* 240, 9–20.

Mokbel, I., Rauzy, E., Meille, J.P., Jose, J. (1998) Low vapor pressures of 12 aromatic hydrocarbons. Experimental and calculated data using a group contribution method. *Fluid Phase Equil.* 147, 271–284.

Morrison, T.J., Billett, F. (1952) The salting out of non-electrolytes. Part II. The effect of variation in non-electrolyte. *J. Chem. Soc.* 3819–3822.

Müller, M., Klein, W. (1992) Comparative evaluation of methods predicting water solubility for organic compounds. *Chemosphere* 25, 769–782.

Müller, M., Kördel, W. (1996) Comparison of screening methods for the estimation of adsorption coefficients on soil. *Chemosphere* 32, 2493–2504.

Munday, E.B., Mullins, J.C., Edie, D.D. (1980) Vapor pressure for toluene, 1-pentanol, 1-butanol, water, and 1-propanol and for the water and 1-propanol system from 273.15 to 323.15 K. *J. Chem. Eng. Data* 25, 191–194.

Myers, H.S., Fenske, M.R. (1955) Measurement and correlation of vapor pressure data for high boiling hydrocarbons. *Ind. Eng. Chem.* 47, 1652–1658.

Nahum, A., Horvath, C. (1980) Evaluation of octanol-water partition coefficients by using high-performance liquid chromatography. *J. Chromatogr.* 192, 315–322.

Nathwani, J.S., Philip, C.R. (1977) Absorption-desorption of selected hydrocarbons in crude oils on soils. *Chemosphere* 6, 157–162.

Neely, W.B. (1980) A method for selecting the most appropriate environmental experiments on a new chemical. In: *Dynamics, Exposure and Hazard Assessment of Toxic Chemicals.* Haque, R., Ed., pp. 287–298, Ann Arbor Sci. Publ., Ann Arbor, MI.

Nielsen, F., Olsen, E., Fredenslund, A. (1994) Henry's law constants and infinite dilution activity coefficients of volatile organic compounds in water by a validated batch air stripping method. *Environ. Sci. Technol.* 28, 2133–2138.

Nirmalakhandan, N.N., Speece, R.E. (1988a) Prediction of aqueous solubility of organic chemicals based on molecular structure. *Environ. Sci. Technol.* 22, 328–338.

Nirmalakhandan, N.N., Speece, R.E. (1988b) QSAR model for predicting Henry's law constant. *Environ. Sci. Technol.* 22, 1349–1357.

Nunes, P., Benville, P.E., Jr. (1979) Uptake and depuration of petroleum hydrocarbons in the Manila clams, Tapes semidecussata Reeve. *Bull. Environ. Contam. Toxicol.* 21, 719–724.

Ogata, M., Miyake, Y. (1978) Disappearance of aromatic hydrocarbons and organic sulfur compounds from fish flesh reared in crude oil suspension. *Water Res.* 12, 1041–1044.

Ogata, M., Fujisawa, K., Ogino, Y., Mano, E. (1984) Partition coefficients as a measure of bioconcentration potential of crude oil compounds in fish and shellfish. *Bull. Environ. Contam. Toxicol.* 33, 561–567.

Ohta, T., Ohyama, T. (1985) A set of rate constants for the reactions of hydroxyl radicals with aromatic hydrocarbons. *Bull. Chem. Soc. Jpn.* 58, 3029–3030.

Olsen, R.L., Davis, A. (1990) Predicting the fate and transport of organic compounds in groundwater. *Hazard. Mat. Control* 3, 40–64.

Osborn, A.G., Douslin, D.R. (1974) Vapor-pressure relations of 15 hydrocarbons. *J. Chem. Eng. Data* 19, 114–117.
Osborn, A.G., Scott, D.W. (1978) Vapor-pressure and enthalpy of vaporization of indan and five methyl-substituted indans. *J. Chem. Thermdyn.* 10, 619–628.
Overberger, J.E., Steele, W.A., Aston, J.G. (1969) The vapor pressure of hexamethylbenzene. The standard entropy of hexamethylbenzene vapor and the barrier to internal rotation. *J. Chem. Thermodyn.* 1, 535–542.
Owens, J.W., Wasik, S.P., DeVoe, H. (1986) Aqueous solubilities and ethalpies of solution of n-alkylbenzenes. *J. Chem. Eng. Data* 31, 47–51.
Pankow, J.F., Isabelle, L.M., Asher, W.E. (1984) Trace organic compounds in rain. 1. Sample design and analysis by adsorption/thermal desorption (ATD). *Environ. Sci. Technol.* 18, 310–318.
Passino, D.R.M., Smith, S.B. (1987) Quantitative structure-activity relationships (QSAR) and toxicity data in hazard assessment. In: *QSAR in Environmental Toxicology-II*. Kaiser, K.L.E., Editor, D. Reidel Publishing Company. pp. 261–270.
Peng, J., Wan, A. (1997) Measurement of Henry's law constants of high volatility organic compounds using a headspace autosampler. *Environ. Sci. Technol.* 31, 2998–3003.
Peng, J., Wan, A. (1998) Effect of ionic strength on Henry's law constant of volatile organic compounds. *Chemosphere* 36, 2731–2740.
Peng, J., Wan, A. (1998) Effect of ionic strength on Henry's constants of volatile organic compounds. *Chemosphere* 36, 2731–2740.
Pennell, K.D., Rhue, R.D., Rao, P.S.C., Johnston, C.T. (1992) Vapor-phase sorption of *p*-xylene and water on soils and clay minerals. *Environ. Sci. Technol.* 26, 756–763.
Perlinger, J.A., Eisenreich, S.J., Capel, P.D. (1993) Application of headspace analysis to the study of sorption of hydrophobic organic chemicals to α-Al_2O_3. *Environ. Sci. Technol.* 27, 928–937.
Perry, R.A., Atkinson, R., Pitts, J.N. (1977) Kinetics and mechanisms of the gas phase reaction of the hydroxyl radicals with aromatic hydrocarbons over temperature range 296–473 K. *J. Phys. Chem.* 81, 296–304.
Phousongphouang, P.T., Arey J. (2002) Rate constants for the gas-phase reactions of a series of alkylnaphthalenes with the OH radicals. *Environ. Sci. Technol.* 36, 1947–1952.
Pierotti, R.A., Liabastre, A.A. (1972) Structure and properties of water solutions. (U.S. Nat. Tech. Inform. Serv. PN Rep. 1972), No. 21163, 113pp.
Pitzer, K.S., Guttman, L., Westrum, Jr., E.F. (1946) The heat capacity, heats of fusion and vaporization, vapor pressure, entropy, vibration frequencies and barrier to internal rotation of styrene. *J. Am. Chem. Soc.* 68, 2209–2212.
Pitzer, K.S., Scott, D.W. (1943) The thermodynamics and molecular structure of benzene and its methyl derivatives. *J. Am. Soc. Chem.* 65, 803–829.
Platford, R.F. (1979) Glyceryl trioleate-water partition coefficients for three simple organic compounds. *Bull. Environ. Contam. Toxicol.* 21, 68–73.
Platford, R.F. (1983) The octanol-water partitioning of some hydrophobic and hydrophilic compounds. *Chemosphere* 12, 1107–1111.
Polak, J., Lu, B.C.Y. (1973) Mutual solubilities of hydrocarbons and water at 0° and 25°C. *Can. J. Chem.* 51, 4018–4023.
Politzki, G.R., Bieniek, D., Lahaniatis, E.S., Sheunert, I., Klein, W., Korte, F. (1982) Determination of vapour pressures of nine organic chemicals on silicagel. *Chemosphere* 11, 1217–1229.
Poulson, S.R., Drever, J.I., Colberg, P.J.S. (1997) Estimation of K_{OC} values for deuterated benzene, toluene, and ethylbenzene, and application to ground water contamination studies. *Chemosphere* 35, 2215–2224.
Poulson, S.R., Harrington, R.R., Drever, J.I. (1999) The solubility of toluene in aqueous salt solutions. *Talanta* 48, 533–641.
Price, L.C. (1973) The Solubility of Hydrocarbons and Petroleum in Water. Ph.D. Thesis., University of California, Riverside, CA.
Price, L.C. (1976) Aqueous solubility of petroleum as applied to its origin and primary migration. *Am. Assoc. Petrol. Geol. Bull.* 60, 213–244.
Ravishankara, A.R., Wagner, S., Fischer, S., Smith, G., Schiff, R., Watson, R.T., Tesi, G., Davis, D.D. (1978) A kinetics study of the reactions of OH with several aromatic and olefinic compounds. *Int. J. Chem. Kinet.* Vol. X, 783–804.
Rapaport, R.A., Eisenreich, S. (1984) Chromatographic determination of octanol-water partition coefficients (K_{OW}'s) for 58 PCB congeners. *Environ. Sci. Technol.* 18, 163–170.
Rekker, R.F. (1977) *The Hydrophobic Fragmental Constants. Its Derivation and Application, a Means of Characterizing Membrane Systems*. Elsevier Sci. Publ. Co., Oxford, England.
Riddick, J.A., Bunger, W.B., Sakano, T.K. (1986) *Organic Solvents. Physical Properties and Methods of Purification*. Fourth edition. Wiley Interscience, New York.
Rintelen, Jr., J.C., Saylor, J.H., Gross, P.M. (1937) The densities and vapor pressures of some alkylbenzenes, aliphatic ketones and *n*-amyl chloride. *J. Am. Chem. Soc.* 39, 1125–1126.
Ritter, S., Hauthal, W.J., Maurer, G. (1995) Octanol/water partition coefficients for environmentally important organic chemicals. *Environ. Sci. Pollut. Res.* 2, 153–160.
Robbins, G.A., Wang, S., Stuart, J.D. (1993) Using the static headspace method to determine Henry's law constants. *Anal. Chem.* 65, 3113–3118.
Rogers, K.S., Cammarata, A. (1969) Superdelocalizability and charge density. A correlation with partition coefficients. *J. Med. Chem.* 12, 692–693.
Rogers, R.D., McFarlane, J.C., Cross, A.J. (1980) Adsorption and desorption of benzene in two soils and montmorillonite clay. *Environ. Sci. Technol.* 14, 457–460.

Rossi, S.S., Thomas, W.H. (1981) Solubility behavior of three aromatic hydrocarbons in distilled water and natural seawater. *Environ. Sci. Technol.* 15, 715–716.

Rutherford, D.W., Chiou, C.T. (1992) Effect of water saturation in soil organic matter on the partition of organic compounds. *Environ. Sci. Technol.* 26, 995–970.

Ryan, J.A., Bell, R.M., Davidson, J.M., O'Connor, G.A. (1988) Plant uptake of non-ionic organic chemicals from soils. *Chemosphere* 17, 2299–2323.

Ryu, S.-A., Park, S.-J. (1999) A rapid determination method of the air/water partition coefficient and its application. *Fluid Phase Equil.* 161, 295–304.

Sabljic, A. (1987a) On the prediction of soil sorption coefficients of organic pollutants from molecular structure: Application of molecular topology model. *Environ. Sci. Technol.* 27, 358–366.

Sabljic, A. (1987b) Nonempirical modeling of environmental distribution and toxicity of major organic pollutants. In: *QSAR in Environmental Toxicology - II*. Kaiser, K.L.E., Ed., pp. 309–332, D. Reidel Publ. Co., Dordrecht, Netherlands.

Sabljic, A., Güsten, H. (1990) Predicting the night-time NO_3 radical reactivity in the troposphere. *Atmos. Environ.* 24A, 73–78.

Sabljic, A., Güsten, H., Verhaar, H., Hermens, J. (1995) QSAR modelling of soil sorption. Improvements and systematics of log K_{OC} vs. log K_{OW} correlations. *Chemosphere* 31, 4489–4514.

Sada, E., Kito, S., Ito, Y. (1975) Solubility of toluene in aqueous salt solutions. *J. Chem. Eng. Data* 20(4), 373–375.

Sanemasa, I., Araki, M., Deguchi, T., Nagai, H. (1981) Solubilities of benzene and alkylbenzenes in water. Methods for obtaining aqueous solutions saturated with vapors in equilibrium with organic liquids. *Chem. Lett.* 2, 255–258.

Sanemasa, I., Araki, M., Deguchi, T., Nagai, H. (1982) Solubility measurement of benzene and the alkylbenzenes in water by making the use of solute vapor. *Bull. Chem. Soc. Jpn.* 53, 1054–1062.

Sanemasa, I., Arakawa, S., Araki, M., Deguchi, T. (1984) The effects of salts on the solubility of benzene, toluene, ethylbenzene and propylbenzene in water. *Bull. Chem. Soc. Jpn.* 57, 1359–1544.

Sangster, J. (1989) Octanol-water partition coefficients of simple organic compounds. *J. Phys. Chem. Ref. Data* 18, 1111–1230.

Sangster, J. (1993) LOGKOW databank. Sangster Research Laboratory, Montreal, Q.C.

Sawamura, S., Nagaoka, K., Machikawa, T. (2001) Effects of pressure and temperature on the solubility of alkylbenzenes in water: Volumetric property of hydrophobic hydration. *J. Phys. Chem. B*, 105, 2429–2436.

Saylor, J.H., Stuckey, J.M., Gross, P.M. (1938) Solubility studies. V. The validity of Henry's law for the calculation of vapor solubilities. *J. Am. Chem. Soc.* 60, 373–415.

Schantz, M.M., Martire, D.E. (1987) Determination of hydrocarbon-water partition coefficients from chromatographic data and based on solution thermodynamics and theory. *J. Chromatogr.* 391, 35–41.

Schoene, S., Steinhanses, J. (1985) Determination of Henry's law constant by automated headspace gas chromatography determination of dissolved gases. *Fresenius Z. Anal. Chem.* 321, 538–543.

Schwarz, F.P. (1977) Determination of temperature dependence of solubilities of polycyclic aromatic hydrocarbons in aqueous solutions by a fluorescence method. *J. Chem. Eng. Data* 22, 273–277.

Schwarz, F.P. (1980) Measurement of the solubilities of slightly soluble organic liquids in water by elution chromatography. *Anal. Chem.* 52, 10–15.

Schwarz, F.P., Miller, J. (1980) Measurement of the solubilities of slightly soluble organic liquids in water by elution chromatography. *Anal. Chem.* 52, 2161–2164.

Schwarzenbach, R.P., Westall, J. (1981) Transport of nonpolar compounds from surface water to groundwater. Laboratory sorption studies. *Environ. Sci. Technol.* 11, 1360–1367.

Scott, D.W., Osborn, A.G. (1979) Representation of vapor-pressure data. *J. Phys. Chem.* 83, 2714–2723.

Seip, H.M., Alstad, J., Carlberg, G.E., Martinsen, K., Skaane, P. (1986) Measurement of mobility of organic compounds in soils. *Sci. Total Environ.* 50, 87–101.

Shaw, D.G., Ed. (1989a) *IUPAC Solubility Data Series* Vol. 37: *Hydrocarbons (C_5–C_7) with Water and Seawater*. Pergamon Press, Oxford, England.

Shaw, D.G., Ed. (1989b) *IUPAC Solubility Data Series* Vol. 38: *Hydrocarbons (C_8–C_{36}) with Water and Seawater*. Pergamon Press, Oxford, England.

Sherblom, P.M., Eganhouse, R.P. (1988) Correlations between octanol-water partition coefficients and reversed-phase high-performance liquid chromatography capacity factors. *J. Chromtogr.* 454, 37–50.

Sherblom, P.M., Gschwend, P.M., Eganhouse, R.P.(1992) Aqueous solubilities, vapor pressures, and 1-octanol-water partition coefficients for C_9-C_{14} linear alkylbenzenes. *J. Chem. Eng. Data* 37, 394–399.

Shiu, W.Y., Ma, K.C. (2000) Temperature dependence of physical-chemical properties of selected chemicals of environmental interest. 1. Mono- and polynuclear aromatic hydrocarbons. *J. Phys. Chem. Ref. Data.* 29, 41–130.

Singh, H.B., Salas, L.J., Smith, A.J., Shigeishi, H. (1981) Measurements of some potentially hazardous organic chemicals in urban environments. *Atmos. Environ.* 15, 601–612.

Smith, A.D., Bharath, A., Mullard, C., Orr, D., McCarthy, L.S., Ozburn, G.W. (1990) Bioconcentration kinetics of some chlorinated benzenes and chlorinated phenols in American flagfish, *Jordanella floridae* (Goode and Bean). *Chemosphere* 20, 379–386.

Söderström, G., Sellström, U., de Wit, C., Tysklind, M. (2004) Photolytic debromination of decabromodiphenyl ether (BDE 209). *Environ. Sci. Technol.* 38, 127–132.

Staudinger, J., Roberts, P.V. (1996) A critical review of Henry's law constant for environmental applications. *Crit. Rev. Environ. Sci. Technol.* 26, 205–297.

Staudinger, J., Roberts, P.V. (2001) A critical compilation of Henry's law constant temperature dependence relations for organic compounds in dilute aqueous solutions. *Chemosphere* 44, 561–576.

Stearns, R.S., Oppenheimer, H., Simon, E., Harkins, W.D. (1947) Solubilization by solutions of long chain colloidal electrolytes. *J. Chem. Phys.* 15, 496–507.

Stephenson, R.M., Malanowski, S. (1987) *Handbook of the Thermodynamic of Organic Compounds.* Elsevier Science Publishing Co. New York, NY.

Stuckey, J.M., Saylor, J.H. (1940) The vapor pressures of some organic compounds. I. *J. Am. Chem. Soc.* 62, 2922–2925.

Stull, D.R. (1947) Vapor pressure of pure substances. Organic compounds. *Ind. Eng. Chem.* 39, 517–540.

Sutton. C., Calder, J.A. (1975) Solubility of alkylbenzenes in distilled water and seawater at 25°C. *J. Chem Eng. Data* 20, 320–322.

Swann, R.L., Laskowski, D.A., McCall, P.J., Vender Kuy, K., Dishburger, J.J. (1983) A rapid method for the estimation of the environmental parameters octanol/water partition coefficient, soil sorption constant, water to air ratio, and water solubility. *Res. Rev.* 85, 17–28.

Swindoll, C.M., Aelion, C.M., Pfaender, F.K. (1987) Inorganic and organic amendment effects of biodegradation of organic pollutants by ground water microorganisms. *Am. Soc. Microbiol. Abst.*, 87th Annual Meeting, pp. 298, Atlanta, GA.

Szabo, G., Prosser, S.L., Bulman, R.A. (1990a) Prediction of the adsorption coefficient (K_{OC}) for soil by a chemically immobilized humic acid column using RP-HPLC. *Chemosphere* 21, 729–740.

Szabo, G., Prosser, S.L., Bulman, R.A. (1990b) Determination of the adsorption coefficient (K_{OC}) of some aromatics for soil by RP-HPLC on two immobilized humic acid phases. *Chemosphere* 21, 777–788.

Szabo, G., Guczi, J., Bulman, R.A. (1995) Examination of silica-salicylic acid and silica-8-hydroxyquinoline HPLC stationary phases for estimation of the adsorption coefficient of soil for some aromatic hydrocarbons. *Chemosphere* 30, 1717–1727.

Tabak, H.H., Quave, S.A., Moshni, C.I., Barth, E.F. (1981) Biodegradability studies with organic priority pollutant compounds. *J. Water Pollut. Control Fed.* 53, 1503–1518.

Taha, A.A., Grisby, R.D., Johnson, J.R., Christian, S.D., Affsprung, H.E. (1966) Monometric apparatus for vapor and solution studies. *J. Chem. Ed.* 43, 432–435.

Tester, D.J., Harker, R.J. (1981) Ground water pollution investigations in the Great Ouse Basin. *Water Pollut. Control* 80, 614–631.

Tewari, Y.B., Martire, D.E., Wasik, S.P., Miller, M.M. (1982a) Aqueous solubilities and octanol-water partition coefficients of binary liquid mixtures of organic compounds at 25°C. *J. Solution Chem.* 11, 435–445.

Tewari, Y.B., Miller, M.M., Wasik, S.P. (1982b) Calculation of aqueous solubilities of organic compounds. *NBS J. Res.* 87, 155–158.

Tewari, Y.B., Miller, M.M., Wasik, S.P., Martire, D.E. (1982c) Aqueous solubility and octanol/water partition coefficient of organic compounds at 25.0°C. *J. Chem. Eng. Data* 27, 451–454.

Trzilova, B., Horska, E. (1988) Biodegradation of amines and alkanes in aquatic environment. *Biologia* (Bratislava) 43, 209–218.

Tse, G., Sandler, S.I. (1994) Determination of infinite dilution activity coefficients and 1-octanol/water partition coefficients of volatile organic pollutants. *J. Chem. Eng. Data* 39, 354–357.

Tsonopoulos, C. (1999) Thermodynamic analysis of the mutual solubilities of normal alkanes and water. *Fluid Phase Equil.* 156, 21–33.

Tsonopoulos, C., Prausnitz, J.M. (1971) Activity coefficients of aromatic solutes in dilute aqueous solutions. *I & EC Fundam.* 10, 593–600.

Tuazon, E.C., Arey, J., Atkinson, R., Aschmann, S.M. (1993) Gas-phase reactions of 2-vinylpyridine and styrene with OH and NO_3 radicals and O_3. *Environ. Sci. Technol.* 27, 1832–1841.

Turner, L.H., Chiew, Y.C., Ahlert, R.C., Kosson, D.S. (1996) Measuring vapor-liquid equilibrium for aqueous-organic systems: Review and a new technique. *Am. Inst. Chem. Eng. J.* 42, 1772–1778.

Udovenko, V.V., Aleksandrova, L.P. (1963) The vapor pressure of three components systems. IV. Formic acid-benzene-water. *Zh. Fiz. Khim.* 37, 52–56.

Ueda, M., Katayama, A., Kuroki, N., Uranhata, T. (1978) Effect of urea on the solubility of benzene and toluene in water. *Prog. Colloid Polym. Sci.* 63, 116–119.

Vaishnav, D.D., Babeu, L. (1987) Comparison of occurrence and rates of chemical biodegradation in natural waters. *Bull. Environ. Contam. Toxicol.* 39, 237–244.

Van der Linden, A.C. (1978) Degradation of oil in the marine environment. *Dev. Biodegrad. Hydrocarbons* 1, 165–200.

Veith, G.D., Austin, N.M., Morris, R.T. (1979) A rapid method for estimating log P for organic chemicals. *Water Res.* 13, 43–47.

Veith, G.D., Macek, K.J., Petrocelli, S.R., Caroll, J. (1980) An evaluation of using partition coefficients and water solubilities to estimate bioconcentration factors for organic chemicals in fish. In: *Aquatic Toxicology.* J.G. Eaton, P.R. Parrish, A.C. Hendricks, Eds., ASTM STP 707, Am. Soc. for Testing and Materials, pp. 116–129.

Vesala, A. (1974) Thermodynamics of transfer nonelectrolytes from light and heavy water. I. Linear free energy correlations of free energy of transfer with solubility and heat of melting of nonelectrolyte. *Acta Chem. Scand.* 28A(8), 839–845.

Vitenberg, A.G., Ioffe, B.V., St. Dimitrova, Z., Butaeva, I.L. (1975) Determination of gas-liquid partition coefficients by means of gas chromatographic analysis. *J. Chromatogr.* 112, 319–327.

Vowles, P.D., Mantoura, R.F.C. (1987) Sediment-water partition coefficients and HPLC retention factors of aromatic hydrocarbons. *Chemosphere* 16, 109–116.

Wahner, A., Zetzsch, C. (1983) Rate constants for the addition of hydroxyl radicals to aromatics (benzene, p-chloroaniline, and o-, m- and p-dichlorobenzene) and the unimolecular decay of the adduct. Kinetics into a quasi-equilibrium. *J. Phys. Chem.* 87, 4945–4951.

Wakeham, S.G., Davis, A.C., Karas, J.L. (1983) Microcosm experiments to determine the fate and persistence of volatile organic compounds in coastal seawater. *Environ. Sci. Technol.* 17, 611–617.

Wallington, T.J., Neuman, D.M., Kurylo, M.J. (1987) Kinetics of the gas phase reaction of hydroxyl radicals with ethane, benzene, and a series of halogenated benzenes over the temperature range 234–438 K. *Int. J. Chem. Kinet.* 19, 725–739.

Walton, B.T., Hendricks, M.S., Anderson, T.A., Griest, W.H., Merriweather, R., Beauchamp, J.J., Francis, C.W. (1992) Soil sorption of volatile and semivolatile organic compounds in a mixture. *J. Environ. Qual.* 21, 552–558.

Wang, L., Wang, X., Xu, O., Tian, L. (1986) Determination of the n-octanol/water partition coefficients of polycyclic aromatic hydrocarbons by HPLC and estimation of their aqueous solubilities. *Huanjing Kexue Xuebao* 6, 491–497.

Wang, L., Zhao, Y., Gao, H. (1992) Predicting aqueous solubility and octanol/water partition coefficients of organic chemicals from molar volume. *Environ. Chem.* (Chinese) 11, 55–70.

Wania, F., Mackay, D. (1996) Tracking the distribution of persistent organic pollutants. *Environ. Sci. Technol.* 30, 390A–396A.

Warner, P.H., Cohen, J.M., Ireland, J.C. (1987) *Determination of Henry's Law Constant of Selected Priority Pollutants.* NTIS PB-87-212684, EPA/600/D-87/229. U.S. EPA, Cincinnati, OH.

Wasik, S.P., Miller, M.M., Tewari, Y.B., May, W.E., Sonnefeld, W.J., DeVoe, H., Zoller, W.H. (1983) Determination of the vapor pressure, aqueous solubility, and octanol/water partition coefficient of hydrophobic substances by coupled generator column/liquid chromatographic methods. *Res. Rev.* 85, 29–42.

Wasik, S.P., Tewari, Y.B., Miller, M.M., Martire, D.E. (1981) *Octanol/Water Partition Coefficients and Aqueous Solubilities of Organic Compounds.* PB82-141797, U.S. EPA, Washington, D.C.

Wasik, S.P., Tewari, Y.B., Miller, M.M. (1982) Measurements of octanol/water partition coefficient by chromatographic method. *J. Res. Natl. Bur. Std.* 87, 311–315.

Watarai, H., Tanaka, M., Suzaki, N. (1982) Determination of partition coefficient of halobenzenes in heptane/water and 1-octanol/water systems and comparison with the scaled particle calculation. *Anal. Chem.* 54, 702–705.

Weast, R.C., Ed. (1972–73) *Handbook of Chemistry and Physics,* 53rd ed. CRC Press, Cleveland.

Weast, R.C., Ed. (1982–83) *Handbook of Chemistry and Physics,* 63rd ed., CRC Press, Florida.

Willingham, C.B., Taylor, W.J., Pignocco, J.M., Rossini, F.D. (1945) Vapor pressure and boiling points of some paraffin, alkylcyclopentane, alkylcyclohexane, and alkylbenzene hydrocarbons. *J. Res. Natl. Bur. Std.* 34, 219–244.

Wilson, B.H., Smith, G.B., Rees, J.F. (1986) Biotransformations of selected alkylbenzenes and halogenated aliphatic hydrocarbons in methanogenic aquifer material: A Microcosm Study. *Environ. Sci. Technol.* 20, 997–1002.

Wilson, J.T., McNabb, J.F., Balkwill, D.L., Ghiorse, W.C. (1983) Enumeration and characterization of bacteria indigenous to a shallow water-table aquifer. *Ground Water* 21, 134–142.

Wilson, J.T., McNabb, J.F., Wilson, R.H., Noonan, M.J. (1983) Biotransformation of selected organic pollutants in ground water. *Dev. Ind. Microbiol.* 24, 225–233.

Winer, A.M., Lloyd, A.C., Darnall, K.R., Pitts, Jr., J.N. (1976) Relative rate constants for the reaction of the hydroxyl radical with selected ketones, chloroethenes, and monoterpene hydrocarbons. *J. Phys. Chem.* 80, 1635–1639.

Witte, F., Urbanik, E., Zetzsch, C. (1986) Temperature dependence of the rate constants for the addition of OH to benzene and to some monosubstituted aromatics (aniline, bromobenzene, and nitrobenzene) and the unimolecular decay of the adducts. Kinetics into a quasi-equilibrium. *J. Phys. Chem.* 90, 3251–3259.

Worley, J.D. (1967) Benzene as a solute in water. *Can. J. Chem.* 45, 2465–2467.

Xing, B., McGill, W.B., Dudas, M.J. (1994) Cross-correlation of polarity curves to predict partition coefficients of nonionic organic contaminants. *Environ. Sci. Technol.* 28, 1929–1933.

Yalkowsky, S.H., Valvani, S.C. (1980) Solubility and Partitioning I: Solubility of nonelectrolytes in water. *J. Pharm. Sci.* 69(8), 912–922.

Yalkowsky, S.H., Valvani, S.C., Mackay, D. (1983) Estimation of the aqueous solubility of some aromatic compounds. *Res. Rev.* 85, 43–55.

Yaws, C.L. (1994) *Handbook of Vapor Pressure,* Vol. 1 C_1 to C_4 Compounds, Vol. 2. C_5 to C_7 Compounds, Vol. 3, C_8 to C_{28} Compounds. Gulf Publishing Co., Houston, Texas.

Yaws, C.L., Yang, J.C., Pan, X. (1991) Henry's law constants for 362 organic compounds in water. *Chem. Eng.* November, 179–185.

Yurteri, C. Ryan, D.F., Callow, J.J., Gurol, J.J. (1987) The effect of chemical composition of water on Henry's law constant. *J. WPCF* 59, 950–956.

Zhang, Z., Pawliszyn, J. (1993) Headspace solid-phase microextraction. *Anal. Chem.* 65, 1843–1852.

Zwolinski, B.J., Wilhoit, R.C. (1971) *Handbook of Vapor Pressures and Heats of Vaporization of Hydrocarbons and Related Compounds.* API-44 TRC Publication No. 101, Texas A & M University, Evans Press, Fort Worth, TX.

Zoeteman, B.C.J., Harmsen, K., Linders, J.B.H. (1980) Persistent organic pollutants in river water and ground water of the Netherlands. *Chemosphere* 9, 231–249.

Zoeteman, B.C.J., De Greef, E., Brinkmann, F.J.J. (1981) Persistency of organic contaminants in groundwater. Lessons from soil pollution incidents in the Netherlands. *Sci. Total Environ.* 21, 187–202.

4 Polynuclear Aromatic Hydrocarbons (PAHs) and Related Aromatic Hydrocarbons

4.1. List of Chemicals and Data Compilations 620
    4.1.1 Polynuclear aromatic hydrocarbons (PAHs) 620
        4.1.1.1. Indan 620
        4.1.1.2 Naphthalene 623
        4.1.1.3 1-Methylnaphthalene 639
        4.1.1.4 2-Methylnaphthalene 646
        4.1.1.5 1,3-Dimethylnaphthalene 651
        4.1.1.6 1,4-Dimethylnaphthalene 653
        4.1.1.7 1,5-Dimethylnaphthalene 655
        4.1.1.8 2,3-Dimethylnaphthalene 657
        4.1.1.9 2,6-Dimethylnaphthalene 659
        4.1.1.10 1-Ethylnaphthalene 661
        4.1.1.11 2-Ethylnaphthalene 665
        4.1.1.12 1,4,5-Trimethylnaphthalene 668
        4.1.1.13 Biphenyl 669
        4.1.1.14 4-Methylbiphenyl 677
        4.1.1.15 4,4′-Dimethylbiphenyl 678
        4.1.1.16 Diphenylmethane 679
        4.1.1.17 Bibenzyl 682
        4.1.1.18 *trans*-Stilbene 685
        4.1.1.19 Acenaphthylene 688
        4.1.1.20 Acenaphthene 691
        4.1.1.21 Fluorene 699
        4.1.1.22 1-Methylfluorene 708
        4.1.1.23 Phenanthrene 709
        4.1.1.24 1-Methylphenanthrene 722
        4.1.1.25 Anthracene 725
        4.1.1.26 2-Methylanthracene 739
        4.1.1.27 9-Methylanthracene 742
        4.1.1.28 9,10-Dimethylanthracene 745
        4.1.1.29 Pyrene 748
        4.1.1.30 Fluoranthene 759
        4.1.1.31 Benzo[*a*]fluorene 767
        4.1.1.32 Benzo[*b*]fluorene 769
        4.1.1.33 Chrysene 771
        4.1.1.34 Triphenylene 777
        4.1.1.35 *o*-Terphenyl 780
        4.1.1.36 *m*-Terphenyl 781

	4.1.1.37	*p*-Terphenyl	783
	4.1.1.38	Naphthacene	785
	4.1.1.39	Benz[*a*]anthracene	788
	4.1.1.40	Benzo[*b*]fluoranthene	796
	4.1.1.41	Benzo[*j*]fluoranthene	799
	4.1.1.42	Benzo[*k*]fluoranthene	800
	4.1.1.43	Benzo[*a*]pyrene	804
	4.1.1.44	Benzo[*e*]pyrene	811
	4.1.1.45	Perylene	814
	4.1.1.46	7,12-Dimethylbenz[*a*]anthracene	818
	4.1.1.47	9,10-Dimethylbenz[*a*]anthracene	820
	4.1.1.48	3-Methylcholanthrene	821
	4.1.1.49	Benzo[*ghi*]perylene	823
	4.1.1.50	Indeno[1,2,3-*cd*]pyrene	826
	4.1.1.51	Dibenz[*a,c*]anthracene	828
	4.1.1.52	Dibenz[*a,h*]anthracene	830
	4.1.1.53	Dibenz[*a,j*]anthracene	834
	4.1.1.54	Pentacene	835
	4.1.1.55	Coronene	837
4.1.2	Chlorinated polynuclear aromatic hydrocarbons		840
	4.1.2.1	2,4″,5-Trichloro-*p*-terphenyl	840
	4.1.2.2	2,4,4″,6-Tetrachloro-*p*-terphenyl	841
4.1.3	Polychlorinated naphthalenes		842
	4.1.3.1	1-Chloronaphthalene	842
	4.1.3.2	2-Chloronaphthalene	845
	4.1.3.3	1,2-Dichloronaphthalene	848
	4.1.3.4	1,4-Dichloronaphthalene	849
	4.1.3.5	1,8-Dichloronaphthalene	851
	4.1.3.6	2,3-Dichloronaphthalene	852
	4.1.3.7	2,7-Dichloronaphthalene	853
	4.1.3.8	1,2,3-Trichloronaphthalene	854
	4.1.3.9	1,3,7-Trichloronaphthalene	855
	4.1.3.10	1,2,3,4-Tetrachloronaphthalene	857
	4.1.3.11	1,2,3,5-Tetrachloronaphthalene	859
	4.1.3.12	1,3,5,7-Tetrachloronaphthalene	860
	4.1.3.13	1,3,5,8-Tetrachloronaphthalene	862
	4.1.3.14	1,2,3,4,6-Pentachloronaphthalene	864
	4.1.3.15	1,2,3,5,7-Pentachloronaphthalene	865
	4.1.3.16	1,2,3,5,8-Pentachloronaphthalene	866
	4.1.3.17	1,2,3,4,5,7-Hexachloronaphthalene	867
	4.1.3.18	1,2,3,4,6,7-Hexachloronaphthalene	868
	4.1.3.19	1,2,3,5,6,7-Hexachloronaphthalene	869
	4.1.3.20	1,2,3,5,7,8-Hexachloronaphthalene	870
	4.1.3.21	1,2,3,4,5,6,7-Heptachloronaphthalene	871
	4.1.3.22	1,2,3,4,5,6,8-Heptachloronaphthalene	872
	4.1.3.23	Octachloronaphthalene	873
4.1.4	Brominated polynuclear aromatic hydrocarbons		875
	4.1.4.1	1-Bromonaphthalene	875
	4.1.4.2	2-Bromonaphthalene	879
	4.1.4.3	1,4-Dibromonaphthalene	882
	4.1.4.4	2,3-Dibromonaphthalene	883
	4.1.4.5	4-Bromobiphenyl	884
	4.1.4.6	4,4′-Dibromobiphenyl	885
	4.1.4.7	2,4,6-Tribromobiphenyl	886
	4.1.4.8	2,2′,5,5′-Tetrabromobiphenyl	887

		4.1.4.9	2,2',4,5,5'-Pentabromobiphenyl	888
		4.1.4.10	2,2',4,4',6,6'-Hexabromobiphenyl	889
		4.1.4.11	Decabromobiphenyl	890
4.2	Summary Tables and QSPR Plots			891
4.3	References			900

4.1. LIST OF CHEMICALS AND DATA COMPILATIONS

4.1.1 POLYNUCLEAR AROMATIC HYDROCARBONS (PAHs)

4.1.1.1. Indan

Common Name: Indan
Synonym: hydroindene, 2,3-dihydroindene, 2,3-dihydro-1H-indene, indane
Chemical Name: indan
CAS Registry No: 496-11-7
Molecular Formula: C_9H_{10}
Molecular Weight: 118.175
Melting Point (°C):
    –51.38   (Lide 2003)
Boiling Point (°C):
    177.97   (Lide 2003)
Density (g/cm³ at 20°C):
    0.9639   (Weast 1982–83; Dean 1985; Lide 2003)
Molar Volume (cm³/mol):
    123.0   (20°C, calculated-density)
    143.7   (calculated-Le Bas method at normal boiling point)
Enthalpy of Evaporation, ΔH_V (kJ/mol):
    49 ± 1.5   (Ambrose & Sprake 1976)
Enthalpy of Fusion, ΔH_{fus} (kJ/mol):
    8.60   (exptl., Chickos et al. 1999)
Entropy of Fusion, ΔS_{fus} (J/mol K):
    38.77, 45.9   (exptl., calculated-group additivity method, Chickos et al. 1999)
Fugacity Ratio at 25°C, F: 1.0

Water Solubility (g/m³ or mg/L at 25°C):
    88.9   (shake flask-GC, Price 1976)
    109.1   (shake flask-fluorescence, Mackay & Shiu 1977)
    100   (recommended-IUPAC, Shaw 1989)

Vapor Pressure (Pa at 25°C or as indicated and reported temperature dependence equations. Additional data at other temperatures designated *, are compiled at the end of this section):
    5523*   (91.68°C, ebulliometry, measured range 91.68–129.05°C, Stull et al. 1961)
    204.0*   (comparative ebulliometry, extrapolated from vapor pressure equation derived from exptl data, Ambrose & Sprake 1976)
    $\log (P/\text{kPa}) = 6.10462 - 1574.160/[(T/K) - 67.079]$; temp range 355.006–482.437 K (vapor pressure eq., ebulliometry, Ambrose & Sprake 1976)
    9585*   (101.124°C, comparative ebulliometry, measured range 101.124–192.408°C, Osborn & Scott 1978)
    $\log (P/\text{atm}) = [1 - 451.051/(T/K)] \times 10^{\{0.859420 - 6.08324 \times 10^{-4} \cdot (T/K) + 4.77502 \times 10^{-7} \cdot (T/K)^2\}}$; temp range 355.01–452.24 K (Cox eq., Chao et al. 1983)
    195.6   (extrapolated-Antoine eq., Boublik et al. 1984)
    $\log (P/\text{kPa}) = 6.1012 - 1571.723/(205.798 + t/°C)$; temp range 81.86–209.3°C (Antoine eq. from reported exptl. data of Ambrose & Sprake 1976, Boublik et al. 1984)
    196.9   (extrapolated-Antoine eq., Stephenson & Malanowski 1987)
    $\log (P_L/\text{kPa}) = 6.11622 - 1580.315/(-66.49 + T/K)$; temp range 374–466 K (Antoine eq., Stephenson & Malanowski 1987)

Henry's Law Constant (Pa m³/mol at 25°C):

Octanol/Water Partition Coefficient, log K_{OW}:
- 3.33 (Hansch & Leo 1979)
- 3.30 (calculated-TSA, Yalkowsky & Valvani 1976)
- 3.57 (calculated-fragment const., Valvani & Yalkowsky 1980)
- 3.29 (calculated-solubility, Mackay et al. 1980)
- 3.31 (calculated-fragment const., Yalkowsky et al. 1983)
- 3.33 (shake flask, Log P Database, Hansch & Leo 1987)
- 3.36 (calculated-molar volume, Wang et al. 1991)
- 3.33 (recommended, Sangster 1993)
- 3.18 (recommended, Hansch et al. 1995)

Octanol/Air Partition Coefficient, log K_{OA}:

Bioconcentration Factor, log BCF:

Sorption Partition Coefficient, log K_{OC}:

Environmental Fate Rate Constants, k, or Half-Lives, $t_{½}$:
  Volatilization:
  Photolysis: photolysis rate $k < 3 \times 10^{-5}$ s^{-1} with $t_{½} > 1$ d (Kwok et al. 1997)
  Oxidation: rate constant k for gas-phase second-order rate constants, k_{OH} for reaction with OH radical, k_{NO_3} with NO$_3$ radical and k_{O_3} with O$_3$ or as indicated; *data at other temperatures see reference:
  $k_{OH} = 9.2 \times 10^{-12}$ cm^3 molecule^{-1} s^{-1} at 295 K (Atkinson 1989)
  k_{OH}(exptl) = $(19 \pm 5) \times 10^{-12}$ cm^3 molecule^{-1} s^{-1}, k_{OH}(calc) = 8.3×10^{-12} cm^3 molecule^{-1} s^{-1} with a calculated lifetime of 8 h; k_{NO_3}(exptl) = $(6.6 \pm 2.0) \times 10^{-15}$ cm^3 molecule^{-1} s^{-1} with a calculated lifetime of 7 d; and k_{O_3}(exptl) $< 3 \times 10^{-19}$ cm^3 molecule^{-1} s^{-1} with a calculated lifetime of > 55 d at 297 ± 2 K (relative rate method; calculated-SAR structure-activity relationship, Kwok et al. 1997)
  Hydrolysis:
  Biodegradation:
  Biotransformation:
  Bioconcentration, Uptake (k_1) and Elimination (k_2) Rate Constants:

Half-Lives in the Environment:
  Air: photolysis $t_{½} > 1$ d; calculated tropospheric lifetimes of 8 h, 7 d and > 55 d due to reactions with OH radical, NO$_3$ radical and O$_3$, respectively, at room temp. (Kwok et al. 1997)

TABLE 4.1.1.1.1
Reported vapor pressures of indan at various temperatures and the coefficients for the vapor pressure equations

log P = A − B/(T/K)   (1)   ln P = A − B/(T/K)   (1a)
log P = A − B/(C + t/°C)   (2)   ln P = A − B/(C + t/°C)   (2a)
log P = A − B/(C + T/K)   (3)
log P = A − B/(T/K) − C·log (T/K) (4)

Stull et al. 1961		Ambrose & Sprake 1976		Osborn & Scott 1978	
ebulliometry		comparative ebulliometry		comparative ebulliometry	
t/°C	P/Pa	t/°C	P/Pa	t/°C	P/Pa
91.68	5523	81.856	4343	101.124	9585
107.38	12086	88.012	5650	104.381	10887
128.56	25246	93.482	7069	107.849	12339
152.36	51766	98.033	6471	111.229	13955
176.03	96999	101.96	9860	114.621	15748

(Continued)

TABLE 4.1.1.1.1 (Continued)

Stull et al. 1961		Ambrose & Sprake 1976		Osborn & Scott 1978	
ebulliometry		comparative ebulliometry		comparative ebulliometry	
t/°C	P/Pa	t/°C	P/Pa	t/°C	P/Pa
178.04	101729	109.194	12920	118.027	17735
129.05	104442	116.391	16722	121.442	19933
		120.345	19150	128.315	25023
mp/K	221.77	123.556	21300	135.235	31177
ΔH_{fus}/(kJ mol^{-1}) = 8.60		128.597	25174	142.204	38565
Antoine eq.		134.718	30586	149.224	47375
eq. 2	P/mmHg	139.494	35453	156.300	57817
A	7.05483	146.528	43667	263.418	70120
B	1625.70	152.479	51799	170.590	84432
C	211.645	158.969	61998	177.812	101325
		164.763	72392	177.812	101325
		171.453	86085	177.811	101325
		176.521	97757	185.085	120790
		177.075	99114	192.408	143240
		177.531	100222		
		25.0	204.0	ΔH_V/(kJ mol^{-1}) = 39.67	
				at bp	
		bp/K	451.12 K	ΔH_V/(kJ mol^{-1}) = 49.03	
				at 298.15 K	
		ΔH_V/(kJ mol^{-1}) = 39.8 at bp			
		ΔH_V/(kJ mol^{-1}) = 49.0			
		at 298.15 K			
		eq. 3	P/kPa		
		A	6.10462		
		B	1574.160		
		C	–67.079		

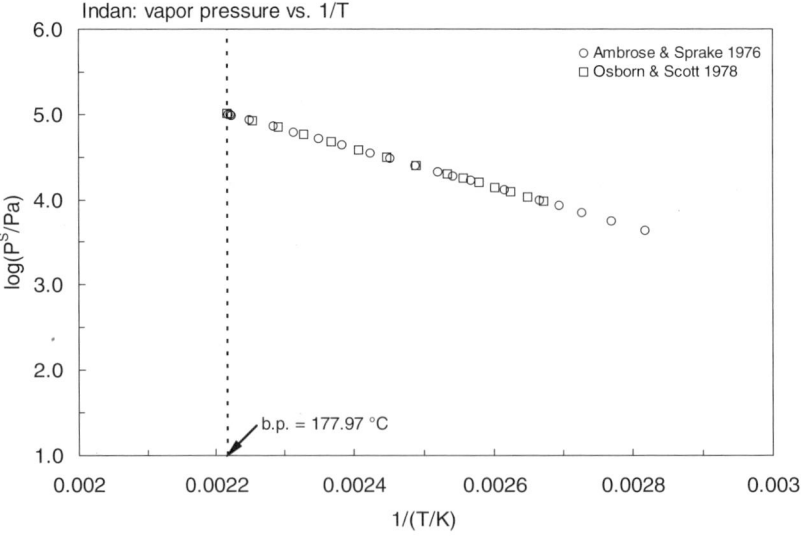

FIGURE 4.1.1.1.1 Logarithm of vapor pressure versus reciprocal temperature for indan.

4.1.1.2 Naphthalene

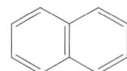

Common Name: Naphthalene
Synonym: naphthene, tar camphor, moth balls
Chemical Name: naphthalene
CAS Registry No: 91-20-3
Molecular Formula: $C_{10}H_8$
Molecular Weight: 128.171
Melting Point (°C):
    80.26   (Lide 2003)
Boiling Point (°C):
    217.9   (Lide 2003)
Density (g/cm³ at 20°C):
    1.0253   (Weast 1983–84)
Molar Volume (cm³/mol):
    125.0   (20°C, calculated-from density)
    133.2   (from density, Bohon & Claussen 1951)
    147.6   (calculated-Le Bas method at normal boiling point)
Enthalpy of Vaporization, ΔH_V (kJ/mol):
Enthalpy of Sublimation, ΔH_{subl} (kJ/mol):
    73.93   (Colomina et al. 1982)
    72.92   (Van Ekeren et al. 1983)
Enthalpy of Fusion, ΔH_{fus} (kJ/mol):
    19.29   (Parks & Huffman 1931)
    19.08   (Wauchope & Getzen 1972; Podoll et al. 1989)
    19.10   (exptl., Chickos et al. 1999)
Entropy of Fusion, ΔS_{fus} (J/mol K):
    54.39   (Casellato et al. 1973)
    54.81   (Ubbelohde 1978)
    53.75, 44.4   (exptl., calculated-group additivity method, Chickos et al. 1999)
Fugacity Ratio at 25°C (assuming ΔS_{fus} = 56 J/mol K), F: 0.287 (mp at 80.26°C)
    0.310   (calculated, Passivirta et al. 1999)

Water Solubility (g/m³ or mg/L at 25°C or as indicated and reported temperature dependence equations. Additional data at other temperatures designated *, are compiled at the end of this section):
    30.0*   (shake flask-gravimetric, measured range 0–25°C, Hilpert 1916)
    31.5   (shake flask-UV, Andrews & Keefer 1949)
    12.5   (shake flask-UV, Klevens 1950)
    34.4*   (shake flask-UV, measured range 2–42°C, Bohon & Claussen 1951)
    30.6   (Stephen & Stephen 1963)
    20.4   (shake flask, Sahyun 1966)
    33.47   (shake flask-GC, Gordon & Thorne 1967)
    38.4   (20°C, shake flask-UV, Eisenbrand & Baumann 1970)
    31.2*   (shake flask-UV, measured range 29–73.4°C, Wauchope & Getzen 1972)
    $R \cdot \ln x = -8690/(T/K) + (0.000408)[(T/K) - 291.15]^2 - 13.4 + 0.0139 \cdot (T/K)$; temp range 29.2–73.4°C (shake flask-UV, Wauchope & Getzen 1972)
    32.17   (shake flask-UV, Vesala 1974)
    31.3   (shake flask-GC, Eganhouse & Calder, 1976)
    22.0   (fluorescence, Schwarz & Wasik 1976)
    31.7   (shake flask-fluorescence, Mackay & Shiu, 1977)
    30.0*   (25°C, shake flask-fluorescence, measured range 8–31°C, Schwarz & Wasik 1977)
    30.25*   (25°C, shake flask-fluorescence, measured range 8.4–31.8°C, Schwarz 1977)

31.69 (generator column-HPLC/UV, measured temp range 5–30°C, May et al. 1978)
S/(mg/kg) = 13.66 + 0.2499·(t/°C) + 0.0189·(t/°C)²; temp range 5–30°C (generator column-HPLC/UV, May et al. 1978)
30.64 (generator column-HPLC/UV, Wasik et al. 1983)
31.94* (25°C, generator column-HPLC/UV, measured range 8.2–27°C, May et al. 1983)
R·ln x = −80.55/(θ/K) + 28.7/[1/(θ/K) − 1/(T/K)] + 0.31·{(θ/K)/(T/K) − 1 − ln [(θ/K)/(T/K)]}, θ = 298.15 K, temp range 8.2–27°C (generator column-HPLC/UV, May et al. 1983)
32.2 (average lit. value, Pearlman et al. 1984)
32.90 (generator column-HPLC/fluorescence, Walters & Luthy 1984)
30.75* (25.2°C, shake flask-UV, Bennet & Canady 1984)
ln x = −1767.4601/R · (T/K) + (17.95209/R) · ln (T/K) + 1; temp range 2–45°C (shake flask-UV, Bennet & Canady 1984)
30.6 (shake flask-HPLC/UV, Fu & Luthy 1985)
31.12 (vapor saturation-GC, Akiyoshi et al. 1987)
31.3, 31.9 (generator column-HPLC/UV, Billington et al. 1988)
31* (recommended, IUPAC Solubility Data Series, Shaw 1989)
33.71* (shake flask-UV, measured range 5–40°C, Perez-Tejeda et al. 1990)
log [S/(mol/dm³)] = −31.24 − 143.5/(T/K) + 4.772·ln (T/K); temp range 5–40°C (shake flask-UV, Perez-Tejeda et al. 1990)
30.6 (generator column-HPLC, Vadas et al. 1991)
29.9 (dialysis tubing equilibration-GC, Etzweiler et al. 1995)
34.8 (generator column-HPLC/fluorescence, De Maagd et al. 1998)
log [S_L/(mol/L)] = 2.992 − 1001/(T/K) (supercooled liquid, Passivirta et al. 1999)
ln x = −1.54117 − 3191.9/(T/K); temp range 5–50°C (regression eq. of literature data, Shiu & Ma 2000)

Vapor Pressure (Pa at 25°C or as indicated and reported temperature dependence equations. Additional data at other temperatures designated * are compiled at the end of this section):

log (P/mmHg) = 7.091 − 3465/(T/K); temp range 87–224 (static isoteniscope method, Mortimer & Murphy 1923)
8.64* (20°C, effusion, measured range 10–30°C Swan & Mack 1925)
log (P/mmHg) = 29.820/(T/K) − 200.682·log (T/K) + 595.642; measured temp range 10–30°C (effusion, Swan & Mack 1925)
log (P/mmHg) = 10.40 − 3429/(T/K); temp range 15–33°C (effusion, Zil'berman-Granovskaya 1940)
133.3* (52.6°C, summary of literature data, temp range 52.6–217.9°C, Stull 1947)
14.26* (manometry-Rodebush gauge, Sears & Hopke 1949)
log (P/mmHg) = −[108.30/(t/°C + 27)] + 1.115; temp range 19–35°C (manometry-Rodebuch gauge, Sears & Hopke 1949)
10.8* (effusion method, measured range 6.7–20.7°C, Bradley & Cleasby 1953)
log (P/cmHg) = 10.597 − 3783/(T/K); temp range 6.7–20.7°C, (Antoine eq., effusion, Bradley & Cleasby 1953)
6815* (126.325°C, manometry, measured range 126.325–218.638°C, Camin & Rossini 1955)
log (P/mmHg) = 6.84577 − 1606.529/(187.227 + t/°C); temp range 126.3–218.6°C (Antoine eq. Camin & Rossini 1955)
log (P/mmHg) = 10.75 − 3616/(T/K); temp range −20 to 10°C (Knudsen effusion method, Hoyer & Peperle 1958)
0.1188* (−13°C, Knudsen effusion, measured range −43 to −13°C, Miller 1963)
46.66* (40.33°C, Hg manometer, measured range 40.33–80.34°C, Fowler et al. 1968)
10.98, 32.95 (manometry, extrapolated solid, supercooled liquid P_L, Fowler et al. 1968)
log (P_S/mmHg) = 9.58102 − 2692.92/(t/°C + 220.651); temp range 40–80°C (Antoine eq., mercury manometer, Fowler et al. 1968)
log (P_L/mmHg) = 7.03382 − 1756.91/(t/°C + 204.931); temp range 81–180°C (Antoine eq., mercury manometer, Fowler et al. 1968)
30.66* (extrapolated-Antoine eq., Zwolinski & Wilhoit 1971)
log (P/mmHg) = 7.01065 − 1733.71/(201.859 + t/°C); temp range 86.581–250.27°C (Antoine eq., Zwolinski & Wilhoit 1971)
11.60 (interpolated-Antoine eq., Weast 1972–73)
log (P/mmHg) = [−0.2185 × 17065.2/(T/K)] + 11.450; temp range 0–80.0°C (Antoine eq., Weast 1972–73)
log (P/mmHg) = [−0.2185 × 12311.6/(T/K)] + 8.413089; temp range 52.6–217.9°C (Antoine eq., Weast 1972–73)
12.26* (Knudsen effusion method, extrapolated from measured data, Radchenko & Kitiagorodskii 1974)

Polynuclear Aromatic Hydrocarbons (PAHs) and Related Aromatic Hydrocarbons

$\log (P/\text{mmHg}) = 11.7041 - 3796.574/(T/K)$; temp range 9.0–23.91°C (Antoine eq., Knudsen effusion, Radchenko & Kitiagorodskii 1974)

10.93*   (gas saturation, Sinke 1974)

11.21*   (Baratron model diaphragm pressure gauge, Ambrose et al. 1975)

$T \cdot \log (P/\text{Pa}) = \tfrac{1}{2} \cdot a_o + \sum a_i E_i(x)$; $a_o = 310.6247$, $a_1 = 791.4937$, $a_2 = -82536$, $a_3 = 0.4043$; temp range: 230–344 K, (Chebyshev polynomial, diaphragm pressure gauge, Ambrose et al. 1975)

$\log (P/\text{Pa}) 13.70 - 3773/(T/K)$ (Antoine eq. derived from exptl data of Ambrose et al. 1975, Wania et al. 1994)

13.5   (effusion method-pressure gauge, DePablo 1976)

10.64*   (gas saturation, interpolated-Clapeyron eq., Macknick & Prausnitz 1979)

$\log (P/\text{mmHg}) = 26.250 - 8575/(T/K)$; temp range 7.15–31.85°C (Clapeyron eq., gas saturation, Macknick & Prausnitz 1979)

11.30*   (effusion method, de Kruif 1980)

$\log (P/\text{Pa}) = 14.187 - 3907/(T/K)$; temp range 253–273 K (torsion effusion, regression, de Kruif 1980)

$\log (P/\text{Pa}) = 14.053 - 3860/(T/K)$; temp range 253–273 K (weighing effusion, regression, de Kruif 1980)

$\log (P/\text{Pa}) = 14.107 - 3886/(T/K)$; temp range 253–273 K (effusion, mean regression, de Kruif 1980)

10.42*   (effusion method, de Kruif et al. 1981)

11.41*   (Knudsen effusion, extrapolated-Antoine eq. from exptl data, Colomina et al. 1982)

$\log (P/\text{Pa}) = 14.01 - 3861.8/(T/K)$; temp range 271.46–284.63 K (Antoine eq., Knudsen effusion, Colomina et al. 1982)

11.33*   (gas saturation-GC, Grayson & Fosbraey 1982)

$\ln (P/\text{Pa}) = 31.8 - 8753/(T/K)$; temp range 302–352 K, (Antoine eq., gas saturation, Grayson & Fosbraey 1982)

$\log (P/\text{atm}) = [1 - 490.988/(T/K)] \times 10^{\{0.832267 - 4.41855 \times 10^{-4} \cdot (T/K) + 2.89627 \times 10^{-7} \cdot (T/K)^2\}}$; (Cox eq., temp range 340.15–751.65 K, Chao et al. 1983)

6.53   (20°C, Mackay et al. 1983)

10.4*   (gas saturation-HPLC/UV, Sonnefeld et al. 1983)

$\log (P/\text{Pa}) = 14.299 - 3960.03/(T/K)$; temp range 10–50°C (Antoine eq., gas saturation, Sonnefeld et al. 1983)

1.63*   (244.19°C, spinning-rotor gauge, measured range 244.19–255.86 K, Van Ekeren et al. 1983)

10.4   (generator column-HPLC, Wasik et al. 1983)

$\log (P_L/\text{kPa}) = 5.93404 - 1579.278/(184.062 + t/°C)$; temp range 126.3–218.6°C (Antoine eq. from reported exptl. data, Boublik et al. 1984)

$\log (P_L/\text{kPa}) = 6.1638 - 1760.018/(215.204 + t/°C)$, temp range 80.3–179.5°C (Antoine eq. from reported exptl. data, Boublik et al. 1984)

22.64, 28.24   (P_{GC} by GC-RT correlation with BP-1 column, Apolane-87 column, Bidleman 1984)

$\log (P_L/\text{mmHg}) = 7.01065 - 1733.71/(201.86 + t/°C)$; temp range 86–250°C (Antoine eq., Dean 1985, 1992)

$\log (P_L/\text{mmHg}) = 6.8181 - 1585.86/(184.82 + t/°C)$; temp range 125–218°C (Antoine eq., Dean 1985, 1992)

10.7*   (25.35°C, gas saturation, temp range 24.85–57.75°C, Sato et al. 1986)

$\ln (P_S/\text{Pa}) = 22.8929 - 4025.35/(T/K - 102.243)$; temp range 298.5–330.9 K (Antoine eq., gas saturation, Sato et al. 1986)

11.27   (interpolated-Antoine eq., Stephenson & Malanowski 1987)

$\log (P_S/\text{kPa}) = 8.70592 - 2619.91/(-52.499 + T/K)$; temp range 310–353 K (Antoine eq.-I., Stephenson & Malanowski 1987)

$\log (P_S/\text{kPa}) = 9.45562 - 3069.145/(-29.892 + T/K)$; temp range 263–353 K (Antoine eq.-II, Stephenson & Malanowski 1987)

$\log (P_S/\text{kPa}) = 11.9681 - 4577.47/(30.394 + T/K)$; temp range not specified (Antoine eq.-III, Stephenson & Malanowski 1987)

$\log (P_L/\text{kPa}) = 6.19487 - 1782.509/(-65.637 + T/K)$; temp range 352–500 K (Antoine eq.-IV, Stephenson & Malanowski 1987)

$\log (P_L/\text{kPa}) = 6.14835 - 1751.644/(-68.319 + T/K)$; temp range 491–565 K (Antoine eq.-V, Stephenson & Malanowski 1987)

$\log (P_L/\text{kPa}) = 6.53231 - 2162.182/(-12.108 + T/K)$; temp range 563–665 K (Antoine eq.-VI, Stephenson & Malanowski 1987)

$\log (P_L/\text{kPa}) = 7.74783 - 4042.567/(227.985 + T/K)$; temp range 661–750 K (Antoine eq.-VII, Stephenson & Malanowski 1987)

11.37*   (pressure gauge, interpolated-Antoine eq., measured range –12.15 to 70.16°C, Sasse et al. 1988)

$\log (P_S/\text{mmHg}) = 10.05263 - 2907.918/(236.459 + t/°C)$; temp range –12.15 to 70.16°C (Antoine eq., pressure gauge, Sasse et al. 1988) log $(P_L/mmHg)$ = 2.25180 – 76.496/(–25.09 + t/°C); temp range 80.16–90.15°C (Antoine eq., pressure gauge, Sasse et al. 1988)

22.65 (PGC, GC-RT correlation with eicosane as reference standard, Hinckley et al. 1990)

41.88, 38.02 (supercooled liquid P_L values converted from literature P_S with different ΔS_{fus} values, Hinckley et al. 1990)

0.7634* (0°C, gas saturation-GC, measured range –30.6 to 0°C, Wania et al. 1994)

log (P/Pa) = 13.93 – 3851/(T/K); temp range –30 to 0°C, (Antoine eq., gas saturation, Wania et al. 1994)

24.0 (supercooled liquid P_L, GC-RT correlation, Donovan 1996)

5.58–12.30; 10.4–14.0; 11.2–14.4; 7.71–17.2; 6.45–8.40 (quoted lit. ranges: effusion method; gas saturation; manometry; calculated; from GC-RT correlation, Delle Site 1997)

11.16 (solid P_S, van der Linde et al. 1998)

40.0; 12.4 (quoted P_L from Hinckley et al. 1990; converted to P_S with fugacity ratio F, Passivirta et al. 1999)

log (P_S/Pa) = 10.90 – 2927/(T/K) (solid, Passivirta et al. 1999)

log (P_L/Pa) = 8.06 – 1923/(T/K) (supercooled liquid, Passivirta et al. 1999)

log (P/Pa) = 13.59 – 3742/(T/K); temp range 5–50°C (regression eq. from literature data, Shiu & Ma 2000)

37.0 (supercooled liquid P_L, GC-RT correlation, Lei et al. 2002)

log (P_L/Pa) = –2930/(T/K) + 11.39; $\Delta H_{vap.}$ = –56.1 kJ·mol^{-1} (GC-RT correlation, Lei et al. 2002)

Henry's Law Constant (Pa m^3/mol at 25°C or as indicated and reported temperature dependence equations. Additional data at other temperatures designated * are compiled at the end of this section):

56.0 (gas stripping, Southworth, 1979)
48.9 (gas stripping, Mackay et al. 1979; Mackay & Shiu 1981)
44.6 (gas stripping, Mackay et al. 1982)
36.5 (20°C, EPICS method, Yurteri et al. 1987)
74.3 (wetted-wall column-GC, Fendinger & Glotfelty 1990)
124 (calculated-vapor-liquid equilibrium (VLE) data, Yaws et al. 1991)
47.1 (headspace solid-phase microextraction (SPME)-GC, Zhang & Pawliszyn 1993)
42.6* (gas stripping-GC, measured range 3.7–35.5°C, Alaee et al. 1996)

ln K_{AW} = 13.95 – 5364.45/(T/K), temp range: 3.7–35.5°C (gas stripping-GC, Alaee et al. 1996)

26.2, 35.5, 48.1, 62.5, 77.7, 108.1 (9.2, 14.5, 20.1, 24.6, 30.5, 34.8°C, seawater with salinity of 35l‰ (0.660M NaCl), gas stripping-GC, Alaee et al. 1996)

44.6 (gas stripping-GC; calculated-P/C, Shiu & Mackay 1997)
45.0 (gas stripping-HPLC/fluo., De Maagd et al. 1998)
57.0 (gas stripping-GC, Altschuh et al. 1999)

log K_{AW} = 6.058 – 2332/(T/K) (van't Hoff eq. derived from literature data, Staudinger & Roberts 2001)

Octanol/Water Partition Coefficient, log K_{OW} at 25°C or as indicated:

3.37 (shake flask, Fujita et al. 1964; Hansch et al. 1973)
3.37, 3.01, 3.45 (Leo et al. 1971)
3.37 (calculated-fragment const., Rekker 1977)
3.395 (shake flask-fluorometry, Krishnamurthy & Wasik 1978)
3.30, 3.01, 3.37, 3.45, 3.59 (quoted, Hansch & Leo 1979)
3.17 (HPLC-RT correlation, Veith et al. 1979a,b)
3.36 (shake flask-UV, concn. ratio, Karickhoff et al. 1979)
3.21 (HPLC-k' correlation, Hanai et al. 1981)
3.18 (HPLC-k' correlation, D'Amboise & Hanai 1981)
3.35 (generator column-HPLC/UV, Wasik et al. 1981 1983)
3.35 (RP-TLC-k' correlation, Bruggeman et al. 1982)
3.45 (HPLC-RT correlation, Hammers et al. 1982)
3.31; 3.35 (shake flask; HPLC correlation, Eadsforth & Moser 1983)
3.36 (HPLC-k' correlation, Hafkenscheid & Tomlinson 1983)
3.35; 3.42 (shake flask; ALPM, Garst & Wilson 1984)
3.57 (HPLC-RV correlation, Garst & Wilson 1984)
3.38 (RP-HPLC correlation, Chin et al. 1986)
3.43 (HPLC-RT correlation, Edsforth 1986)

3.29	(HPLC-RT correlation, Wang et al. 1986)
3.29	(HPLC-RT correlation, de Kock & Lord 1987)
3.30	(average, HPLC-RT correlation, Ge et al. 1987)
3.35	(shake flask-GC, Opperhuizen et al. 1987)
3.23	(HPLC-RT correlation, Minick et al. 1988)
3.29	(RP-HPLC-RT correlation, ODS column with masking agent, Bechalany et al. 1989)
3.35	(recommended, Sangster 1989, 1993)
3.49	(centrifugal partition chromatography, Menges et al. 1990)
3.36	(shake flask-HPLC/UV, Menges & Armstrong 1991)
3.30	(TLC-RT correlation, De Voogt et al. 1990)
3.70	(centrifugal partition chromatography, Berthod et al. 1992)
3.37	(shake flask-UV, pH 7.4, Alcorn et al. 1993)
3.30	(recommended, Hansch et al. 1995)
3.44; 3.68	(26°C; 4°C, quoted, Piatt et al. 1996)
3.47, 3.58; 3.40	(HPLC-k' correlation: ODS column; Diol column; quoted lit. average, Helweg et al. 1997)
3.33	(range 3.24–3.40) (slow stirring method-HPLC/fluo., De Maagd et al. 1998)
3.40	(shake flask-dialysis tubing-HPLC/UV, both phases, Andersson & Schräder 1999)
3.77	(RP-HPLC-RT correlation, short ODP column, Donovan & Pescatore 2002)

Octanol/Air Partition Coefficient, log K_{OA} at 25°C:

5.10	(calculated-K_{OW}/K_{AW}, Wania & Mackay 1996)
5.19	(calculated-S_{oct} and vapor pressure P, Abraham et al. 2001)

Bioconcentration Factor, log BCF at 25°C:

1.64	(mussel *Mytilus edulis*, Lee et al. 1972)
4.11	(bile of rainbow trout, Melancon & Lech 1978)
2.12	(*Daphnia pulex*, Southworth et al. 1978)
2.07	(*Daphnia pulex*, by kinetic estimation, Southworth et al. 1978)
2.63	(fathead minnow, Veith et al. 1979b, 1980)
2.62	(microorganisms-water, calculated from K_{OW}, Mabey et al. 1982)
4.10, 3.84, 4.25	(average, *Selenastrum capricornutum*-dosed singly, dosed simultaneously, Casserly et al. 1983)
2.11; 2.43	(*Chlorella fusca*; calculated-K_{OW}, Geyer et al. 1984)
1.48, 2.10, 3.0	(fish, algae, activated sludge, Freitag et al. 1985)
2.50	(bluegill sunfish, McCarthy & Jimenez 1985)
2.48	(bluegill sunfish with dissolved humic material, McCarthy & Jimenez 1985)

Sorption Partition Coefficient, log K_{OC} at 25°C or as indicated:

3.11	(natural sediment, average sorption isotherms by batch equilibrium technique-UV spec., Karickhoff et al. 1979)
2.38	(22°C, suspended particulates, Herbes et al. 1980)
2.94	(sediment/soil, sorption isotherm by batch equilibrium, Karickhoff 1981)
3.62, 3.87, 4.23	(soil I-very strongly acid sandy soil pH 4.5–5.5, soil II-moderately or slightly acid loamy soil pH 5.6–6.5, soil III-slightly alkaline loamy soil pH 7.1–8.0, OECD 1981)
3.50; 4.43; 3.21	(Speyer soils: pH 7.0, 0.69% OC; pH 5.8, 2.24% OC; pH 7.1, 1.12% OC at 0.15–0.5 mm, batch equilibrium-sorption isotherm, Rippen et al. 1982)
3.11; 3.16	(soils: Alfisol 0.76% OC at pH 7.5, Entisol 1.11% OC at pH 7.9, batch equilibrium-sorption isotherm, Rippen et al. 1982)
3.30	(Offshore Grand Haven sediment, batch equilibrium-sorption isotherm, Voice & Weber 1985)
2.67, 2.77	(Lula aquifer 0.032% OC, Apalachee soil 1.4% OC, batch equilibrium-sorption isotherm, Stauffer & MacIntyre 1986)
2.95	(sediment, calculated, Pavlou 1987)
2.93	(sediment 4.02% OC from Tamar estuary, batch equilibrium-GC, Vowles & Mantoura 1987)
3.27	(calculated-MCI χ, Bahnick & Doucette 1988)
3.02; 2.89	(Aldrich and Fluka humic acid, observed; predicted, Chin et al. 1989)

2.73–3.91   (aquifer materials, Stauffer et al. 1989)
5.00   (sediments average, Kayal & Connell 1990)
3.15, 2.76   (Menlo Park soil, Eustis sand, batch equilibrium-sorption isotherm, Podoll et al. 1989)
3.21, 3.16, 3.10, 3.00 (15, 25, 35, 50°C, Menlo Park soil 1.6% OC, flow sorption equilibrium, Podoll et al. 1989)
2.97, 2.67   (modified, unmodified EPA-6 sediments, batch equilibrium-sorption isotherm, Podoll et al. 1989)
3.11   (soil, RP-HPLC-k′ correlation, Szabo et al. 1990a)
3.29   (sandy surface soil, Wood et al. 1990)
2.97   (dissolved organic matter, Kan & Tomson 1990)
2.98; 2.965, 2.98   (sediment: concn ratio C_{sed}/C_W; concn-based coeff., areal-based coeff. of flux studies of sediment/water boundary layer, Helmstetter & Alden 1994)
3.11   (calculated-MCI $^1\chi$, Sabljic et al. 1995)
3.16, 3.05, 3.06   (RP-HPLC-k′ correlation on different stationary phases, Szabo et al. 1995)
2.75   (HPLC-screening method; Müller & Kördel 1996)
4.06   (range 3.95–4.15); 2.08 (range 2.08–2.11) (4°C, low organic carbon sediment f_{OC} = 0.0002, batch equilibrium; column exptl., Piatt et al. 1996)
3.90   (range 3.81–4.00); 2.11 (range 2.11–2.13) (26°C, low organic carbon sediment f_{OC} = 0.0002, batch equilibrium; column exptl., Piatt et al. 1996)
3.74   (humic acid, HPLC-k′ correlation; Nielsen et al. 1997)
2.42–2.56   (5 soils, 20°C, batch equilibrium-sorption isotherm measured by HPLC/UV, Bayard et al. 1998)
2.61, 2.63, 2.68, 2.77, 2.76, mean 2.69 ± 0.073 (soils: Woodburn soil, Elliot soil, Marlette soil, Piketon soil, Anoka soil, batch equilibrium-sorption isotherms-HPLC-fluorescence, Choiu et al. 1998)
2.91, 2.86, 2.88, 2.87, 2.89, 2.95, 3.07; mean 2.88 ± 0.22 (sediments: Lake Michigan, Mississippi River, Massachusetts Bay, Spectacle Island, Peddocks Island, Port Point Channel, batch equilibrium-sorption isotherms-HPLC-fluorescence, Choiu et al. 1998)
3.11 (3.00–3.19), 2.80 (sediments: Lake Oostvaardersplassen, Lake Ketelmeer, shake flask-HPLC/UV, de Maagd et al. 1998)
3.60; 3.10   (soil, calculated-universal solvation model; quoted exptl., Winget et al. 2000)
3.09–5.51; 2.60–5.0 (range, calculated from sequential desorption of 11 urban soils; lit. range, Krauss & Wilcke 2001)
3.91; 4.12, 4.06, 4.94 (20°C, batch equilibrium, A2 alluvial grassland soil; calculated values of expt 1,2,3-solvophobic approach, Krauss & Wilcke 2001)
2.91, 3.02, 2.71   (soils: organic carbon OC ≥ 0.1%, OC ≥ 0.5%, 0.1 ≤ OC < 0.5%, average, Delle Site 2001)
3.06   (average values for sediments OC ≥ 0.5%, Delle Site 2001)
4.43   (soil humic acid, shake flask-HPLC/UV, Cho et al. 2002)

Environmental Fate Rate Constants, k, or Half-Lives, $t_{1/2}$:

Volatilization/Evaporation: rate of evaporation estimated to be 1.675×10^{-9} mol cm^{-2} h^{-1} at 20°C and air flow rate of 50 L h^{-1} (Gückel et al. 1973);

calculated $t_{1/2}$ = 7.15 h from 1 m depth of water (Mackay & Leinonen 1975; quoted, Haque et al. 1980);

$t_{1/2}$ = 16 h for surface waters for a river 1-m deep, water velocity 0.5 m/s, wind velocity 1m/s (Southworth 1979; Herbes et al. 1980);

evaporation $t_{1/2}$ = 50 h in a river and $t_{1/2}$ = 200 h in a lake when considering current velocity and wind speed in combined with typical reaeration rates for natural bodies of water (Howard 1989);

$t_{1/2}$(exptl) = 28 min and $t_{1/2}$(calc) = 32 min from solution (Mackay et al. 1983).

Photolysis:

$t_{1/2}$ = 71 h calculated for direct photochemical transformation near-surface water and $t_{1/2}$ = 550 d in 5-m deep inland water body at 40°N at midday of midsummer (Zepp & Schlotzhauer 1979)

k = 0.0392 h^{-1} with H_2O_2 under photolysis at 25°C in F-113 solution and with HO- in the gas (Dilling et al. 1988)

k = 0.028 h^{-1} in distilled water with $t_{1/2}$ = 25 h (Fukuda et al. 1988)

k = 6.0×10^{-4} min^{-1} and $t_{1/2}$ = 19.18 h, photodegradation in methanol-water (3:7, v/v) solution with initial concentration of 50.0 ppm by high-pressure mercury lamp or sunlight (Wang et al. 1991);

k(exptl) = 0.000511 min^{-1} with t$_{½}$(calc) = 22.61 h and the predicted k = 0.000303 min^{-1} by QSPR, the pseudo-first-order direct photolysis rate constant of in aqueous solution when irradiated with a 500W medium pressure mercury lamp (Chen et al. 1996)

t$_{½}$(calc) = 15.42 h direct photolysis half-life in atmospheric aerosol (QSPR, Chen et al. 2001).

Oxidation: rate constant k for gas-phase second order rate constants, k$_{OH}$ for reaction with OH radical, k$_{NO_3}$ with NO$_3$ radical and k$_{O_3}$ with O$_3$ or as indicated; *data at other temperatures and/or Arrhenius equation see reference:

k(calc) < 360 M^{-1} h^{-1} for singlet oxygen, k(calc) < 1 M^{-1} h^{-1} for peroxy radical (Mabey et al. 1982)

k$_{OH}$ = (2.42 ± 0.19) × 10^{-11} cm^3 molecule^{-1} s^{-1} with an estimated atmospheric lifetime τ ~ 1d, and k$_{O_3}$ < 2 × 10^{-9} cm^3 molecule^{-1} s^{-1} 294 ± 1 K (relative rate method, Atkinson et al 1984)

k$_{OH}$ = (2.35 ± 0.06) × 10^{-11} cm^3 molecule^{-1} s^{-1} at 298 ± 1 K (relative rate method, Biermann et al. 1985)

k$_{N_2O_5}$ ~ (2 – 3) × 10^{-17} cm^3 molecule^{-1} s^{-1} for reaction with N$_2$O$_5$ at 298 K, the calculated lifetime τ – 10 d due to night-time reaction with N$_2$O$_5$ in atmosphere (relative rate method, Pitts et al. 1985)

k$_{OH}$* = 2.17 × 10^{-11} cm^3 molecule^{-1} s^{-1} at 298 K (recommended, Atkinson 1985)

k$_{OH}$ = (2.59 ± 0.24) × 10^{-11} cm^3 molecule^{-1} s^{-1} with a lifetime of ~11 h, k$_{O_3}$ < 3 × 10^{-19} cm^3 molecule^{-1} s^{-1} at 295 ± 1 K (relative rate method, Atkinson & Aschmann 1986)

k$_{N_2O_5}$ = (1.4 ± 0.2) × 10^{-17} cm^3 molecule^{-1} s^{-1} with N$_2$O$_5$ at 298 K (relative rate method, Atkinson et al. 1987)

k$_{OH}$ = 2.17 × 10^{-11} cm^3 molecule^{-1} s^{-1}, k$_{O_3}$ < 2 × 10^{-19} cm^3 molecule^{-1} s^{-1}, and k$_{N_2O_5}$ = 1.4 × 10^{-17} cm^3 molecule^{-1} s^{-1} with N$_2$O$_5$ at room temp. (Atkinson & Aschmann 1987, 1988)

k$_{OH}$ = (22.8 – 25.9) × 10^{-12} cm^3 molecule^{-1} s^{-1} at 294–298 K (review, Atkinson 1989)

k$_{OH}$* = 2.16 × 10^{-11} cm^3 molecule^{-1} s^{-1} at 298 K (recommended, Atkinson 1989, 1990)

k$_{OH}$(calc) = 26.58 × 10^{-12} cm^3 molecule^{-1} s^{-1} (molecular orbital calculations, Klamt 1996)

k$_{OH}$* = 23 × 10^{-12} cm^3 molecule^{-1} s^{-1} at 298 K, measured range 306–366 K with a calculated atmospheric lifetime of 12 h based on gas-phase OH reaction (Brubaker & Hites 1998)

k$_{OH}$ = (2.39 ± 0.09) × 10^{-11} cm^3 molecule^{-1} s^{-1} at 298 ± 2 K, with a calculated tropospheric lifetime of 5.8 h using a global tropospheric 12-h daytime average OH radical concentration of 2.0 × 10^6 molecule cm^{-3} (relative rate method, Phousongphouang & Arey 2002)

Hydrolysis: not hydrolyzable (Mabey et al. 1982).

Biodegradation:

k = 4 × 10^{-6} g L^{-1} d^{-1} ultimate loss process (Lee & Ryan, 1976)

k = (0.04 – 3.3) × 10^{-6} g L^{-1} d^{-1} (Lee & Anderson 1977)

complete degradation in 8 d in gas-oil contaminated groundwater which was circulated through sand that had been inoculated with groundwater under aerobic conditions (Kappeler & Wuhrmann 1978; quoted, Howard 1989)

k = 0.04 – 3 µg/L d for microorganisms (Callahan et al. 1979)

t$_{½}$ = 1.9 d, in deeper and slowly moving contaminated water (Herbes 1981; Wakeham et al. 1983; quoted, Howard 1989);

half-lives: of 7, 24, 63 and 1700 d in an oil polluted estuarine stream, clean estuarine stream, coastal waters, and in the Gulf stream, respectively (Lee 1977; quoted, Howard 1989)

t$_{½}$(aerobic) = 12 h, based on die-away test data for an oil polluted creek (Walker & Colwell 1976)

t$_{½}$ = 480 h, for an estuarine river (Lee & Ryan 1976).

t$_{½}$(anaerobic) = 25 d at pH 8 and t$_{½}$ = 258 d at pH 5 (Hambrick et al. 1980);

t$_{½}$ = 1.9 d, in deeper and slower moving contaminated water (Herbes 1981; Wakeham et al. 1983)

k = 0.23 h^{-1} (microbial degradation rate constant, Herbes et al. 1980, Hallett & Brecher 1984)

100% degradation within 7 d for an average of three static-flask screening test (Tabak et al. 1981)

k = 3.2 × 10^{-3} h^{-1} with t$_{½}$ = 9 h; k = 7.6 × 10^{-2} h^{-1} with t$_{½}$ = 9 h for mixed bacterial populations in water and sediment from the same stream (NRCC 1983)

k = 0.14 h^{-1} with t$_{½}$ = 5 h; k < 4 × 10^{-4} h^{-1} with t$_{½}$ > 72 d for mixed bacterial populations in oil-contaminated and pristine stream sediments (NRCC 1983)

degraded completely within 1 wk by microbes in non-autoclaved samples of 0.04 mg/L in groundwater from hazardous waste site (Lee et al. 1984)

k = 0.024 d^{-1} with t$_{½}$ = 28 d in groundwater with nutrients and acclimated microbes, k = 0.013 d^{-1} with t$_{½}$ = 53 d in river water with acclimated microbes, and k = 0.018 d^{-1} with t$_{½}$ = 39 d in river water with nutrients and acclimated microbes (Vaishnav & Babeu 1987) k = 0.337 d^{-1} with t$_{½}$ = 2.1 d for Kidman sandy loam and k = 0.308 d^{-1} with t$_{½}$ = 2.2 d for McLarin sandy loam all at –0.33 bar soil moisture (Park et al. 1990)

t$_{½}$(aerobic) = 12–480 h, based on die-away test data and for estuarine river (Howard et al. 1991)

t$_{½}$(anaerobic) = 500–6192 h, based on anaerobic estuarine sediment die-away test data (Howard et al. 1991)

removal rate of 2.4 and 0.43 mg (g of volatile suspended solid d)$^{-1}$, degradation by bacteria from creosote-contaminated marine sediments with nitrate- and sulfate-reducers, respectively, under anaerobic conditions in a fluidized bed reactor (Rockne & Strand 1998)

Biotransformation: estimated rate constant for bacteria, 1×10^{-7} ml cell^{-1} h^{-1} (Mabey et al. 1982).

Bioconcentration, Uptake (k$_1$) or Elimination (k$_2$) Rate Constants:

k$_1$ = 197 h^{-1}; k$_2$ = 1.667 h^{-1} (*Daphnia pulex*, Southworth et al. 1978; quoted, Hawker & Connell 1986)

log k$_2$ = –0.70, –1.70 d^{-1} (fish, calculated-K$_{OW}$, Thomann 1989)

Half-Lives in the Environment:

Air: volatility of 2.28×10^4 s (experimental), 7.7×10^3 s (calculated) for depth of water body of 22.5 m (23°C, Klöpffer et al. 1982);

estimated lifetime of ~1d due to reaction with photochemically produced hydroxyl radical, assuming an average daytime atmospheric OH radical concn of $\sim 1 \times 10^6$ molecule/cm^3 (Atkinson et al. 1984, 1987; quoted, Howard 1989);

calculated lifetime of ~11 h due to reaction with OH radical (Atkinson & Aschmann 1986);

the atmospheric lifetimes of naphthalene and alkyl-substituted naphthalenes due to reaction with OH radical and with N$_2$O$_5$ can be calculated to range from ~4 to 13 h and 20–80 d, respectively (Atkinson & Aschmann 1987);

t$_{½}$ = 2.96–29.6 h based on photooxidation half-life in air (Howard et al. 1991);

calculated atmospheric lifetime of 12 h based on gas-phase OH reactions (Brubaker & Hites 1998).

Surface Water: t$_{½}$ = 16 h (calculated for river water 1 m deep, water velocity 0.5 m/s, wind velocity 1 m/s from air-water partition coefficients (Southworth 1979; Hallett & Brecher 1984)

t$_{½}$ = 71 h of photolysis in near surface water, but t$_{½}$ = 550 d for a depth of 5 m (calculated from surface water in midsummer at 40°N latitude, Zepp & Schlotzhauer 1979);

calculated t$_{½}$ = 7.15 h, based on evaporative loss for a water depth of 1 m at 25°C (Mackay & Leinonen 1975);

an overall t$_{½}$ = 2.3 d in Rhine river based on monitoring data (Zoeteman et al. 1980)

in coastal seawater mesocosm experiments: k = 0.058 d^{-1} with t$_{½}$ = 12 d in winter at 3–7°C during the periods when volatilization dominates, k = 0.061 d^{-1} with t$_{½}$ = 11.3 h with HgCl$_2$ as poison and k = 0.896 d^{-1} corresponding to t$_{½}$ = 0.8 h without poison (Wakeham et al. 1983);

t$_{½}$ = 12–480 h, based on estimated unacclimated aqueous aerobic biodegradation half-life (Howard et al. 1991); photolysis t$_{½}$ = 22.81 h in aqueous solution when irradiated with a 500 W medium pressure mercury lamp (Chen et al. 1996).

Groundwater: estimated t$_{½}$ = 0.6 yr in the Netherlands (Zoeteman et al. 1981);

t$_{½}$ = 24–6193 h based on estimated unacclimated aerobic and anaerobic biodegradation half-lives (Howard et al. 1991).

Sediment: t$_{½}$ = 4.9 h in oil contaminated sediment and t$_{½}$ > 88 d in uncontaminated sediments (Herbes & Schwall 1978; quoted, Howard 1989);

desorption k = 0.031 d^{-1} with t$_{½}$ = 21.3 d from sediment under conditions mimicking marine disposal (Zhang et al. 2000).

Soil: an overall t$_{½}$ = 3.6 months in a solid waste site (Zoeteman et al. 1981);

t$_{½}$ = 0.12–125 d (Sims & Overcash 1983);

t$_{½}$ = 12 d for both 5 mg/kg and 50 mg/kg added (Bulman et al. 1987);

t$_{½}$ > 50 d (Ryan et al. 1988);

t$_{½}$ > 80 d (Howard 1989);

biodegradation k = 0.337 d^{-1} with t$_{½}$ = 2.1 d for Kidman sandy loam soil and k = 0.308 d^{-1} with t$_{½}$ = 2.2 d for McLaurin sandy loam soil (Park et al. 1990);

t$_{½}$ = 398–1152 h, based on soil-die-away test data (Howard et al. 1991); t$_{½}$ = 0.02–46 wk, t$_{½}$ < 2.1 yr (quoted, Ludington soil, Wild et al. 1991).

Biota: elimination half-lives t$_{½}$ = 2 d from Oyster for naphthalenes, t$_{½}$ = 2.0 d from clam *Macoma balthica* quoted, Meador et al. 1995).

TABLE 4.1.1.2.1
Reported aqueous solubilities of naphthalene at various temperatures and empirical temperature dependence equations

$$R \cdot \ln x = -\Delta H_{fus}/(T/K) + (0.000408) \cdot [(T/K) - 291.15]^2 - c + b \cdot (T/K) \quad (1)$$

$$S/(\mu g/kg) = a \cdot t^3 + b \cdot t_{1/2} + c \cdot t + d \quad (2)$$

$$R \cdot \ln x = -30.55/\theta + 28.7(1/\theta - 1/T) + 0.31[\theta/T - 1 - \ln(\theta/T)] \quad (3)$$

$$\ln x = 1767.460/[R \cdot (T/K)] + (17.95209/R) \cdot \ln (T/K) + 1 \quad (4)$$

$$\log (S/\text{mol} \cdot \text{dm}^{-3}) = -31.24 - 143.5/(T/K) + 4.772 \cdot \ln (T/K) \quad (5)$$

1.

Hilpert 1916		Bohon & Claussen 1951		Wauchope & Getzen 1972			
shake flask-gravimetry		shake flask-UV		shake flask-UV			
t/°C	S/g·m⁻³	t/°C	S/g·m⁻³	t/°C	S/g·m⁻³	t/°C	S/g·m⁻³
					exptl.		smoothed
0	19	25	34.4	22.2	28.8	0	13.7
25	30	0	13.74	22.2	29.1	22.2	28.3
		0.4	13.74	24.5	30.8	24.5	30.7
		0.5	13.85	24.5	30.1	25.0	31.2
		0.9	14.59	24.5	30.7	29.9	37.3
		9.4	19.62	29.9	38.1	30.3	37.8
		10	19.42	29.9	38.2	34.5	44.3
		14.9	23.43	29.9	38.3	39.3	53.3
		15.9	24.56	30.3	38.1, 37.6	40.1	55.0
		19.3	28.05	34.5	44.6, 43.8	44.7	66.2
		25.6	36.6	39.3	52.6, 52.8	50.0	82.4
		30.1	42.97	40.1	54.8	50.2	83.1
		30.2	43.87	44.7	66.0	55.6	105
		35.2	54.45	44.7	65.5	64.5	156
		36.0	54.81	44.7	65.3	73.4	239
		42.8	73.49	50.2	78.6	75	258
				55.6	106		
		$\Delta H_{sol}/(\text{kJ mol}^{-1})$		64.5	166	temp dependence eq. 1	
		2°C	4.14	64.5	151	ln x	mole fraction
		7°C	5.77	64.5	157	ΔH_{fus} =	36.36 ± 0.17
		12°C	7.24	73.4	240	$10^2 \cdot b$	1.39 ± 0.05
		17°C	8.49	73.4	247	c	13.4 ± 0.2
		22°C	10.17	73.4	244		
		27°C	12.80				
		32°C	14.23	$\Delta H_{fus}/(\text{kJ mol}^{-1})$ = 36.36			
		37°C	15.69				

(Continued)

TABLE 4.1.1.2.1 (Continued)

2.

Schwarz & Wasik 1977		Schwarz 1977		May 1980		May et al. 1983	
shake flask-fluorescence		shake flask-fluorescence		generator column-HPLC		generator column-HPLC/UV	
t/°C	S/g·m^{-3}	t/°C	S/g·m^{-3}	t/°C	S/g·m^{-3}	t/°C	S/g·m^{-3}
from K_{WA} measurements							
10	18.32	8.4	17.95	8.2	16.91	8.2	16.92
12	19.87	11.1	19.1	11.5	19.23	11.5	19.25
14	21.02	14	21.2	13.4	20.37	13.4	20.39
15	22.17	17.5	24.1	15.1	21.48	15.1	21.50
16	22.56	20.2	26.54	19.3	25.79	19.3	25.81
18	23.97	23.2	28.46	23.4	29.47	23.4	29.50
19	24.61	25	30.25	25.0	31.91	25.0	31.94
20	25.51	26.3	31.79	27.0	34.15	27.0	34.18
22	27.17	29.2	34.36				
24	28.84	31.8	36.28	temp dependence eq. 2		$\Delta_{sol}H$/(kJ mol^{-1}) = 27.4	
25	29.61			S	µg/kg	8–32°C	
26	30.63	ΔH_{sol}/(kJ mol^{-1}) = 22.1		a	0.0013		
28	34.10	for temp 8–32°C		b	–0.0097	data fitted to eq. 3	
30	34.61			c	0.8886		
				d	8.21		
direct measurement							
12	18.58			ΔH_{sol}/(kJ mol^{-1}) = 26.36			
18	24.35						
25	30.0						

3.

Bennet & Canady 1984		Shaw 1989		Perez-Tejeda et al. 1990	
shake flask-UV		recommended values		shake flask-UV	
t/°C	S/g·m^{-3}	t/°C	S/g·m^{-3}	t/°C	S/g·m^{-3}
25	31.1	0	15	5	12.3
1.90	19.2	10	19	10	20.25
10.7	17.2	20	28	15	24.35
15.4	21.66	25	31	20	28.71
21.7	26.74	30	38	25	33.71
25.2	30.75	40	61	35	41.53
30.7	40.12	50	82	35	47.55
35.1	46.36	60	130	40	55.75
39.3	54.85	70	200		
44.9	68.97	75	260	data fitted to eq. 5	

data fitted to eq. 4
ΔH_{sol}/(kJ mol^{-1})
25°C

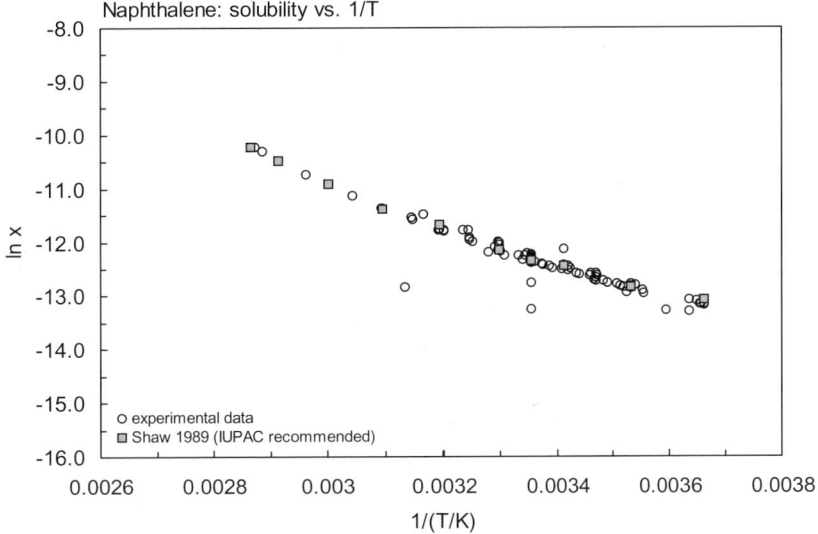

FIGURE 4.1.1.2.1 Logarithm of mole fraction solubility (ln x) versus reciprocal temperature for naphthalene.

TABLE 4.1.1.2.2
Reported vapor pressures of naphthalene at various temperatures and coefficients for vapor pressure equations

$$\log P = A - B/(T/K) \quad (1)$$
$$\log P = A - B/(C + t/°C) \quad (2)$$
$$\log P = A - B/(C + T/K) \quad (3)$$
$$\log P = A - B/(T/K) - C \cdot \log(T/K) \quad (4)$$

$$\ln P = A - B/(T/K) \quad (1a)$$
$$\ln P = A - B/(C + t/°C) \quad (2a)$$

1.

Swan & Mack 1925		Stull 1947		Sears & Hopke 1949		Bradley & Cleasby 1953	
effusion		summary of literature data		manometry		effusion	
t/°C	P/Pa	t/°C	P/Pa	t/°C	P/Pa	t/°C	P/Pa
10	2.32	52.6	133.3	10	2.053	6.7	1.627
20	8.64	74.2	666.6	20	8.533	8.1	1.88
30	23.60	85.8	1333	30	21.87	12.3	2.96
		101.7	2666			12.7	3.13
eq. 4	P/mmHg	119.3	5333	eq. 2	P/mmHg	13.85	3.506
A	595.642	130.2	7999	A	1.115	15.65	4.266
B	29.820	145.5	13332	B	108.3	16.85	4.67
C	200.682	167.7	26664	C	27	17.35	5.09
		193.2	53329			18.7	5.106
		217.9	101325			20.7	7.12
		mp/°C	80.2			eq. 1	P/mmHg
						A	10.597
						B	3783

(Continued)

TABLE 4.1.1.2.2 (Continued)

2.

Camin & Rossini 1955		Miller 1963		Fowler et al. 1968		Zwolinski & Wilhoit 1971	
manometry		Knudsen effusion		Hg manometer		selected values	
t/°C	P/Pa	t/°C	P/Pa	t/°C	P/Pa	t/°C	P/Pa
126.325	6815	–43	0.00148	40.33	46.66	85.581	1333
130.836	8338	–33	0.00746	49.45	98.66	101.788	2666
134.548	9537	–23	0.0332	54.17	148.0	111.451	4000
138.677	10948	–13	0.1188	50.22	244.0	118.688	5333
143.930	13096			61.20	253.3	124.537	6666
155.766	19436	ΔH_{subl}/kJ mol^{-1} = 72.67		65.05	338.6	129.476	7999
161.104	22878	at 25°C		70.18	500.0	137.581	10666
168.540	28550			70.98	521.3	144.146	13332
175.526	34818			75.61	725.3	156.749	19998
183.336	43159			76.7	789.3	166.262	26664
191.702	53748			80.34	931.9	173.966	33331
200.471	67072					180.561	39997
216.319	98125			eq. 2	P/mmHg	191.398	53329
217.237	99729			A	9.58102	200.237	66661
217.848	101112			B	2619.91	207.760	79993
218.638	102925			C	220.651	214.343	93326
						215.569	95992
bp/°C	217.955					216.768	98659
						217.942	101325
eq. 2	P/mmHg						
A	6.84577					eq. 2	P/mmHg
B	1606.529					A	7.01063
C	187.277					B	1733.71
						C	201.859
						bp/°C	217.942
						ΔH_V/(kJ mol^{-1}) =	
						at 25°C	72.68
						at bp	43.26

3.

Sinke 1974		Radchenko & K. 1974		Ambrose 1975		Macknick & Prausnitz 1979	
gas saturation		effusion		diaphragm gauge		gas saturation	
t/°C	P/Pa	t/°C	P/Pa	T/K	P/Pa	t/°C	P/Pa
–53.15	0.00028	9.0	2.357	263.61	0.23	7.15	1.76
–33.15	0.00837	10.7	3.113	267.98	0.40	12.8	3.133
–13.15	0.144	12.5	5.637	273.16	0.74	18.4	5.586
6.85	1.631	14.4	7.546	278.22	1.38	18.85	5.933
25	10.93	16.3	19.59	283.14	2.41	26.4	12.58
26.85	13.092	18.2	31.09	288.01	4.13	31.85	20.53
46.85	79.73	20.0	48.37	288.01	4.17	25	10.64
66.85	386.5	21.95	74.19	293.24	6.93		
80.28	999.8	23.91	112.6	293.25	6.95	eq. 1	P/mmHg
		25	12.26*	298.26	11.35	A	26.25
eq. 1	P/Pa			303.29	18.45	B	8575
A	13.83	ΔH_{subl}/kJ mol^{-1} = 72.72		308.17	28.95		
B	3817	at 25°C		313.24	44.73		

TABLE 4.1.1.2.2 (Continued)

Sinke 1974		Radchenko & K. 1974		Ambrose 1975		Macknick & Prausnitz 1979	
gas saturation		effusion		diaphragm gauge		gas saturation	
t/°C	P/Pa	t/°C	P/Pa	T/K	P/Pa	t/°C	P/Pa
				318.21	68.82		
		eq. 1	P/mmHg	323.14	104.14		
		A	11.7041	328.24	158.41		
		B	3798.574	333.39	237.54		
		*calcd using eq. 1		333.34	238.73		
				333.34	238.47		
				338.10	340.58		
				343.06	488.58		

$\Delta H_{subl}/kJ\ mol^{-1} = 72.5$ at 25°C

4.

de Kruif 1980		de Kruif et al. 1981					
torsion-effusion		diaphragm gauge					
T/K	P/Pa	T/K	P/Pa	T/K	P/Pa	T/K	P/Pa
		solid		cont'd		liquid	
257.21	0.1	274.44	0.87	340.92	421.8	355.23	1069
262.44	0.2	276.40	1.10	348.82	717.5	356.83	1155
265.59	0.3	279.12	1.50	350.44	790.2	359.78	1328
267.88	0.4	280.62	1.78	352.44	914.5	362.77	1526
269.68	0.5	282.73	2.24	344.72	538.0	355.38	1083
271.17	0.6	285.37	3.07	345.76	586.6	372.05	2305
272.44	0.7	285.71	3.11	328.79	159.2	377.33	2899
273.56	0.8	287.98	3.98	324.57	118.5	385.20	4006
274.55	0.9	288.11	4.02	345.92	580.4	381.92	3504
275.44	1.0	290.46	5.16	324.87	121.9	372.45	2351
		290.88	5.35	347.53	650.0	368.07	1929
$\Delta H_{subl}/kJ\ mol^{-1} =$		292.75	6.53	347.57	653.9	363.74	1591
298.15 K	72.6	320.34	83.91	344.72	547.1	353.57	982
253–273 K	74.4	319.87	80.76	344.86	549.1	357.10	1176
		275.50	0.99	352.00	901.3	361.10	1423
		282.12	2.10	319.58	78.38	367.72	1926
		277.86	1.30	319.59	77.9	380.51	3355
		325.41	126.7	325.96	132.1	388.20	4545
		288.51	4.22	341.26	419.9	348.80	3998
		293.83	7.29	349.02	724.1	374.67	2626
		297.40	10.42	346.18	597.6	368.46	1999
		306.67	25.27	352.90	941.6	387.32	4398
		310.71	36.37	338.91	359.7	357.44	1200
		302.92	17.84	346.85	629.9	362.57	1528
		318.58	72.36	350.69	820.2	353.51	995
		282.13	2.10	349.15	751.1	353.33	985
		277.85	1.30	341.72	441.7		
		288.50	4.21	348.6	717.3		

(Continued)

TABLE 4.1.1.2.2 (Continued)

de Kruif 1980		de Kruif et al. 1981					
torsion-effusion		diaphragm gauge					
T/K	P/Pa	T/K	P/Pa	T/K	P/Pa	T/K	P/Pa
		313.96	46.87	349.24	700.9		
		313.88	48.94	345.96	599.0		
		311.49	88.77	353.14	969.3		
		317.68	66.27				
		321.31	89.40	ΔH_{subl}/kJ mol^{-1} =			
		324.05	112.0	298.15	72.5		
		339.29	361.8				

5.

Colomina et al. 1982		Grayson & F. 1982		Sonnefeld et al. 1983		Van Ekeren et al. 1983	
effusion		gas saturation		gas saturation-HPLC		spinning-rotor gauge	
T/K	P/Pa	t/°C	P/Pa	t/°C	P/Pa	T/K	P/Pa
271.46	0.611	28.9	16.38	14.15	3.13	244.19	0.0163
275.08	0.94	42.7	58.33	14.15	3.24	245.31	0.0191
277.10	1.19	50	100.99	14.15	3.27	245.31	0.0192
279.93	1.63	60.2	264.32	19.49	6.0	246.59	0.0233
281.16	1.99	69.6	519.1	19.49	6.19	247.74	0.0268
282.69	2.23	79	970.5	25.05	10.1	247.74	0.0272
283.80	2.54	20	6.75	15.05	10.8	249.18	0.0373
284.63	2.80			25.05	10.4	250.45	0.04
		eq. 1a	P/mmHg	32.1	20.9	252.77	0.0554
ΔH_{subl}/kJ mol^{-1} = 72.8		A	31.80	32.05	21.0	255.86	0.0834
at 25°C		B	8753	32.1	20.8		
						ΔH_{subl}/kJ mol^{-1} =	
eq. 1	P/Pa			eq. 1	P/Pa	248.51	72.92
A	14.01			A	14.299	298.15	72.92
B	3861.80			B	3960.03		

6.

Sato et al. 1986		Sasse et al. 1988				Wania et al. 1994	
gas saturation-electrobalance		gas saturation				gas saturation-GC	
t/°C	P/Pa	t/°C	P/Pa	t/°C	P/Pa	t/°C	P/Pa
		solid		liquid			
24.85	10.7	−12.15	0.1653	80.16	968.85	−30.6	0.014
28.35	14.9	−12.1	0.1653	84.90	1240.4	−25.0	0.0217
30.95	19.2	−9.47	0.2266	84.93	1270.6	−20.1	0.0586
34.15	25.9	0.52	0.8199	87.16	1382.6	−14.9	0.108
41.65	52.8	10.17	2.413	84.23	1395.9	−10.0	0.2122
45.35	71.1	20.11	7.093	90.15	1590.5	−5.0	0.348
48.95	97.5	20.16	7.026			0	0.7634
49.45	101	29.98	18.265	For liquid			
51.85	124	30.03	18.80	eq. 2	P/mmHg	ΔH_{subl}/kJ mol^{-1} = 73.7	
55.35	165	40.04	44.93	A	2.25180	(−30 to 0°C)	
57.75	200	40.04	45.46	B	76.496		
		50.04	106.8	C	−25.09	eq. 1	P/Pa

TABLE 4.1.1.2.2 (Continued)

Sato et al. 1986		Sasse et al. 1988				Wania et al. 1994	
gas saturation-electrobalance		gas saturation				gas saturation-GC	
t/°C	P/Pa	t/°C	P/Pa	t/°C	P/Pa	t/°C	P/Pa
eq. 3	P/Pa	50.04	105.2			A	13.93
A	22.892960.05	60.05	238.2			B	3851
B	4025.3560.12	60.12	232.6				
C	−102.34370.11	70.11	491.0				
		70.16	491.02				
		For solid					
		eq. 2	P/mmHg				
		A	10.05263				
		B	2907.918				
		C	236.459				

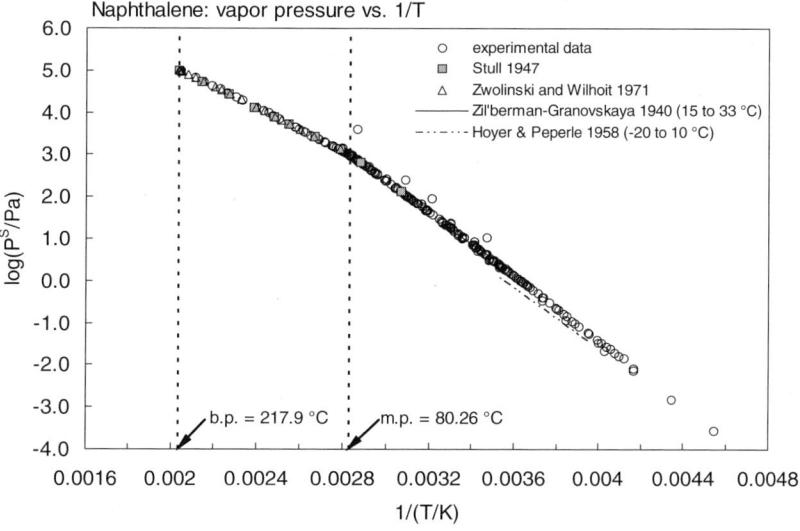

FIGURE 4.1.1.2.2 Logarithm of vapor pressure versus reciprocal temperature for naphthalene.

TABLE 4.1.1.2.3
Reported Henry's law constants of naphthalene at various temperatures

Alaee et al. 1996	
gas stripping-GC	
t/°C	H/(Pa m^3/mol)
3.7	9.65
9.4	15.4
15.3	21.4
15.5	23.1
20	33.2
25	42.6
25.4	45.2
30.2	58.6

(Continued)

TABLE 4.1.1.2.3 (Continued)

Alaee et al. 1996

gas stripping-GC

t/°C	H/(Pa m³/mol)
35.5	79.1

enthalpy of volatilization:
$\Delta H_{vol}/(kJ \cdot mol^{-1}) = 44.6$ at 20°C

$\ln K_{AW} = A - B/(T/K)$

eq. 1	K_{AW}
A	−5364.4
B	13.95

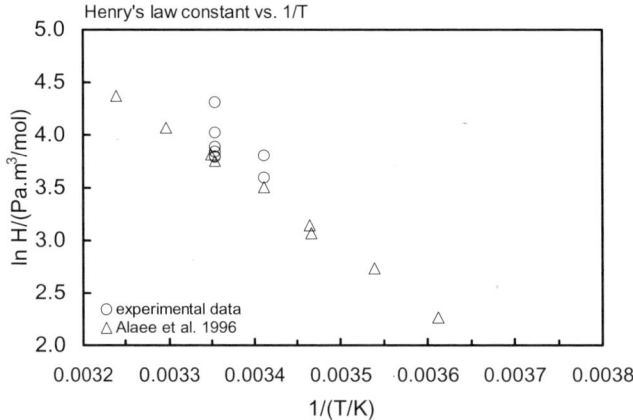

FIGURE 4.1.1.2.3 Logarithm of Henry's law constant versus reciprocal temperature for naphthalene.

4.1.1.3 1-Methylnaphthalene

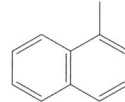

Common Name: 1-Methylnaphthalene
Synonym: α-methylnaphthalene
Chemical Name: 1-methylnaphthalene
CAS Registry No: 90-12-0
Molecular Formula: $C_{11}H_{10}$
Molecular Weight: 142.197
Melting Point (°C):
    –30.43   (Lide 2003)
Boiling Point (°C):
    244.7   (Lide 2003)
Density (g/cm³ at 20°C):
    1.02015, 1.01649   (20°C, 25°C, Dreisbach 1955)
    1.0125   (Dean 1985)
    1.0202   (Lide 2003)
Molar Volume (cm³/mol):
    139.4   (20°C, calculated-density)
    169.8   (calculated-Le Bas method at normal boiling point)
Enthalpy of Vaporization, ΔH_V (kJ/mol):
    60.06, 45.48   (25°, bp, Dreisbach 1955)
Enthalpy of Fusion, ΔH_{fus} (kJ/mol):
    4.853   (Dean 1985)
    4.98, 6.95; 11.92   (–32.45, 34.25°C, total phase change enthalpy, Chickos et al. 1999)
Entropy of Fusion, ΔS_{fus} (J/mol K):
    20.69, 28.82   (–32.45, 34.25°C, Chickos et al. 1999)
    49.3, 44.9   (exptl., calculated-group additivity method, total phase change entropy, Chickos et al. 1999)
Fugacity Ratio at 25°C, F: 1.0

Water Solubility (g/m³ or mg/L at 25°C or as indicated and reported temperature dependence equations. Additional data at other temperatures designated * are compiled at the end of this section):
    25.8   (shake flask-GC, Eganhouse & Calder 1976)
    28.5   (shake flask-fluorescence, Mackay & Shiu 1977)
    29.9*   (shake flask-fluorescence, measured range 8.6–31.7°C, Schwarz & Wasik 1977)
    29.5*   (shake flask-fluorescence, measured range 10–25°C, Schwarz 1977)
    31.7   (generator column-HPLC, Wasik et al. 1983)
    27.02   (average lit. value, Pearlman et al. 1984)
    30.2*   (20°C, shake flask-GC, measured range 20–70°C, Burris & MacIntyre 1986)
    28.0   (recommended, IUPAC Solubility Data Series, Shaw 1989)

Vapor Pressure (Pa at 25°C or as indicated and reported temperature dependence equations. Additional data at other temperatures designated * are compiled at the end of this section):
    7.165*   (manometer, extrapolated-Antoine eq., Camin & Rossini 1955)
    log (P/mmHg) = 7.03592 – 1826.948/(195.002 + t/°C); temp range 142.1–254.4°C (Antoine eq., Hg manometer, Camin & Rossini 1955)
    8.95   (calculated by formula, Dreisbach 1955)
    log (P/mmHg) = 7.06899 – 1852.674/(197.716 + t/°C); temp range 130–305°C (Antoine eq. for liquid state, Dreisbach 1955)
    66.66*   (52.8°C, Hg manometer, measured range 52.8–243.0°C, Myers & Fenske 1955)
    7.19*   (interpolated-Antoine eq., Zwolinski & Wilhoit 1971)

log (P/mmHg) = 7.03592 – 1826.948/(195.002 + t/°C); temp range 107.68–278.32°C (liquid, Antoine eq., Zwolinski & Wilhoit 1971)

8.82*   (gas saturation-GC, Macknick & Prausnitz 1979)
8.84   (extrapolated-Clapeyron eq., Macknick & Prausnitz 1979)

log (P/mmHg) = 20.552 – 6933.2/(T/K); temp range 5.70–38.6°C (Clapeyron eq., gas saturation, Macknick & Prausnitz 1979)

7221*   (151.15°C, differential pressure gauge, measured range 151.15–271.70°C, Wieczorek & Kobayashi 1981)

log (P/atm) = [1 – 517.727/(T/K)] × 10^{0.863323 – 5.26355 × 10^{-4}·(T/K) + 3.85750 × 10^{-7}·(T/K)^2}; temp range 278.85–593.38 K (Cox eq., Chao et al. 1983)

8.84   (interpolated Antoine eq., Boublik et al. 1984)

log (P/kPa) = 6.15971 – 1825.586/(194.848 + t/°C); temp range: 142.1–245.3°C (Antoine eq. from reported exptl. data, Boublik et al. 1984)

7.43, 6.38 (GC-RT correlation with BP-1 column, Apolane column, Bidleman 1984)
7.816   (extrapolated Antoine eq., Dean 1985, 1992)

log (P/mmHg) = 7.03592 – 1826.946/(195.0 + t/°C); temp range 108–278°C (Antoine eq., Dean 1985, 1992)

0.895   (interpolated-Antoine eq., Stephenson & Malanowski 1987)

log (P_L/kPa) = 7.03469 – 3006.467/(T/K); temp range 278–313 K (Antoine eq., Stephenson & Malanowski 1987)

log (P_L/kPa) = 6.15928 – 1826.402/(–72.779 + T/K); temp range 415–526 K (Antoine eq., Stephenson & Malanowski 1987)

8.93*   (pressure gauge, interpolated-Antoine eq. derived from exptl. data, temp range –14.44 to + 115°C, Sasse et al. 1988)

log (P_L/mmHg) = 7.27126 – 2006.862/(212.625 + t/°C); temp range –14.44 to115.1°C, (Antoine eq., pressure gauge, Sasse et al. 1998)

7.43   (P_{GC} by GC-RT correlation with eicosane as reference standard, Hinckley et la. 1990)

log (P/mmHg) = 29.8895 – 3.9535 × 10^3/(T/K) – 7.2253·log (T/K) + 1.1109 × 10^{-11}·(T/K) + 8.9552 × 10^{-7}·(T/K)^2; temp range 243–722 K (vapor pressure eq., Yaws 1994)

8.94, 8.93; 9.50, 8.12, 19.7; 7.42, 5.93 (quoted exptl. values; calculated; GC-RT correlation, Delle Site 1997)

log (P/kPa) = 6.39609 – 2006.662/[(T/K) – 60.525]; temp range 5–50°C (regression eq. from literature data, Shiu & Ma 2000)

6.55; 1.28 (supercooled liquid P_L: calibrated GC-RT correlation, GC-RT correlation, Lei et al. 2002)

log (P_L/Pa) = –3258/(T/K) + 11.74; $\Delta H_{vap.}$ = –62.4 kJ·mol^{-1} (GC-RT correlation, Lei et al. 2002)

8.85*   (24.05°C, transpiration method, measured range 294.1–324.2 K, Verevkin 2003)

ln (P/P°) = 298.831/R – 83537.555/R·(T/K) – (78.6/R)·ln[(T/K)/298.15]; where P° = 101.325 kPa, gas constant R = 8.31451 J·K^{-1}·mol^{-1} (vapor pressure eq. from transpiration measurements, temp range 294.1–324.2 K, Verevkin 2003)

Henry's Law Constant (Pa m^3/mol at 25°C or as indicated and reported temperature dependence equations. Additional data at other temperatures designated * are compiled at the end of this section):

26.3   (gas stripping, Mackay et al. 1979,1982)
26.3   (gas stripping-GC, Mackay et al. 1982)
62.0   (wetted-wall column-GC, Fendinger & Glotfelty 1990)
36.5   (calculated-vapor-liquid equilibrium (VLE) data, Yaws et al. 1991)
24.3   (gas stripping-GC, Shiu & Mackay 1997)
52.1   (gas stripping-GC, Altschuh et al. 1999)
47.8*   (gas stripping-GC; measured range 4.1–31°C, Bamford et al. 1999)

ln K_{AW} = –5821.5/(T/K) + 15.636; ΔH = 48.4 kJ mol^{-1}; measured range 4.1–31°C (gas stripping-GC, Bamford et al. 1999)

Octanol/Water Partition Coefficient, log K_{OW}:

3.87   (shake flask-fluorometry, Krishnamurthy & Wasik 1978)
3.87   (Hansch & Leo 1979)
3.87   (recommended, Sangster 1989, 1994)
3.87   (recommended, Hansch et al. 1995)

Octanol/Air Partition Coefficient, log K_{OA}:

Bioconcentration Factor, log BCF:

Sorption Partition Coefficient, log K_{OC}:
- 3.33, 3.06 (Lula aquifer 0.032% OC, Apalachee soil 1.4% OC, batch equilibrium-sorption isotherm, Stauffer & MacIntyre 1986)
- 3.36 (sediment from Tamar estuary, batch equilibrium-GC, Vowles & Mantoura 1987)
- 2.96–3.83 (aquifer materials, Stauffer et al. 1989)

Environmental Rate Constants, k, or Half-Lives; $t_{1/2}$:
- Volatilization:
- Photolysis: calculated $t_{1/2}$ = 22 h for direct sunlight photolysis of 50% conversion at 40°N latitude of midday in midsummer in near surface water, $t_{1/2}$ = 180 d in 5-m deep inland water and $t_{1/2}$ = 190 d in inland water with a suspended sediment concentration of 20 mg/L partitioning (Zepp & Schlotzhauer 1979);
    - $t_{1/2}$ = 180 d under summer sunlight in surface water (Mill & Mabey 1985);
    - direct photolysis $t_{1/2}$ = 11.14 h (predicted- QSPR) in atmospheric aerosol (Chen et al. 2001).
- Oxidation: rate constant k, for gas-phase second-order rate constants, k_{OH} for reaction with OH radical, k_{NO_3} with NO_3 radical and k_{O_3} with O_3, or as indicated *data at other temperatures see original reference:
    - k_{OH} = (5.30 ± 0.48) × 10^{-11} cm^3 $molecule^{-1}$ s^{-1}, k_{O_3} < 1.3 × 10^{-19} cm^3 $molecule^{-1}$ s^{-1}, $k_{N_2O_5}$ = (3.3 ± 0.7) × 10^{-17} cm^3 $molecule^{-1}$ s^{-1} for reaction with N_2O_5 at 298 ± 2 K (relative rate method, Atkinson & Aschmann 1987, 1988)
    - k_{OH} = 5.30 × 10^{-11} $cm^3 \cdot molecule^{-1}$ s^{-1} at 298 K (Atkinson 1989, 1990)
    - k_{OH}(calc) = 59.77 × 10^{-12} $cm^3 \cdot molecule^{-1}$ s^{-1} (molecular orbital calculations, Klamt 1996)
    - k_{OH} = 5.30 × 10^{-11} $cm^3 \cdot molecule^{-1}$ s^{-1}, (4.09 ± 0.20) × 10^{-11} $cm^3 \cdot molecule^{-1}$ s^{-1} at (298 ± 2)K with a calculated tropospheric lifetime to be 2.9 h using a global tropospheric 12-h daytime average OH radical concentration of 2.0 × 10^6 $molecule \cdot cm^{-3}$ (relative rate method, Phousongphouang & Arey 2002)
- Hydrolysis:
- Biodegradation: k = 0.415 d^{-1} with $t_{1/2}$ = 1.7 d for Kidman sandy loam soil and $t_{1/2}$ = 0.321 d^{-1} with $t_{1/2}$ = 2.2 d for McLaurin sandy loam soil, all at –3.3 bar soil moisture (Park et al. 1990).
- Biotransformation:
- Bioconcentration, Uptake (k_1) and Elimination(k_2) Rate Constants:

Half-Lives in the Environment:
- Air: atmospheric lifetimes of alkyl-substituted naphthalenes due to reaction with OH radical and with N_2O_5 calculated to range from ~4 to 13 h and 20–80 d, respectively (Atkinson & Aschmann 1987);
    - direct photolysis $t_{1/2}$ = 11.14 h (predicted- QSPR) in atmospheric aerosol (Chen et al. 2001); a calculated tropospheric lifetime to be 2.9 h using a global tropospheric 12-h daytime average OH radical concentration of 2.0 × 10^6 $molecule \cdot cm^{-3}$ for the reaction with OH radical (Phousongphouang & Arey 2002).
- Water: computed near-surface $t_{1/2}$ = 22 h and for direct photochemical transformation at latitude 40°N, midday, midsummer, and $t_{1/2}$ = 80 d with no sediment-water partitioning and $t_{1/2}$ = 190 d with sediment-water partitioning for direct photolysis in 5 m deep inland water body integrated over full summer d at latitude 40°N (Zeep & Schlotzhauer 1979);
    - $t_{1/2}$ = 180 d under summer sunlight (Mill & Mabey 1985).
- Soil: biodegradation k = 0.415 d^{-1} with $t_{1/2}$ = 1.7 d for Kidman sandy loam soil, k = 0.321 d^{-1} with $t_{1/2}$ = 2.2 d for McLaurin sandy loam soil (Park et al. 1990).
- Sediment:
- Biota: elimination $t_{1/2}$ = 2 d from Oyster for naphthalenes (quoted, Meador et al. 1995).

TABLE 4.1.1.3.1
Reported aqueous solubilities of 1-methylnaphthalene at various temperatures

Schwarz 1977		Schwarz & Wasik 1977		Burris & MacIntyre 1986	
shake flask-fluorescence		shake flask-fluorescence		shake flask-GC	
t/°C	S/g·m^{-3}	t/°C	S/g·m^{-3}	t/°C	S/g·m^{-3}
8.6	19.91	10	22.75	20	30.2
14.5	22.61	14	28.44	70	87.9
17.1	22.89	20	28.44		
20.0	25.31	25	29.86		
23.0	27.59				
25.0	30.0				
26.1	30.43				
29.2	33.28				
31.7	36.26				
ΔH_{sol}/(kJ mol^{-1}) = 19.1					

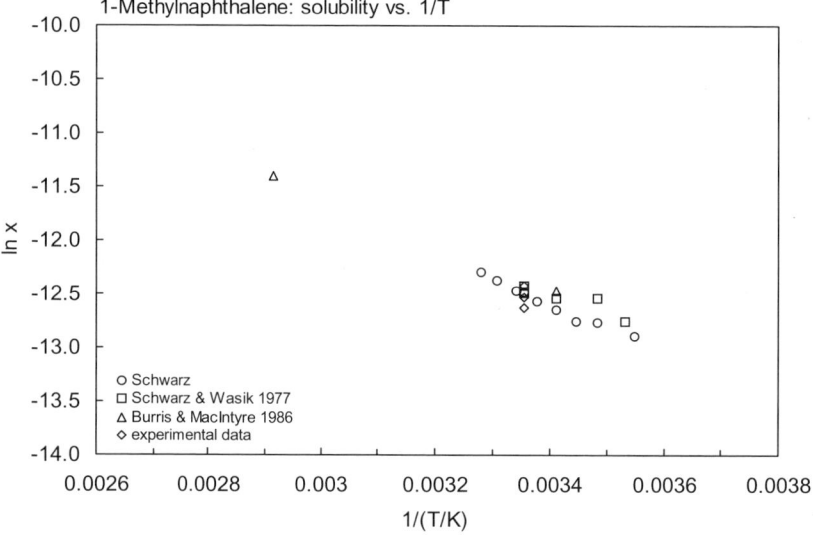

FIGURE 4.1.1.3.1 Logarithm of mole fraction solubility (ln x) versus reciprocal temperature for 1-methylnaphthalene.

TABLE 4.1.1.3.2
Reported vapor pressures of 1-methylnaphthalene at various temperatures and the coefficients for the vapor pressure equations

$$\log P = A - B/(T/K) \quad (1)$$
$$\log P = A - B/(C + t/°C) \quad (2)$$
$$\log P = A - B/(C + T/K) \quad (3)$$
$$\log P = A - B/(T/K) - C \cdot \log (T/K) \quad (4)$$

$$\ln P = A - B/(T/K) \quad (1a)$$
$$\ln P = A - B/(C + t/°C) \quad (2a)$$
$$\ln P/P^o = A - B/(T/K) - C \cdot \ln [(T/K)/298.15] \quad 4(a)$$

1.

Camin & Rossini 1955		Myers & Fenske 1955		Zwolinski & Wilhoit 1971		Macknick & Prausnitz 1979	
ebulliometry		Hg manometer		selected values		gas saturation	
t/°C	P/Pa	t/°C	P/Pa	t/°C	P/Pa	t/°C	P/Pa
142.140	5524	52.8	66.66	107.68	1333	5.70	1.760
153.600	8314	63.5	133.3	123.57	2666	11.40	2.920
157.539	9510	75.8	266.6	133.66	4000	18.10	5.248
161.689	10942	83.0	400.0	141.21	5333	22.15	7.133
167.212	13088	88.9	533.3	147.319	6666	28.85	12.59
179.971	19442	93.3	666.6	152.474	7999	32.35	15.73
185.505	22881	97.1	799.9	160.932	10666	34.90	18.93
193.280	28542	103.8	1067	167.781	13332	38.60	23.46
200.536	34832	108.8	1333	180.927	19998		
208.677	43162	117.6	2000	190.846	26664	eq. 1a	P/mmHg
217.375	53762	124.4	2666	198.908	33331	A	20.552
226.498	67057	134.4	4000	205.750	39997	B	6933.2
246.243	84026	142.2	5333	217.043	53329		
243.177	98081	148.3	6666	226.250	66661		
243.949	99655	153.7	7999	234.084	79993		
244.555	101030	161.9	10666	240.938	93326		
245.326	102757	168.5	13332	242.215	95992		
		182.0	19998	243.463	98659		
eq. 2	P/mmHg	191.6	26664	244.078	99992		
A	7.03592	206.0	39997	244.685	101325		
B	1826.948	217.0	53329				
C	195.002	225.5	66661	eq. 2	P/mmHg		
		233.3	79993	A	7.03592		
$\Delta H_V/(kJ\ mol^{-1}) = 63.82$		240.0	93326	B	1826.948		
at 25°C		243.0	101325	C	195.002		
				bp/°C	244.685		
				$\Delta H_V/(kJ\ mol^{-1}) = 46.0$			
				at bp			

(Continued)

TABLE 4.1.1.3.2 (Continued)

2.

Wieczorek & Kobayashi 81 differential pressure gauge		Sasse et al. 1988 electronic manometer		Verevkin 2003 transpiration method	
t/°C	P/Pa	t/°C	P/Pa	t/°C	P/Pa
151.15	7221	−14.44	0.189	20.95	6.86
158.75	9929	−9.64	0.317	24.05	8.85
166.42	12812	10.14	23386	27.05	11.17
175.40	16969	20.18	6.093	30.15	14.29
182.48	21024	30.15	13.73	33.15	17.51
189.45	25738	40.04	28.40	35.15	20.29
196.03	30891	49.90	56.53	36.15	22.51
202.35	36636	59.89	107	38.05	25.25
210.18	44906	69.98	194	39.05	26.72
217.70	54250	79.94	35.42	42.15	32.68
225.93	66313	89.99	579	45.15	41.63
233.82	79716	100.06	950	47.15	53.96
240.26	92116	110.04	1507	51.05	67.87
243.27	98620	115.10	1888		
244.60	101629			eq. 4a	P/kPa
254.66	126232	eq. 2	P/mmHg	P°	101.325 kPa
263.69	151814	A	7.27126	A	298.831/R
271.70	177915	B	2006.862	B	83537.555/R
		C	212.625	C	78.6/R
data fitted to Chebyshev polynomial temp range 424–593 K		ΔH_V/(kJ mol^{-1}) = 60.58 at 25°C		$R = 8.314$ J K^{-1} mol^{-1} ΔH_V/(kJ mol^{-1}) = 60.7 at 25°C	

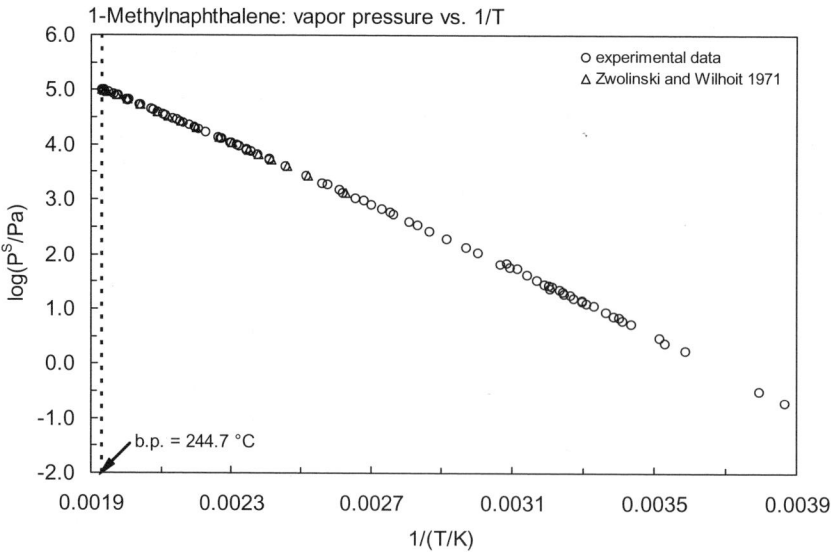

FIGURE 4.1.1.3.2 Logarithm of vapor pressure versus reciprocal temperature for 1-methylnaphthalene.

TABLE 4.1.1.3.3
Reported Henry's law constants of 1-methylnaphthalene at various temperatures and temperature dependence equations

	Bamford et al. 1999	
	gas stripping-GC/MS	
t/°C	H/(Pa m³/mol)	H/(Pa m³/mol)
		average
4.1	9.41, 10.7	10.1
11.0	16.6, 18.2	17.4
18.0	27.8, 30.8	29.2
25.0	44.8, 51.0	47.8
31.0	65.7, 78.2	71.7

$\ln K_{AW} = A - B/(T/K)$
A   15.636
B   5821.5

enthalpy, entropy change:
$\Delta H/(kJ \cdot mol^{-1}) = 48.4 \pm 1.4$
$\Delta S/(J \cdot K^{-1} mol^{-1}) = 130$
at 25°C

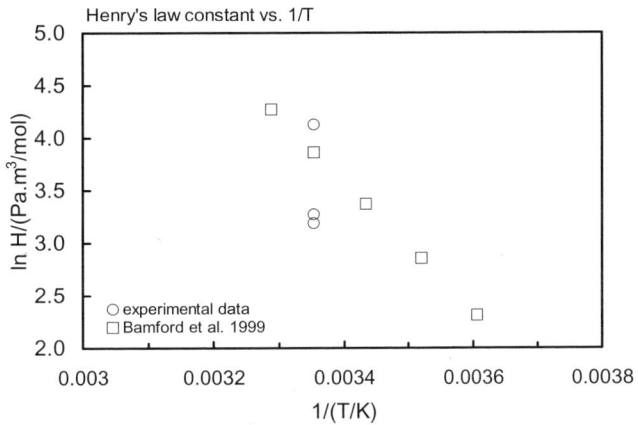

FIGURE 4.1.1.3.3 Logarithm of Henry's law constant versus reciprocal temperature for 1-methylnaphthalene.

4.1.1.4 2-Methylnaphthalene

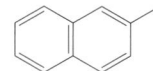

Common Name: 2-Methylnaphthalene
Synonym: β-methylnaphthalene
Chemical Name: 2-methylnaphthalene
CAS Registry No: 91-57-6
Molecular Formula: $C_{11}H_{10}$
Molecular Weight: 142.197
Melting Point (°C):
    34.6   (Weast 1972–73; Lide 2003)
Boiling Point (°C):
    241.1   (Weast 1977; Lide 2003)
Density (g/cm³ at 20°C):
    1.0058   (Weast 1982–83; Lide 2003)
Molar Volume (cm³/mol):
    141.4   (20°C, calculated-density)
    169.8   (calculated-Le Bas method at normal boiling point)
Enthalpy of Vaporization, ΔH_V (kJ/mol):
    59.40, 45.20   (25°, bp, Dreisbach 1955)
Enthalpy of Fusion, ΔH_{fus} (kJ/mol):
    11.924   (Parks & Huffman 1931)
    5.61, 12.13; 17.74   (15.35, 34.25°C; total phase change enthalpy, Chickos et al. 1999)
Entropy of Fusion, ΔS_{fus} (J/mol K):
    39.25   (Tsonopoulos & Prausnitz 1971)
    19.43, 39.43   (15.35, 34.25°C, Chickos et al. 1999)
    58.87, 44.9   (exptl., calculated-group additivity method, total phase change entropy, Chickos et al. 1999)
Fugacity Ratio at 25°C (assuming ΔS_{fus} = 56 J/mol K), F: 0.805 (mp at 34.6°C)

Water Solubility (g/m³ or mg/L at 25°C):
    24.6   (shake flask-GC, Eganhouse & Calder 1976)
    25.4   (shake flask-fluorescence, Mackay & Shiu 1977)
    20.0   (Vozñáková et al. 1978)
    25.6   (average lit. value, Pearlman et al. 1984)
    27.3   (generator column-HPLC, Vadas et al. 1991)

Vapor Pressure (Pa at 25°C or as indicated and reported temperature dependence equations. Additional data at other temperatures designated * are compiled at the end of this section):
    9.07*   (Hg manometer, extrapolated-Antoine eq., Camin & Rossini 1955)
    log (P/mmHg) = 7.06850 – 1840.268/(198.395 + t/°C); temp range 139.1–242°C (Antoine eq., Hg manometer, Camin & Rossini 1955)
    10.67   (calculated by formula, Dreisbach 1955)
    log (P/mmHg) = 7.06850 – 1840.268/(198.395 + t/°C), temp range 130–300°C (Antoine eq. for liquid state, Dreisbach 1955)
    9.07*   (extrapolated-Antoine eq., Zwolinski & Wilhoit 1971)
    log (P/mmHg) = 7.06850 – 1840.268/(198.395 + t/°C); temp range 104.85–274.3°C (Antoine eq., Zwolinski & Wilhoit 1971)
    9.07   (extrapolated-Antoine eq., Boublik et al. 1973)
    5.60   (20°C, Vozñáková et al. 1978)
    8629*   (151.26°C, differential pressure gauge, measured range 151.26–279.81°C, Wieczorek & Kobayashi 1981)
    log (P/atm) = [1– 514.242/(T/K)] × 10^{0.879050 – 5.85793 × 10^{-4}·(T/K) + 4.19235 × 10^{-7}·(T/K)^2}; temp range 378.0–629.32 K (Cox eq., Chao et al. 1983)
    9.08   (extrapolated-Antoine eq., Boublik et al. 1984)

log (P/kPa) = 6.19638 − 1842.831/(198.692 + t/°C); temp range 139.2–241.76°C (Antoine eq. from reported exptl. data, Boublik et al. 1984)
6.31 (GC-RT correlation, supercooled liquid, Bidleman 1984)
9.033 (extrapolated, Antoine eq., Dean 1985, 1992)
log (P/mmHg) = 7.0685 − 1840.268/(198.4 + t/°C); temp range 105–274°C (Antoine eq., Dean 1985, 1992)
9.33 (extrapolated from Antoine eq., Stephenson & Malanowski 1987)
log (P_L/kPa) = 6.21475 − 1858.19/(−72.479 + T/K); temp range 423–515 K (Antoine eq., Stephenson & Malanowski 1987)
log (P/mmHg) = 56.2052 − 5.2563 × 10^3/(T/K) − 16.195·log (T/K) + 8.1583 × 10^{-11}·(T/K) + 3.0253 × 10^{-6}·(T/K)^2; temp range 308–761 K (vapor pressure eq., Yaws 1994)

Henry's Law Constant (Pa m^3/mol at 25°C or as indicated and reported temperature dependence equations. Additional data at other temperatures designated * are compiled at the end of this section):
32.23 (wetted-wall column-GC, Fendinger & Glotfelty 1990)
50.6 (calculated-vapor-liquid equilibrium (VLE) data, Yaws et al. 1991)
20.265* (26°C EPICS-GC, Hansen et al. 1993)
46.0 (gas stripping-HPLC/fluo., De Maagd et al. 1998)
62.0 (gas stripping-GC, Altschuh et al. 1999)
51.3* (gas stripping-GC; Bamford et al. 1999)
ln K_{AW} = −5099.83/(T/K) + 13.23; ΔH = 42.4 kJ mol^{-1}; measured range 4.1–31°C (gas stripping-GC, Bamford et al. 1999)
18620 (20°C, selected value based on Hansen et al. 1993 data, Staudinger & Roberts 1996, 2001)
log K_{AW} = 2.245 − 399/(T/K) (van't Hoff eq. derived from literature data, Staudinger & Roberts 2001)

Octanol/Water Partition Coefficient, log K_{OW}:
3.864 (shake flask-fluorometry, Krishnamurthy & Wasik 1978)
3.86 (Hansch & Leo 1979)
4.11 (shake flask-UV, concn. ratio, Karickhoff et al. 1979)
3.70 (HPLC-k' correlation, Hanai et al. 1981)
4.01 (HPLC-RT correlation, Eadsforth 1986)
4.09 (HPLC-RT correlation, Wang et al. 1986)
4.00 (recommended, Sangster 1989)
3.86 (recommended, Hansch et al. 1995)

Bioconcentration Factor, log BCF:
2.61 (quoted from Davies & Dobbs 1984, Sabljic 1987)
2.65 (calculated-MCI χ, Sabljic 1987)

Sorption Partition Coefficient, log K_{OC}:
3.93 (natural sediments, average sorption isotherms by batch equilibrium technique-UV spec., Karickhoff et al. 1979)
3.40 (sediment 4.02% from Tamar estuary, batch equilibrium-GC, Vowles & Mantoura 1987)
3.719; 3.72, 3.71 (sediment: concn ratio C_{sed}/C_W; concn-based coeff., areal-based coeff. of flux studies of sediment/water boundary layer, Helmstetter & Alden 1994)

Environmental Fate Rate Constants, k, or Half-Lives, $t_{1/2}$:
Volatilization:
Photolysis: $t_{1/2}$ (calc) = 54 h for direct sunlight photolysis at 40°N latitude of midday in midsummer in near surface water, $t_{1/2}$ = 410 d in inland water and $t_{1/2}$ = 440 d in inland water with sediment partitioning (Zepp & Schlotzhauer 1979);
$t_{1/2}$ = 410 d under summer sunlight in surface water (Mill & Mabey 1985);
k = 0.042 h^{-1} with $t_{1/2}$ = 6.4 h in distilled water (Fukuda et al. 1988);
direct photolysis $t_{1/2}$ = 9.23 h predicted - QSPR, in atmospheric aerosol (Chen et al. 2001).
Hydrolysis:

Oxidation: rate constant, k for gas-phase second order rate constants, k_{OH} for reaction with OH radical, k_{NO_3} with NO_3 radical and k_{O_3} with O_3 or as indicated, *data at other temperatures see reference:

$k_{OH} = (5.23 \pm 0.42) \times 10^{-11}$ cm^3 molecule^{-1} s^{-1} with a calculated atmospheric lifetime of ~5 h, and $k_{O_3} < 4 \times 10^{-19}$ cm^3 molecule^{-1} s^{-1} at 295 ± 1 K (relative rate, Atkinson & Aschmann 1986)

$k_{N_2O_5} = (4.2 \pm 0.9) \times 10^{-17}$ cm^3 molecule^{-1} s^{-1} for reaction with N_2O_5 at 298 ± 2 K (relative rate method, Atkinson & Aschmann 1987)

$k_{OH} = 5.23 \times 10^{-12}$ cm^3 molecule^{-1} s^{-1}; $k_{O_3} < 0.4 \times 10^{-19}$ cm^3 molecule^{-1} s^{-1}, $k_{N_2O_5} = 4.2 \times 10^{-17}$ cm^3 molecule^{-1} s^{-1} with N_2O_5 at room temp. (Atkinson & Aschmann 1987, 1988)

$k_{OH} = 52.3 \times 10^{-12}$ cm^3 molecule^{-1} s^{-1} at 298 K (Atkinson 1989, 1990)

k_{OH}(calc) = 57.36×10^{-12} cm^3 molecule^{-1} s^{-1} (molecular orbital calculations, Klamt 1996)

$k_{OH} = (4.86 \pm 0.25) \times 10^{-11}$ cm^3 molecule^{-1} s^{-1} at 298 ± 2 K with a calculated tropospheric lifetime to be 3.4 h using a global tropospheric 12-h daytime average OH radical concn of 2.0×10^6 molecule/cm^3 (relative rate method, Phousongphouang & Arey 2002)

Biodegradation:

Biotransformation:

Bioconcentration, Uptake (k_1) and Elimination (k_2) Rate Constants:

Half-Lives in the Environment:

Air: calculated atmospheric lifetime of ~4 h due to reaction with OH radical (Atkinson & Aschmann 1986) atmospheric lifetimes of naphthalene and alkyl-substituted naphthalenes due to reaction with OH radicals and with N_2O_5 can be calculated to range from ~4 to 13 h and 20–80 d, respectively (Atkinson & Aschmann 1987); direct photolysis $t_{1/2}$ = 9.23 h (predicted- QSPR) in atmospheric aerosol (Chen et al. 2001); a calculated tropospheric lifetime to be 3.4 h using a global tropospheric 12-h daytime average OH radical concn of 2.0×10^6 molecule/cm^3 for the reaction with OH radical (Phousong-phouang & Arey 2002).

Surface water: computed near-surface $t_{1/2}$ = 54 h and for direct photochemical transformation at latitude 40°N, midday, midsummer; $t_{1/2}$ = 410 d with no sediment-water partitioning and $t_{1/2}$ = 440 d with sediment-water partitioning by direct photolysis in a 5-m deep Inland Water Body (Zepp & Schlotzhauer 1979); $t_{1/2}$ = 410 d under summer sunlight (Mill & Mabey 1985);

rate constants and half-lives: k = 0.064 d^{-1}, $t_{1/2}$ = 11 d in spring at 8–16°C, k = 0.687 d^{-1}, $t_{1/2}$ = 1.0 d in summer at 20–22°C, k = 0.054 d^{-1}, $t_{1/2}$ = 13 d in winter at 3–7°C for the periods when volatilization appears to dominate, and k = 0.046 d^{-1}, $t_{1/2}$ = 15.1 d with HgCl$_2$ as poison, and k = 0.954 d^{-1}, $t_{1/2}$ = 0.7 d without poison in September 9–15, in marine mesocosm experiments (Wakeham et al. 1983)

Groundwater:

Sediment:

Soil:

Biota: elimination $t_{1/2}$ = 2 d from Oyster for naphthalenes (quoted, Meador et al. 1995).

TABLE 4.1.1.4.1
Reported vapor pressures of 2-methylnaphthalene at various temperatures and the coefficients for the vapor pressure equations

$\log P = A - B/(T/K)$ (1) $\quad\quad \ln P = A - B/(T/K)$ (1a)

$\log P = A - B/(C + t/°C)$ (2) $\quad\quad \ln P = A - B/(C + t/°C)$ (2a)

$\log P = A - B/(C + T/K)$ (3)

$\log P = A - B/(T/K) - C \cdot \log (T/K)$ (4)

Camin & Rossini 1955		Zwolinski & Wilhoit 1971		Wieczorek & Kobayashi 1981	
ebulliometry		selected values		differential pressure gauge	
t/°C	P/Pa	t/°C	P/Pa	t/°C	P/Pa
139.193	5536	104.85	1333	151.26	8629
145.431	6951	120.68	2666	158.72	10992
150.655	8339	130.73	4000	164.34	13183

TABLE 4.1.1.4.1 (Continued)

Camin & Rossini 1955 ebulliometry		Zwolinski & Wilhoit 1971 selected values		Wieczorek & Kobayashi 1981 differential pressure gauge	
t/°C	P/Pa	t/°C	P/Pa	t/°C	P/Pa
154.676	9539	138.25	5333	169.89	15615
158.689	10942	144.329	6666	175.93	18756
164.155	13100	149.459	7999	181.84	22550
176.722	19433	157.872	10666	189.18	27850
182.322	22882	164.486	13332	196.20	33812
190.033	28551	177.752	19998	202.04	39538
197.234	34820	187.610	26664	209.49	48007
205.329	43163	195.619	33331	217.00	57974
213.963	53782	202.414	39997	225.34	70794
223.026	67089	213.626	53329	232.47	83506
239.613	98125	222.764	66661	239.90	96992
240.336	99734	230.537	79993	246.90	114865
240.957	101114	237.336	93326	254.69	134992
241.760	102929	238.602	95992	263.24	160539
		239.840	98659	271.29	188158
eq. 2	P/mmHg	240.449	99992	279.81	221176
A	7.06850	241.052	101325		
B	1840.268			exptl data fitted to	
C	198.395	eq. 2	P/mmHg	Chebyshev polynomial	
bp/°C	241.052	A	7.06850	temp range 424–639 K	
		B	1840.268		
		C	198.395		
		bp/°C	241.052		
		ΔH_V/(kJ mol^{-1}) = 46.0 at bp			

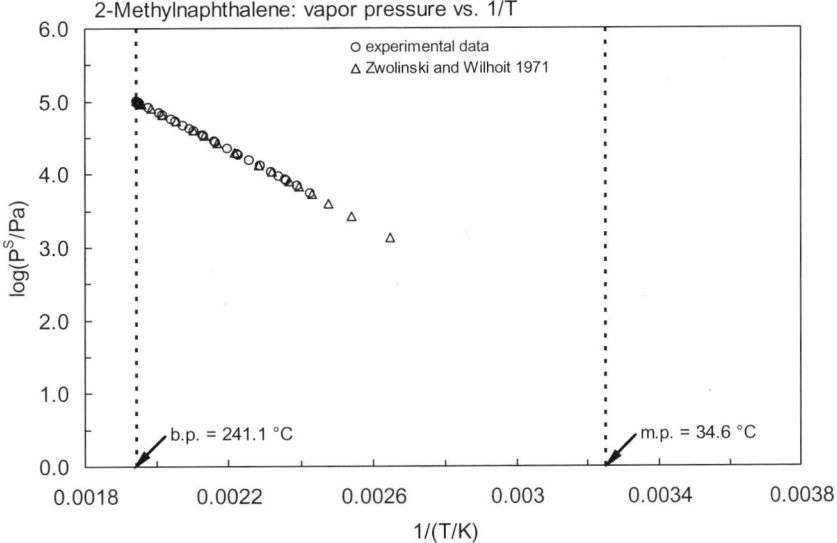

FIGURE 4.1.1.4.1 Logarithm of vapor pressure versus reciprocal temperature for 2-methylnaphthalene.

TABLE 4.1.1.4.2
Reported Henry's law constants of 2-methylnaphthalene at various temperatures and temperature dependence equations

$\ln K_{AW} = A - B/(T/K)$ (1) $\log K_{AW} = A - B/(T/K)$ (1a)

$\ln (1/K_{AW}) = A - B/(T/K)$ (2) $\log (1/K_{AW}) = A - B/(T/K)$ (2a)

$\ln (k_H/\text{atm}) = A - B/(T/K)$ (3)

$\ln [H/(\text{Pa m}^3/\text{mol})] = A - B/(T/K)$ (4) $\ln [H/(\text{atm·m}^3/\text{mol})] = A - B/(T/K)$ (4a)

$K_{AW} = A - B\cdot(T/K) + C\cdot(T/K)^2$ (5)

Hansen et al. 1993		Bamford et al. 1999		
EPICS-GC		gas stripping-GC/MS		
t/°C	H/(Pa m³/mol)	t/°C	H/(Pa m³/mol)	H/(Pa m³/mol)
				average
26.0	20.265	4.1	10.5, 16.2	13.0
35.8	22.900	11.0	18.2, 24.4	21.0
46.0	26.243	18.0	28.8, 38.3	33.2
		25.0	42.1, 62.5	51.3
eq. 4	H/(Pa m³/mol)	31.0	56.2, 95.5	73.3
A	7.0 ± 0.14			
B	1234 ± 44	eq. 1	K_{AW}	
		A	13.2307	
		B	5100	
		enthalpy, entropy change:		
		$\Delta H/(\text{kJ·mol}^{-1}) = 42.4 \pm 4.2$		
		$\Delta S/(\text{J·K}^{-1}\text{ mol}^{-1}) = 110$		
		at 25°C		

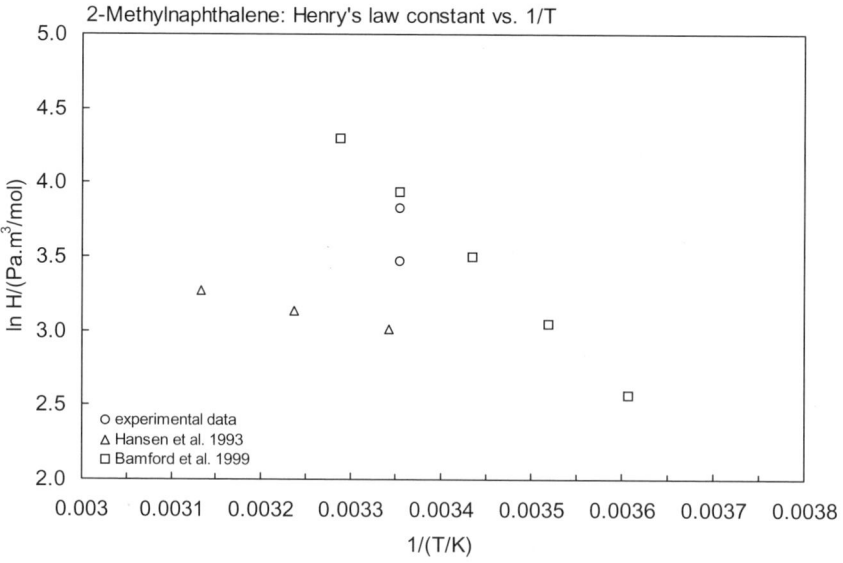

FIGURE 4.1.1.4.2 Logarithm of Henry's law constant versus reciprocal temperature for 2-methylnaphthalene.

4.1.1.5 1,3-Dimethylnaphthalene

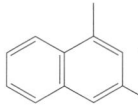

Common Name: 1,3-Dimethylnaphthalene
Synonym:
Chemical Name: 1,3-dimethylnaphthalene
CAS Registry No: 575-41-7
Molecular Formula: $C_{12}H_{12}$
Molecular Weight: 156.223
Melting Point (°C):
    –6.0   (Weast 1982–83, Dean 1985; Lide 2003)
Boiling Point (°C):
    263   (Dreisbach 1955; Weast 1982–83; Dean 1985; Lide 2003)
Density (g/cm³ at 20°C):
    1.0063, 1.0026   (20°C, 25°C, Dreisbach 1955)
    1.0144   (Weast 1982–83; Dean 1985; Lide 2003)
Molar Volume (cm³/mol):
    154.0   (20°C, calculated-density)
    192.0   (calculated-Le Bas method at normal boiling point)
Enthalpy of Vaporization, ΔH_V (kJ/mol):
    69.58, 48.69   (25°C, bp, Dreisbach 1955)
Enthalpy of Fusion, ΔH_{fus} (kJ/mol):
Entropy of Fusion, ΔS_{fus} (J/mol K):
Fugacity Ratio at 25°C, F: 1.0

Water Solubility (g/m³ or mg/L at 25°C):
    8.00   (shake flask-fluorescence, Mackay & Shiu 1977)
    7.81   (average lit. value, Pearlman et al. 1984)

Vapor Pressure (Pa at 25°C and reported temperature dependence equations):
    1.947   (calculated by formula, Dreisbach 1955)
    $\log (P/\text{mmHg}) = 7.0469 - 1845.6/(180.0 + t/°C)$; temp range 150–313°C (Antoine eq. for liquid state, Dreisbach 1955)
    $\log (P/\text{atm}) = [1- 540.353/(T/K)] \times 10^\wedge\{1.72680 - 7.87991 \times 10^{-4}\cdot(T/K) + 42.8535 \times 10^{-7}\cdot(T/K)^2\}$; temp range 400.0–541.0 K (Cox eq., Chao et al. 1983)
    6.950   (interpolated-Antoine eq., Dean 1985, 1992)
    $\log (P/\text{mmHg}) = 7.6347 - 2295.4/(232.4 + t/°C)$; temp range 20–148°C (Antoine eq., Dean 1985, 1992)
    $\log (P/\text{mmHg}) = 7.2698 - 2076.0/(210 + t/°C)$; temp range 148–310°C (Antoine eq., Dean 1985, 1992)

Henry's Law Constant (Pa m³/mol at 25°C):
    71.03   (calculated-vapor-liquid equilibrium (VLE) data, Yaws et al. 1991)

Octanol/Water Partition Coefficient, $\log K_{OW}$:
    4.421   (shake flask-fluorometry, Krishnamurthy & Wasik 1978)
    4.42   (Hansch & Leo 1979)
    4.42   (recommended, Sangster 1993)
    4.42   (recommended, Hansch et al. 1995)

Octanol/Air Partition Coefficient, log K_{OA}:

Bioconcentration Factor, log BCF:

Sorption Partition Coefficient, log K_{OC}:

Environmental Fate Rate Constants, k, or Half-Lives, $t_{½}$:
- Volatilization:
- Photolysis:
- Oxidation: rate constant k, for gas-phase second order rate constants, k_{OH} for reaction with OH radical, k_{NO_3} with NO_3 radical and k_{O_3} with O_3, or as indicated *data at other temperatures see original reference:
  - k_{OH} = (7.49 ± 0.39) × 10^{-11} cm^3 $molecule^{-1}$ s^{-1} at (298 ± 2) K with a calculated tropospheric lifetime ranging from 1.9 to 2.4 h using a global tropospheric 12-h daytime average OH radical concentration of 2.0 × 10^6 molecule cm^{-3} (relative rate method, Phousongphouang & Arey 2002)
- Hydrolysis:
- Biodegradation:
- Biotransformation:
- Bioconcentration, Uptake (k_1) and Elimination (k_2) Rate Constants:

Half-Lives in the Environment:
- Air: calculated tropospheric lifetime ranging from 1.9 to 2.4 h for dimethylnaphthalenes using a global tropospheric 12-h daytime average OH radical concentration of 2.0 × 10^6 molecule cm^{-3} for the reaction with OH radical (Phousongphouang & Arey 2002).
- Surface water:
- Groundwater:
- Sediment:
- Soil:
- Biota: elimination $t_{½}$ = 2 d from Oyster for naphthalenes (quoted, Meador et al. 1995).

4.1.1.6 1,4-Dimethylnaphthalene

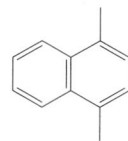

Common Name: 1,4-Dimethylnaphthalene
Synonym:
Chemical Name: 1,4-dimethylnaphthalene
CAS Registry No: 571-58-4
Molecular Formula: $C_{12}H_{12}$
Molecular Weight: 156.223
Melting Point (°C):
    7.60   (Lide 2003)
Boiling Point (°C):
    268   (Dreisbach 1955, Weast 1982–83; Lide 2003)
Density (g/cm³ at 20°C):
    1.0166, 1.0129   (20°C, 25°C, Dreisbach 1955)
    1.0166   (Weast 1982–83)
Molar Volume (cm³/mol):
    153.7   (20°C, calculated-density)
    192.0   (calculated-Le Bas method at normal boiling point)
Enthalpy of Vaporization, ΔH_V (kJ/mol):
    70.315, 48.62   (25°C, bp, Dreisbach 1955)
Enthalpy of Fusion, ΔH_{fus} (kJ/mol):
    10.6   (Chickos et al. 1999)
Entropy of Fusion, ΔS_{fus} (J/mol K):
    37.87, 45.5 (exptl., calculated-group additivity method, Chickos et al. 1999)
Fugacity Ratio at 25°C, F: 1.0

Water Solubility (g/m³ or mg/L at 25°C):
    11.4   (shake flask-fluorescence, Mackay & Shiu 1977)
    11.40   (average lit. value, Pearlman et al. 1984)

Vapor Pressure (Pa at 25°C and reported temperature dependence equations):
    1.55   (calculated by formula, Dreisbach 1955)
    $\log (P/\text{mmHg}) = 7.0527 - 1869.0/(180.0 + t/°C)$; temp range 155–310°C (Antoine eq. for liquid state, Dreisbach 1955)
    $\log (P/\text{atm}) = [1 - 544.363/(T/K)] \times 10\wedge\{1.57594 - 8.55425 \times 10^{-4} \cdot (T/K) + 51.4189 \times 10^{-7} \cdot (T/K)^2\}$; temp range 397.0–544 K (Cox eq., Chao et al. 1983)
    2.27   (calculated-TSA, Amidon & Anik 1981; selected, Ma et al. 1990)
    4.50   (interpolated, Antoine eq., Dean 1985, 1992)
    $\log (P/\text{mmHg}) = 7.6347 - 2345.8/(232.6 + t/°C)$; temp range 20–148°C (Antoine eq., Dean 1985, 1992)
    $\log (P/\text{mmHg}) = 7.2698 - 2076.0/(210 + t/°C)$; temp range 148–310°C (Antoine eq., Dean 1985, 1992)

Henry's Law Constant (Pa m³/mol at 25°C):
    48.84   (calculated-vapor-liquid equilibrium (VLE) data, Yaws et al. 1991)

Octanol/Water Partition Coefficient, log K_{OW}:
    4.372   (shake flask-fluorometry, Krishnamurthy & Wasik 1978)
    4.37   (Hansch & Leo 1979)
    4.37   (recommended, Sangster 1989, 1993)
    4.37   (recommended, Hansch et al. 1995)

Octanol/Air Partition Coefficient, log K_{OA}:

Bioconcentration Factor, log BCF:

Sorption Partition Coefficient, log K_{OC}:

Environmental Fate Rate Constants, k or Half-Lives, $t_{1/2}$:
- Volatilization:
- Photolysis:
- Oxidation: rate constant k, for gas-phase second order rate constants, k_{OH} for reaction with OH radical, k_{NO_3} with NO_3 radical and k_{O_3} with O_3, or as indicated *data at other temperatures see original reference:
    - $k_{OH} = (5.79 \pm 0.36) \times 10^{-11}$ cm^3 molecule^{-1} s^{-1} at (298 ± 2)K with a calculated tropospheric lifetime ranging from 1.9 to 2.4 h using a global tropospheric 12-h daytime average OH radical concentration of 2.0×10^6 molecule cm^{-3} (relative rate method, Phousongphouang & Arey 2002)
- Hydrolysis:
- Biodegradation:
- Biotransformation:
- Bioconcentration, Uptake (k_1) and Elimination (k_2) Rate Constants:

Half-Lives in the Environment:
- Air: calculated tropospheric lifetime ranging from 1.9 to 2.4 h for dimethylnaphthalenes using a global tropospheric 12-h daytime average OH radical concentration of 2.0×10^6 molecule cm^{-3} for the reaction with OH radical (Phousongphouang & Arey 2002).
- Surface water:
- Groundwater:
- Sediment:
- Soil:
- Biota: elimination $t_{1/2}$ = 2 d from Oyster for naphthalenes (quoted, Meador et al. 1995).

4.1.1.7 1,5-Dimethylnaphthalene

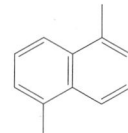

Common Name: 1,5-Dimethylnaphthalene
Synonym:
Chemical Name: 1,5-dimethylnaphthalene
CAS Registry No: 571-61-9
Molecular Formula: $C_{12}H_{12}$
Molecular Weight: 156.223
Melting Point (°C):
    82   (Dreisbach 1955; Weast 1982–83; Lide 2003)
Boiling Point (°C):
    265   (Dreisbach 1955; Weast 1982–83; Lide 2003)
Density (g/cm³ at 20°C):
Molar Volume (cm³/mol):
    192.0   (calculated-Le Bas method at normal boiling point)
Enthalpy of Evaporation, ΔH_V (kJ/mol):
Enthalpy of Fusion, ΔH_{fus} (kJ/mol):
Entropy of Fusion, ΔS_{fus} (J/mol K):
Fugacity Ratio at 25°C (assuming ΔS_{fus} = 56 J/mol K), F: 0.276 (mp at 82°C)

Water Solubility (g/m³ or mg/L at 25°C):
    2.74   (shake flask-GC, Eganhouse & Calder 1976)
    3.38   (shake flask-fluorescence, Mackay & Shiu 1977)
    3.12   (average lit. value, Pearlman et al. 1984)

Vapor Pressure (Pa at 25°C or as indicated and reported temperature dependence equations.):
    log (P/mmHg) = 7.0493 – 1855.0/(180.0 + t/°C); temp range 150–313°C (Antoine eq. for liquid state, Dreisbach 1955)
    1.93; 0.513 (supercooled liquid P_L: calibrated GC-RT correlation; GC-RT correlation, Lei et al. 2002)
    log (P_L/Pa) = –3346/(T/K) + 11.51; $\Delta H_{vap.}$ = –64.1 kJ·mol^{-1} (GC-RT correlation, Lei et al. 2002)

Henry's Law Constant (Pa m³/mol at 25°C):
    35.5   (gas stripping-fluorescence, Mackay et al. 1982)
    61.8   (calculated-vapor-liquid equilibrium (VLE) data, Yaws et al. 1991)
    36.3   (gas stripping, Shiu & Mackay 1997)

Octanol/Water Partition Coefficient, log K_{OW}:
    4.38   (shake flask-fluorometry, Krishnamurthy & Wasik 1978)
    4.38   (Hansch & Leo 1979)
    4.38   (recommended, Sangster 1989, 1993)
    4.38   (recommended, Hansch et al. 1995)

Bioconcentration Factor, log BCF:

Sorption Partition Coefficient, log K_{OC}:

Environmental Fate Rate Constants, k, or Half-Lives, $t_{1/2}$:
    Volatilization:
    Photolysis:
    Hydrolysis:

Oxidation: rate constant k, for gas-phase second order rate constants, k_{OH} for reaction with OH radical, k_{NO_3} with NO_3 radical and k_{O_3} with O_3, or as indicated *data at other temperatures see original reference:
$k_{OH} = (6.01 \pm 0.35) \times 10^{-11}$ cm^3 molecule^{-1} s^{-1} at (298 ± 2) K with a calculated tropospheric lifetime ranging from 1.9 to 2.4 h using a global tropospheric 12-h daytime average OH radical concentration of 2.0×10^6 molecule cm^{-3} (relative rate method, Phousongphouang & Arey 2002)

Biodegradation:
Biotransformation:
Bioconcentration, Uptake (k_1) and Elimination (k_2) Rate Constants:

Half-Lives in the Environment:

Air: a calculated tropospheric lifetime ranging from 1.9 to 2.4 h for dimethylnaphthalenes using a global tropospheric 12-h daytime average OH radical concentration of 2.0×10^6 molecule cm^{-3} for the reaction with OH radical (Phousongphouang & Arey 2002).
Surface water:
Groundwater:
Sediment:
Soil:
Biota: elimination $t_{1/2}$ = 2 d from Oyster for naphthalenes (quoted, Meador et al. 1995).

4.1.1.8   2,3-Dimethylnaphthalene

Common Name: 2,3-Dimethylnaphthalene
Synonym: guaiene
Chemical Name: 2,3-dimethylnaphthalene
CAS Registry No: 581-40-8
Molecular Formula: $C_{12}H_{12}$
Molecular Weight: 156.223
Melting Point (°C):
    105   (Dreisbach 1955; Weast 1972–73; Lide 2003)
Boiling Point (°C):
    268   (Dreisbach 1955; Weast 1982–83; Lide 2003)
Density (g/cm³ at 20°C):
    1.003   (Weast 1982–83; Lide 2003)
Molar Volume (cm³/mol):
    155.8   (20°C, calculated-density)
    192.0   (calculated-Le Bas method at normal boiling point)
Enthalpy of Vaporization, ΔH_V (kJ/mol):
    70.315, 48.97   (25°C, bp, Dreisbach 1955)
Enthalpy of Fusion, ΔH_{fus} (kJ/mol):
    25.10   (Ruelle & Kesselring 1997)
    15.9   (Chickos et al.1999)
Entropy of Fusion, ΔS_{fus} (J/mol K):
    42.06, 45.5   (exptl., calculated-group additivity method, Chickos et al. 1999)
Fugacity Ratio at 25°C (assuming ΔS_{fus} = 56 J/mol K), F: 0.164 (mp at 105°C)

Water Solubility (g/m³ or mg/L at 25°C):
    1.99   (shake flask-GC, Eganhouse & Calder 1976)
    3.0   (shake flask-fluorescence, Mackay & Shiu 1977)
    2.50   (average lit. value, Pearlman et al. 1984)

Vapor Pressure (Pa at 25°C and reported temperature dependence equations):
    1.55   (calculated by formula, Dreisbach 1955)
    $\log (P/\text{mmHg}) = 7.0527 - 1869.0/(180.0 + t/°C)$; temp range 155–315°C (Antoine eq. for liquid state, Dreisbach 1955)
    1.86   (extrapolated-Cox eq., Chao et al. 1983)
    $\log (P/\text{atm}) = [1 - 631.969/(T/K)] \times 10^{\{1.09999 - 10.2378 \times 10^{-4} \cdot (T/K) + 11.3931 \times 10^{-7} \cdot (T/K)^2\}}$; temp range 333.15–408.15 K(Cox eq., Chao et al. 1983)
    0.91   (extrapolated-Antoine eq., Boublik et al. 1984)
    $\log (P/\text{kPa}) = 5.27335 - 1383.083/(141.333 + t/°C)$; temp range 105–135°C (Antoine eq. from reported exptl. data, Boublik et al. 1984)
    1.543   (interpolated, Antoine eq., Dean 1985, 1992)
    $\log (P/\text{mmHg}) = 7.40396 - 2111.9/(201.1 + t/°C)$; temp range 20–155°C (Antoine eq., Dean 1985, 1992)
    $\log (P/\text{mmHg}) = 7.0527 - 1869/(180 + t/°C)$; temp range 155–315°C (Antoine eq., Dean 1985, 1992)
    0.437   (Antoine eq., Stephenson & Malanowski 1987)
    $\log (P_S/\text{kPa}) = 10.635 - 4172.6/(T/K)$; temp range 278–301 K (Antoine eq.-I, Stephenson & Malanowski 1987)
    $\log (P_S/\text{kPa}) = 8.97875 - 2959.733/(-59.936 + T/K)$; temp range 333–373 K (Antoine eq.-II, Stephenson & Malanowski 1987)
    $\log (P_L/\text{kPa}) = 5.18084 - 1544.764/(-116.821 + T/K)$; temp range 378–408 K (Antoine eq.-III, Stephenson & Malanowski 1987)

Henry's Law Constant (Pa m³/mol at 25°C):
- 38.92 (calculated-P/C, Eastcott et al. 1988)
- 92.16, 64.9 (quoted, calculated-bond contribution method, Meylan & Howard 1991)
- 59.9 (calculated-vapor-liquid equilibrium (VLE) data, Yaws et al. 1991)

Octanol/Water Partition Coefficient, log K_{OW}:
- 4.396 (shake flask-fluorometry, Krishnamurthy & Wasik 1978)
- 4.40 (Hansch & Leo 1979)
- 4.40 (recommended, Sangster 1993)
- 4.40 (recommended, Hansch et al. 1995)

Octanol/Air Partition Coefficient, log K_{OA}:
Bioconcentration Factor, log BCF:
Sorption Partition Coefficient, log K_{OC}:

Environmental Fate Rate Constants, k, or Half-Lives, $t_{½}$:
  Volatilization:
  Photolysis:
  Hydrolysis:
  Oxidation: rate constant k, for gas-phase second order rate constants, k_{OH} for reaction with OH radical, k_{NO_3} with NO_3 radical and k_{O_3} with O_3 or as indicated, *data at other temperatures see reference:
    k_{OH} = (7.68 ± 0.48) × 10⁻¹¹ cm³ molecule⁻¹ s⁻¹, with a calculated atmospheric lifetime of ~4 h, and k_{O_3} < 4 × 10⁻¹⁹ cm³ molecule⁻¹ s⁻¹ at 295 ± 1 K (relative rate method, Atkinson & Aschmann 1986)
    $k_{N_2O_5}$ = (5.7 ± 1.9) × 10⁻¹⁷ cm³ molecule⁻¹ s⁻¹ with N_2O_5 at 298 ± 2 K (relative rate method, Atkinson & Aschmann 1987)
    k_{O_3} < 0.4 × 10⁻¹⁸ cm³ molecule⁻¹ s⁻¹, k_{OH} = 7.68 × 10⁻¹¹ cm³ molecule⁻¹ s⁻¹, $k_{N_2O_5}$ = 5.7 × 10⁻¹⁷ cm³ molecule⁻¹ s⁻¹ with N_2O_5 at room temp. (Atkinson & Aschmann 1987, 1988)
    k_{OH} = (76.8 ± 4.8) × 10⁻¹² cm³ molecule⁻¹ s⁻¹ at 295 ± 1 K (Atkinson 1989)
    k_{OH} = 77 × 10⁻¹² cm³ molecule⁻¹ s⁻¹ at 298 K (Atkinson 1990)
    k_{OH} = 100.3 × 10⁻¹² cm³ molecule⁻¹ s⁻¹ (molecular orbital calculations, Klamt 1996)
    k_{OH} = (6.15 ± 0.47) × 10⁻¹¹ cm³ molecule⁻¹ s⁻¹ at (298 ± 2) K using a relative rate method with a calculated tropospheric lifetime ranging from 1.9 to 2.4 h for dimethylnaphthalenes using a global tropospheric 12-h daytime average OH radical concentration of 2.0 × 10⁶ molecule cm⁻³ (Phousongphouang & Arey 2002)
  Biodegradation:
  Biotransformation:
  Bioconcentration, Uptake (k_1) and Elimination (k_2) Rate Constants:

Half-Lives in the Environment:
  Air: calculated atmospheric lifetime of ~4 h due to reaction with OH radical (Atkinson & Aschmann 1986); the atmospheric lifetimes of naphthalene and alkyl-substituted naphthalenes due to reaction with OH radicals and with N_2O_5 can be calculated to range from ~4 to 13 h and 20–80 d, respectively (Atkinson & Aschmann 1987);
    calculated tropospheric lifetime ranging from 1.9 to 2.4 h for dimethylnaphthalenes using a global tropospheric 12-h daytime average OH radical concentration of 2.0 × 10⁶ molecule cm⁻³ for the reaction with OH radical (Phousongphouang & Arey 2002).
  Surface water:
  Groundwater:
  Sediment:
  Soil:
  Biota: elimination $t_{½}$ = 2 d from Oyster for naphthalenes (quoted, Meador et al. 1995).

4.1.1.9 2,6-Dimethylnaphthalene

Common Name: 2,6-Dimethylnaphthalene
Synonym:
Chemical Name: 2,6-dimethylnaphthalene
CAS Registry No: 581-42-0
Molecular Formula: $C_{12}H_{12}$
Molecular Weight: 156.223
Melting Point (°C):
    112   (Lide 2003)
Boiling Point (°C):
    262   (Dreisbach 1955; Dean 1985; Lide 2003)
Density (g/cm³ at 20°C):
    1.003, 0.999   (20°C, 25°C, Dreisbach 1955)
    1.142   (Dean 1985)
Molar Volume (cm³/mol):
    155.8   (calculated-density)
    192.0   (calculated-Le Bas method at normal boiling point)
Enthalpy of Vaporization, ΔH_V (kJ/mol):
    69.45, 48.70   (25°C, bp, Dreisbach 1955)
Enthalpy of Fusion, ΔH_{fus} (kJ/mol):
    24.27   (Tsonopoulos & Prausnitz 1971)
    25.06   (calorimetry, Osborn & Douslin 1975; quoted, Ruelle & Kesselring 1997; Chickos et al. 1999)
Entropy of Fusion, ΔS_{fus} (J/mol K):
    63.18   (Tsonopoulos & Prausnitz 1971)
    65.39, 45.5 (exptl., calculated-group additivity method, total phase change entropy, Chickos et al. 1999)
Fugacity Ratio at 25°C (assuming ΔS_{fus} = 56 J/mol K), F: 0.140 (mp at 112°C)

Water Solubility (g/m³ or mg/L at 25°C):
    1.30   (shake flask-GC, Eganhouse & Calder 1976)
    2.0   (shake flask-fluorescence, Mackay & Shiu 1977)
    1.72   (average lit. value, Pearlman et al. 1984)
    0.997   (generator column-HPLC, Vadas et al. 1991)

Vapor Pressure (Pa at 25°C and reported temperature dependence equations):
    20.41   (calculated by formula, Dreisbach 1955)
    $\log (P/\text{mmHg}) = 7.0460 - 1841.0/(180.0 + t/°C)$; temp range 150–310°C (Antoine eq. for liquid state, Dreisbach 1955)
    0.75   (calculated-TSA, Amidon & Anik 1981)
    1.38   (extrapolated- Cox eq., Chao et al. 1983)
    $\log (P/\text{atm}) = [1 - 687.081/(T/K)] \times 10^{\{1.14901 - 11.9220 \times 10^{-4} \cdot (T/K) + 17.3468 \times 10^{-7} \cdot (T/K)^2\}}$; temp range 328.15–418.15 K (Cox eq., Chao et al. 1983)
    1.41   (extrapolated-Antoine eq., Boublik et al. 1984)
    $\log (P/\text{kPa}) = 5.19014 - 1325.209/(139.781 + t/°C)$; temp range 111–145°C (Antoine eq. from reported exptl. data, Boublik et al. 1984)
    2.036   (interpolated-Antoine eq., Dean 1985, 1992)
    $\log (P/\text{mmHg}) = 7.3968 - 2080.3/(200.8 + t/°C)$; temp range 20–150°C (Antoine eq., Dean 1992)
    $\log (P/\text{mmHg}) = 7.0460 - 1841/(180 + t/°C)$; temp range 150–310°C (Antoine eq., Dean 1985, 1992)
    0.378   (extrapolated-Antoine eq., Stephenson & Malanowski 1987)
    $\log (P_S/\text{kPa}) = 11.290 - 3047.828/(T/K)$; temp range 333–368 K (Antoine eq.-I, Stephenson & Malanowski 1987)

log $(P_S/kPa) = 8.45107 - 2512.509/(-89.765 + T/K)$; temp range 384–418 K (Antoine eq.-II, Stephenson & Malanowski 1987)

log $(P_L/kPa) = 5.18084 - 1320.21/(-133.876 + T/K)$; temp range 384–418 K (Antoine eq.-III, Stephenson & Malanowski 1987)

log $(P/mmHg) = -6.9795 - 2.9488 \times 10^3/(T/K) + 7.4483 \cdot \log (T/K) - 1.15821 \times 10^{-2} \cdot (T/K) + 4.3391 \times 10^{-6} \cdot (T/K)^2$; temp range 383–777 K (vapor pressure eq., Yaws 1994)

Henry's Law Constant (Pa m^3/mol at 25°C):
    121   (calculated-vapor-liquid equilibrium (VLE) data, Yaws et al. 1991)

Octanol/Water Partition Coefficient, log K_{OW}:
    4.313   (shake flask-fluorometry, Krishnamurthy & Wasik 1978)
    4.31   (Hansch & Leo 1979)
    4.38   (calculated-fragment const., Yalkowsky & Valvani 1979, 1980)
    4.31   (recommended, Sangster 1989, 1993)
    4.31   (recommended, Hansch et al. 1995)

Bioconcentration Factor, log BCF:

Sorption Partition Coefficient, log K_{OC}:

Environmental Fate Rate Constants, k or Half-Lives, $t_{1/2}$:
    Volatilization:
    Photolysis: rate constant in distilled water k = 0.045 h^{-1} with $t_{1/2}$ = 15.5 h (Fukuda et al. 1988).
    Hydrolysis:
    Oxidation: rate constant k, for gas-phase second order rate constants, k_{OH} for reaction with OH radical, k_{NO_3} with NO$_3$ radical and k_{O_3} with O$_3$, or as indicated *data at other temperatures see original reference:
        $k_{OH} = (6.65 \pm 0.35) \times 10^{-11}$ cm^3 molecule^{-1} s^{-1} at (298 ± 2) K with a calculated tropospheric lifetime ranging from 1.9 to 2.4 h using a global tropospheric 12-h daytime average OH radical concentration of 2.0×10^6 molecule cm^{-3} (relative rate method, Phousongphouang & Arey 2002)
    Biodegradation:
    Biotransformation:
    Bioconcentration, Uptake (k_1) and Elimination (k_2) Rate Constants:

Half-Lives in the Environment:
    Air: a calculated tropospheric lifetime to be 1.9–2.4 h using a global tropospheric 12-h daytime average OH radical concentration of 2.0×10^6 molecule cm^{-3} for the reaction with OH radical (Phousongphouang & Arey 2002).
    Surface water:
    Groundwater:
    Sediment:
    Soil:
    Biota: elimination $t_{1/2}$ = 2 d from Oyster for naphthalenes (quoted, Meador et al. 1995)

Polynuclear Aromatic Hydrocarbons (PAHs) and Related Aromatic Hydrocarbons

4.1.1.10 1-Ethylnaphthalene

Common Name: 1-Ethylnaphthalene
Synonym: α-ethylnaphthalene
Chemical Name: 1-ethylnaphthalene
CAS Registry No: 1127-76-0
Molecular Formula: $C_{12}H_{12}$
Molecular Weight: 156.223
Melting Point (°C):
    –13.9   (Lide 2003)
Boiling Point (°C):
    258.6   (Lide 2003)
Density (g/cm³ at 20°C):
    1.00816, 1.00446   (20°C, 25°C, Dreisbach 1955)
    1.0082   (Weast 1982–83; Lide 2003)
Molar Volume (cm³/mol):
    155.0   (20°C, calculated-density)
    192.0   (calculated-Le Bas method at normal boiling point)
Enthalpy of Vaporization, ΔH_V (kJ/mol):
    67.42, 46.92   (25°C, bp, Dreisbach 1955)
Enthalpy of Fusion, ΔH_{fus} (kJ/mol):
Entropy of Fusion, ΔS_{fus} (J/mol K):
Fugacity Ratio at 25°C, F: 1.0

Water Solubility (g/m³ or mg/L at 25°C or as indicated and reported temperature dependence equations. Additional data at other temperatures designated * are compiled at the end of this section):
    10.7   (shake flask-fluorescence, Mackay & Shiu 1977)
    10.0*   (shake flask-fluorescence, Schwarz & Wasik 1977)
    10.0*   (shake flask-fluorescence, Schwarz 1977)
    11.58   (generator column-HPLC, Wasik et al. 1981)
    10.31   (average lit. value, Pearlman et al. 1984)

Vapor Pressure (Pa at 25°C and reported temperature dependence equations. Additional data at other temperatures designated *, are compiled at the end of this section):
    133.3*   (70.0°C, summary of literature data, temp range 70.0–258.1°C, Stull 1947)
    3.0   (calculated by formula, Dreisbach 1955)
    $\log (P/\text{mmHg}) = 6.9599 - 1791.4/(180.5 + t/°C)$; temp range 145–310°C (Antoine eq. for liquid state, Dreisbach 1955)
    2.51*   (extrapolated from liquid state, Antoine eq., Zwolinski & Wilhoit 1971)
    $\log (P/\text{mmHg}) = 7.03159 - 1841.320/(185.28 + t/°C)$; temp range 120–292°C (Antoine eq., Zwolinski & Wilhoit 1971)
    $\log (P/\text{atm}) = [1 - 531.480/(T/K)] \times 10^{\{0.923623 - 6.97505 \times 10^{-4} \cdot (T/K) + 5.07450 \times 10^{-7} \cdot (T/K)^2\}}$; temp range 393.15–565.45 K (Cox eq., Chao et al. 1983)
    2.51   (extrapolated-Antoine eq., Stephenson & Malanowski 1987)
    $\log (P_L/\text{kPa}) = 6.15645 - 1841.32/(-87.87 + T/K)$; temp range 393–565 K (Antoine eq., Stephenson & Malanowski 1987)
    $\log (P/\text{mmHg}) = 7.5650 - 3.7597 \times 10^3/(T/K) + 2.6035 \cdot \log (T/K) - 1.1581 \times 10^{-2} \cdot (T/K) + 5.1365 \times 10^{-6} \cdot (T/K)^2$; temp range 259–776 K (vapor pressure eq., Yaws 1994)

Henry's Law Constant (Pa m^3/mol at 25°C):
- 14.8 (calculated-P/C, Eastcott et al. 1988)
- 36.7 (calculated-vapor-liquid equilibrium (VLE) data, Yaws et al. 1991)
- 69.4 (gas stripping-GC, Altschuh et al. 1999)

Octanol/Water Partition Coefficient, log K_{OW}:
- 4.39 (calculated-fragment const., Yalkowsky & Valvani 1979, 1980)
- 4.38 (calculated-fragment const., Yalkowsky et al. 1983)
- 4.42 (calculated-solvatochromic parameters and V_I, Kamlet et al. 1988)
- 4.40 (recommended, Sangster 1989, 1994)
- 4.44 (calculated-molar volume, Wang et al. 1992)
- 4.8016 (calculated-UNIFAC group contribution, Chen et al. 1993)
- 4.39 (recommended, Hansch et al. 1995)

Octanol/Air Partition Coefficient, log K_{OA}:

Bioconcentration Factor, log BCF:

Sorption Partition Coefficient, log K_{OC}:
- 3.77 (sediment, HPLC-k' correlation, Vowles & Mantoura 1987)
- 3.89 (HPLC-capacity factor correlation, Hodson & Williams 1988)
- 3.78 (calculated-MCI $^1\chi$, Sabljic et al. 1995)

Environmental Fate Rate Constants, k, or Half-Lives, $t_{1/2}$:

Volatilization:

Photolysis:

Oxidation: rate constant k, for gas-phase second order rate constants, k_{OH} for reaction with OH radical, k_{NO_3} with NO_3 radical and k_{O_3} with O_3, or as indicated *data at other temperatures see original reference:
$k_{OH} = (3.64 \pm 0.41) \times 10^{-11}$ cm^3 molecule^{-1} s^{-1} at (298 ± 2)K with a calculated tropospheric lifetime to be 3.8 h using a global tropospheric 12-h daytime average OH radical concentration of 2.0×10^6 molecule cm^{-3} (relative rate method, Phousongphouang & Arey 2002)

Hydrolysis:

Biodegradation:

Biotransformation:

Bioconcentration, Uptake (k_1) and Elimination (k_2) Rate Constants:

Half-Lives in the Environment:

Air: a calculated tropospheric lifetime to be 3.8 h using a global tropospheric 12-h daytime average OH radical concentration of 2.0×10^6 molecule cm^{-3} for the reaction with OH radical (Phousongphouang & Arey 2002).

Surface water:

Groundwater:

Sediment:

Soil:

Biota: elimination $t_{1/2}$ = 2 d from Oyster for naphthalenes (quoted, Meador et al. 1995).

TABLE 4.1.1.10.1
Reported aqueous solubilities and vapor pressures of 1-ethylnaphthalene at various temperatures

Aqueous solubility				Vapor pressure			
Schwarz 1977		Schwarz & Wasik 1977		Stull 1947		Zwolinski & Wilhoit 1971	
shake flask-fluorescence		shake flask-fluorescence		summary of literature data		selected values	
t/°C	S/g·m^{-3}	t/°C	S/g·m^{-3}	t/°C	P/Pa	t/°C	P/Pa
8.6	8.124	10	8.1	70.0	133.3	120.0	1333
11.1	8.124	14	8.1	101.4	666.6	136.04	2666
14.0	8.28	20	10	116.8	1333	146.22	4000
17.1	8.593	25	10	133.8	2666	153.85	5333
20.0	8.436			152.0	5333	160.01	6666
23.0	8.593			164.1	7999	165.22	7999
25.0	10.0			180.0	13332	173.76	10666
26.1	9.842			204.6	26664	180.67	13332
31.7	11.72			230.8	53329	193.94	19998
				258.1	101325	203.96	26664
ΔH_{sol}/(kJ mol^{-1}) = 14.27						212.10	33331
				mp/°C	−27	219.01	39997
						230.41	53329
						239.71	66661
						247.62	79993
						254.54	93326
						255.83	95992
						257.09	98659
						258.94	101325
						log P = A − B/(C + t/°C)	
							P/mmHg
						A	7.03159
						B	1841.320
						C	185.28
						bp/°C	258.33
						ΔH_V/(kJ mol^{-1}) =	
						at 25°C	–
						at bp	48.1

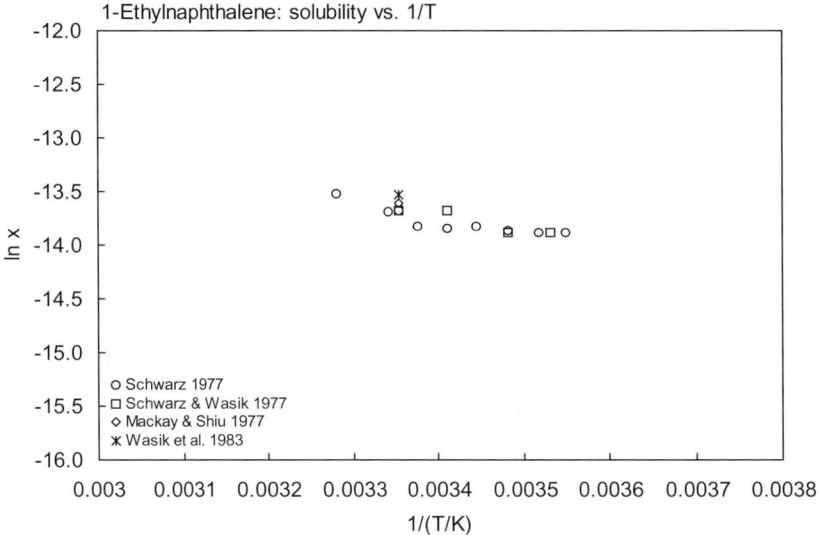

FIGURE 4.1.1.10.1 Logarithm of mole fraction solubility (ln x) versus reciprocal temperature for 1-ethylnaphthalene.

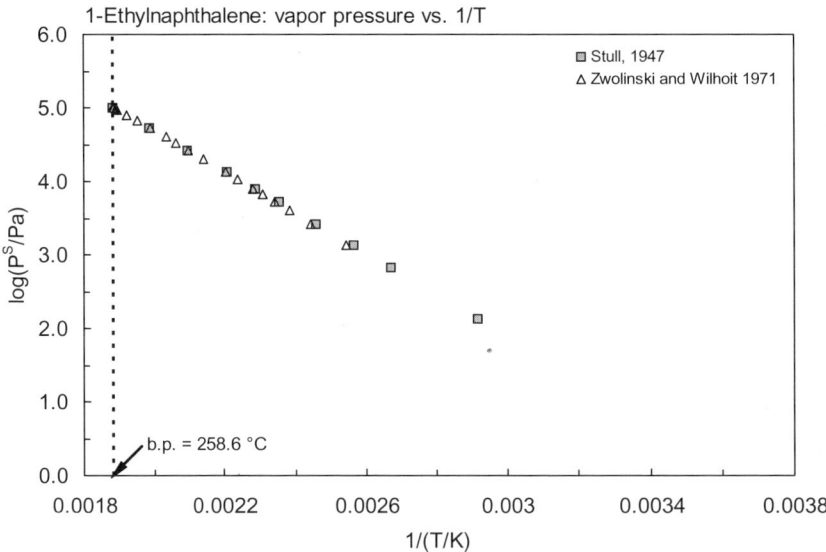

FIGURE 4.1.1.10.2 Logarithm of vapor pressure versus reciprocal temperature for 1-ethylnaphthalene.

4.1.1.11 2-Ethylnaphthalene

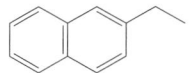

Common Name: 2-Ethylnaphthalene
Synonym: β-ethylnaphthalene
Chemical Name: 2-ethylnaphthalene
CAS Registry No: 939-27-5
Molecular Formula: $C_{12}H_{12}$
Molecular Weight: 156.223
Melting Point (°C):
    –7.4   (Lide 2003)
Boiling Point (°C):
    258   (Lide 2003)
Density (g/cm³ at 20°C):
    0.9922   (Weast 1982–83; Lide 2003)
Molar Volume (cm³/mol):
    157.4   (20°C, calculated from density)
    192.0   (calculated-Le Bas method at normal boiling point)
Enthalpy of Vaporization, ΔH_V (kJ/mol):
    66.99, 47.33   (25°, bp, Dreisbach 1955)
    64.7   (Lei et al. 2002)
Enthalpy of Fusion, ΔH_{fus} (kJ/mol):
Entropy of Fusion, ΔS_{fus} (J/mol K):
Fugacity Ratio at 25°C), F: 1.0

Water Solubility (g/m³ or mg/L at 25°C):
    7.97   (shake flask-GC, Eganhouse & Calder 1976)
    7.97   (average lit. value, Pearlman et al. 1984)

Vapor Pressure (Pa at 25°C or as indicated and reported temperature dependence equations. Additional data at other temperatures designated * are compiled at the end of this section):
    3.76   (calculated by formula, Dreisbach 1955)
    log (P/mmHg) = 8.0819 – 1886.0/(191.0 + t/°C); temp range 145–300°C (Antoine eq. for liquid state, Dreisbach 1955)
    3.24*   (extrapolated from liq. state, Antoine eq., Zwolinski & Wilhoit 1971; quoted, Mackay & Shiu 1981)
    log (P/mmHg) = 7.07566 – 1880.73/(191.41 + t/°C); temp range 119.14–291.9°C (Antoine eq., Zwolinski & Wilhoit 1971)
    4.21*   (extrapolated exptl. data, Macknick & Prausnitz 1979; quoted, Mackay & Shiu 1981)
    log (P/mmHg) = 21.485 – 7435.9/(T/K); temp range 13.05–45.1°C (Clapeyron eq., gas saturation, Macknick & Prausnitz 1979)
    log (P/atm) = [1–531.189/(T/K)] × 10^{0.871612 – 5.23140 × 10⁻⁴·(T/K) + 3.70623 × 10⁻⁷·(T/K)²}; temp range 286.2–565.05 K (Cox eq., Chao et al. 1983)
    4.21   (interpolated-Antoine eq.-I, Stephenson & Malanowski 1987)
    log (P_L/kPa) = 7.46683 – 3232.791/(T/K); temp range 286–319 K (Antoine eq.-I, Stephenson & Malanowski 1987)
    log (P_L/kPa) = 6.20056 – 1880.73/(–82.74 + T/K); temp range 393–565 K (Antoine eq.-II, Stephenson & Malanowski 1987)
    3.71*   (pressure gauge in vacuum cell, interpolated-Antoine eq. derived from exptl. data, temp range, –4.65–125°C, Sasse et al. 1988)
    log (P_L/mmHg) = 6.83511 – 1799.779/(189.505 + t/°C); temp range –4.65 to 125.09°C (Antoine eq., pressure gauge, Sasse et al. 1988)
    log (P/kPa) = 7.46683 – 3232.79/(T/K); temp range 5–50°C (regression eq. from literature data, Shiu & Ma 2000)
    2.56; 0.633   (supercooled liquid P_L: calibrated GC-RT correlation; GC-RT correlation, Lei et al. 2002)
    log (P_L/Pa) = –3381/(T/K) + 11.75; $\Delta H_{vap.}$ = –64.7 kJ·mol⁻¹ (GC-RT correlation, Lei et al. 2002)

Henry's Law Constant (Pa m³/mol at 25°C):
- 82.2 (calculated-P/C, Mackay & Shiu 1981)
- 63.2 (calculated-vapor-liquid equilibrium (VLE) data, Yaws et al. 1991)
- 54.5 (gas stripping-GC, Altschuh et al. 1999)

Octanol/Water Partition Coefficient, log K_{OW}:
- 4.377 (shake flask-fluorometry, Krishnamurthy & Wasik 1978)
- 4.43 (HPLC-k′ correlation, Vowles & Mantoura 1987)
- 4.38 (recommended, Sangster 1989, 1994)
- 4.38 (recommended, Hansch et al. 1995)
- 4.00 (HPLC-k′ correlation, Ritter et al. 1995)

Octanol/Air Partition Coefficient, log K_{OA}:

Bioconcentration Factor, log BCF:

Sorption Partition Coefficient, log K_{OC}:
- 3.76 (sediment 4.02% OC from Tamar estuary, batch equilibrium-GC, Vowles & Mantoura 1987)
- 3.76 (calculated-MCI $^1\chi$, Sabljic et al. 1995)

Environmental Fate Rate Constants, k, or Half-Lives, $t_{1/2}$:

Volatilization:

Photolysis: k = 0.038 h⁻¹ in distilled water with $t_{1/2}$ = 18.4 h (Fukuda et al. 1988).

Hydrolysis:

Oxidation: rate constant k, for gas-phase second order rate constants, k_{OH} for reaction with OH radical, k_{NO_3} with NO_3 radical and k_{O_3} with O_3, or as indicated *data at other temperatures see original reference:
k_{OH} = (4.02 ± 0.55) × 10⁻¹¹ cm³ molecule⁻¹ s⁻¹ at (298 ± 2) K with a calculated tropospheric lifetime to be 3.5 h using a global tropospheric 12-h daytime average OH radical concentration of 2.0 × 10⁶ molecule cm⁻³ (relative rate method, Phousongphouang & Arey 2002)

Biodegradation:

Biotransformation:

Bioconcentration, Uptake (k_1) and Elimination (k_2) Rate Constants:

Half-Lives in the Environment:

Air:

Surface water: $t_{1/2}$ = 18.4 h in distilled water (Fukuda et al. 1988).

Groundwater:

Sediment:

Soil:

Biota: elimination $t_{1/2}$ = 2 d from Oyster for naphthalenes (quoted, Meador et al. 1995).

TABLE 4.1.1.11.1
Reported vapor pressures of 2-ethylnaphthalene at various temperatures and the coefficients for the vapor pressure equations

$$\log P = A - B/(T/K) \quad (1)$$
$$\log P = A - B/(C + t/°C) \quad (2)$$
$$\log P = A - B/(C + T/K) \quad (3)$$
$$\log P = A - B/(T/K) - C \cdot \log(T/K) \quad (4)$$
$$\ln P = A - B/(T/K) \quad (1a)$$
$$\ln P = A - B/(C + t/°C) \quad (2a)$$

Zwolinski & Wilhoit 1971		Macknick & Prausnitz 1979		Sasse et al. 1988	
selected values		gas saturation		electronic manometer	
t/°C	P/Pa	t/°C	P/Pa	t/°C	P/Pa
119.14	1333	13.05	1.533	−4.65	0.164
135.28	2666	17.90	2.213	0.26	0.30

TABLE 4.1.1.11.1 (Continued)

Zwolinski & Wilhoit 1971		Macknick & Prausnitz 1979		Sasse et al. 1988	
selected values		gas saturation		electronic manometer	
t/°C	P/Pa	t/°C	P/Pa	t/°C	P/Pa
145.52	4000	22.90	3.40	10.17	0.880
153.19	5333	29.50	6.213	20.08	2.426
159.38	6666	34.85	9.799	20.16	2.40
164.61	7999	39.40	13.16	30.11	6.05
173.19	10666	45.10	20.0	40.01	13.33
180.13	13332			49.93	27.33
193.45	19998	eq. 1a	P/mmHg	59.92	54.26
203.49	26664	A	21.485	69.96	103.6
211.65	33331	B	7435.9	79.98	187.9
218.57	39997			90.01	326.8
230.00	53329			100.08	551.8
239.31	66661			110.08	898.1
247.22	79993			120.02	1420
254.15	93326			125.09	1771
255.44	94659				
255.83	95992			eq. 2	P/mmHg
256.70	98659			A	6.83541
257.32	99992			B	1799.779
257.93	101325			C	189.505
eq. 2	P/mmHg				
A	7.07566				
B	1880.73				
C	191.41				
bp/°C	257.93				
$\Delta H_V/(kJ\ mol^{-1}) = 48.1$ at bp					

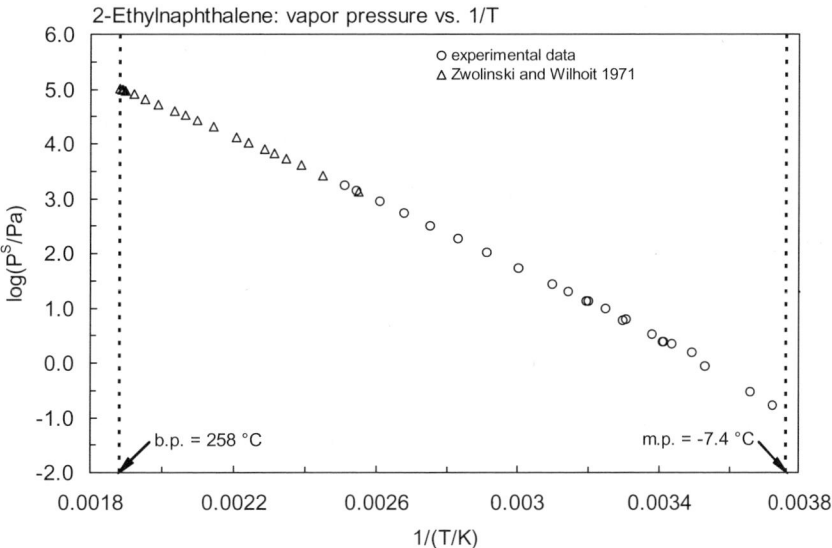

FIGURE 4.1.1.11.1 Logarithm of vapor pressure versus reciprocal temperature for 2-ethylnaphthalene.

4.1.1.12 1,4,5-Trimethylnaphthalene

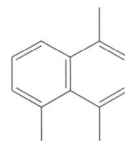

Common Name: 1,4,5-Trimethylnaphthalene
Synonym:
Chemical Name: 1,4,5-trimethylnaphthalene
CAS Registry No: 2131-41-1
Molecular Formula: $C_{13}H_{14}$
Molecular Weight: 170.250
Melting Point (°C):
    63   (Lide 2003)
Boiling Point (°C):
    285   (Zwolinski & Wilhoit 1971)
Density (g/cm³ at 20°C):
Molar Volume (cm³/mol):
    169.0   (calculated-density, liquid molar volume, Lande & Banerjee 1981)
    214.2   (calculated-Le Bas method at normal boiling point)
Enthalpy of Fusion, ΔH_{fus} (kJ/mol):
Entropy of Fusion, ΔS_{fus} (J/mol K):
Fugacity Ratio at 25°C (assuming ΔS_{fus} = 56 J/mol K), F: 0.424 (mp at 63°C)

Water Solubility (g/m³ or mg/L at 25°C):
    2.10   (shake flask-fluorescence, Mackay & Shiu 1977)
    2.04   (average lit. value, Pearlman et al. 1984)

Vapor Pressure (Pa at 25°C or as indicated and reported temperature dependence equations):
    0.681   (supercooled liquid P_L, Chao et al. 1983)
    $\log (P_L/atm) = [1-558.187/(T/K)] \times 10^{\wedge}\{0.998467 - 7.03095 \times 10^{-4} \cdot (T/K) + 1.71801 \times 10^{-7} \cdot (T/K)^2\}$; temp range 415.15–558.15 K (Cox eq., Chao et al. 1983)

Henry's Law Constant (Pa m³/mol at 25°C):
    23.50   (calculated-P/C, Eastcott et al. 1988)

Octanol/Water Partition Coefficient, log K_{OW}:
    4.90   (calculated-fragment const., Yalkowsky & Valvani 1979, 1980)
    4.79   (calculated-solubility and mp, Mackay et al. 1980)
    4.94   (calculated-solvatochromic parameters and intrinsic molar volume V_I, Kamlet et al. 1988)
    5.10 ± 0.50   (recommended, Sangster 1989)
    4.91   (calculated-molar volume, Wang et al. 1992)
    5.6829   (calculated-UNIFAC group contribution, Chen et al. 1993)
    4.79   (recommended, Sangster 1993)
    4.90   (recommended, Hansch et al. 1995)

Octanol/Air Partition Coefficient, log K_{OA}:
Bioconcentration Factor, log BCF:
Sorption Partition Coefficient, log K_{OC}:

Environmental Fate Rate Constants, k, or Half-Lives, $t_{1/2}$:

Half-Lives in the Environment:
    Biota: elimination $t_{1/2}$ = 2 d from Oyster for naphthalenes (quoted, Meador et al. 1995).

4.1.1.13 Biphenyl

Common Name: Biphenyl
Synonym: diphenyl, phenylbenzene
Chemical Name: biphenyl
CAS Registry No: 92-52-4
Molecular Formula: $C_{12}H_{10}$
Molecular Weight: 154.207
Melting Point (°C):
  68.93 (Lide 2003)
Boiling Point (°C):
  256.1 (Lide 2003)
Density (g/cm³ at 20°C):
  0.866 (Weast 1982–1983)
  1.04 (Lide 2003)
Molar Volume (cm³/mol):
  148.3 (20°C, calculated-density)
  184.6 (calculated-Le Bas method at normal boiling point)
Enthalpy of Fusion, ΔH_{fus} (kJ/mol):
  18.58 (Parks & Huffman 1931)
  18.66 (exptl., Chickos et al. 1999)
Entropy of Fusion, ΔS_{fus} (J/mol K):
  51.05 (Miller et al. 1984)
  54.81, 59.2 (exptl., calculated-group additivity method, Chickos et al. 1999)
Fugacity Ratio at 25°C (assuming ΔS_{fus} = 56 J/mol K), F: 0.371 (mp at 68.93°C)
  0.35 (Mackay et al. 1980, 1983; Shiu & Mackay 1986; Shiu et al. 1987)
  0.381 (calculated, ΔS_{fus} = 54 J/mol K, Passivirta et al. 1999)

Water Solubility (g/m³ or mg/L at 25°C or as indicated and reported temperature dependence equations. Additional data at other temperatures designated * are compiled at the end of this section):
  5.94 (shake flask-UV, Andrews & Keefer 1949)
  7.48* (shake flask-UV, measured range 0.4–42.8°C, Bohon & Claussen 1951)
  3.87 (shake flask-UV, Sahyun 1966)
  7.08* (shake flask-UV, measured range 0–64.5°C, Wauchope & Getzen 1972)
  $R \cdot \ln x = -4520/(T/K) + 4.08 \times 10^{-4} \cdot [(T/K) - 298.15]^2 - 20.8 + 0.0273 \cdot (T/K)$, temp range 24.6–73.4°C
    (shake flask-UV measurements, Wauchope & Getzen 1972)
  7.45 (shake flask-GC, Eganhouse & Calder 1976)
  7.0 (shake flask-fluorescence, Mackay & Shiu 1977)
  8.50 (shake flask-nephelometry, Hollifield 1979)
  7.51 (shake flask-LSC, Banerjee et al. 1980)
  8.09 (TLC-RT correlation, Bruggeman et al. 1982)
  6.71 (generator column-GC/ECD, Miller et al. 1984, 1985; quoted, Hawker 1989b)
  7.09 (recommended, Pearlman et al. 1984)
  7.05 (vapor saturation-UV, Akiyoshi et al. 1987)
  6.5 (29°C, shake flask-GC/FID; Stucki & Alexander 1987)
  7.20, 7.55 (generator column-HPLC/UV, Billington et al. 1988)
  10.67 (calculated average of HPLC-RI, Brodsky & Ballschmiter 1988)
  7.2* (recommended, IUPAC Solubility Data Series, Shaw 1989)
  $\log [S_L/(mol/L)] = 1.872 - 973.4/(T/K)$ (supercooled liquid, Passivirta et al. 1999)
  $\ln x = -1.5792 - 3669.26/(T/K)$, temp range 5–50°C (regression eq. of literature data, Shiu & Ma 2000)
  5.37, 5.32 (generator column-GC/ECD, different flow rates, Oleszek-Kudlak et al. 2004)

Vapor Pressure (Pa at 25°C or as indicated and reported temperature dependence equations. Additional data at other temperatures designated * are compiled at the end of this section):

7933*   (162.5°C, isoteniscope-manometer, measured range 162.5–255.2°C, Chipman & Peltier 1929)
104*   (69.20°C, temp range 69.20–271.2°C, Cunningham 1930; quoted, Boublik et al. 1984)
133.3*   (70.6°C, summary of literature data, temp range 70.6–254.0°C, Stull 1947)
1.30   (effusion method, Bright 1951)
log (P/mmHg) = 10.38 – 3799/(T/K); temp range 4.0–34.5°C (Antoine eq., effusion, Bright 1951)
0.031   (manometry, Augood et al. 1953; selected, Bidleman 1984)
1.273*   (effusion method, Bradley & Cleasby 1953; selected, Bidleman 1984; Neely 1983; Erickson 1986)
log (P/cmHg) = 11.282 – 4263/(T/K); temp range 15.05–40.55°C (Antoine eq., Bradley & Cleasby 1953)
log (P/mmHg) = [–0.2185 × 12910.0/(T/K)] + 8.218583; temp range 70.6–254.9°C (Antoine eq., Weast 1972–73)
7.60   (selected P_L, Mackay & Wolkoff 1973; Mackay & Leinonen 1975; Mackay et al. 1982; Bopp 1983)
1.41*   (effusion method, interpolated-Antoine eq., measured range 24.9–50.33°C, Radchenko & Kitiagorodskii 1974; selected, Bidleman 1984)
log (P/mmHg) = 12.6789 – 4367.436/(T/K); temp range 24.9–50.33°C (Antoine eq., Knudsen effusion, Radchenko & Kitiagorodskii 1974)
16.0*   (53.05°C, gas saturation-GC, measured range 53.05–81.05°C, Sharma & Palmer 1974)
2040*   (123.0°C, pressure transducer, measured range 123.0–327.55°C, Nasir et al. 1980)
1.40   (HPLC-RT correlation, Swann et al. 1983)
log (P/atm) = [1– 528.437/(T/K)] × 10^{0.821410 – 2.73337 × 10^{-4}·(T/K) + 1.02285 × 10^{-7}·(T/K)^2}; temp range: 342.35–673.15 K (Cox eq., Chao et al. 1983)
5.608   (P_L supercooled liquid converted from literature P_S with ΔS_{fus} Bidleman 1984)
3.35, 3.41 (P_{GC} by GC-RT correlation with octadecane as reference standard, different columns, BP-1 column, Apolane-87 column, Bidleman 1984)
log (P/kPa) = 6.36895 – 1997.558/(202.608 + t/°C); temp range 69.2–271.1°C (Antoine eq. from reported exptl. data, Boublik et al. 1984)
1.19 ± 0.03; 1.03, 1.29, 0.579, 0.969 (gas saturation-GC; quoted lit. values Burkhard et al. 1984, 1985b)
1.15*   (24.7°C, gas saturation-GC/FID, measured range 5.2–24.7°C, Burkhard et al. 1984)
log (P/Pa) = 14.840 – 4402.1/(T/K); temp range 5.2–24.7°C (gas saturation data, Clapeyron eq., Burkhard et al. 1984)
0.423, 0.703, 0.594 (calculated-MW, GC-RI correlation, calculated-MCI χ, Burkhard et al. 1985a)
2.03   (supercooled liquid P_L, GC-RI correlation, Burkhard et al. 1985b)
log (P/mmHg) = 7.24541 – 1998.725/(202.733 + t/°C); temp range 69–271°C (Antoine eq., Dean 1985, 1992)
5.61; 6.62 (supercooled liquid P_L, quoted lit.; GC-RT correlation, Foreman & Bidleman 1985)
2.43; 6.90 (selected P_S; supercooled liq. P_L, Shiu & Mackay 1986; Shiu et al. 1987; Sklarew & Girvin 1987)
1.443; 1.23 (P_S, interpolated - Antoine equations; Stephenson & Malanowski 1987)
log (P_S/kPa) = 11.71929 – 4143.054/(T/K); temp range 297–324 K (Antoine eq.-I, Stephenson & Malanowski 1987)
log (P_S/kPa) = 28.5175 – 21141.5/(374.85 + T/K); temp range 283–342 K (Antoine eq.-II, Stephenson & Malanowski 1987)
log (P_L/kPa) = 6.37526 – 1794.8/(–74.85 + T/K); temp range 390–563 K (Antoine eq.-III, Stephenson & Malanowski 1987)
3.35   (P_{GC} by GC-RT correlation with eicosane as reference standard, Hinckley et al. 1990)
log(P/mmHg) = 53.0479 – 5.3509 × 10^3/(T/K) – 14.955·log (T/K) + 2.1039 × 10^{-9}·(T/K) + 2.4345 × 10^{-6}·(T/K)^2; temp range 342–789 K (vapor pressure eq., Yaws 1994)
0.422–2.54; 2.03–7.04   (quoted range of lit. P_S values; lit. P_L values, Delle Site 1997)
5.31; 2.02 (quoted supercooled liquid P_L from Hinckley et al. 1990; converted to solid P_S with fugacity ratio F, Passivirta et al. 1999)
log (P_S/Pa) = 11.05 – 3201/(T/K) (solid, Passivirta et al. 1999)
log (P_L/Pa) = 8.20 – 2228/(T/K) (supercooled liquid, Passivirta et al. 1999)
log (P/kPa) = 14.840 – 4402.1/(T/K); temp range 5–50°C (regression eq. from literature data, Shiu & Ma 2000)
3.63; 0.822 (supercooled liquid P_L, calibrated GC-RT correlation; GC-RT correlation, Lei et al. 2002)
log (P_L/Pa) = –3265/(T/K) + 11.51; $\Delta H_{vap.}$ = –62.5 kJ·mol^{-1} (GC-RT correlation, Lei et al. 2002)

Henry's Law Constant (Pa m³/mol at 25°C or as indicated and reported temperature dependence equations):
- 41.34 (gas stripping-GC, Mackay et al. 1979)
- 30.4 (gas stripping-GC, Mackay et al. 1980)
- 11.55 (gas stripping-GC, Warner et al. 1987)
- 19.57 (wetted-wall column-GC, Fendinger & Glotfelty 1990)
- 86.5 (calculated-vapor-liquid equilibrium (VLE) data, Yaws et al. 1991)
- 31.20 (gas stripping-GC, Shiu & Mackay 1997)
- $\log [H/(Pa\ m^3/mol)] = 6.33 - 1255/(T/K)$ (Passivirta et al. 1999)

Octanol/Water Partition Coefficient, log K_{OW}:
- 3.16 (shake flask-UV, Rogers & Cammarata 1969)
- 4.09 (shake flask, Leo et al. 1971; Hansch & Leo 1979)
- 4.04 (shake flask, Hansch et al. 1973)
- 4.17, 4.09, 3.16, 4.04 (Neely et al. 1974; Hansch & Leo 1979)
- 3.95 (HPLC-k' correlation, Rekker & De Kort 1979)
- 3.75 (HPLC-RT correlation, Veith et al. 1979a)
- 4.04 (shake flask-HPLC, Banerjee et al. 1980)
- 3.88 (lit. average, Kenaga & Goring 1980; Freitag et al. 1985)
- 4.10 (RP-TLC-k' correlation, Bruggeman et al. 1982)
- 4.08 (HPLC-k' correlation, Hammers et al. 1982)
- 3.70 (HPLC-RT correlation, Woodburn 1982; Woodburn et al, 1984)
- 3.16–4.09, 3.91 (shake flask, range, average, Eadsforth & Moser, 1983)
- 3.91–4.15, 4.05 (HPLC, range, average, Eadsforth & Moser 1983)
- 4.03 (HPLC-k' correlation, Hafkenscheid & Tomlinson 1983)
- 3.93 (HPLC correlation; Harnisch et al. 1983)
- 3.76 (generator column-GC/ECD, Miller et al. 1984,1985)
- 3.89 (generator column-HPLC, Woodburn et al. 1984)
- 3.79 (RP-HPLC-RT correlation, Rapaport & Eisenreich 1984)
- 4.11–4.13 (HPLC-RV correlation, quoted exptl., Garst 1984)
- 4.10 (HPLC-RV correlation, Garst & Wilson 1984)
- 4.05 (HPLC-RT correlation, Edsforth 1986)
- 3.81 (shake flask-GC, Menges & Armstrong 1991)
- 4.13 (HPLC-RT correlation, Wang et al. 1986)
- 3.63 (HPLC-k' correlation, De Kock & Lord 1987)
- 3.89 (generator column-GC, Doucette & Andren 1987, 1988)
- 4.14, 4.06, 4.00, 3.94 (RP-HPLC-RI correlation, Brodsky & Ballschmiter 1988)
- 3.69 (HPLC-RT correlation, Doucette & Andren 1988)
- 3.75 (HPLC-RT correlation, Sherblom & Eganhouse 1988)
- 4.008; 4.10 (slow stirring-GC; calculated-π const., De Bruijn et al. 1989; De Bruijn & Hermens 1990)
- 3.98 (recommended, Sangster 1989, 1993)
- 4.29 (dual-mode centrifugal partition chromatography, Gluck & Martin 1990)
- 4.26 (HPLC-k' correlation, Noegrohati & Hammers 1992)
- 4.01 (recommended, Hansch et al. 1995)

Octanol/Air Partition Coefficient, log K_{OA} at 25°C or as indicated:
- 6.92, 6.09; 6.09 (0, 20°C, multi-column GC-k' correlation; calculated at 20°C, Zhang et al. 1999)
- 6.15 (calculated-S_{oct} and vapor pressure P, Abraham et al. 2001)

Bioconcentration Factor, log BCF:
- 2.64 (trout, calculated-k_1/k_2, Neely et al. 1974)
- 3.12 (rainbow trout, Veith et al. 1979; Veith & Kosian 1983)
- 2.53 (fish, flowing water, Kenaga & Goring 1980; Kenaga 1980)
- 2.73, 2.45, 3.41 (algae, fish, activated sludge, Freitag et al. 1985; selected, Halfon & Reggiani 1986)

Sorption Partition Coefficient, log K_{OC}:

- 3.15 (soil, Kenaga 1980)
- 3.0, 3.27 (Aldrich humic acid, reversed phase separation, Landrum et al. 1984)
- 3.57, 3.77 (humic materials in aqueous solutions: RP-HPLC-LSC, equilibrium dialysis, Lake Erie water with 9.6 mg/L DOC: Landrum et al. 1984)
- 5.58, 4.04 (humic materials in aqueous solutions: RP-HPLC-LSC, equilibrium dialysis, Huron River with 7.8 mg/L DOC, Landrum et al. 1984)
- 5.68, 5.34, 5.23, 3.57 (humic materials in natural water: Huron River 6.7% DOC spring, Grand River 10.7% DOC spring, Lake Michigan 4.7% DOC spring, Lake Erie 9.6% DOC spring, RP-HPLC separation method, Landrum et al. 1984)
- 3.52, 2.94 (Apison soil 0.11% OC, Dormont soil 1.2% OC, batch equilibrium, Southworth & Keller 1986)
- 3.40 (calculated, soil, Chou & Griffin 1987)
- 3.04, 3.32, 3.26, 3.04, 3.08 (5 soils: clay loam/kaolinite, light clay/montmorillonite, light clay/montmorillite, sandy loam/allophane, clay loam/allophane, batch equilibrium-sorption isotherm, Kishi et al. 1990)
- 4.20; 3.30 (soil, calculated-universal solvation model; quoted lit., Winget et al. 2000)
- 3.03, 3.12 (soils: organic carbon OC $\geq$ 0.1%, OC $\geq$ 0.5%, average values, Delle Site 2001)

Sorption Partition Coefficient, log K_P:

- 2.146 (lake sediment, calculated-K_{OW}, f_{OC}, Formica et al. 1988)

Environmental Fate Rate Constants, k and Half-Lives, $t_{1/2}$:

Volatilization/Evaporation: $t_{1/2}$ = 7.52 d evaporation from water depth of 1 m (Mackay & Leinonen 1975), rate of volatilization k = 0.92 g m^{-2} h^{-1} (Metcalfe et al. 1988)

Photolysis: k = 5.1 × 10^{-4} h^{-1} to 7.4 × 10^{-3} h^{-1} with H_2O_2 under photolysis at 25°C in F-113 solution and with HO- in the gas (Dilling et al. 1988);

photodegradation k = 5.1 × 10^{-4} min^{-1} and $t_{1/2}$ = 22.61 h in methanol-water (3:7, v/v) with initial concentration of 16.2 ppm by high pressure mercury lamp or sunlight (Wang et al. 1991).

Oxidation: rate constant k, for gas-phase second order rate constants, k_{OH} for reaction with OH radical, k_{NO_3} with NO_3 radical and k_{O_3} with O_3 or as indicated, *data at other temperatures see reference:

k_{OH} = (8.06 ± 0.77) × 10^{-12} cm^3 molecule^{-1} s^{-1} with an estimated lifetime of ~3 d, and k_{O_3} < 2.0 × 10^{-19} cm^3 molecule^{-1} s^{-1} at 294 ± 1 K (relative rate method, Atkinson et al. 1984)

k_{OH} = (8.5 ± 0.8) × 10^{-12} cm^3 molecule^{-1} s^{-1} at 295 K (relative rate method, Atkinson & Aschmann 1985)

k_{OH} = (7 ± 2) × 10^{-12} cm^3 molecule^{-1} s^{-1} at 298 K (recommended, Atkinson 1985)

k_{OH}(calc) = 7.9 × 10^{-12} cm^3 molecule^{-1} s^{-1}, k_{OH}(obs.) = (5.8 – 8.2) × 10^{-12} cm^3 molecule^{-1} s^{-1} with a calculated tropospheric lifetime of 3 d (Atkinson 1987a)

k_{OH}(calc) = 7.1 × 10^{-12} cm^3 molecule^{-1} s^{-1}, k_{OH}(obs.) = 7.0 × 10^{-12} cm^3 molecule^{-1} s^{-1} (SAR structure-activity relationship, Atkinson 1987b)

k_{O_3} < 2 × 10^{-19} cm^3 molecule^{-1} s^{-1}; k_{OH} = 7.0 × 10^{-12} cm^3 molecule^{-1} s^{-1}; $k_{N_2O_5}$ < 2.0 × 10^{-19} cm^3 molecule^{-1} s^{-1} for reaction with N_2O_5 at room temp. (Atkinson & Aschmann 1988)

k_{OH}* = 7.2 × 10^{-12} cm^3 molecule^{-1} s^{-1} at 298 K (recommended, Atkinson 1989)

k_{OH}(calc) = 6.44 × 10^{-12} cm^3 molecule^{-1} s^{-1} (molecular orbital calculations, Klamt 1993)

k_{OH}(exptl) = 7.2 × 10^{-12} cm^3 molecule^{-1} s^{-1}, k_{OH}(calc) = 6.7 × 10^{-12} cm^3 molecule^{-1} s^{-1} with a calculated tropospheric lifetime of 2.0 d (Kwok et al. 1995)

Hydrolysis:

Biodegradation: 100% degraded by activated sludge in 47 h cycle (Monsanto Co. 1972);

k = 109 yr^{-1} in the water column and k = 1090 yr^{-1} in the sediment, microbial degradation pseudo first-order rate constant (Wong & Kaiser 1975; selected, Neely 1981);

k = 9.3–9.8 nmol L^{-1} d^{-1} with an initial biphenyl concentration of 4.4–4.7 µmol/L, and k = 3.2 nmol L^{-1} d^{-1} with initial concentration of 2.9 µmol/L, rate of biodegradation in water from Port Valdez (estimated, Reichardt et al. 1981)

$t_{1/2}$ = 1.5 d, estimated by using water die-away test (Bailey et al. 1983)

$t_{1/2}$(aq. aerobic) = 36–168 h, based on river die-away test data and activated sludge screening test data (Howard et al. 1991)

$t_{1/2}$(aq. anaerobic) = 144–672 h, based on estimated unacclimated aqueous aerobic biodegradation half-life (Howard et al. 1991) removal rate of 5.3 and 0.52 mg (g of volatile suspended solid d)$^{-1}$, degradation by bacteria from creosote-contaminated marine sediments with nitrate- and sulfate-reducers, respectively, under anaerobic conditions in a fluidized bed reactor (Rockne & Strand 1998)

Biotransformation

Bioconcentration, Uptake (k_1) and Elimination (k_2) Rate Constants or Half-Lives:
$k_1 = 6.79$ h^{-1}; $k_2 = 0.0155$ h^{-1} (trout muscle, Neely et al. 1974; Neely 1979)
$k_1 = 6.8$ h^{-1}; $1/k_2 = 65$ h (trout, quoted, Hawker & Connell 1985)
log $k_1 = 2.21$ d^{-1}; log $1/k_2 = 0.43$ d (fish, Connell & Hawker 1988)
log $k_2 = -0.43$ d^{-1} (fish, quoted, Thomann 1989)

Half-Lives in the Environment:

Air: calculated lifetime of ~3 d due to reaction with OH radical, assuming an average daytime atmospheric OH radical concn of ~1 × 10^6 molecule/cm^3 (Atkinson et al. 1984);

estimated atmospheric lifetime of ~2.7 d due to reaction with the OH radical for a 24-h average OH radical concn of 5 × 10^5 cm^{-3} (Atkinson & Aschmann 1985);

calculated tropospheric lifetime of 9 d due to the rate constants of gas-phase reaction with OH radical (Atkinson 1987);

$t_{1/2} = 7.8$–110 h, based on photooxidation half-life in air (Howard et al. 1991);

tropospheric lifetime of 2.0 d based on the experimentally determined rate constant for gas phase reaction with OH radical for biphenyl (Kwok et al. 1995).

Surface water: $t_{1/2}$ ~1.5 d in river water (Bailey et al. 1983);

$t_{1/2} = 36$–168 h, based on unacclimated aqueous aerobic biodegradation half-life (Howard et al. 1991);

photolysis $t_{1/2} = 19.18$ min in aqueous solution when irradiated with a 500 W medium pressure mercury lamp (Chen et al. 1996).

Groundwater: $t_{1/2} = 72$–336 h, based on unacclimated aqueous aerobic biodegradation half-life (Howard et al. 1991)

Sediment:

Soil: $t_{1/2} = 36$–168 h, based on unacclimated aqueous aerobic biodegradation half-life (Howard et al. 1991)

Biota: estimated $t_{1/2} = 29$ h from fish in simulated ecosystem (Neely 1980).

TABLE 4.1.1.13.1
Reported aqueous solubilities of biphenyl at various temperatures

Bohon & Claussen 1951		Wauchope & Getzen 1972				Shaw 1989	
shake flask-UV		shake flask-UV				IUPAC recommended	
t/°C	S/g·m^{-3}	t/°C	S/g·m^{-3}	t/°C	S/g·m^{-3}	t/°C	S/g·m^{-3}
			experimental		smoothed		
0.40	2.83	24.6	7.13	0	2.64	0	2.72
2.4	2.97	24.6	7.29	24.6	6.96	10	4.1
5.2	3.38	24.6	7.35	25	7.08	20	6.3
7.6	3.64	29.9	8.77	29.9	8.73	25	7.2
10	4.06	29.9	8.64	30.3	8.88	30	9.1
12.6	4.58	29.9	8.95	38.4	12.7	40	14.4
14.9	5.11	30.3	8.55	40.1	13.8	50	22
15.9	5.27	30.3	8.54	47.5	19.5	69	37
25	7.48	30.3	8.48	50	22.0		
25.6	7.78	38.4	13.2	50.1	22.1		
30.1	9.64	38.4	13.3	50.2	22.2		
30.4	9.58	38.4	13.5	54.7	27.7		
33.3	11.0	40.1	13.1	59.2	34.8		
34.9	11.9	40.1	13.4	60.5	37.2		
36	12.5	40.1	13.4	64.5	45.9		
42.8	17.2	47.5	18.8				

(Continued)

TABLE 4.1.1.13.1 (*Continued*)

Bohon & Claussen 1951		Wauchope & Getzen 1972				Shaw 1989	
shake flask-UV		shake flask-UV				IUPAC recommended	
t/°C	S/g·m⁻³	t/°C	S/g·m⁻³	t/°C	S/g·m⁻³	t/°C	S/g·m⁻³
		47.5	19.0	temp dependence eq. 1			
for supercooled liquid:		47.5	18.7	ln x	mole fraction		
ΔH_{sol}/(kJ mol⁻¹) =		50.1	20.6	ΔH_{fus}	18.9 ± 0.50		
at 275 K	7.03	50.1	21.6	$10^2 \cdot b$	2.73 ± 0.12		
280 K	10.13	50.1	21.8	c	20.8 ± 0.4		
285 K	11.25	50.2	20.7				
290 K	12.55	50.2	21.8				
295 K	13.43	54.7	28.3				
300 K	15.02	54.7	28.8				
305 K	18.58	59.2	36.4				
310 K	21.42	59.2	36.3				
315 K	21.09	59.2	36.0				
		60.5	40.4				
		64.5	43.7, 44.7				
		64.5	46.5				
		ΔH_{fus}/(kJ mol⁻¹) = 18.91					

Empirical temperature dependence equations:

Wauchope & Getzen (1972): $R \cdot \ln x = -[H_{fus}/(T/K)] + (0.000408)[(T/K) - 291.15]^2 - c + b \cdot (T/K)$ (1)

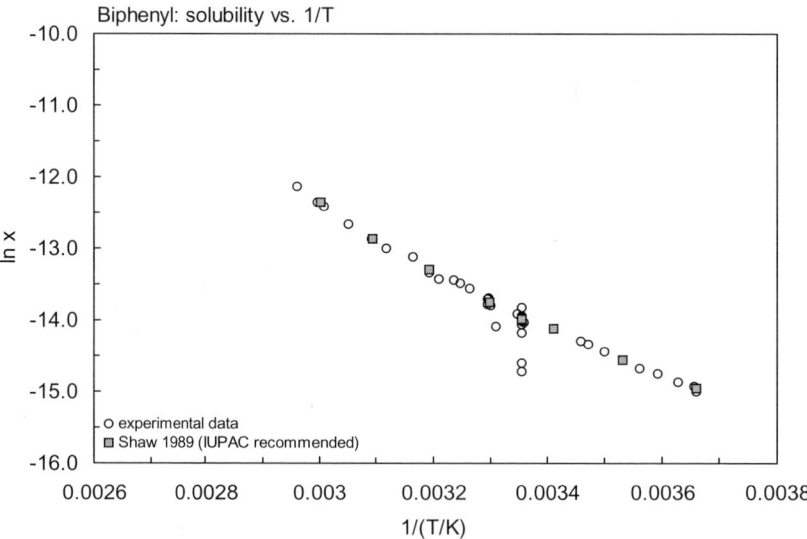

FIGURE 4.1.1.13.1 Logarithm of mole fraction solubility (ln x) versus reciprocal temperature for biphenyl.

TABLE 4.1.1.13.2
Reported vapor pressures of biphenyl at various temperatures and the coefficients for the vapor pressure equations

$$\log P = A - B/(T/K) \quad (1)$$
$$\log P = A - B/(C + t/°C) \quad (2)$$
$$\log P = A - B/(C + T/K) \quad (3)$$
$$\log P = A - B/(T/K) - C \cdot \log (T/K) \quad (4)$$
$$\log P = A - B/(T/K) - C \cdot \log (T/K)^2 \quad (5)$$
$$\ln P = A - B/(T/K) \quad (1a)$$
$$\ln P = A - B/(C + t/°C) \quad (2a)$$

1.

Chipman & Peltier 1929		Stull 1947		Bright 1951		Bradley & Cleasby 1953	
isoteniscope-manometer		summary of literature data		effusion		effusion	
t/°C	P/Pa	t/°C	P/Pa	t/°C	P/Pa	t/°C	P/Pa
162.5	7933	70.6	133.3	Data presented in graph		15.05	0.416
172.3	10959	101.8	666.6	25	0.579	20.7	0.7786
177.7	12799	117.0	1333		(interpolated)	24.7	1.2252
183.5	15705	134.2	2666	eq. 1	P/mmHg	24.0	1.1825
191.6	19972	152.5	5333	A	10.38	24.1	1.184
198.75	24691	165.2	7999	B	3799	27.05	1.600
293.8	28504	180.7	13332	temp range 4.9–34.5°C		29.15	2.053
211.25	34677	204.2	26664			32.45	2.973
220.05	43756	229.4	53329	ΔH_{subl}/(kJ mol^{-1}) = 72.80		35.05	3.866
229.8	56329	254.0	101325			37.9	5.160
238.2	68901					40.55	6.693
247.7	85580	mp/°C	69.5			23.05	1.027
253.7	98019					36.5	1.533
255.2	101178					31.25	2.546
						35.9	4.133
bp/°C	266.25						
ΔH_V/(kJ mol^{-1}) = 44.99						eq. 1	P/mmHg
	at bp					A	11.282
						B	4262
eq. 5	P/mmHg					temp range 15–41°C	
A	7.0220						
B	1723						
C	245700						
temp range 162–322°C							

2.

Radchenko & K. 1974		Nasir et al. 1980		Burkhard et al. 1984		Cunningham 1930	
Knudsen effusion		pressure transducer		gas saturation-GC		in Boublik et al. 1984	
t/°C	P/Pa	t/°C	P/Pa	t/°C	P/Pa	t/°C	P/Pa
24.9	1.433	123.0	2040	5.2	0.106	69.20	104
31.75	2.976	143.81	4773	14.9	0.361	93.3	413
33.7	3.734	164.69	9962	24.7	1.15	148.7	4833
35.5	4.538	181.28	16447			160.0	7239
37.6	5.726	200.87	28599	eq. 1	P/Pa	171.1	10548
39.57	6.913	223.66	51518	A	14.840	182.2	15031
41.52	8.26	245.65	86254	B	4402.1	193.3	21098

(Continued)

TABLE 4.1.1.13.2 (*Continued*)

2.

Radchenko & K. 1974		Nasir et al. 1980		Burkhard et al. 1984		Cunningham 1930	
Knudsen effusion		pressure transducer		gas saturation-GC		in Boublik et al. 1984	
t/°C	P/Pa	t/°C	P/Pa	t/°C	P/Pa	t/°C	P/Pa
43.48	10.26	257.91	111343			204.4	28958
45.45	12.35	274.09	154493			215.6	39093
47.4	15.49	296.14	235345			226.7	51986
50.0	19.46	315.19	329300			237.8	68051
		327.55	400175			248.9	88252
eq. 1	P/mmHg					255.3	101353
A	12.6789	data fitted to				260.0	112384
B	4367.436	Chebyshev polynomial				271.2	142032
for temp range 24.9–50°C							
						eq. 2	P/kPa
						A	6.36895
Sharma & Palmer 1974						B	1997.558
gas saturation-GC						C	202.608
t/°C	P/Pa					bp/°C	255.208
53.05	16.0						
61.05	34.66						
71.95	92.0						
81.05	220.0						

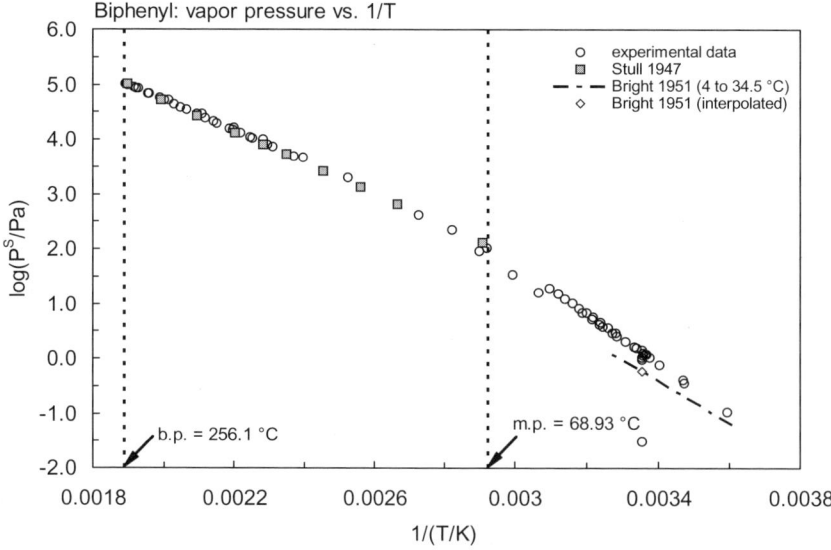

FIGURE 4.1.1.13.2 Logarithm of vapor pressure versus reciprocal temperature for biphenyl.

4.1.1.14 4-Methylbiphenyl

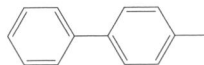

Common Name: 4-Methylbiphenyl
Synonym: 4-phenyltoluene
Chemical Name: 4-methylbiphenyl
CAS Registry No: 644-08-6
Molecular Formula: $C_{13}H_{12}$
Molecular Weight: 168.234
Melting Point (°C):
    49.5   (Lide 2003)
Boiling Point (°C):
    267.5   (2003)
Density (g/cm³ at 20°C):
    1.015   (27°C, Weast 1982–83; Lide 2003)
Molar Volume (cm³/mol):
    165.7   (27°C, calculated-density)
    206.8   (calculated-Le Bas method at normal boiling point)
Enthalpy of Evaporation, ΔH_V (kJ/mol):
Enthalpy of Fusion, ΔH_{fus} (kJ/mol):
Entropy of Fusion, ΔS_{fus} (J/mol K):
Fugacity Ratio at 25°C (assuming ΔS_{fus} = 56 J/mol K), F: 0.575 (mp at 49.5°C)

Water Solubility (g/m³ or mg/L at 25°C or as indicated and reported temperature dependence equations.):
    1.834, 4.05, 7.03 (4.9, 25, 40°C, generator column-HPLC/GC, Doucette & Andren 1988a)
    S/(mol/L) = 9.18×10^{-6} exp(0.038·t/°C) (generator column-GC/ECD, temp range 4.9–40°C, Doucette & Andren 1988a); or
    log x = –1436/(T/K) – 1.541; temp. range 4.9–40°C (generator column-GC/ECD, Doucette & Andren 1988a)

Vapor Pressure (Pa at 25°C):

Henry's Law Constant (Pa m³/mol at 25°C):

Octanol/Water Partition Coefficient, log K_{OW}:
    4.63   (generator column-HPLC/GC, calculated-group contribution, TSA, Doucette & Andren 1987)
    4.63   (recommended, Sangster 1989, 1994)
    4.63   (recommended, Hansch et al. 1995)

Octanol/Air Partition Coefficient, log K_{OA}:

Bioconcentration Factor, log BCF:

Sorption Partition Coefficient, log K_{OC}:

Environmental Fate Rate Constants, k or Half-Lives, $t_{½}$:

Half-Lives in the Environment:

4.1.1.15 4,4′-Dimethylbiphenyl

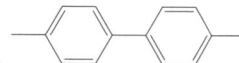

Common Name: 4,4′-Dimethylbiphenyl
Synonym: 4,4′-dimethyl-1,1′-biphenyl
Chemical Name: 4,4′-dimethylbiphenyl
CAS Registry No: 613-33-2
Molecular Formula: $C_{14}H_{14}$
Molecular Weight: 182.261
Melting Point (°C):
    125   (Weast 1982–83; Ruelle & Kesselring 1997; Lide 2003)
Boiling Point (°C):
    295   (Weast 1982–83; Lide 2003)
Density (g/cm³ at 20°C):
Molar Volume (cm³/mol):
    194.0   (Ruelle & Kesselring 1997)
    229.0   (calculated-Le Bas method at normal boiling point)
Enthalpy of Evaporation, ΔH_V (kJ/mol):
Enthalpy of Fusion, ΔH_{fus} (kJ/mol):
Entropy of Fusion, ΔS_{fus} (J/mol K):
Fugacity Ratio at 25°C (assuming ΔS = 56 J/mol K), F: 0.104 (mp at 125°C)

Water Solubility (g/m³ or mg/L at 25°C or as indicated and reported temperature dependence equations.):
    0.0687, 0.175, 0.441 (4.9, 25, 40°C, generator column-GC, Doucette & Andren 1988a)
    $S/(mol/L) = 2.90 \times 10^{-7} \exp(0.052 \cdot t/°C)$ (generator column-GC/ECD, temp range 4–40°C, Doucette & Andren 1988a)
    $\log x = -1913/(T/K) - 1.288$; temp. range 4.9–40°C (generator column-GC/ECD, Doucette & Andren 1988a)

Vapor Pressure (Pa at 25°C):

Henry's Law Constant (Pa m³/mol at 25°C):
    0.931   (calculated-P/C)

Octanol/Water Partition Coefficient, log K_{OW}:
    5.09   (generator column-GC/ECD, Doucette & Andren 1987)
    5.09   (recommended, Sangster 1989, 1994)
    5.09   (recommended, Hansch et al. 1995)

Octanol/Air Partition Coefficient, log K_{OA}:

Bioconcentration Factor, log BCF:

Sorption Partition Coefficient, log K_{OC}:

Environmental Fate Rate Constants, k or Half-Lives, $t_{1/2}$:

Half-Lives in the Environment:

4.1.1.16 Diphenylmethane

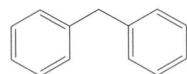

Common Name: Diphenylmethane
Synonym: diphenyl methane, 1,1′-methylenebis-benzene
Chemical Name: diphenylmethane
CAS Registry No: 101-81-5
Molecular Formula: $C_{13}H_{12}$
Molecular Weight: 168.234
Melting Point (°C):
    25.4   (Lide 2003)
Boiling Point (°C):
    265   (Lide 2003)
Density (g/cm³ at 20°C):
    1.00592, 1.00192   (20°C, 25°C. Dreisbach 1955)
    1.001   (26°C Lide 2003)
Molar Volume (cm³/mol):
    168.1   (27°C, from density, Stephenson & Malanowski 1987)
    167.2   (20°C, calculated-density)
    206.8   (calculated-Le Bas method at normal boiling point)
Enthalpy of Evaporation, ΔH_V (kJ/mol):
Enthalpy of Sublimation, ΔH_{subl} (kJ/mol):
    64.02   (Bright 1951)
    66.845, 45.34   (25°C, bp, Dreisbach 1955)
Enthalpy of Fusion, ΔH_{fus} (kJ/mol):
    18.28   (Dreisbach 1955)
    18.58   (Parks & Huffman 1931; Chickos et al. 1999)
Entropy of Fusion, ΔS_{fus} (J/mol K):
    61.92   (Stephenson & Malanowski 1987)
    62.34, 62.1 (exptl., calculated-group additivity method, Chickos et al. 1999)
Fugacity Ratio at 25°C (assuming ΔS_{fus} = 56 J/mol K), F: 0.991 (mp at 25.4°C)

Water Solubility (g/m³ or mg/L at 25°C):
    14.10   (shake flask/UV, Andrews and Keefer 1949)
    16.40   (Deno & Berkheimer 1960)
    3.76   (Lu et al. 1978)
    3.00   (shake flask-nephelometry, Hollifield 1979)
    16.19   (lit. mean, Pearlman et al. 1984)

Vapor Pressure (Pa at 25°C and reported temperature dependence equations. Additional data at other temperatures designated *, are compiled at the end of this section):
    133.3*   (76.0°C, summary of literature data, temp range 76.0–264.5°C, Stull 1947)
    1.09   (effusion method, interpolated from reported Antoine eq., Bright 1951)
    log (P/mmHg) = 9.12 – 3341/(T/K); temp range 5.1–26.5°C (Antoine eq., effusion, Bright 1951)
    2.266   (calculated by formula, Dreisbach 1955)
    log (P/mmHg) = 7.16125 – 1944.42/(190.0 + t/°C); temp range 150–310°C (Antoine eq. for liquid state, Dreisbach 1955)
    3904*   (151.49°C, static-differential pressure gauge, measured range 151.49–336.32°C, Wieczorek & Kobayashi 1980)
    0.0452   (extrapolated-Antoine eq., Boublik et al. 1984)
    log (P/kPa) = 5.94201 – 1668.355/(186.212 + t/°C); temp range 217.5–282.2°C (Antoine eq. from reported exptl. data, Boublik et al. 1984)

13100*   (457.95 K, vapor-liquid equilibrium, measured range 457.95–581.85 K, Klara et al. 1987)
0.052   (extrapolated-Antoine eq., Dean 1985, 1992)
log (P/mmHg) = 6.291 – 1261/(105 + t/°C); temp range 217–282°C (Antoine eq., Dean 1985, 1992)
0.0885   (interpolated-Antoine eq., Stephenson & Malanowski 1987)
log (P_L/kPa) = 5.8765 – 1707.9/(–101.15 + T/K); temp range 295–383 K (Antoine eq.-I, Stephenson & Malanowski 1987)
log (P_L/kPa) = 6.28615 – 1944.42/(–83.15 + T/K); temp range 423–583 K (Antoine eq.-II, Stephenson & Malanowski 1987)
log (P/mmHg) = 50.8894 – 5.2749 × 10^3/(T/K) – 14.246·log (T/K) – 4.2994 × 10^{-10}·(T/K) + 2.4197 × 10^{-6}·(T/K)^2; temp range 298–768 K (vapor pressure eq., Yaws 1994)
1.456*   (22.25°C, transpiration-GC, measured range –0.95 to 22.25°C, Verevkin 1999)

Henry's Law Constant (Pa m^3/mol at 25°C):
   0.931 (calculated-P/C, Mackay et al. 1992)

Octanol/Water Partition Coefficient, log K_{OW}:
   4.14   (Hansch & Leo 1979)
   4.22   (HPLC-RT correlation, Burkhard et al. 1985)
   4.33   (HPLC-RT correlation, Eadsforth 1986)
   4.14   (recommended, Sangster 1989,1994)
   4.14   (recommended, Hansch et al. 1995)

Octanol/Air Partition Coefficient, log K_{OA}:

Bioconcentration Factor, log BCF:

Sorption Partition Coefficient, log K_{OC}:

Environmental Fate Rate Constants, k, or Half-Lives, $t_{1/2}$:

Half-Lives in the Environment:

TABLE 4.1.1.16.1
Reported vapor pressures of diphenylmethane at various temperatures and the coefficients for the vapor pressure equations

log P = A – B/(T/K)   (1)   ln P = A – B/(T/K)   (1a)
log P = A – B/(C + t/°C)   (2)   ln P = A – B/(C + t/°C)   (2a)
log P = A – B/(C + T/K)   (3)
log P = A – B/(T/K) – C·log (T/K)   (4)

Stull 1947		Bright 1951		Wieczorek & Kobayashi 1980		Verevkin 1999	
summary of literature data		effusion method		static-differential pressure gauge		transpiration-GC	
t/°C	P/Pa	t/°C	P/Pa	t/°C	P/Pa	t/°C	P/Pa
76.0	133.3	Data presented in graph		151.49	3904	–0.95	0.07946
107.4	666.6			151.77	3929	5.15	0.1584
122.8	1333			158.06	4942	10.15	0.3261
138.8	2666	eq. 1	P/mmHg	164.06	6182	15.05	0.6196
157.8	5333	A	9.12	170.31	7799	18.65	0.9556
170.2	7999	B	3341	176.25	9487	22.25	1.456
186.3	13332	temp range 5.1–26.5°C		182.32	11487		
210.7	26664			189.57	14279	eq. 1a	P/Pa
237.5	53329	ΔH_{subl}/(kJ mol^{-1}) = 64.015		196.01	17435	A	36.43
264.5	101325			202.69	21342	B	10639

TABLE 4.1.1.16.1 (Continued)

Stull 1947 summary of literature data		Bright 1951 effusion method		Wieczorek & Kobayashi 1980 static-differential pressure gauge		Verevkin 1999 transpiration-GC	
t/°C	P/Pa	t/°C	P/Pa	t/°C	P/Pa	t/°C	P/Pa
				209.7	26436	ΔH_{subl}/(kJ mol^{-1}) = 88.48	
mp/°C	26.5			217.16	32381	at 284.3 K	
				225.99	41037	ΔH_{subl}/(kJ mol^{-1}) = 87.63	
		Klara et al. 1987		232.67	58832	at 298.15 K	
		vapor-liquid equilibrium		247.79	70500		
		T/K	P/Pa	255.23	83501	30.25	3.099
		457.95	13100	263.79	100021	35.05	4.653
		488.05	31000	264.36	100938	40.05	7.019
		527.25	82720	272.29	119269	44.95	10.19
		560.95	168900	280.24	137819	55.05	21.63
		581.85	250200	288.0	1377.37	60.05	30.55
				295.97	162025	65.15	47.05
		eq. 3	P/kPa	303.88	186511	70.15	67.41
		A	15.1413	311.45	218741		
		B	5078.0	320.63	253090	eq. 1a	P/Pa
		C	53.89	336.32	298026	A	27.43
						B	7981.2
				fitted to Chebyshev Polynomial		ΔH_V/(kJ mol^{-1}) = 66.36	
						at 323.3 K	
						ΔH_V/(kJ mol^{-1}) = 67.83	
						at 298.15 K	

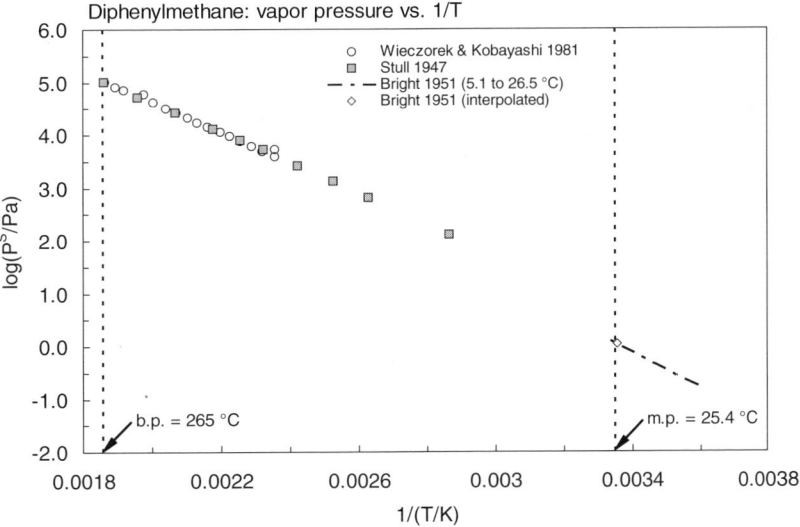

FIGURE 4.1.1.16.1 Logarithm of vapor pressure versus reciprocal temperature for diphenylmethane.

4.1.1.17 Bibenzyl

Common Name: Bibenzyl
Synonym: 1,2-Diphenylethane, dibenzyl, 1,1′-(1,2-ethanediyl) bis-benzene
Chemical Name: 1,2-diphenylethane
CAS Registry No: 103-29-7
Molecular Formula: $C_{14}H_{14}$
Molecular Weight: 182.261
Melting Point (°C):
    52.5   (Lide 2003)
Boiling Point (°C):
    284   (Lide 2003)
Density (g/cm³ at 20°C):
    0.9780   (25°C Lide 2003)
Molar Volume (cm³/mol):
    190.2   (60°C, calculated-density, Stephenson & Malanowski 1987)
    186.4   (25°C, calculated-density)
    229.0   (calculated-Le Bas method at normal boiling point)
Enthalpy of Fusion, ΔH_{fus} (kJ/mol):
    23.43   (Parks & Huffman 1931)
    30.54   (Stephenson & Malanowski 1987)
    22.73   (Chickos et al. 1999)
Entropy of Fusion, ΔS_{fus} (J/mol K):
    94.14   (Stephenson & Malanowski 1987)
Fugacity Ratio at 25°C (assuming $\Delta S = 56$ J/mol K), F: 0.537 (mp at 52.5°C)

Water Solubility (g/m³ or mg/L at 25°C):
    4.37   (shake flask-UV, Andrews & Keefer 1950b)
    4.37   (quoted, Pearlman et al. 1984)
    1.89; 0.44   (generator column-HPLC/UV; HPLC-RT correlation, Swann et al. 1983)

Vapor Pressure (Pa at 25°C and reported temperature dependence equations. Additional data at other temperatures designated *, are compiled at the end of this section):
    133.3*   (86.8°C, summary of literature data, temp range 86.8–284.0°C, Stull 1947)
    0.198   (effusion method, interpolated from reported Antoine eq., Bright 1951)
    $\log (P/mmHg) = 9.86 - 3783/(T/K)$; temp range 17.1–44.2°C (Antoine eq., effusion, Bright 1951)
    17.1*   (60°C, inclined piston, measured range 60–140°C, Osborn & Scott 1980)
    $\log (P/atm) = [1 - 547.288/(T/K)] \times 10^{\wedge}\{0.914704 - 6.08831 \times 10^{-4} \cdot (T/K) + 5.11258 \times 10^{-7} \cdot (T/K)^2\}$; temp range 333.15–413.15 K, (Cox eq., Chao et al. 1983)
    0.406   (interpolated-Antoine eq., Stephenson & Malanowski 1987)
    $\log (P_S/kPa) = 11.319 - 4386/(T/K)$, temp range 286–308 K, (Antoine eq.-I, Stephenson & Malanowski 1987)
    $\log (P_L/kPa) = 6.93271 - 2636.21/(-22.009 + T/K)$; temp range 369–557 K (Antoine eq.-II, Stephenson & Malanowski 1987)
    $\log (P/mmHg) = 48.5573 - 5.2841 \times 10^3/(T/K) - 13.41 \cdot \log (T/K) - 1.0073 \times 10^{-9} \cdot (T/K) + 2.1338 \times 10^{-6} \cdot (T/K)^2$; temp range 324–780 K (vapor pressure eq., Yaws 1994)
    0.734; 0.249   (supercooled liquid P_L, calibrated GC-RT correlation; GC-RT correlation, Lei et al. 2002)
    $\log (P_L/Pa) = -3522/(T/K) + 11.67$; $\Delta H_{vap.} = -67.4$ kJ·mol⁻¹ (GC-RT correlation, Lei et al. 2002)

Henry's Law Constant (Pa m³/mol at 25°C):

Octanol/Water Partition Coefficient, log K_{OW}:

    4.79, 4.82   (Hansch & Leo 1979)
    4.76   (quoted, HPLC-k' correlation, Hammers et al. 1982)
    3.67   (HPLC-RT correlation, Swann et al. 1983)
    4.60   (HPLC-RT correlation, Webster et al. 1985)
    4.71   (HPLC-RT correlation, Eadsworth 1986)
    4.70 ± 0.20   (recommended, Sangster 1989)
    4.79   (recommended, Sangster 1993)
    4.79   (recommended, Hansch et al. 1995)

Octanol/Air Partition Coefficient, log K_{OA}:

Bioconcentration Factor, log BCF:

Sorption Partition Coefficient, log K_{OC}:

Environmental Fate Rate Constants, k or Half-Lives, $t_{1/2}$:

Half-Lives in the Environment:

TABLE 4.1.1.17.1
Reported vapor pressures of bibenzyl at various temperatures and the coefficients for the vapor pressure equations

$$\log P = A - B/(T/K) \quad (1)$$
$$\log P = A - B/(C + t/°C) \quad (2)$$
$$\log P = A - B/(C + T/K) \quad (3)$$
$$\log P = A - B/(T/K) - C \cdot \log(T/K) \quad (4)$$
$$\ln P = A - B/(T/K) \quad (1a)$$
$$\ln P = A - B/(C + t/°C) \quad (2a)$$

Stull 1947		Bright 1951		Osborn & Scott 1980	
summary of literature data		effusion method		inclined piston	
t/°C	P/Pa	t/°C	P/Pa	t/°C	P/Pa
86.8	133.3	Data presented in graph		60	17.1
119.8	666.6			65	24.1
136.0	1333			70	34.0
153.7	2666	eq. 1	P/mmHg	75	47.7
173.7	5333	A	9.56	80	65.9
186.0	7999	B	3783	85	89.7
202.8	13332	temp range 17.1–44.2°C		90	121.2
227.8	26664			95	161.9
255.0	53329	ΔH_{subl}/(kJ mol^{-1}) = 72.38		100	213.6
284.0	101325			105	280.1
				110	364.2
mp/°C	51.5			115	469.8
				120	602.4
				125	761.8
				130	961.1
				135	1202.7
				140	1498
				data fitted to a 4–constant vapor pressure eq.	

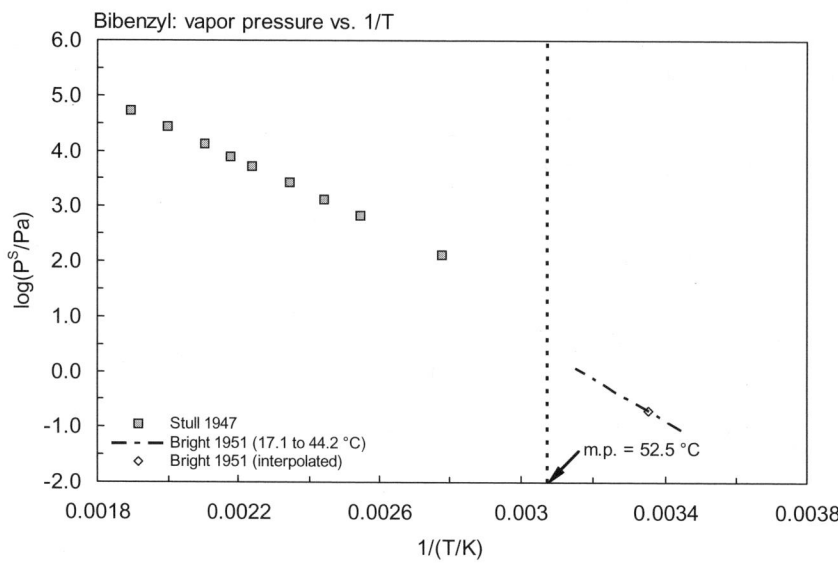

FIGURE 4.1.1.17.1 Logarithm of vapor pressure versus reciprocal temperature for bibenzyl.

4.1.1.18 *trans*-Stilbene

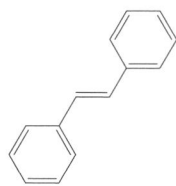

Common Name: *trans*-1,2-Diphenylethene
Synonym: *trans*-stilbene, *trans*-diphenylethylene, E-stilbene
Chemical Name: *trans*-1,2-diphenylethene
CAS Registry No: 103-30-0
Molecular Formula: $C_{14}H_{12}$
Molecular Weight: 180.245
Melting Point (°C):
    124.2   (Lide 2003)
Boiling Point (°C):
    307   (Lide 2003)
Density (g/cm³ at 20°C):
    0.9707   (Weast 1982–83; Lide 2003)
Molar Volume (cm³/mol):
    185.0   (Ruelle & Kesselring 1997)
    185.7   (20°C, calculated-density)
    221.6   (calculated-Le Bas method at normal boiling point)
Enthalpy of Sublimation, ΔH_{subl} (kJ/mol):
    100.7   (Van Ekeren et al. 1983)
Enthalpy of Fusion, ΔH_{fus} (kJ/mol):
    30.125   (Stephenson & Malanowski 1987)
    27.40   (Ruelle & Kesselring 1997; Chickos et al. 1999)
Entropy of Fusion, ΔS_{fus} (J/mol K):
    76.73   (Stephenson & Malanowski 1987)
    68.81, 69.7   (exptl., calculated-group additivity method, Chickos et al. 1999)
Fugacity Ratio at 25°C (assuming ΔS_{fus} = 56 J/mol K), F: 0.106 (mp at 124.2°C)

Water Solubility (g/m³ or mg/L at 25°C):
    0.29   (shake flask-UV, Andrews & Keefer 1950)

Vapor Pressure (Pa at 25°C or as indicated and reported temperature dependence equations. Additional data at other temperatures designated * are compiled at the end of this section):
    133.3*   (113.2°C, summary of literature data, temp range 113.2–306.5°C, Stull 1947)
    0.00764*   (manometer-spinning rotor friction gauge, torsion mass loss effusion, measured range 297.45–364.5 K, Van Ekeren et al. 1983)
    0.0274   (34.65°C, effusion-quartz crystal microbalance, Offringa et al. 1983)
    0.00647   (interpolated-Antoine eq., Stephenson & Malanowski 1987)
    $\log (P_S/kPa) = 12.25604 - 5201.358/(T/K)$; temp range 298–343 K (Antoine eq.-I, solid, Stephenson & Malanowski 1987)
    $\log (P_L/kPa) = 6.97928 - 2610.05/(-54.759 + T/K)$; temp range 419–580 K (Antoine eq.-II, liquid, Stephenson & Malanowski 1987)
    $\log (P/mmHg) = 68.6303 - 6.3776 \times 10^3/(T/K) - 21.015 \cdot \log (T/K) + 5.7813 \times 10^{-3} \cdot (T/K) + 1.8334 \times 10^{-12} \cdot (T/K)^2$; temp range 397–820 K (vapor pressure eq., Yaws 1994)

Henry's Law Constant (Pa m³/mol at 25°C):

Octanol/Water Partition Coefficient, log K_{OW}:
    4.81   (Hansch & Leo 1979)

4.81 (recommended, Sangster 1989)
4.81 (recommended, Hansch et al. 1995)

Octanol/Air Partition Coefficient, log K_{OA}:
 7.48 (calculated-S_{oct} and vapor pressure P, Abraham et al. 2001)

Bioconcentration Factor, log BCF:

Sorption Partition Coefficient, log K_{OC}:

Environmental Fate Rate Constants, k, or Half-Lives, $t_{½}$:

Half-Lives in the Environment:

TABLE 4.1.1.18.1
Reported vapor pressures of *trans*-stilbene at various temperatures and the coefficients for the vapor pressure equations

$\log P = A - B/(T/K)$ (1) $\ln P = A - B/(T/K)$ (1a)
$\log P = A - B/(C + t/°C)$ (2) $\ln P = A - B/(C + t/°C)$ (2a)
$\log P = A - B/(C + T/K)$ (3)
$\log P = A - B/(T/K) - C \cdot \log(T/K)$ (4)

Stull 1947		Van Ekeren et al. 1983					
Summary of literature data		Spinning rotor fraction gauge		Torsion mass loss effusion		Static method	
t/°C	P/Pa	t/°C	P/Pa	t/°C	P/Pa	t/°C	P/Pa
113.2	133.3	24.3	0.00704	44.97	0.100	69.17	1.43
145.8	666.6	24.3	0.00699	50.78	0.200	69.45	1.44
161.0	1333	24.3	0.00715	54.27	0.300	71.25	1.83
179.8	2666	24.43	0.0693	56.80	0.400	71.52	1.80
199.0	5333	27.24	0.0102	60.43	0.600	72.2	1.94
211.5	7999	27.24	0.0104	64.14	0.900	73.70	2.25
227.4	13332	27.24	0.0105	65.11	1.00	74.18	2.39
251.7	26664	30.65	0.0151	69.45	1.44	75.73	2.81
287.3	53329	30.65	0.015	71.25	1.83	75.83	2.80
306.5	101325	32.03	0.0196	71.25	1.80	75.86	2.83
		32.03	0.0195	72.2	1.94	77.15	3.10
mp/°C	124	34.96	0.0287	73.7	2.25	78.27	3.55
		34.96	0.0288	74.18	2.39	8.06	4.34
		34.96	0.0289	75.73	2.81	80.39	4.39
		37.33	0.0386	75.86	2.83	81.79	5.07
		37.33	0.0387	76.88	3.10	83.36	5.84
		40.61	0.0584	78.27	3.55	85.04	6.86
		40.61	0.0585	80.79	4.39	86.35	7.71
		43.32	0.0792	81.79	5.07	86.38	7.74
		25	0.00765	83.36	5.84	86.51	7.90
				85.04	6.86	87.35	8.47
				86.35	7.71	89.74	10.57
				86.51	7.90	90.43	11.22
				87.35	8.47	91.35	12.20
				89.74	10.57		
				90.43	11.22		
				91.35	12.20		
				ΔH_{fus} = 100.17 kJ/mol			
				at 331.64 K			

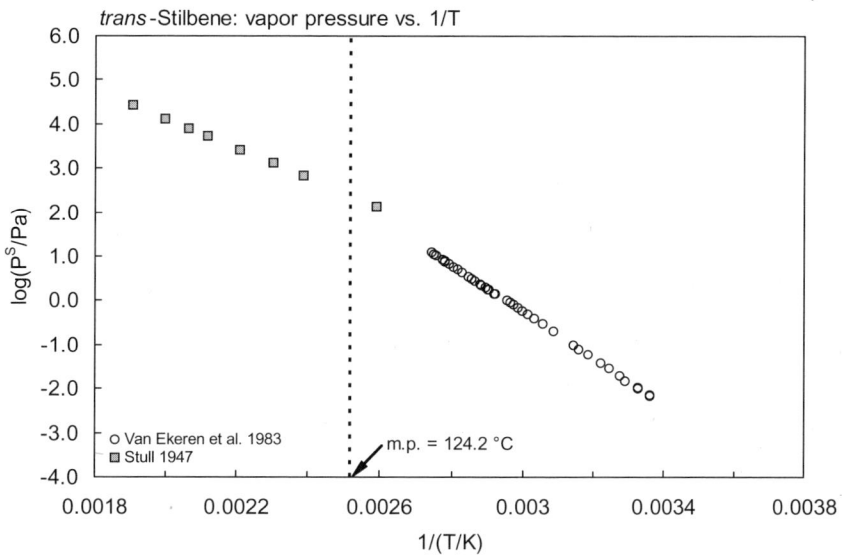

FIGURE 4.1.1.18.1 Logarithm of vapor pressure versus reciprocal temperature for *trans*-stilbene.

4.1.1.19 Acenaphthylene

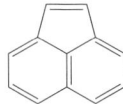

Common Name: Acenaphthylene
Synonym:
Chemical Name: acenaphthylene
CAS Registry No: 208-96-8
Molecular Formula: $C_{12}H_8$
Molecular Weight: 152.192
Melting Point (°C):
 91.8 (Lide 2003)
Boiling Point (°C):
 280 (Lide 2003)
Density (g/cm³ at 20°C):
 0.899 (Dean 1985)
 0.8987 (17°C, Lide 2003)
Molar Volume (cm³/mol):
 141.2 (Ruelle & Kesselring 1997)
 167.1 (17°C, calculated-density)
 165.7 (calculated-Le Bas method at normal boiling point)
Enthalpy of Fusion, ΔH_{fus} (kJ/mol):
 10.96 (Ruelle & Kesselring 1997)
 1.4, 6.95, 10.96; 12.36 (–156.55, 88.45, 88.85°C; total phase change enthalpy, Chickos et al. 1999)
Entropy of Fusion, ΔS_{fus} (J/mol K):
 30.3 (Passivirta et al. 1999)
 42.4, 37.8 (exptl., calculated-group additivity method, total phase change entropy, Chickos et al. 1999)
Fugacity Ratio at 25°C (assuming ΔS_{fus} = 56 J/mol K), F: 0.221 (mp at 91.8°C)
 0.458 (calculated, ΔS_{fus} = 30.3 J/mol K, Passivirta et al. 1999)

Water Solubility (g/m³ or mg/L at 25°C and reported temperature dependence equations):
 3.93 (misquoted from Mackay & Shiu 1977)
 16.1 (generator column-HPLC/fluorescence, Walters & Luthy 1984)
 log $[S_L/(mol/L)]$ = 1.315 – 573.5/(T/K) (supercooled liquid, Passivirta et al. 1999)

Vapor Pressure (Pa at 25°C or as indicated and reported temperature dependence equations. Additional data at other temperatures designated * are compiled at the end of this section):
 0.893* (gas saturation-HPLC/fluo./UV, Sonnefeld et al. 1983)
 log (P/Pa) = 12.768 – 3821.55/(T/K); temp range 10–50°C (Antoine eq., Sonnefeld et al. 1983)
 0.893 (generator column-HPLC, Wasik et al. 1983)
 1.105 (interpolated, Antoine eq., Stephenson & Malanowski 1987)
 log (P_S/kPa) = 9.500 – 3714/(T/K); temp range 286–318 K (Antoine eq., Stephenson & Malanowski 1987)
 0.90 (selected, Mackay et al. 1992, 1996; quoted, Shiu & Mackay 1997)
 0.90; 1.97 (quoted solid P_S from Mackay et al. 1992; converted to supercooled liquid P_L with fugacity ratio F, Passivirta et al. 1999)
 log (P_S/Pa) = 11.11 – 3201/(T/K) (solid, Passivirta et al. 1999)
 log (P_L/Pa) = 9.53 – 2751/(T/K) (supercooled liquid, Passivirta et al. 1999)

Henry's Law Constant (Pa m³/mol at 25°C or as indicated and reported temperature dependence equations. Additional data at other temperatures designated * are compiled at the end of this section):
 11.55 (gas stripping-GC, Warner et al. 1987)
 11.40 (wetted-wall column-GC, Fendinger & Glotfelty 1990)
 12.7* (gas stripping-GC, measured range 4.1–31°C, Bamford et al. 1999)

ln K_{AW} = –6278.6/(T/K) + 15.757; ΔH = 52.2 kJ mol^{-1}; measured range 4.1–31°C (gas stripping-GC, Bamford et al. 1999)

log [H/(Pa m^3/mol)] = 8.22 – 2178/(T/K) (Passivirta et al. 1999)

Octanol/Water Partition Coefficient, log K_{OW}:
- 4.07 (calculated as per Leo et al. 1971)
- 3.94 (Yalkowsky & Valvani 1979)
- 3.72 (calculated-fragment const., Mabey et al. 1982)
- 4.08 (selected, Mills et al. 1982)
- 4.06 (calculated-molar refraction MR, Yoshida et al. 1983)
- 3.90 (calculated-MCI χ as per Rekker & De Kort 1979, Ruepert et al. 1985)
- 3.55 (HPLC-RT correlation, Chin et al. 1986)
- 4.07–4.10; 4.08 (quoted lit. range; lit. mean, Meadors et al. 1995)
- 4.00; 3.67 (quoted lit.; calculated, Passivirta et al. 1999)

Octanol/Air Partition Coefficient, log K_{OA}:

Bioconcentration Factor, log BCF:
- 3.0 (microorganisms-water, calculated-K_{OW}, Mabey et al. 1982)
- 2.58 (Isnard & Lambert 1988)

Sorption Partition Coefficient, log K_{OC} at 25°C or as indicated:
- 3.83, 3.75 (soil, RP-HPLC correlation on CIHAC, on PIHAC, Szabo et al. 1990b)
- 4.91–5.21; 3.60–3.80 (range, calculated from sequential desorption of 11 urban soils; lit. range, Krauss & Wilcke 2001)
- 4.96; 5.05, 5.14, 5.45 (20°C, batch equilibrium, A2 alluvial grassland soil; calculated values of expt 1,2,3-solvophobic approach, Krauss & Wilcke 2001)

Environmental Fate Rate Constants, k or Half-Lives, $t_{½}$:

Volatilization:

Photolysis: not environmentally significant (Mabey et al. 1982);
$t_{½}$ = 0.7 h on silica gel, $t_{½}$ = 2.2 h on alumina and $t_{½}$ = 44 h on fly ash for different atmospheric particulate substrates determined in the rotary photoreactor (appr. 25 µg/g on substrate) (Behymer & Hites 1985); direct photolysis $t_{½}$ = 9.08 h (predicted-QSPR) in atmospheric aerosol (Chen et al. 2001).

Photodegradation k = 3 × 10^{-5} s^{-1} in surface water during the summertime at mid-latitude (Fasnacht & Blough 2002)

Oxidation: rate constant k, for gas-phase second order rate constants, k_{OH} for reaction with OH radical, k_{NO_3} with NO$_3$ radical and k_{O_3} with O$_3$ or as indicated, *data at other temperatures see reference:

k = 4 × 10^7 M^{-1} h^{-1} for singlet oxygen and k = 5 × 10^3 M^{-1} h^{-1} for peroxy radical (calculated, Mabey et al. 1982)

k_{O_3} ~ 5.50 × 10^{-16} cm^3·molecule^{-1}·s^{-1}, k_{OH} = (11.0 ± 0.1) × 10^{-11} cm^3 molecule^{-1} s^{-1} and k_{NO_3} = (54 ± 0.8) × 10^{-12} cm^3 molecule^{-1} s^{-1} at 296 ± 2 K (relative rate methods, Atkinson & Aschmann 1988)

k_{OH} = 11.0 × 10^{-11} cm^3 molecule^{-1} s^{-1} at 296 K (Atkinson 1989)

k_{OH} = 12.4 × 10^{-11} cm^3 molecule^{-1} s^{-1} at 296 ± 2 K with a atmospheric lifetime of 1.1 h assuming an average ambient 12-h daytime OH radical concn of 2 × 10^6 molecule/cm^3; k_{O_3} = 1.60 × 10^{-16} cm^3·molecule^{-1}·s^{-1} at 296 K with lifetime of 2.5 h assuming ambient O$_3$ concn of 7 × 10^{11} molecule/cm^3 (relative rate method, Reisen & Arey 2002)

Hydrolysis: not hydrolyzable (Mabey et al. 1982; Howard et al. 1991).

Biodegradation: > 98% degradation within 7 d, based on domestic sewer for an average of three static-flask screening test (Tabak et al. 1981);
aerobic $t_{½}$ = 1020–1440 h, based on soil column study data (Kincannon & Lin 1985; quoted, Howard et al. 1991);
anaerobic $t_{½}$ = 4080–5760 h, based on estimated unacclimated aqueous aerobic biodegradation half-life (Howard et al. 1991).

Biotransformation: k = 3 × 10^{-9} mL cell^{-1} h^{-1}, estimated rate constant for bacteria (Mabey et al. 1982).

Bioconcentration, Uptake (k_1) and Elimination (k_2) Rate Constants:

Half-Lives in the Environment:

Air: $t_{1/2}$ = 0.191–1.27 h, based on photooxidation half-life in air (Atkinson 1987; quoted, Howard et al. 1991). Atmospheric lifetime of 1.1 h and 2.5 h due to reaction with OH and O_3 at 296 K, respectively (Reisens & Arey 2002)

Surface water: $t_{1/2}$ = 1020–1440 h, based on estimated unacclimated aqueous aerobic biodegradation half-life (Kincannon & Lin 1985; quoted, Howard et al. 1991).

Groundwater: $t_{1/2}$ = 2040–2880 h, based on estimated unacclimated aqueous aerobic biodegradation half-life (Howard et al. 1991).

Sediment:

Soil: $t_{1/2}$ = 1020–1440 h, based on soil column study data (Kincannon & Lin 1985; quoted, Howard et al. 1991); $t_{1/2}$ > 50 d (Ryan et al. 1988).

Biota: elimination $t_{1/2}$ = 1 d from rainbow trout (quoted, Meador et al. 1995).

TABLE 4.1.1.19.1
Reported vapor pressures and Henry's law constants of acenaphthlyene at various temperatures and the coefficients for the vapor pressure equations

Vapor pressure		Henry's law constant		
Sonnefeld et al. 1983		Bamford et al. 1999		
gas saturation-HPLC		gas stripping-GC/MS		
t/°C	P/Pa	t/°C	H/(Pa m³/mol)	H/(Pa m³/mol)
				average
11.20	0.206	4.1	1.98, 2.87	2.38
11.20	0.205	11.0	3.76. 4.86	4.27
11.20	0.216	18.0	6.67, 8.33	7.46
20.56	0.590	25.0	10.9, 14.6	12.7
20.56	0.585	31.0	16.2, 23.7	19.6
20.56	0.588			
30.40	1.50	ln K_{AW} = A – B/(T/K)		
30.40	1.54	A	15.7566	
30.40	1.52	B	6278.6	
39.05	3.30			
39.05	3.41	enthalpy, entropy change:		
39.05	3.27	ΔH/(kJ·mol⁻¹) = 52.2 ± 3.3		
39.05	3.34	ΔS/(J·K⁻¹ mol⁻¹) = 131		
25.0	0.89			at 25°C
log P/Pa = A – B/(T/K)				
A	12.768			
B	3821.55			

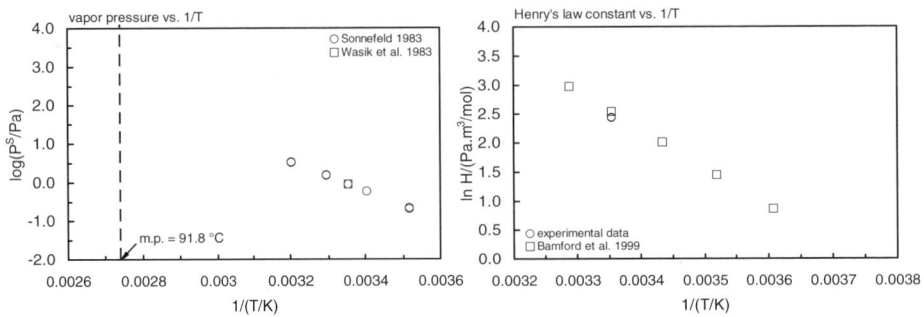

FIGURE 4.1.1.19.1 Logarithm of vapor pressure and Henry's law constant versus reciprocal temperature for acenaphthylene.

4.1.1.20 Acenaphthene

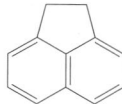

Common Name: Acenaphthene
Synonym: 1,8-hydroacenaphthylene, ethylenenaphthalene, periethylenenaphthalene, 1,2-dihydro-acenaphthalene
Chemical Name: 1,8-hydroacenaphthylene
CAS Registry No: 83-32-9
Molecular Formula: $C_{12}H_{10}$
Molecular Weight: 154.207
Melting Point (°C):
    93.4   (Lide 2003)
Boiling Point (°C):
    279   (Weast 1982–82; Lide 2003)
Density (g/cm³ at 20°C):
    1.069   (95°C, Dean 1985)
    1.222   (Lide 2003)
Molar Volume (cm³/mol):
    126.2   (20°C, calculated-density)
    173.1   (calculated-Le Bas method at normal boiling point)
Enthalpy of Sublimation, ΔH_{subl} (kJ/mol):
    84.68   (Radchenko & Kitiagorodskii 1974)
Enthalpy of Fusion, ΔH_{fus} (kJ/mol):
    21.88   (differential calorimetry, Wauchope & Getzen 1972)
    21.46   (calorimetry, Osborn & Douslin 1975; Chickos et al. 1999)
Entropy of Fusion, ΔS_{fus} (J/mol K):
    60.25   (Wauchope & Getzen 1972)
    59.83   (Casellato et al. 1973)
    56.90   (Ubbelohde 1978)
    58.55, 41.09 (exptl., calculated-group additivity method, total phase change entropy, Chickos et al. 1999)
Fugacity Ratio at 25°C (assuming ΔS_{fus} = 56 J/mol K), F: 0.213 (mp at 93.4°C)
    0.197   (calculated, Passivirta et al. 1999)

Water Solubility (g/m³ or mg/L at 25°C or as indicated and reported temperature dependence equations. Additional data at other temperatures designated * are compiled at the end of this section):
    6.14   (Deno & Berkheimer 1960)
    3.88*   (shake flask-UV, measured range 0–74.7°C, Wauchope & Getzen 1972)
    $R \cdot \ln x = -5230/(T/K) + 4.08 \times 10^{-4} \cdot [(T/K) - 291.15]^2 - 17.1 + 0.0186 \cdot (T/K)$; temp range 22.2–73.4°C (shake flask-UV measurements, Wauchope & Getzen 1972)
    3.59   (shake flask-UV, Vesala 1974)
    3.47   (shake flask-GC, Eganhouse & Calder 1976)
    3.93   (shake flask-fluorescence, Mackay & Shiu 1977)
    7.37   (shake flask-LSC, Banerjee et al. 1980)
    2.42   (shake flask-GC, Rossi & Thomas 1981)
    4.47   (average lit. value, Pearlman et al. 1984)
    4.16   (generator column-HPLC/fluorescence, Walters & Luthy 1984)
    3.8*   (recommended, IUPAC Solubility Data Series, Shaw 1989)
    3.88   (shake flask-HPLC, Haines & Sandler 1995)
    $\log [S_L/(mol/L)] = 2.505 - 1127/(T/K)$ (supercooled liquid S_L, Passivirta et al. 1999)
    $\ln x = 0.684974 - 4541.77/(T/K)$; temp range 5–50°C (regression eq. of literature data, Shiu & Ma 2000)

Vapor Pressure (Pa at 25°C or as indicated and reported temperature dependence equations. Additional data at other temperatures designated * are compiled at the end of this section):

    2560*   (147.2°C, static isoteniscope method, measured range 147.2–287.8°C, Mortimer & Murphy 1923)

    log (P/mmHg) = 8.033 – 2835/(T/K); temp range 147.2–287.8°C (Antoine eq. static isoteniscope method, Mortimer & Murphy 1923)

    666.6*   (114.8°C, summary of literature data, temp range 114.8–277.5°C, Stull 1947)

    0.207   (Hoyer & Peperle 1958)

    log (P/mmHg) = 11.50 – 4264/(T/K); temp range –15 to 30°C (Knudsen effusion method, Hoyer & Peperle 1958)

    4.02   (extrapolated-Antoine eq., Weast 1972–73)

    log (P/mmHg) = [–0.2185 × 13078.5/(T/K)] + 8.069478; temp range 114.8–277.5°C (Antoine eq., Weast 1972–73)

    3.07   (extrapolated-Antoine eq., liquid state P_L, Boublik et al. 1973)

    8.622*   (54.1°C, effusion method, measured range 54.1–83.45°C, Radchenko & Kitiagorodskii 1974)

    log (P/mmHg) = 12.2930 – 4422.921/(T/K); temp range 54.10–83.15°C (Antoine eq., Knudsen effusion, Radchenko & Kitiagorodskii 1974)

    0.373*   (manometry-extrapolated, measured range 65–140°C, Osborn & Douslin 1975)

    0.287*   (gas saturation-HPLC/fluo./UV, Sonnefeld et al. 1983)

    log (P/Pa) = 14.669 – 4535.39/(T/K); temp range 10–50°C (Antoine eq., gas saturation, Sonnefeld et al. 1983)

    0.287   (gas saturation/generator column-HPLC, Wasik et al. 1983)

    3.03, 1.48   (extrapolated-Antoine eq., Boublik et al. 1984)

    log (P_L/kPa) = 6.84571 – 2527.716/(244.912 + t/°C); temp range 147.2–287.8°C (Antoine eq. from reported exptl. data of Mortimer & Murphy 1923, Boublik et al. 1984)

    log (P_L/kPa) = 6.38504 – 2102.491/(203.124 + t/°C); temp range 95–140°C (Antoine eq. from reported exptl. data of Osborn & Douslin 1966, Boublik et al. 1984)

    3.07   (extrapolated from liquid state P_L, Dean 1985)

    log (P/mmHg) = 7.72819 – 2534.234/(245.576 + t/°C); temp range 147–187°C (Antoine eq., Dean 1885, 1992)

    log (P/mmHg) = 8.033 – 2834.99/(T/K); temp range 147–288°C (Antoine eq., Dean 1985, 1992)

    0.237, 0.319   (19.95°C, 26.85°C, gas saturation-GC, Sato al. 1986)

    0.427*   (gas saturation, interpolated-Antoine eq., temp range 20–169°C, Sato al. 1986)

    ln (P/Pa) = 22.9288 – 5183.86/(T/K – 80.153); temp range: 293.1–342 K (Antoine eq., gas saturation, Sato et al. 1986)

    0.311   (interpolated-Antoine eq., Stephenson & Malanowski 1987)

    log (P_S/kPa) = 10.883 – 4290.5/(T/K); temp range 290–311 K (Antoine eq.-I, Stephenson & Malanowski 1987)

    log (P_S/kPa) = 9.4944 – 3248.008/(–48.055 + T/K); temp range 338–366 K (Antoine eq.-II, Stephenson & Malanowski 1987)

    log (P_L/kPa) = 6.3519 – 2082.356/(–71.578 + T/K); temp range 368–413 K (Antoine eq.-III, Stephenson & Malanowski 1987)

    log (P_L/kPa) = 7.30401 – 2975/(10.674 + T/K); temp range 388–552 K (Antoine eq.-IV, Stephenson & Malanowski 1987)

    0.336, 0.211, 0.383; 0.287; 0.375; 0.377, 0.122, 0.306, 0.862 (quoted lit. values: effusion method; gas saturation-HPLC; calculated, Delle Site 1997)

    0.30; 1.52   (quoted solid P_S from Mackay et al. 1992; converted to supercooled liquid P_L with fugacity ratio F, Passivirta et al. 1999)

    log (P_S/Pa) = 11.20 – 3492/(T/K) (solid, Passivirta et al. 1999)

    log (P_L/Pa) = 8.13 – 2367/(T/K) (supercooled liquid, Passivirta et al. 1999)

    log (P/kPa) = 10.883 – 4290.5/(T/K); temp range 5–50°C (regression eq. from literature data, Shiu & Ma 2000)

    1.52; 0.428   (supercooled liquid P_L, calibrated GC-RT correlation; GC-RT correlation, Lei et al. 2002)

    log (P_L/Pa) = –3337/(T/K) + 11.37; ΔH_V = –63.9 kJ·mol^{-1} (GC-RT correlation, Lei et al. 2002)

Henry's Law Constant (Pa m^3/mol at 25°C or as indicated and reported temperature dependence equations. Additional data at other temperatures designated * are compiled at the end of this section):

    14.79   (gas stripping-GC, Mackay et al. 1979)

    15.7   (gas stripping-GC, Mackay & Shiu 1981; Mackay et al. 1982)

    24.42   (gas stripping-GC, Warner et al. 1987)

6.45 (wetted-wall column-GC, Fendinger & Glotfelty 1990)
9.17 (headspace solid-phase microextraction (SPME)-GC, Zhang & Pawliszyn 1993)
16.20 (gas stripping-fluorescence, Shiu & Mackay 1997)
18.5* (gas stripping-GC, measured range 4.1–31°C, Bamford et al. 1999)
ln K_{AW} = –6242.48/(T/K) + 16.0, ΔH = 51.9 kJ mol^{-1}; measured range 4.1–31°C, (gas stripping-GC, Bamford et al. 1999)
log [H/(Pa m^3/mol)] = 5.63 – 1240/(T/K) (Passivirta et al. 1999)

Octanol/Water Partition Coefficient, log K_{OW} at 25°C or as indicated:
3.92 (shake flask-LSC, Veith et al. 1979, 1980)
3.92; 4.49 (shake flask-GC; RP-HPLC-RT correlation; Veith et al. 1980)
3.92 (23°C, shake flask, Banerjee et al. 1980)
3.92 (recommended, Sangster 1989, 1993)
3.92 (recommended, Hansch et al. 1995)

Octanol/Air Partition Coefficient, log K_{OA} at 25°C:
6.31 (calculated-S_{oct} and vapor pressure P, Abraham et al. 2001)

Bioconcentration Factor, log BCF:
2.59 (bluegill sunfish, Veith et al. 1979, 1980)
2.59 (bluegill sunfish, Barrows et al. 1980)
2.59 (bluegill sunfish, Davies & Dobbs, 1984)

Sorption Partition Coefficient, log K_{OC} at 25°C or as indicated:
5.38 (sediments average, Kayal & Connell 1990)
3.79 (RP-HPLC correlation on CIHAC, Szabo 1990b)
3.59 (RP-HPLC correlation on PIHAC, Szabo 1990b; quoted, Pussemier et al. 1990)
3.58; 3.79 (HPLC-screening method; calculated-PCKOC fragment method, Müller & Kördel 1996)
3.40–5.33; 3.80–5.40 (range, calculated from sequential desorption of 11 urban soils; lit. range, Krauss & Wilcke 2001)
4.79; 4.89, 4.31, 4.20 (20°C, batch equilibrium, A2 alluvial grassland soil; calculated values of expt 1,2,3-solvophobic approach, Krauss & Wilcke 2001)

Environmental Fate Rate Constants, k or Half-Lives, $t_{1/2}$:
Volatilization:
Photolysis: half-lives on different atmospheric substrates determined in the rotary photoreactor (appr. 25 µg/g on substrate): $t_{1/2}$ = 2.0 h on silica gel, $t_{1/2}$ = 2.2 h on alumina and $t_{1/2}$ = 44 h on fly ash (Behymer & Hites 1985); k = 0.23 h^{-1} in distilled water with $t_{1/2}$ = 3 h (Fukuda et al. 1988);
direct photolysis $t_{1/2}$ = 7.67 h (predicted- QSPR) in atmospheric aerosol (Chen et al. 2001).
Hydrolysis: not hydrolyzable (Mabey et al. 1982).
Oxidation: rate constant k, for gas-phase second order rate constants, k_{OH} for reaction with OH radical, k_{NO_3} with NO$_3$ radical and k_{O_3} with O$_3$ or as indicated, *data at other temperatures see reference:
k < 3600 M^{-1} h^{-1} for singlet O$_2$, k = 8000 M^{-1} h^{-1} for peroxy radical at 25°C (Mabey et al. 1982)
k_{OH}(exptl) = 1.03 × 10^{-10} cm^3 molecule^{-1} s^{-1}, k_{OH}(calc) = 1.49 × 10^{-11} cm^3 molecule^{-1} s^{-1} (Atkinson et al. 1988)
k_{O_3} < 5.0 × 10^{-19} cm^3 molecule^{-1} s^{-1}, k_{OH} = (1.03 ± 0.13) × 10^{-10} cm^3 molecule^{-1} s^{-1} and k_{NO_3} = (4.6 ± 2.6) × 10^{-13} cm^3 molecule^{-1} s^{-1} at 296 ± 2 K (relative rate methods, Atkinson & Aschmann 1988)
k_{OH} = (103 – 58.4) × 10^{-12} cm^3 molecule^{-1} s^{-1} at 296–300 K (Atkinson 1989)
k_{OH}(calc) = 84.03 × 10^{-12} cm^3 molecule^{-1} s^{-1} (molecular orbital calculations, Klamt 1996)
k_{OH}* = 58 × 10^{-12} cm^3 molecule^{-1} s^{-1} at 298 K, measured range 325–365 K with a calculated atmospheric lifetime of 4.9 h based on gas-phase OH reaction (Brubaker & Hites 1998)
k_{OH} = 8.9 × 10^{-11} cm^3 molecule^{-1} s^{-1} at 296 ± 2 K with a atmospheric lifetime of 1.8 h assuming an average ambient 12-h daytime OH radical concn of 2 × 10^6 molecule/cm^3 (Reisen & Arey 2002)
Biodegradation: significant degradation within 7 d for a domestic sewer test (Tabak et al. 1981);
aerobic $t_{1/2}$ = 295–2448 h, based on aerobic soil column test data (Kincannon & Lin 1985; quoted, Howard et al. 1991);

anaerobic $t_{1/2}$ = 1180–9792 h, based on estimated unacclimated aqueous aerobic biodegradation half-life (Howard et al. 1991).

Biotransformation: 3×10^{-9} mL cell^{-1} h^{-1}, estimated bacterial transformation rate constant (Mabey et al. 1982).

Bioconcentration, Uptake (k_1) and Elimination (k_2) Rate Constants:

Sorption (k_1)-Desorption (k_2) Rate constants: desorption rate constant of 0.018 d^{-1} with $t_{1/2}$ = 38.5 d from sediment under conditions mimicking marine disposal (Zhang et al. 2000).

Half-Lives in the Environment:

Air: $t_{1/2}$ = 0.879–8.79 h, based on estimated photooxidation half-life in air (Howard et al. 1991); calculated atmospheric lifetime of 4.9 h based on gas-phase OH reactions (Brubaker & Hites 1998).

Surface water: $t_{1/2}$ = 3–300 h, based on photolysis half-life in water (Howard et al. 1991).

Groundwater: $t_{1/2}$ = 590–4896 h, based on estimated unacclimated aqueous aerobic biodegradation half-life (Howard et al. 1991).

Sediment: desorption $t_{1/2}$ = 38.5 d from sediment under conditions mimicking marine disposal (Zhang et al. 2000).

Soil: $t_{1/2}$ = 295–2448 h, based on aerobic soil column test data (Kincannon & Lin 1985; quoted, Howard et al. 1991);

$t_{1/2}$ > 50 d (Ryan et al. 1988).

Biota: $t_{1/2}$ < 1.0 d in the tissue of bluegill sunfish (Veith et al. 1980).

TABLE 4.1.1.20.1
Reported aqueous solubilities of acenaphthene at various temperatures

Wauchope & Getzen 1972				Shaw 1989	
shake flask-UV				IUPAC recommended	
t/°C	S/g·m^{-3}	t/°C	S/g·m^{-3}	t/°C	S/g·m^{-3}
experimental		smoothed data			
22.2	3.57	0	1.45	0	1.5
30.0	4.76	22.2	3.46	20	3.2
30.0	4.60	25.0	3.88	25	3.8
30.0	4.72	30.0	4.80	30	4.8
34.5	6.00	34.5	5.83	40	7.4
34.5	5.68	39.3	7.20	50	9.2
34.5	5.75	44.7	9.20	60	19
39.3	6.80	50	11.9	70	32
39.3	7.10	50.1	11.9	75	43
39.3	7.00	55.6	15.6		
44.7	9.40	64.5	24.3		
44.7	9.40	65.2	25.2		
44.7	9.30	69.8	32.1		
50.1	12.5	71.9	35.9		
50.1	12.4	73.4	39.0		
50.1	12.4	74.7	41.8		
55.6	15.8	75.0	42.5		
55.6	16.3				
55.6	15.9				
64.5	25.9	temp dependence eq. 1			
64.5	27.8	ln x	mole fraction		
65.2	23.7	ΔH_{fus}	21.88 ± 0.21		
65.2	23.4	$10^2 \cdot b$	1.86 ± 0.11		
65.2	22.8	c	20.8 ± 0.4		
69.8	30.1				
69.8	34.3				

TABLE 4.1.1.20.1 (Continued)

Wauchope & Getzen 1972				Shaw 1989	
shake flask-UV				IUPAC recommended	
t/°C	S/g·m^{-3}	t/°C	S/g·m^{-3}	t/°C	S/g·m^{-3}
69.8	33.6				
71.9	35.2				
73.4	39.1				
79.4	40.1				
74.7	40.8				
74.7	39.3				
ΔH_{fus}/(kJ mol^{-1}) = 21.88					

Empirical temperature dependence equations:

Wauchope & Getzen (1972): $R \cdot \ln x = -[H_{fus}/(T/K)] + (0.000408)[(T/K) - 291.15]^2 - c + b \cdot (T/K)$ (1)

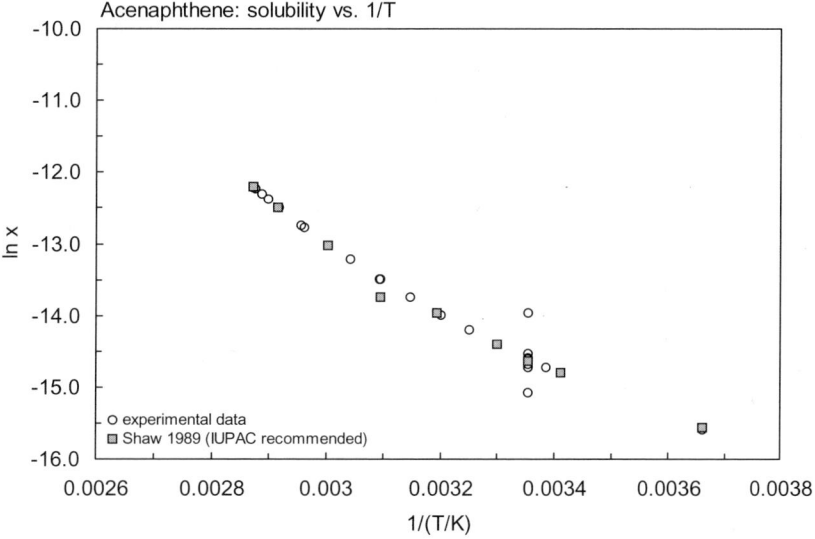

FIGURE 4.1.1.20.1 Logarithm of mole fraction solubility (ln x) versus reciprocal temperature for acenaphthene.

TABLE 4.1.1.20.2
Reported vapor pressures of acenaphthene at various temperatures and the coefficients for the vapor pressure equations

$$\log P = A - B/(T/K) \quad (1)$$
$$\log P = A - B/(C + t/°C) \quad (2)$$
$$\log P = A - B/(C + T/K) \quad (3)$$
$$\log P = A - B/(T/K) - C \cdot \log (T/K) \quad (4)$$
$$\ln P = A - B/(T/K) \quad (1a)$$
$$\ln P = A - B/(C + t/°C) \quad (2a)$$

1.

Mortimer & Murphy 1923		Stull 1947		Hoyer & Peperle 1958		Radchenko & K. 1974	
static isoteniscope method		summary of literature data		effusion		effusion method	
t/°C	P/Pa	t/°C	P/Pa	t/°C	P/Pa	t/°C	P/Pa
147.2	2560	114.8	666.6	data presented as		54.1	8.622
182.4	8479	131.2	1333	eq. 1	P/mmHg	57.85	10.576
182.4	8479	148.7	2666	A	11.50	58.95	13.180
210.2	19732	168.2	5333	B	4264	61.3	16.545
210.4	19865	181.2	7999	temp range −15–35°C		64.55	20.438
227.2	31277	197.5	13332	ΔH_{sub}/(kJ mol^{-1}) = 83.26		67.25	26.757
233.2	36264	222.1	26664			69.5	31.304
246.2	49943	250.0	53329			71.8	37.77
246.6	50383	277.5	101325			73.65	47.22
247.0	51049					76.0	57.23
252.4	57929	mp/°C	55			78.05	67.66
252.5	57955					80.5	82.045
264.4	76047					83.45	130.30
264.4	76460						
275.3	97779					eq. 1	P/mmHg
275.4	97779					A	12.2930
275.4	97779					B	4222.924
286.8	124030						
287.0	124296						
287.8	125723						
eq. 1	P/mmHg						
A	8.033						
B	2835						
temp range 147–288°C							

2.

Osborn & Douslin 1975		Sonnefeld et al. 1983		Sato et al. 1986	
inclined-piston manometry		gas saturation-HPLC		gas saturation-electrobalance	
t/°C	P/Pa	t/°C	P/Pa	t/°C	P/Pa
solid					
65.0	25.865	10.87	0.048	19.95	0.237
70.0	30.797	10.87	0.0504	26.85	0.519
75.0	46.796	10.87	0.0515	30.85	0.792
80.0	70.526	20.45	0.167	35.75	1.30
85.0	104.92	20.45	0.161	37.95	1.60
90.0	153.45	20.45	10.66	43.35	3.21
92.0	185.05	30.15	0.539	45.35	4.21
92.5	195.05	30.15	0.512	48.35	5.76

TABLE 4.1.1.20.2 (*Continued*)

Osborn & Douslin 1975		Sonnefeld et al. 1983		Sato et al. 1986	
inclined-piston manometry		gas saturation-HPLC		gas saturation-electrobalance	
t/°C	P/Pa	t/°C	P/Pa	t/°C	P/Pa
liquid		32.15	0.580	51.85	7.68
95.0	214.91	38.9	1.35	58.05	9.72
100.0	281.04	38.9	1.32	60.25	12.0
105.0	364.23	38.9	1.32	63.65	15.6
110.0	468.35	25	0.287	68.85	23.5
115.0	597.94				
120.0	755.39	eq. 1	P/Pa	eq. 3	P/Pa
125.0	948.97	A	14.385	A	22.9288
130.0	1184	B	4616.07	B	5183.86
135.0	1469			C	−80.153
140.0	1809				

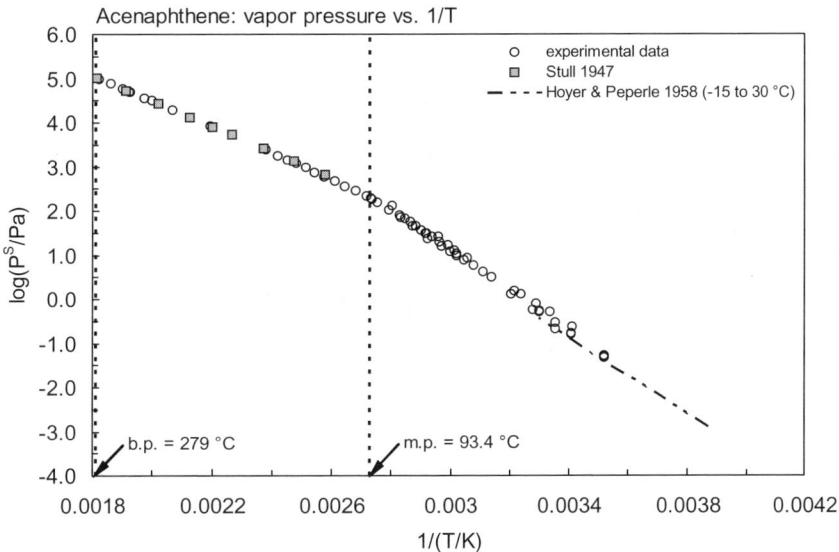

FIGURE 4.1.1.20.2 Logarithm of vapor pressure versus reciprocal temperature for acenaphthene.

TABLE 4.1.1.20.3
Reported Henry's law constants of acenaphthene at various temperatures

	Bamford et al. 1999	
	gas stripping–GC/MS	
t/°C	H/(Pa m³/mol)	H/(Pa m³/mol)
		average
4.1	3.27, 3.79	3.52
11.0	5.98, 6.62	6.29
18.0	10.5, 11.4	10.9
25.0	17.5, 19.6	18.5
31.0	26.5, 30.8	28.6

$\ln K_{AW} = A - B/(T/K)$

	K_{AW}
A	15.997
B	6242.5

enthalpy, entropy change:
$\Delta H/(kJ \cdot mol^{-1}) = 51.9 \pm 1.3$
$\Delta S/(J \cdot K^{-1} \cdot mol^{-1}) = 133$
at 25°C

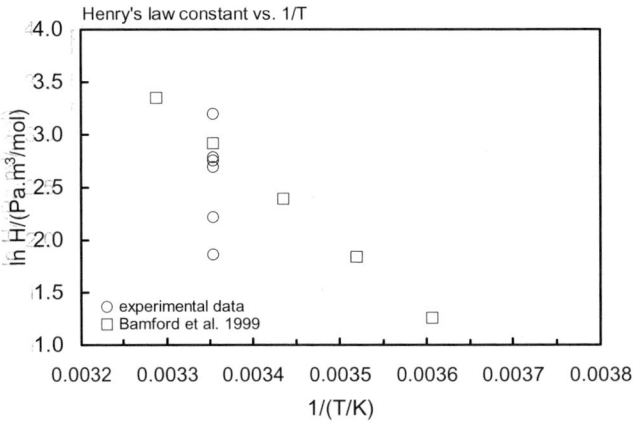

FIGURE 4.1.1.20.3 Logarithm of Henry's law constant versus reciprocal temperature for acenaphthene.

4.1.1.21 Fluorene

Common Name: Fluorene
Synonym: 2,3-benzindene, diphenylenemethane, 9H-fluorene
Chemical Name: diphenylenemethane
CAS Registry No: 86-73-7
Molecular Formula: $C_{13}H_{10}$
Molecular Weight: 166.218
Melting Point (°C):
    114.77   (Lide 2003)
Boiling Point (°C):
    295   (Dean 1985; Lide 2003)
Density (g/cm³ at 20°C):
    1.203   (0°C, Lide 2003)
Molar Volume (cm³/mol):
    138.0   (calculated-density, liquid molar volume, Lande & Banerjee 1981)
    187.9   (calculated-Le Bas method at normal boiling point)
Enthalpy of Fusion, ΔH_{fus} (kJ/mol):
    19.54   (Wauchope & Getzen 1972)
    19.58   (Osborn & Douslin 1975; Ruelle & Kesselring 1997; Chickos et al. 1999)
Entropy of Fusion, ΔS_{fus} (J/mol K):
    50.63   (Wauchope & Getzen 1972)
    48.53   (Casellato et al. 1973)
    50.48, 51.0   (exptl., calculated-group additivity method, Chickos et al. 1999)
    50.5   (Passivirta et al. 1999)
Fugacity Ratio at 25°C (assuming ΔS_{fus} = 56 J/mol K), F: 0.132 (mp at 114.77°C)
    0.161   (calculated, ΔS_{fus} = 50.5 J/mol K, Passivirta et al. 1999)

Water Solubility (g/m³ or mg/L at 25°C or as indicated and reported temperature dependence equations. Additional data at other temperatures designated * are compiled at the end of this section):
    1.90   (Pierotti et al. 1959)
    1.66   (shake flask, binding to bovine serum albumin, Sahyun 1966)
    1.90*   (shake flask-UV, measured range 24.6–73.4°C, Wauchope & Getzen 1972)
$R \cdot \ln x = -4670/(T/K) + 4.08 \times 10^{-4} \cdot [(T/K) - 291.15]^2 - 24.2 + 0.0309 \cdot (T/K)$; temp range 24.6–73.4°C (shake flask-UV measurements, Wauchope & Getzen 1972)
    1.98   (shake flask-fluorescence, Mackay & Shiu 1977)
    1.68*   (generator column-HPLC, measured range 6.6–31°C, May et al. 1978)
$S/(\mu g/kg) = 324.0 + 5.413 \cdot (t/°C) + 0.8059 \cdot (t/°C)^2 + 0.0025 \cdot (t/°C)^3$; temp range 4–29°C (generator column-HPLC/UV, May et al. 1978)
    1.62*   (24°C, generator column-HPLC, measured range 279.75–304.25 K, May et al. 1983)
    1.68   (generator column-HPLC, Wasik et al. 1983)
    1.83   (average lit. value, Pearlman et al. 1984)
    1.90   (generator column-HPLC/fluorescence, Walters & Luthy 1984)
    1.96   (generator column-HPLC/UV, Billington et al. 1988)
    1.9*   (recommended, IUPAC Solubility Data Series, Shaw 1989)
    2.23   (generator column-HPLC, Vadas et al. 1991)
$\log [S_L/(mol/L)] = 1.664 - 1024/(T/K)$ (supercooled liquid, Passivirta et al. 1999)
$\ln x = 0.82861 - 4824/(T/K)$; temp range 5–50°C (regression eq. of literature data, Shiu & Ma 2000)

Vapor Pressure (Pa at 25°C or as indicated and reported temperature dependence equations. Additional data at other temperatures designated * are compiled at the end of this section):

2400* (161.0°C, static isoteniscope method, measured range 161.0–300.4°C, Mortimer & Murphy 1923)

log $(P/mmHg)$ = 8.059 – 2957/(T/K); temp range 161–300.4°C (Antoine eq., static isoteniscope method, Mortimer & Murphy 1923)

666.6* (129.3°C, summary of literature data, temp range 129.3–295.0°C, Stull 1947)

0.087* (effusion method, measured range 33.3–49.55°C, Bradley & Cleasby 1953)

log $(P/cmHg)$ = 10.325 – 4324/(T/K); temp range 33.3–49.55°C (Antoine eq., Bradley & Cleasby 1953)

1.66 (extrapolated-Antoine eq., liquid state P_L, Weast 1972-73)

log $(P/mmHg)$ = [–0.2185 × 13682.8/(T/K)] + 8.18894; temp range 129.3–295°C (Antoine eq., Weast 1972–73)

1.13 (extrapolated-Antoine eq., liquid state P_L, Boublik et al. 1973)

1.133 (extrapolated-Antoine eq., supercooled liquid P_L, Dean 1985)

0.127* (static method-manometry, measured range 75.8–114°C, Osborn & Douslin 1975)

0.0946 (Irwin 1982)

0.080* (gas saturation-HPLC/UV, measured range 10–50°C, Sonnefeld et al. 1983)

log (P/Pa) = 14.385 – 4616.07/(T/K); temp range 10–50°C (Antoine eq., Sonnefeld et al. 1983)

0.080 (generator column-HPLC, Wasik et al. 1983)

0.473, 0.380 (P_{GC} by GC-RT correlation with eicosane as reference standard, different GC columns, Bidleman 1984)

log (P/kPa) = 2.88490 – 2635.371/(243.022 + t/°C); temp range 161–300.4°C (Antoine eq. from reported exptl. data, Boublik et al. 1984)

0.403 (Yamasaki et al. 1984)

log $(P/mmHg)$ = 7.7619 – 2637.1/(243.2 + t/°C); temp range 161–300°C (Antoine eq., Dean 1985, 1992)

0.0875* (gas saturation, interpolated-Antoine eq. derived from exptl. data, temp range 34–72°C, Sato et al. 1986)

ln (P/Pa) = 17.0935 – 2815.52/(T/K – 153.984); temp range 307.7–347.5 K (Antoine eq., gas saturation, Sato et al. 1986)

0.088 (extrapolated-Antoine eq.-I, Stephenson & Malanowski 1987)

log (P_S/kPa) = 10.449 – 4324/(T/K); temp range 306–323 K (Antoine eq.-I, Stephenson & Malanowski 1987)

log (P_S/kPa) = 10.04542 – 4122.908/(T/K); temp range 348–388 K (Antoine eq.-II, Stephenson & Malanowski 1987)

log (P_L/kPa) = 8.31368 – 4133.08/(86.582 + T/K); temp range 402–568 K (Antoine eq.-III, Stephenson & Malanowski 1987)

0.0850*, 0.566 (pressure gauge in vacuum cell: solid P_S, supercooled liquid P_L, extrapolated for 25°C from reported Antoine eq., measured temp range 30.03–154.81°C, Sasse et al. 1988)

log $(P_S/mmHg)$ = 11.64431 – 4268.644/(262.656 + t/°C); temp range: 30.03–100.08°C (Antoine eq., pressure gauge, Sasse et al. 1988)

log $(P_L/mmHg)$ = 7.74839 – 2641.73/(230.963 + t/°C); temp range 110.06–154.83°C (Antoine eq., pressure gauge, Sasse et al. 1988)

0.474 (P_{GC} by GC-RT correlation with eicosane as reference standard, Hinckley et al. 1990)

0.793, 0.652 (supercooled P_L converted from literature Ps with different ΔS_{fus} values, Hinckley et al. 1990)

log $(P/mmHg)$ = 53.9382 – 5.322 × 10^3/(T/K) – 16.059·log (T/K) + 4.5696 × 10^{-3}·(T/K) + 8.1430 × 10^{-13}·(T/K)^2; temp range 388–870 K (vapor pressure eq., Yaws 1994)

0.407 (supercooled liquid P_L, calculated from Yamasaki et al. 1984, Finizio et al. 1997)

0.0575, 0.0885; 0.080; 0.0851 (quoted exptl.: effusion method; gas saturation; manometry, Delle Site 1997)

0.0792, 0.243; 0.00594, 0.00477 (quoted lit.; calculated; GC-RT correlation, Delle Site 1997)

0.72; 0.116 (quoted supercooled liquid P_L from Hinckley et al. 1990; converted to solid P_S with fugacity ratio F, Passivirta et al. 1999)

log (P_S/Pa) = 11.27 – 3638/(T/K) (solid, Passivirta et al. 1999)

log (P_L/Pa) = 8.63 – 2614/(T/K) (supercooled liquid, Passivirta et al. 1999)

log (P/Pa) = 14.385 – 4616.07/(T/K); temp range 5–50°C (regression eq. from literature data, Shiu & Ma 2000)

0.526; 0.194 (supercooled liquid P_L: calibrated GC-RT correlation; GC-RT correlation, Lei et al. 2002)

log (P_L/Pa) = –3492/(T/K) + 11.43; ΔH_{vap} = –66.9 kJ·mol^{-1} (GC-RT correlation, Lei et al. 2002)

0.086*   (25.05°C, transpiration method, measured range 288.7–359.2 K, Verevkin 2004)
ln (P/Pa) = 298.47/R − 95086.65/[R(T/K)] − (30.2/R)·ln [(T/K)/298.15]; temp range 288.7–359.2 K (transpiration method, Verevkin 2004)

Henry's Law Constant (Pa m^3/mol at 25°C or as indicated and reported temperature dependence equations. Additional data at other temperatures designated * are compiled at the end of this section):
- 7.75   (batch stripping, Mackay & Shiu 1981)
- 10.13   (batch stripping, Mackay et al. 1982)
- 11.85   (batch stripping, Warner et al. 1987)
- 6.45   (wetted-wall column, Fendinger & Glotfelty 1990)
- 9.75   (gas stripping-fluorescence, Shiu & Mackay 1997)
- 6.50   (gas stripping-HPLC/fluorescence, De Maagd et al. 1998)
- 9.81*   (gas stripping-GC; measured range 4.1–31°C, Bamford et al. 1999)

ln K_{AW} = −5869.62/(T/K) + 14.193; ΔH = 48.8 kJ mol^{-1}; measured range 4.1–31°C (gas stripping-GC, Bamford et al. 1999)

log (H/(Pa m^3/mol)) = 6.97 − 1590/(T/K) (Passivirta et al. 1999)

Octanol/Water Partition Coefficient, log K_{OW}:
- 4.18   (Hansch & Leo 1979)
- 4.12   (Chou & Jurs 1979)
- 4.18   (HPLC-k' correlation, Rekker & De Kort 1979)
- 3.91   (HPLC-k' correlation, Hanai et al. 1981)
- 4.18   (RP-TLC-k' correlation, Bruggeman et al. 1982)
- 4.18   (shake flask-UV, Yalkowsky et al. 1983b)
- 4.23   (HPLC-RT correlation, Rapaport et al. 1984)
- 4.18   (shake flask-GC, Haky & Leja 1986)
- 4.10   (RP-HPLC-RT correlation, Chin et al. 1986)
- 4.23   (HPLC-RT correlation, Wang et al. 1986)
- 4.13   (TLC-RT correlation, De Voogt et al. 1990)
- 4.18   (recommended, Sangster 1993)
- 4.18   (recommended, Hansch et al. 1995)
- 4.32 ± 0.19, 3.68 ± 0.62   (HPLC-k' correlation: ODS column; Diol column, Helweg et al. 1997)

Octanol/Air Partition Coefficient, log K_{OA} at 25°C or as indicated and reported temperature dependence equation. Additional data at other temperatures designated * are compiled at the end of this section:
- 6.68   (calculated, Finizio et al. 1997)
- 6.79*; 6.59   (generator column-GC; calculated-C_O/C_A, measured range 0–40°C, Harner & Bidleman 1998)

log K_{OA} = −7.74 + 4332/(T/K); temp range 0–40°C (generator column-GC. Harner & Bidleman 1998)
- 6.83, 6.79   (calculated-S_{oct} and vapor pressure P, quoted lit., Abraham et al. 2001)

Bioconcentration Factor, log BCF:
- 3.67   (microorganisms-water, calculated-K_{OW}, Mabey et al. 1982)
- 2.70   (*Daphnia magna*, Newsted & Giesy 1987)

Sorption Partition Coefficient, log K_{OC}:
- 3.95; 3.87   (Aldrich and Fluka humic acids, observed; predicted, Chin et al. 1989)
- 5.47   (sediments average, Kayal & Connell 1990)
- 3.76   (RP-HPLC correlation, Pussemier et al. 1990)
- 4.15, 4.21   (RP-HPLC correlation on CIHAC, on PIHAC, Szabo 1990b)
- 4.68   (humic acid, HPLC-k' correlation, Nielsen et al. 1997)
- 3.24–5.75; 4.10–5.50   (range, calculated from sequential desorption of 11 urban soils; lit. range, Krauss & Wilcke 2001)
- 4.81; 4.93, 4.24, 4.63 (20°C, batch equilibrium, A2 alluvial grassland soil; calculated values of expt 1,2,3-solvophobic approach, Krauss & Wilcke 2001)

3.95 (Askov soil, a Danish Agricultural soil, Sverdrup et al. 2002)

3.93–6.19 (sediment/water, initial-final values of 5–100 d contact time, gas-purge technique-HPLC/fluo., ten Hulscher et al. 2003)

Environmental Fate Rate Constants, k, or Half-Lives, $t_{1/2}$:

Volatilization:

Photolysis: half-lives on different atmospheric particulate substrates determined in rotary photoreactor (approx. 25 µg/g on substrate): $t_{1/2}$ = 110 h on silica gel, $t_{1/2}$ = 62 h on alumina and $t_{1/2}$ = 37 h on fly ash (Behymer & Hites 1985);

photolysis rate k < 2 × 10^{-5} s^{-1} with $t_{1/2}$ > 1.6 d (Kwok et al. 1997):

direct photolysis $t_{1/2}$ = 7.69 h (predicted- QSPR) in atmospheric aerosol (Chen et al. 2001)

Photodegradation k = 9.0 × 10^{-7} s^{-1} in surface water during the summertime at mid-latitude (Fasnacht & Blough 2002).

Oxidation: rate constant k, for gas-phase second order rate constants, k_{OH} for reaction with OH radical, k_{NO_3} with NO_3 radical and k_{O_3} with O_3, or as indicated *data at other temperatures and/or the Arrhenius expression see reference:

k(calc) < 360 M^{-1} h^{-1} for singlet oxygen and k = 3 × 10^3 M^{-1} h^{-1} for peroxy radical (Mabey et al. 1982)

k_{OH}(calc) = 13.0 × 10^{-12} cm^3 $molecule^{-1}$ s^{-1} at 300 K (SAR structure-activity relationship, Arey et al. 1989, Atkinson 1989)

k_{OH}(calc) = 9.90 × 10^{-12} cm^3 $molecule^{-1}$ s^{-1} (molecular orbital calculation, Klamt 1996)

k_{OH}(exptl) = (16 ± 5) × 10^{-12} cm^3 $molecule^{-1}$ s^{-1}, k_{OH}(calc) = 9.2 × 10^{-12} cm^3 $molecule^{-1}$ s^{-1} with a calculated lifetime τ = 9 h; k_{NO_3}(exptl) = (3.5 ± 1.2) × 10^{-14} cm^3 $molecule^{-1}$ s^{-1} with a calculated lifetime τ = 1.3 d; and k_{O_3}(exptl) < 2 × 10^{-19} cm^3 $molecule^{-1}$ s^{-1} with a calculated lifetime τ > 82 d at 297 ± 2 K (relative rate method; calculated-SAR structure-activity relationship, Kwok et al. 1997)

k_{OH}* = 23 × 10^{-12} cm^3 $molecule^{-1}$ s^{-1} at 298 K, measured range 306–366 K with a calculated atmospheric lifetime of 22 h based on gas-phase OH reactions (Brubaker & Hites 1998)

Hydrolysis: no hydrolyzable groups (Howard et al. 1991).

Biodegradation: significant degradation with gradual adaptation within 7 d for an average of three static-flask screening test (Tabak et al. 1981);

nonautoclaved groundwater samples of approx. 0.06 mg/L are degraded at rates of about 30% per week by microbes (Lee et al. 1984);

$t_{1/2}$(aq. aerobic) = 768–1440 h, based on aerobic soil die-away test data (Coover & Sims 1987; quoted, Howard et al. 1991);

$t_{1/2}$(aq. anaerobic) = 3072–5760 h, based on estimated unacclimated aqueous aerobic biodegradation half-life (Howard et al. 1991).

Biotransformation: estimated rate constant for bacteria, 3 × 10^{-9} mL $cell^{-1}$ h^{-1} (Mabey et al. 1982).

Bioconcentration, Uptake (k_1) and Elimination (k_2) Rate Constants:

k_1 = 12.3 mg g^{-1} h^{-1}; k_2 = 0.051 h^{-1} (freshwater oligochaete from sediment, Van Hoof et al. 2001)

Half-Lives in the Environment:

Air: $t_{1/2}$ = 6.81–68.1 h, based on reported rate constant for reaction with hydroxyl radical in air (Howard et al. 1991);

photolysis $t_{1/2}$ > 1.6 d; calculated tropospheric lifetimes of 9 h, 1.3 d and > 82d due to reactions with OH radical, NO_3 radical and O_3, respectively, at room temp. (Kwok et al. 1997);

calculated atmospheric lifetime of 22 h based on gas-phase reactions with OH radical (Brubaker & Hites 1998).

Surface water: $t_{1/2}$ = 768–1440 h, based on aerobic soil die-away test data (Howard et al. 1991).

Groundwater: $t_{1/2}$ = 1536–2880 h, based on estimated unacclimated aqueous aerobic biodegradation half-life (Howard et al. 1991).

Sediment:

Soil: $t_{1/2}$ = 768–1440 h, based on aerobic soil die-away test data (Howard et al. 1991);

$t_{1/2}$ > 50 d (Ryan et al. 1988).

Biota: elimination $t_{1/2}$ = 7 d from rainbow trout (quoted, Meador et al. 1995).

TABLE 4.1.1.21.1
Reported aqueous solubilities of fluorene at various temperatures

Wauchope & Getzen 1972				May 1980, 1983		Shaw 1989	
shake flask-UV				generator column-HPLC		IUPAC recommended	
t/°C	S/g·m^{-3}	t/°C	S/g·m^{-3}	t/°C	S/g·m^{-3}	t/°C	S/g·m^{-3}
	experimental		smoothed				
24.6	1.93	0	0.66	6.60	0.7184	0	0.70
24.6	1.87	24.6	1.86	13.2	0.9673	25	1.90
24.6	1.88	25	1.90	18.0	1.203	30	2.40
29.9	2.41	29.9	2.37	24.0	1.616	40	3.80
29.9	2.33	30.3	2.41	27.0	1.845	50	6.30
29.9	2.34	38.4	3.53	31.1	2.248	60	10
30.3	2.10	40.1	3.84			70	19
30.3	2.25	47.5	5.54				
30.3	2.23	50	6.29	temp dependence eq. 2			
38.4	3.72	50.1	6.32	S	µg/kg		
38.4	3.73	50.2	6.35	a	0.0185		
40.1	3.88	54.7	8.02	b	0.4543		
40.1	3.84, 3.85	59.2	10.2	c	22.76		
47.5	5.59, 5.62	60.5	10.9	d	543.3		
47.5	5.68	65.1	14.1				
50.1	6.31, 6.42	70.7	19.2	ΔH_{sol}/(kJ mol^{-1}) = 32.97			
50.1	6.54	71.9	20.6	measured between 5–30°C			
50.2	6.27	73.4	22.5				
54.7	8.31, 8.41	75	24.7				
54.7	8.56						
59.2	10.5						
60.5	10.7	temp dependence eq. 1					
60.5	11.0	ln x	mole fraction				
60.5	11.6	ΔH_{fus}	19.54 ± 0.13				
65.1	14.2	10^2·b	3.09 ± 0.1				
65.1	14.1	c	24.2 ± 0.3				
70.7	18.5						
70.7	18.9						
71.9	18.8						
73.4	21.5						

ΔH_{fus}/(kJ mol^{-1}) = 19.54

Empirical temperature dependence equations:

Wauchope & Getzen (1972): R·ln x = −[H$_{fus}$/(T/K)] + (0.000408)[(T/K) − 291.15]^2 −c + b·(T/K)　　(1)

May et al. (1978):− S/(µg/kg) = a·t^3 + b·t$_{½}$ +c·t + d　　(2)

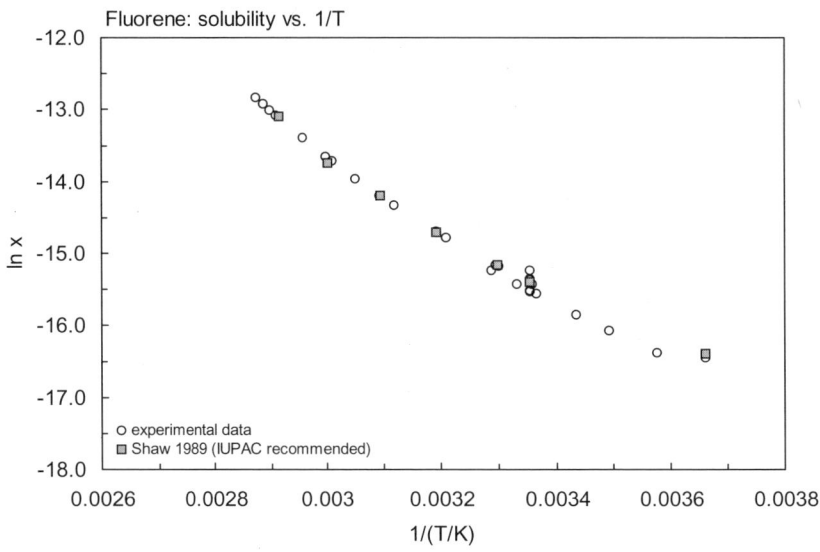

FIGURE 4.1.1.21.1 Logarithm of mole fraction solubility (ln x) versus reciprocal temperature for fluorene.

TABLE 4.1.1.21.2
Reported vapor pressures of fluorene at various temperatures and the coefficients for the vapor pressure equations

$$\log P = A - B/(T/K) \quad (1) \qquad \ln P = A - B/(T/K) \quad (1a)$$
$$\log P = A - B/(C + t/°C) \quad (2) \qquad \ln P = A - B/(C + t/°C) \quad (2a)$$
$$\log P = A - B/(C + T/K) \quad (3)$$
$$\log P = A - B/(T/K) - C \cdot \log (T/K) \quad (4)$$
$$\ln P = A/R - B/[R(T/K)] - (C/R) \cdot \ln [(T/K)/298.15]; \; R - \text{gas constant} \quad (5)$$

1.

Mortimer & Murphy 1923		Stull 1947		Bradley & Cleasby 1953		Osborn & Douslin 1975	
isoteniscope-Hg manometer		summary of literature data		effusion		manometry	
t/°C	P/Pa	t/°C	P/Pa	t/°C	P/Pa	t/°C	P/Pa
161.0	2400	129.3	666.6	33.3	0.2186	75.0	15.065
202.5	9266	146.0	1333	37.2	0.3333	80.0	22.264
203.0	9399	164.2	2666	40.3	0.4573	85.0	32.263
240.0	27398	185.2	5333	45.0	0.7239	90.0	47.462
241.4	27784	197.8	7999	49.25	1.0906	95.0	68.26
276.6	64995	214.7	13332	34.85	0.2600	100.0	99.19
277.1	65421	240.3	26664	38.45	0.3746	105.0	141.45
295.6	98499	268.6	53329	42.45	0.5546	110.0	199.98
295.7	98499	295.0	101325	47.75	0.9439	114.0	259.04
295.7	98499			49.55	1.1106		
299.8	107231	mp/°C	113				
300.4	108298			eq. 1	P/mmHg	triple point	387.943 K
				A	10.325		
				B	4324	$\Delta H_{subl}/(kJ\,mol^{-1}) = 81.76$	
eq. 1	P/mmHg			temp range 33–50°C			at bp

TABLE 4.1.1.21.2 (Continued)

Mortimer & Murphy 1923		Stull 1947		Bradley & Cleasby 1953		Osborn & Douslin 1975	
isoteniscope-Hg manometer		summary of literature data		effusion		manometry	
t/°C	P/Pa	t/°C	P/Pa	t/°C	P/Pa	t/°C	P/Pa
A	8.059					ΔH_{fus}/(kJ mol^{-1}) = 19.58	
B	2957						
temp range 161–300°C						ΔH_V/(kJ mol^{-1}) = 62.17 at bp	

2.

Sonnefeld et al. 1983		Sato et al. 1986		Sasse et al. 1988		Verevkin 2004	
gas saturation–HPLC		gas saturation–electrobalance		electronic manometry		transpiration	
t/°C	P/Pa	t/°C	P/Pa	t/°C	P/Pa	T/K	P/Pa
				solid			
10.5	0.0132	34.55	0.297	30.03	0.157	288.8	0.028
10.5	0.0133	36.35	0.360	39.92	0.184	290.4	0.034
10.5	0.013	41.15	0.620	49.88	1.241	292.4	0.044
20.4	0.0425	47.45	1.21	59.92	3.426	289.6	0.031
20.4	0.0438	49.85	1.53	60.03	3.533	296.2	0.067
20.4	0.0451	52.35	1.95	69.93	8.746	298.2	0.086
30.0	0.147	55.45	2.60	79.96	20.67	300.2	0.112
29.97	0.153	57.35	3.07	79.98	20.80	302.2	0.132
29.97	0.146	58.75	3.88	89.99	46.26	304.3	0.166
38.85	0.387	63.25	4.78	89.99	46.40	306.3	0.218
38.85	0.384	64.55	5.88	100.04	97.86	308.3	0.268
38.85	0.382	67.25	7.38	100.08	98.93	310.2	0.328
38.9	0.387	69.75	9.13			313.3	0.449
38.9	0.393	71.25	10.2	eq. 2	P/mmHg	314.3	0.507
25.0	0.080	74.35	13.0	A	11.64431	315.3	0.545
				B	4268.664	317.3	0.666
eq. 1	P/Pa	eq. 3	P/Pa	C	262.656	320.3	0.930
A	14.385	A	17.0935	temp range: 30–100.08°C		323.4	1.245
B	4616.07	B	2815.52			326.5	1.645
		C	−153.984	liquid		329.3	2.223
				110.06	212.5	332.3	3.103
				120.04	352.0	335.3	3.867
				120.04	353.0	338.2	4.838
				129.97	568.6	341.2	6.552
				139.95	894.3	344.2	8.417
				149.86	1369	347.1	11.190
				154.81	1683	350.3	14.497
						353.2	18.302
				eq. 2	P/mmHg	356.3	23.900
				A	7.94839	359.2	30.456
				B	2641.73		
				C	230.963	ΔH_{subl}/(kJ mol^{-1}) = 86.08	
				temp range: 110–155°C		eq. 5	P/Pa
						A	298.47
						B	95086.65
						C	30.2

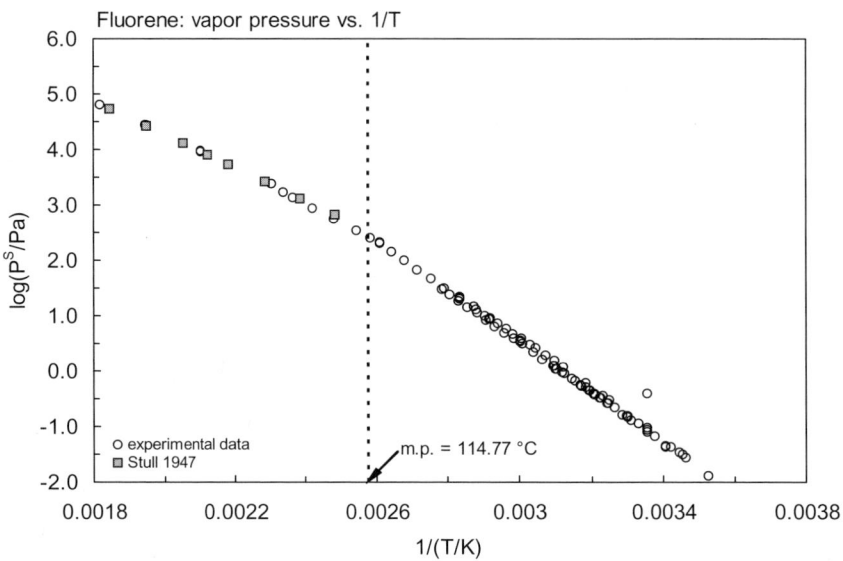

FIGURE 4.1.1.21.2 Logarithm of vapor pressure versus reciprocal temperature for fluorene.

TABLE 4.1.1.21.3
Reported Henry's law constants and octanol–air partition coefficients of fluorene at various temperatures and temperature dependence equations

Henry's law constant			log K_{OA}	
Bamford et al. 1999			Harner & Bidleman 1998	
gas stripping-GC/MS			generator column-GC/FID	
t/°C	H/(Pa m³/mol)	H/(Pa m³/mol)	t/°C	log K_{OA}
		average		
4.1	1.96, 2.14	2.05	0	8.134
11.0	3.44, 3.65	3.54	10	7.501
18.0	5.81, 6.12	5.96	20	7.130
25.0	9.49, 10.1	9.81	30	6.516
31.0	14.1, 15.4	14.8	40	6.093
			25(exptl)	6.79
ln K_{AW} = A − B/(T/K)			25(calc)	6.59
eq. 1	K_{AW}			
A	14.193		log K_{OA} = A + B/(T/K)	
B	5870		A	−7.74
			B	4332
enthalpy, entropy change:				
ΔH/(kJ·mol⁻¹) = 48.8 ± 0.8			enthalpy of phase change	
ΔS/(J·K⁻¹ mol⁻¹) = 118			$ΔH_{OA}$/(kJ mol⁻¹) = 82.9	
at 25°C				

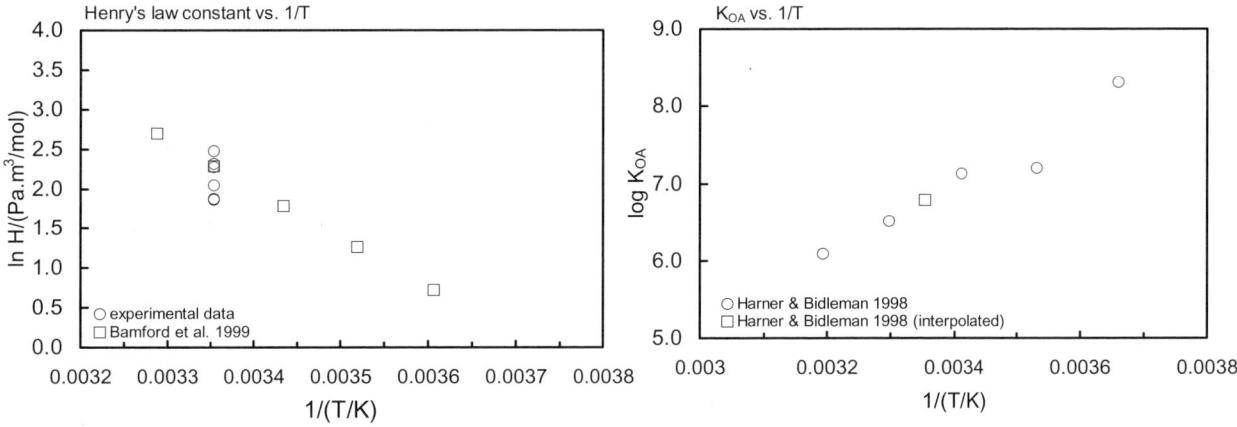

FIGURE 4.1.1.21.3 Logarithm of Henry's law constant and K_{OA} versus reciprocal temperature for fluorene.

4.1.1.22 1-Methylfluorene

Common Name: 1-Methylfluorene
Synonym:
Chemical Name: 1-methylfluorene
CAS Registry No: 1730-37-6
Molecular Formula: $C_{14}H_{12}$
Molecular Weight: 180.245
Melting Point (°C):
    87   (Lide 2003)
Boiling Point (°C):
    318   (Weast 1982)
Density (g/cm³ at 20°C):
Molar Volume (cm³/mol):
    177.1   (Ruelle & Kesselring 1997)
    210.1   (calculated-Le Bas method at normal boiling point)
Enthalpy of Fusion, ΔH_{fus} (kJ/mol):
Entropy of Fusion, ΔS_{fus} (J/mol K):
Fugacity Ratio at 25°C (assuming ΔS_{fus} = 56 J/mol K), F: 0.246 (mp at 87°C)

Water Solubility (g/m³ or mg/L at 25°C):
    1.092, 4.867   (measured, supercooled liquid value, Miller et al. 1985)

Vapor Pressure (Pa at 25°C and the reported temperature dependence equations):
    0.136; 0.0708   (supercooled liquid P_L: calibrated GC-RT correlation; GC-RT correlation, Lei et al. 2002)
    $\log (P_L/Pa) = -3711/(T/K) + 11.58$; $\Delta H_{vap.} = -71.1$ kJ·mol⁻¹ (GC-RT correlation, Lei et al. 2002)
    0.032   (24.95°C, transpiration method, measured range 298.1–375.3 K, Verevkin 2004)
    $\ln (P_S/Pa) = 311.78/R - 101590.4/[R(T/K)] - (35.1/R)\cdot\ln [(T/K)/298.15]$; temp range 298.1–359.2 K (solid, transpiration method, Verevkin 2004)
    $\ln (P_L/Pa) = 330.39/R - 104778.1/[R(T/K)] - (87.5/R)\cdot\ln [(T/K)/298.15]$; temp range 361.2–375.3 K (liquid, transpiration method, Verevkin 2004)

Henry's Law Constant (Pa m³/mol):

Octanol/Water Partition Coefficient, log K_{OW}:
    4.97   (calculated, Miller et al. 1985)
    4.63   (calculated-solvatochromic parameters and V_I, Kamlet et al. 1988)
    5.7640   (calculated-UNIFAC group contribution, Chen et al. 1993)
    4.97   (recommended, Hansch et al. 1995)

Octanol/Air Partition Coefficient, log K_{OA}:

Bioconcentration Factor, log BCF:

Sorption Partition Coefficient, log K_{OC}:

Environmental Fate Rate Constants, k, or Half-Lives, $t_{½}$:

Half-Lives in the Environment:

4.1.1.23 Phenanthrene

Common Name: Phenanthrene
Synonym: o-diphenyleneethylene, phenanthren
Chemical Name: phenanthrene
CAS Registry No: 85-01-8
Molecular Formula: $C_{14}H_{10}$
Molecular Weight: 178.229
Melting Point (°C):
    99.24   (Lide 2003)
Boiling Point (°C):
    340   (Dean 1985; Lide 2003)
Density (g/cm³ at 20°C):
    1.174   (Dean 1985)
    0.980   (4°C, Weast 1982–83; Lide 2003)
Molar Volume (cm³/mol):
    182.0   (calculated-density, liquid molar volume, Lande & Banerjee 1981)
    199.2   (calculated-Le Bas method at normal boiling point)
Enthalpy of Fusion, ΔH_{fus} (kJ/mol):
    18.62   (Parks & Huffman 1931; Tsonopoulos & Prausnitz 1971; Fu & Luthy 1985)
    16.28   (differential calorimetry, Wauchope & Getzen 1972)
    16.44   (calorimetry, Osborn & Douslin 1975)
    0.22, 16.46; 15.58   (74.35, 99.25°C; total phase change enthalpy, Chickos et al. 1999)
Entropy of Fusion, ΔS_{fus} (J/mol K):
    42.68   (Wauchope & Getzen 1972)
    45.19   (Casellato et al. 1973)
    50.63   (Ubbelohde 1978)
    47.70   (De Kruif 1980)
    44.83, 44.2   (exptl., calculated-group additivity method, total phase change entropy, Chickos et al. 1999)
Fugacity Ratio at 25°C (assuming ΔS_{fus} = 56 J/mol K), F: 0.187 (mp at 99.24°C)

Water Solubility (g/m³ or mg/L at 25°C or as indicated and reported temperature dependence equations. Additional data at other temperatures designated * are compiled at the end of this section):
    1.65   (shake flask-nephelometry, Davis & Parker 1942)
    1.60   (27°C, nephelometry, Davis et al. 1942)
    0.994   (shake flask-UV, Andrews & Keefer 1949)
    1.60   (shake flask-UV, Klevens 1950)
    1.18   (Pierotti et al. 1959)
    0.71   (shake flask, binding to bovine serum albumin, Sahyun 1966)
    1.60   (shake flask-UV/fluorescence, Barone et al. 1967)
    2.67   (20°C, shake flask-UV, Eisenbrand & Baumann 1970)
    1.18*   (shake flask-UV, measured range 24.6–73.4°C, Wauchope & Getzen 1972)
    $R \cdot \ln x = -3890/(T/K) + 4.08 \times 10^{-4} \cdot [(T/K) - 291.15]^2 - 27.9 + 0.0374 \cdot (T/K)$; temp range 24.6–73.4°C
        (shake flask-UV measurements, Wauchope & Getzen 1972)
    3.03, 2.85 (20°C, HPLC-relative retention correlation, different stationary and mobile phases, Locke 1974)
    1.21   (shake flask-UV, Vesala 1974)
    1.07   (shake flask-GC, Eganhouse & Calder 1976)
    1.29   (shake flask-fluorescence, Mackay & Shiu 1977)
    1.002   (Rossi 1977; Neff 1979)
    1.151*   (shake flask-UV, measured range 8.4–31.8°C, Schwarz 1977)

1.002*   (generator column-HPLC/UV, measured range 4–29°C, May et al. 1978, 1983)

$S/(\mu g/kg) = 324.0 + 5.413 \cdot (t/°) + 0.8059 \cdot (t/°C)^2 + 0.0025 \cdot (t/°C)^3$; temp range 4–29°C (generator column-HPLC/UV, May et al. 1978)

0.955*   (24.3°C, generator column-HPLC, measured range 4.0–29.9°C, May 1980)
0.816   (quoted, Verschueren 1983)
1.0   (generator column-HPLC/UV, Wasik et al. 1983)
1.28   (average lit. value, Pearlman et al. 1984)
1.29   (generator column-HPLC/fluorescence, Walters & Luthy 1984)
1.10*   (generator column-HPLC/UV, measured range 4.6–25.3°C, Whitehouse 1984)
1.69   (29°C, shake flask-GC/FID, Stucki & Alexander 1987)
0.0446   (vapor saturation-UV, Akiyoshi et al. 1987)
1.08   (generator column-HPLC, Billington et al. 1988)
1.10*   (recommended, IUPAC Solubility Data Series, Shaw 1989)
1.0   (generator column-HPLC, Vadas et al. 1991)
1.03   (dialysis tubing equilibration-GC, Etzweiler et al. 1995)
0.823   (generator column-HPLC/fluorescence, De Maagd et al. 1998)
1.20   (microdroplet sampling and multiphoton ionization-based fast-conductivity technique MPI-FC, Gridin et al. 1998)

$\log S_L$ (mol/L) $= 0.930 - 861.6/(T/K)$ (supercooled liquid, Passivirta et al. 1999)

$\ln x = -2.546051 - 4053/(T/K)$; temp range 5–50°C (regression eq. of literature data, Shiu & Ma 2000)

Vapor Pressure (Pa at 25°C or as indicated and reported temperature dependence equations. Additional data at other temperatures designated * are compiled at the end of this section):

7773*   (230°C, isoteniscope-Hg, measured range 230–340°C, Nelson & Senseman 1922)
3626*   (203.6°C, isoteniscope-Hg manometer, measured range 203.6–346.8°C, Mortimer & Murphy 1923)

$\log (P/mmHg) = 7.771 - 2990/(T/K)$; temp range 203.6–346.8°C (Antoine eq., static isoteniscope method, Mortimer & Murphy 1923)

133.3*   (118.2°C, summary of literature data, temp range 118.2–340.2°C, Stull 1947)
0.0997   (effusion method, Inokuchi et al. 1952)
0.0227*   (effusion method, Bradley & Cleasby 1953)

$\log (P/cmHg) = 10.388 - 4519/(T/K)$; temp range 36.7–49.65°C (Antoine eq., Bradley & Cleasby 1953)

$\log (P/mmHg) = 16.0 - 5008/(T/K)$; temp range 0–60°C (Knudsen effusion method, Hoyer & Peperle 1958)

0.464   (extrapolated from Antoine eq. of liquid state P_L, Weast 1972–73)

$\log (P/mmHg) = [-0.2185 \times 14184.0/(T/K)] + 7.936781$; temp range 118.2–340°C (Antoine eq., Weast 1972–73)

0.159   (extrapolated from Antoine eq. of liquid state P_L, Boublik et al. 1973)
30.4*   (100.0°C, inclined-piston manometry, measured range 100.0–150.0°C, Osborn & Douslin 1975)
0.0187   (lit. average-interpolated, API 1979)
0.0263*   (gas saturation, Macknick & Prausnitz 1979)
0.0267   (extrapolated-Clapeyron eq., Macknick & Prausnitz 1979)

$\log (P/mmHg) = 26.648 - 10484/(T/K)$; temp range 51.6–90.3°C (Clapeyron eq., gas saturation, Macknick & Prausnitz 1979)

0.018*   (effusion, De Kruif 1980)
0.0161*   (gas saturation-HPLC/UV, Sonnefeld et al. 1983)

$\log (P/Pa) = 14.852 - 4962.77/(T/K)$; temp range 10–50°C (Antoine eq., Sonnefeld et al. 1983)

0.016   (generator column-HPLC/UV, Wasik et al. 1983)
0.111, 0.0688 (P_{GC} by GC-RT correlation with eicosane as reference standard, different GC columns, Bidleman 1984)
0.134   (supercooled liquid P_L converted from literature P_S, Bidleman 1984)

$\log (P_L/kPa) = 6.61335 - 2593.134/(224.402 + t/°C)$; temp range 203.6–346.8°C (Antoine eq. from reported exptl. data, Boublik et al. 1984)

$\log (P_L/kPa) = 6.01392 - 2039.351/(168.569 + t/°C)$; temp range 230–340°C (Antoine eq. from reported exptl. data, Boublik et al. 1984)

$\log (P/mmHg) = 7.26082 - 2379.04/(203.76 + t/°C)$; temp range 176–379°C (Antoine eq., Dean 1985, 1992)

0.070   (Yamasaki et al. 1984)
0.0149   (selected, Howard et al. 1986)

0.012* (gas saturation, extrapolated-Antoine eq. from exptl. data, temp range 49–74°C, Sato et al. 1986)
ln (P/Pa) = 20.3950 − 2931.20/(T/K − 139.743); temp range: 322.9–347.8 K (Antoine eq., gas saturation, Sato et al. 1986)
0.025 (interpolated-Antoine eq., Stephenson & Malanowski 1987)
log (P_S/kPa) = 10.305 − 4444/(T/K); temp range 296–315 K (Antoine eq.-I, Stephenson & Malanowski 1987)
log (P_S/kPa) = 10.70162 − 4554.38/(T/K); temp range 313–363 K (Antoine eq.-II, Stephenson & Malanowski 1987)
log (P_L/kPa) = 6.64812 − 2513.134/(−65.345 + T/K); temp range 373–423 K (Antoine eq., Stephenson & Malanowski 1987)
log (P_L/kPa) = 7.17186 − 3235.19/(12.908 + T/K); temp range 391–613 K (Antoine eq., Stephenson & Malanowski 1987)
0.0127, 0.0827 (literature mean P_S, supercooled liquid P_L, Bidleman & Foreman 1987)
0.111, 0.0556 (P_{GC}, GC-RT correlation with different reference standards, Hinckley et al. 1990)
0.134, 0.10 (supercooled liquid P_L converted from literature P_S with different ΔS_{fus} values, Hinckley et al. 1990)
log (P_L/Pa) = 11.46 − 3716/(T/K) (GC-RT correlation, Hinckley et al. 1990)
log (P/mmHg) = 50.2858 − 5.7409 × 10^3/(T/K) − 13.935·log (T/K) − 8.852 × 10^{-10}·(T/K) + 2.11343 × 10^{-6}·(T/K)^2; temp range 372–869 K (vapor pressure eq., Yaws 1994)
0.0708 (supercooled liquid P_L, calculated from Yamasaki et al. 1984, Finizio et al. 1997)
0.0251, 0.0227, 0.018, 0.0186; 0.0267, 0.0161 (quoted exptl.: effusion method, gas saturation, Delle Site 1997)
0.0288, 0.0227, 0.0181; 0.0196, 0.0122, 0.0173 (quoted lit.: calculated; from GC-RT correlation, Delle Site 1997)
0.0197* (Knudsen effusion/thermogravimetry technique, extrapolated Clausius-Clapeyron eq., Oja & Suuberg 1998)
log (P/Pa) = 34.387 − 11423/(T/K); temp range: 303–333 K (Clausius-Clapeyron eq., Knudsen effusion, Oja & Suuberg 1998)
0.115; 0.0306 (quoted supercooled liquid P_L from Hinckley et al. 1990; converted to solid P_S with fugacity ratio F, Passivirta et al. 1999)
log (P_S/Pa) = 11.38 − 3842/(T/K) (solid, Passivirta et al. 1999)
log (P_L/Pa) = 9.07 − 2982/(T/K) (supercooled liquid, Passivirta et al. 1999)
0.0163 ± 0.004 (gas saturation-HPLC/fluorescence; de Seze et al. 2000)
log (P/Pa) = 14.852 − 4962.77/(T/K); temp range 5–50°C (regression eq. from literature data, Shiu & Ma 2000)
0.0799; 0.0475 (supercooled liquid P_L, calibrated GC-RT correlation; GC-RT correlation, Lei et al. 2002)
log (P_L/Pa) = −3768/(T/K) + 11.54; $\Delta H_{vap.}$ = −72.2 kJ·mol^{-1} (GC-RT correlation, Lei et al. 2002)
0.0202 (solid P_S, gas saturation-GC/MS, Mader & Pankow 2003)
0.0966 (supercooled liquid P_L, calculated from P_S assuming ΔS_{fus} = 56 J/mol K, Mader & Pankow 2003)

Henry's Law Constant (Pa m^3/mol at 25°C or as indicated and reported temperature dependence equations. Additional data at other temperatures designated * are compiled at the end of this section):
5.55 (gas stripping-GC, Southworth 1979)
3.981 (gas stripping-GC, Mackay et al. 1979; Mackay & Shiu 1981; Mackay et al. 1982)
3.65 (gas stripping-GC, Mackay & Shiu 1981; Mackay et al. 1982)
2.38 (wetted-wall column-GC, Fendinger & Glotfelty 1990; quoted, Shiu & Mackay 1997; Shiu et al. 1999)
3.97 (headspace solid-phase microextraction (SPME)-GC, Zhang & Pawliszyn 1993)
4.68* (gas stripping-GC, measured range 5.9–34.7°C, Alaee et al. 1996)
ln K_{AW} = 6.0314 − 3524.18/(T/K); temp range 5.9–34.7°C (gas stripping-GC, Alaee et al. 1996)
3.61 (gas stripping-GC, Shiu & Mackay 1997)
2.90 (gas stripping-HPLC/fluo., De Maagd et al. 1998)
4.29* (gas stripping-GC; measured range 4.1–31°C, Bamford et al. 1999)
ln K_{AW} = −5689.2/(T/K) + 12.75, ΔH = 47.3 kJ mol^{-1}; measured range 4.1–31°C (gas stripping-GC, Bamford et al. 1999)
log (H/(Pa m^3/mol)) = 8.14 − 2120/(T/K) (Passivirta et al. 1999)
3.85 (20°C, selected from reported experimentally measured data, Staudinger & Roberts 2001)
log K_{AW} = 2.417 − 1530/(T/K) (van't Hoff eq. derived from literature data, Staudinger & Roberts 2001)

Octanol/Water Partition Coefficient, log K_{OW} at 25°C or as indicated:
- 4.46 (Hansch & Fujita 1964; Leo et al. 1971; Hansch et al. 1973; Hansch & Leo 1979)
- 4.66 (calculated-molecular connectivity indices MCI, Kier et al. 1971)
- 4.67 (calculated-fragment const., Rekker 1977)
- 4.57 (shake flask-UV, concn ratio, Karickhoff et al. 1979)
- 4.45 (HPLC-k′ correlation, McDuffie 1981)
- 4.63 (RP-TLC-k′ correlation, Bruggeman et al. 1982)
- 4.53 (HPLC-k′ correlation, Hammers et al. 1982)
- 4.52; 4.31 (shake flask; HPLC correlation, Eadsforth & Moser 1983)
- 4.46 (HPLC-k′ correlation, Hafkenscheid & Tomlinson 1983)
- 4.28 (HPLC-k′ correlation, Haky & Young 1984)
- 4.39 (RP-HPLC-RT correlation, Chin et al. 1986)
- 4.50 (HPLC-RT correlation, Wang et al. 1986)
- 4.56 (shake flask/slow stirring-GC, De Bruijn et al. 1989)
- 4.52 ± 0.15 (recommended, Sangster 1989, 1993)
- 4.374 ± 0.034, 4.562 ± 0.006 (shake flask/slow stirring-GC/HPLC, interlaboratory studies, Brooke et al. 1990)
- 4.30 (centrifugal partition chromatography, Berthod et al. 1992)
- 4.46 (shake flask-UV spectroscopy, pH 7.4, Alcron et al. 1993)
- 4.46 (recommended, Hansch et al. 1995)
- 4.53, 4.83 (26°C, 4°C, quoted, Piatt et al. 1996)
- 4.48 ± 0.19, 4.54 ± 0.61 (HPLC-k′ correlation: ODS column, Diol column, Helweg et al. 1997)
- 4.57, 4.49–4.64 (shake flask/slow stirring-HPLC/fluo., mean value, De Maagd et al. 1998)
- 4.60 (shake flask-dialysis tubing-HPLC/UV, both phases, Andersson & Schräder 1999)
- 4.50; 4.65, 4.52 (shake flask-SPME solid-phase micro-extraction; quoted lit. values, Paschke et al. 1999)

Octanol/Air Partition Coefficient, log K_{OA} at 25°C or as indicated and reported temperature dependence equation:
- 7.45 (calculated, Finizio et al. 1997)
- 7.57*; 7.41 (generator column-GC; calculated-concn ratio C_O/C_A, measured range 0–40°C, Harner & Bidleman 1998)
- log K_{OA} = –5.62 + 3942/(T/K); temp range 0–40°C, ΔH_{OA} = 63.3 kJ/mol (generator column-GC, Harner & Bidleman 1998)
- 7.52 (calculated-S_{oct} and vapor pressure P, Abraham et al. 2001)
- 7.88 (solid-phase microextraction SPME-GC, Treves et al. 2001)

Bioconcentration Factor, log BCF:
- 3.42 (fathead minnow, Carlson et al. 1978)
- 2.51 (*Daphnia pulex*, Southworth et al. 1978)
- 2.57 (kinetic estimation, Southworth et al. 1978)
- 3.80 (mixed microbial population, Steen & Karickhoff 1981)
- 4.28 (*P. hoyi*, Eadie et al. 1982)
- 3.67 (microorganisms-water, Mabey et al. 1982)
- 4.38, 4.03, 4.57 (average, *Selenastum capricornutum*-dosed singly, dosed simultaneously, Casserly et al. 1983)
- 3.25 (*Chlorella fusca*; Geyer et al. 1984)
- 2.51 (fish, Govers et al. 1984)
- 2.97, 3.25, 3.25 (activated sludge, algae, fish, Freitag et al. 1985)
- 4.18; 4.28 (*P. hoyi* of Lake Michigan interstitial waters; of high sediment study site, Landrum et al. 1985)
- 2.51 (*Daphnia magna*, Newsted & Giesy 1987)
- 3.21 (10–20°C, *h. limbata*, rate constant ratio k_1/k_2, Landrum & Poore 1988)
- 4.45; 3.77; 3.43 (4°C, *P. hoyi*; *S. heringianus*; *Mysis relicta*, quoted, Landrum & Poore 1988)
- 0.756, 1.487 (*Polychaete sp, Capitella capitata*, Bayona et al. 1991)

Sorption Partition Coefficient, log K_{OC} at 25°C or as indicated:

- 4.36 (natural sediment, average of sorption isotherms by batch equilibrium-UV spec., Karickhoff et al. 1979)
- 4.08 (sediment/soil, sorption isotherm by batch equilibrium technique, Karickhoff 1981)
- 4.60 (fluorescence quenching interaction with AB humic acid, Gauthier et al. 1986)
- 4.28 (sediment from Tamar estuary, batch equilibrium-GC, Vowles & Mantoura 1987)
- 4.00 (Aldrich and Fluka humic acids, Chin et al. 1989)
- 6.12 (sediments average, Kayal & Connell 1990)
- 4.22, 4.28 (RP-HPLC correlation on CIHAC, on PIHAC stationary phases, Szabo 1990b)
- 4.42 (sandy surface soil, batch equilibrium-sorption isotherm, Wood et al. 1990)
- 4.07 (Quarry dark sand, batch equilibrium-sorption isotherm, Magee et al. 1991)
- 4.64 (DOM derived from soil, fluorescence quenching method, Magee et al. 1991)
- 4.42, 4.30 (marine porewater organic colloids; marine sediment, Fort Point Channel FPC 25–29 cm, Chin & Gschwend 1992)
- 4.17; 4.18, 4.17 (sediment: concn ratio C_{sed}/C_W; concn-based coeff., areal-based coeff. of flux studies of sediment/water boundary layer, Helmstetter & Alden 1994)
- 4.50 (Rotterdam Harbor sediment 4.6% OC, batch sorption equilibrium, Hegeman et al. 1995)
- 4.37 (Speyer soil 1.08% OC, batch sorption equilibrium, Ou et al. 1995)
- 6.07, 7.03, 6.39; 4.12 (marine sediments: Fort Point Channel, Spectacle Island, Peddocks Island; quoted lit., McGroddy & Farrington 1995)
- 5.77 (marine sediments: Fort Point Channel FPC 25–29 cm, McGroddy & Farrington 1995)
- 4.28, 4.12, 4.23 (RP-HPLC-k′ correlation on different stationary phases, Szabo et al. 1995)
- 4.09; 4.32 (HPLC-screening method; calculated-PCKOC fragment method, Müller & Kördel 1996)
- 4.18 (range 4.13–4.20); 3.56 (range 3.54–3.56) (4°C, low organic carbon sediment f_{OC} = 0.0002, batch equilibrium; column exptl., Piatt et al. 1996)
- 4.13 (range 4.06–4.19); 3.48 (range 3.47–3.48) (26°C, low organic carbon sediment f_{OC} = 0.0002, batch equilibrium; column exptl., Piatt et al. 1996)
- 4.65; 4.81 ± 0.16 (humic acid, HPLC-k′ correlation; quoted lit., Nielsen et al. 1997)
- 2.42–2.56 (5 soils, 20°C, batch equilibrium-sorption isotherm measured by HPLC/UV, Bayard et al. 1998)
- 4.27, 4.27, 4.12, 4.27, 4.10; mean 4.12 ± 0.088 (soils: Woodburn soil, Elliot soil, Marlette soil, Piketon soil, Anoka soil, batch equilibrium-sorption isotherms-HPLC-fluorescence, Choiu et al. 1998)
- 4.38, 4.45, 4.53, 4.33, 4.42, 4.62, 4.64 ± 0.087 (sediments: Lake Michigan, Mississippi River, Massachusetts Bay, Spectacle Island, Peddocks Island, Port Point Channel, batch equilibrium-sorption isotherms-HPLC-fluorescence, Choiu et al. 1998)
- 4.48 (4.46–4.50), 4.22 (4.20–4.23) (sediments: Lake Oostvaardersplassen, Lake Ketelmeer, shake flask-HPLC/UV, de Maagd et al. 1998)
- 3.67; 3.29, 4.04, 3.27, 4.37, 4.21 (calculated-K_{OW}; HPLC-screening method with different LC-columns, Szabo et al. 1999)
- 4.31–6.02 (range, calculated from sequential desorption of 11 urban soils; Krauss & Wilcke 2001)
- 5.34; 5.23, 4.82, 4.98 (20°C, batch equilibrium, A2 alluvial grassland soil; calculated values of expt 1,2,3-solvophobic approach, Krauss & Wilcke 2001)
- 4.39 - algae, 4.66 - degraded algae, 3.33 - cellulose, 4.72 - collagen, 4.50 - cuticle, 4.18 - lignin, 4.67 - humic acid, 4.56 - oxidized humic acid, 4.64 - Green River kerogen, 4.88 - Pula kerogen (aliphatic-rich sedimentary organic matter, batch experiments, Salloum et al. 2002)
- 4.28 (Askov soil - a Danish agricultural soil, Sverdrup et al. 2000)
- 4.03, 4.08 (soils: organic carbon OC ≥ 0.1%, OC ≥ 0.5%, average, Delle Site 2001)
- 4.34 (average values for sediment OC ≥ 0.5%, Delle Site 2001)
- 4.70 (soil humic acid, shake flask-HPLC/UV, Cho et al. 2002)
- 4.66–4.90 (sediment/water, initial-final values of 5–100 d contact time, gas-purge technique-HPLC/fluo., ten Hulscher et al. 2003)
- 5.29–5.92, 5.98 (NIST SRM diesel particulate matter: flocculation-based batch equilibrium method with 59-d equilibration time, air-bridge equilibrium with 123-d equilibration time, Nguyen et al. 2004)

4.30–5.3, 5.2–5.5 (sediments of 5 lakes with OC ranges from 0.12–21.03%, solute concn at 1 mg/L, at 1 ng/L, sorption isotherms, Cornelissen et al. 2004)

4.80, 4.80, 4.50, 4.84 (sediment free of BC "black carbon": Lake Varparanta OC 0.12%, Lake Kuorinka OC 1.39%, Lake Höytiäinen OC 3.3%, Lake Ketelmeer OC 5.51%, equilibrium sorption isotherm, Cornelissen et al. 2004)

Environmental Fate Rate Constants, k, or Half-Lives, $t_{1/2}$:

Volatilization: half-lives from solution: 97, 108 min (exptl., calculated, Mackay et al. 1983).

Photolysis: calculated $t_{1/2}$ = 8.4 h for direct sunlight photolysis at 40°N latitude of midday in midsummer, near surface water and $t_{1/2}$ = 59 d (inland water) and $t_{1/2}$ = 69 d for inland water with sediment partitioning in a 5-m deep inland water body (Zepp & Schlotzhauer 1979)

$t_{1/2}$ = 3 h, atmospheric and aqueous half-life, based on measured aqueous photolysis quantum yields and calculated for midday summer sunlight at 40°N latitude and $t_{1/2}$ = 25 h after adjusting for approximate winter sunlight intensity (Howard et al. 1991);

half-lives on different atmospheric particulate substrates (appr. 25 μg/g on substrate): $t_{1/2}$ = 150 h on silica gel, $t_{1/2}$ = 45 h on alumina and $t_{1/2}$ = 49 h on fly ash (Behymer & Hites 1985);

$t_{1/2}$ = 59 d under sunlight (Mill & Mabey 1985);

k = 0.11 h^{-1} with $t_{1/2}$ = 6.3 h in distilled water (Fukuda et al. 1988);

photodegradation k = 6.53 × 10^{-3} min and $t_{1/2}$ = 1.78 h in methanol-water (2:3, v/v) solution with an initial concn of 5.0 ppm under high pressure mercury lamp or sunlight (Wang et al. 1991)

k(expt) = 0.00653 min^{-1} pseudo-first-order direct photolysis rate constant with the calculated $t_{1/2}$ = 1.78 h and the predicted k = 0.00165 min^{-1} calculated by QSPR, in aqueous solution when irradiated with a 500-W medium pressure mercury lamp (Chen et al. 1996);

direct photolysis $t_{1/2}$ = 4.62 h (observed), $t_{1/2}$ = 3.89 h (predicted- QSPR) in atmospheric aerosol (Chen et al. 2001);

photochemical degradation under atmospheric conditions: k = (4.49 ± 0.68) × 10^{-5} s^{-1} and $t_{1/2}$ = (4.29 ± 0.57)h in diesel particulate matter, k = (2.11 ± 0.04) × 10^{-5} s^{-1} and $t_{1/2}$ = (9.1 ± 0.19)h in diesel particulate matter/soil mixture, and $t_{1/2}$ = 17 – 4.97 h in various soil components using a 900-W photo-irradiator as light source; k = (3.17 ± 0.06) × 10^{-6} s^{-1} and $t_{1/2}$ = (60.63 ± 1.33)h in diesel particulate matter using a 300-W light source (Matsuzawa et al. 2001)

Photodegradation k = 0.09 × 10^{-4} s^{-1} in surface water during the summertime at mid-latitude (Fasnacht & Blough 2002)

Oxidation: rate constant k, for gas-phase second order rate constants, k_{OH} for reaction with OH radical, k_{NO_3} with NO$_3$ radical and k_{O_3} with O$_3$ or as indicated, *data at other temperatures see reference:

k (aquatic fate rate) = 0.01 L M^{-1} s^{-1} with $t_{1/2}$ = 8 × 10^6 d (Callahan et al. 1979)

k (calc) < 360 M^{-1} h^{-1} for singlet oxygen and k < 36 M^{-1} h^{-1} for peroxy radical (Mabey et al. 1982)

k = (1.33 – 1.57) × 10^4 dm^3 mol^{-1} s^{-1} over the pH range 1–7 for the reaction with O$_3$ in water at 25°C, $t_{1/2}$ = 0.44 s in presence of 10^{-4} M ozone at pH 7 (Butković et al. 1983)

k_{OH} = 3.4 × 10^{-11} cm^3 molecule^{-1} s^{-1} at 298 ± 1 K, k_{OH} = 2.8 × 10^{-11} cm^3 molecule^{-1} s^{-1} at 319 ± 1 K (relative rate method, Biermann et al. 1985)

k_{OH}* = 3.4 × 10^{-11} cm^3 molecule^{-1} s^{-1} at 298 K (review, Atkinson 1989)

k_{OH} = 25.7 × 10^{-12} cm^3 molecule^{-1} s^{-1} (molecular orbital calculations, Klamt 1996)

k_{OH}* = 27 × 10^{-12} cm^3 molecule^{-1} s^{-1} at 298 K, measured range 346–386 K with a calculated atmospheric lifetime of 11 h based on gas-phase OH reactions (Brubaker & Hites 1998)

Hydrolysis: not hydrolyzable (Mabey et al. 1982); no hydrolyzable groups (Howard et al. 1991).

Biodegradation: 100% degradation within 7 d for a domestic sewage of an average of three static-flask screening test (Tabak et al. 1981);

$t_{1/2}$(aerobic) = 384–4800 h, based on aerobic soil die-away test data (Coover & Sims 1987; quoted, Howard et al. 1991);

k = 0.0447 d^{-1} with $t_{1/2}$ = 16 d for Kidman sandy loam and k = 0.0196 d^{-1} with $t_{1/2}$ = 35 d for McLarin sandy loam all at –0.33 bar soil moisture (Park et al. 1990);

$t_{1/2}$(anaerobic) = 1536–19200 h, based on estimated unacclimated aqueous aerobic biodegradation half-life (Howard et al. 1991);

$t_{1/2}$ = 4 d in inorganic solution and $t_{1/2}$ = 11 d in Kendaia soil (Manilal & Alexander 1991)

removal rate of 1.10 and 0.12 mg (g of volatile suspended solid d)$^{-1}$, degradation by bacteria from creosote-contaminated marine sediments with nitrate- and sulfate-reducers respectively under anaerobic conditions in a fluidized bed reactor (Rockne & Strand 1998)

first-order k = 0.033 to 0.139 L mg^{-1} d^{-1} for a marine PAH-degrading enrichment without sediment, the degradation rate was 2.1 to 3.5 times faster with sediment present (Poeton et al. 1999)

Biotransformation: for bacteria, 1.6×10^{-7} mL cell^{-1} h^{-1} (Paris et al. 1980; quoted, Mabey et al. 1982)

Bioconcentration, Uptake (k_1) and Elimination (k_2) Rate Constants:

k_1 = 203 h^{-1}; k_2 = 0.543 h^{-1} (*Daphnia pulex*, Southworth et al. 1978)

log k_1 = 2.31 h^{-1}; log k_2 = –0.27 h^{-1} (*Daphnia pulex*, as per the correlation of Mackay & Hughes 1984, Hawker & Connell 1986)

k_1 = 129.0 mL g^{-1} h^{-1}; k_2 = 0.0046 h^{-1} (4°C, *p. hoyi*, Landrum 1988; quoted, Landrum & Poore 1988)

k_1 = 52.5 h^{-1}; k_2 = 0.032 h^{-1} (10–20°C, *H. limbata*, Landrum & Poore 1988)

k_1 = 94.0 h^{-1}; k_2 = 0.016 h^{-1} (4°C, *S. heringianus*, quoted, Landrum & Poore 1988)

k_1 = 32.0 h^{-1}; k_2 = 0.012 h^{-1} (4°C, *Mysis relicta*, quoted, Landrum & Poore 1988)

k_1 = 8.8 mg g^{-1} h^{-1}; k_2 = 0.045 h^{-1} (freshwater oligochaete from sediment, Van Hoof et al. 2001)

Half-Lives in the Environment:

Air: $t_{1/2}$ = 6 h with a steady-state concn of tropospheric ozone of 2×10^{-9} M in clean air (Butković et al. 1983)

$t_{1/2}$ = 2.01–20.1 h, based on photooxidation half-life in air (Howard et al. 1991);

calculated atmospheric lifetime of 11 h based on gas-phase OH reactions (Brubaker & Hites 1998).

Surface water: computed near-surface of a water body, $t_{1/2}$ = 8.4 h for direct photochemical transformation at latitude 40°N, midday, midsummer with $t_{1/2}$ = 59 d (no sediment-water partitioning), $t_{1/2}$ = 69 d (with sediment-water partitioning) on direct photolysis in a 5-m deep inland water body (Zepp & Schlotzhauer 1979);

$t_{1/2}$ = 0.44 s in presence of 10^{-4} M ozone at pH 7 (Butković et al. 1983);

calculated $t_{1/2}$ = 59 d under sunlight for summer at 40°N latitude (Mill & Mabey 1985);

$t_{1/2}$ = 3–25 h, based on aqueous photolysis half-life (Howard et al. 1991);

photolysis $t_{1/2}$ = 1.78 h in aqueous solution when irradiated with a 500 W medium pressure mercury lamp (Chen et al. 1996).

Groundwater: $t_{1/2}$ = 768–9600 h, based on estimated unacclimated aqueous aerobic biodegradation half-life (Howard et al. 1991).

Sediment: reduction $t_{1/2}$(est.) = 1196 h, $t_{1/2}$(exptl) = 825 h for chemical available phenanthrene and $t_{1/2}$ = 151 h for bioavailable phenanthrene for amphipod, *P. hoyi* in Lake Michigan sediments at 4°C. The average uptake clearance from sediment was (0.041 ± 0.023)g of dry sediment·g^{-1} of organism·h^{-1}, and the rate constants to become biologically unavailable was (0.0055 ± 0.003) h^{-1} resulting a bioavailable $t_{1/2}$ = 126 h (Landrum 1989);

desorption $t_{1/2}$ = 8.6 d from sediment under conditions mimicking marine disposal (Zhang et al. 2000).

Soil: $t_{1/2}$ = 2.5–26 d (Sims & Overcash 1983; quoted, Bulman et al. 1987);

$t_{1/2}$ = 9.7 d for 5 mg/kg treatment and $t_{1/2}$ = 14 d for 50 mg/kg (Bulman et al. 1987);

biodegradation k = 0.0447 d^{-1} with $t_{1/2}$ = 16 d in Kidman sandy loam soil and k = 0.0196 d^{-1} with $t_{1/2}$ = 35 d in McLaurin sandy loam soil (Park et al. 1990);

biodegradation $t_{1/2}$ = 11 d in Kendaia soil (Manilal & Alexander 1991);

$t_{1/2}$ = 384–4800 h, based on aerobic soil die-away test data (Howard et al. 1991);

$t_{1/2}$ > 50 d (Ryan et al. 1988);

$t_{1/2}$ = 0.4–26 wk, 5.7 yr (quoted, Luddington soil, Wild et al. 1991).

Biota: depuration $t_{1/2}$ = 40.9 h in *S. heringianus* (Frank et al. 1986);

elimination $t_{1/2}$ = 9 d from rainbow trout, $t_{1/2}$ = 8.4 d from clam *Mya arenaria*, $t_{1/2}$ = 1.9 d from mussel *Mytilus edulis*; $t_{1/2}$ = 7 d from polychaete *Abarenicola pacifica*, $t_{1/2}$ = 3.4 d from Oyster, $t_{1/2}$ = 0.9 d from shrimp, $t_{1/2}$ = 4.8 d from polychaete *Nereis virens*, $t_{1/2}$ = 6.1 d from clam *Mercenario mercenaria* (Meador et al. 1995).

TABLE 4.1.1.23.1
Reported aqueous solubilities of phenanthrene at various temperatures

1.

Wauchope & Getzen 1972				Schwarz 1977		May et al. 1978a	
shake flask-UV				shake flask-fluorescence		generator column-HPLC	
t/°C	S/g·m^{-3}	t/°C	S/g·m^{-3}	t/°C	S/g·m^{-3}	t/°C	S/g·m^{-3}
	experimental		smoothed				
24.6	1.12	0	0.39	8.4	0.501	8.5	0.423
24.6	1.11	24.8	1.16	11.1	0.5507	10.0	0.468
29.9	1.49	25	1.18	14.0	0.640	12.5	0.512
29.9	1.49	29.9	1.49	17.5	0.784	15.0	0.601
30.3	1.47	30.3	1.52	20.2	0.881	21.0	0.816
30.3	1.48	38.4	2.27	23.3	1.085	24.3	0.995
38.4	2.44, 2.45	40.1	2.47	25.0	1.151	29.9	1.277
40.1	2.27, 2.28	47.5	3.63	29.3	1.372		
40.1	2.25	50	4.14	34.6	1.627	temp dependence eq. 2	
47.5	3.87, 3.88	50.1	4.16			S	µg/kg
47.5	3.87	50.2	4.19	ΔH_{sol}/(kJ mol^{-1}) = 36.32		a	0.0025
50.1	4.30, 4.38	54.7	5.34			b	0.8059
50.1	4.32	59.2	6.85			c	5.413
50.2	4.08, 4.04	60.5	7.30			d	324
50.2	4.11	65.1	9.60				
54.7	5.66, 5.64	70.7	13.3				
54.7	5.63	71.9	14.2			**May et al. 1978b**	
59.2	7.17, 7.19	73.4	15.6			generator column-HPLC	
59.2	7.21	75	17.2			t/°C	S/g·m^{-3}
60.5	7.20, 7.60						
65.1	9.80	temp dependence eq. 1				25	1.002
65.1	9.70	ln x	mole fraction			29	1.220
65.1	9.80	ΔH_{fus}	16.28 ± 0.08				
70.7	12.4	10^2·b	3.74 ± 0.13			temp dependence eq. 2	
70.7	12.6	c	27.9 ± 0.4			S	µg/kg
70.7	12.4					a	0.0025
71.9	18.8					b	0.8059
73.4	21.5					c	5.413
						d	324

ΔH_{fus}/(kJ mol^{-1}) = 16.28

ΔH_{sol}/(kJ mol^{-1}) = 34.81
for temp range 5–30°C

Empirical temperature dependence equations:

Wauchope & Getzen (1972): R·ln x = −[H$_{fus}$/(T/K)] + (0.000408)[(T/K) − 291.15]^2 −c + b·(T/K) (1)

May et al. (1978):− S/(µg/kg) = a·t^3 + b·t$_{1/2}$ +c·t + d (2)

TABLE 4.1.1.23.1 (Continued)

2.

May 1980		May et al. 1983		Whitehouse 1984		Shaw 1989	
generator column-HPLC		generator column-HPLC		generator column-HPLC/UV		IUPAC recommended	
t/°C	S/g·m⁻³	t/°C	S/g·m⁻³	t/°C	S/g·m⁻³	t/°C	S/g·m⁻³
4.0	0.361	4.0	0.361	4.6	0.358	0	0.40
8.5	0.423	8.5	0.423	8.8	0.437	10	0.50
10	0.468	10.0	0.468	12.9	0.556	20	0.85
12.5	0.512	12.5	0.512	17.0	0.720	25	1.10
15	0.601	15.5	0.602	21.1	0.880	30	1.40
20	0.787	20.0	0.788	25.3	1.10	40	2.50
21	0.816	21.0	0.817			50	4.10
24.3	0.955	24.3	0.956			50	7.20
29.9	1.227	29.9	1.188			70	13.0
						75	17.0

temp dependence eq. 2
S   μg/kg
a   0.0025
b   0.8059
c   5.412
d   324

$\Delta H_{sol}/(kJ\ mol^{-1}) = 34.81$
for temp range 5–30°C

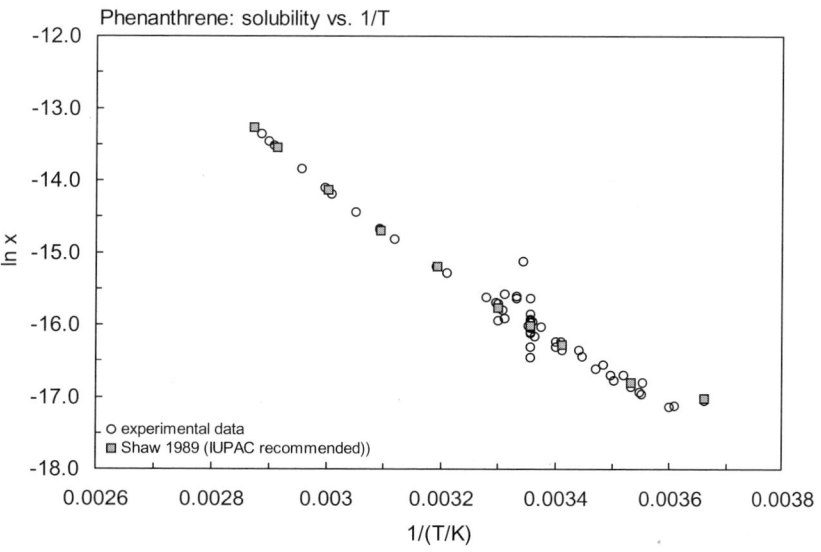

FIGURE 4.1.1.23.1 Logarithm of mole fraction solubility (ln x) versus reciprocal temperature for phenanthrene.

TABLE 4.1.1.23.2
Reported vapor pressures of phenanthrene at various temperatures and the coefficients for the vapor pressure equations

$\log P = A - B/(T/K)$ (1) $\ln P = A - B/(T/K)$ (1a)

$\log P = A - B/(C + t/°C)$ (2) $\ln P = A - B/(C + t/°C)$ (2a)

$\log P = A - B/(C + T/K)$ (3) $\ln P = A - B/(C + T/K)$ (3a)

$\log P = A - B/(T/K) - C \cdot \log(T/K)$ (4)

1.

Nelson & Senseman 1922		Mortimer & Murphy 1923		Stull 1947		Bradley & Cleasby 1953	
isoteniscope-Hg manometer		isoteniscope-Hg manometer		summary of literature data		effusion	
t/°C	P/Pa	t/°C	P/Pa	t/°C	P/Pa	t/°C	P/Pa
230	7773	203.6	3626	118.2	133.3	36.7	0.0853
235	9079	233.8	8999	154.3	666.6	39.85	0.1187
240	10532	246.0	12572	173.0	1333	42.0	0.160
245	12146	271.5	24265	193.7	2666	46.7	0.2466
250	13919	271.5	24398	215.8	5333	48.8	0.2973
255	15892	293.1	39957	229.0	7999	48.8	0.3000
260	18065	293.2	39970	249.0	13332	39.15	0.1080
265	20492	293.2	39983	277.1	26664	42.1	0.1480
270	23145	306.4	53222	308.0	53329	44.62	0.1933
275	26091	324.5	53289	340.2	101325	46.7	0.2426
280	18331	324.9	77954			49.65	0.3213
285	32891	325.4	78487	mp/°C	99.5		
290	36810	337.1	79007			eq. 2	P/mmHg
300	45823	337.1	98792			A	10.388
310	56529	345.1	115244			B	4519
320	69169	245.7	116110				
330	83913	346.8	117844				
340	100925						

2.

Hoyer & Peperle 1958		Osborn & Douslin 1975		Macknick & Prausnitz 1979		de Kruif 1980	
effusion		inclined-piston manometry		gas saturation-GC		effusion	
t/°C	P/Pa	t/°C	P/Pa	t/°C	P/Pa	t/°C	P/Pa
data presented in		100.0	30.4	51.60	0.465	38.96	0.1
eq. 1	P/mmHg	105.0	41.20	57.70	0.813	45.29	0.2
A	16.0	110.0	54.66	61.85	1.321	49.11	0.3
B	5998	115.0	73.59	67.35	2.120	51.88	0.4
for temp range 0–60°C		120.0	95.99	71.80	3.093	54.06	0.5
		125.0	124.79	78.90	5.653	55.86	0.6
ΔH_{subl}/(kJ mol^{-1}) = 95.90		130.0	161.05	83.40	8.892	57.40	0.7
		135.0	207.18	90.30	14.52	58.75	0.8
		140.0	264.91			9.95	0.9
		145.0	335.30	eq. 1a	P/mmHg	61.02	1.0
		150.0	420.76	A	26.648	25.0	0.18
				B	10.484		
		triple pt.				ΔH_{subl}/(kJ mol^{-1}) = 90.5	
		tp/K	372.385				

TABLE 4.1.1.23.2 (*Continued*)

Hoyer & Peperle 1958		Osborn & Douslin 1975		Macknick & Prausnitx 1979		de Kruif 1980	
effusion		inclined-piston manometry		gas saturation-GC		effusion	
t/°C	P/Pa	t/°C	P/Pa	t/°C	P/Pa	t/°C	P/Pa
		ΔH_V/(kJ mol^{-1}) =					
		at 398.15 K	68.58				
		at tp	70.79				
		ΔH_{subl}/(kJ mol^{-1}) = 87.24					
		at tp					
		ΔH_{fus}/(kJ mol^{-1}) = 16.44					

3.

Sonnefeld et al. 1983		Sato et al. 1986		Oja & Suuberg 1998	
generator column-HPLC		gas saturation-electrobalance		Knudsen effusion	
t/°C	P/Pa	T/K	P/Pa	t/°C	P/Pa
10.35	0.00219	322.9	0.340	30.34	0.0357
10.35	0.00222	324.6	0.420	37.22	0.0909
10.35	0.00211	328.5	0.655	40.77	0.140
10.35	0.00238	333.0	1.05	44.79	0.227
18.85	0.00738	335.2	1.34	40.1	0.323
18.85	0.00731	337.8	1.73	59.78	0.998
18.85	0.00749	339.5	2.03		
29.5	0.0262	343.2	2.98	eq. 1a	P/Pa
29.5	0.0271	346.0	3.75	A	11.423
29.5	0.0268	347.8	4.42	B	33.387
38.65	0.0899				
38.65	0.0917	eq. 3a	P/Pa	ΔH_{subl}/(kJ mol^{-1}) = 95.0	
3865	0.0889	A	20.3950		
38.65	0.0849	B	3931.20		
38.80	0.0863	C	−139.743		
38.80	0.0844				
38.88	0.0902	ΔH_{subl}/(kJ mol^{-1}) = 96.5			
38.90	0.0902				
38.90	0.0922				
38.90	0.0906				
25.0	0.0161				
eq. 1	P/Pa				
A	14.852				
B	4962.77				
ΔH_{subl}/(kJ mol^{-1}) = 95.0					

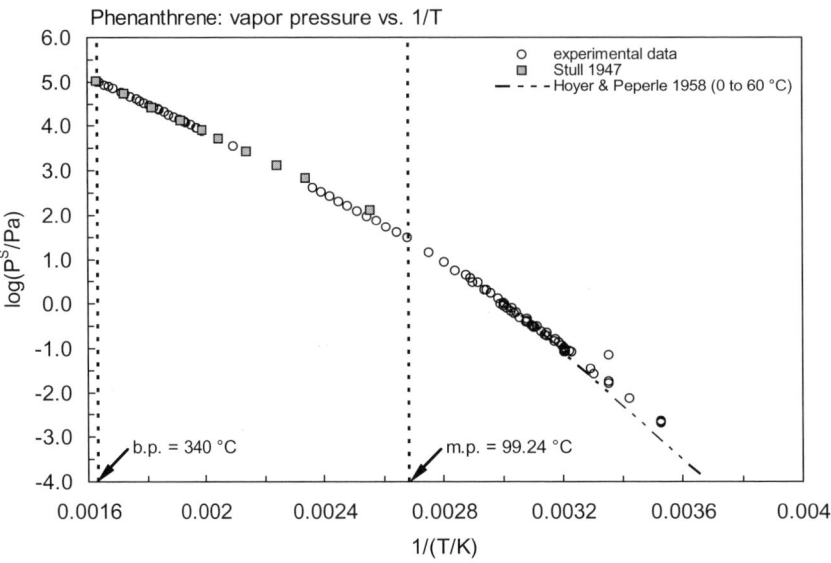

FIGURE 4.1.1.23.2 Logarithm of vapor pressure versus reciprocal temperature for phenanthrene.

TABLE 4.1.1.23.3
Reported Henry's law constants and octanol-air partition coefficients of phenanthrene at various temperatures and temperature dependence equations

$\ln K_{AW} = A - B/(T/K)$ (1)	$\log K_{AW} = A - B/(T/K)$ (1a)
$\ln (1/K_{AW}) = A - B/(T/K)$ (2)	$\log (1/K_{AW}) = A - B/(T/K)$ (2a)
$\ln (k_H/\text{atm}) = A - B/(T/K)$ (3)	
$\ln [H/(\text{Pa m}^3/\text{mol})] = A - B/(T/K)$ (4)	$\ln [H/(\text{atm·m}^3/\text{mol})] = A - B/(T/K)$ (4a)
$K_{AW} = A - B·(T/K) + C·(T/K)^2$ (5)	

Henry's law constant					log K_{OA}	
Alaee et al. 1996		Bamford et al. 1999			Harner & Bidleman 1998	
gas stripping-GC		gas stripping-GC/MS			generator column-GC/FID	
t/°C	H/(Pa m³/mol)	t/°C	H/(Pa m³/mol)	H/(Pa m³/mol)	t/°C	log K_{OA}
				average		
5.9	1.81	4.1	0.88, 1.00	0.94	0	8.808
10.4	2.99	11.0	1.53, 1.67	1.60	10	8.267
15.0	3.06	18.0	2.55, 2.70	2.65	20	7.898
20.2	3.66	25.0	4.08, 4.51	4.29	30	7.418
25.7	4.73	31.0	5.97, 6.82	6.38	40	6.926
30.2	5.54				25(exptl)	7.57
34.7	7.90	$\ln K_{AW} = A - B/(T/K)$			25(calc)	7.41
		A	14.1293			
$\ln K_{AW} = A - B/(T/K)$		B	5689.2		$\log K_{OA} = A + B/(T/K)$	
A	6.0313				A	−5.62
B	3524.2	enthalpy, entropy change:			B	3942
		$\Delta H/(\text{kJ·mol}^{-1}) = 47.3 \pm 1.2$				
enthalpy of volatilization:		$\Delta S/(\text{J·K}^{-1} \text{ mol}^{-1}) = 118$			enthalpy of phase change	
$\Delta H_{vol}/(\text{kJ·mol}^{-1}) = 29.3$ at 20°C		at 25°C			$\Delta H_{OA}/(\text{kJ mol}^{-1}) = 75.5$	

Polynuclear Aromatic Hydrocarbons (PAHs) and Related Aromatic Hydrocarbons

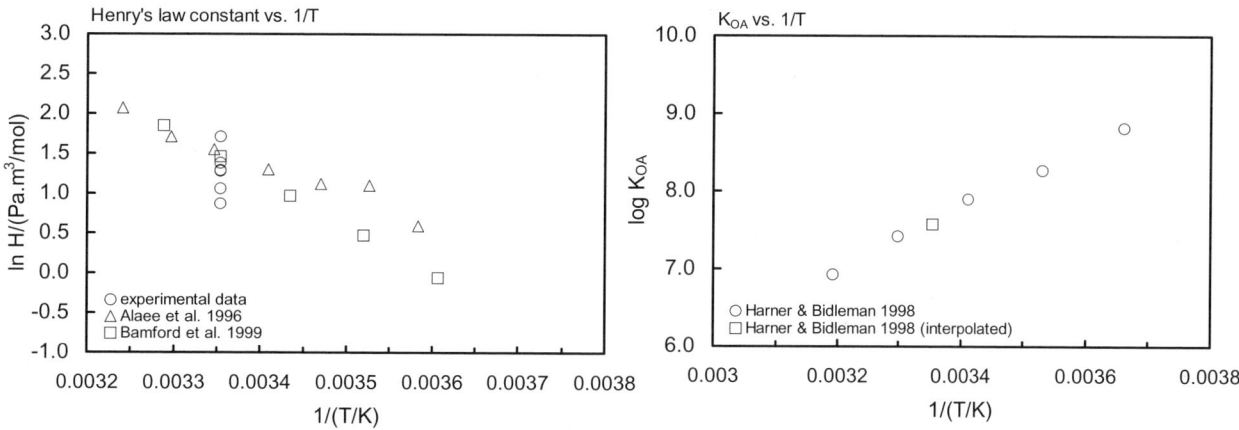

FIGURE 4.1.1.23.3 Logarithm of Henry's law constant and K_{OA} versus reciprocal temperature for phenanthrene.

4.1.1.24 1-Methylphenanthrene

Common Name: 1-Methylphenanthrene
Synonym:
Chemical Name: 1-methylphenanthrene
CAS Registry No: 832-69-6
Molecular Formula: $C_{15}H_{12}$
Molecular Weight: 192.256
Melting Point (°C):
    123   (Weast 1982–83; Lide 2003)
Boiling Point (°C):
    354   (Lide 2003)
Density (g/cm³ at 20°C):
Molar Volume (cm³/mol):
    188.0   (Ruelle & Kesselring 1997)
    221.4   (calculated-Le Bas method at normal boiling point)
Enthalpy of Fusion, ΔH_{fus} (kJ/mol):
Entropy of Fusion, ΔS_{fus} (J/mol K):
Fugacity Ratio at 25°C (assuming ΔS_{fus} = 56 J/mol K), F: 0.109 (mp at 123°C)

Water Solubility (g/m³ or mg/L at 25°C or as indicated. Additional data at other temperatures designated *, are compiled at the end of this section):

    0.255*   (24.1°C, generator column-HPLC/UV, measured range 6.6–29.9°C, May et al. 1978a, 1983)
    0.269   (generator column-HPLC, May et al. 1978b)
    $S/(\mu g/kg) = 55.42 + 6.8016 \cdot (t/°) - 0.1301 \cdot (t/°C)^2 + 0.0080 \cdot (t/°C)^3$; temp range 6.6–29.9°C (generator column-HPLC/UV, May et al. 1978)

Vapor Pressure (Pa at 25°C and the reported temperature dependence equation):

    0.0186; 0.0160   (supercooled liquid P_L, calibrated GC-RT correlation; GC-RT correlation, Lei et al. 2002)
    $\log (P_L/Pa) = -3987/(T/K) + 11.64$; $\Delta H_{vap.} = -76.3$ kJ·mol⁻¹ (GC-RT correlation, Lei et al. 2002)

Henry's Law Constant (Pa m³/mol at 25°C or as indicated and reported temperature dependence equations. Additional data at other temperatures designated * are compiled at the end of this section):

    5.0*   (gas stripping-GC, measured range 4.1–31°C, Bamford et al. 1999)
    $\ln K_{AW} = -4257.88/(T/K) + 8.0587$, $\Delta H = 35.4$ kJ mol⁻¹; measured range 4.1–31°C (gas stripping-GC, Bamford et al. 1999)

Octanol/Water Partition Coefficient, log K_{OW}:

    5.08   (HPLC-RT correlation, Wang et al. 1986)
    5.08   (recommended, Sangster 1989, 1994)
    5.08   (recommended, Hansch et al. 1995)
    5.10–5.20; 5.15 (quoted lit. range; lit. mean, Meador et al. 1995)

Octanol/Air Partition Coefficient, log K_{OA}:

Bioconcentration Factor, log BCF:

Sorption Partition Coefficient, log K_{OC}:

Environmental Fate Rate Constants, k or Half-Lives, $t_{1/2}$:
    Volatilization:
    Photolysis: photodegradation of 5 ppm initial concentration in methanol-water (3:7, v/v) by high pressure mercury lamp or sunlight with a rate constant k = 1.84 × 10⁻³ min⁻¹ and $t_{1/2}$ = 6.27 h (Wang et al. 1991); the

pseudo-first-order direct photolysis k(exptl) = 0.00184 min^{-1} with the calculated $t_{1/2}$ = 6.27 h and the predicted k = 0.0026 min^{-1} calculated by QSPR method in aqueous solution when irradiated with a 500 W medium pressure mercury lamp (Chen et al. 1996);

direct photolysis $t_{1/2}$ = 3.10 h (predicted-QSPR) in atmospheric aerosol (Chen et al. 2001).

Hydrolysis:
Oxidation:
Biodegradation:
Biotransformation:
Bioconcentration, Uptake (k_1) and Elimination (k_2) Rate Constants:

Half-Lives in the Environment:

Air: direct photolysis $t_{1/2}$ = 3.10 h (predicted-QSPR) in atmospheric aerosol (Chen et al. 2001).

Surface water: photolysis $t_{1/2}$ = 6.27 h in aqueous solution when irradiated with a 500 W medium pressure mercury lamp (Chen et al. 1996).

Groundwater:
Sediment:
Soil:
Biota: elimination $t_{1/2}$ = 6.7 d from Oyster, 6.0 d from clam *Mercenario mercenaria* (quoted, Meador et al. 1995).

TABLE 4.1.1.24.1
Reported aqueous solubilities and Henry's law constant of 1-methylphenanthrene at various temperature and the empirical temperature dependence equations

$$S/(\mu g/kg) = a \cdot t^3 + b \cdot t^2 + c \cdot t + d \quad (1)$$

Aqueous solubility						Henry's law constant	
May et al. 1978a		May et al. 1978b		May 1980, 1983		Bamford et al. 1999	
generator column-HPLC		generator column-HPLC		generator column-HPLC		gas stripping-GC/MS	
t/°C	S/g·m^{-3}	t/°C	S/g·m^{-3}	t/°C	S/g·m^{-3}	t/°C	H/(Pa m^3/mol)
6.6	0.0952	25.0	0.269	6.6	0.0952	4.1	1.42, 1.75
8.9	0.114			8.9	0.114		av. 1.58
14.0	0.147	temp dependence eq. 1		14.0	0.147	11.0	2.20, 2.53
19.2	0.193	S	µg/kg	19.2	0.193		av. 2.36
24.1	0.255	a	0.0080	24.1	0.255	18.0	3.26, 3.69
26.9	0.304	b	0.1301	26.9	0.304		av. 3.47
29.9	0.355	c	6.8016	29.9	0.355	25.0	4.62, 5.42
		d	55.42				av. 5.00
data of May et al. 1978a				data of May 1980 fitted to		31.0	6.09, 7.52
temp dependence eq. 1		ΔH_{sol}/(kJ mol^{-1}) = 30.08		temp dependence eq. 1			av. 6.77
S	µg/kg	measured between 5–30°C		S	µg/kg		
a	0.0080			a	0.0074	ln K_{AW} = A – B/(T/K)	
b	0.1301			b	–0.0858	A	8.0587
c	6.8016			c	5.785	B	4257.88
d	55.42			d	62.9		
						enthalpy, entropy change:	
						ΔH/(kJ·mol^{-1}) = 35.4 ± 1.9	
						ΔS/(J·K^{-1} mol^{-1}) = 67	
						at 25°C	

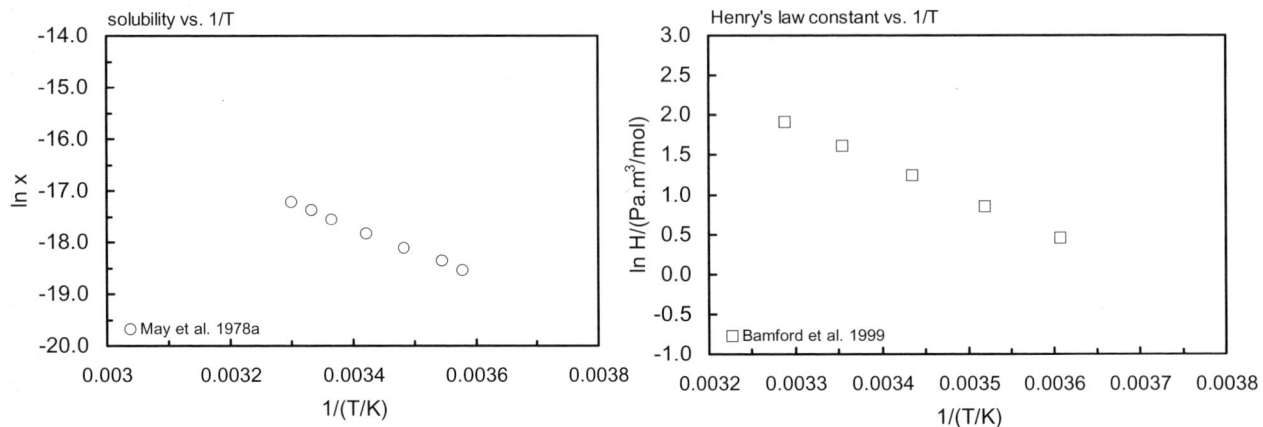

FIGURE 4.1.1.24.1 Logarithm of mole fraction solubility and Henry's law constant versus reciprocal temperature for 1-methylphenanthrene.

4.1.1.25 Anthracene

Common Name: Anthracene
Synonym: paranaphthalene, green oil, tetra olive NZG
Chemical Name: anthracene
CAS Registry No: 120-12-7
Molecular Formula: $C_{14}H_{10}$
Molecular Weight: 178.229
Melting Point (°C):
    215.76   (Lide 2003)
Boiling Point (°C):
    339.9   (Lide 2003)
Density (g/cm³ at 20°C):
    1.25   (27°C, Dean 1985)
    1.28   (25°C, Lide 2003)
Molar Volume (cm³/mol):
    139   (calculated-density, liquid molar volume, Lande & Banerjee 1981)
    196.7   (calculated-Le Bas method at normal boiling point)
Enthalpy of Fusion, ΔH_{fus} (kJ/mol):
    28.83   (Parks & Huffman 1931)
    28.87   (Tsonopoulos & Prausnitz 1971; Ruelle & Kesselring 1997)
    29.37   (Chickos et al. 1999)
Entropy of Fusion, ΔS_{fus} (J/mol K):
    58.99   (Tsonopoulos & Prausnitz 1971; Wauchope & Getzen 1972; Ubbelohde 1978)
    58.58   (Casellato et al. 1973)
    55.65   (De Kruif 1980)
    60.08, 44.2   (exptl., calculated-group additivity method, Chickos et al. 1999)
Fugacity Ratio at 25°C (assuming ΔS_{fus} = 56 J/mol K), F: 0.0134 (mp at 215.76°C)
    0.0101   (calculated, ΔS_{fus} = 58.6 J/mol K, Passivirta et al. 1999)

Water Solubility (g/m³ or mg/L at 25°C or as indicated and reported temperature dependence equations. Additional data at other temperatures designated * are compiled at the end of this section):
    0.075   (27°C, shake flask-nephelometry, Davis et al. 1942)
    0.075   (shake flask-UV, Klevens 1950)
    0.075   (Pierotti et al. 1959; Weimer & Prausnitz 1965)
    0.112   (shake flask, binding to bovine serum albumin-UV, Sahyun 1966)
    0.080   (shake flask-UV/fluorescence, Barone et al. 1967)
    0.043   (20°C, shake flask-UV, Eisenbrand & Baumann 1970)
    0.040   (shake flask-UV, Eisenbrand & Baumann 1970)
    0.171, 0.0392 (20°C, HPLC-relative retention correlation, different stationary or mobile phases, Locke 1974)
    0.075*   (extrapolated value, shake flask-UV, measured range 25–74.7°C, Wauchope & Getzen 1972)
    $R \cdot \ln x = -6930/(T/K) + 4.08 \times 10^{-4} \cdot [(T/K) - 291.15]^2 - 19.3 + 0.0181 \cdot (T/K)$, temp range 35.4–73.4°C (shake flask-UV measurements, Wauchope & Getzen 1972)
    0.030   (fluorescence/UV, Schwarz & Wasik 1976)
    0.073   (shake flask-fluorescence, Mackay & Shiu 1977)
    0.0446   (Rossi 1977; Neff 1979)
    0.041*   (shake flask-UV, measured range 8.6–31.3°C, Schwarz 1977)
    0.074   (Lu et al. 1978)
    0.0446*   (generator column-HPLC/UV, measured range 5.2–29.3°C, May et al. 1978)
    0.0434*   (24.6°C, generator column-HPLC, measured range 5.2–28.7°C, May 1980)

S/(μg/kg) = 8.21 + 0.8861·(t/°C) + 0.0097·(t/°C)^2 + 0.0013·(t/°C)^3; temp range 5.2–29.3°C (generator column-HPLC/UV, May et al. 1978)
0.033   (20°C, generator column-fluorescence, Hashimoto et al. 1982)
0.0434*   (24.6°C, generator column-HPLC, measured range 5.6–29.3°C, May et al. 1983)
0.030, 0.051   (generator column-HPLC/UV, Swann et al. 1983)
0.04257*   (generator column-spectrofluorimetry, measured range 10–30°C, Velapoldi et al. 1983)
0.0446   (generator column-HPLC/UV, Wasik et al. 1983)
0.066   (average lit. value, Pearlman et al. 1984)
0.0698   (generator column-HPLC/fluorescence, Walters & Luthy 1984)
0.0442*   (generator column-HPLC/UV, measured range 4.6–25.3°C, Whitehouse 1984)
0.0446   (vapor saturation-UV, Akiyoshi et al. 1987)
0.0443, 0.034   (generator column-HPLC/UV, Billington et al. 1988)
0.041   (20°C, shake flask/UV, ring test, Kishi & Hashimoto 1989)
0.062*   (recommended, IUPAC Solubility Data Series, Shaw 1989)
0.070   (23°C, shake flask-HPLC/UV/fluorescence, Pinal et al. 1991)
0.058   (generator column-HPLC, Vadas et al. 1991)
0.0488   (dialysis tubing equilibration-GC, Etzweiler et al. 1995)
0.043   (shake flask-HPLC, Haines & Sandler 1996)
0.0796; 0.138, 0.0743 (quoted, exptl.; calculated-molar volume, mp and mobile order thermodynamics, Ruelle & Kesselring 1997)
0.093   (generator column-HPLC/fluorescence, De Maagd et al. 1998)
0.070   (microdroplet sampling and multiphoton ionization-based fast-conductivity technique MPI-FC, Gridin et al. 1998)
log [S_L/(mol/L)] = 1.679 – 1509/(T/K) (supercooled liquid, Passivirta et al. 1999)
ln x = –1.43611 – 5307.35/(T/K); temp range 5–50°C (regression eq. of literature data, Shiu & Ma 2000)
0.0434*   (24.61°C, generator column-HPLC/fluo., Reza et al. 2002)
ln x = (0.50 ± 0.45) + [(–5876 ± 135)/(T/K)]; temp range 282.09–323.07 K (Reza et al. 2002)
0.0438*   (generator column-HPLC/UV, measured range 0–50°C, Dohányosová et al. 2003)
ln x = –33.7547 + 14.5018/τ + C ln τ, τ = (T/K)/298.15 K; temp range 0–50°C (generator column-HPLC/UV, temp range 0–50°C, Dohányosová et al. 2003)

Vapor Pressure (Pa at 25°C or as indicated and reported temperature dependence equations. Additional data at other temperatures designated * are compiled at the end of this section):
5680*   (220°C, isoteniscope-Hg, measured range 220–342°C, Nelson & Senseman 1922)
6399*   (223.3°C, isoteniscope-Hg manometer, measured range 223.3–340.5°C, Mortimer & Murphy 1923)
log (P_S/mmHg) = 10.972 – 4595/(T/K); temp range 100–210°C (Antoine eq., static isoteniscope method, Mortimer & Murphy 1923)
log (P_L/mmHg) = 7.910 – 3093/(T/K); temp range 223.2–340.5°C (Antoine eq., static isoteniscope method, Mortimer & Murphy 1923)
133.3*   (145.0°C, summary of literature data, temp range 145.0–342.0°C, Stull 1947)
0.001014   (static method-Rodebush gauge, Sears & Hopke 1949)
log (P/mmHg) = 12.0072 – 5102.6/(T/K); temp range 378–398 K (Rodebush gauge, Sears & Hopke 1949)
3.60×10^{-3} (effusion method, Inokuchi et al. 1952; quoted, Bidleman 1984)
8.31×10^{-4}* (effusion method, Bradley & Cleasby 1953)
log (P/cmHg) = 11.638 – 5320/(T/K); temp range 65.7–80.4°C (Antoine eq., Bradley & Cleasby 1953)
1.04×10^{-3} (fluorescence spectroscopy, Stevens 1953)
log (P/mmHg) = 12.002 – 5102/(T/K); temp range 396–421 K (fluorescence, Stevens 1953)
log (P/mmHg) = 11.15 – 5401/(T/K); temp range 30–100°C, (Knudsen effusion method, Hoyer & Peperle 1958)
8.62×10^{-4} (effusion method, Kelley & Rice 1964)
log (P/mmHg) = 12.068 – 5145/(T/K); temp range 69–86°C (effusion method, Kelley & Rice 1964)
3.87×10^{-7} (Wakayama & Inokuchi 1967)
0.0024*   (effusion method-interpolated, measured range 290.1–358 K, Wiedemann & Vaughan 1969, Wiedemann 1972)

log (P/mmHg) = 10.0216 − 4397.60/(T/K); temp range 290.1–358 K (Knudsen method, Wiedemann & Vaughan 1969, Wiedemann 1972)

log (P/mmHg) = 7.67401 − 2819.63/(247.02 + t/°C); temp range: 175.5–380°C (liquid state, Antoine eq., Zwolinski & Wilhoit 1971)

log (P/mmHg) = [−0.2185 × 16823.6/(T/K)] + 8.70760; temp range 100–600°C (Antoine eq., Weast 1972–73)

1.113×10^{-3}*   (Knudsen effusion weight-loss method, extrapolated Malaspina et al. 1973)

log (P/torr) = 12.616 − 5277/(T/K); temp range 352.7–432.3 K (Knudsen method, Malaspina et al. 1973)

1.47×10^{-5} (effusion method, Murray et al. 1974)

0.026   (20°C, Radding et al. 1976)

9.04×10^{-4} (effusion method, Taylor & Crooks 1976)

5.59×10^{-3} (gas saturation, Power et al. 1977)

1.41×10^{-3}* (gas saturation, extrapolated-Clapeyron eq., measured range 85.25–119.95°C, Macknick & Prausnitz 1979)

log (P/mmHg) = 26.805 − 11402/(T/K); temp range 85.25–119.95°C (Clapeyron eq., gas saturation, Macknick & Prausnitz 1979)

7.50×10^{-4}* (effusion methods, extrapolated, measured range 64.44–87.74°C, De Kruif 1980)

4.90×10^{-4} (calculated-TSA, Amidon & Anik 1981)

1.83×10^{-3}* (gas saturation, extrapolated-Antoine eq., measured range 50–85°C, Grayson & Fosbraey 1982)

1.44×10^{-3}* (gas saturation, extrapolated-Antoine eq. measured range 80–125°C, Bender et al. 1983)

ln (P/Pa) = 31.620 − 11378/(T/K); temp range 353.6–398.6 K (Antoine eq., Bender et al. 1983)

8.0×10^{-4}* (gas saturation-HPLC/fluo./UV, Sonnefeld et al. 1983)

log (P/Pa) = 12.977 − 4791.87/(T/K); temp range 10–50°C (Antoine eq., Sonnefeld et al. 1983)

8.0×10^{-4} (gas saturation-HPLC, Wasik et al. 1983)

0.10, 0.0638 (P_{GC} by GC-RT correlation with eicosane as reference standard, different GC columns, Bidleman 1984)

0.0865   (supercooled liquid P_L, converted from literature P_S with ΔS_{fus} Bidleman 1984)

log (P_L/kPa) = 6.53182 − 2550.737/(221.756 + t/°C); temp range 223.2–340.5°C (Antoine eq. from reported exptl. data, Boublik et al. 1984)

log (P_L/kPa) = 6.66266 − 2659.55/(230.119 + t/°C); temp range 220–310°C (Antoine eq. from reported exptl. data, Boublik et al. 1984)

log (P/mmHg) = 8.91 − 3761/(T/K); temp range: 100–160°C (Antoine eq., Dean 1985, 1992)

log (P/mmHg) = 7.67401 − 2819.63/(247.02 + t/°C); temp range: 176–380°C (Antoine eq., Dean 1985, 1992)

1.06×10^{-3}* (gas saturation-GC, Rordorf 1985)

8.05×10^{-4}* (extrapolated-Clausius-Clapeyron eq. on gas saturation data, Hansen & Eckert 1986)

log (P/mPa) = 17.88 − 5359/(T/K); temp range 313–363 K (Clausius-Clapeyron eq., Hansen & Eckert 1986)

1.14×10^{-3} (extrapolated-Antoine eq., Stephenson & Malanowski 1987)

log (P_S/kPa) = 10.58991 − 4903.3/(−1.58 + T/K); temp range 299–439 K (solid, Antoine eq.-I, Stephenson & Malanowski 1987)

log (P_S/kPa) = 11.76139 − 5315.532/(T/K); temp range 313–363 K, (solid, Antoine eq.-II, Stephenson & Malanowski 1987)

log (P_S/kPa) = 10.75544 − 4947.751/(T/K); temp range 363–393 K (solid, Antoine eq.-III, Stephenson & Malanowski 1987)

log (P_L/kPa) = 7.47799 − 3612.44/(−44.906 + T/K); temp range 504–615 K (liquid, Antoine eq.-IV, Stephenson & Malanowski 1987)

5.73×10^{-4}; 5.60×10^{-3} (literature mean solid P_S, supercooled liquid P_L, Bidleman & Foreman 1987)

0.086, 0.0940 (supercooled P_L, converted from literature P_S with different ΔS_{fus} values, Hinckley et al. 1990)

0.100, 0.0689 (P_{GC} by GC-RT correlation with different reference standards, Hinckley et al. 1990)

log P_L/Pa = 11.18 − 3642/(T/K) (GC-RT correlation, Hinckley et al. 1990)

log (P/mmHg) = −120.0992 + 4.478/(T/K) + 52.574·log (T/K) − 4.7697×10^{-2}·(T/K) + 1.5020×10^{-5}·(T/K)^2; temp range 489–873 K (vapor pressure eq., Yaws 1994)

0.0162; 7.64×10^{-4}, 0.0617 (liquid P_L by GC-RT correlation; quoted P_S, converted to P_L, Donovan 1996)

0.0661   (supercooled liquid P_L, calculated from Yamasaki et al. 1984, Finizio et al. 1997)

0.000144–0.00313; 0.000804–0.00511; 0.000683–0.00484 (quoted exptl. values measured by: effusion, gas saturation; manometry, Delle Site 1997)

0.0049–0.00125; 0.000925–0.00129   (quoted lit. values by: calculation; from GC-RT relation, Delle Site 1997)

8.69×10^{-4}* (Knudsen effusion, extrapolated-Antoine eq. derived from exptl. data, temp range 30–60°C, Oja & Suuberg 1998)

log (P/Pa) = 33.281 – 12024/(T/K); temp range 318–363 K (Antoine eq., Knudsen effusion, Oja & Suuberg 1998)

9.01×10^{-2}; 9.08×10^{-4} (quoted supercooled liquid P_L from Hinckley et al. 1990; converted to solid P_S with fugacity ratio F, Passivirta et al. 1999)

log (P_S/Pa) = 11.66 – 4380/(T/K) (solid, Passivirta et al. 1999)

log (P_L/Pa) = 8.39 – 2872/(T/K) (supercooled liquid value, Passivirta et al. 1999)

log (P/Pa) = 12.977 – 4791.89/(T/K); temp range 5–50°C (regression eq. from literature data, Shiu & Ma 2000)

0.0724; 0.0442 (supercooled liquid P_L, calibrated GC-RT correlation; GC-RT correlation, Lei et al. 2002)

log (P_L/Pa) = –3780/(T/K) + 11.54; $\Delta H_{vap.}$ = –72.4 kJ·mol^{-1} (GC-RT correlation, Lei et al. 2002)

0.0014 (solid P_S, gas saturation-GC/MS, Mader & Pankow 2003)

0.0922 (supercooled liquid P_L, calculated from P_S assuming ΔS_{fus} = 56 J/mol K, Mader & Pankow 2003)

ln (P/Pa) = 34.261 – 12339/(T/K); temp range 313–363 K (regression eq. of Hansen & Eckert 1986 data, Li et al. 2004)

ln (P/Pa) = (34.199 ± 0.641) – (12332 ± 229)/(T/K); temp range 348–368 K (Knudsen effusion technique, Li et al. 2004)

Henry's Law Constant (Pa m^3/mol at 25°C or as indicated and reported temperature dependence equations. Additional data at other temperatures designated * are compiled at the end of this section):

6.59 (gas stripping-GC, Southworth 1977, 1979)
7.19 (gas stripping-GC, Mackay & Shiu 1981)
1.96 (wetted-wall column/GC, Fendinger & Glotfelty 1990)
3.30 (gas stripping-GC, Friesen et al. 1993)
8.68 (headspace solid-phase microextraction (SPME)-GC, Zhang & Pawliszyn 1993)
4.94* (gas stripping-GC, measured range 4.1–31°C, Alaee et al. 1996)
ln K_{AW} = 4.6774 – 3235.5/(T/K); temp range: 5.4–35.3°C (gas stripping-GC, Alaee et al. 1996)
5.64* (gas stripping-GC, measured range 4.1–31°C, Bamford et al. 1999)
ln K_{AW} = –5629.06/(T/K) + 12.75, ΔH = 46.8 kJ mol^{-1}, measured range 4.1–31°C (gas stripping-GC, Bamford et al. 1999)
log [H/(Pa m^3/mol)] = 6.91 – 1363/(T/K), (Passivirta et al. 1999)
4.58 (20°C, selected from reported experimentally measured data, Staudinger & Roberts 2001)
log K_{AW} = 2.065 – 1404/(T/K) (van't Hoff eq. derived from literature data, Staudinger & Roberts 2001)

Octanol/Water Partition Coefficient, log K_{OW}:

4.45 (Hansch & Fujita 1964; Leo et al. 1971; Hansch & Leo 1979)
4.67 (calculated-fragment const., Rekker 1977)
4.45 (calculated from Leo 1975, Southworth et al. 1978)
4.54 (shake flask-UV, concn. ratio, Karickhoff et al. 1979)
3.45 (HPLC-RT correlation, Veith 1979a)
4.34 (Kenaga & Goring 1980)
4.49 (HPLC-k′ correlation, McDuffie 1981)
4.38 (HPLC-k′ correlation, Hanai et al. 1981)
4.63 (RP-TLC-k′ correlation, Bruggeman et al. 1982)
4.20 (HPLC-k′ correlation, D'Amboise 1982)
4.45 (HPLC-k′ correlation, Hafkenscheid & Tomlinson 1983)
4.45 (RP-HPLC-RT correlation, Rapaport et al. 1984)
4.45 (shake flask-GC, Haky & Leja 1986)
4.51 (HPLC-RT correlation, Eadsforth 1986)
4.63 (HPLC-RT correlation, Wang et al. 1986)
4.80 (HPLC-RT correlation, De Kock & Lord 1987)
4.50 ± 0.15 (recommended, Sangster 1989, 1993)
4.57 (TLC-RT correlation, De Voogt et al. 1990)
4.45 (recommended, Hansch et al. 1995)
4.53 ± 0.19, 4.55 ± 0.61 (HPLC-k′ correlation: ODS column, Diol column, Helweg et al. 1997)

4.68   (range 4.55–4.79) (shake flask-HPLC/fluo., De Maagd et al. 1998)
5.34   (RP-HPLC-RT correlation, short ODP column, Donovan & Pescatore 2002)

Octanol/Air Partition Coefficient, log K_{OA} at 25°C:
7.30   (calculated-K_{OW}/K_{AW}, Wania & Mackay 1996)
7.34   (calculated, Finizio et al. 1997)
7.55   (calculated-S_{oct} and vapor pressure P, Abraham et al. 2001)

Bioconcentration Factor, log BCF:
3.08, 2.68   (*Daphnia, pimephales*, Southworth 1977)
2.88   (*Daphnia pulex*, Herbes & Risi 1978)
2.96   (*Daphnia pulex*, Southworth et al. 1978)
3.08   (kinetic estimation, Southworth et al. 1978)
3.89   (algae, Geyer et al. 1981)
4.22   (*P. hoyi*, Eadie et al. 1982)
3.67   (microorganisms-water, Mabey et al. 1982)
2.83   (bluegill sunfish, Spacie et al. 1983)
3.83   (activated, sludge, Freitag et al. 1984)
3.89   (algae, Geyer et al. 1984)
2.21   (goldfish, Ogata et al. 1984)
2.96, 3.89, 3.83   (fish, algae, activated sludge, Freitag et al. 1985)
2.99   (*Daphnia magna*, Newsted & Giesy 1987)
0.820, 1.373   (*Polychaete sp., Capitella capitata*, Bayona et al. 1991)

Sorption Partition Coefficient, log K_{OC}:
4.42   (natural sediment, average of sorption isotherms by batch equilibrium-UV spec., Karickhoff et al. 1979)
3.74   (22°C, suspended particulates, Herbes et al. 1980)
4.204   (sediment/soil, sorption isotherm by batch equilibrium technique, Karickhoff 1981)
4.20   (soil, shake flask-UV, Karickhoff 1981)
3.95, 4.46   (Aldrich humic acid 9.4 mg/L DOC, RP-HPLC separation, equilibrium dialysis, Landrum et al. 1984)
3.95, 4.73   (humic materials in aqueous solutions: RP-HPLC-LSC, equilibrium dialysis, Lake Erie water with 9.6 mg/L DOC: Landrum et al. 1984)
4.87, 5.70   (humic materials in aqueous solutions: RP-HPLC-LSC, equilibrium dialysis, Huron River with 7.8 mg/L DOC, Landrum et al. 1984)
3.81, 4.87, 4.62, 4.20   (humic materials in natural water: Huron River 6.1% DOC spring, Huron river 6.7% DOC winter, Grand River 10.7% DOC spring, Lake Michigan 5.5% DOC spring, Lake Erie 9.6% DOC spring, RP-HPLC separation method, Landrum et al. 1984)
4.20   (soil, shake flask-LSC, Nkedi-Kizza 1985)
4.93   (fluorescence quenching interaction with AB humic acid, Gauthier et al. 1986)
4.38   (HPLC-k' correlation, Hodson & Williams 1988)
4.21 ± 0.11 (Aldrich and Fluka humic acids, observed; Chin et al. 1989)
4.11   (soil-fine sand 0.2% OC, dynamic soil column studies, Enfield et al. 1989)
5.76   (sediments average, Kayal & Connell 1990)
4.41   (RP-HPLC correlation, Pussemier et al. 1990)
4.53, 4.42   (RP-HPLC correlation on CIHAC, on PIHAC stationary phases, Szabo et al. 1990b)
4.34, 4.38, 4.32   (RP-HPLC-k' correlation on different stationary phases, Szabo et al. 1995)
4.65   (humic acid, HPLC-k' correlation; Nielsen et al. 1997)
4.62 (4.60–4.64)   (sediment from Lake Oostvaardersplassen, shake flask-HPLC/UV, de Maagd et al. 1998)
4.40; 4.30   (soil, calculated-universal solvation model; quoted exptl., Winget et al. 2000)
4.36–5.97; 4.20–6.90 (range, calculated from sequential desorption of 11 urban soils; lit. range, Krauss & Wilcke 2001)
5.31; 5.33, 5.12, 4.92 (20°C, batch equilibrium, A2 alluvial grassland soil; calculated values of expt 1,2,3-solvophobic approach, Krauss & Wilcke 2001)
4.34; 4.26; 3.95–5.70 (calculated-K_{OW}; calculated-solubility; quoted lit. range, Schlautman & Morgan 1993a)

4.375 at pH 4, 4.42 at pH 7, 4.39 at pH 10 in 0.001 M NaCl; 4.415 at pH 4, 4.37 at pH 7, 4.30 at pH 10 in 0.01 M NaCl; 4.50 at pH 4, 4.24 at pH 7, 4.27 at pH 10 in 0.1 M NaCl; 4.38 at pH 4, 4.40 at pH 7, 4.12 at pH 10 in 1 mM Ca^{2+} in 0.1 M total ionic strength solutions (shake flask/fluorescence, humic acid; Schlautmam & Morgan 1993a)

4.28 at pH 4, 4.18 at pH 7, 4.24 at pH 10 in 0.001 M NaCl; 4.285 at pH 4, 4.15 at pH 7, 4.22 at pH 10 in 0.01 M NaCl; 4.23 at pH 4, 4.12 at pH 7, 4.20 at pH 10 in 0.1 M NaCl; 4.21 at pH 4, 4.19 at pH 7, 4.24 at pH 10 in 1 mM Ca^{2+} in 0.1 M total ionic strength solutions (shake flask/fluorescence, fulvic acid; Schlautmam & Morgan 1993a)

Environmental Fate Rate Constants, k, or Half-Lives, $t_{1/2}$:

Volatilization: removal rate constants from the water column at 25°C in midsummer sunlight were: $k = 0.002$ h^{-1} in deep, slow, somewhat turbid water; $k = 0.001$ h^{-1} in deep, slow muddy water; $k = 0.002$ h^{-1} in deep slow, clear water; $k = 0.042$ h^{-1} in shallow, fast, clear water; and $k = 0.179$ h^{-1} in very shallow, fast, clear water (Southworth 1977);

aquatic $t_{1/2}$ = 18–300 h (Callahan et al. 1979);

calculated $t_{1/2}$ = 62 h for a river 1-m deep with water velocity of 0.5 m/s and wind velocity of 1 m/s (Southworth 1979; quoted, Herbes et al. 1980; Hallett & Brecher 1984).

Photolysis: removal rate constants from the water at 25°C in midsummer sunlight were: $k = 0.004$ h^{-1} in deep, slow somewhat turbid water; $k < 0.001$ h^{-1} in deep, slow, muddy water; $k = 0.018$ h^{-1} in deep, slow, clear water; $k = 0.086$ h^{-1} in shallow, fast, clear water; and $k = 0.238$ h^{-1} in very shallow, fast, clear water (Southworth 1977)

24-h photolytic $t_{1/2}$ ~1.6 h in summer and $t_{1/2}$ = 4.8 h in winter at 35°N latitude (Southworth 1977)

direct sunlight $k = 0.15$ h^{-1} in winter at 35°N latitude (Callahan et al. 1979)

$t_{1/2}$(calc) = 0.75 h near surface water for direct sunlight photolysis at 40°N latitude of midday in midsummer (quoted, Herbes et al. 1980; Harris 1982)

$t_{1/2}$ = 4.5 d in inland water, and $t_{1/2}$ = 5.2 d in inland water with sediment partitioning and $t_{1/2}$ = 0.75 h for direct photochemical transformation near water surface (Zepp & Schlotzhauer 1979)

atmospheric and aqueous photolysis $t_{1/2}$ = 0.58 h, based on measured aqueous photolysis rate constant for midday summer sunlight at 35°N latitude (Southworth 1979; quoted, Howard et al. 1991) and adjusted for approximate winter sunlight intensity (Lyman et al. 1982; quoted, Howard et al. 1991)

half-lives on different atmospheric particulate substrates (appr. 25 µg/g on substrate): $t_{1/2}$ = 2.9 h on silica gel, $t_{1/2}$ = 0.5 h on alumina and $t_{1/2}$ = 48 h on flyash (Behymer & Hites 1985)

$t_{1/2}$ = 4.5 d for summer at 40°N latitude under sunlight in surface water (Mill & Mabey 1985)

$k = 0.66$ h^{-1} in distilled water with $t_{1/2}$ = 1.0 h (Fukuda et al. 1988)

photodegradation $k = 0.023$ min^{-1} and $t_{1/2}$ = 0.50 h for initial concentration of 5 ppm in methanol-water (1:1, v/v) solution by high pressure mercury lamp or sunlight (Wang et al. 1991)

pseudo-first-order direct photolysis rate constants, k(exptl) = 0.023 min^{-1} with the calculated $t_{1/2}$ = 0.50 h, and the predicted $k = 0.030$ min^{-1} calculated by QSPR method in aqueous solution when irradiated with a 500 W medium pressure mercury lamp (Chen et al. 1996)

$k = 0.0503$–0.0521 min^{-1} in natural water system by UV and sunlight (Yu et al. 1999)

direct photolysis $t_{1/2}$(calc) = 3.10 h, predicted by QSPR, in atmospheric aerosol (Chen et al. 2001)

Photodegradation $k = 4.7 \times 10^{-4}$ s^{-1} in surface water during the summertime at mid-latitude (Fasnacht & Blough 2002)

Oxidation: rate constant k, for gas-phase second order rate constants, k_{OH} for reaction with OH radical, k_{NO_3} with NO_3 radical and k_{O_3} with O_3 or as indicated, *data at other temperatures see reference:

photooxidation $t_{1/2}$ = 1111–38500 h, based on measured rate constant for reaction with hydroxyl radical in water (Radding et al. 1976; quoted, Howard et al. 1991)

k(aquatic) fate rate of 50 L mol^{-1} s^{-1} with $t_{1/2}$ = 1600 d (Callahan et al. 1979)

k(calc) = 5.0×10^8 M^{-1} h^{-1} for singlet O_2 and 2.2×10^5 M^{-1} h^{-1} for peroxy radical (Mabey et al. 1982)

k(aq.) = 3.3×10^{-4} s^{-1} with $t_{1/2}$ = 0.6 h under natural sunlight conditions (NRCC 1983)

k_{OH} = 110×10^{-12} cm^3 $molecule^{-1}$ s^{-1} at 325 ± 1 K (relative rate technique for propene, (Biermann et al. 1985; Atkinson 1989)

photooxidation $t_{1/2}$ = 0.501–5.01 h, based on estimated rate constant for reaction with hydroxyl radical in air (Atkinson 1987)

k_{OH}(calc) = 203×10^{-12} cm^3 $molecule^{-1}$ s^{-1} (molecular orbital calculations, Klamt 1996)

$k = 3.5 \times 10^{-5}$ s^{-1}, indirect total photoreaction rate constant in surface waters (Mill 1999)

$k_{OH}^* = 190 \times 10^{-12}$ cm^3 molecule^{-1} s^{-1} at 298 K, measured range 306–366 K with a calculated atmospheric lifetime of 1.5 h based on gas-phase OH reaction (Brubaker & Hites 1998)

Hydrolysis: $k < 0.001$ h^{-1} at 25°C (Southworth 1977); not hydrolyzable (Mabey et al. 1982); no hydrolyzable groups (Howard et al. 1991).

Biodegradation:

$k = 0.061$ h^{-1} for microbial degradation in Third Creek water incubated 18 h at 25°C; removal rate constants from water column at 25°C in midsummer sunlight were: $k = 0.060$ h^{-1} in deep, slow, somewhat turbid water; $k = 0.030$ h^{-1} in deep, slow, muddy water; $k = 0.061$ h^{-1} in deep, slow, clear water; $k = 0.061$ h^{-1} in shallow, fast, clear water; and $k = 0.061$ h^{-1} in very shallow, fast, clear water (Southworth 1977)

$k = 0.035$ h^{-1} for microbial degradation (Herbes et al. 1980; quoted, Hallett & Brecher 1984)

significant degradation in 7 d with rapid adaptation for an average of three static-flask screening test (Tabak et al. 1981)

$k = 2.5 \times 10^{-3}$ h^{-1} with $t_{1/2} = 12$ d and $k = 2.5 \times 10^{-4}$ h^{-1} with $t_{1/2} = 115$ d for mixed bacterial populations in oil-contaminated and pristine stream sediments (Herbes & Schwall 1978, NRCC 1983)

$t_{1/2}$(aq. aerobic) = 1200–11040 h, based on aerobic soil die-away test data (Coover & Sims 1987; Sims 1990; quoted, Howard et al. 1991)

$k = 0.0052$ d^{-1} with $t_{1/2} = 134$ d for Kidman sandy loam and $k = 0.0138$ d^{-1} with $t_{1/2} = 50$ d for McLarin sandy loam all at –0.33 bar soil moisture (Park et al. 1990)

$t_{1/2}$(aq. anaerobic) = 4800–44160 h, based on estimated unacclimated aqueous aerobic biodegradation half-life (Howard et al. 1991).

Biotransformation: aquatic fate rate, $k < 0.0612$ h^{-1} with $t_{1/2} > 11.3$ h (Callahan et al. 1979); estimated rate constant for bacteria of 3×10^{-9} mL cell^{-1} h^{-1} (Mabey et al. 1982).

Bioconcentration, Uptake (k_1) and Elimination (k_2) Rate Constants:

log $k_1 = 2.89$ h^{-1}; log $k_2 = 0.0043$ h^{-1} (*Daphnia pulex*, Herbes & Risi 1978)

$k_1 = 702$ h^{-1}; $k_2 = 0.589$ h^{-1} (*Daphnia pulex*, Southworth et al. 1978)

$k_1 = (1.73 \times 10^{-3} - 36)$h^{-1}; $k_2 = 0.040$ h^{-1} (average, bluegill sunfish, Spacie et al. 1983)

log $k_1 = 2.85$ h^{-1}; log $k_2 = -0.23$ h^{-1} (*Daphnia pulex*, correlated to Mackay & Hughes 1984, Hawker & Connell 1986)

$k_1 = 1.46, 16.9$ h^{-1}; $k_2 = (1.58–1.88) \times 10^{-3}$ h^{-1} (rainbow trout, Linder et al. 1985)

$k_1 = 87.2$ h^{-1}; $k_2 = 0.019$ h^{-1} (4°C, *S. heringianus*, Frank et al. 1986)

$k_1 = 131.1$ mL g^{-1} h^{-1}; $k_2 = 0.0033$ h^{-1} (4°C, *P. hoyi*, Landrum 1988)

log $k_2 = .2, -0.01$ d^{-1} (fish, calculated-K_{OW}, Thomann 1989)

log $k_2 = -0.96$ d^{-1} (oyster, calculated-K_{OW}, Thomann 1989)

$k_1 = 1.8–2.3$ mg g^{-1} h^{-1}; $k_2 = 0.045$ h^{-1} (freshwater oligochaete from sediment, Van Hoof et al. 2001)

Half-Lives in the Environment:

Air: $t_{1/2} = 0.58–1.7$ h, based on photolysis half-life in water (Howard et al. 1991);

half-lives under simulated atmospheric conditions: $t_{1/2} = 0.20$ h for simulated sunlight, $t_{1/2} = 0.15$ h for simulated sunlight + ozone with concn of 0.2 ppm, $t_{1/2} = 1.23$ h for dark reaction ozone with concn of 0.2 ppm (Katz et al. 1979; quoted, Bjørseth & Olufsen 1983);

calculated atmospheric lifetime of 1.5 h based on gas-phase OH reactions (Brubaker & Hites 1998).

Surface water: half-lives for removal from water column at 25°C in midsummer sunlight were, $t_{1/2} = 10.5$ h for deep, slow, somewhat turbid water; $t_{1/2} = 21.6$ h for deep, slow, muddy water; $t_{1/2} = 8.5$ h for deep, slow, clear water; $t_{1/2} = 3.5$ h for shallow, fast, clear water; and $t_{1/2} = 1.4$ h for very shallow, fast, clear water (Southworth 1977, Herbes et al. 1980);

computed near-surface $t_{1/2} = 0.75$ h of a water body and for direct photochemical transformation at latitude 40°N, midday, midsummer and half-lives: $t_{1/2} = 4.5$ d for no sediment-water partitioning, $t_{1/2} = 5.2$ d with sediment-water partitioning and for direct photolysis in a 5-m deep inland water body (Zepp & Schlotzhauer 1979);

$t_{1/2} = 0.58–1.7$ h, based on photolysis half-life in water (Howard et al. 1991);

$t_{1/2} = 4.5$ d at 40°N under summer sunlight (Mill & Mabey 1985);

photolysis $t_{1/2} = 0.50$ h in aqueous solution when irradiated with a 500 W medium pressure mercury lamp (Chen et al. 1996);

indirect photoreaction $t_{1/2} = 5.5$ h in surface waters (Mill 1999);

photolysis $t_{1/2}$ = 13.3–13.80 min at 15°C in natural water system by UV and sunlight illumination (Yu et al. 1999).

Groundwater: $t_{1/2}$ = 2400–22080 h, based on estimated unacclimated aqueous aerobic biodegradation half-life (Howard et al. 1991).

Sediment: reduction $t_{1/2}$ = 600 h for chemical available anthracene and $t_{1/2}$ = 77 h for bioavailable anthracene for amphipod, *P. hoyi* in Lake Michigan sediments at 4°C. The uptake clearance from sediment was (0.024 ± 0.002)g of dry sediment·g^{-1} of organism·h^{-1}, and the rate constants to become biologically unavailable were (0.009 ± 0.002)h^{-1} corresponding to $t_{1/2}$ = 77 h (Landrum 1989).

Soil: $t_{1/2}$ = 3.3 – 175 d (Sims & Overcash 1983; quoted, Bulman et al. 1987);

$t_{1/2}$ = 17 d for 5 mg/kg treatment and $t_{1/2}$ = 45 d for 50 mg/kg treatment (Bulman et al. 1987);

degradation rate constant k = 0.0052 d^{-1} with $t_{1/2}$ = 134 d for Kidman sandy loam soil and k = 0.138 d^{-1} with $t_{1/2}$ = 50 d for McLauren sandy loam soil (Park et al. 1990);

$t_{1/2}$ = 1200–11040 h, based on aerobic soil die-away test data (Howard et al. 1991);

$t_{1/2}$ = 0.5–26 wk, 7.9 yr (quoted, Luddington soil, Wild et al. 1991).

Biota: $t_{1/2}$ = 17 h in bluegill sunfish (Spacie et al. 1983);

with depuration $t_{1/2}$ = 37.75 h in *s. heringianus* (Frank et al. 1986);

elimination $t_{1/2}$ = 7 d from rainbow trout, $t_{1/2}$ = 1.9 d from mussel *Mytilus edulis* (quoted, Meador et al. 1995).

TABLE 4.1.1.25.1
Reported aqueous solubilities of anthracene at various temperatures and the reported empirical temperature dependence equations

$$R \cdot \ln x = -[\Delta H_{fus}/(T/K)] + (0.000408)[(T/K) - 291.15]^2 - c + b \cdot (T/K) \quad (1)$$

$$S/(\mu g/kg) = a \cdot t^3 + b \cdot t^2 + c \cdot t + d \quad (2)$$

$$\ln x = A - B/T(K) \quad (3)$$

$$\ln x = A + B/(T/K) + C \cdot \ln (T/K) \quad (4)$$

$$\ln x = A + B/\tau + C \ln \tau, \text{ where } \tau = T/T_o, T_o = 298.15 \text{ K} \quad (4a)$$

1.

Wauchope & Getzen 1972				Schwarz 1977		May et al. 1978a	
shake flask-UV				shake flask-fluorescence		generator column-HPLC	
t/°C	S/g·m^{-3}	t/°C	S/g·m^{-3}	t/°C	S/g·m^{-3}	t/°C	S/g·m^{-3}
experimental		smoothed*					
35.4	0.125	0	0.022	8.60	0.0233	5.2	0.0127
35.4	0.122	25	0.075	11.1	0.0244	10.0	0.0175
35.4	0.119	35.4	0.123	12.2	0.0257	14.1	0.0222
39.3	0.152	39.3	0.159	14.0	0.0274	18.3	0.0291
39.3	0.151	44.7	0.214	15.5	0.0296	22.4	0.0372
39.3	0.148	47.5	0.249	19.2	0.0323	24.6	0.0434
44.7	0.208	50	0.286	20.3	0.0396	28.7	0.0557
44.7	0.210	50.1	0.288	23.3	0.0417		
44.7	0.206	54.7	0.372	25.0	0.0410		
47.5	0.279	59.2	0.481	26.2	0.0476	temp dependence eq. 2	
50.1	0.301	64.5	0.66	28.5	0.0579	S	μg/kg
50.1	0.297	65.1	0.68	31.3	0.0695	a	0.0013
50.1	0.302	69.8	0.90			b	–0.0097
54.7	0.391	70.7	0.95			c	0.8861
54.7	0.389	71.9	1.02	$\Delta H_{sol}/(kJ\ mol^{-1})$ = 34.81		d	8.21
54.7	0.402	74.7	1.21				
59.2	0.480	75.0	1.23				
59.2	0.488						
59.2	0.525						
64.5	0.72	temp dependence eq. 1					

TABLE 4.1.1.25.1 (Continued)

Wauchope & Getzen 1972		Schwarz 1977		May et al. 1978a	
shake flask-UV		shake flask-fluorescence		generator column-HPLC	
t/°C	S/g·m⁻³	t/°C	S/g·m⁻³	t/°C	S/g·m⁻³
64.5	0.62, 0.64	ln x	mole fraction		
65.1	0.67, 0.64	ΔH_{fus}	29.0 ± 0.29		
69.8	0.92	$10^2 \cdot b$	1.81 ± 0.15		
70.7	0.90, 0.97	c	19.3 ± 0.5		
70.7	0.96				
71.9	0.91	$\Delta H_{fus}/(kJ\ mol^{-1}) = 29.0$			
74.7	1.19, 1.13				
74.7	1.26				

2.

May et al. 1978b		May 1980		May et al. 1983		Velapoldi et al. 1983	
generator column-HPLC		generator column-HPLC		generator column-HPLC		generator column-fluo.	
t/°C	S/g·m⁻³	t/°C	S/g·m⁻³	t/°C	S/g·m⁻³	t/°C	S/g·m⁻³
25	0.0446	5.2	0.0127	5.20	0.0127	10	0.0177
29	0.0579	10.0	0.0175	10.0	0.0175	15	0.02282
		14.1	0.0222	14.1	0.0222	20	0.03061
temp dependence eq. 2		18.3	0.0281	18.3	0.0291	25	0.04257
S	µg/kg	22.4	0.0372	22.4	0.0372	30	0.06123
a	0.0013	24.6	0.0434	24.6	0.0434		
b	−0.0097	28.7	0.0557	28.7	0.0557		
c	0.8861			9.70	0.0162	eq. 4	x
d	8.21			16.6	0.0251	A	−1078.056
		temp dependence eq. 2		23.2	0.0378	B	41884.5
$\Delta H_{sol}/(kJ\ mol^{-1}) = 43.76$		S	µg/kg	29.3	0.0572	C	161.175
for temp range 5–30°C		a	0.0013				
		b	−0.0097			$\Delta H_{sol}/(kJ\ mol^{-1}) = 51.3$	
		c	0.8886				
		d	8.21				

3.

Whitehouse 1984		Shaw 1989		Reza et al. 2002		Dohányosová et al. 2003	
generator column-HPLC/UV		IUPAC recommended		generator column-HPLC/fluo		generator column-HPLC/UV	
t/°C	S/g·m⁻³	t/°C	S/g·m⁻³	t/°C	S/g·m⁻³	t/°C	S/g·m⁻³
4.6	0.00961	0	0.022	8.94	0.0155	0.30	0.00963
8.8	0.0129	10	0.021	11.57	0.0183	5.0	0.0124
12.9	0.0177	20	0.034	13.39	0.0198	10	0.0169
17.0	0.0237	25	0.062*	15.88	0.0246	15	0.0227
21.1	0.0323	30	0.080*	22.54	0.0287	20	0.0320
25.3	0.0442	40	0.16*	24.61	0.0434	25	0.0438
		50	0.29*	27.10	0.0505	30	0.0584
		60	0.51*	28.20	0.0540	35	0.0784
		70	0.91*	29.12	0.0565	40	0.106
				30.53	0.0633	45	0.145

(Continued)

TABLE 4.1.1.25.1 (Continued)

Whitehouse 1984		Shaw 1989		Reza et al. 2002		Dohányosová et al. 2003	
generator column-HPLC/UV		IUPAC recommended		generator column-HPLC/fluo		generator column-HPLC/UV	
t/°C	S/g·m⁻³	t/°C	S/g·m⁻³	t/°C	S/g·m⁻³	t/°C	S/g·m⁻³
			*tentative	34.87	0.0890	50	0.190
				39.91	0.1157		
				44.90	0.1569	eq. 4a	x
				49.20	0.2123	A	−33.7647
						B	14.5018
				eq. 3	x	C	32.7269
				A	0.050 ± 0.45		
				B	5876 ± 135	$\Delta H_{sol}/(kJ\ mol^{-1}) = 45.2 \pm 0.3$	
				temp range 282–323 K		at 298.15 K.	
						mp/K	489
						$\Delta H_{sol}/(kJ\ mol^{-1}) = 29.37$	

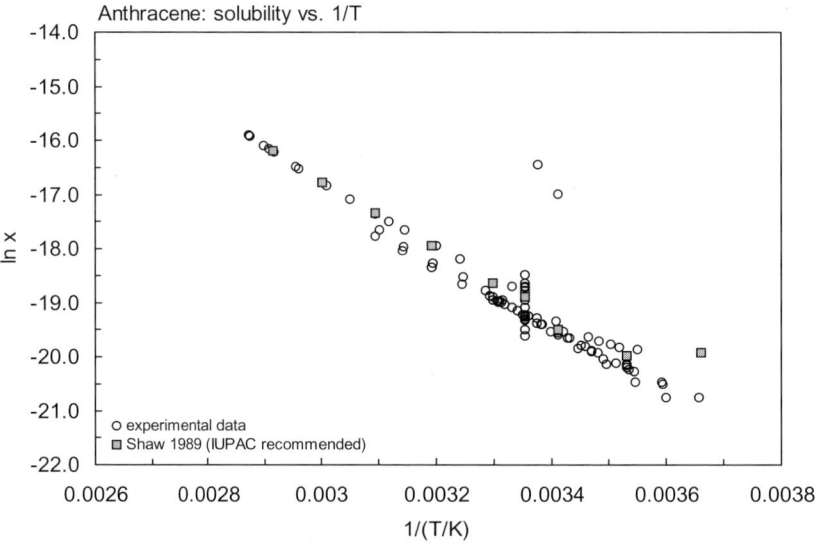

FIGURE 4.1.1.25.1 Logarithm of mole fraction solubility (ln x) versus reciprocal temperature for anthracene.

TABLE 4.1.1.25.2
Reported vapor pressures of anthracene at various temperatures and the coefficients for the vapor pressure equations

$$\log P = A - B/(T/K) \quad (1) \qquad \ln P = A - B/(T/K) \quad (1a)$$
$$\log P = A - B/(C + t/°C) \quad (2) \qquad \ln P = A - B/(C + t/°C) \quad (2a)$$
$$\log P = A - B/(C + T/K) \quad (3)$$
$$\log P = A - B/(T/K) - C \cdot \log(T/K) \quad (4)$$

1.

Nelson & Senseman 1922		Mortimer & Murphy 1923		Stull 1947		Sears & Hopke 1949	
isoteniscope-Hg manometer		isoteniscope-Hg manometer		summary of literature data		Rodebush gauge	
t/°C	P/Pa	t/°C	P/Pa	t/°C	P/Pa	t/°C	P/Pa
220	5680	223.3	6399	145.0	133.3	Data presented by	
225	6599	225.0	7333	173.5	666.6		
230	6306	244.4	11452	187.2	1333	eq. 1	P/mmHg
235	8813	244.6	11532	201.9	2666	A	12.002
240	10133	259.4	16972	217.5	5333	B	5102.0
245	11612	259.8	17105	231.8	7999	measured range 105–125°C	
250	13279	260.3	17359	250.0	13332	$\Delta H_{fus}/(kJ\,mol^{-1}) = 97.70$	
255	15132	282.0	29277	279.0	26664		
260	17212	282.1	29304	210.2	53329	25	.00103
265	19532	299.0	44050	342.0	101325	extrapolated	
270	22105	300.0	44170				
275	14971	312.8	57715	mp/°C	217.5		
280	28131	313.2	58155				
285	31624	313.4	58262				
290	35450	317.4	76980				
300	44263	327.9	77780				
310	54729	328.0	77860				
320	67048	340.6	98952				
330	81380	340.5	98950				
342	101325						
		eq. 1	P/mmHg				
bp/°C	342	A	7.910				
		B	3093				
		temp range 232–340°C					
		eq. 2	P/mmHg				
		A	10.972				
		B	4584				
		temp range 100–350°C					

2.

Bradley & Cleasby 1953		Kelley & Rice 1964		Malaspina et al. 1973		Power et al. 1977	
effusion		effusion		Knudsen effusion		gas saturation	
t/°C	P/Pa	t/°C	P/Pa	t/°C	P/Pa	t/°C	P/Pa
65.7	0.115	data represented by		79.55	0.612	30	0.00937
69.91	0.167			88.25	1.373	30	0.00913
73.35	0.259	eq. 1	P/mmHg	98.05	3.506	50	0.0928

(Continued)

TABLE 4.1.1.25.2 (*Continued*)

Bradley & Cleasby 1953		Kelley & Rice 1964		Malaspina et al. 1973		Power et al. 1977	
effusion		effusion		Knudsen effusion		gas saturation	
t/°C	P/Pa	t/°C	P/Pa	t/°C	P/Pa	t/°C	P/Pa
79.95	0.489	A	12.068	105.95	6.413	70	0.582
67.1	0.140	B	5145	113.75	12.17	70	0.5826
68.75	0.157	measured range 69–86°C		123.75	29.33	100	7.738
71.25	0.208	$\Delta H_{fus}/(kJ\,mol^{-1}) = 93.91$		135.25	61.06	100	7.771
73.2	0.223			142.35	106.32		
80.4	0.524	reported extrapolated data		149.75	189.31		
		95	1.667	159.15	483.95	eq. 2	P/mmHg
		100	2.546			A	36.40
eq. 1	P/mmHg	105	3.880			B	8634
A	11.638					C	238.6
B	5320			eq. 1	P/mmHg		
				A	12.616		
				B	5277		

3.

Macknick & Prausnitz 1979		de Kruif 1980		Grayson & Fosbraey 1982		Sonnefeld et al. 1983	
gas saturation-GC		teorsion-, weighing effusion		gas saturation-GC		gas saturation-HPLC	
t/°C	P/Pa	t/°C	P/Pa	t/°C	P/Pa	t/°C	P/Pa
85.25	0.892	64.44	0.10	20.0	0.00102	12.3	1.64×10^{-4}
90.15	1.36	71.18	0.20	50.1	0.035	12.3	1.57×10^{-4}
95.65	2.13	75.20	0.30	60.0	0.0849	12.3	1.60×10^{-4}
100.7	3.32	78.12	0.40	70.0	0.267	19.2	3.76×10^{-4}
104.7	4.59	80.41	0.50	75.0	0.359	19.2	3.83×10^{-4}
111.9	8.04	82.31	0.60	80.5	0.588	19.2	3.54×10^{-4}
116.4	11.41	83.93	0.70			19.2	3.72×10^{-4}
119.93	14.67	85.35	0.80			25.0	8.43×10^{-4}
		86.61	0.90	eq. 1	P/Pa	25.0	8.17×10^{-4}
		87.74	1.00	A	30.5	25.0	8.19×10^{-4}
eq. 1	P/mmHg			B	10968	25.0	3.30×10^{-4}
A	26.805	25.0	0.00875			30.1	1.46×10^{-3}
B	111402		extrapolated			30.1	1.54×10^{-3}
						30.1	1.51×10^{-3}
		$\Delta H_{sub}/(kJ\,mol^{-1}) = 100.4$				34.93	2.59×10^{-3}
						34.93	2.63×10^{-3}
						34.93	2.69×10^{-3}
						eq. 1	P/Pa
						A	12.977
						B	4891.87

TABLE 4.1.1.25.2 (Continued)

4.

Bender et al. 1983		Rordorf 1985		Hansen & Eckert 1986		Oja & Suuberg 1998	
gas saturation		gas saturation-GC		gas saturation-IR		Knudsen effusion	
t/°C	P/Pa	t/°C	P/Pa	t/°C	P/Pa	t/°C	P/Pa
80.45	0.55	25	0.00106	40	0.0058	27.7	0.00114
82.05	0.67	50	0.023	45	0.0105	39.7	0.00575
85.35	0.864	75	0.322	50	0.0193	47.6	0.0162
90.5	1.418	100	3.17	55	0.039	54.6	0.0355
94.5	2.019	125	23.4	60	0.0675	60.1	0.062
100.8	3.426	150	136.2	65	0.0987	72.7	0.204
105.65	4.99			70	0.1688	74.1	0.258
110.4	7.17	$\Delta H_{subl}/(kJ\,mol^{-1}) = 98.75$		75	0.3056		
115.35	10.01			80	0.5252	eq. 1a	P/Pa
115.65	10.58	av. selected literature value		85	0.9247	A	33.281
110.857	15.30	25	0.00108	90	1.244	B	12024
125.45	21.30	50	0.0243				
		75	0.344			$\Delta H_{subl}/(kJ\,mol^{-1}) = 100.0$	
eq. 1	P/Pa	100	3.38	eq. 1	P/mPa		
A	31.620	125	24.5	A	17.88		
B	1138	150	139.7	B	5359		
		$\Delta H_{subl}/(kJ\,mol^{-1}) = 98.79$		$\Delta H_{subl}/(kJ\,mol^{-1}) = 102.6$			

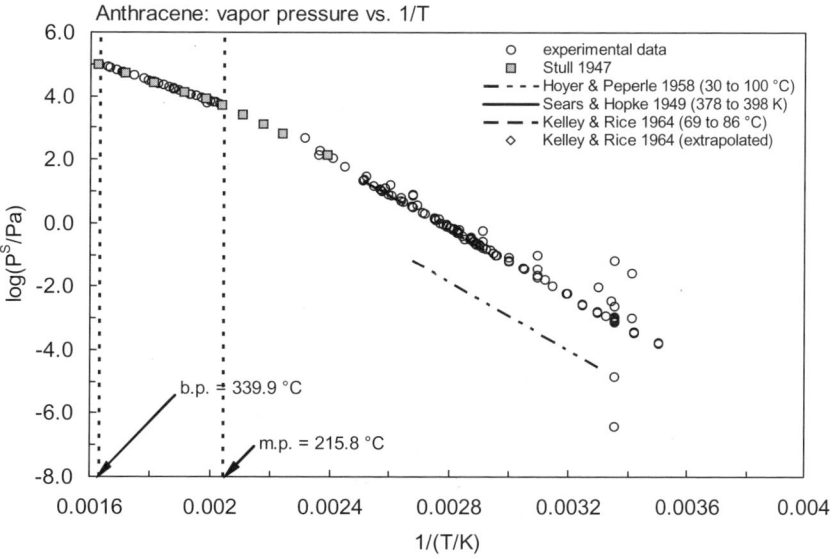

FIGURE 4.1.1.25.2 Logarithm of vapor pressure versus reciprocal temperature for anthracene.

TABLE 4.1.1.25.3
Reported Henry's law constants of anthracene at various temperatures and temperature dependence equations

$\ln K_{AW} = A - B/(T/K)$	(1)		$\log K_{AW} = A - B/(T/K)$	(1a)
$\ln (1/K_{AW}) = A - B/(T/K)$	(2)		$\log (1/K_{AW}) = A - B/(T/K)$	(2a)
$\ln (k_H/atm) = A - B/(T/K)$	(3)			
$\ln [H/(Pa\ m^3/mol)] = A - B/(T/K)$	(4)		$\ln [H/(atm \cdot m^3/mol)] = A - B/(T/K)$	(4a)
$K_{AW} = A - B \cdot (T/K) + C \cdot (T/K)^2$	(5)			

Alaee et al. 1996		Bamford et al. 1999		
gas stripping-GC		gas stripping-GC/MS		
t/°C	H/(Pa m³/mol)	t/°C	H/(Pa m³/mol)	H/(Pa m³/mol)
				average
5.4	2.76	4.1	1.14, 1.38	1.25
10.1	3.12	11.0	1.99, 2.26	2.12
14.8	3.44	18.0	3.31, 3.70	3.50
20.6	3.91	25.0	5.26, 6.06	5.64
25.0	4.94	31.0	7.60, 9.18	8.36
30.2	8.05			
34.3	8.94	eq. 1	K_{AW}	
		A	12.75	
eq. 1	K_{AW}	B	5629	
A	4.680			
B	3235.5	enthalpy, entropy change:		
		$\Delta H/(kJ \cdot mol^{-1}) = 46.8 \pm 1.7$		
enthalpy of volatilization:		$\Delta S/(J \cdot K^{-1}\ mol^{-1}) = 106$		
$\Delta H_{vol}/(kJ \cdot mol^{-1}) = 26.9$		at 25°C		

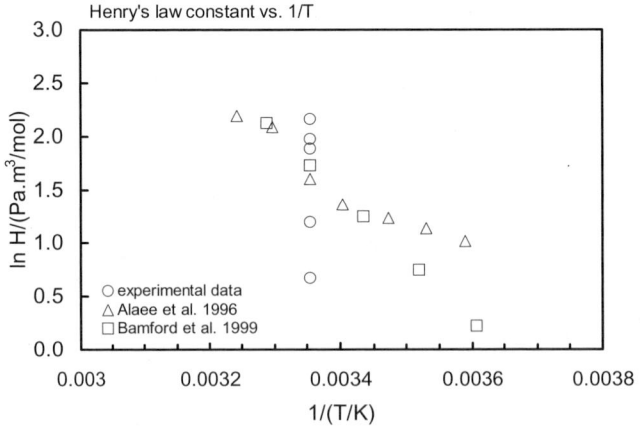

FIGURE 4.1.1.25.3 Logarithm of Henry's law constant versus reciprocal temperature for anthracene.

4.1.1.26 2-Methylanthracene

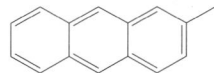

Common Name: 2-Methylanthracene
Synonym:
Chemical Name: 2-methylanthracene
CAS Registry No: 613-12-7
Molecular Formula: $C_{15}H_{12}$
Molecular Weight: 192.256
Melting Point (°C):
    209   (Weast 1982–83; Lide 2003)
Boiling Point (°C):
    359   (sublimation, Bjørseth 1983)
Density (g/cm³ at 20°C):
    1.80   (0°C, Lide 2003)
Molar Volume (cm³/mol):
    106   (calculated-density, liquid molar volume, Lande & Banerjee 1981)
    218.9   (calculated-Le Bas method at normal boiling point)
Enthalpy of Fusion, ΔH_{fus} (kJ/mol):
Entropy of Fusion, ΔS_{fus} (J/mol K):
Fugacity Ratio at 25°C (assuming ΔS_{fus} = 56 J/mol K), F: 0.0157 (mp at 209°C)

Water Solubility (g/m³ or mg/L at 25°C or as indicated and reported temperature dependence equations. Additional data at other temperatures designated * are compiled at the end of this section):
    0.039   (shake flask-fluorescence, Mackay & Shiu 1977)
    0.0219*   (generator column-HPLC/UV, measured range 6.3–31.1°C, May et al. 1978a)
    0.0213   (generator column-HPLC/UV, measured range 5–30°C, May et al. 1978b)
    $S/(\mu g/kg) = 324.0 + 5.413 \cdot (t/°) + 0.8059 \cdot (t/°C)^2 + 0.0025 \cdot (t/°C)^3$; temp range 5–30°C (generator column-HPLC/UC, May et al. 1978b)
    0.0191*   (23.1°C, generator column-HPLC/fluo., temp range 278.25–302.45 K, May et al. 1983)
    0.031   (average lit. value, Pearlman et al. 1984)
    0.0225*   (generator column-HPLC/UV, measured range 4.6–25.3°C, Whitehouse 1984)
    0.03*   (tentative value, IUPAC Solubility Data Series, Shaw 1989)
    $\ln x = -1.841995 - 4616.86/(T/K)$, temp range 5–50°C (regression eq. of literature data, Shiu & Ma 2000)
    0.0262*   (generator column-HPLC/UV, measured range 0–50°C, Dohányosová et al. 2003)
    $\ln x = -42.7975 + 22.9752/\tau + C \ln \tau$, $\tau = T/298.15$ K; temp range 0–50°C (generator column-HPLC/UV, Dohányosová et al. 2003)

Vapor Pressure (Pa at 25°C and reported temperature dependence equations):
    0.0207   (supercooled liquid P_L, calibrated GC-RT correlation, Lei et al. 2002)
    $\log (P_L/Pa) = -3976/(T/K) + 11.65$; $\Delta H_{vap.} = -76.1$ kJ·mol⁻¹ (GC-RT correlation, Lei et al. 2002)

Henry's Law Constant (Pa m³/mol):

Octanol/Water Partition Coefficient, log K_{OW} at 25°C and reported temperature dependence equations:
    5.15   (calculated-fragment const., Yalkowsky & Valvani 1979,1980)
    5.00   (shake flask-UV, Alcorn et al. 1993)
    5.00   (recommended, Sangster 1993)

4.97; 4.70 (calibrated GC-RT correlation; GC-RT correlation, Lei et al. 2000)

log K_{OW} = 1.093 + 1154.2/(T/K); temp range 5–55°C (temperature dependence HPLC-k' correlation, Lei et al. 2000)

Octanol/Air Partition Coefficient, log K_{OA}:

Bioconcentration Factor, log BCF:

Sorption Partition Coefficient, log K_{OC}:

Environmental Fate Rate Constants, k or Half-Lives, $t_{1/2}$:
  Volatilization:
  Photolysis:
  Hydrolysis:
  Oxidation:

Half-Lives in the Environment:
  Biota: elimination $t_{1/2}$ = 2 d from rainbow trout (quoted, Meador et al. 1995).

TABLE 4.1.1.26.1
Reported aqueous solubilities of 2-methylanthracene at various temperature and the empirical temperature dependence equations

$S/(\mu g/kg) = a \cdot t^3 + b \cdot t^2 + c \cdot t + d$ (1)

$\ln x = A + B/\tau + C \ln \tau$ (2)   where $\tau = T/T_o$ and T_o = 298.15 K

May et al. 1978a, May 1983		Whitehouse 1984		Shaw 1989b		Doháiyosová et al. 2003	
generator column-HPLC		generator column-HPLC/UV		IUPAC "tentative" values		generator column-HPLC	
t/°C	S/g·m⁻³	t/°C	S/g·m⁻³	t/°C	S/g·m⁻³	t/°C	S/g·m⁻³
6.3	0.00706	4.6	–	5	0.006	0.30	0.00567
9.1	0.00848	8.8	0.00754	10	0.009	5.0	0.00738
10.8	0.00943	12.9	0.00969	20	0.016	20	0.0106
13.9	0.0111	17.0	0.0123	25	0.03	15	0.0142
18.3	0.0145	21.1	0.0161	30	0.03	20	0.0192
23.1	0.0191	25.3	0.0225			25	0.0262
27.0	0.0242					30	0.0352
31.1	0.0321					35	0.048
						40	0.0674
temp dependence eq. 1						45	0.0943
S	μg/kg					50	0.125
a	0.0011						
b	–0.0306					eq. 2	x
c	0.8180					A	–42.7975
d	2.78					B	22.9752
ΔH_{sol}/(kJ mol⁻¹) = 39.08						C	41.7206
measured between 5–30°C							
						ΔH_{sol}/(kJ mol⁻¹) = 46.15±0.3 0.3	
							at 298.15 K
						mp/K	479
						ΔH_{fus}/(kJ mol⁻¹) = 24.06	

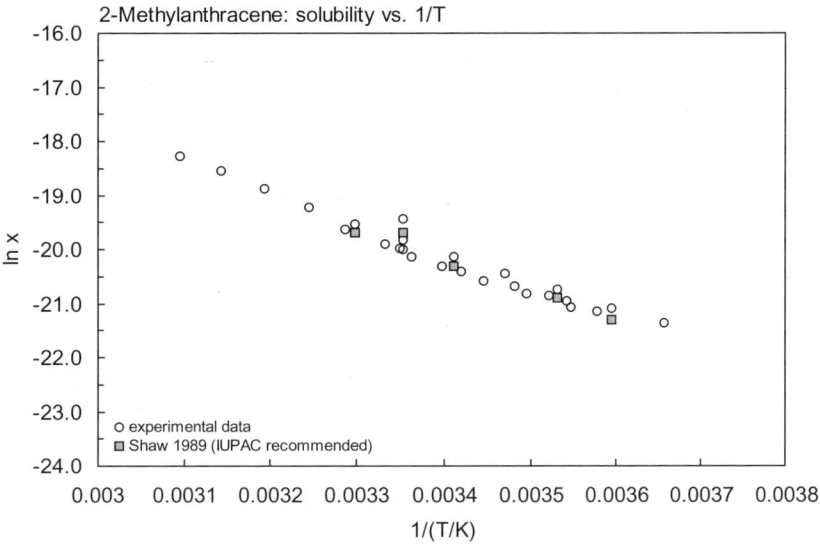

FIGURE 4.1.1.26.1 Logarithm of mole fraction solubility (ln x) versus reciprocal temperature for 2-methylanthracene.

4.1.1.27 9-Methylanthracene

Common Name: 9-Methylanthracene
Synonym:
Chemical Name: 9-methylanthracene
CAS Registry No: 779-02-2
Molecular Formula: $C_{15}H_{12}$
Molecular Weight: 192.256
Melting Point (°C):
    81.5   (Weast 1982–83; Lide 2003)
Boiling Point (°C):
    196   (12 mm Hg, Weast 1982–83; Lide 2003)
Density (g/cm^3 at 20°C):
    1.065   (99°C, Lide 2003)
Molar Volume (cm^3/mol):
    181   (calculated-density, liquid molar volume, Lande & Banerjee 1981)
    218.9   (calculated-Le Bas method at normal boiling point)
Enthalpy of Fusion, ΔH_{fus} (kJ/mol):
Entropy of Fusion, ΔS_{fus} (J/mol K):
Fugacity Ratio at 25°C (assuming ΔS_{fus} = 56 J/mol K), F: 0.279 (mp at 81.5°C)

Water Solubility (g/m^3 or mg/L at 25°C or as indicated and reported temperature dependence equations. Additional data at other temperatures designated * are compiled at the end of this section):
    0.261   (shake flask-fluorescence, Mackay & Shiu 1977)
    0.269   (average lit. value, Pearlman et al. 1984)
    0.530   (generator column-HPLC, Vadas et al. 1991)
    0.0376*   (generator column-HPLC/UV, measured range 0–50°C, Dohányosová et al. 2003)
    $\ln x = -76.9798 + 59.8386/\tau + C \ln \tau$, $\tau = T/298.15$ K, temp range 0–50°C (generator column-HPLC/UV, Dohányosová et al. 2003)

Vapor Pressure (Pa at 25°C and reported temperature dependence equations):
    0.00224   (extrapolated-Antoine eq., Stephenson & Malanowski 1987)
    $\log (P_L/\text{kPa}) = 11.683 - 5168/(T/K)$; temp range 354–402 K (Antoine eq., Stephenson & Malanowski 1987)

Henry's Law Constant (Pa m^3/mol):

Octanol/Water Partition Coefficient, log K_{OW}:
    5.12   (calculated-π const., Southworth et al. 1978)
    5.07   (shake flask-UV, concn. ratio, Karickhoff et al. 1979)
    5.15   (calculated-fragment const., Valvani & Yalkowsky 1980; Yalkowsky & Valvani 1979,1980)
    5.14   (average lit. value, Yalkowsky et al. 1983)
    5.61   (HPLC-RT correlation; Burkhard et al. 1985)
    5.10   (HPLC-RT correlation, Wang et al. 1986)
    5.07 ± 0.20   (recommended, Sangster 1989, 1993)
    5.07   (recommended, Hansch et al. 1995)

Octanol/Air Partition Coefficient, log K_{OA}:

Bioconcentration Factor, log BCF:
    3.66; 3.59   (*Daphnia pulex*; kinetic estimation, Southworth et al. 1978)

3.75 (calculated-K_{OW}, Mackay 1982)
3.94 (calculated-MCI χ, Sabljic 1987b)
3.683, 3.778 (calculated-MCI χ, calculated-K_{OW}, Lu et al. 1999)

Sorption Partition Coefficient, log K_{OC}:
4.81 (natural sediment, average of isotherms by batch equilibrium-UV spec., Karickhoff et al. 1979)
4.50 (calculated-molecular connectivity indices χ, Sabljic 1984)
4.81 (calculated-MCI $^1\chi$, Sabljic et al. 1995)

Environmental Fate Rate Constants, k, or Half-Lives, $t_{1/2}$:
Volatilization:
Photolysis:
$t_{1/2}$(calc) = 0.13 h for direct photochemical transformation near water surface and $t_{1/2}$ = 0.78 d for no sediment-water partitioning; and $t_{1/2}$ = 1.2 d with sediment-water partitioning (Zepp & Scholtzhauer 1979)
$t_{1/2}$ = 0.79 d for summer at 40°N latitude under sunlight in surface water (Mill & Mabey 1985)
photodegradation k = 0.163 min^{-1} and $t_{1/2}$ = 0.07 h in methanol-water (2:3, v/v) solution for initial concentration of 5 ppm by high pressure mercury lamp or sunlight (Wang et al. 1991)
the pseudo-first-order direct photolysis k(exptl) = 0.0163 min^{-1} with calculated $t_{1/2}$ = 0.07 h and the predicted k(calc) = 0.00343 min^{-1} calculated by QSPR in aqueous solution when irradiated with a 500 W medium pressure mercury lamp (Chen et al. 1996)
direct photolysis $t_{1/2}$ = 1.85 h predicted by QSPR in atmospheric aerosol (Chen et al. 2001)
Oxidation: $t_{1/2}$ = 10 h for photosensitized oxygenation with singlet oxygen at near-surface natural water, 40°N, midday, midsummer (Zepp & Schlotzhauer 1979)
Hydrolysis:
Biodegradation:
Biotransformation:
Bioconcentration, Uptake (k_1) and Elimination (k_2) Rate Constants:
k_1 = 561 h^{-1}; k_2 = 0.144 h^{-1} (*Daphnia pulex*, Southworth et al. 1978)
log k_1 = 2.75 h^{-1}; log k_2 = –0.84 h^{-1} (*Daphnia pulex*, correlated as per Mackay & Highes 1984, Hawker & Connell 1986)

Half-Lives in the Environment:
Air: direct photolysis $t_{1/2}$ = 1.85 h predicted by QSPR in atmospheric aerosol (Chen et al. 2001)
Surface water: computed $t_{1/2}$ = 0.13 h at near-surface of a water body, for direct photochemical transformation, and $t_{1/2}$ = 0.79 d for direct photolysis in a 5-m deep inland water body with no sediment-water partitioning, $t_{1/2}$ = 1.2 d with sediment-water partitioning to top cm bottom sediment; and $t_{1/2}$ = 10 h for photosensitized oxygenation with singlet oxygen at near-surface natural water, 40°N, midday, midsummer (Zepp & Schlotzhauer 1979);
$t_{1/2}$ = 0.79 d for summer at 40°N latitude under sunlight (Mill & Mabey 1985);
photolysis $t_{1/2}$ = 0.07 h in aqueous solution when irradiated with a 500 W medium pressure mercury lamp (Chen et al. 1996).
Groundwater:
Sediment:
Soil:
Biota: elimination $t_{1/2}$ = 4 d from rainbow trout (quoted, Meador et al. 1995).

TABLE 4.1.1.27.1
Reported aqueous solubilities of 9-methylanthracene at various temperature and the empirical temperature dependence equations

$$\ln x = A + B/\tau + C \ln \tau, \text{ where } \tau = T/T_o \text{ and } T_o = 298.15 \text{ K} \quad (1)$$

Doháyosová et al. 2003

Generator column-HPLC

t/°C	S/g·m^{-3}
0.30	0.113
5.0	0.137
20	0.174
15	0.228
20	0.286
25	0.376
30	0.508
35	0.699
40	0.953
45	1.270
50	1.770
eq. 1	mole fraction
A	−76.9798
B	59.8386
C	76.7066
$\Delta H_{sol}/(\text{kJ mol}^{-1}) = 41.8 \pm 0.2$	
	at 298.15 K
mp/K	348
$\Delta H_{fus}/(\text{kJ mol}^{-1}) = 16.95$	

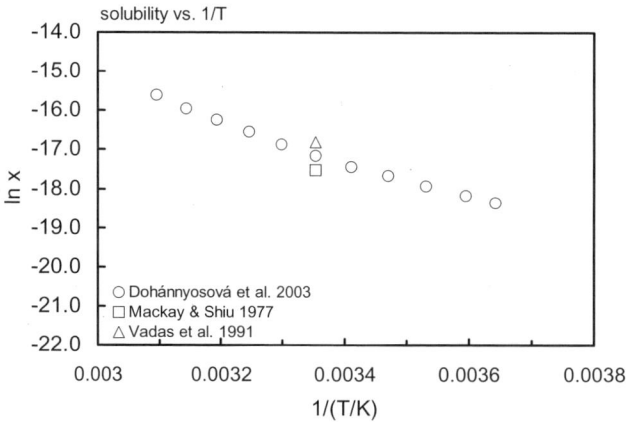

FIGURE 4.1.1.27.1 Logarithm of mole fraction solubility (ln x) versus reciprocal temperature for 9-methylanthracene.

4.1.1.28 9,10-Dimethylanthracene

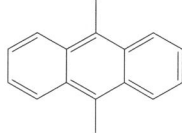

Common Name: 9,10-Dimethylanthracene
Synonym:
Chemical Name: 9,10-dimethylanthracene
CAS Registry No: 781-43-1
Molecular Formula: $C_{16}H_{14}$
Molecular Weight: 206.282
Melting Point (°C):
    183.6   (Lide 2003)
Boiling Point (°C):
    360   (Lide 2003)
Density (g/cm³ at 20°C):
Molar Volume (cm³/mol):
    241.1   (calculated-Le Bas method at normal boiling point)
Enthalpy of Fusion, ΔH_{fus} (kJ/mol):
Entropy of Fusion, ΔS_{fus} (J/mol K):
Fugacity Ratio at 25°C (assuming ΔS_{fus} = 56 J/mol K), F: 0.0278 (at mp = 183.6°C)

Water Solubility (g/m³ or mg/L at 25°C or as indicated and reported temperature dependence equations. Additional data at other temperatures designated * are compiled at the end of this section):
    0.056   (shake flask-fluorescence, Mackay & Shiu 1977)
    0.0129*   (generator column-HPLC/UV, measured range 0–50°C, Dohányosová et al. 2003)
    $\ln x = -73.2594 + 52.6685/\tau + C \ln \tau$, τ = T/298.15 K; temp range 0–50°C (generator column-HPLC/UV, Dohányosová et al. 2003)

Vapor Pressure (Pa at 25°C and reported temperature dependence equations):
    1.53×10^{-4} (extrapolated-Antoine eq., Stephenson & Malanowski 1987)
    $\log (P_S/\text{kPa}) = 11.266 - 5391/(T/K)$; temp range 381–434 K (Antoine eq., Stephenson & Malanowski 1987)

Henry's Law Constant (Pa m³/mol):

Octanol/Water Partition Coefficient, log K_{OW}:
    5.69   (HPLC-RT correlation, Wang et al. 1986)
    5.69   (recommended, Sangster 1989, 1993)
    5.69   (recommended, Hansch et al. 1995)

Octanol/Air Partition Coefficient, log K_{OA}:

Bioconcentration Factor, log BCF:

Sorption Partition Coefficient, log K_{OC}:

Environmental Fate Rate Constants, k or Half-Lives, $t_{1/2}$:
    Volatilization:
    Photolysis: direct photochemical transformation $t_{1/2}$(calc) = 0.35 h, computed near-surface water, latitude 40°N, midday, midsummer (Zepp & Schlotzhauer 1979)
        photodegradation in methanol-water (2:3, v/v) solution for initial concentration of 5 ppm by high pressure mercury lamp or sunlight with a rate constant k = 0.0633 min⁻¹ and $t_{1/2}$ = 0.18 h (Wang et al. 1991)

pseudo-first-order direct photolysis k (exptl) = 0.0633 min^{-1} with the calculated t$_{1/2}$ = 0.18 h and the predicted k(calc) = 0.0379 min^{-1} calculated by QSPR metnod in aqueous solution when irradiated with a 500 W medium pressure mercury lamp (Chen et al. 1996)

direct photolysis t$_{1/2}$ = 1.17 h predicted by QSPR method in atmospheric aerosol (Chen et al. 2001).

Oxidation: t$_{1/2}$ = 1.5 h for photosensitized oxygenation with singlet oxygen at near-surface natural water, 40°N, midday, midsummer (Zepp & Schlotzhauer 1979)

Hydrolysis:

Biodegradation:

Biotransformation:

Bioconcentration, Uptake (k$_1$) and Elimination (k$_2$) Rate Constants:

Half-Lives in the Environment:

Air: direct photolysis t$_{1/2}$ = 1.17 h predicted by QSPR method in atmospheric aerosol (Chen et al. 2001).

Surface water: photolysis t$_{1/2}$ = 0.35 h near surface water, 40°N; midday, midsummer and photosensitized oxygenation t$_{1/2}$ = 1.5 h at near surface water, 40°N, midday, midsummer (Zepp & Schlotzhauer 1979).

photolysis t$_{1/2}$ = 0.18 h in aqueous solution when irradiated with a 500 W medium pressure mercury lamp (Chen et al. 1996).

Groundwater:

Sediment:

Soil:

Biota:

TABLE 4.1.1.28.1
Reported aqueous solubilities of 9,10-dimethylanthracene at various temperature and the empirical temperature dependence equations

$$\ln x = A + B/\tau + C \ln \tau, \text{ where } \tau = T/T_o \text{ and } T_o = 298.15 \text{ K} \quad (1)$$

Doháynosová et al. 2003

generator column-HPLC

t/°C	S/g·m^{-3}
0.30	0.000391
5.0	0.00382
20	0.00534
15	0.00698
20	0.00932
25	0.0129
30	0.0186
35	0.0252
40	0.0351
45	0.0501
50	0.0728
eq. 1	mole fraction
A	−73.2594
B	52.6685
C	71.9873

ΔH$_{sol}$/(kJ mol^{-1}) = 47.9 ± 0.3
at 298.15 K
mp/K   455
ΔH$_{fus}$/(kJ mol^{-1}) = 23.46

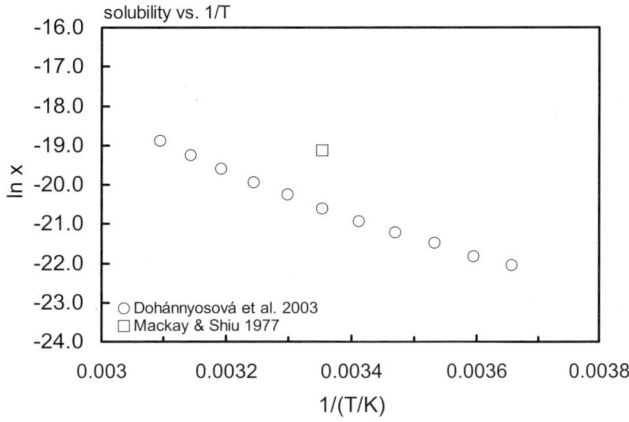

FIGURE 4.1.1.28.1 Logarithm of mole fraction solubility (ln *x*) versus reciprocal temperature for 9, 10-dimethylanthracene.

4.1.1.29 Pyrene

Common Name: Pyrene
Synonym: benzo[*def*]phenanthrene
Chemical Name: pyrene
CAS Registry No: 129-00-0
Molecular Formula: $C_{16}H_{10}$
Molecular Weight: 202.250
Melting Point (°C):
    150.62   (Lide 2003)
Boiling Point (°C):
    404   (Lide 2003)
Density (g/cm³ at 20°C):
    1.271   (23°C, Weast 1982–83; Lide 2003)
Molar Volume (cm³/mol):
    159.0   (calculated-density, liquid molar volume, Lande & Banerjee 1981)
    213.8   (calculated-Le Bas method at normal boiling point)
Enthalpy of Fusion, ΔH_{fus} (kJ/mol):
    17.11   (Ruelle & Kesselring 1997)
    0.29, 17.36; 17.65   (–152.35, 150.65°C; total phase change enthalpy, Chickos et al. 1999)
Entropy of Fusion, ΔS_{fus} (J/mol K):
    35.98   (Wauchope & Getzen 1972)
    40.17   (Casellato et al. 1973)
    54.8   (Hinckley et al. 1990)
    40.97   (150.65°C, Chickos et al. 1999)
    43.36, 43.8   (exptl., calculated-group additivity method, total phase change entropy, Chickos et al. 1999)
Fugacity Ratio at 25°C (assuming ΔS_{fus} = 56 J/mol K), F: 0.0585 (mp at 150.62°C)
    0.128   (calculated, ΔS_{fus} = 40.3 J/mol K, Passivirta et al. 1999)
Water Solubility (g/m³ or mg/L at 25°C or as indicated and reported temperature dependence equations. Additional data at other temperatures designated * are compiled at the end of this section):
    0.165   (27°C, shake flask-nephelometry, Davis et al. 1942)
    0.175   (shake flask-UV, Klevens 1950)
    0.148   (Pierotti et al. 1959)
    1.56   (shake flask-UV/fluorescence, Barone et al. 1967)
    0.105   (20°C, shake flask-UV, Eisenbrand & Baumann 1970)
    0.148*   (shake flask-UV, measured range 22.2–74.7°C, Wauchope & Getzen 1972)
    $R \cdot \ln x = -3660/(T/K) + 4.08 \times 10^{-4} \cdot [(T/K) - 291.15]^2 - 38.1 + 0.0478 \cdot (T/K)$; temp range 22.2–73.4°C (shake flask-UV measurements, Wauchope & Getzen 1972)
    0.105, 0.133, 0.107, 0.069   (HPLC-relative retention correlation, different stationary and mobile phases, Locke 1974)
    0.171   (shake flask-fluorescence/UV, Schwarz & Wasik 1976)
    0.135   (shake flask-fluorescence, Mackay & Shiu 1977)
    0.132   (Rossi 1977; Neff 1979)
    0.1295*   (shake flask-fluorescence, measured range 12.2–31.3°C, Schwarz 1977)
    0.132*   (generator column-HPLC/UV, measured range 5–30°C, May et al. 1978b)
    $S/(\mu g/kg) = 50.2 - 1.051 \cdot (t/°C) + 0.2007 \cdot (t/°C)^2 - 0.0011 \cdot (t/°C)^3$; temp range 5–30°C (generator column-HPLC/UV, May et al. 1978b)
    0.032   (shake flask-nephelometry, Hollifield 1979)
    0.135   (shake flask-LSC, Means et al. 1979; 1980)
    0.130   (shake flask-GC/UV, Rossi & Thomas 1981)

0.136*   (25.5°C, generator column-HPLC/UV, measured range 277.85–303.05 K, May et al. 1983)
0.129   (average lit. value, Pearlman et al. 1984)
0.133   (generator column-HPLC/fluorescence, Walters & Luthy 1984)
0.135   (RP-HPLC-RT correlation, Chin et al. 1986)
0.118   (generator column-HPLC/UV, Billington et al. 1988)
0.132*   (recommended, IUPAC Solubility Data Series, Shaw 1989)
0.150   (shake flask-LSC, Eadie et al. 1990)
0.107   (generator column-HPLC, Vadas et al. 1991)
0.132, 0.050 (solid S_S at 26°C); 2.61, 1.01 (supercooled liquid S_L at 4°C) (quoted, Piatt et al. 1996)
0.131; 0.077, 0.422   (quoted, exptl.; calculated-molar volume, mp and mobile order thermodynamics, Ruelle & Kesselring 1997)
0.150   (microdroplet sampling and multiphoton ionization-based fast-conductivity technique MPI-FC, Gridin et al. 1998)
log $[S_L/(mol/L)]$ = 0.316 – 895.4/(T/K); (supercooled liquid, Passivirta et al. 1999)
ln x = –4.007476 – 4252.03/(T/K), temp range 5–50°C (regression eq. of literature data, Shiu & Ma 2000)
0.1331*   (25.55°C, generator column-HPLC/fluorescence, measured range 0.75–32.08°C, Reza et al. 2002)
ln x = (–1.30 ± 0.56) + (–5059 ± 165)/(T/K); temp range 281.9–305.23 K (Reza et al. 2002)

Vapor Pressure (Pa at 25°C or as indicated and reported temperature dependence equations. Additional data at other temperatures designated * are compiled at the end of this section):
0.00339   (effusion method, Inokuchi et al. 1952)
0.000882*   (effusion method, extrapolated from Antoine eq., Bradley & Cleasby 1953)
log (P/cmHg) = 10.270 – 4904/(T/K); temp range 71.75–85.25°C (Antoine eq., Bradley & Cleasby 1953)
347*   (200.4°C, temp range 200.4–394.7°C, Tsypkina 1955; quoted, Boublik et al. 1984)
0.00033   (effusion method, Hoyer & Peperle 1958; quoted, Mabey et al. 1982; Tsai et al. 1991)
log (P/mmHg) = 12.0 – 5248/(T/K), temp range 25–90°C, (Knudsen effusion method, Hoyer & Peperle 1958)
0.2118*   (348 K, Knudsen effusion, measured range 348–419 K, Malaspina et al. 1974)
12.4*   (125°C, inclined-piston gauge, measured range 125–185°C, Smith et al. 1980)
0.00091   (effusion method, Pupp et al. 1974; quoted, Bidleman 1984)
0.00027   (lit. average-interpolated, API 1979; quoted, Wasik et al. 1983)
0.00088   (extrapolated from Antoine eq., Amidon & Anik 1981)
0.00060*   (gas saturation-HPLC/fluo./UV, Sonnefeld et al. 1983)
log (P/Pa) = 12.748 – 4760.73/(T/K); temp range 10–50°C (solid, Antoine eq., Sonnefeld et al. 1983)
0.0006   (generator column-HPLC/fluo./UV, Wasik et al. 1983)
0.0113, 0.0049   (P_{GC} by GC-RT correlation with eicosane as reference standard, different GC columns, Bidleman 1984)
0.0158   (supercooled liquid P_L, converted from literature P_S with ΔS_{fus} Bidleman 1984)
log (P_L/kPa) = 4.75092 – 1127.529/(16.02 + t/°C); temp range 200–394°C (Antoine eq. from reported exptl. data, Boublik et al. 1984)
0.00442   (Yamasaki et al. 1984)
log (P_L/mmHg) = 5.6184 – 1122.0/(15.2 + t/°C); temp range 200–395°C (Antoine eq., Dean 1985, 1992)
0.00033   (selected, Howard et al. 1986; quoted, Banerjee et al. 1990)
0.00055   (interpolated Antoine eq.-I, Stephenson & Malanowski 1987)
log (P_S/kPa) = 10.75452 – 5072.78/(T/K); temp range 298–401 K (Antoine eq.-I, Stephenson & Malanowski 1987)
log (P_S/kPa) = 11.35032 – 5286.784/(T/K); temp range 360–419 K (Antoine eq.-II, Stephenson & Malanowski 1987)
log (P_L/kPa) = 5.5106 – 1743.57/(–170.83 + T/K), temp range 513–668 K, (Antoine eq.-III, Stephenson & Malanowski 1987)
0.000413, 0.00973   (lit. mean, supercooled liquid value P_L, Bidleman & Foreman 1987)
0.000293*   (pressure gauge, extrapolated-Antoine eq. derived exptl. data, temp 80–140°C, Sasse et al. 1988)
log (P_S/mmHg) = 8.654859 – 2967.129/(182.314 + t/°C); temp range 80.11–139.97°C (solid, Antoine eq., pressure gauge, Sasse et al. 1988)
log (P_L/mmHg) = 5.62672 – 1553.755/(112.964 + t/°C); temp range 139.93–194.16°C (liquid, Antoine eq., pressure gauge, Sasse et al. 1988)

0.010, 0.014   (quoted P_L, supercooled liquid P_L, GC-RT correlation, Hinckley 1989)
0.0158, 0.0144   (supercooled P_L, converted from literature P_S with different ΔS_{fus} values, Hinckley et al. 1990)
0.0113, 0.00752   (P_{GC} by GC-RT correlation with different reference standards, Hinckley et al. 1990)
$\log (P_L/Pa) = 11.92 - 4104/(T/K)$ (GC-RT correlation, Hinckley et al. 1990)
$\log (P/mmHg) = 70.7671 - 6.9413 \times 10^3/(T/K) - 21.79 \cdot \log (T/K) + 6.0727 \times 10^{-3} \cdot (T/K) + 1.5767 \times 10^{-12} \cdot (T/K)^2$; temp range 424–926 K (vapor pressure eq., Yaws 1994)
0.00446   (supercooled liquid P_L, calculated from Yamasaki et al. 1984, Finizio et al. 1997)
0.000334–0.00306; 0.000601; 0.000293 (quoted exptl.: effusion, gas saturation, manometry, Delle Site 1997)
0.000173, 0.00088; 0.000247, 0.000572, 0.000713 (quoted lit., calculated; from GC-RT correlation, Delle Site 1997)
0.00044*   (Knudsen effusion, extrapolated-Antoine eq. derived from exptl. data, temp range 35–125°C, Oja & Suuberg 1998)
$\log (P/Pa) = 33.856 - 12400/(T/K)$; temp range 308–398 K (Clausius-Clapeyron eq., Knudsen effusion, Oja & Suuberg 1998)
1.51×10^{-2}; 1.94×10^{-3} (quoted supercooled liquid P_L from Hinckley et al. 1990; converted to solid P_S with fugacity ratio F, Passivirta et al. 1999)
$\log (P_S/Pa) = 11.60 - 4263/(T/K)$ (solid, Passivirta et al. 1999)
$\log (P_L/Pa) = 9.49 - 3370/(T/K)$ (supercooled liquid, Passivirta et al. 1999)
0.00073 ± 0.00033   (gas saturation-HPLC/fluorescence, de Seze et al. 2000)
$\log (P/Pa) = 12.748 - 4760.73/(T/K)$; temp range 5–50°C (regression eq. from literature data, Shiu & Ma 2000)

Henry's Law Constant (Pa m^3/mol 25°C or as indicated and reported temperature dependence equations. Additional data at other temperatures designated * are compiled at the end of this section):

1.89   (gas stripping, Southworth 1979)
1.10   (gas stripping, Mackay & Shiu 1981)
1.21   (gas stripping-fluorescence, Shiu & Mackay 1997)
2.0, 0.92   (gas stripping-HPLC/fluo., De Maagd et al. 1998)
0.496   (wetted wall column-GC, Altschuh et al. 1999)
1.71*   (gas stripping-GC; measured range 4.1–31°C, Bamford et al. 1999)
$\ln K_{AW} = -5159.97/(T/K) + 10.103$, $\Delta H = 42.9$ kJ mol^{-1}, measured range 4.1–31°C (gas stripping-GC, Bamford et al. 1999)
$\log [H/(Pa\ m^3/mol)] = 9.17 - 2475/(T/K)$ (Passivirta et al. 1999)

Octanol/Water Partition Coefficient, log K_{OW}:

4.90   (calculated-π const., Southworth et al. 1978)
5.32   (calculated-fragment const., Callahan et al. 1979)
4.88   (Hansch & Leo 1979)
5.18   (shake flask-UV, concn. ratio, Karickhoff et al. 1979)
5.09   (shake flask-LSC, Means et al. 1979, 1980)
5.22   (calculated-f const., Yalkowsky & Valvani 1980, Yalkowsky et al. 1983)
5.03   (HPLC-k' correlation, Hanai et al. 1981)
5.05   (HPLC-k' correlation, McDuffie 1981)
5.22   (RP-TLC-k' correlation, Bruggeman et al. 1982)
4.50   (HPLC-k' correlation, D'Amboise & Hanai 1982)
4.88   (HPLC-k' correlation, Hammers et al. 1982)
4.88   (HPLC-k' correlation, Hafkenscheid & Tomlinson 1983)
4.96   (HPLC-RT correlation, Rapaport 1984)
5.52   (HPLC-RT/MS correlation, Burkhard et al. 1985)
4.80   (Hansch & Leo 1985)
4.97   (RP-HPLC-RT correlation, Chin et al. 1986)
4.95   (Leo 1986)
5.00 ± 0.20 (recommended, Sangster 1989, 1993)
4.88   (recommended, Hansch et al. 1995)
5.08; 5.39   (26°C; 4°C, Piatt et al. 1996)
4.84 ± 0.19, 5.14 ± 0.62 (HPLC-k' correlation: ODS column; Diol column, Helweg et al. 1997)
4.77   (shake flask-dialysis tubing-HPLC/UV, both phases, Andersson & Schräder 1999)
4.79   (shake flask-SPME solid-phase micro-extraction, Paschke et al. 1999)

Polynuclear Aromatic Hydrocarbons (PAHs) and Related Aromatic Hydrocarbons

Octanol/Air Partition Coefficient, log K_{OA} at 25°C or as indicated and reported temperature dependence equations. Additional data at other temperatures designated * are compiled at the end of this section:

8.60 (calculated-K_{OW}/K_{AW}, Wania & Mackay 1996)
8.61 (calculated, Finizio et al. 1997)
8.80*; 8.49 (generator column-GC; calculated-C_O/C_A, measured range 0–40°C, Harner & Bidleman 1998)
log K_{OA} = –4.56 + 3985/(T/K); temp range 0–40°C (generator column-GC, Harner & Bidleman 1998)
8.75 (calculated-S_{oct} and vapor pressure P, Abraham et al. 2001)

Bioconcentration Factor, log BCF:

3.43 (*Daphnia pulex*, Southworth et al. 1978)
4.38 (mixed microbial population, Steen & Karickhoff 1981)
4.65 (*P. hoyi*, Eadie et al. 1982)
4.56, 4.22, 4.75 (average, *Selenustrum capricornutum*-dosed singly, dosed simultaneously, Casserly et al. 1983)
2.66 (goldfish, shake flask-GC, concn. ratio, Ogata et al. 1984)
3.43 (*Daphnia pulex*, Mackay & Hughes 1984)
3.43 (*Daphnia magna*, Newsted & Giesy 1987)
3.65, 3.81, 2.35 (mussel, clam, shrimp, Gobas & Mackay 1989)
2.85, 2.70 (*Polychaete, Shrimo-hepatopancreas*, Gobas & Mackay 1989)
0.716, 1.124 (*Polychaete sp, Capitella capitata*, Bayona et al. 1991)

Sorption Partition Coefficient, log K_{OC}:

4.92 (natural sediments, sorption isotherms by batch equilibrium technique-UV spec., Karickhoff et al. 1979)
4.81 (average value of soil and sediment, shake flask-LSC, sorption isotherms, Means et al. 1979)
4.92 (Kenaga & Goring 1980)
4.80 (average value of 12 soil/sediment samples, sorption isotherms by shake flask-LSC, Means et al. 1980)
4.78, 4.80 (soil/sediment: calculated-K_{OW}, regress of K_P versus substrate properties, Means et al. 1980)
4.826 (sediment/soil, sorption isotherm by batch equilibrium technique, Karickhoff 1981)
3.11, 3.46 (sediment suspensions, Karickhoff & Morris 1985)
5.23; 5.08 (fluorescence quenching interaction with AB humic acid; AB fulvic acid, Gauthier et al. 1986)
4.46–4.81; 4.94–5.51; 4.73–5.02 (marine humic acids; soil humic acids; soil fulvic acids, fluorescence quenching technique, Gauthier et al. 1987)
5.02 (dissolved humic materials, Aldrich humic acid, fluorescence quenching technique, Gauthier et al. 1987)
5.13 (sediment, batch equilibrium-GC, Vowles & Mantoura 1987)
4.88 (soil-fine sand 0.2% OC, dynamic soil column studies, Enfield et al. 1989)
5.65 (LSC, Eadie et al. 1990)
6.51 (sediments average, Kayal & Connell 1990)
4.83 (RP-HPLC-RT correlation, Pussemier et al. 1990)
4.82, 4.77 (RP-HPLC-RT correlation on CIHAC, on PIHAC stationary phases, Szabo et al. 1990b)
6.50 (Baltic Sea particulate field samples, concn distribution-GC/MS, Broman et al. 1991)
5.05, 5.00, 4.88; 4.71 (marine porewater organic colloids: Fort Point Channel FPC 7–9 cm, FPC 15–17 cm, FPC 25–29 cm; Spectacle Island 14–16 cm, Chin & Gschwend 1992)
5.20, 5.18, 4.99; 5.23 (marine sediments: Fort Point Channel FPC 7–9 cm, FPC 15–17 cm, FPC 25–29 cm; Spectacle Island 14–16 cm, Chin & Gschwend 1992)
4.78; 4.78, 4.78 (sediment: concn ratio C_{sed}/C_W; concn-based coeff., area-based coeff. of flux studies of sediment/water boundary layer, Helmstetter & Alden 1994)
5.50, 6.61, 6.06 (marine sediments: Fort Point Channel, Spectacle Island, Peddocks Island, McGroddy & Farrington 1995)
5.51, 5.34, 5.31; 7.43 (marine sediments: Fort Point Channel FPC 7–9 cm, FPC 15–17 cm, FPC 25–29 cm; Spectacle Island 14–16 cm, McGroddy & Farrington 1995)
4.64 (Aldrich humic acid, Ozretich et al. 1995)
4.80, 4.81, 4.72 (RP-HPLC-k' correlation on different stationary phases, Szabo et al. 1995)
4.81 (range 4.73–4.66); 4.22 (range 4.20–4.22) (4°C, low organic carbon sediment f_{OC} = 0.0002, batch equilibrium; column exptl., Piatt et al. 1996)

4.62 (range 4.56–4.67); 4.0 (range 3.98–4.00) (26°C, low organic carbon sediment f_{OC} = 0.0002, batch equilibrium; column exptl., Piatt et al. 1996)

4.42–2.56 (5 soils, 20°C, batch equilibrium-sorption isotherm measured by HPLC/UV, Bayard et al. 1998)

4.99, 4.98, 4.96, 4.97, 4.97 (soils: Woodburn soil, Elliot soil, Marlette soil, Piketon soil, Anoka soil, batch equilibrium-sorption isotherms-HPLC-fluorescence, Choiu et al. 1998)

5.14, 5.22, 5.23, 5.12, 5.04, 5.24, 5.45; mean 4.98 ± 0.009 (sediments: Lake Michigan, Mississippi River, Massachusetts Bay, Spectacle Island, Peddocks Island, Port Point Channel, batch equilibrium-sorption isotherms-HPLC-fluorescence, Choiu et al. 1998)

3.47, 4.60, 3.53, 4.78, 4.61; mean 5.18 ± 0.056 (HPLC-screening method with different LC-columns, Szabo et al. 1999)

4.66, 4.78 (soils: organic carbon OC ≥ 0.1%, OC ≥ 0.5%, average, Delle Site 2001)

4.88, 4.90 (sediments: organic carbon OC ≥ 0.1%, OC ≥ 0.5%, average, Delle Site 2001)

5.47–6.68; 4.60–6.80 (range, calculated from sequential desorption of 11 urban soils; lit. range, Krauss & Wilcke 2001)

5.90; 5.89, 5.60, 5.56 (20°C, batch equilibrium, A2 alluvial grassland soil; calculated values of expt 1,2,3- solvophobic approach, Krauss & Wilcke 2001)

4.96; 4.70; 4.46–5.74 (calculated-K_{OW}; calculated-solubility; quoted lit. range, Schlautman & Morgan 1993a)

4.50 at pH 4, 4.37 at pH 7, 4.33 at pH 10 in 0.001M NaCl; 4.35 at pH 4, 4.20 at pH 7, 4.245 at pH 10 in 0.01M NaCl; 4.35 at pH 4, 4.15 at pH 7, 4.15 at pH 10 in 0.1M NaCl; 4.33 at pH 4, 4.29 at pH 7, 4.15 at pH 10 in 1mM Ca^{2+} in 0.1M total ionic strength solutions (shake flask/fluorescence, humic acid; Schlautmam & Morgan 1993a)

4.19 at pH 4, 3.89 at pH 7, 3.92 at pH 10 in 0.001M NaCl; 4.15 at pH 4, 3.88 at pH 7, 3.86 at pH 10 in 0.01M NaCl; 4.08 at pH 4, 3.81 at pH 7, 3.785 at pH 10 in 0.1M NaCl; 4.08 at pH 4, 3.90 at pH 7, 4.06 at pH 10 in 1mM Ca^{2+} in 0.1M total ionic strength solutions (shake flask/fluorescence, fulvic acid; Schlautmam & Morgan 1993a)

5.52 (soil humic acid, shake flask-HPLC/UV, Cho et al. 2002)

4.66 (Askov soil, a Danish agriculture soil, Sverdrup et al. 2002)

5.35–6.33 (field contaminated sediment, initial-final values of 5–100 d contact time, gas-purge technique- HPLC/fluorescence, ten Hulscher et al. 2003)

Environmental Fate Rate Constants, k, or Half-Lives, $t_{1/2}$:

Volatilization: sublimation rate constant of 1.1×10^{-4} s^{-1} was measured as loss from glass surface at 24°C at an air flow rate of 3 L/min (Cope & Kalkwarf 1987)

Photolysis: calculated $t_{1/2}$ = 4.2 d for direct sunlight photolysis in midday of midsummer at 40°N for inland water, and $t_{1/2}$ = 5.9 d for inland water with sediment partitioning (Zepp & Schlotzhauer 1979)

k = 1.014 h^{-1} (Zepp 1980)

$t_{1/2}$ = 0.68 hm atmospheric and aqueous photolysis half-life, based on measured aqueous photolysis quantum yields calculated for midday summer sunlight at 40°N latitude (Zepp & Schlotzhauer 1979; quoted, Harris 1982; Howard et al. 1991) and $t_{1/2}$ = 2.04 h after adjusting for approximate winter sunlight intensity (Lyman et al. 1982; quoted, Howard et al. 1991)

half-lives: $t_{1/2}$ = 21 h on silica gel, $t_{1/2}$ = 31 h on alumina and $t_{1/2}$ = 46 h on fly ash on different atmospheric particulate substrates (approximate 25 µg/g on substrate) (Behymer & Hites 1985);

$t_{1/2}$ = 4.2 d for summer sunlight photolysis in surface water (Mill & Mabey 1995)

k < 1.05×10^{-4} m/s at 24°C with [O_3] = 0.16 ppm and light intensity of 1.3 kW/m^2 on glass surface of(Cope & Kalkwarf 1987)

photodegradation $t_{1/2}$ = 1 h in summer to days in winter by sunlight for adsorption on airborne particulates (Valerio et al. 1991);

photolysis $t_{1/2}$ = 0.68 h in water, based on direct photolysis in sunlight at midday, midsummer, latitude 40°N (Zepp 1991)

k(exptl) = 0.00362 min^{-1} for pseudo-first-order direct photolysis, with $t_{1/2}$ = 3.18 h, and the predicted k(calc) = 0.00382 min^{-1} by QSPR in aqueous solution when irradiated with a 500 W medium pressure mercury lamp (Chen et al. 1996)

$t_{1/2}$(obs.) = 2.63 h, $t_{1/2}$(calc) = 2.56 h by QSPR in atmospheric aerosol (Chen et al. 2001);

k = (2.08 ± 0.13) × 10^{-5} s^{-1} and $t_{1/2}$ = (9.24 ± 0.53)h in diesel particulate matter, photochemical degradation under atmospheric conditions, k = (1.88 ± 0.16) × 10^{-5} s^{-1} and $t_{1/2}$ = (10.22 ± 0.95)h in diesel particulate matter/soil mixture, and $t_{1/2}$ = 0.80 to 1.59 h in various soil components using a 900-W photo-irradiator as light source; k = (2.61 ± 0.53) × 10^{-7} s^{-1} and $t_{1/2}$ = (737.55 ± 124.49)h in diesel particulate matter using a 300 W light source (Matsuzawa et al. 2001)

Photodegradation k = 3.9 × 10^{-4} s^{-1} in surface water during the summertime at mid-latitude (Fasnacht & Blough 2002)

Oxidation: rate constant k, for gas-phase second order rate constants, k_{OH} for reaction with OH radical, k_{NO_3} with NO$_3$ radical and k_{O_3} with O$_3$ or as indicated, *data at other temperatures see reference:

k = 5 × 10^8 M^{-1} h^{-1} for singlet oxygen and 2.2 × 10^4 M^{-1} h^{-1} for peroxy radical (Mabey et al. 1982)

k = (3.4–5.3) × 10^4 dm^3 mol^{-1} s^{-1} over pH range 1–7, with $t_{1/2}$ = 0.18 s in presence of 10^{-4} M ozone at pH 7 for the reaction with O$_3$ in water at 25°C (Butković et al. 1983)

k_{OH} = 5.0 × 10^{-11} cm^3 molecule^{-1} s^{-1} and $k_{N_2O_5}$ ≈ –5.6 × 10^{-17} cm^3 molecule^{-1} s^{-1} for reaction with N$_2$O$_5$ at 296 ± 2 K (relative rate method, Atkinson et al. 1990)

photooxidation $t_{1/2}$ = 0.802–8.02 h, based on estimated rate constant for reaction with hydroxyl radical in air (Atkinson 1987; quoted, Howard et al. 1991)

Hydrolysis: not hydrolyzable (Mabey et al. 1982); no hydrolyzable groups (Howard et al. 1991).

Biodegradation: significant degradation within 7 d for a domestic sewage 28-d test for an average of three static-flask screening (Tabak et al. 1981)

$t_{1/2}$(aq. aerobic) = 5040–45600 h, based on aerobic soil die-away test data at 10–30°C (Coover & Sims 1987; Sims 1990; quoted, Howard et al. 1991)

k = 0.29 h^{-1} in atmosphere (Dragoscu & Friedlander 1989; quoted, Tsai et al. 1991)

k = 0.0027 d^{-1} with $t_{1/2}$ = 260 d for Kidman sandy loam and k = 0.0035 d^{-1} with $t_{1/2}$ = 199 d for McLarin sandy loam all at –0.33 bar soil moisture (Park et al. 1990)

$t_{1/2}$(anaerobic) = 20160–182400 h, based on estimated unacclimated aqueous aerobic biodegradation half-life (Howard et al. 1991).

Biotransformation: estimated to be 1 × 10^{-10} mL cell^{-1} h^{-1} for bacteria (Mabey et al. 1982)

Bioconcentration, Uptake (k_1) and Elimination (k_2) Rate Constants:

k_1 = 1126 h^{-1}; k_2 = 0.343 h^{-1} (*Daphnia pulex*, Southworth et al. 1978)

log k_1 = 3.05 h^{-1}; log k_2 = –0.46 h^{-1} (*Daphnia pulex*, correlated as per Mackay & Hughes 1984, Hawker & Connell 1986)

k_1 = 113.0 h^{-1}; k_2 = 0.017 h^{-1} (4°C, *S. heringianus*, Frank et al. 1986)

k_2 = 0.017 h^{-1} (*S. heringianus*, Frank et al. 1986)

k_1 = 199.2 mL g^{-1}·h^{-1}; k_2 = 0.0012 h^{-1} (4°C, *P. hoyi*, Landrum 1988)

k_1 = 3.4–5.3 mg g^{-1} h^{-1}; k_2 = 0.022 h^{-1} (freshwater oligochaete from sediment, Van Hoof et al. 2001)

Half-Lives in the Environment:

Air: $t_{1/2}$ = 0.68–2.04 h, based on estimated sunlight photolysis half-life in water (Zepp & Stotzhauer 1979; Lyman et al. 1982; quoted, Howard et al. 1991);

half-lives under simulated atmospheric conditions: $t_{1/2}$ = 4.20 h under simulated sunlight, $t_{1/2}$ = 2.75 h under simulated sunlight + ozone (0.2 ppm), $t_{1/2}$ = 15.72 h under dark reaction ozone (0.2 ppm) (Katz et al. 1979; quoted, Bjørseth & Olufsen 1983);

$t_{1/2}$ = 2.5 h with a steady-state concn of tropospheric ozone of 2 × 10^{-9} M in clean air (Butković et al. 1983);

photooxidation $t_{1/2}$ = 0.802–8.02 h, based on estimated rate constant for reaction with hydroxyl radical in air (Atkinson 1987; quoted, Howard et al. 1991).

Surface water: computed near-surface $t_{1/2}$ = 0.58 h for direct photochemical transformation at latitude 40°N, midday, midsummer, $t_{1/2}$ = 4.2 h for direct photolysis in a 5-m deep inland water body with no sediment-water partitioning and $t_{1/2}$ = 5.9 d with sediment-water partitioning (Zepp & Schlotzhauer 1979);

$t_{1/2}$ = 0.68–2.04 h, based on estimated sunlight photolysis half-life in water (Lyman et al. 1982; quoted, Howard et al. 1991);

$t_{1/2}$ = 1.8 s in presence of 10^{-4} M ozone at pH 7 (Butković et al. 1983);

$t_{1/2}$ = 4.2 d for summer at 40°N latitude under sunlight (Mill & Mabey 1985);

$t_{1/2}$ = 0.68 h, based on direct photolysis in sunlight at midday, mid-summer and 40°N latitude (quoted, Zepp 1991);

photolysis $t_{1/2}$ = 3.18 h in aqueous solution when irradiated with a 500 W medium pressure mercury lamp (Chen et al. 1996).

Groundwater: $t_{1/2}$ = 10080–91200 h, based on estimated unacclimated aqueous aerobic biodegradation half-life (Howard et al. 1991).

Sediment: reduction $t_{1/2}$ = 547 h for chemical available pyrene and $t_{1/2}$ = 298 h for bioavailable pyrene for amphipod, *P. hoyi* in Lake Michigan sediments at 4°C. The uptake clearance from sediment, k = (0.019–0.015)g of dry sediment·g^{-1} of organism·h^{-1}, and the rate constants to become biologically unavailable were k = 0.0019 h^{-1} for 10-d aged sediment corresponding to a t$_{½}$ = 365 d and k = 0.0030 h^{-1} for nonaged sediment corresponding to a t$_{½}$ = 231 h (Landrum 1989).

Soil: t$_{½}$ = 3–35 h (Sims & Overcash 1983; quoted, Bulman et al. 1987);
t$_{½}$ = 58 d for 5 mg/kg treatment and t$_{½}$ = 48 d for 50 mg/kg treatment (Bulman et al. 1987);
t$_{½}$ = 5040–45600 h, based on aerobic soil die-away test data at 10–30°C (Coover & Sims 1987; Sims 1990; quoted, Howard et al. 1991);
t$_{½}$ > 50 d (Ryan et al. 1988);
degradation rate constant, k = 0.0027 d^{-1} with t$_{½}$ = 260 d for Kidman sandy loam soil and k = 0.0035 d^{-1} with t$_{½}$ = 199 d for McLaurin sandy loam soil (Park et al. 1990);
t$_{½}$ = 500 d in soil (Jury et al. 1990);
t$_{½}$ = 0.4 to more than 90 wk, 8.5 yr (quoted, Luddington soil, Wild et al. 1991).

Biota: depuration t$_{½}$ = 40.8 h in *s. heringianus* (Frank et al. 1986);
elimination t$_{½}$ = 4.1–5.5 d from mussel *Mytilus edulis*; t$_{½}$ = 10.3 d from clam *Mya areneria*, t$_{½}$ = 14.4 d from polychaete *Abarenicola pacifica*, t$_{½}$ = 6.7 d from Oyster, t$_{½}$ = 0.80 d from shrimp, t$_{½}$ = 3.6 d from clam *Mercenario mercenaria* (quoted, Meador et al. 1995).

TABLE 4.1.1.29.1
Reported aqueous solubilities of pyrene at various temperatures

1.

Wauchope & Getzen 1972				Schwarz 1977		May et al. 1978a	
shake flask-UV				shake flask-fluorescence		generator column-HPLC	
t/°C	S/g·m^{-3}	t/°C	S/g·m^{-3}	t/°C	S/g·m^{-3}	t/°C	S/g·m^{-3}
experimental		smoothed					
22.2	0.129	0	0.049	12.2	0.055	25	0.132
22.2	0.128	22.2	0.130	15.5	0.069	29	0.162
22.2	0.124	25	0.148	17.4	0.079		
34.5	0.228	34.5	0.235	20.3	0.092	temp dependence eq. 2	
34.5	0.235	44.7	0.399	23.0	0.117	S	µg/kg
44.7	0.397	50	0.532	23.3	0.118	a	−0.0011
44.7	0.395	50.1	0.534	25.0	0.129	b	0.2007
44.7	0.405	55.6	0.730	26.2	0.144	c	−1.051
50.1	0.558	56.0	0.74	26.7	0.145	d	50.2
50.1	0.576	60.7	0.97	28.5	0.164		
50.1	0.556	65.2	1.27	31.3	0.188	ΔH$_{sol}$/(kJ mol^{-1}) = 35.44	
55.6	0.75	71.9	1.90			for temp range 5–30°C	
55.6	0.75	74.7	2.26	ΔH$_{sol}$/(kJ mol^{-1}) = 47.70			
55.6	0.77	75.0	2.31				
56.0	0.74						
60.7	0.96	temp dependence eq. 1					
60.7	0.95	ln x	mole fraction				
60.7	0.90	ΔH$_{fus}$	15.3 ± 0.084				
65.2	1.27	10^2·b	4.78 ± 0.09				
65.2	1.29	c	38.1 ± 0.3				
71.9	1.83						
71.9	1.86						
71.9	1.89						
74.7	2.21						
ΔH$_{fus}$/(kJ mol^{-1}) = 15.3							

Empirical temperature dependence equations:

Wauchope & Getzen (1972): R·ln x = −[H$_{fus}$/(T/K)] + (0.000408)[(T/K) − 291.15]^2 −c + b·(T/K)   (1)

May et al. (1978):- S/(µg/kg) = a·t^3 + b·t^2 +c·t + d   (2)

Polynuclear Aromatic Hydrocarbons (PAHs) and Related Aromatic Hydrocarbons

TABLE 4.1.1.29.1 (Continued)

2.

May 1980		May et al. 1983		Shaw 1989		Reza et al. 2002	
generator column-HPLC		generator column-HPLC		IUPAC recommended		generator column-HPLC/fluo	
t/°C	S/g·m^{-3}	t/°C	S/g·m^{-3}	t/°C	S/g·m^{-3}	t/°C	S/g·m^{-3}
4.7	0.0492	4.7	0.0492	0	0.050	8.54	0.0474
9.5	0.0585	9.5	0.0586	20	0.090	10.39	0.0566
14.3	0.0720	14.3	0.0721	25	0.132	13.5	0.0635
18.7	0.0933	18.7	0.0934	30	0.175	14.46	0.0694
21.2	0.109	21.2	0.1091	40	0.30	15.7	0.0804
25.5	0.136	25.5	0.1361	50	0.50	18.05	0.0871
29.9	0.170	29.9	0.1701	60	0.90	21.53	0.1087
				70	1.70	25.55	0.1331
temp dependence eq. 2				75	2.30	27.36	0.1505
S	µg/kg					29.66	0.1686
a	−0.0011					21.28	0.1931
b	0.2007						
c	−1.051					ln x = A − B/T(K)	
d	50.2					A	−1.30
						B	5059
ΔH_{sol}/(kJ mol^{-1}) = 35.44 for temp range 5–30°C						temp range 282–305 K	

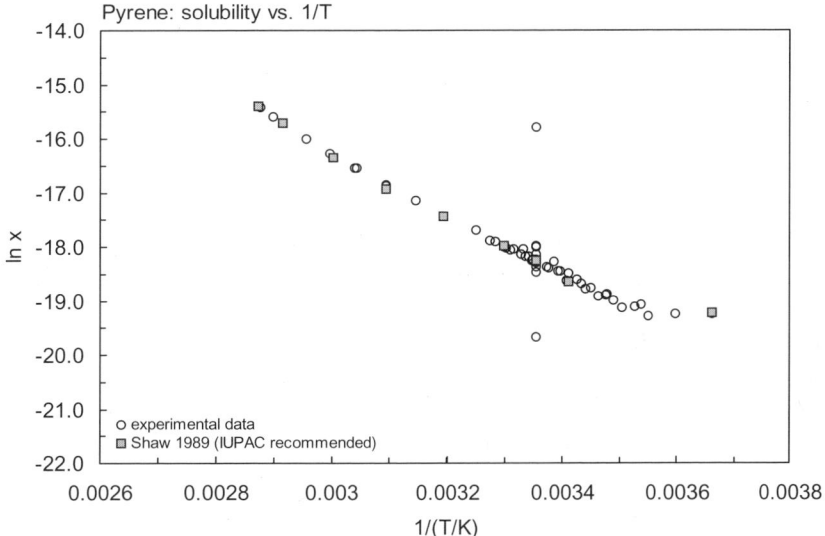

FIGURE 4.1.1.29.1 Logarithm of mole fraction solubility (ln x) versus reciprocal temperature for pyrene.

TABLE 4.1.1.29.2
Reported vapor pressures of pyrene at various temperatures and the coefficients for the vapor pressure equations

$\log P = A - B/(T/K)$ (1) $\quad\quad\quad$ $\ln P = A - B/(T/K)$ (1a)

$\log P = A - B/(C + t/°C)$ (2) $\quad\quad\quad$ $\ln P = A - B/(C + t/°C)$ (2a)

$\log P = A - B/(C + T/K)$ (3)

$\log P = A - B/(T/K) - C \cdot \log (T/K)$ (4)

1.

Bradley & Cleasby 1953		Tsypkina 1955		Hoyer & Peperle 1958		Malaspina et al. 1974	
effusion		in Boublik et al. 1984		effusion		Knudsen effusion	
t/°C	P/Pa	t/°C	P/Pa	t/°C	P/Pa	T/K	P/Pa
68.9	0.1147	200.4	347	data presented by		348	0.2118
74.15	0.1880	220.8	920	eq. 1	P/mmHg	361	0.508
78.1	0.263	242.7	2440	A	12.00	364	0.717
78.1	0.274	256.4	4373	B	5348	369	1.034
81.7	0.2746	270.0	6493	temp range 25–90°C		377	2.016
82.65	0.3893	277.0	7999			384	3.871
85.0	0.4093	288.7	10866	$\Delta H_{subl}/(kJ\,mol^{-1}) = 100.5$		393	7.934
71.75	0.1440	293.0	12599			402	15.20
75.85	0.2226	306.0	17932			411	31.01
82.7	0.2880	316.0	22665			419	55.42
82.25	0.4053	394.7	101324				
85.0	0.5066					for solid pyrene:	
		bp/°C	394.707			eq. 1	P/atm
eq. 1	P/mmHg					A	8.848
A	10.270					B	5091
B	4904						
						$\Delta H_{subl}/(kJ\,mol^{-1}) = 101.04$	
$\Delta H_{subl}/(kJ\,mol^{-1}) = 93.90$						at 298.15 K	

2.

Smith et al. 1980		Sonnefeld et al. 1983		Sasse et al. 1988		Oja & Suuberg 1998	
inclined-piston gauge		gas saturation-HPLC		electronic manometry		Knudsen effusion	
t/°C	P/Pa	t/°C	P/Pa	t/°C	P/Pa	t/°C	P/Pa
	solid			solid		46.95	0.00863
125	12.4	10.5	9.20×10^{-5}	80.11	0.30	56.99	0.0243
130	17.5	10.5	9.21×10^{-5}	90.10	0.764	37.98	0.0209
135	24.7	10.5	9.53×10^{-5}	100.11	1.853	68.03	0.0852
140	34.7	20.9	3.35×10^{-4}	110.14	4.360	72.99	0.164
145	48.0	20.9	3.35×10^{-4}	120.08	9.213	83.99	0.428
149	62.1	20.9	3.39×10^{-4}	120.10	9.40	93.02	0.945
150	66.1	30.0	1.10×10^{-4}	129.99	19.20		
	liquid	30.0	1.09×10^{-4}	134.98	26.80		
125	17.1*	30.0	1.03×10^{-4}	139.97	37.07	eq. 1a	P/Pa
130	22.5*	39.34	3.31×10^{-3}	liquid		A	33.856
135	29.9*	39.34	3.47×10^{-3}	139.93	40.80	B	12400
140	39.5*	39.34	3.25×10^{-3}	144.90	53.06		
145	51.3*	39.45	3.41×10^{-3}	149.82	68.66	$\Delta H_{sub}/(kJ\,mol^{-1}) = 103.10$	

TABLE 4.1.1.29.2 (Continued)

2.

Smith et al. 1980		Sonnefeld et al. 1983		Sasse et al. 1988		Oja & Suuberg 1998	
inclined-piston gauge		gas saturation-HPLC		electronic manometry		Knudsen effusion	
t/°C	P/Pa	t/°C	P/Pa	t/°C	P/Pa	t/°C	P/Pa
149	63.3*	39.45	3.20×10^{-3}	159.75	112.5		
150	66.7*	25.0	6.0×10^{-4}	169.61	178.8		
152	78.3			179.45	279.4		
155	86.5	eq. 2	P/Pa	189.25	405.7		
160	110.7	A	12.748	194.16	490.1		
165	141.6	B	4760.73				
170	178.9	temp range 10–50°C		for solid			
175	224.6			eq. 2	P/mmHg		
180	280.4	ΔH_{subl}/(kJ mol^{-1}) = 91.20		A	8.654859		
185	345.3			B	2967.129		
				C	182.314		
*supercooled liquid values				temp range: 80.1–139.97°C			
				ΔH_{subl}/(kJ mol^{-1}) = 97.70			
reported vapor pressure eq.							
see foot note			eq. 2	P/mmHg			
				A	5.62672		
ΔH_{subl}/(kJ mol^{-1}) = 103.9			B	1553.755			
				C	112.964		
				temp range: 139.97–194°C			

note: $\ln(P_L/P_S) = 268.6187 - 699.31/(T/K) - 45.6846 \cdot \ln(T/K) + 0.057217(T/K)$; in which P_L and P_S are the vapor pressure of supercooled liquid and crystal phase, respectively, at temperature T.

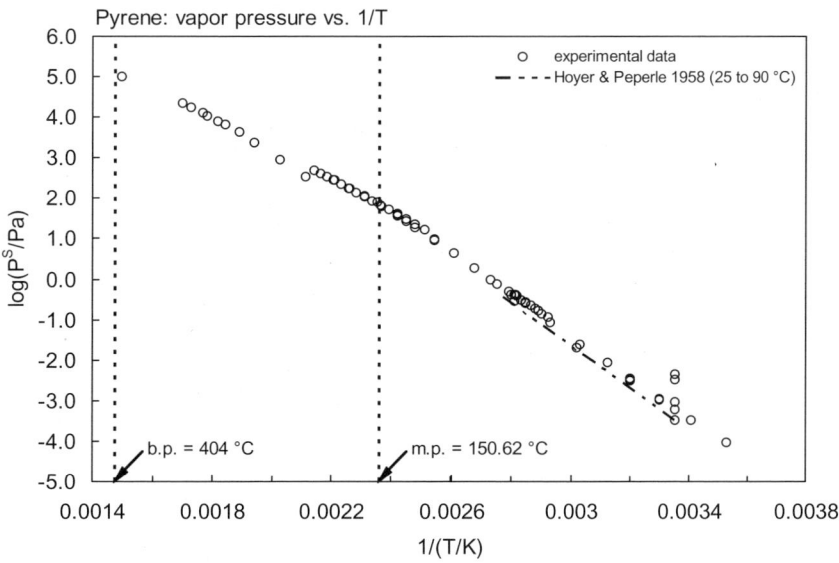

FIGURE 4.1.1.29.2 Logarithm of vapor pressure versus reciprocal temperature for pyrene.

TABLE 4.1.1.29.3
Reported Henry's law constants and octanol-air partition coefficients of pyrene at various temperatures and temperature dependence equations

	Henry's law constant			log K_{OA}	
	Bamford et al. 1999			Harner & Bidleman 1998	
	gas stripping-GC/MS			generator column-GC/FID	
t/°C	H/(Pa m³/mol)	H/(Pa m³/mol)		t/°C	log K_{OA}
		average			
4.1	0.37, 0.49	0.43		0	9.966
11.0	0.63, 0.76	0.69		10	9.528
18.0	1.02, 1.19	1.10		20	9.155
25.0	1.54, 1.89	1.71		30	8.647
31.0	2.15, 2.80	2.45		40	8.121
				25(exptl)	8.80
ln K_{AW} = A − B/(T/K)				25(calc)	8.49
A	10.1034				
B	5160			log K_{OA} = A + B/(T/K)	
				A	−4.56
enthalpy, entropy change:				B	3985
ΔH/(kJ·mol⁻¹) = 42.9 ± 2.3					
ΔS/(J·K⁻¹ mol⁻¹) = 84				enthalpy of phase change	
	at 25°C			ΔH_{OA}/(kJ mol⁻¹) = 76.3	

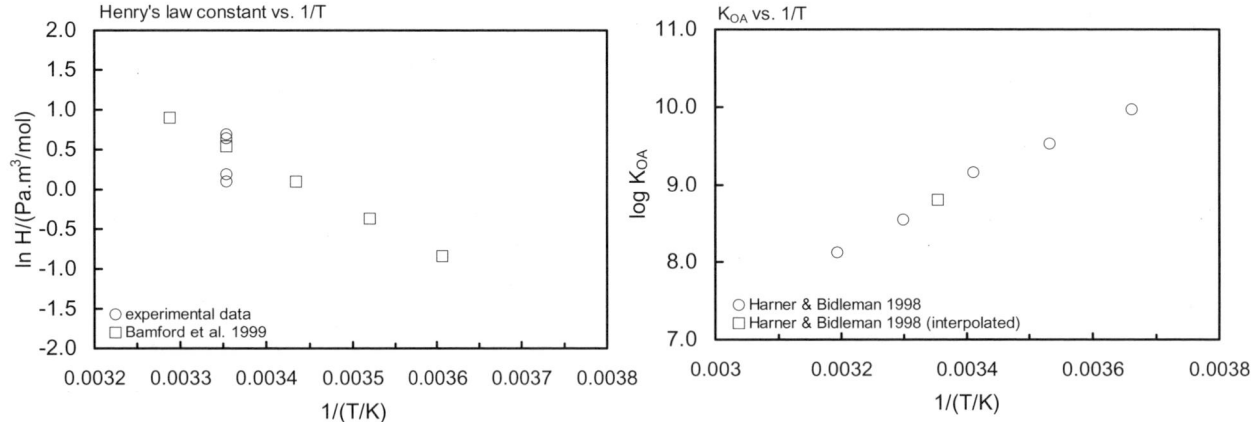

FIGURE 4.1.1.29.3 Logarithm of Henry's law constant and K_{OA} versus reciprocal temperature for pyrene.

4.1.1.30 Fluoranthene

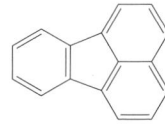

Common Name: Fluoranthene
Synonym: idryl, 1,2-benzacenaphthene, benzo[j,k]fluorene, benz[a]acenaphthylene, fluoranthrene
Chemical Name: 1,2-benzacenaphthene
CAS Registry No: 206-44-0
Molecular Formula: $C_{16}H_{10}$
Molecular Weight: 202.250
Melting Point (°C):
    110.19   (Lide 2003)
Boiling Point (°C):
    384   (Lide 2003)
Density (g/cm³ at 20°C):
    1.252   (0°C, Weast 1982–83, Dean 1985; Lide 2003)
Molar Volume (cm³/mol):
    162   (calculated-density, liquid molar volume, Lande & Banerjee 1981)
    217.3   (calculated-Le Bas method at normal boiling point)
Enthalpy of Fusion, ΔH_{fus} (kJ/mol):
    18.87   (Ruelle & Kesselring 1997)
    18.74   (exptl., Chickos et al. 1999)
Entropy of Fusion, ΔS_{fus} (J/mol K):
    49.37   (Casellato et al. 1973; quoted, Yalkowsky 1981)
    47.70   (differential scanning calorimetry, Hinckley et al. 1990)
    48.89, 36.5   (exptl., calculated-group additivity method, total phase change entropy, Chickos et al. 1999)
    49.6   (Passivirta et al. 1999)
Fugacity Ratio at 25°C (assuming ΔS_{fus} = 56 J/mol K), F: 0.146 (mp at 110.19°C)
    0.191   (calculated, ΔS_{fus} = 49.6 J/mol K, Passivirta et al. 1999)

Water Solubility (g/m³ or mg/L at 25°C or as indicated and reported temperature dependence equations. Additional data at other temperatures designated * are compiled at the end of this section):
    0.240   (27°C, shake flask-nephelometry, Davis et al. 1942)
    0.265   (shake flask-UV, Klevens 1950)
    0.240   (20°C, shake flask-UV, Eisenbrand & Baumann 1970)
    0.236   (fluorescence/UV, Schwarz & Wasik 1976)
    0.260   (shake flask-fluorescence, Mackay & Shiu 1977)
    0.206   (Rossi 1977; Neff 1979)
    0.206*   (generator column-HPLC/UV, measured range 8.1–29.9°C, May et al. 1978a,b)
$S/(\mu g/kg) = 50.4 + 4.322 \cdot (t/°C) - 0.1047 \cdot (t/°C)^2 + 0.0072 \cdot (t/°C)^3$; temp range 5–30°C (generator column-HPLC/UC, May et al. 1978b)
    0.120   (shake flask-nephelometry, Hollifield 1979)
    0.218   (OECD 1979/1980; quoted, He et al. 1995)
    0.275, 0.373   (15, 25°C, generator column/elution method, average values of 6–7 laboratories, OECD 1981)
    0.200   (20°C, quoted, Schmidt-Bleek et al. 1982)
    0.190   (20°C, generator column-fluorescence, Hashimoto et al. 1982)
    0.203*   (24.6°C, generator column-HPLC, measured range 281.25–303.05 K, May et al. 1983)
    0.243   (average lit. value, Pearlman et al. 1984)
    0.199   (generator column-HPLC/fluorescence, Walters & Luthy 1984)
    1.43   (RP-HPLC-RT correlation, Chin et al. 1986)
    0.283   (vapor saturation-UV, Akiyoshi et al. 1987)
    0.240   (recommended, Shaw 1989)
    0.222   (generator column-HPLC/fluorescence, Kishi & Hashimoto 1989)

0.373   (average value of Japan, OECD tests, Kishi & Hashimoto 1989)
0.166   (shake flask-fluorescence, Kishi & Hashimoto 1989)
0.265   (shake flask-HPLC/UV/fluorescence, Pinal et al. 1991)
0.177   (generator column-HPLC, Vadas et al. 1991)
0.248   (generator column-HPLC/UV, Yu & Xu 1993)
0.207   (generator column-HPLC/fluorescence, De Maagd et al. 1998)
0.2289 ± 0.0008   (shake flask-SPME (solid-phase micro-extraction)-GC, Paschke et al. 1999)
log $[S_L/(mol/L)]$ = 0.779 − 987.5/(T/K) (supercooled liquid, Passivirta et al. 1999)
ln x = −1.796327 − 4772.17/(T/K); temp range 5–50°C (regression eq. of literature data, Shiu & Ma 2000)

Vapor Pressure (Pa at 25°C or as indicated and reported temperature dependence equations. Additional data at other temperatures designated * are compiled at the end of this section):

653*   (197.0°C, temp range 197.0–384.2°C, Tsypkina 1955; quoted, Boublik et al. 1984)
6.67 × 10^{-4} (effusion method, Hoyer & Peperle 1958)
log (P/mmHg) = 12.67 − 5357/(T/K); temp range 25–85°C (Knudsen effusion method, Hoyer & Peperle 1958)
0.00123*   (gas saturation-HPLC/fluo.or UV, Sonnefeld et al. 1983)
log (P/Pa) = 11.901 − 4416.56/(T/K); temp range 10–50°C (Antoine eq., gas saturation, Sonnefeld et al. 1983)
0.00124   (generator column-HPLC/fluo., Wasik et al. 1983)
1.79   (supercooled liquid P_L, extrapolated from Antoine eq., Boublik et al. 1984)
0.0154, 0.0067 (P_{GC} by GC-RT correlation with eicosane as reference standard, different GC columns, Bidleman 1984)
0.00861   (supercooled liquid P_L, converted from literature P_S with ΔS_{fus} Bidleman 1984)
0.000125   (extrapolated-Antoine eq., Boublik et al. 1984)
log (P/kPa) = 5.45017 − 1717.489/(114.025 + t/°C); temp range 197–384.2°C (Antoine eq. from reported exptl. data, Boublik et al. 1984)
1.65 × 10^{-4} (extrapolated, Antoine eq., Dean 1985, 1992)
log (P/mmHg) = 6.373 − 1756/(118 + t/°C)l;temp range 197–384°C (Antoine eq., Dean 1985, 1992)
0.00105   (extrapolated-Antoine eq.-I, Stephenson & Malanowski 1987)
log (P_S/kPa) = 11.96071− 5348.06/(T/K); temp range 298–383 K (Antoine eq.-I, Stephenson & Malanowski 1987)
log (P_L/kPa) = 6.67549 − 2957.01/(−24.15 + T/K); temp range 503–658 K (Antoine eq.-II, Stephenson & Malanowski 1987)
0.00068, 0.0056 (lit. solid P_S, supercooled liquid P_L, Bidleman & Foreman 1987)
0.992   (WERL Treatability database, quoted, Ryan et al. 1988)
0.00861, 0.00635 (supercooled P_L, converted from literature P_S with different ΔS_{fus} values, Hinckley et al. 1990)
0.0155, 0.00955 (P_{GC} by GC-RT correlation with different reference standards, Hinckley et al. 1990)
log (P_L/Pa) = 11.35 − 4040/(T/K) (GC-RT correlation, Hinckley et al. 1990)
log (P/mmHg) = 70.6802 − 6.484 × 10^3/(T/K) − 22.241·log (T/K) + 7.2184 × 10^{-3}·(T/K) − 6.3035 × 10^{-13}·(T/K)^2; temp range 383–905 K (vapor pressure eq., Yaws 1994)
0.00692 (supercooled liquid P_L, calculated from Yamasaki et al. 1984, Finizio et al. 1997)
0.00168, 0.000672; 0.00124 (quoted exptl. values, effusion; gas saturation, Delle Site 1997)
0.00218, 0.000939, 0.000889 (quoted lit. values, from GC-RT correlation, Delle Site 1997)
7.48 × 10^{-3}; 1.43 × 10^{-3} (supercooled liquid P_L from Hinckley et al. 1990; converted to solid P_S with fugacity ratio F, Passivirta et al. 1999)
log (P_S/Pa) = 11.62 − 4310/(T/K) (solid, Passivirta et al. 1999)
log (P_L/Pa) = 9.03 − 3323/(T/K) (supercooled liquid, Passivirta et al. 1999)
log (P/Pa) = 11.901 − 4415.56/(T/K); temp range 5–50°C (regression eq. from literature data, Shiu & Ma 2000)
0.00598   (supercooled liquid P_L, calibrated GC-RT correlation, Lei et al. 2002)
log (P_L/Pa) = −4141/(T/K) + 11.66; $\Delta H_{vap.}$ = −79.3 kJ·mol^{-1} (GC-RT correlation, Lei et al. 2002)

Henry's Law Constant (Pa m^3/mol 25°C or as indicated and reported temperature dependence equations. Additional data at other temperatures designated * are compiled at the end of this section):

0.65*   (gas stripping-HPLC/fluorescence, measured range 10–55°C, ten Hulscher et al. 1992)
1.10   (gas stripping-HPLC/fluorescence, De Maagd et al. 1998)
1.96*   (gas stripping-GC, measured range 4.1–31°C, Bamford et al. 1999)
ln K_{AW} = −4654.8/(T/K) + 8.42, ΔH = 38.7 kJ mol^{-1}; measured range 4.1–31°C (gas stripping-GC, Bamford et al. 1999)
log [H/(Pa m^3/mol)] = 8.23 − 2336/(T/K) (Passivirta et al. 1999)

1.96 (quoted from Bamford et al. 1999; Dachs & Eisenreich 2000)
0.602 (20°C, selected from reported experimentally measured values, Staudinger & Robers 1996, 2001)
log K_{AW} = 6.175 − 2868/(T/K) (van't Hoff eq. derived from literature data, Staudinger & Roberts 2001)

Octanol/Water Partition Coefficient, log K_{OW}:
5.22 (RP-TLC-k' correlation, Bruggeman et al. 1982)
4.47 (HPLC-k' correlation, Harnisch et al. 1983)
4.84 (RP-HPLC-RT correlation, Chin et al. 1986)
4.85 (HPLC-RT correlation, Wang et al. 1986)
5.16 (shake flask/slow stirring-GC, De Bruijn et al. 1989)
5.20 (recommended, Sangster 1989, 1993)
5.17 (TLC-RT correlation, De Voogt et al. 1990)
5.148 ± 0.077, 5.155 ± 0.015 (shake flask/slow stirring-GC/HPLC, interlaboratory studies, Brooke et al. 1990)
5.00 (shake flask-UV spec., Alcorn et al. 1993)
5.16 (recommended, Hansch et al. 1995)
5.23 (5.12–5.31) (shake flask/slow stirring-HPLC/fluorescence., De Maagd et al. 1998)
5.16 (shake flask-SPME solid-phase micro-extraction; Paschke et al. 1999)
0.602 (20°C, selected from literature experimentally measured data, Staudinger & Roberts 2001)
log K_{AW} = 5.485 − 2682/(T/K), (van't Hoff eq. derived from literature data, Staudinger & Roberts 2001)

Octanol/Air Partition Coefficient, log K_{OA} at 25°C or as indicated and reported temperature dependence equations. Additional data at other temperatures designated * are compiled at the end of this section:
8.60 (calculated-K_{OW}/K_{AW}, Wania & Mackay 1996)
8.60 (calculated, Finizio et al. 1997)
8.80*; 8.60 (generator column-GC; calculated-C_O/C_A, measured range 0–40°C, Harner & Bidleman 1998)
log K_{OA} = −5.94 + 4417/(T/K); temp range 20–40°C (generator column-GC, Harner & Bidleman 1998)
8.61 (calculated-S_{oct} and vapor pressure P, Abraham et al. 2001)

Bioconcentration Factor, log BCF:
3.18 (calculated as per Kenaga & Goring 1979, Eadie et al. 1982)
4.90 (*P. hoyi*, Eadie et al. 1982)
4.08 (microorganisms-water, calculated from K_{OW}, Mabey et al. 1982)
3.24 (*Daphnia magna*, Newsted & Giesy 1987)
0.756, 1.079 (*Polychaete sp, Capitella capitata*, Bayona et al. 1991)

Sorption Partition Coefficient, log K_{OC} at 25°C or as indicated:
6.38 (sediments average, Kayal & Connell 1990)
4.74, 4.62 (RP-HPLC correlation on CIHAC, on PIHAC stationary phases, Szabo et al. 1990b)
6.30 (Baltic Sea particulate field samples, concn distribution-GC/MS, Broman et al. 1991)
4.816; 4.81, 4.82 (sediment: concn ratio C_{sed}/C_W; concn-based coeff., areal-based coeff. of flux studies of sediment/water boundary layer, Helmstetter & Alden 1994)
4.51, 5.05, 4.16 (sediments from Brown's Lake, Hamlet City Lake, WES reference soil, shake flask-LSC, Brannon et al. 1995)
6.56, 6.66, 6.08 (marine sediments: Fort Point Channel, Spectacle Island, Peddocks Island, McGroddy & Farrington 1995)
4.62 (calculated-MCI $^1\chi$, Sabljic et al. 1995)
5.25 (10°C), 5.22, 5.12 (20°C), 5.05 (35°C), 4.89, 4.96 (45°C) (log K_{DOC} - dissolved organic material from lake, gas-purge technique-HPLC/fluorescence, Lüers & ten Hulscher 1996)
5.40 (20°C, log K_{POC} - particulate organic material from lake, Lüers & ten Hulscher 1996)
4.81, 4.65, 4.80, 4.82; 4.813; 4.727 (4 soils with different organic carbon content f_{OC}, adsorption equilibrium-shake flask-HPLC; calculated-K_{OW}; calculated-S, He et al. 1996)
5.32 (5.29–5.35), 4.89 (4.89–4.90) (sediments: Lake Oostvaardersplassen, Lake Ketelmeer, shake flask-HPLC/UV, de Maagd et al. 1998)
4.62, 4.03; 3.40, 4.49, 3.55, 4.53, 4.56 (quoted lit., calculated-K_{OW}; HPLC-screening method with different LC-columns, Szabo et al. 1999)

5.32–6.59; 4.60–6.70 (range, calculated from sequential desorption of 11 urban soils; lit. range, Krauss & Wilcke 2001)

5.83; 6.79, 5.53, 5.52 (20°C, batch equilibrium, A2 alluvial grassland soil; calculated values of expt 1,2,3-solvophobic approach, Krauss & Wilcke 2001)

4.91, 4.65 (average values for sediments, soils, organic carbon OC ≥ 0.5%, Delle Site 2001)

4.62 (Askov soil, a Danish agricultural soil, Sverdrup et al. 2002)

5.21–6.60 (field contaminated sediment, initial-final values of 5–100 d contact time, gas-purge technique-HPLC/fluorescence, ten Hulscher et al. 2003)

Environmental Fate Rate Constants, k, or Half-Lives, $t_{1/2}$:

Volatilization:

Photolysis: direct photochemical transformation $t_{1/2}$(calc) = 21 h, computed near-surface water, latitude 40°N, midday, midsummer and photolysis $t_{1/2}$ = 160 d and 200 d in 5-m deep inland water body without and with sediment-water partitioning, respectively, to top cm of bottom sediment over full summer day, 40°N (Zepp & Schlotzhauer 1979)

$t_{1/2}$ = 21 h, atmospheric and aqueous photolysis half life, based on measured sunlight photolysis rate constant in water adjusted for midday summer sunlight at 40°N latitude and $t_{1/2}$ = 63 h after adjusting for approximate winter sunlight intensity (Howard et al. 1991);

$t_{1/2}$ = 160 d under summer sunlight in surface water (Mill & Mabey 1985);

half-lives on different atmospheric particulate substrates (appr. 25 μg/g on substrate): $t_{1/2}$ = 74 h on silica gel, $t_{1/2}$ = 23 h on alumina and $t_{1/2}$ = 44 h on fly ash (Behymer & Hites 1985);

direct photolysis $t_{1/2}$(obs.) = 3.61 h, $t_{1/2}$(calc) = 4.78 h, by QSPR in atmospheric aerosol (Chen et al. 2001);

k = (1.76 ± 0.13) × 10^{-5} s^{-1} and $t_{1/2}$ = (10.93 ± 0.75)h in diesel particulate matter, photochemical degradation under atmospheric conditions, k = (2.97 ± 0.40) × 10^{-5} s^{-1} and $t_{1/2}$ = (6.48 ± 1.03)h in diesel particulate matter/soil mixture, and $t_{1/2}$ = 1.6 to 4.15 h in various soil components using a 900-W photo-irradiator as light source; k = (8.69 ± 0.29) × 10^{-6} s^{-1} and $t_{1/2}$ = (22.16 ± 0.77) h in diesel particulate matter using a 300-W light source (Matsuzawa et al. 2001)

Photodegradation k = 5.0 × 10^{-6} s^{-1} in surface water during the summertime at mid-latitude (Fasnacht & Blough 2002)

Oxidation: rate constant, k, and for gas-phase second-order rate constants, k_{OH} for reaction with OH radical, k_{NO_3} with NO_3 radical and k_{O_3} with O_3, or as indicated *data at other temperatures and/or the Arrhenius expression see reference:

k (calc) < 3600 M^{-1} h^{-1} for singlet oxygen and < 360 M^{-1} h^{-1} for peroxy radical (Mabey et al. 1982)

k_{OH} = 5.0 × 10^{-11} cm^3 $molecule^{-1}$ s^{-1} and $k_{N_2O_5}$ ≈ 1.8 × 10^{-17} cm^3 $molecule^{-1}$ s^{-1} for reaction with N_2O_5 at 296 ± 2 K (relative rate method, Atkinson et al. 1990)

photooxidation half-life of 2.02–20.2 h, based on estimated rate constant for reaction with hydroxyl radical in air (Atkinson 1987; quoted, Howard et al. 1991)

k_{OH}* = 11 × 10^{-12} cm^3 $molecule^{-1}$ s^{-1} at 298 K, measured range 306–366 K with a calculated atmospheric lifetime of 26 h based on gas-phase OH reactions (Brubaker & Hites 1998)

Hydrolysis: not hydrolyzable (Mabey et al. 1982); no hydrolyzable groups (Howard et al. 1991).

Biodegradation: aquatic k = 2.2 × 10^{-3} μmol h^{-1} mg^{-1} with bacterial protein (Barnsley 1975; quoted, Callahan et al. 1979)

significant with gradual degradation for a domestic sewer test for an average three static-flask screening (Tabak et al. 1981)

$t_{1/2}$(aerobic) = 3360–10560 h, based on aerobic soil die-away test data at 10–30°C (Coover & Sims 1987; quoted, Howard et al. 1991)

k = 0.19 h^{-1} in atmosphere (Dragoescu & Friedlander 1989)

k = 0.0018 d^{-1} with $t_{1/2}$ = 377 d for Kidman sandy loam and k = 0.0026 d^{-1} with $t_{1/2}$ = 268 d for McLarin sandy loam all at –0.33 bar soil moisture (Park et al. 1990)

$t_{1/2}$(anaerobic) = 13440–42240 h, based on estimated unacclimated aqueous aerobic biodegradation half-life (Howard et al. 1991)

first order k = 0.132 to 0.162 L mg^{-1} d^{-1} for a marine PAH-degrading enrichment without sediment, the degradation rate was 2.1 to 5.3 times faster with sediment present (Poeton et al. 1999)

Biotransformation: estimated rate constant for bacteria, 1 × 10^{-10} mL $cell^{-1}$ h^{-1} (Mabey et al. 1982).

Bioconcentration, Uptake (k_1) and Elimination (k_2) Rate Constants:

k_1 = 4.1–6.1 mg g^{-1} h^{-1}; k_2 = 0.026 h^{-1} (freshwater oligochaete from sediment, Van Hoof et al. 2001)

$k_2 = 0.15$ h^{-1} in water with corresponding $t_{1/2} = 4.7$ h and $k_2 = 0.14$ h^{-1} in sediment with corresponding $t_{1/2} = 4.9$ h for copepods *S. knabeni* in 24-h experiments (Lotufo 1998)

$k_2 = 0.17$ h^{-1} in sediment with corresponding $t_{1/2} = 4.2$ h and $k_2 = 0.09$ h^{-1} in sediment with corresponding $t_{1/2} = 7.4$ h for copepods *Coullana* sp. in 24-h experiments (Lotufo 1998)

Half-Lives in the Environment:

Air: $t_{1/2} = 2.02–20.2$ h, based on estimated sunlight photolysis half-life in water (Howard et al. 1991); calculated atmospheric lifetime of 26 h based on gas-phase OH reactions (Brubaker & Hites 1998).

Surface water: computed near-surface $t_{1/2} = 21$ h for photochemical transformation of a water body (latitude 40°N, midday, midsummer), $t_{1/2} = 160$ d for direct photolysis in a 5-m deep inland water body with no sediment-water partitioning and $t_{1/2} = 200$ d with sediment-water partitioning (Zepp & Schlotzhauer 1979);

$t_{1/2} = 21–63$ h, based on photolysis half-life in water (Howard et al. 1991);

$t_{1/2} = 160$ d for summer sunlight at 40°N latitude (Mill & Mabey 1985).

Groundwater: $t_{1/2} = 6720–21120$ h, based on estimated unacclimated aqueous aerobic biodegradation half-life (Howard et al. 1991).

Sediment: desorption $t_{1/2} = 8.3$ d from sediment under conditions mimicking marine disposal (Zhang et al. 2000).

Soil: $t_{1/2} = 44–182$ d (Sims & Overcash 1983; quoted, Bulman et al. 1987);

$t_{1/2} = 39$ d for 5 mg/kg treatment and 34 d for 50 mg/kg treatment (Bulman et al. 1987);

biodegradation rate constant $k = 0.0018$ d^{-1} with $t_{1/2} = 377$ d for Kidman sandy loam soil, and $k = 0.0026$ d^{-1} with $t_{1/2} = 268$ d for McLaurin sandy loam soil (Park et al. 1990);

$t_{1/2} = 3360–10560$ h, based on aerobic soil die-away test data at 10–30°C (Howard et al. 1991);

$t_{1/2} > 50$ d (Ryan et al. 1988);

$t_{1/2} = 17.961$ wk, 7.8 yr (Luddington soil, Wild et al. 1991).

Biota: $t_{1/2} = 5$ d depuration half life by oysters (Lee et al. 1978);

elimination $t_{1/2} = 6$ d from rainbow trout, $t_{1/2} = 2.0–29.8$ d from mussel *Mytilus edulis*; $t_{1/2} = 8.4$ d from clam *Mya arenaria*, $t_{1/2} = 5.8$ d from polychaete *Abarenicola pacifica*, $t_{1/2} = 5.9$ d from Oyster, $t_{1/2} = 0.8$ d from shrimp (Meador et al. 1995)

depuration $t_{1/2} \sim 4.8$ h in sediment and water for copepod *S. knabeni*, $t_{1/2} = 4.2$ h in sediment and $t_{1/2} = 7.4$ h in water for copepod *Coullana* sp. (Lotufo 1998)

TABLE 4.1.1.30.1
Reported aqueous solubilities of fluoranthene at various temperatures and reported temperature dependence equation

$$S/(\mu g/kg) = a \cdot t^3 + b \cdot t^2 + c \cdot t + d \quad (1)$$

May 1980, 1983		May et al. 1978b	
generator column-HPLC		generator column-HPLC/fluo.	
t/°C	S/g·m^{-3}	t/°C	S/g·m^{-3}
8.1	0.082	25	0.206
13.2	0.107	29	0.264
19.7	0.1483		
24.6	0.2027	temp dependence eq. 1	
29.9	0.2793	S	µg/kg
		a	0.0072
temp dependence eq. 1		b	–0.1047
given in May et al. 1980		c	4.322
S	µg/kg	d	50.4
a	0.0072		
b	–0.1047	ΔH_{sol}/(kJ mol^{-1}) = 39.83	
c	4.322	measured between 5–30°C	
d	50.4		
ΔH_{sol}/(kJ mol^{-1}) = 39.83			
measured between 5–30°C			

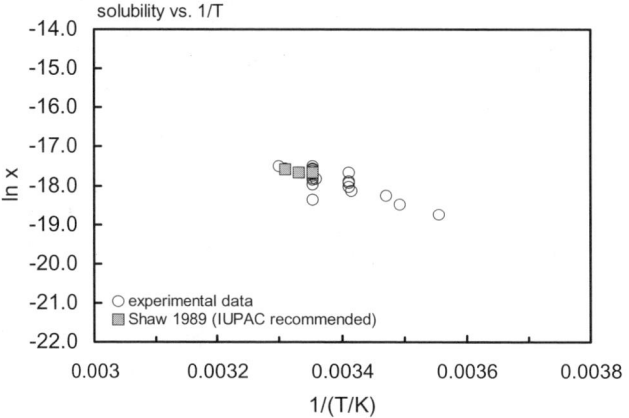

FIGURE 4.1.1.30.1 Logarithm of mole fraction solubility (ln x) versus reciprocal temperature for fluoranthene.

TABLE 4.1.1.30.2
Reported vapor pressures of fluoranthene at various temperatures and the coefficients for the vapor pressure equations

$$\log P = A - B/(T/K) \quad (1) \qquad \ln P = A - B/(T/K) \quad (1a)$$
$$\log P = A - B/(C + t/°C) \quad (2) \qquad \ln P = A - B/(C + t/°C) \quad (2a)$$
$$\log P = A - B/(C + T/K) \quad (3)$$
$$\log P = A - B/(T/K) - C \cdot \log(T/K) \quad (4)$$

Tsypkina 1955		Hoyer & Peperle 1958		Sonnefeld et al. 1983	
		effusion		generator column-HPLC	
t/°C	P/Pa	t/°C	P/Pa	t/°C	P/Pa
197.0	653	data presented in eq.		10.88	2.17×10^{-4}
209.0	1053	eq. 1	P/mmHg	10.88	2.05×10^{-4}
228.5	2586	A	12.67	10.88	2.15×10^{-4}
238.1	3786	B	5357	10.88	2.57×10^{-4}
247.7	5386	for temp range 25–85°C		20.25	7.07×10^{-4}
255.0	6586			20.25	7.39×10^{-4}
261.3	7919	ΔH_{sub}(kJ/mol) = 100		20.25	7.25×10^{-4}
270.9	10319			20.25	7.63×10^{-4}
281.5	13386			29.79	2.03×10^{-3}
305.0	20318			29.79	1.99×10^{-3}
314.5	27384			29.80	2.17×10^{-3}
382.9	99058			29.80	2.20×10^{-3}
384.2	101325			38.9	5.81×10^{-3}
				38.9	5.67×10^{-3}
				38.9	5.58×10^{-3}
				38.9	5.77×10^{-3}
				38.9	5.46×10^{-3}
				38.9	5.45×10^{-3}
				25.0	1.23×10^{-3}
				eq. 2	P/Pa
				A	11.901
				B	4415.56
				temp range 10–50°C	

ref. Tsypkina, O.YA. Zh. Prikl. Khim 28, 185 (1955).

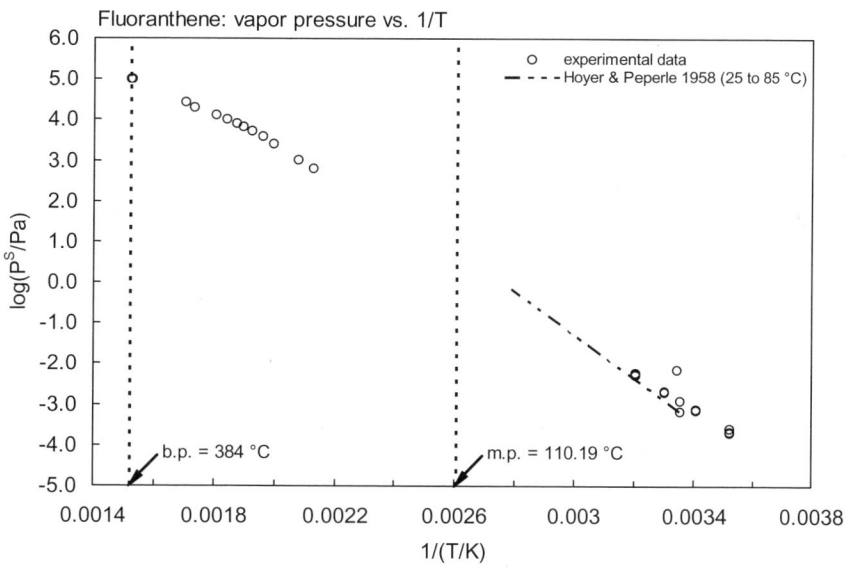

FIGURE 4.1.1.30.2 Logarithm of vapor pressure versus reciprocal temperature for fluoranthene.

TABLE 4.1.1.30.3
Reported Henry's law constants and octanol-air partition coefficients of fluoranthene at various temperatures and temperature dependence equations

Henry's law constant					log K_{OA}	
ten Hulscher et al. 1992		Bamford et al. 1999			Harner & Bidleman 1998	
gas stripping-HPLC/fluo.		gas stripping-GC/MS			generator column-GC/FID	
t/°C	H/(Pa m³/mol)	t/°C	H/(Pa m³/mol)	H/(Pa m³/mol)	t/°C	log K_{OA}
				average		
10.0	0.26	4.1	0.37, 0.49	0.56	0	-
20.0	0.64	11.0	0.79, 0.95	0.87	10	-
35.0	1.63	18.0	1.21, 1.43	1.32	20	11.124
40.1	2.38	25.0	1.76, 2.18	1.96	30	8.652
45.0	5.84	31.0	2.36, 3.14	2.72	40	8.161
55.0	6.23				25(exptl)	8.88
					25(calc)	8.60
ln K_{AW} = –ΔH/RT + ΔS/R		ln K_{AW} = A − B/(T/K)					
R = 8.314 Pa m³ mol⁻¹ K⁻¹		A	8.4195			
ΔS/R	22.16	B	4654.8		log K_{OA} = A + B/(T/K)	
ΔH/R	6855.9				A	–5.94
					B	4417
		enthalpy, entropy change:					
enthalpy of volatilization:		ΔH/(kJ·mol⁻¹) = 38.7 ± 2.5			enthalpy of phase change	
ΔH/(kJ·mol⁻¹) = 57 ± 5		ΔS/(J·K⁻¹ mol⁻¹) = 70			ΔH_{OA}/(kJ mol⁻¹) =	
entropy of volatilization, ΔS		at 25°C				
TΔS/(kJ·K⁻¹ mol⁻¹) = 54 ± 5					
at 20°C					

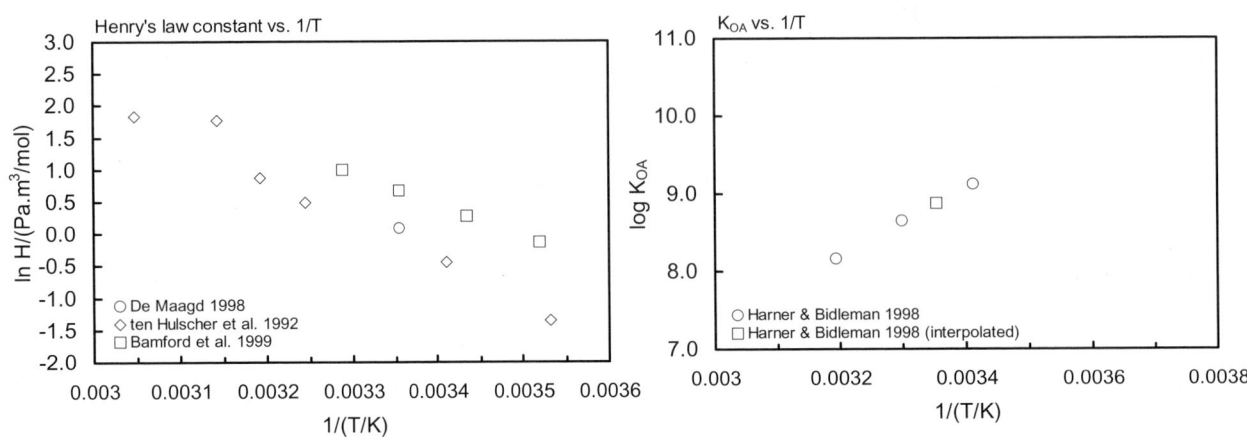

FIGURE 4.1.1.30.3 Logarithm of Henry's law constant and K_{OA} versus reciprocal temperature for fluoranthene.

4.1.1.31 Benzo[a]fluorene

Common Name: Benzo[a]fluorene
Synonym: 1,2-benzofluorene, 11H-benzo[a]fluorene, chrysofluorene
Chemical Name: benzo[a]fluorene, 1,2-benzofluorene
CAS Registry No: 238-84-6
Molecular Formula: $C_{17}H_{12}$
Molecular Weight: 216.227
Melting Point (°C):
    189.5   (Lide 2003)
Boiling Point (°C):
    405   (Lide 2003)
Density (g/cm³ at 20°C):
Molar Volume (cm³/mol):
    200.9   (Ruelle & Kesselring 1997)
    239.5   (calculated-Le Bas method at normal boiling point)
Enthalpy of Fusion, ΔH_{fus} (kJ/mol):
    3.8, 18.4; 22.2 (126.75, 189.65°C; total phase change enthalpy, Chickos et al. 1999)
Entropy of Fusion, ΔS_{fus} (J/mol K):
    39.76   (exptl., Chickos et al. 1999)
    49.26, 50.9 (exptl., calculated-group additivity method, total phase change entropy, Chickos et al. 1999)
Fugacity Ratio at 25°C (assuming ΔS_{fus} = 56 J/mol K), F: 0.0243 (mp at 189.5°C)

Water Solubility (g/m³ or mg/L at 25°C):
    0.045   (shake flask-fluorescence, Mackay & Shiu 1977)
    0.045   (average lit. value, Pearlman et al. 1984)

Vapor Pressure (Pa at 25°C and reported temperature dependence equation):
    0.00136   (supercooled liquid P_L, calibrated GC-RT correlation, Lei et al. 2002)
    $\log (P_L/Pa) = -4373/(T/K) + 11.80$; $\Delta H_{vap.} = -83.7$ kJ·mol⁻¹ (GC-RT correlation, Lei et al. 2002)

Henry's Law Constant (Pa m³/mol at 25°C or as indicated):
    2.70*   (gas stripping-GC, measured range 4.1–31°C, Bamford et al. 1999)
    $\ln K_{AW} = -4113.54/(T/K) + 6.976$, $\Delta H = 34.2$ kJ mol⁻¹, measured range 4.1–31°C (gas stripping-GC, Bamford et al. 1999)

Octanol/Water Partition Coefficient, log K_{OW}:
    5.68   (HPLC-RT correlation, Wang et al. 1986)
    5.40   (recommended, Sangster 1989, 1993)
    6.5387   (calculated-UNIFAC group contribution, Chen et al. 1993)
    5.68   (recommended, Hansch et al. 1995)

Octanol/Air Partition Coefficient, log K_{OA}:

Bioconcentration Factor, log BCF:

Sorption Partition Coefficient, log K_{OC}:

Environmental Fate Rate Constants, k, or Half-Lives, $t_{1/2}$:

Half-Lives in the Environment:

Biota: elimination $t_{1/2}$ = 10.5 d from Oyster, $t_{1/2}$ = 4.2 d from clam *Mercenario mercenaria* (quoted, Meador et al. 1995).

TABLE 4.1.1.31.1
Reported Henry's law constants of benzo[a]fluorene at various temperatures

	Bamford et al. 1999	
	gas stripping-GC/MS	
t/°C	H/(Pa m³/mol)	H/(Pa m³/mol)
		average
4.1	0.67, 1.16	0.88
11.0	1.08, 1.57	1.30
18.0	1.61., 2.23	1.89
25.0	2.19, 3.33	2.70
31.0	2.74, 4.78	3.62

ln K_{AW} = A − B/(T/K)
A   6.9762
B   4113.5
enthalpy, entropy change:
$\Delta H/(kJ \cdot mol^{-1})$ = 34.2 ± 4.9
$\Delta S/(J \cdot K^{-1} \, mol^{-1})$ = 58
at 25°C

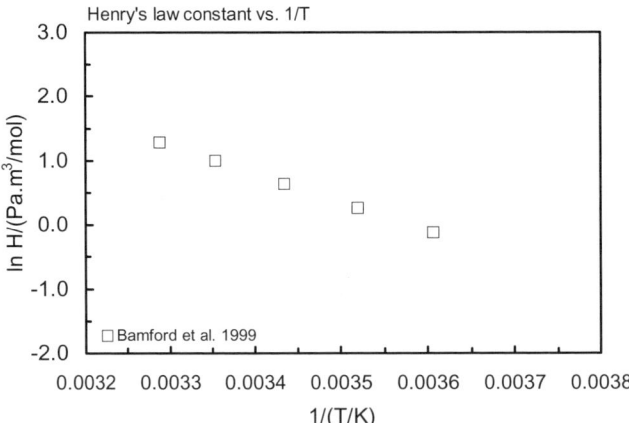

FIGURE 4.1.1.31.1 Logarithm of Henry's law constant versus reciprocal temperature for benzo[a]fluorene.

4.1.1.32 Benzo[b]fluorene

Common Name: Benzo[b]fluorene
Synonym: 2,3-benzofluorene, 11H-benzo[b]fluorene, isonaphthofluorene
Chemical Name: benzo[b]fluorene
CAS Registry No: 243-17-4
Molecular Formula: $C_{17}H_{12}$
Molecular Weight: 216.227
Melting Point (°C):
    212   (Lide 2003)
Boiling Point (°C):
    401   (Lide 2003)
Density (g/cm³ at 20°C):
Molar Volume (cm³/mol):
    200.9   (Ruelle & Kesselring 1997)
    239.5   (calculated-Le Bas method at normal boiling point)
Enthalpy of Fusion, ΔH_{fus} (kJ/mol):
    23.4   (exptl., Chickos et al. 1999)
Entropy of Fusion, ΔS_{fus} (J/mol K):
    47.78, 50.9   (exptl., calculated-group additivity method, Chickos et al. 1999)
Fugacity Ratio at 25°C (assuming ΔS_{fus} = 56 J/mol K), F: 0.0146 (mp at 212°C)

Water Solubility (g/m³ or mg/L at 25°C):
    0.0020   (shake flask-fluorescence, Mackay & Shiu 1977)
    0.002   (average lit. value, Pearlman et al. 1984)

Vapor Pressure (Pa at 25°C or as indicated and reported temperature dependence equations. Additional data at other temperatures designated * are compiled at the end of this section):
    7.37×10^{-6}*   (gas saturation, extrapolated-Antoine eq. derived from exptl. data, temp range 71–125°C, Oja & Suuberg 1998)
    log (P/Pa) = 36.325 – 14354/(T/K); temp range 344–398 K (Antoine eq., Knudsen effusion, Oja & Suuberg 1998)
    0.00107   (supercooled liquid P_L, calibrated GC-RT correlation, Lei et al. 2002)
    log (P_L/Pa) = –4423/(T/K) + 11.86; $\Delta H_{vap.}$ = –84.7 kJ·mol⁻¹ (GC-RT correlation, Lei et al. 2002)

Henry's Law Constant (Pa m³/mol):

Octanol/Water Partition Coefficient, log K_{OW}:
    5.77   (HPLC-RT correlation, Wang et al. 1986)
    5.75   (recommended, Sangster 1989, 1994)
    5.77   (recommended, Hansch et al. 1995)

Octanol/Air Partition Coefficient, log K_{OA}:

Bioconcentration Factor, log BCF:

Sorption Partition Coefficient, log K_{OC}:

Environmental Fate Rate Constants, k or Half-Lives, $t_{1/2}$:

Half-Lives in the Environment:
    Biota: elimination $t_{1/2}$ = 10.5 d from Oyster, $t_{1/2}$ = 4.3 d from clam *Mercenario mercenaria* (quoted, Meador et al. 1995).

TABLE 4.1.1.32.1
Reported vapor pressures of benzo[b]fluorene at various temperatures and the coefficients for the vapor pressure equations

$$\log P = A - B/(T/K) \quad (1) \qquad \ln P = A - B/(T/K) \quad (1a)$$
$$\log P = A - B/(C + t/°C) \quad (2) \qquad \ln P = A - B/(C + t/°C) \quad (2a)$$
$$\log P = A - B/(C + T/K) \quad (3)$$
$$\log P = A - B/(T/K) - C \cdot \log (T/K) \quad (4)$$

Oja & Suuberg 1998	
Knudsen effusion	
t/°C	P/Pa
70.92	0.00454
83.14	0.0188
97.55	0.0936
124.4	1.230
mp/K	484–486
eq. 1a	P/Pa
A	36.325
B	14354
for temp range 344–398 K	
ΔH_{sub}/(kJ/mol) = 119.3	

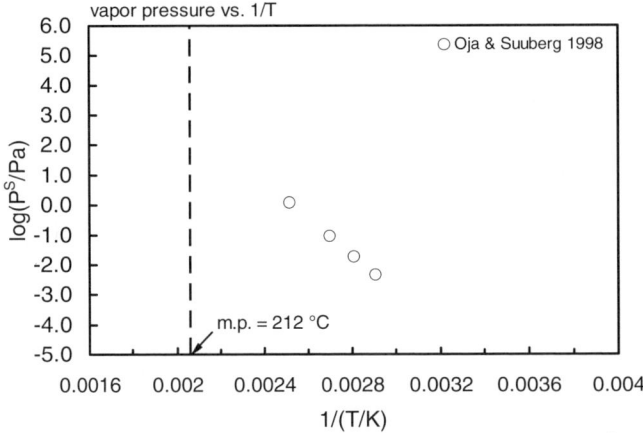

FIGURE 4.1.1.32.1 Logarithm of vapor pressure versus reciprocal temperature for benzo[b]fluorene.

4.1.1.33 Chrysene

Common Name: Chrysene
Synonym: 1,2-benzophenanthrene, benzo(a)phenanthrene, 1,2,5,6-dibenzonaphthalene
Chemical Name: chrysene
CAS Registry No: 218-01-9
Molecular Formula: $C_{18}H_{12}$
Molecular Weight: 228.288
Melting Point (°C):
    255.5   (Lide 2003)
Boiling Point (°C):
    448   (Weast 1975; Lide 2003)
Density (g/cm³ at 20°C):
    1.274   (Weast 1982–83; Lide 2003)
Molar Volume (cm³/mol):
    179.2   (20°C, calculated-density)
    250.8   (calculated-Le Bas method at normal boiling point)
Enthalpy of Fusion, ΔH_{fus} cal/mol:
    26.153   (Ruelle & Kesselring 1997)
    3.22, 26.15; 29.37 (239.05, 28.25°C, total phase change enthalpy, Chickos et al. 1999)
Entropy of Fusion, ΔS_{fus} (J/mol K):
    49.37   (Casellato et al. 1973)
    62.34   (Ubbelohde 1978)
    49.21   (Chickos et al. 1999)
    55.5, 44.1   (exptl., calculated-group additivity method, total phase change entropy, Chickos et al. 1999)
Fugacity Ratio at 25°C (assuming ΔS_{fus} = 56 J/mol K), F: 0.00548 (mp at 255.5°C)
    0.00976   (calculated, Passivirta et al. 1999)

Water Solubility (g/m³ or mg/L at 25°C or as indicated and reported temperature dependence equations. Additional data at other temperatures designated * are compiled at the end of this section):
    0.0015   (27°C, shake flask-nephelometry, Davis et al. 1942)
    0.006   (shake flask-UV, Klevens 1950)
    0.0015   (Weimer & Prausnitz 1965)
    0.0041, 0.0014   (HPLC-relative retention correlation, different stationary and mobile phases, Locke 1974)
    0.002   (shake flask-fluorescence, Mackay & Shiu 1977)
    0.0018   (Rossi 1977; Neff 1979; quoted, Eadie et al. 1982)
    0.0018*   (generator column-HPLC/UV, measured range 6.5–29°C May et al. 1978a,b)
    S/(μg/kg) = 0.609 + 0.0144·(t/°C) + 0.0024·(t/°C)²; temp range 5–30°C (generator column-HPLC/UV, May et al. 1978)
    0.017   (shake flask-nephelometry, Hollifield 1979)
    0.00189*   (25.3°C, generator column-HPLC, measured range 279.65–301.85 K, May et al. 1983)
    0.0018*   (average lit. value, Pearlman et al. 1984)
    0.00327   (generator column-HPLC/fluorescence, Walters & Luthy 1984)
    0.00102, 0.0012   (generator column-HPLC/UV, Billington et al. 1988)
    0.0019   (recommended, Shaw 1989)
    0.0016   (generator column-HPLC, Vadas et al. 1991)
    0.0015   (generator column-HPLC/fluorescence, De Maagd et al. 1998)
    $\log [S_L/(mol/L)] = -0.323 - 1369/(T/K)$ (supercooled liquid, Passivirta et al. 1999)

Vapor Pressure (Pa at 25°C or as indicated and reported temperature dependence equations. Additional data at other temperatures designated * are compiled at the end of this section):

5.7×10^{-7}* (effusion method, extrapolated, De Kruif 1980)

8.4×10^{-7} (effusion method, Hoyer & Peperle 1958; quoted, Mabey et al. 1982)

$\log (P/mmHg) = 13.07 - 6340/(T/K)$; temp range 80–145°C (Knudsen effusion method, Hoyer & Peperle 1958)

6.08×10^{-7} (extrapolated-Antoine eq., Stephenson & Malanowski 1987)

$\log (P_S/kPa) = 11.445 - 6160/(T/K)$; temp range 358–463 K (Antoine eq., Stephenson & Malanowski 1987)

$\log (P/mmHg) = -50.1566 - 3.4381 \times 10^3/(T/K) + 25.178 \cdot \log (T/K) - 2.462 \times 10^{-2} \cdot (T/K) + 7.1044 \times 10^{-6} \cdot (T/K)^2$; temp range 531–979 K (vapor pressure eq., Yaws 1994)

2.29×10^{-4} (supercooled liquid P_L, calculated from Yamasaki et al. 1984, quoted, Finizio et al. 1997)

5.70×10^{-7}; 5.84×10^{-5} (quoted solid P_S from Mackay et al. 1992; converted to supercooled liquid P_L with fugacity ratio F, Passivirta et al. 1999)

$\log (P_S/Pa) = 12.24 - 5507/(T/K)$ (solid, Passivirta et al. 1999)

$\log (P_L/Pa) = 9.66 - 4139/(T/K)$ (supercooled liquid, Passivirta et al. 1999)

$\log (P/Pa) = 14.848 - 6189/(T/K)$; temp range 5–50°C (regression eq. from literature data, Shiu & Ma 2000)

1.70×10^{-4} (supercooled liquid P_L, calibrated GC-RT correlation, Lei et al. 2002)

$\log (P_L/Pa) = -4679/(T/K) + 11.92$; $\Delta H_{vap.} = -89.6$ kJ·mol^{-1} (GC-RT correlation, Lei et al. 2002)

2.11×10^{-6} (solid P_S, gas saturation-GC/MS, Mader & Pankow 2003)

4.03×10^{-4} (supercooled liquid P_L, calculated from P_S assuming $\Delta S_{fus} = 56$ J/mol K, Mader & Pankow 2003)

Henry's Law Constant (Pa m^3/mol at 25°C or as indicated and reported temperature dependence equations. Additional data at other temperatures designated * are compiled at the end of this section):

0.107 (headspace solid-phase microextraction (SPME)-GC, Zhang & Pawliszyn 1993)

0.53* (gas stripping-GC, measured range 4.1–31°C, Bamford et al. 1999)

$\ln K_{AW} = -12136.2/(T/K) + 32.235$, $\Delta H = 100.9$ kJ mol^{-1}, measured range 4.1–31°C (gas stripping-GC, Bamford et al. 1999)

$\log (H/(Pa\ m^3/mol)) = 9.98 - 2770/(T/K)$ (Passivirta et al. 1999)

0.53 (quoted, Dachs & Eisenreich 2000)

Octanol/Water Partition Coefficient, log K_{OW}:

5.79 (HPLC-k' correlation, Hanai et al. 1981)

5.91 (RP-TLC-k' correlation, Bruggeman et al. 1982)

5.73 (HPLC-RT correlation, Wang et al. 1986)

5.61 ± 0.40 (recommended, Sangster 1989)

5.84 (TLC retention time correlation, De Voogt et al. 1990)

5.50 (shake flask-UV, Alcorn et al. 1993)

5.86 (recommended, Sangster 1993)

5.73 (recommended, Hansch et al. 1995)

5.81 (range 6.63–5.94) (shake flask/slow stirring-HPLC/fluo., De Maagd et al. 1998)

5.78 (shake flask-dialysis tubing-HPLC/UV, both phases, Andersson & Schröader 1999)

Octanol/Air Partition Coefficient, log K_{OA}:

10.40 (calculated-K_{OW}/K_{AW}, Wania & Mackay 1996)

10.44 (calculated, Finizio et al. 1997)

Bioconcentration Factor, log BCF:

4.31 (*P. hoyi*, Eadie et al. 1982)

4.72 (microorganisms-water, calculated from K_{OW}, Mabey et al. 1982)

3.785 (*Daphnia magna*, Newsted & Giesy 1987)

1.17, 0.792 (*Polychaete sp, Capitella capitata*, Bayona et al. 1991)

Sorption Partition Coefficient, log K_{OC}:

6.27 (sediments average, Kayal & Connell 1990)

6.9 (Baltic Sea particulate field samples, concn distribution-GC/MS, Broman et al. 1991)

4.0 (predicted dissolved log K_{OC}, Broman et al. 1991)

5.79 (5.74–5.83), 5.40 (5.35–5.50) (sediments: Lake Oostvaarderplassen, Lake Ketelmeer, shake flask-HPLC/UV, de Maagd et al. 1998)

5.52–7.38; 4.90–7.80 (for chrysene + triphenylene, range, calculated from sequential desorption of 11 urban soils; lit. range, Krauss & Wilcke 2001)

5.92; 6.12, 5.77, 6.14 (for chrysene + triphenylene, 20°C, batch equilibrium, A2 alluvial grassland soil; calculated values of expt 1,2,3-solvophobic approach, Krauss & Wilcke 2001)

Environmental Fate Rate Constants, k or Half-Lives, $t_{1/2}$:

Volatilization:

Photolysis: calculated $t_{1/2}$ = 4.4 h of direct sunlight photolysis for 50% conversion at 40°N latitude of midday in midsummer: 4.4 h in near-surface water; (Herbes et al. 1980)

direct photochemical transformation $t_{1/2}$(calc) = 4.4 h, computed near-surface water, latitude 40°N, midday, midsummer and photolysis $t_{1/2}$ = 13 d and 68 d in 5-m deep inland water body without and with sediment-water partitioning, respectively, to top cm of bottom sediment over full summer day, 40°N (Zepp & Schlotzhauer 1979)

$t_{1/2}$ = 13 d in 5-m deep inland water and $t_{1/2}$ = 68 d in inland water with sediment partitioning (Zepp & Schlotzhauer 1979)

half-lives on different atmospheric particulate substrates (appr. 25 µg/g on substrate): $t_{1/2}$ = 100 h on silica gel, $t_{1/2}$ = 78 h on alumina and $t_{1/2}$ = 38 h on fly ash (Behymer & Hites 1985)

first order daytime decay constants:: k = 0.0056 min^{-1} for soot particles loading of 1000–2000 ng/mg and k = 0.0090 min^{-1} with 30–350 ng/mg loading (Kamens et al. 1988)

photodegradation k = 7.07 × 10^{-3} min with $t_{1/2}$ = 1.63 h in ethanol-water (1:1, v/v) solution for initial concentration of 5.0 ppm by high pressure mercury lamp or sunlight (Wang et al. 1991)

k(exptl) = 0.00707 min^{-1} with the calculated $t_{1/2}$ = 1.63 h and the predicted k = 0.0114 min^{-1} calculated by QSPR in aqueous solution when irradiated with a 500 W medium pressure mercury lamp (Chen et al. 1996)

k = 1.01–1.30 min^{-1} in natural water system by UV and sunlight (Yu et al. 1999)

$t_{1/2}$(obs) = 1.58 h, $t_{1/2}$(calc) = 1.34 h predicted by QSPR in atmospheric aerosol (Chen et al. 2001)

photochemical degradation k = (1.60 ± 0.08) × 10^{-5} s^{-1} and $t_{1/2}$ = (11.99 ± 0.53)h in diesel particulate matter under atmospheric conditions; k = (2.29 ± 0.22) × 10^{-5} s^{-1} and $t_{1/2}$ = (8.41 ± 0.91)h in diesel particulate matter/soil mixture, and $t_{1/2}$ = 1.69 – 8.82 h in various soil components using a 900-W photo-irradiator as light source; k = (4.76 ± 0.40) × 10^{-7} s^{-1} and $t_{1/2}$ = (405.26 ± 37.27)h in diesel particulate matter using a 300-W light source (Matsuzawa et al. 2001)

Photodegradation k = 9.0 × 10^{-5} s^{-1} in surface water during the summertime at mid-latitude (Fasnacht & Blough 2002)

Oxidation: $t_{1/2}$ = 2.6 h for photosensitized oxygenation with singlet oxygen at near-surface natural water, 40°N, midday, midsummer (Zepp & Schlotzhauer 1979)

k > 1 × 10^6 M^{-1} h^{-1} for singlet oxygen and k = 1 × 10^3 M^{-1} h^{-1} for peroxy radical (Mabey et al. 1982); photooxidation $t_{1/2}$ = 0.802–8.02 h in air, based on estimated rate constant for reaction with hydroxyl radical in air (Howard et al. 1991)

Hydrolysis: not hydrolyzable (Mabey et al. 1982); no hydrolyzable groups (Howard et al. 1991).

Biodegradation: significant degradation with gradual adaptation within 7 d for a domestic sewer 28 d test for an average of three static-flask screening (Tabak et al. 1981)

aerobic $t_{1/2}$ = 8904–24000 h, based on aerobic soil dieaway test data (Howard et al. 1991)

rate constants k = 0.0019 d^{-1} with $t_{1/2}$ = 371 d for Kidman sandy loam and k = 0.0018 d^{-1} with $t_{1/2}$ = 387 d for McLarin sandy loam all at –0.33 bar soil moisture (Park et al. 1990)

anaerobic $t_{1/2}$ = 35616–96000 h, based on estimated unacclimated aqueous aerobic biodegradation half-life (Howard et al. 1991).

Biotransformation: estimated to be 1 × 10^{-10} mL cell^{-1} h^{-1} for bacteria (Mabey et al. 1982).

Bioconcentration, Uptake (k_1) and Elimination (k_2) Rate Constants:

k_1 = 0.35–0.71 mg g^{-1} h^{-1}; k_2 = 0.15 h^{-1} (freshwater oligochaete from sediment, Van Hoof et al. 2001)

Half-Lives in the Environment:

Air: $t_{1/2}$ = 0.802–8.02 h, based on estimated photooxidation half-life in air (Howard et al. 1987);

$t_{1/2}$ = 1.3 h for adsorption on wood soot particles in an outdoor Teflon chamber with an estimated rate constant k = 0.0092 min^{-1} at 1 cal cm^{-2} min^{-1}, 10 g/m^3 H$_2$O and 20°C (Kamens et al. 1988).

Surface water: photolysis $t_{1/2}$ = 4.4 h near surface water, $t_{1/2}$ = 13 d and 68 d in 5-m deep water body without and with sediment-water partitioning in full summer day, 40°N; photosensitized oxygenation $t_{1/2}$ = 2.6 h at near surface water, 40°N, midday, midsummer (Zepp & Schlotzhauer 1979)

$t_{1/2}$ ~ 4.4–13 h, based on photolysis half-life in water (Howard et al. 1991);

photolysis $t_{1/2}$ = 1.63 h in aqueous solution when irradiated with a 500 W medium-pressure mercury lamp (Chen et al. 1996);

photolysis $t_{1/2}$ = 533–693 min at 18°C in natural water system by UV and sunlight illumination (Yu et al. 1999).

Groundwater: $t_{1/2}$ = 17808–48000 h, based on estimated unacclimated aqueous aerobic biodegradation half-life (Howard et al. 1991).

Sediment: desorption $t_{1/2}$ = 31.9 d from sediment under conditions mimicking marine disposal (Zhang et al. 2000).

Soil: $t_{1/2}$. > 5.5 d (Sims & Overcash 1983; quoted, Bulman et al. 1987);

$t_{1/2}$ = 328 d for 5 mg/kg treatment and $t_{1/2}$ = 224 d for 50 mg/kg treatment (Bulman et al. 1987);

biodegradation rate constant l = 0.0019 d^{-1} with $t_{1/2}$ = 371 d for Kidman sandy loam soil, and k = 0.0018 h^{-1} with $t_{1/2}$ = 387 d for McLaurin sandy loam soil (Park et al. 1990);

$t_{1/2}$ = 8904–24000 h, based on aerobic soil die-away test data (Howard et al. 1991);

$t_{1/2}$ > 50 d (Ryan et al. 1988).

Biota: elimination $t_{1/2}$ = 5.0–14.2 d from mussel *Mytilus edulis*; $t_{1/2}$ = 15.1 d from Oyster, $t_{1/2}$ = 4.3 d from clam *Mercenario mercenaria*, $t_{1/2}$ = 3.3 d from clam *Macoma balthica* (quoted, Meador et al. 1995).

TABLE 4.1.1.33.1
Reported aqueous solubilities of chrysene at various temperatures and reported temperature dependence equations

$$S/(\mu g/kg) = a \cdot t^3 + b \cdot t^2 + c \cdot t + d \quad (1)$$

May et al. 1978b		May 1980, May et al. 1983	
generator column-HPLC		generator column-HPLC	
t/°C	S/g·m^{-3}	t/°C	S/g·m^{-3}
25	0.0018	6.5	0.00071
29	0.0022	11.0	0.00080
		20.4	0.0014
		24.0	0.00168
temp dependence eq. 1		25.3	0.00189
S	μg/kg	28.7	0.00221
a	0		
b	0.0024	temp dependence eq. 1 given	
c	–0.0144	in May 1980	
d	0.69	S	μg/kg
		a	0
ΔH_{sol}/(kJ mol^{-1}) = 41.25		b	0.0024
measured between 5–30°C		c	–0.0144
		d	0.69
		ΔH_{sol}/(kJ mol^{-1}) = 41.25	
		measured between 5–30°C	

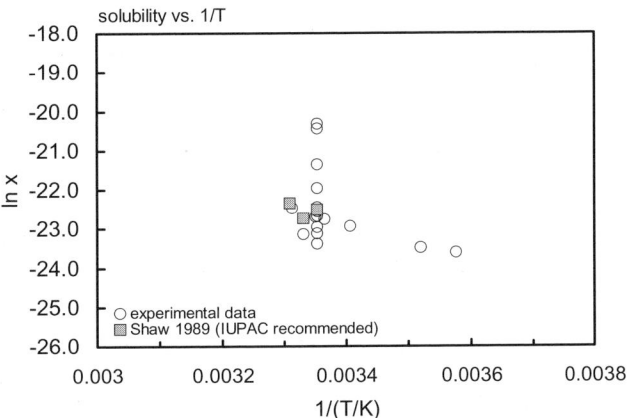

FIGURE 4.1.1.33.1 Logarithm of mole fraction solubility (ln x) versus reciprocal temperature for chrysene.

TABLE 4.1.1.33.2
Reported vapor pressures and Henry's law constants of chrysene at various temperatures and temperature dependence equations

Vapor pressure				Henry's law constant		
Hoyer & Peperle 1958		de Kruif 1980		Bamford et al. 1999		
effusion		torsion-, weighing effusion		gas stripping-GC/MS		
t/°C	P/Pa	t/°C	P/Pa	t/°C	H/(Pa m³/mol)	H/(Pa m³/mol)
						average
data presented by equation		117.31	0.1	4.1	0.01, 0.03	0.02
log P/mmHg = A − B/(T/K)		124.84	0.2	11.0	0.05, 0.09	0.07
A	13.07	129.38	0.3	18.0	0.15, 0.25	0.19
B	6340	132.67	0.4	25.0	0.38, 0.73	0.53
temp range 80–145°C		135.26	0.5	31.0	0.78, 1.86	1.20
		137.39	0.6			
ΔH_{subl}/(kJ/mol) = 118.8		139.22	0.7	ln K_{AW} = A − B/(T/K)		
		140.81	0.8	A	32.235	
		142.23	0.9	B	12136	
		143.51	1.0			
		25.0	5.7 × 10⁻⁷	enthalpy, entropy change:		
			extrapolated)	ΔH/(kJ·mol⁻¹) = 100.9 ± 7.7		
		ΔH_{subl}/(kJ/mol) = 118.5		ΔS/(J·K⁻¹ mol⁻¹) = 268		
				at 25°C		

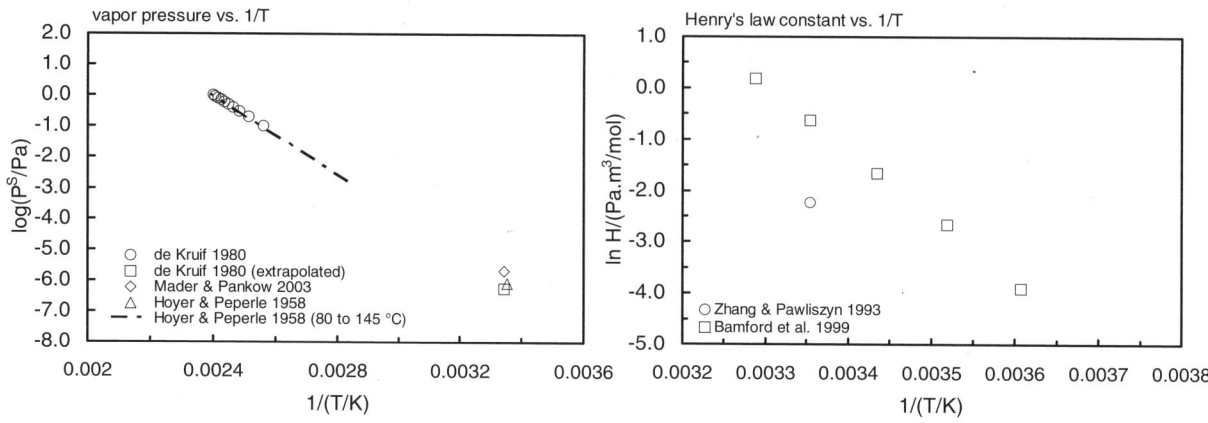

FIGURE 4.1.1.33.2 Logarithm of vapor pressure and Henry's law constant versus reciprocal temperature for chrysene.

4.1.1.34 Triphenylene

Common Name: Triphenylene
Synonym: 9,10-benzophenanthrene, isochrysene, 1,2,3,4-dibenznaphthalene
Chemical Name: triphenylene
CAS Registry No: 217-59-4
Molecular Formula: $C_{18}H_{12}$
Molecular Weight: 228.288
Melting Point (°C):
    197.8   (Lide 2003)
Boiling Point (°C):
    425   (Weast 1982–83; Dean 1985; Pearlman et al. 1984; Budavari 1989; Lide 2003)
Density (g/cm³ at 20°C):
    1.302   (Dean 1985; Budavari 1989)
Molar Volume (cm³/mol):
    211.8   (Ruelle & Kesselring 1997)
    250.8   (calculated-Le Bas method at normal boiling point)
Enthalpy of Fusion, ΔH_{fus} (kJ/mol):
    24.74   (exptl., Chickos et al. 1999)
Entropy of Fusion, ΔS_{fus} (J/mol K):
    52.53, 44.1 (exptl., calculated-group additivity method, Chickos et al. 1999)
Fugacity Ratio at 25°C (assuming ΔS_{fus} = 56 J/mol K), F: 0.0202 (mp at 197.8°C)

Water Solubility (g/m³ or mg/L at 25°C or as indicated and reported temperature dependence equations. Additional data at other temperatures designated * are compiled at the end of this section):
    0.0388   (27°C, nephelometry, Davis et al. 1942)
    0.043   (shake flask-UV, Klevens 1950)
    0.043   (shake flask-fluorescence, Mackay & Shiu 1977)
    0.0049*   (20.5°C, generator column-HPLC, measured range 281.15–301.35 K, May et al. 1983)
    0.041   (lit. mean, Pearlman et al. 1984)
    0.0307   (generator column-HPLC/fluorescence, Walters & Luthy 1984)
    0.041   (vapor saturation-UV, Akiyoshi et al. 1987)

Vapor Pressure (Pa at 25°C or as indicated and reported temperature dependence equations. Additional data at other temperatures designated * are compiled at the end of this section):
    $\log (P/\text{mmHg}) = 12.89 - 6154/(T/K)$; temp range: 65–125°C (Knudsen effusion method, Hoyer & Peperle 1958)
    2.30×10^{-6}*   (effusion, De Kruif 1980)
    3.85×10^{-7} (extrapolated-Antoine eq.-I, Stephenson & Malanowski 1987)
    $\log (P_S/\text{kPa}) = 9.435 - 5620/(T/K)$; temp range: 363–468 K (Antoine eq.-I, Stephenson & Malanowski 1987)
    1.17×10^{-2} (extrapolated-Antoine eq.-II, supercooled liquid P_L, Stephenson & Malanowski 1987)
    $\log (P_L/\text{kPa}) = 6.8974 - 3527/(T/K)$; temp range: 600–720 K (Antoine eq.-II, Stephenson & Malanowski 1987)
    2.39×10^{-4} (supercooled liquid P_L, GC-RT correlation, Lei et al. 2002)
    $\log (P_L/\text{Pa}) = -4624/(T/K) + 11.89$; $\Delta H_{vap.} = -88.5$ kJ·mol^{-1} (GC-RT correlation, Lei et al. 2002)

Henry's Law Constant (Pa m³/mol):

Octanol/Water Partition Coefficient, log K_{OW} at 25°C and the reported temperature dependence equations:
    5.45   (shake flask-UV, Karickhoff et al. 1979)
    5.49   (HPLC-RT correlation, Wang et al. 1986)
    5.84   (TLC retention time correlation, De Voogt et al. 1990)

5.49 (recommended, Sangster 1993)
5.49 (recommended, Hansch et al. 1995)
5.15; 4.83 (calibrated GC-RT correlation; GC-RT correlation, Lei et al. 2000)
log K_{OW} = 1.313 + 1138.55/(T/K); temp range 5–55°C (temperature dependence HPLC-k′ correlation, Lei et al. 2000)
6.27 (RP-HPLC-RT correlation, short ODP column, Donovan & Pescatore 2002)

Octanol/Air Partition Coefficient, log K_{OA}:

Bioconcentration Factor, log BCF:
3.96 (*Daphnia magna*, Newsted & Giesy 1987)

Sorption Partition Coefficient, log K_{OC}:
6.90 (Baltic Sea particulate field samples, concn distribution-GC/MS, Broman et al. 1991)
4.0 (predicted dissolved log K_{OC}, Broman et al. 1991)
5.52–7.38; 4.90–7.80 (for chrysene + triphenylene, range, calculated from sequential desorption of 11 urban soils; lit. range, Krauss & Wilcke 2001)
5.92; 6.12, 5.77, 6.14 (for chrysene + triphenylene, 20°C, batch equilibrium, A2 alluvial grassland soil; calculated values of expt 1,2,3-solvophobic approach, Krauss & Wilcke 2001)

Environmental Fate Rate Constants, k or Half-Lives, $t_{1/2}$:

Half-Lives in the Environment:

Biota: elimination $t_{1/2}$ = 2 d from rainbow trout, $t_{1/2}$ = 4.4 d from clam *Mya arenaria*, $t_{1/2}$ = 8.0d from mussel *Mytilus edulis*, $t_{1/2}$ = 14.4 d from polychaete *Abarenicola pacifica*, $t_{1/2}$ = 21.7 d from Oyster, $t_{1/2}$ = 2.4 d from shrimp (quoted, Meador et al. 1995).

TABLE 4.1.1.34.1
Reported aqueous solubilities and vapor pressures of triphenylene at various temperatures and reported temperature dependence equation

$$S/(\mu g/kg) = a \cdot t^3 + b \cdot t^2 + c \cdot t + d \quad (1)$$

Aqueous solubility		Vapor pressure			
May 1980, May et al. 1983		Hoyer & Peperle 1958		de Kruif 1980	
generator column-HPLC		effusion		torsion, weighing effusion	
t/°C	S/g·m⁻³	t/°C	P/Pa	t/°C	P/Pa
8.0	0.00299	data presented by		107.02	0.1
12.0	0.00393	log P/mmHg = A − B/(T/K)		114.36	0.2
14.8	0.00339	A	12.89	118.39	0.3
20.5	0.00489	B	6154	134.52	0.4
27.3	0.00765	temp range 65–125°C		122.0	0.5
28.2	0.00811			126.61	0.6
		ΔH_{subl}/(kJ mol⁻¹) = 114.6		128.38	0.7
temp dependence eq. 1				129.94	0.8
given in May 1980				131.32	0.9
S	μg/kg			132.56	1.0
a	−0.0002			25.0	2.3 × 10⁻⁶
b	0.0250				extrapolated
c	−0.4250				
d	4.89			ΔH_{subl}/(kJ mol⁻¹) = 115.2	
ΔH_{sol}/(kJ mol⁻¹) = 41.25					
measured between 5–30°C					

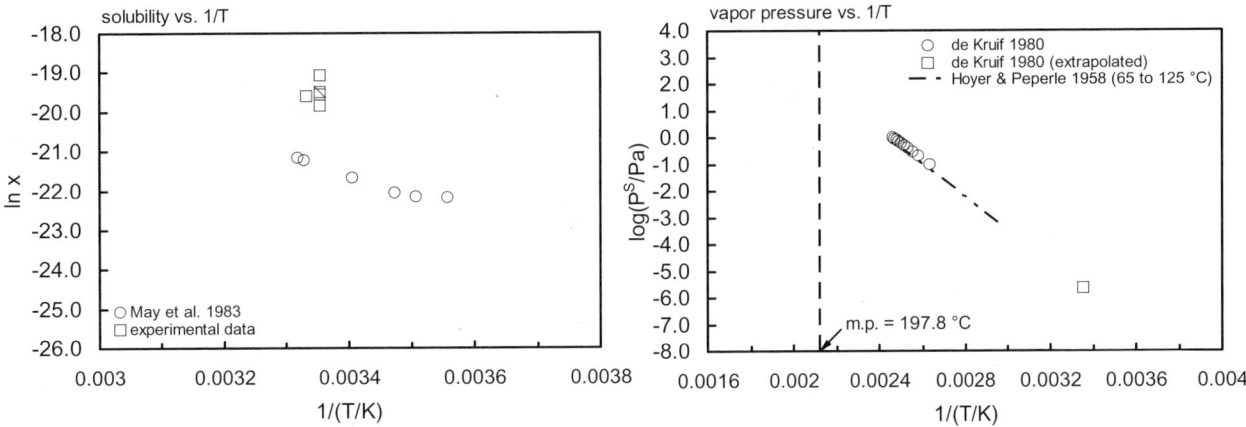

FIGURE 4.1.1.34.1 Logarithm of mole fraction solubility and vapor pressure versus reciprocal temperature for triphenylene.

4.1.1.35 *o*-Terphenyl

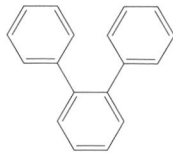

Common Name: *o*-Terphenyl
Synonym: 1,2-diphenyl benzene
Chemical Name:
CAS Registry No: 84-15-1
Molecular Formula: $C_{18}H_{14}$, $1,2\text{-}(C_6H_5)_2C_6H_4$
Molecular Weight: 230.304
Melting Point (°C):
    56.2 (Lide 2003)
Boiling Point (°C):
    332 (Weast 1982–83; Stephenson & Malanowski 1987; Lide 2003)
Density (g/cm^3):
Molar Volume (cm^3/mol):
    225.4 (93°C, Stephenson & Malanowski 1987)
    273.2 (calculated-Le Bas method at normal boiling point)
Enthalpy of Fusion, ΔH_{fus} (kJ/mol):
    17.2 (Chickos et al. 1999)
Entropy of Fusion, ΔS_{fus} (J/mol K):
    52.3, 73.9 (exptl., calculated-group additivity method, Chickos et al. 1999)
Fugacity Ratio at 25°C (assuming ΔS_{fus} = 56 J/mol K), F: 0.494 (mp at 56.2°C)

Water Solubility (g/m^3 or mg/L at 25°C):
    1.24 (vapor saturation-spectrophotometry, Akiyoshi et al. 1987)

Vapor Pressure (Pa at 25°C and reported temperature dependence equations):
    $\log (P/kPa) = 6.29308 - [2160.24/(-106.38 + (T/K))]$; temp range 462–650 K (Antoine eq., liquid phase, Stephenson & Malanowski 1987)
    $\log (P/mmHg) = -8.0641 - 4.0928 \times 10^3/(T/K) + 9.1076 \cdot \log (T/K) - 1.6326 \times 10^{-2} \cdot (T/K) + 6.0467 \times 10^{-6} \cdot (T/K)^2$; temp range 329–891 K (vapor pressure eq., Yaws 1994)

Henry's Law Constant (Pa·m^3/mol):
Octanol/Water Partition Coefficient, log K_{OW}:
Octanol/Air Partition Coefficient, log K_{OA}:
Bioconcentration Factor, log BCF or log K_B:
Sorption Partition Coefficient, log K_{OC}:

Environmental Fate Rate Constants, k and Half-Lives, $t_{1/2}$:

Half-Lives in the Environment:
    Air:
    Surface water: a first order reduction process in river water with an estimated $t_{1/2}$ = 0.6 d for terphenyl in Rhine River, isomer unspecified (Zoeteman et al. 1980).
    Ground water:
    Sediment:
    Soil:
    Biota:

4.1.1.36 *m*-Terphenyl

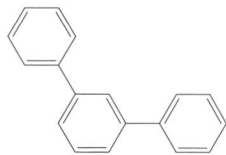

Common Name: *m*-Terphenyl
Synonym: 1,3-diphenyl benzene
Chemical Name:
CAS Registry No: 92-06-8
Molecular Formula: $C_{18}H_{14}$, 1,3-$(C_6H_5)_2C_6H_4$
Molecular Weight: 230.304
Melting Point (°C):
    87   (Lide 2003)
Boiling Point (°C):
    363   (Lide 2003)
Density (g/cm³ at 20°C):
    1.199   (Lide 2003)
Molar Volume (cm³/mol):
    227   (93°C, Stephenson & Malanowski 1987)
    192.1   (20°C, calculated-density)
    273.2   (calculated-Le Bas method at normal boiling point)
Enthalpy of Fusion, ΔH_{fus} (kJ/mol):
    22.59   (Chickos et al. 1999)
Entropy of Fusion, ΔS_{fus} (J/mol K):
    62.76, 73.9   (exptl., calculated-group additivity method, Chickos et al. 1999)
Fugacity Ratio at 25°C (assuming ΔS_{fus} = 56 J/mol K), F: 0.246 (mp at 87°C)

Water Solubility (g/m³ or mg/L at 25°C or as indicated and reported temperature dependence equations. Additional data at other temperatures designated * are compiled at the end of this section):
    1.51   (vapor saturation-spectrophotometry, Akiyoshi et al. 1987)
    0.0305*   (24.99°C, generator column-HPLC/fluorescence, Reza et al. 2002)
    $\ln x = (-2.62 \pm 0.91) + (-5134 \pm 271)/(T/K)$; temp range 278–323.13 K (Reza et al. 2002)

Vapor Pressure (Pa at 25°C and reported temperature dependence equations):
    $\log (P/kPa) = 6.48808 - [2445.98/(-102.76 + (T/K))]$; temp range 462–691 K (Antoine eq., liquid phase, Stephenson & Malanowski 1987)
    $\log (P/mmHg) = -14.7175 - 4.3577 \times 10^3/(T/K) + 11.935 \cdot \log (T/K) - 1.8441 \times 10^{-2} \cdot (T/K) + 6.437 \times 10^{-6} \cdot (T/K)^2$; temp range 360–925 K (vapor pressure eq., Yaws 1994)

Henry's Law Constant (Pa·m³/mol):
Octanol/Water Partition Coefficient, log K_{OW}:
Octanol/Air Partition Coefficient, log K_{OA}:
Bioconcentration Factor, log BCF or log K_B:
Sorption Partition Coefficient, log K_{OC}:
Environmental Fate Rate Constants, k and Half-Lives, $t_{1/2}$:

Half-Lives in the Environment:
    Air:
    Surface water: a first order reduction process in river water with an estimated $t_{1/2}$ = 0.6 d for terphenyl in Rhine River, isomer unspecified (Zoeteman et al. 1980).

Ground water:
Sediment:
Soil:
Biota:

TABLE 4.1.1.36.1
Reported aqueous solubilities of m-terphenyl at various temperatures

Reza et al. 2002			
generator column-HPLC			
t/°C	S/g·m^{-3}	t/°C	S/g·m^{-3}
4.85	0.009	40.07	0.0741
7.64	0.0115	45.07	0.095
11.8	0.012	49.98	0.1159
14.98	0.0183	$\ln x = A - B/(T/K)$	
19.95	0.0219	A	−2.62
24.99	0.0305	B	5134
30.0	0.0409	for temp range 287–323 K	
35.0	0.0535		

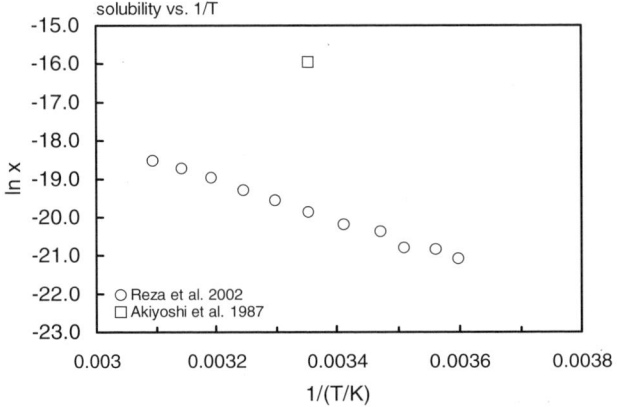

FIGURE 4.1.1.36.1 Logarithm of mole fraction solubility (ln x) versus reciprocal temperature for *m*-terphenyl.

4.1.1.37 *p*-Terphenyl

Common Name: *p*-Terphenyl
Synonym: 1,4-diphenyl benzene
Chemical Name: *p*-terphenyl
CAS Registry No: 92-94-4
Molecular Formula: $C_{18}H_{14}$, $1,4\text{-}(C_6H_5)_2C_6H_4$
Molecular Weight: 230.304
Melting Point (°C):
    213.9   (Lide 2003)
Boiling Point (°C):
    376   (Lide 2003)
Density (g/cm³):
Molar Volume (cm³/mol):
    262.0   (315.6°C, Stephenson & Malanowski 1987)
    273.2   (calculated-Le Bas method at normal boiling point)
Enthalpy of Fusion, ΔH_{fus} (kJ/mol):
    0.3, 35.3; 35.6   (–79.55, 213.85°C; total phase change enthalpy, Chickos et al. 1999)
Entropy of Fusion, ΔS_{fus} (J/mol K):
    74.1, 73.9   (exptl., calculated-group additivity method, total phase change entropy, Chickos et al. 1999)
Fugacity Ratio at 25°C (assuming ΔS_{fus} = 56 J/mol K), F: 0.014 (mp at 213.9°C)

Water Solubility (g/m³ or mg/L at 25°C):
    1.80   (vapor saturation-spectrophotometry, Akiyoshi et al. 1987)

Vapor Pressure (Pa at 25°C and reported temperature dependence equations):
    4.86×10^{-6} (extrapolated from solid, Stephenson & Malanowski 1987)
    $\log (P_S/\text{kPa}) = 12.515 - 6210/(T/K)$; temp range 338–431 K (Antoine eq.-I, solid phase, Stephenson & Malanowski 1987)
    1.78×10^{-5} (P_L extrapolated from liquid state, Stephenson & Malanowski 1987)
    $\log (P_L/\text{kPa}) = 6.16107 - [2125.84/(-145.29 + (T/K))]$; temp range 499–700 K (Antoine eq.-II, liquid phase, Stephenson & Malanowski 1987)
    $\log (P/\text{mmHg}) = -39.6342 - 3.2661 \times 10^3/(T/K) + 21.08 \cdot \log(T/K) - 2.2574 \times 10^{-2} \cdot (T/K) + 6.902 \times 10^{-6} \cdot (T/K)^2$; temp range 485–926 K (vapor pressure eq., Yaws 1994)
    5.40×10^{-4} (supercooled liquid P_L, GC-RT correlation, Lei et al. 2002)
    $\log (P_L/\text{Pa}) = -4135/(T/K) + 10.60$; $\Delta H_{vap.} = -79.2$ kJ·mol⁻¹ (GC-RT correlation, Lei et al. 2002)

Henry's Law Constant (Pa·m³/mol):

Octanol/Water Partition Coefficient, log K_{OW}:
    6.03, 5.88   (HPLC-RV correlation, Garst 1984)
    6.03 ± 0.50   (recommended, Sangster 1989)

Octanol/Air Partition Coefficient, log K_{OA}:

Bioconcentration Factor, log BCF or log K_B:

Sorption Partition Coefficient, log K_{OC}:

Environmental Fate Rate Constants, k and Half-Lives, $t_{1/2}$:

Half-Lives in the Environment:
- Air:
- Surface water: a first order reduction process in river water with an estimated $t_{1/2}$ = 0.6 d for terphenyl in Rhine River, isomer unspecified (Zoeteman et al. 1980).
- Ground water:
- Sediment:
- Soil:
- Biota:

4.1.1.38 Naphthacene

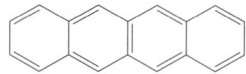

Common Name: Naphthacene
Synonym: benz[*b*]anthracene, 2,3-benzanthracene, tetracene
Chemical Name: benz[*b*]anthracene
CAS Registry No: 92-24-0
Molecular Formula: $C_{18}H_{12}$
Molecular Weight: 228.288
Melting Point (°C):
    357   (Lide 2003)
Boiling Point (°C):
    450   (sublimation, Bjørseth 1983; Lide 2003)
Density (g/cm³ at 20°C):
Molar Volume (cm³/mol):
    211.8   (Ruelle & Kesselring 1997)
    250.8   (calculated-Le Bas method at normal boiling point)
Enthalpy of Fusion, ΔH_{fus} (kJ/mol):
    38.64   (Ruelle & Kesselring 1997)
Entropy of Fusion, ΔS_{fus} (J/mol K):
Fugacity Ratio at 25°C (assuming ΔS_{fus} = 56 J/mol K), F: 0.000553 (mp at 357°C)

Water Solubility (g/m³ or mg/L at 25°C or as indicated):
    0.0010   (27°C, shake flask-nephelometry, Davis et al. 1942)
    0.0015   (approximate, shake flask-UV, Klevens 1950)
    0.0036   (shake flask-UV, Eisenbrand & Baumann 1970)
    0.00057   (shake flask-fluorescence, Mackay & Shiu 1977)
    0.044   (shake flask-nephelometry, Hollifield 1979)
    0.00103   (lit. mean, Pearlman et al. 1984)

Vapor Pressure (Pa at 25°C or as indicated and reported temperature dependence equations. Additional data at other temperatures designated * are compiled at the end of this section):
    7.30×10^{-9}*   (effusion method, De Kruif 1980)
    3.70×10^{-8} (extrapolated-Antoine eq., Stephenson & Malanowski 1987)
    log $(P_S/kPa) = 11.505 - 6540/(T/K)$; temp range 376–489 K (Antoine eq., Stephenson & Malanowski 1987)
    2.31×10^{-8}*   (gas saturation, extrapolated-Antoine eq. derived from exptl. data, temp range 113–199°C,
                    Oja & Suuberg 1998)
    log $(P/Pa) = 33.594 - 15151/(T/K)$; temp range 368–472 K (Antoine eq., Knudsen effusion, Oja & Suuberg 1998)

Henry's Law Constant (Pa m³/mol):

Octanol/Water Partition Coefficient, log K_{OW}:
    5.90   (shake flask-UV, concn. ratio, Karickhoff et al., 1979)
    6.02   (HPLC-k' correlation, McDuffie 1981)
    5.76   (HPLC-RT correlation, Wang et al. 1986)
    5.76.   (recommended, Sangster 1989, 1993)
    5.84   (TLC retention time correlation, De Voogt et al. 1990)
    5.90   (shake flask-HPLC, De Voogt et al. 1990)
    5.90   (recommended, Hansch et al 1995)

Octanol/Air Partition Coefficient, log K_{OA}:

Bioconcentration Factor, log BCF:

Sorption Partition Coefficient, log K_{OC}:
- 5.81 (sediment, batch equilibrium-sorption isotherms by GC/UV, Karickhoff et al. 1979)

Environmental Fate Rate Constants, k or Half-Lives, $t_{1/2}$:

Volatilization:

Photolysis: direct photochemical transformation $t_{1/2}$ = 0.034 h, computed near-surface water, latitude 40°N, midday, midsummer and photolysis $t_{1/2}$ = 0.20 d and 0.95 d in 5-m deep inland water body without and with sediment-water partitioning, respectively, to top cm of bottom sediment over full summer day, 40°N (Zepp & Schlotzhauer 1979)

photodegradation k = 0.051 min^{-1} and $t_{1/2}$ = 0.23 h in ethanol-water (1:1, v/v) solution for initial concentration of 5.0 ppm by high pressure mercury lamp or sunlight (Wang et al. 1991)

pseudo-first-order direct photolysis k(exptl) = 0.051 min^{-1} with the calculated k = 0.051 min^{-1} and $t_{1/2}$ = 0.23 h and the predicted k(calc) = 0.0355 min^{-1} calculated by QSPR in aqueous solution when irradiated with a 500 W medium pressure mercury lamp (Chen et al. 1996)

direct photolysis $t_{1/2}$ = 0.92 h predicted by QSPR in atmospheric aerosol (Chen et al. 2001).

Oxidation: $t_{1/2}$ = 2.6 h for photosensitized oxygenation with singlet oxygen at near-surface natural water, 40°N, midday, midsummer (Zepp & Schlotzhauer 1979)

Hydrolysis:

Biodegradation:

Biotransformation:

Bioconcentration, Uptake (k_1) and Elimination (k_2) Rate Constants:

Half-Lives in the Environment:

Air: direct photolysis $t_{1/2}$ = 0.92 h predicted by QSPR method in atmospheric aerosol (Chen et al. 2001).

Surface water: photolysis $t_{1/2}$ = 0.034 h near surface water, $t_{1/2}$ = 0.20 d and 0.95 d in 5-m deep water body without and with sediment-water partitioning in full summer day, 40°N; photosensitized oxygenation $t_{1/2}$ = 2.6 h at near surface water, 40°N, midday, midsummer (Zepp & Schlotzhauer 1979).

photolysis $t_{1/2}$ = 0.23 h in aqueous solution when irradiated with a 500W medium pressure mercury lamp (Chen et al. 1996).

TABLE 4.1.1.38.1
Reported vapor pressures of naphthacene at various temperatures and the coefficients for the vapor pressure equations

$$\log P = A - B/(T/K) \quad (1)$$
$$\log P = A - B/(C + t/°C) \quad (2)$$
$$\log P = A - B/(C + T/K) \quad (3)$$
$$\log P = A - B/(T/K) - C \cdot \log (T/K) \quad (4)$$

$$\ln P = A - B/(T/K) \quad (1a)$$
$$\ln P = A - B/(C + t/°C) \quad (2a)$$

de Kruif 1980		Oja & Suuberg 1998	
torsion-, weighing effusion		Knudsen effusion	
t/°C	P/Pa	t/°C	P/Pa
147.35	0.1	113.4	0.00344
155.46	0.2	128.78	0.0199
160.35	0.3	139.56	0.0535
163.89	0.4	145.44	0.0496
166.67	0.5	160.69	0.322
168.97	0.6	172.88	0.584
170.94	0.7	198.99	4.81
172.65	0.8		
173.98	0.9	mp/K	630
175.65	1.0		

TABLE 4.1.1.38.1 (Continued)

de Kruif 1980		Oja & Suuberg 1998	
torsion-, weighing effusion		Knudsen effusion	
t/°C	P/Pa	t/°C	P/Pa
25.0	9.7 × 10⁻⁹	eq. 1a	P/Pa
	extrapolated	A	35.594
ΔH_{subl}/(kJ mol⁻¹) = 155.0		for temp range 386–472 K	
		ΔH_{subl}/(kJ mol⁻¹) = 126.1	

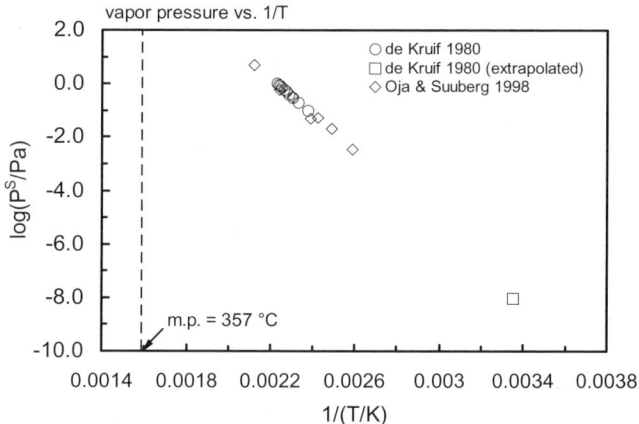

FIGURE 4.1.1.38.1 Logarithm of vapor pressure versus reciprocal temperature for naphthacene.

4.1.1.39 Benz[a]anthracene

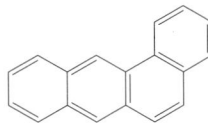

Common Name: Benz[a]anthracene
Synonym: 1,2-benzanthracene, 2,3-benzophenanthrene, naphthanthracene, BaA, B(a) a, tetraphene
Chemical Name: 1,2-benzanthracene
CAS Registry No: 56-55-3
Molecular Formula: $C_{18}H_{12}$
Molecular Weight: 228.288
Melting Point (°C):
    160.5   (Lide 2003)
Boiling Point (°C):
    438   (Lide 2003)
Density (g/cm³ at 20°C):
    1.2544   (Mailhot & Peters 1988)
Molar Volume (cm³/mol):
    211.8   (Ruelle & Kesselring 1997; Passivirta et al. 1999)
    248.3   (calculated-Le Bas method at normal boiling point)
Enthalpy of Fusion, ΔH_{fus} (kJ/mol):
    21.38   (Ruelle & Kesselring 1997; Chickos et al. 1999)
Entropy of Fusion, ΔS_{fus} (J/mol K):
    49.23, 44.1   (exptl., calculated-group additivity method, Chickos et al. 1999)
    49.2   (Passivirta et al. 1999)
Fugacity Ratio at 25°C (assuming ΔS_{fus} = 56 J/mol K), F: 0.0468 (mp at 160.5°C)
    0.040   (calculated, assuming ΔS_{fus} = 56 J/mol K, Mackay et al. 1980)
    0.0661   (calculated, ΔS_{fus} = 49.2 J/mol K, Passivirta et al. 1999)

Water Solubility (g/m³ or mg/L at 25°C or as indicated and reported temperature dependence equations. Additional data at other temperatures designated * are compiled at the end of this section):
    0.011   (27°C, shake flask-nephelometry, Davis & Parker 1942)
    0.010   (shake flask-UV, Klevens 1950)
    0.014   (shake flask-fluorescence, Mackay & Shiu 1977)
    0.0094, 0.0122   (25, 29°C, generator column-HPLC/UV, May et al. 1978b)
    $S/(\mu g/kg) = 1.74 + 0.1897 \cdot (t/°C) + 0.0031 \cdot (t/°C)^2 + 0.0003 \cdot (t/°C)^3$, temp range 5–30°C (generator column-HPLC/UV, May et al. 1978b, May 1980)
    0.044   (shake flask-nephelometry, Hollifield 1979)
    0.00837*   (generator column-HPLC, measured range 6.9–29.7°C, May 1980)
    0.0086*   (generator column-HPLC, measured range 6.9–29.7°C, May et al. 1983)
    0.00935*   (generator column-fluo., measured range 10–30°C, Velapoldi et al. 1983)
    0.011   (average lit. value, Pearlman et al. 1984)
    0.0168   (generator column-HPLC/fluorescence, Walters & Luthy 1984)
    0.00854   (generator column-HPLC/UV, measured range 3.7–25.0°C, Whitehouse 1984)
    0.011   (recommended, IUPAC Solubility Data Series, Shaw 1989)
    0.0146   (shake flask-HPLC, Haines & Sandler 1995)
    0.0130   (generator column-HPLC/fluorescence, De Maagd et al. 1998)
    $\log [S_L/(mol/L)] = -0.326 - 1119/(T/K)$ (supercooled liquid, Passivirta et al. 1999)
    $\ln x = -3.060466 - 5354.51/(T/K)$, temp range 5–50°C (regression eq. of literature data, Shiu & Ma 2000)

Vapor Pressure (Pa at 25°C or as indicated and reported temperature dependence equations. Additional data at other temperatures designated * are compiled at the end of this section):
    2.93×10^{-6} (20°C, Hoyer & Peperle 1958)
    $\log (P/mmHg) = 13.68 - 6250/(T/K)$; temp range 60–120°C (Knudsen effusion method, Hoyer & Peperle 1958)

2.17 × 10⁻⁵ (solid, extrapolated from Antoine eq., Kelley & Rice 1964; quoted, Bidleman 1984)
log (P/mmHg) = 11.528 – 5461/(T/K); temp range: 104–127°C (effusion method, Kelley & Rice 1964)
3.87 × 10⁻⁷ (effusion method, Wakayama & Inokuchi 1967)
1.47 × 10⁻⁵ (solid, effusion method, extrapolated-Antoine eq., Murray et al. 1974)
log (P/mmHg) = 10.045 – 5929/(T/K); temp range: 330–390 K (effusion method, Murray et al. 1974)
6.67 × 10⁻⁷ (20°C, effusion, Pupp et al. 1974)
7.30 × 10⁻⁶* (effusion method, De Kruif 1980)
2.71 × 10⁻⁵* (gas saturation-HPLC/fluo./UV, Sonnefeld et al. 1983)
log (P/Pa) = 9.684 – 4246.51/(T/K); temp range 10–50°C (solid, Antoine eq., Sonnefeld et al. 1983)
0.00107, 0.0003 (P_{GC} by GC-RT correlation, different GC columns, Bidleman 1984)
0.000543 (supercooled liquid P_L, converted from literature P_S with ΔS_{fus} Bidleman 1984)
2.49 × 10⁻⁴ (Yamasaki et al. 1984)
4.10 × 10⁻⁶ (selected, Howard et al. 1986)
1.51 × 10⁻⁵, 2.17 × 10⁻⁵ (extrapolated-Antoine eq., Stephenson & Malanowski 1987)
log (P_S/kPa) = 12.0507 – 5925/(T/K); temp range 330–390 K (solid, Antoine eq.-I, Stephenson & Malanowski 1987)
log (P_S/kPa) = 10.653 – 5461/(T/K); temp range 377–400 K (solid, Antoine eq.-II., Stephenson & Malanowski 1987)
5.43 × 10⁻⁴ (supercooled P_L, converted from literature P_S, Hinckley et al. 1990)
0.00107, 3.23 × 10⁻⁴ (P_{GC} by GC-RT correlation with different reference standards, Hinckley et al. 1990)
log P_L/Pa = 12.63 – 4742/(T/K) (GC-RT correlation, Hinckley et al. 1990)
2.51 × 10⁻⁴ (supercooled liquid P_L, calculated from Yamasaki et al. 1984, Finizio et al. 1997)
(4.11–281) × 10⁻⁷; 2.76 × 10⁻⁵ (P_S, quoted exptl., effusion; gas saturation, Delle Site 1997)
3.39 × 10⁻⁵; 5.29 × 10⁻⁵, 1.48 × 10⁻⁵, 2.57 × 10⁻⁵ (P_S, quoted lit., calculated; GC-RT correlation, Delle Site 1997)
5.47 × 10⁻⁴; 3.59 × 10⁻⁵ (quoted supercooled liquid P_L from Hinckley et al. 1990; converted to solid P_S with fugacity ratio F, Passivirta et al. 1999)
log (P_S/Pa) = 11.91 – 4858/(T/K) (solid, Passivirta et al. 1999)
log (P_L/Pa) = 9.34 – 3760/(T/K) (supercooled liquid, Passivirta et al. 1999)
log (P/Pa) = 9.683 – 4246.51/(T/K); temp range 5–50°C (regression eq. from literature data, Shiu & Ma 2000)

Henry's Law Constant (Pa m³/mol at 25°C or as indicated and reported temperature dependence equations. Additional data at other temperatures designated * are compiled at the end of this section):

0.813   (gas stripping-GC, Southworth 1979)
0.102   (headspace solid-phase microextraction (SPME)-GC, Zhang & Pawliszyn 1993)
1.22*   (gas stripping-GC; measured range 4.1–31°C, Bamford et al. 1999)
ln K_{AW} = –7986.53/(T/K) + 19.124, ΔH = 66.4 kJ mol⁻¹, measured range 4.1–31°C (gas stripping-GC, Bamford et al. 1999)
log (H/(Pa m³/mol)) = 9.67 – 2641/(T/K) (Passivirta et al. 1999)

Octanol/Water Partition Coefficient, log K_{OW} at 25°C and the reported temperature dependence equations:
   5.61   (Radding et al. 1976)
   5.66   (Leo 1986; quoted, Schüürmann & Klein 1988)
   5.79   (HPLC-RT correlation, Wang et al. 1986)
   5.91   (recommended, Sangster 1989, 1993)
   5.84   (TLC retention time correlation, De Voogt et al. 1990)
   5.79   (recommended, Hansch et al. 1995))
   5.54 ± 0.19, 5.50 ± 0.64 (HPLC-k' correlation: ODS column; Diol column, Helweg et al. 1997)
   5.91   (range 5.74–6.04) (shake flask/slow stirring-HPLC/fluorescence, De Maagd et al. 1998)
   5.75   (shake flask-SPME solid-phase micro-extraction, Paschke et al. 1999)
   5.33; 4.98 (calibrated GC-RT correlation; GC-RT correlation, Lei et al. 2000)
   log K_{OW} = 1.238 + 1216.89/(T/K); temp range 5–55°C (temperature dependence HPLC-k' correlation, Lei et al. 2000)

Octanol/Air Partition Coefficient, log K_{OA}:
   9.50   (calculated-K_{OW}/K_{AW}, Wania & Mackay 1996)
   9.54   (calculated, Finizio et al. 1997)

Bioconcentration Factor, log BCF:
- 4.56 (Smith et al. 1978; Steen & Karickhoff 1981)
- 4.0 (*Daphnia pulex*, Southworth et al. 1978)
- 4.0 (fathead minnow, Veith et al. 1979)
- 4.56, 5.0 (bacteria, Baughman & Paris 1981)
- 4.39 (activated sludge, Freitag et al. 1984)
- 4.0 (*Daphnia pulex*, correlated as per Mackay & Hughes 1984, Howell & Connell 1986)
- 4.39, 3.50, 2.54 (activated sludge, algae, fish, Freitag et al. 1985)
- 4.01 (*Daphnia magna*, Newsted & Giesy 1987)
- 4.303, 4266 (calculated-molecular connectivity indices, calculated-K_{OW}, Lu et al. 1999)

Sorption Partition Coefficient, log K_{OC}:
- 4.52 (22°C, suspended particulates, Herbes et al. 1980)
- 6.30 (sediments average, Kayal & Connell 1990)
- 7.30 (Baltic Sea particulate field samples, concn distribution-GC/MS, Broman et al. 1991)
- 5.62 (humic acid, HPLC-k' correlation; Nielsen et al. 1997)
- 5.77 (5.73–5.80), 5.47 (5.44–5.50) (sediments: Lake Oostvaardersplassen, Lake Ketelmeer, shake flask-HPLC/UV, de Maagd et al. 1998)
- 5.20 (soil, calculated-universal solvation model; Winget et al. 2000)
- 5.63–7.53; 4.50–6.70 (range, calculated from sequential desorption of 11 urban soils; lit. range, Krauss & Wilcke 2001)
- 5.11; 6.33, 5.84, 6.18 (20°C, batch equilibrium, A2 alluvial grassland soil; calculated values of expt 1,2,3-solvophobic approach, Krauss & Wilcke 2001)

Environmental Fate Rate Constants, k or Half-Lives, $t_{1/2}$:

Volatilization: aquatic fate rate $k = 8 \times 10^3$ h^{-1} with $t_{1/2} \sim 90$ h (Callahan et al. 1979);
  half-lives predicted by one compartment model: $t_{1/2} > 1000$ h in stream, eutrophic pond or lake and oligotrophic lake (Smith et al. 1978);
  calculated $t_{1/2} = 500$ h for a river of 1-m deep with water velocity of 0.5 m/s and wind velocity of 1 m/s (Southworth 1979; quoted, Herbes et al. 1980; Hallett & Brecher 1984).

Photolysis: aquatic fate rate $k \sim 6 \times 10^{-5}$ s^{-1} with $t_{1/2} = 10$–50 h (Callahan et al. 1979)
  $t_{1/2} = 20$ h in stream, $t_{1/2} = 50$ h in eutrophic pond or lake and $t_{1/2} = 10$ h in oligotrophic lake, predicated by one compartment model (Smith et al. 1978)
  direct photochemical transformation $t_{1/2}$(calc) = 0.59 h, computed near-surface water, latitude 40°N, midday, midsummer and photolysis $t_{1/2} = 3.7$ d and 9.2 d in 5-m deep inland water body without and with sediment-water partitioning, respectively, to top cm of bottom sediment over full summer day, 40°N (Zepp & Schlotzhauer 1979)
  $t_{1/2} = 0.58$ h in aquatics (quoted of EPA Report 600/7-78-074, Haque et al. 1980)
  $t_{1/2} = 0.2$ d for early day in March (Mill et al. 1981);
  $k = 1.93$ h^{-1} (Zepp 1980; quoted, Mill & Mabey 1985)
  $k = 13.4 \times 10^{-5}$ s^{-1} in early March with $t_{1/2} = 5$ h in pure water at 366 nm, in sunlight at 23–28°C and $k = 2.28 \times 10^{-5}$ s^{-1} at 313 nm with 1% acetonitrile in filter-sterilized natural water (Mill et al. 1981);
  $k = 1.39$ h^{-1} for summer midday at 40°N latitude (quoted, Mabey et al. 1982)
  $t_{1/2} = 1$–3 h, atmospheric and aqueous photolysis half-life, based on measured photolysis rate constant for midday March sunlight on a cloudy day (Smith et al. 1978; quoted, Harris 1982; Howard et al. 1991) and adjusted for approximate summer and winter sunlight intensity (Lyman et al. 1982; quoted, Howard et al. 1991)
  half-lives on different atmospheric particulate substrates (approx. 25 µg/g on substrate): $t_{1/2} = 4.0$ h on silica gel, $t_{1/2} = 2.0$ h on alumina and $t_{1/2} = 38$ h on fly ash (Behymer & Hites 1985)
  first order daytime decay $k = 0.0125$ min^{-1} for soot particles loading of 1000–2000 ng/mg and $k = 0.0250$ min^{-1} for soot particles loading of 30–350 ng/mg (Kamens et al. 1988)
  photodegradation $k = 0.0251$ min^{-1} with $t_{1/2} = 0.46$ h in ethanol-water (2:3, v/v) solution for initial concentration of 12.5 ppm by high pressure mercury lamp or sunlight (Wang et al. 1991)
  pseudo-first-order direct photolysis k(exptl) = 0.0251 min^{-1} with the calculated $t_{1/2} = 0.46$ h and the predicted $k = 0.0245$ min^{-1} calculated by QSPR method in aqueous solution when irradiated with a 500 W medium pressure mercury lamp (Chen et al. 1996)

direct photolysis $t_{1/2}$(obs) = 0.94 h, $t_{1/2}$(calc) = 0.89 h predicted by QSPR method in atmospheric aerosol (Chen et al. 2001)

Photodegradation k = 5.0×10^{-4} s^{-1} in surface water during the summertime at mid-latitude (Fasnacht & Blough 2002)

Oxidation: half-lives predicted by one compartment model: $t_{1/2}$ = 38 h in stream, eutrophic pond or lake and oligotrophic lake based on peroxy radical concentration of 10^{-9} M (Smith et al. 1978)

aquatic fate rate k = 5×10^3 M^{-1} s^{-1} with $t_{1/2}$ = 38 h (Callahan et al. 1979);

$t_{1/2}$ = 6400 h for photosensitized oxygenation with singlet oxygen at near-surface natural water, 40°N, midday, midsummer (Zepp & Schlotzhauer 1979)

k = 5×10^8 M^{-1} h^{-1} for singlet oxygen and 2×10^4 M^{-1} h^{-1} for peroxy radical (Mabey et al. 1982)

k = 3.3×10^{-4} h^{-1} with $t_{1/2}$ = 0.6 h under natural sunlight conditions; k(aq.) = 5.0×10^3 M^{-1} h^{-1} with $t_{1/2}$ = 1.6 d for free-radical oxidation in air-saturated water (NRCC 1983)

photooxidation $t_{1/2}$ = 0.801–8.01 h, based on estimated rate constant for reaction with hydroxyl radical in air (Howard et al. 1991);

photooxidation $t_{1/2}$ = 77–3850 h in water, based on measured rate constant for reaction with hydroxyl radical in water (Howard et al. 1991)

Hydrolysis: not hydrolyzable (Mabey et al. 1982); no hydrolyzable groups (Howard et al. 1991).

Biodegradation: not observed during enrichment procedures (Smith et al. 1978)

no significant degradation in 7 d for an average of three static-flask screening test (Tabak et al. 1981)

k = 3.3×10^{-3} h^{-1} with $t_{1/2}$ = 208 h for mixed bacterial populations in stream sediment (NRCC 1983)

k = 1.0×10^{-4} h^{-1} with $t_{1/2}$ = 288 d; k = 4.0×10^{-6} h^{-1} with $t_{1/2}$ = 20 yr for mixed bacterial populations in oil-contaminated and pristine stream sediments (NRCC 1983)

k = 0.0026 d^{-1} with $t_{1/2}$ = 261 d for Kidman sandy loam and k = 0.0043 d^{-1} with $t_{1/2}$ = 162 d for McLarin sandy loam all at –0.33 bar soil moisture (Park et al. 1990)

$t_{1/2}$(aq.aerobic) = 2448–16320 h, based on aerobic soil dieaway test data at 10–30°C (Howard et al. 1991)

$t_{1/2}$(aq. anaerobic) = 9792–65280 h, based on estimated unacclimated aqueous aerobic biodegradation half-life (Howard et al. 1991).

Biotransformation: rate constant estimated to be 1×10^{-10} mL cell^{-1} h^{-1} for bacteria (Mabey et al. 1982).

Bioconcentration, Uptake (k_1) and Elimination (k_2) Rate Constants:

k_1 = 669 h^{-1}; k_2 = 0.144 h^{-1} (*Daphnia pulex*, Southworth et al. 1978)

log k_1 = 2.83 h^{-1}; log k_2 = –0.84 h^{-1} (*Daphnia pulex*, correlated as per Mackay & Hughes 1984, Hawker & Connell 1986)

k_1 = 138.6 mL g^{-1} h^{-1}; k_2 = 0.0022 h^{-1} (4°C, *P. hoyi*, Landrum 1988)

k_1 = 0.72–1.4 mg g^{-1} h^{-1}; k_2 = 0.0096 h^{-1} (freshwater oligochaete from sediment, Van Hoof et al. 2001)

Half-Lives in the Environment:

Air: $t_{1/2}$ = 1–3 h, based on estimated photolysis half-life in air (Howard et al. 1991);

$t_{1/2}$ = 4.20 h under simulated sunlight, $t_{1/2}$ = 1.35 h in simulated sunlight + ozone (0.2 ppm), $t_{1/2}$ = 2.88 h in dark reaction ozone (0.2 ppm), under simulated atmospheric conditions (Katz et al. 1979)

$t_{1/2}$ = 0.4 h for adsorption on soot particles in an outdoor Teflon chamber with an estimated rate constant k = 0.0265 min^{-1} at 1 cal cm^{-2} min^{-1}, 10 g/m^3 H$_2$O and 20°C (Kamens et al. 1988).

Surface water: photolysis $t_{1/2}$ = 0.59 h near surface water, $t_{1/2}$ = 3.7 d and 9.2 d in 5-m deep water body without and with sediment-water partitioning in full summer day, 40°N; photosensitized oxygenation $t_{1/2}$ = 2.6 h at near surface water, 40°N, midday, midsummer (Zepp & Schlotzhauer 1979)

$t_{1/2}$ = 0.20 d under summer sunlight (Mill & Mabey 1985);

$t_{1/2}$ = 1–3 h, based on estimated photolysis half-life in water, Howard et al. 1991);

photolysis $t_{1/2}$ = 0.46 h (reported in units of minutes) in aqueous solution when irradiated with a 500 W medium pressure mercury lamp (Chen et al. 1996).

Groundwater: $t_{1/2}$ = 4896–32640 h, based on estimated unacclimated aqueous aerobic biodegradation half-life (Howard et al. 1991).

Sediment: The uptake clearance from sediment was k = (0.005 ± 0.001)g of dry sediment·g^{-1} of organism·h^{-1}, and the elimination rate constants k = (0.0014 ± 0006)h^{-1} for amphipod, *P. hoyi* in Lake Michigan sediments at 4°C (Landrum 1989);

desorption $t_{1/2}$ = 11.1 d from sediment under conditions mimicking marine disposal (Zhang et al. 2000).

Soil: $t_{1/2}$ = 4–6250 d (Sims & Overcash 1983; quoted, Bulman et al. 1987);

$t_{1/2}$ = 240 d for 5 mg/kg treatment and 130 d for 50 mg/kg treatment (Bulman et al. 1987);
biodegradation k = 0.0026 d^{-1} with $t_{1/2}$ = 261 d for Kidman sandy loam soil, and k = 0.0043 d^{-1} with $t_{1/2}$ = 162 d for McLaurin sandy loam soil (Park et al. 1990);
$t_{1/2}$ ~ 2448–16320 h, based on aerobic die-away test data at 10–30°C (Howard et al. 1991);
$t_{1/2}$ > 50 d (Ryan et al. 1988).

Biota: depuration $t_{1/2}$ = 9 d by oysters (Lee et al. 1978);
elimination $t_{1/2}$ = 4.3–17.8 d from mussel *Mytilus edulis*; $t_{1/2}$ = 7–15.4 d from Oyster, $t_{1/2}$ = 8.0 d from clam *Mercenario mercenaria* (quoted, Meador et al. 1995).

TABLE 4.1.1.39.1
Reported aqueous solubilities of benz[a]anthracene at various temperatures and reported temperature dependence equation

$$S/(\mu g/kg) = a \cdot t^3 + b \cdot t^2 + c \cdot t + d \quad (1)$$

$$\ln x = A + B/(T/K) + C \cdot \ln(T/K) \quad (2)$$

May 1980		May et al. 1978b		May et al. 1983		Velapoldi et al. 1983	
generator column-HPLC		generator column-HPLC		generator column-HPLC		generator column-fluo.	
t/°C	S/g·m^{-3}	t/°C	S/g·m^{-3}	t/°C	S/g·m^{-3}	t/°C	S/g·m^{-3}
6.9	0.00299	25	0.0094	6.9	0.00299	10	0.00342
10.7	0.00378	29	0.0.0122	10.7	0.00378	15	0.00475
14.3	0.00479			11.0	0.00361	20	0.00669
19.3	0.00633			14.7	0.00558	25	0.00935
23.1	0.00837	temp dependence eq. 1		18.1	0.00634	30	0.01297
29.7	0.0127	S	µg/kg	19.3	0.00801		
		a	0.0003	23.6	0.00838		
temp dependence eq. 1		b	–0.0031	25.0	0.00862	eq. 2	mole fraction
S	µg/kg	c	0.1897	29.5	0.0124	A	–83.75982
a	0.0003	d	1.74	29.7	0.0127	B	41884.5
b	–0.0031					C	161.175
c	0.1897	ΔH_{sol}/(kJ mol^{-1}) = 44.81					
d	1.74	measured between 5–30°C				ΔH_{sol}/(kJ mol^{-1}) = 49.0	
						at 25°C	
ΔH_{sol}/(kJ mol^{-1}) = 44.81							
measured between 5–30°C							

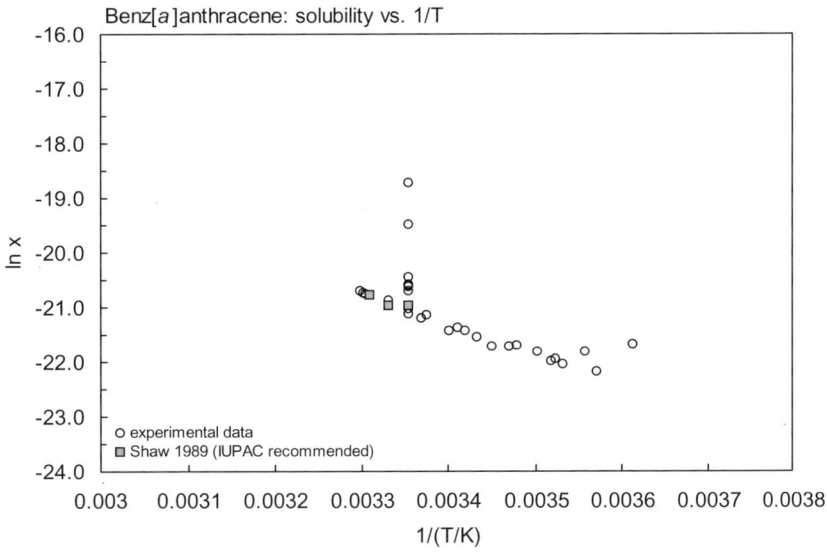

FIGURE 4.1.1.39.1 Logarithm of mole fraction solubility (ln x) versus reciprocal temperature for benz[a]anthracene.

TABLE 4.1.1.39.2
Reported vapor pressures of benz[a]anthracene at various temperatures and the coefficients for the vapor pressure equations

$$\log P = A - B/(T/K) \quad (1)$$
$$\log P = A - B/(C + t/°C) \quad (2)$$
$$\log P = A - B/(C + T/K) \quad (3)$$
$$\log P = A - B/(T/K) - C \cdot \log(T/K) \quad (4)$$

$$\ln P = A - B/(T/K) \quad (1a)$$
$$\ln P = A - B/(C + t/°C) \quad (2a)$$

Kelley & Rice 1964		Murray et al. 1972		de Kruif 1980		Sonnefeld et al. 1983	
effusion-electrobalance		Knudsen effusion		torsion-, effusion method		generator column-HPLC	
t/°C	P/Pa	t/°C	P/Pa	t/°C	P/Pa	t/°C	P/Pa
data represented by		data presented by graph and		98.07	0.1	13.81	8.05 × 10⁻⁵
		eq. 2	P/atm	105.21	0.2	13.81	6.06 × 10⁻⁵
eq. 1	P/mmHg	A	10.045	109.51	0.3	13.81	1.13 × 10⁻⁵
A	11.528	B	5925	112.62	0.4	25.1	2.66 × 10⁻⁵
B	5461	temp range 330–390 K		115.06	0.5	25.1	2.56 × 10⁻⁵
measured range 104–127°C				117.09	0.6	25.1	2.81 × 10⁻⁵
$\Delta H_{subl}/(kJ\,mol^{-1}) = 104.56$				118.82	0.7	40.12	1.39 × 10⁻⁴
		$\Delta H_{subl}/(kJ\,mol^{-1}) = 113.5$		120.32	0.8	40.12	1.41 × 10⁻⁴
mp/°C	160–161.5			121.66	0.9	40.12	1.36 × 10⁻⁴
				122.87	1.0	40.75	1.31 × 10⁻⁴
				25.0	7.3 × 10⁻⁶	40.85	1.17 × 10⁻⁴
					extrapolated	40.85	1.27 × 10⁻⁴
						40.85	1.21 × 10⁻⁴
Hoyer & Peperle 1958				$\Delta H_{sub}/(kJ\,mol^{-1}) = 113$		49.56	3.87 × 10⁻⁴
effusion method						49.56	3.85 × 10⁻⁴
t/°C	P/Pa					49.56	3.88 × 10⁻⁴
						34.93	2.69 × 10⁻⁴
data presented by equation.						25.0	2.80 × 10⁻⁵

(*Continued*)

TABLE 4.1.1.39.2 (Continued)

Kelley & Rice 1964		Murray et al. 1972		de Kruif 1980		Sonnefeld et al. 1983	
effusion-electrobalance		Knudsen effusion		torsion-, effusion method		generator column-HPLC	
t/°C	P/Pa	t/°C	P/Pa	t/°C	P/Pa	t/°C	P/Pa
eq. 1	P/mmHg					eq. 1	P/Pa
A	13.68					A	9.684
B	6250					B	4246.51
for temp range 60–120°C						$\Delta H_{subl}/(kJ\ mol^{-1}) = 51.83$	
						for temp range 10–50°C	

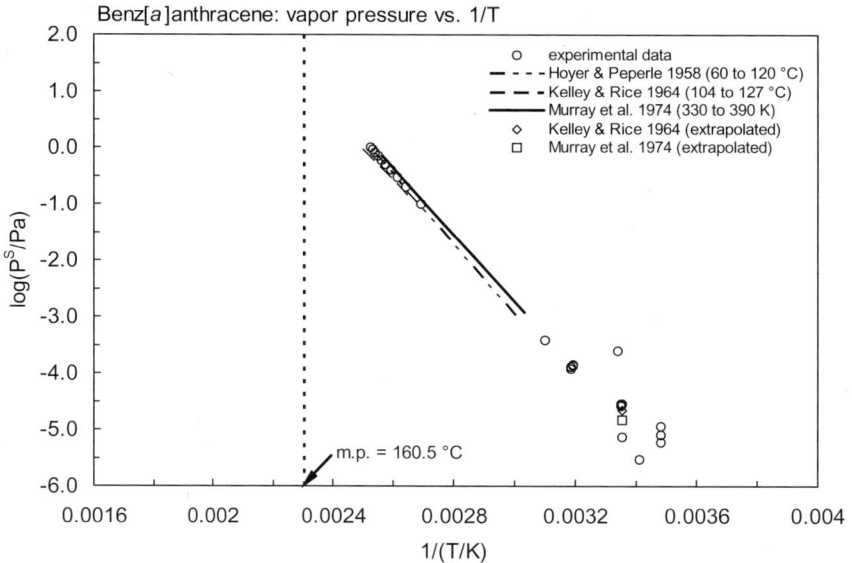

FIGURE 4.1.1.39.2 Logarithm of vapor pressure versus reciprocal temperature for benz[a]anthracene.

TABLE 4.1.1.39.3
Reported Henry's law constants of benz[a]anthracene at various temperatures and temperature dependence equations

$$\ln K_{AW} = A - B/(T/K) \quad (1)$$
$$\ln (1/K_{AW}) = A - B/(T/K) \quad (2)$$
$$\ln (k_H/atm) = A - B/(T/K) \quad (3)$$
$$\ln [H/(Pa\ m^3/mol)] = A - B/(T/K) \quad (4)$$
$$K_{AW} = A - B \cdot (T/K) + C \cdot (T/K)^2 \quad (5)$$

$$\log K_{AW} = A - B/(T/K) \quad (1a)$$
$$\log (1/K_{AW}) = A - B/(T/K) \quad (2a)$$

$$\ln [H/(atm \cdot m^3/mol)] = A - B/(T/K) \quad (4a)$$

	Bamford et al. 1999	
	gas stripping-GC/MS	
t/°C	H/(Pa m³/mol)	H/(Pa m³/mol)
		average
4.1	0.10, 0.22	0.15
11.0	0.24, 0.41	0.31
18.0	0.50, 0.79	0.63
25.0	0.91, 1.64	1.22
31.0	1.43, 3.13	2.11

$\ln K_{AW} = A - B/(T/K)$

TABLE 4.1.1.39.3 (*Continued*)

t/°C	Bamford et al. 1999 gas stripping-GC/MS H/(Pa m³/mol)	H/(Pa m³/mol)
A	19.124	
B	7986.5	
enthalpy, entropy change:		
$\Delta H/(kJ \cdot mol^{-1}) = 66.4 \pm 6.9$		
$\Delta S/(J \cdot K^{-1}\, mol^{-1}) = 159$		
at 25°C		

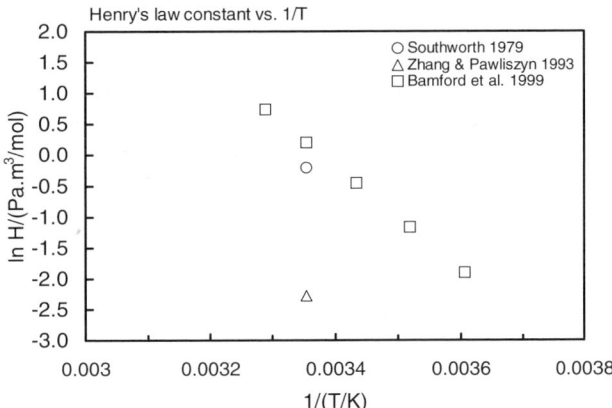

FIGURE 4.1.1.39.3 Logarithm of Henry's law constant versus reciprocal temperature for benz[a]anthracene.

4.1.1.40 Benzo[b]fluoranthene

Common Name: Benzo[b]fluoranthene
Synonym: 2,3-benzofluoranthene, 3,4-benzofluoranthene, benz[e]acephenanthrylene, B[b]F
Chemical Name: 2,3-benzofluoranthene
CAS Registry No: 205-99-2
Molecular Formula: $C_{20}H_{12}$
Molecular Weight: 252.309
Melting Point (°C):
    168   (Bjørseth 1983; Pearlman et al. 1984; Lide 2003)
Boiling Point (°C):
    481   (Bjørseth 1983)
Density (g/cm³ at 20°C):
Molar Volume (cm³/mol):
    222.8   (Ruelle & Kesselring 1997; Passivirta et al. 1999)
    268.9   (calculated-Le Bas method at normal boiling point)
Enthalpy of Fusion, ΔH_{fus} (kJ/mol):
Entropy of Fusion, ΔS_{fus} (J/mol K):
    56.5   (Passivirta et al. 1999)
Fugacity Ratio at 25°C (assuming ΔS_{fus} = 56 J/mol K), F: 0.0395 (mp at 168°C)

Water Solubility (g/m³ or mg/L at 25°C or as indicated and reported temperature dependence equations):
    0.0015   (generator column-HPLC/fluorescence, Wise et al. 1981)
    0.0015   (average lit. value, Pearlman et al. 1984)
    0.00109   (generator column-HPLC/fluo., De Maagd et al. 1998)
    $\log [S_L/(mol/L)] = -0.351 - 1303/(T/K)$ (supercooled liquid, Passivirta et al. 1999)

Vapor Pressure (Pa at 25°C or as indicated and reported temperature dependence equations):
    6.67×10^{-5} (20°C, estimated, Callahan et al. 1979)
    2.12×10^{-5} (Yamasaki et al. 1984)
    5.0×10^{-8}; 1.30×10^{-6} (quoted solid P_S from Mackay et al. 1992; converted to supercooled liquid P_L with fugacity ratio F, Passivirta et al. 1999)
    $\log (P_S/Pa) = 12.43 - 5880/(T/K)$ (solid, Passivirta et al. 1999)
    $\log (P_L/Pa) = 9.48 - 4578/(T/K)$ (supercooled liquid, Passivirta et al. 1999)
    7.55×10^{-6} (supercooled liquid P_L, calibrated GC-RT correlation, Lei et al. 2002)
    $\log (P_L/Pa) = -4682/(T/K) + 10.58$, $\Delta H_{vap.} = -89.7$ kJ·mol^{-1} (GC-RT correlation, Lei et al. 2002)

Henry's Law Constant (Pa m³/mol at 25°C or as indicated and reported temperature dependence equations.):
    0.051   (20°C, gas stripping-HPLC/fluorescence, measured range 10–55°C, ten Hulscher et al. 1992)
    $\log [H/(Pa\ m^3/mol)] = 9.83 - 3274/(T/K)$ (Passivirta et al. 1999)
    0.0485   (20°C, selected from reported experimentally measured values, Staudinger & Roberts 1996, 2001)
    $\log K_{AW} = 2.955 - 2245/(T/K)$, (van't Hoff eq. derived from literature data, Staudinger & Roberts 2001)

Octanol/Water Partition Coefficient, $\log K_{OW}$:
    5.78   (HPLC-RT correlation, Wang et al. 1986)
    5.78   (recommended, Sangster 1989, 1993)

Polynuclear Aromatic Hydrocarbons (PAHs) and Related Aromatic Hydrocarbons

Octanol/Air Partition Coefficient, log K_{OA}:

Bioconcentration Factor, log BCF:
- 5.15 (microorganisms-water, Mabey et al. 1982)
- 4.00 (*Daphnia magna*, Newsted & Giesy 1987)
- 0.959, 0.230 (*Polychaete sp, Capitella capitata*, Bayona et al. 1991)

Sorption Partition Coefficient, log K_{OC} at 25°C or as indicated:
- 6.182; 6.00, 6.18 (sediment: concn ratio C_{sed}/C_W; concn-based coeff., areal-based coeff. of flux studies of sediment/water boundary layer, Helmstetter & Alden 1994)
- 5.45 (log K_{DOC} - Aldrich humic acid, RP-HPLC, Ozretich et al. 1995)
- 6.57 (10°C), 6.55, 6.61 (20°C), 6.26 (35°C), 6.44, 6.45 (45°C) (log K_{DOC}, dissolved organic material from lake, gas-purge technique-HPLC/fluorescence, Lüers & ten Hulscher 1996)
- 6.20 (20°C, log K_{DOC}, particulate organic material from lake, Lüers & ten Hulscher 1996)
- 6.15–8.02; 5.70–7.50 (range, calculated from sequential desorption of 11 urban soils; lit. range, Krauss & Wilcke 2001, for benzo[$b + j + k$]fluoranthenes)
- 5.91; 6.50, 6.26, 6.68 (20°C, batch equilibrium, A2 alluvial grassland soil; calculated values of expt 1,2,3-solvophobic approach, Krauss & Wilcke 2001, for benzo[$b + j + k$]fluoranthenes)

Environmental Fate Rate Constants, k or Half-Lives, $t_{1/2}$:

Volatilization:

Photolysis: atmospheric and aqueous $t_{1/2}$ = 8.7–720 h, based on measured rate of photolysis in heptane irradiated with light > 290 nm (Howard et al. 1991);
- first order daytime decay rate constants: k = 0.0065 min^{-1} for 1000–2000 ng/mg soot particles loading and k = 0.0090 min^{-1} with 30–350 ng/mg loading (Kamens et al. 1988);
- $t_{1/2}$(obs.) = 4.31 h, $t_{1/2}$(calc) = 1.49 h predicted by QSPR in atmospheric aerosol (Chen et al. 2001)

Photodegradation k = 3 × 10^{-5} s^{-1} in surface water during the summertime at mid-latitude (Fasnacht & Blough 2002).

Oxidation: rate constant k = 4 × 10^7 M^{-1} h^{-1} for singlet oxygen and k = 5 × 10^3 M^{-1} h^{-1} for peroxy radical (Mabey et al. 1982);
- photooxidation $t_{1/2}$ = 1.43–14.3 h, based on estimated rate constant for reaction with hydroxyl radical in air (Howard et al. 1991).

Hydrolysis: not hydrolyzable (Mabey et al. 1982; no hydrolyzable groups (Howard et al. 1991).

Biodegradation:
- aerobic $t_{1/2}$ = 8640–14640 h, based on aerobic soil die-away test data (Coover & Sims 1987; quoted, Howard et al. 1991);
- k = 0.0024 d^{-1} with $t_{1/2}$ = 294 d for Kidman sandy loam and k = 0.0033 d^{-1} with $t_{1/2}$ = 211 d for McLarin sandy loam all at –0.33 bar soil moisture (Park et al. 1990);
- $t_{1/2}$(aq. anaerobic) = 34560–58560 h, based on estimated unacclimated aqueous aerobic degradation half-life (Howard et al. 1991).

Biotransformation: estimated to be 3 × 10^{-12} mL cell^{-1} h^{-1} for bacteria (Mabey et al. 1982).

Bioconcentration, Uptake (k_1) and Elimination (k_2) Rate Constants:
- k_1 = 0.11–0.38 mg g^{-1} h^{-1}; k_2 = 0.0029 h^{-1} (freshwater oligochaete from sediment, Van Hoof et al. 2001)

Sorption-Desorption Rate constants: desorption rate constant k = 0.016 d^{-1} with $t_{1/2}$ = 42.4 d from sediment under conditions mimicking marine disposal (Zhang et al. 2000).

Half-Lives in the Environment:

Air: $t_{1/2}$ = 1.43–14.3 h, based on estimated photooxidation half-life in air (Howard et al. 1991)
- half-lives under simulated atmospheric conditions: simulated sunlight – $t_{1/2}$ = 8.70 h, simulated sunlight + ozone (0.2 ppm) $t_{1/2}$ = 4.20 h, dark reaction ozone (0.2 ppm) $t_{1/2}$ = 52.70 h (Katz et al. 1979; quoted, Bjørseth & Olufsen 1983);
- $t_{1/2}$ = 1.3 h for adsorption on soot particles in an outdoor Teflon chamber with an estimated rate constant k = 0.0091 min^{-1} at 1 cal cm^{-2} min^{-1}, 10 g/m^3 H$_2$O and 20°C (Kamens et al. 1988).

Surface water: $t_{1/2}$ = 8.7–720 h, based on estimated aqueous photolysis half-life (Lane & Katz 1977; Muel & Saguem 1985; quoted, Howard et al. 1991).

Groundwater: $t_{1/2}$ = 17280–29280 h, based on estimated unacclimated aqueous aerobic biodegradation half-life (Howard et al. 1991).
Sediment: desorption $t_{1/2}$ = 42.4 d from sediment under conditions mimicking marine disposal (Zhang et al. 2000).
Soil: biodegradation rate constant k = 0.0024 d^{-1} with $t_{1/2}$ = 294 d for Kidman sandy loam soil, and k = 0.0033 d^{-1} with $t_{1/2}$ = 211 d for McLaurin sandy loam soil (Park et al. 1990);
$t_{1/2}$ = 8640–14640 h, based on aerobic die-away test data (Coover & Sims 1987; quoted, Howard et al. 1991);
$t_{1/2}$ = 42 wk, 9.0 yr (quoted, Luddington soil, Wild et al. 1991).
Biota: elimination $t_{1/2}$ = 5.7–16.9 d from mussel *Mytilus edulis*; $t_{1/2}$ = 7.7 d from Oyster (isomer unspecified), $t_{1/2}$ = 3.9 d from clam *Mercenario mercenaria* (isomer unspecified) (quoted, Meador et al. 1995).

4.1.1.41 Benzo[*j*]fluoranthene

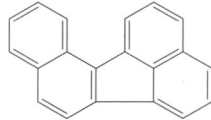

Common Name: Benzo[*j*]fluoranthene
Synonym: 7,8-benzofluoranthene, 10,11-fluoranthene
Chemical Name: benzo[*j*]fluoranthene
CAS Registry No: 205-82-3
Molecular Formula: $C_{20}H_{12}$
Molecular Weight: 252.309
Melting Point (°C):
    166   (Bjørseth 1983; Pearlman et al. 1984; Lide 2003)
Boiling Point (°C):
    480   (Bjørseth 1983)
Density (g/cm³ at 20°C):
Molar Volume (cm³/mol):
    222.8   (Ruelle & Kesselring 1997)
    268.9   (calculated-Le Bas method at normal boiling point)
Enthalpy of Fusion, ΔH_{fus} (kJ/mol):
Entropy of Fusion, ΔS_{fus} (J/mol K):
Fugacity Ratio at 25°C (assuming ΔS_{fus} = 56 J/mol K), F: 0.0414 (mp at 166°C)

Water Solubility (g/m³ or mg/L at 25°C):
    0.0025   (generator column-HPLC/fluorescence, Wise et al. 1981)
    0.0025   (average lit. value, Pearlman et al. 1984)

Vapor Pressure (Pa at 25°C):

Henry's Law Constant (Pa m³/mol):

Octanol/Water Partition Coefficient, log K_{OW}:
    6.44   (calculated-MCI χ as per Rekker & De Kort 1979, Ruepert et al. 1985)
    6.40   (Bayona et al. 1991)

Octanol/Air Partition Coefficient, log K_{OA}:

Bioconcentration Factor, log BCF:
    0.914; –0.222   (*Polychaete sp, Capitella capitata*, Bayona et al. 1991)

Sorption Partition Coefficient, log K_{OC}:
    6.15–8.02; 5.70–7.50 (range, calculated from sequential desorption of 11 urban soils; lit. range, Krauss & Wilcke 2001, for benzo[*b* + *j* + *k*]fluoranthenes)
    5.91; 6.50, 6.26, 6.68 (20°C, batch equilibrium, A2 alluvial grassland soil; calculated values of expt 1,2,3-solvophobic approach, Krauss & Wilcke 2001, for benzo[*b* + *j* + *k*]fluoranthenes)

Environmental Fate Rate Constants, k or Half-Lives, $t_{1/2}$:

Half-Lives in the Environment:
    Biota: elimination $t_{1/2}$ = 7.7 d from Oyster (isomer unspecified), $t_{1/2}$ = 3.9 d from clam *Mercenario mercenaria* (isomer unspecified) (quoted, Meador et al. 1995).

4.1.1.42 Benzo[k]fluoranthene

Common Name: Benzo[k]fluoranthene
Synonym: 8,9-benzofluoranthene, 11,12-benzofluoranthene, B[k]F
Chemical Name: 8,9-benzofluoranthene
CAS Registry No: 207-08-9
Molecular Formula: $C_{20}H_{12}$
Molecular Weight: 252.309
Melting Point (°C):
    217   (Weast 1977; Bjørseth 1983; Stephenson & Malanowski 1987; Lide 2003)
Boiling Point (°C):
    480   (Stephenson & Malanowski 1987; Lide 2003)
Density (g/cm³ at 20°C):
Molar Volume (cm³/mol):
    222.8   (Ruelle & Kesselring 1997; Passivirta et al. 1999)
    268.9   (calculated-Le Bas method at normal boiling point)
Enthalpy of Fusion, ΔH_{fus} (kJ/mol):
Entropy of Fusion, ΔS_{fus} (J/mol K):
    56.6   (Passivirta et al. 1999)
Fugacity Ratio at 25°C (assuming ΔS_{fus} = 56 J/mol K), F: 0.0131 (mp at 217°C)
    0.0126   (calculated, Passivirta et al. 1999)

Water Solubility (g/m³ or mg/L at 25°C and reported temperature dependence equations):
    0.0008   (generator column-HPLC/UV, Wise et al. 1981)
    0.00081   (average lit. value, Pearlman et al. 1984)
    0.00109   (generator column-HPLC/fluorescence, De Maagd et al. 1998)
    $\log [S_L/(mol/L)] = -0.351 - 1448/(T/K)$ (supercooled liquid, Passivirta et al. 1999)

Vapor Pressure (Pa at 25°C or as indicated and reported temperature dependence equations):
    1.28×10^{-8} (20°C, Radding et al. 1976)
    6.70×10^{-5} (20°C, Mabey et al. 1982)
    2.07×10^{-5} (Yamasaki et al. 1984)
    5.20×10^{-8}, 4.93×10^{-6} (20°C, lit. mean solid P_S, supercooled liquid value P_L, Bidleman & Foreman 1987)
    1.29×10^{-7} (extrapolated, Antoine eq., Stephenson & Malanowski 1987)
    $\log (P_S/kPa) = 12.8907 - 6792/(T/K)$; temp range 363–430 K (Antoine eq., Stephenson & Malanowski 1987)
    2.09×10^{-5} (supercooled liquid P_L, calculated from Yamasaki et al. 1984, Finizio et al. 1997)
    5.20×10^{-8}; 4.14×10^{-6} (quoted solid P_S from Mackay et al. 1992; converted to supercooled liquid P_L with fugacity ratio F, Passivirta et al. 1999)
    $\log (P_S/Pa) = 12.43 - 5874/(T/K)$ (solid, Passivirta et al. 1999)
    $\log (P_L/Pa) = 9.48 - 4427/(T/K)$ (supercooled liquid, Passivirta et al. 1999)
    8.96×10^{-6} (supercooled liquid P_L, calibrated GC-RT correlation, Lei et al. 2002)
    $\log (P_L/Pa) = -4623/(T/K) + 10.46$; $\Delta H_{vap.} = -88.5$ kJ·mol⁻¹ (GC-RT correlation, Lei et al. 2002)

Henry's Law Constant (Pa m³/mol at 25°C or as indicated and reported temperature dependence equations. Additional data at other temperatures designated * are compiled at the end of this section):
    0.111   (15°C, calculated, Baker & Eisenreich 1990)
    0.043*   (20°C, gas stripping-HPLC/fluorescence, measured range 10–55°C, ten Hulscher et al. 1992)
    $\log (H/(Pa\ m^3/mol)) = 9.83 - 2979/(T/K)$ (Passivirta et al. 1999)
    0.0422   (20°C, selected from reported experimentally measured values, Staudinger & Roberts 1996, 2001)
    $\log K_{AW} = 3.498 - 2421/(T/K)$ (van't Hoff eq. derived from literature data, Staudinger & Roberts 2001)

Octanol/Water Partition Coefficient, log K_{OW}:
- 6.84 (calculated-fragment const., Callahan et al. 1979)
- 6.06 (calculated-f const., Mabey et al. 1982)
- 6.44 (calculated-MCI χ as per Rekker & De Kort 1979)
- 6.40 (Bayona et al. 1991)
- 6.50 (calculated-S and mp, Capel et al. 1991)
- 7.20 (calculated-K_{OC}, Broman et al. 1991)
- 6.00 (selected, Mackay et al. 1992; quoted, Finizio et al. 1997)
- 6.30 (computed-expert system SPARC, Kollig 1995)
- 6.50–6.85; 6.73 (quoted lit. range; lit. mean, Meador et al. 1995)
- 6.11 (range 5.86–6.28) (shake flask/slow stirring-HPLC/fluo., De Maagd et al. 1998)
- 5.94; 6.16 (quoted lit.; calculated, Passivirta et al. 1999)

Octanol/Air Partition Coefficient, log K_{OA}:
- 11.19 (calculated, Finizio et al. 1997)

Bioconcentration Factor, log BCF:
- 5.15 (microorganisms-water, calculated from K_{OW}, Mabey et al. 1982)
- 4.12 (*Daphnia magna*, Newsted & Giesy 1987)

Sorption Partition Coefficient, log K_{OC} at 25°C or as indicated:
- 5.99 (sediments average, Kayal & Connell 1990)
- 7.00 (Baltic Sea particulate field samples, concn distribution-GC/MS, Broman et al. 1991)
- 6.80 (10°C), 6.74, 6.89 (20°C), 6.46 (35°C), 6.44, 6.45 (45°C) (log K_{DOC} - dissolved organic material from lake, gas-purge technique- HPLC/fluorescence, Lüers & ten Hulscher 1996)
- 6.30 (20°C, log K_{POC} - particulate organic material from lake, Lüers & ten Hulscher 1996)
- 6.04 (5.93–6.12), 5.47 (5.39–5.54) (sediments: Lake Oostvaardersplassen, Lake Ketelmeer, shake flask-HPLC/UV, de Maagd et al. 1998)
- 6.15–8.02; 5.70–7.50 (range, calculated from sequential desorption of 11 urban soils; lit. range, Krauss & Wilcke 2001, for benzo[$b + j + k$]fluoranthenes)
- 5.91; 6.50, 6.26, 6.68 (20°C, batch equilibrium, A2 alluvial grassland soil; calculated values of expt 1, 2, 3-solvophobic approach, Krauss & Wilcke 2001, for benzo[$b + j + k$]fluoranthenes)

Environmental Fate Rate Constants, k or Half-Lives, $t_{1/2}$:

Volatilization:

Photolysis: atmospheric and aqueous photolysis $t_{1/2}$ = 3.8–499 h, based on measured rate of photolysis in heptane under November sunlight and adjusted by ratio of sunlight photolysis half-lives in water versus heptane (Howard et al. 1991);

first-order daytime decay constants: k = 0.0047 min^{-1} for soot particles loading of 1000–2000 ng/mg and k = 0.0013 min^{-1} with 30–350 ng/mg loading (Kamens et al. 1988);

direct photolysis $t_{1/2}$(obs) = 0.88 h, $t_{1/2}$(calc) = 0.80 h predicted by QSPR in atmospheric aerosol (Chen et al. 2001)

Photodegradation k = 3 × 10^{-5} s^{-1} in surface water during the summertime at mid-latitude (Fasnacht & Blough 2002)

Oxidation: rate constant k = 4 × 10^7 M^{-1} h^{-1} for singlet oxygen and k = 5 × 10^3 M^{-1} h^{-1} for peroxy radical (Mabey et al. 1982);

photooxidation $t_{1/2}$ = 1.1–11 h, based on estimated rate constant for reaction with hydroxyl radical in air (Howard et al. 1991).

Hydrolysis: not hydrolyzable (Mabey et al. 1982);
no hydrolyzable groups (Howard et al. 1991).

Biodegradation:
aerobic $t_{1/2}$ = 21840–51360 h, based on aerobic soil die-away test data (Howard et al. 1991);

$t_{1/2}$(aq. anaerobic) = 87360–205440 h, based on estimated unacclimated aqueous aerobic biodegradation half-life (Howard et al. 1991).

Biotransformation: estimated to be 3×10^{-12} mL cell^{-1} h^{-1} for bacteria (Mabey et al. 1982).
Bioconcentration, Uptake (k_1) and Elimination (k_2) Rate Constants:

Half-Lives in the Environment:
Air: $t_{½}$ = 1.1–11 h, based on estimated photooxidation half-life in air (Howard et al. 1991);
$t_{½}$ = 14.10 h in simulated sunlight: $t_{½}$ = 3.90 h in simulated sunlight + ozone (0.2 ppm), $t_{½}$ = 34.90 h in dark reaction ozone (0.2 ppm) u) under simulated atmospheric conditions (Katz et al. 1979);
$t_{½}$ = 0.8 h for adsorption on soot particles in an outdoor Teflon chamber with an estimated rate constant k = 0.0138 min^{-1} at 1 cal cm^{-2} min^{-1} and 10 g/m^3 H$_2$O at 20°C (Kamens et al. 1988).
Surface water: $t_{½}$ = 3.8–499 h, based on photolysis half-life in water (Howard et al. 1991).
Groundwater: $t_{½}$ = 42680–102720 h, based on estimated unacclimated aqueous aerobic biodegradation half-life (Howard et al. 1991).
Sediment: desorption $t_{½}$ = 23.2 d from sediment under conditions mimicking marine disposal (Zhang et al. 2000).
Soil: $t_{½}$ = 21840–51360 h, based on aerobic soil die-away test data Howard et al. 1991);
$t_{½}$ > 50 d (Ryan et al. 1988);
mean $t_{½}$ = 8.7 yr for Luddington soil (Wild et al. 1991).
Biota: elimination $t_{½}$ = 11.9 d from mussel *Mytilus edulis*; $t_{½}$ = 7.7 d from Oyster (isomer unspecified), $t_{½}$ = 3.9 d from clam *Mercenario mercenaria* (isomer unspecified) (quoted, Meador et al. 1995).

TABLE 4.1.1.42.1
Reported Henry's law constants of benzo[k]fluoranthene at various temperatures and temperature dependence equations

$\ln K_{AW} = A - B/(T/K)$ (1)	$\log K_{AW} = A - B/(T/K)$ (1a)
$\ln (1/K_{AW}) = A - B/(T/K)$ (2)	$\log (1/K_{AW}) = A - B/(T/K)$ (2a)
$\ln (k_H/atm) = A - B/(T/K)$ (3)	
$\ln [H/(Pa\ m^3/mol)] = A - B/(T/K)$ (4)	$\ln [H/(atm \cdot m^3/mol)] = A - B/(T/K)$ (4a)
$K_{AW} = A - B \cdot (T/K) + C \cdot (T/K)^2$ (5)	

ten Hulscher et al. 1992

gas stripping-HPLC/fluorescence

t/°C	H/(Pa m^3/mol)
10.0	0.022
20.0	0.043
35.0	0.107
40.1	0.138
45.0	0.198
55.0	0.403

$\ln K_{AW} = -\Delta H_{vol}/RT + \Delta S_{vol}/R$
R = 8.314 Pa m^3 mol^{-1} K^{-1}

ΔS_{vol}	16.41
ΔH_{vol}	5893.7

enthalpy of volatilization:
$\Delta H_{vol}/(kJ \cdot mol^{-1}) = 49 \pm 1.9$
entropy of volatilization, ΔS
$T\Delta S_{vol}/(kJ \cdot mol^{-1}) = 40 \pm 4$
at 20°C

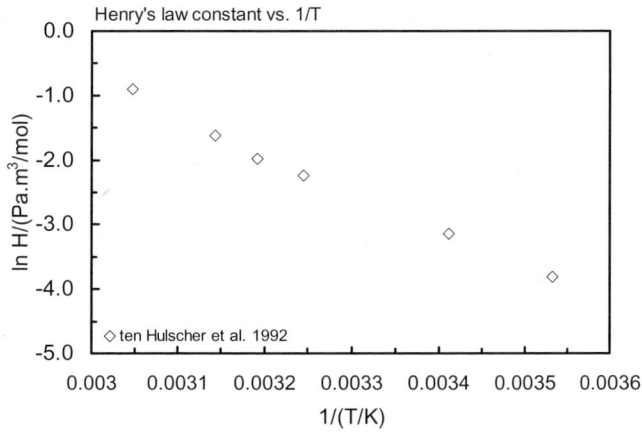

FIGURE 4.1.1.42.1 Logarithm of Henry's law constant versus reciprocal temperature for benzo[*k*]fluoranthrene.

4.1.1.43 Benzo[a]pyrene

Common Name: Benzo[a]pyrene
Synonym: BaP, B(a)P, 3,4-benzopyrene
Chemical Name: benzo[a]pyrene
CAS Registry No: 50-32-8
Molecular Formula: $C_{20}H_{12}$
Molecular Weight: 252.309
Melting Point (°C):
    181.1   (Lide 2003)
Boiling Point (°C):
    495   (Stephenson & Malanowski 1987; Dean 1992)
Density (g/cm³ at 20°C):
Molar Volume (cm³/mol):
    222.8   (Ruelle & Kesselring 1997; Passivirta et al. 1999)
    262.9   (calculated-Le Bas method at normal boiling point)
Enthalpy of Fusion, ΔH_{fus} (kJ/mol):
    17.324   (Ruelle & Kesselring 1997)
    8.49, 17.32; 25.61 (117.05, 181.05°C; total phase change enthalpy, Chickos et al. 1999)
Entropy of Fusion, ΔS_{fus} (J/mol K):
    38.5   (differential scanning calorimetry, Hinckley et al. 1990)
    21.77, 38.13 (117.05, 181.05°C, Chickos et al. 1999)
    42.35, 43.7   (exptl., calculated-group additivity method, total phase change entropy, Chickos et al. 1999)
    38.2   (Passivirta et al. 1999)
Fugacity Ratio at 25°C (assuming ΔS_{fus} = 56 J/mol K), F: 0.0294 (mp at 181.1°C)
    0.0903   (calculated, Passivirta et al. 1999)

Water Solubility (g/m³ or mg/L at 25°C or as indicated and reported temperature dependence equations. Additional data at other temperatures designated * are compiled at the end of this section):
    0.004   (27°C, nephelometry, Davis et al. 1942)
    0.0043   (shake flask-UV/fluorescence, Barone et al. 1967)
    0.0061   (average, Barone et al. 1967)
    0.0005   (20°C, shake flask-UV, Eisenbrand & Baumann 1970)
    0.000038, 0.0000606, 0.000038, 0.0000505 (HPLC-relative retention correlation, different stationary and mobile phases, Locke 1974)
    0.00121   (Haque & Schmedding 1975)
    0.0038   (shake flask-fluorescence, Mackay & Shiu 1977)
    0.0002   (Rossi 1977; Neff 1979)
    0.0012   (generator column-HPLC, Wise et al. 1981)
    0.00162*   (generator column-HPLC, measured range 10–30°C, May et al. 1983)
    0.00158*   (generator column-spectrofluorimetry, measured range 10–30°C, Velapoldi et al. 1983)
    0.0038   (selected value, Pearlman et al. 1984)
    0.00154*   (generator column-HPLC/UV, measured 8.0–25.0°C, Whitehouse 1984)
    0.0016   (generator column-HPLC/UV, Billington et al. 1988)
    0.000504   (shake flask-LSC, Eadie et al. 1990)
    0.00472   (shake flask-fluorescence, Haines & Sandler 1995)
    0.00182   (generator column-HPLC/fluorescence, De Maagd et al. 1998)
    0.00622 ± 0.00023; 0.0038, 0.0018 ± 0.0003 (shake flask-SPME (solid-phase micro-extraction)-GC; quoted lit. values; Paschke et al. 1999)

log S_L (mol/L) = –1.310 – 906.6/(T/K) (supercooled liquid, Passivirta et al. 1999)
ln x = –2.59638 – 6046.87/(T/K); temp range 5–50°C (regression eq. of literature data, Shiu & Ma 2000)

Vapor Pressure (Pa at 25°C and reported temperature dependence equations):
7.32×10^{-7} (effusion method, extrapolated, Murray et al. 1974)
log (P/mmHg) = 9.601 – 6181/(T/K); temp range 358–431 K (Knudsen effusion method, Murray et al. 1974)
1.12×10^{-4}, 1.50×10^{-5} (P_{GC}, GC-RT correlation with different GC columns, Bidleman 1984)
2.35×10^{-5} (supercooled liquid P_L, converted from literature P_S with ΔS_{fus} Bidleman 1984)
1.22×10^{-5} (Yamasaki et al. 1984)
3.2×10^{-7}, 1.23×10^{-5} (lit. mean, supercooled liq. value, Bidleman & Foreman 1987)
7.51×10^{-7} (extrapolated-Antoine eq., Stephenson & Malanowski 1987)
log (P_S/kPa) = 11.6067 – 6181/(T/K); temp range 358–431 K (Antoine eq., Stephenson & Malanowski 1987)
1.12×10^{-4}, 7.24×10^{-5} (P_{GC}, GC-RT correlation with different reference standards, Hinckley et al. 1990)
2.35×10^{-5}, 7.28×10^{-6} (supercooled liquid P_L, converted from literature P_S with different ΔS_{fus} Hinckley et al. 1990)
log (P_L/Pa) = 11.59 – 4989/(T/K) (liquid phase, GC-RT correlation, Clausius-Clapeyron eq., Hinckley et al. 1990)
1.17×10^{-5} (supercooled liquid P_L, calculated from Yamasaki et al 1984, Finizio et al. 1997)
7.51×10^{-7}, 7.45×10^{-7} (quoted exptl., effusion, Delle Site 1997)
7.01×10^{-7}; 3.51×10^{-6}, 4.73×10^{-7}, 2.25×10^{-7} (quoted lit., calculated-UNIFAC; GC-RT correlation, Delle Site 1997)
1.54×10^{-5}; 1.39×10^{-6} (quoted supercooled liquid P_L from Hinckley et al. 1990; converted to solid P_S with fugacity ratio F, Passivirta et al. 1999)
log (P_S/Pa) = 12.17 – 5371/(T/K) (solid, Passivirta et al. 1999)
log (P_L/Pa) = 10.71 – 4465/(T/K) (supercooled liquid, Passivirta et al. 1999)
5.90×10^{-6} (supercooled liquid P_L, calibrated GC-RT correlation, Lei et al. 2002)
log (P_L/Pa) = –4755/(T/K) + 10.72; $\Delta H_{vap.}$ = –91.0 kJ·mol^{-1} (GC-RT correlation, Lei et al. 2002)

Henry's Law Constant (Pa·m^3/mol at 25°C or as indicated and reported temperature dependence equations. Additional data at other temperatures designated * are compiled at the end of this section):
0.009   (15°C, calculated, Baker & Eisenreich 1990)
0.0079   (10°C, estimated, McLachlan et al. 1990)
0.034*   (20°C, gas stripping-HPLC/fluorescence, measured range 10–55°C, ten Hulscher et al. 1992)
0.074   (wetted wall column-GC, Altschuh et al. 1999)
log [H/(Pa·m^3/mol)] = 12.02 – 3558/(T/K) (Passivirta et al. 1999)
0.035   (20°C, selected from reported experimentally measured values, Staudinger & Roberts 1996, 2001)
log K_{AW} = 1.732 – 1927/(T/K) (van't Hoff eq. derived from literature data, Staudinger & Roberts 2001)

Octanol/Water Partition Coefficient, log K_{OW}:
6.04   (Radding et al. 1976)
6.31   (Smith et al. 1978)
5.99, 5.78 (calculated-fragment const., Hansch & Leo 1979)
6.34   (Steen & Karickhoff 1981)
6.50   (RP-TLC-k' correlation, Bruggeman et al. 1982)
6.20   (shake flask-GC, Hanai et al. 1982)
5.85–5.12; 5.88–6.04; 5.99, 6.00 ± 0.1 (23°C, shake flask- concentration ratio/UV spec.; shake flask-HPLC/UV; exptl. mean value, recommended value; Mallon & Harris 1984)
6.74, 7.77, 7.99 (HPLC-RT correlation, Sarna et al. 1984)
6.42   (HPLC-RT correlation, Rapaport et al. 1984)
5.97   (Hansch & Leo 1985)
6.78   (HPLC-RT correlation, Webster et al. 1985)
6.04   (HPLC-RT correlation, Wang et al. 1986)
6.44   (TLC retention time correlation, De Voogt et al. 1990)
6.35   (recommended, Sangster 1993)
5.97   (recommended, Hansch et al. 1995)

6.02 ± 0.19, 6.14 ± 0.71 (HPLC-k′ correlation: ODS column; Diol column, Helweg et al. 1997)
6.13 (5.91–6.28) (slow stirring-HPLC/fluorescence, De Maagd et al. 1998)
6.27 (shake flask-SPME solid-phase micro-extraction, Paschke et al. 1999)

Octanol/Air Partition Coefficient, log K_{OA}:
10.80 (calculated-K_{OW}/K_{AW}, Wania & Mackay 1996)
10.77 (calculated, Finizio et al. 1997)

Bioconcentration Factor, log BCF:
1.09; 2.22; 3.45 (steady-state, bluegills; midge larva; periphyton, Leversee et al. 1981)
4.74 (*P. hoyi*, Eadie et al. 1982)
3.90 (*Daphnia magna*, McCarthy 1983)
3.69, 4.45 (*Lepomis macrochirus*, bluegill sunfish, Spacie et al., 1983)
4.00 (activated sludge, Freitag et al. 1984)
3.42 (bluegills, McCarthy & Jimenez 1985)
2.35, 2.45 (bluegills-with dissolved humic material, McCarthy & Jimenez 1985)
2.68, 3.52, 4.0 (fish, algae, activated sludge, Freitag et al. 1985)
3.51 (worms, Frank et al. 1986)
6.95, 6.51 (*P. hoyi* of Lake Michigan interstitial waters, Landrum et al. 1985)
3.34 (*P. hoyi* of Government Pond of Grand Haven in Michigan, Landrum et al. 1985)
2.69 (Gobas et al. 1987)
4.11 (*Daphnia magna*, Newsted & Giesy 1987)
3.77 (10–20°C, *H. limbata*, Landrum & Poore 1988)
4.61, 3.86, 3.87 (4°C, *P. hoyi*, *S. heringianus*, *Mysis relicta*, Landrum & Poore 1988)
4.69, 3.93 (calculated for amphipods and mysids, Evans & Landrum 1989)
3.22–3.59; 2.96–3.32 (*Daphnia magna* in natural waters with humic substances, measured range; predicted range, Kokkonen et al. 1989)
1.140, –0.155 (*Polychaete sp, Capitella capitata*, Bayona et al. 1991)
6.22, 6.04; 3.68, 5.01, 4.90, 5.15 (oligochaetes; chironomid larvae, Bott & Standley 2000)

Sorption Partition Coefficient, log K_{OC} at 25°C or as indicated:
5.95 (Aldrich humic acid, RP-HPLC separation, Landrum et al. 1984)
4.59, 4.72, 4.26 (humic materials in natural water: Huron River 6.1% OC winter, Grand River 10.7% DOC spring, Lake Michigan 5.5% DOC spring, RP-HPLC separation method, Landrum et al. 1984)
6.66 (LSC, Eadie et al. 1990)
6.26 (sediments average, Kayal & Connell 1990)
8.30 (Baltic Sea particulate field samples, concn distribution-GC/MS, Broman et al. 1991)
7.0 (Rotterdam Harbor sediment 4.6% OC, batch sorption equilibrium, Hegeman et al. 1995)
6.00, 6.28, 6.17; 5.81 (marine sediments: Fort Point Channel, Spectacle Island, Peddocks Island; quoted lit., McGroddy & Farrington 1995)
5.93 (Aldrich humic acid, Ozretich et al. 1995)
6.54 (10°C), 6.46, 6.60 (20°C), 6.14 (35°C), 6.07, 6.09 (45°C) (log K_{DOC} - dissolved organic material from lake, gas-purge technique-HPLC/fluorescence, Lüers & ten Hulscher 1996)
6.30 (20°C, log K_{POC} - particulate organic material from lake, Lüers & ten Hulscher 1996)
6.27; 6.30 (humic acid, HPLC-k′ correlation; quoted lit., Nielsen et al. 1997)
5.72, 5.89, 5.51 (pH 5, 6.5, 8, humic acid from sediments of River Arno, De Paolis & Kukkonen 1997)
4.81, 4.87, 4.49 (pH 5, 6.5, 8, fulvic acid from sediments of River Arno, De Paolis & Kukkonen 1997)
5.54, 5.59, 5.37 (pH 5, 6.5, 8, HA + FA extracted from sediments of River Arno, De Paolis & Kukkonen 1997)
5.51, 5.74, 5.68 (pH 5, 6.5, 8, HA extracted from sediments of Tyrrenhian Sea, De Paolis & Kukkonen 1997)
4.93, 4.84, 4.85 (pH 5, 6.5, 8, FA extracted from sediments of Tyrrenhian Sea, De Paolis & Kukkonen 1997)
5.66, 5.46, 5.60 (pH 5, 6.5, 8, HA + FA from sediments of Tyrrenhian Sea, De Paolis & Kukkonen 1997)
5.22, 5.46, 5.60 (pH 5, 6.5, 8, HA extracted from water of River Arno, De Paolis & Kukkonen 1997)
4.67, 4.80, 4.45 (pH 5, 6.5, 8, FA extracted from water of River Arno, De Paolis & Kukkonen 1997)
5.21, 5.29, 5.18 (pH 5, 6.5, 8, HA + FA extracted from water of River Arno, De Paolis & Kukkonen 1997)
4.62, 4.52, 4.61 (pH 5, 6.5, 8, FA extracted from water of Tyrrenhian Sea, De Paolis & Kukkonen 1997)

5.99 (5.92–6.04), 5.53 (5.43–5.61) (sediments: Lake Oostvaardersplassen, Lake Ketelmeer, shake flask-HPLC/UV, de Maagd et al. 1998)

5.25–6.18 (Lake Michigan sediment, Kukkonen & Landrum 1998)

5.48–5.69, 5.56, 5.55, 5.49, 5.30 (log K_{DOC}: humic acid from Lake Hohlohsee in Black Forest, soil leachate, fulvic acid from brown coal-derived production effluent, fulvic acid from groundwater, fulvic acid from effluent of a waste water plant near Karlsruhe, Haitzer et al. 1999)

5.53 (Clay Creek sediment with organic matter content 0.45%, Bott & Standley. 2000)

6.23 (sediment: organic carbon OC -0.5%, average, Delle Site 2001)

6.39–8.17; 6.30–8.50 (range, calculated from sequential desorption of 11 urban soils; lit. range, Krauss & Wilcke 2001)

3.12; 6.67, 6.58, 6.79 (20°C, batch equilibrium method, A2 alluvial grassland soil; calculated values of expt 1,2,3-solvophobic approach, Krauss & Wilcke 2001)

6.15; 6.34 (Plym river sediment; plym sea sediment, batch equilibrium-LSC, Turner & Rawling 2002)

6.49; 6.31 (Carnon river sediment; Carnon sea sediment, batch equilibrium-LSC, Turner & Rawling 2002)

Environmental Fate Rate Constants, k or Half-Lives, $t_{1/2}$:

Volatilization: aquatic fate rate of 300 h^{-1} with $t_{1/2}$ = 22 h (Callahan et al. 1979);

half-lives predicted by one compartment model: $t_{1/2}$ = 140 h in river water, $t_{1/2}$ = 350 h in eutrophic pond, $t_{1/2}$ = 700 h in eutrophic lake and oligotrophic lake (Smith et al. 1978);

calculated $t_{1/2}$ = 1500 h for a river of 1-m deep with water velocity of 0.5 m s^{-1} and wind velocity of 1 m/s (Southworth 1979; Herbes et al. 1980);

sublimation rate constant from glass surface of < 1 × 10^{-5} s^{-1} was measured at 24°C at an airflow rate of 3 L/min (Cope & Kalkwarf 1987).

Photolysis: photolysis $t_{1/2}$ = 2 h in methanol solution when irradiated at 254 nm (Lu et al. 1977);

k = 0.58 h^{-1} for winter at midday at 40°N latitude (Smith et al. 1978);

direct photochemical transformation $t_{1/2}$(calc) = 0.54 h, computed near-surface water, latitude 40°N, midday, midsummer and photolysis $t_{1/2}$ = 3.2 d and 13 d in 5-m deep inland water body without and with sediment-water partitioning, respectively, to top cm of bottom sediment over full summer day, 40°N (Zepp & Schlotzhauer 1979)

k(aq.) = 2.8 × 10^{-4} s^{-1} with $t_{1/2}$ = 1–2 h (Callahan et al. 1979)

photolytic $t_{1/2}$(aq) = 0.53 h (quoted of EPA Report 600/7-78-074, Haque et al. 1980)

k = 1.30 h^{-1} (Zepp 1980)

half-lives predicted by one compartment model: $t_{1/2}$ = 3.0 h in river water based on the photolysis rates estimated for summer sunlight, $t_{1/2}$ = 7.5 h in eutrophic pond or eutrophic lake, and $t_{1/2}$ = 1.5 h in oligotrophic lake (Smith et al. 1978; quoted, Harris 1982)

k = 2.8 × 10^{-4} s^{-1} with $t_{1/2}$ = 1–2 h (Callahan et al., 1979)

calculated direct photolysis k = 3.86 × 10^{-4} s^{-1} in late January with $t_{1/2}$ = 0.69 h in pure water at 366 nm and in sunlight at 23–28°C and k = 1.05 × 10^{-5} s^{-1} in mid-December with $t_{1/2}$ = 1.1 h at 313 nm with 1–20% acetonitrile as cosolvent in filter-sterilized natural water (Mill et al. 1981)

$t_{1/2}$ = 0.37–1.1 h, based on estimated photolysis half-life in air (Howard et al. 1991)

sunlight photolysis $t_{1/2}$ = 0.045 d for mid-December (Mill & Mabey 1985)

half-lives on different atmospheric particulate substrates (approx. 25 µg/g on substrate):$t_{1/2}$ = 4.7 h on silica gel, $t_{1/2}$ = 1.4 h on alumina and $t_{1/2}$ = 31 h on fly ash (Behymer & Hites 1985)

ozonation rate constant k < 6.1 × 10^{-4} m/s was measured at 24°C with [O_3] = 0.16 ppm and light intensity of 1.3 kW/m^2 (Cope & Kalkwarf 1987)

first order daytime decay k = 0.0090 min^{-1} for soot particles loading of 1000–2000 ng/mg and k = 0.0211 min^{-1} with 30–350 ng/mg loading (Kamens et al. 1988)

photodegradation half-life was found ranging from 1 h in summer to days in winter (Valerio et al. 1991)

photodegradation k = 0.0322 min^{-1} and $t_{1/2}$ = 0.35 h in ethanol-water (3:7, v/v) solution for initial concentration of 2.5 ppm by high pressure mercury lamp or sunlight (Wang et al. 1991)

pseudo-first-order direct photolysis k(exptl) = 0.0322 min^{-1} with the calculated $t_{1/2}$ = 0.35 h and the predicted $t_{1/2}$ = 0.0416 min^{-1} calculated by QSPR method, in aqueous solution when irradiated with a 500 W medium pressure mercury lamp (Chen et al. 1996)

direct photolysis $t_{1/2}$(obs) = 0.50 h, $t_{1/2}$(calc) = 0.57 h predicted by QSPR method in atmospheric aerosol (Chen et al. 2001) photochemical degradation under atmospheric conditions: $k = (1.18 \pm 0.50) \times 10^{-4}$ s^{-1} and $t_{½} = (1.63 \pm 0.48)$h in diesel particulate matter, rate constant $k = (3.09 \pm 0.23) \times 10^{-5}$ s^{-1} and $t_{½} = (6.22 \pm 0.51)$h in diesel particulate matter/soil mixture, and $t_{½} = 0.35$ to 1.62 h in various soil components using a 900-W photo-irradiator as light source; rate constant $k = (2.92 \pm 0.20) \times 10^{-5}$ s^{-1} and $t_{½} = (6.59 \pm 0.49)$h in diesel particulate matter using a 300-W light source (Matsuzawa et al. 2001)

Photodegradation $k = 2.1 \times 10^{-3}$ s^{-1} in surface water during the summertime at mid-latitude (Fasnacht & Blough 2002)

Oxidation: rate constant k, for gas-phase second order rate constants, k_{OH} for reaction with OH radical, k_{NO_3} with NO$_3$ radical and k_{O_3} with O$_3$ or as indicated, *data at other temperatures see reference:

$t_{½} > 340$ h in river water, eutrophic pond or lake and oligotrophic lake, half-lives predicted by one compartment model (Smith et al. 1978)

k(aquatic fate rate) = 1680 M^{-1} s^{-1}, with half-life of 96 h (Callahan et al. 1979)

$t_{½} = 1500$ h for photosensitized oxygenation with singlet oxygen at near-surface natural water, 40°N, midday, midsummer (Zepp & Schlotzhauer 1979)

$k = 5 \times 10^8$ M^{-1} h^{-1} for singlet oxygen and 2×10^4 M^{-1} h^{-1} for peroxy radical (Mabey et al. 1982)

$k = 0.62 \times 10^4$ dm^3 mol^{-1} s^{-1} for the reaction with O$_3$ in water at pH 7 and 25°C with $t_{½} = 1.0$ s in presence of 10^{-4} M of ozone at pH 7 (Butković et al. 1983)

k(aq.) = 3.6×10^{-4} h^{-1} with $t_{½} = 0.5$ h under natural sunlight conditions; k(aq.) = 1.9×10^3 M^{-1} h^{-1} with $t_{½} = 4.3$ d free-radicals oxidation in air-saturated water (NRCC 1983)

photooxidation $t_{½} = 0.428–4.28$ h, based on estimated rate constant for reaction with hydroxyl radical in air (Atkinson 1987; quoted, Howard et al. 1991)

$k_{HO.}$(calc) = 1×10^{10} M^{-1} s^{-1} with hydroxyl radical in aqueous solutions (Haag & Yao 1992)

Hydrolysis: not hydrolyzable (Mabey et al. 1982; Howard et al. 1991).

Biodegradation:

$t_{½} > 10000$ h (quoted, Smith et al. 1978)

$k = 0.2–0.9$ μmol^{-1} mg^{-1} for bacterial protein (Callahan et al. 1979)

$k = 3.4 \times 10^{-4}$ h^{-1} with $t_{½} = 83$ h for mixed bacterial populations in stream sediment (NRCC 1983)

$k < 3 \times 10^{-5}$ h^{-1} with $t_{½} > 2.5$ yr; $k < 3.0 \times 10^{-5}$ h^{-1} with $t_{½} > 2.5$ yr for mixed bacterial populations in oil-contaminated and pristine stream sediments (NRCC 1983)

$k = 3.5 \times 10^{-5}$ h^{-1} estimated in water and soil (Ryan & Cohen 1986)

$t_{½}$(aq. aerobic) = 57 d to 1.45 yr at 10–30°C, soil die-away test (Coover & Sims 1987; quoted, Howard et al. 1991); $k = 0.0022$ d^{-1} with $t_{½} = 309$ d for Kidman sandy loam and $k = 0.0030$ d^{-1} with $t_{½} = 229$ d for McLarin sandy loam all at –0.33 bar soil moisture (Park et al. 1990)

$t_{½}$(aq. anaerobic) = 228 d to 5.8 yr, based on estimated unacclimated aqueous aerobic biodegradation half-life (Coover & Sims 1987; quoted, Howard et al. 1991).

Biotransformation: estimated to be 3×10^{-12} mL cell^{-1} h^{-1} for bacteria (Mabey et al. 1982).

Bioconcentration, Uptake (k_1) and Elimination (k_2) Rate Constants:

$k_1 = 49$ h^{-1}; $k_2 = 0.010$ h^{-1} (bluegill sunfish, Spacie et al. 1983)

$k_1 = 131.1$ mL g^{-1} h^{-1}; $k_2 = 0.0033$ h^{-1} (4°C, *P. hoyi*, Landrum 1988)

$k_1 = 81.3$ h^{-1}; $k_2 = 0.014$ h^{-1} (10–20°C, *H. limbata*, Landrum & Poore 1988)

$k_1 = 16.8$ h^{-1}; $k_2 = 0.0016$ h^{-1} (4°C, *P. hoyi*, Landrum & Poore 1988)

$k_1 = 87.8$ h^{-1}; $k_2 = 0.012$ h^{-1} (4°C, *S. heringianus*, quoted, Landrum & Poore 1988)

$k_1 = 112.0$ h^{-1}; $k_2 = 0.013$ h^{-1} (4°C, *Mysis relicta*, quoted, Landrum & Poore 1988)

$k_1 = 75.9$ mL g^{-1} h^{-1} (*Pontoporeia hoyi*, Evans & Landrum 1989)

$k_1 = 39.9$ mL g^{-1} h^{-1} (*Mysis relicta*, Evans & Landrum 1989)

$k_2 = 0.0017$ h^{-1} (*Amphipods*, Evans & Landrum 1989)

$k_2 = 0.0047$ h^{-1} (*Mysids*, Evans & Landrum 1989)

$k_1 = 0.11–0.36$ mg g^{-1} h^{-1}; $k_2 = 0.0032$ h^{-1} (freshwater oligochaete from sediment, Van Hoof et al. 2001)

Half-Lives in the Environment:

Air: $t_{½} = 0.37–1.1$ h, based on estimated photolysis half-life (Lyman et al. 1982; quoted, Howard et al. 1991); half-lives under simulated atmospheric conditions: $t_{½} = 5.30$ h in simulated sunlight, $t_{½} = 0.58$ h in simulated sunlight + ozone (0.2 ppm), $t_{½} = 0.62$ h in dark reaction ozone (0.2 ppm) (Katz et al. 1979; quoted, Bjørseth & Olufsen 1983);

$t_{½} = 14$ h with a steady-state concn of tropospheric ozone of 2×10^{-9} M in clean air (Butković et al. 1983);

photooxidation $t_{1/2}$ = 0.428–4.28 h, based on estimated rate constant for reaction with hydroxyl radical in air (Atkinson 1987; quoted, Howard et al. 1991);

$t_{1/2}$ = 0.5 h for adsorption on soot particles in an outdoor Teflon chamber with an estimated k = 0.0234 min^{-1} at 1 cal cm^{-2} min^{-1}, 10 g m^{-3} H$_2$O and 20°C (Kamens et al. 1988).

Surface Water: $t_{1/2}$ = 2 h in methanol solution irradiated at 254 nm (Lu et al. 1977);

half-lives predicted by one compartment model: $t_{1/2}$ > 340 h in river water, eutrophic pond or lake and oligotrophic lake (Smith et al. 1978);

very slow, not an important process (Callahan et al. 1979);

computed near-surface half-life for direct photochemical transformation of a natural water body $t_{1/2}$ = 0.54 h at latitude 40°N, midday, midsummer, and direct photolysis, $t_{1/2}$ = 3.2 d (no sediment-water partitioning) and $t_{1/2}$ = 13 d (with sediment-water partitioning) in a 5-m deep inland water body (Zepp & Schlotzhauer 1979);

$t_{1/2}$ = 0.37–1.1 h, based on photolysis half-life in water (Lyman et al. 1982; quoted, Howard et al. 1991);

$t_{1/2}$ = 1.0 s in presence of 10^{-4} M of ozone at pH 7 (Butković et al. 1983);

$t_{1/2}$ = 4.3 d free-radical oxidation in air-saturated water (NRCC 1983);

$t_{1/2}$ = 0.045 d under mid-December sunlight (Mill & Mabey 1985);

photolysis $t_{1/2}$ = 0.35 h in aqueous solution when irradiated with a 500 W medium pressure mercury lamp (Chen et al. 1996).

Groundwater: $t_{1/2}$ = 2736–25440 h, based on estimated unacclimated aqueous aerobic biodegradation half-life (Howard et al. 1991).

Sediment: uptake clearance from sediment k = (0.0023 ± 0.001)g of dry sediment·g^{-1} of organism·h^{-1} for amphipod, *P. hoyi* in Lake Michigan sediments at 4°C (Landrum 1989);

desorption $t_{1/2}$ = 19.5 d from sediment under conditions mimicking marine disposal (Zhang et al. 2000).

Soil: $t_{1/2}$ > 2 d (Sims & Overcash 1983; quoted, Bulman et al. 1987);

$t_{1/2}$ = 347 d for 5 mg/kg treatment and $t_{1/2}$ = 218 d for 50 mg/kg treatment (Bulman et al. 1987);

biodegradation k = 0.002 d^{-1} with $t_{1/2}$ = 309 d for Kidman sandy loam soils and k = 0.0030 d^{-1} with $t_{1/2}$ = 229 d for Mclaurin sandy loam soils (Park et al. 1990);

$t_{1/2}$ = 1368–12720 h, based on aerobic soil dieaway test data at 10–30°C (Groenewegen & Stolp 1976; Coover & Sims 1987; quoted, Howard et al. 1991);

$t_{1/2}$ > 50 d (Ryan et al. 1988);

$t_{1/2}$ = 0.3 to > 300 wk, 8.2 yr (literature, Luddington soil, Wild et al. 1991).

Biota: depuration $t_{1/2}$ = 18 d by oysters (Lee et al. 1978; quoted, Verschueren 1983);

$t_{1/2}$ = 67 h in bluegill sunfish (Spacie et al. 1983);

depuration $t_{1/2}$ = 52 h in *s. heringianus* (Frank et al. 1986);

calculated half-lives in different tissues of sea bass: $t_{1/2}$ = 12.4 d for fat, $t_{1/2}$ = 6.5 d for kidney, $t_{1/2}$ = 5.1 d for kidney, $t_{1/2}$ = 5.1 d for intestine, $t_{1/2}$ = 4.8 d for gallbladder, $t_{1/2}$ = 4.5 d for spleen, $t_{1/2}$ = 2.9 d for muscle, $t_{1/2}$ = 2.4 d for whole body, $t_{1/2}$ = 2.3 d for gonads, $t_{1/2}$ = 2.3 d for gills, and $t_{1/2}$ = 2.2 d for liver (Lemaire et al. 1990);

elimination $t_{1/2}$ = 4.8–16 d from mussel *Mytilus edulis*; $t_{1/2}$ = 7 d from polychaete *Abarenicola pacifica*, $t_{1/2}$ = 21.7 d from Oyster, $t_{1/2}$ = 8.0 d from clam *Mercenario mercenaria*, $t_{1/2}$ = 8 d from clam *Rangia cuneata* (quoted, Meador et al. 1995);

$t_{1/2}$ = 15–17d for blue mussel *Mytilus edulis* in 32-d exposure laboratory studies (Magnusson et al. 2000).

TABLE 4.1.1.43.1
Reported aqueous solubilities and Henry's law constants of benzo[a]pyrene at various temperatures

$\ln x = A + B/(T/K) + C \cdot \ln(T/K)$ (1)

Aqueous solubility						Henry's law constant	
May et al. 1983		Velapoldi et al. 1983		Whitehouse 1984		ten Hulscher et al. 1992	
generator column-HPLC		generator column-fluorescence		generator column-HPLC		gas stripping-HPLC/fluorescence	
t/°C	S/g·m^{-3}	t/°C	S/g·m^{-3}	t/°C	S/g·m^{-3}	t/°C	Pa m^3/mol
10	0.00056	10	0.00061	8.0	0.00066	10.0	0.022
15	0.00080	15	0.00082	12.4	0.00077	20.0	0.034
20	0.00114	20	0.00113	16.7	0.00094	35.0	0.074
25	0.00162	25	0.00158	20.9	0.00116	40.1	0.092
30	0.00229	30	0.00224	25.0	0.00154	45.0	0.110
						55.0	0.239
		eq. 1	x				
		A	–677.4109			$\ln K_{AW} = -\Delta H/RT + \Delta S/R$	
		B	23963.0			R = 8.314 Pa m^3 mol^{-1} K^{-1}	
		C	100.767			$\Delta S/R$	11.90
						$\Delta H/R$	4690.88
		ΔH_{sol}/(kJ mol^{-1}) = 50.6				enthalpy of volatilization:	
		at 25°C				ΔH/(kJ·mol^{-1}) = 39 ± 3	
						entropy of volatilization, ΔS:	
						$T\Delta S$/(kJ·mol^{-1}) = 29 ± 3	
						at 20°C	

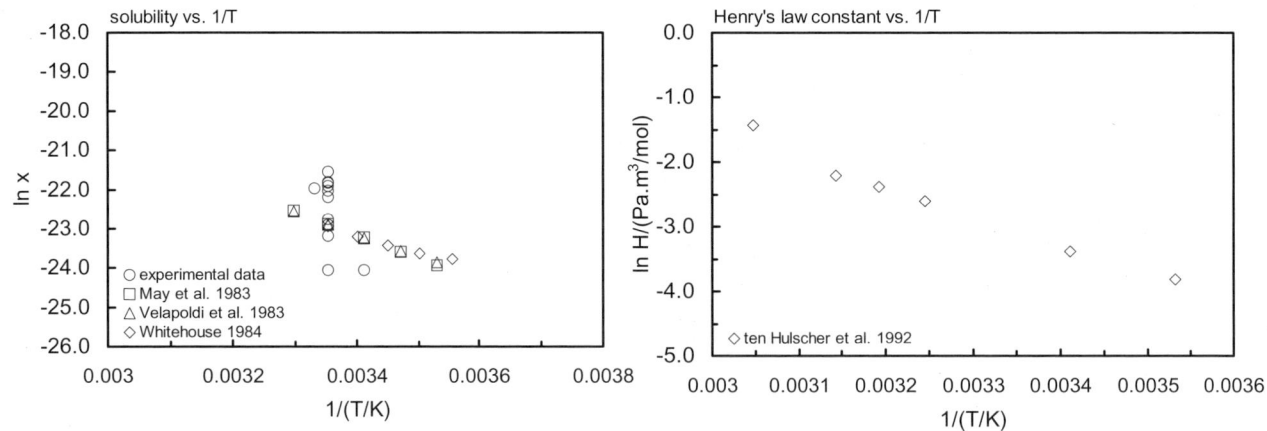

FIGURE 4.1.1.43.1 Logarithm of mole fraction solubility and Henry's law constant versus reciprocal temperature for benzo[a]pyrene.

4.1.1.44 Benzo[e]pyrene

Common Name: Benzo[e]pyrene
Synonym: B[e]P, 4,5-benzopyrene
Chemical Name: 4,5-benzopyrene
CAS Registry No: 192-97-2
Molecular Formula: $C_{20}H_{12}$
Molecular Weight: 252.309
Melting Point (°C):
    181.4   (Lide 2003)
Boiling Point (°C):
    311   (Lide 2003)
Density (g/cm³ at 20°C):
Molar Volume (cm³/mol):
    222.8   (Ruelle & Kesselring 1997; Passivirta et al. 1999)
    262.9   (calculated-Le Bas method at normal boiling point)
Enthalpy of Fusion, ΔH_{fus} (kJ/mol):
    16.57   (exptl., Chickos et al. 1999)
Entropy of Fusion, ΔS_{fus} (J/mol K):
    42.30   (differential scanning calorimetry, Hinckley et al. 1990)
    36.5   (Passivirta et al. 1999)
Fugacity Ratio at 25°C (assuming ΔS_{fus} = 56 J/mol K), F: 0.0292 (mp at 181.4°C)
    0.010   (calculated, Passivirta et al. 1999)

Water Solubility (g/m³ or mg/L at 25°C or as indicated and reported temperature dependence equations. Additional data at other temperatures designated * are compiled at the end of this section):
    0.0035   (27°C, shake flask-nephelometry, Davis et al. 1942)
    0.00732   (shake flask-UV/fluorescence, Barone et al. 1967)
    0.00014, 0.000172, 0.0000252 (HPLC-relative retention correlation, different stationary and phases, Locke 1974)
    0.0050*   (23°C, shake flask-fluorescence, measured range 8.6–31.7°C, Schwarz 1977)
    0.00732, 0.004; 0.0063 (quoted values; lit. mean, Pearlman et al. 1984)
    $\log S_L$ (mol/L) = $-1.398 - 866.8/(T/K)$ (supercooled liquid, Passivirta et al. 1999)
    $\ln x = -11.8754 - 2916.84/(T/K)$; temp range 5–50°C (regression eq. of literature data, Shiu & Ma 2000)

Vapor Pressure (Pa at 25°C and reported temperature dependence equations):
    7.40×10^{-7} (Knudsen effusion method, extrapolated, Murray et al. 1974)
    $\log (P/mmHg) = 9.736 - 6220/(T/K)$; temp range 359–423 K (Knudsen effusion method, Murray et al. 1974)
    7.58×10^{-7} (extrapolated-Antoine eq., Stephenson & Malanowski 1987)
    $\log (P_S/kPa) = 11.7417 - 6220/(T/K)$; temp range 358–423 K (Antoine eq., Stephenson & Malanowski 1987)
    3.20×10^{-7}, 1.28×10^{-5} (20°C, literature solid P_S, converted to supercooled liquid P_L with ΔS_{fus} Bidleman & Foreman 1987)
    8.59×10^{-5} (P_{GC}, GC-RT correlation with p,p = –DDT as reference standard, Hinckley et al. 1990)
    2.53×10^{-5}, 1.02×10^{-5} (supercooled liquid values P_L, converted from literature P_S with different ΔS_{fus} values, Hinckley et al. 1990)
    $\log (P_L/Pa) = 11.11 - 4803/(T/K)$ (GC-RT correlation, Hinckley et al. 1990)
    1.29×10^{-5} (supercooled liquid values P_L, calculated from Yamasaki et al. 1984, Finizio et al. 1997)
    1.78×10^{-5}; 1.78×10^{-6} (quoted supercooled liquid P_L from Hinckley et al. 1990; converted to solid P_S with fugacity ratio F, Passivirta et al. 1999)
    $\log (P_S/Pa) = 12.15 - 5333/(T/K)$ (solid, Passivirta et al. 1999)

log (P_L/Pa) = 10.14 – 4467/(T/K) (supercooled liquid, Passivirta et al. 1999)

Henry's Law Constant (Pa m^3/mol at 25°C and reported temperature dependence equations):
- 0.02 (calculated-P/C, Mackay et al. 1992)
- log $[H/(Pa\ m^3/mol)]$ = 11.64 – 3660/(T/K) (Passivirta et al. 1999)

Octanol/Water Partition Coefficient, log K_{OW}:
- 6.44 (calculated-MCI χ as per Rekker & De Kort 1979, Ruepert et al. 1985)
- 7.40 (calculated-K_{OC}, Broman et al. 1991)
- 6.44 (TLC retention time correlation, De Voogt et al. 1990)
- 6.44 (recommended, Sangster 1993)
- 6.10 (quoted, Meador et al. 1995)
- 5.68 (calculated, Passivirta et al. 1999)

Octanol/Air Partition Coefficient, log K_{OA}:
- 11.13 (calculated, Finizio et al. 1997)

Bioconcentration Factor, log BCF:

Sorption Partition Coefficient, log K_{OC}:
- 7.20 (Baltic Sea particulate field samples, concn distribution-GC/MS, Broman et al. 1991)
- 4.00 (predicted dissolved log K_{OC}, Broman et al. 1991)
- 6.11–7.90; 7.20–8.30 (range, calculated from sequential desorption of 11 urban soils; lit. range, Krauss & Wilcke 2001)
- 5.84; 6.12, 6.11, 6.62 20°C, batch equilibrium, A2 alluvial grassland soil; calculated values of expt 1,2,3-solvophobic approach, Krauss & Wilcke 2001)

Environmental Fate Rate Constants or Half-Lives:

Bioconcentration, Uptake (k_1) and Elimination (k_2) Rate Constants:
k_1 = 0.13–0.36 mg g^{-1} h^{-1}; k_2 = 0.0031 h^{-1} (freshwater oligochaete from sediment, Van Hoof et al. 2001)

Half-Lives in the Environment:

Air: half-lives under simulated atmospheric conditions: $t_{1/2}$ = 21.10 h in simulated sunlight, $t_{1/2}$ = 5.38 h in simulated sunlight + ozone (0.2 ppm), $t_{1/2}$ = 7.6 h in dark reaction ozone (0.2 ppm) (Katz et al. 1979; quoted, Bjørseth & Olufsen 1983).
Surface water:
Groundwater:
Sediment:
Soil:
Biota: elimination $t_{1/2}$ = 6.9–14.4 d from mussel *Mytilus edulis*; $t_{1/2}$ = 30.1 d from Oyster, $t_{1/2}$ = 4.7 d from clam *Mercenario mercenaria* (quoted, Meador et al. 1995).

TABLE 4.1.1.44.1
Reported aqueous solubilities of benzo[e]pyrene at various temperatures

	Schwarz 1977
	shake flask-fluorescence
t/°C	S/g·m^{-3}
8.6	3.25 × 10^{-3}
14.0	3.58 × 10^{-3}
17.0	4.44 × 10^{-3}
17.5	3.94 × 10^{-3}

TABLE 4.1.1.44.1 (*Continued*)

Schwarz 1977

shake flask-fluorescence

t/°C	S/g·m^{-3}
20.2	4.79 × 10^{-3}
23.2	5.35 × 10^{-3}
23.0	5.07 × 10^{-3}
29.2	6.38 × 10^{-3}
29.2	6.48 × 10^{-3}
31.7	6.81 × 10^{-3}

ΔH_{sol}/(kJ mol^{-1}) = 25.56

at 25°C

4.1.1.45 Perylene

Common Name: Perylene
Synonym: peri-dinaphthalene
Chemical Name: perylene
CAS Registry No: 198-55-0
Molecular Formula: $C_{20}H_{12}$
Molecular Weight: 252.309
Melting Point (°C):
    277.76   (Lide 2003)
Boiling Point (°C):
    503   (Pearlman et al. 1984)
Density (g/cm³ at 25°C):
    1.35   (Lide 2003)
Molar Volume (cm³/mol):
    186.9   (25°C, calculated-density)
    222.8   (Ruelle & Kesselring 1997; Passivirta et al. 1999)
    262.9   (calculated-Le Bas method at normal boiling point)
Enthalpy of Fusion, ΔH_{fus} (kJ/mol):
    23.51   (quoted, Tsonopoulos & Prausnitz 1971)
    31.753   (Ruelle & Kesselring 1997)
    31.88   (exptl., Chickos et al. 1999)
Entropy of Fusion, ΔS_{fus} (J/mol K):
    42.68   (quoted, Tsonopoulos & Prausnitz 1971)
    57.4   (Passivirta et al. 1999)
    67.87, 43.7   (exptl., calculated-group additivity method, Chickos et al. 1999)
Fugacity Ratio at 25°C (assuming ΔS_{fus} = 56 J/mol K), F: 0.00331 (mp at 277.76°C)
    0.00268   (calculated, Passivirta et al. 1999)

Water Solubility (g/m³ or mg/L at 25°C or as indicated and reported temperature dependence equations):
    0.0005   (27°C, shake flask-nephelometry Davis et al. 1942)
    < 0.0005   (Weimer & Prausnitz 1965)
    0.00011   (20°C, shake flask-UV, Eisenbrand & Baumann 1970)
    0.0004   (shake flask-fluorescence, Mackay & Shiu 1977)
    0.0003   (average lit. value, Pearlman et al. 1984)
    0.0007   (microdroplet sampling and multiphoton ionization-based fast-conductivity technique MPI-FC, Gridin et al. 1998)
    log $[S_L/(mol/L)]$ = –0.306 – 1662/(T/K) (supercooled liquid, Passivirta et al. 1999)

Vapor Pressure (Pa at 25°C and the reported temperature dependence equations. Additional data at other temperatures designated * are compiled at the end of this section):
    log (P/mmHg) = 13.95 – 7260/(T/K); temp range 110–180°C (Knudsen effusion method, Hoyer & Peperle 1958)
    5.31 × 10⁻⁹ (extrapolated-Antoine eq., Stephenson & Malanowski 1987)
    log (P_S/kPa) = 13.075 – 7260/(T/K); temp range 383–453 K (Antoine eq.-I, Stephenson & Malanowski 1987)
    log (P_S/kPa) = 12.9379 – 7210/(T/K); temp range 383–516 K (Antoine eq.-II, Stephenson & Malanowski 1987)
    7.00 × 10⁻⁷   (quoted, Riederer 1990)
    1.84 × 10⁻⁸* (gas saturation, extrapolated-Antoine eq. derived from exptl. data, temp range 118–210°C, Oja & Suuberg 1998)
    ln (P/Pa) = 35.702 – 15955/(T/K); temp range 391–424 K (Clausius-Clapeyron eq., Knudsen effusion, Oja & Suuberg 1998)

1.40 × 10⁻⁸; 5.22 × 10⁻⁶ (quoted solid P_S from Mackay et al. 1992; converted to supercooled liquid P_L with fugacity ratio F, Passivirta et al. 1999)

$\log (P_S/Pa) = 12.53 - 6074/(T/K)$ (solid, Passivirta et al. 1999)

$\log (P_L/Pa) = 9.53 - 4414/(T/K)$ (supercooled liquid, Passivirta et al. 1999)

4.88 × 10⁻⁶ (supercooled liquid P_L, calibrated GC-RT correlation, Lei et al. 2002)

$\log (P_L/Pa) = -4694/(T/K) + 10.43$; $\Delta H_{vap.} = -89.9$ kJ·mol⁻¹ (GC-RT correlation, Lei et al. 2002)

Henry's Law Constant (Pa m³/mol at 25°C and the report temperature dependence equations):

0.440 (calculated-P/C, Riederer 1990)

$\log [H/(Pa\ m^3/mol)] = 9.84 - 2752/(T/K)$, (Passivirta et al. 1999)

Octanol/Water Partition Coefficient, log K_{OW}:

6.30, 5.10	(HPLC-RV predicted, Brooke et al. 1986)
5.30	(HPLC-RV measured, Brooke et al. 1986)
5.82	(HPLC-RT correlation, Wang et al. 1986)
6.25	(recommended, Sangster 1989, 1994)
6.40	(Bayona et al. 1991)
5.82	(recommended, Hansch et al. 1995)
6.50	(shake flask-dialysis tubing-HPLC/UV, both phases, Andersson & Schräder 1999)
6.25; 6.53	(quoted lit.; calculated, Passivirta et al. 1999)

Octanol/Air Partition Coefficient, log K_{OA}:

Bioconcentration Factor, log BCF:

3.86, 3.73	(*Daphnia pulex*, kinetic estimation, Southworth et al. 1978)
4.36	(activated sludge, Freitag et al. 1984)
3.30, 4.36, < 1.0	(algae, activated sludge, fish, Klein et al. 1984)
3.85	(*Daphnia pulex*, correlated as per Mackay & Hughes 1984, Hawker & Connell 1986)
3.30, 4.36, < 1.0	(algae, activated sludge, fish, Freitag et al. 1985)
3.86	(*Daphnia magna*, Newsted & Giesy 1987)
1.196, –0.398	(*Polychaete sp, Capitella capitata*, Bayona et al. 1991)

Sorption Partition Coefficient, log K_{OC} at 25°C or as indicated:

6.39–7.93 (range, calculated from sequential desorption of 11 urban soils; lit. range, Krauss & Wilcke 2001)

5.88; 6.73, 6.63, 6.76 (20°C, batch equilibrium, A2 alluvial grassland soil; calculated values of expt 1,2,3- solvophobic approach, Krauss & Wilcke 2001)

6.30; 5.89; 4.15–6.38 (calculated-K_{OW}; calculated-solubility; quoted lit. range, Schlautman & Morgan 1993a)

6.05 at pH 4, 5.98 at pH 7, 5.71 at pH 10 in 0.001 M NaCl; 6.01 at pH 4, 5.95 at pH 7, 5.29 at pH 10 in 0.01 M NaCl; 5.98 at pH 4, 5.67 at pH 7, 4.86 at pH 10 in 0.1 M NaCl; 5.97 at pH 4, 5.61 at pH 7, 4.78 at pH 10 in 1 mM Ca^{2+} in 0.1M total ionic strength solutions (shake flask/fluorescence, dissolved humic substances-humic acid; Schlautmam & Morgan 1993)

5.17 at pH 4 in 0.001 M NaCl; 5.14 at pH 4 in 0.01 M NaCl; 5.08 at pH 4 in 0.1 M NaCl; 5.11 at pH 4 in 1 mM Ca^{2+} in 0.1 M total ionic strength solutions (shake flask/fluorescence, dissolved humic substances-fulvic acid; Schlautmam & Morgan 1993a, b)

5.82 at pH 4, < 4.49 at pH 7, < 4.18 at pH 10 in 0.001 M NaCl; 5.65 at pH 4, < 4.46 at pH 7, < 3.85 at pH 10 in 0.01 M NaCl; 5.67 at pH 4, < 4.17 at pH 7, < 3.78 at pH 10 in 0.1 M NaCl; 5.74 at pH 4, 5.02 at pH 7, 4.43 at pH 10 in 1 mM Ca^{2+} in 0.1 M total ionic strength solutions (shake flask/fluorescence, adsorbed humic substances-humic acid; Schlautmam & Morgan 1993b)

< 3.48 at pH 4 in 0.001 M NaCl; < 3.30 at pH 4 in 0.01 M NaCl; < 3.48 at pH 4 in 0.1 M NaCl; < 3.48 at pH 4 in 1 mM Ca^{2+} in 0.1 M total ionic strength solutions (shake flask/fluorescence, adsorbed humic substances-fulvic acid; Schlautmam & Morgan 1993b)

6.00 (soil humic acid, shake flask-HPLC/UV, Cho et al. 2002)

Environmental Fate Rate Constants, k or Half-Lives, $t_{1/2}$:

Volatilization: sublimation $k < 1 \times 10^{-5}$ s^{-1} from glass surface was measured at 24°C at an airflow rate of 3 L/min (Cope & Kalkwarf 1987).

Photolysis:

half-lives on different atmospheric particulate substrates (appr. 25 µg/g on substrate): $t_{1/2}$ = 3.9 h on silica gel, $t_{1/2}$ = 1.2 h on alumina and $t_{1/2}$ = 35 h on fly ash (Behymer & Hites 1985)

ozonation $k < 4.7 \times 10^{-5}$ m/s was measured from glass surface at 24°C with $[O_3]$ = 0.16 ppm and light intensity of 1.3 kW/m^2 (Cope & Kalkwarf 1987)

photodegradation k = 0.0152 min^{-1} and $t_{1/2}$ = 0.78 h in ethanol-water (2:3, v/v) solution for initial concentration of 5.0 ppm by high pressure mercury lamp or sunlight (Wang et al. 1991)

k(expt) = 0.0152 min^{-1} the pseudo-first-order rate constant with the calculated $t_{1/2}$ = 0.78 h and the predicted k = 0.0406 min^{-1}, calculated by QSPR in aqueous solution when irradiated with a 500 W medium pressure mercury lamp (Chen et al. 1996);

$t_{1/2}$ = 0.53 h (predicted- QSPR) in atmospheric aerosol (Chen et al. 2001)

Photodegradation $k = 4.4 \times 10^{-4}$ s^{-1} in surface water during the summertime at mid-latitude (Fasnacht & Blough 2002)

Hydrolysis:

Oxidation:

Biodegradation:

Biotransformation:

Bioconcentration, Uptake (k_1) and Elimination (k_2) Rate Constants:

k_1 = 752 h^{-1}; k_2 = 0.139 h^{-1} (*Daphnia pulex*, Southworth et al. 1978)

log k_1 = 2.88 h^{-1}; log k_2 = –0.86 h^{-1} (*Daphnia pulex*, correlated as per Mackay & Hughes 1984, Hawker & Connell 1986)

k_1 = 0.12–0.38 mg g^{-1} h^{-1}; k_2 = 0.0034 h^{-1} (freshwater oligochaete from sediment, Van Hoof et al. 2001)

Half-Lives in the Environment:

Air: direct photolysis $t_{1/2}$ = 0.53 h (predicted- QSPR) in atmospheric aerosol (Chen et al. 2001).

Surface water: photolysis $t_{1/2}$ = 0.78 h (reported in units of minutes) in aqueous solution when irradiated with a 500 W medium pressure mercury lamp (Chen et al. 1996).

Groundwater:

Sediment:

Soil:

Biota: elimination half-lives: $t_{1/2}$ = 2.0 d from rainbow trout, $t_{1/2}$ = 26.2 d from clam *Mya arenaria*, $t_{1/2}$ = 6.3–13.3 d from mussel *Mytilus edulis*; $t_{1/2}$ = 9.2 d from Oyster, $t_{1/2}$ = 1.2 d from shrimp, $t_{1/2}$ = 5.7 d from polychaete *Nereis virens*, $t_{1/2}$ = 8.0 d from clam *Mercenario mercenaria* (quoted, Meador et al. 1995).

TABLE 4.1.1.45.1
Reported vapor pressures of perylene at various temperatures and the coefficients for the vapor pressure equations

log P = A – B/(T/K)	(1)	ln P = A – B/(T/K) (1a)
log P = A – B/(C + t/°C)	(2)	ln P = A – B/(C + t/°C) (2a)
log P = A – B/(C + T/K)	(3)	
log P = A – B/(T/K) – C·log (T/K)	(4)	

Hoyer & Peperle 1958		Oja & Suuberg 1998	
effusion		Knudsen effusion	
t/°C	P/Pa	t/°C	P/Pa
data presented by equation		124.10	0.0114
eq. 1	P/mmHg	127.37	0.0164
A	13.95	131.82	0.0249
B	7260	135.96	0.0373

TABLE 4.1.1.45.1 (*Continued*)

Hoyer & Peperle 1958		Oja & Suuberg 1998	
effusion		Knudsen effusion	
t/°C	P/Pa	t/°C	P/Pa
for temp range 110–180°C		141.87	0.0638
mp/°C	270.5–273.5	mp/K	551
ΔH_{sub}/(kJ/mol) = 140.1		eq. 1a	P/Pa
		A	35.702
		B	15955
		for temp range 391–424 K	
		ΔH_{sub}/(kJ/mol) = 132.6	

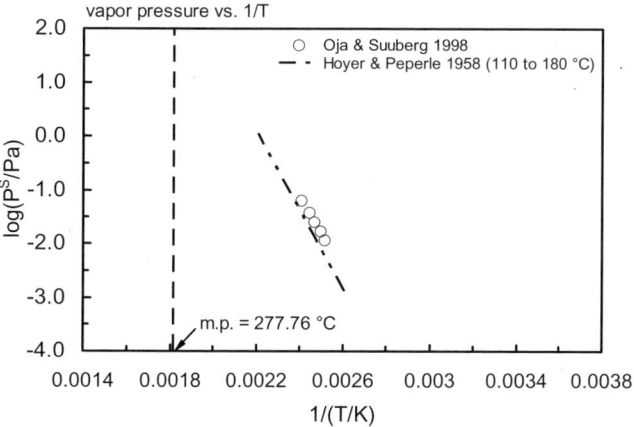

FIGURE 4.1.1.45.1 Logarithm of vapor pressure versus reciprocal temperature for perylene.

4.1.1.46 7,12-Dimethylbenz[a]anthracene

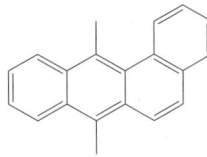

Common Name: 7,12-Dimethylbenz[a]anthracene
Synonym: 7,12-dimethylbenz[a]anthracene, 9,10-dimethyl-1,2-benzanthracene, 7,12-dimethylbenzanthracene
Chemical Name: 7,12-dimethylbenz[a]anthracene
CAS Registry No: 57-97-6
Molecular Formula: $C_{20}H_{16}$
Molecular Weight: 256.341
Melting Point (°C):
    122.5   (Lide 2003)
Boiling Point (°C):
Density (g/cm³ at 20°C):
Molar Volume (cm³/mol):
    245.8   (Ruelle & Kesselring 1997)
    292.7   (calculated-Le Bas method at normal boiling point)
Enthalpy of Vaporization, ΔH_V (kJ/mol):
    104.56   (Kelley & Rice 1964)
Enthalpy of Fusion, ΔH_{fus} (kJ/mol):
    22.09   (Kelley & Rice 1964)
Entropy of Fusion, ΔS_{fus} (J/mol K):
Fugacity Ratio at 25°C (assuming ΔS_{fus} = 56 J/mol K), F: 0.111 (mp at 122.5°C)

Water Solubility (g/m³ or mg/L at 25°C or as indicated):
    0.043   (27°C, shake flask-nephelometry, Davis et al. 1942)
    0.061   (shake flask-fluorescence, Mackay & Shiu 1977)
    0.053   (24°C, shake flask-nephelometry, Hollifield 1979)
    0.025   (24°C, shake flask-LSC, Means et al. 1979)
    0.0244   (shake flask-LSC, Means et al. 1980b)
    0.043, 0.061, 0.053; 0.054 (quoted lit. values; lit. mean, Pearlman et al. 1984)

Vapor Pressure (Pa at 25°C and reported temperature dependence equations):
    3.73×10^{-7} (solid vapor pressure, extrapolated, effusion method, Kelley & Rice 1974)
    3.70×10^{-6} (extrapolated, supercooled liquid value P_L, Kelley & Rice 1974)
    $\log (P_S/\text{mmHg}) = 15.108 - 7051/(T/K)$; temp range: 106–122°C (effusion method, Kelley & Rice 1964)
    $\log (P_L/\text{mmHg}) = 12.232 - 5987/(T/K)$; temp range: 122–135°C (effusion method, Kelley & Rice 1964)
    3.84×10^{-7} (extrapolated-Antoine eq.-I, Stephenson & Malanowski 1987)
    $\log (P_S/\text{kPa}) = 14.233 - 7051/(T/K)$; temp range 379–396 K (Antoine eq.-I, Stephenson & Malanowski 1987)
    6.78×10^{-6} (extrapolated-Antoine eq.-II, Stephenson & Malanowski 1987)
    $\log (P_S/\text{kPa}) = 10.70417 - 5629.911/(T/K)$; temp range 379–390 K (Antoine eq.-II, Stephenson & Malanowski 1987)
    $\log (P_L/\text{kPa}) = 11.357 - 5897/(T/K)$, temp range 396–408 K (Antoine eq.-III, Stephenson & Malanowski 1987)
    6.38×10^{-6} (supercooled liquid P_L, calibrated GC-RT correlation, Lei et al. 2002)
    $\log (P_L/\text{Pa}) = -4643/(T/K) + 10.38$; $\Delta H_{vap.} = -88.9$ kJ·mol^{-1} (GC-RT correlation, Lei et al. 2002)

Henry's Law Constant (Pa m³/mol):

Octanol/Water Partition Coefficient, log K_{OW}:
    5.98   (shake flask-LSC, concn. ratio, Means et al. 1979)
    5.80   (shake flask-LSC, Means et al. 1980b)

6.16	(UNIFAC activity coeff., Banerjee & Howard 1988)
5.80	(recommended, Sangster 1989)
5.89	(recommended, Sangster 1993)
5.80	(recommended, Hansch et al. 1995)

Octanol/Air Partition Coefficient, log K_{OA}:

Bioconcentration Factor, log BCF:

Sorption Partition Coefficient, log K_{OC}:

    5.68   (average of 3 soil/sediment samples, sorption isotherms by shake flask-LSC, Means et al. 1979)

    5.37   (average of 12 soil/sediment samples, sorption isotherms by shake flask-LSC, Means et al. 1980b)

Environmental Fate Rate Constants, k or Half-Lives, $t_{1/2}$:

    Volatilization:

    Photolysis:

    Hydrolysis: no hydrolyzable groups (Howard et al. 1991).

    Oxidation: photooxidation $t_{1/2}$ = 0.32–3.2 h estimated, based on estimated rate constant for the reaction with hydroxyl radicals in air (Atkinson 1987; quoted, Howard et al. 1991); photooxidation $t_{1/2}$(aq.) = 1,57 – 157 yr estimated, based on measured rate constant for the reaction with singlet oxygen in benzene (Stevens et al. 1974; quoted, Howard et al. 1991).

    Biodegradation: biodegradation k = 0.0339 d^{-1} with $t_{1/2}$ = 20 d for Kidman sandy loam soil and k = 0.0252 d^{-1} with $t_{1/2}$ = 28 d for McLaurin sandy loam soil (Park et al. 1990);

    $t_{1/2}$(aq. aerobic) ~ 480–672 h, based on aerobic soil die-away test data (Sims 1990; quoted, Howard et al. 1991);

    $t_{1/2}$(aq. anaerobic) ~ 1920–2688 h, based on estimated unacclimated aqueous aerobic biodegradation half-life (Howard et al. 1991).

    Biotransformation:

    Bioconcentration, Uptake (k_1) and Elimination (k_2) Rate Constants:

Half-Lives in the Environment:

    Air: $t_{1/2}$ = 0.32–3.2 h, based on estimated photooxidation half-life in air (Atkinson 1987; quoted, Howard et al. 1991).

    Surface water: $t_{1/2}$ = 480–672 h, based on aerobic soil die-away test data (Howard et al. 1991).

    Groundwater: $t_{1/2}$ = 960–1344 h, based on estimated unacclimated aqueous aerobic biodegradation half-life (Howard et al. 1991).

    Sediment:

    Soil: biodegradation rate constant k = 0.0339 d^{-1} with $t_{1/2}$ = 20 d for Kidman sandy loam soil and k = 0.0252 d^{-1} with $t_{1/2}$ = 28 d for McLaurin sandy loam soil (Park et al. 1990);

    $t_{1/2}$ ~ 480–672 h, based on aerobic soil die-away test data (Howard et al. 1991).

4.1.1.47 9,10-Dimethylbenz[a]anthracene

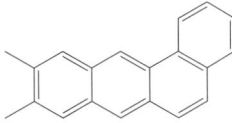

Common Name: 9,10-Dimethylbenz[a]anthracene
Synonym:
Chemical Name: 9,10-dimethylbenz[a]anthracene
CAS Registry No: 56-56-4
Molecular Formula: $C_{20}H_{16}$
Molecular Weight: 256.341
Melting Point (°C):
    122   (Yalkowsky et al. 1983; Ruelle & Kesselring 1997)
Boiling Point (°C):
Density (g/cm³ at 20°C):
Molar Volume (cm³/mol):
    245.8   (Ruelle & Kesselring 1997)
    292.7   (calculated-Le Bas method at normal boiling point)
Entropy of Fusion, ΔS_{fus} (J/mol K):
    54.81   (Kelley & Rice 1974)
Fugacity Ratio at 25°C (assuming ΔS_{fus} = 56 J/mol K), F: 0.112 (mp at 122°C)

Water Solubility (g/m³ or mg/L at 25°C or as indicated):
    0.0435   (27°C, shake flask-nephelometry, Davis et al. 1942)
    0.0435   (recommended, Shaw 1989)

Vapor Pressure (Pa at 25°C and reported temperature dependence equation):
    log (P/mmHg) = 15.108 – 7051/(T/K); temp range: 106–135°C (effusion method, Kelley & Rice 1964)
    log (P/mmHg) = 12.232 – 5897/(T/K) (liquid, effusion method, Kelley & Rice 1964)

Henry's Law Constant (Pa m³/mol):

Octanol/Water Partition Coefficient, log K_{OW}:
    6.93   (calculated-fragment const., Yalkowsky et al. 1983)

Octanol/Air Partition Coefficient, log K_{OA}:

Bioconcentration Factor, log BCF:

Sorption Partition Coefficient, log K_{OC}:

Environmental Fate Rate Constants, k or Half-Lives, $t_{½}$:

Half-Lives in the Environment:

4.1.1.48 3-Methylcholanthrene

Common Name: 3-Methylcholanthrene
Synonym: 20-methylcholanthrene, 1,2-dihydro-3-methyl-benz[*j*]aceanthrylene
Chemical Name: 3-methylcholanthrene
CAS Registry No: 56-49-5
Molecular Formula: $C_{21}H_{16}$
Molecular Weight: 268.352
Melting Point (°C):
    180   (Bjørseth 1983)
Boiling Point (°C):
Density (g/cm³ at 20°C):
    1.28   (Lide 2003)
Molar Volume (cm³/mol):
    247.8   (Ruelle & Kesselring 1997)
    209.6   (20°C, calculated-density)
    296.0   (calculated-Le Bas method at normal boiling point)
Enthalpy of Fusion, ΔH_{fus} (kJ/mol):
Entropy of Fusion, ΔS_{fus} (J/mol K):
Fugacity Ratio at 25°C (assuming ΔS_{fus} = 56 J/mol K), F: 0.0301 (mp at 180°C)
    0.003   (Mackay et al. 1980)

Water Solubility (g/m³ or mg/L at 25°C):
    0.0015   (Weimer & Prausnitz 1965)
    0.0029   (shake flask-fluorescence, Mackay & Shiu 1977)
    0.00323   (shake flask-liquid scintillation counting, Means et al. 1980)

Vapor Pressure (Pa at 25°C and reported temperature dependence equations):
    log (P/mmHg) = 13.168 − 6643/(T/K); temp range 128–152°C (effusion method, Kelley & Rice 1964)
    1.03×10^{-7} (extrapolated-Antoine eq., Stephenson & Malanowski 1987)
    log (P_S/kPa) = 12.293 − 6643/(T/K); temp range 401–425 K (Antoine eq., Stephenson & Malanowski 1987)
    1.48×10^{-6} (supercooled liquid P_L, calibrated GC-RT correlation, Lei et al. 2002)
    log (P_L/Pa) = −4901/(T/K) + 10.61; ΔH_{vap} = −93.8 kJ·mol⁻¹ (GC-RT correlation, Lei et al. 2002)

Henry's Law Constant (Pa m³/mol):

Octanol/Water Partition Coefficient, log K_{OW}:
    6.42   (shake flask-LSC, Means et al. 1980)
    7.11   (calculated-f const., Valvani & Yalkowsky 1980)
    6.69   (calculated-S and mp, Mackay et al. 1980)
    7.11   (Hansch & Leo 1985)
    6.45, 7.07   (calculated-UNIFAC, calculated-fragment const., Banerjee & Howard 1988)
    6.75 ± 0.50 (recommended, Sangster 1989)
    6.45   (recommended, Sangster 1993)
    6.42   (recommended, Hansch et al. 1995)

Octanol/Air Partition Coefficient, log K_{OA}:

Bioconcentration Factor, log BCF:
    4.12   (*Daphnia magna*, McCarthy et al. 1985)

Sorption Partition Coefficient, log K_{OC}:
    6.09   (soil/sediment, sorption isotherm by batch equilibrium-LSC, Means et al. 1980b)
    6.25   (average of 14 soil/sediment samples, sorption isotherm by shake flask-LSC, Means et al. 1980)
    6.09, 6.10   (calculated-regression of K_p versus substrate properties, calculated.-K_{OW}, Means et al. 1980)
    4.02   (soil, calculated-K_{OW}, Briggs 1981)
    6.18   (soil, calculated-K_{OW}, Means et al. 1982)
    5.07   (soil, calculated-K_{OW}, Chiou et al. 1983)

Environmental Fate Rate Constants, k or Half-Lives, $t_{1/2}$:
    Volatilization:
    Photolysis:
    Hydrolysis: no hydrolyzable groups (Howard et al. 1991).
    Oxidation: photooxidation $t_{1/2}$ = 0.317–3.17 h, based on estimated rate constant for reaction with hydroxyl radical in air (Atkinson 1987; quoted, Howard et al. 1991).
    Biodegradation: aerobic $t_{1/2}$ = 14616–33600 h, based on mineralization half-life in fresh water and estuarine ecosystems (Heitkamp 1988; quoted, Howard et al. 1991);
        anaerobic $t_{1/2}$ = 58464–134400 h, based on estimated unacclimated aqueous aerobic biodegradation half-life (Howard et al. 1991).
    Biotransformation:
    Bioconcentration, Uptake (k_1) and Elimination (k_2) Rate Constants:

Half-Lives in the Environment:
    Air: $t_{1/2}$ = 0.317–3.17 h, based on estimated photooxidation half-life in air (Atkinson 1987; quoted, Howard et al. 1991).
    Surface water: $t_{1/2}$ = 14616–33600 h, based on mineralization half-life in fresh water and estuarine ecosystems (Heitkamp 1988; quoted, Howard et al. 1991).
    Groundwater: $t_{1/2}$ = 29232–672000 h, based on estimated unacclimated aqueous aerobic biodegradation half-life (Howard et al. 1991).
    Sediment:
    Soil: $t_{1/2}$ = 14616–33600 h, based on estimated mineralization half-life in fresh water and estuarine ecosystems (Heitkamp 1988; quoted, Howard et al. 1991).
    Biota:

4.1.1.49 Benzo[*ghi*]perylene

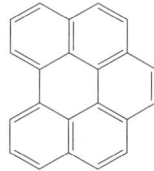

Common Name: Benzo[*ghi*]perylene
Synonym: 1,12-benzoperylene, benzoperylene
Chemical Name: 1,12-benzoperylene
CAS Registry No: 191-24-2
Molecular Formula: $C_{22}H_{12}$
Molecular Weight: 276.330
Melting Point (°C):
    272.5   (Lide 2003)
Boiling Point (°C):
    525   (Pearlman et al. 1984)
Density (g/cm³ at 20°C):
Molar Volume (cm³/mol):
    233.8   (Ruelle & Kesselring 1997; Passivirta et al. 1999)
    277.5   (calculated-Le Bas method at normal boiling point)
Enthalpy of Fusion, ΔH_{fus} (kJ/mol):
    17.365   (Ruelle & Kesselring 1997)
    17.37   (exptl., Chickos et al. 1999)
Entropy of Fusion, ΔS_{fus} (J/mol K):
    31.34, 43.2   (exptl., calculated-group additivity method, Chickos et al. 1999)
    31.4   (Passivirta et al. 1999)
Fugacity Ratio at 25°C (assuming ΔS_{fus} = 56 J/mol K), F = 0.00373 (mp at 272.5°C)
    0.039   (calculated, ΔS_{fus} = 31.4 J/mol K, Passivirta et al. 1999)

Water Solubility (g/m³ or mg/L at 25°C or as indicated and reported temperature dependence equations):
    0.00026   (shake flask-fluorescence, Mackay & Shiu 1977)
    0.0007   (generator column-HPLC/fluo., Wise et al. 1981)
    0.00083   (quoted, Pearlman et al. 1984)
    0.000137   (generator column-HPLC/fluo., De Maagd et al. 1998)
    $\log [S_L/(mol/L)] = -2.073 - 908.7/(T/K)$ (supercooled liquid, Passivirta et al. 1999)

Vapor Pressure (Pa at 25°C or as indicated and reported temperature dependence equations):
    1.39×10^{-8} (Knudsen effusion method, Murray et al. 1974)
    $\log (P/atm) = 9.519 - 6674/(T/K)$; temp range: 389–468 K (Knudsen effusion method, Murray et al. 1974)
    1.33×10^{-8} (20°C, estimated, Callahan et al. 1979)
    6.69×10^{-7} (Yamasaki et al. 1984)
    1.38×10^{-8}; 7.51×10^{-9} (extrapolated-Antoine eq. I, II, Stephenson & Malanowski 1987)
    $\log (P_S/kPa) = 11.5247 - 6674/(T/K)$; temp range 389–468 K (Antoine eq.-I, Stephenson & Malanowski 1987)
    $\log (P_S/kPa) = 10.945 - 6580/(T/K)$; temp range 391–513 K (Antoine eq.-II, Stephenson & Malanowski 1987)
    7.20×10^{-6}; 1.84×10^{-6} (quoted solid P_S from Mackay et al. 1992; converted to supercooled liquid P_L with fugacity ratio F, Passivirta et al. 1999)
    $\log (P_S/Pa) = 12.40 - 5824/(T/K)$ (solid, Passivirta et al. 1999)
    $\log (P_L/Pa) = 10.76 - 4915/(T/K)$ (supercooled liquid, Passivirta et al. 1999)
    4.28×10^{-7} (supercooled liquid P_L, calibrated GC-RT correlation, Lei et al. 2002)
    $\log (P_L/Pa) = -5018/(T/K) + 10.46$; $\Delta H_{vap.} = -96.1$ kJ·mol⁻¹ (GC-RT correlation, Lei et al. 2002)

Henry's Law Constant (Pa m³/mol at 25°C or as indicated and reported temperature dependence equations. Additional data at other temperatures designated * are compiled at the end of this section):

    0.027*   (20°C, gas stripping-HPLC/fluorescence, measured range 10–55°C, ten Hulscher et al. 1992)

    $\log (H/(Pa\ m^3/mol)) = 12.83 - 4006/(T/K)$ (Passivirta et al. 1999)

    0.0278 (20°C, selected from reported experimentally measured values, Staudinger & Roberts 1996, 2001)

    $\log K_{AW} = -0.651 - 1258/(T/K)$ (van't Hoff eq. derived from literature data, Staudinger & Roberts 2001)

Octanol/Water Partition Coefficient, log K_{OW}:

    7.10   (RP-TLC-k' correlation, Bruggeman et al. 1982)
    7.05   (HPLC-RT correlation, Rapaport et al. 1984)
    6.63   (HPLC-RT correlation, Wang et al. 1986)
    6.90   (recommended, Sangster 1989, 1993)
    7.04   (TLC retention time correlation, De Voogt et al. 1990)
    6.63   (recommended, Hansch et al. 1995)
    6.22 (range 5.95–6.38) (shake flask/slow stirring-HPLC/fluo., De Maagd et al. 1998)

Octanol/Air Partition Coefficient, log K_{OA}:

Bioconcentration Factor, log BCF:

    5.54   (microorganisms-water, calculated from K_{OW}, Mabey et al. 1982)
    4.45   (*Daphnia magna*, Newsted & Giesy 1987)

Sorption Partition Coefficient, log K_{OC} at 25°C or as indicated:

    6.70   (Baltic Sea particulate field samples, concn distribution-GC/MS, Broman et al. 1991)

    7.215 (10°C), 7.08, 6.93 (20°C), 6.68 (35°C), 6.46, 6.51 (45°C) (log K_{DOC} - dissolved organic material from lake, gas-purge technique-HPLC/fluorescence, Lüers & ten Hulscher 1996)

    6.80   (20°C, log K_{POC} - particulate organic material from lake, Lüers & ten Hulscher 1996)

    6.82–8.25; 6.20–9.20 (range, calculated from sequential desorption of 11 urban soils; lit. range, Krauss & Wilcke 2001)

    5.87; 6.84, 6.82, 7.26 (20°C, batch equilibrium, A2 alluvial grassland soil; calculated values of expt 1,2,3- solvophobic approach, Krauss & Wilcke 2001)

Environmental Fate Rate Constants, k or Half-Lives, $t_{1/2}$:

    Volatilization:

    Photolysis: half-lives on different atmospheric particulate substrates (approx. 25 μg/g on substrate): $t_{1/2}$ = 7.0 h on silica gel, $t_{1/2}$ = 22 h on alumina and $t_{1/2}$ = 29 h on fly ash (Behymer & Hites 1985);

        first order daytime photodegradation rate constants for adsorption on wood soot particles in an outdoor Teflon chamber: k = 0.0077 min^{-1} with 1000–2000 ng/mg loading and k = 0.0116 min^{-1} with 30–350 ng/mg loading (Kamens et al. 1988);

        direct photolysis $t_{1/2}$(obs) = 0.89 h, $t_{1/2}$(calc) = 0.86 h predicted by QSPR method in atmospheric aerosol (Chen et al. 2001).

    Hydrolysis: no hydrolyzable groups (Howard et al. 1991).

    Oxidation: rate constants k < 60 M^{-1} h^{-1} for singlet oxygen and k < 6 M^{-1} h^{-1} for peroxy radical (Mabey et al. 1982); photooxidation $t_{1/2}$ = 0.321–3.21 h, based on estimated rate constant for reaction with hydroxyl radical in air (Atkinson 1987; quoted, Howard et al. 1991).

    Biodegradation: aerobic $t_{1/2}$ = 14160–15600 h, based on aerobic soil die-away test data at 10–30°C (Coover & Sims 1987; quoted, Howard et al. 1991); anaerobic $t_{1/2}$ = 56640–62400 h, based on aerobic soil die-away test data at 10–30°C (Coover & Sims 1987; quoted, Howard et al. 1991).

    Biotransformation: estimated to be 3×10^{-12} mL cell^{-1} h^{-1} for bacteria (Mabey et al. 1982).

    Bioconcentration, Uptake (k_1) and Elimination (k_2) Rate Constants:

        k_1 = 0.076–0.21 mg g^{-1} h^{-1}; k_2 = 0.0012–0.0014 h^{-1} (freshwater oligochaete from sediment, Van Hoof et al. 2001)

Half-Lives in the Environment:

Air: $t_{1/2}$ = 0.321–3.21 h, based on estimated photooxidation half-life in air (Atkinson 1987; quoted, Howard et al. 1991); $t_{1/2}$ = 0.6 h for adsorption on wood soot particles in an outdoor Teflon chamber with an estimated first order rate constant k = 0.0179 min^{-1} at 1 cal cm^{-2} min^{-1}, 10 g/m^3 H$_2$O and 20°C (Kamens et al. 1988).

Surface water: $t_{1/2}$ = 14160–15600 h, based on aerobic soil die-away test data at 10–30°C (Coover & Sims 1987; quoted, Howard et al. 1991).

Groundwater: $t_{1/2}$ = 28320–31200 h, based on aerobic soil die-away test data at 10–30°C (Coover & Sims 1987; quoted, Howard et al. 1991).

Sediment:

Soil: $t_{1/2}$ = 14160–15600 h, based on aerobic soil die-away test data (Coover & Sims 1987; quoted, Howard et al. 1991);

$t_{1/2}$ > 50 d (Ryan et al. 1988);

mean $t_{1/2}$ = 9.1 yr for Luddington soil (Wild et al. 1991).

Biota: elimination $t_{1/2}$ = 12.4 d from Oyster, $t_{1/2}$ = 4.8 d from clam *Mercenario mercenaria* (quoted, Meador et al. 1995).

TABLE 4.1.1.49.1
Reported Henry's law constants of benzo[ghi]perylene at various temperatures

ten Hulscher et al. 1992

gas stripping-HPLC/fluo.

t/°C	H/(Pa m³/mol)
10.0	0.019
20.0	0.027
35.0	0.052
40.1	0.054
45.0	0.066
55.0	0.087

enthalpy of volatilization:
ΔH_{vol}/(kJ·mol^{-1}) = 26.1±1.0
entropy of volatilization, ΔS
$T\Delta S_{vol}$/(kJ·mol^{-1}) = 15.9±1.3
                             at 20°C
ln K_{AW} = –ΔH_{vol}/RT + ΔS_{vol}/R
R = 8.314 Pa m^3 mol^{-1} K^{-1}

4.1.1.50 Indeno[1,2,3-*cd*]pyrene

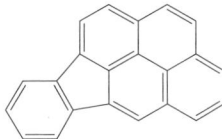

Common Name: Indeno[1,2,3-*cd*]pyrene
Synonym: 2,3-*o*-phenylenepyrene,
Chemical Name: indeno[1,2,3-*cd*]pyrene
CAS Registry No: 193-39-5
Molecular Formula: $C_{22}H_{12}$
Molecular Weight: 276.330
Melting Point (°C):
    162   (Lide 2003)
Boiling Point (°C):
Density (g/cm³):
Molar Volume (cm³/mol):
    233.8   (Ruelle & Kesselring 1997, Passivirta et al. 1999)
    283.5   (calculated-Le Bas method at normal boiling point)
Enthalpy of Fusion, ΔH_{fus} (kJ/mol):
    21.51   (Chickos et al. 1999)
Entropy of Fusion, ΔS_{fus} (J/mol K):
    49.41, 36.0 (exptl., calculated-group additivity method, Chickos et al. 1999)
    56.5   (Passivirta et al. 1999)
Fugacity Ratio at 25°C (assuming ΔS_{fus} = 56 J/mol K), F: 0.0453 (mp at 162°C)
    0.043   (Passivirta et al. 1999)

Water Solubility (g/m³ or mg/L at 25°C and the reported temperature dependence equations):
    0.00019   (generator column-HPLC/UV, Wise et al. 1981)
    0.000438; 0.0023, 0.000191 (quoted, exptl.; calculated-molar volume, mp and mobile order thermodynamics, Ruelle & Kesselring 1997)
    $\log [S_L/(mol/L)] = -0.758 - 1631/(T/K)$ (supercooled liquid, Passivirta et al. 1999)

Vapor Pressure (Pa at 25°C and the reported temperature dependence equations):
    1.33×10^{-7} (estimated, Callahan et al. 1979)
    1.00×10^{-8}; 2.32×10^{-7} (quoted solid P_S from Mackay et al. 1992; converted to supercooled liquid P_L with fugacity ratio F, Passivirta et al. 1999)
    $\log (P_S/Pa) = 12.56 - 6126/(T/K)$ (solid, Passivirta et al. 1999)
    $\log (P_L/Pa) = 9.60 - 4839/(T/K)$ (supercooled liquid, Passivirta et al. 1999)

Henry's Law Constant (Pa·m³/mol at 25°C and the reported temperature dependence equations. Additional data at other temperatures designated * are compiled at the end of this section):
    0.029*   (20°C, gas stripping-HPLC/fluorescence, measured range 10–55°C, ten Hulscher et al. 1992)
    $\log [H/(Pa\ m^3/mol)] = 10.36 - 3208/(T/K)$ (Passivirta et al. 1999)
    0.0285   (20°C, selected from reported experimentally measured values, Staudinger & Roberts 1996, 2001)
    $\log K_{AW} = 0.033 - 1455/(T/K)$ (van't Hoff eq. derived from literature data, Staudinger & Roberts 2001)

Octanol/Water Partition Coefficient, $\log K_{OW}$:
    7.66   (calculated-π substituent const., Callahan et al. 1979)
    8.20   (calculated, Broman et al. 1991)
    6.72   (calculated, Passivirta et al. 1999)

Octanol/Air Partition Coefficient, log K_{OA}:

Bioconcentration Factor, log BCF or log K_B:

Sorption Partition Coefficient, log K_{OC}:
- 8.00 (Baltic Sea particulate field samples, concn distribution-GC/MS, Broman et al. 1991)
- 6.93 (10°C), 6.88, 6.84 (20°C), 6.42 (35°C), 6.32, 6.31 (45°C) (log K_{DOC} - dissolved organic material from lake, gas-purge technique-HPLC/fluorescence, Lüers & ten Hulscher 1996)
- 6.80 (20°C, log K_{POC} - particulate organic material from lake, Lüers & ten Hulscher 1996)

Environmental Fate Rate Constants, k and Half-Lives, $t_{½}$:

Bioconcentration and Uptake and Elimination Rate Constants (k_1 and k_2):
k_1 = 0.067–0.20 mg g^{-1} h^{-1}; k_2 = 0.0010–0.0013 h^{-1} (freshwater oligochaete from sediment, Van Hoof et al. 2001)

Half-Lives in the Environment:

TABLE 4.1.1.50.1
Reported Henry's law constants of indeno[123-cd]pyrene at various temperatures

	ten Hulscher et al. 1992
	gas stripping-HPLC/fluo.
t/°C	H/(Pa m³/mol)
10.0	0.018
20.0	0.029
35.0	0.057
40.1	0.061
45.0	0.077
55.0	0.105

enthalpy of volatilization:
$\Delta H/(kJ \cdot mol^{-1})$ = 30.0 ± 1.1
entropy of volatilization, ΔS:
$T\Delta S/(J \cdot K^{-1} mol^{-1})$ = 19.9±1.3
at 20°C

ln K_{AW} = $-\Delta H/RT + \Delta S/R$
R = 8.314 Pa m³ mol^{-1} K^{-1}

4.1.1.51 Dibenz[a,c]anthracene

Common Name: Dibenz[a,c]anthracene
Synonym: 1,2:3,4-Dibenzanthracene, naphtho-2',3':9,10-phenanthrene
Chemical Name: dibenz[a,c]anthracene
CAS Registry No: 215-58-7
Molecular Formula: $C_{22}H_{14}$
Molecular Weight: 278.346
Melting Point (°C):
    205 (Bjørseth 1983; Lide 2003)
Boiling Point (°C):
    518 (Weast 1982–83)
Density (g/cm³ at 20°C):
Molar Volume (cm³/mol):
    299.9 (calculated-Le Bas method at normal boiling point)
Enthalpy of Fusion, ΔH_{fus} (kJ/mol):
    25.82 (exptl., Chickos et al. 1999)
Entropy of Fusion, ΔS_{fus} (J/mol K):
    46.65, 44.0 (exptl., calculated-group additivity method, Chickos et al. 1999)
Fugacity Ratio at 25°C (assuming ΔS_{fus} = 56 J/mol K), F: 0.0171 (mp at 205°C)

Water Solubility (g/m³ or mg/L at 25°C):
    0.0016 (generator column-HPLC/UV, Billington et al. 1988)

Vapor Pressure (Pa at 25°C or as indicated and reported temperature dependence equations. Additional data at other temperatures designated * are compiled at the end of this section):
    1.30×10^{-9}* (effusion method, extrapolated, De Kruif 1980)
    log (P/Pa) = 16.25 – 7322.2/(T/K); temp range: 152–179°C (torsion-effusion, de Kruif 1980)
    log (P/Pa) = 16.011 – 7207.2/(T/K); temp range: 152–179°C (weighing-effusion, de Kruif 1980)
    log (P/Pa) = 16.131 – 7265/(T/K); temp range: 152–179°C (mean, de Kruif 1980)
    3.44×10^{-7} (supercooled liquid P_L, calibrated GC-RT correlation, Lei et al. 2002)
    log (P_L/Pa) = –5094/(T/K) + 10.62; $\Delta H_{vap.}$ = –97.5 kJ·mol⁻¹ (GC-RT correlation, Lei et al. 2002)

Henry's Law Constant (Pa m³/mol):

Octanol/Water Partition Coefficient, log K_{OW}:
    7.19 (calculated, Miller et al. 1985)
    7.11 (calculated-MCI χ as per Rekker & De Kort 1979, Ruepert et al. 1985)
    7.19 (recommended, Sangster 1989, 1993)
    7.11 (TLC retention time correlation, De Voogt et al. 1990)
    8.0068 (calculated-UNIFAC group contribution, Chen et al. 1993)
    6.17 (recommended, Hansch et al. 1995)
    6.40 ± 0.19, 6.48 ± 0.78 (HPLC-k' correlation: ODS column; Diol column, Helweg et al. 1997)

Octanol/Air Partition Coefficient, log K_{OA}:

Bioconcentration Factor, log BCF:

Sorption Partition Coefficient, log K_{OC}:
  6.54   (humic acid, HPLC-k' correlation, Nielsen et al. 1997)

Environmental Fate Rate Constants, k or Half-Lives, $t_{1/2}$:

Half-Lives in the Environment:
  Air: half-lives under simulated atmospheric conditions: $t_{1/2}$ – 9.20 h in simulated sunlight, $t_{1/2}$ = 4.60 h in simulated sunlight + ozone (0.2 ppm), $t_{1/2}$ = 3.82 h in dark reaction ozone (0.2 ppm) (Katz et al. 1979; quoted, Bjørseth & Olufsen 1983).

TABLE 4.1.1.51.1
Reported vapor pressures of dibenz[a,c]anthracene at various temperatures and the coefficients for the vapor pressure equations

$\log P = A - B/(T/K)$ (1) $\ln P = A - B/(T/K)$ (1a)
$\log P = A - B/(C + t/°C)$ (2) $\ln P = A - B/(C + t/°C)$ (2a)
$\log P = A - B/(C + T/K)$ (3)
$\log P = A - B/(T/K) - C \cdot \log (T/K)$ (4)

de Kruif 1980

torsion-, weighing effusion

t/°C	P/Pa
151.65	0.1
158.52	0.2
163.08	0.3
166.38	0.4
168.97	0.5
171.11	0.6
172.94	0.7
174.53	0.8
175.95	0.9
177.22	1.0
298.15	1.3×10^{-9}
	extrapolated

$\Delta H_{subl}/(kJ\ mol^{-1}) = 138.0$

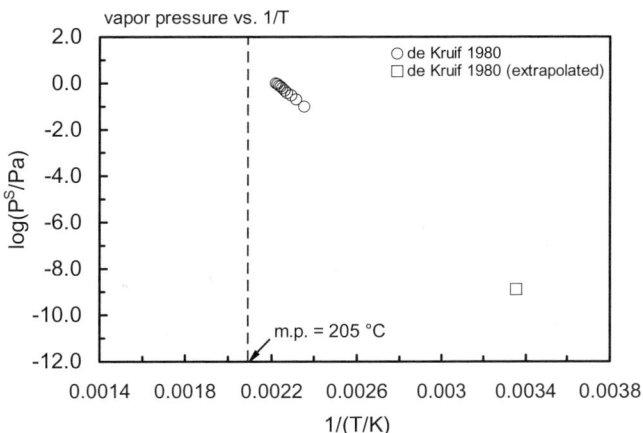

FIGURE 4.1.1.51.1 Logarithm of vapor pressure versus reciprocal temperature for dibenz[a,c]anthracene.

4.1.1.52 Dibenz[a,h]anthracene

Common Name: Dibenz[a,h]anthracene
Synonym: DB[a,h]A, 1,2,5,6-dibenzanthracene, 1,2:5,6-dibenzanthracene
Chemical Name: 1,2:5,6-dibenzanthracene
CAS Registry No: 53-70-3
Molecular Formula: $C_{22}H_{14}$
Molecular Weight: 278.346
Melting Point (°C):
   269.5   (Lide 2003)
Boiling Point (°C):
   524   (Weast 1977)
Density (g/cm³ at 20°C):
Molar Volume (cm³/mol):
   252.6   (Ruelle & Kesselring 1997; Passivirta et al. 1999)
   299.9   (calculated-Le Bas method at normal boiling point)
Enthalpy of Fusion, ΔH_{fus} (kJ/mol):
   31.165   (Ruelle & Kesselring 1997)
   31.16   (exptl., Chickos et al. 1999)
Entropy of Fusion, ΔS_{fus} (J/mol K):
   58.26, 44.0   (exptl., calculated-group additivity method, Chickos et al. 1999)
   57.3   (Passivirta et al. 1999)
Fugacity Ratio at 25°C (assuming ΔS_{fus} = 56 J/mol K), F: 0.00399 (mp at 269.5°C)
   0.00389   (calculated, Passivirta et al. 1999)

Water Solubility (g/m³ or mg/L at 25°C or as indicated and the reported temperature dependence equations):
   0.0005   (27°C, shake flask-nephelometry, Davis et al. 1942)
   0.0006   (shake flask-UV, Klevens 1950)
   0.0025   (shake flask-LSC, Means et al. 1980b)
   0.00056 (lit. mean, Pearlman et al. 1984)
   $\log [S_L/(mol/L)] = -1.409 - 1631/(T/K)$ (supercooled liquid, Passivirta et al. 1999)

Vapor Pressure (Pa at 25°C or as indicated and reported temperature dependence equations. Additional data at other temperatures designated * are compiled at the end of this section):
   1.33×10^{-8} (20°C, estimated, Callahan et al. 1979)
   3.70×10^{-10}* (effusion method, De Kruif 1980)
   $\log (P/Pa) = 16.049 - 7395.4/(T/K)$; temp range: 163–189°C (torsion-effusion, de Kruif 1980)
   $\log (P/Pa) = 15.876 - 7312/(T/K)$; temp range: 163–189°C (weighing-effusion, de Kruif 1980)
   $\log (P/Pa) = 15.962 - 7730/(T/K)$; temp range: 163–189°C (mean, de Kruif 1980)
   4.25×10^{-10} (extrapolated-Antoine eq., Stephenson & Malanowski 1987)
   $\log (P_S/kPa) = 12.515 - 7420/(T/K)$; temp range 403–513 K (Antoine eq., Stephenson & Malanowski 1987)
   3.70×10^{-10}; 9.31×10^{-8} (quoted solid P_S from Mackay et al. 1992; converted to supercooled liquid P_L with fugacity ratio F, Passivirta et al. 1999)
   $\log (P_S/Pa) = 12.82 - 5824/(T/K)$ (solid, Passivirta et al. 1999)
   $\log (P_L/Pa) = 9.82 - 5002/(T/K)$ (supercooled liquid, Passivirta et al. 1999)
   2.51×10^{-7} (supercooled liquid P_L, calibrated GC-RT correlation, Lei et al. 2002)
   $\log (P_L/Pa) = -5193/(T/K) + 10.82$; $\Delta H_{vap.} = -99.4$ kJ·mol⁻¹ (GC-RT correlation, Lei et al. 2002)

Henry's Law Constant (Pa m^3/mol at 25°C and reported temperature dependence equations):
- 0.0074 (calculated-P/C, Mabey et al. 1982)
- 0.0076 (calculated-P/C, Eastcott et al. 1988)
- log [H/(Pa m^3/mol)] = 11.23 − 3371/(T/K), (Passivirta et al. 1999)

Octanol/Water Partition Coefficient, log K_{OW}:
- 6.50 (shake flask-LSC, Means et al. 1980b)
- 6.88 (HPLC-RT/MS, Burkhard et al. 1985)
- 5.80 (Hansch & Leo 1985)
- 6.75 ± 0.40 (recommended, Sangster 1989, 1993)
- 7.11 (TLC retention time correlation, De Voogt et al. 1990)
- 6.60 (shake flask-UV, pH 7.4, Alcorn et al. 1993)
- 6.50 (recommended, Hansch et al. 1995)
- 6.54 ± 0.19, 6.60 ± 0.78 (HPLC-k′ correlation: ODS column; Diol column, Helweg et al. 1997)

Octanol/Air Partition Coefficient, log K_{OA}:

Bioconcentration Factor, log BCF:
- 5.84 (microorganisms-water, calculated from K_{OW}, Mabey et al. 1982)
- 4.63 (activated sludge, Freitag et al. 1984)
- 3.38, 4.63, 1.0 (algae, activated sludge, fish, Freitag et al. 1985)
- 4.00 (*Daphnia magna*, Newsted & Giesy 1987)

Sorption Partition Coefficient, log K_{OC}:
- 6.31 (average of 14 soil/sediment samples, equilibrium sorption isotherms by shake flask-LSC, Means et al. 1980b)
- 6.22, 6.18 (calculated-regression of k_p versus substrate properties, calculated-K_{OW}, Means et al. 1980b)
- 6.22; 6.11, 5.30, 5.62 (quoted; calculated-K_{OW}, calculated-S and mp, calculated-S, Karickhoff 1981)
- 6.52 (calculated-K_{OW}, Mabey et al. 1982)
- 5.20 (calculated, Pavlou 1987)
- 6.31; 6.44; 3.75–5.77 (soil, quoted exptl.; calculated-MCI $^1\chi$, calculated-K_{OW} range, Sabljic 1987a,b)
- 5.77 (soil, calculated-K_{OW} based on model of Karickhoff et al. 1979, Sabljic 1987b)
- 5.66 (soil, calculated-K_{OW} based on model of Means et al. 1982, Sabljic 1987b)
- 4.60 (soil, calculated-K_{OW} based on model of Chiou et al. 1983, Sabljic 1987b)
- 4.61 (soil, calculated-K_{OW} based on model of Kenaga 1980, Sabljic 1987b)
- 3.75 (soil, calculated-K_{OW} based on model of Briggs 1981, Sabljic 1987b)
- 6.22 (calculated-MCI $^1\chi$, Sabljic et al. 1995)
- 6.44 (humic acid, HPLC-k′ correlation, Nielsen et al. 1997)
- 6.00; 6.30 (soil, calculated-universal solvation model; quoted exptl., Winget et al. 2000)
- 6.76–8.42; 5.80–8.50 (range, calculated from sequential desorption of 11 urban soils; lit. range, Krauss & Wilcke 2001)
- 6.03; 7.0, 6.76, 7.32 (20°C, batch equilibrium, A2 alluvial grassland soil; calculated values of expt 1,2,3-solvophobic approach, Krauss & Wilcke 2001)

Environmental Fate Rate Constants, k or Half-Lives, $t_{1/2}$:

Volatilization:

Photolysis: atmospheric and aqueous photolysis $t_{1/2}$ = 782 h, based on measured rate of photolysis in heptane under November sunlight (Muel & Saguim 1985; quoted, Howard et al. 1991) and $t_{1/2}$ = 6 h after adjusting the ratio of sunlight photolysis in water versus heptane (Smith et al. 1978; Muel & Saguem 1985; quoted, Howard et al. 1991);

pseudo-first-order direct photolysis rate constant k(exptl) = 0.014 min^{-1} with the calculated $t_{1/2}$ = 0.83 h and the predicted k = 0.0216 min^{-1} calculated by QSPR method in aqueous solution when irradiated with a 500 W medium pressure mercury lamp (Chen et al. 1996);

direct photolysis $t_{1/2}$(obs.) = 0.31 h, $t_{1/2}$ = 0.38 h predicted by QSPR method in atmospheric aerosol (Chen et al. 2001).

Oxidation: rate constant k = 5 × 10⁸ M⁻¹ h⁻¹ for singlet oxygen and k = 1.5 × 10⁴ M⁻¹ h⁻¹ for peroxy radical (Mabey et al. 1982);
   photooxidation $t_{1/2}$ = 0.428–4.28 h, based on estimated rate constant for reaction with hydroxyl radical in air (Atkinson 1987; quoted, Howard et al. 1991).
Hydrolysis: not hydrolyzable (Mabey et al. 1982); no hydrolyzable groups (Howard et al. 1991).
Biodegradation: aerobic $t_{1/2}$ = 8664–22560 h, based on aerobic soil die-away test data (Coover & Sims 1987; Sims 1990; quoted, Howard et al. 1991);
   k = 0.0019 d⁻¹ with $t_{1/2}$ = 361 d for Kidman sandy loam and k = 0.0017 d⁻¹ with $t_{1/2}$ = 420 d for McLarin sandy loam all at –0.33 bar soil moisture (Park et al. 1990).
Biotransformation: estimated to be 3 × 10⁻¹² mL cell⁻¹ h⁻¹ for bacteria (Mabey et al. 1982).
Bioconcentration, Uptake (k_1) and Elimination (k_2) Rate Constants:

Half-Lives in the Environment:

Air: half-lives under simulated atmospheric conditions: $t_{1/2}$ = 9.6 h in simulated sunlight, $t_{1/2}$ = 4.8 h in simulated sunlight + ozone (0.2 ppm), $t_{1/2}$ = 2.71 h in dark reaction ozone (0.2 ppm) (Katz et al. 1979; quoted, Bjørseth & Olufsen 1983);
   $t_{1/2}$ = 0.428–4.28 h, based on estimated photooxidation half-life in air (Atkinson 1987; quoted, Howard et al. 1991).
Surface water: $t_{1/2}$ = 6–782 h, based on sunlight photolysis half-life in water (Smith et al. 1978; Muel & Saguem 1985; quoted, Howard et al. 1991);
   photolysis $t_{1/2}$ = 0.83 h in aqueous solution when irradiated with a 500 W medium-pressure mercury lamp (Chen et al. 1996).
Groundwater: $t_{1/2}$ = 17328–45120 h, based on estimated unacclimated aqueous aerobic biodegradation half-life (Howard et al. 1991).
Sediment:
Soil: biodegradation rate constant k = 0.0019 d⁻¹ with $t_{1/2}$ = 361 d for Kidman sandy loam soil and k = 0.117 d⁻¹ with $t_{1/2}$ = 420 d for McLaurin sandy loam soil (Park et al. 1990);
   $t_{1/2}$ ~ 8664–22560 h, based on aerobic soil dieaway test data (Coover & Sims 1987; Sims 1990; quoted, Howard et al. 1991);
   mean $t_{1/2}$ = 20.607 wk (quoted, Wild et al. 1991).
Biota:

TABLE 4.1.1.52.1
Reported vapor pressures of dibenz[a,h]anthracene at various temperatures and the coefficients for the vapor pressure equations

$\log P = A - B/(T/K)$ (1) $\quad \ln P = A - B/(T/K)$ (1a)

$\log P = A - B/(C + t/°C)$ (2) $\quad \ln P = A - B/(C + t/°C)$ (2a)

$\log P = A - B/(C + T/K)$ (3)

$\log P = A - B/(T/K) - C \cdot \log(T/K)$ (4)

de Kruif 1980	
torsion-, weighing effusion	
t/°C	P/Pa
160.38	0.1
168.21	0.2
172.93	0.3
176.33	0.4
179.01	0.5
181.22	0.6
183.11	0.7
184.76	0.8

TABLE 4.1.1.52.1 (*Continued*)

de Kruif 1980

torsion-, weighing effusion

t/°C	P/Pa
186.22	0.9
187.54	1.0
25.0	3.7×10^{-10}
	extrapolated

$\Delta H_{subl}/(kJ\ mol^{-1}) = 140.0$

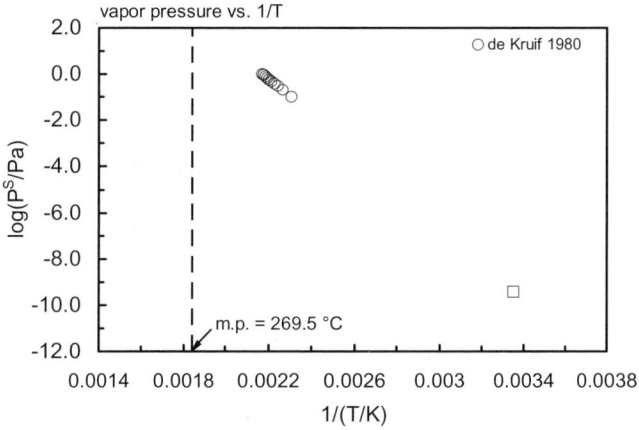

FIGURE 4.1.1.52.1 Logarithm of vapor pressure versus reciprocal temperature for dibenz[a,h]anthracene.

4.1.1.53 Dibenz[*a,j*]anthracene

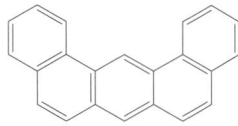

Common Name: Dibenz[*a,j*]anthracene
Synonym: 1,2:7,8-dibenzanthracene, 1,2:7,8-dibenzanthracene, a,a′-dibenzanthracene, dinaphthanthracene
Chemical Name: dibenz[*a,j*]anthracene
CAS Registry No: 58-70-3
Molecular Formula: $C_{22}H_{14}$
Molecular Weight: 278.346
Melting Point (°C):
    197.5   (Lide 2003)
Boiling Point (°C):
Density (g/cm³ at 20°C):
Molar Volume (cm³/mol):
    222.8   (Ruelle & Kesselring 1997)
    299.9   (calculated-Le Bas method at normal boiling point)
Enthalpy of Fusion, ΔH_{fus} (kJ/mol):
Entropy of Fusion, ΔS_{fus} (J/mol K):
Fugacity Ratio at 25°C (assuming ΔS = 56 J/mol K), F: 0.0203 (mp at 197.5°C)

Water Solubility (g/m³ or mg/L at 25°C or as indicated):
    0.012   (27°C, shake flask-nephelometry, Davis et al. 1942; quoted, Shaw 1989)
    0.012   (quoted, Yalkowsky et al. 1983; Pearlman et al. 1984)
    0.000041, 0.00022 (calculated-molar volume, mp and mobile order thermodynamics, Ruelle & Kesselring 1997)

Vapor Pressure (Pa at 25°C):

Henry's Law Constant (Pa m³/mol):

Octanol/Water Partition Coefficient, log K_{OW}:
    7.19   (calculated-fragment const., Yalkowsky et al. 1983)
    7.11   (calculated-MCI χ as per Rekker & De Kort 1979, Ruepert et al. 1985)
    7.11   (TLC retention time correlation, De Voogt et al. 1990)
    7.11   (quoted and recommended, Sangster 1993)
    6.54 ± 0.19, 6.44 ± 0.75 (HPLC-k′ correlation: ODS column; Diol column, Helweg et al. 1997; quoted, Nielsen et al. 1997)

Octanol/Air Partition Coefficient, log K_{OA}:
Bioconcentration Factor, log BCF:

Sorption Partition Coefficient, log K_{OC}:
    6.58   (humic acid, HPLC-k′ correlation, Nielsen et al. 1997)

Environmental Fate Rate Constants, k or Half-Lives, $t_{1/2}$:
Half-Lives in the Environment:

4.1.1.54 Pentacene

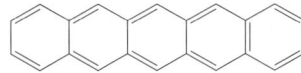

Common Name: Pentacene
Synonym: 2,3,6,7-dibenzanthracene, 2,3:6,7-dibenzanthracene
Chemical Name: pentacene
CAS Registry No: 135-48-8
Molecular Formula: $C_{22}H_{14}$
Molecular Weight: 278.346
Melting Point (°C):
    270–271 (Weast 1982–83)
Boiling Point (°C):
    290–300 (sublimation, Weast 1982–83)
Density (g/cm³ at 20°C):
Molar Volume (cm³/mol):
    299.9   (calculated-Le Bas method at normal boiling point)
Enthalpy of Fusion, ΔH_{fus} (kJ/mol):
    35.19   (Chickos et al. 1999)
Entropy of Fusion, ΔS_{fus} (J/mol K):
    55.22, 44.0 (exptl., calculated-group additivity method, Chickos et al. 1999)
Fugacity Ratio at 25°C (assuming ΔS_{fus} = 56 J/mol K), F: 0.00395 (mp at 270°C)

Water Solubility (g/m³ or mg/L at 25°C):

Vapor Pressure (Pa at 25°C or as indicated and reported temperature dependence equations. Additional data at other temperatures designated * are compiled at the end of this section):
    1.0×10^{-13}* (effusion method, De Kruif 1980)
    log (P_S/kPa) = 12.725 – 8260/(T/K); temp range 444–566 K (Antoine eq., Stephenson & Malanowski 1987)
    1.19×10^{-12}* (gas saturation, extrapolated-Antoine eq. derived from exptl. data, temp range 170–210°C, Oja & Suuberg 1998)
    log (P/Pa) = 35.823 – 18867/(T/K); temp range 443–483 K (Antoine eq., Knudsen effusion, Oja & Suuberg 1998)

Henry's Law Constant (Pa m³/mol):

Octanol/Water Partition Coefficient, log K_{OW}:
    7.19   (calculated-f const., Miller et al. 1985)
    7.19   (recommended, Sangster 1989, 1993)
    7.11   (TLC retention time correlation, De Voogt et al. 1990)
    8.0068 (calculated-UNIFAC group contribution, Chen et al. 1993)
    7.19   (recommended, Sangster 1993)

Octanol/Air Partition Coefficient, log K_{OA}:

Bioconcentration Factor, log BCF:

Sorption Partition Coefficient, log K_{OC}:

Environmental Fate Rate Constants, k or Half-Lives, $t_{1/2}$:

Half-Lives in the Environment:

TABLE 4.1.1.54.1
Reported vapor pressures of pentacene at various temperatures and the coefficients for the vapor pressure equations

$$\log P = A - B/(T/K) \quad (1)$$
$$\log P = A - B/(C + t/°C) \quad (2)$$
$$\log P = A - B/(C + T/K) \quad (3)$$
$$\log P = A - B/(T/K) - C \cdot \log (T/K) \quad (4)$$

$$\ln P = A - B/(T/K) \quad (1a)$$
$$\ln P = A - B/(C + t/°C) \quad (2a)$$

de Kruif 1980		Oja & Suuberg 1998	
torsion-, weighing effusion		Knudsen effusion	
t/°C	P/Pa	t/°C	P/Pa
220.51	0.1	171.33	0.00161
229.8	0.2	184.8	0.00413
235.4	0.3	185.99	0.00421
239.45	0.4	190.98	0.00760
242.63	0.5	200.29	0.0171
245.27	0.6	203.1	0.0157
247.51	0.7	212.6	0.0515
249.47	0.8		
251.22	0.9	mp/K	> 573
252.79	1.0		
25.0	1.0×10^{-13}	eq. 1a	P/Pa
	extrapolated	A	35.823
		B	18823
$\Delta H_{subl}/(kJ\ mol^{-1})$ = 155.0		for temp range 443–483 K	
		$\Delta H_{subl}/(kJ\ mol^{-1})$ = 155.9	

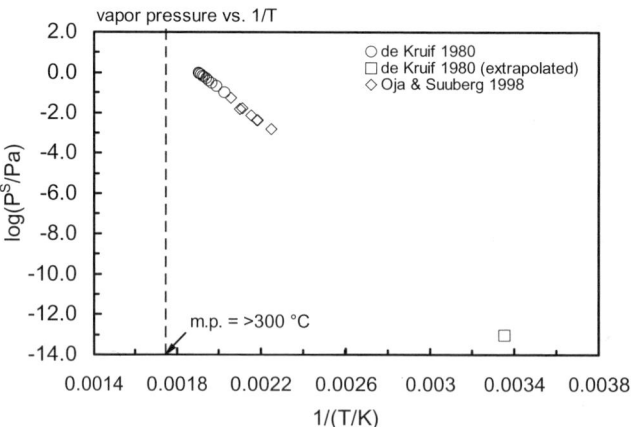

FIGURE 4.1.1.54.1 Logarithm of vapor pressure versus reciprocal temperature for pentacene.

4.1.1.55 Coronene

Common Name: Coronene
Synonym: hexabenzobenzene
Chemical Name: coronene
CAS Registry No: 191-07-1
Molecular Formula: $C_{24}H_{12}$
Molecular Weight: 300.352
Melting Point (°C):
  437.4   (Lide 2003)
Boiling Point (°C):
  525   (Weast 1982–83; Stephenson & Malanowski 1987; Lide 2003)
Density (g/cm³ at 25°C):
  1.371   (Lide 2003)
Molar Volume (cm³/mol):
  244.8   (Ruelle & Kesselring 1997)
  219.1   (25°C, calculated-density)
  292.1   (calculated-Le Bas method at normal boiling point)
Enthalpy of Fusion, ΔH_{fus} (kJ/mol):
  19.202   (Ruelle & Kesselring 1997)
  19.2   (exptl., Chickos et al. 1999)
Entropy of Fusion, ΔS_{fus} (J/mol K):
  27.02, 42.8 (exptl., calculated-group additivity method, Chickos et al. 1999)
Fugacity Ratio at 25°C (assuming ΔS_{fus} = 56 J/mol K), F: 8.99×10^{-5} (mp at 437.4°C)

Water Solubility (g/m³ or mg/L at 25°C):
  0.00014   (shake flask-fluorescence, Mackay & Shiu 1977)
  0.00014   (average lit. value, Pearlman et al. 1984)
  0.00010   (generator column-HPLC/UV, Billington et al. 1988)

Vapor Pressure (Pa at 25°C or as indicated and reported temperature dependence equations. Additional data at other temperatures designated * are compiled at the end of this section):
  log (P/mmHg) = 12.62 – 7675/(T/K); temp range 160–240°C (Knudsen effusion method, Hoyer & Peperle 1958)
  1.95×10^{-10} (Knudsen effusion method, extrapolated, Murray et al. 1974)
  log (P/atm) = 9.110 – 7100/(T/K); temp range 427–510 K (Knudsen effusion method, Murray et al. 1974)
  2.00×10^{-10} (extrapolated-Antoine eq.-I, Stephenson & Malanowski 1987)
  log (P_S/kPa) = 11.1157 – 7100/(T/K); temp range 427–510 K (solid, Antoine eq.-I, Stephenson & Malanowski 1987)
  log (P_S/kPa) = 8.886 – 5764/(T/K); temp range not specified (solid, Antoine eq.-II, Stephenson & Malanowski 1987)
  log (P_L/kPa) = 8.318 – 5362/(T/K); temp range not specified (liquid, Antoine eq.-III, Stephenson & Malanowski 1987)
  2.89×10^{-10}* (gas saturation, extrapolated-Antoine eq. derived from exptl. data, temp range 148–231°C, Oja & Suuberg 1998)
  log (P/Pa) = 31.72 – 16006/(T/K); temp range 421–504 K (Antoine eq., Knudsen effusion, Oja & Suuberg 1998)
  2.55×10^{-8} (supercooled liquid P_L, calibrated GC-RT correlation, Lei et al. 2002)
  log (P_L/Pa) = –5446/(T/K) + 10.67; $\Delta H_{vap.}$ = –104.2 kJ·mol⁻¹ (GC-RT correlation, Lei et al. 2002)

Henry's Law Constant (Pa m³/mol at 25°C):

Octanol/Water Partition Coefficient, log K_{OW}:

7.64	(average lit. value, Yalkowsky et al. 1983)
7.64	(calculated-MCI χ as per Rekker & De Kort 1979, Ruepert et al. 1985)
8.20, 6.70	(HPLC-RV correlation, different mobile phases, Brooke et al. 1986)
5.40	(shake flask/slow stirring-GC, Brooke et al. 1986)
6.50	(recommended, Sangster 1989, 1993)
7.64	(TLC retention time correlation, De Voogt et al. 1990)
8.0	(calculated-K_{OC}, Broman et al. 1991)
5.40, 6.70	(Hansch et al. 1995)

Bioconcentration Factor, log BCF:

Sorption Partition Coefficient, log K_{OC}:

7.80	(Baltic Sea particulate field samples, concn distribution-GC/MS, Broman et al. 1991)
5.0	(predicted dissolved log K_{OC}, Broman et al. 1991)

Environmental Fate Rate Constants, k or Half-Lives, $t_{1/2}$:

Half-Lives in the Environment:

Soil: mean $t_{1/2}$ = 16.5 yr for Luddington soil (Wild et al. 1991).

TABLE 4.1.1.55.1
Reported vapor pressures of coronene at various temperatures and the coefficients for the vapor pressure equations

$$\log P = A - B/(T/K) \quad (1)$$
$$\log P = A - B/(C + t/°C) \quad (2)$$
$$\log P = A - B/(C + T/K) \quad (3)$$
$$\log P = A - B/(T/K) - C \cdot \log(T/K) \quad (4)$$

$$\ln P = A - B/(T/K) \quad (1a)$$
$$\ln P = A - B/(C + t/°C) \quad (2a)$$

Hoyer & Peperle 1958		Murray et al. 1974		Oja & Suuberg 1998	
effusion		Knudsen effusion		Knudsen effusion	
t/°C	P/Pa	t/°C	P/Pa	t/°C	P/Pa
data presented by equation		data presented by graph and		147.9	0.00191
eq. 1	P/mmHg	eq. 1	P/atm	163.6	0.00686
A	12.62	A	9.110	178.1	0.0236
B	7676	B	7100	196.9	0.0895
for temp range 160–240°C		for temp range 427–510 K		109.9	0.222
				210.0	0.309
ΔH_{sub}/(kJ/mol) = 147.0		25.0	2.0 × 10⁻¹⁰ extrapolated	231.11	0.927
				mp/K	711
		mp/°C	~440		
				eq. 1a	P/Pa
		ΔH_{sub}/(kJ/mol) = 135.9		A	31.72
		at av. temp measurements		B	16006
				temp range 421–504 K	
				ΔH_{sub}/(kJ/mol) = 133.1	

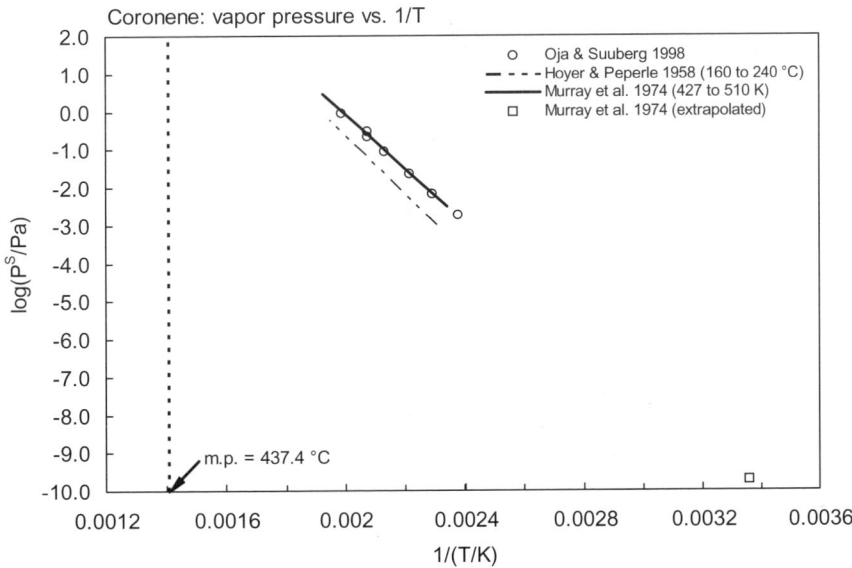

FIGURE 4.1.1.55.1 Logarithm of vapor pressure versus reciprocal temperature for coronene.

4.1.2 Chlorinated polynuclear aromatic hydrocarbons

4.1.2.1 2,4″,5-Trichloro-p-terphenyl

Common Name: 2,4″,5-Trichloro-*p*-terphenyl
Synonym:
Chemical Name:
CAS Registry No: 61576-93-0
Molecular Formula: $C_{18}H_{11}Cl_3$
Molecular Weight: 333.639
Melting Point (°C):
    92   (Dickhut et al. 1994)
Boiling Point (°C):
Density (g/cm³):
Molar Volume (cm³/mol):
    335.9   (calculated-Le Bas method at normal boiling point)
Heat of Fusion, ΔH_{fus} (kJ/mol):
Entropy of Fusion, ΔS_{fus} (J/mol K):
Fugacity Ratio at 25°C (assuming ΔS_{fus} = 56 J/mol K), F: 0.22 (mp at 92°C)

Water Solubility (g/m³ or mg/L at 25°C or as indicated):
    3.8×10^{-4}; 3.37×10^{-4} (exptl. mean by generator column-GC method; calculated-TSA, Dickhut et al. 1994)
    1.01×10^{-4}, 6.56×10^{-4}, 8.56×10^{-4} (5, 25, 30°C, generator column-GC, Dickhut et al. 1994)

Vapor Pressure (Pa at 25°C):

Henry's Law Constant (Pa·m³/mol):

Octanol/Water Partition Coefficient, log K_{OW}:

Octanol/Air Partition Coefficient, log K_{OA}:

Bioconcentration Factor, log BCF or log K_B:

Sorption Partition Coefficient, log K_{OC}:

Environmental Fate Rate Constants, k and Half-Lives, $t_{1/2}$:

Half-Lives in the Environment:

4.1.2.2 2,4,4″,6-Tetrachloro-p-terphenyl

Common Name: 2,4,4″,6-Tetrachloro-*p*-terphenyl
Synonym:
Chemical Name:
CAS Registry No:
Molecular Formula: $C_{18}H_{11}Cl_4$
Molecular Weight: 368.084
Melting Point (°C):
    114   (Dickhut et al. 1994)
Boiling Point (°C):
Density (g/cm³):
Molar Volume (cm³/mol):
    356.8   (calculated-Le Bas method at normal boiling point)
Heat of Fusion, ΔH_{fus} (kJ/mol):
Entropy of Fusion, ΔS_{fus} (J/mol K):
Fugacity Ratio at 25°C (assuming ΔS_{fus} = 56 J/mol K), F: 0.134 (m.p at 114°C)

Water Solubility (g/m³ or mg/L at 25°C or as indicated):
    1.79×10^{-4}; 7.91×10^{-5} (exptl. mean by generator column-GC method; calculated-TSA, Dickhut et al. 1994)
    5.91×10^{-5}, 1.74×10^{-4}, 4.07×10^{-4} (5, 25, 40°C, generator column-GC, Dickhut et al. 1994)

Vapor Pressure (Pa at 25°C):

Henry's Law Constant (Pa·m³/mol):

Octanol/Water Partition Coefficient, log K_{OW}:

Octanol/Air Partition Coefficient, log K_{OA}:

Bioconcentration Factor, log BCF or log K_B:

Sorption Partition Coefficient, log K_{OC}:

Environmental Fate Rate Constants, k and Half-Lives, $t_{1/2}$:

Half-Lives in the Environment:

4.1.3 Polychlorinated naphthalenes

4.1.3.1 1-Chloronaphthalene

Common Name: 1-Chloronaphthalene
Synonym: PCN-1, α-chloronaphthalene
Chemical Name: 1-chloronaphthalene
CAS Registry No: 90-13-1
Molecular Formula: $C_{10}H_7Cl$
Molecular Weight: 162.616
Melting Point (°C):
    –2.5   (Lide 2003)
Boiling Point (°C):
    259   (Lide 200)
Density (g/cm³):
    1.1938   (20°C, Weast 1982–83; Windholz 1983; Budavari 1989)
    1.1976, 1.1938 (15°C, 20°C, Riddick et al. 1986)
    1.188   (25°C, Lide 2003)
Molar Volume (cm³/mol):
    136.2   (20°C, calculated-density)
    168.5   (calculated-Le Bas method at normal boiling point)
Enthalpy of Fusion, ΔH_{fus} (kJ/mol):
Entropy of Fusion, ΔS_{fus} (J/mol K):
Fugacity Ratio at 25°C, F: 1.0

Water Solubility (g/m³ or mg/L at 25°C):
    22.4   (shake flask-fluorescence, Mackay & Shiu 1981)
    19.0; 8.93   (quoted; calculated-molecular connectivity indices, Nirmalakhandan & Speece 1989)
    2.87   (quoted, Crookes & Howe 1993, Alcock et al. 1999)
    6.75; 36.3   (quoted exptl value; calculated-molar volume, Wang et al. 1992)
    19.1; 25.8   (quoted; calculated-group contribution method, Kühne et al. 1995)
    19.1; 25.2   (quoted; calculated-molar volume, mp and mobile order thermodynamics, Ruelle & Kesselring 1997)

Vapor Pressure (Pa at 25°C or as indicated and reported temperature dependence equations. Additional data at other temperatures designated * are compiled at the end of this section):
    133.3*   (80.6°C, static-Hg manometer, measured range 80.6–269.3°C, Kahlbaum 1898)
    133.3*   (80.6°C, summary of literature data, temp range 80.6–269.3°C, Stull 1947)
    3.055   (extrapolated from Antoine eq., temp range 353–533 K, Stephenson & Malanowski 1987)
    $\log (P/kPa) = 6.15143 - [1861.65/(T/K - 83.337)]$; temp range: 353–533 K (Antoine eq., Stephenson & Malanowski 1987)
    2.133   (estimated, Crookes & Howe 1993)
    1.2×10^{-4} (estimated, Alcock et al. 1999)
    3.597, 3.84 (calibrated GC-RT correlation, GC-RT correlation, P_L supercooled liquid values, Lei et al. 1999)
    $\log (P_L/Pa) = -3058/(T/K) + 10.81$ (HPLC-RT correlation, Lei et al. 1999)
    5.588   (supercooled liquid P_L, regression with GC-RT from literature, Lei et al. 1999)
    $\log (P_L/Pa) = -3054/(T/K) + 9.97$; (regression with GC-RT from literature, supercooled liquid, Lei et al. 1999)
    6.89*   (23.15°C, transpiration method, measured range 289.1–332.3 K, Verevkin 2003)

ln (P/P°) = 299.001/R − 83941.481/R·(T/K) − (73.5/R)·ln[(T/K)/298.15], where P° = 101.325 kPa, gas constant R = 8.31451 J·K^{-1}·mol^{-1} (vapor pressure eq. from transpiration measurement, temp range 289.1–332.3 K, Verevkin 2003)

Henry's Law Constant (Pa·m³/mol at 25°C):
- 35.5 (gas stripping-GC, Mackay & Shiu 1981)
- 36.3 (gas stripping-GC, Shiu & Mackay 1997)

Octanol/Water Partition Coefficient, log K_{OW} at 25°C and the reported temperature dependence equations:
- 3.80 (HPLC-k′ correlation, Hanai et al. 1981)
- 4.08 (calculated-fragment constant, Yalkowsky et al. 1983)
- 3.90 (shake flask, Opperhuizen 1987)
- 4.08 (estimated, Abernethy & Mackay 1987)
- 3.80 (calculated-molar volume, Wang et al. 1992)
- 4.24 (recommended, Hansch et al. 1995)
- 4.0; 3.97 (calibrated HPLC-RT correlation; HPLC-RT correlation, Lei et al. 2000)
- log K_{OW} = 0.841 + 940.09/(T/K), temp range 5–55°C (temperature dependence HPLC-k′ correlation, Lei et al. 2000)
- 4.06 (GC-RT correlation, Hackenberg et al. 2003)

Octanol/Air Partition Coefficient, log K_{OA} at 25°C or as indicated and reported temperature dependence equations:
- 6.39, 6.10, 5.52, 5.30, 5.13 (10, 20, 30, 40, 50°C, GC-RT correlation, Su et al. 2002)
- log K_{OA} = 58300/(2.303·RT) − 4.40; temp range 10–50°C (GC-RT correlation, Su et al. 2002)

Bioconcentration Factor, log BCF or log K_B:
- 2.28 (*Cyprinus carpio*, for monochloronaphthalenes, Matsuo 1984; quoted, Crookes & Howe 1993)

Sorption Partition Coefficient, log K_{OC}:
- 2.97 (estimated for mono-chloronaphthalenes, Crookes & Howe 1993)

Environmental Fate Rate Constants, k and Half-Lives, $t_{1/2}$:

Half-Lives in the Environment:

TABLE 4.1.3.1.1
Reported vapor pressures of 1-chloronaphthalene at various temperatures and the coefficients for the vapor pressure equations

$$\log P = A − B/(T/K) \quad (1)$$
$$\log P = A − B/(C + t/°C) \quad (2)$$
$$\log P = A − B/(C + T/K) \quad (3)$$
$$\log P = A − B/(T/K) − C·\log(T/K) \quad (4)$$

$$\ln P = A − B/(T/K) \quad (1a)$$
$$\ln P = A − B/(C + t/°C) \quad (2a)$$

$$\ln P/P° = A − B/(T/K) − C·\ln[(T/K)/298.15] \quad (4a)$$

Kahlbaum 1898		Stull 1947		Verevkin 2003	
static-Hg manometer		summary of lit. data		transpiration-GC	
t/°C	P/Pa	t/°C	P/Pa	t/°C	P/Pa
80.6	133.3	80.6	133.3	15.95	3.57
104.8	666.6	104.8	666.6	5.55	1.47
118.6	1333	118.6	1333	7.15	1.75
134.7	2666	134.4	2666	13.85	3.12
140.3	2333	153.2	5333	20.05	5.29
145.1	3999	165.6	7999	23.15	6.89
149.4	4666	180.4	13332	26.05	8.43
153.0	5333	204.2	26664	30.25	11.51

(*Continued*)

TABLE 4.1.3.1.1 (*Continued*)

Kahlbaum 1898		Stull 1947		Verevkin 2003	
static-Hg manometer		summary of lit. data		transpiration-GC	
t/°C	P/Pa	t/°C	P/Pa	t/°C	P/Pa
159.3	6666	230.8	53329	32.15	14.07
171.4	9999	269.3	101325	35.35	18.30
180.4	13332			38.35	24.71
204.2	26664	mp/°C	−20	40.35	28.10
218.3	39997			41.15	29.56
230.8	53329			44.05	35.53
240.5	66661			47.05	46.62
248.6	79993			50.15	55.93
255.5	93326			53.15	69.72
269.3	101325			56.15	84.66
				59.15	105.54
				eq. 4a	P/kPa
				P°	101.325 kPa
				A	299.011/R
				B	83941.481/R
				C	73.5/R
				$R = 8.314$ J K^{-1} mol^{-1}	
				ΔH_V/(kJ mol^{-1}) = 62.03	
				at 298.15 K	

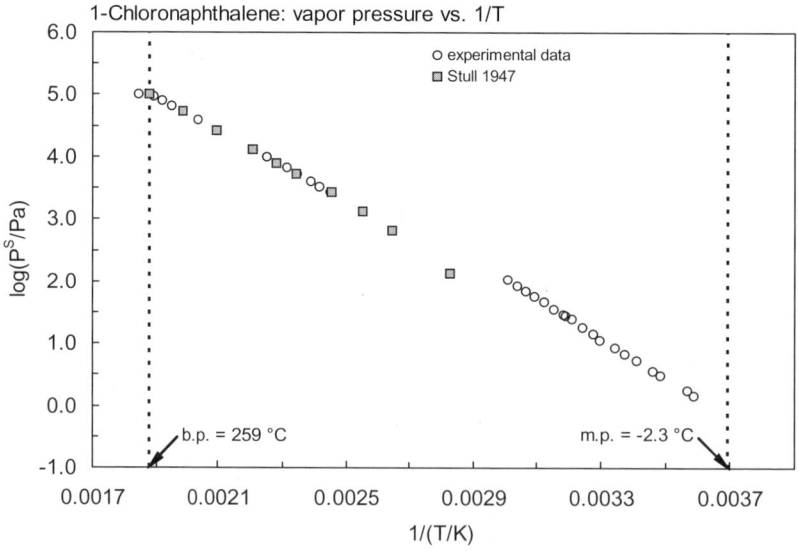

FIGURE 4.1.3.1.1 Logarithm of vapor pressure versus reciprocal temperature for 1-chloronaphthalene.

Polynuclear Aromatic Hydrocarbons (PAHs) and Related Aromatic Hydrocarbons

4.1.3.2 2-Chloronaphthalene

Common Name: 2-Chloronaphthalene
Synonym: PCN-2, β-chloronaphthalene
Chemical Name: 2-chloronaphthalene
CAS Registry No: 91-58-7
Molecular Formula: $C_{10}H_7Cl$
Molecular Weight: 162.616
Melting Point (°C):
    58   (Lide 2003)
Boiling Point (°C):
    256   (Weast 1982–83; Windholz 1983; Budavari 1989; Järnberg et al. 1994; Lide 2003)
Density (g/cm³):
    1.1377   (71°C, Weast 1982–83; Lide 2003)
Molar Volume (cm³/mol):
    142.9   (71°C, calculated from density, Stephenson & Malanowski 1987)
    168.5   (calculated-Le Bas method at normal boiling point)
Enthalpy of Fusion, ΔH_{fus} (kJ/mol):
    3.346   (Ruelle & Kesselring 1997)
Entropy of Fusion, ΔS_{fus} (J/mol K):
Fugacity Ratio at 25°C (assuming ΔS_{fus} = 56 J/mol K), F: 0.474 (mp at 58°C)

Water Solubility (g/m³ or mg/L at 25°C):
    11.7   (shake flask-fluorescence, Mackay & Shiu 1981)
    0.924   (shake flask, Opperhuizen et al. 1985, 1986)
    8.93   (calculated-molecular connectivity indices, Nirmalakhandan & Speece 1989)
    16.3   (calculated-group contribution method, Kühne et al. 1995)
    7.80   (calculated-molar volume, mp and mobile order thermodynamics, Ruelle & Kesselring 1997)

Vapor Pressure (Pa at 25°C or as indicated and reported temperature dependence equations. Additional data at other temperatures designated * are compiled at the end of this section):
    5.34   (supercooled liquid value, extrapolated from Antoine eq., temperature range 400–435 K, Stephenson & Malanowski 1987)
    $\log (P_L/\text{kPa}) = 7.8608 - [3021.2/(T/K)]$; temp range 400–435 K (Antoine eq., Stephenson & Malanowski 1987)
    3.679, 3.84 (supercooled liquid values P_L: calibrated GC-RT correlation, GC-RT correlation, Lei et al. 1999)
    $\log (P_L/\text{Pa}) = -3054/(T/K) + 10.81$ (GC-RT correlation, supercooled liquid, Lei et al. 1999)
    2.526   (supercooled liquid P_L, regression with GC-RT data from literature, Lei et al. 1999)
    $\log (P_L/\text{Pa}) = -3054/(T/K) + 9.97$ (regression with GC-RT data from literature, Lei et al. 1999)
    2.301*   (24.15°C, transpiration method, measured range 280.2–330.7 K, Verevkin 2003)
    $\ln (P/P^\circ) = 301.255/R - 87496.950/R \cdot (T/K) - (39.5/R) \cdot \ln[(T/K)/298.15]$, where P° = 101.325 kPa, gas constant R = 8.31451 J·K⁻¹·mol⁻¹ (vapor pressure eq. from transpiration measurement, solid, temp range 280.2–330.7 K, Verevkin 2003)
    53.71*   (59.05°C, transpiration method, measured range 332.2–362.2 K, Verevkin 2003)
    $\ln (P/P^\circ) = 294.501/R - 84197.803/R \cdot (T/K) - (73.5/R) \cdot \ln[(T/K)/298.15]$, where P° = 101.325 kPa, gas constant R = 8.31451 J·K⁻¹·mol⁻¹ (vapor pressure eq. from transpiration measurements, liquid, temp range 332.2–362.2 K, Verevkin 2003)

Henry's Law Constant (Pa·m³/mol at 25°C):
    31.9   (gas stripping-GC, Mackay & Shiu 1981)
    33.5   (gas stripping-GC, Shiu & Mackay 1997)

Octanol/Water Partition Coefficient, log K_{OW} at 25°C and the reported temperature dependence equations:
- 4.80 (calculated-fragment constant, Yalkowsky et al. 1983)
- 4.19 (HPLC-RT correlation, Opperhuizen et al. 1985, 1986)
- 4.08 (estimated, Abernethy & Mackay 1987)
- 3.98 (shake flask, Opperhuizen 1987)
- 4.07 (selected, Isnard & Lambert 1988; 1989)
- 4.6024 (calculated-UNIFAC group contribution, Chen et al. 1993)
- 4.14 (selected, Hansch et al. 1995)
- 3.90; 3.91 (calibrated HPLC-RT correlation; HPLC-RT correlation, Lei et al. 2000)
- log K_{OW} = 0.821 + 924.42/(T/K), temp range 5–55°C (temperature dependence HPLC-k′ correlation, Lei et al. 2000)

Octanol/Air Partition Coefficient, log K_{OA} at 25°C or as indicated and reported temperature dependence equations. Additional data at other temperatures designated * are compiled at the end of this section:
- 6.36, 6.08, 5.50, 5.28, 5.11 (10, 20, 30, 40, 50°C, GC-RT correlation, Su et al. 2002)
- log K_{OA} = 58000/(2.303·RT) – 4.40; temp range 10–50°C (GC-RT correlation, Su et al. 2002)

Bioconcentration Factor, log BCF or log K_B:
- 2.28 (*Cyprinus carpio*, for monochloronaphthalenes, Matsuo 1981)
- 3.63 (guppies, Opperhuizen et al. 1985)
- 3.63, 4.81 (whole fish, fish lipid, Gobas et al. 1987)
- 4.52 (guppy, lipid-weight based, Gobas et al. 1989)
- 3.63; 3.06 (quoted means; calculated-K_{OW} and S_0, Banerjee & Baughman 1991)
- 3.63 (*Poecilia reticulata*, under static and semi-static conditions, quoted, Devillers et al. 1996)
- 2.496, 2.721 (calculated-MCI χ, calculated-K_{OW}, Lu et al. 1999)

Sorption Partition Coefficient, log K_{OC}:
- 2.97 (estimated for monochloronaphthalenes, Crookes & Howe 1993)

Environmental Fate Rate Constants, k and Half-Lives, $t_{½}$:

Volatilization:

Photolysis:

Photooxidation:

Hydrolysis: laboratory determined hydrolysis rate constant k = (9.5 ± 2.8) × 10^{-6} h^{-1} at neutral conditions, calculated $t_{½}$ = 8.3 yr at pH 7 (Ellington et al. 1988).

Biodegradation:

Biotransformation:

Bioconcentration and Uptake and Elimination Rate Constants (k_1 and k_2):
- k_1 = 7.3 × 10^2 d^{-1}; k_2 = 3.1 × 10^{-1} d^{-1} (guppy, Opperhuizen et al. 1985; quoted, Connell & Hawker 1988)
- log k_1 = 2.83 d^{-1}; log k_2 = –0.51 d^{-1} (guppy, Gobas et al. 1989)

Half-Lives in the Environment:

Air:

Surface water:

Ground water:

Sediment:

Soil: experimentally measured abiotic disappearance $t_{½}$ = 11.3 d in two different soil types, a Captina silt loam (Typic Fragiudult) and McLaurin sandy loam (Typic Paleudults) (Anderson et al. 1991).

Biota: $t_{½}$ = 2.3 d (female guppies, Opperhuizen et al. 1985, quoted, Crookes & Howe 1993)

TABLE 4.1.3.2.1
Reported vapor pressures of 2-chloronaphthalene at various temperatures and the coefficients for the vapor pressure equations

$\log P = A - B/(T/K)$ (1) $\qquad \ln P = A - B/(T/K)$ (1a)

$\log P = A - B/(C + t/°C)$ (2) $\qquad \ln P = A - B/(C + t/°C)$ (2a)

$\log P = A - B/(C + T/K)$ (3)

$\log P = A - B/(T/K) - C \cdot \log (T/K)$ (4) $\qquad \ln P/P^o = A - B/(T/K) - C \cdot \ln [(T/K)/298.15]$ (4a)

Verevkin 2003			
transpiration-GC			
t/°C	P/Pa	t/°C	P/Pa
	solid		liquid
7.05	0.342	59.05	53.71
10.15	0.494	62.05	64.43
15.05	0.881	65.05	77.53
17.05	1.103	68.05	95.98
19.05	1.394	71.05	111.91
21.05	1.716	74.05	134.91
24.15	2.301	77.05	161.0
27.15	3.162	80.05	190.2
29.05	3.854	83.05	223.41
30.15	4.361	85.05	252.43
32.05	5.167	89.05	310.78
34.05	6.297		
36.05	7.569	ΔH_V/(kJ mol^{-1}) = 62.3 ± 1.1	
38.05	9.386	at 298.15 K	
40.05	10.941		
42.05	13.466	eq. 4a	P/kPa
44.04	15.786	P^o	101.325 kPa
46.05	18,715	A	294.501/R
48.05	22.307	B	84197.803/R
50.05	26.990	C	73.5/R
52.05	31.158	R = 8.314 J K^{-1} mol^{-1}	
54.05	36.801		
56.05	43.420		
57.55	49.194		
eq. 4a	P/kPa		
P^o	101.325 kPa		
A	301.255/R		
B	87496.95/R		
C	39.5/R		
ΔH_{subl}/(kJ mol^{-1}) = 75.72			
at 298.15 K			

4.1.3.3 1,2-Dichloronaphthalene

Common Name: 1,2-Dichloronaphthalene
Synonym: PCN-3
Chemical Name: 1,2-dichloronaphthalene
CAS Registry No: 2050-69-3
Molecular Formula: $C_{10}H_6Cl_2$
Molecular Weight: 197.061
Melting Point (°C):
    36   (Lide 2003)
Boiling Point (°C):
    295.6   (Lide 2003)
Density (g/cm^3): 1.3147 (49°C, Weast 1982–83; Lide 2003)
Molar Volume (cm^3/mol):
    156   (Ruelle & Kesselring 1997)
    189.4   (calculated-Le Bas method at normal boiling point)
Enthalpy of Fusion, ΔH_{fus} (kJ/mol):
Entropy of Fusion, ΔS_{fus} (J/mol K):
Fugacity Ratio at 25°C (assuming ΔS_{fus} = 56 J/mol K), F: 0.78 (mp at 36°C)

Water Solubility (g/m^3 or mg/L at 25°C):
    0.137   (generator column-GC/ECD, Opperhuizen 1987)
    4.31   (calculated-molar volume, mp and mobile order thermodynamics, Ruelle & Kesselring 1997)

Vapor Pressure (Pa at 25°C and reported temperature dependence equations):
    0.344; 0.333   (supercooled liquid P_L: calibrated GC-RT correlation; GC-RT correlation, Lei et al. 1999)
    log (P_L/Pa) = -3172/(T/K) + 10.18; (GC-RT correlation, supercooled liquid, Lei et al. 1999)
    0.301   (supercooled liquid P_L, regression with GC-RT from literature, Lei et al. 1999)
    log (P_L/Pa) = -3172/(T/K) + 10.11 (regression with GC-RT from literature, Lei et al. 1999)

Henry's Law Constant (Pa·m^3/mol):

Octanol/Water Partition Coefficient, log K_{OW} at 25°C and the reported temperature dependence equations:
    4.40   (HPLC-RT correlation, Opperhuizen et al. 1985)
    4.42   (shake flask, Opperhuizen 1987; quoted, Sangster 1993, Crookes & Howe 1993, Hansch et al. 1995)
    4.66   (selected, Alcock et al. 1999)
    4.60; 4.45   (calibrated HPLC-RT correlation; HPLC-RT correlation, Lei et al. 2000)
    log K_{OW} = 1.064 + 1060.21/(T/K), temp range 5–55°C (temperature dependence HPLC-k′ correlation, Lei et al. 2000)
    4.69   (GC-RT correlation, Hackenberg et al. 2003)

Octanol/Air Partition Coefficient, log K_{OA} at 25°C or as indicated and reported temperature dependence equations:
    6.93   (generator column-GC, Harner & Bidleman 1998)
    7.35, 7.01, 6.44, 6.13, 5.91 (10, 20, 30, 40, 50°C, GC-RT correlation, Su et al. 2002)
    log K_{OA} = 66000/(2.303·RT) – 4.800; temp range 10–50°C (GC-RT correlation, Su et al. 2002)
    6.89; 7.01   (calibrated GC-RT correlation; GC-RT correlation, Wania et al. 2002)

Bioconcentration Factor, log BCF or log K_B:
    3.40   (fish, Opperhuizen et al. 1985)

Sorption Partition Coefficient, log K_{OC}:

Environmental Fate Rate Constants, k and Half-Lives, $t_{1/2}$:

Half-Lives in the Environment:

4.1.3.4 1,4-Dichloronaphthalene

Common Name: 1,4-Dichloronaphthalene
Synonym: PCN-5
Chemical Name: 1,4-dichloronaphthalene
CAS Registry No: 1825-31-6
Molecular Formula: $C_{10}H_6Cl_2$
Molecular Weight: 197.061
Melting Point (°C):
    67.5   (Lide 2003)
Boiling Point (°C):
    288   (Lide 2003)
Density (g/cm³):
Molar Volume (cm³/mol):
    156.0   (Ruelle & Kesselring 1997)
    189.4   (calculated-Le Bas method at normal boiling point)
Enthalpy of Fusion, ΔH_{fus} (kJ/mol):
Entropy of Fusion, ΔS_{fus} (J/mol K):
Fugacity Ratio at 25°C (assuming ΔS_{fus} = 56 J/mol K), F: 0.383 (mp at 288°C)

Water Solubility (g/m³ or mg/L at 25°C):
    0.314   (generator column-GC/ECD, Opperhuizen et al. 1985)
    0.314   (generator column-GC/ECD, Opperhuizen et al. 1987)
    4.02   (calculated-TSA, Dickhut et al. 1994)
    1.98   (calculated-molar volume, mp and mobile order thermodynamics, Ruelle & Kesselring 1997)

Vapor Pressure (Pa at 25°C and reported temperature dependence equations):
    0.173   (estimated, Crookes & Howe 1993)
    9.98×10^{-6} (estimated, Alcock et al. 1999)
    0.428; 0.416 (supercooled liquid P_L: calibrated GC-RT correlation; GC-RT correlation, Lei et al. 1999)
    $\log (P_L/Pa) = -3067/(T/K) + 9.92$ (GC-RT correlation, supercooled liquid, Lei et al. 1999)
    0.353   (supercooled liquid P_L, regression with GC-RT from literature, Lei et al. 1999)
    $\log (P_L/Pa) = -3067/(T/K) + 10.17$ (regression with GC-RT from literature, Lei et al. 1999)

Henry's Law Constant (Pa·m³/mol):

Octanol/Water Partition Coefficient, log K_{OW} at 25°C and the reported temperature dependence equations:
    4.88   (HPLC-RT correlation, Opperhuizen et al. 1985)
    4.66   (shake flask, Opperhuizen 1987; quoted, Gobas et al. 1987; 1989; Clark et al. 1990; Sangster 1993; Crookes & Howe 1993; Hansch et al. 1995; Devillers et al. 1996; Alcock et al. 1999)
    4.79   (calculated, Oliver & Niimi 1984; Oliver 1987)
    4.80; 4.57   (calibrated HPLC-RT correlation; HPLC-RT correlation, Lei et al. 2000)
    $\log K_{OW} = 1.269 + 1049.8/(T/K)$, temp range 5–55°C (temperature dependence HPLC-k′ correlation, Lei et al. 2000)
    4.56   (GC-RT correlation, Hackenberg et al. 2003)

Octanol/Air Partition Coefficient, log K_{OA} at 25°C or as indicated and reported temperature dependence equations:
    6.93   (generator column-GC/MS, Harner & Bidleman 1998)
    7.52, 7.13, 6.72, 6.38, 6.13 (10, 20, 30, 40, 50°C, generator column-GC/MS, Harner & Bidleman 1998)

log K_{OA} = –3.97 + 3248/(T/K), temp range: 10–50°C (generator column-GC/MS, Harner & Bidleman 1998)
    6.78; 6.91 (calibrated GC-RT correlation; GC-RT correlation, Wania et al. 2002)

Bioconcentration Factor, log BCF or log K_B:
    3.75   (*Oncorhynchus mykiss*, Oliver & Niimi 1984)
    3.36   (female guppies, Opperhuizen et al. 1985)
    3.80   (Opperhuizen et al. 1985)
    4.04   (Opperhuizen et al. 1985)
    3.36, 4.54 (guppies: whole fish, fish lipid, Gobas et al. 1987)
    3.75   (rainbow trout, Oliver & Niimi 1984; Oliver 1987)
    5.18   (guppy, lipid-weight based, Gobas et al. 1989)
    3.36; 4.63 (quoted means; calculated-K_{OW} and S_0, Banerjee & Baughman 1991)
    3.75   (*Oncorhynchus mykiss*, under flow-through condition, quoted Devillers et al. 1996)
    3.36   (*Poecilia reticulata*, under static and semi-static conditions, quoted, Devillers et al. 1996)

Sorption Partition Coefficient, log K_{OC}:

Environmental Fate Rate Constants, k and Half-Lives, $t_{1/2}$:
    Volatilization:
    Photolysis:
    Photooxidation: rate constant k = 5.8 × 10^{-12} cm^3 molecule^{-1} s^{-1} for the gas-phase reactions with OH radical at 298 ± 2 K (Atkinson 1989).
    Hydrolysis:
    Biodegradation:
    Biotransformation:
    Bioconcentration and Uptake and Elimination Rate Constants (k_1 and k_2):
        k_1 = 1.2 × 10^3 d^{-1}; k_2 = 1.1 × 10^{-1} d^{-1} (guppies, Opperhuizen et al. 1985)
        log k_1 = 3.04 d^{-1}; log k_2 = –0.96 d^{-1} (guppy, Gobas et al. 1989)

Half-Lives in the Environment:
    Biota: $t_{1/2}$ = 6.2 d (guppies, Opperhuizen et al. 1985; quoted, Crookes & Howe 1993)

4.1.3.5 1,8-Dichloronaphthalene

Common Name: 1,8-Dichloronaphthalene
Synonym: PCN-9
Chemical Name: 1,8-dichloronaphthalene
CAS Registry No: 2050-74-0
Molecular Formula: $C_{10}H_6Cl_2$
Molecular Weight: 197.061
Melting Point (°C):
    89   (Weast 1982-83; Lide 2003)
Boiling Point (°C):
    sublimation (Lide 2003)
Density (g/cm³):
Molar Volume (cm³/mol):
    156.0   (Ruelle & Kesselring 1997)
    189.4   (calculated-Le Bas method at normal boiling point)
Enthalpy of Fusion, ΔH_{fus} (kJ/mol):
Entropy of Fusion, ΔS_{fus} (J/mol K):
Fugacity Ratio at 25°C (assuming ΔS_{fus} = 56 J/mol K), F: 0.236 (mp at 89°C)

Water Solubility (g/m³ or mg/L at 25°C):
    0.315   (generator column-GC/ECD, Opperhuizen et al. 1985)
    0.059   (generator column-GC/ECD, Opperhuizen et al. 1987)
    0.309   (Isnard & Lambert 1988, 1989; quoted, Crookes & Howe 1993)
    1.27   (calculated-molar volume mp and mobile order thermodynamics, Ruelle & Kesselring 1997)

Vapor Pressure (Pa at 25°C and the reported temperature dependence equation):
    0.198   (supercooled liquid P_L, regression with GC-RT from literature, Lei et al. 1999)
    $\log (P_L/Pa) = -3169/(T/K) + 9.93$ (regression with GC-RT from literature, Lei et al. 1999)

Henry's Law Constant (Pa·m³/mol):

Octanol/Water Partition Coefficient, log K_{OW}:
    4.41   (HPLC-RT, Opperhuizen et al. 1985)
    4.19   (shake flask-GC, Opperhuizen 1987; quoted, Sangster 1993; Hansch et al. 1995)
    5.4348   (calculated-UNIFAC, Chen et al. 1993)
    4.85   (GC-RT correlation, Hackenberg et al. 2003)

Octanol/Air Partition Coefficient, log K_{OA}:

Bioconcentration Factor, log BCF or log K_B:
    3.79   (guppies, Opperhuizen et al. 1985)
    3.79. 4.96   (guppies: whole fish, fish lipid, Gobas et al. 1987)
    4.95   (guppy, lipid-weight based, Gobas et al. 1989)
    3.79   (*Poecilia reticulata*, under static and semi-static conditions, quoted, Devillers et al. 1996)

Sorption Partition Coefficient, log K_{OC}:

Environmental Fate Rate Constants, k and Half-Lives, $t_{½}$:
    Bioconcentration and Uptake and Elimination Rate Constants (k_1 and k_2):
        $k_1 = 9.8 \times 10^2$ d^{-1}; $k_2 = 1.6 \times 10^{-1}$ d^{-1} (guppies, Opperhuizen et al. 1985)
        $\log k_1 = 2.97$ d^{-1}; $\log k_2 = -0.80$ d^{-1} (guppy, Gobas et al. 1989)

Half-Lives in the Environment:
    Biota: elimination $t_{½}$ = 4.3 d (guppies, Opperhuizen et al. 1985; Crookes & Howe 1993)

4.1.3.6 2,3-Dichloronaphthalene

Common Name: 2,3-Dichloronaphthalene
Synonym: PCN-10
Chemical Name: 2,3-dichloronaphthalene
CAS Registry No: 2050-75-1
Molecular Formula: $C_{10}H_6Cl_2$
Molecular Weight: 197.061
Melting Point (°C):
   120   (Weast 1982-83; Ruelle & Kesselring 1997; Lide 2003)
Boiling Point (°C):
Density (g/cm³):
Molar Volume (cm³/mol):
   156.0   (Ruelle & Kesselring 1997)
   189.4   (calculated-Le Bas method at normal boiling point)
Enthalpy of Fusion, ΔH_{fus} (kJ/mol):
Entropy of Fusion, ΔS_{fus} (J/mol K):
Fugacity Ratio at 25°C (assuming ΔS_{fus} = 56 J/mol K), F: 0.117 (mp at 120°C)

Water Solubility (g/m³ or mg/L at 25°C):
   0.0862   (generator column-GC/ECD, Opperhuizen et al. 1985; 1987)
   0.623   (calculated-molar volume, mp and mobile order thermodynamics, Ruelle & Kesselring 1997)

Vapor Pressure (Pa at 25°C and the reported temperature dependence equation):
   0.333   (supercooled liquid P_L, regression with GC-RT from literature, Lei et al. 1999)
   $\log (P_L/Pa) = -3169/(T/K) + 10.15$ (regression with GC-RT from literature, Lei et al. 1999)

Henry's Law Constant (Pa·m³/mol):

Octanol/Water Partition Coefficient, log K_{OW}:
   4.71   (HPLC-RT correlation, Opperhuizen et al. 1985)
   4.51   (shake flask, Opperhuizen 1987)
   4.78   (GC-RT correlation, Hackenberg et al. 2003)

Octanol/Air Partition Coefficient, log K_{OA}:

Bioconcentration Factor, log BCF or log K_B:
   4.04   (guppies, Opperhuizen et al. 1985)
   4.04, 5.22 (guppies: whole fish, fish lipid, Gobas et al. 1987)
   5.08   (guppy, lipid-weight based, Gobas et al. 1989)
   4.04   (*Poecilia reticulata*, under static and semi-static conditions, quoted, Devillers et al. 1996)

Sorption Partition Coefficient, log K_{OC}:

Environmental Fate Rate Constants, k and Half-Lives, $t_{½}$:
   Bioconcentration and Uptake and Elimination Rate Constants (k_1 and k_2):
      $k_1 = 1.6 \times 10^3$ d^{-1}; $k_2 = 1.4 \times 10^{-1}$ d^{-1} (guppies, Opperhuizen et al. 1985)
      log k_1 = 3.05 d^{-1}; log k_2 = –0.85 d^{-1} (guppy, Gobas et al. 1989)

Half-Lives in the Environment:
   Biota: elimination $t_{½}$ = 5.1 d (guppies, Opperhuizen et al. 1985; quoted, Crookes & Howe 1993)

4.1.3.7 2,7-Dichloronaphthalene

Common Name: 2,7-Dichloronaphthalene
Synonym: PCN-12
Chemical Name: 2,7-dichloronaphthalene
CAS Registry No: 2198-77-8
Molecular Formula: $C_{10}H_6Cl_2$
Molecular Weight: 197.061
Melting Point (°C):
    115   (Lide 2003)
Boiling Point (°C):
Density (g/cm³):
Molar Volume (cm³/mol):
    156.0   (Ruelle & Kesselring 1997)
    189.4   (calculated-Le Bas method at normal boiling point)
Enthalpy of Fusion, ΔH_{fus} (kJ/mol):
Entropy of Fusion, ΔS_{fus} (J/mol K):
Fugacity Ratio at 25°C (assuming ΔS_{fus} = 56 J/mol K), F: 0.131 (mp at 115°C)

Water Solubility (g/m³ or mg/L at 25°C):
    0.236   (generator column-GC/ECD, Opperhuizen et al. 1985)
    0.235   (reported as 2,8-dichloronaphthalene, generator column-GC/ECD, Opperhuizen 1987)
    0.699   (calculated-molar volume mp and mobile order thermodynamics, Ruelle & Kesselring 1997)

Vapor Pressure (Pa at 25°C and reported temperature dependence equation):
    0.344   (supercooled liquid P_L, regression with GC-RT from literature, Lei et al. 1999)
    $\log (P_L/Pa) = -3169/(T/K) + 10.16$ (regression with GC-RT from literature, Lei et al. 1999)

Henry's Law Constant (Pa·m³/mol):

Octanol/Water Partition Coefficient, log K_{OW}:
    4.81   (HPLC-RT correlation, Opperhuizen et al. 1985)
    4.56   (shake flask-GC, Opperhuizen 1987)

Octanol/Air Partition Coefficient, log K_{OA} at 25°C and reported temperature dependence equation:
    7.28, 6.95, 6.38, 6.08, 5.85 (10, 20, 30, 40, 50°C, GC-RT correlation, Su et al. 2002)
    $\log K_{OA} = 65400/(2.303 \cdot RT) - 4.80$; temp range 10–50°C (GC-RT correlation, Su et al. 2002)

Bioconcentration Factor, log BCF or log K_B:
    4.04   (guppies, Opperhuizen et al. 1985; Crookes & Howe 1993, Lu et al. 1999)
    4.04   (guppies: whole fish, fish lipid, Gobas et al. 1987)
    5.11   (guppy, lipid-weight based, Gobas et al. 1989)

Sorption Partition Coefficient, log K_{OC}:

Environmental Fate Rate Constants, k and Half-Lives, $t_{1/2}$:
    Bioconcentration and Uptake and Elimination Rate Constants (k_1 and k_2):
        $k_1 = 1.6 \times 10^3$ d^{-1}; $k_2 = 1.4 \times 10^{-1}$ d^{-1} (guppies, Opperhuizen et al. 1985)
        $\log k_1 = 3.08$ d^{-1}; $\log k_2 = -0.85$ d^{-1} (guppy, Gobas et al. 1989)

Half-Lives in the Environment:
    Biota: depuration $t_{1/2}$ = 5.1 d (guppies, Opperhuizen et al. 1985; quoted, Crookes & Howe 1993)

4.1.3.8 1,2,3-Trichloronaphthalene

Common Name: 1,2,3-Trichloronaphthalene
Synonym: PCN-13
Chemical Name: 1,2,3-trichloronaphthalene
CAS Registry No:50402-52-3
Molecular Formula: $C_{10}H_5Cl_3$
Molecular Weight: 231.506
Melting Point (°C):
    81–84   (Järnberg et al. 1994)
Boiling Point (°C):
Density (g/cm³):
Molar Volume (cm³/mol):
    210.3   (calculated-Le Bas method at normal boiling point)
Enthalpy of Vaporization, ΔH_V (kJ/mol):
Enthalpy of Sublimation, ΔH_{subl} (kJ/mol):
Enthalpy of Fusion, ΔH_{fus} (kJ/mol):
Entropy of Fusion, ΔS_{fus} (J/mol K):
Fugacity Ratio at 25°C (assuming ΔS_{fus} = 56 J/mol K), F:

Water Solubility (g/m³ or mg/L at 25°C):

Vapor Pressure (Pa at 25°C and reported temperature dependence equation):
    0.071; 0.0652(supercooled liquid P_L: calibrated GC-RT correlation; GC-RT correlation, Lei et al. 1999)
    log (P_L/Pa) = – 3551/(T/K) + 10.76 (GC-RT correlation, supercooled liquid, Lei et al. 1999)
    0.0791   (supercooled liquid P_L, regression with GC-RT data from literature, Lei et al. 1999)
    log (P_L/Pa) = –3485/(T/K) + 10.59 (regression with GC-RT from literature, Lei et al. 1999)

Henry's Law Constant (Pa·m³/mol at 25°C):

Octanol/Water Partition Coefficient, log K_{OW}:

Octanol/Air Partition Coefficient, log K_{OA} at 25°C and reported temperature dependence equation:
    8.24, 7.85, 7.30, 6.91, 6.63 (10, 20, 30, 40, 50°C, GC-RT correlation, Su et al. 2002)
    log K_{OA} = 73200/(2.303·RT) – 5.20; temp range 10–50°C (GC-RT correlation, Su et al. 2002)
    7.66; 7.72 (calibrated GC-RT correlation; GC-RT correlation, Wania et al. 2002)

Bioconcentration Factor, log BCF or log K_B:
Sorption Partition Coefficient, log K_{OC}:

Environmental Fate Rate Constants, k and Half-Lives, $t_{1/2}$:
Half-Lives in the Environment:

4.1.3.9 1,3,7-Trichloronaphthalene

Common Name: 1,3,7-Trichloronaphthalene
Synonym: PCN-21
Chemical Name: 1,3,7-trichloronaphthalene
CAS Registry No: 55720-37-1
Molecular Formula: $C_{10}H_5Cl_3$
Molecular Weight: 231.506
Melting Point (°C):
    113   (Crookes & Howe 1993)
Boiling Point (°C): 274
Density (g/cm³):
Molar Volume (cm³/mol):
    168.9   (Ruelle & Kesselring 1997)
    210.3   (calculated-Le Bas method at normal boiling point)
Enthalpy of Fusion, ΔH_{fus} (kJ/mol):
Entropy of Fusion, ΔS_{fus} (J/mol K):
Fugacity Ratio at 25°C (assuming ΔS_{fus} = 56 J/mol K), F: 0.137 (mp at 113°C)

Water Solubility (g/m³ or mg/L at 25°C):
    0.0644   (generator column-GC/ECD, Opperhuizen et al. 1985, 1987)
    0.049   (Opperhuizen et al. 1986)
    2.85   (calculated-molar volume, mp and mobile order thermodynamics, Ruelle & Kesselring 1997)

Vapor Pressure (Pa at 25°C and reported temperature dependence equation):
    0.127   (estimated, Crookes & Howe 1993)
    7.10×10^{-6} (estimated, Alcock et al. 1999)
    0.114   (supercooled liquid P_L, regression with GC-RT from literature, Lei et al. 1999)
    $\log (P_L/\text{Pa}) = -3485/(T/K) + 10.74$ (regression with GC-RT from literature, Lei et al. 1999)
    $0.0778 - P_S$, $0.359 - P_L$ (estimated for trichloronaphthalenes, Kaupp & McLachlan 1999)

Henry's Law Constant (Pa·m³/mol):

Octanol/Water Partition Coefficient, log K_{OW}:
    5.59   (HPLC-RT correlation, Opperhuizen et al. 1985)
    5.60   (selected, Opperhuizen et al. 1986)
    5.35   (shake flask, Opperhuizen 1987)
    5.08   (GC-RT correlation, Hackenberg et al. 2003)

Octanol/Air Partition Coefficient, log K_{OA}:

Bioconcentration Factor, log BCF or log K_B:
    3.94   (*Cyprinus carpio*, for trichloronaphthalenes, Matsuo 1981)
    4.43   (guppies, Opperhuizen et al. 1985)
    4.43, 5.61 (guppies: whole fish, fish lipid, Gobas et al. 1987)
    5.96   (guppy, lipid-weight based, Gobas et al. 1989)
    4.08   (calculated-K_{OW} and solubility, Banerjee & Baughman 1991)
    4.43   (*Poecilia reticulata*, under static and semi-static conditions, Devillers et al. 1996)

Sorption Partition Coefficient, log K_{OC}:

Environmental Fate Rate Constants, k and Half-Lives, $t_{1/2}$:
    Bioconcentration and Uptake and Elimination Rate Constants (k_1 and k_2):
        $k_1 = 2.3 \times 10^3$ d^{-1}; $k_2 = 8.4 \times 10^{-2}$ d^{-1} (guppy, Opperhuizen et al. 1985)
        $k_1 = 1.7 \times 10^3$ d^{-1} (estimated, Opperhuizen et al. 1985)
        $\log k_1 = 3.14$ d^{-1}; $\log k_2 = -1.64$ d^{-1} (guppy, Gobas et al. 1989)

Half-Lives in the Environment:
    Biota: $t_{1/2} = 8.3$ d (guppies, Opperhuizen et al. 1985; quoted, Crookes & Howe 1993)

4.1.3.10 1,2,3,4-Tetrachloronaphthalene

Common Name: 1,2,3,4-Tetrachloronaphthalene
Synonym: PCN-27
Chemical Name: 1,2,3,4-tetrachloronaphthalene
CAS Registry No: 20020-02-4
Molecular Formula: $C_{10}H_4Cl_4$
Molecular Weight: 265.951
Melting Point (°C):
    199   (Lide 2003)
Boiling Point (°C):
Density (g/cm³):
Molar Volume (cm³/mol):
    181.8   (Ruelle & Kesselring 1997)
    231.2   (calculated-Le Bas method at normal boiling point)
Enthalpy of Fusion, ΔH_{fus} (kJ/mol):
Entropy of Fusion, ΔS_{fus} (J/mol K):
Fugacity Ratio at 25°C (assuming ΔS_{fus} = 56 J/mol K), F: 0.0196 (mp at 199°C)

Water Solubility (g/m³ or mg/L at 25°C):
    0.00426   (generator column-GC/ECD, Opperhuizen et al. 1985)
    0.0042   (generator column-GC/ECD, Opperhuizen 1987)
    0.0172   (calculated-AQUAFAC, Myrdal et al. 1995)
    0.016   (calculated-molar volume, mp and mobile order thermodynamics, Ruelle & Kesselring 1997)

Vapor Pressure (Pa at 25°C and reported temperature dependence equations):
    0.0197; 0.0173 (supercooled liquid P_L: calibrated GC-RT correlation; GC-RT correlation, Lei et al. 1999)
    log (P_L/Pa) = –3825/(T/K) + 11.12; (GC-RT correlation, supercooled liquid, Lei et al. 1999)
    0.0162(supercooled liquid P_L, regression with GC-RT from literature, Lei et al. 1999)
    log (P_L/Pa) = –3825/(T/K) + 10.96 (regression with GC-RT from literature, Lei et al. 1999)
    0.00536 – P_S; 0.0975 – P_L (estimated for tetrachloronaphthalenes, Kaupp & McLachlan 1999)

Henry's Law Constant (Pa·m³/mol):

Octanol/Water Partition Coefficient, log K_{OW} at 25°C and the reported temperature dependence equations:
    5.94   (HPLC-RT correlation, Opperhuizen et al. 1985)
    5.90   (HPLC-RT correlation, Opperhuizen et al. 1985)
    5.75   (shake flask-GC, Opperhuizen 1987)
    6.30; 5.76   (calibrated HPLC-RT correlation; HPLC-RT correlation, Lei et al. 2000)
    log K_{OW} = 1.832 + 1347.46/(T/K), temp range 5–55°C (temperature dependence HPLC-k' correlation, Lei et al. 2000)
    5.91   (GC-RT correlation, Hackenberg et al. 2003)

Octanol/Air Partition Coefficient, log K_{OA} at 25°C or as indicated and reported temperature dependence equation:
    9.03, 8.59, 8.05, 7.59, 7.26 (10, 20, 30, 40, 50°C, GC-RT correlation, Su et al. 2002)
    log K_{OA} = 79500/(2.303·RT) – 5.60; temp range 10–50°C (GC-RT correlation, Su et al. 2002)
    8.30; 8.29   (calibrated GC-RT correlation; GC-RT correlation, Wania et al. 2002)

Bioconcentration Factor, log BCF or log K_B:

    3.94   (*Cyprinus carpio*, Matsuo 1981)
    3.71   (*Oncorhynchus mykiss*, Oliver & Niimi 1984)
    4.52   (guppies, Opperhuizen et al. 1985)
    4.50   (fish, Opperhuizen et al. 1985)
    4.66, 5.71   (whole fish, fish lipid, Gobas et al. 1987)
    3.71   (rainbow trout, mean value, Oliver & Niimi 1985)
    5.96   (guppy, lipid-weight based, Gobas et al. 1989)

Sorption Partition Coefficient, log K_{OC}:

Environmental Fate Rate Constants, k and Half-Lives, $t_{1/2}$:

    Bioconcentration and Uptake and Elimination Rate Constants (k_1 and k_2):
        $k_1 = 3.3 \times 10^3$ d^{-1}; $k_2 = 9.9 \times 10^{-2}$ d^{-1} (guppy, Opperhuizen et al. 1985)
        $k_1 = 1.3 \times 10^3$ d^{-1} (estimated, Opperhuizen et al. 1985)
        log $k_1 = 3.70$ d^{-1}; log $k_2 = -1.08$ d^{-1} (guppy, Gobas et al. 1989)

Half-Lives in the Environment:

    Biota: $t_{1/2} = 7$ d (guppies, Opperhuizen et al. 1985; quoted, Crookes & Howe 1993)

4.1.3.11 1,2,3,5-Tetrachloronaphthalene

Common Name: 1,2,3,5-Tetrachloronaphthalene
Synonym: PCN-28
Chemical Name: 1,2,3,5-tetrachloronaphthalene
CAS Registry No: 53555-63-8
Molecular Formula: $C_{10}H_4Cl_4$
Molecular Weight: 265.951
Melting Point (°C):
    141   (Järnberg et al. 1994)
Boiling Point (°C):
Density (g/cm^3):
Molar Volume (cm^3/mol):
    231.2   (calculated-Le Bas method at normal boiling point)
Enthalpy of Vaporization, ΔH_V (kJ/mol):
Enthalpy of Sublimation, ΔH_{subl} (kJ/mol):
Enthalpy of Fusion, ΔH_{fus} (kJ/mol):
Entropy of Fusion, ΔS_{fus} (J/mol K):
Fugacity Ratio at 25°C (assuming ΔS_{fus} = 56 J/mol K), F: 0.0728 (mp at 141°C)

Water Solubility (g/m^3 or mg/L at 25°C):

Vapor Pressure (Pa at 25°C and reported temperature dependence equation):
    0.0203; 0.0179   (supercooled liquid P_L: calibrated GC-RT correlation; GC-RT correlation, Lei et al. 1999)
    log (P_L/Pa) = $-3836/(T/K) + 11.17$; (GC-RT correlation, supercooled liquid, Lei et al. 1999)
    0.0205   (supercooled liquid P_L, regression with GC-RT data from literature, Lei et al. 1999)
    log (P_L/Pa) = $-3800/(T/K) + 11.06$ (regression with GC-RT data from literature, Lei et al. 1999)

Henry's Law Constant (Pa·m^3/mol at 25°C):

Octanol/Water Partition Coefficient, log K_{OW}:
    5.78   (GC-RT correlation, Hackenberg et al. 2003)

Octanol/Air Partition Coefficient, log K_{OA} at 25°C or as indicated and reported temperature dependence equation:
    8.98, 8.55, 8.00, 7.55, 7.22 (10, 20, 30, 40, 50°C, GC-RT correlation, Su et al. 2002)
    log K_{OA} = $79100/(2.303 \cdot RT) - 5.60$; temp range 10–50°C (GC-RT correlation, Su et al. 2002)
    8.29; 8.28 (calibrated GC-RT correlation; GC-RT correlation, Wania et al. 2002)

Bioconcentration Factor, log BCF or log K_B:
Sorption Partition Coefficient, log K_{OC}:

Environmental Fate Rate Constants, k and Half-Lives, $t_{1/2}$:
Half-Lives in the Environment:

4.1.3.12 1,3,5,7-Tetrachloronaphthalene

Common Name: 1,3,5,7-Tetrachloronaphthalene
Synonym: PCN-42
Chemical Name: 1,3,5,7-tetrachloronaphthalene
CAS Registry No: 53555-64-9
Molecular Formula: $C_{10}H_4Cl_4$
Molecular Weight: 265.951
Melting Point (°C):
    179   (Crookes & Howe 1993)
Boiling Point (°C):
Density (g/cm^3):
Molar Volume (cm^3/mol):
    181.8   (Ruelle & Kesselring 1997)
    231.2   (calculated-Le Bas method at normal boiling point)
Enthalpy of Fusion, ΔH_{fus} (kJ/mol):
Entropy of Fusion, ΔS_{fus} (J/mol K):
Fugacity Ratio at 25°C (assuming ΔS_{fus} = 56 J/mol K), F: 0.0308 (mp at 179°C)

Water Solubility (g/m^3 or mg/L at 25°C):
    0.00426   (generator column-GC/ECD, Opperhuizen et al. 1985)
    0.0040   (generator column-GC/ECD, Opperhuizen 1987)
    0.0237   (calculated-molar volume, mp and mobile order thermodynamics, Ruelle & Kesselring 1997)

Vapor Pressure (Pa at 25°C and reported temperature dependence equation):
    0.0480   (estimated, Crookes & Howe 1993)
    2.70×10^{-6} (estimated, Alcock et al. 1999)
    0.0415   (supercooled liquid P_L, regression with GC-RT from literature, Lei et al. 1999)
    $\log (P_L/Pa) = -3800/(T/K) + 11.36$ (regression with GC-RT from literature, Lei et al. 1999)
    0.00536 – P_S; 0.0975 – P_L (estimated for tetrachloronaphthalenes, Kaupp & McLachlan 1999)

Henry's Law Constant (Pa·m^3/mol):

Octanol/Water Partition Coefficient, log K_{OW}:
    6.38   (HPLC-RT correlation, Opperhuizen et al. 1985)
    6.40   (Opperhuizen 1986)
    6.19   (shake flask, Opperhuizen 1987; selected, Hansch et al. 1995)
    5.54   (GC-RT correlation, Hackenberg et al. 2003)

Octanol/Air Partition Coefficient, log K_{OA} at 25°C or as indicated and reported temperature dependence equation:
    8.39   (estimated value for tetrachloronaphthalenes, Kaupp & McLachlan 1999)
    8.58, 8.18, 7.62, 7.21, 6.90 (10, 20, 30, 40, 50°C, GC-RT correlation, Su et al. 2002)
    $\log K_{OA} = 75000/(2.303 \cdot RT) - 5.40$; temp range 10–50°C (GC-RT correlation, Su et al. 2002)

Bioconcentration Factor, log BCF or log K_B:
    4.53   (guppies, Opperhuizen et al. 1985)
    4.50   (guppies, Opperhuizen et al. 1985)
    5.06, 5.71   (guppies: whole fish, fish lipid, Gobas et al. 1987)

5.81 (guppy, lipid-weight based, Gobas et al. 1989)
4.53; 4.37 (quoted means; calculated-K_{OW} and S, Banerjee & Baughman 1991)
4.701, 4.961 (calculated-MCI χ, calculated-K_{OW}, Lu et al. 1999)

Sorption Partition Coefficient, log K_{OC}:

Environmental Fate Rate Constants, k and Half-Lives, $t_{1/2}$:

Bioconcentration and Uptake and Elimination Rate Constants (k_1 and k_2):
$k_1 = 7.5 \times 10^2$ d^{-1}; $k_2 = 2.2 \times 10^{-2}$ d^{-1} (guppies, Opperhuizen et al. 1985)
log $k_1 = 2.97$ d^{-1}; log $k_2 = -1.66$ d^{-1} (guppy, Gobas et al. 1989)

Half-Lives in the Environment:

Biota: $t_{1/2} = 30$ d (guppies, Opperhuizen et al. 1985)

4.1.3.13 1,3,5,8-Tetrachloronaphthalene

Common Name: 1,3,5,8-Tetrachloronaphthalene
Synonym: PCN-43
Chemical Name: 1,3,5,8-tetrachloronaphthalene
CAS Registry No: 31604-28-1
Molecular Formula: $C_{10}H_4Cl_4$
Molecular Weight: 265.951
Melting Point (°C):
    131   (Crookes & Howe 1993; Järnberg et al. 1994)
Boiling Point (°C):
Density (g/cm³):
Molar Volume (cm³/mol):
    181.8   (Ruelle & Kesselring 1997)
    231.2   (calculated-Le Bas method at normal boiling point)
Enthalpy of Fusion, ΔH_{fus} (kJ/mol):
Entropy of Fusion, ΔS_{fus} (J/mol K):
Fugacity Ratio at 25°C (assuming ΔS_{fus} = 56 J/mol K), F: 0.0912 (mp at 131°C)

Water Solubility (g/m³ or mg/L at 25°C):
- 0.00825   (generator column-GC/ECD, Opperhuizen et al. 1985)
- 0.0030   (Opperhuizen 1986)
- 0.0082   (generator column-GC/ECD, Opperhuizen 1987)
- 0.0716   (calculated-molar volume, mp and mobile order thermodynamics, Ruelle & Kesselring 1997)

Vapor Pressure (Pa at 25°C and reported temperature dependence equation):
- 0.0208   (supercooled liquid P_L, regression with GC-RT from literature, Lei et al. 1999)
- log (P_L/Pa) = $-3800/(T/K) + 11.07$ (regression with GC-RT from literature, Lei et al. 1999)
- 0.00536 – P_S; 0.0975 – P_L (estimated for tetrachloronaphthalenes, Kaupp & McLachlan 1999)

Henry's Law Constant (Pa·m³/mol):

Octanol/Water Partition Coefficient, log K_{OW}:
- 5.96   (HPLC-RT correlation, Opperhuizen et al. 1985; selected, Sangster 1993)
- 6.00   (Opperhuizen et al. 1985; Opperhuizen 1986)
- 5.76   (shake flask, Opperhuizen 1987; selected, Hansch et al. 1995)
- 5.81   (selected, Gobas et al. 1987, 1989)
- 5.78   (GC-RT correlation, Hackenberg et al. 2003)

Octanol/Air Partition Coefficient, log K_{OA} at 25°C or as indicated and reported temperature dependence equations:
- 8.39   (value for tetrachloronaphthalenes, Kaupp & McLachlan 1999)
- 8.98, 8.55, 8.00, 7.55, 7.22 (10, 20, 30, 40, 50°C, GC-RT correlation, Su et al. 2002)
- log K_{OA} = $79100/(2.303 \cdot RT) - 5.60$; temp range 10–50°C (GC-RT correlation, Su et al. 2002)

Bioconcentration Factor, log BCF or log K_B:
- 4.40   (guppies, Opperhuizen et al. 1985)
- 4.69, 5.57   (guppies: whole fish, fish lipid, Gobas et al. 1987)
- 5.62   (guppy, lipid-weight based, Gobas et al. 1989)
- 4.701, 4.582 (calculated-MCI χ, calculated-K_{OW}, Lu et al. 1999)

Sorption Partition Coefficient, log K_{OC}:

Environmental Fate Rate Constants, k and Half-Lives, $t_{1/2}$:

Bioconcentration and Uptake and Elimination Rate Constants (k_1 and k_2):
- $k_1 = 1.2 \times 10^3$ d^{-1}; $k_2 = 4.5 \times 10^{-2}$ d^{-1} (guppies, Opperhuizen et al. 1985)
- $k_1 = 1.4 \times 10^3$ d^{-1} (fish, Opperhuizen 1986)
- log k_1 = 3.10 d^{-1}; log k_2 = –1.35 d^{-1} (guppy, Gobas et al. 1989)

Half-Lives in the Environment:

Biota: $t_{1/2}$ = 15.5 d (guppies, Opperhuizen et al. 1985)

4.1.3.14 1,2,3,4,6-Pentachloronaphthalene

Common Name: 1,2,3,4,6-Pentachloronaphthalene
Synonym: PCN-50
Chemical Name: 1,2,3,4,6-pentachloronaphthalene
CAS Registry No: 67922-25-2
Molecular Formula: $C_{10}H_3Cl_5$
Molecular Weight: 300.396
Melting Point (°C):
    147   (Crookes & Howe 1993; Järnberg et al. 1994)
Boiling Point (°C):
Density (g/cm³):
Molar Volume (cm³/mol):
    252.1   (calculated-Le Bas method at normal boiling point)
Enthalpy of Vaporization, ΔH_V (kJ/mol):
Enthalpy of Sublimation, ΔH_{subl} (kJ/mol):
Enthalpy of Fusion, ΔH_{fus} (kJ/mol):
Entropy of Fusion, ΔS_{fus} (J/mol K):
Fugacity Ratio at 25°C (assuming ΔS_{fus} = 56 J/mol K), F: 0.0635 (mp at 147°C)

Water Solubility (g/m³ or mg/L at 25°C):

Vapor Pressure (Pa at 25°C):
    0.00562; 0.00475   (supercooled liquid P_L: calibrated GC-RT correlation; GC-RT correlation, Lei et al. 1999)
    log (P_L/Pa) = −4123/(T/K) + 11.58; (GC-RT correlation, supercooled liquid, Lei et al. 1999)
    0.0055   (supercooled liquid P_L, regression with GC-RT data from literature, Lei et al. 1999)
    log (P_L/Pa) = −4116/(T/K) + 11.55 (regression with GC-RT data from literature, Lei et al. 1999)

Henry's Law Constant (Pa·m³/mol at 25°C):

Octanol/Water Partition Coefficient, log K_{OW} at 25°C and the reported temperature dependence equations:
    7.0; 6.27   (calibrated HPLC-RT correlation; HPLC-RT correlation, Lei et al. 2000)
    log K_{OW} = 2.166 + 1446.69/(T/K), temp range 5–55°C (temperature dependence HPLC-k′ correlation, Lei et al. 2000)

Octanol/Air Partition Coefficient, log K_{OA} at 25°C or as indicated and the reported temperature dependence equations:
    8.91   (generator column-GC/MS, Harner & Bidleman 1998)
    9.73, 9.20, 8.63, 8.11, 7.75 (10, 20, 30, 40, 50°C, generator column-GC/MS, Harner & Bidleman 1998)
    log K_{OA} = −6.63 + 4629/(T/K), temp range: 10–50°C (generator column-GC/MS, Harner & Bidleman 1998)
    8.92; 8.85   (calibrated GC-RT correlation; GC-RT correlation, Wania et al. 2002)

Bioconcentration Factor, log BCF or log K_B:

Sorption Partition Coefficient, log K_{OC}:

Environmental Fate Rate Constants, k and Half-Lives, $t_{1/2}$:

Half-Lives in the Environment:

4.1.3.15 1,2,3,5,7-Pentachloronaphthalene

Common Name: 1,2,3,5,7-Pentachloronaphthalene
Synonym: PCN-52
Chemical Name: 1,2,3,5,7-pentachloronaphthalene
CAS Registry No: 53555-65-0
Molecular Formula: $C_{10}H_3Cl_5$
Molecular Weight: 300.396
Melting Point (°C):
    171   (Crookes & Howe 1993; Järnberg et al. 1994)
Boiling Point (°C):
    313   (estimated, Crookes & Howe 1993)
Density (g/cm³):
Molar Volume (cm³/mol):
    252.1   (calculated-Le Bas method at normal boiling point)
Enthalpy of Fusion, ΔH_{fus} (kJ/mol):
Entropy of Fusion, ΔS_{fus} (J/mol K):
Fugacity Ratio at 25°C (assuming ΔS_{fus} = 56 J/mol K), F: 0.0369 (mp at 171°C)

Water Solubility (g/m³ or mg/L at 25°C):
    0.0073   (estimated, Crookes & Howe 1993)

Vapor Pressure (Pa at 25°C and reported temperature dependence equation):
    4.266×10^{-3} (estimated, Crookes & Howe 1993)
    2.40×10^{-6} (estimated, Alcock et al. 1999)
    0.00696; 0.00593 (supercooled liquid P_L: calibrated GC-RT correlation; GC-RT correlation, Lei et al. 1999)
    log (P_L/Pa) = $-4082/(T/K)$ + 11.53 (GC-RT correlation, supercooled liquid, Lei et al. 1999)
    0.00798   (supercooled liquid P_L, regression with GC-RT data from literature, Lei et al. 1999)
    log (P_L/Pa) = $-4082/(T/K)$ + 11.71 (regression with GC-RT data from literature, Lei et al. 1999)
    0.0–133 – P_S; 0.0277 – P_L (estimated for pentachloronaphthalenes, Kaupp & McLachlan 1999)

Henry's Law Constant (Pa·m³/mol):

Octanol/Water Partition Coefficient, log K_{OW}:
    5.46   (selected, Hawker 1990)
    6.87   (estimated, Crookes & Howe 1993, quoted, Alcock et al. 1999)
    6.87   (quoted, Falandysz et al. 1997)

Octanol/Air Partition Coefficient, log K_{OA} at 25°C or as indicated and reported temperature dependence equations:
    8.73   (generator-column-GC, Harner & Bidleman 1998)
    9.50, 9.04, 8.47, 7.97, 7.63 (10, 20, 30, 40, 50°C, generator column-GC/MS, Harner & Bidleman 1998)
    log K_{OA} = $-6.02 + 4394/(T/K)$; temp range: 10–50°C (generator column-GC/MS, Harner & Bidleman 1998)
    9.00   (estimated, Kaupp & McLachlan 1999)
    8.82; 8.76   (calibrated GC-RT correlation; GC-RT correlation, Wania et al. 2002)

Bioconcentration Factor, log BCF or log K_B:
    4.23   (selected, Hawker 1990)

Sorption Partition Coefficient, log K_{OC}:
    4.36   (estimated, Crookes & Howe 1993)

Environmental Fate Rate Constants, k and Half-Lives, $t_{1/2}$:

Half-Lives in the Environment:

4.1.3.16 1,2,3,5,8-Pentachloronaphthalene

Common Name: 1,2,3,5,8-Pentachloronaphthalene
Synonym: PCN-53
Chemical Name: 1,2,3,5,8-pentachloronaphthalene
CAS Registry No: 150224-24-1
Molecular Formula: $C_{10}H_3Cl_5$
Molecular Weight: 300.396
Melting Point (°C):
    174–176 (Järnberg et al. 1994)
Boiling Point (°C):
Density (g/cm³):
Molar Volume (cm³/mol):
    252.1   (calculated-Le Bas method at normal boiling point)
Enthalpy of Vaporization, ΔH_V (kJ/mol):
Enthalpy of Sublimation, ΔH_{subl} (kJ/mol):
Enthalpy of Fusion, ΔH_{fus} (kJ/mol):
Entropy of Fusion, ΔS_{fus} (J/mol K):
Fugacity Ratio at 25°C (assuming ΔS_{fus} = 56 J/mol K), F:

Water Solubility (g/m³ or mg/L at 25°C):

Vapor Pressure (Pa at 25°C or as indicated and reported temperature dependence equations):
    0.00394; 0.00329   (supercooled liquid P_L: calibrated GC-RT correlation; GC-RT correlation, Lei et al. 1999)
    log (P_L/Pa) = –4204/(T/K) + 11.70 (GC-RT correlation, supercooled liquid, Lei et al. 1999)
    0.00428   (supercooled liquid P_L, regression with GC-RT data from literature, Lei et al. 1999)
    log (P_L/Pa) = –4116/(T/K) + 11.44 (regression with GC-RT data from literature, Lei et al. 1999)

Henry's Law Constant (Pa·m³/mol at 25°C):

Octanol/Water Partition Coefficient, log K_{OW} at 25°C and the temperature dependence equations:
    6.80; 6.13   (calibrated HPLC-RT correlation; HPLC-RT correlation, Lei et al. 2000)
    log K_{OW} = 2.305 + 1431.02/(T/K), temp range 5–55°C (temperature dependence HPLC-k′ correlation, Lei et al. 2000)
    6.46   (GC-RT correlation, Hackenberg et al. 2003)

Octanol/Air Partition Coefficient, log K_{OA} at 25°C or as indicated and reported temperature dependence equations:
    9.13   (generator column-GC/MS, Harner & Bidleman 1998)
    9.97, 9.44, 8.86, 8.34, 7.96 (10, 20, 30, 40, 50°C, generator column-GC/MS, Harner & Bidleman 1998)
    log K_{OA} = –6.59 + 4684/(T/K); temp range: 10–50°C (generator column-GC/MS, Harner & Bidleman 1998)
    9.10; 9.01   (calibrated GC-RT correlation; GC-RT correlation, Wania et al. 2002)

Bioconcentration Factor, log BCF or log K_B:

Sorption Partition Coefficient, log K_{OC}:

Environmental Fate Rate Constants, k and Half-Lives, $t_{1/2}$:

Half-Lives in the Environment:

4.1.3.17 1,2,3,4,5,7-Hexachloronaphthalene

Common Name: 1,2,3,4,5,7-Hexachloronaphthalene
Synonym: PCN-64
Chemical Name: 1,2,3,4,5,7-hexachloronaphthalene
CAS Registry No: 67927-67-4
Molecular Formula: $C_{10}H_2Cl_6$
Molecular Weight: 334.842
Melting Point (°C):
    194   (Crookes & Howe 1993)
    164–166   (Järnberg et al. 1994)
Boiling Point (°C):
    331   (estimated, Crookes & Howe 1993)
Density (g/cm³):
Molar Volume (cm³/mol):
    273.0   (calculated-Le Bas method at normal boiling point)
Enthalpy of Fusion, ΔH_{fus} (kJ/mol):
Entropy of Fusion, ΔS_{fus} (J/mol K):
Fugacity Ratio at 25°C (assuming ΔS_{fus} = 56 J/mol K), F: 0.022 (mp at 194°C)

Water Solubility (g/m³ or mg/L at 25°C):
    0.00011   (estimated, Crookes & Howe 1993)

Vapor Pressure (Pa at 25°C and reported temperature dependence equation):
    9.47×10^{-4} (estimated, Crookes & Howe 1993)
    5.30×10^{-8} (estimated, Alcock et al. 1999)
    0.00134   (supercooled liquid P_L, regression with GC-RT data from literature, Lei et al. 1999)
    $\log (P_L/Pa) = -4432/(T/K) + 11.99$ (regression with GC-RT data from literature, Lei et al. 1999)
    0.000257 – P_S; 0.00809 – P_L (estimated for hexachloronaphthalenes, Kaupp & McLachlan 1999)

Henry's Law Constant (Pa·m³/mol):

Octanol/Water Partition Coefficient, log K_{OW}:
    7.58   (estimated, Crookes & Howe 1993, quoted, Alcock et al. 1999)

Octanol/Air Partition Coefficient, log K_{OA} at 25°C or as indicated and the reported temperature dependence equations:
    9.80   (generator-column-GC, Harner & Bidleman 1998)
    10.07, 9.57, 8.95, 8.54 (20, 30, 40, 50°C, generator column-GC/MS, Harner & Bidleman 1998)
    $\log K_{OA} = -6.77 + 4393/(T/K)$; temp range: 20–50°C (generator column-GC/MS, Harner & Bidleman 1998)
    10.02   (value for hexachloronaphthalenes, Kaupp & McLachlan 1999)

Bioconcentration Factor, log BCF or log K_B:

Sorption Partition Coefficient, log K_{OC}:

Environmental Fate Rate Constants, k and Half-Lives, $t_{1/2}$:

Half-Lives in the Environment:

4.1.3.18 1,2,3,4,6,7-Hexachloronaphthalene

Common Name: 1,2,3,4,6,7-Hexachloronaphthalene
Synonym: PCN-66
Chemical Name: 1,2,3,4,6,7-hexachloronaphthalene
CAS Registry No: 103426-96-6
Molecular Formula: $C_{10}H_2Cl_6$
Molecular Weight: 334.842
Melting Point (°C):
    205-206 (Järnberg et al. 1994)
Boiling Point (°C):
Density (g/cm^3):
Molar Volume (cm^3/mol):
    273.0   (calculated-Le Bas method at normal boiling point)
Enthalpy of Vaporization, ΔH_V (kJ/mol):
Enthalpy of Sublimation, ΔH_{subl} (kJ/mol):
Enthalpy of Fusion, ΔH_{fus} (kJ/mol):
Entropy of Fusion, ΔS_{fus} (J/mol K):
Fugacity Ratio at 25°C (assuming ΔS_{fus} = 56 J/mol K), F:

Water Solubility (g/m^3 or mg/L at 25°C):

Vapor Pressure (Pa at 25°C and the reported temperature dependence equations):
    0.0015; 0.00121 (supercooled liquid P_L: calibrated GC-RT correlation; GC-RT correlation, Lei et al. 1999)
    $\log (P_L/Pa) = -4411/(T/K) + 11.97$ (GC-RT correlation, supercooled liquid, Lei et al. 1999)
    0.00157(supercooled liquid P_L, regression with GC-RT from literature, Lei et al. 1999)
    $\log (P_L/Pa) = -4432/(T/K) + 12.06$ (regression with GC-RT from literature, Lei et al. 1999)

Henry's Law Constant (Pa·m^3/mol at 25°C):

Octanol/Water Partition Coefficient, log K_{OW} at 25°C and the temperature dependence equations:
    7.70; 6.79   (calibrated HPLC-RT correlation; HPLC-RT correlation, Lei et al. 2000)
    $\log K_{OW} = 2.489 + 1556.37/(T/K)$; temp range 5–55°C (temperature dependence HPLC-k' correlation, Lei et al. 2000)
    6.77   (GC-RT correlation, Hackenberg et al. 2003)

Octanol/Air Partition Coefficient, log K_{OA} at 25°C or as indicated and reported temperature dependence equations:
    9.70   (generator column-GC/MS, Harner & Bidleman 1998)
    10.58, 10.01, 9.46, 8.84, 8.42 (10, 20, 30, 40, 50°C, generator column-GC/MS, Harner & Bidleman 1998)
    $\log K_{OA} = -7.09 + 5003/(T/K)$; temp range: 10–50°C (generator column-GC/MS, Harner & Bidleman 1998)
    9.58; 9.45   (calibrated GC-RT correlation; GC-RT correlation, Wania et al. 2002)

Bioconcentration Factor, log BCF or log K_B:

Sorption Partition Coefficient, log K_{OC}:

Environmental Fate Rate Constants, k and Half-Lives, $t_{1/2}$:

Half-Lives in the Environment:

4.1.3.19 1,2,3,5,6,7-Hexachloronaphthalene

Common Name: 1,2,3,5,6,7-Hexachloronaphthalene
Synonym: PCN-67
Chemical Name: 1,2,3,5,6,7-hexachloronaphthalene
CAS Registry No: 103426-97-7
Molecular Formula: $C_{10}H_2Cl_6$
Molecular Weight: 334.842
Melting Point (°C):
    234–235 (Järnberg et al. 1994)
Boiling Point (°C):
Density (g/cm^3):
Molar Volume (cm^3/mol):
    273.0   (calculated-Le Bas method at normal boiling point)
Enthalpy of Fusion, ΔH_{fus} (kJ/mol):
Entropy of Fusion, ΔS_{fus} (J/mol K):
Fugacity Ratio at 25°C (assuming ΔS_{fus} = 56 J/mol K), F:

Water Solubility (g/m^3 or mg/L at 25°C):

Vapor Pressure (Pa at 25°C):
    0.00150; 0.00121   (supercooled liquid P_L: calibrated GC-RT correlation; GC-RT correlation, Lei et al. 1999)
    log (P_L/Pa) = – 4411/(T/K) + 11.97 (GC-RT correlation, supercooled liquid, Lei et al. 1999)
    0.00157   (supercooled liquid P_L, regression with GC-RT data from literature, Lei et al. 1999)
    log (P_L/Pa) = –4432/(T/K) + 12.06 (regression with GC-RT data from literature, Lei et al. 1999)

Henry's Law Constant (Pa·m^3/mol at 25°C):

Octanol/Water Partition Coefficient, log K_{OW}:

Octanol/Air Partition Coefficient, log K_{OA} at 25°C and reported temperature dependence equation:
    9.70   (generator column-GC.MS, Harner & Bidleman 1998)
    10.58, 10.01, 9.46, 8.84, 8.42 (10, 20, 30, 40, 50°C, generator column-GC/MS, Harner & Bidleman 1998)
    log K_{OA} = –7.09 + 5003/(T/K); temp range: 10–50°C (generator column-GC/MS, Harner & Bidleman 1998)
    9.58; 9.45   (calibrated GC-RT correlation; GC-RT correlation, Wania et al. 2002)

Bioconcentration Factor, log BCF or log K_B:
Sorption Partition Coefficient, log K_{OC}:

Environmental Fate Rate Constants, k and Half-Lives, $t_{½}$:
Half-Lives in the Environment:

4.1.3.20 1,2,3,5,7,8-Hexachloronaphthalene

Common Name: 1,2,3,5,7,8-Hexachloronaphthalene
Synonym: PCN-69
Chemical Name: 1,2,3,5,7,8-hexachloronaphthalene
CAS Registry No: 103426-94-4
Molecular Formula: $C_{10}H_2Cl_6$
Molecular Weight: 334.842
Melting Point (°C):
    148–149 (Järnberg et al. 1994)
Boiling Point (°C):
Density (g/cm^3):
Molar Volume (cm^3/mol):
    273.0   (calculated-Le Bas method at normal boiling point)
Enthalpy of Fusion, ΔH_{fus} (kJ/mol):
Entropy of Fusion, ΔS_{fus} (J/mol K):
Fugacity Ratio at 25°C (assuming ΔS_{fus} = 56 J/mol K), F:

Water Solubility (g/m^3 or mg/L at 25°C):

Vapor Pressure (Pa at 25°C and the reported temperature dependence equations):
    0.00124; 0.0010   (supercooled liquid P_L: calibrated GC-RT correlation; GC-RT correlation, Lei et al. 1999)
    log (P_L/Pa) = – 4441/(T/K) + 11.99 (GC-RT correlation, supercooled liquid, Lei et al. 1999)
    0.00134(supercooled liquid P_L, regression with GC-RT from literature, Lei et al. 1999)
    log (P_L/Pa) = –4432/(T/K) + 11.99 (regression with GC-RT from literature, Lei et al. 1999)

Henry's Law Constant (Pa·m^3/mol at 25°C):

Octanol/Water Partition Coefficient, log K_{OW} at 25°C and the reported temperature dependence equations:
    7.50; 6.69 (calibrated HPLC-RT correlation; HPLC-RT correlation, Lei et al. 2000)
    log K_{OW} = 2.413 + 1535.48/(T/K); temp range 5–55°C (temperature dependence HPLC-k′ correlation, Lei et al. 2000)
    6.87   (GC-RT correlation, Hackenberg et al. 2003)

Octanol/Air Partition Coefficient, log K_{OA} at 25°C or as indicated and reported temperature dependence equation:
    9.83   (generator column-GC.MS, Harner & Bidleman 1998)
    10.09, 9.62, 8.99, 8.57 (20, 30, 40, 50°C, generator column-GC/MS, Harner & Bidleman 1998)
    log K_{OA} = –6.64 + 4909/(T/K); temp range: 20–50°C (generator column-GC, Harner & Bidleman 1998)
    9.67; 9.53 (calibrated GC-RT correlation; GC-RT correlation, Wania et al. 2002)

Bioconcentration Factor, log BCF or log K_B:
Sorption Partition Coefficient, log K_{OC}:

Environmental Fate Rate Constants, k and Half-Lives, $t_{1/2}$:
Half-Lives in the Environment:

4.1.3.21 1,2,3,4,5,6,7-Heptachloronaphthalene

Common Name: 1,2,3,4,5,6,7-Heptachloronaphthalene
Synonym: PCN-73
Chemical Name: 1,2,3,4,5,6,7-heptachloronaphthalene
CAS Registry No: 58863-14-2
Molecular Formula: $C_{10}HCl_7$
Molecular Weight: 369.287
Melting Point (°C):
Boiling Point (°C):
Density (g/cm³):
Molar Volume (cm³/mol):
    293.9   (calculated-Le Bas method at normal boiling point)
Enthalpy of Fusion, ΔH_{fus} (kJ/mol):
Entropy of Fusion, ΔS_{fus} (J/mol K):
Fugacity Ratio at 25°C (assuming ΔS_{fus} = 56 J/mol K), F:

Water Solubility (g/m³ or mg/L at 25°C):
    0.00062 (generator column-GC/ECD, Opperhuizen 1985; quoted, Opperhuizen 1986)

Vapor Pressure (Pa at 25°C and reported temperature dependence equations):
    $5.44 \times 10^{-4} - P_S$; $0.00258 - P_L$ (estimated for heptachloronaphthalenes, Kaupp & McLachlan 1999)
    2.93×10^{-4}, 2.78×10^{-4} (supercooled liquid P_L: calibrated GC-RT correlation; GC-RT correlation, Lei et al. 1999)
    $\log (P_L/Pa) = -4745/(T/K) + 12.38$ (GC-RT correlation, supercooled liquid, Lei et al. 1999)
    0.000278 (supercooled liquid P_L, regression with GC-RT from literature, Lei et al. 1999)
    $\log (P_L/Pa) = -4745/(T/K) + 12.37$ (regression with GC-RT from literature, Lei et al. 1999)

Henry's Law Constant (Pa·m³/mol):

Octanol/Water Partition Coefficient, log K_{OW} at 25°C and the reported temperature dependence equations:
    8.20   (HPLC-RT, Opperhuizen et al. 1985; quoted, Opperhuizen 1986)
    7.69   (calculated-fragment const., Burreau et al. 1997)
    8.20; 7.18   (calibrated HPLC-RT correlation; HPLC-RT correlation, Lei et al. 2000)
    $\log K_{OW} = 2.638 + 1660.82/(T/K)$; temp range 5–55°C (HPLC-k' correlation, Lei et al. 2000)
    7.33   (GC-RT correlation, Hackenberg et al. 2003)

Octanol/Air Partition Coefficient, log K_{OA} at 25°C or as indicated and reported temperature dependence equation:
    11.52, 10.96, 10.44, 9.75, 9.28 (10, 20, 30, 40, 50°C, GC-RT correlation, Su et al. 2002)
    $\log K_{OA} = 99500/(2.303 \cdot RT) - 6.80$; temp range 10–50°C (GC-RT correlation, Su et al. 2002)

Bioconcentration Factor, log BCF or log K_B:
Sorption Partition Coefficient, log K_{OC}:

Environmental Fate Rate Constants, k and Half-Lives, $t_{1/2}$:
Half-Lives in the Environment:

4.1.3.22 1,2,3,4,5,6,8-Heptachloronaphthalene

Common Name: 1,2,3,4,5,6,8-Heptachloronaphthalene
Synonym: PCN-74
Chemical Name: 1,2,3,4,5,6,8-heptachloronaphthalene
CAS Registry No: 58863-15-3
Molecular Formula: $C_{10}HCl_7$
Molecular Weight: 369.287
Melting Point (°C):
    194   (Crookes & Howe 1993; Järnberg et al. 1994)
Boiling Point (°C):
    348   (estimated, Crookes & Howe 1993)
Density (g/cm³):
Molar Volume (cm³/mol):
    293.9   (calculated-Le Bas method at normal boiling point)
Enthalpy of Fusion, ΔH_{fus} (kJ/mol):
Entropy of Fusion, ΔS_{fus} (J/mol K):
Fugacity Ratio at 25°C (assuming ΔS_{fus} = 56 J/mol K), F: 0.022 (mp at 194°C)

Water Solubility (g/m³ or mg/L at 25°C):
    0.00004   estimated, Crookes & Howe 1993; quoted, Alcock et al. 1999)

Vapor Pressure (Pa at 25°C and reported temperature dependence equation):
    3.73×10^{-4} (estimated, Crookes & Howe 1993)
    2.10×10^{-8} (estimated, Alcock et al. 1999)
    2.46×10^{-4} (supercooled liquid P_L, regression with GC-RT from literature, Lei et al. 1999)
    log (P_L/Pa) = -4748/(T/K) + 12.31 (regression with GC-RT from literature, Lei et al. 1999)
    5.44×10^{-4} – P_S; 0.00258 – P_L (estimated for heptachloronaphthalenes, Kaupp & McLachlan 1999)

Henry's Law Constant (Pa·m³/mol):

Octanol/Water Partition Coefficient, log K_{OW}:
    8.30   (estimated, Crookes & Howe 1993, quoted, Alcock et al. 1999)
    8.50; 7.46 (calibrated HPLC-RT correlation; HPLC-RT correlation, Lei et al. 2000)

Octanol/Air Partition Coefficient, log K_{OA} at 25°C or as indicated and reported temperature dependence equation:
    11.56, 10.99, 10.47, 9.79, 9.31 (10, 20, 30, 40, 50°C, GC-RT correlation, Su et al. 2002)
    log K_{OA} = 99800/(2.303·RT) – 6.80; temp range 10–50°C (GC-RT correlation, Su et al. 2002)

Bioconcentration Factor, log BCF or log K_B:
Sorption Partition Coefficient, log K_{OC}:

Environmental Fate Rate Constants, k and Half-Lives, $t_{\frac{1}{2}}$:
Half-Lives in the Environment:

4.1.3.23 Octachloronaphthalene

Common Name: Octachloronaphthalene
Synonym: PCN-75
Chemical Name: octachloronaphthalene
CAS Registry No: 2234-13-1
Molecular Formula: $C_{10}Cl_8$
Molecular Weight: 403.731
Melting Point (°C):
    197.5   (Lide 2003)
Boiling Point (°C):
    365   (estimated, Crookes & Howe 1993)
Density (g/cm³):
Molar Volume (cm³/mol):
    314.8   (calculated-Le Bas method at normal boiling point)
    233.7   (Ruelle & Kesselring 1997)
Enthalpy of Fusion, ΔH_{fus} (kJ/mol):
Entropy of Fusion, ΔS_{fus} (J/mol K):
Fugacity Ratio at 25°C (assuming ΔS_{fus} = 56 J/mol K), F: 0.0203 (mp at 197.5°C)

Water Solubility (g/m³ or mg/L at 25°C):
    0.000078, 0.00008   (generator column-GC/ECD, Opperhuizen 1986)
    0.00008   (generator column-GC/ECD, Opperhuizen 1987)
    0.00008; 0.00027   (quoted; calculated-molar volume, mp and mobile order thermodynamics, Ruelle & Kesselring 1997)

Vapor Pressure (Pa at 25°C and reported temperature dependence equations):
    1.33×10^{-4} (estimated, Crookes & Howe 1993)
    7.50×10^{-9} (estimated, Alcock et al. 1999)
    $7.61 \times 10^{-5}, 5.60 \times 10^{-5}$ (supercooled liquid P_L: calibrated GC-RT correlation; GC-RT correlation, Lei et al. 1999)
    log (P_L/Pa) = $-5021/(T/K) + 12.72$ (Antoine eq., GC-RT correlation, supercooled liquid, Lei et al. 1999)
    6.84×10^{-5} (supercooled liquid P_L, regression with GC-RT from literature, Lei et al. 1999)
    log (P_L/Pa) = $-5021/(T/K) + 12.82$ (regression with GC-RT from literature, Lei et al. 1999)
    $1.60 \times 10^{-5} - P_S$; $8.71 \times 10^{-4} - P_L$ (estimated for heptachloronaphthalenes, Kaupp & McLachlan 1999)

Henry's Law Constant (Pa·m³/mol):

Octanol/Water Partition Coefficient, log K_{OW} at 25°C and the reported temperature dependence equations.
    6.50   (calculated, Kaiser 1983)
    8.40   (HPLC-RT correlation, Opperhuizen et al. 1985)
    8.50   (Opperhuizen 1986)
    6.42   (shake flask, Opperhuizen)
    7.90   (calculated, Banerjee & Baughman 1991)
    8.50, 7.47   (calibrated HPLC-k' correlation, HPLC-k' correlation, Lei et al. 2000)
    log K_{OW} = $2.998 + 1660.82/(T/K)$, temp range 5–55°C (temperature dependence HPLC-k' correlation, Lei et al. 2000)
    7.70   (GC-RT correlation, Hackenberg et al. 2003)

Octanol/Air Partition Coefficient, log K_{OA} at 25°C or as indicated and the reported temperature dependence equations:
    12.39, 11.78, 11.27, 10.51, 9.98 (10, 20, 30, 40, 50°C, GC-RT correlation, Su et al. 2002)
    log K_{OA} = 10600/(2.303·RT) – 7.20; temp range 10–50°C (GC-RT correlation, Su et al. 2002)

Bioconcentration Factor, log BCF or log K_B:
    2.50   (*Oncorhynchus mykiss*, Oliver & Niimi 1985; quoted, Crookes & Howe 1993
    2.52   (rainbow trout, mean value, Oliver & Niimi 1985)
    5.0   (calculated-K_{OW} and S_0, Banerjee & Baughman 1991)
    2.52   (*Oncorhynchus mykiss*, under flow-through condition, quoted, Devillers et al. 1996)

Sorption Partition Coefficient, log K_{OC}:
    5.38   (estimated, Crookes & Howe 1993)

Environmental Fate Rate Constants, k and Half-Lives, $t_{1/2}$:

Volatilization:

Photolysis:

Photooxidation: $t_{1/2}$ = 1608–16082 h based on estimated rate constant for the reaction with hydroxyl radical in air (Atkinson 1987; quoted, Howard et al. 1991).

Hydrolysis:

Biodegradation: aerobic biodegradation $t_{1/2}$ = 4320–8760 h, based on essentially no biodegradation observed for hexachlorobenzene in soil die-away tests (Griffin & Chou 1981; quoted, Howard et al. 1991);
anaerobic biodegradation $t_{1/2}$ = 17280–35040 h, based on essentially no biodegradation observed for hexachlorobenzene in soil die-away tests (Griffin & Chou 1981; quoted, Howard et al. 1991).

Biotransformation:

Bioconcentration and Uptake and Elimination Rate Constants (k_1 and k_2):

Half-Lives in the Environment:

Air: photooxidation $t_{1/2}$ = 1608–16082 h in air based on estimated rate constant for the reaction with hydroxyl radical in air (Atkinson 1987; quoted, Howard et al. 1991).

Surface water: $t_{1/2}$ = 4320–8760 h, based on essentially no biodegradation observed for hexachlorobenzene in soil die-away tests (Griffin & Chou 1981; quoted, Howard et al. 1991).

Ground water: $t_{1/2}$ = 8640–17520 h, based on essentially no biodegradation observed for hexachloro-benzene in soil die-away tests (Griffin & Chou 1981; quoted, Howard et al. 1991).

Sediment:

Soil: $t_{1/2}$ = 4320–8760 h, based on essentially no biodegradation observed for hexachlorobenzene in soil die-away tests (Griffin & Chou 1981; quoted, Howard et al. 1991).

Biota:

4.1.4 BROMINATED POLYNUCLEAR AROMATIC HYDROCARBONS

4.1.4.1 1-Bromonaphthalene

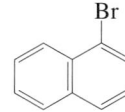

Common Name: 1-Bromonaphthalene
Synonym: α-bromonaphthalene
Chemical Name: 1-bromonaphthalene
CAS Registry No: 90-11-9
Molecular Formula: $C_{10}H_7Br$
Molecular Weight: 207.067
Melting Point (°C):
    6.1   (Lide 2003)
Boiling Point (C):
    281   (Lide 2003)
Density (g/cm³ at 20°C):
    1.4834   (Dean 1992)
    1.4785   (Lide 2003)
Molar Volume (cm³/mol):
    140.6   (30°C, calculated from density, Stephenson & Malanowski 1987)
    140.0   (20°C, calculated-density)
    170.9   (calculated-Le Bas method at normal boiling point)
Dissociation Constant pK_a:
Enthalpy of Vaporization, ΔH_V (kJ/mol):
    39.33   (at normal bp, Hon et al. 1976)
Enthalpy of Fusion, ΔH_{fus} (kJ/mol):
Entropy of Fusion, ΔS_{fus} (J/mol K):
Fugacity Ratio at 25°C, F: 1.0

Water Solubility (g/m³ or mg/L at 25°C):
    9.95   (Yalkowsky et al. 1983)
    7.72   (calculated-molecular connectivity indices, Nirmalakhandan & Speece 1989)
    9.08, 9.80, 13.35, and 18.98 (4, 10, 25, and 40°C, generator column-UV spec., Dickhut et al. 1994)
    25.3   (calculated-TSA, Dickhut et al. 1994)
    14.0   (calculated-group contribution method, Kühne et al. 1995)
    23.3   (calculated-molar volume, mp and mobile order thermodynamics, Ruelle & Kesselring 1997)

Vapor Pressure (Pa at 25°C or as indicated and reported temperature dependence equations. Additional data at other temperatures designated * are compiled at the end of this section):
    266.6*   (97.9°C, static-Hg manometer, measured range 97.9–281.1°C, Kahlbaum 1898)
    133*   (84.2°C, summary of literature data, temp range 84.2–281.1°C, Stull 1947)
    $\log (P/mmHg) = [-0.2185 \times 13274.9/(T/K)] + 8.131285$; temp range 84.2–281.1°C, (Antoine eq., Weast 1972–73)
    18681*   (196.27°C, ebulliometry, measured range 97.9–285.92°C, Hon et al. 1976)
    $\log (P/mmHg) = 5.38175 - 929.64/(91.06 + t/°C)$; temp range: 196.27–285.93°C (Antoine eq., twin ebulliometry, Hon et al. 1976)
    0.713*, 1.07* (torsion effusion, measured range 295–359 K, Urbani et al. 1980)
    $\log (P/kPa) = (6.90 - 2950/(T/K)$, temp range 295–359 K (torsion and Knudsen effusion, Urbani et al. 1980)
    $\log (P/mmHg) = 7.00350 - 1927.05/(186.0 + t/°C)$; temp range: liquid (Antoine eq., Dean 1985, 1992)
    $\log (P_L/kPa) = 6.56365 - 2303.73/[-48.841 + (T/K)]$; temp range 357–555 K (liquid, Antoine eq.-I, Stephenson & Malanowski 1987)

log (P_L/kPa) = 4.50679 − 929.871/[182.045 + (T/K)], temp range 469–559 K, (liquid, Antoine eq.-II, Stephenson & Malanowski 1987)

2.754*   (30.15°C, transpiration method, measured range 303.3–336.3 K, Verevkin 2003)

ln (P/P°) = 299.001/R − 8.3941.481/R·(T/K) − (73.5/R)·ln[(T/K)/298.15], where P° = 101.325 kPa, gas constant R = 8.31451 J·K^{-1}·mol^{-1} (vapor pressure eq. from transpiration measurement, temp range 303.3–336.3 K, Verevkin 2003)

Henry's Law Constant (Pa·m^3/mol):

Octanol/Water Partition Coefficient, log K_{OW}:

4.35   (calculated-fragment const., Yalkowsky et al. 1983)

Octanol/Air Partition Coefficient, log K_{OA}:

Bioconcentration Factor, log BCF or log K_B:

Sorption Partition Coefficient, log K_{OC}:

Environmental Fate Rate Constants, k or Half-Lives, $t_{½}$:

Half-Lives in the Environment:

TABLE 4.1.4.1.1
Reported vapor pressures of 1-bromonaphthalene at various temperatures and the coefficients for the vapor pressure equations

log P = A − B/(T/K)   (1)   ln P = A − B/(T/K)   (1a)
log P = A − B/(C + t/°C)   (2)   ln P = A − B/(C + t/°C)   (2a)
log P = A − B/(C + T/K)   (3)
log P = A − B/(T/K) − C·log (T/K) (4)

1.

Kahlbaurm 1898		Stull 1947		Hon et al. 1976		Urbani et al. 1980	
static method-manometer		summary of literature data		ebulliometry		torsion effusion	
t/°C	P/Pa	t/°C	P/Pa	t/°C	P/Pa	t/°C	P/Pa
						Run A.01	
97.9	266.6	84.2	133.3	196.27	18681	25	0.713
117.5	666.6	117.5	666.6	213.14	28206	28	0.950
134.0	1333	133.6	1333	228.08	39235	30	1.37
151.3	2666	150.2	2666	234.45	44740	33	1.31
157.2	3333	170.2	5333	244.51	54504	34	1.43
162.3	3999	183.5	7999	245.13	55150	40	2.85
166.7	4666	198.8	13332	250.71	61192	42	3.05
170.5	5333	224.2	26664	256.24	67574	45	3.62
177.1	6666	252.2	53329	262.78	75759	46	3.72
189.6	9999	281.1	101325	275.35	93151	47	4.16
198.8	13332			279.49	99510	48	4.51
223.8	26664	mp/°C	5.5	281.40	102474	49	4.75
239.7	39997			284.04	106737	50	4.99
252.0	53329			285.92	109863	52	5.82
261.6	66661					58	7.78
269.8	79993			bp/°C	280.56	61	8.94
277.2	93326					66	12.8

TABLE 4.1.4.1.1 (Continued)

Kahlbaurm 1898		Stull 1947		Hon et al. 1976		Urbani et al. 1980	
static method-manometer		summary of literature data		ebulliometry		torsion effusion	
t/°C	P/Pa	t/°C	P/Pa	t/°C	P/Pa	t/°C	P/Pa
281.1	101325			eq. 2	P/mmHg	74	20.2
				A	5.38157	77	24.3
				B	929.64	86	42.0
				C	91.06		
				ΔH_V/(kJ mol^{-1}) = 39.33 at bp			

2.

Urbani et al. 1980 (Continued)		Knudsen effusion		Verevkin 2003	
torsion effusion				transpiration-GC	
t/°C	P/Pa	t/°C	P/Pa	t/°C	P/Pa
Run A.02		42	4.95	30.15	2.754
50	8.33	49	9.26	33.15	3.466
53	8.54	50	8.97	36.15	4.388
55	10.4	54	11.6	39.14	5.529
58	12.5	60	16.1	42.15	7.095
58	11.7	65	18.4	45.15	8.644
67	17.3	70	28.3	48.15	10.97
Run A.04				51.25	13.66
26	1.07	For Knudsen effusion:		54.15	16.64
34	1.72	eq. 1	P/kPa	57.15	20.95
36	2.32	A	6.33 ± 0.33	60.15	25.50
38	2.61	B	2710 ± 110	63.15	30.34
43	3.86				
50	6.18			ΔH_V/(kJ mol^{-1}) = 63.91	
51	6.21			at 298.15 K	
55	7.00				
60	8.58				
63	10.3			eq. 4a	P/kPa
68	12.9			P°	101.325 kPa
70	17.5	Overall temp dependence eq.		A	303.761/R
82	38.7	eq. 1	P/kPa	B	89574.863/R
for torsion effusion:		A	6.90 ± 1.1	C	78.8/R
eq. 1	P/kPa	B	2950 ± 300	R = 8.314 J K^{-1} mol^{-1}	
A	6.96 ± 0.18				
B	2980 ± 57	ΔH_V/(kJ mol^{-1}) = 56 ± 6			

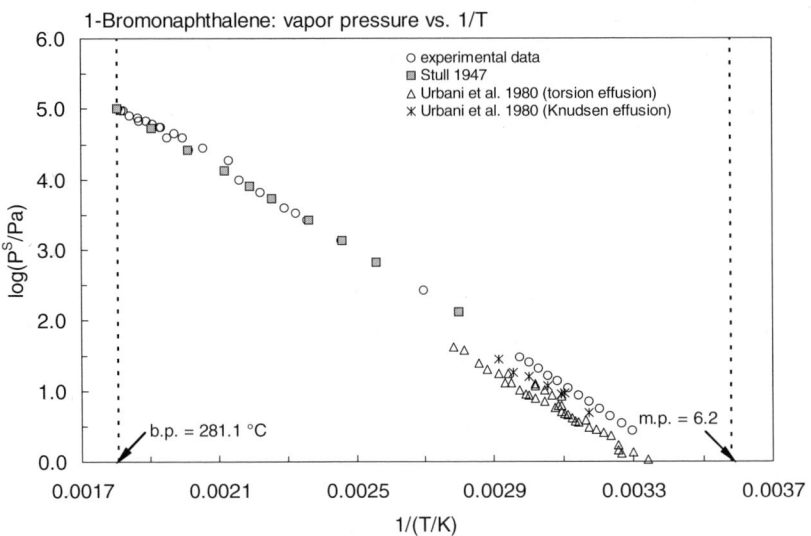

FIGURE 4.1.4.1.1 Logarithm of vapor pressure versus reciprocal temperature for 1-bromonaphthalene.

4.1.4.2 2-Bromonaphthalene

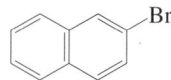

Common Name: 2-Bromonaphthalene
Synonym: β-bromonaphthalene
Chemical Name: 2-bromonaphthalene
CAS Registry No: 580-13-2
Molecular Formula: $C_{10}H_7Br$
Molecular Weight: 207.067
Melting Point (°C):
    55.9   (Lide 2003)
Boiling Point (°C):
    281.5   (Lide 2003)
Density (g/cm³ at 20°C):
Molar Volume (cm³/mol):
    146.3   (Ruelle & Kesselring 1997)
    170.9   (calculated-Le Bas method at normal boiling point)
Dissociation Constant pK_a:
Enthalpy of Fusion, ΔH_{fus} (kJ/mol):
    11.97   (Ruelle & Kesselring 1997)
Entropy of Fusion, ΔS_{fus} (J/mol K):
Fugacity Ratio at 25°C (assuming ΔS_{fus} = 56 J/mol K), F: 0.498 (mp at 55.9°C)

Water Solubility (g/m³ or mg/L at 25°C or as indicated):
    8.27   (Yalkowsky et al. 1983)
    7.72   (calculated-molecular connectivity indices, Nirmalakhandan & Speece 1989)
    3.80, 8.04, 15.76 (4, 25, 40°C, generator column-UV spec., Dickhut et al. 1994)
    10.1   (calculated-TSA, Dickhut et al. 1994)
    9.04   (calculated-group contribution method, Kühne et al. 1995)
    8.06   (calculated-molar volume, mp and mobile order thermodynamics, Ruelle & Kesselring 1997)

Vapor Pressure (Pa at 25°C or as indicated and reported temperature dependence equations. Additional data at other temperatures designated * are compiled at the end of this section):
    0.501*   (25.15°C, solid I, transpiration method-GC, measured range 280.4–318.3 K, Verevkin 2003)
    $\ln(P/P°) = 314.110/R - 94001.596/R \cdot (T/K) - (41.6/R) \cdot \ln[(T/K)/298.15]$, where P° = 101.325 kPa, gas constant R = 8.31451 J·K⁻¹·mol⁻¹ (solid I, vapor pressure eq. from transpiration measurement, temp range 280.4–318.3 K, Verevkin 2003)
    7.34*   (46.05°C, solid-II, transpiration method-GC, measured range 319.2–328.2 K, Verevkin 2003)
    $\ln(P/P°) = 302.672/R - 90417.272/R \cdot (T/K) - (41.6/R) \cdot \ln[(T/K)/298.15]$, where P° = 101.325 kPa, gas constant R = 8.31451 J·K⁻¹·mol⁻¹ (solid II, vapor pressure eq. from transpiration measurements, temp range 319.2–328.2 K, Verevkin 2003)
    18.75*   (57.05°C, liquid, transpiration method-GC, measured range 330.2–360.2 K, Verevkin 2003)
    $\ln(P/P°) = 303.761/R - 89574.863/R \cdot (T/K) - (78.8/R) \cdot \ln[(T/K)/298.15]$, where P° = 101.325 kPa, gas constant R = 8.31451 J·K⁻¹·mol⁻¹ (liquid, vapor pressure eq. from transpiration measurements, temp range 330.2–360.2 K, Verevkin 2003)

Henry's Law Constant (Pa·m³/mol):

Octanol/Water Partition Coefficient, log K_{OW}:
    4.35   (calculated-fragment const., Yalkowsky et al. 1983)

Octanol/Air Partition Coefficient, log K_{OA}:

Bioconcentration Factor, log BCF or log K_B:

Sorption Partition Coefficient, log K_{OC}:

Environmental Fate Rate Constants, k or Half-Lives, $t_{½}$:

Half-Lives in the Environment:

TABLE 4.1.4.2.1
Reported vapor pressures of 2-bromonaphthalene at various temperatures and the coefficients for the vapor pressure equations

$\log P = A - B/(T/K)$ (1) $\quad\quad$ $\ln P = A - B/(T/K)$ (1a)

$\log P = A - B/(C + t/°C)$ (2) $\quad\quad$ $\ln P = A - B/(C + t/°C)$ (2a)

$\log P = A - B/(C + T/K)$ (3)

$\log P = A - B/(T/K) - C \cdot \log(T/K)$ (4) $\quad\quad$ $\ln P/P^o = A - B/(T/K) - C \cdot \ln[(T/K)/298.15]$ (4a)

Verevkin 2003

transpiration-GC

t/°C	P/Pa	t/°C	P/Pa	t/°C	P/Pa
solid I		solid II		liquid	
7.25	0.105	46.05	77.34	57.05	18.75
10.35	0.152	47.15	8.04	60.15	23.55
15.15	0.287	48.05	8.68	63.15	28.58
20.05	0.501	49.15	9.73	66.15	34.83
25.05	0.886	51.05	11.47	69.15	41.76
28.15	1.239	51.15	11.59	72.15	52.31
31.15	1.709	53.15	13.60	75.15	61.76
34.15	2.285	54.65	15.68	78.05	75.34
37.15	3.107	55.05	16.20	81.05	88.25
40.05	4.219			85.05	103.89
41.15	4.588			87.05	125.40
43.15	5.703				
45.15	6.686			eq. 4a	P/kPa
		eq. 4a	P/kPa	P^o	101.325 kPa
eq. 4a	P/kPa	P^o	101.325 kPa	A	303.761/R
P^o	101.325 kPa	A	302.672/R	B	89574.863/R
A	314.110/R	B	90417.272/R	C	78.8/R
B	94001.596/R	C	41.6/R	R = 8.314 J K^{-1} mol^{-1}	
C	41.6/R	R = 8.314 J K^{-1} mol^{-1}			
R = 8.314 J K^{-1} mol^{-1}				ΔH_V/(kJ mol^{-1}) = 66.08	
		ΔH_{subl}/(kJ mol^{-1}) = 78.01		at 298.15 K	
ΔH_{subl}/(kJ mol^{-1}) = 81.60		at 298.15 K			
at 298.15 K					

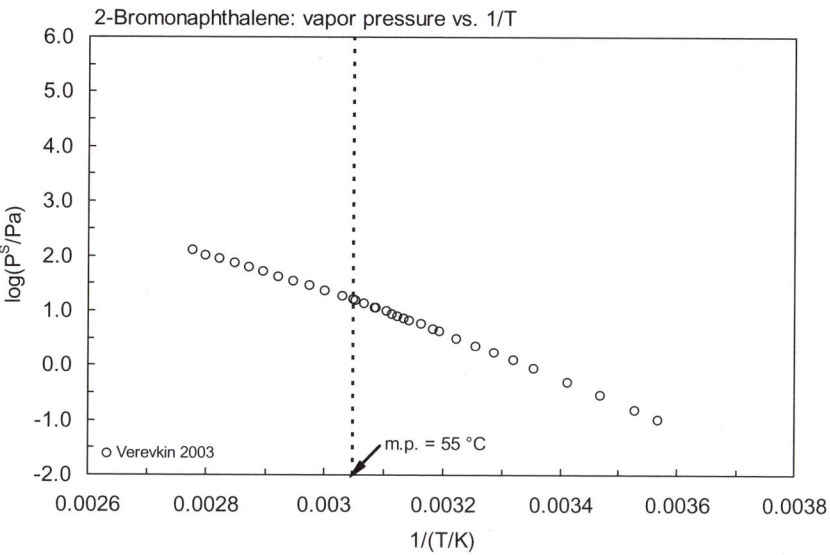

FIGURE 4.1.4.2.1 Logarithm of vapor pressure versus reciprocal temperature for 2-bromonaphthalene.

4.1.4.3 1,4-Dibromonaphthalene

Common Name: 1,4-Dibromonaphthalene
Synonym:
Chemical Name: 1,4-dibromonaphthalene
CAS Registry No: 83-53-4
Molecular Formula: $C_{10}H_6Br_2$
Molecular Weight: 285.963
Melting Point (°C):
    83   (Lide 2003)
Boiling Point (°C):
    310   (Lide 2003)
Density (g/cm³ at 20°C):
Molar Volume (cm³/mol):
    194.2   (calculated-Le Bas method at normal boiling point)
Dissociation Constant pK_a:
Enthalpy of Fusion, ΔH_{fus} (kJ/mol):
Entropy of Fusion, ΔS_{fus} (J/mol K):
Fugacity Ratio at 25°C (assuming ΔS_{fus} = 56 J/mol K), F: 0.27 (mp at 83°C)

Water Solubility (g/m³ or mg/L at 25°C or as indicated):
    0.125, 0.35, and 0.866 (4, 25, and 40°C, generator column-GC/ECD, Dickhut et al. 1994)
    1.92   (calculated-TSA, Dickhut et al. 1994)

Vapor Pressure (Pa at 25°C):

Henry's Law Constant (Pa·m³/mol):

Octanol/Water Partition Coefficient, log K_{OW}:

Octanol/Air Partition Coefficient, log K_{OA}:

Bioconcentration Factor, log BCF or log K_B:

Sorption Partition Coefficient, log K_{OC}:

Environmental Fate Rate Constants, k or Half-Lives, $t_{½}$:

Half-Lives in the Environment:

4.1.4.4 2,3-Dibromonaphthalene

Common Name: 2,3-Dibromonaphthalene
Synonym:
Chemical Name: 2,3-dibromonaphthalene
CAS Registry No:
Molecular Formula: $C_{10}H_6Br_2$
Molecular Weight: 285.963
Melting Point (°C):
    140   (Lide 2003)
Boiling Point (°C):
Density (g/cm³ at 20°C):
Molar Volume (cm³/mol):
    194.2   (calculated-Le Bas method at normal boiling point)
Dissociation Constant pK_a:
Enthalpy of Fusion, ΔH_{fus} (kJ/mol):
Entropy of Fusion, ΔS_{fus} (J/mol K):
Fugacity Ratio at 25°C (assuming ΔS_{fus} = 56 J/mol K), F: 0.0744 (mp at 140°C)

Water Solubility (g/m³ or mg/L at 25°C):
    0.0554, 0.138, and 0.352 (4, 25, and 40°C, generator column-GC/ECD, Dickhut et al. 1994)
    0.432   (calculated-TSA, Dickhut et al. 1994)

Vapor Pressure (Pa at 25°C):

Henry's Law Constant (Pa·m³/mol):

Octanol/Water Partition Coefficient, log K_{OW}:

Octanol/Air Partition Coefficient, log K_{OA}:

Bioconcentration Factor, log BCF or log K_B:

Sorption Partition Coefficient, log K_{OC}:

Environmental Fate Rate Constants, k or Half-Lives, $t_{1/2}$:

Half-Lives in the Environment:

4.1.4.5 4-Bromobiphenyl

Common Name: 4-Bromobiphenyl
Synonym:
Chemical Name: 4-bromobiphenyl
CAS Registry No: 92-66-0
Molecular Formula: $C_{12}H_9Br$
Molecular Weight: 233.103
Melting Point (°C):
    91.5   (Lide 2003)
Boiling Point (°C):
    310   (Weast 1983–84; Stephenson & Malanowski 1987, Lide 2003)
Density (g/cm³ at 25°C):
    0.9327   (Weast 1983–84, Lide 2003)
Molar Volume (cm³/mol):
    176.1   (Ruelle & Kesselring 1997)
    250.0   (25°C, calculated-density)
    207.9   (calculated-Le Bas method at normal boiling point)
Dissociation Constant pK_a:
Enthalpy of Fusion, ΔH_{fus} (kJ/mol):
    26.86   (Doucette & Andren 1988)
Entropy of Fusion, ΔS_{fus} (J/mol K):
Fugacity Ratio at 25°C (assuming ΔS_{fus} = 56 J/mol K), F: 0.223 (mp at 91.5°C)

Water Solubility (g/m³ or mg/L at 25°C or as indicated and reported temperature dependence equations.):
    0.235, 0.653, 0.874 (4.9, 25, 40°C, generator column-GC, Doucette & Andren 1988a)
    $S/(mol/L) = 9.36 \times 10^{-7} \exp(0.037 \cdot t/°C)$ (generator column-GC/ECD, temp range 4.9–40°C, Doucette & Andren 1988a); or
    $\log x = -1436/(T/K) - 1.541$, temp range 4.9–40°C (generator column-GC/ECD, Doucette & Andren 1988a)
    0.546   (calculated-TSA, Dickhut et al. 1994)
    0.256   (calculated-molar volume, mp and mobile order thermodynamics, Ruelle & Kesselring 1997)

Vapor Pressure (Pa at 25°C and reported temperature dependence equations):
    $\log (P_L/kPa) = 2.24643 - 2174.97/[(T/K) - 70.067]$; (Antoine eq., liquid state, temp range 371–583 K, Stephenson & Malanowski 1987)

Henry's Law Constant (Pa·m³/mol):

Octanol/Water Partition Coefficient, $\log K_{OW}$:
    4.96; 4.89, 5.00 (generator column-GC; calculated-group contribution method, estimated-TSA and K_{OW}, Doucette & Andren 1987)
    4.96; 4.89, 4.83, 5.10, 5.14, 5.10 (generator column-GC; calculated-π const., HPLC-RT correlation, calculated-MW, calculated-MCI χ, calculated-TSA and K_{OW}, Doucette & Andren 1988b)
    4.96   (recommended, Sangster 1993)
    4.95   (Hansch et al. 1995)

Octanol/Air Partition Coefficient, $\log K_{OA}$:

Bioconcentration Factor, $\log BCF$ or $\log K_B$:

Sorption Partition Coefficient, $\log K_{OC}$:

Environmental Fate Rate Constants, k or Half-Lives, $t_{1/2}$:

Half-Lives in the Environment:

4.1.4.6 4,4′-Dibromobiphenyl

Br—⟨⟩—⟨⟩—Br

Common Name: 4,4′-Dibromophenyl
Synonym: PBB-15, 4,4′-dibromo-1,1′-biphenyl
Chemical Name: 4,4′-dibromophenyl
CAS Registry No: 92-86-4
Molecular Formula: $C_{12}H_8Br_2$
Molecular Weight: 312.000
Melting Point (°C):
    164   (Ruelle & Kesselring 1997, Lide 2003)
Boiling Point (°C):
    357.5   (Lide 2003)
Density (g/cm³):
Molar Volume (cm³/mol):
    192.2   (Ruelle & Kesselring 1997)
    231.2   (calculated-Le Bas method at normal boiling point)
Enthalpy of Fusion, ΔH_{fus} (kJ/mol):
Entropy of Fusion, ΔS_{fus} (J/mol K):
Fugacity Ratio at 25°C (assuming ΔS_{fus} = 56 J/mol K), F: 0.0433 (mp at 164°C)

Water Solubility (g/m³ or mg/L at 25°C or as indicated and reported temperature dependence equations):
    0.00574   (generator column-GC, Gobas et al. 1988)
    0.104   (quoted, Chessells et al. 1992)
    0.0164   (calculated-MCI χ, Ruelle & Kesselring et al. 1997)

Vapor Pressure (Pa at 25°C):

Henry's Law Constant (Pa·m³/mol):

Octanol/Water Partition Coefficient, log K_{OW}:
    5.72; 5.75, 5.57 (generator column-GC; calculated-group contribution method, estimated-TSA and K_{OW}, Doucette & Andren 1987)
    5.72; 5.75, 5.85, 6.14, 5.81, 5.68 (generator column-GC; calculated-π const., HPLC-RT correlation, calculated-MW, calculated-MCI χ, calculated-TSA and K_{OW}, Doucette & Andren 1988b)
    5.72   (HPLC-RT correlation, Gobas et al. 1988,)
    5.72   (recommended, Sangster 1993)
    4.67   (calculated-UNIFAC, Chen et al. 1993)
    5.72   (recommended, Hansch et al. 1995)

Octanol/Air Partition Coefficient, log K_{OA}:

Bioconcentration Factor, log BCF or log K_B:
    5.43   (guppy, lipid weight-based, Gobas et al. 1989)
    4.19; 5.43   (flowing water system-whole weight of fish; lipid content, quoted, Lu et al. 1999)
    3.825, 4.365 (calculated-MCI χ, K_{OW}, Lu et al. 1999)

Sorption Partition Coefficient, log K_{OC}:

Environmental Fate Rate Constants, k and Half-Lives, $t_{1/2}$:
    Bioconcentration and Uptake and Elimination Rate Constants (k_1 and k_2):
        log k_1 = 3.35 d^{-1} (guppy, Gobas et al. 1989)
        log k_2 = –0.91 d^{-1} (guppy, Gobas et al. 1989)

Half-Lives in the Environment:

4.1.4.7 2,4,6-Tribromobiphenyl

Common Name: 2,4,6-Tribromobiphenyl
Synonym: PBB-30
Chemical Name: 2,4,6-tibromobiphenyl
CAS Registry No: 59080-33-0
Molecular Formula: $C_{12}H_7Br_3$
Molecular Weight: 390.896
Melting Point (°C):
    64   (Ruelle & Kesselring 1997)
Boiling Point (°C):
Density (g/cm^3):
Molar Volume (cm^3/mol):
    208.3   (Ruelle & Kesselring 1997)
    254.5   (calculated-Le Bas method at normal boiling point)
Enthalpy of Fusion, ΔH_{fus} (kJ/mol):
Entropy of Fusion, ΔS_{fus} (J/mol K):
Fugacity Ratio at 25°C (assuming ΔS_{fus} = 56 J/mol K), F: 0.414 (mp at 64°C)

Water Solubility (g/m^3 or mg/L at 25°C):
    0.016   (generator column-GC, Gobas et al. 1988)
    0.0131   (quoted, Chessells et al. 1992)
    0.054   (calculated-MCI χ, Ruelle & Kesselring 1997)

Vapor Pressure (Pa at 25°C):

Henry's Law Constant (Pa·m^3/mol):

Octanol/Water Partition Coefficient, log K_{OW}:
    6.03   (HPLC-RT correlation, Gobas et al. 1988, 1989)
    6.03; 4.78   (quoted; calculated-UNIFAC, Chen et al. 1993)

Octanol/Air Partition Coefficient, log K_{OA}:

Bioconcentration Factor, log BCF or log K_B:
    5.06   (guppy, lipid weight-based, Gobas et al. 1989)
    3.97; 5.06   (flowing water system-whole weight of fish; lipid content, quoted, Lu et al. 1999)
    4.408, 4.645 (calculated-MCI χ, K_{OW}, Lu et al. 1999)

Sorption Partition Coefficient, log K_{OC}:

Environmental Fate Rate Constants, k and Half-Lives, $t_{1/2}$:
    Bioconcentration and Uptake and Elimination Rate Constants (k_1 and k_2):
        log k_1 = 3.05 d^{-1}; log k_2 = –0.83 d^{-1} (guppy, Gobas et al. 1989)

Half-Lives in the Environment:

4.1.4.8 2,2′,5,5′-Tetrabromobiphenyl

Common Name: 2,2′,5,5′-Tetrabromobiphenyl
Synonym: PBB-52
Chemical Name: 2,2′,5,5′-tetrabromobiphenyl
CAS Registry No: 59080-37-4
Molecular Formula: $C_{12}H_6Br_4$
Molecular Weight: 469.792
Melting Point (°C):
    144   (Ruelle & Kesselring 1997)
Boiling Point (°C):
Density (g/cm³):
Molar Volume (cm³/mol):
    224.4   (Ruelle & Kesselring 1997)
    277.8   (calculated-Le Bas at normal boiling point)
Enthalpy of Fusion, ΔH_{fus} (kJ/mol):
Entropy of Fusion, ΔS_{fus} (J/mol K):
Fugacity Ratio at 25°C (assuming ΔS_{fus} = 56 J/mol K), F: 0.068 (mp at 144°C)

Water Solubility (g/m³ or mg/L at 25°C):
    0.00429   (generator column-GC, Gobas et al. 1988)
    0.0545   (quoted, Chessells et al. 1992)
    0.00409   (calculated-AQUAFAC, Myrdal et al. 1995)
    0.00246   (calculated-molar volume, mp and mobile order thermodynamics, Ruelle & Kesselring 1997)

Vapor Pressure (Pa at 25°C):

Henry's Law Constant (Pa·m³/mol):

Octanol/Water Partition Coefficient, log K_{OW}:
    6.50   (HPLC-RT correlation, Gobas et al. 1988,)
    4.88   (calculated-UNIFAC group contribution, Chen et al. 1993)

Bioconcentration Factor, log BCF or log K_B:
    6.16   (guppy, lipid weight-based, Gobas et al. 1989)
    4.62; 6.16   (flowing water system-whole weight of fish; lipid content, quoted, Lu et al. 1999)
    5.097; 6.16   (calculated-MCI χ, K_{OW}, Lu et al. 1999)

Sorption Partition Coefficient, log K_{OC}:

Environmental Fate Rate Constants, k and Half-Lives, $t_{1/2}$:
    Bioconcentration and Uptake and Elimination Rate Constants (k_1 and k_2):
        log k_1 = 2.96 d^{-1}; log k_2 = –2.02 d^{-1} (guppy, Gobas et al. 1989)

Half-Lives in the Environment:

4.1.4.9 2,2′,4,5,5′-Pentabromobiphenyl

Common Name: 2,2′,4,5,5′-Pentabromobiphenyl
Synonym:
Chemical Name: 2,2′,4,5,5′-pentabromobiphenyl
CAS Registry No: 6788-96-4
Molecular Formula: $C_{12}H_5Br_5$
Molecular Weight: 548.688
Melting Point (°C):
    157   (Dickhut et al. 1994; Ruelle & Kesselring 1997)
Boiling Point (°C):
Density (g/cm³ at 20°C):
Molar Volume (cm³/mol):
    301.1   (calculated-Le Bas method at normal boiling point)
Dissociation Constant pK_a:
Enthalpy of Fusion, ΔH_{fus} (kJ/mol):
    45.44   (Doucette & Andren 1988)
Entropy of Fusion, ΔS_{fus} (J/mol K):
Fugacity Ratio at 25°C (assuming ΔS_{fus} = 56 J/mol K), F: 0.0507 (mp at 157°C)

Water Solubility (g/m³ or mg/L at 25°C or as indicated and reported temperature dependence equations.):
    1.032×10^{-4}, 4.42×10^{-4}, 9.82×10^{-4} (4.9, 25, 40°C, generator column-GC/ECD, Doucette & Andren 1988a)
    S/(mol/L) = 1.52×10^{-10} exp(0.063·t/°C) (generator column-GC, temp range 4–40°C, Doucette & Andren 1988a)
    log x = –2374/(T/K) – 2.373, temp range 4–40°C (generator column-GC, Doucette & Andren 1988a)

Vapor Pressure (Pa at 25°C):

Henry's Law Constant (Pa·m³/mol):

Octanol/Water Partition Coefficient, log K_{OW}:
    77.10   (generator column-GC Doucette & Andren 1987)
    77.10; 8.76 (generator column-GC; HPLC-RT correlation, Doucette & Andren 1988b)
    7.10   (recommended, Hansch et al. 1995)

Octanol/Air Partition Coefficient, log K_{OA}:

Bioconcentration Factor, log BCF or log K_B:

Sorption Partition Coefficient, log K_{OC}:

Environmental Fate Rate Constants, k or Half-Lives, $t_{1/2}$:

Half-Lives in the Environment:

4.1.4.10 2,2',4,4',6,6'-Hexabromobiphenyl

Common Name: 2,2',4,4',6,6'-Hexabromobiphenyl
Synonym: 2,2',4,4',6,6'-HBB
Chemical Name: 2,2',4,4',6,6'-hexabromobiphenyl
CAS Registry No: 59261-08-4
Molecular Formula: $C_{12}H_4Br_6$
Molecular Weight: 627.584
Melting Point (°C):
    176   (Gobas et al. 1988; Ruelle & Kesselring 1997)
Boiling Point (°C):
Density (g/cm³):
Molar Volume (cm³/mol):
    256.6   (Ruelle & Kesselring 1997)
    324.4   (calculated-Le Bas method at normal boiling point)
Enthalpy of Fusion, ΔH_{fus} (kJ/mol):
Entropy of Fusion, ΔS_{fus} (J/mol K):
Fugacity Ratio at 25°C (assuming ΔS_{fus} = 56 J/mol K), F: 0.033 (mp at 176°C)

Water Solubility (g/m³ or mg/L at 25°C):
    6.23×10^{-4} (generator column-GC, Gobas et al. 1988)
    0.0210   (lit. mean, Chessells et al. 1992)
    1.04×10^{-4} (calculated-molar volume, mp and mobile order thermodynamics, Ruelle & Kesselring 1997)

Vapor Pressure (Pa at 25°C):
    8.033×10^{-6} (for hexabrominated biphenyl, GC-RT correlation, Watanabe & Tatsukawa 1989)
    4.52×10^{-10} (quoted, Pijnenburg et al. 1995)

Henry's Law Constant (Pa·m³/mol):

Octanol/Water Partition Coefficient, log K_{OW}:
    7.50   (for hexabrominated biphenyl, reversed phase-HPLC-RT correlation, Watanabe & Tatsukawa 1989)
    7.20   (HPLC-RT correlation, Gobas et al. 1987, 1989)
    7.20; 5.09   (quoted; calculated-UNIFAC group contribution, Chen et al. 1993)

Octanol/Air Partition Coefficient, log K_{OA}:

Bioconcentration Factor, log BCF or log K_B:
    5.85   (guppy, lipid weight-based, Gobas et al. 1989)
    4.26   (calculated-K_{OW}, Chessells et al. 1992)

Sorption Partition Coefficient, log K_{OC}:

Environmental Fate Rate Constants, k and Half-Lives, $t_{1/2}$:
    Bioconcentration and Uptake and Elimination Rate Constants (k_1 and k_2):
        log k_1 = 2.51 d⁻¹; log k_2 = –2.15 d⁻¹ (guppy, Gobas et al. 1989)

Half-Lives in the Environment:

4.1.4.11 Decabromobiphenyl

Common Name: Decabromobiphenyl
Synonym: PBB-209
Chemical Name:
CAS Registry No: 13654-09-6
Molecular Formula: $C_{12}Br_{10}$
Molecular Weight: 943.168
Melting Point (°C):
Boiling Point (°C):
Density (g/cm^3):
Molar Volume (cm^3/mol):
   417.6   (calculated-Le Bas method at normal boiling point)
Heat of Fusion, ΔH_{fus} (kJ/mol):
Entropy of Fusion, ΔS_{fus} (J/mol K):
Fugacity Ratio at 25°C (assuming ΔS_{fus} = 56 J/mol K), F:

Water Solubility (g/m^3 or mg/L at 25°C):

Vapor Pressure (Pa at 25°C):
   > 1.33×10^{-9} (GC-RT correlation, Watanabe & Tatsukawa 1989)
   < 7.4×10^{-4} (quoted, Pijnenburg et al. 1995)

Henry's Law Constant (Pa·m^3/mol):

Octanol/Water Partition Coefficient, log K_{OW}:
   8.58; 12.63, 9.36   (generator column-GC; calculated-group contribution method, estimated-TSA and K_{OW}, Doucette & Andren 1987)
   8.58; 10.42, 13.87, 8.46, 8.69, 7.10 (generator column-GC; calculated-π const., HPLC-RT correlation, calculated-MW, calculated-MCI χ, calculated-TSA and K_{OW}, Doucette & Andren 1988b)
   8.60   (reversed phase-HPLC-RT correlation, Watanabe & Tatsukawa 1989)

Octanol/Air Partition Coefficient, log K_{OA}:

Bioconcentration Factor, log BCF or log K_B:

Sorption Partition Coefficient, log K_{OC}:

Environmental Fate Rate Constants, k and Half-Lives, $t_{1/2}$:

Half-Lives in the Environment:

4.2 SUMMARY TABLES AND QSPR PLOTS

TABLE 4.2.1
Summary of physical properties of polynuclear aromatic hydrocarbons (PAHs)

Compound	CAS no.	Molecular formula	Molecular weight, MW g/mol	m.p. °C	b.p. °C	Fugacity ratio, F at 25°C*	Molar volume, V_M cm³/mol MW/ρ at 20°C	Le Bas
Indan	496-11-7	C_9H_{10}	118.175	−51.38	177.97	1	123.0	143.7
Naphthalene	91-20-3	$C_{10}H_8$	128.171	80.26	217.9	0.287	125.0	147.6
1-Methyl-	90-12-0	$C_{11}H_{10}$	142.197	−30.43	244.7	1	139.4	169.8
2-Methyl-	91-57-6	$C_{11}H_{10}$	142.197	34.6	241.1	0.805	141.4	169.8
1,2-Dimethyl-	573-98-8	$C_{12}H_{12}$	156.223	0.8	266.5	1		192.0
1,3-Dimethyl-	575-41-7	$C_{12}H_{12}$	156.223	−6	263	1	154.0	192.0
1,4-Dimethyl-	571-58-4	$C_{12}H_{12}$	156.223	7.6	268	1	153.7	192.0
1,5-Dimethyl-	571-61-9	$C_{12}H_{12}$	156.223	82	265	0.276		192.0
2,3-Dimethyl-	581-40-8	$C_{12}H_{12}$	156.223	105	268	0.164	155.8	192.0
2,6-Dimethyl-	581-42-0	$C_{12}H_{12}$	156.223	112	262	0.140	155.8	192.0
1-Ethyl-	1127-76-0	$C_{12}H_{12}$	156.223	−10.9	258.6	1	155.0	192.0
2-Ethyl-	939-27-5	$C_{12}H_{12}$	156.223	−7.4	258	1	157.4	192.0
1,4,5-Trimethyl-	2131-41-1	$C_{13}H_{14}$	170.250	63	285	0.424		214.2
Biphenyl	92-52-4	$C_{12}H_{10}$	154.207	68.93	256.1	0.371	148.3	184.6
4-Methyl-	644-08-6	$C_{13}H_{12}$	168.234	49.5	267.5	0.575		206.8
4,4'-Dimethyl-	613-33-2	$C_{14}H_{14}$	182.261	125	295	0.104		229.0
Diphenylmethane	101-81-5	$C_{13}H_{12}$	168.234	25.4	265	0.991	167.2	206.8
Bibenzyl	103-29-7	$C_{14}H_{14}$	182.261	52.5	284	0.537		229.0
trans-Stilbene	103-30-0	$C_{14}H_{12}$	180.245	124.2	307	0.106	185.7	221.6
Acenaphthylene	208-96-8	$C_{12}H_8$	150.192	91.8	280	0.221		165.7
Acenaphthene	83-32-9	$C_{12}H_{10}$	154.207	93.4	279	0.213	126.2	173.1
Fluorene	86-73-7	$C_{13}H_{10}$	166.218	114.77	295	0.132		187.9
1-Methylfluorene	1730-37-6	$C_{14}H_{12}$	180.245	87	318	0.246		210.1
Phenanthrene	85-01-8	$C_{14}H_{10}$	178.229	99.24	340	0.187		199.2
1-Methyl-	832-69-9	$C_{15}H_{12}$	192.256	123	354	0.109		221.4
Anthracene	120-12-7	$C_{14}H_{10}$	178.229	215.76	339.9	0.0134		196.7
2-Methyl-	613-12-7	$C_{15}H_{12}$	192.256	209		0.0157		218.9
9-Methyl-	779-02-2	$C_{15}H_{12}$	192.256	81.5	359	0.279		218.9
9,10-Dimethyl-	781-43-1	$C_{16}H_{14}$	206.282	183.6	360	0.0278		241.1

(Continued)

TABLE 4.2.1 (*Continued*)

Compound	CAS no.	Molecular formula	Molecular weight, MW g/mol	m.p.°C	b.p.°C	Fugacity ratio, F at 25°C*	Molar volume, V_M cm³/mol MW/ρ at 20°C	Le Bas
Pyrene	129-00-0	$C_{16}H_{10}$	202.250	150.62	404	0.0585		213.8
Fluoranthene	206-44-0	$C_{16}H_{10}$	202.250	110.19	384	0.146		217.3
Benzo[a]fluorene	238-84-6	$C_{17}H_{12}$	216.227	189.5	405	0.0243		239.5
Benzo[b]fluorene	243-17-4	$C_{17}H_{12}$	216.227	212	401	0.0146		239.5
Chrysene	218-01-9	$C_{18}H_{12}$	228.288	255.5	448	0.00548	179.2	250.8
Triphenylene	217-59-4	$C_{18}H_{12}$	228.288	197.8	425	0.0202		250.8
o-Terphenyl	84-15-1	$C_{18}H_{14}$	230.304	56.2	332	0.494		273.2
m-Terphenyl	92-06-8	$C_{18}H_{14}$	230.304	87	363	0.246	192.1	273.2
p-Terphenyl	92-94-4	$C_{18}H_{14}$	230.304	213.9	376	0.0140		273.2
Naphthacene	92-24-0	$C_{18}H_{12}$	228.288	357	sublim	0.00055		250.8
Benz[a]anthracene	56-55-3	$C_{18}H_{12}$	228.288	160.5	438	0.0468		248.3
Benzo[b]fluoranthene	205-99-2	$C_{20}H_{12}$	252.309	168	481	0.0395		268.9
Benzo[j]fluoranthene	205-82-3	$C_{20}H_{12}$	252.309	166	480	0.0414		268.9
Benzo[k]fluoranthene	207-08-9	$C_{20}H_{12}$	252.309	217	480	0.0131		268.9
Benzo[a]pyrene	50-32-8	$C_{20}H_{12}$	252.309	181.1	495	0.0294		262.9
Benzo[e]pyrene	192-97-2	$C_{20}H_{12}$	252.309	181.4	311	0.0292		262.9
Perylene	198-55-0	$C_{20}H_{12}$	252.309	277.76	503	0.00331		262.9
7,12-DMBA	57-97-6	$C_{20}H_{16}$	256.341	122.5		0.111		292.7
9,10-DMBA	56-56-35	$C_{20}H_{16}$	256.341	122		0.112		292.7
3-MCA	56-49-5	$C_{21}H_{16}$	268.352	180		0.0301	209.6	296.0
Benzo[ghi]perylene	191-24-2	$C_{22}H_{12}$	276.330	272.5		0.00373		277.5
Indeno[1,2,3-c,d]pyrene	193-39-5	$C_{22}H_{12}$	276.330	162		0.0453		283.5
Dibenz[a,c]anthracene	215-58-7	$C_{22}H_{14}$	278.346	205		0.0171		299.9
Dibenz[a,h]anthracene	53-70-3	$C_{22}H_{14}$	278.346	269.5	524	0.00399		299.9
Dibenz[a,j]anthracene	224-41-9	$C_{22}H_{14}$	278.346	197.5		0.0203		299.9
Pentacene	135-48-8	$C_{22}H_{14}$	278.346	> 300 dec				299.9
Coronene	191-07-1	$C_{24}H_{12}$	300.352	437.4	525	0.00009		292.1
2,4'',5-Trichloro-p-terphenyl	61576-93-0	$C_{18}H_{11}Cl_3$	333.639	92		0.220		335.9
2,4',4'',6-Tetrachloro-p-terphenyl		$C_{18}H_{10}Cl_4$	368.084	114		0.134		356.8
1-Chloronaphthalene	90-13-1	$C_{10}H_7Cl$	162.616	-2.5	259	1	136.2	168.5
2-Chloronaphthalene	91-58-7	$C_{10}H_7Cl$	162.616	58	256	0.474		168.5
1,2-Dichloronaphthalene	2050-69-3	$C_{10}H_6Cl_2$	197.061	36	296.5	0.780		189.4
1,4-Dichloronaphthalene	1825-31-6	$C_{10}H_6Cl_2$	197.061	67.5	288	0.383		189.4
1,8-Dichloronaphthalene	2050-74-0	$C_{10}H_6Cl_2$	197.061	89	sublim	0.236		189.4

Name	CAS	Formula	MW	mp (°C)	bp (°C)		Solubility	
2,3-Dichloronaphthalene	2050-75-1	$C_{10}H_6Cl_2$	197.061	120			0.117	189.4
2,7-Dichloronaphthalene	2198-77-8	$C_{10}H_6Cl_2$	197.061	115			0.131	189.4
1,2,3-Trichloronaphthalene	50402-52-3	$C_{10}H_5Cl_3$	231.506	81–84			0.273	210.3
1,3,7-Trichloronaphthalene	55720-37-1	$C_{10}H_5Cl_3$	231.506	113	274		0.137	210.3
1,2,3,4-Tetrachloronaphthalene	20020-02-4	$C_{10}H_4Cl_4$	265.951	199			0.0196	231.2
1,2,3,5-Tetrachloronaphthalene	53555-63-8	$C_{10}H_4Cl_4$	265.951	141			0.0728	231.2
1,3,5,7-Tetrachloronaphthalene	53555-64-9	$C_{10}H_4Cl_4$	265.951	179			0.0308	231.2
1,3,5,8-Tetrachloronaphthalene	31604-28-1	$C_{10}H_4Cl_4$	265.951	131			0.0912	231.2
1,2,3,4,6-Pentachloronaphthalene	67922-25-2	$C_{10}H_3Cl_5$	300.396	147			0.0635	252.1
1,2,3,5,7-Pentachloronaphthalene	53555-65-0	$C_{10}H_3Cl_5$	300.396	171	313		0.0369	252.1
1,2,3,5,8-Pentachloronaphthalene	150224-24-1	$C_{10}H_3Cl_5$	300.396	174–176			0.0340	252.1
1,2,3,4,5,7-Hexachloronaphthalene	67927-67-4	$C_{10}H_2Cl_6$	334.842	194	331		0.0220	273.0
1,2,3,4,6,7-Hexachloronaphthalene	103426-96-6	$C_{10}H_2Cl_6$	334.842	205–206			0.0340	273.0
1,2,3,5,6,7-Hexachloronaphthalene	103426-97-7	$C_{10}H_2Cl_6$	334.842	234–235			0.00880	273.0
1,2,3,5,7,8-Hexachloronaphthalene	103426-94-4	$C_{10}H_2Cl_6$	334.842	148–149			0.0614	273.0
1,2,3,4,5,6,7-Heptachloro-	58863-14-2	$C_{10}HCl_7$	369.287	194	348		0.0220	293.9
1,2,3,4,5,6,8-Heptachloro-	58863-15-3	$C_{10}HCl_7$	369.287	197.5	365		0.0203	293.9
Octachloronaphthalene	2234-13-1	$C_{10}Cl_8$	403.731	197.5		140.0		314.8
1-Bromonaphthalene	90-11-9	$C_{10}H_7Br$	207.067	6.1	281		1	170.9
2-Bromonaphthalene	580-13-1	$C_{10}H_7Br$	207.067	55.9	281.5		0.498	170.9
1,4-Dibromonaphthalene	83-53-4	$C_{10}H_6Br_2$	285.963	83	310		0.270	194.2
2,3-Dibromonaphthalene	13214-70-5	$C_{10}H_6Br_2$	285.963	140			0.0744	194.2
4-Bromobiphenyl	92-66-0	$C_{12}H_9Br$	233.103	91.5	310		0.223	207.9
4,4'-Dibromobiphenyl	92-86-4	$C_{12}H_8Br_2$	312.000	164	357.5		0.0433	231.2
2,4,6-Tribromobiphenyl	59080-33-0	$C_{12}H_7Br_3$	390.896	64			0.414	254.5
2,2',5,5'-Tetrabromobiphenyl	59080-37-4	$C_{12}H_6Br_4$	469.792	144			0.0680	277.8
2,2',4,5,5'-Pentabromobiphenyl	6788-96-4	$C_{12}H_5Br_5$	548.688	157			0.0507	301.1
2,2',4,4',6,6'-Hexabromobiphenyl	59261-08-4	$C_{12}H_4Br_6$	627.584	176			0.0330	324.4
Decabromobiphenyl	13654-09-6	$C_{12}Br_{10}$	943.168					417.6

Note:
3-MCA   3-Methylcholanthrene
7,12-DMBA   7,12-Dimethylbenz[a]anthracene
9,10-DMBA   9,10-Dimethylbenz[a]anthracene

* Assuming ΔS_{fus} = 56 J/mol K.

TABLE 4.2.2
Summary of selected physical-chemical properties of polynuclear aromatic hydrocarbons (PAHs) at 25°C

Compound	Selected properties			Solubility			log K_{OW}	Henry's law constant $H/(Pa \cdot m^3/mol)$
	Vapor pressure							
	P^S/Pa	P_L/Pa	$S/(g/m^3)$	$C^S/(mmol/m^3)$	$C_L/(mmol/m^3)$		log K_{OW}	calculated P/C
Indan	197	197	100	846.2	846.2		3.33	232.8
Naphthalene	10.4	36.24	31	241.9	842.7		3.37	43.00
1-Methyl-	8.84	8.84	28	196.9	196.9		3.87	44.89
2-Methyl-	9.0	11.2	25	175.8	218.4		3.86	51.19
1,2-Dimethyl-	0.87	0.87					4.31	
1,3-Dimethyl-			8	51.21	51.21		4.42	
1,4-Dimethyl-	2.27	2.27	11.4	72.97	72.97		4.37	31.11
1,5-Dimethyl-			3.1	19.84	71.90		4.38	
2,3-Dimethyl-	1.0	6.10	2.5	16.00	97.58		4.40	62.49
2,6-Dimethyl-	1.4	10.0	1.7	10.88	77.73		4.31	128.7
1-Ethyl-	2.51	2.51	10.1	64.65	64.65		4.40	38.82
2-Ethyl-	4.0	4.0	8.0	51.21	51.21		4.38	78.11
1,4,5-Trimethyl-	0.681	1.61	2.1	12.33	29.09		5.00	55.21
Biphenyl	1.3	3.50	7.0	45.39	122.4		3.90	28.64
4-Methyl-			4.05	24.07	41.87		4.63	
4,4'-Dimethyl-			0.175	0.960	9.232		5.09	
Diphenylmethane	0.0885	0.0893	16	95.10	95.10		4.14	0.931
Bibenzyl	0.406	0.756	4.37	23.98	44.65		4.70	16.93
trans-Stilbene	0.065	0.613	0.29	1.609	15.18		4.81	40.40
Acenaphthylene	0.9	4.14	16.1	107.2	485.0		4.00	8.396
Acenaphthene	0.3	1.41	3.80	24.64	115.7		3.92	12.17
Fluorene	0.09	0.682	1.90	11.43	85.60		4.18	7.873
1-Methyl-			1.09	6.047	24.58		4.97	
Phenanthrene	0.02	0.107	1.10	6.172	33.00		4.57	3.240
1-Methyl-			0.27	1.404	12.88		5.14	
Anthracene	0.001	0.0746	0.045	0.252	18.84		4.54	3.961
2-Methyl-			0.03	0.156	9.939		5.15	
9-Methyl-	0.00224	0.00803	0.261	1.358	4.866		5.07	1.650
9,10-Dimethyl-	1.53×10^{-4}	5.50×10^{-3}	0.056	0.271	9.765		5.25	0.564
Pyrene	0.0006	0.0119	0.132	0.652	12.89		5.18	0.919
Fluoranthene	0.00123	8.42×10^{-3}	0.26	1.286	8.805		5.22	0.957
Benzo[a]fluorene			0.045	0.208	8.564		5.40	

Benzo[b]fluorene	5.70×10^{-7}	0.002	0.00925	0.634	5.75
Chrysene	2.30×10^{-9}	0.002	0.00876	1.599	5.60
Triphenylene	1.07×10^{-4}	0.043	0.188	9.325	5.49
p-Terphenyl	4.86×10^{-9} 1.21×10^{-4}	0.0180	0.0782	5.583	6.03
Naphthacene	7.30×10^{-9} 3.47×10^{-4}	0.0006	0.00263	4.779	5.76	0.065
Benz[a]anthracene	2.80×10^{-5} 1.33×10^{-5}	0.011	0.0482	1.030	5.91	0.012
Benzo[b]fluoranthene	5.98×10^{-4}	0.0015	0.00595	0.150	5.80	0.062
Benzo[j]fluoranthene		0.0025	0.0099	0.239		2.77×10^{-3}
Benzo[k]fluoranthene	5.20×10^{-8} 3.97×10^{-9}	0.0008	0.00317	0.242	6.00	0.581
Benzo[a]pyrene	7.00×10^{-7} 2.38×10^{-5}	0.0038	0.0151	0.512	6.04	0.016
Benzo[e]pyrene	7.40×10^{-7} 2.53×10^{-5}	0.004	0.0159	0.543		0.046
Perylene	1.40×10^{-8} 4.23×10^{-9}	0.0004	0.00159	0.479	6.25	0.047
7,12-DMBA	3.84×10^{-8} 3.45×10^{-7}	0.0500	0.195	1.757	6.00	8.83×10^{-3}
9,10-DMBA	3.73×10^{-7} 3.33×10^{-9}	0.0435	0.170	1.543	6.00	1.97×10^{-4}
3-MCA	1.03×0^{-9} 3.42×10^{-5}	0.0019	0.00708	0.235	6.42	2.20×10^{-3}
Benzo[ghi]perylene	2.25×10^{-5}	0.00026	0.000941	0.252	6.50	0.145
Indeno[1,2,3-c,d]pyrene						
Dibenz[a,c]anthracene	1.30×10^{-9} 7.84×10^{-8}	0.0016	0.00575	0.336		2.26×10^{-4}
Dibenz[a,h]anthracene	3.70×10^{-10} 9.27×10^{-8}	0.0006	0.00216	0.540	6.75	1.72×10^{-4}
Dibenz[a,j]anthracene		0.012	0.0431	2.210		
Pentacene	1.0×10^{-10}					
Coronene	2.0×10^{-10} 2.22×10^{-9}	0.00014	0.000466	5.179	6.75	4.29×10^{-4}

Abbreviations:
3-MCA   3-Methylcholanthrene
7,12-DMBA   7,12-Dimethylbenz[a]anthracene
9,10-DMBA   9,10-Dimethylbenz[a]anthracene

TABLE 4.2.3
Suggested half-life classes of polynuclear aromatic hydrocarbons (PAHs) in various environment compartments at 25°C

Compound	Air class	Water class	Soil class	Sediment class
Indan	2	4	6	7
Naphthalene	2	4	6	7
1-Methyl-	2	4	6	7
2,3-Dimethyl-	2	4	6	7
1-Ethyl-	2	4	6	7
1,4,5-Trimethyl-	2	4	6	7
Biphenyl	3	4	5	6
Acenaphthene	3	5	7	8
Fluorene	3	5	7	8
Phenanthrene	3	5	7	8
Anthracene	3	5	7	8
Pyrene	4	6	8	9
Fluoranthene	4	6	8	9
Chrysene	4	6	8	9
Benz[a]anthracene	4	6	8	9
Benzo[k]fluoranthene	4	6	8	9
Benzo[a]pyrene	4	6	8	9
Perylene	4	6	8	9
Dibenz[a,h]anthracene	4	6	8	9

where,

Class	Mean half-life (hours)	Range (hours)
1	5	< 10
2	17 (~ 1 day)	10–30
3	55 (~ 2 days)	30–100
4	170 (~ 1 week)	100–300
5	550 (~ 3 weeks)	300–1,000
6	1700 (~ 2 months)	1,000–3,000
7	5500 (~ 8 months)	3,000–10,000
8	17000 (~ 2 years)	10,000–30,00
9	55000 (~ 6 years)	> 30,000

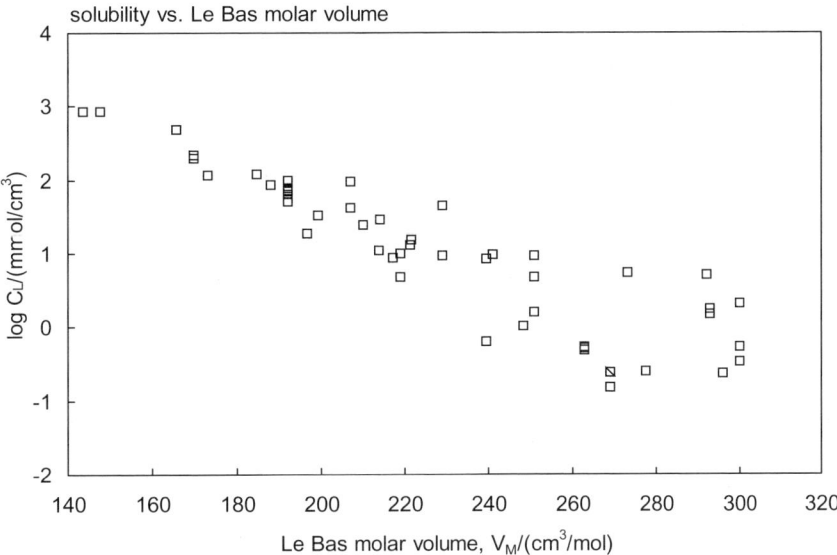

FIGURE 4.2.1 Molar solubility (liquid or supercooled liquid) versus Le Bas molar volume for polynuclear aromatic hydrocarbons.

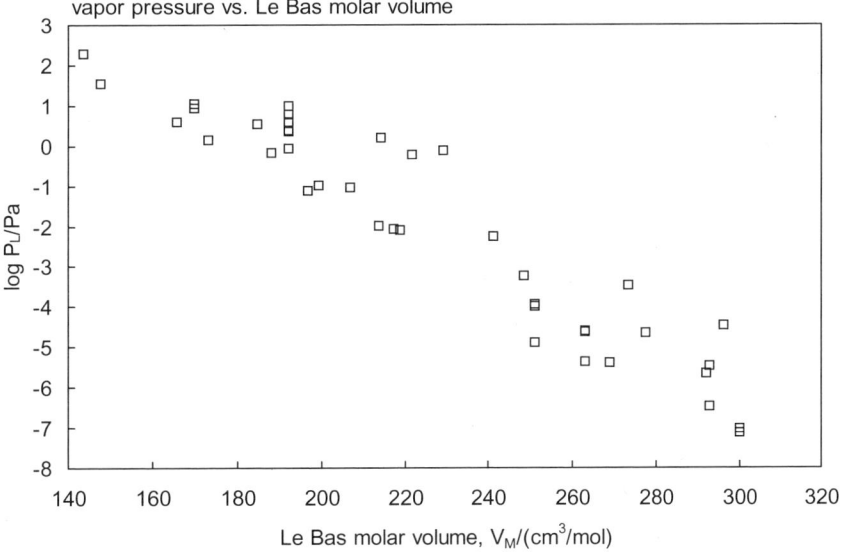

FIGURE 4.2.2 Vapor pressure (liquid or supercooled liquid) versus Le Bas molar volume for polynuclear aromatic hydrocarbons.

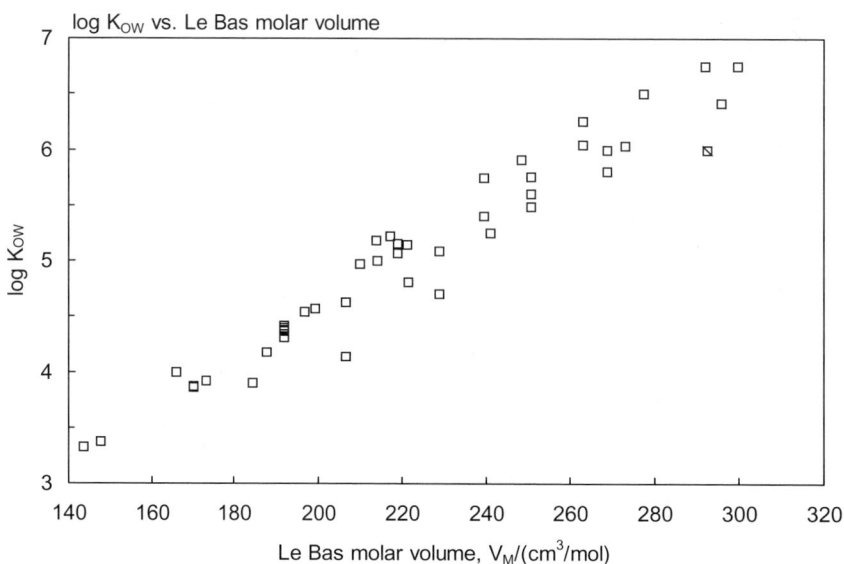

FIGURE 4.2.3 Octanol-water partition coefficient versus Le Bas molar volume for polynuclear aromatic hydrocarbons.

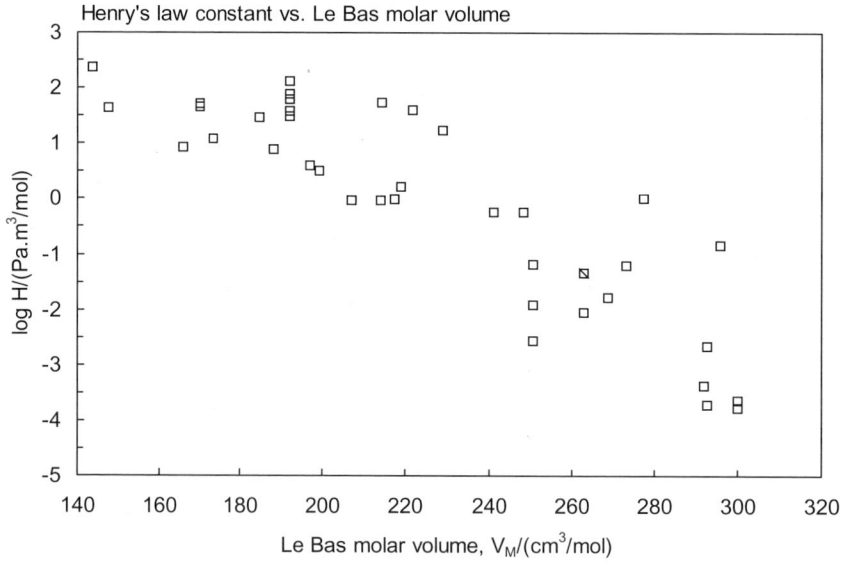

FIGURE 4.2.4 Henry's law constant versus Le Bas molar volume for polynuclear aromatic hydrocarbons.

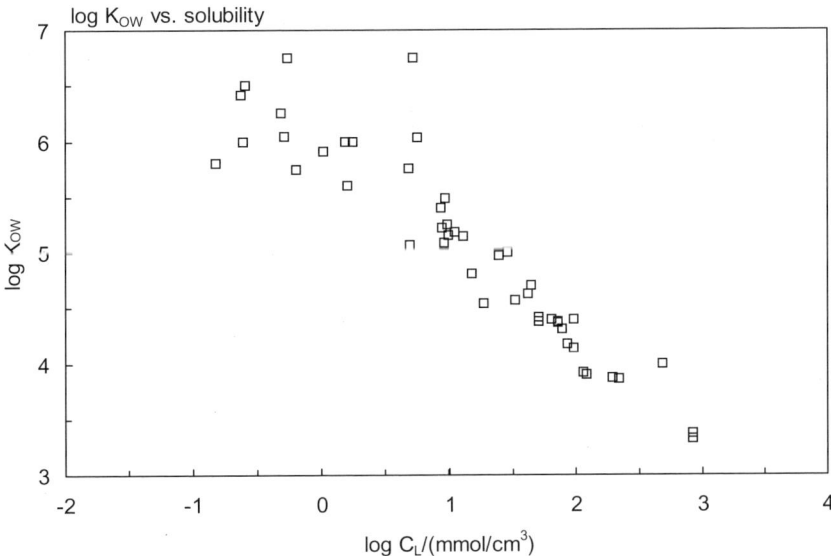

FIGURE 4.2.5 Octanol-water partition coefficient versus molar solubility (liquid or supercooled liquid) for polynuclear aromatic hydrocarbons.

4.3 REFERENCES

Abernethy, S., Mackay, D. (1987) A discussion of correlations for narcosis in aquatic species. In: *QSAR in Environmental Toxicology* - II, Kaiser, K.L.E., Ed., pp. 1–16, D. Reidel Publ. Co., Dordrecht, Netherlands.

Abraham, M.H., Le J., Acree, Jr., W.E., Carr, P.W., Dallas, A.J.(2001) The solubility of gases and vapours in dry octan-1-ol at 298 K.*Chemosphere* 44, 855–863.

Akiyoshi, M., Deguchi, T., Sanemasa, I. (1987) The vapor saturation method for preparing aqueous solutions of solid aromatic hydrocarbons. *Bull. Chem. Soc. Jpn.* 60, 3935–3939.

Alaee, M., Whittal, R.M., Strachan, W.M., J. (1996) The effect of water temperature and composition on Henry's law constant for various PAHs. *Chemosphere* 32, 1153–1164.

Alcock, R.E., Sweetman, A., Jones, K.C. (1999) Assessment of organic contaminant fate in waste water treatment plants. *Chemosphere* 38, 2247–2262.

Alcorn, C.J., Simpson, R.J., Leahy, D.E., Peters, T.J. (1993) Partition and distribution coefficients of solutes and drugs in brush border membrane vesicles. *Biochem Pharmacol.* 45, 1775–1782.

Altschuh, J., Brüggemann, Santl, H., Eichinger, G., Piringer, O.G. (1999) Henry's law constants for a diverse set of organic chemicals: Experimental determination and comparison of estimation methods. *Chemosphere* 39, 1871–1887.

Ambrose, M., Lawrenson, I.J., Sprake, C.H.S. (1975) The vapour pressure of naphthalene. *J. Chem. Thermodyn.* 7, 1173–1176.

Ambrose, D., Sprake, H.S. (1976) The vapour pressure of indane. *J. Chem. Thermodyn.* 8, 601–602.

Amidon, G.L., Anik, S.T. (1980) Hydrophobicity of polycyclic aromatic compounds. Thermodynamic partitioning analysis. *J. Phys. Chem.* 84, 970–974.

Anderson, T.A., Beauchamp, J.J., Walton, B.T. (1991) Organic chemicals in the environment. *J. Environ. Qual.* 20, 420–424.

Andersson, J.T., Schräder, W. (1999) A method for measuring 1-octanol-water partition coefficients. *Anal. Chem.* 71, 3610–3614.

Andrews, L.J., Keefer, R.M. (1949) Cation complexes of compounds containing carbon-carbon double bonds. IV. The argentation of aromatic hydrocarbons. *J. Am. Chem. Soc.* 71, 3644–3647.

Andrews, L.J., Keefer, R.M., (1950) Cation complexes of compounds containing carbon-carbon double bonds. VI. The argentation of substituted benzene. *J. Am. Chem. Soc.* 72, 3110–3116.

API (1979) *Monograph Series 707, Naphthalene; 708, Anthracene and Phenanthrene; and 709, A Ring Condensed Series.* American Petroleum Institute, Washington, D.C.

Arey, J., Atkinson, R., Zielinska, B., McElroy, P.A. (1989) Diurnal concentrations of volatile polycyclic aromatic hydrocarbons and nitroarenes during a photochemical air pollution episode in Glendora, California. *Environ. Sci. Technol.* 23, 321–327.

Atkinson, R. (1985) Kinetics and mechanisms of the gas phase reactions of hydroxyl radicals with organic compounds under atmospheric conditions. *Chem. Rev.* 85, 69–201.

Atkinson, R. (1987) Structure-activity relationship for the estimation of rate constants for the gas phase reactions of OH radicals with organic compounds. *Int'l J. Chem. Kinetics* 19, 799–828.

Atkinson, R. (1989) Kinetics and Mechanisms of the gas-phase reactions of the hydroxyl radical with organic compounds. *J. Phys. Chem. Data* Monograph No.1.

Atkinson, R. (1990) Gas phase tropospheric chemistry of organic compounds: a review. *Atmos. Environ.* 24A, 1–41.

Atkinson, R., Aschmann, S.M. (1985) Rate constants for the gas-phase reaction of hydroxyl radicals with biphenyl and the monochlorobiphenyls at 295 ± 1 K. *Environ. Sci. Technol.* 19, 462–464.

Atkinson, R., Aschmann, S.M. (1986) Kinetics of the reactions of naphthalene, 2-methylnaphthalene and 2,3-dimethylnaphathalene with OH radicals and with O_3 at 295 ± 1 K. *Intl. J. Chem. Kinetics* 18, 569–573.

Atkinson, R., Aschmann, S.M. (1987) Kinetics of the gas-phase reactions of alkylnaphthalenes with O_3, N_2O_5 and OH radicals at 298 ± 2 K. *Atmos. Environ.* 21, 2323–2326.

Atkinson, R., Aschmann, S.M. (1988) Kinetics of the reactions of acenaphthene and acenaphthylene and structurally-related aromatic compounds with OH and NO_3 radicals, N_2O_5 and O_3 at 296 ± 2 K. *Int. J. Chem. Kinet.* 20, 513–539.

Atkinson, R., Aschmann, S.M., Pitts Jr., J.N. (1984) Kinetics of the reactions of naphthalene and biphenyl with OH radicals and O_3 at 294 ± 1 K. *Environ. Sci. Technol.* 18, 110–113.

Atkinson, R., Arey, J., Zielinska, B., Aschmann, S.M. (1987) Kinetics and products of the gas-phase reactions of OH radicals and N_2O_5 with naphthalene and biphenyl. *Environ. Sci. Technol.* 21, 1014–1022.

Atkinson, R., Carter, W.P.L. (1984) Kinetics and mechanisms of gas-phase ozone reaction with organic compounds under atmospheric conditions. *Chem. Rev.* 84, 437–470.

Atkinson, R., Tuazon, E.C., Arey, J. (1990) Reaction of naphthalene in N_2O_5–NO_3–NO_2–air mixtures. *Int. J. Chem. Kinet.* 22, 1071–1082.

Augood, D.R., Hey, D.H., Williams, G.H. (1953) Homolytic aromatic substitution. Part III. Ratio of isomerides formed in the phenylation of chlorobenzene. Competitive experiments on the phenylation of *p*-dichlorobenzene and 1,3,5-trichlorobenzene. Partial rate factors of chlorobenzene. *J. Chem. Soc.* (London), pp. 44–50.

Bahnick, D.A., Doucette, W.J. (1988) Use of molecular indices to estimate soil sorption coefficients for organic chemicals. *Chemosphere* 17(9), 1703–1715.

Bailey, R.E., Gonslor, S.J., Rhinehart, W.L. (1983) Biodegradation of the monochlorobiphenyls and biphenyl in river water. *Environ. Sci. Technol.* 17, 617–621.

Baker, J.E., Eisenreich, S.J. (1990) Concentrations and fluxes of polycyclic aromatic hydrocarbons and polychlorinated biphenyls across the air-water interface of Lake Superior. *Environ. Sci. Technol.* 24, 342–352.

Bamford, H.A., Poster, D.L., Baker, J.E. (1999) Temperature dependence of Henry's law constants of thirteen polycyclic aromatic hydrocarbons between 4°C and 31°C . *Environ. Toxicol. Chem.* 18,1905–1912.

Banerjee, S., Baughman, G.L. (1991) Bioconcentration factors and lipid solubility. *Environ. Sci. Technol.* 25(3), 536–539.

Banerjee, S., Howard, P. (1988) Improved estimation of solubility and partitioning through correction of UNIFAC-derived activity coefficients. *Environ. Sci. Technol.* 22(7), 839–841.

Banerjee, S., Howard, P., Lande, S.S. (1990) General structure-vapor pressure relationships for organics. *Chemosphere* 21(10–11), 1173–1180.

Banerjee, S., Yalkowsky, S.H., Valvani, S.C. (1980) Water solubility and octanol/water partition coefficients of organics. Limitations of the solubility-partition coefficient correlation. *Environ. Sci. Technol.* 14, 1227–1229.

Barnsley, E.A. (1975) The bacterial degradation of fluoranthene and benzo(a) pyrene. *Can. J. Microbiol.* 21, 1004–1008.

Barone, G., Crescenzi, V., Liquori, A.M., Quadrifoglio, F. (1967) Solubilization of polycyclic aromatic hydrocarbons in poly(methacrylic acid) aqueous solutions. *J. Phys. Chem.* 71, 2341–2345.

Barrows M.E., Petrocelli, S.R., Macek, K.J., Carroll, J.J. (1980) Bioconcentration and elimination of selected water pollutants by bluegill sunfish (*Lepomis macrochirus*). In: *Dynamic, Exposure, Hazard Assessment of Toxic Chemicals.* Haque, R. Editor. Ann Arbor Press, pp. 379–392, Ann Arbor, Michigan.

Baughman, G.L., Paris, D.F. (1981) Microbial bioconcentration of organic pollutants from aquatic systems-a critical review. *CRC Critical Reviews in Microbiology.* pp. 205–228.

Baulch, D.L., Campbell, I.M., Saunders, S.M., Louie, P.K.K. (1989) *J. Chem. Soc. Faraday Trans.2.* 85, 1919.

Bayard, R., Barna, L., Mahjoub, B., Gourdon, R. (1998) Investigation of naphthalene sorption in soils and soil fractions using batch and column assays. *Environ. Toxicol. Chem.* 17, 2383–2390.

Bayona, J.M., Fernandez, P., Porte, C., Tolosa, I., Valls, M., Albaiges, J. (1991) Partitioning of urban wastewater organic microcontaminants among coastal compartments. *Chemosphere* 23(3), 313–326.

Bechalany, A., Röthlisberger, T., El Tayler, N., Testa, B. (1989) Comparison of various non-polar stationary phases used for assessing lipophilicity. *J. Chromatog.* 473, 115–124.

Behymer, T.D., Hites, R.A. (1985) Photolysis of polycyclic aromatic hydrocarbons adsorbed on simulated atmospheric particulates. *Environ. Sci. Technol.* 19(10), 1004–1006.

Bender, R., Bieling, V., Maurer, G. (1983) The vapour pressures of solids: anthracene, hydroquinone, and resorcinol. *J. Chem. Thermodyn.* 15, 585–594.

Bennet, D., Canady, J. (1984) Thermodynamics of solution of naphthalene in various water-ethanol mixtures. *J. Am. Chem. Soc.* 106, 910–915.

Berthod, A., Menges, R.A., Armstrong, D.W. (1992) Direct octanol-water partition coefficient determination using co-current chromatography. *J. Liq. Chromatogr.* 15, 2769–2785.

Bidleman T.F. (1984) Estimation of vapor pressures for nonpolar organic compounds by capillary gas chromatography. *Anal. Chem.* 56, 2490–2496.

Bidleman T.F., Foreman, W.T. (1987) vapor-particle partitioning of semivolatile organic compounds. In: *Sources, and Fates of Aquatic Pollutants.* Hites, R.A., Eisenreich, S.J., Editors, *Advances in Chemistry Series* 216, American Chemical Society, Washington, D.C.

Biermann, H.W., MacLeod, H., Atkinson, R., Winer, A.M., Pitts Jr., J.N. (1985) Kinetics of the gas-phase reactions of the hydroxyl radical with naphthalene, phenanthrene and anthracene. *Environ. Sci. Technol.* 19, 244–248.

Billington, J.W., Huang, G.L., Szeto, F., Shiu, W.Y., Mackay, D. (1988) Preparation of sparingly soluble organic substances: I. Single component systems. *Environ. Toxicol. Chem.* 7, 117–124.

Bjørseth, A. Editor (1983) *Handbook of Polycyclic Aromatic Hydrocarbons.* Marcel Dekker, N.Y. and Basel.

Bjørseth, A., Olufsen, B.S. (1983) Long-range transport of polycyclic aromatic Hydrocarbons. In: *Handbook of Polycyclic Aromatic Hydrocarbons.* Bjørseth, A. Editor, pp. 50–524, Marcel Dekker, N.Y. and Basel.

Bohon, R.L., Claussen, W.F. (1951) The solubility of aromatic hydrocarbons in water. *J. Am. Chem. Soc.* 73, 1571–1578.

Bopp, R.F. (1983) Revised parameters for modelling the transport of PCB compounds across the air-water interface. *J. Geophys. Res.* 88, 2521–2529.

Bott, T.L., Standley, L.J. (2000) Transfer of benzo[*a*]pyrene and 2,2′,5,5′-tetrachlorobiphenyl from bacteria and algae to sediment-associated freshwater invertebrates. *Environ. Sci. Technol.* 34, 4936–4942.

Boublik, T., Fried, V., Hala, E. (1973) *The Vapour Pressures of Pure Substances.* Elsevier, Amsterdam.

Boublik, T., Fried, V., Hala, E. (1984) *The Vapour Pressures of Pure Substances.* 2nd revised ed., Elsevier, Amsterdam.

Bradley, R.S., Cleasby, T.G. (1953) The vapor pressure and lattice energy of some aromatic ring compounds. *J. Chem. Soc* 1690–1692.

Brannon, J.N., Pennington, J.C., McFarland, V.A., Hayes, C. (1995) The effects of sediment contact on K_{OC} of nonpolar organic contaminants. *Chemosphere* 31, 3465–3473.

Briggs, G.G. (1981) Theoretical and experimental relationships between soil adsorption, octanol-water partition coefficients, water solubilities, bioconcentration factors, and the Parachor. *J. Agric Food Chem.* 29, 1050–1059.

Bright, N.F.H. (1951) The vapor pressure of diphenyl, dibenzyl and diphenylmethane. *J. Chem. Soc.* Part I, 624–625.

Brodsky, J., Ballschmiter, K. (1988) Reversed phase liquid chromatography of PCBs as a basis for the calculation of water solubility and log K_{OW} for polychlorobiphenyls. *Fresenius Z. Anal. Chem.* 331, 295–301.

Broman, D., Naf, C., Rolff, C., Zebuhr, Y. (1991) Occurrence and dynamics of polychlorinated dibenzo-*p*-dioxins and dibenzofurans and polycyclic aromatic hydrocarbons in the mixed surface layer of remote coastal and offshore waters of the Baltic. *Environ. Sci. Technol.* 25, 1850–1864.

Brooke, D.N., Dobbs, A.J., Williams, N. (1986) Octanol/water partition coefficients (P) : Measurements, estimation, and interpartition particularly for chemicals with P > 10^5. *Ecotoxicol. Environ. Saf.* 11, 251–260.

Brooke, D., Nielsen, I., de Bruijn, J., Hermens, J. (1990) An interlaboratory evaluation of the stir-flask method for the determination of octanol-water partition coefficients (log P_{OW}) . *Chemosphere* 21, 119–133.

Brubaker, Jr. W.W., Hites, R.A. (1998) OH reaction kinetics of polycyclic aromatic hydrocarbons and polychlorinated dibenzo-*p*-dioxins and dibenzofurans. *J. Phys. Chem. A* 102, 915–921.

Bruggeman, W.A., Van Der Steen, J., Hutzinger, O. (1982) Reversed-phase thin-layer chromatography of polynuclear aromatic hydrocarbons and chlorinated biphenyls. Relationship with hydrophobicity as measured by aqueous solubility and octanol-water partition coefficient. *J. Chromatogr.* 238, 335–346.

Budavari, S., Editor (1989) The Merck Index. *An Encyclopedia of Chemicals, Drugs and Biologicals.* 11th edition, Merck & Co., Rahway, New Jersey, U.S.A.

Bulman, T.L., Lesage, S., Fowlie, P., Webber, M.D. (1987) The fate of polynuclear aromatic hydrocarbons in soil. In: *Oil in Fresh Water: Chemistry, Biology, Countermeasure Technology.* Vandermeulan, J.H., Hurley, S.E., Editors, Pergamon Press, New York.

Burkhard, L.P. (1984) *Physical-chemical Properties of the Polychlorinated Biphenyls: Measurement, Estimation, and Application to Environmental Systems.* Ph.D. Thesis, University of Wisconsin-Madison.

Burkhard, L.P., Andren, A.W., Armstrong, D.E. (1985) Estimation of vapor pressures for polychlorinated biphenyls: A comparison of eleven predictive methods. *Environ. Sci. Technol.* 19, 500–507.

Burkhard, L.P., Armstrong, D.E., Andren, A.W. (1985) Henry's law constants for the polychlorinated biphenyls. *Environ. Sci. Technol.* 19, 590–596.

Burkhard, L.P., Kuehl, D.W., Veith, G.D. (1985) Evaluation of reverse phase liquid chromatography/mass spectrometry for estimation of *n*-octanol/water partition coefficients for organic chemicals. *Chemosphere* 14(10), 1551–1560.

Burris, D.R., MacIntyre, W.G. (1986) A thermodynamic study of solutions of liquid hydrocarbon mixtures in water. *Geochim. Cosmochim. Acta* 50, 1545–1549.

Butković, V., Klasinc, L., Orhanović, Turk, J., Güsten, H. (1983) Reaction rates of polynuclear aromatic hydrocarbons with ozone in water. *Environ. Sci. Technol.* 17, 546–548.

Callahan, M.A., Slimak, M.W., Gabel, N.W., May, I.P., Fowler, C.F., Freed, J.R., Jennings, P., Durfee, R.L., Whitmore, F.C., Maestri, B., Mabey, W.R., Holt, B.R., Gould, C. (1979) *Water Related Environmental Fate of 129 Priority Pollutants.* EPA-440-4-79-029a,b.

Camin, D.L., Rossini, F.D. (1955) Physical properties of fourteen American Petroleum Institute research hydrocarbons, C_9 to C_{15}. *J. Phys. Chem.* 59, 1173–1179.

Capel, P.D., Leuenberger, C., Giger, W. (1991) Hydrophobic organic chemicals in urban fog. *Atmos. Environ.* 25A(7), 1335–1346.

Carlson, R.M., Carlson, R.E., Kopperman, H.L. (1975) Determination of partition coefficients by liquid chromatography. *J. Chromatogr.* 107, 219–223.

Carlson, A.R., Kosian, P.A. (1987) Toxicity of chlorobenzenes to fathead minnows (*pimephales promelas*) . *Arch. Environ. Contam. Toxicol.* 16, 129–135.

Casellato, F., Vecchi, C., Grielli, A., Casu, B. (1973) Differential calorimetric study of polycyclic aromatic hydrocarbons. *Thermochim. Acta* 6, 361–368.

Casserly, D.M., Davis, E.M., Downs, T.D., Guthrie, R.K. (1983) Sorption of organics by *selenastrum capricornutum*. *Water Res.* 17(11), 1591–1594.

Chao, J., Lin, C.T., Chung, T.H. (1983) vapor pressure of coal chemicals. *J. Phys. Chem. Ref. Data* 12, 1033–1063.

Chen, F., Holten-Andersen, J., Tyle, H. (1993) New developments of the UNIFAC model for environmental applications. *Chemosphere* 26, 1325–1354.

Chen, J., Quan, X., Yan, Y., Yang, F., Peijnenburg, W.J.G.M. (2001) Quantitative structure-property relationship studies on direct photolysis of selected polycyclic aromatic hydrocarbons in atmospheric aerosol. *Chemosphere* 42, 263–270.

Chen, J.W., Kong, L.R., Zhu, C.M., Huang, Q.G., Wang, L.S. (1996) Correlation between photolysis rate constants of polycyclic aromatic hydrocarbons and frontier molecular orbital energy. *Chemosphere* 33, 1143–1150.

Chessells, M., Hawker, D.W., Connell, D.W. (1992) Influence of solubility in lipid on bioconcentration of hydrophobic compounds. *Ecotoxicol. Environ. Saf.* 23, 260–273.

Chickos, J.S., Acree, Jr., W.E., Liebman, J.F. (1999) Estimating solid-liquid phase change enthalpies and entropies. *J. Phys. Chem. Ref. Data* 28, 1535–1673.

Chin, Y.-P., Gschwend, P.M. (1992) Partitioning of polycyclic aromatic hydrocarbons to marine porewater organic colloids. *Environ. Sci. Technol.* 26, 1621–1626.

Chin, Y.-P., Weber, Jr., W.J. (1989) Estimating the effects of dispersed organic polymers on the sorption contaminants by natural solid, 1. A predictive thermodynamic humic substance-organic solute interaction model. *Environ. Sci. Technol.* 23, 978–984.

Chin, Y.P., Weber Jr., W.J., Voice, T.C. (1986) Determination of partition coefficient and water solubilities by reversed phase chromatography. II. Evaluation of partitioning and solubility models. *Water Res.* 20, 1443–1450.

Chiou, C.T., McGroddy, S.E., Kile, D. (1998) Partition characteristics of polycyclic aromatic hydrocarbons on soils and sediments. *Environ. Sci. Technol.* 32, 264–269.

Chiou, C.T., Porter, P.E., Schmedding, D.W. (1983) Partition equilibria of nonionic organic compounds between soil organic matter and water. *Environ. Sci. Technol.* 17, 227–231.

Chipman, J., Peltier, S.B. (1929) Vapor pressure and heat of vaporization of diphenyl. *Ind. Eng. chem.* 21, 1106–1108.

Cho, H.-H., Park, J.-W., Liu, C.K. (2002) Effect of molecular structures on the solubility enhancement of hydrophobic organic compounds by environmental amphiphiles. *Environ. Toxicol. Chem.* 21, 999–1003.

Chou, S.F.J., Griffin, R.A. (1987) Solubility and soil mobility of polychlorinated biphenyls. In: *PCBs and the Environment*. Waid, J.S., Editor, CRC Press, Inc., Boca Raton, Florida. pp. 101–120.

Chou, J.T., Jurs, P.C. (1979) Computation of partition coefficients from molecular structures by a fragment addition method. In: *Physical Chemical Properties of Drugs*. Medical Research Series, Vol. 10, Yalkowsky, S.H., Sindula, A.A., Valvani, S.C., Editors, Marcel Dekker, New York. pp. 163–199.

Clark, K.E., Gobas, F.A.P.C., Mackay, D. (1990) Model of organic chemical uptake and clearance by fish from food and water. *Environ. Sci. Technol.* 24(8), 1203–1213.

Colomina, M., Jimenez, P., Turrion, C. (1982) Vapor pressures and enthalpies of sublimation of naphthalene and benzoic acid. *J. Chem. Thermodynamics* 14, 779–784.

Connell, D.W., Hawker, D.W. (1988) Use of polynomial expressions to describe the bioconcentration of hydrophobic chemicals by fish. *Ecotoxicol. Environ. Safety* 16, 242–257.

Coover, M.P., Sims, R.C.C. (1987) The effects of temperature on polycyclic aromatic hydrocarbon persistence in an unacclimated agricultural soil. *Haz. Waste Haz. Mat.* 4, 69–82.

Cope, V.W., Kalkwarf, D.R. (1987) Photooxidation of selected polycyclic aromatic hydrocarbons and pyrenequinones coated on glass surfaces. *Environ. Sci. Technol.* 21(7), 643–648.

Cornelissen, G., Kukulska, Z. Kalaitzidis, S., Christanis, K., Gustafsson, Ö. (2004) Relations between environmental black carbon sorption and geochemical sorbent characteristics. *Environ. Sci. Technol.* 38, 3632–3640.

Crookes, M.J., Howe, P.D. (1993) *Environmental Hazard Assessment: Halogenated naphthalenes*. Toxic Substances Division, Directorate for Air, Climate and Toxic Substances, Department of the Environment, Build Research Establishment, Garston, Watford, WD2 7JR.

Cunningham, G.B. (1930) *Power* 72, 374-. (-from Boublik et al.1984).

Dachs, J., Eisenreich, S.J. (2000) Adsorption onto aerosol soot carbon dominates gas-particle partitioning of polycyclic aromatic hydrocarbons. *Environ. Sci. Technol.* 34, 3690–3697.

D'Amboise, M., Hanai, T. (1982) Hydrophobicity and retention in reverse phase liquid chromatography. *J. Liq. Chromatogr.* 229–244.

Davies, R.P., Dobbs, A.J. (1984) The prediction of bioconcentration in fish. *Water Res.* 18 (10), 1253–1262.

Davis, W.W., Krahl, M.E., Clowes, H.A. (1942) Solubility of carcinogenic and related hydrocarbons in water. *J. Am. Chem. Soc.* 64, 108–110.

Davis, W.W., Parker, Jr., T.V. (1942) A nephelometric method for determination of solubilities of extremely low order. *J. Am. Chem. Soc.* 64, 101.

Dean, J.D., Editor (1985) *Lange's Handbook of Chemistry*. 12th ed., McGraw-Hill, New York.

De Bruijn, J., Busser, F., Seinen, W., Hermens, J. (1989) Determination of octanol/water partition coefficients for hydrophobic organic chemicals with the "slowing-stirring" method. *Environ. Toxicol. Chem.* 8, 499–512.

De Bruijn, J., Hermens, J. (1990) Relationships between octanol/water partition coefficients and total molecular surface area and total molecular volume of hydrophobic organic chemicals. *Quant. Strut.-Act. Relat.* 9, 11–21.

De Kock, A.C., Lord, D.A. (1987) A simple procedure for determining octanol-water partition coefficients using reverse phase high performance liquid chromatography (RPHPLC) . *Chemosphere* 16, 133–142.

De Kruif, C.G. (1980) Enthalpies of sublimation and vapor pressures of 11 polycyclic hydrocarbons. *J. Chem. Thermodyn.* 12, 243–248.

De Kruif, C.G., Kuipers, T., Van Miltenburg, J.C., Schaake, R.C.F., Stevens, G. (1981) The vapour pressure of solid and liquid naphthalene. *J. Chem. Thermodyn.* 13, 1081–1086.

Delle Site, A. (1997) The vapor pressure of environmentally significant organic chemicals: A review of methods and data at ambient temperature. *J. Phys. Chem. Ref. Data* 26, 157–193.

Delle Site, A. (2001) Factors affecting sorption of organic compounds in natural sorbent/water systems and sorption coefficients for selected pollutants. A review. *J. Phys. Chem. Ref. Data* 30, 187–439.

De Maagd, P.G.-J., ten Hulscher, T.E.M., Van den Heuvel, H., Opperhuizen, A., Sijm, D.T.H.M. (1998) Physicochemical properties of polycyclic aromatic hydrocarbons: aqueous solubilities, *n*-octanol/water partition coefficients, and Henry's law constants. *Environ. Toxicol. Chem.* 17, 252–257.

De Maagd, P.G.-J., Sinnige, T.L., Schrap, M., Opperhuizen, A., Sijm, T.H.M. (1998) Sorption coefficients of polycyclic aromatic hydrocarbons for two lake sediments: Influence of the bactericide sodium azide. *Environ. Toxicol. Chem.* 17, 1977–1907.

De Pablo, R.S. (1976) Determination of saturated vapor pressure in range 10^{-1}–10^{-4} torr by effusion method. *J. Chem. Eng. Data* 21, 141–143.

De Paolis, F., Kukkonen, J. (1997) Binding of organic pollutants to humic and fulvic acids: Influence of pH and the structure of humic material. Chemosphere 34, 1693–1704.

De Seze, G., Valsaraj, K.T., Reible, D.D., Thibodeaux, L.J. (2000) Sediment-air equilibrium partitioning of semi-volatile hydrophobic organic compounds. Part 2. Saturated vapor pressures, and the effects of sediment moisture content and temperature on the partitioning of polyaromatic hydrocarbons. *Sci. Total Environ.* 253, 27–44.

Devillers, J., Bintein, S., Domine, D. (1996) Comparison of BCF models based on log P. *Chemosphere* 33, 1047–1065.

De Voogt, P., Van Zijl, G.A., Govers, H., Brinkman, U.A.T. (1990) Reversed-phase TLC and structure-activity relationships of polycyclic (hetero) aromatic hydrocarbons. *J. Planar Chromatog.-Mod. TLC* 3, 24–33.

Deno, N.C., Berkheimer, H.E. (1960) Phase equilibria molecular transport thermodynamics: Activity coefficients as a function of structure and media. *J. Chem. Eng. Data* 5, 1–5.

Dickhut, R.M., Miller, K.E., Andren, A.W. (1994) Evaluation of total molecular surface area for predicting air water partitioning properties of hydrophobic aromatic chemicals. *Chemosphere* 29, 283–297.

Dilling, W.L., Gonsior, S.J., Boggs, G.U., Mendoza, C.G. (1988) Organic photochemistry. 20. A method for estimating gas-phase rate constants for reactions of hydroxyl radicals with organic compounds from their relative rates of reaction with hydrogen peroxide under photolysis in 1,1,2-trichlorotrifluoroethane solution. *Environ. Sci. Technol.* 22, 1447–1453.

Dohányosová, P., Dohnal, V., Fenclová, D. (2003) Temperature dependence of aqueous solubility of anthracenes: accurate determination by a new generator column apparatus. *Fluid Phase Equil.* 214, 151–167.

Donovan, S.F. (1996) New method for estimating vapor pressure by the use of gas chromatography. *J. Chromatogr. A*, 749, 123–129.

Donovan, S.F., Pescatore, M.C. (2002) Method for measuring the logarithm of the octanol-water partition coefficient by using short octadecyl-poly-poly(vinyl alcohol) high-performance liquid chromatography columns. *J. Chromatog. A*, 952, 47–61.

Doucette, W.J., Andren, A.W. (1987) Correlation of octanol/water partition coefficients and total molecular surface area for highly hydrophobic aromatic compounds. *Environ. Sci. Technol.* 21, 621–624.

Doucette, W.J., Andren, A.W. (1988) Aqueous solubility of biphenyl, furan, and dioxin congeners. *Chemosphere* 17, 243–252.

Doucette, W.J., Andren, A.W. (1988) Estimation of octanol/water partition coefficients: Evaluation of six methods for highly hydrophobic aromatic hydrocarbons. *Chemosphere* 17, 345–359.

Dreisbach, R.R. (1955) *Physical Properties of Chemical Compounds. Advances in Chem. Series* 15, Am. Chem. Soc., Washington DC.

Dragoescu, C., Friedlander, S. (1989) Dynamics of the aerosol products of incomplete combustion in urban atmospheres. *Aerosol Sci. Technol.* 10, 249–257.

Eadie, B.J., Landrum, P.F., Faust, W. (1982) Polycyclic aromatic hydrocarbons in sediments, pore water and the *amphipod Pontoporeia hoyi* from Lake Michigan. *Chemosphere* 11(9), 847–858.

Eadie, B.J., Morehead, N.R., Landrum, P.F. (1990) Three-phase partitioning of hydrophobic organic compounds in Great Lakes waters. *Chemosphere* 20, 161–178.

Eadsforth, C.V. (1986) Application of reverse-phase HPLC for the determination of partition coefficients. *Pestic, Sci.* 17, 311–325.

Eadsforth, C.V., Moser, P. (1983) Assessment of reverse phase chromatographic methods for determining partition coefficients. *Chemosphere* 12, 1459–1475.

Eastcott, L., Shiu, W.Y., Mackay, D. (1988) Environmentally relevant physical-chemical properties of hydrocarbons: a review of data and development of simple correlations. *Oil & Chem. Pollut.* 4, 191–216.

Eganhouse, R.P., Calder, J.A. (1976) The solubility of medium molecular weight aromatic hydrocarbons and the effects of hydrocarbon co-solutes and salinity. *Geochim. Cosmochim. Acta* 40, 555–561.

Eisenbrand, J., Baumann, K. (1970) Über die bestimmung der wasserlöslichkeit von coronen, fluoranthen, perylen, picen, tetracen und triphenylen und über die bildung wasserlöslicher komplexe dieser kohlenwasserstoffe mit coffein. *Zeitschrift für Lebensmittel Untersuchung und Forschung* 144, 312–317.

Ellington, J. J., Stancil, Jr., F. E., Payne, W. D., Trusty, C. D. (1988) *Interim Protocol for Measurement Hydrolysis Rate Constants in Aqueous Solutions.* USEPA, EPA/600/3-88/014, Athens, Georgia.

Enfield, C.G. Bengtsson, G., Lindqvist, R. (1989) Influence of macromolecules on chemical transport. *Environ. Sci. Technol.* 23, 1278–1286

Erickson, M.D. (1986) *Analytical Chemistry of PCB's.* Ann Arbor Science, Butterworth Publishers, Stoneham, MA.

Etzweiler, F., Senn, E., Schmidt, H.W.H. (1995) Method for measuring aqueous solubilities of organic compounds. *Anal. Chem.* 67, 655–658.

Evans, M.S., Landrum, P.F. (1989) Toxicokinetics of DDE, benzo(a) pyrene, and 2,4,5,2′,4′,5′-hexachlorobiphenyl in *Pontoporeia hoyi* and *Mysis relicta. J. Great Lakes Res.* 15, 589–600.

Falandysz, J., Strandberg, L., Bergqvist, P.-A., Strandberg, B., Rappe, C. (1997) Spatial distribution and bioaccumulation of polychlorinated naphthalenes (PCN) in mussel and fish from the Gulf of Gdansk, Baltic Sea. *Sci. Total Environ.* 203, 93–104.

Fasnacht, M.P., Blough, N.V. (2002) Aqueous photodegradation of polycyclic aromatic hydrocarbons. *Environ. Sci. Technol.* 36, 4364–4389.

Fendinger, N.J., Goltfelty, D.E. (1988) A laboratory method for the experimental determination of air-water Henry's law constants for several pesticides. *Environ. Sci. Technol.* 22, 1289–1293.

Fendinger, N.J., Goltfelty, D.E. (1990) Henry's law constants for selected pesticides, PAHs and PCBs. *Environ. Toxicol. Chem.* 9, 731–735.

Finizio, A., Mackay, D., Bidleman, T.F., Harner, T. (1997) Octanol-air partition coefficient as a predictor of partitioning of semi-volatile organic chemicals to aerosols. *Atoms. Environ.* 31(15), 2289–2296.

Foreman, W.T., Bidleman, T.F. (1985) Vapor pressure estimates of individual polychlorinated biphenyls and commercial fluids using gas chromatographic retention data. *J. Chromatogr.* 330, 203–216.

Formica, S.J., Baron, J.A., Thibodeaux, L.J., Valsaraj, K.T. (1988) PCB transport into lake sediments. Conceptual model and laboratory simulation. *Environ. Sci. Technol.* 22, 1435–1440.

Fowler, L., Trump, W.N., Vogler, C.E. (1968) Vapor pressure of naphthalene. *J. Chem. Eng. Data* 13(2), 209–210.

Frank, A.P., Landrum, P.F., Eadie, B.J. (1986) Polycyclic aromatic hydrocarbon rates of uptake, depuration, and biotransformation by Lake Michigan *Stylodrilus heringianus*. *Chemosphere* 15, 317–330.

Freitag, D., et al. (1984) Environmental hazard profile-test results as related structures and translation into the environment. In: *QSAR in Environmental Toxicology*. Kaiser, K.L.E., Editor, pp. 111–136, D. Reidel Publishing Company, Dordrecht, The Netherlands.

Freitag, D., Ballhorn, L., Geyer, H., Korte, F. (1985) Environmental hazard profile of organic chemicals. An experimental method for the assessment of the behaviour of chemicals in the ecosphere by simple laboratory tests with C-14 labelled chemicals. *Chemosphere* 14, 1589–1616.

Friesen, K.J., Fairchild, W.L., Loewen, M.D., Lawrence, S.G., Holoka, M.H., Muir, D.C.G. (1993) Evidence for particle-mediated transport of 2,3,7,8-tetrachlorodibenzofuran during gas sparging of natural water. *Environ. Toxicol. Chem.* 12, 2037–2044.

Fu, J.-K., Luthy, R.G. (1985) Aromatic compound solubility in solvent/water mixtures. *J. Environ. Eng.* 112, 328–346.

Fu, J.-K., Luthy, R.G. (1985) *Pollutant Sorption to Soils and Sediments in Organic/Aqueous Solvent Systems*. NTIS P–85-242535. EPA/600/3–85/050.

Fujita, T., Iwasa, J., Hansch, C. (1964) A new substituent constant, "pi" derived from partition coefficients. *J. Am. Chem. Soc.* 86, 5175–5180.

Fukuda, K., Inagaki, Y., Maruyama, T., Kojima, H.I., Yoshida, T. (1988) On the photolysis of akylated naphthalenes in aquatic systems. *Chemosphere* 17(4), 651–659.

Garst, J.E. (1984) Accurate, wide-range, automated high-performance liquid chromatographic method for the estimation of octanol/water partition coefficients. II: Equilibrium in partition coefficient measurements, additivity of substituent constants, and correlation of biological data. *J. Pharm. Sci.* 73, 1623–1629.

Garst, J.E., Wilson, W.C. (1984) Accurate, wide-range, automated high-performance liquid chromatographic method for the estimation of octanol/water partition coefficients. I: Effect of chromatographic conditions and procedure variables on accuracy and reproducibility of the method. *J. Pharm. Sci.* 73, 1616–1622.

Gauthier, T.D., Shane, E.C., Guerin, W.F., Seltz, W.R., Grant, C.L. (1986) Fluorescence quenching method for determining equilibrium constants for polycyclic aromatic hydrocarbons binding to dissolved humic materials. *Environ. Sci. Technol.* 20(11), 1162–1166.

Gauthier, T.D., Selfz, W.R., Grant, C.L. (1987) Effects of structural and compositional variations of dissolved humic materials on pyrene K_{OC} values. *Environ. Sci. Technol.* 21, 243–248.

Ge, J., Liu, W., Dong, S. (1987) Determination of partition coefficient with chemically bonded omega-hydroxysilica as HPLC column packing. *Sepu* 5, 182–185.

Geyer, H., Politzki, G., Freitag, D. (1984) Prediction of ecotoxicological behaviour of chemicals: Relationship between n-octanol/water partition coefficient and bioaccumulation of organic chemicals by *alga chlorella*. *Chemosphere* 13, 269–284.

Geyer, H., Viswanathan, R., Freitag, D., Korte, F. (1981) Relationship between water solubility of organic chemicals and their bioaccumulation by the *Alga Chlorella*. *Chemosphere* 10, 1307–1313.

Gluck, S.J., Martin, E.J. (1990) Extended octanol-water partition coefficient determination by dual-mode centrifugal partition chromatography. *J. Liq. Chromatogr.* 13, 3559–3570.

Gobas, F.A.P.C., Clark, K.E., Shiu, W.Y., Mackay, D. (1989) Bioconcentration of polybrominated benzenes and biphenyls and related superhydrophobic chemicals in fish: role of bioavailability and elimination into the feces. *Environ. Toxicol. Chem.* 8, 231–245.

Gobas, F.A.P.C., Shiu, W.Y., Mackay, D. (1987) Factors determining partitioning of hydrophobic organic chemicals in aquatic organisms. In: *QSAR in Environmental Toxicology* II. Kaiser, K.L.E., Editor, D. Reidel Publ. Co., Dordrecht, The Netherlands.

Gobas, F.A.P.C., Lahittete, J.M., Garofalo, G., Shiu, W.Y., Mackay, D. (1988) A novel method for measuring membrane-water partition coefficients of hydrophobic organic chemicals: Comparison with 1-octanol-water partitioning. *J. Pharm. Sci.* 77, 265–272.

Gordon, J.E., Thorne, R.L. (1967) Salt effects on the activity coefficient of naphthalene in mixed aqueous electrolyte solutions. I. Mixtures of two salts. *J. Phys. Chem.* 71, 4390–4399.

Govers, H., Ruepert, C., Aiking, H. (1984) Quantitative structure-activity relationships for polycyclic aromatic hydrocarbons: Correlation between molecular connectivity, physico-chemical properties, bioconcentration and toxicity in *Daphnia pulex*. *Chemosphere* 13, 227–236.

Grayson, B.T., Fosbraey, L.A. (1982) Determination of the vapor pressure of pesticides. *Pestic. Sci.* 13, 269–278.

Gridin, V.V., Litani-Barzilai, I., Kadosh, M., Schechter, I. (1998) Determination of aqueous solubility and surface adsorption of polycyclic aromatic hydrocarbons by laser multiphoton ionization. *Anal. Chem.* 70, 2685–2692.

Griffin, R.A., Chou, S.J. (1981) *Attenuation of polybrominated biphenyls and hexachlorobenzene in Earth Materials*. Final Report, U.S. EPA-600/2-81-191. Urbana, Il: Illinois State Geological Survey, pp. 60.

Groenewegen, D., Stolp, H. (1975) Microbial degradation of polycyclic aromatic hydrocarbons. *Erdoel Kohle, Erdgas, Pentrachem. Bremst. Chem.* 28(4), 206.

Groenewegen, D., Stolp, H. (1976) Microbial breakdown of polycyclic aromatic hydrocarbons. *Zentralbl. Bakteriol. Parasitenkd. Infekitionskr. Hyg. Abt:* 1, Orig., Reihe, B. 162, 225–232.

Gückel, W., Synnatschke, G., Rittig, R. (1973) A method for determining the volatility of active ingredients used in plant protection. *Pest. Sci.* 4, 137–147.

Haag, W.R., Yao, C.C.D. (1992) Rate constants for reaction of hydroxyl radicals with several drinking water contaminants. *Environ. Sci. Technol.* 26, 1005–1013.

Hackenberg, R., Schütz, A., Ballschmiter, K. (2003) High-resolution gas chromatography retention data as basis for the estimation of K_{OW} values using PCB congeners as secondary standards. *Environ. Sci. Technol.* 37, 2274–2279.

Hafkenscheid, T.L., Tomlinson, E. (1983) Correlations between alkane/water and octan-1-ol/water distribution coefficients and isocratic reversed-phase liquid chromatographic capacity factors of acids, bases and neutrals. *Int'l J. Pharmaceutics* 16, 225–239.

Haines, R.L.S., Sandler, S.L. (1995) Aqueous solubilities and infinite dilution activity coefficients of several polycyclic aromatic hydrocarbons. *J. Chem. Eng. Data* 40, 835–836.

Haitzer, M., Höss, S., Traunspurger, W., Steinberg, C. (1999) Relationship between concentration of dissolved organic matter (DOM) and the effect of DOM on the bioconcentration of benzo[a]pyrene. *Aqua. Toxicol.* 45, 147–158.

Haky, J.E., Leja, B. (1986) Evaluation of octanol-water partition coefficients using capillary gas chromatography with cold on-column injection. *Anal. Lett.* 19, 123–134.

Haky, J.E., Young, A.M. (1984) Evaluation of a simple HPLC correlation method for the determination of the octanol-water partition coefficients of organic compounds. *J. Liq. Chromatogr.* 7(4), 675–689.

Halfon & Reggiani (1986) On ranking chemicals for environmental hazard. *Environ. Sci. Technol.* 20, 1173–1179.

Hallett, D.J., Brecher, R.W. (1984) Cycling of polynuclear aromatic hydrocarbons in the Great Lakes ecosystem. In: *Toxic Contaminants in the Great Lakes. Advances in Environment Sciences and Technology.* Nriagu, J.O., Simmons, M.S., Editors, John Wiley & Sons, New York, New York. pp. 213–237.

Hambrick, G.A., Delaune, R.D., Patrick, W.H., Jr. (1980) Effects of estuarine sediment, pH and oxidation-reduction potential on microbial hydrocarbon degradation. *Appl. Environ. Microbiol.* 40, 365–9.

Hammers, W.E., Meurs, G.J., De Ligny, C.L. (1982) Correlations between liquid chromatographic capacity ratio data on Lichrosorb RP-18 and partition coefficients in the octanol-water system. *J. Chromatogr.* 247, 1–13.

Hanai, T., Hubert, J. (1982) Hydrophobicity and chromatographic behaviour of aromatic acids found in urine. *J. Chromatogr.* 239, 527–536.

Hanai, T., Tran, C., Hubert, J. (1981) An approach to the prediction of retention times in liquid chromatography. *J. High Resolution Chromatography & Chromatography Communication* (J. HRC & CC) 4, 454–460.

Hansch, C., Fujita, T. (1964) ρ-σ-π Analysis: method for the correlation of biological activity and chemical structure. *J. Am. Chem. Soc.* 86, 1616–1626.

Hansch, C., Leo, A. (1979) *Substituent Constants for Correlation Analysis in Chemistry and Biology.* Wiley, N.Y.

Hansch, C., Leo, A. (1985) *Medchem Project Issue No. 26.* Pomona College, Claremont, CA.

Hansch, C., Leo, A. (1987) *Medchem Project.* Pomona College, Claremont, California.

Hansch. C., Leo, A.J., Hoekman, D. (1995) *Exploring QSAR, Hydrophobic, Electronic, and Steric Constants.* ACS Professional Reference Book, American Chemical Society, Washington, DC.

Hansch, C., Leo, A., Unger, S.H., Kim, K.H., Nikaitani, D., Lien, E.J. (1973) Aromatic substituent constants for structure-activity correlations. *J. Med. Chem.* 16, 1207.

Hansen, K.C., Zhou, Z., Yaws, C.L., Aminabhavi, T.J. (1993) Determination of Henry's law constants of organics in dilute aqueous solutions. *J. Chem. Eng. Data* 38, 546–550.

Hansen, P.C., Eckert, C.A. (1986) An improved transpiration method for the measurement of very low pressures. *J. Chem. Eng. Data* 31, 1–3.

Haque, R., Falco, J., Cohen, S., Riordan, C. (1980) Role of transport and fate studies in the exposure, assessment and screening of toxic chemicals. In: *Dynamics, Exposure and Hazard Assessment of Toxic Chemicals.* R. Haque, Editor. Ann Arbor Science Publishers Inc., Ann Arbor, Michigan.

Haque, R., Schmedding, D. (1975) A method of measuring the water solubility of hydrophobic chemicals. *Bull. Environ. Contam. Toxicol.* 14, 13–18.

Harner, T., Bidleman, T.F. (1998) Measurement of octanol-air partition coefficients for polycyclic aromatic hydrocarbons and polychlorinated naphthalenes. *J. Chem. Eng. Data* 43, 40–46.

Harnisch, M., Möckel, H.J., Schule, G. (1983) Relationship between log P_{OW} shake-flask values and capacity factors derived from reversed-phase high-performance liquid chromatography for n-alkylbenzenes and some OECD reference substances. *J. Chromatogr.* 282, 315–332.

Harris, J.C. (1982) Rate of aqueous photolysis. Chapter 8, In: *Handbook of Chemical Property Estimation Methods.* Lyman, W.J., Reehl, W.F., Rosenblatt, D.H., Editors, McGraw-Hill, New York.

Hashimoto, Y., Tokura, K., Ozaki, K., Strachan, W.M.J. (1982) A comparison of water solubilities by the flask and micro-column methods. *Chemosphere* 11(10), 991–1001.

Hawker, D.W. (1989) The relationship between octan-1-ol/water partition coefficient and aqueous solubility in terms of solvatochromic parameters. *Chemosphere* 19(10/11), 1585–1593.

Hawker, D.W. (1990) Description of fish bioconcentration factors in terms of solvatochromic parameters. *Chemosphere* 20, 467–477.

Hawker, D.W., Connell, D.W. (1985) Relationships between partition coefficient, uptake rate constant, clearance rate constant and time to equilibration for bioaccumulation. *Chemosphere* 14, 1205–1219.

Hawker, D.W., Connell, D.W. (1986) Bioconcentration of lipophilic compounds by some aquatic organisms. *Ecotox. Environ. Safety* 11, 184–197.

Hawker, D.W., Connell, D.W. (1988) Influence of partition coefficient of lipophilic compounds on bioconcentration kinetics with fish. *Water Res.* 22, 701–707.

He, Y., Yediler, A., Sun, T., Kettrup, A. (1995) Adsorption of fluoranthene on soil and lava: Effects of the common carbon contents of adsorbents and temperature. *Chemosphere* 30, 141–150.

Hegeman, W.J.M., van der Weijden, C.H., Loch, J.P.G. (1995) Sorption of benzo[a]pyrene and phenanthrene on suspended harbor sediment as a function of suspended sediment concentration and salinity: A laboratory study using the cosolvent partition coefficient. *Environ. Sci. Technol.* 29, 363–371.

Heitkamp, M.A. (1988) Environmental and biological factors affecting the biodegradation and detoxification of polycyclic aromatic hydrocarbons. *Diss. Abstr. Int'l. B.* 48, 1926.

Helmstetter, M.F., Alden III, R.W. (1994) Release rates of polynuclear aromatic hydrocarbons from natural sediments and their relationship to solubility and octanol-water partitioning. *Arch. Environ. Contam. Toxicol.* 26, 282–291.

Helweg, C., Nielson, T., Hansen, P.E. (1997) Determination of octanol-water partition coefficients of polar polycyclic aromatic (NPAC) by high performance liquid chromatography. *Chemosphere* 34, 1673–1684.

Herbes, S.E. (1981) Rates of microbial transformation of polycyclic aromatic hydrocarbons in water and sediments in the vicinity of coal-coking waste water discharge. *Appl. Environ. Microbiol.* 41, 20–28.

Herbes, S.E., Risi, G.F. (1978) Metabolic alteration and excretion of anthracene by *Daphnia pulex*. *Bull. Environ. Contam. Toxicol.* 19, 147–155.

Herbes, S.E., Schwall, L.R. (1978) Microbial transformation of polycyclic aromatic hydrocarbons in pristine and petroleum contaminated sediments. *Appl. Environ. Microbiol.* 35, 306–316.

Herbes, S.E., Southworth, G.R., Shaeffer, D.L., Griest, W.H., Maskarinec, M.P. (1980) Critical pathways of polycyclic aromatic hydrocarbons in aquatic environments. In: *The Scientific Basis of Toxicity Assessment.* Witschi, H. Editor, pp. 113–128, Elsevier/North-Holland Biomedical Press, Amsterdam.

Hilpert, S. (1916) The solubility of naphthalene in ammonia. A possible cause for naphthalene stoppages. *Angew. Chem.* 29, 57–59.

Hinckley, D.A. (1989) Vapor Pressures, Henry's Law Constants and Air-Sea Gas Exchange of Selected Organochlorine Pollutants. Ph.D. Thesis, University of South Carolina.

Hinckley, D.A., Bidleman, T.F., Foreman, W.T. (1990) Determination of vapor pressures for nonpolar and semipolar organic compounds from gas chromatographic retention data. *J. Chem. Eng. Data* 35, 232–237.

Hodson, J., Williams, N.A. (1988) The estimation of the adsorption (K_{OC}) for soils by high performance liquid chromatography. *Chemosphere* 17(1), 67–77.

Hollifield, H.C. (1979) Rapid nephelometric estimate of water solubility of highly insoluble organic chemicals of environmental interest. *Bull. Environ. Contam. Toxicol.* 23, 579–586.

Hon, H.C., Singh, R.P., Kuochadker, A.P. (1976) Vapor pressure-boiling point measurements of five organic substances by twin ebulliometry. *J. Chem. Eng. Data* 21, 430–431.

Howard, P.H., Editor (1989) *Handbook of Environmental Fate and Exposure Data for Organic Chemicals.* Volume I, Lewis Publishers, Inc., Chelsea, Michigan.

Howard, P.H., Boethling, R.S., Jarvis, W.F., Meylan, W.M., Michalenco, E.M., Editors (1991) *Handbook of Environmental Degradation Rates.* Lewis Publishers, Inc., Chelsea, Michigan.

Howard, P.H., Hueber, A.E., Mulesky, B.C., Crisman, J.S., Meylan, W., Crosbie, E., Gray, D.A., Sage, G.W., Howard, K.P., LaMacchia, A., Boethling, R., Troast, R. (1986) Biology, biodegradation and fate/exposure: New files on microbial degradation and toxicity as well as environmental fate/exposure of chemicals. *Environ. Toxicol. Chem.* 5, 977–988.

Hoyer, H., Peperle, W. (1958) Dampfdruckmessungen an organischen substanzen und ihre sublimationswarmen. *Z. Elektrochem.* 62, 61–66.

Inokuchi, H., Shiba, S., Handa, T., Akamatsu, H. (1952) Heats of sublimation of condensed polynuclear aromatic hydrocarbons. *Bull. Chem. Soc. Japan* 25, 299–302.

Irwin, K.C. (1982) SRI International unpublished analysis.

Isnard, P., Lambert, S. (1988) Estimating bioconcentration factors from octanol-water partition coefficient and aqueous solubility. *Chemosphere* 17, 21–34.

Isnard, P., Lambert, S. (1989) Aqueous solubility and *n*-octanol/water partition coefficient correlations. *Chemosphere* 18, 1837–1853.

Järnberg, U., Asplund, L., Jakobsson, E. (1994) Gas chromatographic retention behaviour of polychlorinated naphthalenes on nonpolar, polarizable, polar and smectic capillary columns. *J. Chromatog. A* 683, 385–396.

Jury, W.A., Russo, D., Streile, G., El Abd, H. (1990) Evaluation of volatilization by organic chemicals residing below the soil surface. *Water Resources Res.* 26, 13–26.

Kaiser, K.L.E. (1983) A non-linear function for the calculation of partition coefficients of aromatic compounds with multiple chlorine substitution. *Chemosphere* 12, 1159–1165.

Kamens, R.M., Guo, Z., Fulcher, J.N., Bell, D.A. (1988) Influence of humidity, sunlight, and temperature on daytime decay of polyaromatic hydrocarbons on atmospheric soot particles. *Environ. Sci. Technol.* 22, 103–108.

Kamlet, M.J., Doherty, R.M., Carr, P.W., Mackay, D., Abraham, M.H., Taft, R.W. (1988) Linear solvation energy relationships. 44. Parameter estimation rules that allow accurate prediction of octanol/water partition coefficients and other solubility and toxicity properties of polychlorinated biphenyls and polycyclic aromatic hydrocarbons. *Environ. Sci. Technol.* 22, 503–509.

Kan, A.T., Tomson, M.B. (1990) Ground water transport of hydrophobic organic compounds in the presence of dissolved organic matter. *Environ. Sci. Technol.* 9, 253–263.

Kappeler, T., Wuhrmann, K. (1978) Microbial degradation of water soluble fraction of gas oil. *Wat. Res.* 12, 327–333.

Karickhoff, S.W. (1981) Semi-empirical estimation of sorption of hydrophobic pollutants on natural sediments and soils. *Chemosphere* 10, 833–846.

Karickhoff, S.W., Brown, D.S., Scott, T.A. (1979) Sorption of hydrophobic pollutants on natural water sediments. *Water Res.* 13, 241–248.

Karickhoff, S.W., Morris, K.R. (1985) Sorption dynamics of hydrophobic pollutants in sediment suspensions. *Environ. Toxicol. Chem.* 4, 469–479.

Katz, M., Chan, C., Tosine, H., Sakuma, T. (1979) Relative rates of photochemical and biological oxidation (*in vitro*) of polynuclear aromatic hydrocarbons. In: *Polynuclear Aromatic Hydrocarbons*. Jones, P.W., Leher, P., Eds., pp. 171–189, Ann Arbor Science Publishers, Ann Arbor, MI.

Kaupp, K., McLachlan, M.S. (1999) Gas/particle partitioning of PCDDs/Fs, PCBs, PCNs and PAHs. *Chemosphere* 38, 3411–3421.

Kayal, S.I., Connell, D.W. (1990) Partitioning of unsubstituted polycyclic aromatic hydrocarbons between surface sediments and the water column in the Brisbane River estuary. *Aust. J. Mar. Freshwater Res.* 41, 443–456.

Kelley, J.D., Rice, F.O. (1964) The vapor pressures of some polynuclear aromatic hydrocarbons. *J. Phys. Chem.* 68, 3794–3796.

Kenaga, E.E. (1980a) Predicted bioconcentration factors and soil sorption coefficients of pesticides and other chemicals. *Ecotoxicol. Environ. Safety* 4, 26–38.

Kenaga, E.E. (1980b) Correlation of bioconcentration factors of chemicals in aquatic and terrestrial organisms with their physical and chemical properties. *Environ. Sci. Technol.* 14, 553–556.

Kenaga, E.E., Goring, C.A.I. (1980) Relationship between water solubility, solubility, soil sorption, octanol-water partitioning, and concentration of chemicals in biota. In: *Aquatic Toxicology.* Eaton, J.G., Parrish, P.R., Hendrick, A.C. (Eds.) Am. Soc. for Testing and Materials, STP 707, pp. 78–115, Philadelphia.

Kier, L.B., Hall, L.H., Murray, W.J., Randic, M. (1971) Molecular connectivity I:Relationship to nonspecific local anesthesia. *J. Pharm. Sci.* 64, 1971–1981.

Kincannon, D.F., Lin, Y.S. (1985) Microbial degradation of hazardous wastes by land treatment. In: *Proc. Indust. Waste Conf.* 40, 607–619.

Kishi, H., Hashimoto, Y. (1989) Evaluation of the procedures for the measurements of water solubility and n-octanol/water partition coefficient of chemicals results of a ring test in Japan. *Chemosphere* 18(9/10), 1749–1759.

Kishi, H., Kogure, N., Hashimoto, Y. (1990) Contribution of soil constituents in adsorption coefficient of aromatic compounds, halogenated alicyclic and aromatic compounds to soil. *Chemosphere* 21, 867–876.

Klamt, A. (1993) Estimation of gas-phase hydroxyl radical rate constants of organic compounds from molecular orbital calculations. *Chemosphere* 26, 1273–1289.

Klamt, A. (1996) Estimation of gas-phase hydroxyl radical rate constants of oxygenated compounds based on molecular orbital calculations. *Chemosphere* 32, 717–726.

Klara, C.M., Mohamed, R.S., Dempsey, D.M., Holder G.D. (1987) Vapor-liquid equilibria for the binary systems of benzene/toluene, diphenylmethane/toluene, *m*-cresol/1,2,3,4-tetrahydronaphthalene and quinoline/benzene. *J. Chem. Eng. Data* 32, 143–147.

Klein, W., Geyer, H., Freitag, D., Rohleder, H. (1984) Sensitivity of schemes for ecotoxicological hazard ranking of chemicals. *Chemosphere* 13, 203–211.

Klevens, H.B. (1950) Solubilization of polycyclic hydrocarbons. *J. Phys. Colloid Chem.* 54, 283–298.

Klöpffer, W., Rippen, G., Frische, R. (1982) Physicochemical properties as useful tools for predicting the environmental fate of organic chemicals. *Ecotoxicol. Environ. Saf.* 6, 294–301.

Kollig, H.P. (1995) *Environmental Fate Constants for Additional 27 Organic Chemicals under Consideration for EPA's Hazardous Water Identification Projects.* EPA/600/R-95/039. Environmental Research Laboratory, Office of Research and Development, U.S. EPA, Athens, GA.

Krauss, M., Wilcke, W. (2001) Predicting soil-water partitioning of polycyclic aromatic hydrocarbons and polychlorinated biphenyls by desorption with methanol-water mixtures at different temperatures. *Environ. Sci. Technol.* 35, 2319–2325.

Krishnamurthy, T., Wasik, S.P. (1978) Fluorometric determination of partition coefficient of naphthalene homologues in octanol-water mixtures. *J. Environ. Sci. Health* A13(8), 595–602.

Kühne, R., Ebert, R.-U., Kleint, F., Schmidt, G., Schüürmann, G. (1995) Group contribution methods to estimate water solubility of organic chemicals. *Chemosphere* 30, 2061–2077.

Kukkonen, J., Landrum, P.F. (1998) Effect of particle-xenobiotic contact time on bioavailability of sediment-associated benzo(*a*)pyrene to benthic amphipod, *Diporeia* spp. *Aqua. Toxicol.* 42, 229–242.

Kwok, E.S.C., Atkinson R. (1995) Estimation of hydroxyl radical reaction rate constants for gas-phase organic compounds using a structure-reactivity relationship: An update. *Atmos. Environ.* 29, 1685–1695.

Kwok, E.S.C., Atkinson, R., Arey, J. (1997) Kinetic of the gas-phase reactions of indan, indene, fluorene, and 9,10-dihydroanthracene with OH radicals, NO_3 radicals, and O_3. *Int. J. Chem. Kinet.* 29, 299–309.

Lande, S.S., Banerjee, S. (1981) Predicting aqueous solubility of organic nonelectrolytes from molar volume. *Chemosphere* 10, 751–759.

Landrum, P.F. (1988) Toxicokinetics of organic xenobiotics in the amphipod *Pontoporeia hoyi*: role of physiological and environmental variables. *Aqua. Toxicol.* 12, 245–271.

Landrum, P.F. (1989) Bioavailability and toxicokinetics of polycyclic aromatic hydrocarbons sorbed to sediments for the amphipod *Pontoporeia hoyi. Environ. Sci. Technol.* 23, 588–595.

Landrum,, R.F., Nihart, S.R., Edie, B.J., Gardner, W.S. (1984) Reverse-phase separation method for determining pollutant binding to Aldrich humic acid and dissolved organic carbon of natural waters. *Environ. Sci. Technol.* 18, 187–192.

Landrum, P.F., Poore, R. (1988) Toxicokinetics of selected xenobiotics in *Hexagenia limbata. J. Great Lakes Res.* 14(4), 427–437.

Landrum, P.F., Reinhold, M.D., Nihart, S.R., Eadie, B.J. (1985) Predicting the bioavailability of organic xenobiotics to *Pontoporeia hoyi* in the presence of humic and fulvic materials and natural dissolved organic matter. *Environ. Toxicol. & Chem.* 4, 459–467.

Lane, D.A., Katz, M. (1977) The photomodification of benzo(*a*)pyrene, benzo(*b*)fluoranthene, and benzo(*k*)fluoranthene under simulated atmospheric conditions. *Adv. Environ. Sci. Technol.* 8, 137–154.

Lee, R.F. (1977) Oil Spill Conference, Am. Petrol. Inst. pp. 611–616.

Lee, R.F., Anderson, J.W. (1977) Fate and effect of naphthalenes: controlled ecosystem pollution experiment. *Bull. Mar. Sci.* 27, 127.

Lee, R.F., Ryan, C. (1976) Biodegradation of petroleum hydrocarbons by marine microbes. In: Proc. of the Third International Biodegradation Symposium of 1975, pp. 119–125.

Lee, R.F., Gardner, W.S., Anderson, J.W., Blaylock, J.W. (1978) Fate of polycyclic aromatic hydrocarbons in controlled ecosystem enclosures. *Environ. Sci. Technol.* 12, 832–838.

Lee, R.F., Sauerheber, R., Benson, A.A. (1972) Petroleum hydrocarbons: Uptake and discharge by the marine mussel *Mytilus edulis. Science* 177, 344–345.

Lee, M.D., Wilson, J.T., Ward, C.H. (1984) Microbial degradation of selected aromatics in a hazardous waste site. *Devel. Indust. Microbiol.* 25, 557–565.

Lei, Y.D., Chankalal, R., Chan, A., Wania, F. (2002) Supercooled liquid vapor pressures of the polycyclic aromatic hydrocarbons. *J. Chem. Eng. Data* 47, 801–806.

Lei, Y.D., Wania, F., Shiu, W.Y. (1999) Vapor pressures of the polychlorinated naphthalenes. *J. Chem. Eng. Data* 44, 577–582.

Lei, Y.D., Wania, F., Shiu, W.Y., Boocock, G.B. (2000) HPLC-based method for estimating the temperature dependence of *n*-octanol-water partition coefficients. *J. Chem. Eng. Data* 45, 738–742.

Lemaire, P., Mathieu, A., Carriere, S., Drai, P., Giudicelli, J., Lafaurie, M. (1990) The uptake mechanism and biological half-life of benzo(*a*) pyrene in different tissues of sea bass, *Dicentrarchus labrax. Ecotoxicol. Environ. Saf.* 20, 223–233.

Leo, A.J. (1975) Calculation of partition coefficients useful in the evaluation of relative hazards of various chemicals in the environment. In: *Symposium on Structure-Activity Correlations in Studies of Toxicity and Bioconcentration with Aquatic Organisms*. G.D. Veith and D.E. Konasewich, Editors, International Joint Commission, Ontario, Canada.

Leo, A.J. (1986) *CLOGP-3.42 Medchem. Software, Medicinal Chemistry Project*, Pomona College, Claremont, CA.

Leo, A., Hansch, C., Elkins, D. (1971) Partition coefficients and their uses. *Chemical Reviews* 71, 525–616.

Leversee, G.J., Giesy, J.P., Landrum, P.F., Bartell, S., Gerould, S. Bruno, M., Spacie, A., Bowling, J., Haddock, J., Fannin, T. (1981) Disposition of benzo(*a*) pyrene in aquatic systems components: *Periphyton, Chironomids, Daphania*, Fish. In: *Polynuclear Aromatic Hydrocarbons: Chemical Analysis and Biological Fate*. Cooke, M., Dennis, A.J., Eds., pp. 357–367. Battelle Press, Columbus, Ohio.

Li, X.-W., Shibata, E., Kasai, E., Nakamura, T. (2004) Vapor pressures and enthalpies of sublimation of 17 polychlorinated dibenzo-*p*-dioxins, and five polychlorinated dibenzofurans. *Environ. Toxicol. Chem.* 23, 348–354.

Linder, G., Bergman, H.L., Meyer, J.S. (1985) Anthracene bioconcentration in rainbow trout during single-compound and complex mixture exposures. *Environ. Toxicol. Chem.* 4, 549–558.

Locke, D. (1974) Selectivity in reversed-phase liquid chromatography using chemically bonded stationary phases. *J. Chromatogr. Sci.* 12, 433–437.

Lotufo, G.R. (1998) Bioaccumulation of sediment-associated fluoranthene in benthic copdpods: uptake, elimination and biotransformation. *Aqua. Toxicol.* 44, 1–15.

Lu, P.Y., Metcalf, R.L., Carlson, E.M. (1978) Environmental fate of five radiolabelled coal conversion by-products evaluated in a laboratory model ecosystems. *Environ. Health Perspect.* 24, 201.

Lu, P.Y., Metcalf, R.L., Plummmer, N., Mandel, D. (1977) The environmental fate of three carcinogens: benzo[*a*]pyrene, benzidine, and vinyl chloride evaluated in laboratory model ecosystems. *Arch. Environ. Contam. Toxicol.* 6, 129–142.

Lu, X., Tao, S., Cao, J., Dawson, R.W. (1999) Prediction of fish bioconcentration factors of nonpolar organic pollutants based on molecular connectivity indices. *Chemosphere* 39, 987–999.

Lüers, F., ten Hulscher, Th.E.M. (1996) Temperature effect on the partitioning of polycyclic aromatic hydrocarbons between natural organic carbon and water. *Chemosphere* 33, 643–657.

Lyman, W.J., Reehl, W.F., Rosenblatt, D.H., Editors (1982) *Handbook on Chemical Property Estimation Methods, Environmental Behavior of Organic Compounds*. McGraw-Hill, New York, pp. 960.

Ma, K.C., Shiu, W.Y., Mackay, D. (1990) A Critically Reviewed Compilation of Physical and Chemical and Persistence Data for 110 Selected EMPPL Substances. A report prepared for the Ontario Ministry of Environment, Water Resources Branch, Toronto, Ontario.

Mabey, W., Mill, T. (1978) Critical review of hydrolysis of organic compounds in water under environmental conditions. *J. Phys. Chem. Ref. Data* 7, 838–415.

Mabey, W., Smith, J.H., Podoll, R.T., Johnson, H.L., Mill, T., Chou, T.W., Gate, J., Waight-Partridge, I., Jaber, H., Vandenberg, D. (1982) *Aquatic Fate Process for Organic Priority Pollutants*. EPA Report, No. 440/4-81-14.

Mackay, D. (1982) Correlation of bioconcentration factors. *Environ. Sci. Technol.* 16, 274–278.

Mackay, D., Hughes, A.I. (1984) Three parameter equation describing the uptake of organic compounds by fish. *Environ. Sci. Technol.* 18, 439–444.

Mackay, D., Leinonen, P.J. (1975) Rate of evaporation of low-solubility contaminants from water bodies to atmosphere. *Environ. Sci. Technol.* 9, 1178–1180.

Mackay, D., Shiu, W.Y. (1977) Aqueous solubility of polynuclear aromatic hydrocarbons. *J. Chem. Eng. Data* 22, 399–402.

Mackay, D. Shiu, W.Y., Chau, E. (1983) Calculation of diffusion resistance controlling volatilization rates of organic contaminants from water. *Can. J. Fish. Aqua. Sci.* 40, 295–303.

Mackay, D., Shiu, W.Y., Ma, K.C. (1992a) *Illustrated Handbook of Physical-Chemical Properties and Environmental Fate for Organic Chemicals. Vol. 1, Monoaromatic Hydrocarbons, Chlorobenzenes and PCBs.* Lewis Publishers, Inc., Chelsea, Michigan.

Mackay, D., Shiu, W.Y., Ma, K.C. (1992b) *Illustrated Handbook of Physical-Chemical Properties and Environmental Fate for Organic Chemicals. Vol. 2, Polynuclear Aromatic Hydrocarbons, Polychlorinated Dioxins, and Dibenzofurans.* Lewis Publishers, Inc., Chelsea, Michigan.

Mackay, D., Shiu, W.Y., Sutherland, R.P. (1979) Determination of air-water Henry's law constants for hydrophobic pollutants. *Environ. Sci. Technol.* 13, 333–337.

Mackay, D., Bobra, A.M., Shiu, W.Y., Yalkowsky, S.H. (1980) Relationships between aqueous solubility and octanol-water partition coefficient. *Chemosphere* 9, 701–711.

Mackay, D., Bobra, A.M., Chan, D.W., Shiu, W.Y. (1982) Vapor pressure correlation for low-volatility environmental chemicals. *Environ. Sci. Technol.* 16, 645–649.

Mackay, D., Shiu, W.Y., Bobra, A., Billington, J., Chau, E., Yeun, A., Ng, C., Szeto, F. (1982) *Volatilization of Organic Pollutants from Water*. EPA600/3-82-019. National Technical Information Service, Springfield, Virginia.

Mackay, D., Shiu, W.Y. (1981) A critical review of Henry's law constants for chemicals of environmental interest. *J. Phys. Chem. Ref. Data* 10, 1175–1199.

Mackay, D., Wolkoff, A.W. (1973) Rate of evaporation of low-solubility contaminants from water bodies to atmosphere. *Environ. Sci. Technol.* 7, 611–614.

Macknick, A.B., Prausnitz, J.M. (1979) Vapor pressures of high-molecular-weight hydrocarbons. *J. Chem. Eng. Data* 24, 175–178.

Mader, B.T., Pankow, J.F. (2003) Vapor pressures of the polychlorinated dibenzodioxins (PCDDs) and the polychlorinated dibenzofurans (PCDFs) . *Atmos. Environ.* 37, 3103–3114.

Magee, B.R., Lion, LW., Lemley, A. (1991) Transport of dissolved organic macromolecules and their effect on the transport of phenanthrene in porous media. *Environ. Sci. Technol.* 25, 323–331.

Magnusson, K., Ekelund, R., Ingebrigtsen, K., Granmo, Å., Brandt, I. (2000) Tissue disposition of benzo[a]pyrene in blue mussel (*Mytilus edulis*) and effect of algal concentration on metabolism and depuration. *Environ. Toxicol. Chem.* 19, 2683–2690.

Mailhot, H., Peters, R.H. (1988) Empirical relationships between the 1-octanol/water partition coefficient and nine physicochemical properties. *Environ. Sci. Technol.* 22, 1479–1488.

Malaspina, L., Bardi, G., Gigli, R. (1974) Simultaneous determination by Knudsen-effusion microcalorimetric technique of the vapor pressure and enthalpy of vaporization of pyrene and 1,3,5-triphenylbenzene. *J. Chem. Thermodyn.* 6, 1053–1064.

Malaspina, L., Gigli, R., Bardi, G. (1973) Microcalorimetric determination of the enthalpy of sublimation of benzoic acid and anthracene. *J. Chem. Phys.* 59, 387–394.

Mallon, B.J., Harris, F. (1984) Octanol-water partition coefficient of benzo[a]pyrene, measurement, calculation and environmental implication. *Bull. Environ. Contam. Toxicol.* 32, 316–323.

Manilal, V.B., Alexander, M. (1991) Factors affecting the microbial degradation of phenanthrene in soil. *Appl. Microbiol. Biotechnol.* 35, 401–405.

Matsuo, J. (1981) i/o* - Characters to describe bioconcentration factors of chlorobenzenes and naphthalenes - meaning of the sign of the coefficients of $\Sigma i/\Sigma o$ in the correlating equations. *Chemosphere* 10, 1073–1078.

Matsuzawa, S., Masser-Ali, L., Garrigues, P. (2001) Photolytic behavior of polycyclic aromatic hydrocarbons in diesel particulate matter deposited on the ground. *Environ. Sci. Technol.* 36, 3139–3143.

May, W.E. (1980) The solubility behavior of polycyclic aromatic hydrocarbons in aqueous systems. In: *Petroleum in the Marine Environment*, Petrakis, L., Weiss, F.T., Eds., *Advances in Chemistry Series* No. 85, pp. 143–192, Am. Chem. Soc., Washington, D.C.

May, W.E., Brown, J.M., Chesler, S.N., Guenther, F., Hilpert, L.R., Hertz, H.S., Wise, S.A. (1979) Development of an aqueous polynuclear aromatic hydrocarbon standard reference material. In: *Polynuclear Aromatic Hydrocarbons.* Jones, P.W., Leber, P., Editors, pp. 411–418, Ann Arbor Science Publishers, Inc., Ann Arbor, Michigan.

May, W.E., Wasik, S.P., Freeman, D.H. (1978a) Determination of aqueous solubility of polynuclear aromatic hydrocarbons by coupled column liquid chromatographic technique. *Anal. Chem.* 50, 175–179.

May, W.E, Wasik, S.P., Freeman, D.H. (1978b) Determination of solubility behaviour of some polycyclic aromatic hydrocarbons in water. *Anal. Chem.* 50, 997–1000.

May, W.E., Wasik, S.P., Miller, M.M., Tewari, Y.B., Brown-Thomas, J.M., Goldberg, R.N. (1983) Solution thermodynamics of some slightly soluble hydrocarbons in water. *J. Chem. Eng. Data* 28, 197–200.

McCarthy, J.F. (1983) Role of particulate organic matter in decreasing accumulation of polynuclear aromatic hydrocarbons by *Daphnia magna. Arch. Environ. Contam. Toxicol.* 12, 559–568.

McCarthy, J.F., Jimenez, B.D. (1985) Reduction in bioavailability to bluegills of polycyclic aromatic hydrocarbons bound to dissolved humic material. *Environ. Toxicol. Chem.* 4, 511–521.

McCarthy, J.F., Jimenez, B.D., Barbee, T. (1985) Effect of dissolved humic material on accumulation of polycyclic aromatic hydrocarbons: structure-activity relationships. *Aqua. Toxicol.* 7, 15–24.

McDuffie, D. (1981) Estimation of octanol/water partition coefficients for organic pollutants using reverse-phase HPLC. *Chemosphere* 10, 73–83.

McGroddy, S.E., Farrington, J.W. (1995) Sediment porewater partitioning of polycyclic aromatic hydrocarbons in three cores from Boston Harbor, Massachusetts. *Environ. Sci. Technol.* 29, 1542–1550.

Mclachlan, M., Mackay, D., Jones, P.H. (1990) A conceptual model of organic chemical volatilization at waterfalls. *Environ. Sci. Technol.* 24, 252–257.

Meador, J.P., Stein, J.E., Reichert, W.L., Varanasi, U. (1995) Bioaccumulation of polycyclic aromatic hydrocarbons by marine organisms. *Rev. Environ. Contam. Toxicol.* 143, 79–165.

Means, J.C., Hassett, J.J., Wood, S.G., Banwart, W.L. (1979) Sorption properties of energy-related pollutants and sediments. In: *Polynuclear Aromatic Hydrocarbons.* P.W. Jones and P. Leber Editors, pp. 329–340, Ann Arbor Science Publishers, Ann Arbor, Michigan.

Means, J.C., Wood, S.G., Hassett, J.J., Banwart, W.L. (1980) Sorption of polynuclear aromatic hydrocarbons by sediments and soils. *Environ. Sci. Technol.* 14, 1524–1528.

Means, J.C., Wood, S.G., Hassett, J.J., Banwart, W.L. (1982) Sorption of amino-and carboxy-substituted polynuclear aromatic hydrocarbons by sediments and soils. *Environ. Sci. Technol.* 16, 93–98.

Melancon, M.J. Jr., Lech, J.J. (1978) Distribution and elimination of naphthalene and 2-methylnaphthalene in rainbow trout during short and long-term exposures. *Arch. Environ. Contam. Toxicol.* 7, 207.

Menges, R.A., Armstrong, D.W. (1991) Use of a three-phase model with hydroxypropyl-β-cyclodextrin for the direct determination of large octanol-water and cyclodextrin-water partition coefficients. *Anal. Chim. Acta* 255, 157–162.

Menges, R.A., Bertrand, G.L., Armstrong, D.W. (1990) Direct measurement of octanol-water partition coefficients using centrifugal partition chromatography with a back-flushing technique. *J. Liq. Chromatogr.* 13, 3061–3077.

Metcalfe, D.E., Zukovs, G., Mackay, D., Paterson, S. (1988) Polychlorinated biphenyls (PCBs): physical and chemical property data. In: *Hazards, Decontamination and Replacement of PCB: A Comprehensive Guide.* Crine, J.P., Ed., pp. 3–33, Plenum Press, New York.

Meylan, W.M., Howard, P.H. (1991) Bond contribution method for estimating Henry's law constants. *Environ. Toxicol. Chem.* 10, 1283–1293.

Mill, T. (1999) Predicting photoreaction rates in surface waters. *Chemosphere* 38, 1379–1390.

Mill, T., Mabey, W. (1985) Photodegradation in water. In: *Environmental Exposure from Chemicals.* Vol. 1. Neely, W. B., Blau, G. E., Eds., pp. 175–216, CRC Press, Inc., Boca Raton, Florida.

Mill, T., Mabey, W.R., Lan, B.Y., Baraze, A. (1981) Photolysis of polycyclic aromatic hydrocarbons in water. *Chemosphere* 10, 1281–1290.

Mill, T., et al. (1982) *Laboratory Protocols for Evaluating the Fate of Organic Chemicals in Air and Water.* p. 255 US-EPA-600/3-82-022.

Miller, G.A. (1963) Vapor pressure of naphthalene. Thermodynamic consistency with proposed frequency assignments. *J. Chem. Eng. Data* 8, 69–72.

Miller, M.M., Ghodbane, S., Wasik, S.P., Tewari, Y.B., Martire, D.E. (1984) Aqueous solubilities, octanol/water partition coefficients and entropies of melting of chlorinated benzenes and biphenyls. *J. Chem. Eng. Data* 29, 184–190.

Miller, M.M., Wasik, S.P., Huang, G.L., Shiu, W.Y., Mackay, D. (1985) Relationships between octanol-water partition coefficient and aqueous solubility. *Environ. Sci. Technol.* 19, 522–529.

Mills, W.B., Dean, J.D., Porcella, D.B., Gherini, S.A., Huson, R.J.M., Frick, W.E., Rupp, G.L. (1982) *Water Quality Assessment: A Screening Procedure for Toxic and Conventional Pollutants.* Part 1. U.S. EPA Report EPA-600/6-82-004a.

Minick, D.J., Frenz, J.H., Patrick, M.A., Brent, D.A. (1988) A comprehensive method for determining hydrophobicity constants by reversed-phase high-performance liquid chromatography. *J. Med. Chem.* 31, 1923–1933.

Monsanto Co. (1972) Presentation to the interdepartmental task force on PCB. May 15, 1972, Washington, DC.

Mortimer, F.S., Murphy, R.V. (1923) The vapor pressures of some substances found in coal tar. *Ind. Eng. Chem.* 15, 1140–1142.

Muel, B., Saguem, S. (1985) Determination of 23 polycyclic hydrocarbons in atmospheric particulate matter of the Paris area and photolysis by sunlight. *Int'l. J. Environ. Anal. Chem.* 19, 111–131.

Müller, M., Kördel, W. (1996) Comparison of screening methods for the estimation of adsorption coefficients on soil. *Chemosphere* 32, 2493–2504.

Murray, J.M., Pottie, R.F., Pupp, C. (1974) The vapor pressures and enthalpies of sublimation of five polycyclic aromatic hydrocarbons. *Can. J. Chem.* 52, 557–563.

Myers, H.S., Fenske, M.R. (1955) Measurement and correlation of vapor pressure data for high boiling hydrocarbons. *Ind. Eng. Chem.* 47, 1652–1658.

Myrdal, P.B., Manka, A.M., Yalkowsky, S.H. (1995) AQUAFAC 3. Aqueous functional group activity coefficients; application to the estimation of aqueous solubility. *Chemosphere* 30, 1619–1637.

Nasir, P., Hwang, S.C., Kobayashi, R. (1980) Development of an apparatus to measure vapor pressures at high temperatures and its application to three higher-boiling compounds. *J. Chem. Eng. Data* 25, 298–301.

Neely, W.B. (1979) Estimating rate constants for the uptake and clearance of chemicals by fish. *Environ. Sci. Technol.* 13, 1506–1508.

Neely, W.B. (1980) A method for selecting the most appropriate environmental experiments on a new chemical. In: *Dynamics, Exposure and Hazard Assessment of Toxic Chemicals.* Haque, R., Editor, pp. 287–296, Ann Arbor Science Publishers, Ann Arbor, Michigan.

Neely, W.B. (1981) Complex problems-simple solutions. *Chemtech.* 11, 249–252.

Neely, W.B. (1983) Reactivity and environmental persistence of PCB isomers. In: *Physical Behavior of PCBs in the Great Lakes.* Mackay, D., Paterson, S., Eisenreich, S.J., Editors, pp. 71–88. Ann Arbor Science Publishers, Ann Arbor, Michigan.

Neely, W.B., Branson, D.R., Blau, G.E. (1974) Partition coefficient to measure bioconcentration potential of organic chemicals in fish. *Environ. Sci. Technol.* 8, 1113–1115.

Neff, J.M. (1979) *Polycyclic Aromatic Hydrocarbons in the Aquatic Environment.* Applied Sci. Publisher, London, England, 262pp.

Nelson, O.A., Senseman, C.E. (1922) Vapor pressure determinations on naphthalene, anthracene, phenanthrene, and anthraquinone between their melting and boiling points. *Ind. Eng. Chem.* 14, 58–62.

Newsted, J.L., Giesy, J.P. (1987) Predictive models for photoinduced acute toxicity of polycyclic aromatic hydrocarbons to *Daphnia magna, Strauss (Cladocera, crustacea)* . *Environ. Toxicol. Chem.* 6, 445–461.

Nguyen, T.H., Sabbah, I. Ball, W.P. (2004) Sorption nonlinearity for organic contaminants with diesel soot: method development and isotherm interpretation. *Environ. Sci. Technol.* 38, 3593–3603.

Nielson, T., Siigur, S., Helweg, C., Jørgensen, O., Hansen, O.E., Kirso, U. (1997) Sorption of polycyclic aromatic compounds to humic acid as studied by high-performance liquid chromatography. *Environ Sci. Technol.* 37, 1102–1108.

Nirmalakhandan, N.N., Speece, R.E. (1989) Prediction of aqueous solubility of organic chemicals based on molecular structure. 2. Application to PNAs, PCBs, PCDDs, etc. *Environ. Sci. Technol.* 23, 708–713.

Nkedi-Kizza, P., Rao, P.S.C., Hornsby, A.G. (1985) Influence of organic cosolvent on sorption of hydrophobic organic chemicals by soils. *Environ. Sci. Technol.* 19, 975–979.

Noegrohati, S., Hammers, W.E. (1992) Regression models for octanol-water partition coefficients, and for bioconcentration in fish. *Toxicol. Environ. Chem.* 34, 155–173.

NRCC (1983) Polycyclic aromatic hydrocarbons in the aquatic environment: Formation, sources, fate and effects on aquatic biota. NRCC/CNRC, Ottawa, Canada.

OECD (1981) *OECD Guidelines for Testing of Chemicals.* Section 1: Physical-Chemical Properties. Organization for Economic Co-operation and Development, Paris.

Offringa, J.C.A., de Kruif, C.G., Van Ekeren, P.J., Jacobs, M.H.G. (1983) Measurement of the evaporation coefficient and saturation vapor pressure of *trans*-diphenylethene using a temperature-controlled vacuum quartz-crystal microbalance. *J. Chem. Thermodyn.* 15, 681–690.

Ogata, M., Fujisawa, K., Ogino, Y., Mano, E. (1984) Partition coefficients as a measure of bioconcentration potential of crude oil compounds in fish and shellfish. *Bull. Environ. Contam. Toxicol.* 33, 561–567.

Oja, V., Suuberg, E.M. (1998) Vapor pressures and enthalpies of sublimation of polycyclic aromatic hydrocarbons and their derivatives. *J. Chem. Eng. Data* 43, 486–492.

Oleszek-Kudlak, S., Shibata, E., Nakamura, T. (2004) The effects of temperature and inorganic salts on the aqueous solubility of selected chlorobenzenes. *J. Chen. Eng. Data* 49, 570–575.

Oliver, B.G. (1987a) Biouptake of chlorinated hydrocarbons from laboratory-spiked and field sediments by oligochaete worms. *Environ. Sci. Technol.* 21, 785–790.

Oliver, B.G. (1987b) Fate of some chlorobenzenes from the Niagara River in Lake Ontario. In: *Sources and Fates of Aquatic Pollutants.* Hite, R.A., Eisenreich, S.J., Eds., pp. 471–489, Advances in Chemistry Series 216, Am. Chem. Soc., Washington, D.C.

Oliver, B.G., Niimi, A.J. (1984) Rainbow trout bioconcentration of some halogenated aromatics from water at environmental concentrations. *Environ. Toxicol. Chem.* 3, 271–277.

Oliver, B.G., Niimi, A.J. (1985) Bioconcentration factors of some halogenated organics for rainbow trout: Limitations in their use for prediction of environmental residues. *Environ. Sci. Technol.* 19, 842–849.

Opperhuizen, A. (1986) Bioconcentration of hydrophobic chemicals in fish. In: *Aquatic Toxicology and Environmental Fate*: Nineth Volume. ASTM STP 921. Poston, T.M., Purdy, R., Eds., pp. 304–315. American Society for Testing and Materials, Philadelphia.

Opperhuizen, A. (1987) Relationships between octan-1-ol/water partition coefficients, aqueous activity coefficients and reversed phase HPLC capacity factors of alkylbenzenes, chlorobenzenes, chloronaphthalenes and chlorobiphenyls. *Toxicol. Environ. Chem.* 15, 349–364.

Opperhuizen, A., Sinnige, T.L., van der Steen, J.M.D., Hutzinger O. (1987) Differences between retentions of various classes of aromatic hydrocarbons in reversed-phase high-performance liquid chromatography. Implications of using retention data for characterizing hydrophobicity. *J. Chromatogr.* 388, 51–64.

Opperhuizen, A., van der Velde, E.W., Gobas, F.A.P.C., Liem, D.A.K., van der Steen, J.M. (1985) Relationship between bioconcentration in fish and steric factors of hydrophobic chemicals. *Chemosphere* 14, 1871–189.

Osborn, A.G., Douslin, D.R. (1975) Vapor pressures and derived enthalpies of vaporization for some condensed-ring hydrocarbons. *J. Chem. Eng. Data* 20, 229–231.

Osborn, A.G., Scot, D.W. (1978) Vapor-pressure and enthalpy of vaporization of indan and five methyl-substituted indans. *J. Chem. Thermodyn.* 10, 619–628.

Osborn, A.G., Scott, D.W. (1980) Vapor pressures of 17 miscellaneous organic compounds. *J. Chem. Thermodyn.* 12, 429–438.

Ou, Z.Q., Yediler, A., He, Y.W., Kettrup, A., Sun, T.H. (1995) Effects of linear alkylbenzene sulfonate (LAS) on the adsorption behaviour of phenanthrene on soils. *Chemosphere* 31, 313–325.

Paris, D.F., Steen, W.C., Barnett, J.T., Bates, E.H. (1980) Kinetics of degradation of xenobiotics by microorganisms. *Paper ENVR-21, 180th National Meeting of American Chemical Society,* San Francisco.

Park, K.S., Sims, R.C., Dupont, R.R., Doucette, W.J., Matthews, J.E. (1990) Fate of PAH compounds in two soil types: Influence of volatilization, abiotic loss and biological activity. *Environ. Toxicol. Chem.* 9, 187–195.

Parks, G.S., Huffman, H.M. (1931) Some fusion and transition data for hydrocarbons. *Ind. Eng. Chem.* 23, 1138–1139.

Paschke, A., Popp, P., Schüürmann, G. (1999) Solubility and partitioning studies with polycyclic aromatic hydrocarbons using an optimized SPME procedure. *Fresenius J. Anal. Chem.* 363, 426–428.

Paschke, A., Popp, P., Schüürmann, G. (1999) Water solubility and octanol/water-partitioning of hydrophobic chlorinated organic substances. determined by using SPME/GC. *Fresenius J. Anal. Chem.* 360, 52–57.

Passivirta, J., Sinkkonen, S., Mikkelson, P., Rantio, T., Wania, F. (1999) Estimation of vapor pressures, solubilities and Henry's law constants of selected persistent organic pollutants as functions of temperature. *Chemosphere* 39, 811–832.

Pavlou, S.P. (1987) The use of equilibrium partition approach in determining safe levels of contaminants in marine sediments. p. 388–412. In: *Fate and Effects of Sediments-Bound Chemicals in Aquatic Systems.* Dickson, K.L., Maki, A.W., Brungs, W.A., Editors. Proceedings of the Sixth Pellston Workshop, Florissant, Colorado, August 12–17, 1984. SETAC Special Publ. Series, Ward, C.H., Walton, B.T., Eds., Pergamon Press, N.Y.

Pearlman, R.S., Yalkowsky, S.H., Banerjee, S. (1984) Water solubilities of polynuclear aromatic and heteroaromatic compounds. *J. Phys. Chem. Ref. Data* 13(2), 555–562.

Perez-Tejeda, P., Yanes, C., Maestre, A. (1990) Solubility of naphthalene in water and alcohol solutions at various temperatures. *J. Chem. Eng. Data* 35, 244–246.

Phousongphouang, P.T., Arey J. (2002) Rate constants for the gas-phase reactions of a series of alkylnaphthalenes with the OH radicals. *Environ. Sci. Technol.* 36, 1947–1952.

Piatt, J.J., Backhus, D.A., Capel, P.D., Eisenreich, S.J. (1996) Temperature-dependent sorption of naphthalene, phenanthrene, and pyrene to low organic carbon aquifer sediments. *Environ. Sci. Technol.* 30, 751–760.

Pierotti, C., Deal, C., Derr, E. (1959) Activity coefficient and molecular structure. *Ind. Eng. Chem. Fundam.* 51, 95–101.

Pinal, R., Lee, L.S., Rao, P.S.C. (1991) Prediction of the solubility of hydrophobic compounds in nonideal solvent mixtures. *Chemosphere* 22, 939–951.

Pitts, Jr., J.N., Atkinson, R., Sweetman, J.A., Zielinska, B. (1985) The gas-phase reaction of naphthalene with N_2O_5 t0 form nitronaphthalenes. *Atmos. Environ.* 19, 701–705.

Podoll, R.T., Irwin, K.C., Parish, H.J. (1989) Dynamic studies of naphthalene sorption on soil from aqueous solution. *Chemosphere* 18, 2399–2412.

Poeto, T.S., Stensel, H.D., Strand, S.E. (1999) Biodegradation of polyaromatic hydrocarbons by marine bacteria: effect of solid phase on degradation kinetics. *Water. Res.* 33, 868–880.

Power, W.H., Woodworth, C.L., Loughary, W.G. (1977) Vapor pressure determination by gas chromatography in the microtorr range-anthracene and triethylene glycol di-2-ethyl butyrate. *J. Chromatogr. Sci.* 15, 203–207.

Price, L.C. (1976) Aqueous solubility of petroleum as applied to its origin and primary migration. *Am. Assoc. Petrol. Geol. Bull.* 60, 213–244.

Pupp, C., Lao, R.C., Murray, J.J., Pottie, R.F. (1974) Equilibrium vapor concentrations of some polycyclic aromatic hydrocarbons, arsenic trioxide (As_4O_6) and selenium dioxide, and the collection efficiencies of these air pollutants. *Atoms. Environ.* 8, 915–925.

Pussemier, L., Szabo, G., Bulman, R.A. (1990) Prediction of the soil adsorption coefficient K_{OC} for aromatic pollutants. *Chemosphere* 21(10–11), 1199–1212.

Radchenko, L.G., Kitiagorodskii, A.I. (1974) Vapor pressure and heat of sublimation of naphthalene, biphenyl, octafluoronaphthalene, decafluorobiphenyl, acenaphthene and α-nitronaphthalene. *Zhur. Fiz. Khim.* 48, 2702–2704.

Radding, S.B., Mill, T., Gould, C.W., Lin, D.H., Johnson, H.L., Bomberger, D.C., Fojo, C.V. (1976) *The Environmental Fates of Selected Polycyclic Aromatic Hydrocarbons.* U.S. Environmental Protection Agency Report No. EPDA-560/5-75-009.

Radding, S.B., Liu, D.H., Johnson, H.L., Mill, T. (1977) *Review of the Environmental Fate of Selected Chemicals*. USEPA report No. EPA-560/5-77-003.

Rapaport, R.A., Eisenreich, S.J. (1984) Chromatographic determination of octanol-water partition coefficients (K_{OW}'s) for 58 polychlorinated biphenyl congeners. *Environ. Sci. Technol.* 18, 163–170.

Reza, J., Trejo, A., Vera-Ávila, L.E. (2002) Determination of the temperature dependence of water solubilities of polycyclic aromatic hydrocarbons by a generator column-*on-line* solid-phase extraction-liquid chromatographic method. *Chemosphere* 47, 933–945.

Reichardt, P.B., Chadwick, B.L., Cole, M.A., Robertson, B.R., Button, D.K. (1981) Kinetic study of biodegradation of biphenyl and its monochlorinated analogues by a mixed marine microbial community. *Environ. Sci. Technol.* 15, 75–79.

Reisen, F., Arey, J. (2002) Reaction of hydroxyl radicals and ozone with acenaphthene and acenaphthylene. *Environ. Sci. Technol.* 36, 4302–4311.

Rekker, R.F. (1977) *The Hydrophobic Fragmental Constant. Its Derivation and Application. A Means of Characterizing Membrane Systems.* Elsevier Sci. Publishers Co., New York.

Rekker, R.F., De Kort, N.N. (1979) The hydrophobic fragment constant: an extension to a 1000 data point set. *Eur. J. Med. Chem.-Chim. Ther.* 14(6), 479–488.

Riddick, J.A., Bunger, W.B., Sakano, T.K. (1986) *Organic Solvents: Physical Properties and Methods of Purification*. 4th Ed. J. Wiley & Sons, New York, N.Y.

Riederer, M.(1990) Estimating partitioning and transport of organic chemicals in the foliage/atmosphere system: Discussion of a fugacity-based model. *Environ. Sci. Technol.* 24, 829–837.

Rippen, G., Ilgenstein, M., Klöpffer, W., Poreniski, H.J. (1982) Screening of the adsorption behavior of new chemicals: natural soils and model adsorbents. *Ecotox. Environ. Saf.* 6, 236–245.

Ritter, S., Hauthal, W.H., Maurer, G. (1995) Octanol/water partition coefficients for environmentally important organic compounds. *Environ. Sci. Pollut. Res.* 2(3), 153–160.

Rockne, K.J., Strand, S.E. (1998) Biodegradation of bicyclic and polycyclic aromatic hydrocarbons in anaerobic enrichments. *Environ. Sci. Technol.* 32, 3962–3967.

Rogers, K.S., Cammarata, A. (1969) Superdelocalizability and charge density. A correlation with partition coefficients. *J. Med. Chem.* 12, 692.

Rordorf, B.F. (1985) Thermodynamics and thermal properties of polychlorinated compounds: the vapor pressures and flow-tube kinetics of ten dibenzo-*p*-dioxins. *Chemosphere* 14, 885–892.

Rossi, S.S. (1977) Bioavailability of petroleum hydrocarbons from water, sediments and detritus to the marine annelid, *neanthes arenaceodentata*. *Proc. Oil Spill Conf.*, pp. 621–625. Am. Petrol. Inst., Washington DC.

Rossi, S.S., Thomas, W.H. (1981) Solubility behavior of three aromatic hydrocarbons in distilled water and natural seawater. *Environ. Sci. Technol.* 15, 715–716.

Ruelle, P., Kesselring, U.W. (1997) Aqueous solubility prediction of environmentally important chemicals form the mobile order thermodynamics. *Chemosphere* 34, 275–298.

Ruepert, C., Grinwis, A., Govers, H. (1985) Prediction of partition coefficients of unsubstituted polycyclic aromatic hydrocarbons from C_{18} chromatographic and structural properties. *Chemosphere* 14, 279–291.

Ryan, J.A., Bell, R.M., Davidson, J.M., O'Connor, G.A. (1988) Plant uptake of non-ionic organic chemicals from soils. *Chemosphere* 17, 2299–2323.

Ryan, P.A., Cohen, Y. (1986) Multimedia transport of particle-bound organics: benzo(*a*)pyrene test case. *Chemosphere* 15, 21–47.

Sabljic, A. (1984) Predictions of the nature and strength of soil sorption of organic pollutants by molecular topology. *J. Agric. Food Chem.* 32, 243–246.

Sabljic, A. (1987a) On the prediction of soil sorption coefficients of organic pollutants from molecular structure: application of molecular topology model. *Environ. Sci. Technol.* 21, 358–366.

Sabljic, A. (1987b) Nonempirical modeling of environmental distribution and toxicity of major organic pollutants. In: *QSAR in Environmental Toxicology - II*. Kaiser, K.L.E., Ed., pp 309–322, D. Reidel Publ. Co., Dordrecht, The Netherlands.

Sabljic, A., Güsten, H., Verhaar, H., Hermens, J. (1995) QSAR modelling of soil sorption, improvement and systematics of log K_{OC} vs. log K_{OW} correlations. *Chemosphere* 31, 4489–4514.

Sahyun, M.R.V. (1966) Binding of aromatic compounds to bovine serum albumin. *Nature* 209, 613–614.

Sallounm, M.J., Chefetz, B., Hatcher, P.G. (2002) Phenanthrene sorption by aliphatic-rich natural organic matter. *Environ. Sci. Technol.* 36, 1953–1958.

Sangster, J. (1989) Octanol-water partition coefficients of simple organic compounds. *J. Phys. Chem. Ref. Data* 18, 1111–1230.

Sangster, J. (1993) LOGKOW, A Databank of Evaluated Octanol-Water Partition Coefficients. 1st Edition, Montreal, Quebec, Canada.

Sarna, L.P., Hodge, P.E., Webster, G.R.B. (1984) Octanol-water partition coefficients of chlorinated dioxins and dibenzofurans by reversed-phase HPLC using several C_{18} columns. *Chemosphere* 13, 975–983.

Sasse, K., Jose, J., Merlin, J.-C. (1988) A static apparatus for measurement of low vapor pressures. Experimental results on high molecular-weight hydrocarbons. *Fluid Phase Equil.* 42, 287–304.

Sato, N., Inomata, H., Arai, K., Saito, S. (1986) Measurements of vapor pressures for coal-related aromatic compounds by gas saturation method. *J. Chem. Eng. Jpn.* 19, 145–147.

Schlautman, M.A., Morgan, J.J. (1993a) Effects of aqueous chemistry on the binding of polycyclic aromatic hydrocarbons by dissolved humic materials. *Environ. Sci. Technol.* 27, 961–969.

Schlautman, M.A., Morgan, J.J. (1993b) Binding of a fluorescent hydrophobic organic probe by dissolved humic substances and organically-coated aluminum oxide surfaces. *Environ. Sci. Technol.* 27, 2523–2532.

Schmidt-Bleek, F., Haberland, W., Klein, A.W., Caroli, S. (1982) Steps towards environmental hazard assessment of new chemicals (including a hazard ranking scheme, based upon Directive 79/831/EEC) . *Chemosphere* 11, 383–415.

Schüürmann, G., Klein, W. (1988) Advances in bioconcentration prediction. *Chemosphere* 17(8), 1551–1574.

Schwarz, F.P. (1977) Determination of temperature dependence of solubilities of polycyclic aromatic hydrocarbons in aqueous solutions by a fluorescence method. *J. Chem. Eng. Data* 22, 273–277.

Schwarz, F.P., Wasik, S.P. (1976) Fluorescence measurements of benzene, naphthalene, anthracene, pyrene, fluoranthene, and benzo[*a*]pyrene in water. *Anal. Chem.* 48, 524–528.

Schwarz, F.P., Wasik, S.P. (1977) A fluorescence method for the measurement of the partition coefficients of naphthalene, 1-methyl-naphthalene, and 1-ethylnaphthalene in water. *J. Chem. Eng. Data* 22, 270–273.

Sears, G.W., Hopke, E.R. (1949) Vapor pressures of naphthalene, anthracene, and hexachlorobenzene in a low pressure region. *J. Am. Chem. Soc.* 71, 1632–1634.

Sharma, R.K., Palmer, H.B. (1974) Vapor pressures of biphenyl near fusion temperature. *J. Chem. Eng. Data* 19, 6–8.

Shaw, D.G., Editor (1989) IUPAC Solubility Data Series, Vol. 38: *Hydrocarbons (C_8 -C_{36}) with Water and Seawater*. Pergamon Press, Oxford, England.

Sherblom, P.M., Eganhouse, R.P. (1988) Correlations between octanol-water partition coefficients and reversed-phase high-performance liquid chromatography capacity factors. *J. Chromtogr.* 454, 37–50.

Shiu, W.Y., Gobas, F.A.P.C., Mackay, D. (1987) Physical-chemical properties of three congeneric series of chlorinated hydrocarbons. In: *QSAR in Environmental Toxicology-II*. Kaiser, K.L.E., Editor, D. Reidel Publishing Company. pp. 347–362.

Shiu, W.Y., Ma, K.C. (2000) Temperature dependence of physical-chemical properties of selected chemicals of environmental interest. 1. Mono- and polynuclear aromatic hydrocarbons. *J. Phys. Chem. Ref. Data* 29, 41–130.

Shiu, W.Y., Mackay, D. (1986) A Critical review of aqueous solubilities, vapor pressures, Henry's law constants and octanol-water partition coefficients of the polychlorinated biphenyls. *J. Phys. Chem. Data* 15, 911–929.

Shiu, W.Y., Mackay, D. (1997) Henry's law constants of selected aromatic hydrocarbons, alcohols, and ketones. *J. Chem. Eng. Data* 42, 27–30.

Sims, R.C.C. (1990) Fate of PAH compounds in soil loss mechanisms. *Environ. Toxicol. Chem.* 9, 187-.

Sims, R.C.C., Overcash, M.R. (1983) Fate of polynuclear aromatic compounds (PNA's) in soil plant systems. *Residue Rev.* 88.

Sinke, G.C. (1974) A method for measurement of vapor pressures of organic compounds below 0.1 torr, naphthalene as a reference substance. *J. Chem. Thermodyn.* 6, 311–316.

Sklarew, D.S., Girvin, D.C. (1987) Attenuation of polychlorinated biphenyls in soil. *Rev. Environ. Contam. Toxicol.* 98, 1–41.

Smith, J.H., Mabey, W.R., Bahonos, N., Holt, B.R., Lee, S.S., Chou, T.W., Venberger, D.C., Mill, T. (1978) *Environmental Pathways of Selected Chemicals in Fresh Water Systems: Part II. Laboratory Studies*. Interagency Energy-Environment Research Program Report. EPA-600/7-78-074. Environmental Research Laboratory Office of Research and Development. US EPA, Athens, GA 30605, pp. 304.

Smith, N.K., Stewart, Jr., R.C., Osborn, A.G. (1980) Pyrene: vapor pressures, enthalpy of combustion, and chemical thermodynamic properties. *J. Chem. Thermodyn.* 21, 919–926.

Sonnefeld, W.J., Zoller, W.H., May, W.E. (1983) Dynamic coupled-column liquid chromatographic determination of ambient temperature vapor pressures of polynuclear aromatic hydrocarbons. *Anal. Chem.* 5, 275–280.

Southworth, R.G. (1977) Transport and transformations of anthracene in natural waters. In: *Aquatic Toxicology*. ASTM ATP 667, Marking, L.L., Kimerle, R.A., Editors, American Society for Testing and Materials, pp. 359–380, Philadelphia.

Southworth, G.R. (1979) The role of volatilization in removing polycyclic aromatic hydrocarbons from aquatic environments. *Bull. Environ. Contam. Toxicol.* 21, 507–514.

Southworth, G.R., Beauchamp, J.J., Schmieders, P.K. (1978) Bioaccumulation potential of polycyclic aromatic hydrocarbons in *Daphnia Pulex*. *Water Res.* 12, 973–977.

Southworth, G.R., Keller, J.L. (1986) *Water Air Soil Pollut.* 28, 239.

Spacie, A., Landrum, R.F., Leversee, G.J. (1983) Uptake, depuration and biotransformation of anthracene and benzo-*a*-pyrene in bluegill sunfish. *Ecotox. Environ. Saf.* 7, 330–341.

Staudinger, J., Roberts, P.V. (1996) A critical review of Henry's law constant for environmental applications. *Crit. Rev. Environ. Sci. Technol.* 26, 205–297.

Staudinger, J., Roberts, P.V. (2001) A critical compilation of Henry's law constant temperature dependence relations for organic compounds in dilute aqueous solutions. *Chemosphere* 44, 561–576.

Stauffer, T.B., MacIntyre, W.G. (1986) Sorption of low-polarity organic compounds on oxide minerals and aquifer material. *Environ. Toxicol. Chem.* 5, 949–955.

Stauffer, T.B., MacIntyre, W.G., Wickman, D.C. (1989) Sorption of nonpolar organic chemicals on low-carbon-content aquifer materials. *Environ. Toxicol. Chem.* 8, 845–852.

Steen, W.C., Karickhoff, S.W. (1981) Biosorption of hydrophobic organic pollutants by mixed microbial populations. *Chemosphere* 10, 27–32.

Stephen, H., Stephen, D., Editors (1963) *Solubility of Inorganic and Organic Compounds.* Macmillan Co., New York.

Stephenson, R.M., Malanowski, A. (1987) *Handbook of the Thermodynamics of Organic Compounds.* Elsevier, N.Y.

Stevens, B., Perez, S.R., Ors, J.A. (1974) Photoperoxidation of unsaturated organic molecules O_2 delta G acceptor properties and reactivity. *J. Am. Chem. Soc.* 96, 6846–6850.

Stevens, N. (1953) Vapour pressures and the heats of sublimation of anthracene and of 9:10-diphenyl anthracene. *J. Chem. Soc. (Lond.)* 2973–2974.

Stucki, G., Alexander, M. (1987) Role of dissolution rate and solubility in biodegradation of aromatic compounds. *Appl. Environ. Microbiol.* 53, 292–297.

Stull, D.R. (1947) Vapor pressure of pure substances. Organic compounds. *Ind. Eng. Chem.* 39, 517–560.

Stull, D.R., Sinke, G.C., McDonald, R.A., Hatton, W.E., Hildenbrand, D.L. (1961) Thermodynamic properties of indane and indene. *Pure and Applied Chem.* 2, 315–322.

Su, Y., Lei, Y.D., Daly, G.L., Wania, F. (2002) Determination of octanol-air partition coefficient (K_{OA}) values for chlorobenzenes and polychlorinated naphthalenes from gas chromatographic retention times. *J. Chem. Eng. Data* 47, 449–455.

Sverdrup, L.E., Jensen, J., Kelley, A.E., Krogh, P.J., Stenersen, J. (2002) Effects of eight polycyclic aromatic compounds on the survival and reproduction on *Enchytraeus crypticus* (Oligochaeta, Clitellata) . *Environ. Toxicol. Chem.* 21, 109–114.

Swan, T.H., Mack, Jr., E. (1925) Vapor pressures of organic crystals by an effusion method. *J. Am. Chem. Soc.* 47, 2112–2116.

Swann, R.L., Laskowski, D.A., McCall, P.J., Vander Kuy, K., Dishburger, H.J. (1983) A rapid method for the estimation of the environmental parameters octanol/water partition coefficient, soil sorption constant, water to air ratio, and water solubility. *Residue Rev.* 85, 17–28.

Szabo, G., Guczi, J., Bulman, R.A. (1995) Examination of silica-salicylic acid and silica-8-hydroxyquinoline HPLC stationary phases for estimation of the adsorption coefficient of soil for some aromatic hydrocarbons. *Chemosphere* 30, 1717–1727.

Szabo, G., Guczi, J., Ködel, W., Zsolnay, A., Major, V., Keresztes, P. (1999) Comparison of different HPLC stationary phases for determination of soil-water distribution coefficient, K_{OC}, values of organic chemicals in RP-HPLC system. *Chemosphere* 39, 431–442.

Szabo, G., Prosser, S.L., Bulman, R.A. (1990a) Prediction of the adsorption coefficient (K_{OC}) for soil by a chemically immobilized humic acid column using RP-HPLC. *Chemosphere* 21, 729–739.

Szabo, G., Prosser, S.L., Bulman, R.A. (1990b) Determination of the adsorption coefficient (K_{OC}) of some aromatics for soil by RP-HPLC on two immobilized humic acid phases. *Chemosphere* 21 777–788.

Tabak, H.H., Quave, S.A., Mashni, C.I., Barth, E.F. (1981) Biodegradability studies with organic priority pollutant compounds. *J. Water Pollut. Control. Fed.* 53, 1503–1518.

Taylor, J.W., Crookes, R.J. (1976) Vapour pressure and enthalpy of sublimation of 1,3,5,7-tetranitro-1,3,5,7-tetra-azacyclo-octane. *J. Chem. Soc. Farad. Trans.* 72, 723–729.

Ten Hulscher, T.E.M., van der Velde, L.E., Bruggeman, W.A. (1992) Temperature dependence of Henry's law constants for selected chlorobenzenes, polychlorinated biphenyls and polycyclic aromatic hydrocarbons. *Environ. Toxicol. Chem.* 11, 1595–1603.

Ten Hulscher, T.E.M., Vrind, B., Van den Heuvel, H., van Noort, P., Govers, H. (2003) Influence of desorption and contact time on sediment of spiked polychlorinated biphenyls and polycyclic aromatic hydrocarbons: relationship with in situ distribution. *Environ. Toxicol. Chem.* 22, 1208–1211.

Thomann, R.V. (1989) Bioaccumulation model of organic chemical distribution in aquatic food chains. *Environ. Sci. Technol.* 23, 699–707.

Treves, K., Shragina, L., Rudich, Y. (2001) Measurement of octanol-air partition coefficients using solid-phase microextraction (SPME) - application to hydroxy alkyl nitrates. *Atmos. Environ.* 35, 5843–5854.

Tsai, W., Cohen, Y., Sakugawa, H., Kaplan, I.R. (1991) Dynamic partitioning of semivolatile organics in gas/particle/rain phases during rain scavenging. *Environ. Sci. Technol.* 25, 2012–2023.

Tsonopoulos, C., Prausnitz, J.M. (1971) Activity coefficients of aromatic solutes in dilute aqueous solutions. *Ind. Eng. Chem. Fundam.* 10, 593–600.

Tsypkina, O.Ya. (1955) Effect of vacuum on the separation of polycyclic compounds of coal-tar resins by rectification. *Zh. Prikl. Khim.* 28, 185-192. *J. Appl. Chem.* (U.S.S.R.) 28, 167–172.

Turner, A., Rawling, M.C. (2002) Sorption of benzo[a]pyrene to sediment contaminated by acid mine drainage: contrasting particle concentration-dependencies in river water and seawater. *Water Res.* 36, 2011–2019.

Ubbelohde, A.R. (1978) *The Molten State of Matter.* p.148, Wiley, New York, New York.

Urbani, M., Gigli, R., Piacente, V. (1980) Vaporization study of α-bromonaphthalene. *J. Chem. Eng. Data* 25, 97–100.

Vadas, G.G., MacIntyre, W.G., Burris, D.R. (1991) Aqueous solubility of liquid hydrocarbon mixtures containing dissolved solid components. *Environ. Toxicol. Chem.* 10, 633–639.

Vaishnav, D.D., Babeu, L. (1987) Comparison of occurence and rates of chemical biodegradation in natural waters. *Bull. Environ. Contam. Toxicol.* 39, 237–244.

Valerio, F., Pala, M., Brescianini, C., Lazaazrotto, A., Balducci, D. (1991) Effect of sunlight and temperature on concentration of pyrene and benzo[a]pyrene adsorbed on airborne particulate. *Toxicol. Environ. Chem.* 31–32, 113–118.

Valvani, S.C., Yalkowsky, S.H. (1980) Solubility and partitioning in drug design. In: *Physical Chemical Properties of Drug. Med. Res. Series.* volume 10., pp. 201–229, Yalkowsky, S.H., Sinkula, A.A., Valvani, S.C., Editors, Marcel Dekker, New York.

Van der Linde, P.R., Blok, J.G., Oonk, A.J. (1998) Naphthalene as a reference substance for vapour pressure measurements looked upon from an unconventional point of view. *J. Chem. Thermodyn.* 30, 909–917.

Van Ekeren, P.J., Jacobs, M.H.G., Offringa, J.C.A., De Kruif, C.G. (1983) Vapour-pressure measurements on *trans*-diphenylethene and naphthalene using a spinning-rotor friction gauge. *J. Chem. Thermodyn.* 15, 409–417.

Van Hoof, P.L., Kukkonen, J.V.K., Landrum, P.F. (2001) Impact of sediment manipulation on the bioaccumulation of polycyclic aromatic hydrocarbons from field-contaminated and laboratory-dosed sediments by an oligochaete. *Environ. Toxicol. Chem.* 20, 1752–1761.

Veith, G.D., Austin, N.M., Morris, R.T. (1979a) A rapid method for estimation log P for organic chemicals. *Water Res.* 13, 43–47.

Veith, G.D., Defor, D.L. Bergstedt, B.V. (1979b) Measuring and estimating the bioconcentration factor of chemicals in fish. *J. Fish Res. Board Can.* 26, 1040–1048.

Veith, G.D., Kosian, P. (1983) Estimating bioconcentration potential from octanol/water partition coefficients. In: *Physical Behavior of PCBs in the Great Lakes.* Mackay, D., Paterson, S., Eisenreich, S.J., Simmons, M.S., Editors, pp. 269–282, Ann Arbor Science Publishers, Ann Arbor, Michigan.

Velapoldi, R.A., White, P.A., May, W.E., Eberhardt, K.R. (1983) Spectrofluorimetric determination of polycyclic aromatic hydrocarbons in aqueous effluents from generator columns. *Anal. Chem.* 55, 1896–1901.

Verevkin, S.P. (1999) Thermochemical properties of diphenylalkanes. *J. Chem. Eng. Data* 44, 175–179.

Verevkin, S.P. (2003) Vapor pressures and enthalpies of vaporization of a series of 1- and 2-halogenated naphthalene. *J. Chem. Thermodyn.* 35, 1237–1251.

Verevkin, S.P. (2004) Vapor pressure measurements on fluorene and methyl-fluorenes. *Fluid Phase Equil.* 225, 145–152.

Verschueren, K. (1983) *Handbook of Environmental Data on Organic Chemicals*, 2nd Edition, Van Nostrand Reinhold Co., New York.

Vesala, A. (1974) Thermodynamics of transfer of nonelectrolytes from light to heavy water. I. Linear free energy correlations of free energy of transfer with solubility and heat of melting of a nonelectrolyte. *Acta Chemica Scand.* A28, 839–845.

Voice, T.C., Weber Jr., W.J. (1985) Sorbent concentration effects in liquid/solid partitioning. *Environ. Sci. Technol.* 19(9), 789–796.

Vowles, P.D., Mantoura, R.F.C. (1987) Sediment-water partition coefficients and HPLC retention factors of aromatic hydrocarbons. *Chemosphere* 16, 109–116.

Vozňáková, Z., Popl, M., Berka, M. (1978) Recovery of aromatic hydrocarbons from water. *J. Chromatogr. Sci.* 16, 123–127.

Wakayama, N., Inokuchi, H. (1967) Heats of sublimation of polycyclic aromatic hydrocarbons and their molecular packings. *Bull. Chem. Soc. Japan* 40, 2267–2271.

Wakeham S.G., Davis, A.C., Karas, J. (1983) Mesocosm experiments to determine the fate and persistence of volatile organic compounds in coastal seawater. *Environ. Sci. Technol.* 17, 611–617.

Walker, J.D., Colwell, R.R. (1976) Measuring the potential activity of hydrocarbon-degrading bacteria. *Appl. Environ. Microbiol.* 31, 187–197.

Walters, R.W., Luthy, R.G. (1984) Equilibrium adsorption of polycyclic aromatic hydrocarbons from water onto activated carbon. *Environ. Sci. Technol.* 18(6), 395–403.

Wang, L., Kong, L., Chang, C. (1991) Photodegradation of 17 PAHs in methanol (or acetonitrile) -water solution. *Environ. Chem.* 10(2), 15–20.

Wang, L., Wang, X., Xu, O., Tian, L. (1986) Determination of the *n*-octanol/water partition coefficients of polycyclic aromatic hydrocarbons by HPLC and estimation of their aqueous solubilities. *Huanjing Kexue Xuebao* 6, 491–497.

Wang, L., Zhao, Y., Hong, G. (1992) Predicting aqueous solubility and octanol/water partition coefficients of organic chemicals from molar volume. *Environ. Chem.* (China) 11, 55–70.

Wania, F., Lei, Y.D., Harner, T. (2002) Estimating octanol-air partition coefficients of nonpolar semivolatile organic compounds form gas chromatographic retention times. *Anal. Chem.* 74, 3478–3483.

Wania, F., Mackay, D. (1996) Tracking the distribution of persistent organic pollutants. *Environ. Sci. Technol.* 30, 390A–396A.

Wania, F., Shiu, W.-Y., Mackay, D. (1994) Measurement of the vapor pressure of several low-volatility organochlorine chemicals at low temperatures with a gas saturation method. *J. Chem. Eng. Data* 39, 572–577.

Warner, P.H., Cohen, J.M., Ireland, J.C. (1987) *Determination of Henry's Law Constants of Selected Priority Pollutants.* NTIS PB87-212684, EPA/600/D-87/229, U.S. Environmental Protection Agency, Cincinnati, Ohio.

Wasik, S.P., Miller, M.M., Teware, Y.B., May, W.E., Sonnefeld, W.J., DeVoe, H., Zoller, W.H. (1983) Determination of the vapor pressure, aqueous solubility, and octanol/water partition coefficient of hydrophobic substances by coupled generator column/liquid chromatographic methods. *Residue Rev.* 85, 29–42.

Wasik, S.P., Tewari, Y.B., Miller, M.M., Martire, D.E. (1981) *Octanol/Water Partition Coefficients and Aqueous Solubilities of Organic Compounds.* NBSIR 81-2406., U.S. Department of Commerce, National Bureau of Standards, Washington, D.C.

Watanabe, I., Tatsukawa, R. (1989) Anthropogenic brominated aromatics in the Japanese environment. In: *Proceedings: Workshop on Brominated Aromatic Flame Retardants.* pp. 63–70. Skokloster, Sweden, 24–26 October, 1989.

Wauchope, R.D., Getzen, F.W. (1972) Temperature dependence of solubilities in water and heats of fusion of solid aromatic hydrocarbons. *J. Chem. Eng. Data* 17, 38–41.

Weast, R. (1972–73) *Handbook of Chemistry and Physics.* 53rd ed., CRC Press, Cleveland.

Weast, R. (1983–84) *Handbook of Chemistry and Physics.* 64th ed., CRC Press, Boca Raton, Florida.

Webster, G.R.B., Friesen, K.J., Sarna, L.P., Muir, D.C.G. (1985) Environmental fate modelling of chlorodioxins: Determination of physical constants. *Chemosphere* 14, 609–622.

Weimer, R.F., Prausnitz, J.M. (1965) Complex formation between carbon tetrachloride and aromatic hydrocarbons. *J. Chem. Phys.* 42, 3643–3644.
Whitehouse, B.G. (1984) The effects of temperature and salinity on the aqueous solubility of polynuclear aromatic hydrocarbons. *Mar. Chem.* 14, 319–332.
Wieczorek, S.A., Kobayashi, R. (1980) Vapor pressure measurements of diphenylmethane, thianaphthene, and bicyclohexyl at elevated temperatures. *J. Chem. Eng. Data* 25, 302–305.
Wieczorek, S.A., Kobayashi, R. (1981) Vapor pressure measurements of 1-methylnaphthalene, 2-methylnaphthalene, and 9.10-dehydrophenanthrene at elevated temperatures. *J. Chem. Eng. Data* 26, 8–11.
Wiedemann, H.G. (1972) Application of thermogravimetry for vapor pressure determination. *Thermochim. Acta* 3, 355–366.
Wiedemann, H.G., Vaughan, H.P. (1969) Application of thermogravimetry for vapor pressure determinations. In: *Proc. Toronto Symp. Therm. Anal.* (3rd), pp. 233–249.
Wild, S.R., Berrow, M.L., Jones, K.C. (1991) The persistence of polynuclear aromatic hydrocarbons (PAHs) in sewage sludge amended agricultural soils. *Environ. Pollut.* 72(2), 141–157.
Windholtz, M., Budavari, S., Blumetti, R.F., Otterbein, E.S., Editors (1983) *The Merck Index*, 10th edition, Merck and Co., Rahway, New Jersey.
Winget, P., Cramer, C.J., Truhlar, D.G. (2000) Prediction of soil sorption coefficients using a universal solvation model. *Environ. Sci. Technol.* 34, 4733–4740.
Wise, S.A., Bonnett, W.J., Guenther, F.R., May, W.E. (1981) A relationship between reversed phase C_{18} liquid chromatographic retention and the shape of polycyclic aromatic hydrocarbons. *J. Chromatographic Sci.* 19, 457–465.
Wong, P.T.S., Kaiser, K.L.E. (1975) Bacterial degradation of polychlorinated biphenyls. II. Rate studies. *Bull. Contam. Toxiocol.* 3, 249.
Wood, A.L., Bouchard, D.C., Brusseau, M.L., Rao, P.S.C. (1990) Cosolvent effects on sorption and mobility of organic contaminants in soil. *Chemosphere* 21, 575–587.
Woodburn, K.B. (1982) Measurement and Application of the Octanol/Water Partition Coefficient for Selected Polychlorinated Biphenyls. M.Sc. Thesis, University of Wisconsin, Madison, Wisconsin.
Woodburn, K.B., Doucette, W.J., Andren, A.W. (1984) Generator column determination of octanol/water partition coefficients for selected polychlorinated biphenyl congeners. *Environ. Sci. Technol.* 18, 457–459.
Woodburn, K.B., Lee, L.S., Rao, P.S.C., Delfino, J.J. (1989) Comparison of sorption energetics for hydrophobic organic chemicals by synthetic and natural sorbents from methanol/water solvent mixtures. *Environ. Sci. Technol.* 23, 407–412.
Yalkowsky, S.H. (1981) Solubility and partitioning V: Dependence of solubility on melting point. *J. Pharmaceutical Sci.* 70, 971–973.
Yalkowsky, S.H., Valvani, S.C. (1979) Solubilities and partitioning 2. Relationships between aqueous solubilities, partition coefficients, and molecular surface areas of rigid aromatic hydrocarbons. *J. Chem. Eng. Data* 24, 127–129.
Yalkowsky, S.H., Valvani, S.C. (1980) Solubility and Partitioning. I:Solubility of nonelectrolytes in water. *J. Pharm. Sci.* 69, 912–922.
Yalkowsky, S.H., Valvani, S.C., Mackay, D. (1983a) Estimation of the aqueous solubility of some aromatic compounds. *Residue Rev.* 85, 43–55.
Yalkowsky, S.H., Valvani, S.C., Roseman, T.J. (1983b) Solubility and partitioning. VI: Octanol solubility and octanol-water partition coefficients. *J. Pharm. Sci.* 72, 866–870.
Yamasaki, H., Kuwata, K., Kuge, Y. (1984) Determination of vapor pressure of polycyclic aromatic hydrocarbons in the supercooled liquid phase and their adsorption on airborne particulate matter. *Nippon Kagaka Kaish.* 8, 1324–1329.
Yaws, C.L. (1994) *Handbook of Vapor Pressure*, Vol. 1 C_1 to C_4 Compounds, Vol. 2. C_5 to C_7 Compounds, Vol. 3, C_8 to C_{28} Compounds. Gulf Publishing Co., Houston, Texas.
Yaws, C.L., Yang, J.C., Pan, X. (1991) Henry's law constants for 362 organic compounds in water. *Chem. Eng.* November, 179–185.
Young, S. (1989) XLVIII. On the vapour-pressures and specific volumes of similar compounds of elements in relation to the position of those elements in the periodic Table. Part I. *J. Chem. Soc.* 55, 486–521.
Yu, G., Xu, X. (1993) Investigation of aqueous solubility of nitro-PAH by dynamic coupled-column HPLC. *Chemosphere* 24, 1699–1993.
Yu, W., Lin, F., Lin, Z., Xu, Z., Tang, Y. (1999) Photolysis of anthracene and chrysene in aquatic systems. *Chemosphere* 38, 1273–1278.
Yuteri, C., Ryan, D.F., Callow, J.J., Gurol, M.D. (1987) The effect of chemical composition of water on Henry's law constant. *J. Water Pollut. Control Fed.* 59, 950–956.
Zepp, R.G. (1980) In: *Dynamics, Exposure and Hazard Assessment of Toxic Chemicals.* pp. 69–110. Haque, R., Editor, Ann Arbor Science, Ann Arbor, Michigan.
Zepp, R.G. (1991) Photochemical fate of agrochemicals in natural waters. In: *Pesticide Chemistry.* Advances in International Research, Development, and Legislation. pp. 329–345, Frechse, H., Editor, VCH, Weinheim.
Zepp, R.G., Scholtzhauer, P.F. (1979) Photoreactivity of selected aromatic hydrocarbons in water. In: *Polynuclear Aromatic Hydrocarbons.* Jones, P.W., Leber, P., Editors, pp. 141–58. Ann Arbor Sci. Publ. Inc., Ann Arbor, Michigan.
Zhang, Y., Wu, R.S.S., Hong, H.-S., Poon, K.-F., Lam, M.H.W. (2000) Field study on desorption rates of polynuclear aromatic hydrocarbons from contaminated marine sediment. *Environ. Toxicol. Chem.* 19, 2431–2435.
Zhang, Z., Pawliszyn, J. (1993) Headspace solid-phase microextraction. *Anal. Chem.* 65, 1843–1852.
Zil'berman-Granovskaya, A.A. (1940) Measurement of small vapor pressures. I. Vapor pressures of naphthalene, camphor and glycerol. *J. Phys. Chem.* (U.S.S.R.) 14, 759–767.
Zoeteman, B.C.J., Harmsen, K., Linders, J.B.H. (1980) Persistent organic pollutants in river water and ground water of the Netherlands. *Chemosphere* 9, 231–249.

Zoeteman, B.C.J., De Greef, E., Brinkmann, F.J.J. (1981) Persistency of organic contaminants in groundwater, lessons from soil pollution incidents in the Netherlands. *Sci. Total Environ.* 21, 187–202.

Zwolinski, B.J., Wilhoit, R.C. (1971) *Handbook of Vapor Pressures and Heats of Vaporization of Hydrocarbons and Related Compounds.* American Petroleum Institute Project 44, API 44-TRC Publications in Science and Engineering, Texas A & M University, College Station, Texas.